W9-CPA-525

WINDOWS

TO ALGEBRA AND GEOMETRY

AN INTEGRATED APPROACH

Roland E. Larson

Laurie Boswell

Timothy D. Kanold

Lee Stiff

D.C. Heath and Company
Lexington, Massachusetts / Toronto, Ontario

About the Cover

Each chapter has a theme which is used in the chapter introduction, in several examples and exercises throughout the chapter, and for the last page of the Chapter Review. Water resources is the theme for Chapter 7 (see pages 292, 293, 307, 331, and 340) and it is also the theme of the cover.

Acknowledgements

Editorial Development Jane Bordzol, Anne M. Collier, Rita Campanella, Tamara Gorman, Susan E. Kipp, Albert S. Jacobson, James O'Connell, George J. Summers
Marketing Jo DiGiustini
Advertising Phyllis Lindsay
Design Jane Bigelow-Orner, Robert Botsford, Carmen Johnson
Production Pamela Tricca

D. C. Heath is committed to publishing educational materials that accurately and fairly reflect the diversity of all peoples; that promote a better understanding of one another; that acknowledge the contributions of all groups; and that avoid stereotypes, ridicule, and bias. Our instructional materials foster an appreciation of differences in culture, religion, age, gender, ability, and socio-economic background. Our products promote respect for the intrinsic worth of all individuals and prepare people to live and work together in a diverse world. D. C. Heath believes that in order to flourish in a changing world, we must value diversity.

Published simultaneously in Canada

Printed in the United States of America

International Standard Book Number 0-669-37690-6

1 2 3 4 5 6 7 8 9 10 –RRD– 00 99 98 97 96 95

ABOUT THE AUTHORS

Roland E. Larson is a professor of mathematics at the Behrend College of Pennsylvania State University at Erie. He is a member of NCTM and author of many well-known high school and college mathematics textbooks, including D. C. Heath's *Calculus, Pre-Calculus,* and *Algebra-Geometry* series. He is a pioneer of interactive textbooks, and his calculus textbook is published in a CD-ROM version.

Laurie Boswell is a mathematics teacher at Profile Junior-Senior High School in Bethlehem, New Hampshire. She is active in NCTM and local mathematics associations and is a frequent convention speaker. A 1986 recipient of the Presidential Award for Excellence in Mathematics Teaching, she is also the 1992 Tandy Technology Scholar and the 1991 recipient of the Richard Balomenos Mathematics Education Service Award, presented by the New Hampshire Association of Teachers of Mathematics. She is also an author of D. C. Heath's *Geometry.*

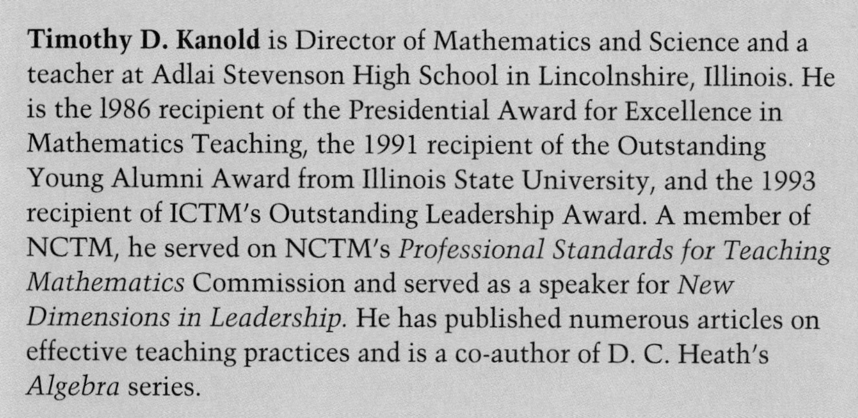

Timothy D. Kanold is Director of Mathematics and Science and a teacher at Adlai Stevenson High School in Lincolnshire, Illinois. He is the 1986 recipient of the Presidential Award for Excellence in Mathematics Teaching, the 1991 recipient of the Outstanding Young Alumni Award from Illinois State University, and the 1993 recipient of ICTM's Outstanding Leadership Award. A member of NCTM, he served on NCTM's *Professional Standards for Teaching Mathematics* Commission and served as a speaker for *New Dimensions in Leadership.* He has published numerous articles on effective teaching practices and is a co-author of D. C. Heath's *Algebra* series.

Lee Stiff is an associate professor of mathematics education in the College of Education and Psychology of North Carolina State University at Raleigh and has taught mathematics at the high school and middle school levels. He was a member of both the NCTM Board of Directors and the writing team of NCTM's *Professional Standards for Teaching Mathematics.* He is the 1992 recipient of the W. W. Rankin Award for Excellence in Mathematics Education presented by the North Carolina Council of Teachers of Mathematics. He is an author of D. C. Heath's *Algebra-Geometry* series and a contributing author to *Heath Mathematics CONNECTIONS..*

REVIEWERS AND CONTRIBUTORS

Linda Bailey
Putman City Schools
Oklahoma City, OK

David S. Bradley
Thomas Jefferson Junior High School
Kearns, UT

Blanche S. Brownley
Mathematics, Science, Technology Initiative
District of Columbia Public Schools

John A. Carter
West Chicago Community High School
West Chicago, IL

Susan Currier
Pioneer Valley Regional School
Northfield, MA

Anthony C. Dentino
Plainfield School District
Plainfield, NJ

Susan E. Ewart
Centreville Middle School
Centreville, MD

John Paul Fox
Clinton Middle School
Columbus, OH

Gregory J. Fry and Keith Tuominen
White Bear Lake High School
White Bear Lake, MN

Sue D. Garriss
Millbrook Senior High School
Raleigh, NC

Linda Gojak
Hawken School
Lyndhurst, OH

Leigh M. Graham
Awtrey Middle School
Cobb County, GA

Judy Hall
St. John School
Seattle, WA

Sandra A. Hinker
Marshfield Junior High School
Marshfield, WI

Audrey M. Johnson
Luke O'Toole Elementary School
Chicago, IL

Kathy Johnson
East Paulding Middle School
Dallas, GA

Lieshen Johnson
Schimelpfenig Middle School
Plano, TX

Scott Hemingway Killam
Carter Junior High School
Arlington, TX

Charlene M. Kincaid
Gulf Breeze High School
Gulf Breeze, FL

Marsha W. Lilly
Alief Independent School District
Alief, TX

Christine S. Losq
The Whole Math™ Project
Palo Alto, CA

Veronica G. Meeks
Western Hills High School
Fort Worth, TX

Joyce E. Nitz
Kellogg Middle School
Portland, OR

Eileen Paul and Ellen Silbert
Adlai E. Stevenson High School
Lincolnshire, IL

Clementine Sherman
Dade County Public Schools
Miami, FL

Robyn Silbey
Math Specialist
Montgomery County Public Schools, MD

Tomas M. Tobiasen
Parsippany–Troy Hills Township Schools
Parsippany, NJ

Ricardo Torres
M. B. Lamar Middle School
Laredo, TX

Linda Tucci
Rice Avenue Middle School
Girard, PA

Marianne Weber
Middle School Mathematics Consultant
Chesterfield, MO

Betsy L. Wiens
Washburn Rural Middle School
Topeka, KS

To the Students

Mathematics evolved over thousands years in many stages. The early stages were concerned with using mathematics to answer questions about real life. We wrote this book in much the same way. We centered the concepts around the real-life use of mathematics.

As more and more mathematics was discovered, people began to collect and categorize the different rules, formulas, and properties. This took place independently in many different parts of the world: Africa, Asia, Europe, North America, and South America. The mathematics that we use today is a combination of the work of literally thousands of people.

As you study our book, be sure you understand the value and purpose of what you are learning. Knowing how and why a concept is used helps you master it. That's why we begin each lesson explaining what you should learn and why you should learn it.

Remember, math is not a spectator sport It's a valuable tool you can use in everyday life, and the more you use it the more useful it becomes!

Roland E. Larson

Laurie Boswell

Timothy D. Kanold

Lee Stiff

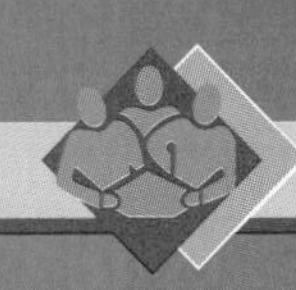

EXPLORING AND INVESTIGATING MATHEMATICS

Mathematics is more fun and more understandable when you can play around with mathematical ideas and discover how things work before you are told the formal definitions and rules. There are many opportunities in this text to explore mathematical ideas through hands-on investigations and using technology. The full-page investigations that precede lessons are listed below. In addition, shorter investigations are built into many of the lessons. Also listed below are the full-page technology activities.

REAL-LIFE APPLICATIONS

Look through this list for things that interest you.Then find out how they are linked to mathematics.

People

Places

United States Facts

World Facts

TABLE OF CONTENTS

CHAPTER 1

Exploring Patterns

Integrating Algebra and Geometry
In all chapters, algebra and geomety are integrated wherever possible to strengthen important skills, develop concepts, and pave the way for future mathematics.

C Integrates Coordinate Geometry D Integrates Data Analysis or Probability
P/F Integrates Patterns and Functions

CHAPTER 2

Investigations in Algebra

Using Problem Solving to Model Math in Real Life

Windows introduces you to a problem-solving plan for writing algebraic models—the major problem-solving strategy of the program. By introducing the plan early, problem solving becomes an integral part of every lesson.

CHAPTER 3

Modeling Integers

Introducing the Coordinate Plane

Windows introduces the coordinate plane beginning in Lesson 3.8. It is a pictorial model to help you discover the patterns and relationships of algebra and geometry. Introducing the coordinate plane early prepares you for future study of mathematics.

CHAPTER 4

Exploring the Language of Algebra

CHAPTER 5

Exploring Data and Graphs

Emphasis on Communication
Communicating about Mathematics—a feature found in every lesson—offers opportunities for you to share your ideas about the lesson and to do cooperative learning activities.

C Integrates Coordinate Geometry D Integrates Data Analysis or Probability
P/F Integrates Patterns and Functions

CHAPTER 6

Exploring Number Theory

Connections Across Content Strands

Windows provides connections within mathematics—between data analysis and probability, patterns and functions, algebra, geometry and coordinate geometry—and to other disciplines, such as social sciences, physical sciences, music, and many more.

CHAPTER 7

Rational Numbers and Percents

CHAPTER 8

Proportion, Percent, and Probability

Emphasis on Hands-On Investigations
Frequent, hands-on use of models in lesson investigations gives you the chance to experiment with your classmates and discover—by inductive reasoning—important mathematics concepts for yourselves.

CHAPTER 9

Real Numbers and Inequalities

C Integrates Coordinate Geometry D Integrates Data Analysis or Probability
P/F Integrates Patterns and Functions

CHAPTER 10

Geometry Concepts and Spatial Thinking

CHAPTER 11

Congruence, Similarity, and Transformations

Career Interviews
Career Interviews, which feature people of various backgrounds who use math on the job, demonstrate the utility of math in daily life.

CHAPTER 12

Measurements in Geometry

Milestones
Through anecdotes from the history of math, **Milestones** highlight the develeopement of mathematics over the centuries, with time lines that show other important events of a period.

CHAPTER 13

Exploring Linear Equations

C Integrates Coordinate Geometry **D** Integrates Data Analysis or Probability
P/F Integrates Patterns and Functions

CHAPTER 14

Exploring Data and Polynomials

Student Handbook

CHAPTER 1

Exploring Patterns

LESSONS

Rocky IV is a smart robot that can explore places that are too dangerous for humans. It is designed to withstand the temperatures on Mars that get as low as −190°F.

Real Life
Building a Rover
Distance (in feet)
7
6
5
4
3
2
1
3.3 ft / min
1
2
Time (in minutes)
Because of size and weight limits in modern space exploration, this Mars rover will be under 2 feet long, and weigh less than 17 pounds. It recognizes patterns and makes decisions by using a computer program based on a guess-and-check strategy.
In this chapter, you will study number patterns and ways to present data, such as this graph of distance, rate, and time for Rocky IV.
NASA
JPL

1.1 Number Patterns

What you should learn:

Goal 1 How to use numbers to identify and measure objects

Goal 2 How to recognize and describe number patterns

Why you should learn it:

You can use numbers to identify and measure real-life objects, such as the numbers in a zip code or the waist size of a pair of jeans.

Nebraska's Zip Codes

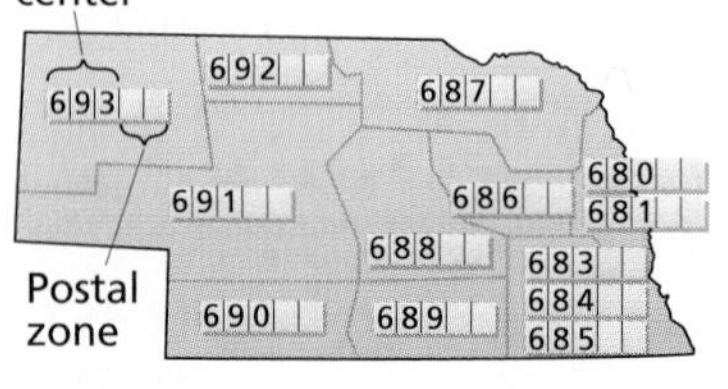

Zip codes identify the state, sectional center, and postal zone of an address.

Goal 1 Identifying and Measuring

Mathematics is not just the study of numbers. Mathematics is also the study of how numbers are *used* in real life. For instance, numbers can be used as zip codes to identify addresses. Numbers can also be used to measure distances on a highway.

Highway signs indicate the distance to a city or town.

LESSON INVESTIGATION

Investigating Number Uses

Group Activity With the other students in your group, make 4 different lists of numbers that are used for identification. Then make 4 different lists of numbers used for measurement. (Each list should contain at least five numbers.)

Share the numbers from one of your group's lists with the other groups in your class. Ask the groups if they can tell what the numbers identify or measure. For instance, the numbers

10199, 20066, 60607, 75260, and 94188

are the postmaster zip codes for New York City; Washington D.C.; Dallas; San Francisco; and Chicago.

Goal 2 Describing Number Patterns

The most common shoe sizes for women are 7, $7\frac{1}{2}$, 8, and $8\frac{1}{2}$. The most common shoe sizes for men are $9\frac{1}{2}$, 10, and $10\frac{1}{2}$.

Sometimes when you see numbers in real life, the numbers form a pattern. For instance, when you go into a shoe store, you might see numbers representing shoe sizes.

$5, 5\frac{1}{2}, 6, 6\frac{1}{2}, 7, 7\frac{1}{2}, 8, 8\frac{1}{2}, 9, 9\frac{1}{2}, 10, 10\frac{1}{2}$ *Shoe Sizes*

An ordered list of numbers is called a **sequence.** The pattern for the sequence above is that each number is $\frac{1}{2}$ more than the preceding number.

Example *Describing Number Patterns*

Describe a pattern for each sequence. Then use the pattern to write the next three numbers in the sequence.

a. 4, 8, 12, 16, ?, ?, ?

b. 128, 64, 32, 16, ?, ?, ?

Solution

a. One pattern for this sequence is that each number is 4 more than the preceding number. The next three numbers are shown below.

4, 8, 12, 16, 20, 24, 28

16 + 4 → 20; 20 + 4 → 24; 24 + 4 → 28

b. One pattern for this sequence is that each number is half the preceding number. The next three numbers are shown below.

128, 64, 32, 16, 8, 4, 2

$\frac{1}{2} \cdot 16$ → 8; $\frac{1}{2} \cdot 8$ → 4; $\frac{1}{2} \cdot 4$ → 2 ■

Communicating about MATHEMATICS

SHARING IDEAS about the Lesson

Describing Patterns In Example 1, only one pattern was described for each list of numbers. Often, however, a list of numbers can be described by two or more patterns. For instance, suppose that you are given the numbers

2, 3, 5, ?, ?, ?, ?, ?.

Describe several possible patterns for this list. For each pattern, write the next five numbers in the list.

EXERCISES

Guided Practice

CHECK for Understanding

P **1.** *Writing* In your own words, describe what mathematics is.

2. State several examples of how numbers are used to identify objects in real life.

3. State several examples of how numbers are used to measure objects in real life. Name some units of measure, such as centimeters or pounds.

4. State an example of a real-life number sequence. Describe its pattern.

Independent Practice

In Exercises 5–14, describe the pattern. Then list the next 3 numbers.

5. 1, 3, 5, 7, [?], [?], [?]

6. 5, 10, 15, 20, [?], [?], [?]

7. 1, 3, 6, 10, [?], [?], [?]

8. 60, 57, 53, 48, [?], [?], [?]

9. $\frac{1}{2}, \frac{2}{3}, \frac{3}{4}, \frac{4}{5}$, [?], [?], [?]

10. $\frac{2}{3}, \frac{4}{5}, \frac{6}{7}, \frac{8}{9}$, [?], [?], [?]

11. $2, \frac{7}{2}, 5, \frac{13}{2}$, [?], [?], [?]

12. 100, 81, 64, 49, [?], [?], [?]

13. 2, 6, 18, 54, [?], [?], [?]

14. 4096, 1024, 256, 64, [?], [?], [?]

In Exercises 15–20, describe the pattern. Then list the next 3 letters.

15. A, C, E, G, [?], [?], [?]

16. A, Y, C, W, [?], [?], [?]

17. Z, A, Y, B, X, C, [?], [?], [?]

18. A, N, B, O, [?], [?], [?]

19. O, T, T, F, F, S, [?], [?], [?]

20. T, F, S, E, T, [?], [?], [?]

Number Sense **In Exercises 21 and 22, write the first 6 numbers in the sequence.**

21. The first number is 50. Each succeeding number is 3 less than the preceding number.

22. The first number is 1 and the second number is 2. Each succeeding number is the sum of the two preceding numbers.

Visualizing Patterns **In Exercises 23 and 24, draw the next 3 figures in the pattern.**

23.

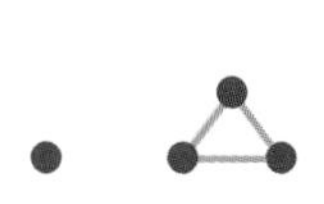
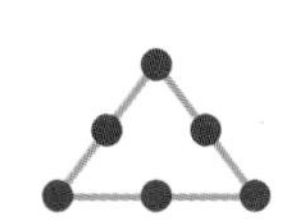
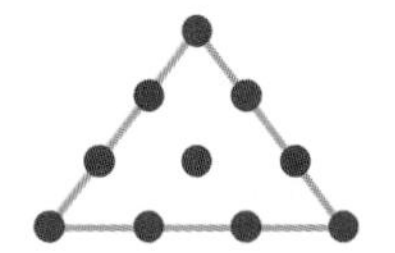

24.

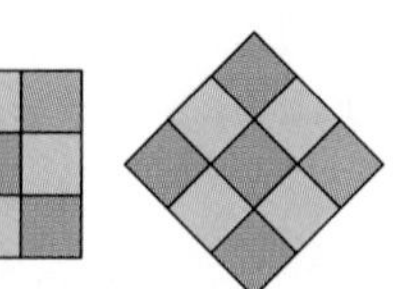
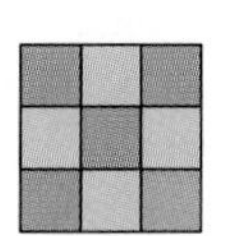
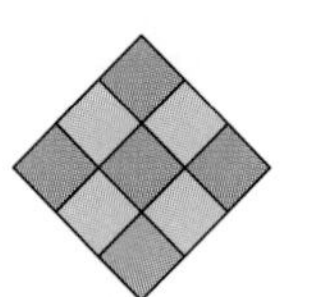

25. *Race Times* You run a 400-meter race with a time of 2:39.4. Your friend's time is 2:41.8. Explain how these numbers measure your times. Who won the race?

26. *Long Jump* You and a friend are in a standing long-jump contest. You jump 4 feet 9 inches and your friend jumps 4 feet 11 inches. Who won the contest? Explain.

27. *Vacation Travel* You and your family travel to an amusement park. As you near the park, you begin seeing signs that state the number of miles to the park. What can you say about the numbers on the signs as you get closer and closer to the park?

28. *Amusement Park* Imagine that you are at an amusement park. Describe several ways that numbers are used to identify objects at the park and measure objects at the park.

In 1994, approximately 255 million people visited a theme park in North America. The most popular of these was the Magic Kingdom at Walt Disney World in Florida.

Integrated Review

Making Connections within Mathematics

Measurement Sense **In Exercises 29–34, rewrite the measure as indicated. (Measurement tables are found on pages 680 and 681.)**

29. 1.5 feet = ? inches

30. 1.5 hours = ? minutes

31. 1.5 meters = ? centimeters

32. 1.5 kilometers = ? meters

33. 30 inches = ? feet

34. 90 seconds = ? minutes

Exploration and Extension

Making a Table **In Exercises 35 and 36, imagine that you have completed a homework assignment. Without a penalty, your score would be 94. Complete the table to show what your score would be with penalties.**

35. 1 day late: Lose 5 points.
 2 days late: Lose 10 points.
 3 days late: Lose 15 points.

Days Late	0	1	2	3	4	5
Score	94	89	?	?	?	?

36. 1 day late: Lose 4 points.
 2 days late: Lose 4 + 8 points.
 3 days late: Lose 4 + 8 + 12 points.

Days Late	0	1	2	3	4	5
Score	94	90	?	?	?	?

37. *It's Up to You* If you were the teacher, which penalty system would you use: the one described in Exercise 35 or in Exercise 36? Explain.

LESSON INVESTIGATION 1.2

Computation Practice

Materials Needed: toothpicks

Grape Game can be played with two or more people. Players take turns at placing a toothpick along three hexagons in a row. The hexagons must be in a horizontal or diagonal row. Each player than mentally performs the operations in each of the three hexagons and adds the results together. Players should agree on the correct sum.

$6+2$

$14-1$, 3×5

2×7, $6\div 3$, $7-2$

$6-3$, $5+4$, 8×2, $6+4$

$4+4$, $8+4$, $6+5$, $5+2$, 2×4

$6+7$, $8+7$, 3×2, 2×2, $12+1$, $10+5$

$13+1$, $8\div 4$, $10\div 2$, $8+9$, $10+4$, $3-1$, $8-3$

$9\div 3$, $12-3$, $9+7$, $7+3$, $5-2$, 3×3, $20-4$, 5×2

$13-1$, $9+2$, $8-1$, $10-2$, 3×4, $16-5$, $6+1$

$12\div 2$, $6-2$, $11+2$, $18-3$, $9-3$, $8-4$

$11+6$, $15-1$, $8-6$, $3+2$, $19-2$

$7-4$, $11-2$, $8+8$, $11-1$

6×2, $7+4$, $10-3$

$3+3$, $9-5$

$18-1$

1. Describe a quick way to find the sum of four hexagons in a row.
2. Can you discover a quick way to find the sum of three hexagons in a row?
3. Describe a quick way to find the sum of seven hexagons that form a "flower."

1.2 Number Operations

What you should learn:

Goal 1 How to use the four basic number operations

Goal 2 How to use multiplication models

Why you should learn it:

Many real-life uses of numbers require number operations, such as multiplication to find the area of a rectangle.

Hindu-Arabic Numerals

Mathematical symbols that are used today came from many different cultures. For instance, the numerals 0, 1, 2, 3, . . . , 9 were invented in India and Arabia more than 2000 years ago. The fraction bar was first introduced by the Arabic author al-Hassar in about A.D. 1202.

Goal 1 Using Number Operations

There are four basic number operations: **addition, subtraction, multiplication,** and **division.** Each of these operations can be described verbally or symbolically.

The symbol "$+$" signifies addition and the symbol "$-$" signifies subtraction. Multiplication can be specified by "$\times$", or by "$\cdot$", or by parentheses. Division can be specified by "$\div$", or by "/", or by a fraction bar, as in $\frac{1}{2}$.

Example 1 *Finding Sums and Differences*

a. *Addition:*

The **sum** of 5 and 13 is 18. — *Verbal description*

$5 + 13 = 18$ — *Symbolic description*

b. *Subtraction:*

The **difference** of 9 and 6 is 3. — *Verbal description*

$9 - 6 = 3$ — *Symbolic description* ■

Example 2 *Finding Products and Quotients*

a. *Multiplication:*

The **product** of 3 and 5 is 15. — *Verbal description*

$3 \times 5 = 15$ — *Symbolic description*

$3 \cdot 5 = 15$ — *Symbolic description*

$3(5) = 15$ — *Symbolic description*

b. *Division:*

The **quotient** of 14 and 7 is 2. — *Verbal description*

$14 \div 7 = 2$ — *Symbolic description*

$14/7 = 2$ — *Symbolic description*

$\frac{14}{7} = 2$ — *Symbolic description* ■

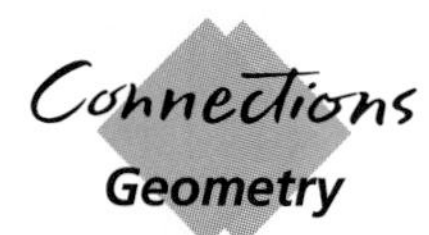

Goal 2 Using Multiplication Models

A **model** is something that helps you visualize or understand an actual process or object. For instance, drawing a family tree helps you understand how your cousins are related to you. The next example shows how area can be used as a model for multiplication.

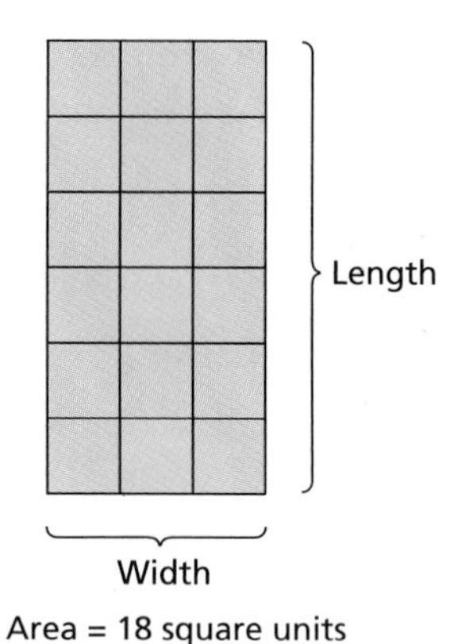

Area = 18 square units

Example 3 *Using Area as a Multiplication Model*

Show how to use area to model the product of 3 and 6.

Solution One way to model the product is to use squares to form a rectangle that is 3 units wide and 6 units long. Each square has an area of 1 square unit. By counting the squares, you can see that the rectangle has an area of 18 square units. Because the area of a rectangle is the product of its width and length,

$$(\text{Area of rectangle}) = (\text{Width}) \times (\text{Length}),$$

you can see that $3 \times 6 = 18$. ■

Communicating about MATHEMATICS

SHARING IDEAS about the Lesson

Multiplication Models Each of the multiplication models shown below represents a multiplication sentence. Work with a partner. Write the sentence verbally and symbolically. Describe other ways to model multiplication.

A.

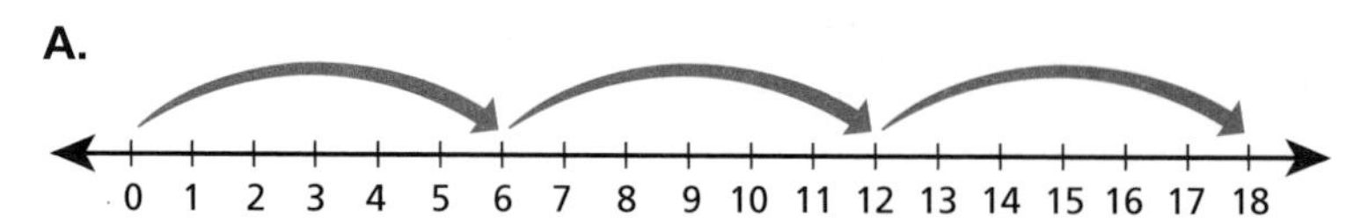

B.

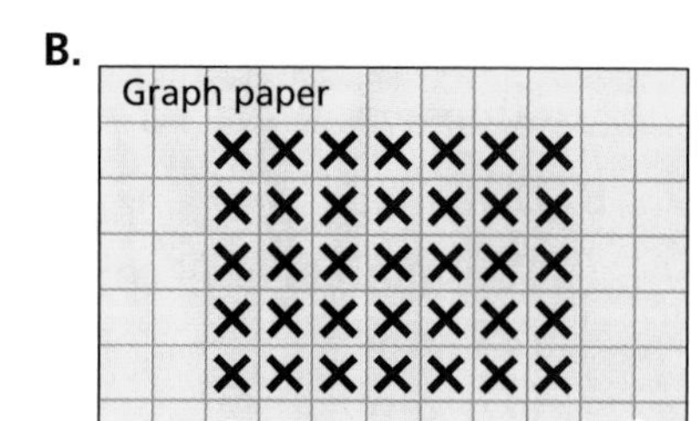

C.

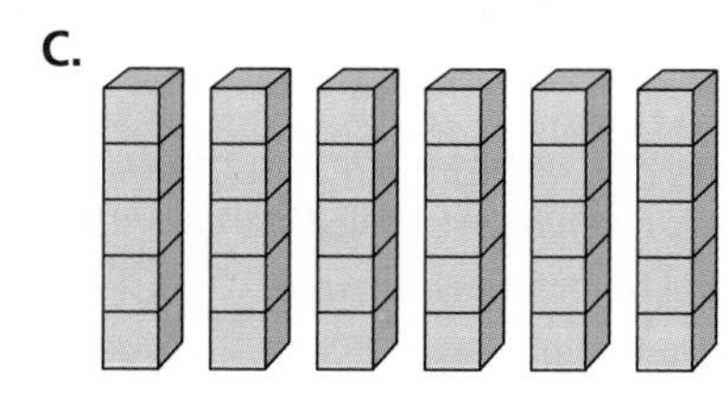

EXERCISES

Guided Practice

CHECK for Understanding

1. State the four basic number operations.
2. State the symbol or symbols that represent each number operation.
3. What is a model? Describe a model that can be used to represent multiplication.
4. *Problem Solving* Describe a real-life example of a number operation.

Independent Practice

Computation Sense **In Exercises 5-12, write a verbal description of the number sentence.**

5. $6 \times 8 = 48$ **6.** $25 \div 5 = 5$ **7.** $3 + 14 = 17$ **8.** $9(7) = 63$

9. $111 - 56 = 55$ **10.** $\frac{12}{4} = 3$ **11.** $2 \cdot 54 = 108$ **12.** $\frac{132}{11} = 12$

Computation Sense **In Exercises 13–28, find the sum or difference.**

13. $659 + 23$ **14.** $350 + 211$ **15.** $746 - 27$ **16.** $858 - 349$

17. $75 + 40 + 98$ **18.** $352 + 67 + 20$ **19.** $10.9 - 8.6$ **20.** $112.7 - 72.9$

21. $316.41 + 589.02$ **22.** $4203.9 + 123.12$ **23.** $0.248 - 0.097$ **24.** $2.385 - 0.597$

25. $\frac{5}{6} + \frac{1}{6}$ **26.** $\frac{3}{8} + \frac{1}{8}$ **27.** $\frac{9}{12} - \frac{5}{12}$ **28.** $\frac{6}{13} - \frac{3}{13}$

Number Sense **In Exercises 29–44, find the product or quotient.**

29. 16×7 **30.** 21×14 **31.** $527 \div 31$ **32.** $1435 \div 35$

33. $(4.7)(8.9)$ **34.** $(13.2)(5.1)$ **35.** $3 \cdot \frac{4}{9}$ **36.** $7 \cdot \frac{1}{8}$

37. $\frac{256}{32}$ **38.** $\frac{1024}{64}$ **39.** 321×156 **40.** 497×38

41. $76.97 \div 4.3$ **42.** $145.2 \div 33$ **43.** $1977/15$ **44.** $2125/34$

In Exercises 45–48, write a number sentence for each model.

45.

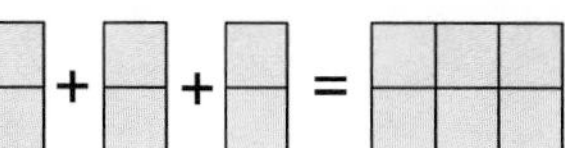

46.

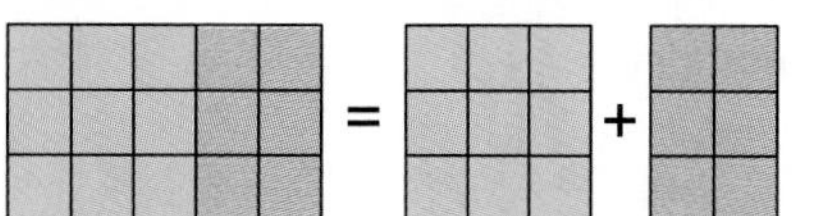

47.

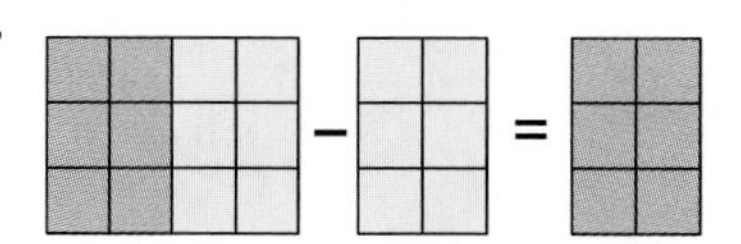

48.

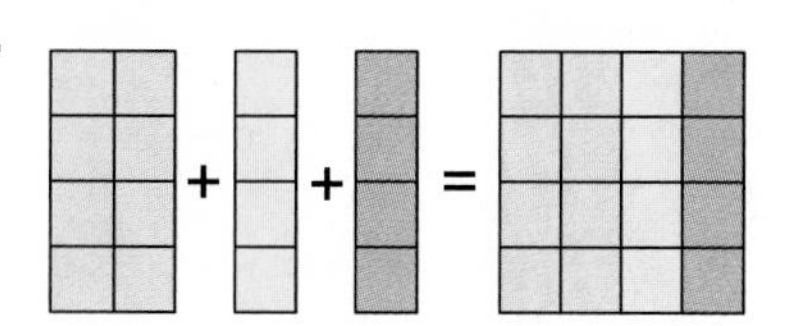

This symbol indicates exercises where you should choose the appropriate method of calculation: mental math, paper and pencil, or calculator.

CDs and Cassettes **In Exercises 49–52, use the graph.** *(Source: Sound Scan)*

49. In 1993, how many more CDs were sold than cassettes?

50. How many more CDs were sold in 1993 than in 1992?

51. How many CDs and cassettes were sold in 1993?

52. *It's Up to You* What do you predict will happen in years to come with regard to CDs and cassettes?

53. *Collecting CDs* You have a CD storage unit that holds 5 stacks of 14 CDs. How many CDs does it hold?

54. *Number Sense* Complete the table. Is the result of the sentence odd or even?

CDs OUTSELL CASSETTES

Number sold in millions

	1992	1993
CDs	248	304
Cassettes	300	268

First number	Operator	Second number	Result
Odd	+	Even	?
Odd	+	Odd	?
Odd	×	Even	?
Odd	×	Odd	?

Integrated Review

Making Connections within Mathematics

Fraction Sense **In Exercises 55–58, simplify the fraction.**

55. $\frac{2}{4}$ **56.** $\frac{6}{3}$ **57.** $\frac{12}{8}$ **58.** $\frac{6}{20}$

Decimal Sense **In Exercises 59–62, write the fraction as a decimal.**

59. $\frac{1}{2}$ **60.** $\frac{1}{4}$ **61.** $\frac{2}{3}$ **62.** $\frac{2}{5}$

Exploration and Extension

Guess, Check, and Revise **In Exercises 63–66, find the largest number that can be made using each of the digits 2, 4, 6, 8, and 9 only once. Then find the smallest.**

63. ? ? ? − ? ?

64. ? ? ? + ? ?

65. ? ? ? × ? ?

66. ? ? ? ÷ ? ?

1.3 Powers and Square Roots

What you should learn:

How to use powers

How to use square roots

Why you should learn it:

You can use powers and square roots to solve real-life problems, such as finding the area of a square room.

Goal 1 Using Powers

The squares shown below have sides whose lengths are 1, 2, 3, 4, and 5. The areas of the squares are

$1 \times 1, 2 \times 2, 3 \times 3, 4 \times 4,$ and $5 \times 5.$

These areas can also be written as **powers.** For instance, 1×1 can be written as 1^2, 2×2 can be written as 2^2, and so on.

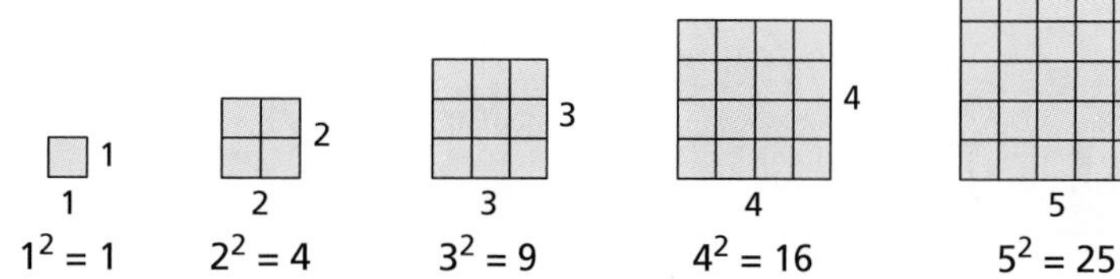

Study Tip...

When you encounter new symbols in mathematics, be sure you can state the symbols verbally. For instance, the statement

$2^5 = 32$

is read as "2 raised to the 5th power is 32."

A power has two parts, a **base** and an **exponent.** For instance, the base of 4^2 is 4 and the exponent of 4^2 is 2.

Base → 4^2 ← Exponent

Power

Any natural number (1, 2, 3, . . .) can be used as an exponent.

$4^2 = 4 \times 4 = 16$	*4 raised to the 2nd power or 4 squared*
$5^3 = 5 \times 5 \times 5 = 125$	*5 raised to the 3rd power or 5 cubed*
$3^4 = 3 \times 3 \times 3 \times 3 = 81$	*3 raised to the 4th power*
$6^1 = 6$	*6 raised to the 1st power*

Example 1 *Raising Numbers to Powers*

Technology
Using a Calculator

Use a calculator to evaluate the powers.

a. 1.5^2 **b.** 4.2^3 **c.** 6^5

Solution

	Calculator Steps	Display	Written Result
a.	1.5 [x^2]	2.25	$1.5^2 = 2.25$
b.	4.2 [y^x] 3 [=]	74.088	$4.2^3 = 74.088$
c.	6 [y^x] 5 [=]	7776	$6^5 = 7776$

■

Goal 2 Using Square Roots

The **square root** of a number is denoted by the symbol $\sqrt{\ }$, which is called a **radical** or **square root symbol.** When you square the square root of a number, you obtain the original number. For instance, $\sqrt{9} = 3$ because $3^2 = 9$.

Real Life
Architecture

Example 2 *Designing a Room*

You are designing a bedroom for an apartment. You want the room to be square and have an area of 90.25 square feet. How long should each side of the room be? What is the perimeter?

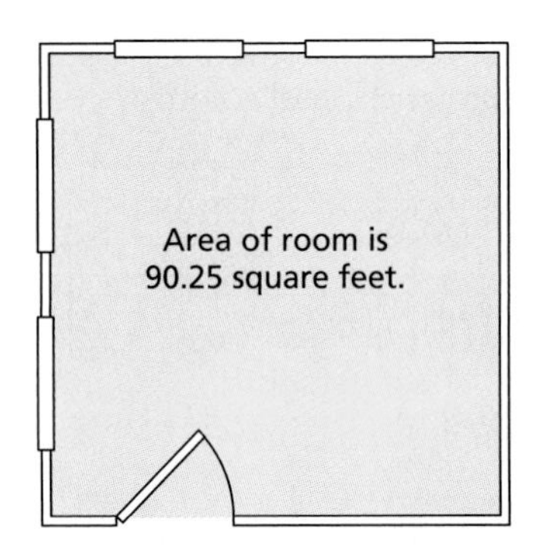

The perimeter of a room is the sum of the lengths of the four sides.

Solution You need to find a number whose square is 90.25.

$$(\text{Side})^2 = 90.25$$

The solution of this equation is the square root of 90.25.

$$\text{Side} = \sqrt{90.25}$$

With a calculator, you can obtain the following.

Calculator Steps	Display	Written Result
90.25 $\boxed{\sqrt{x}}$	9.5	$\sqrt{90.25} = 9.5$

Each side of the room should be 9.5 feet long. The perimeter is the sum of the lengths of the four sides, which is 4(9.5) or 38 feet. ■

Communicating about MATHEMATICS

SHARING IDEAS about the Lesson

Rounding Numbers The square root of a **perfect square** can be written as an exact decimal. Two examples of perfect squares are $\sqrt{49} = 7$ and $\sqrt{1.44} = 1.2$. The square root of a number that is not a perfect square can only be approximated by a decimal. Here are two examples.

$$\sqrt{2} \approx 1.4142136 \quad \text{and} \quad \sqrt{5.5} \approx 2.3452079$$

(The symbol $\approx$ means "is approximately equal to.") Rounded to two decimal places, these square roots are 1.41 and 2.35. Round each of the following to two decimal places. Explain how to check your answers.

A. $\sqrt{6}$ **B.** $\sqrt{10}$ **C.** $\sqrt{6.5}$ **D.** $\sqrt{140.3}$

EXERCISES

Guided Practice

CHECK for Understanding

1. Fill in the blanks. A power has two parts, a [?] and an [?].

2. State a verbal description of the number sentence $3^4 = 81$.

In Exercises 3–5, find the value of the expression.

3. $\sqrt{16}$ **4.** $\sqrt{49}$ **5.** $\sqrt{81}$

6. State a verbal description of the number sentence $\sqrt{36} = 6$.

Independent Practice

In Exercises 7–10, write a verbal description of the number sentence.

7. $6^4 = 1296$ **8.** $2.9^2 = 8.41$ **9.** $\sqrt{1.21} = 1.1$ **10.** $\sqrt{225} = 15$

In Exercises 11–16, write each expression as a power. Then use a calculator to find the value of the power.

11. 12×12 **12.** $8 \times 8 \times 8 \times 8$ **13.** $(3.4)(3.4)(3.4)$

14. $(9.7)(9.7)(9.7)(9.7)(9.7)$ **15.** $\frac{1}{5} \cdot \frac{1}{5} \cdot \frac{1}{5} \cdot \frac{1}{5}$ **16.** $\frac{2}{3} \cdot \frac{2}{3} \cdot \frac{2}{3} \cdot \frac{2}{3} \cdot \frac{2}{3}$

In Exercises 17–22, find the value of the expression using a calculator. Round your result to two decimal places.

17. $\sqrt{169}$ **18.** $\sqrt{441}$ **19.** $\sqrt{117}$

20. $\sqrt{372}$ **21.** $\sqrt{5.5}$ **22.** $\sqrt{8.26}$

Guess and Check **In Exercises 23–28, find the number that is represented by $\triangle$.**

23. $\triangle \cdot \triangle \cdot \triangle = 512$ **24.** $\triangle \cdot \triangle \cdot \triangle \cdot \triangle = 625$ **25.** $\triangle \cdot \triangle = 4.41$

26. $\triangle \cdot \triangle \cdot \triangle = 42.875$ **27.** $\sqrt{\triangle} = 9$ **28.** $\sqrt{\triangle} = 17$

Guess and Check **In Exercises 29–32, replace each [?] with >, <, or =. (The symbol < means "is less than" and the symbol > means "is greater than.")**

29. 2^3 [?] 3^2 **30.** 2^4 [?] 4^2 **31.** 4^3 [?] 3^4 **32.** 5^2 [?] 2^5

33. *Estimation* Without using a calculator, predict which is greater 10^2 or 2^{10}. Then use a calculator to check your answer.

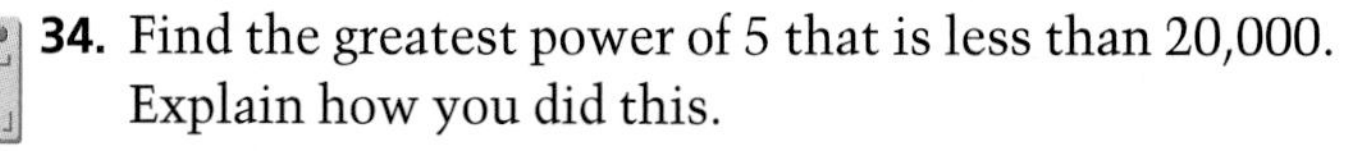

34. Find the greatest power of 5 that is less than 20,000. Explain how you did this.

Reasoning **In Exercises 35–38, determine whether the statement is true or false. Explain your reasoning.**

35. $5^3 = 5 \times 5 \times 5$

36. $5^3 = 2^3 + 3^3$

37. $\sqrt{5} = \sqrt{2} + \sqrt{3}$

38. $\sqrt{5} = \sqrt{2+3}$

39. ***Geometry*** Each edge of the cube shown at the right is 7 inches. The surface area of a cube is the sum of the areas of the faces.

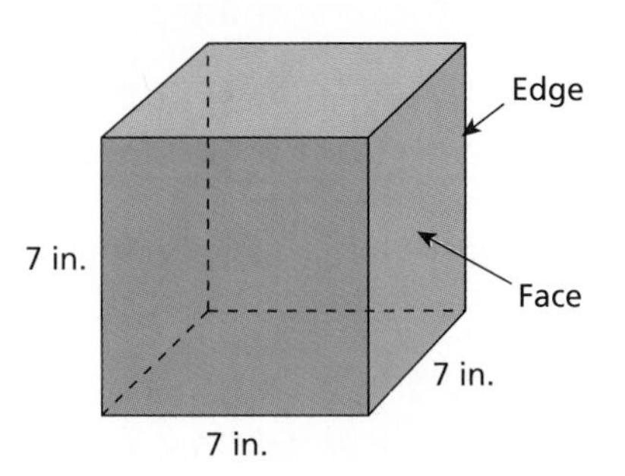

a. How many faces does a cube have?

b. What is the area of one face of the cube?

c. Find the surface area of the cube. Explain how you obtained your answer.

d. If you double the length of each edge of the cube, does its surface area double? Explain your reasoning.

40. ***Geometry*** Volume is a measure of the amount of space that an object occupies. A formula for the volume of a cube is

$$\text{Volume} = (\text{Edge})^3.$$

Use the cube in Exercise 39 to answer the following questions.

a. What is the length, width, and height of the cube?

b. Write an expression that gives the volume of the cube.

c. What is the volume?

41. ***Designing a Room*** You are designing a computer classroom. You want the classroom to be square and have an area of 420.25 square feet.

a. Find the length and width of the classroom.

b. Find the perimeter of the classroom.

c. If you double the area of the classroom, does its perimeter double? Explain your reasoning.

42. The classroom in Exercise 41 must be divided into nine equal square work areas.

a. Sketch a simple diagram of the classroom.

b. What is the area of each work area?

c. Explain how you can find the side length of each work area. Compare your explanation to others in the class. Do you all agree?

d. What is the side length of a work area?

I. M. Pei is an American architect known for his imaginative and creative designs. Born in China, Pei became a United States citizen in 1954. Some of his works include the John Hancock Building and the John F. Kennedy Library. Both are in Boston, Massachusetts.

Integrated Review

Making Connections within Mathematics

Estimation **In Exercises 43–48, round the number to two decimal places.**

43. 2.5914 **44.** 11.3496 **45.** 318.067

46. 42.8934 **47.** 26.1966 **48.** 285.095

49. *How Many Buses?* 216 students and 9 teachers are boarding buses to go on a field trip. Each bus holds 50 people. How many buses are needed?

50. *Fundraiser* Your band club is selling T-shirts to raise money for a trip. The profit from each shirt is $9.00. Your club needs to raise $5000. How many shirts must be sold?

Exploration and Extension

51. Complete the table.

Expression	$\sqrt{6400}$	$\sqrt{640}$	$\sqrt{64}$	$\sqrt{6.4}$	$\sqrt{0.64}$	$\sqrt{0.064}$
Value	?	?	?	?	?	?

52. *Finding a Pattern* Describe a pattern of the values you obtained in the table.

53. *Making a Prediction* Use the pattern to predict the value of $\sqrt{64{,}000}$ and $\sqrt{0.0064}$. Verify your answers by using a calculator.

Mixed REVIEW

In Exercises 1–6, describe the pattern. Then list the next 3 numbers (1.1)

1. 2, 4, 6, 8, ?, ?, ?

2. 30, 27, 24, 21, ?, ?, ?

3. 1, 6, 11, 16, ?, ?, ?

4. 3, 1, $\frac{1}{3}$, $\frac{1}{9}$, ?, ?, ?

5. 1, 3, 7, 15, ?, ?, ?

6. 2, 5, 14, 41, ?, ?, ?

In Exercises 7–14, perform the operation. (1.2)

7. $724 + 693$ **8.** $415 + 219$ **9.** $532 - 421$ **10.** $864 - 179$

11. 84×31 **12.** $(243)(16)$ **13.** $4472 \div 52$ **14.** $2327 \div 13$

In Exercises 15–20, use a calculator to find the value of the expression. Round your answer to two decimal places. (1.3)

15. 5^7 **16.** 7^5 **17.** $\left(\frac{1}{3}\right)^3$

18. $\sqrt{3}$ **19.** $\sqrt{22}$ **20.** $\sqrt{45.28}$

USING A CALCULATOR

Problem Solving: Guess, Check, and Revise

Problem Solving

Guess, Check, Revise

Guess a reasonable solution based on data in the problem.

↓

Check the guess.

↓

Revise the guess and continue until a correct solution is found.

Problem solving is a process for discovering relationships. It is not simply a process for finding answers. In this text, problem solving is something you will do while you learn new skills. To solve problems successfully, you should consider using different strategies, such as *making a table, looking for a pattern, using a model, drawing a diagram,* and *guessing, checking, and revising.*

Example **Designing a Testing Ground**

You are part of a team that is designing a moon rover. You are in a desert location and want to design a testing ground for the rover. The testing ground should be square and have an area of 10 square kilometers. You have only a simple calculator—one that doesn't have a square root key. Explain how to use the calculator to find the dimensions of the testing ground.

Solution Because the area of the square testing ground is

$$\text{Area} = (\text{Side})^2$$

you need to find a number whose square is 10. With your calculator, you can try **guessing, checking,** and **revising.**

Guess	Calculator Steps	Display	Conclusion
3.1	3.1 [×] 3.1 [=]	9.61	Too small
3.2	3.2 [×] 3.2 [=]	10.24	Too large
3.15	3.15 [×] 3.15 [=]	9.9225	Too small
3.16	3.16 [×] 3.16 [=]	9.9856	Too small
3.17	3.17 [×] 3.17 [=]	10.0489	Too large

From these guesses and revisions, you can see that each side should be between 3.16 kilometers and 3.17 kilometers. ■

Exercises

1. Use a guess, check, and revise strategy to find the dimensions of a square testing ground that has an area of 5 square kilometers. Are the dimensions half the dimensions of the testing ground described in the Example? Explain your reasoning.

2. You are designing a cubical storage container for the moon rover. The storage container should have a volume of 2 cubic feet. Find the dimensions of the container. (*Hint:* Use the fomula $\text{Volume} = (\text{Edge})^3$).

1.4 Order of Operations

What you should learn:

How to use order of operations

How to use order of operations on a calculator

Why you should learn it:

When more than one operation is used in an expression, it is important to know the order in which the operations must be performed.

Goal 1 Using the Order of Operations

One of your goals as you study this book is to learn to read and write about numbers. One way to avoid confusion when you are communicating mathematical ideas is to establish an **order of operations.**

LESSON INVESTIGATION

Investigating Order of Operations

Group Activity Individually perform the following operations. Then compare your results with those obtained by other members in your group.

1. $3 \times 8 + 2 = \boxed{?}$ **2.** $2 + 8 \times 3 = \boxed{?}$

3. $12 \div 4 \times 3 = \boxed{?}$ **4.** $3 \div 12 = \boxed{?}$

5. $40 - 4^2 \times 2 = \boxed{?}$ **6.** $2 \times 3^2 = \boxed{?}$

7. $16 + 12 - 20 = \boxed{?}$ **8.** $20 - 12 + 16 = \boxed{?}$

Because the value of an expression can change by performing the operations in different orders, we give a different priority to each of the operations. First priority is given to exponents, second priority is given to multiplication and division, and third priority is given to addition and subtraction.

Numerical Expression

A **numerical expression** is a collection of numbers, operations, and grouping symbols. When you perform the operations to obtain a single number or value, you are **evaluating** the expression.

Example 1 *Priority of Operations*

a. $2 + \underbrace{8 \times 3} = 2 + 24$ *First priority: multiplication*
$= 26$ *Second priority: addition*

b. $4 \times \underbrace{3^2} = 4 \times 9$ *First priority: exponent*
$= 36$ *Second priority: multiplication*

Operations having the same priority, such as multiplication and division *or* addition and subtraction, are performed from left to right. This is called the **Left-to-Right Rule.**

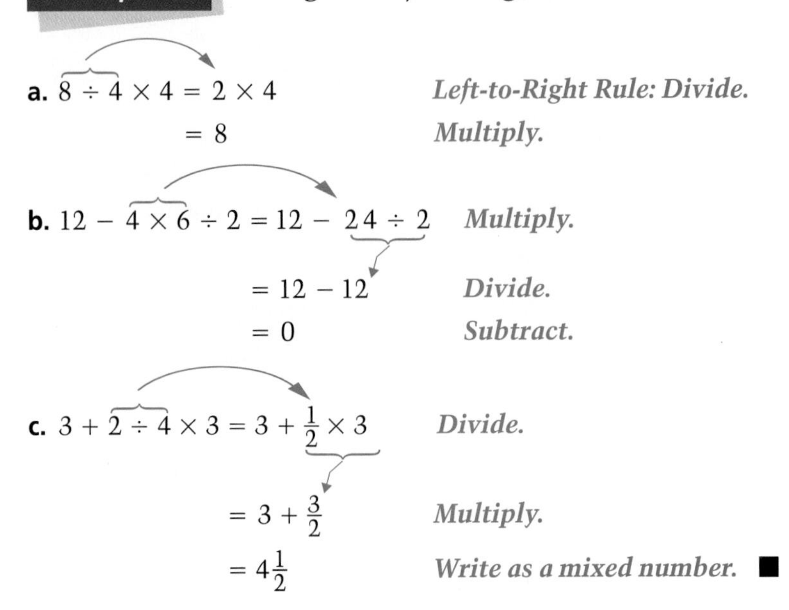

Example 2 ***Using the Left-to-Right Rule***

a. $8 \div 4 \times 4 = 2 \times 4$ *Left-to-Right Rule: Divide.*

$= 8$ *Multiply.*

b. $12 - 4 \times 6 \div 2 = 12 - 24 \div 2$ *Multiply.*

$= 12 - 12$ *Divide.*

$= 0$ *Subtract.*

c. $3 + 2 \div 4 \times 3 = 3 + \frac{1}{2} \times 3$ *Divide.*

$= 3 + \frac{3}{2}$ *Multiply.*

$= 4\frac{1}{2}$ *Write as a mixed number.* ■

When you want to change the established order of operations *or* simply want to make an expression clearer, you should use parentheses or other **grouping symbols.** The most common grouping symbols are parentheses ( ) and brackets [ ]. Braces { } are also sometimes used as grouping symbols.

Expressions within grouping symbols must be evaluated first. Here is an example.

$(3 + 4) \times 2 = 7 \times 2$ *Add within parentheses.*

$= 14$ *Multiply.*

The established rules for order of operations are summarized below.

Order of Operations

To evaluate an expression involving more than one operation, use the following order.

1. First do operations that occur within grouping symbols.
2. Then evaluate powers.
3. Then do multiplications and divisions from left to right.
4. Finally do additions and subtractions from left to right.

Goal 2 Evaluating Expressions with a Calculator

Many calculators use the Order of Operations used in this book, but some do not. Try using your calculator to see whether it gives a result that is listed in the next example.

Technology
Using a Calculator

Example 3 *Order of Operations on a Calculator*

When you enter 6 [+] 10 [÷] 2 [−] 3 [=] on your calculator, does it display 8 or 5? Using the established order of operations, it should display 8.

$$6 + 10 \div 2 - 3 = 6 + 5 - 3 = 11 - 3 = 8$$

If it displays 5, it would have performed the operations as

$$[(6 + 10) \div 2] - 3 = [16 \div 2] - 3 = 8 - 3 = 5.$$

Example 4 *Finding an Area of a Region*

Write an expression that represents the area of the region at the left. Then evaluate the expression.

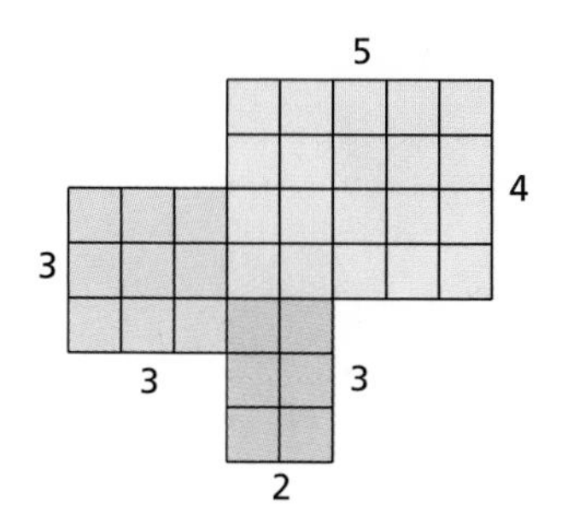

Solution The region at the left can be divided into two rectangles and one square.

Verbal Model Area of region = Area of rectangle + Area of rectangle + Area of square

$$\text{Area} = 2 \times 3 + 4 \times 5 + 3^2$$
$$= 6 + 20 + 9$$
$$= 35 \text{ square units}$$

■

Communicating about MATHEMATICS

Cooperative Learning

SHARING IDEAS about the Lesson

Using Multiplication Models Work with a partner. Copy the figure in Example 4 on graph or dot paper. Divide it to form a 2-by-7 rectangle, a 3-by-4 rectangle, and a 3-by-3 square. Then write an expression for the area and evaluate the expression. Do you obtain the same result as in Example 4? Repeat this process with other regions. What can you conclude?

EXERCISES

Guided Practice

CHECK for Understanding

1. Why is it important to learn to communicate mathematics? Give a real-life example of communicating mathematics.
2. State the established order of operations. Why is it important to have an established order of operations?
3. Evaluate the expression.

   a. $18 - 4 \times 3$ **b.** $48 \div 6 \times 3$ **c.** $12 + 4^2 - 3 \times (5 - 2)$

4. *Reasoning* Insert parentheses in order to make the number sentence true.

   a. $3 \times 4 + 8 - 2 = 34$ **b.** $7 - 3 \div 2 \times 8 + 2 = 20$

Independent Practice

Computation Sense **In Exercises 5–18, evaluate the expression without using a calculator.**

5. $7 + 12 \div 6$
6. $12 - 3 \times 4$
7. $5 \cdot 3 + 2^2$
8. $5^2 - 8 \div 2$
9. $11 + 4 \div 2 \times 9$
10. $21 - 1 \cdot 2 \div 4$
11. $8 + 14 - 2 + 4 \cdot 2^3$
12. $30 - 3^3 + 8 \cdot 3 \div 12$
13. $(9 + 7) \div 4 \times 2$
14. $6 \div (17 - 11) \cdot 14$
15. $4 \cdot (5)(5) + 13$
16. $16 \div 4(2) \times 9$
17. $3[16 - (3 + 7) \div 5]$
18. $[1 + 3(9 + 12)] - 4^3$

In Exercises 19–26, use a calculator to evaluate the expression.

19. $29 + 16 \div 8 \cdot 25$
20. $36 + 16 - 50 \div 25$
21. $18 \cdot 3 \div 3^3$
22. $10 + 5^3 - 25$
23. $149 - (2^8 - 40) \div 6$
24. $20 - (3^5 \div 27) \cdot 2$
25. $22 + (34 \cdot 2)^2 \div 8 + 59$
26. $85 - (4 \cdot 2)^2 - 3 \cdot 7$

Reasoning **In Exercises 27–34, decide whether the number sentence is true or false according to the established order of operations. If it is false, insert parentheses to make it true.**

27. $6 + 21 \div 3 = 9$
28. $21 - 8 \cdot 2 = 26$
29. $6 \cdot 3 - 2 \cdot 5 = 8$
30. $24 - 12 - 4 \cdot 2 = 8$
31. $6 + 3^2 \div 3 = 5$
32. $8^2 - 1 \cdot 3 - 5 = 56$
33. $7 + 7 \cdot 2 + 6 = 63$
34. $36 \div 9 - 6 \div 2 = 6$

In Exercises 35–38, write a numerical expression for the phrase. Then evaluate your expression.

35. 9 added to the quotient of 24 and 8

36. 6 minus the difference of 18 and 15

37. 36 divided by the sum of 6 and 12

38. 42 divided by the product of 1 and 7

39. *School Clothes* You go to the mall to shop for school clothes. You purchase 2 pairs of jeans for $25 each, 3 shirts for $20 each, and 2 pairs of shoes for $25 each. Write an expression that represents your total cost. How much money did you spend?

40. *School Supplies* You go to your school supply store to buy school supplies. You have $40. You purchase 5 folders for $1 each, 3 notebooks for $1.20 each, 2 packs of pencils for $.80 each, and a calculator for $21.50. All prices include tax. Write an expression that represents the amount of money you have left. How much money do you have left?

Alberta, Canada ***In 1993, the world's largest mall was the West Edmonton Mall in Alberta, Canada. It has 839 stores, including 11 major department stores.***

Integrated Review

Making Connections within Mathematics

Geometry **In Exercises 41 and 42, write an expression that represents the area of the region. Then evaluate the expression.**

41.

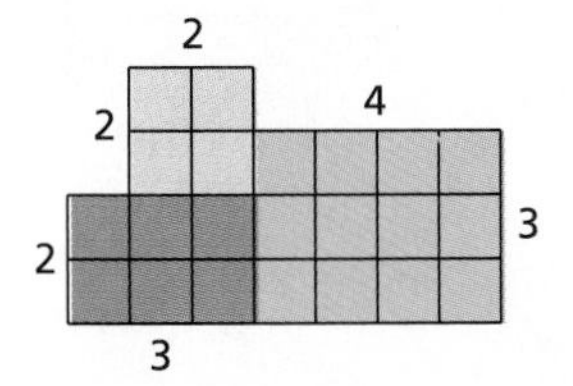

42.

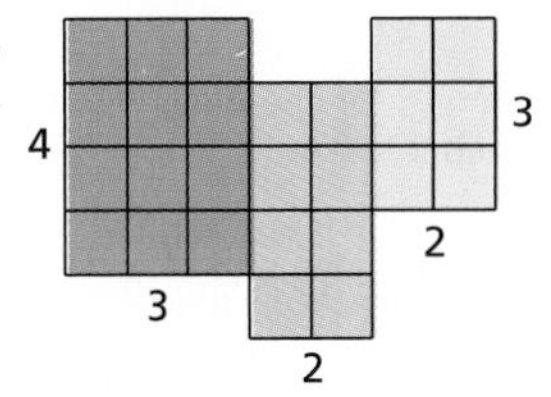

Exploration and Extension

Twenty-Four Game **In Exercises 43–46, use the numbers to play a game of 24. The object of this game is to use the established order of operations to write an expression whose value is 24. You can reorder the numbers, but you cannot use grouping symbols. For instance, the numbers 3, 8, 9, and 9 can be written as $8 \cdot 3 + 9 - 9 = 24$.**

43. 1, 8, 4, 8

44. 1, 5, 9, 2

45. 3, 6, 2, 3

46. 2, 7, 8, 8

47. Play the Twenty-Four Game with a friend. Each person should create 4 sets of 4 numbers. (You must be able to get 24 with each set.) Then exchange sets, write the expressions, and check your results.

Mid-Chapter SELF-TEST

Take this test as you would take a test in class. The answers to the exercises are given in the back of the book.

1. List several ways that numbers are used to describe objects in real life. **(1.1)**

In Exercises 2–5, describe the pattern. Then list the next 3 numbers. (1.1)

2. 3, 6, 9, ?, ?, ?

3. 90, 81, 72, ?, ?, ?

4. 1, 4, 9, ?, ?, ?

5. $1, \frac{1}{2}, \frac{1}{3},$?, ?, ?

In Exercises 6 and 7, draw the next three figures in the pattern. (1.1)

6.

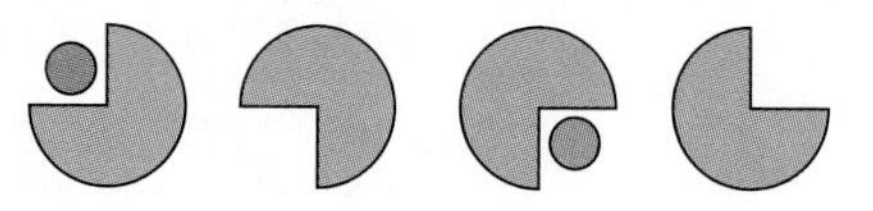

7.

In Exercises 8–13, write a verbal description of the expression and evaluate the expression. (1.2, 1.3, 1.4)

8. 12×4

9. $(176)(12)$

10. $15 \div 3$

11. $\frac{369}{41}$

12. 3^3

13. $\sqrt{256}$

In Exercises 14 and 15, imagine that you are in a music store and decide to buy 2 CDs and 3 cassettes. Each CD costs $14 and each cassette costs $8 including tax. (1.4)

14. How much money did you spend?

15. You brought $70 to the music store. After buying the CDs and cassettes and lending $6 to a friend, how much will you have left?

In Exercises 16–18, consider a square checkerboard that has an area of 400 square inches. (1.4)

16. What is the length of each side of the checkerboard?

17. The checkerboard has 8 small squares on each side. What are the dimensions of each small square?

18. How many small squares are on the checkerboard? What is the area of each?

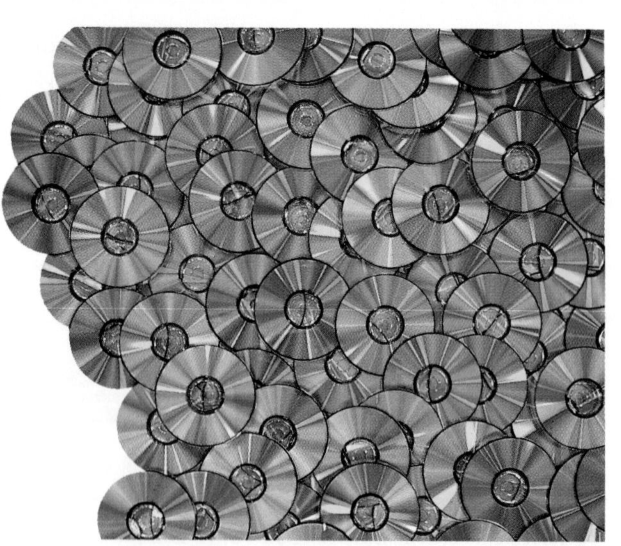

In 1991, consumers in the United States spent about $5.3 billion on CDs and about $4 billion on cassettes. (Source: Recording Industry Association of America)

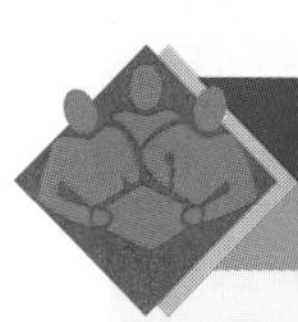

LESSON INVESTIGATION 1.5

Finding Patterns

Materials Needed: Pencil, graph paper, or dot paper

Example

The 5 figures at the right form a pattern.

a. Find the perimeter of each figure.

b. Describe the 6th figure. Without drawing it, predict its perimeter. Then draw the figure to confirm your result.

c. What is the perimeter of the 41st figure?

d. Explain how the perimeter of each figure is related to its figure number.

1 1
Figure 1

Figure 2

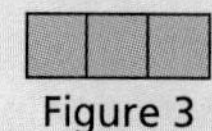
Figure 3

Figure 4

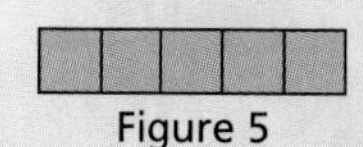
Figure 5

Solution

a. The perimeters of the figures are as follows:

Figure	1	2	3	4	5
Perimeter	4	6	8	10	12
Pattern	$2 + 2(1)$	$2 + 2(2)$	$2 + 2(3)$	$2 + 2(4)$	$2 + 2(5)$

b. The sixth figure is a 1-by-6 rectangle. From the pattern in the table, its perimeter should be $2 + 2(6)$ or 14. The 6th figure is shown at the right. Notice that its perimeter is $1 + 6 + 1 + 6$ or 14.

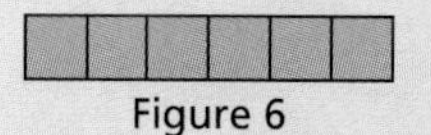
Figure 6

c. From the pattern in the table, the perimeter of the 41st figure is $2 + 2(41)$ or 84.

d. To find the perimeter of any figure, multiply the figure number by 2 and add 2.

Exercises

In Exercises 1 and 2, repeat the steps described in the Example.

1. 1 1 1

Figure 1

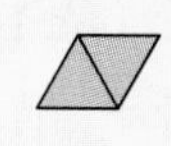
Figure 2

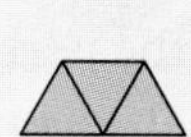
Figure 3

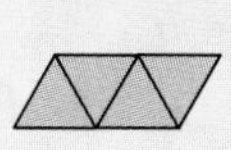
Figure 4

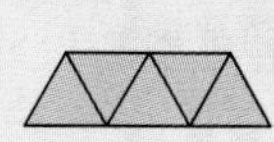
Figure 5

2.

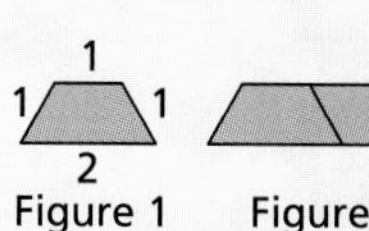

Figure 1 Figure 2

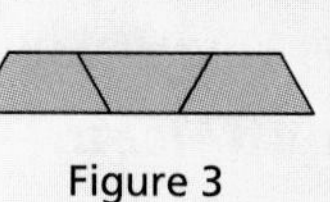
Figure 3

Figure 4

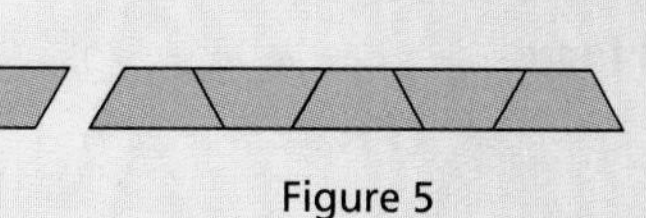
Figure 5

1.5 Introduction to Algebra: Variables in Expressions

What you should learn:

Goal 1 How to evaluate variable expressions

Goal 2 How to use variable expressions in formulas to model real-life situations

Why you should learn it:

Learning to use variable expressions is a key part of learning algebra. Variable expressions allow you to model real-life situations, such as finding the distance traveled by a moon rover.

Goal 1 Evaluating Variable Expressions

A **variable** is a letter that is used to represent one or more numbers. The numbers are the **values of the variable.** An **algebraic expression** is a collection of numbers, variables, operations, and grouping symbols. Here are some examples.

Algebraic Expression	Meaning
$5n$	5 times n
$4x^2$	4 times the square of x
$2a + bc$	2 times a, plus b times c

Replacing each variable in an algebraic expression by a number is called **substituting** in the expression. The number obtained by simplifying the expression is the **value of the expression.** To **evaluate** an algebraic expression, use the following flowchart.

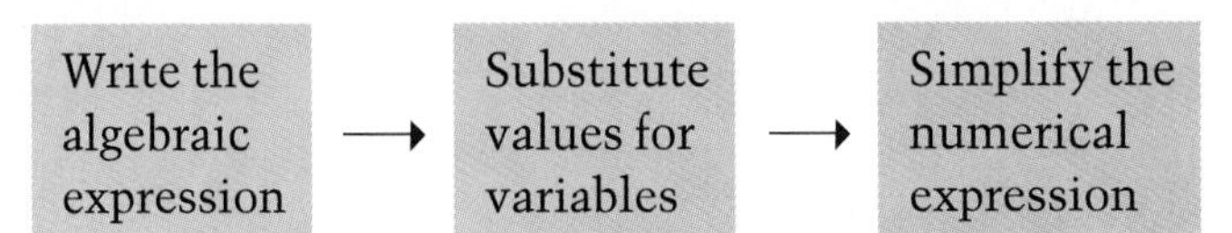

Write the algebraic expression → Substitute values for variables → Simplify the numerical expression

Parts of expressions have special names.

Sum: $4x$ and 5 are the **terms** of $4x + 5$.
Product: 7 and n are the **factors** of $7n$.
Quotient: $3a$ is the **numerator** and b is the **denominator** of $\frac{3a}{b}$.

Need to Know

When substituting a number for a variable, you must replace each occurrence of the variable in the expression. For example, you can evaluate $3x + 2x^2$ when $x = 4$ as follows.

Expression
$3x + 2x^2$
↓
Substitute 4 for x.
$3(4) + 2(4^2)$
↓
Simplify.
$12 + 32 = 44$

Example 1 *Evaluating an Algebraic Expression*

The expression $2 + 2n$ represents the perimeter of a 1-by-n rectangle. (See the Lesson Investigation on page 23.) Evaluate this expression when $n = 37$. What does the result represent?

Solution To evaluate the expression, substitute 37 for n.

$2 + 2n$	*Write the expression.*
$= 2 + 2(37)$	*Substitute 37 for n.*
$= 2 + 74$	*Simplify 2(37).*
$= 76$	*Value of the expression*

The result represents the perimeter of a 1-by-37 rectangle. ■

Goal 2 Modeling Real-Life with Formulas

Algebraic expressions are often used to model real-life quantities. Here are two examples.

Quantity	Labels	Formula
Area of a Rectangle	A = area l = length w = width	$A = lw$
Distance	d = distance r = rate or speed t = time	$d = rt$

Each of these **formulas** (or algebraic models) can be written as a **verbal model.** For instance, $A = lw$ can be written as "the area of a rectangle is the product of its length and width."

Problem Solving
Verbal Model

Example 2 *Finding a Distance*

As an astronaut on the moon, you are driving your rover at a speed of 18 miles per hour. How far will you travel in 75 minutes?

Solution Because the speed is stated in miles per *hour,* you should begin by writing 75 minutes as 1.25 hours.

Verbal Model Distance = Rate • Time

Distance = d (miles)
Speed = r (miles per hour)
Time = t (hours)

$d = r \cdot t$ *Write algebraic model.*
$= (18)(1.25)$ *Substitute for r and t.*
$= 22.5$ *Simplify.*

In 75 minutes, you will travel 22.5 miles. ■

Communicating about MATHEMATICS

SHARING IDEAS about the Lesson

Extending the Example Repeat Example 2 for the following speeds and times. Describe the process you used.

A. 16 mph for 90 min. **B.** 22 mph for 50 min.

EXERCISES

Guided Practice

CHECK for Understanding

1. What is an algebraic expression? Give an example and state the expression in words.

2. What does it mean to *evaluate* an expression?

In Exercises 3–6, using $4 + n$ with $n = 3$, identify the following.

3. The variable

4. The value of the variable

5. The expression

6. The value of the expression

In Exercises 7 and 8, evaluate the expression for $a = 5$ and $b = 3$.

7. $3a + b$

8. $(2b^2 + 7) \div a$

Independent Practice

In Exercises 9–20, evaluate the expression for $x = 4$.

9. $5 + x$ **10.** $32 \div x$ **11.** $12x$ **12.** $x \cdot 3x$

13. $3x^2 + 9$ **14.** $2x^2 \cdot 3x$ **15.** $(x + 3)6$ **16.** $(x - 2) \div 4$

17. $(9 - x)^2$ **18.** $(7 - x)^3$ **19.** $(8 - x + 8) \div x$ **20.** $x^2 - 3 \cdot x$

In Exercises 21–32, evaluate the expression for $a = 2$ and $b = 7$.

21. $b - a$ **22.** ab **23.** $3b - a$ **24.** $5a + 2b$

25. $3a^2 \cdot b$ **26.** $(4b) \div (2a)$ **27.** $(24a - 6) \div b$ **28.** $b(9 - a)$

29. $(b - a)^3$ **30.** $(5 + b)^2 + a$

31. $(a + b) \div (b - 2a)$ **32.** $6(b - a) \div (3a)$

In Exercises 33–36, evaluate the expression for $x = 5$, $y = 8$, and $z = 9$.

33. $x + y - z$ **34.** $x + (z - y)$ **35.** $z \div (y - x) + z$ **36.** $y(z - x) + x$

In Exercises 37–40, match the algebraic expression with its verbal description.

a. $a + 8$ **b.** $x - 8$ **c.** $n \div 8$ **d.** $8y$

37. The difference of a number and 8

38. The product of a number and 8

39. The quotient of a number and 8

40. The sum of a number and 8

41. *Reasoning* If the expression $4x + 2$ has a value of 18, what is the value of x?

42. *Reasoning* If the expression $5n - 3$ has a value of 47, what is the value of n?

43. *Pteranodon* You are driving to Wild Animal Park, near San Diego, California. You see the sign shown below.

a. What does 22 miles represent? Does it mean that the driving distance to Wild Animal Park is 22 miles? Explain.

b. Suppose it takes a Pteranodon 120 minutes to reach Wild Animal Park. How fast would the Pteranodon be flying?

Pteranodons, flying reptiles, became extinct millions of years ago. Most species of Pteranodons had wingspans that ranged from 36 to 40 feet.

Moon Rover **In Exercises 44 and 45, imagine that you are an astronaut driving a moon rover.**

44. You travel at a speed of 20 miles per hour. How many minutes does it take you to travel 30 miles?

45. You travel at a speed of 25 miles per hour. How far will you travel in 105 minutes?

46. *Going to a Movie* You are treating some friends to a movie and popcorn. Tickets cost \$6.00 per person. Popcorn costs \$2.00 per bag. Let f represent the number of friends.

a. Write an expression for the total amount you will spend.

b. If $f = 2$, how much did it cost to treat you and your friends?

Integrated Review

Making Connections within Mathematics

Computation Sense **In Exercises 47–54, evaluate the expression.**

47. $6(5 - 3)$
48. $24 \div (8 - 4)$
49. $(9 + 7) - 3^2$
50. $4^3 - (8 + 9)$
51. $1 + 11 \cdot 3^2$
52. $12 - 2^2 \cdot 3$
53. $6(2 \div 4)$
54. $8(6 \div 8) + 5$

Exploration and Extension

Guess, Check, and Revise **In Exercises 55–60, find a value of the variable so that the values of the expressions are the same.**

55. $2x$ and $2 + x$
56. $4n + 6$ and $6n$
57. $8m - 3$ and $6m + 1$
58. $x + 4$ and $2x + 1$
59. $4n - 8$ and $2n - 4$
60. $4y + 3$ and $8y - 9$

1.6 Exploring Data: Tables and Graphs

What you should learn:

Goal 1 How to use tables to organize data

Goal 2 How to use graphs to model data visually

Why you should learn it:

You can use tables and graphs to help you see relationships among such data collections as survey results.

Goal 1 Using Tables to Organize Data

The word *data* is plural and it means facts or numbers that describe something. A collection of data is easier to understand when it is organized in a table or graph. There is no "best way" to organize data, but there are many good ways. One way to organize data is with a table. For instance, on page 23 the information about the perimeters of several rectangles was organized in a table. In that table, suppose you represent the figure number as n. Then the table could look like this.

Figure number, n	1	2	3	4	5
Perimeter, $2 + 2n$	4	6	8	10	12

Example 1 Constructing a Table

Real Life
Football

From 1985 through 1994, the winning and losing scores at the Super Bowl were as follows: ***(Source: The Sporting News)***

38 to 16 (1985), 46 to 10 (1986), 39 to 20 (1987),
42 to 10 (1988), 20 to 16 (1989), 55 to 10 (1990),
20 to 19 (1991), 37 to 24 (1992), 52 to 17 (1993),
30 to 13 (1994)

Represent this data by a table.

Solution One way to represent the data by a table is shown below. Can you think of another way to represent the data?

Year	1985	1986	1987	1988	1989
Winning Score	38	46	39	42	20
Losing Score	16	10	20	10	16

Year	1990	1991	1992	1993	1994
Winning Score	55	20	37	52	30
Losing Score	10	19	24	17	13

Notice how the table helps to organize the data. ■

The Super Bowl is a football game between the winning teams of the NFC and AFC. The first Super Bowl was held in 1967.

Goal 2 Using Graphs to Organize Data

There are many ways to organize data graphically. Two types of graphs are shown in the next example.

Example 2 *Drawing Graphs*

Draw a bar graph and a line graph that represent the Super Bowl data given in Example 1.

Solution

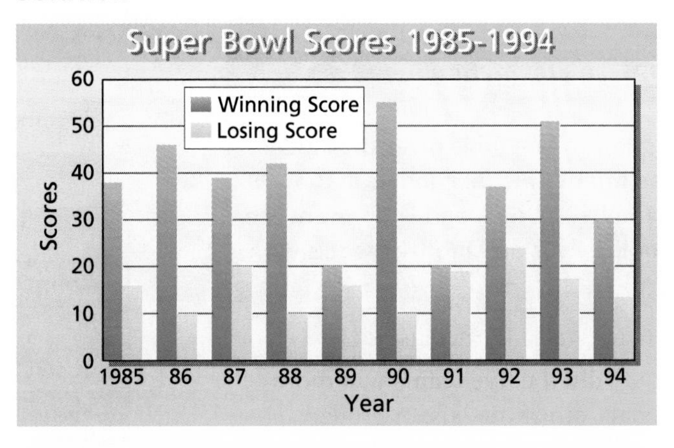

■

Communicating about MATHEMATICS

SHARING IDEAS about the Lesson

Its Up to You Which Super Bowl in Example 2 do you think was the most exciting? Why? How are the two graphs similar? Different?

EXERCISES

Guided Practice

CHECK for Understanding

1. Name some ways that you can organize data.
2. How could you organize the following real-life data?
   a. The number of students eating school lunches each day last week
   b. The number of hours all students spent studying last night

Independent Practice

Olympic Medals **The table gives the medal count for the top five medal winners in the 1992 Summer Olympics. Let *G, S,* and *B* represent the number of gold, silver, and bronze medals won by a team.**

1992 Summer Olympics

Country	Gold	Silver	Bronze
Unified Team	45	38	29
United States	37	34	37
Germany	33	21	28
China	16	22	16
Cuba	14	6	11

3. Which team won 112 medals?
4. As a group, did the five teams win more gold medals, silver medals, or bronze medals?
5. For which team is it true that $G - 5 = B$?
6. For which team is it true that $B - 5 = S$?
7. Draw a bar graph or a line graph to represent the number of gold and silver medals won by each team. Which graph do you think is more appropriate? Explain.

Javelin Throw **The bar graph shows the winning distances of the javelin throw in the Olympics.**

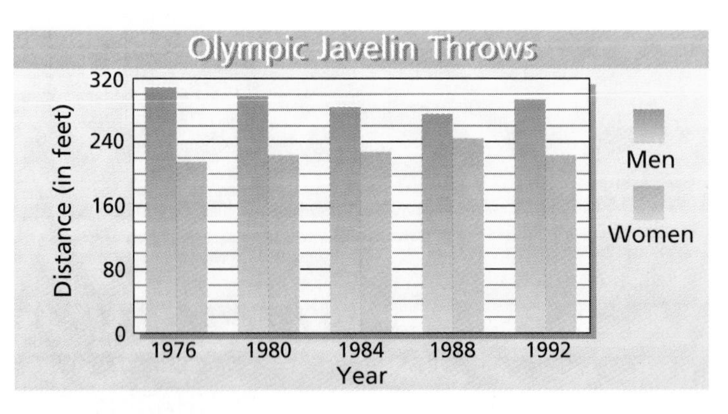

8. How far did the men's champion throw the javelin in 1980?
9. In what year was the difference between the men's and women's throws the least? The greatest?
10. Would a table better represent this data? Explain why or why not.
11. *Perimeter and Area* The side lengths of seven squares are 1, 2, 3, 4, 5, 6, and 7. Create a table that shows the perimeter and area of each square.
12. *Perimeter and Area* The widths and lengths of seven rectangles are n and $n + 2$, where n is equal to 1, 2, 3, 4, 5, 6, and 7. Create a table that shows the perimeter and area of each rectangle.

13. *Average Temperature* The average monthly temperatures (in degrees Fahrenheit) for Cleveland, Ohio, (C) and Seattle, Washington, (S) are listed in the table. Represent this data with a bar graph and a line graph. ***(Source: U. S. National Oceanic and Atmospheric Administration)***

Month	J	F	M	A	M	J
C	33°	35°	45°	58°	69°	78°
S	45°	50°	53°	58°	65°	69°

Month	J	A	S	O	N	D
C	82°	80°	74°	63°	49°	38°
S	75°	74°	69°	60°	51°	47°

The greatest recorded temperature variation in a 24-hour period occurred in Browning, Montana, in 1916. In January, the temperature fell 100°F from 44°F to −56°F.

14. *Estimation* Use the table in Exercise 13 to estimate the average annual temperature in Cleveland. Explain how you obtained your estimate.

15. *Estimation* Use the table in Exercise 13 to estimate the average annual temperature in Seattle. Explain how you obtained your estimate.

16. *It's Up to You* Explain what the line graph at the right could represent.

17. *Writing* Describe some advantages and disadvantages of organizing data with a table and with a graph.

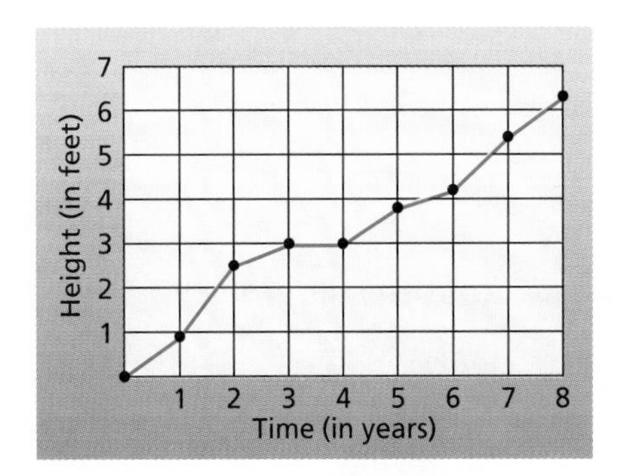

Integrated Review

Making Connections within Mathematics

Decimal Sense **In Exercises 18–23, find the difference.**

18. $107.2 - 84.3$

19. $256.8 - 174.1$

20. $4.28 - 3.69$

21. $8.14 - 5.47$

22. $14.823 - 11.602$

23. $72.073 - 56.941$

Exploration and Extension

24. *Research Project* Use your school's library or some other reference source to collect data on a topic. You can research any topic you want. After collecting your data, organize it in a table, bar graph, or line graph.

25. *Writing* Write a short paper discussing your research.

Mixed REVIEW

In Exercises 1–6, perform the operation. Round the result to two decimal places. (1.2, 1.3)

1. $6(14.2)$ **2.** $8.45 \cdot 5$ **3.** $13 \div 9$

4. $19/2$ **5.** $\frac{6.47}{5}$ **6.** $\sqrt{155}$

In Exercises 7–10, use the established order of operations to evaluate $72 - (8 + 4) \times 6$. (1.3)

7. Which operation is performed first?

8. Which operation is performed second?

9. Which operation is performed third?

10. What is the value of the expression?

In Exercises 11–18, evaluate the expression for $x = 4$, $y = 5$, and $z = 2$. (1.4)

11. $2y$ **12.** $4z - x$ **13.** $(x + y)^2$ **14.** $5x(x - 4)$

15. $x + y \times z$ **16.** $2x \div z^2$ **17.** $(y - z)^2 - x$ **18.** $(y - x) \div z$

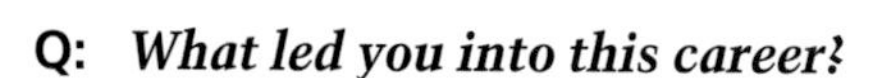

Robotics Engineer

Larry Chao-Hsiung Li is a robotics engineer for NASA (the National Aeronautics and Space Administration). He works in a laboratory designing robots for use in current and future space missions.

Q: ***What led you into this career?***

A: In the summer between my junior and senior years of college, I started working for NASA as part of a cooperative education program with the NASA Johnson Space Center in Houston, TX. After I graduated, I started working there full time and I have been there ever since.

Q: ***Do you use geometric reasoning in your work?***

A: Yes, if we want a robot to move its arm to a certain spot, we have to change the x-, y-, and z-coordinates into angles of joint motor rotation and amounts of movement. This requires using sines, cosines, arctangents, and many other trig functions.

Q: ***What would you like to tell kids who are in school about math or geometry?***

A: If you want to work in a really fun job you need to learn math! Math is really being used today!

Perimeter and Area

Materials Needed: Square tiles or graph paper

Example ***Comparing Area and Perimeter***

How many different rectangular shapes can be formed with 12 square tiles? Form each rectangle and record its dimensions, area, and perimeter in a table.

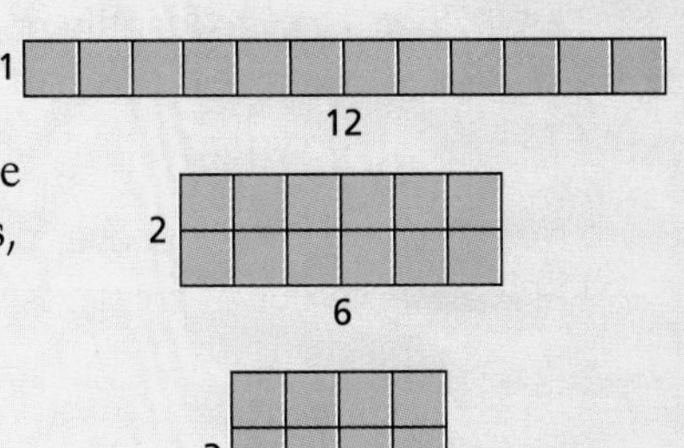

Solution In the figure at the right, you can see that there are only three different shapes of rectangles that can be formed. The dimensions, areas, and perimeters of these rectangles are shown in the following table.

Dimensions	1-by-12	2-by-6	3-by-4
Area	12 square units	12 square units	12 square units
Perimeter	26 units	16 units	14 units

■

Exercises

In Exercises 1–6, decide how many different rectangular shapes can be formed with the tiles. Record the dimensions, area, and perimeter of each rectangle in a table.

1. 20 tiles **2.** 24 tiles **3.** 25 tiles

4. 18 tiles **5.** 36 tiles **6.** 37 tiles

7. If you double the number of tiles, do you double the number of rectangles that can be formed? Explain.

8. How should you arrange a given number of tiles to obtain the greatest perimeter? The least perimeter?

In Exercises 9 and 10, write the number sentence that is suggested by the geometric model. For instance, the model shown below suggests the number sentence $4 \times (4 + 2) = 4 \times 4 + 4 \times 2$.

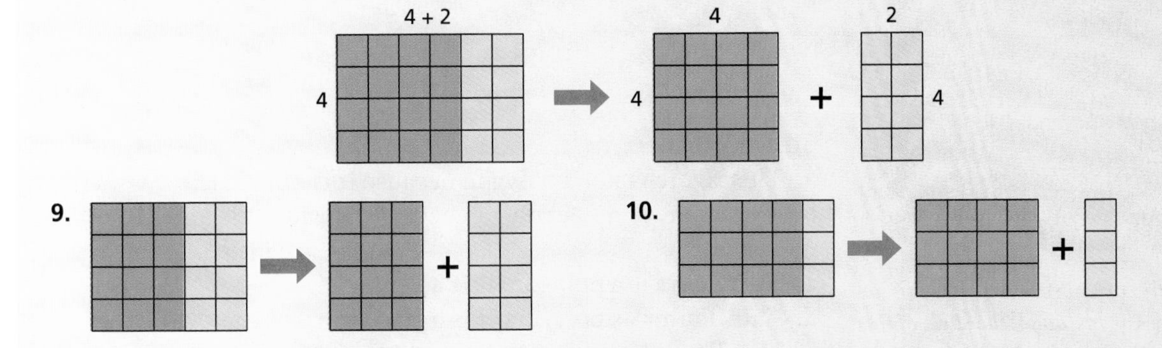

1.7 Exploring Patterns in Geometry

What you should learn:

Goal 1 How to identify polygons and parts of polygons

Goal 2 How to discover properties of polygons

Why you should learn it:

Knowing the names of geometric figures, such as polygons, helps you communicate mathematical ideas.

Goal 1 Identifying Polygons

Geometry is the study of shapes and their measures. One of the most common geometric shapes is a polygon.

> **Polygons**
>
> A **polygon** is a closed figure that is made up of straight line segments that intersect at their endpoints. Each line segment is a **side** of the polygon, and each endpoint is a **vertex.** (The plural of vertex is *vertices.*) A polygon has the same number of vertices as it has sides. Polygons have special names, depending on their number of sides.
>
> 3 : Triangle
> 4 : Quadrilateral
> 5 : Pentagon
> 6 : Hexagon
> 7 : Heptagon
> 8 : Octagon
> 9 : Nonagon
> 10 : Decagon
> n : n-gon

Polygons occur frequently in nature. For instance, the shell of this desert tortoise has pentagons as part of its pattern.

Example 1 *Identifying Polygons*

State whether the figure is a polygon. If it is, name it.

a. 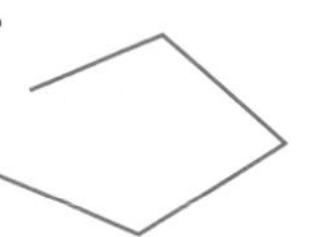

b. 

c. 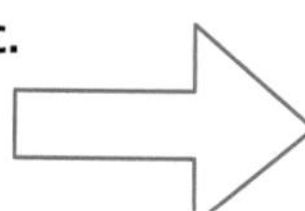

d. 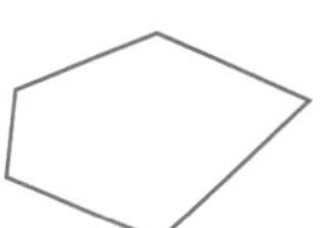

e. 

f.

Solution

a. This figure is not a polygon because it is not closed.
b. This figure is not a polygon because its sides are not all straight.
c. This figure is a polygon. It is a heptagon.
d. This figure is a polygon. It is a pentagon.
e. This figure is a polygon. It is a hexagon.
f. This figure is a polygon. It is a quadrilateral. ■

Goal 2 Discovering Properties of Polygons

A segment that connects two vertices of a polygon and is not a side is a **diagonal** of the polygon.

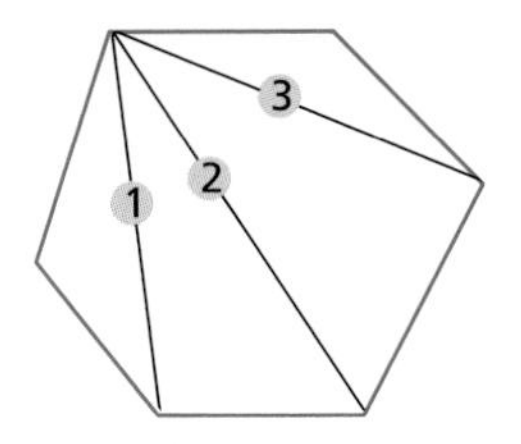

From each vertex of a hexagon you can draw 3 diagonals.

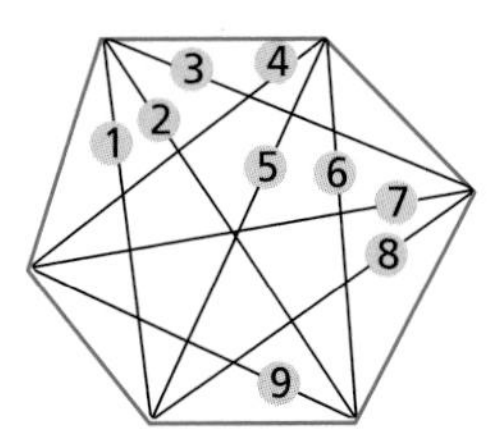

In a hexagon, you can draw a total of 9 different diagonals.

LESSON INVESTIGATION

Investigating Diagonals of Polygons

Group Activity Determine the number of diagonals that can be drawn from each vertex of a triangle, a quadrilateral, and a pentagon. Then determine the total number of diagonals that can be drawn in each polygon. Record your results in a table. What patterns do you observe?

Type of Polygon	n	Diagonals from Each Vertex	Total Number of Diagonals
Triangle	3	?	?
Quadrilateral	4	?	?
Pentagon	5	?	?

Example 2 *Problem Solving: Verbal and Algebraic Models*

In the investigation, you may have discovered the following.

Verbal Model

Total diagonals = Number of vertices × Diagonals from each vertex ÷ 2

Total number of diagonals = T
Number of vertices = n
Number of diagonals from each vertex = $n - 3$

Algebraic Model

$T = n \times (n - 3) \div 2 = \frac{n(n-3)}{2}$ ■

Communicating about MATHEMATICS

SHARING IDEAS about the Lesson

Using an Algebraic Model Use the algebraic model in Example 2 to find the total number of diagonals of a heptagon and an octagon. Confirm your results by drawing a diagram.

EXERCISES

Guided Practice

CHECK for Understanding

1. In your own words, state the definition of a polygon.
2. State the name of each polygon in GEOMETRY.
3. How many diagonals are there in the "O" of GEOMETRY?
4. *Problem Solving* Give an example of a verbal model and an algebraic model.

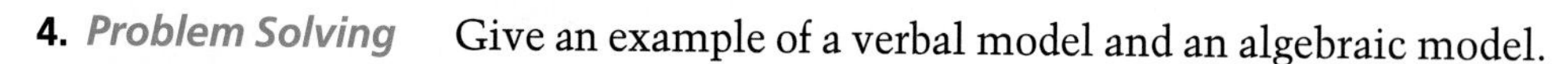

Independent Practice

In Exercises 5–8, name the polygon.

5. 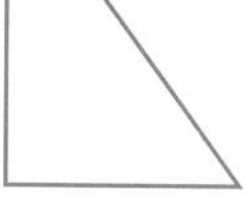

6. 

7. 

8. 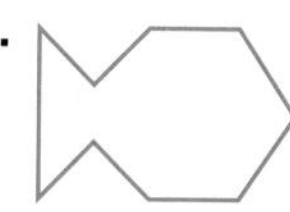

In Exercises 9–12, decide whether the figure is a polygon. If it is, name it. If it is not, explain why.

9. 

10. 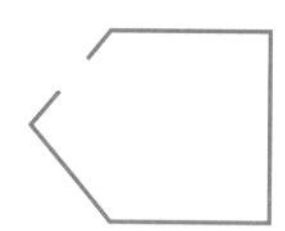

11. 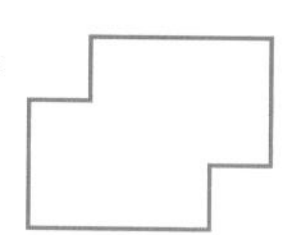

12. 

13. Draw a hexagon with 2 sides equal in measure.
14. Draw an octagon with 4 sides equal to one length and the remaining 4 sides equal to a different length.

Interior Angles **In Exercises 15–18, a polygon is regular if all its side lengths are equal and all its interior angle measures are equal. Each polygon shown at the right is regular.**

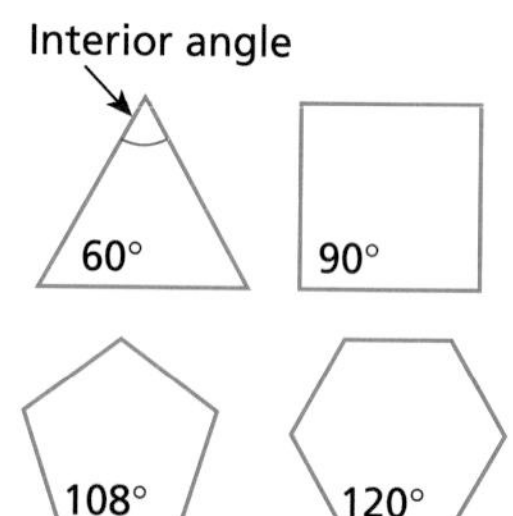

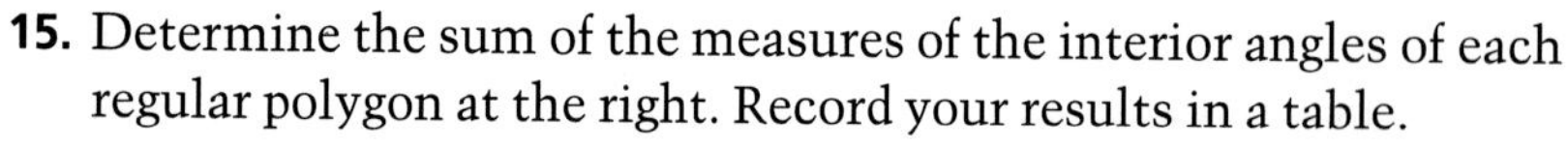

15. Determine the sum of the measures of the interior angles of each regular polygon at the right. Record your results in a table.
16. One of the algebraic models below is a formula for determining the measure of an interior angle of a regular n-gon. Based on your results in Exercise 15, which one is it?

    a. $I = \frac{180°}{n}$ **b.** $I = \frac{360°}{n}$ **c.** $I = \frac{(n-2)(180°)}{n}$ **d.** $I = \frac{(n+1)(180°)}{12}$

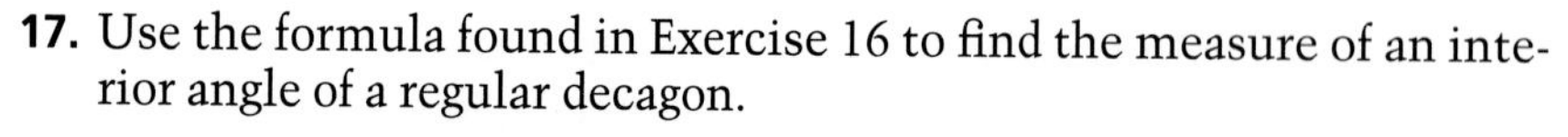

17. Use the formula found in Exercise 16 to find the measure of an interior angle of a regular decagon.
18. Use the formula found in Exercise 16 to find the measure of an interior angle of a regular octagon.

Area of a Triangle **The table below lists the base, height, and area of several triangles. Use the table to answer Exercises 19 and 20.**

Triangle	Base	Height	(Base)(Height)	Area
1	6	4	24	12
2	3	2	6	3
3	10	5	50	25

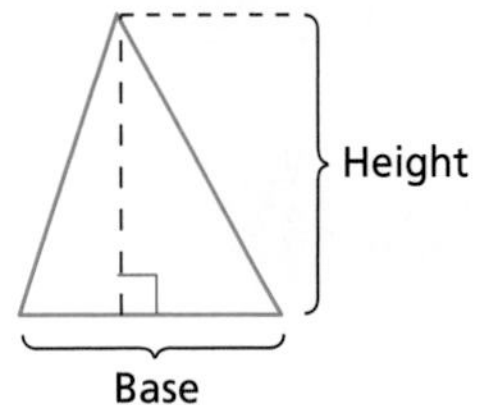

19. Write a verbal model and an algebraic model for the area of a triangle.

20. Find the area of a triangle whose base is 16 inches and height is 8 inches.

Reasoning **In Exercises 21–24, decide whether the outer edge of the sign is a polygon. If it is not, explain why it isn't.**

21. 

22. 

23. 

24. 

Integrated Review

Making Connections within Mathematics

Perimeter **In Exercises 25 and 26, find the perimeter of the polygon when $a = 9$.**

25. 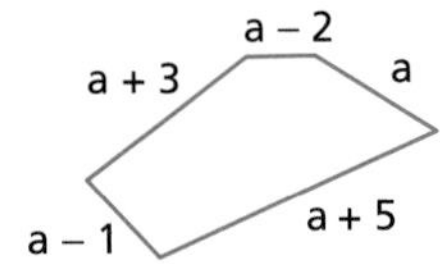

26. 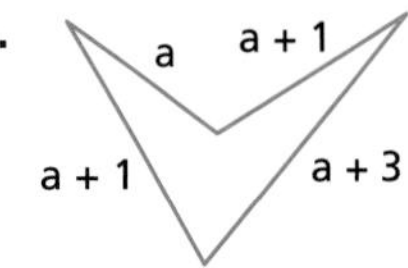

Language Skills **In Exercises 27–32, match each word with its definition.**

a. Government by 7 persons
b. Being between 90 and 100 years old
c. A period of 4 years
d. An athletic contest consisting of 5 events
e. Occurring every 8 years
f. A period of 10 years

27. Quadrennium
28. Octennial
29. Pentathlon
30. Decade
31. Heptarchy
32. Nonagenarian

Exploration and Extension

Convex or Nonconvex **A polygon is convex if a segment joining any two interior points lies completely within the polygon. In Exercises 33–36, decide whether the polygon is convex.**

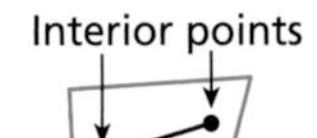

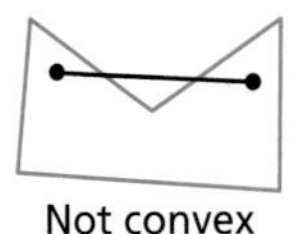

33. 

34. 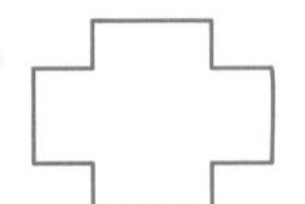

35. 

36. 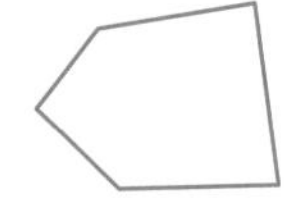

1.8 Exploring Patterns with Technology

What you should learn:

Goal 1 How to use a calculator to discover number patterns

Goal 2 How to use diagrams to discover number patterns in real-life situations

Why you should learn it:

Many real-life situations, such as setting up a carnival game, have patterns that can be described with algebraic models. Knowing the patterns can help you understand the real-life situation.

Goal 1 Discovering Number Patterns

In Lesson 1.7, you learned that polygons have special names, depending on their number of sides. Numbers also have special names, such as *whole numbers* (0, 1, 2, 3, . . .), *natural numbers* (1, 2, 3, 4, . . .), decimal numbers, and fractions.

Sequences of numbers often have patterns. Sometimes, a calculator or a computer can help you discover the pattern.

Example 1 *Finding a Pattern for Products*

Use a calculator to calculate the product of 89 and the first nine natural numbers. Describe the pattern.

Solution You are asked to evaluate $89n$, when n is a natural number from 1 to 9. The results are shown in the table.

n	1	2	3	4	5	6	7	8	9
$89n$	89	178	267	356	445	534	623	712	801

From the table, you can see that the hundreds digit *increases* by 1, and the tens and the units digits *decrease* by 1.

Example 2 *Finding a Pattern for Decimals*

Evaluate fractions of the form $\frac{1}{n}$ as a decimal for several values of n. Does the decimal repeat or terminate? What values of n produce terminating decimals?

Solution

Fraction	$\frac{1}{2}$	$\frac{1}{3}$	$\frac{1}{4}$	$\frac{1}{5}$	$\frac{1}{6}$	$\frac{1}{7}$	$\frac{1}{8}$
Decimal	0.5	$0.\overline{3}$	0.25	0.2	$0.1\overline{6}$	$0.\overline{142857}$	0.125

From the table, it appears that the fraction $\frac{1}{n}$ terminates if n is a product of 2's and 5's. For instance, $8 = 2 \cdot 2 \cdot 2$. This makes sense when you think that the decimal system is based on the number 10, which is the product of 2 and 5. ■

Repeating and Terminating Decimals

The decimals for $\frac{1}{2}$ and $\frac{1}{4}$ terminate.

$\frac{1}{2} = 0.5$, $\frac{1}{4} = 0.25$

The decimals for $\frac{1}{3}$ and $\frac{1}{6}$ repeat.

$\frac{1}{3} = 0.333\ldots$, $\frac{1}{6} = 0.1666\ldots$

Repeating decimals are indicated by a bar, as in $\frac{1}{3} = 0.\overline{3}$.

Draw a diagram that shows the facts from the problem.

↓

Use the diagram to visualize the action of the problem.

↓

Use arithmetic or algebra to find a solution. Then check the solution against the facts.

A 3-row stack of wooden milk bottles has 6 bottles. How many bottles are in an n-row stack?

Goal 2 Number Patterns in Real Life

Sometimes the facts of a problem are easier to understand when you picture them.

Example 3 *Modeling Triangular Numbers*

Your school is having a carnival to raise money for band uniforms. You are in charge of a booth in which people try to knock over a stack of wooden bottles with a baseball. A "3-row" stack has 6 bottles. How many bottles does an "n-row" stack have?

Solution Begin by drawing some examples.

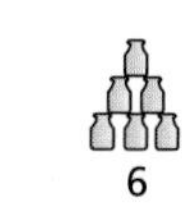

1 3 6 10 15 21

These numbers are called **triangular** numbers. To help discover their pattern, try simplifying the drawing. By doubling the number of squares in each "stack," you can see that the nth triangular number corresponds to half the area of an n-by-$(n + 1)$ rectangle.

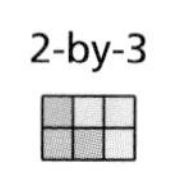
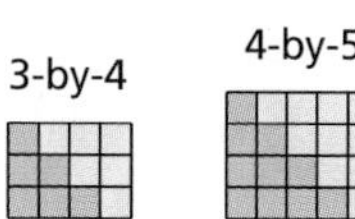
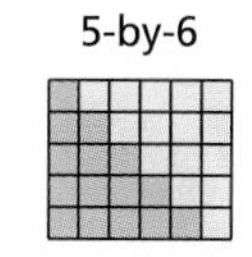
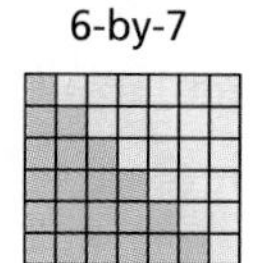

1-by-2 2-by-3 3-by-4 4-by-5 5-by-6 6-by-7

The algebraic model for triangular numbers is as follows.

Verbal Model nth triangular number = Width of rectangle × Length of rectangle ÷ 2

nth triangular number = T

Width of rectangle = n

Length of rectangle = $n + 1$

Algebraic Model $T = n \times (n+1) \div 2 = \frac{n(n+1)}{2}$ ■

Communicating about MATHEMATICS

SHARING IDEAS about the Lesson

Using an Algebraic Model Use the algebraic model in Example 3 to find the 7th and 8th triangular numbers. Confirm your results by drawing diagrams.

EXERCISES

Guided Practice

CHECK for Understanding

1. What is the first whole number? What is the first natural number?
2. Evaluate fractions of the form $\frac{n}{11}$ as a decimal for values of n from 1 through 4. Describe any patterns.

Independent Practice

In Exercises 3–8, create a table showing your calculations.

3. Calculate the quotient of 192 and each of the first 4 natural numbers.
4. Calculate the product of 75 and each of the first 9 whole numbers.
5. Evaluate fractions of the form $\frac{n}{3}$ as a decimal for the values of n from 1 through 9.
6. Evaluate fractions of the form $\frac{2}{n}$ as a decimal for the values of n from 1 through 9. What values of n produce repeating decimals?
7. Calculate $\frac{n^2}{2}$ for the first 7 whole numbers.
8. Calculate $\frac{n}{n+1}$ for the first 5 natural numbers.

In Exercises 9–12, use a calculator to evaluate the expressions. Then describe the pattern.

9.	10.	11.	12.
8(2) + 2	77(1443)	$\frac{1}{5} + \frac{1}{8}(400)$	$\frac{5}{6} + 9\left(\frac{1}{3}\right)$
8(23) + 3	154(1443)	$\frac{2}{5} + \frac{1}{8}(400)$	$\frac{5}{6} + 9\left(\frac{2}{3}\right)$
8(234) + 4	231(1443)	$\frac{3}{5} + \frac{1}{8}(400)$	$\frac{5}{6} + 9\left(\frac{3}{3}\right)$
8(2,345) + 5	308(1443)	$\frac{4}{5} + \frac{1}{8}(400)$	$\frac{5}{6} + 9\left(\frac{4}{3}\right)$
8(23,456) + 6	385(1443)	$\frac{5}{5} + \frac{1}{8}(400)$	$\frac{5}{6} + 9\left(\frac{5}{3}\right)$
8(234,567) + 7	462(1443)	$\frac{6}{5} + \frac{1}{8}(400)$	$\frac{5}{6} + 9\left(\frac{6}{3}\right)$

Constant Function **In Exercises 13–18, write the keystrokes that will produce the sequence on your calculator. Then write the next four numbers in the sequence. For instance, the sequence 3, 5, 7, 9 can be produced on some calculators by these keystrokes. 3 [+] 2 [=] [=] [=]**

13. 1, 9, 17, 25, [?] [?] [?] [?]
14. 100, 87, 74, 61, [?] [?] [?] [?]
15. 5, 15, 45, 135, [?] [?] [?] [?]
16. 1025, 205, 41, 8.2, [?] [?] [?] [?]
17. 1008, 100.8, 10.08, [?] [?] [?] [?]
18. 12.3, 24.6, 49.2, [?] [?] [?] [?]

This symbol indicates exercises where you should choose the appropriate method of calculations: mental math, paper and pencil, or calculator.

Modeling Cubic Numbers **The cubic numbers can be modeled as follows.**

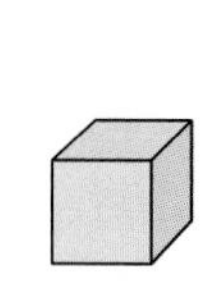
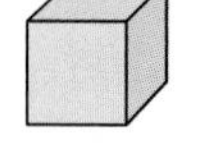
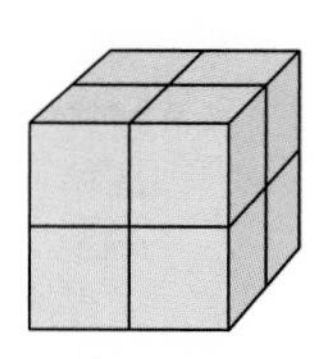
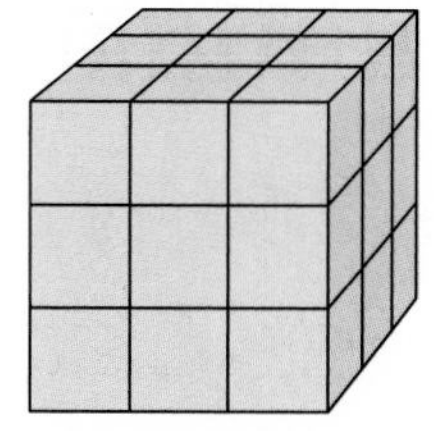

1 cubic unit 8 cubic units 27 cubic units

19. Draw a model for the next cubic number. What is the number?

20. Write a verbal and algebraic model that gives the cubic numbers.

Integrated Review

Making Connections within Mathematics

Sequences **In Exercises 21–28, find the next 4 numbers in the sequence.**

21. 2, 5, 8, 11, [?] [?] [?] [?]

22. 55, 50, 45, 40, [?] [?] [?] [?]

23. 1, 2, 4, 8, [?] [?] [?] [?]

24. 1, 4, 9, 16, [?] [?] [?] [?]

25. $\frac{1}{2}, \frac{2}{3}, \frac{3}{4}, \frac{4}{5},$ [?] [?] [?] [?]

26. $\frac{2}{3}, \frac{2}{4}, \frac{2}{5}, \frac{2}{6},$ [?] [?] [?] [?]

27. 1, 5, 2, 6, [?] [?] [?] [?]

28. 100, 81, 64, 49, [?] [?] [?] [?]

Exploration and Extension

Fibonacci Sequence **The sequence 1, 1, 2, 3, 5, 8, 13, . . . is called a Fibonacci Sequence. It was developed by the Italian mathematician Leonardo Fibonacci (1175–1250).**

29. Describe the pattern of the sequence.

30. State the next three numbers in the sequence.

31. *Group Investigation* In groups of four, you will investigate one of the many special proprieties of the Fibonacci Sequence. To begin, pick any two numbers such that the second is larger than the first. Each member of the group is to develop a Fibonacci-like sequence from the two numbers. Then, using the sequence, each member of the group is to compute the ratio of the consecutive terms and record the results in a table. For example, 12 and 20 would yield 12, 20, 32, 52, ... Describe the pattern.

Term	12	20	32	52
Quotient	$\frac{20}{12} = 1.6$	$\frac{32}{20} = 1.6$	$\frac{52}{32} = 1.625$	$\frac{84}{52} \approx 1.615$

The numbers of overlapping clockwise and counterclockwise spirals in the heads of sunflowers are consecutive terms in the Fibonacci sequence like 34 and 55, or 55 and 89.

Chapter Summary

What did you learn?

Skills

1. Use numbers to identify and measure objects. **(1.1)**
2. Recognize and describe number patterns. **(1.1, 1.8)**
3. Use the four basic number operations. **(1.2)**
   - Use multiplication models. **(1.2)**
4. Use powers and square roots. **(1.3)**
5. Use order of operations. **(1.4)**
   - Use order of operations on a calculator. **(1.4)**
6. Use variables in expressions. **(1.5)**
   - Evaluate algebraic expressions. **(1.5)**
7. Identify polygons and parts of polygons. **(1.7)**

Problem-Solving Strategies

8. Use verbal and algebraic models to solve real-life problems. **(1.5–1.8)**
9. Solve problems by using strategies such as guess, check, and revise and draw a diagram. **(1.8)**

Exploring Data

10. Organize and display data
    - with a table. **(1.6–1.8)**
    - with a graph. **(1.6)**

Why did you learn it?

Number patterns and geometric patterns occur frequently in real-life situations. Often these patterns can be modeled by using algebraic expressions or formulas. Being able to write and use models for real-life situations helps people succeed in their occupations. For instance, accountants use formulas to compute taxes. Medical technicians use formulas to analyze samples. Engineers use formulas to measure the strength of a structure. More importantly, whatever your work turns out to be, you will be able to perform more successfully if you can use formulas to model real life.

How does it fit into the bigger picture of mathematics?

Hundreds of years ago, mathematics was divided into two major branches: algebra and geometry. These branches used to have little overlap—equations and algebraic expressions belonged to algebra, and graphs, curves, and lines belonged to geometry. Today, however, algebra and geometry have a lot of overlap. For instance, you have seen that some geometrical patterns can be described by algebraic expressions.

In this course you will be preparing for future mathematical studies—not only studies in algebra and geometry, but also studies in statistics, probability, and trigonometry.

Chapter REVIEW

Number Patterns **In Exercises 1–6, describe the pattern. Then list the next 3 numbers. (1.1, 1.3)**

1. 15, 30, 45, 60 **2.** 100, 94, 88, 82 **3.** 2, 6, 12, 20

4. 3, 6, 18, 72 **5.** 1, 8, 27, 64 **6.** $1, \sqrt{3}, \sqrt{5}, \sqrt{7}$

Sequences **In Exercises 7 and 8, write the first 6 numbers in the sequence. (1.1)**

7. The first number is 1. Each succeeding number is 9 more than the preceding number.

8. The first two numbers are 1 and 2. Each succeeding number is the product of the two preceding numbers.

Reasoning **In Exercises 9–12, without researching, match the state with its abbreviation. (1.1)**

a. AR **b.** AL **c.** AZ **d.** AK

9. Alaska **10.** Alabama **11.** Arkansas **12.** Arizona

13. Describe how you obtained your answers in Exercises 9–12. **(1.1)**

14. Describe the pattern, $z1, y4, x7, w10, \ldots$. **(1.1)**

15. ***Migration*** Geese usually migrate south for the winter months. A flock of geese migrated 552 miles in 16 hours. How fast did the geese fly? **(1.1, 1.5)**

16. ***Migration*** Ducks, like geese, usually migrate south for the winter months. Ducks fly an average speed of 40 mph. If a flock of ducks migrates 18 hours in a day, how far will they travel? **(1.1, 1.5)**

Snow geese can fly for many hours without stopping. Some flocks have flown 1700 miles in less than $2\frac{1}{2}$ days.

Computation Sense **In Exercises 17–24, evaluate the expression. (1.2)**

17. $21.86 + 53.09$ **18.** 4.28×1.7 **19.** $542 - 291$ **20.** $1146/6$

21. $972 \div 18$ **22.** $\frac{585}{15}$ **23.** $(78)(64)$ **24.** $109 - 33$

Number Sense **In Exercises 25 and 26, write a number sentence for the model. (1.2)**

25.

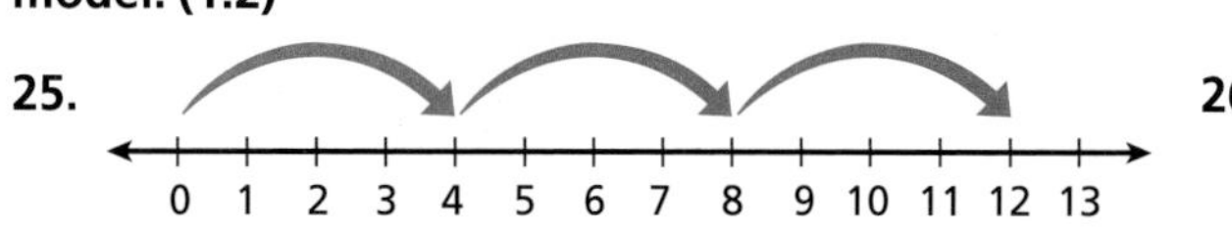

26.

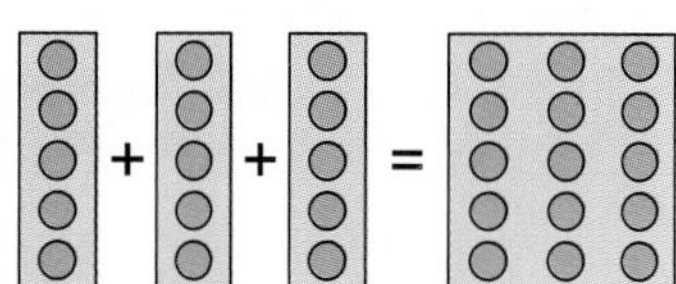

Computation Sense **In Exercises 27–32, write the expression as a power. Then evaluate. (1.3)**

27. 13×13 **28.** $11 \times 11 \times 11 \times 11$ **29.** $\left(\frac{1}{2}\right)\left(\frac{1}{2}\right)\left(\frac{1}{2}\right)$

30. $(3.2)(3.2)(3.2)(3.2)(3.2)$ **31.** $\frac{5}{9} \cdot \frac{5}{9} \cdot \frac{5}{9}$ **32.** $1.01 \cdot 1.01 \cdot 1.01 \cdot 1.01$

This symbol indicates exercises where you should choose the appropriate method of calculation: mental math, paper and pencil, or calculator.

In Exercises 33–36, use a calculator to evaluate the expression. Round to 2 decimal places. (1.3)

33. $\sqrt{289}$ **34.** $\sqrt{729}$ **35.** $\sqrt{2.07}$ **36.** $\sqrt{5.63}$

37. *Surveying* A square plot of land contains 65,946 square feet. What are the lengths of the sides of the plot? **(1.3)**

38. *Writing* In your own words, write the Left-to-Right Rule. **(1.4)**

39. *Writing* In your own words, describe the established order of operations. **(1.4)**

In Exercises 40–48, evaluate the expression. Verify your answer using a calculator. (1.3, 1.4)

40. $4 + 7 - 3$ **41.** $21 - 13 + 8$ **42.** $5 + 18 \div 6$

43. $8 \cdot 4 - 2^3$ **44.** $7^2 - 14 \times 3$ **45.** $(8 - 2) \times 12 \div 3^2$

46. $(2^3 + 1) \div (12 - 9)$ **47.** $(5 - 3)[(8 - 2)^2 - 4^2]$ **48.** $(3^3 + 15) \div \sqrt{49}$

In Exercises 49–60, evaluate the expression for $m = 8$ and $n = 4$. (1.3, 1.4, 1.5)

49. $m + 5$ **50.** $n - 1$ **51.** $3n - 6$ **52.** $2m + 7$

53. $n \div m$ **54.** $m \div n$ **55.** $2n - m$ **56.** $\frac{1}{2}m + 3n$

57. $n^2(m - n)$ **58.** $m^2 + \sqrt{n}$ **59.** $3m \div (2n^2 - m)$ **60.** $n(m - n) + mn$

Distance, Rate, Time **In Exercises 61 and 62, use the formula $d = rt$. (1.5)**

61. Find d for $r = 35$ miles per hour and $t = 7$ hours.

62. Find r for $d = 76$ miles and $t = 1.5$ hours.

U. S. Patents **The number of patents (in thousands) granted by the United States for the years 1987 to 1991 is listed in the table. Use the table to answer Exercises 63–66.** (*Source: U. S. Patent and Trademark Office*) **(1.6)**

Year	Total Number Granted	Individuals	Corporations	U.S. Government
1987	83.0	15.3	66.7	1.0
1988	77.9	14.3	62.9	0.7
1989	95.5	18.0	76.7	0.9
1990	90.4	17.3	72.1	1.0
1991	96.5	18.1	77.3	1.2

63. In what year were the most patents issued?

64. How many individuals were granted a patent in 1989?

65. In 1990, how many more patents were issued to corporations than to individuals?

66. Represent this data with a bar graph or a line graph.

See footnote on page 43.

Population **In Exercises 67–70, use the graph at the right. (1.6)**

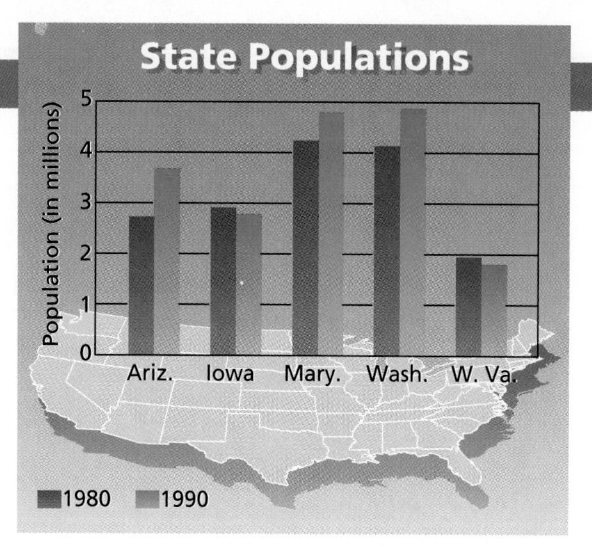

67. In which states did the population increase between 1980 and 1990?

68. Which state had the highest population in 1980?

69. Estimate the 1990 population of West Virginia.

70. Represent the data in the graph with a table.

Reasoning **In Exercises 71–76, decide whether the figure is a polygon. If it is, name it. If it is not, explain why. (1.7)**

71.

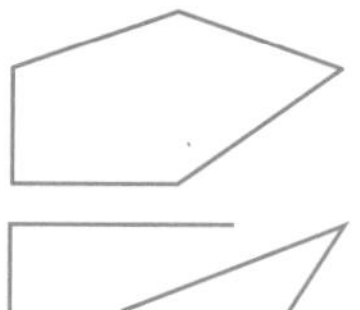

72.

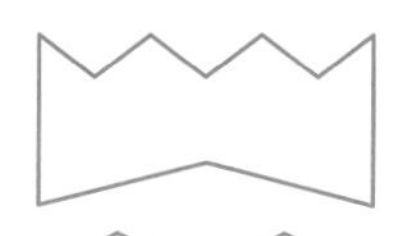

73.

74.

75.

76.

Use the following information to answer Exercises 77 and 78. (1.4, 1.5, 1.6, 1.7)

Area of a Trapezoid A trapezoid is a quadrilateral that has exactly one pair of parallel sides. The table below lists the bases, b_1 and b_2, the height, and the area of several trapezoids. (The variable b_1 is read "b sub 1," and the variable b_2 is read "b sub 2.")

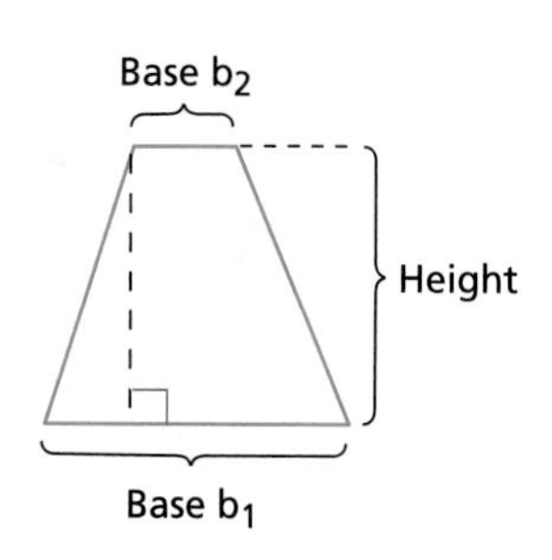

Trapezoid	b_1	b_2	h	$h(b_1 + b_2)$	Area
1	8	6	6	84	42
2	4	2	3	18	9
3	10	5	4	60	30

77. Write a verbal model and an algebraic model for the area of a trapezoid.

78. Find the area of a trapezoid for which $b_1 = 12$, $b_2 = 8$, and $h = 7$.

In Exercises 79 and 80, use a calculator to create a table. Describe the pattern. (1.1, 1.8)

79. The product of 105 and the first 9 natural numbers

80. The sum of $9 + 9n$ using the first 9 natural numbers

In Exercises 81 and 82, give the keystrokes for your calculator that produce the sequence. Then write the next three numbers. (1.8)

81. 15309, 5103, 1701, 567, [?] [?] [?]

82. 100, 90, 81, 72.9, [?] [?] [?]

Manufacturing Rovers **In Exercises 83–85, use the following information.**

You own a business that makes rovers. Last month, your business spent $1 million to build 72 rovers. Each rover sold for $45,000.

83. A model for your profit is

Profit = Income − Expenses .

Use this verbal model to write an algebraic model for the profit.

84. What was your profit last month?

85. Which of the following would increase your profit?

- **a.** Reduce your expenses.
- **b.** Sell more rovers.
- **c.** Increase the price of a rover.

David R. Scott and James B. Irwin were the first astronauts to travel in a moon rover.

Driving Rovers **In Exercises 86–88, imagine that you are an astronaut driving a moon rover. You are traveling at a speed of 22 miles per hour.**

86. How many minutes will it take you to travel 33 miles?

87. How far will you travel in 30 minutes?

88. If the rover gets 13 miles per gallon, how many gallons of fuel will it take to travel 80 miles?

Transporting Rovers **In Exercises 89–91, use the following information.**

On the highway you see a tractor-trailer transporting all-terrain rovers, as shown below. Each rover weighs 4,500 pounds. Without its cargo, the tractor-trailer weighs 33,000 pounds.

89. Write a verbal model for the total weight of the tractor-trailer and the rovers.

90. Write an algebraic model that represents the total weight of the tractor-trailer and the rovers.

91. What is the total weight?

Chapter TEST

In Exercises 1–6, evaluate the expression for $a = 8$, $b = 3$, $c = 5$.

1. $a \cdot c$
2. b^2
3. $\frac{a}{4}$
4. $3c^3$
5. $4(a + 1)^2$
6. $\sqrt{(a+1)}$

In Exercises 7 and 8, describe the pattern. Then list the next 3 numbers.

7. 1, 5, 9, 13, ?, ?, ?
8. 1, 4, 9, 16, ?, ?, ?

In Exercises 9–12, evaluate the expression.

9. $3 + 5 \cdot 2 + 4$
10. $10 - 5 \div 5 + 4$
11. $(6 + 2) \cdot 2 + 3^2$
12. $14 \div (9 - 2) + 2^3$

13. Calculate the product of 321 and the first 7 whole numbers. Record the results in a table. Describe the pattern.

In Exercises 14 and 15, state whether the figure is a polygon. If it is, name it.

14. 

15. 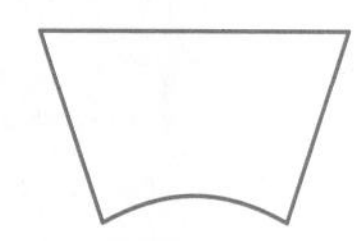

16. The side lengths of 5 squares are 1, 2, 3, 4, and 5. Complete the table by finding the perimeter and area of each square.

Side Length	1	2	3	4	5
Perimeter	?	?	?	?	?
Area	?	?	?	?	?

In Exercises 17–20, use the bar graph at the right. ***(Source: National Basketball Association)***

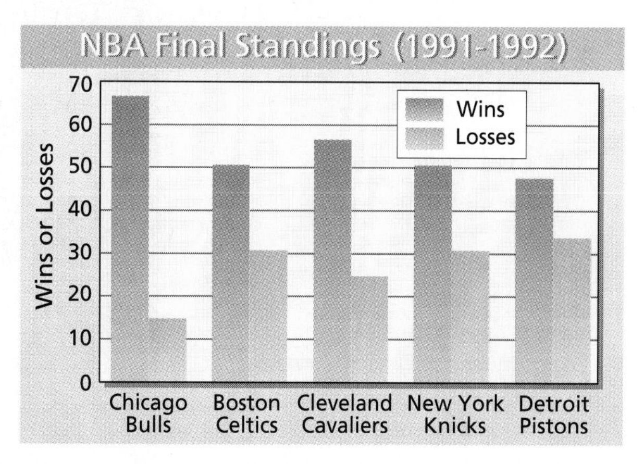

17. About how many games did the Detroit Pistons win?
18. Which two teams had the same number of wins and losses?
19. Which team had the greatest difference between wins and losses?
20. Represent this data with a table.
21. *Gold Prices* The average price of gold (in dollars per fine ounce) for the years 1987 through 1992 is listed in the table. Represent this data with a bar graph and a line graph. ***(Source: U.S. Bureau of Mines)***

Year	1987	1988	1989	1990	1991	1992
Average Price	448	438	383	385	363	350

CHAPTER

2

Investigations in Algebra

LESSONS

Apparel and accessory stores in the United States generate sales in excess of $95 million. However, each individual store must take in enough money to pay for the goods it is selling, to cover its operating costs (rent, insurance, advertising, salaries, etc.), and to pay off any outstanding debts or loans.

Increases in Retail Prices

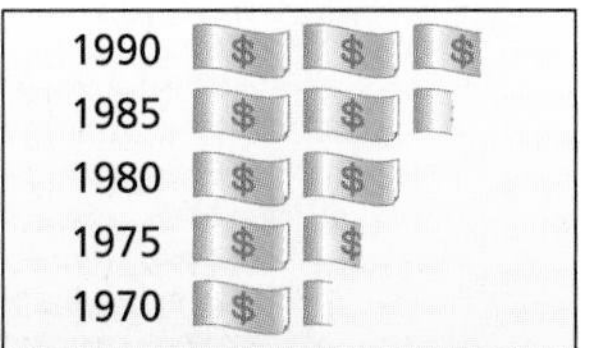

Each [$] represents the retail price of an item in 1960

The amount you pay for any item in a store is called the **retail price.** The graph at the left shows how clothing prices have changed over the years. The amount the store paid for the item is called the **wholesale price.** The **markup** is the amount the store adds to the wholesale price to get the retail price. So, in words,

retail price = wholesale price + markup.

In this chapter, you will learn how to assign variables and write verbal and algebraic models to describe real-life situations as part of a problem-solving plan.

LESSON INVESTIGATION 2.1

Algebraic Expressions

Materials Needed: algebra tiles or graph paper

In this investigation you will use algebra tiles, like those at the right, to represent algebraic expressions. The smaller tile is a 1-by-1 square whose area is 1 square unit. It represents the number 1. The larger tile is a 1-by-x rectangle whose area is x square units. It represents the variable x.

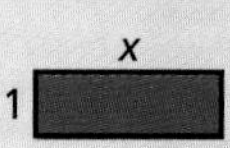

Example — *Using Algebra Tiles*

The expression $x + 3$ can be represented with algebra tiles as shown at the right. Use algebra tiles to represent the expression $2(x + 3)$. Then rearrange the tiles to represent an equivalent expression.

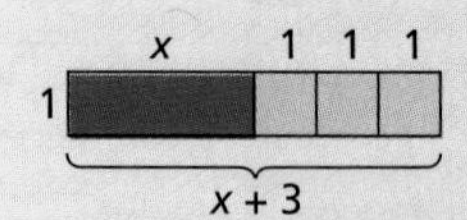

Solution As shown at the right, you can represent the expression $2(x + 3)$ by doubling the number of tiles used to represent $x + 3$. These tiles can be rearranged to represent the expression $2x + 6$.

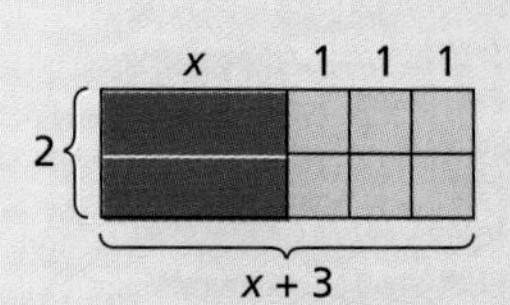

Thus, $2(x + 3)$ and $2x + 6$ are equivalent expressions. ■

Exercises

In Exercises 1–4, model the expression with algebra tiles. Make a sketch of the model.

1. $x + 3$ **2.** $2x + 4$ **3.** $3x + 1$ **4.** $2x + 5$

In Exercises 5–8, match the algebra tiles with *two* of the expressions.

a. $2(3x + 1)$ **b.** $3(2x + 3)$ **c.** $4(x + 1)$ **d.** $2(2x + 1)$

e. $6x + 2$ **f.** $4x + 4$ **g.** $4x + 2$ **h.** $6x + 9$

5.

6.

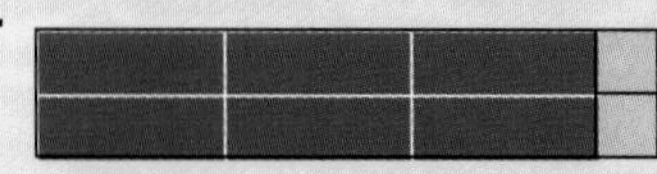

7.

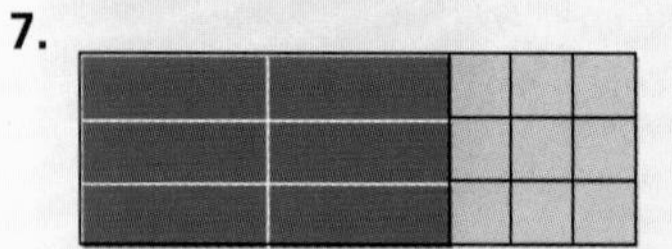

8.

2.1 The Distributive Property

What you should learn:

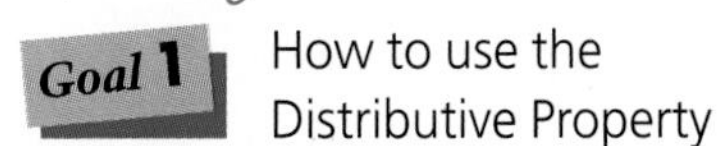

Goal 1 How to use the Distributive Property

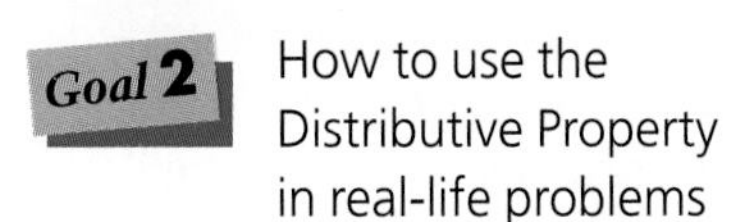

Goal 2 How to use the Distributive Property in real-life problems

Why you should learn it:

You can use the Distributive Property to model real-life situations, such as ordering clothes for a clothing retailer.

Goal 1 Using the Distributive Property

Two expressions that have one or more variables are **equivalent** if they have the same values when numbers are substituted for the variables. For instance, the expressions $2(3x)$ and $6x$ are equivalent.

$2(3x) = 6x$ *$2(3x)$ and $6x$ are equivalent expressions.*

In the Lesson Investigation on page 50, you used algebra tiles to show that $2(x + 3)$ and $2x + 6$ are equivalent expressions. Another way to show these expressions are equivalent is with the **Distributive Property.**

The Distributive Property

Let a, b, and c be numbers or variable expressions.

$a(b + c) = ab + ac$ and $ab + ac = a(b + c)$

Example 1 *Using the Distributive Property*

a. $2(x + 3) = 2(x) + 2(3)$ *Apply Distributive Property.*

$= 2x + 6$ *Simplify.*

b. $3(2x + 4) = 3(2x) + 3(4)$ *Apply Distributive Property.*

$= 6x + 12$ *Simplify.*

c. $5(2) + 5(4) = 5(2 + 4)$ *Apply Distributive Property.*

$= 5(6)$ *Simplify.*

$= 30$ *Simplify.*

d. $x(x + 4) = x(x) + x(4)$ *Apply Distributive Property.*

$= x^2 + 4x$ *Simplify.* ■

In part **d,** notice that $x(4)$ is usually written as $4x$. This is a use of the **Commutative Property of Multiplication.**

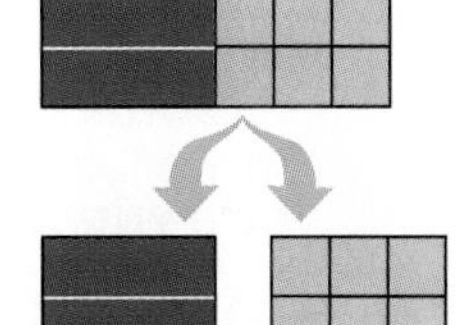

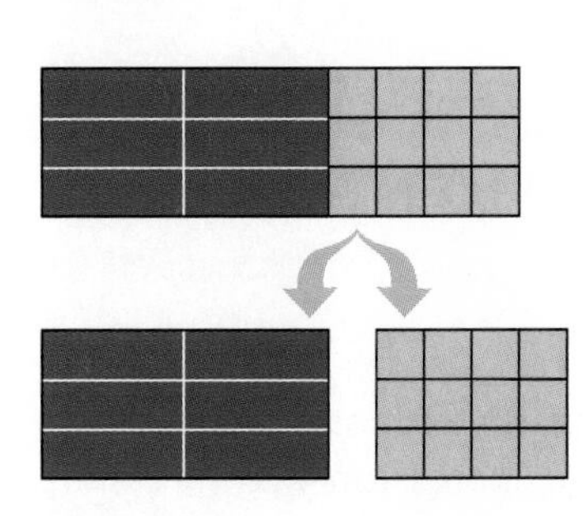

Algebra tiles can be used to model the Distributive Property. Which parts of Example 1 are illustrated by the algebra tiles shown above?

Goal 2 Solving Real-Life Problems

The Distributive Property is usually stated with a sum involving only two terms. However, it also applies to sums involving three or more terms.

$$a(b + c + d) = ab + ac + ad \quad \textit{Sum with 3 terms}$$
$$a(b + c + d + e) = ab + ac + ad + ae \quad \textit{Sum with 4 terms}$$

Example 2 *Using the Distributive Property*

Real Life
Fashion Buying

You are a fashion buyer for a clothing retailer that has 5 stores. You are attending a fashion show and decide to order 10 dresses of Style A, 12 dresses of Style B, and 15 dresses of Style C for *each* of the 5 stores. Use the Distributive Property to find the total number of dresses ordered.

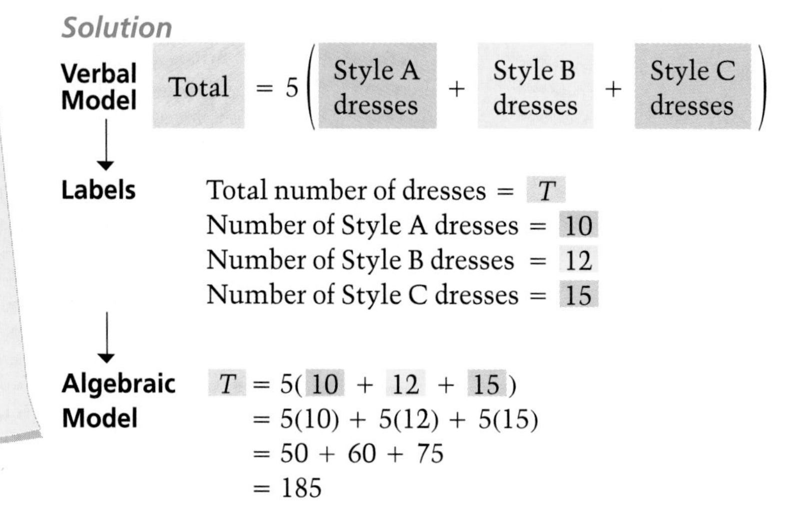

Study Tip...
Problem Solving *One of your goals in this chapter is to learn how to organize your problem-solving solutions using a verbal model, labels, and an algebraic model. You will study this in more detail in Lesson 2.7.*

You are ordering 185 dresses. You can check this result by writing $5(10 + 12 + 15) = 5(37) = 185$. ■

Communicating about MATHEMATICS

Cooperative Learning

SHARING IDEAS about the Lesson

Using the Distributive Property Discuss with others in your group different ways to evaluate the expression $4(3x + 2)$ when $x = 5$. Which way do you prefer? Explain your reasoning.

EXERCISES

Guided Practice

CHECK for Understanding

1. State the Distributive Property.

2. Which are correct applications of the Distributive Property?

a. $2(3 + 5) = 2(3) + 5$ **b.** $4(y + 7) = 4(y) + 4(7)$

c. $16(x + 1) = 16x + 16$ **d.** $2(a + 6) = 2a + 6$

3. Decide whether the statement is true. Explain your reasoning.

a. $3(5x) \stackrel{?}{=} 15x$ **b.** $4(9x) \stackrel{?}{=} 13x$ **c.** $2(5 + 6) \stackrel{?}{=} 21$

4. Give an example of the Commutative Property of Multiplication.

Independent Practice

Modeling Expressions **In Exercises 5–8, write the dimensions of the rectangle and an expression for its area. Then use the Distributive Property to rewrite the expression.**

5.
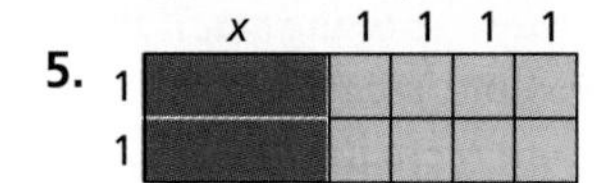

6.
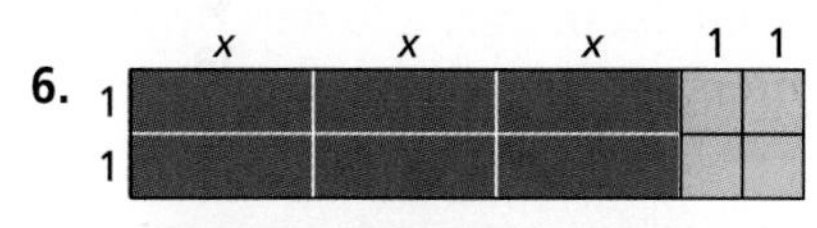

7.
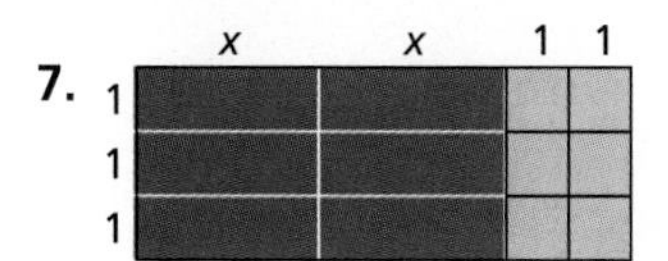

8.
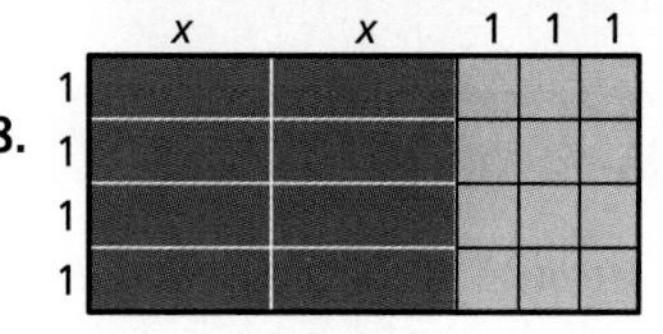

In Exercises 9–12, use the Distributive Property to write an equivalent expression. Illustrate your result with an algebra tile sketch.

9. $4(x + 2)$ **10.** $2(x + 1)$ **11.** $2(5x + 3)$ **12.** $5(2x + 3)$

In Exercises 13–32, use the Distributive Property to rewrite the expression.

13. $9(8 + 7)$ **14.** $11(10 + 5)$ **15.** $4(x + 9)$ **16.** $16(z + 3)$

17. $5(y + 20)$ **18.** $8(4 + q)$ **19.** $1(x + 12)$ **20.** $1(t + 42)$

21. $a(b + 4)$ **22.** $p(q + 2)$ **23.** $r(s + t)$ **24.** $m(n + k)$

25. $4(6 + 10 + 12)$ **26.** $3(5+8+9)$ **27.** $y(3 + z)$ **28.** $a(b + c + 4)$

29. $12(s + t + w)$ **30.** $b(e + f + g)$ **31.** $6(m + n + r + t)$ **32.** $z(x + 4 + y)$

It's Up to You **In Exercises 33–35, use a calculator to evaluate the expression two ways. Which way do you prefer? Why?**

33. $3(1.21+5.48)$ **34.** $10(6.81+9.06)$ **35.** $525(11.19+27.60)$

Geometry **In Exercises 36 and 37, use the rectangle at the right.**

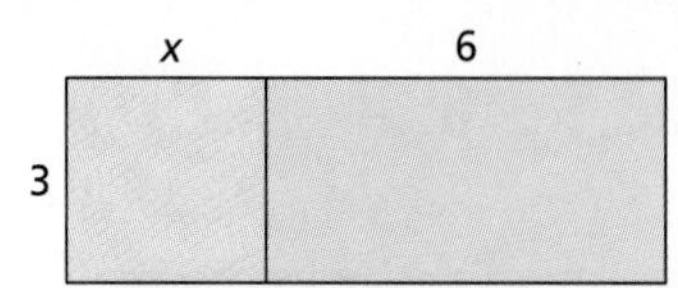

36. Write an expression for the area of the rectangle.

37. Rewrite your expression using the Distributive Property.

38. *Geometry* Which of the following is a correct formula for the perimeter of a rectangle? Explain.

a. $P = 2 + (\text{length} \times \text{width})$

b. $P = 2(\text{length} + \text{width})$

c. $P = (\text{length} \times \text{width} \times 2)$

39. *Business* You own a small business that has 3 employees. You pay one employee \$1800 a month, the second \$1500 a month, and the third \$1300 a month.

a. Write a verbal model that represents how much you pay all 3 employees in the year.

b. Use the model in Part **a** to determine how much you pay your employees in a year.

Nearly two-thirds of all Americans get their first job in a small business. The fastest growing small businesses in 1993 were restaurants, computer and data services, and amusement and recreation services.

Integrated Review

Making Connections within Mathematics

Evaluating Expressions **In Exercises 40–51, evaluate the expression for $a = 2, b = 5, c = 8$, and $d = 10$.**

40. $2(a + 1)$	**41.** $7(d + 6)$	**42.** $a(b + 3)$	**43.** $c(a + 9)$
44. $3(a + c)$	**45.** $12(b + d)$	**46.** $a(c + d)$	**47.** $b(c + a)$
48. $3(a + b + c)$	**49.** $8(b + c + d)$	**50.** $a(b + c + d)$	**51.** $d(c + b + a)$

Exploration and Extension

Mental Math **In Exercises 52–55, use mental math with the Distributive Property to evaluate the product. For instance, 15(405) can be written as 15(400 + 5). Using mental math, the result is 6000 + 75 or 6075.**

52. 3(522)	**53.** 8(612)	**54.** 11(409)	**55.** 21(203)

56. *You Be the Teacher* Your friend applies the Distributive Property to multiplication and determines $2(3 \times 5) = 2(3) \times 2(5)$. Write a paragraph explaining how you would convince your friend that this is incorrect.

2.2

Simplifying by Adding Like Terms

What you should learn:

How to simplify expressions by adding like terms

How to add like terms to simplify expressions in geometry

Why you should learn it:

Evaluating a simplified expression is usually easier than evaluating one that has not been simplified.

Goal 1 Adding Like Terms

Two or more terms in an expression are **like terms** if they have the same variables, raised to the same powers.

Expression	Like Terms
$3x + x + 2$	$3x$ and x
$5y + 5 + 4$	5 and 4
$3y^2 + 4y + y^2 + y$	$3y^2$ and y^2, $4y$ and y

LESSON INVESTIGATION

Investigating Addition of Like Terms

Group Activity The expression $3x + 2x + 2$ can be modeled with algebra tiles, as shown at the left. After collecting like tiles, you can see that there are five x's and two 1's. This means that the expression $5x + 2$ is equivalent to $3x + 2x + 2$.

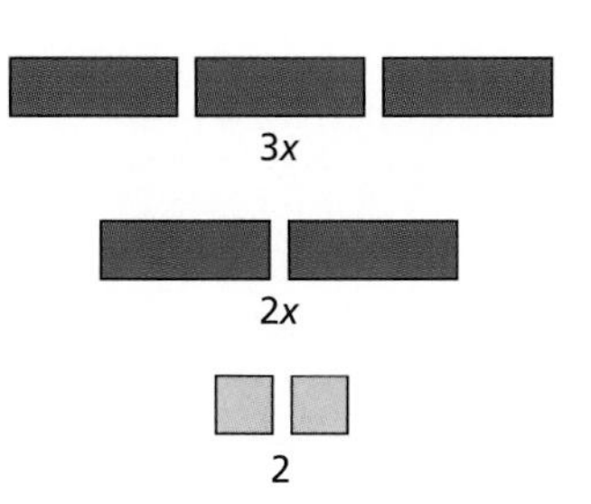

Model each of the following with algebra tiles, collect like tiles, and write the simplified expression.

a. $2x + 3 + 4x$ **b.** $x + 2x + 3x + 5$ **c.** $4x + 3 + x$

This procedure is called **adding like terms** (or **collecting like terms**). The rewritten algebraic expression is said to be **simplified.**

The Distributive Property can be used to add like terms.

Need to Know

In part **b** of Example 1, notice that you can reorder the terms of an expression. That is, $3b + 2 + 5b$ can be rewritten as $3b + 5b + 2$. This procedure is justified by the **Commutative Property of Addition.**

Example 1 *Simplifying by Adding Like Terms*

a. $2x + 5x + 1 = (2 + 5)x + 1$ *Distributive Property*
$= 7x + 1$ *Simplify.*

b. $3b + 2 + 5b = 3b + 5b + 2$ *Commutative Property*
$= (3 + 5)b + 2$ *Distributive Property*
$= 8b + 2$ *Simplify.* ■

Goal 2 Simplifying Expressions in Geometry

Example 2 *Simplifying before Evaluating*

Write an expression that represents the perimeter of the figure. Then evaluate the perimeter when x is 1, 2, 3, 4, and 5. Organize your results in a table and in a graph. Describe the pattern as the values of x increase by 1.

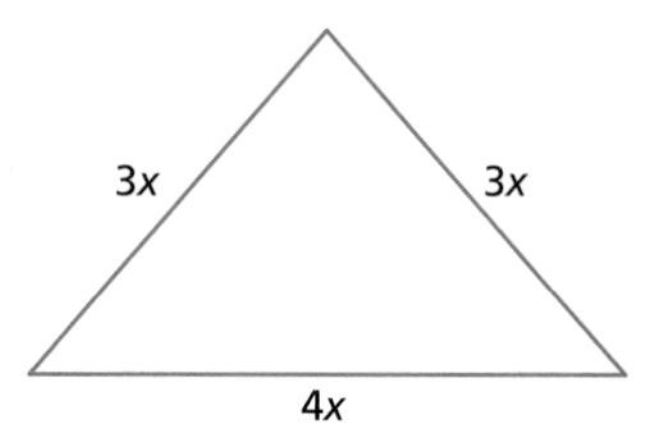

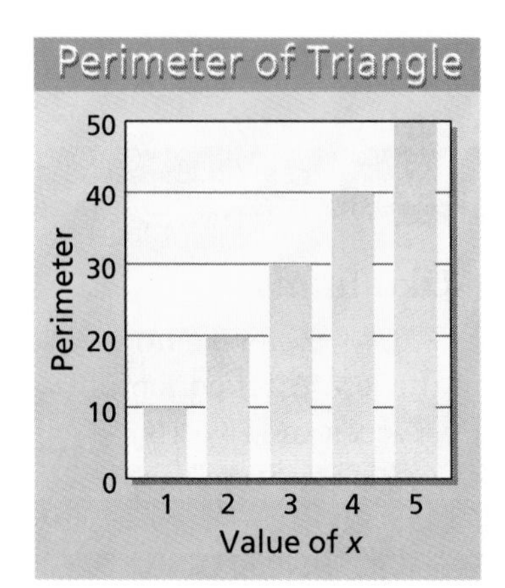

Solution To find an expression for the perimeter, add the lengths of the three sides.

$$\text{Perimeter} = 3x + 3x + 4x \quad \textit{Add the side lengths.}$$
$$= 10x \quad \textit{Add like terms.}$$

Next, evaluate the expression $10x$ when x is 1, 2, 3, 4, and 5. The results are shown in the table and in the bar graph.

x	1	2	3	4	5
Perimeter	10	20	30	40	50

From the table or the graph, you can see that the perimeter increases by 10 each time x increases by 1. ■

Communicating about MATHEMATICS

SHARING IDEAS about the Lesson

A Perimeter Pattern Write an expression that represents the perimeter of the figure. Then evaluate the perimeter when x is 1, 2, 3, 4, and 5. Organize your results in a table or in a graph. Describe the pattern as the values of x increase by 1.

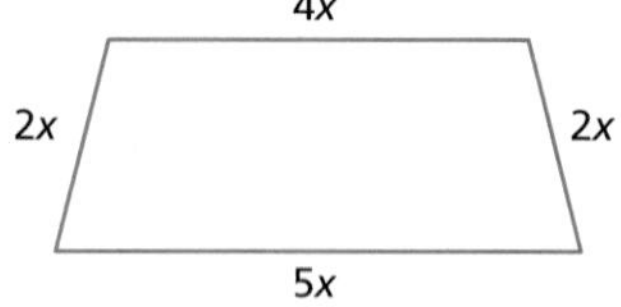

EXERCISES

Guided Practice

CHECK for Understanding

1. What is meant by like terms? State an example.
2. State an example of adding like terms. Illustrate your result with an algebra tile sketch.
3. Describe in your own words the Commutative Property of Addition.
4. When is it helpful to add like terms?

In Exercises 5–8, decide whether the expression can be simplified. Explain.

5. $3x + x$ **6.** $2a + 7$ **7.** $2r + 5r^2$ **8.** $5 + 2(x + 8)$

Independent Practice

In Exercises 9–20, simplify the expression.

9. $2a + a$ **10.** $5b + 7b$ **11.** $3x + 6x + 9$

12. $8x + 4x + 12$ **13.** $9y + 10 + 3y$ **14.** $y + 5 + 13y$

15. $3a + 2b + 5a$ **16.** $6x + x + 2y$ **17.** $5 + r + 2s + 13r$

18. $p + 9q + 9 + 14p$ **19.** $2x + 4y + 3z + 17z$ **20.** $a + 2b + 2a + b + 2c$

21. *Its Up to You* Write an expression that has four terms and simplifies to $16x + 5$.

22. *Reasoning* Explain the steps used to simplify $2x + 3 + 5x$.

In Exercises 23–34, simplify the expression.

23. $b + b^2 + 2b$ **24.** $5x + 6x + x^3$ **25.** $x^2 + x^2$

26. $2a^3 + a^3$ **27.** $3(x + 3) + 4x$ **28.** $8(y + 2) + y + 4$

29. $3(a + b) + 3(b + a)$ **30.** $5(x + y) + 2(y + x)$ **31.** $xy + 15xy$

32. $24rs + 12rs$ **33.** $x^2y + x^2y$ **34.** $6(r^2s + s^2) + r^2s$

In Exercises 35–40, simplify the expression. Then evaluate when $x = 2$ and $y = 5$.

35. $2x + 3x + y$ **36.** $y + 4y + 8x$ **37.** $4(x + y) + x$

38. $(x + y)4 + 7x$ **39.** $xy + x^2 + xy$ **40.** $6xy + x^2 + x^2$

Geometry **In Exercises 41 and 42, write an expression for the perimeter of each polygon. Are the expressions equivalent? Explain.**

41.

x + 3
x
2x
x + 1

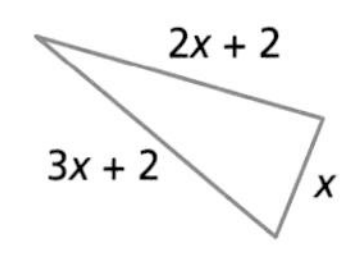

42.

x + 3
x
x
x + 2
x + 2

x + 2
2x + 1
x + 1
x + 1

Geometry **In Exercises 43 and 44, write an expression for the perimeter. Find the perimeter when x is 1, 2, 3, 4, and 5. Represent your results with a table or a bar graph. Then describe the pattern.**

43. 6x, 3x, 3x, 6x

44.

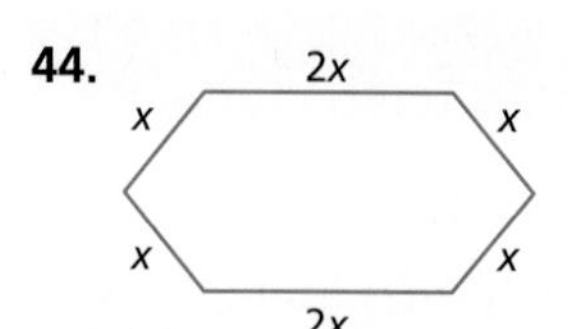

45. *Baby-sitting* You have a baby-sitting job. You work after school for 3 hours on Wednesday, Thursday, and Friday, and 9 hours on Saturday. You earn x dollars an hour.

a. Write an expression that represents your weekly earnings.

b. Suppose you earn \$3.25 an hour. How much money would you earn?

46. *Arcade* While you are at the arcade, you play your favorite video games. You play 2 different video games; one that costs \$0.25 and the other that costs \$0.50. You play each video game x number of times.

a. Write an expression for the total amount of money spent.

b. If you play each game 5 times, how much do you spend?

The circuit boards of arcade video games are made up of chips that operate by a tiny computer called a microprocessor.

Integrated Review

Making Connections within Mathematics

Modeling Perimeter **In Exercises 47 and 48, write a model for the perimeter. Then use your model to find the perimeter of the 10th and 15th figures in the sequence.**

47. 1st 2nd 3rd 4th

1, 1, 1

48. 1st 2nd 3rd 4th

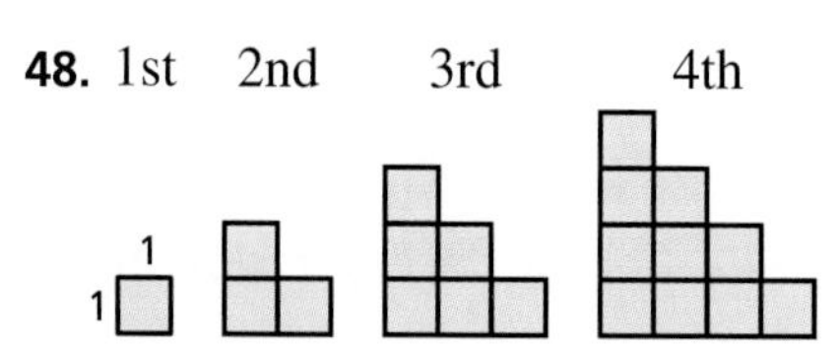

Exploration and Extension

49. *You Be the Teacher* Imagine that you are teaching this class. You decide to check your students on the material they have learned up to this point in the book. Write a test that you think covers the important concepts and skills. The test should have 20 questions with 2 questions from each lesson.

2.3 Solving Equations: Mental Math

What you should learn:

Goal 1 How to check that a number is a solution of an equation

Goal 2 How to use mental math to solve an equation

Why you should learn it:

Many real-life situations can be modeled with equations. Learning how to solve equations helps you solve real-life problems.

Goal 1 Checking Solutions

An **equation** states that two expressions are equivalent. Some equations, called **identities,** are true for all values of the variables they contain. Here are two examples.

$3 + 2^2 = 7$ *Identity: Always true*

$2(x + 3) = 2x + 6$ *Identity: True for all x*

Other equations, called **conditional equations,** are not true for all values of the variables they contain.

$x + 1 = 4$ *Conditional equation: true only for x = 3*

$3x = 12$ *Conditional equation: true only for x = 4*

Finding the values of a variable that make a conditional equation true is called **solving the equation,** and these values of the variable are **solutions** of the equation. Two equations are **equivalent** if they have the same solutions. You can check that a number is a solution by substituting in the original equation.

Example 1 *Checking Possible Solutions*

Which of the following numbers are solutions of the equation $4x - 3 = 5$?

a. $x = 2$ **b.** $x = 3$

Solution

a.

$4x - 3 = 5$ *Write original equation.*

$4(2) - 3 \stackrel{?}{=} 5$ *Substitute 2 for x.*

$8 - 3 \stackrel{?}{=} 5$ *Simplify.*

$5 = 5$ *x = 2 is a solution.* ✓

b.

$4x - 3 = 5$ *Write original equation.*

$4(3) - 3 \stackrel{?}{=} 5$ *Substitute 3 for x.*

$12 - 3 \stackrel{?}{=} 5$ *Simplify.*

$9 \neq 5$ *x = 3 is not a solution.* ⊗

Need to Know

Problem Solving
Checking a solution is an important part of solving equations. Throughout this course, you can learn to be a better problem solver if you develop the habit of always **checking your solutions.**

In 1991, Sidney Swartz, the chief executive officer of Timberland, asked his son Jeffrey to help restructure the shoe and boot manufacturing company. The company installed a team approach to manufacturing, and the company's sales increased dramatically. (Source: Timberland)

Goal 2 Solving Equations with Mental Math

As you study algebra, you will learn many techniques for solving equations. Some equations are simple enough that they can be solved mentally. For instance, to solve the equation

$$x + 2 = 8$$

you can think "what number can be added to 2 to produce 8?" The solution is $x = 6$. You can check this solution by writing $6 + 2 = 8$.

Example 2 *Solving Equations with Mental Math*

Solve the following equations.

a. $x + 4 = 10$ **b.** $x - 12 = 18$ **c.** $3x = 15$ **d.** $\frac{x}{4} = 5$

Solution

Equation	Stated as a Question	Solution
a. $x + 4 = 10$	*What number can be added to 4 to obtain* 10?	$x = 6$
b. $x - 12 = 18$	*What number can* 12 *be subtracted from to obtain* 18?	$x = 30$
c. $3x = 15$	*What number can be multiplied by* 3 *to obtain* 15?	$x = 5$
d. $\frac{x}{4} = 5$	*What number can be divided by* 4 *to obtain* 5?	$x = 20$

■

Communicating about MATHEMATICS

SHARING IDEAS about the Lesson

Real Life **Retail Sales**

Modeling a Real-Life Situation In 1992, *Timberland* sold \$290 million worth of shoes and boots. In 1993, the company sold \$425 million worth of shoes and boots. How much more did the company sell in 1993 than in 1992?

1992 sales	+	Increase in sales	=	1993 sales

Let x represent the increase in sales. Then rewrite the verbal model as an equation and solve the equation.

EXERCISES

Guided Practice

CHECK for Understanding

1. Explain the difference between an identity and a conditional equation.
2. Write an example of an identity.
3. Write an example of a conditional equation.
4. Explain how to check a solution of an equation.

Independent Practice

Checking Solutions **In Exercises 5–8, match the equation with a solution.**

a. 2 **b.** 3 **c.** 4 **d.** 5

5. $5x + 7 = 22$ **6.** $10 - 2y = 2$ **7.** $n^2 - 4 = 21$ **8.** $\frac{20}{x^2} = 5$

Mental Math **In Exercises 9–12, write the equation as a question. Then solve it mentally.**

9. $z + 8 = 14$ **10.** $7x = 42$ **11.** $y - 18 = 16$ **12.** $\frac{9}{x} = 3$

Mental Math **In Exercises 13–16, write the question as an equation. Then solve it mentally.**

13. What number can be subtracted from 33 to obtain 24?

14. What number can be added to 7 to obtain 19?

15. What number can be multiplied by 8 to obtain 56?

16. What number can be divided by 9 to obtain 5?

Mental Math **In Exercises 17–20, decide whether $r = 4$ is a solution of the equation. If it isn't, use mental math to find the solution.**

17. $5r = 20$ **18.** $19 - r = 15$ **19.** $\frac{24}{r} = 8$ **20.** $3r + r = 16$

Reasoning **In Exercises 21–24, decide whether the equations have the same solutions. Explain why.**

21. a. $x - 15 = 8$ **b.** $15 - x = 8$

22. a. $x + 4 = 17$ **b.** $4 + x = 17$

23. a. $3x = 12$ **b.** $\frac{x}{12} = 3$

24. a. $x \div 3 = 6$ **b.** $3 \div x = 6$

Mental Math **In Exercises 25–36, solve the equation using mental math. Check your solution.**

25. $p + 7 = 18$ **26.** $r + 11 = 14$ **27.** $27 + n = 41$ **28.** $x - 13 = 2$

29. $y - 20 = 32$ **30.** $81 - y = 76$ **31.** $11x = 55$ **32.** $2t = 18$

33. $21 = 3m$ **34.** $\frac{x}{4} = 9$ **35.** $\frac{s}{12} = 4$ **36.** $\frac{26}{y} = 2$

In Exercises 37–39, decide whether the equation is an identity or a conditional equation. Explain your reasoning.

37. $4(x + 7) = 4x + 28$ **38.** $3 + x = 10$ **39.** $9x = 72$

Growing Pumpkins **In Exercises 40–43, use algebra to answer the question. Then, use the graph to check your answer.** ***(Source: World Pumpkin Confederation)***

40. The 1987 winner weighed 212 pounds less than the 1990 winner. How much did the 1990 winner weigh?

41. The 1989 winner weighed $1\frac{1}{4}$ times as much as the 1987 winner. How much did the 1989 winner weigh?

42. The 1992 winner weighed $10\frac{1}{2}$ pounds more than the 1990 winner. How much did the 1992 winner weigh?

43. The 1988 winner's weight divided by the 1992 winner's weight is $\frac{3}{4}$. How much did the 1988 winner weigh?

Mental Math **In Exercises 44–47, solve the equation using mental math. Then write another equation that has the same solution.**

44. $x + 15 = 29$ **45.** $7x = 70$ **46.** $31 - r = 24$ **47.** $\frac{t}{12} = 4$

Integrated Review

Making Connections within Mathematics

Mental Math **In Exercises 48–50, use mental math to evaluate the expression.**

48. $7 \cdot 2 + 3 \cdot 6$ **49.** $24 \div 3 - (2)(3)$ **50.** $3[(7 - 4)^2 + 1]$

Mental Math **In Exercises 51–56, simplify the equation. Then solve it mentally.**

51. $6x + 3x = 18$ **52.** $4x + 8x = 24$ **53.** $5x + 4x = 11 + 7$

54. $17x = 9 + 25$ **55.** $3x + 4x + 2x = 0$ **56.** $8x + 7x + 3x = 9$

Exploration and Extension

57. *Geometry* **Which two rectangles have the same perimeter for all values of x? Explain your reasoning.**

a.

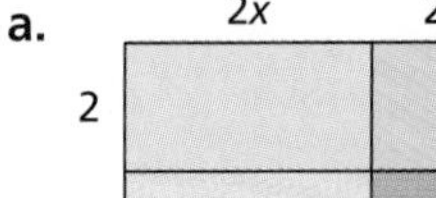

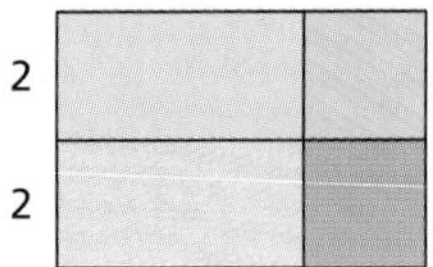

b.

x 3

4

c.

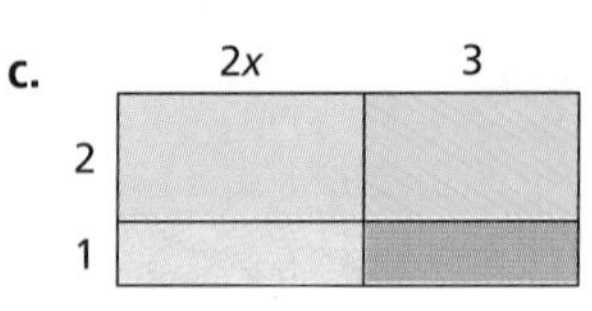

Mixed REVIEW

In Exercises 1–8, evaluate the expression when $x = 6$. (1.5)

1. $3x + 1$ **2.** $\frac{x}{3} + 4$ **3.** $x^2 - 8$ **4.** $\frac{x}{2} - 3$

5. $(x - 1) \cdot 2$ **6.** $2 \cdot (x \div 3)$ **7.** $\frac{1}{4} \cdot x^2$ **8.** $(x^2 - 12) \div 4$

In Exercises 9–14, simplify the expression and evaluate for $r = 4$ and $s = 5$. (1.5, 2.1, 2.2)

9. $16r + 2 - 12r$ **10.** $2r^2 + r + r^2$ **11.** $4s + 3r + 2s$

12. $3(r + s) - 2r$ **13.** $2(r^2 + s) + r^2$ **14.** $\frac{1}{5}r + s + 3s$

In Exercises 15–20, use a calculator to evaluate the expression. (2.1)

15. $616(1.8 + 2.5)$ **16.** $412(2.02 - 1.64)$ **17.** $951(8.25 \cdot 4)$

18. $827(3.03 \div 3)$ **19.** $\frac{1}{3}(3.14 \cdot 15)$ **20.** $\frac{1}{5}(330 \div 3)$

Milestones THE CHANGING FACE OF MONEY

700 BC | 400 BC | 100 BC | 200 AD | 500 AD | 800 AD | 1100 AD | 1400 AD | 1700 AD | 2000 AD

stater 650 B.C. — Greek drachma 269 B.C. — Latin aureus 87 B.C. — Chinese paper money 600's — Banks, Italy 1200-1600 — U.S Mint 1792 — Universal Credit Card, 1950

In order to get the things you need you must find someone who has them and is willing to trade for things that you have. This process was made much easier in about 650 B.C., when a convenient item was created that was widely accepted in trade for anything, the coin. The first known metal coin, the *stater* (meaning standard) was made by King Croesus in Lydia, a kingdom on the western coast of modern Turkey. It was made of 0.3 ounces of gold and stamped with a lion's head. It had a standard size, weight, and purity. For centuries, the value of coins was based on the amount of precious metal in them. Between A.D. 700 and 900, the Chinese introduced paper money which had value only because people trusted that the issuer would take it back in return for its face value in coins. In the Middle Ages, banks (from the Italian word for bench) were established to help merchants establish credit to buy goods.

Tribes in the Pacific Northwest used fish-hooks as currency.

In our day world trading demands that the money of one country be exchanged for the money of another. Newspapers publish the rates of exchange between currencies daily.

- *Use the chart to determine how many English pounds you would get in exchange for 100 U.S. dollars.*
- *How many U.S. dollars could you get for 100 Japanese yen?*

EXCHANGE RATES (8/26/94)

Currency	In U.S. dollars	per U.S. dollars
British pound	1.5312	0.6531
Canadian dollar	0.7302	1.369
Greek drachma	0.0042	239.15
Japanese yen	0.0099	100.45
Mexican peso	0.2980	3.3554
Turkish lira	0.00003	32745.2

LESSON INVESTIGATION 2.4
Solving Equations

Materials Needed: algebra tiles

In this investigation, you will use algebra tiles to model and solve equations involving addition. Remember that each smaller tile represents the number 1 and each larger tile represents the variable *x*.

Example — *Using Algebra Tiles*

Model the equation $x + 3 = 6$ with algebra tiles. Then use the tiles to solve the equation. Finally, check your solution.

Solution

Original equation:
$x + 3 = 6$

To isolate x, remove (subtract) 3 tiles from both sides.

Solution is $x = 3$.

You can check that $x = 3$ is a solution as follows.

$x + 3 = 6$	*Write original equation.*
$3 + 3 \stackrel{?}{=} 6$	*Substitute 3 for x.*
$6 = 6$	*$x = 3$ is a solution.* ✔

■

Exercises

In Exercises 1 and 2, an equation has been modeled and solved with algebra tiles. Write the equation and its solution. Then check the solution.

1.

2.

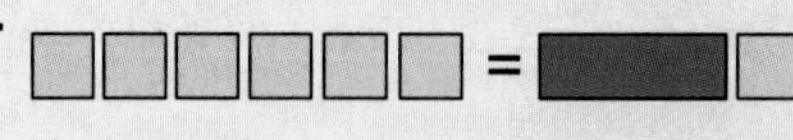

In Exercises 3–6, use algebra tiles to model and solve the equation.

3. $x + 4 = 7$ **4.** $2 + x = 9$ **5.** $6 = x + 4$ **6.** $5 = 1 + x$

7. *You Be the Teacher* Suppose you are teaching a friend how to use algebra tiles to solve an equation. How would you explain the process to your friend?

2.4 Solving Equations: Addition or Subtraction

What you should learn:

Goal 1 How to use addition or subtraction to solve an equation

Goal 2 How to use equations as algebraic models to solve real-life problems

Why you should learn it:

You can use addition or subtraction to solve equations that model real-life situations, such as buying a basketball.

Goal 1 Using Addition or Subtraction

The Lesson Investigation on page 64 shows how algebra tiles can be used to model and solve an equation. Another way to model an equation is to use a scale. Your goal is to find the value of x. Whatever you do to one side of the scale, you must also do to the other so that the scale remains in balance.

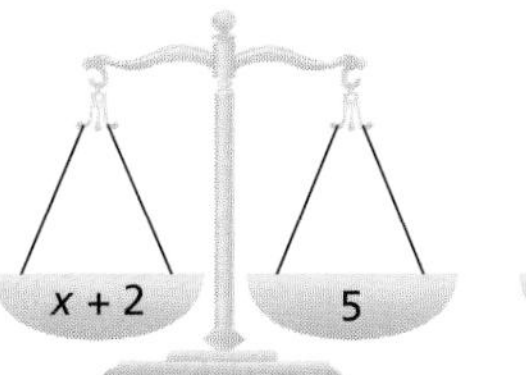

Original equation: *$x + 2 = 5$.*

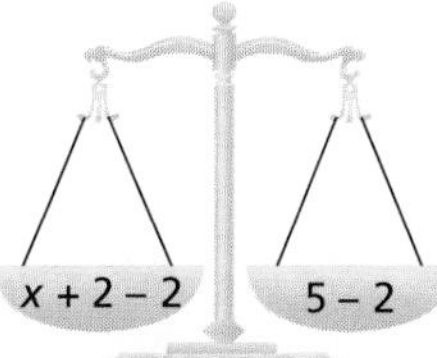

Subtract 2 from both sides. *Scale stays in balance.*

Simplify both sides. *Solution is $x = 3$.*

Addition and Subtraction Properties of Equality

Adding the same number to both sides of an equation or subtracting the same number from both sides of an equation produces an equivalent equation.

Study Tip...

*Some people like to use a vertical format to show the steps of a solution. For instance, in part **a** of Example 1, you could write the following.*

$$\begin{array}{r} x-31= \ \ 14 \\ +31=+31 \\ \hline x= \ \ 45 \end{array}$$

Example 1 *Solving Equations*

Solve the equations **a.** $x - 31 = 14$ and **b.** $214 = y + 112$.

Solution Each equation that is part of a written solution is called a **step** of the solution.

a.

$x - 31 = 14$	*Rewrite original equation.*
$x - 31 + 31 = 14 + 31$	*Add 31 to both sides.*
$x = 45$	*Simplify.*

The solution is 45. Check this in the original equation.

b.

$214 = y + 112$	*Rewrite original equation.*
$214 - 112 = y + 112 - 112$	*Subtract 112 from both sides.*
$102 = y$	*Simplify.*

The solution is 102. Check this in the original equation. ■

Goal 2 Modeling Real Life with Equations

Example 2 *Using an Equation as a Real-Life Model*

Real Life
Retail Sales

For your birthday, your grandmother sent you a \$12.50 gift certificate for a sporting-goods store. You decide to use the certificate to buy a basketball that cost \$14.59 (including tax). How much extra money do you need to pay?

Solution

Verbal Model ↓ | Gift certificate | + | Extra money | = | Price of Basketball |

Labels ↓

Value of gift certificate = 12.50 (dollars)
Extra money = x (dollars)
Price of basketball = 14.59 (dollars)

Algebraic Model

$$12.50 + x = 14.59$$
$$12.50 + x - 12.50 = 14.59 - 12.50$$
$$x + 12.50 - 12.50 = 14.59 - 12.50$$
$$x = 2.09$$

You need an extra \$2.09. You can check this result by noting that $12.50 + 2.09 = 14.59$. ■

Study Tip...
Problem Solving *Even though you could use mental math to answer the question in Example 2, it is important to learn to use algebra to answer the question. Later, you will encounter problems that are too complicated to be solved with mental math. Then, your knowledge of algebra will really pay off!*

In the 3rd step of the solution, note that the terms 12.50 and x were reordered. This is justified by the **Commutative Property of Addition.** This and other properties of addition and multiplication are listed below.

Property

Commutative Property of Addition: $a + b = b + a$

Commutative Property of Multiplication: $ab = ba$

Associative Property of Addition: $a + (b + c) = (a + b) + c$

Associative Property of Multiplication: $a(bc) = (ab)c$

Communicating about MATHEMATICS

SHARING IDEAS about the Lesson

Real-Life Modeling Think of a real-life situation that can be modeled with an equation. Then solve the equation.

EXERCISES

Guided Practice

CHECK for Understanding

1. Give an example of the Addition Property of Equality.
2. Explain why the equations $x + 5 = 8$ and $x = 3$ are equivalent.

In Exercises 3–6, explain how to solve the equation.

3. $x + 24 = 38$
4. $y - 16 = 53$
5. $152 = r + 72$
6. $185 = s - 68$

In Exercises 7–10, state the property that is demonstrated.

7. $3 \cdot (8 \cdot 4) = (3 \cdot 8) \cdot 4$
8. $(3)(8) = (8)(3)$
9. $4 + 7 = 7 + 4$
10. $8 + (9 + 7) = (8 + 9) + 7$

Independent Practice

In Exercises 11–14, copy and complete the solution.

11.
$$\begin{aligned} x - 34 &= 52 \\ x - 34 + \boxed{?} &= 52 + \boxed{?} \\ x &= \boxed{?} \end{aligned}$$

12.
$$\begin{aligned} 76 &= y - 29 \\ 76 + \boxed{?} &= y - 29 + \boxed{?} \\ \boxed{?} &= y \end{aligned}$$

13.
$$\begin{aligned} r + 62 &= 111 \\ r + 62 - \boxed{?} &= 111 - \boxed{?} \\ r &= \boxed{?} \end{aligned}$$

14.
$$\begin{aligned} 279 &= t + 194 \\ 279 - \boxed{?} &= t + 194 - \boxed{?} \\ \boxed{?} &= t \end{aligned}$$

In Exercises 15–26, solve the equation. Then check your solution.

15. $n + 17 = 98$
16. $m + 39 = 81$
17. $x - 61 = 78$
18. $z - 129 = 200$
19. $356 = y - 219$
20. $445 = t - 193$
21. $736 = x + 598$
22. $907 = s + 316$
23. $n + 1.7 = 3.9$
24. $11.31 = 5.31 + y$
25. $7.49 = m - 5.86$
26. $q - 12.42 = 9$
27. Explain the steps you used to solve Exercise 20.
28. Explain the steps you used to solve Exercise 21.

In Exercises 29–34, use a calculator to solve the equation.

29. $r + 217.46 = 598.07$
30. $952.7 = s + 420.38$
31. $1.397 = x - 1.973$
32. $y - 4.85 = 13.01$
33. $s + 1024 = 9785$
34. $5826 = r - 2290$

35. *Geometry* Two angles are **supplementary** if the sum of their measures is 180°. An angle whose measure is 74° is supplementary to an angle whose measure is m. Find m.

Computation Sense **In Exercises 36–43, write an equation that represents the statement. Then solve the equation.**

36. The sum of x and 49 is 165.

37. The sum of r and 2.4 is 7.2.

38. The difference of y and 5.8 is 12.2.

39. The difference of n and 40 is 38.

40. 89 is the sum of a number and 37.

41. 173 is the sum of a number and 93.

42. 141 is the difference of a number and 503.

43. 317 is the difference of a number and 723.

Downhill Skiing **In Exercises 44 and 45, use the information given in the caption of the photo.**

44. How much higher is the summit than the base?

45. You are skiing on a beginner trail. The head of the trail is 1530 feet higher than the base elevation. At what elevation does the beginner slope start?

46. *Buying Skis* After purchasing a pair of used skis for $89.99 (including tax), you have $5.63 left. How much money did you have before buying the skis?

Vail, Colorado, is a popular Rocky Mountain ski resort. The base elevation of the resort is 8200 feet. The summit elevation is 11,450 feet.

Integrated Review

Making Connections within Mathematics

Measurement Sense **In Exercises 47–50, complete the statement using <, >, or =.**

47. 7 feet [?] 82 inches

48. 2.5 miles [?] 13,200 feet

49. 320 cm [?] 3.2 m

50. 6500 mm [?] 65 m

Mental Math **In Exercises 51–56, use mental math to solve the equation.**

51. $x + \frac{1}{2} = 1$

52. $y + \frac{3}{5} = 1$

53. $z - \frac{3}{4} = 1$

54. $t - \frac{5}{2} = 1$

55. $4 - s = 2.5$

56. $3.3 - r = 2.6$

Exploration and Extension

57. *Commutative Property?* Does subtraction have a commutative property? That is, does $a - b = b - a$? Explain.

58. *Associative Property?* Does subtraction have an associative property? That is, does $a - (b - c) = (a - b) - c$? Explain.

59. *It's Up to You* Decide whether the commutative and associative properties are true for division. Explain.

Mid-Chapter SELF-TEST

Take this test as you would take a test in class. The answers to the exercises are given in the back of the book.

In Exercises 1 and 2, simplify the expression. (2.1)

1. $5(x + 3)$

2. $2(a + 2b + 4)$

In Exercises 3 and 4, match the rectangle with its perimeter. (There may be more than one correct match.) (2.1)

a. $2x + 2y$

b. xy

c. $x + y$

d. $2(x + y)$

e. $2(x + 1)$

f. x

g. $2x + 2$

h. $x + 1$

3. y x

4. x 1

5. You are buying a \$2.50 notebook and a \$1.20 book cover for each of your 6 classes. What property allows you to compute your total cost as $6(2.50 + 1.20)$ or as $6(2.50) + 6(1.20)$? What is your total cost? **(2.1)**

In Exercises 6–8, simplify the expression. (2.2)

6. $2a + 10a$

7. $2x + 8 + x$

8. $7(x + 3) + 2x$

9. Simplify $2(3x + 4) + x$. Then evaluate the expression when $x = 4$. **(2.2)**

10. List all of the expressions that are equivalent to $3(2x + 1) + 3x$. **(2.2)**

a. $6x + 3 + 3x$

b. $5x + 3 + 3x$

c. $9x + 3$

11. Write expressions for the perimeter and area of the rectangle. **(2.1, 2.2)**

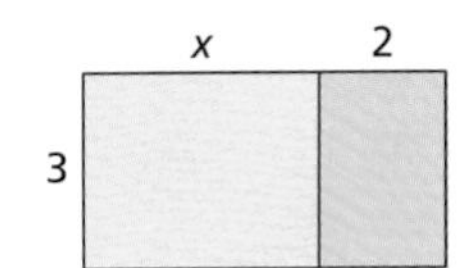

In Exercises 12–14, use mental math to solve the equation. (2.3)

12. $3x = 39$

13. $\frac{n}{4} = 20$

14. $\frac{1}{2}x = 17$

In Exercises 15–17, solve the equation. Show your work. (2.4)

15. $x + 13 = 28$

16. $19 = m - 4$

17. $7 + y = 11$

In Exercises 18–20, use the bar graph at the right. (2.4)
(Source: Video Store)

18. *Fantasia* has sold 3.6 million more units than *Bambi*. How many million units did *Bambi* sell?

19. *101 Dalmatians* sold 0.9 million units less than *E.T.* How many million units did *E.T.* sell?

20. *101 Dalmatians* sold 1.2 million units more than *Batman*. How many million units did *Batman* sell?

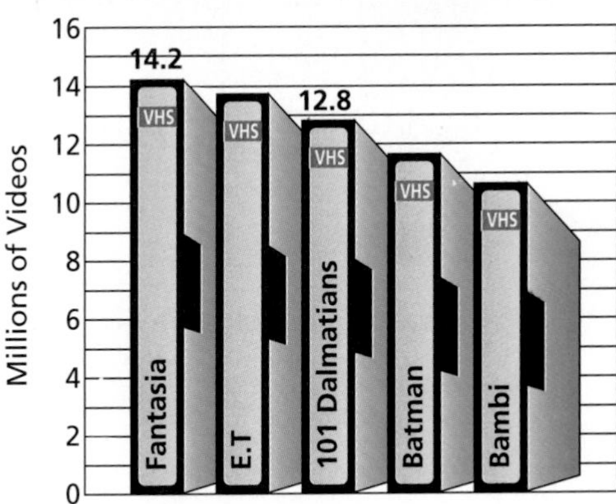

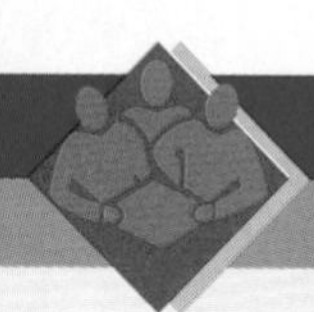

LESSON INVESTIGATION 2.5
Solving Equations

Materials Needed: algebra tiles

In this investigation, you will use algebra tiles to model and solve equations involving multiplication.

Example ***Using Algebra Tiles***

Model the equation $2x = 6$ with algebra tiles. Then use the tiles to solve the equation. Finally, check your solution.

Solution

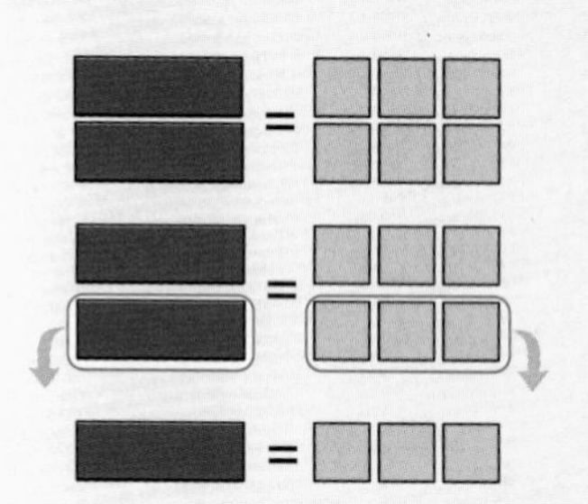

Original equation: $2x = 6$

To isolate x, divide each side into 2 groups and remove one group from each side.

Solution is $x = 3$.

You can check that $x = 3$ is a solution as follows.

$2x = 6$	*Write original equation.*
$2(3) \stackrel{?}{=} 6$	*Substitute 3 for x.*
$6 = 6$	*$x = 3$ is a solution.* ✔

■

Exercises

In Exercises 1 and 2, an equation has been modeled and solved with algebra tiles. Write the equation and its solution. Then check the solution.

1.

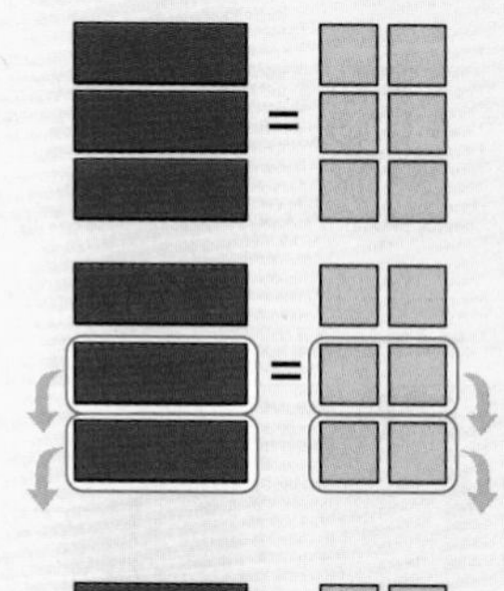

2.

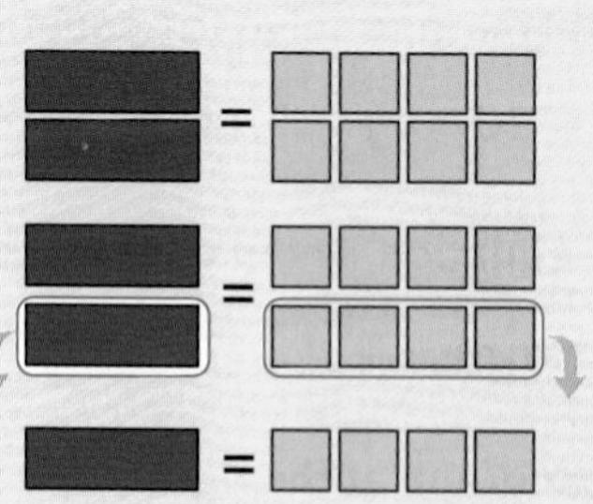

In Exercises 3–6, use algebra tiles to model and solve the equation.

3. $2x = 10$ **4.** $3n = 9$ **5.** $4y = 12$ **6.** $2b = 12$

2.5 Solving Equations: Multiplication or Division

What you should learn:

How to use multiplication or division to solve an equation

How to use equations to solve real-life problems

Why you should learn it:

You can use multiplication or division to solve equations that model real-life situations, such as finding the hourly rate for baby-sitting.

Goal 1 Using Multiplication or Division

In Lesson 2.4 you learned how to use addition or subtraction to solve an equation. In this lesson you will learn how to use multiplication or division to solve an equation.

Multiplication and Division Properties of Equality

Multiplying both sides of an equation by the same nonzero number or dividing both sides of an equation by the same nonzero number produces an equivalent equation.

Example 1 *Solving Equations*

Solve the equations **a.** $5x = 20$ and **b.** $12 = \frac{n}{4}$.

Solution

a. $5x = 20$ — *Rewrite original equation.*

$\frac{5x}{5} = \frac{20}{5}$ — *Divide both sides by 5.*

$x = 4$ — *Simplify.*

The solution is 4. Check this in the original equation.

b. $12 = \frac{n}{4}$ — *Rewrite original equation.*

$4 \cdot 12 = 4 \cdot \frac{n}{4}$ — *Multiply both sides by 4.*

$48 = n$ — *Simplify.*

The solution is 48. Check this in the original equation. ■

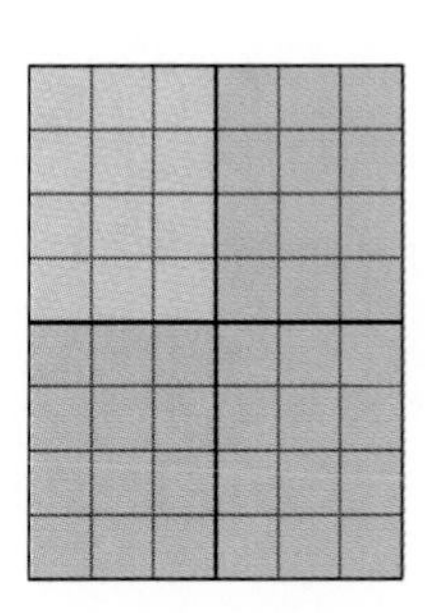

This is an area model for part b of Example 1. One-fourth of the area is 12. The entire area is 48.

In Example 1, notice that you can simplify a fraction that has a **common** factor in its numerator and denominator. Here are two other examples.

Fraction	Factor	Divide	Simplify
$\frac{6}{4}$	$\frac{2 \cdot 3}{2 \cdot 2}$	$\frac{\cancel{2} \cdot 3}{\cancel{2} \cdot 2}$	$\frac{3}{2}$
$\frac{3x}{3}$	$\frac{3 \cdot x}{3}$	$\frac{\cancel{3} \cdot x}{\cancel{3}}$	$\frac{x}{1}$ or x

Goal 2 Modeling Real Life with Equations

Example 2 *Using an Equation as a Real-Life Model*

Real Life
Baby-Sitting

In 1991, 2000 eighth-grade students were surveyed about after-school work. The 4 most common types of work were lawn work, restaurant work, newspaper deliveries, and baby-sitting. (Source: Teacher Magazine, May/June 1991)

You are baby-sitting for a neighbor. You arrive at 3:30 P.M. and leave at 8:00 P.M. Your neighbor pays you \$13.50. How much did you get paid per hour?

Solution From 3:30 to 8:00 is a total of 4.5 hours.

Verbal Model Total time • Hourly rate = Total pay

↓

Labels Total time = 4.5 (hours)
Hourly rate = x (dollars per hour)
Total pay = 13.5 (dollars)

↓

Algebraic Model $4.5 \cdot x = 13.5$

$$\frac{4.5x}{4.5} = \frac{13.5}{4.5}$$

$$x = 3$$

You got paid \$3 per hour. You can check this by multiplying to see that $4.5(3) = 13.5$.

Problem Solving: Checking Solutions Using Unit Analysis
When you are solving real-life problems that involve division, be sure to check that your units of measure make sense. For instance, you can check that the units of measure in Example 2 make sense as follows.

$$4.5 \text{ ~~hours~~} \cdot \frac{3 \text{ dollars}}{\text{~~hour~~}} = 13.5 \text{ dollars}$$

Communicating about MATHEMATICS

SHARING IDEAS about the Lesson

Unit Analysis Each of the following expressions was taken from a real-life situation. Simplify the expression, stating the units for the simplified result.

A. $\frac{50 \text{ miles}}{\text{hour}} \cdot 3 \text{ hours}$ **B.** $\frac{80 \text{ kilometers}}{\text{hour}} \cdot 2.5 \text{ hours}$

C. $\frac{2.5 \text{ dollars}}{\text{pound}} \cdot 4 \text{ pounds}$ **D.** $\frac{28 \text{ miles}}{\text{gallon}} \cdot 10 \text{ gallons}$

For each of the above expressions, describe a real-life situation that the expression could represent.

EXERCISES

Guided Practice

CHECK for Understanding

1. In your own words, state the Multiplication and Division Properties of Equality.

2. Give an example of how the Multiplication Property of Equality can be used.

3. Give an example of how the Division Property of Equality can be used.

In Exercises 4 and 5, state whether you would multiply or divide to solve the equation.

4. $6x = 54$

5. $\frac{x}{3} = 12$

6. **Problem Solving** Describe a real-life situation that can be modeled with an equation.

Independent Practice

In Exercises 7–10, write the equation as a verbal sentence. Then solve the equation.

7. $2x = 4$ **8.** $3x = 21$ **9.** $\frac{b}{2} = 3$ **10.** $\frac{a}{3} = 3$

In Exercises 11–34, solve the equation. Check your solution.

11. $4x = 16$ **12.** $12y = 144$ **13.** $56 = 7n$ **14.** $6s = 48$

15. $6 = \frac{x}{5}$ **16.** $\frac{m}{2} = 2$ **17.** $\frac{t}{8} = 9$ **18.** $5y = 100$

19. $2z = 50$ **20.** $10a = 240$ **21.** $\frac{b}{20} = 2$ **22.** $16 = \frac{x}{4}$

23. $\frac{n}{4} = 25$ **24.** $\frac{m}{3} = 33$ **25.** $5x = 625$ **26.** $7s = 175$

27. $6.3 = 3y$ **28.** $5t = 6.5$ **29.** $524 = \frac{a}{1}$ **30.** $\frac{y}{6} = 345$

31. $\frac{z}{3.2} = 8$ **32.** $\frac{t}{7.4} = 6$ **33.** $4.8b = 36$ **34.** $9.6x = 72$

35. Sketch an area model for $\frac{x}{3} = 36$.

36. Sketch an area model for $\frac{n}{5} = 20$.

In Exercises 37–44, use a calculator to solve the equation.

37. $456x = 1368$ **38.** $824x = 1648$ **39.** $23x = 966$ **40.** $55x = 3025$

41. $\frac{x}{9} = 1025$ **42.** $\frac{x}{8} = 624$ **43.** $\frac{x}{136} = 17$ **44.** $\frac{x}{189} = 19$

Geometry **In Exercises 45–48, find the width of the rectangle.**

45.

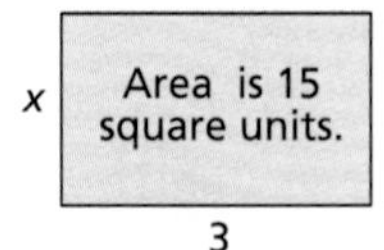

46.

x Area is 27 square units.

9

47.

x Area is 38 square units.

2

48.

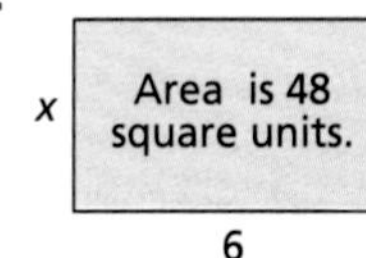

In Exercises 49–54, write an equation that represents the sentence. Then solve the equation.

49. The number of football players f times 4 equals 28 players.

50. The product of the number of dancers d and 4 is 100 dancers.

51. The number of bicycles b divided by 12 equals 2 bicycles.

52. The quotient of the number of telephones t and 6 is 10 telephones.

53. 5 comic books times x dollars is \$3.75.

54. The number of board games g times 20 dollars is 100 dollars.

55. *Basketball Court* The area of a basketball court is 4700 square feet. The width of a basketball court is 50 feet.

a. Write an equation that represents the area of a basketball court.

b. Solve the equation to find the length of a basketball court.

Basketball **In Exercises 56 and 57, use the following information.**

You play on your school's basketball team. This season your team played 24 games. You averaged 15.75 points per game.

56. Use the number of games played and your points-per-game average to write a verbal model that represents the points you scored this season.

57. Write an algebraic model that represents your total points scored this season. Then solve the equation to find your total.

58. *Rollerblading* You rollerblade 5 days a week. Each day you rollerblade the same distance in miles. How many miles a day do you rollerblade if you rollerblade a total of 20.5 miles in 5 days?

In the 1992–93 season, Shaquille O'Neal, a 7-foot-1-inch center for the Orlando Magic, finished among the top 10 in scoring for his first season. O'Neal scored 23.4 points per game, ranking eighth.

Integrated Review

Making Connections within Mathematics

Computing with Measures **In Exercises 59–62, perform the indicated operation.**

59. 1 hr 50 min + 37 min

60. 20 lb − (3 lb 4 oz + 1 lb 14 oz)

61. $3\frac{1}{2}$ yd ÷ 6

62. 126 gal ÷ 8

Exploration and Extension

Measurement Sense **In Exercises 63 and 64, perform the indicated conversion.**

63. Convert 4.5 dollars per pound to cents per ounce.

64. Convert 55 miles per hour to feet per second.

USING A SPREADSHEET

Make a Table

A **spreadsheet** is a computer program that creates tables. The following example shows how a spreadsheet can be used to help solve a real-life problem.

Example *Creating a Wage Table*

Make a table that shows your total pay for working from 1 to 8 hours at hourly rates ranging from $2.00 per hour to $4.50 per hour.

Solution The table is shown below.

Hourly Rate

Number of Hours Worked	**$2.00**	**$2.50**	**$3.00**	**$3.50**	**$4.00**	**$4.50**
1.0	$2.00	$2.50	$3.00	$3.50	$4.00	$4.50
1.5	$3.00	$3.75	$4.50	$5.25	$6.00	$6.75
2.0	$4.00	$5.00	$6.00	$7.00	$8.00	$9.00
2.5	$5.00	$6.25	$7.50	$8.75	$10.00	$11.25
3.0	$6.00	$7.50	$9.00	$10.50	$12.00	$13.50
3.5	$7.00	$8.75	$10.50	$12.25	$14.00	$15.75
4.0	$8.00	$10.00	$12.00	$14.00	$16.00	$18.00
4.5	$9.00	$11.25	$13.50	$15.75	$18.00	$20.25
5.0	$10.00	$12.50	$15.00	$17.50	$20.00	$22.50
5.5	$11.00	$13.75	$16.50	$19.25	$22.00	$24.75
6.0	$12.00	$15.00	$18.00	$21.00	$24.00	$27.00
6.5	$13.00	$16.25	$19.50	$22.75	$26.00	$29.25
7.0	$14.00	$17.50	$21.00	$24.50	$28.00	$31.50
7.5	$15.00	$18.75	$22.50	$26.25	$30.00	$33.75
8.0	$16.00	$20.00	$24.00	$28.00	$32.00	$36.00

You can use the table to find your total pay. For instance, if you work 5.5 hours at $3.50 per hour, then your total pay is $19.25 ■

Exercises

1. Make a table that shows the distance traveled for several different times and speeds.
2. Make a table that shows the areas of several rectangles of different widths and heights.

2.6 Modeling Verbal Expressions

What you should learn:

Goal 1 How to translate verbal phrases into algebraic expressions

Goal 2 How to model real-life situations with algebraic expressions

Why you should learn it:

To use algebra to solve real-life problems, you must translate verbal phrases into algebraic expressions.

Goal 1 Translating Verbal Phrases

When you are translating verbal phrases into algebraic expressions, look for words that indicate a number operation.

	Verbal Phrase	Algebraic Expression
Addition:	The *sum* of 5 and a number	$5 + x$
	Nine *more than* a number	$n + 9$
	A number *plus* 2	$y + 2$
Subtraction	The *difference* of 8 and a number	$8 - n$
	Ten *less than* a number	$y - 10$
	Twelve *minus* a number	$12 - x$
Multiplication:	The *product* of 3 and a number	$3x$
	Seven *times* a number	$7y$
	A number *multiplied* by 4	$4n$
Division:	The *quotient* of a number and 3	$\frac{x}{3}$
	Four *divided* by a number	$\frac{4}{n}$

Order is important in subtraction and division, but not for addition and multiplication. For instance, "ten less than a number" is written as $y - 10$, not $10 - y$. On the other hand, "the sum of 5 and a number" can be written as $5 + x$ or $x + 5$.

Example 1 gives you some examples of phrases that contain two number operations.

Study Tip...
Problem Solving *When you are modeling a verbal phrase, you can usually choose any letter to represent the variable. Although common choices are x, y, and n, it is sometimes helpful to choose a letter that helps remind you what the variable represents. For instance, you might choose t to represent time or A to represent area.*

Example 1 *Translating Verbal Phrases*

a. Three more than twice a number can be written as

$2x + 3$. *Label: x represents a number.*

b. The sum of a number and 3 times another number can be written as

$n + 3m$. *Labels: n is a number, m is another number.*

c. One number times the sum of 2 and another number can be written as

$x(2 + y)$. *Labels: x is a number, y is another number.* ■

Goal 2 Modeling Real-Life Phrases

Problem Solving When you are modeling a real-life situation, we suggest that you use three steps.

Write a verbal model. → Assign labels to the model. → Write the algebraic model.

In 1974, long-playing albums were the most popular form of storing music. In 1984, cassettes were the most popular. In 1994, compact discs were the most popular. What form do you think will be most popular in the year 2004? (Source: Record Industry Association of America)

Example 2 *Modeling a Real-Life Situation*

You are buying some cassettes and some compact discs. Each cassette costs $12 and each compact disc costs $15. Write an algebraic expression that represents your total cost. Then use the expression to find the cost of **a.** 3 cassettes and 2 compact discs and **b.** 2 cassettes and 3 compact discs.

Solution

Verbal Model [Cost per cassette] • [Number of cassettes] + [Cost per disc] • [Number of discs]

Labels
Cost per cassette = 12 (dollars per cassette)
Number of cassettes = c (cassettes)
Cost per disc = 15 (dollars per disc)
Number of discs = d (discs)

Algebraic Model $12 \cdot c + 15 \cdot d$

a. The cost of 3 cassettes and 2 discs is as follows.

$$\text{Total cost} = 12c + 15d \quad \textit{Algebraic expression}$$
$$= 12(3) + 15(2) \quad \textit{Substitute 3 for c and 2 for d.}$$
$$= \$66 \quad \textit{Simplify.}$$

b. The cost of 2 cassettes and 3 discs is as follows.

$$\text{Total cost} = 12c + 15d \quad \textit{Algebraic expression}$$
$$= 12(2) + 15(3) \quad \textit{Substitute 2 for c and 3 for d.}$$
$$= \$69 \quad \textit{Simplify.}$$

■

Communicating about MATHEMATICS

SHARING IDEAS about the Lesson

Guess, Check, and Revise In Example 2, suppose your total cost is $120. How many different combinations of cassettes and discs could you have purchased? Explain.

EXERCISES

Guided Practice

CHECK for Understanding

In Exercises 1–4, write the phrase as an algebraic expression.

1. The temperature, decreased by 20°
2. Five miles per hour more than the speed limit
3. $0.25 per minute, plus $1.40
4. Five dollars times the number of people, less $15
5. **Problem Solving** Describe the steps for modeling a real-life situation.
6. **Problem Solving** Describe the two kinds of labels that are used to develop an algebraic model.

Independent Practice

In Exercises 7–12, match the verbal phrase with its algebraic expression.

a. $20y$ **b.** $20 + n$ **c.** $\frac{m}{8} - 4$ **d.** $2s + 8$ **e.** $x - 6$ **f.** $6 - x$

7. The sum of a number and 20
8. Eight more than twice a number
9. The difference of 6 and a number
10. 6 less than a number
11. The product of a number and 20
12. 4 less than the quotient of a number and 8

In Exercises 13–26, translate the verbal phrase into an algebraic expression.

13. The difference of 32 and a number
14. 72 plus a number
15. A number divided by 23
16. The quotient of a number and 7
17. 9 more than 12 times a number
18. The sum of 1 and a number multiplied by 2
19. 4 less than 3 times a number
20. 8 minus the product of 5 and a number
21. 18 times the product of a number and 5
22. The sum of 3 and a number multiplied by 11
23. The quotient of a number and 2 more than another number
24. The difference of 10, and one number divided by another number
25. The product of one number and the sum of another number and 13
26. The product of one number minus 4 and another number plus 4

In Exercises 27–34, write as an algebraic expression.

27. Four more miles than yesterday
28. Your salary plus $572
29. 3 times as much money as your sister
30. Half as much money as your brother
31. 3 less runs than the Pirates scored
32. 7 years younger than your cousin
33. Number of passengers times 3
34. Number of students divided by 3

In Exercises 35–38, write an algebraic expression that represents the phrase. Let a represent your age now.

35. Your age eight years ago

36. 10 times your age

37. Three-fourths your age

38. Your age 17 years from now

39. *Collectors' Items* You are buying some comic books and baseball cards. Each comic book costs $1.50 and each pack of baseball cards costs $1.

a. Write an algebraic expression that represents your total cost. Let c be the number of comics and b be the number of packs of baseball cards.

b. Find the cost of 3 comic books and 4 packs of baseball cards.

c. Find the cost of 4 comic books and 2 packs of baseball cards.

40. *Playing Pool* You and two friends are playing pool. The cost of renting the pool table is $7 for the first hour and $3 for each additional half hour.

a. Write an algebraic expression that represents the total cost. Let h be the number of additional half hours.

b. You and your friends are sharing the total cost equally. Write an expression that represents *your* cost.

c. You play for $2\frac{1}{2}$ hours. Find *your* cost.

Basketball **In Exercises 41–43, use the graph and algebra to answer the questions. Then use the graph to check your answer.** *(Source: American Basketball Council)*

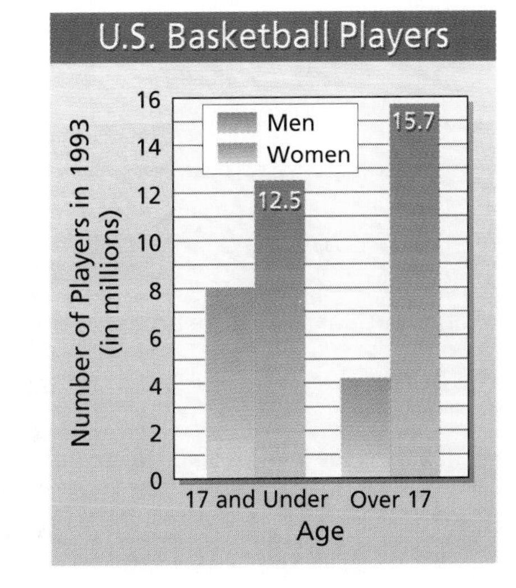

41. Let m and w represent the numbers (in millions) of men and women (17 years old and younger) who played basketball. These two numbers are related by the equation $m - 4.5 = w$. How many women (17 years old and younger) played basketball?

42. Let M and W represent the numbers (in millions) of men and women (over 17) who played basketball. These two numbers are related by the equation $M - 11.5 = W$. How many women (over 17) played basketball?

43. What is the total number of players?

44. *Currency Exchange* On February 7, 1994, the exchange rate between U. S. currency and Canadian currency was 1.25 Canadian dollars per 1 U. S. dollar. Use this information to complete the table.

U.S. Dollars	$1	$2	$5	$10	$$x$	?	?	?
Canadian Dollars	$1.25	$2.50	?	?	?	$7.50	$10	$1.25x$

Integrated Review

Making Connections within Mathematics

Algebraic Expressions **In Exercises 45–50, simplify the expression. Then evaluate the expression when $a = 5$.**

45. $6a + 10a - 65$ **46.** $7a + 4a + 3$ **47.** $9a - 5a + 4a$
48. $20a - 13a - 2a$ **49.** $3(a + 2) + 8(a + 7)$ **50.** $10(a + 2) + 2(a + 1)$

Exploration and Extension

Computation Puzzles **In Exercises 51 and 52, solve the puzzle. In each puzzle, you can use each natural number from 1 through 9 exactly once.**

51.

	+		×		=16
×	■	+	■	+	■
	+		−		=3
+	■	÷	■	−	■
	−		+		=9
=14	■	=5	■	=8	■

52.

	−		+		=9
−	■	×	■	+	■
	+		÷		=7
×	■	−	■	−	■
	+		÷		=4
=3	■	=15	■	=6	■

Mixed REVIEW

In Exercises 1–4, state an operation that can be used to solve the equation. (2.4, 2.5)

1. $c + 2 = 6$ **2.** $5b = 75$ **3.** $\frac{a}{3} = 2$ **4.** $d - 4 = 10$

In Exercises 5–12, solve the equation. (2.4, 2.5)

5. $8n = 32$ **6.** $m - 12 = 20$ **7.** $3n = 2$ **8.** $y + 6 = 10$
9. $x - 4 = 6$ **10.** $\frac{t}{8} = 2$ **11.** $y + 2 = 2$ **12.** $\frac{z}{12} = 3$

In Exercises 13–20, simplify the equation. Then solve it mentally. (1.5, 2.3)

13. $2f + 3f = 10$ **14.** $12r - 3r = 81$
15. $6g = 10 + 2$ **16.** $\frac{t}{2} = 4 + 1$
17. $2s + 3s + s = 18$ **18.** $8p + 2p - p = 9$
19. $4h - 3h + 2h = 9$ **20.** $20q + 9q - 12q = 51$

2.7 Real-Life Modeling with Equations

What you should learn:

How to translate verbal sentences into algebraic equations

How to model real-life situations with algebraic equations

Why you should learn it:

You can use algebraic equations to solve real-life problems, such as deciding how many pairs of jeans to order for a clothing store.

Goal 1 Translating Verbal Sentences

A phrase does not usually contain a verb, but a sentence must contain a verb. In Lesson 2.6, you learned that many phrases can be modeled as algebraic expressions. In this lesson, you will learn how to model sentences as algebraic equations. Here are examples of an expression and an equation.

Verbal Phrase	**Algebraic Expression**
The cost of several cassettes at \$12 each	$12x$
Verbal Sentence	**Algebraic Equation**
The cost of several cassettes at \$12 each is \$60.	$12x = 60$

Notice that you can solve an equation, but you cannot solve an expression. For instance, the solution of the equation $12x = 60$ is $x = 5$, but it doesn't make sense to try to "solve" the expression $12x$.

Problem Solving
Writing a Model

Example 1 *Modeling and Solving a Word Problem*

The price of one television set is \$283. It costs \$147 less than another set. What is the price of the more expensive set?

Solution

Verbal Model ↓

Smaller price = Larger price − 147

Labels ↓

Smaller price = 283 (dollars)
Larger price = x (dollars)

Algebraic Model

$283 = x - 147$ *Algebraic equation*

$283 + 147 = x - 147 + 147$ *Add 147 to each side.*

$430 = x$ *Simplify.*

The price of the more expensive set is \$430. You can check this by observing that \$283 is \$147 less than \$430. ■

Goal 2 Modeling Real-Life Sentences

Real Life
Retail Sales

Example 2 *Modeling a Real-Life Situation*

You work in a clothing store and are responsible for the jeans department. To earn a bonus in November–December, you need to sell at least 1.25 times as many pairs of jeans as you sold last November–December. The graph shows this year's sales, and you sold exactly as many pairs as was needed to make a bonus. How many pairs did you sell last November–December?

Solution From the graph at the left, you can estimate that this November the store sold about 450 pairs of jeans, and this December the store sold about 650 pairs of jeans.

Verbal Model $1.25 \cdot$ [Number sold last year] $=$ [Number sold this year]

↓

Labels Number sold last year $= n$ (pairs)

Number sold this year $= 450 + 650 = 1100$ (pairs)

↓

Algebraic Model

$$1.25 \cdot n = 1100$$

$$\frac{1.25n}{1.25} = \frac{1100}{1.25}$$

$$n = 880$$

You sold 880 pairs of jeans last November–December. ■

Communicating about MATHEMATICS

Cooperative Learning

▶ SHARING IDEAS about the Lesson

Modeling Real Life Work with a partner. Describe a real-life problem that fits the verbal model. Then assign labels to the model, write the equation, and solve the equation. Finally, explain how the solution of the equation helps you solve the real-life problem.

A. [Miles driven this week] $+ 72 =$ [Miles driven last week]

B. [Number of tires] $\cdot$ [\$110 per tire] $=$ [Total cost]

EXERCISES

Guided Practice

CHECK for Understanding

In Exercises 1–4, state whether the quantity is an expression or an equation. Solve or simplify.

1. $3 + x = 19$ **2.** $5x - 3x + 6$ **3.** $2x = 18$ **4.** $\frac{x}{3} = 7$

In Exercises 5 and 6, write an algebraic equation that represents the verbal sentence. Then solve the equation.

5. The number of cars decreased by 21 is 84.

6. The cost of 10 T-shirts at x dollars each is $75.

Independent Practice

In Exercises 7–12, match the sentence with an equation.

a. $7 = x + 5$ **b.** $\frac{x}{7} = 5$ **c.** $x - 7 = 5$ **d.** $7 = \frac{x}{5}$ **e.** $7x = 35$ **f.** $5x = 35$

7. The difference of x and 7 is 5.

8. 5 times x equals 35.

9. 7 is the sum of x and 5.

10. 7 equals x divided by 5.

11. The quotient of x and 7 is 5.

12. The product of x and 7 is 35.

In Exercises 13–18, write an algebraic equation that represents the verbal sentence. Then solve the equation.

13. The number of dogs increased by 9 is 20.

14. 16 equals the number of tennis shoes decreased by 3.

15. The number of pencils divided by 12 is 4.

16. The cost of 3 sweaters at x dollars each is $90.75.

17. y miles divided by 45 miles per hour equals 5 hours.

18. 128 baseball cards is the sum of 65 baseball cards and x baseball cards.

In Exercises 19–24, write a verbal sentence that represents the equation.

19. $f + 15 = 33$ **20.** $90 = s - 3$ **21.** $7c = 56$

22. $\frac{b}{9} = 8$ **23.** $11 + a = 23$ **24.** $21 = d - 18$

In Exercises 25 and 26, use a verbal model, labels, and an algebraic model to answer the question.

25. One number is 251 more than another number. The larger number is 420. What is the smaller number?

26. The product of a number and 38 is 912. Find the missing number.

Geometry **In Exercises 27 and 28, use the verbal model to write an algebraic equation. Then use the guess, check, and revise method to solve the equation.**

27. Perimeter = 2(Width + Length)

8

w

Perimeter is 25 inches.

28. Area = Width • Length

$w + 2$

w

Area is 24 square units.

The Fashion Business **In Exercises 29 and 30, use the following information.**

The company you work for designs a variety of clothes. Last week's sales (in units) are shown below. You expect the total sales for this week to be 500 units more than last week.

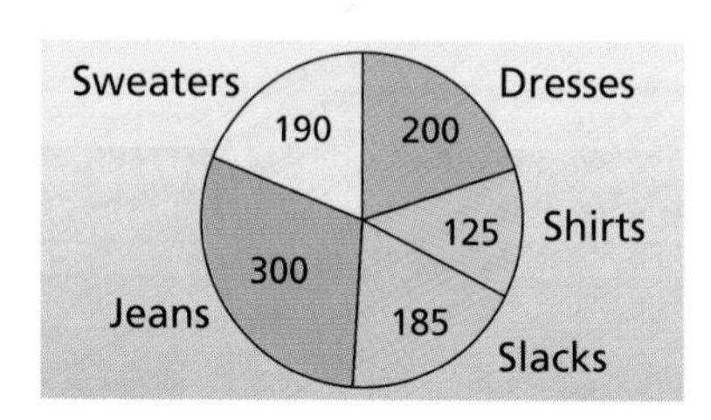

29. In the following equation, what does n represent?
$n = (200 + 125 + 185 + 190 + 300) + 500$

30. How many of the 500 additional would you expect to be sweaters? Explain your reasoning.

Many fashion designers use computers to design clothing.

Integrated Review

Making Connections within Mathematics

Equation Sense **In Exercises 31–38, solve the equation.**

31. $x + 9 = 13$ **32.** $x - 21 = 19$ **33.** $12x = 48$ **34.** $\frac{x}{8} = 10$

35. $x + 3.4 = 6.6$ **36.** $x - 5.5 = 6.6$ **37.** $7.2x = 36$ **38.** $\frac{x}{8.2} = 20$

Exploration and Extension

Mental Math **In Exercises 39–42, write an algebraic equation that represents the verbal sentence. Use mental math to solve the equation.**

39. 2 times x, increased by 6, is 12.

40. The difference of 7 times x and 6 is 8.

41. 5 times the sum of x and 4 is 25.

42. 8 times the difference of x and 2 is 8.

2.8 A Problem-Solving Plan

What you should learn:

Goal 1 How to use a systematic problem-solving plan

Goal 2 How to use other problem-solving strategies such as "solving a simpler problem"

Why you should learn it:

Learning to be a systematic problem solver will help you in many real-life occupations, such as being a sales representative.

Goal 1 Using a Problem-Solving Plan

In Lessons 2.6 and 2.7, you studied the three steps for algebraic modeling. These steps can be used as part of your **general problem-solving plan.**

A Problem-Solving Plan

1. Ask yourself what you need to know to solve the problem. Then **write a verbal model** that will give you what you need to know.
2. **Assign labels** to each part of your verbal model.
3. Use the labels to **write an algebraic model** based on your verbal model.
4. **Solve** the algebraic model.
5. **Answer** the original question.
6. **Check** that your answer is reasonable.

Example 1 *Using a Problem-Solving Plan*

Real Life
Sales Bonus

You are a sales representative for a publisher. You receive a salary plus a bonus, which is one-twentieth of the amount by which you exceed your previous year's sales. Last year your sales totaled \$456,000. This year you earned a bonus of \$9300. What were your sales this year?

Solution

Verbal Model $\frac{1}{20} \cdot$ Amount over last year's sales $=$ Bonus

Labels Amount over last year's sales $= x$ (dollars)
Bonus $= 9300$ (dollars)

Algebraic Model

$$\frac{1}{20} \cdot x = 9300$$
$$20 \cdot \frac{1}{20} \cdot x = 20 \cdot 9300$$
$$x = 186{,}000$$

You sold \$186,000 more than last year. Thus, your sales for this year were 456,000 + 186,000 or \$642,000. ■

Study Tip...
Problem Solving *An important part of your problem-solving plan is to be sure that you have actually* ***answered the question*** *posed in the real-life problem. For instance, in Example 1, the solution of the equation (x = 186,000) is not the answer to the question. The answer to the question is 456,000 + 186,000 or \$642,000.*

Goal 2 Using Other Problem-Solving Strategies

In this (and every) lesson, don't be afraid to try a variety of problem-solving strategies. For instance, you might try "guess, check, and revise" or "solving a simpler problem."

Real Life
Planning a Career

Eight years after working with only a high school diploma, Deloris McClam went back to school. She earned her bachelor's degree in 1987. By 1993, her salary was over $60,000. She maintains that her degree has made a significant difference in her career. (Source: Black Enterprise)

Example 2 *Decision Making*

It is 1990, and you have just graduated from high school. Explain how to use the table to decide whether a college degree is a good financial investment. ***(Source: U.S. Bureau of Census)***

Education	Median Income (1990)
9th to 12th Grade, No Diploma	$5,904
High school graduate	$12,924
Some College, No Degree	$15,360
Associate Degree	$20,064
Bachelor's Degree	$25,392
Master's Degree	$33,864
Doctorate Degree	$46,260
Professional Degree	$59,532

Solution **Simplify the Problem.** This is a complicated question, with many unknowns, such as the cost of college, the rate of inflation, and the actual salaries you will receive. However, to obtain a general sense of the answer, assume that a 4-year degree will cost $30,000. Assume that you work for 45 years with a high school degree or work for 41 years with a college degree.

High School Degree $45(\$12{,}924) = \$581{,}580$

College Degree $41(\$25{,}392) - \$30{,}000 = \$1{,}011{,}072$

Although this analysis is overly simple, it still appears clear that a college degree is a good financial investment. ■

Communicating about MATHEMATICS

SHARING IDEAS about the Lesson

It's Up to You Suppose you have just obtained a bachelor's degree. Use the above table to decide whether a master's degree (requiring 2 years) is a good financial investment.

EXERCISES

Guided Practice

CHECK for Understanding

1. Order the steps for a general problem-solving plan.
   a. Assign labels.
   b. Answer the original question.
   c. Solve the algebraic model.
   d. Write a verbal model.
   e. Check your solution.
   f. Write an algebraic model.
2. When using the problem-solving plan, why is it important to *answer the original question*?

Independent Practice

Getting an A **In Exercises 3–7, consider the following question.**

Your history grade is based on five 100-point tests. To earn an A, you need a total of 460 points. Your first four test scores are 89, 85, 92, and 97. What is the minimum score you need to earn an A?

3. Write a verbal model that relates the total points needed, the number of points obtained, and the final test score.
4. Assign labels to the three parts of your model.
5. Use the labels to translate your verbal model into an algebraic model.
6. Solve the algebraic model.
7. Your final test score is 98 points. Did you get an A?

Taking a Hike **In Exercises 8–12, consider the following question.**

You are taking a three-day hiking and camping trip. The trail is 29 miles long. On the first day you hike 8 miles. On the second day you hike 11 miles. How much farther do you need to go?

8. Write a verbal model that relates the trail length, the number of miles traveled, and the distance left to hike.
9. Assign labels to the three parts of your model.
10. Use the labels to translate your verbal model into an algebraic model.
11. Solve the algebraic model.
12. Explain why your solution is reasonable.

The Appalachian National Scenic Trail extends about 2000 miles from Mt. Katahdin in Maine to Springer Mountain in Georgia.

Earning a Commission **In Exercises 13–17, consider the following question.**

You are an assistant manager at a computer software store. Your annual earnings consist of your base salary and your sales commission. Your commission is one-twentieth of the amount of your annual sales. Your base salary is $16,500 and you sell $150,000 of merchandise this year. How much money did you make?

13. Write a verbal model that relates your sales commission, your commission rate, and your annual sales.

14. Assign labels to each part of your model.

15. Use the labels to translate your verbal model into an algebraic model.

16. Solve the algebraic model.

17. Answer the original question and check your answer.

Necessary Information **In Exercises 18 and 19, decide which information is necessary to solve the problem. Then solve the problem.**

18. Every morning, except Sunday, your sister gets up at 5:30 A.M. and travels 7 miles to the skating rink to practice. Each practice lasts $2\frac{1}{2}$ hours. How many miles does she travel each week in order to practice?

19. Your Spanish club is selling T-shirts. Each shirt sells for $9.50. The club needs to raise $2,000 for a trip. You sell 17 T-shirts and have 3 left to sell. How much money have you raised?

Integrated Review

Making Connections within Mathematics

Mental Math **In Exercises 20–23, evaluate the expression.**

20. $\frac{1}{3}(84)$ **21.** $\frac{1}{5}(375)$ **22.** $\frac{1}{20}(72{,}000)$ **23.** $\frac{1}{15}(81{,}000)$

Geometry **In Exercises 24–27, find the length of each side of the square.**

24. Area is 225 square inches.

25. Area is 441 square centimeters.

26. Area is 96.43 square meters.

27. Area is 51.98 square feet.

Exploration and Extension

Is Your Answer Reasonable? **In Exercises 28–30, decide if the answer seems reasonable. Explain your reasoning.**

28. A problem asks you to find the time it takes to drive a car from Washington, D.C., to Portland, Oregon, and your answer is 12 hours.

29. A problem asks you to find the area of your bedroom, and your answer is 650 square feet.

30. A problem asks you to find the average age of the students in your class, and your answer is 13.46.

2.9 Exploring Variables and Inequalities

What you should learn:

How to solve simple inequalities

How to use inequalities as algebraic models to solve real-life problems

Why you should learn it:

You can use inequalities to solve real-life problems, such as describing the number of yards needed for a first down.

Goal 1 Solving Simple Inequalities

Some sentences are better modeled with **inequalities** instead of equations. An inequality is formed when an **inequality symbol** is placed between two expressions. Here are four kinds of inequality symbols.

Inequality Symbol	Meaning
$<$	is less than
$\leq$	is less than or equal to
$>$	is greater than
$\geq$	is greater than or equal to

A **solution** of an inequality is a number that produces a true statement when it is substituted for the variable in the inequality. For instance, 2 is one of the many solutions of $x < 5$ because $2 < 5$ is a true statement.

Finding all solutions of an inequality is called **solving the inequality.** You can do this in much the same way you solve an equation. That is, you can add or subtract the same number from each side, or you can multiply or divide both sides by the same positive number.

Reading Inequalities

The inequality in part c is read as "16 is greater than x." This inequality can also be written as $x < 16$, which is read as "x is less than 16."

Example 1 *Solving Inequalities*

a.

$x + 4 \leq 6$	*Original inequality*
$x + 4 - 4 \leq 6 - 4$	*Subtract 4 from each side.*
$x \leq 2$	*Solution of inequality*

b.

$3x \geq 12$	*Original inequality*
$\frac{3x}{3} \geq \frac{12}{3}$	*Divide both sides by 3.*
$x \geq 4$	*Solution of inequality*

c.

$14 > x - 2$	*Original inequality*
$14 + 2 > x - 2 + 2$	*Add 2 to each side.*
$16 > x$	*Solution of inequality*

■

Goal 2 Modeling Real Life with Inequalities

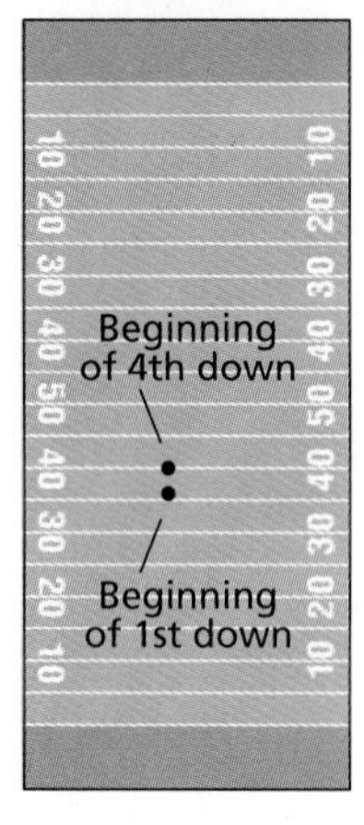

Example 2 *Modeling a Real-Life Situation*

Your team begins its first down on its own 36th yard line. By the fourth down, the team is beginning on its own 40th yard line. How many additional yards must the team gain to be awarded a new first down?

Solution Going into its fourth down, the team has gained 4 yards. By the end of the fourth down, the team needs to have gained at least 10 yards to be awarded a new first down.

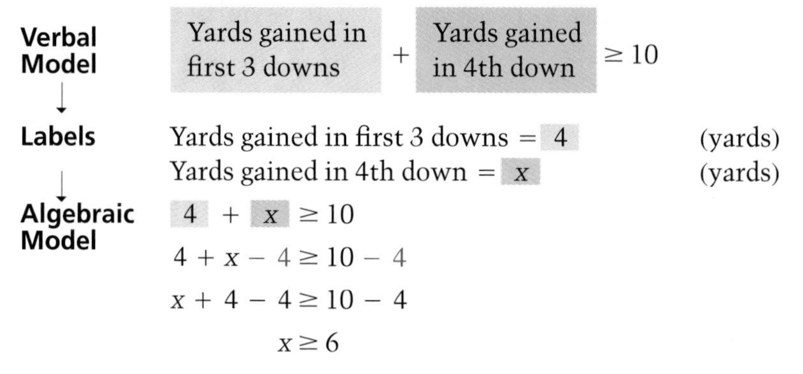

Verbal Model

Yards gained in first 3 downs + Yards gained in 4th down ≥ 10

Labels

Yards gained in first 3 downs = 4 (yards)

Yards gained in 4th down = x (yards)

Algebraic Model

$$4 + x \geq 10$$
$$4 + x - 4 \geq 10 - 4$$
$$x + 4 - 4 \geq 10 - 4$$
$$x \geq 6$$

The team must gain at least 6 yards in its fourth down. ■

Communicating about MATHEMATICS

SHARING IDEAS about the Lesson

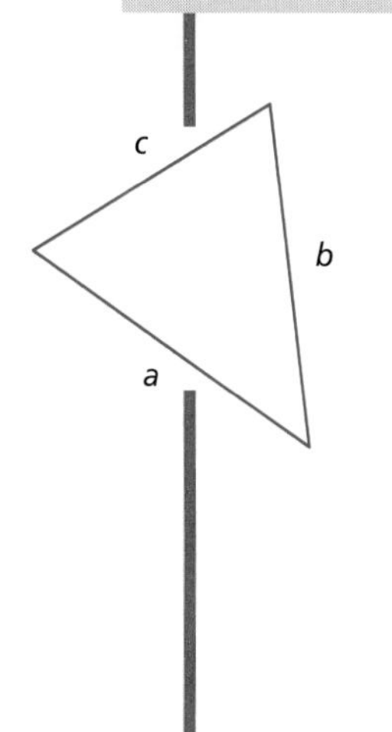

The Triangle Inequality A theorem in geometry states the sum of the lengths of any two sides of a triangle must be greater than the length of the third side of the triangle. For instance, in the triangle at the left, you can write the inequalities

$$a + b > c, \quad a + c > b, \quad \text{and} \quad b + c > a.$$

In the following triangles, what does the Triangle Inequality allow you to say about the value of x?

A.

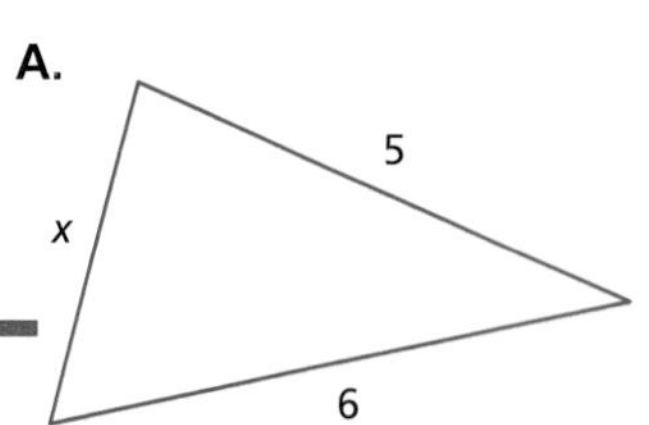

B.

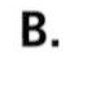

EXERCISES

Guided Practice

CHECK for Understanding

1. Write the four kinds of inequality symbols and their meanings.
2. *True or False?* An inequality can have many solutions.

In Exercises 3–6, state whether the number is a solution of the inequality $x - 5 < 11$.

3. 11
4. 16
5. 15
6. 30

In Exercises 7–9, solve the inequality.

7. $x + 4 \leq 7$
8. $3x \geq 10$
9. $9 < x - 5$
10. **Problem Solving** State an example of an inequality used in a real-life situation.

Independent Practice

In Exercises 11–16, state two solutions of the inequality.

11. $x < 4$
12. $y \geq 12.3$
13. $45 < x$
14. $100 \geq a$
15. $t < 2\frac{1}{2}$
16. $y \geq 0$

In Exercises 17–34, solve the inequality.

17. $x + 5 < 11$
18. $s - 4 > 9$
19. $7y \leq 42$
20. $\frac{x}{8} \geq 11$
21. $22 \leq b + 22$
22. $16 \geq s - 3$
23. $56 < 14t$
24. $45 > 5m$
25. $x + 25 \geq 26$
26. $n - 34 \leq 16$
27. $17y \geq 68$
28. $\frac{x}{2} \leq 52$
29. $x + 3.4 > 5.8$
30. $y - 13.7 < 5.4$
31. $8.9k \geq 17.8$
32. $\frac{a}{2.5} \leq 4.2$
33. $3.8 < \frac{x}{5.5}$
34. $138.6 > 5.5y$

In Exercises 35–40, write an inequality that represents the sentence. Then solve the inequality.

35. c plus 5 is greater than or equal to 19.36.
36. The difference of b and 7 is less than 24.
37. x times 2 is less than 42.
38. The product of y and 3 is greater than 39.
39. 20 is greater than or equal to m divided by 6.
40. 15 is less than or equal to the quotient of x and 5.

In Exercises 41–46, write a sentence that represents the inequality.

41. $d + 11 < 52$ **42.** $f - 5 \geq 29$ **43.** $3h \leq 60$

44. $\frac{p}{36} > 2$ **45.** $17 > c - 31$ **46.** $23 \leq e + 9$

Number Sense **In Exercises 47–54, state whether the inequality $x \leq 5.6$ is true for the value of x.**

47. 5 **48.** 5.65 **49.** 5.61 **50.** 5.60

51. 5.59 **52.** 5.7 **53.** 5.06 **54.** 0.56

55. ***Bike Racing*** You are a member of the "Hot Wheelers" bicycling team. Your team competes in a relay race with 4 other teams. The race consists of 20 miles of semi-rugged terrain. The table at the right shows the times that each team took to finish.

a. Who was the first place team?

b. Convert the first place team's time from minutes to hours.

c. Write verbal and algebraic models to show how many fewer minutes the last place team would need to place first.

d. Write verbal and algebraic models to show how many fewer minutes the Cruisin' Kids would need to tie or beat the Brave Bikers.

e. Using the formula $D = r \cdot t$, find the average speed of the Hot Wheelers.

Team Name	Time
Hot Wheelers	69 minutes
Bikin' Buddies	85 minutes
Cruisin' Kids	76 minutes
Brave Bikers	71 minutes
Flyin' Friends	81 minutes

Integrated Review

Making Connections within Mathematics

Reasoning **In Exercises 56–61, complete the statement with $<$, $>$, or $=$.**

56. $6 \cdot 9$? $13 \cdot 4$ **57.** $124 - 14$? $50 \cdot 2$ **58.** $\frac{120}{6}$? $\frac{342}{18}$

59. $3(x)$? $3(x + 1)$ **60.** $4x + 4x$? $5x + 3x + 2$ **61.** $6(x + 2)$? $6x + 2$

Exploration and Extension

Triangle Inequality **In Exercises 62–65, decide whether the numbers could be lengths of the sides of a triangle. The *Triangle Inequality* states that the sum of the lengths of *any* two sides must be greater than the length of the third side.**

62. 4, 5, 6 **63.** 2, 3, 6 **64.** 4, 6, 12 **65.** 8, 9, 11

2 Chapter Summary

What did you learn?

Skills

1. Use the Distributive Property to simplify expressions. **(2.1)**
2. Simplify expressions by adding like terms. **(2.2)**
3. Use substitution to check a solution of an equation. **(2.3)**
4. Solve equations
   - using mental math. **(2.3)**
   - using addition or subtraction. **(2.4)**
   - using multiplication or division. **(2.5)**
5. Write algebraic models.
   - Translate verbal phrases as algebraic expressions. **(2.6)**
   - Translate verbal sentences as algebraic equations. **(2.7)**
   - Translate verbal sentences as algebraic inequalities. **(2.9)**
6. Solve inequalities.

Problem-Solving Strategies

7. Write equations or inequalities as algebraic models of real-life situations. **(2.4–2.9)**
8. Use other problem-solving strategies such as guess, check, and revise and solve a simpler problem. **(2.1–2.9)**
9. Use a general problem-solving plan to solve real-life problems. **(2.8–2.9)**

Exploring Data

10. Use tables and graphs to interpret and organize data. **(2.2, 2.7–2.9)**

Why did you learn it?

In this chapter, you saw how companies can use mathematics to help plan for future sales or order merchandise. You also saw how a sales representative can use mathematics to determine his or her bonus, how you can use mathematics to help determine the financial value of further education, and how a football team can use mathematics to determine the number of yards needed to make a first down. As you continue studying this book, you will see many other examples that point out that algebra and geometry are not just useful to engineers and scientists. They can be useful to anyone—in a wide variety of real-life situations.

How does it fit into the bigger picture of mathematics?

In this chapter, you studied the three basic building blocks of algebra: expressions, equations, and inequalities. You were also introduced to techniques that can be used to solve simple equations and inequalities. Most of the equations and inequalities in this chapter can be solved with mental math. In addition to using mental math, be sure that you can also solve equations and inequalities using systematic techniques, such as adding the same number to both sides. The value of these techniques will become clear in later chapters when you study more difficult equations.

Chapter REVIEW

In Exercises 1–4, use the Distributive Property to rewrite the expression. (2.1)

1. $3(9+10)$
2. $7(5+8)$
3. $4(x+6)$
4. $2(y+12)$

In Exercises 5–16, simplify the expression. (2.1, 2.2)

5. $3m+8m$
6. $12x+5x$
7. $18x^2+12x+12x^2$
8. $5r^2+2r+r^2+9r^2$
9. $x^2+x+3x+7x^2$
10. $2m+4m+7m+9m^2$
11. $16w+10+4w+12$
12. $9+r+47+39r$
13. $3(a+4b)+4(3a+b)$
14. $9(x+2y)+2(9x+y)$
15. $3(x+y)+2x+y$
16. $4(s+t)+5s$

Mental Math **In Exercises 17–20, decide whether the given value is a solution. (2.3, 2.9)**

17. $4x+7=43;\ x=9$
18. $12m-16=36;\ m=4$
19. $\frac{s}{2}-24<8;\ s=64$
20. $\frac{36}{t}+10\geq 16;\ t=6$

In Exercises 21–29, use mental math to solve the equation. (2.3)

21. $17+p=25$
22. $q+11=35$
23. $y-17=34$
24. $16-z=9$
25. $7m=63$
26. $64=16m$
27. $\frac{x}{8}=7$
28. $\frac{n}{5}=6$
29. $\frac{72}{r}=18$

Raising the Vasa **In Exercises 30–32, use the following information. (2.3, 2.4)**

The *Vasa,* a Swedish war vessel, was one of the largest warships of its time. However, in 1628, the *Vasa* sank on its maiden voyage. It never fought a battle. In 1961, restoration crews raised it from the bottom of the Stockholm Harbor. It took 18 years to completely restore the hull of the vessel.

30. How many years was the *Vasa* underwater?
31. How long ago did the *Vasa* sink?
32. In what year was the restoration on the hull of the vessel completed?

A Swedish Warship ***This model of the Vasa is one-tenth the original size. It took four years to build.***

In Exercises 33–40, write an algebraic model for the phrase or sentence. (2.6, 2.7, 2.9)

33. The sum of a number and 71
34. The difference of 15 and a number
35. Eight students receive an equal share.
36. Three times more than your friend has
37. The difference of a number and 16 is 19.
38. Twelve plus another number is 51.
39. The product of a number and 7 is 84.
40. The quotient of 35 and a number is 7.

In Exercises 41–44, write an algebraic equation or inequality for the verbal sentence. Then solve. (2.7, 2.9)

41. The quotient of a number and 6 is equal to 3.

42. The product of a number and 8 is equal to 72.

43. The difference of a number and 15 is less than or equal to 3.

44. The product of a number and the sum of 7 and 8 is greater than 45.

In Exercises 45–56, solve the equation. Then check your solution. (2.3, 2.4, 2.5)

45. $x + 34 = 82$ **46.** $m + 92 = 131$ **47.** $37 = y - 15$

48. $z - 39 = 61$ **49.** $14.2 + t = 29.1$ **50.** $19.68 = s - 14.58$

51. $20x = 420$ **52.** $18x = 711$ **53.** $\frac{r}{12} = 12$

54. $\frac{x}{1} = 1$ **55.** $\frac{n}{2.2} = 3.3$ **56.** $5.5m = 100.1$

In Exercises 57–65, solve the inequality. (2.9)

57. $x + 5 > 14$ **58.** $71 \le x + 49$ **59.** $y - 23 < 14.4$

60. $r - 1.1 \ge 3.02$ **61.** $3x > 18$ **62.** $12x \le 144$

63. $\frac{x}{8} < 5$ **64.** $\frac{x}{2} \ge 54$ **65.** $\frac{x}{4.4} > 12.5$

Geometry **In Exercises 66 and 67, write simplified expressions for the perimeter and area of the region. (2.1, 2.2)**

66.

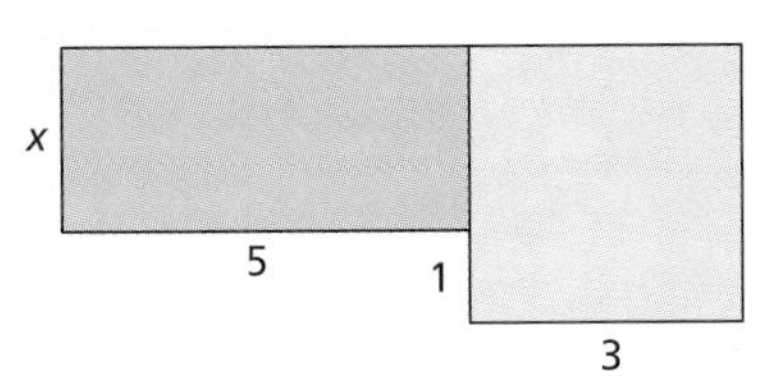

67.

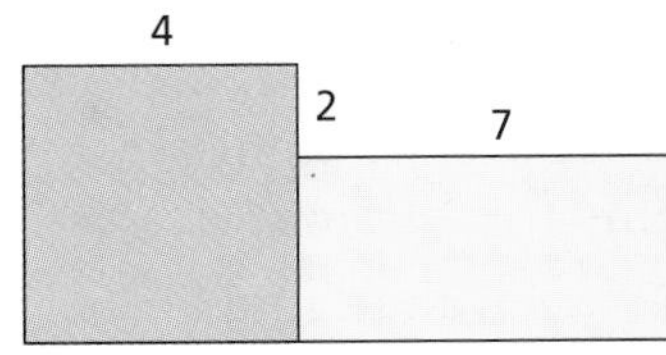

Savings Plan **In Exercises 68–72, consider the following question. (2.8)**

You are saving your money to buy a portable CD player. You have a paper route and you earn \$14.50 a week. The portable CD player costs \$135 including tax. You spend \$5 a week and save the rest of your money. In how many weeks will you have enough money to buy the CD player?

68. Write a verbal model that relates the price of the portable CD player, the amount you save each week, and the number of weeks you need to save your money.

69. Assign labels to each part of your model.

70. Use the verbal model to write an algebraic model.

71. Solve the algebraic model.

72. Answer the question and check your answer.

Fashion Sales **In Exercises 73–78, use the following information.**

The Gap is a popular retail clothing company. It specializes in jeans, sweatsuits, shirts, sweaters, and other casual apparel. The table gives information about *The Gap*. ***(Source: The Gap, Inc.)***

Year	1990	1991	1992	1993	1994
Sales (millions)	\$1,934	\$2,519	\$2,960	\$3,300	\$3,800
Number of Stores	1,092	1,216	1,307	1,420	1,570
Profit (millions)	\$144.5	\$229.9	\$210.7	\$243	\$282

73. What was the total profit for 1990 through 1994?

74. Profit is the difference between sales and cost. Write an algebraic model for this statement.

75. Use the algebraic model to find *The Gap's* cost for each year given in the table.

76. Let S represent the number of stores in 1990 and let R represent the number of stores in 1994. Write an equation that relates S and R.

77. Let P represent the profit (in millions of dollars). During which years did the profit satisfy the inequality $P \geq 210$?

78. Let P represent the profit (in millions of dollars) and let S represent the number of stores. The expression $\frac{P}{S}$ represents the average profit per store. For which year was this average profit greatest?

79. *Research Project* Research a career in the field of fashion or sports uniform design.

a. Write your results. Include the name of the profession, job requirements, and the environment of the job.

b. List several ways that mathematics is used in the profession.

c. If you know a person who has the profession you have researched, interview the person and write about the results.

Clothing Patterns **In Exercises 80 and 81, estimate the area of the pattern piece. Each square is 6 inches by 6 inches.**

80.

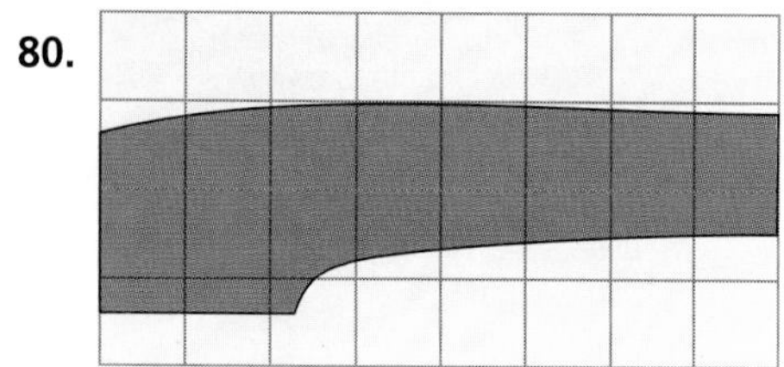

81.

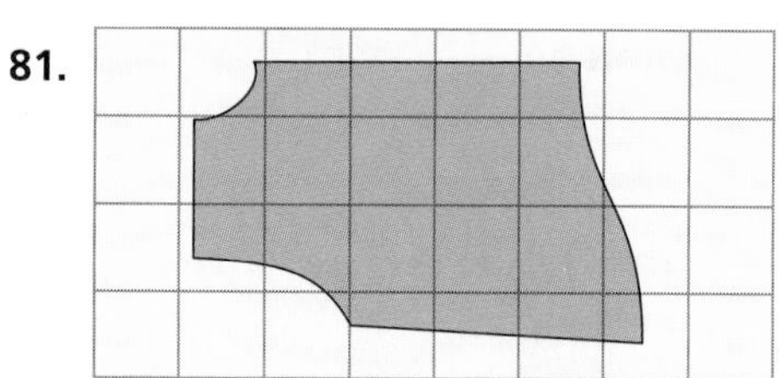

In Exercises 1 and 2, apply the Distributive Property.

1. $7(a+2)$

2. $3(b+2c+3)$

In Exercises 3–6, solve the equation. Show your work.

3. $3x=15$

4. $y-6=0$

5. $p+2=9$

6. $\frac{1}{4}q=3$

In Exercises 7–9, solve the inequality. Show your work.

7. $x-2>4$

8. $10\geq 3+y$

9. $7+z<14$

10. Write simplified expressions for the perimeter and area of the rectangle.

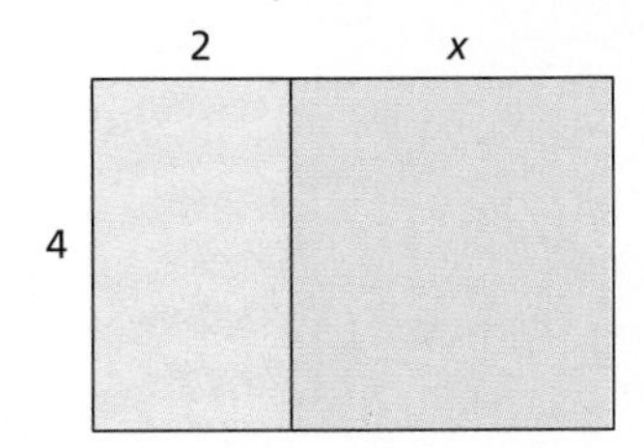

11. ***Business*** You are the manager of two sporting goods stores. The weekly cost to operate Store 1 is \$3500 and the weekly cost to operate Store 2 is \$5600.

a. Write a verbal model that represents the yearly costs of operating the two stores.

b. Use the model in Part **a** to determine the yearly cost of operating both stores.

In Exercises 12–14, you buy *n* juice boxes. Each juice box costs \$0.50.

12. Write an algebraic expression for the total amount you spent.

13. You spent \$3.00. Write an equation that can be solved to find the number of juice boxes you bought.

14. Solve the equation to find the number of juice boxes.

In Exercises 15 and 16, simplify by adding like terms.

15. $3a+6b+2a$

16. $12p+4q+2q+3p$

In Exercises 17–19, write an algebraic equation or inequality that represents the verbal sentence. Then solve.

17. A number decreased by 14 is at least 36.

18. The product of a number and 12 equals 60.

19. A number divided by 21 is 15.

In Exercises 20 and 21, write an expression for the perimeter. Find the perimeter when *x* is 1, 2, 3, and 4. Represent your results in a table.

20.

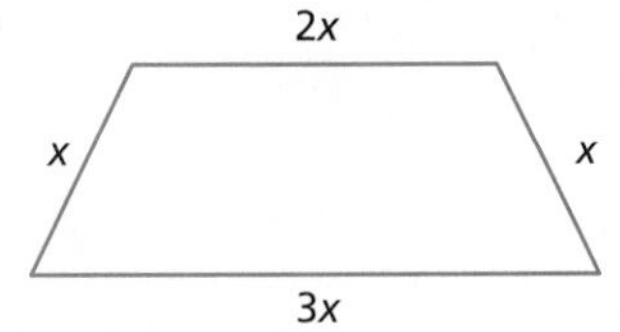

21.

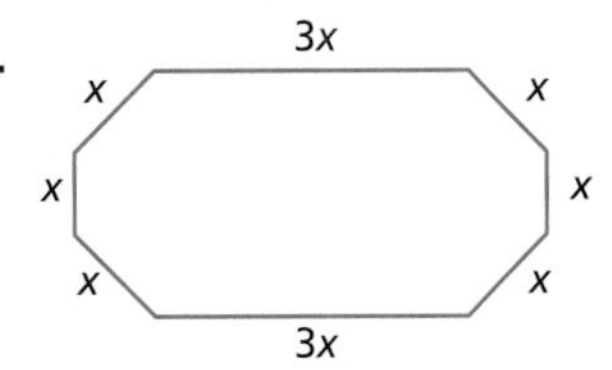

CHAPTER 3

Modeling Integers

LESSONS

Sports medicine has helped today's coaches learn how to protect their athletes from permanent damage and from injuries from sprains and breaks. Of all athletes, dancers have been proven to be the most fit and the most flexible. So, many athletic coaches have their teams trained in dance movements to improve their endurance and flexibility.

To test your own flexibility, mark a line on the floor and place a yardstick across the line as shown. Sit with your feet about 5 inches apart, heels on the line. With your knees straight, *slowly* bend forward from your waist as far as you can. Mark the point at which you touch the stick. Measure the distance from this point to the line. Use this chart to determine your score.

Distance from Line			
< −7″ Needs help	−7″ to 2″ Fair	2″ to 7″ Good	> 7″ Excellent

In this test, the yardstick is used as a kind of number line. The number line is just one of the models for positive and negative numbers that are used to explore integers in this chapter.

3.1 Integers and Absolute Value

What you should learn:

Goal 1 How to model integers on a number line

Goal 2 How to find the absolute value of a number

Why you should learn it:

You can use integers to model real-life situations, such as temperatures that are below zero.

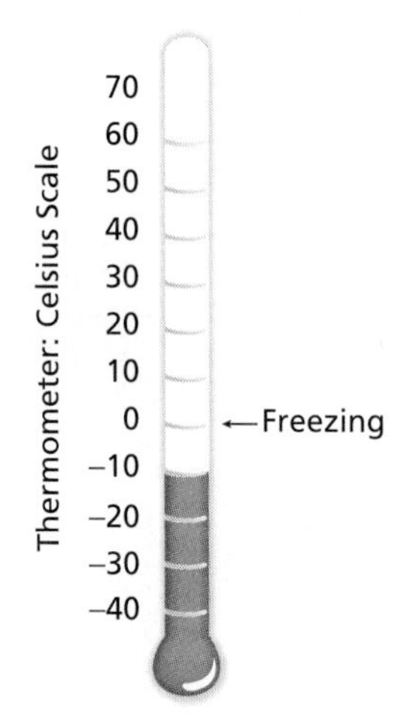

This thermometer shows a temperature of $-10°C$. This is read as "negative 10 degrees Celsius" or "10 degrees below zero Celsius." On a vertical number line such as this the negative direction is downward and the positive direction is upward.

Goal 1 Integers and the Number Line

Many real-life situations can be modeled with whole numbers. Some situations, such as temperatures, are more easily modeled with an expanded set of numbers called *integers*.

Integers

The following numbers are called **integers.**

$\ldots, \underbrace{-3, -2, -1}_{\text{Negative}}, \underbrace{0}_{\text{Zero}}, \underbrace{1, 2, 3, \ldots}_{\text{Positive}}$

Integers can be visually modeled on a number line, as follows.

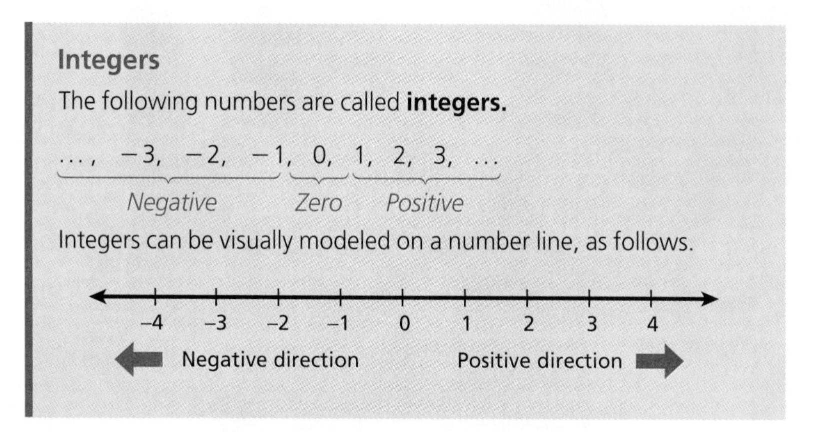

Here are some things you need to know about integers.

1. A negative integer such as -5 is read as "negative 5." Although the negative sign "$-$" is the same sign that is used for subtraction, it does not mean the same thing.
2. A positive integer such as 12 can be written with a positive sign as $+12$. It is more common, however, to omit the positive sign.
3. If a and b are integers, then the inequality $a < b$ means that a lies *to the left* of b on the number line.

Example 1 *Plotting Integers on the Number Line*

Draw a number line and plot the integers -6, -2, and 3.

Solution To *plot* the integer on the number line, draw a dot at the point that represents the integer. Note that -6 is to the left of -2, which means that $-6 < -2$.

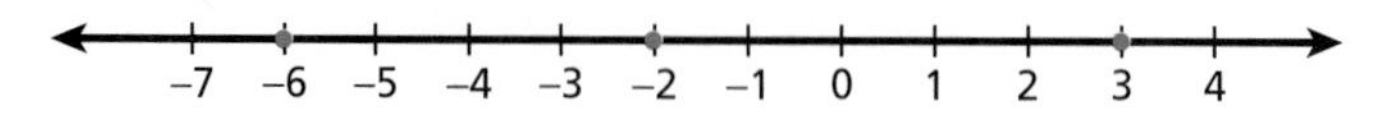

Goal 2 Finding the Absolute Value of a Number

Elevations are measured in distances above or below sea level. The lowest elevation in the United States is Death Valley. Its elevation is -282 feet. Because $|-282| = 282$, this can also be described as 282 feet below sea level.

The **absolute value** of a number is the distance between the number and 0. Absolute values are written with two vertical rules, $|\ \ |$, called **absolute value signs**. Because distance cannot be negative, it follows that the absolute value of a number cannot be negative.

Example 2 *Finding Absolute Values*

Find the absolute value of **a.** -4 and **b.** 3.

Solution

a. On a number line, the distance between -4 and 0 is 4. This means that $|-4| = 4$.

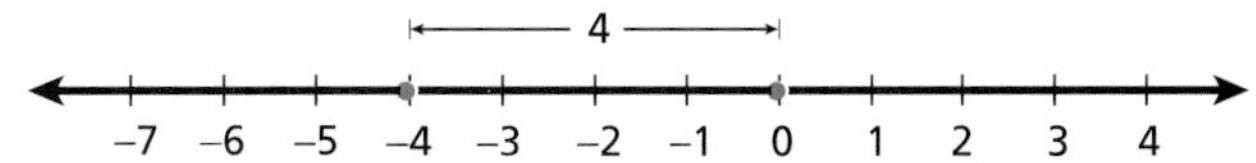

b. On a number line, the distance between 3 and 0 is 3. This means that $|3| = 3$.

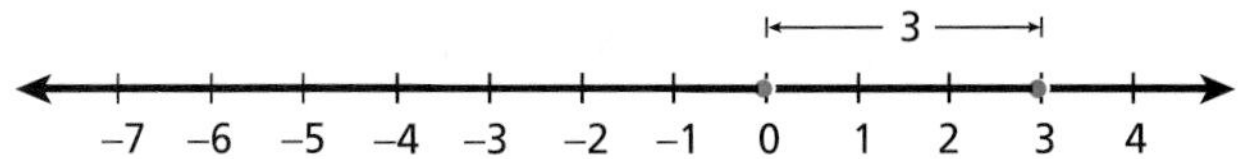

Two numbers that have the same absolute value but opposite signs are called **opposites.** For instance, -4 and 4 are opposites. Zero is its own opposite.

Communicating about MATHEMATICS

SHARING IDEAS about the Lesson

Reasoning Work with a partner. Decide whether the statement is true or false. In each case, explain your reasoning.

A. The absolute value of a negative integer is a positive integer.

B. The absolute value of any integer is positive.

C. The absolute value of -6 is greater than the absolute value of -4.

D. -6 is greater than -4.

E. Zero is the only integer that is its own opposite.

F. If $a \le b$, then $|a| \le |b|$.

EXERCISES

Guided Practice

CHECK for Understanding

In Exercises 1–6, use the following set of numbers.

$-4, -3, -2, -1, 0, 1, 2, 3, 4$

1. Which are integers?

2. Which are whole numbers?

3. Which are natural numbers?

4. Which is the smallest positive integer?

5. Which is the greatest negative integer?

6. Which is neither positive nor negative?

7. Give an example of two numbers that are opposites.

8. State two values of x that make $|x| = 5$ true.

9. On a number line, which direction is positive? Which is negative?

10. *Problem Solving* Describe an example (other than temperature) of how negative integers are used in real life.

Independent Practice

In Exercises 11–16, draw a number line and plot the integers.

11. $0, 4, -3$ **12.** $-1, 2, -6$ **13.** $-5, -3, 0$ **14.** $-4, 2, 3$ **15.** $-7, -8, -5$ **16.** $0, -2, 7$

In Exercises 17–22, compare the integers using the symbols $<$ or $>$.

17. 0 ? 4

18. -2 ? 1

19. 0 ? -3

20. 4 ? -6

21. $|-1|$? -2

22. $|-14|$? $|-13|$

In Exercises 23–30, write the opposite and the absolute value of the integer.

23. 1 **24.** -4 **25.** -3 **26.** 3

27. 20 **28.** -32 **29.** -100 **30.** 144

Number Sense **In Exercises 31–38, write the integer that represents the situation.**

31. 250 feet below sea level

32. An elevation of 5050 feet

33. A gain of 25 yards

34. $100 deposit in a checking account

35. 17 degrees below zero

36. A profit of $40

37. A loss of 15 pounds

38. A gain of 6 hours

In Exercises 39–44, order the integers from least to greatest.

39. $0, -6, 5, -3, 4$

40. $-1, -10, 1, -4, -6$

41. $-1, 2, -2, 0, -4$

42. $-9, -11, 11, 1, -7$

43. $6, 4, -4, -5, -2$

44. $-7, -8, -9, -6, -5$

Cities in Illinois **In Exercises 45–50, use the number line to estimate the distance in miles between the two cities.**

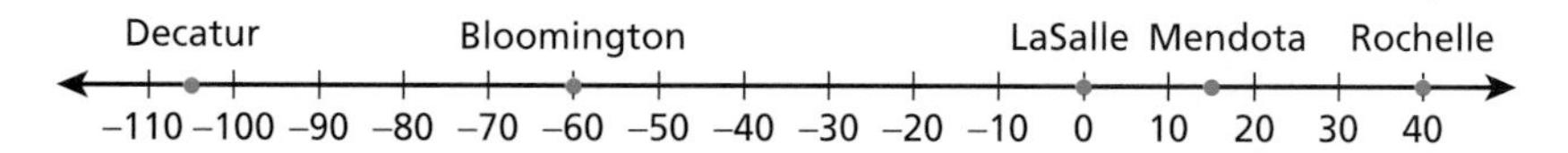

45. LaSalle and Rochelle

46. LaSalle and Bloomington

47. Mendota and Bloomington

48. Decatur and LaSalle

49. Bloomington and Decatur

50. Rochelle and Decatur

Swimming **In Exercises 51–53, use the following information.**

You are at a swimming pool that is 3 feet deep at the shallow end and 14 feet deep at the deep end. There are three diving boards: a low diving board (3 feet above the water), a middle diving board (10 feet above the water), and a high diving board (16 feet above the water).

51. Draw a vertical number line showing the heights and depths in feet.

52. You dive off the middle diving board to a depth of 9 feet below the surface. How far did you dive vertically?

53. You dive off the high diving board and touch the bottom of the deep end. How far did you dive vertically?

Diving platforms used in competitions can be as high as 10 meters (33 feet) above the water.

Integrated Review

Making Connections within Mathematics

Mental Math **In Exercises 54–62, solve the equation.**

54. $x + 13 = 20$

55. $m - 15 = 6$

56. $3x = 36$

57. $\frac{x}{12} = 5$

58. $n - 17 = 9$

59. $12y = 96$

60. $14 = x - 4$

61. $17 = 6 + b$

62. $34 = 2x$

Exploration and Extension

63. *Logical Reasoning* Find and plot on a number line four different integers a, b, c, and d such that

- d is neither negative nor positive,
- a and b are the same distance from d,
- c is the least positive integer and three units to the left of b.

LESSON INVESTIGATION 3.2

Modeling Addition

Materials Needed: number counters, mat, pencil, paper

In this investigation, you will use number counters to model integer addition.

Sample 1: Finding the Sum of Two Positive Integers

3 (+ + +) 2 (+ +)

Choose 3 black counters to represent positive 3 and 2 black counters to represent positive 2.

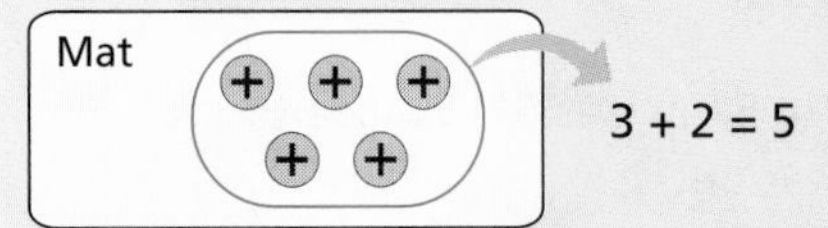

Place the counters on the mat and count the total number of black counters.

Sample 2: Finding the Sum of Two Negative Integers

–4 (– – – –) –1 (–)

Choose 4 red counters to represent negative 4 and 1 red counter to represent negative 1.

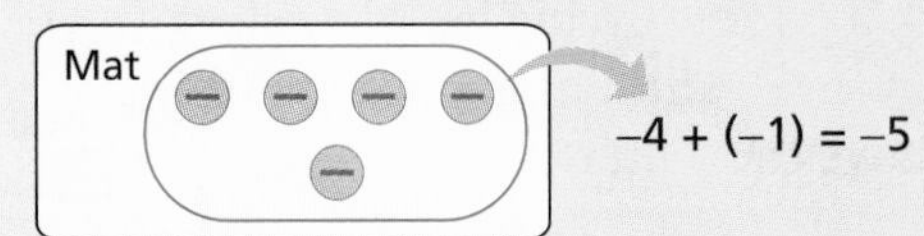

Place the counters on the mat and count the total number of red counters.

Sample 3: Finding the Sum of Positive and Negative Integers

–7 (– – – – – – –)

4 (+ + + +)

Choose 7 red counters to represent negative 7 and 4 black counters to represent positive 4.

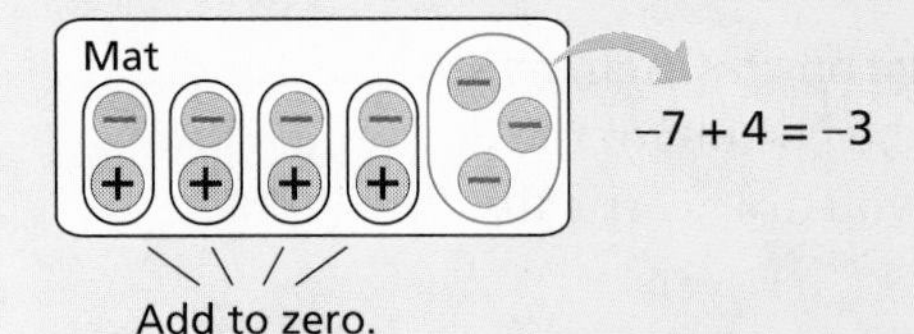

Add to zero.

Place the counters on the mat and group pairs of black and red counters. Each pair has a sum of zero. The remaining counters represent the sum.

Exercises

In Exercises 1–12, use number counters to find the sum.

1. $3 + 4$ **2.** $-1 + (-3)$ **3.** $-2 + (-5)$ **4.** $-4 + 4$

5. $3 + (-3)$ **6.** $6 + (-4)$ **7.** $-7 + 2$ **8.** $-5 + 8$

9. $6 + (-3)$ **10.** $5 + (-7)$ **11.** $8 + (-8)$ **12.** $4 + (-9)$

13. Write a general statement about the sum of two positive integers.

14. Write a general statement about the sum of two negative integers.

15. Can the sum of a positive integer and a negative integer be positive? Can it be negative? How can you predict when a sum will be positive or negative?

16. Can the sum of two integers be zero? What can you say about such integers?

3.2 Adding Two Integers

What you should learn:

How to use absolute values to add two integers

How to use integer addition to solve real-life problems

Why you should learn it:

You can use integer addition to solve real-life problems, such as finding the temperature.

Goal 1 Adding Two Integers

The *Investigation* on page 104 shows how number counters can be used to add two integers. In this lesson, you will learn rules that can be used to add two integers.

Adding Two Integers

Rule	Examples
1. To add two integers with the *same sign*, add their absolute values and write the common sign.	$3 + 4 = 7$ $-2 + (-6) = -8$
2. To add two integers with *opposite signs*, subtract the smaller absolute value from the larger absolute value and write the sign of the integer with the greater absolute value.	$-2 + 5 = 3$ $1 + (-7) = -6$

Example 1 *Adding Integers with the Same Sign*

a. The sum of two positive integers is positive.

$4 + 5 = 9$

b. The sum of two negative integers is negative.

$-12 + (-3) = -15$ ■

Example 2 *Adding Integers with Opposite Signs*

a. The sum of 5 and -8 is negative because -8 has a greater absolute value than 5.

Write sign of -8. *Subtract 5 from 8.*

$5 + (-8) = -3$

b. The Sum of Opposites The sum of 6 and -6 is zero because both integers have the same absolute value.

$6 + (-6) = 6 - 6$ *Subtract 6 from 6.*

$= 0$ *The sum is zero.*

The sum of any two opposites is zero. ■

Zero Property of Addition

When zero is added to an integer, the sum is the same integer. Here are two examples of this property.

$3 + 0 = 3$

$0 + (-5) = -5$

Goal 2 Solving Real-Life Problems

Example 3 *Finding a Temperature*

Real Life Meteorology

In northern cities like Chicago, the temperature can change rapidly. For instance, in an hour, the temperature can drop from 20°C (68°F) to 10°C (50°F).

The thermometers at the right show the temperatures (in degrees Celsius) at 1 P.M., 2 P.M., and 3 P.M.

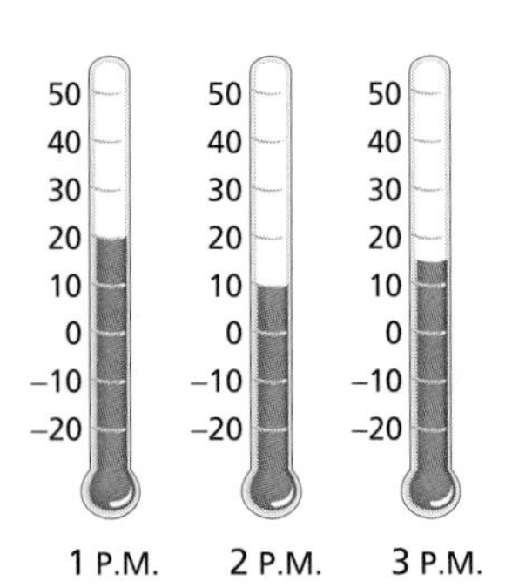

a. Use the thermometers to approximate the temperatures.

b. Write an addition equation that relates the temperatures at 1 P.M. and 2 P.M.

c. Write an equation that relates the temperatures at 2 P.M. and 3 P.M.

Solution

a. At 1 P.M. the temperature is 20°C. At 2 P.M. the temperature is 10°C. At 3 P.M. the temperature is 15°C.

b. From 1 P.M. to 2 P.M. the temperature *dropped* 10 degrees. This can be represented by adding −10 degrees to the 1 P.M. temperature.

1 P.M. temperature + Temperature drop of 10° = 2 P.M. temperature

$20 + (-10) = 10$

c. From 2 P.M. to 3 P.M. the temperature *rose* 5 degrees. This can be represented by adding 5 degrees to the 2 P.M. temperature.

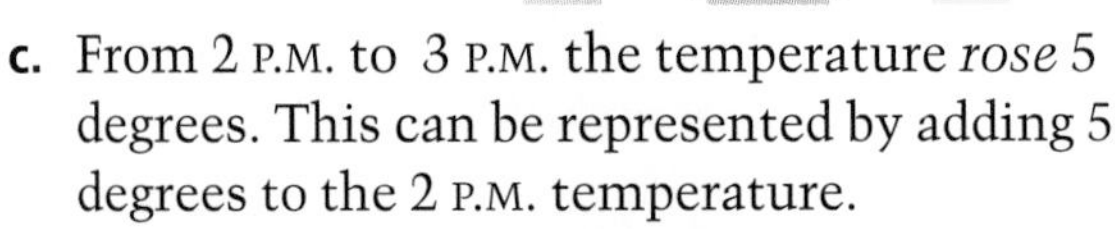

2 P.M. temperature + Temperature rise of 5° = 3 P.M. temperature

$10 + 5 = 15$ ■

Communicating about MATHEMATICS

SHARING IDEAS about the Lesson

Problem Solving Explain how each real-life situation can be modeled by integer addition.

A. You owe your uncle \$25. You pay back \$15.

B. You owe your sister \$15. You borrow another \$10.

C. You owe your mom \$40. You pay all of it back.

EXERCISES

Guided Practice

CHECK for Understanding

In Exercises 1–4, find the sum. Write your conclusion as an equation.

1. $4 + 3$ **2.** $2 + (-2)$ **3.** $7 + (-5)$ **4.** $-7 + (-5)$

In Exercises 5–8, write the equation modeled by the number counters.

5. 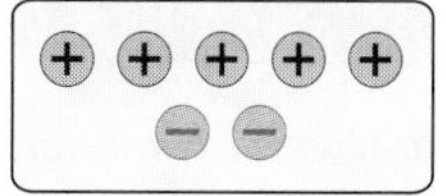

6. 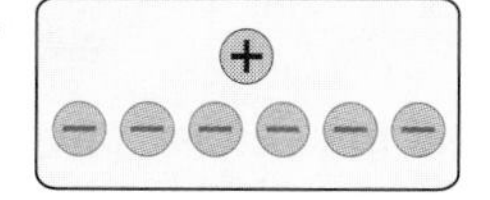

7. 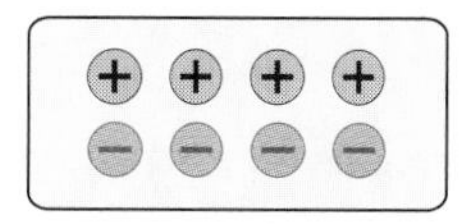

8. 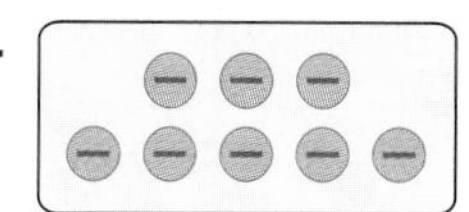

9. Discuss the two rules used in this lesson to add integers. Which rule should be used to find the sum of 8 and -11? Explain.

10. *Computation Sense* From the number line below, state whether $a + b$ is negative, zero, or positive. Explain.

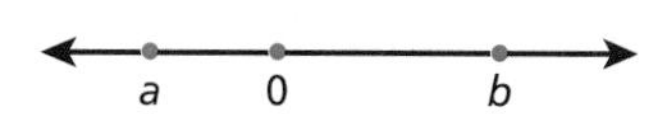

Independent Practice

Mental Math **In Exercises 11–26, find the sum. Write your conclusion as an equation.**

11. $11 + 15$ **12.** $-8 + (-2)$ **13.** $-13 + (-13)$ **14.** $10 + 24$

15. $10 + (-10)$ **16.** $-8 + 8$ **17.** $-13 + 13$ **18.** $24 + (-24)$

19. $13 + 0$ **20.** $-7 + 0$ **21.** $0 + 15$ **22.** $0 + (-33)$

23. $2 + (-9)$ **24.** $39 + (-21)$ **25.** $-16 + 12$ **26.** $-17 + 13$

Addition Patterns **In Exercises 27-30, complete the statements. Then describe the pattern.**

27.
$4 + 3 = \boxed{?}$
$4 + 2 = \boxed{?}$
$4 + 1 = \boxed{?}$
$4 + 0 = \boxed{?}$
$4 + (-1) = \boxed{?}$

28.
$-2 + (-3) = \boxed{?}$
$-2 + (-2) = \boxed{?}$
$-2 + (-1) = \boxed{?}$
$-2 + 0 = \boxed{?}$
$-2 + 1 = \boxed{?}$

29.
$3 + (-7) = \boxed{?}$
$3 + (-5) = \boxed{?}$
$3 + (-3) = \boxed{?}$
$3 + (-1) = \boxed{?}$
$3 + 1 = \boxed{?}$

30.
$-6 + (-4) = \boxed{?}$
$-6 + (-5) = \boxed{?}$
$-6 + (-6) = \boxed{?}$
$-6 + (-7) = \boxed{?}$
$-6 + (-8) = \boxed{?}$

Equation Sense **In Exercises 31–34, find three sets of values of x and y that make the equation true. (There are many correct answers.)**

31. $x + y = 7$ **32.** $x + y = -2$ **33.** $x + y = -8$ **34.** $x + y = 10$

Mental Math **In Exercises 35–38, use mental math to solve the equation.**

35. $4 + x = 7$ **36.** $6 + n = 5$ **37.** $-2 + m = -5$ **38.** $-3 + y = 0$

Modeling Real Life **In Exercises 39–42, match the equation with the real-life situation. Then solve the equation for x and explain what x represents in the problem.**

a. *Elevator Ride* You enter an elevator on the 5th floor. The elevator goes down 3 floors.

b. *Football* Your team is on its own 35-yard line. The running back rushes for a 10-yard gain.

c. *Temperature* The temperature is 35°F. The temperature drops 10 degrees.

d. *Borrowed Money* Your sister owes you $5. She pays $3 of it back to you.

39. $-5 + 3 = x$

40. $5 - 3 = x$

41. $35 + 10 = x$

42. $35 + (-10) = x$

Emmitt Smith won the National Football League rushing title several times. In 1992, he rushed for 1713 yards.

Problem Solving **In Exercises 43 and 44, write a real-life situation that can be represented by the equation.**

43. $12 + (-15) = x$

44. $-15 + 40 = x$

Number Line Model **In Exercises 45 and 46, find a pair of integers whose sum is − 1. (Use the integers labeled *a*, *b*, *c*, and *d*.)**

45. a b c d

−4 −3 −2 −1 0 1 2 3 4

46. a b c d

−4 −3 −2 −1 0 1 2 3 4

Integrated Review

Making Connections within Mathematics

Computation Sense **In Exercises 47–54, simplify the expression.**

47. $|9 + 7|$

48. $|15 + 6|$

49. $|11 + (-2)|$

50. $|5 + (-8)|$

51. $|-7 + 4|$

52. $|-3 + 10|$

53. $|-2 + (-6)|$

54. $|-10 + (-1)|$

Exploration and Extension

Dinosaurs **In Exercises 55–57, find the period in which the dinosaur lived.**

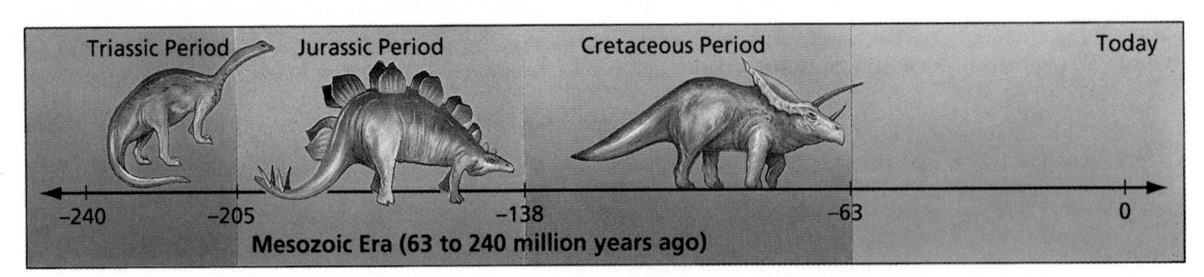

55. *Stegasaurus:* 122 million years before the end of the Cretaceous Period

56. *Torosaurus:* 117 million years after the beginning of the Jurassic Period

57. *Plateosaurus:* 72 million years before the end of the Jurassic Period

3.3 Adding Three or More Integers

What you should learn:

Goal 1 How to add three or more integers

Goal 2 How to simplify expressions by adding like terms

Why you should learn it:

You can use integer addition to solve real-life problems, such as finding the total number of yards gained in football.

Goal 1 Adding Three or More Integers

One way to model the sum of three or more integers is with a number line. On the number line, adding a positive number is represented by *movement to the right* and adding a negative number is represented by *movement to the left.*

Example 1 *Modeling Addition with a Number Line*

The sum of 3, −5, and 4 can be modeled as shown below.

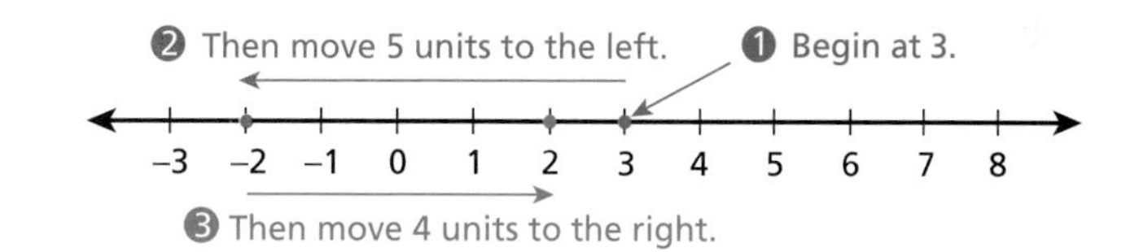

Begin at 3, then move 5 units to the left, then move 4 units to the right. Because you end at 2, you can conclude that $3 + (-5) + 4 = 2$. ■

Technology
Using a Calculator

Example 2 *Using a Calculator to Add Integers*

Use a calculator to find $-6 + 5 + (-7) + 1$.

Solution

Calculator Steps	Display	Conclusion
6 [+/-] [+] 5 [+] 7 [+/-] [+] 1 [=]	−7	The sum is −7. ■

Real Life
Football

Example 3 *Finding the Number of Yards Gained*

On its first down, your team gained 7 yards. On its second down, it lost 13 yards. On its third down, it gained 14 yards. On its fourth down, it gained 1 yard. Did your team gain 10 yards and thus earn a first down?

Solution The total number of yards gained is $7 + (-13) + 14 + 1$ or 9.

Thus, the team did not earn a first down. ■

Goal 2 Adding Like Terms

In the expression $-5x + 2$, the number -5 is the **coefficient** of x. When you add like terms, you can add the coefficients. Here are two examples.

Expression	Apply Distributive Property.	Simplify.
$4n + (-6n) + 3$	$[4 + (-6)]n + 3$	$-2n + 3$
$-5x + 7x + (-3x)$	$[-5 + 7 + (-3)]x$	$-x$

Example 4 *Evaluating an Expression*

Simplify the expression $-15x + 40x + (-16x)$. Then evaluate the expression when **a.** $x = 2$ and **b.** $x = 4$.

Solution

$$-15x + 40x + (-16x) = [-15 + 40 + (-16)]x$$
$$= 9x \quad \textit{Simplify.}$$

a. When $x = 2$, the value of the expression is $9x = 9(2) = 18$.

b. When $x = 4$, the value of the expression is $9x = 9(4) = 36$. ■

Need to Know

Coefficients of -1 and 1 are usually implied (rather than written). For instance, the coefficent of x is 1 and the coefficient of $-x$ is -1.

Communicating about MATHEMATICS

Real Life **Business**

SHARING IDEAS about the Lesson

Profit or Loss? You own an art business whose monthly profits and losses are shown in the graph. Did your business have an overall profit or an overall loss for the year? Explain.

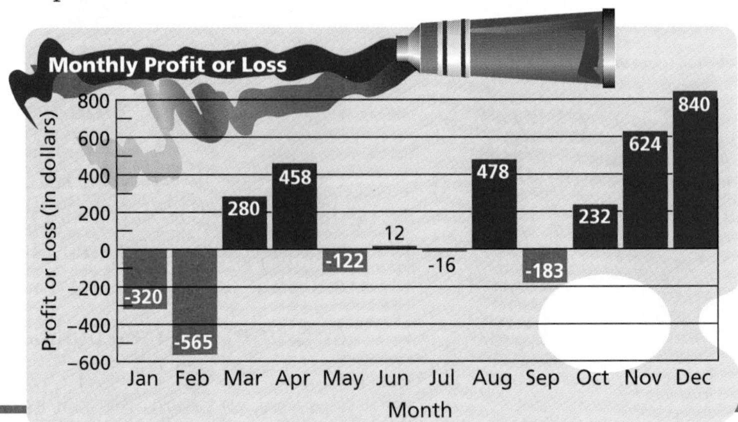

EXERCISES

Guided Practice

CHECK for Understanding

In Exercises 1 and 2, use a number line to illustrate the movement. Then write your conclusion as an equation.

1. Begin at 4. Then move 2 units to the right. Then move 6 units to the left.

2. Begin at 2. Then move 5 units to the left. Then move 8 units to the right.

3. What are the terms in $-3x+5x+7$? Simplify the expression.

4. *Computation Sense* From the number line, state whether $a+b+c$ is negative, positive, or zero. Explain.

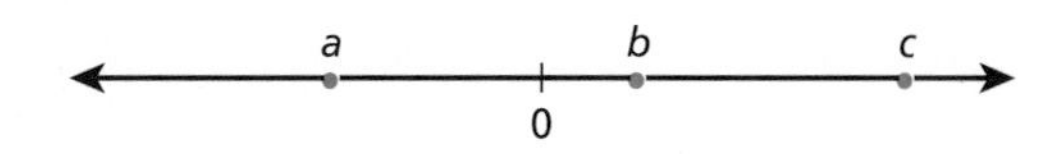

Independent Practice

In Exercises 5–16, find the sum. Write your conclusion as an equation.

5. $4 + (-5) + 6$ **6.** $3 + (-9) + 13$ **7.** $-7 + 1 + (-8)$

8. $-6 + 2 + (-15)$ **9.** $-8 + 12 + (-1)$ **10.** $-10 + 16 + (-4)$

11. $-12 + (-4) + (-8)$ **12.** $-11 + (-7) + (-3)$ **13.** $5 + (-6) + (-13)$

14. $4 + (-8) + 9 + (-2)$ **15.** $-7 + (-6) + 2 + (-7)$ **16.** $-12 + (-4) + 20$

In Exercises 17–22, use a calculator to find the sum.

17. $-36 + 49 + (-2) + 15$ **18.** $-23 + 112 + (-9) + 13$ **19.** $19 + (-39) + (-51)$

20. $92 + (-20) + (-101)$ **21.** $84 + (-89) + (-40)$ **22.** $111 + 105 + (-99)$

Reasoning **In Exercises 23 and 24, decide whether the sum is positive or negative. Explain how you can make your decision *without* actually finding the sum.**

23. $-237 + 122 + 69$ **24.** $-142 + 127 + 89$

In Exercises 25–36, simplify the expression. Then evaluate it when $x = 2$.

25. $-2x + 5x + 9$ **26.** $8x + (-6x) + 3$ **27.** $-2x + 10x + (-7x)$

28. $9x + 13x + (-10x)$ **29.** $13x + (-11x) + x$ **30.** $2x + (-9x) + x$

31. $-3x + 2x + 26x$ **32.** $-4x + 9x + 6x$ **33.** $-7x + 8 + 17x$

34. $-8x + 10 + 12x$ **35.** $5x + 3 + (-8) + (-3x)$ **36.** $9x + 6 + (-18) + (-6x)$

In Exercises 37–42, complete the statement using $>$, $<$, or $=$.

37. -4 ? $6 + (-2)$ **38.** $3 + (-4)$? $6 + (-9)$ **39.** $-5 + 5$? $-5 + (-5)$

40. $-5 + (-3)$? -9 **41.** -9 ? $-7 + 16$ **42.** $4 + (-2)$? $-6 + 8$

First Downs **In Exercises 43 and 44, decide whether the football team earns a first down. Explain your reasoning.**

43. On its first down, the team gains 6 yards. On its second down, it loses 3 yards. On its third down, it loses 4 yards. On its fourth down, it gains 8 yards.

44. On its first down, the team gains 8 yards. On its second down, it gains 1 yard. On its third down, it loses 5 yards. On its fourth down, it gains 7 yards.

Flying a Plane **In Exercises 45–47, imagine that you are flying a Boeing 747 jet airliner. You are instructed by air traffic control to change your elevation, then change it again. What is your final elevation? In which exercise must the airplane have changed directions? Explain.**

45. You are flying at an altitude of 31,000 feet. You lower the airplane 2,000 feet, then raise the airplane 4,000 feet.

46. You are flying at an altitude of 40,000 feet. You raise the airplane 4,000 feet. Then you lower the airplane 6,000 feet.

47. You are flying at an altitude of 35,000 feet. You raise the airplane 5,000 feet. Then you lower the airplane 6,000 feet.

Commercial airplanes whose direction is north or east must fly at "odd" altitudes. Those whose direction is south or west must fly at "even" altitudes.

Integrated Review

Making Connections within Mathematics

Number Sense **In Exercises 48–51, complete the statement using $>$, $<$, or $=$.**

48. $|3 + (-2)|$? $|3| + |-2|$

49. $|-4 + (-5)|$? $|-4| + |-5|$

50. $|-4 + 7|$? $|-4| + |7|$

51. $|-5 + 3|$? $|-5| + |3|$

Exploration and Extension

52. ***Finding an Exam Average*** You have the following exam scores in math.

84, 91, 82, 89, 93, 84, 87, 94

Here is one way to find your average without using a calculator. Begin by guessing the average, say 85. Then, determine the amounts by which your scores are above or below 85, as shown at the right. Find the average of these amounts. Add that average to 85 to determine your actual average.

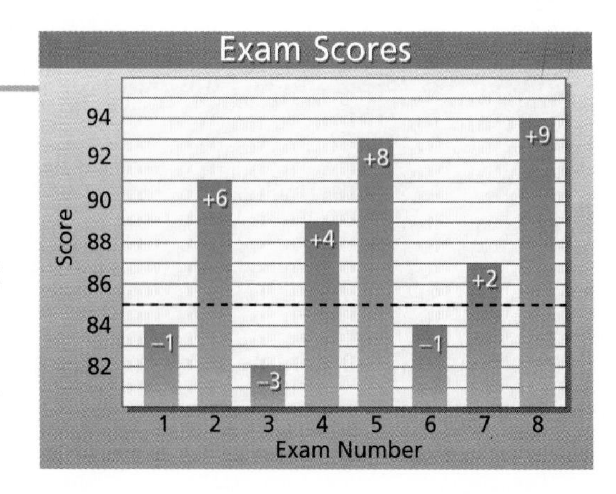

Mixed REVIEW

In Exercises 1–8, evaluate the expression when $n = 3$. (1.3–1.5)

1. $6n - 12$ **2.** $n^2 - 4$ **3.** $13n + 16$ **4.** $1.2n + 1.5$

5. $6 \times (n - 1)$ **6.** $\frac{1}{4}(n \times 8)$ **7.** $n^2 \div 3$ **8.** $14 \times (n + 2) - 15$

In Exercises 9–11, complete the statement using $>$, $<$, or $=$. (3.1)

9. $\frac{1}{8}(32)\ \boxed{?}\ \frac{1}{4}(16)$ **10.** $16 + (-8)\ \boxed{?}\ 10 + (-3)$ **11.** $-12 + (-2)\ \boxed{?}\ -17 + 5$

12. Are the equations $x - 12 = 4$ and $x = 17$ equivalent? Explain. **(2.4)**

In Exercises 13–20, solve the equation. Check your solution. (2.4)

13. $x - 2 = 14$ **14.** $y + 5 = 6$ **15.** $14 + z = 34$ **16.** $a - 11 = 22$

17. $3b = 6$ **18.** $16c = 32$ **19.** $\frac{1}{2}p = 5$ **20.** $10q = \frac{1}{2}$

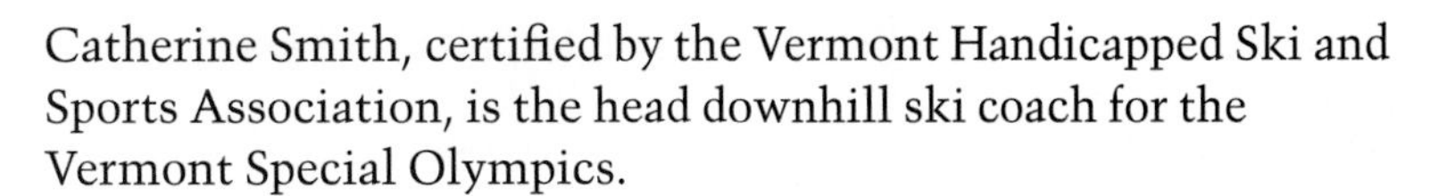

Downhill Ski Coach

Catherine Smith, certified by the Vermont Handicapped Ski and Sports Association, is the head downhill ski coach for the Vermont Special Olympics.

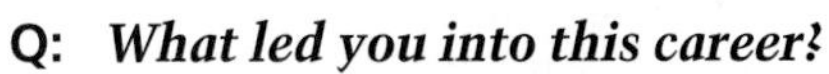

Q: ***What led you into this career?***

A: My brother has Down's syndrome and required special instruction when learning to ski.

Q: ***What is your favorite part of your job?***

A: Working with the athletes. Our athletes have a unique perspective—they care as much about the success of their peers as they do about their own success.

Q: ***What math do you use in your job?***

A: To maintain balance and stability, skiers have to learn how to position their skis at different angles in order to turn, speed, and stop. I draw a diagram in the snow of a clock and use the hand positions on the clock to describe the position of the skis.

Q: ***What would you like to tell kids who are in school about math?***

A: Don't get uptight about math. It's easy when you relate it to things you know and it is fun! There are a lot of uses for it and math is one of the most important skills you'll use when you get any type of job.

LESSON INVESTIGATION 3.4
Modeling Subtraction

Materials Needed: number counters, mat, pencil, paper

In this investigation, you will use number counters to model integer subtraction.

Sample 1: Finding the Difference of Two Positive Integers

The difference $5 - 3 = 2$ can be modeled as follows.

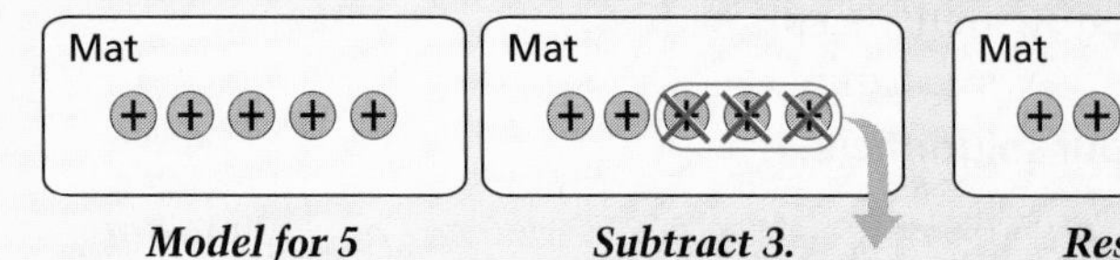

Model for 5 *Subtract 3.* *Result is 2.*

Sample 2: Finding the Difference of Two Positive Integers

The difference $4 - 7 = -3$ can be modeled as follows.

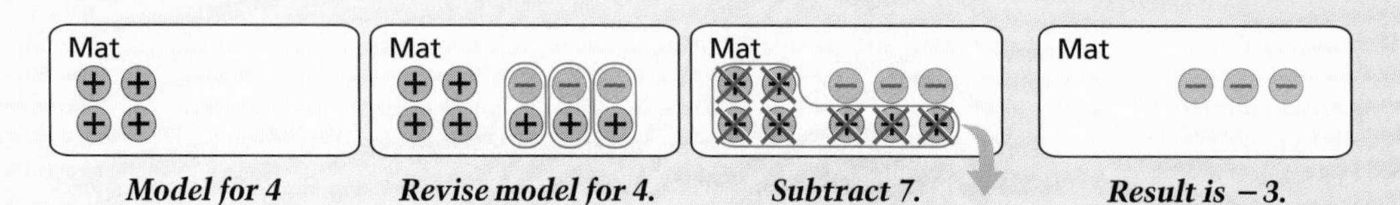

Model for 4 *Revise model for 4.* *Subtract 7.* *Result is −3.*

Sample 3: Finding the Difference of Two Negative Integers

The difference $-5 - (-3) = -2$ can be modeled as follows:

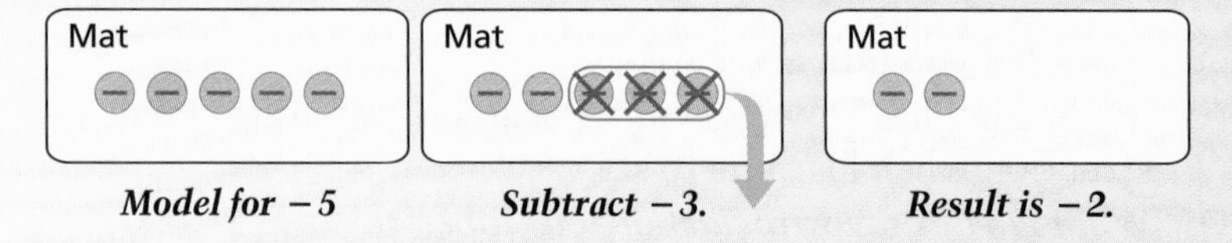

Model for −5 *Subtract −3.* *Result is −2.*

Sample 4: Finding the Difference of a Positive and a Negative Integer

The difference $1 - (-4) = 5$ can be modeled as follows.

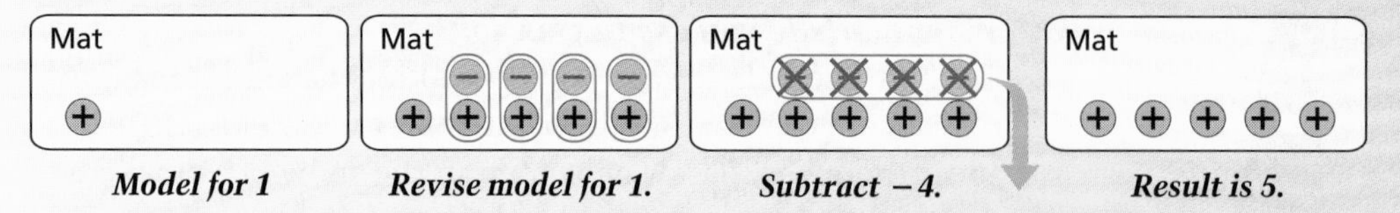

Model for 1 *Revise model for 1.* *Subtract −4.* *Result is 5.*

Exercises

In Exercises 1–8, use number counters to find the difference.

1. $6 - 4$ **2.** $3 - 2$ **3.** $4 - 6$ **4.** $5 - 8$

5. $-2 - (-4)$ **6.** $-1 - (-2)$ **7.** $1 - (-5)$ **8.** $-6 - 2$

9. ***Think About It*** Can the difference of a positive integer and a negative integer be positive? Can it be negative? Explain.

3.4 Subtracting Integers

What you should learn:

Goal 1 How to use opposites to subtract integers

Goal 2 How to simplify expressions involving subtraction

Why you should learn it:

You can use integer subtraction to simplify algebraic expressions, such as $5x-(-4x)+3$.

Goal 1 Subtracting Integers

The *Investigation* on page 114 shows how counters can be used to subtract integers. In this lesson, you will use opposites to subtract integers. The number you obtain from subtracting one integer from another is the **difference** of the integers.

LESSON INVESTIGATION

Investigating Integer Subtraction

Group Activity Complete the following. Compare your results with others in your group. How do the results of the two columns compare? Use your results to describe a rule for subtracting integers.

$5-3=$?	$5+(-3)=$?
$5-4=$?	$5+(-4)=$?
$5-5=$?	$5+(-5)=$?
$5-6=$?	$5+(-6)=$?
$5-7=$?	$5+(-7)=$?

Need to Know

If an integer b is positive, then its opposite $-b$ is negative. For instance, the opposite of 17 is -17.

If an integer b is negative, then its opposite $-b$ is positive. For instance, the opposite of -5 is 5.

Notice that the integer $-b$ is *not necessarily negative.* If b is itself negative, then $-b$ is positive.

Subtracting Integers

To subtract an integer b from an integer a, add the opposite of b to a.

$$a-b=a+(-b)$$

Example 1 *Subtracting Integers*

a. $5-7=5+(-7)=-2$ *Opposite of 7 is -7.*

b. $-6-8=-6+(-8)=-14$ *Opposite of 8 is -8.*

c. $-9-(-9)=-9+(9)=0$ *Opposite of -9 is 9.*

d. $-5-(-1)=-5+1=-4$ *Opposite of -1 is 1.*

e. $13-12=13+(-12)=1$ *Opposite of 12 is -12.* ■

Goal 2 Simplifying Expressions

The *terms* of an algebraic expression are separated by addition, not subtraction. To recognize the terms of an expression involving subtraction, you can rewrite the expression as a sum. Here are two examples.

Expression	Rewrite as Sum	Terms
$2x - 4$	$2x + (-4)$	$2x$ and -4
$-3x - 2x + 5$	$-3x + (-2x) + 5$	$-3x$, $-2x$, and 5

The *Distributive Property,* $a(b + c) = ab + ac$, also applies to subtraction. The "subtraction form" of the property is

$a(b - c) = ab - ac$. *Distributive Property*

The next example shows how this property can be used.

Example 2 *Simplifying Expressions*

Simplify the expression $11x - 2x + 3$. Then evaluate the expression when **a.** $x = 5$ and **b.** $x = 8$.

Solution

$11x - 2x + 3 = (11 - 2)x + 3$ *Distributive Property*

$= 9x + 3$ *Simplify.*

a. When $x = 5$ the value of the expression is

$9x + 3 = 9(5) + 3 = 48$.

b. When $x = 8$ the value of the expression is

$9x + 3 = 9(8) + 3 = 75$. ■

Communicating about MATHEMATICS

SHARING IDEAS about the Lesson

Connections English

Comparing Mathematics and English A friend of yours says "There isn't no way I will finish on time." Your friend meant to say that he will not be able to finish on time, but what does his statement really mean?

Your friend's statement contains a *double negative.* How is his statement related to the mathematical equation $3 - (-4) = 7$?

EXERCISES

Guided Practice

CHECK for Understanding

1. Explain how to evaluate $5 - (-2)$.
2. Apply the Distributive Property to $r(s - t)$.

Reasoning **In Exercises 3–6, decide whether the statement is true for all values of x, some values of x, or no values of x. Explain.**

3. $5x - 2x - 12 = 5x + (-2x) + (-12)$
4. The opposite of x is 0.
5. $3(x - 4) = 3x - (-12)$.
6. The opposite of x is negative.

Independent Practice

In Exercises 7–22, find the difference. Write your conclusion as an equation.

7. $11 - 6$
8. $19 - 17$
9. $13 - 18$
10. $5 - 9$
11. $23 - (-8)$
12. $2 - (-4)$
13. $-10 - 7$
14. $-3 - 3$
15. $-5 - (-5)$
16. $-16 - (-8)$
17. $-5 - 5$
18. $-16 - 8$
19. $0 - 27$
20. $0 - 13$
21. $0 - (-61)$
22. $0 - (-43)$

In Exercises 23–28, evaluate the expression when $a = 5$ and when $a = -5$.

23. $a - 1$
24. $1 - a$
25. $a - 6$
26. $6 - a$
27. $a - a$
28. $a + a$

In Exercises 29–32, rewrite the expression as a sum. Then identify the terms of the expression.

29. $3x - 2x + 16$
30. $7x - 9x - 5$
31. $7a - 5b$
32. $4 - 2n + 4m$

In Exercises 33–44, simplify the expression.

33. $9x - 6x - 17$
34. $18n - 12n + 4$
35. $-11y - (-15y) - 2$
36. $-20x - (-30x) + 10$
37. $b - (-2b)$
38. $3x - (-3x)$
39. $-2a - 3a - 4$
40. $-13x - 13x - 13$
41. $4m - 6m + 8$
42. $16y - 20y + 24$
43. $-14x - (-10x)$
44. $-30x - (-19x)$

In Exercises 45–50, use a calculator to evaluate the expression.

45. $-8 - 8 - (-8)$
46. $8 - (-8) - 8$
47. $21 - (-62) - 43$
48. $43 - (-21) - 62$
49. $-84 - 72 - (-100)$
50. $-69 - (-96) - 56$
51. *Technology* Explain the difference between the [+/-] key and the [−] key on a calculator.

Computation Sense **In Exercises 52–55, find values for x and y so that the statement is true. (There are many correct answers.)**

52. x is positive, y is positive, and $x - y$ is positive.

53. x is positive, y is positive, and $x - y$ is negative.

54. x is negative, y is negative, and $x - y$ is negative.

55. x is negative, y is negative, and $x - y$ is positive.

Science **In Exercises 56–58, use the table, which gives the low and high surface temperatures of four planets in degrees Fahrenheit.**

Planet	Mercury	Earth	Mars	Pluto
Low	− 279	− 129	− 225	− 387
High	801	136	63	− 369

56. Find the difference between the high temperature and the low temperature of each planet.

57. Find the difference between the low temperatures of Earth and Pluto.

58. Find the difference between the high temperatures of Mercury and Mars.

Of the other eight planets in our solar system, Mars has a temperature range that is most similar to the range on Earth.

Integrated Review

Making Connections within Mathematics

Opposites **In Exercises 59–62, copy the number line and plot the opposite of a.**

59. 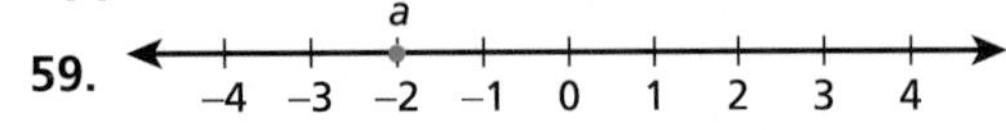

60. 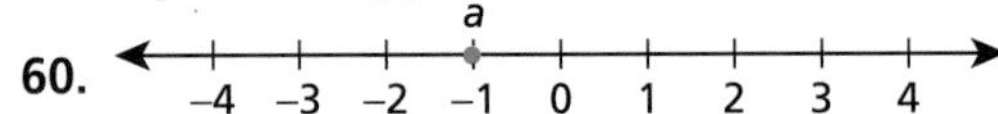

61. 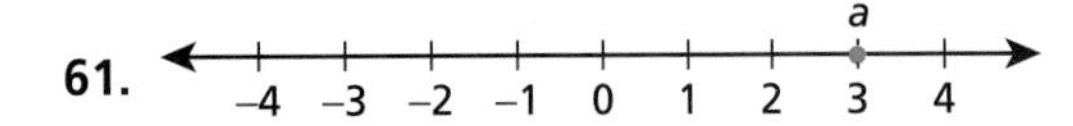

62. 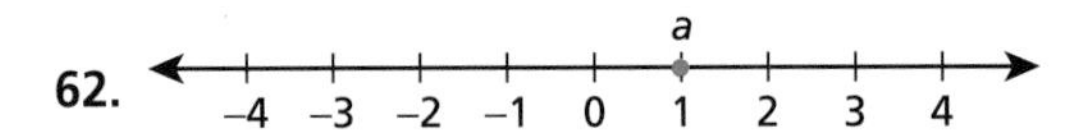

Equation Sense **In Exercises 63–66, solve the equation. Check your solution.**

63. $t + 34 = 71$ **64.** $r - 85 = 97$ **65.** $36n = 324$ **66.** $\frac{s}{28} = 5$

Exploration and Extension

In Exercises 67–70, evaluate the expression.

67. $5 - 3$ **68.** $3 - 5$ **69.** $-7 - 12$ **70.** $12 - (-7)$

71. *Making a Conjecture* Use the results of Exercises 67–70 to make a conjecture about the values of $a - b$ and $b - a$. Test your conjecture with several values of a and b.

USING A CALCULATOR

Negative Numbers

Technology
Using a Calculator

All scientific and graphing calculators have a key that allows you to enter a negative number. For example, to enter −9 on a scientific or graphing calculator, use the following keystrokes.

	Keystrokes	Display
Scientific calculator:	9 [+/-]	−9
Graphing calculator:	[(−)] 9	−9

Make sure you do not confuse these keys with the subtraction key. They are not the same!

Example *Adding and Subtracting Integers*

Use a scientific or graphing calculator to evaluate the following expressions.

a. $-9 + 4 + (-11)$ **b.** $7 + (-10) - (-5)$

Solution

Use the keystrokes that correspond to your calculator.

Scientific calculator

a. Keystrokes:

9 [+/-] [+] 4 [+] 11 [+/-] [=]

The display should show −16.

b. Keystrokes:

7 [+] 10 [+/-] [−] 5 [+/-] [=]

The display should show 2.

Graphing calculator

a. Keystrokes:

[(−)] 9 [+] 4 [+] [(−)] 11 [ENTER]

The display should show −16.

b. Keystrokes:

7 [+] [(−)] 10 [−] [(−)] 5 [ENTER]

The display should show 2.

Exercises

In Exercises 1–6, use a scientific or graphing calculator to evaluate the expression.

1. $-12 + 16 + (-6)$

2. $8 - (-17) - 30$

3. $37 + 16 + (-53)$

4. $83 + (-62) - 31$

5. $154 - 23 + (-76) - (-2)$

6. $148 - (-16) + (-121)$

7. Using your calculator, evaluate $31 + 14$ and $31 - (-14)$. What is the result of each expression? Is $a + b$ always equal to $a - (-b)$ when a and b are integers? Investigate.

Mid-Chapter SELF-TEST

Take this test as you would take a test in class. The answers to the exercises are given in the back of the book.

In Exercises 1–3, draw a number line and plot the numbers. (3.1)

1. $-2, 1, -4$ **2.** $4, -1, 0$ **3.** $2, -2, 0$

4. Write the opposite and absolute value of 7. **(3.1)**

5. Write the opposite and absolute value of -5. **(3.1)**

6. Order the integers $-2, 4, 3$, and -3 from least to greatest. **(3.1)**

In Exercises 7–9, use mental math to solve the equation. (3.2)

7. $2 + a = 10$ **8.** $-5 + b = 5$ **9.** $3 + c = -2$

In Exercises 10–13, evaluate the expression when $a = -2$ and $b = -8$. (3.2)

10. $|a| + |b|$ **11.** $|b| - |a|$ **12.** $|a + b|$ **13.** $|a - b|$

In Exercises 14 and 15, use the bar graph at the right. The graph shows the changes in weights (in pounds) of eight racehorses during a ten-week training program. (3.3)

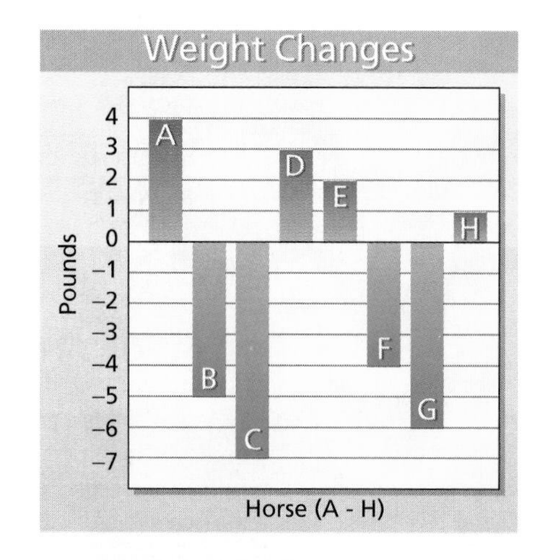

14. Find the sum of the changes in weights of the eight horses.

15. Which horse's weight changed the most? Explain your reasoning.

In Exercises 16 and 17, simplify the expression. Then evaluate the expression when $a = 6$ and $b = 2$. (3.3)

16. $4a + 2a - 6b$ **17.** $7a - 2b - 3a$

In Exercises 18–20, consider the following. (3.3)

You are a golfer in the PGA tour. Each tournament consists of 4 rounds of 18 holes. You have completed 2 tournaments with the indicated scores. (Each score lists your strokes above or below par.) To find your final score, add the scores for the 4 rounds.

First Tournament: $-2, 1, 2, -2$
Second Tournament: $3, -2, -2, 1$

18. Find the final score for your first tournament.

19. Find the final score for your second tournament.

20. Your opponent scored $3, -2, -2$, and 1 in the first tournament. Whose score was better?

In 1991, about 1.7 million teenagers played golf in the United States. (Source: National Sporting Goods Association)

LESSON INVESTIGATION 3.5

Modeling Multiplication

Materials Needed: number counters, mat, pencil, paper

In this investigation, you will use number counters to model integer multiplication.

Sample 1: Finding the Product of a Positive and a Negative Integer

To model the product $3\times(-2)$, you can *put in* 2 red counters, 3 times. The result shows that $3\times(-2)=-6$.

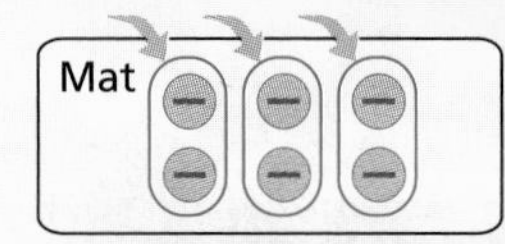

Put in* -2*, 3 times.

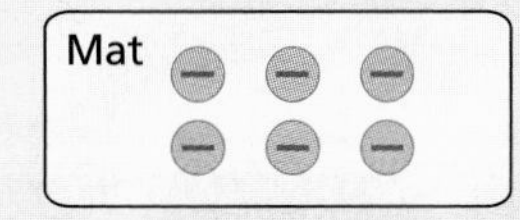

Result is* -6*.

Sample 2: Finding the Product of a Negative and a Positive Integer

To model the product -3×2, you can *take out* 2 black counters, 3 times. Because there are no black counters to take out, begin by modeling 0 with 6 pairs of red and black counters. The result shows that $-3\times 2=-6$.

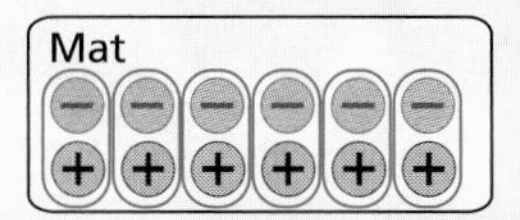

Model for 0

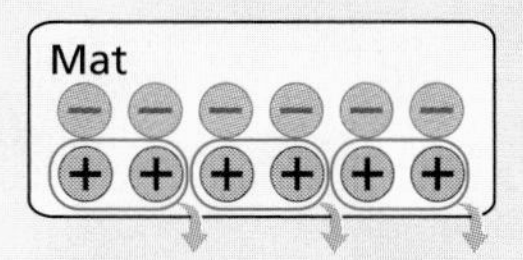

Take out 2, 3 times.

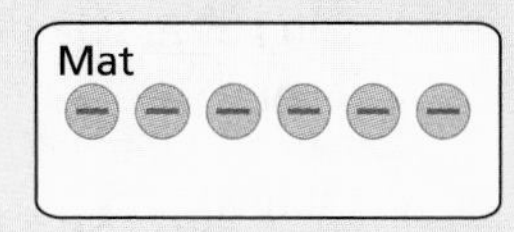

Result is* -6*.

Sample 3: Finding the Product of Two Negative Integers

To model the product $-3\times(-2)$, you can *take out* 2 red counters, 3 times. Because there are no red counters to take out, begin by modeling 0 with 6 pairs of red and black counters. The result shows that $-3\times(-2)=6$.

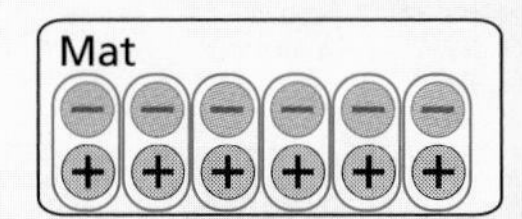

Model for 0

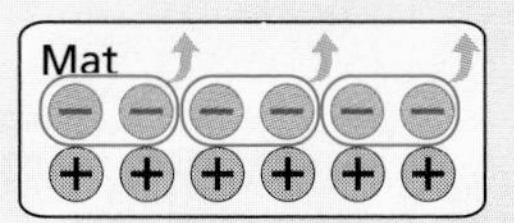

Take out* -2*, 3 times.

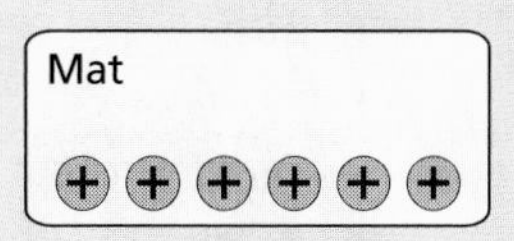

Result is 6.

Exercises

In Exercises 1–8, use number counters to find the product.

1. 2×4 **2.** 3×2 **3.** $2\times(-5)$ **4.** $4\times(-3)$

5. -3×3 **6.** -2×2 **7.** $-2\times(-4)$ **8.** $-3\times(-4)$

9. What can you say about the product of a negative integer and a positive integer? Can the result be positive?

10. Write a general statement about the product of two negative integers.

3.5 Multiplying Integers

What you should learn:

Goal 1 How to multiply integers

Goal 2 How to use integer multiplication to model real-life problems

Why you should learn it:

You can use integer multiplication to solve real-life problems, such as finding the wind-chill factor.

Goal 1 Multiplying Integers

In this lesson, you will learn how to find products of one or more negative factors. Here is an example.

$$3(-3) = (-3) + (-3) + (-3) = -9$$

LESSON INVESTIGATION

Investigating Integer Multiplication

Group Activity Complete the following. What patterns can you observe?

$(3)(3) = \boxed{?}$	$(3)(-2) = \boxed{?}$
$(2)(3) = \boxed{?}$	$(2)(-2) = \boxed{?}$
$(1)(3) = \boxed{?}$	$(1)(-2) = \boxed{?}$
$(0)(3) = \boxed{?}$	$(0)(-2) = \boxed{?}$
$(-1)(3) = \boxed{?}$	$(-1)(-2) = \boxed{?}$
$(-2)(3) = \boxed{?}$	$(-2)(-2) = \boxed{?}$
$(-3)(3) = \boxed{?}$	$(-3)(-2) = \boxed{?}$

Need to Know

When there are no grouping symbols, the order of operations for powers and for negative signs is that the power is evaluated *before* the negative sign.

a. $-3^2 = -(3)(3) = -9$
b. $(-3)^2 = (-3)(-3) = 9$

This order of operations also applies to expressions that have variables. For instance, the value of $-x^2$ when $x = -4$ is

$-(-4)(-4) = -16.$

Multiplying Integers

1. The product of two positive numbers is positive.
2. The product of two negative numbers is positive.
3. The product of a positive and a negative number is negative.

Example 1 ***Multiplying Integers***

a. $4(3) = 12$ *Product is positive.*
b. $5(-2) = -10$ *Product is negative.*
c. $(-4)(6) = -24$ *Product is negative.*
d. $(-3)(-11) = 33$ *Product is positive.* ■

Goal 2 Modeling Real-Life Situations

The rules for multiplying positive and negative integers also apply to fractions and decimals. For example, $\frac{1}{2} \cdot (-4) = -2$.

Real Life *Temperature*

Example 2 *Writing and Using a Model*

To convert from a Celsius temperature to a Fahrenheit temperature, multiply the Celsius temperature by $\frac{9}{5}$ and add 32. Write an algebraic model for this relationship. Then use the model to find the Fahrenheit temperature that corresponds to **a.** $-20°C$ and **b.** $-30°C$.

Solution

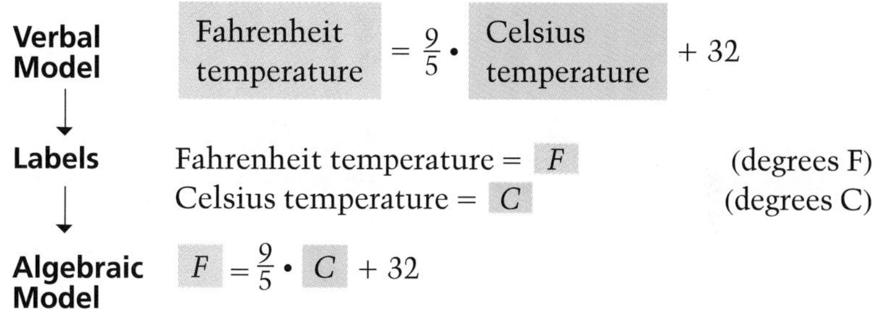

a. The Fahrenheit temperature corresponding to $-20°C$ is

$$
\begin{aligned}
F &= \tfrac{9}{5}C + 32 && \textit{Write model.} \\
&= \tfrac{9}{5}(-20) + 32 && \textit{Substitute } -20 \textit{ for } C. \\
&= -36 + 32 && \textit{Simplify.} \\
&= -4°F. && \textit{Simplify.}
\end{aligned}
$$

b. The Fahrenheit temperature corresponding to $-30°C$ is

$$
\begin{aligned}
F &= \tfrac{9}{5}C + 32 && \textit{Write model.} \\
&= \tfrac{9}{5}(-30) + 32 && \textit{Substitute } -30 \textit{ for } C. \\
&= -54 + 32 && \textit{Simplify.} \\
&= -22°F. && \textit{Simplify.}
\end{aligned}
$$

■

Wind-Chill Factor

Wind Speed (miles per hour)	Actual Air Temperature (°F) 30	20	10	0	−10	−20
5	27	16	7	−5	−15	−26
10	16	3	−9	−22	−34	−46
15	9	−5	−18	−31	−45	−58
20	4	−10	−24	−39	−53	−67
25	1	−15	−29	−44	−59	−74
30	−2	−18	−33	−49	−64	−79
35	−4	−20	−35	−52	−67	−82
40	−5	−21	−37	−53	−69	−84
45	−6	−22	−38	−54	−70	−85

Apparent Temperature (°F)

When the wind is blowing, you feel colder than the actual temperature given by a thermometer. The temperature that you feel is called the wind-chill factor.

Communicating about MATHEMATICS

SHARING IDEAS about the Lesson

Finding the Wind-Chill Factor The temperature is $-10°C$. The wind speed is 40 miles per hour. How cold does it feel on the Fahrenheit scale? Refer to the table above.

EXERCISES

Guided Practice

CHECK for Understanding

In Exercises 1–3, write the sum as a product. Then simplify.

1. $(-4) + (-4) + (-4)$ **2.** $(-x) + (-x) + (-x)$ **3.** $(-8) + (-8) + (-8) + (-8)$

4. *Computation Sense* Use the number line to decide whether ab, ac, and bc are positive or negative.

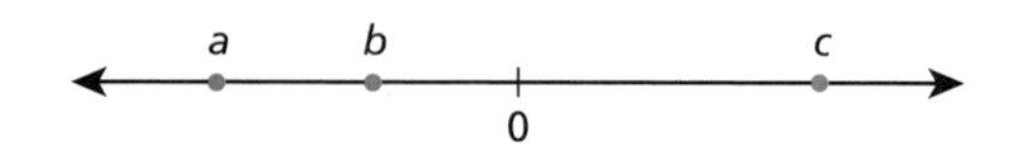

5. *Logical Reasoning* If x is not zero, which expression must be positive: x^2, $2x$, $-4x$.

6. Which expression is equal to $(-4)(-4)$: -4^2 or $(-4)^2$?

Independent Practice

In Exercises 7–18, find the product. Write your conclusion as an equation.

7. $3 \cdot 9$ **8.** $7(8)$ **9.** $-4 \cdot (-6)$ **10.** $-10 \cdot (-2)$

11. $5 \cdot (-11)$ **12.** $-8 \cdot 6$ **13.** $(-7)(-9)$ **14.** $(-4)(-12)$

15. $(-10)(3)$ **16.** $(-1)(54)$ **17.** $(-20)(0)$ **18.** $0 \cdot (-4)$

In Exercises 19–22, simplify the expression.

19. $-7 \cdot x$ **20.** $6 \cdot (-y)$ **21.** $(-14)(-a)$ **22.** $(-b)(-25)$

In Exercises 23–26, evaluate the expression when $a = 8$ and $b = -2$.

23. ab **24.** ab^2 **25.** $-b$ **26.** $-a^2$

In Exercises 27–30, find the product. Sample: $-3 \cdot (-2) \cdot (-4) = -24$

27. $1 \cdot (-9) \cdot 3$ **28.** $6(10)(-2)$ **29.** $4 \cdot (-2) \cdot (-8)$ **30.** $11(-3)(-3)$

31. Is the product of three negative numbers positive or negative?

32. Is the product of four negative numbers positive or negative?

In Exercises 33–40, use a calculator to find the product.

33. $-1.4 \cdot (-6)$ **34.** $-8 \cdot (-1.5)$ **35.** $\frac{7}{8}(-24)$ **36.** $\frac{5}{6}(-42)$

37. $(-542)(-15)$ **38.** $(-712)(-21)$ **39.** $(43)(-2)(-4)$ **40.** $(-36)(7)(-4)$

Mental Math **In Exercises 41–48, use mental math to solve the equation.**

41. $2x = -4$ **42.** $3m = -9$ **43.** $-6b = 12$ **44.** $-8n = 24$

45. $-3a = 27$ **46.** $-7y = 42$ **47.** $1.5m = -3$ **48.** $-2x = -3.2$

Geometry **In Exercises 49–51, use the rectangle at the right. The rectangle has an area of 15 square units. The base of the rectangle rests on a number line.**

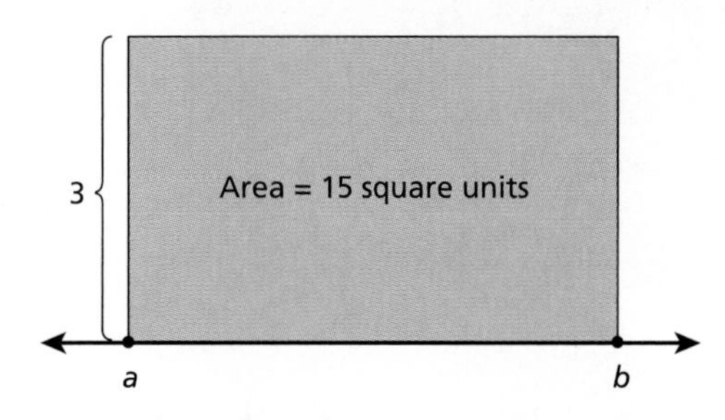

49. If $a = -4$, what is b?

50. If $b = -6$, what is a?

51. If $b = 3$, what is a?

Temperature Scales **In Exercises 52 and 53, use the following.**

Temperatures are sometimes measured on the Kelvin scale.

- To convert from Fahrenheit to Celsius, subtract 32 from the Fahrenheit temperature and multiply by $\frac{5}{9}$.
- To convert from Celsius to Kelvin, add 273.15 to the Celsius temperature.

52. Write algebraic models for converting from Fahrenheit to Celsius and from Celsius to Kelvin. (See page 123 for a model for converting from Celsius to Fahrenheit.)

53. Copy and complete the table.

Scale	Fahrenheit	Celsius	Kelvin
Temperature	212°F	?	373.15K
Temperature	32°F	0°C	?
Temperature	−40°F	?	?
Temperature	14°F	?	?

Helsinki University of Technology
The lowest temperature ever reached is two billionths of a degree above absolute zero. It was achieved at the Low Temperature Laboratory at the Helsinki University of Technology in Finland.

Integrated Review

Making Connections within Mathematics

Number Sense **In Exercises 54 and 55, choose values of *a* and *b* so that both statements are true.**

54. ab is positive and $a + b$ is negative.

55. ab is negative and $a + b$ is positive.

Exploration and Extension

Matrices **In Exercises 56–58, multiply using the following information.**

$$\begin{bmatrix} 3 & -2 \\ -1 & 4 \end{bmatrix}$$

The array at the right is a **matrix.** To multiply a matrix by a number, multiply each element of the matrix by the number.

$$-2\begin{bmatrix} 3 & -2 \\ -1 & 4 \end{bmatrix} = \begin{bmatrix} (-2)(3) & (-2)(-2) \\ (-2)(-1) & (-2)(4) \end{bmatrix} = \begin{bmatrix} -6 & 4 \\ 2 & -8 \end{bmatrix}$$

56. $-1\begin{bmatrix} 1 & -3 \\ 0 & -1 \end{bmatrix}$

57. $-3\begin{bmatrix} 1 & -5 \\ 4 & -2 \end{bmatrix}$

58. $2\begin{bmatrix} 8 & 6 \\ -7 & -4 \end{bmatrix}$

3.6 Dividing Integers

What you should learn:

Goal 1 How to divide integers

Goal 2 How to use integer division to model real-life problems

Why you should learn it:

You can use integer division to solve real-life problems involving averages, such as finding the average weight loss of a wrestling team.

Goal 1 Dividing Integers

To perform division by hand, use **long division** as shown below. Notice that you can check your result by multiplying.

Divisor → 26, Quotient → 14, Dividend → 364

$$26\overline{)364}\quad\text{(quotient } 14\text{)}$$

		Check:
26	*Multiply: (1)(26)*	26
104	*Subtract*	× 14
104	*Multiply: (4)(26)*	104
0	*Subtract*	26
		364

LESSON INVESTIGATION

Investigating Integer Division

Group Activity Solve the following division problems. Explain how to use the multiplication check illustrated above to determine the sign of the quotient.

a. $24 \div (-6)$ **b.** $-24 \div (-6)$ **c.** $(-24) \div 6$

Discuss the results with members of your group. What can you conclude about the rules for division of integers?

Zero in Divison

The number 0 has special rules regarding division.

1. You cannot divide a number by 0. Expressions with 0 divisors, such as $4 \div 0$, are meaningless.
2. When 0 is divided by a nonzero number, the result is 0. For instance, $0 \div (-4) = 0$.

Dividing Integers

1. The quotient of two positive numbers is positive.
2. The quotient of two negative numbers is positive.
3. The quotient of a positive and a negative number is negative.

Example 1 *Dividing Integers*

a. $\frac{12}{-2} = -6$ *Quotient is negative.*

b. $\frac{-20}{4} = -5$ *Quotient is negative.*

c. $\frac{-36}{-6} = 6$ *Quotient is positive.* ■

Goal 2 Modeling Real-Life Problems

To find the average (or **mean**) of n numbers, add the numbers and divide the result by n. For instance, the average of 24, 37, 21, and 42 is

$$\text{Average} = \frac{24 + 37 + 21 + 42}{4} = \frac{124}{4} = 31.$$

There are 13 weight classes in high school wrestling. The classes range from 103 pounds to a heavyweight class of no more than 275 pounds.

Example 2 *Finding an Average Weight Change*

You are coaching your school's wrestling team. During the months of December and January, the 15 members of the team recorded the following weight gains or losses (in pounds).

$$-5, 0, -7, -3, 2, -4, -6, -1, 0, -3, -4, -4, -2, -5, -3$$

What was the average weight gain or loss per team member?

Solution

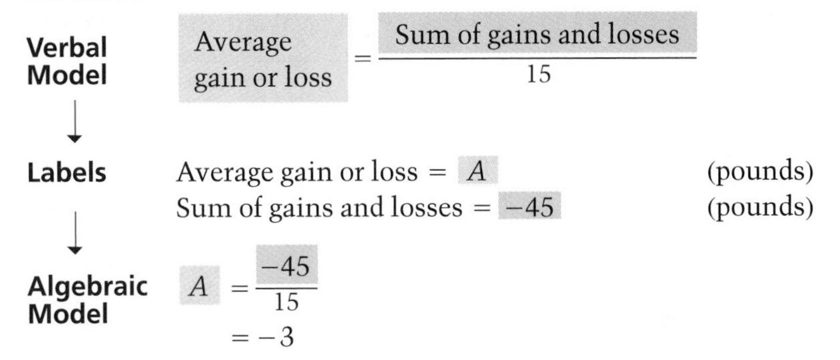

Verbal Model $\quad$ Average gain or loss $= \dfrac{\text{Sum of gains and losses}}{15}$

↓

Labels $\quad$ Average gain or loss $= A$ (pounds)

Sum of gains and losses $= -45$ (pounds)

↓

Algebraic Model

$$A = \frac{-45}{15} = -3$$

The average was a loss of 3 pounds per team member. ■

Communicating about MATHEMATICS

Cooperative Learning

SHARING IDEAS about the Lesson

It's Up to You Work with a partner to create a single list of weight changes that satisfies all of the following.

- 10 people are on the wrestling team.
- 4 people gained, 5 lost, and 1 stayed the same.
- The average weight change was a loss of 3 pounds.

EXERCISES

Guided Practice

CHECK for Understanding

In Exercises 1–4, state whether the quotient is negative or positive.

1. $216 \div 9$ **2.** $-28 \div 4$ **3.** $\frac{48}{-6}$ **4.** $\frac{-522}{-9}$

5. Explain how to check that $\frac{-6}{-2} = 3$.

Computation Sense **In Exercises 6–8, decide whether $\frac{a}{b}$ is positive or negative.**

6.

7.

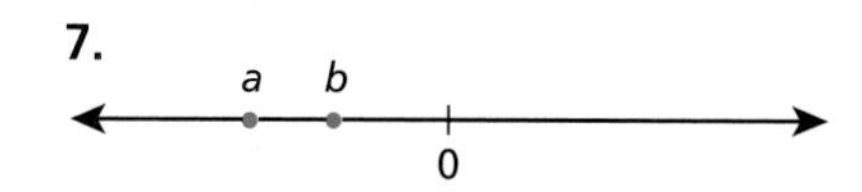

8.

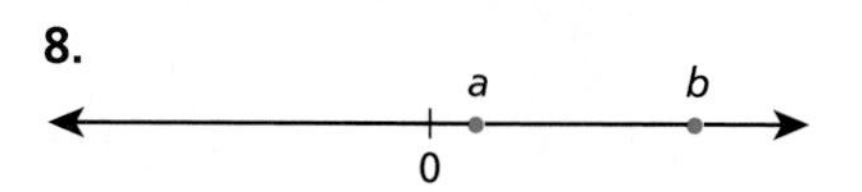

Independent Practice

Computation Sense **In Exercises 9–24, evaluate the expression. Check your result by multiplying.**

9. $\frac{54}{2}$ **10.** $\frac{100}{5}$ **11.** $\frac{-90}{15}$ **12.** $\frac{-130}{26}$

13. $384 \div (-12)$ **14.** $376 \div (-8)$ **15.** $-954 \div (-18)$ **16.** $-1058 \div (-46)$

17. $\frac{0}{-75}$ **18.** $\frac{0}{1000}$ **19.** $-1020 \div 30$ **20.** $-300 \div 12$

21. $1568 / (-16)$ **22.** $242 / (-11)$ **23.** $-1659 / (-21)$ **24.** $-3621 / (-71)$

Division Patterns **In Exercises 25–28, complete the statements. Then describe the pattern.**

25.
$-2 \div 2 = \boxed{?}$
$-4 \div 2 = \boxed{?}$
$-6 \div 2 = \boxed{?}$
$-8 \div 2 = \boxed{?}$
$-10 \div 2 = \boxed{?}$

26.
$-1 \div (-1) = \boxed{?}$
$-2 \div (-2) = \boxed{?}$
$-3 \div (-3) = \boxed{?}$
$-4 \div (-4) = \boxed{?}$
$-5 \div (-5) = \boxed{?}$

27.
$0 \div (-1) = \boxed{?}$
$0 \div (-2) = \boxed{?}$
$0 \div (-3) = \boxed{?}$
$0 \div (-4) = \boxed{?}$
$0 \div (-5) = \boxed{?}$

28.
$64 \div (-32) = \boxed{?}$
$32 \div (-16) = \boxed{?}$
$16 \div (-8) = \boxed{?}$
$8 \div (-4) = \boxed{?}$
$4 \div (-2) = \boxed{?}$

In Exercises 29–32, evaluate the expression when $x = -2$, $y = 3$, and $z = -4$.

29. $\frac{-2x}{4}$ **30.** $\frac{xz}{2}$ **31.** $\frac{xy}{z}$ **32.** $\frac{-yz}{x}$

Mental Math **In Exercises 33–40, use mental math to solve the equation.**

33. $\frac{b}{7} = -7$ **34.** $\frac{a}{4} = -5$ **35.** $\frac{x}{-4} = 6$ **36.** $\frac{z}{-8} = 5$

37. $\frac{m}{-2} = -24$ **38.** $\frac{n}{-3} = -13$ **39.** $\frac{63}{n} = -7$ **40.** $\frac{81}{p} = -3$

In Exercises 41–46, find the average of the numbers.

41. 35, 38, 34, 39, 32, 38

42. 22, 19, 21, 20, 18, 22, 25

43. $-4, -2, 0, 1, -3, 1, 2, -5, 3, -3$

44. $3, 3, -4, 6, 2, 1, -2, -1, 2, -4, 5$

45. $-6, -9, -4, -2, -7, -6, -8$

46. $-20, -16, -12, -17, -10$

Coaching a Speed Skater **In Exercises 47–49, use the following information.**

In five trial runs on a 500 m track, your skater has times of 44.21 seconds, 45.02 seconds, 44.78 seconds, 45.10 seconds, and 44.13 seconds.

47. Find the average trial time of the skater.

48. Another skater, after five trial runs, had an average trial time of 45.07 seconds. What could her trial times have been?

49. Is it possible that the second skater had a faster trial time than the first? Explain.

By 1994, Bonnie Blair had won 5 Olympic gold medals.

Integrated Review

Making Connections within Mathematics

Logical Reasoning **In Exercises 50–53, decide whether the statement is *always, sometimes,* or *never* true.**

50. The sum of a positive number and a negative number is positive.

51. The difference of a negative number and a positive number is negative.

52. The product of a positive number and a negative number is positive.

53. The quotient of a positive number and a negative number is equal to -1.

Computation Sense **In Exercises 54–56, evaluate the expression.**

54. $(24 - 16) \div 2$

55. $(36 + 18) \div (-6)$

56. $(10 - 24) \div (-5 - 2)$

Equation Sense **In Exercises 57–64, solve the equation.**

57. $m + 21 = 44$

58. $p + 16 = 57$

59. $r - 76 = 12$

60. $47 = t - 19$

61. $12m = 60$

62. $8s = 72$

63. $\frac{x}{6} = 12$

64. $\frac{y}{10} = 80$

Geometry **In Exercises 65–68, write an expression that does not contain parentheses for the area of the rectangle.**

65.

66.

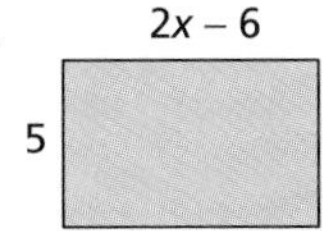

67.

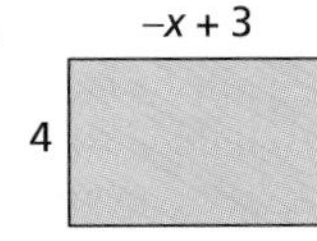

68.

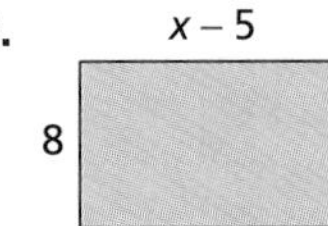

Exploration and Extension

Finding Reciprocals **In Exercises 69–72, find the reciprocal of the number. The reciprocal of a nonzero number a is $\frac{1}{a}$. For instance, the reciprocal of −2 is $\frac{1}{-2}$ or $-\frac{1}{2}$.**

69. −4 **70.** 5 **71.** 7 **72.** −3

73. ***Writing a Conjecture*** Multiply several numbers by their reciprocals. Then write a conjecture about the product of a number and its reciprocal.

Reasoning **In Exercises 74 and 75, solve the puzzle by replacing each $\boxed{?}$ with a single digit.**

74.

$$\begin{array}{r} 2\,\boxed{?} \\ \boxed{?}\,5\,\overline{)\,3\,\boxed{?}\,0} \\ -\,\boxed{?}\,\boxed{?} \\ \hline \boxed{?}\,0 \\ -\,6\,0 \\ \hline 0 \end{array}$$

75.

$$\begin{array}{r} \boxed{?}\,\boxed{?} \\ 3\,\boxed{?}\,\overline{)\,\boxed{?}\,\boxed{?}\,\boxed{?}} \\ -\,\boxed{?}\,6 \\ \hline \boxed{?}\,2\,\boxed{?} \\ -\,3\,\boxed{?}\,\boxed{?} \\ \hline 0 \end{array}$$

Mixed REVIEW

In Exercises 1–8, evaluate the expression. (3.2–3.6)

1. $9 + 3$ **2.** $4 - 2$ **3.** $6 + (-3)$ **4.** $8 - (-2)$

5. $8 \times (-4)$ **6.** $(-3) \times (-5)$ **7.** $-84 \div 4$ **8.** $(-7)^2$

In Exercises 9 and 10, translate the verbal sentence into an algebraic equation. (2.7)

9. The cost of 3 burritos is \$4.20

10. The depth of 4 fathoms is 24 feet.

In Exercises 11–14, describe the pattern and list the next three numbers in the sequence. (1.1, 3.5, 3.6)

11. $2, -4, 8, -16, \boxed{?}\ \boxed{?}\ \boxed{?}$

12. $-1, 3, -9, 27, \boxed{?}\ \boxed{?}\ \boxed{?}$

13. $-2, 1, -\frac{1}{2}, \frac{1}{4}, \boxed{?}\ \boxed{?}\ \boxed{?}$

14. $-78125, 15625, -3125, 625, \boxed{?}\ \boxed{?}\ \boxed{?}$

In Exercises 15–20, evaluate the expression. (1.4, 3.2–3.6)

15. $3 + (-6) \times (-4)$ **16.** $-4 \times 6 \div 3$ **17.** $(3 - 6)^2$

18. $8 - (-2) \times 4$ **19.** $15 \times \left(\frac{3}{5} - \frac{2}{5}\right)$ **20.** $6.2 - 4.5 \div 1.5$

3.7 Problem Solving Using Integers

What you should learn:

How to use properties of equality to solve equations involving integers

How to use integer operations to model real-life problems

Why you should learn it:

You can use integer operations to solve real-life problems, such as finding the profit or loss from a dance.

Goal 1 Solving Equations

In Lessons 2.4 and 2.5, you learned how to use properties of equality to solve equations. These properties also apply to negative and positive integers. For instance, to solve the equation $x - 5 = -7$, you can add 5 to each side of the equation.

Example 1 *Using Addition or Subtraction*

Solve the equations **a.** $x - 5 = -7$ and **b.** $-12 = n + 3$.

Solution

a.

$x - 5 = -7$	*Rewrite original equation.*
$x - 5 + 5 = -7 + 5$	*Add 5 to each side.*
$x = -2$	*Simplify.*

The solution is -2. Check this in the original equation.

b.

$-12 = n + 3$	*Rewrite original equation.*
$-12 - 3 = n + 3 - 3$	*Subtract 3 from each side.*
$-15 = n$	*Simplify.*

The solution is -15. Check this in the original equation. ■

Inverse Operations

To help you decide which operation to perform to solve an equation, remember that addition and subtraction are **inverse operations** and multiplication and division are inverse operations. For instance, in the equation $-12 = n + 3$, notice that the side of the equation with *n* involves addition. To isolate *n*, you should perform the inverse (or opposite) operation. That is, you should subtract 3 from each side.

Example 2 *Using Multiplication or Division*

Solve the equations **a.** $3y = -18$ and **b.** $\frac{m}{-2} = 15$.

Solution

a.

$3y = -18$	*Rewrite original equation.*
$\frac{3y}{3} = \frac{-18}{3}$	*Divide each side by 3.*
$y = -6$	*Simplify.*

The solution is -6. Check this in the original equation.

b.

$\frac{m}{-2} = 15$	*Rewrite original equation.*
$-2 \cdot \frac{m}{-2} = -2 \cdot 15$	*Multiply each side by -2.*
$m = -30$	*Simplify.*

The solution is -30. Check this in the original equation. ■

Goal 2 Modeling Real-Life Problems

Example 3 *Planning a Dance*

Real Life
Budgeting

Your class is sponsoring a school video dance. Your expenses will be \$750 for a disc jockey, \$75 for security, and \$40 for advertisements. You will charge \$6 per person. Will your class make a profit if 125 people attend? If 250 people attend?

Many current songs are accompanied by videos. The first commercially successful home video recorder was introduced in 1975.

Solution The profit is the difference between your total income (from ticket sales) and your total expenses.

Verbal Model Profit = Ticket price • Number of people − Expenses

Labels
Profit = P (dollars)
Ticket price = 6 (dollars per person)
Number of people attending = n (people)
Expenses = 750 + 75 + 40 = 865 (dollars)

Algebraic Model $P = 6 \cdot n - 865$

If 125 people attend, then the profit is

$$P = 6(125) - 865 = -115$$

which means that your class had a loss of \$115.

If 250 people attend, then the profit is

$$P = 6(250) - 865 = 635$$

which means that your class had a profit of \$635. ■

Communicating about MATHEMATICS

SHARING IDEAS about the Lesson

Real Life
Budgeting

Work Backward You are on the school yearbook staff. The printing company that is producing the yearbooks will charge \$2400, plus \$6 per book. You expect to sell 300 yearbooks. Use the following verbal model to decide how much you can reasonably charge for each yearbook. (Your answer depends on the amount of profit you want to make.)

Profit = Number of books • Price per book − (2400 + 6 • Number of books)

EXERCISES

Guided Practice

CHECK for Understanding

1. *Writing* In your own words, state the four properties of equality.

In Exercises 2–5, state the property of equality that can be used to solve the equation. Then solve the equation.

2. $x - 4 = -8$
3. $-6 = y + 8$
4. $\frac{a}{-5} = 7$
5. $-5b = 35$
6. *Problem Solving* Describe a real-life situation in which you can use integer operations.

Independent Practice

In Exercises 7–12, decide whether the value of the variable is a solution of the equation. If not, find the solution.

7. $x - 7 = 3; x = 10$
8. $t + 7 = -10; t = -17$
9. $9 = s + 5; s = 14$
10. $-42 = -14b; b = 3$
11. $\frac{m}{-2} = 12; m = -24$
12. $\frac{n}{-6} = -8; n = -48$

In Exercises 13–28, solve the equation. Check your solution.

13. $x + 2 = -11$
14. $y + 1 = 5$
15. $x - 9 = 15$
16. $-17 = p - 13$
17. $q + 12 = 3$
18. $r - 6 = -2$
19. $72 = -6x$
20. $-15t = -60$
21. $\frac{y}{-7} = 9$
22. $-5 = \frac{s}{-11}$
23. $2x = -34$
24. $\frac{a}{20} = -4$
25. $b + 5.6 = -8.4$
26. $y - 3.8 = 5.2$
27. $2 = -4z$
28. $\frac{c}{-6.1} = -9$

In Exercises 29–32, write an algebraic equation for the sentence. Then solve the equation and write your conclusion as a sentence.

29. The difference of x and 20 is -4.
30. -10 is the sum of y and 25.
31. 51 is the product of a and -3.
32. The quotient of t and -6 is -14.

In Exercises 33–38, use a calculator to solve the equation. Then check the solution.

33. $-1088 = y + 129$
34. $m - 364 = -1980$
35. $-486s = 7776$
36. $-555t = -8325$
37. $-56 = \frac{p}{-23}$
38. $\frac{q}{67} = -31$

In Exercises 39–44, match the equation with its solution.

a. 9 b. -13 c. -8 d. 7 e. -12 f. -16

39. $a - 4 + 9 = -8$
40. $t - 12 + 3 = -2$
41. $x + 6 - 7 = -13$
42. $-27 = 4x - 7x$
43. $y + 3y = -32$
44. $\frac{p}{2} = 13 - 21$

Hot-Air Balloon Ride **In Exercises 45 and 46, imagine that you are taking a hot-air balloon ride.**

45. You are flying at an altitude of x feet. You descend 6,891 feet to an altitude of 18,479 feet. Which of the following models will correctly determine your original altitude? Solve the correct model and use the result to determine your original altitude.

a. $x - 6{,}891 = 18{,}479$

b. $18{,}479 - x = 6{,}891$

46. You are flying at an altitude of 19,653 feet. You descend 8,905 feet, rise 9,842 feet, descend 14,450 feet, and descend another 6,140 feet. Write an algebraic model that represents your final altitude. What is your final altitude?

The altitude record in a hot-air balloon is 64,996 feet over Laredo, Texas, achieved by Per Lindstrand in 1988.

Rock-N-Roll **In Exercises 47 and 48, use the following information.**

Your school is sponsoring a rock-n-roll concert. The expenses include \$800 for the band, \$20 for posters, \$200 for refreshments, and \$70 for security. The tickets cost \$5 per person.

47. Use the verbal model Profit = Income − Expenses to write an algebraic model for the profit. Let P represent the profit and let n represent the number of tickets sold.

48. How many people must attend to make a profit of \$250?

Integrated Review

Making Connections within Mathematics

Endangered Species **In Exercises 49–52, use the following equations and the information at the right.** ***(Source: U.S. Fish and Wildlife Service)***

$C + 17 = B \qquad F - B = -5$

$B - M = \frac{1}{2}C \qquad I + M = C + 10$

49. How many species of birds are endangered?

50. How many species of fish are endangered?

51. How many species of mammals are endangered?

52. How many species of insects are endangered?

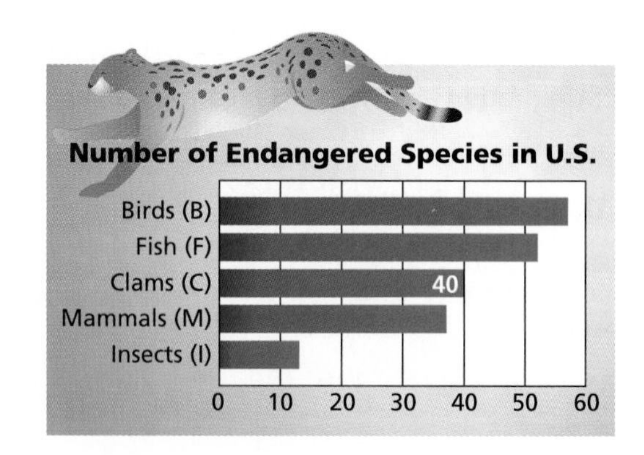

Exploration and Extension

53. *Research Project* Use some reference source to find information about a world record. Use your findings to write a story problem.

3.8 Exploring Patterns in the Coordinate Plane

What you should learn:

How to plot points in a coordinate plane

How to use a coordinate plane to represent data graphically

Why you should learn it:

Representing data in a coordinate plane helps you discover relationships and patterns between two variables.

Points in a Coordinate Plane

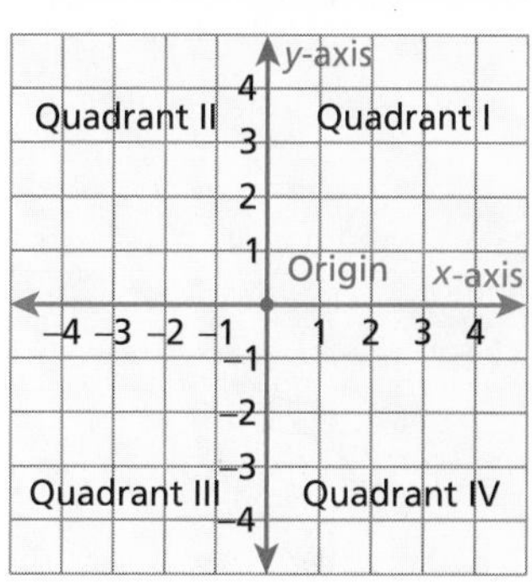

A **coordinate plane** has two number lines that intersect at a right angle. The point of intersection is the **origin.** The horizontal number line is usually called the **x-axis,** and the vertical number line is usually called the **y-axis.** (The plural of axis is *axes.*)

The two axes divide the coordinate plane into four parts called **quadrants.**

Each point in a coordinate plane can be represented by an **ordered pair** of numbers, (x, y). The first number is the **x-coordinate,** and it gives the position of the point relative to the x-axis. The second number is the **y-coordinate,** and it gives the position of the point relative to the y-axis.

(x, y)

x-coordinate ↗ ↖ y-coordinate

Locating the point in the coordinate plane that corresponds to an ordered pair is called **plotting the point.**

Example 1 *Plotting Points in a Coordinate Plane*

Plot the points $A(4, 3)$, $B(4, -2)$, $C(-4, -2)$, and $D(-4, 3)$. Then find the perimeter and area of rectangle $ABCD$.

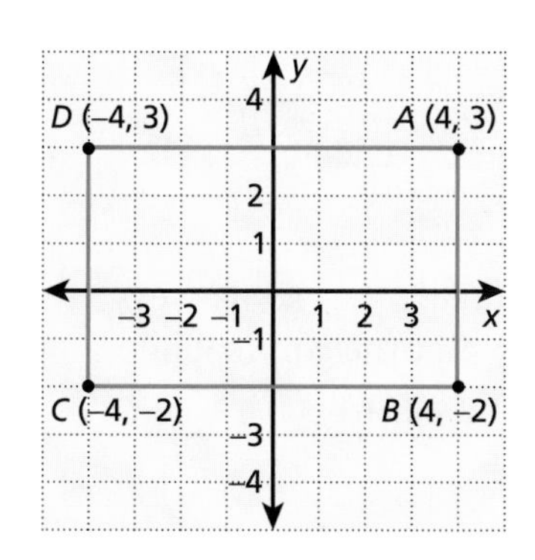

Solution The points are plotted in the coordinate plane at the left. The sides of the rectangle are denoted by the line segments

$\overline{AB}$, $\overline{BC}$, $\overline{CD}$, and $\overline{AD}$. *Sides of rectangle ABCD*

The lengths of the sides are

$AB = 5$, $BC = 8$, $CD = 5$, and $AD = 8$. *Lengths*

The rectangle has a perimeter of $5 + 8 + 5 + 8$ or 26 units, and an area of $5(8)$ or 40 square units. ■

Goal 2 Representing Data Graphically

An ordered pair (x, y) is a **solution** of an equation involving x and y if the equation is true when the values of x and y are substituted into the equation. Most equations in two variables have many solutions. For instance, 3 solutions of $x + y = 4$ are shown below.

Equation	Solution	Check
$x + y = 4$	$(1, 3)$	$1 + 3 = 4$
$x + y = 4$	$(-3, 7)$	$-3 + 7 = 4$
$x + y = 4$	$(0, 4)$	$0 + 4 = 4$

Example 2 *Representing Data*

Construct a table that shows several solutions of the equation $2 + x = y$. Then plot the corresponding points and describe the graphical pattern.

Solution Begin by choosing an x-value, such as $x = -3$. Substitute -3 for x in the equation to obtain

$$2 + (-3) = y \quad \text{or} \quad y = -1.$$

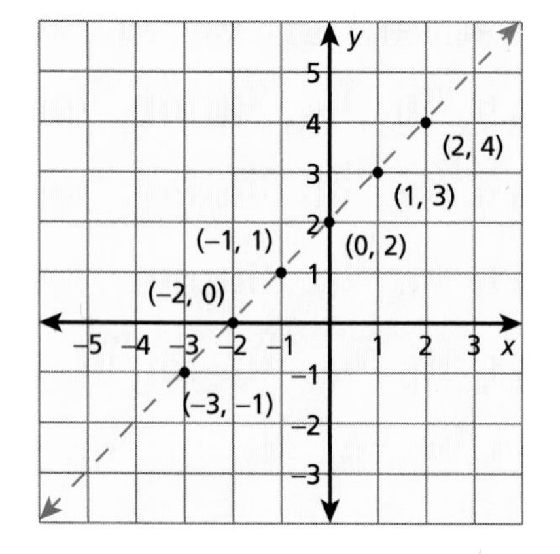

This implies that $(-3, -1)$ is a solution of the equation. Other solutions are shown in the table.

x	-3	-2	-1	0	1	2
y	-1	0	1	2	3	4
(x, y)	$(-3, -1)$	$(-2, 0)$	$(-1, 1)$	$(0, 2)$	$(1, 3)$	$(2, 4)$

After plotting the points in a coordinate plane, you can see that all 6 points lie on a line. If you try finding other solution points of the equation, you will discover that they lie on the same line. ■

Communicating about MATHEMATICS

SHARING IDEAS about the Lesson

Extending the Example Construct a table that shows several solutions of the equation. Then plot the corresponding points and describe the graphical pattern.

$x + y = 3$

EXERCISES

Guided Practice

CHECK for Understanding

1. Explain how to construct and label a coordinate plane.
2. Explain how to decide whether an ordered pair (x, y) is a solution of an equation.

In Exercises 3–10, use the coordinate plane at the right to state the x- and y-coordinate of the point. Then state the quadrant in which the point lies.

3. A 4. B 5. C 6. D
7. E 8. F 9. G 10. H

Independent Practice

In Exercises 11–16, match the ordered pair with its corresponding point in the coordinate plane. Identify the quadrant in which the point lies.

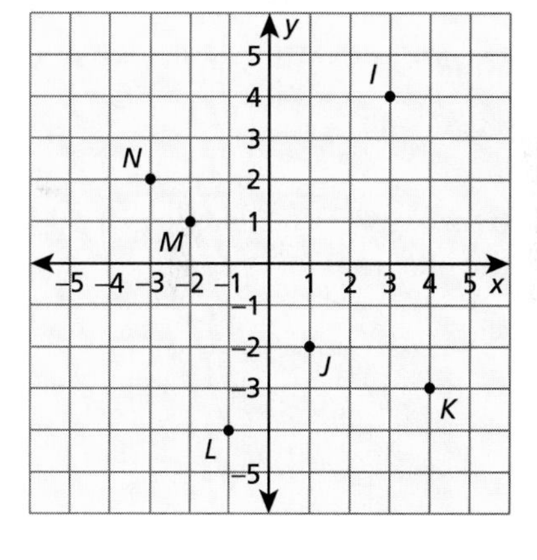

11. $(1, -2)$ 12. $(-2, 1)$ 13. $(3, 4)$
14. $(-3, 2)$ 15. $(-1, -4)$ 16. $(4, -3)$

In Exercises 17–24, plot the points on a single coordinate plane. Determine the quadrant in which each point lies.

17. $A(-6, -2)$ 18. $B(-1, -5)$
19. $C(3, 7)$ 20. $D(7, 3)$
21. $E(-1, 0)$ 22. $F(0, 4)$
23. $G(-4, 5)$ 24. $H(2, -6)$

In Exercises 25–28, determine the quadrant in which (x, y) lies.

25. $x < 0$ and $x = y$ 26. $x > 0$ and $y = -x$ 27. $y > 0$ and $y = -x$ 28. $y > 0$ and $x = y$

Coordinate Geometry **In Exercises 29 and 30, plot the points to form the vertices of a rectangle. Find the area and perimeter of the rectangle.**

29. $A(1, 3), B(-2, 3), C(1, -4), D(-2, -4)$ 30. $A(-4, 4), B(1, 4), C(1, -1), D(-4, -1)$

Equation Sense **In Exercises 31–33, show that the ordered pair is a solution of the equation. Then find three other solutions.**

31. $3 + x = y$; $(7, 10)$ 32. $y - 5 = x$; $(-8, -3)$ 33. $x + y = 6$; $(8, -2)$

Visualizing Patterns **In Exercises 34–39, construct a table that lists several solutions of the equation. Then plot the points and describe the pattern.**

34. $y = 4 - x$

35. $-5x = y$

36. $2x - y = 1$

37. $\frac{1}{2}x + 3 = y$

38. $x + y = -2$

39. $3x - 2 = y$

Selling Jeans **In Exercises 40–43, use the following.**

You own a clothing store. You pay \$20 for a pair of jeans. The number of pairs you sell in a month depends on the price you charge customers, as shown in the graph.

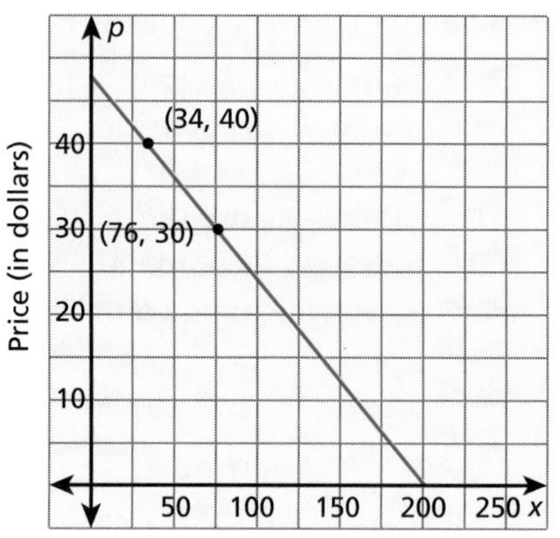

40. When the price is \$30, how many pairs will you sell?

41. When the price is \$40, how many pairs will you sell?

42. Your profit is given by

$$\text{Profit} = px - 20x.$$

What is your profit when $p = \$30$? When $p = \$40$?

43. *It's Up to You* Can stores always make a greater profit by charging more? Explain your reasoning.

Integrated Review

Making Connections within Mathematics

Equation Sense **In Exercises 44–47, solve the equation.**

44. $x + 15 = -3$

45. $n - 5 = -12$

46. $10m = -1$

47. $\frac{p}{-1} = 5$

Geometry **In Exercises 48–50, write an expression for the area of the blue region.**

48.

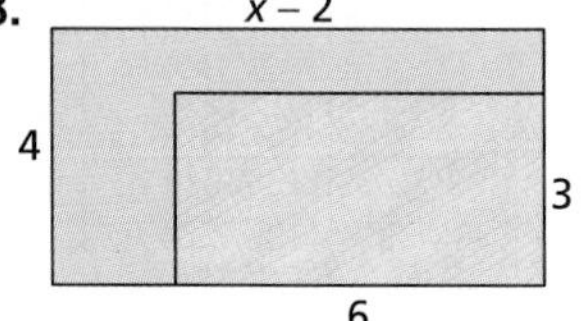

49.

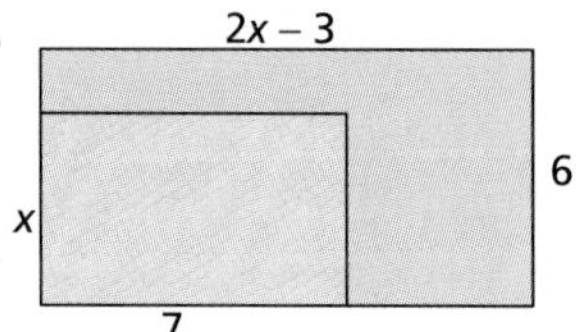

50.

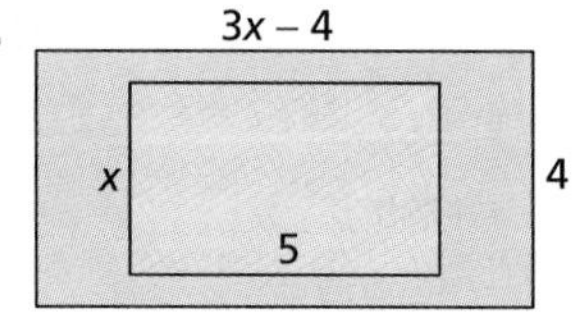

Exploration and Extension

A Coordinate Sequence **You are given a point on the coordinate plane and a procedure that produces a sequence. In Exercises 51 and 52,**

a. find the sequence,

b. plot the points on a coordinate plane, and

c. describe the results.

51. $(0, 0)$; add 1 to the x-coordinate and add 1 to the y-coordinate.

52. $(3, 1)$; subtract 3 from the x-coordinate and add 2 to the y-coordinate.

3

Chapter Summary

What did you learn?

Skills

1. Plot integers on a number line. **(3.1)**
2. Find the absolute value of an integer. **(3.1)**
3. Perform operations with integers.
- Add two integers. **(3.2)**
- Add three or more integers. **(3.3)**
- Subtract integers. **(3.4)**
- Multiply integers. **(3.5)**
- Divide integers. **(3.6)**

4. Simplify expressions with integer coefficients. **(3.3, 3.4)**
5. Solve equations involving integers. **(3.7)**
6. Plot points in a coordinate plane. **(3.8)**

Problem-Solving Strategies

7. Use integers to solve real-life problems. **(3.1–3.8)**

Exploring Data

8. Use a coordinate plane to organize and interpret data. **(3.8)**

Why did you learn it?

In prehistoric times, humans led simple lives. They could solve most of their real-life problems with counting numbers, such as 1, 2, 3, and so on. As life became more and more complex, humans found that they needed other types of numbers. In this chapter, you have seen that many real-life quantities can be modeled with negative numbers. For instance, a weight change of -4 pounds means that you have lost 4 pounds. A "profit" of $-\$260$ means that you had a loss of $260. A temperature of $-25°$F means that the temperature is 25 degrees below zero.

How does it fit into the bigger picture of mathematics?

Some people think that mathematics consists of dozens of unrelated rules that have to be memorized. With that outlook, mathematics is much more difficult than it has to be. Throughout this course, we hope that you will look for connections between different parts of mathematics. For instance, in this chapter you learned that the equation-solving techniques you studied in Chapter 2 can also be applied to equations involving integers. Later, you will learn that the same techniques can be applied to equations with variables on both sides and to equations that contain fractions or decimals. Rather than remembering dozens of rules for solving equations, you will learn that you only need one basic rule: *performing the same operation to both sides of an equation produces an equivalent equation.*

Chapter REVIEW

In Exercises 1–4, plot the integers on a number line. (3.1)

1. 1, −4, 0, −7 **2.** 2, 3, −2, −6 **3.** 0, −1, −2, 5 **4.** −4, −3, 3, 4

In Exercises 5 and 6, order the integers from least to greatest. (3.1)

5. 5, −7, 8, −3, 0 **6.** 0, −1, 12, −5, −13

Number Sense **In Exercises 7–10, write the integer associated with the phrase. (3.1)**

7. A loss of $365
8. 5 units to the left
9. Up 17 units
10. 875 feet above sea level

In Exercises 11–19, evaluate the expression. (3.2, 3.3, 3.4)

11. $17 + (-16)$ **12.** $25 + (-35)$ **13.** $-27 + (-15)$
14. $-63 + 12$ **15.** $-10 - 31$ **16.** $-5 - |-11|$
17. $49 - |-51|$ **18.** $-10 - (-11) + (-12)$ **19.** $10 + 11 - (-12)$

In Exercises 20–28, simplify the expression. (3.3, 3.4)

20. $3x + (-x) + 10$ **21.** $5x + (-2x) - 6$ **22.** $-7x + (-8x) + 15$
23. $-12x + 16x + (-8)$ **24.** $-3x - (-12x) + 2$ **25.** $-4x - (-9x) - 6$
26. $-6x - 6x - 6$ **27.** $12x - (-12x) + 12$ **28.** $8x - (-10x) + (-7x)$

An Investment **In Exercises 29–31, use the following information. (3.3)**

You have two $250 investments. At the end of each quarter you receive a statement that shows the change in your balance for that quarter.

Time (in months)	3	6	9	12
Investment 1	$18	−$15	$7	$20
Investment 2	$4	$4	$4	$4

29. Which investment earned more after 6 months?
30. Which investment earned more after 12 months?
31. One of these investments is a stock purchase and one is a bank savings account. Which is which? Explain your reasoning.

Computation Sense **In Exercises 32–35, *a* is positive and *b* is negative. Decide whether the expression is positive or negative. (3.1, 3.5, 3.6)**

32. ab **33.** $-\frac{a}{b}$ **34.** $-5ab$ **35.** $\frac{6a}{-2b}$

In Exercises 36–43, evaluate the expression. (3.5, 3.6)

36. $9(8)$ **37.** $(-3)(7)$ **38.** $5(-9)$ **39.** $(-6)(-4)$
40. $14 \div (-2)$ **41.** $-52 \div 13$ **42.** $\frac{-96}{-12}$ **43.** $\frac{-144}{4}$

44. *Averages* Find the average of the numbers. **(3.6)**
−274, −266, −290, −281, −275, −300, −259, −262, −270, −283

Batting Average **In Exercises 45 and 46, use the information given in the caption of the photo. (3.3)**

45. During a baseball season, you had 28 hits in 56 times at bat. What is your batting average?

46. During a baseball season, you played 16 games. Your hit totals in each game were

1, 4, 0, 2, 2, 1, 3, 0, 0, 2, 4, 3, 1, 1, 0, 4.

What was your average number of hits per game?

A baseball player's batting average is found by dividing the number of hits by the number of times at bat.

In Exercises 47–50, write an algebraic model for the statement. Then solve the model. (3.7)

47. The difference of a number and -8 is 7.

48. The sum of a number and -9 is -21.

49. The product of a number and 4 is -28.

50. The quotient of a number and -15 is 3.

In Exercises 51–58, solve the equation. Check your solution. (3.3, 3.4, 3.7)

51. $x + 9 = -6$ **52.** $y + 15 = 8$ **53.** $m - 8 = 6$ **54.** $-14 = n - 27$

55. $-14s = 42$ **56.** $10t = -80$ **57.** $\frac{r}{-7} = -8$ **58.** $\frac{p}{5} = 11$

In Exercises 59–62, plot the point and name its quadrant. (3.8)

59. $(-5, 1)$ **60.** $(-1, -6)$ **61.** $(4, 5)$ **62.** $(6, -3)$

63. *Coordinate Geometry* In a coordinate plane, plot the points $A(7, 1)$, $B(7, -5)$, $C(-1, -5)$, and $D(-1, 1)$. What type of figure is $ABCD$? Find its perimeter and area. **(3.8)**

Equation Sense **In Exercises 64–66, determine whether the ordered pair is a solution of the equation. (3.8)**

64. $x - y = -2$; $(-3, 1)$ **65.** $x + 7 = y$; $(-4, -3)$ **66.** $y = -4x$; $(3, -12)$

Equation Sense **In Exercises 67–70, construct a table that lists several solutions of the equation. Then plot the points and describe the pattern. (3.8)**

67. $x + 1 = y$ **68.** $y = x$ **69.** $y = -2x$ **70.** $x + y = 1$

Logical Reasoning **In Exercises 71–74, decide whether the statement is sometimes, always, or never true. Explain. (3.4)**

71. The opposite of n is $-n$.

72. The opposite of n is negative.

73. The absolute value of $-n$ is n.

74. $x(y - z) = xy + (-xz)$

FOCUS ON *Coaching an Athlete*

Real Life Connections

Gymnastics **In Exercises 75 and 76, use the following information.**

Gymnasts are judged by how well they perform their routine. A score of 10.00 is a perfect score. In women's gymnastics, seven judges score each performance. To determine a gymnast's score, the head judge eliminates the highest and lowest scores of the other six judges and averages the remaining four scores. (The head judge's score is usually not used.)

75. Listed below are the judges' scores (excluding the head judge's) for the gold, silver, and bronze medalists in an international women's gymnastics competition. Which gymnast won the gold?

Gymnast 1: 9.85, 9.30, 9.70, 9.65, 9.35, 9.50
Gymnast 2: 9.80, 9.60, 9.45, 9.30, 9.50, 9.25
Gymnast 3: 9.10, 9.45, 9.95, 9.70, 9.55, 9.65

76. One gymnast finished with a final score of 9.2. What could her scores have been? (There are many correct answers.)

In gymnastics competitions, judges award a maximum of 3.4 points for the level of difficulty of the routine, 2.6 points for the combination of movements, and 4.0 points for originality and performance.

Word Puzzle **In Exercises 77–96, solve the equation. Then use the results to decode the message.**

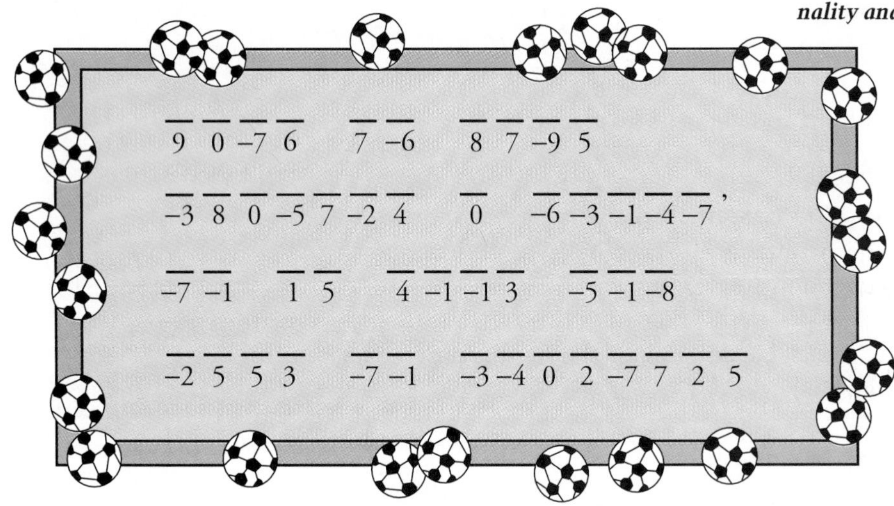

77. $Y + 17 = 12$

78. $4 = 0 + G$

79. $N - 3 = -5$

80. $S - (-4) = -2$

81. $-5I = -35$

82. $8D = 24$

83. $\frac{U}{4} = -2$

84. $\frac{H}{-1} = -6$

85. $6 + K = -3$

86. $C - 7 = -5$

87. $79B = 79$

88. $-7 + M = 2$

89. $\frac{R}{-4} = 1$

90. $16P = -48$

91. $3 - O = 4$

92. $E + 11 = 16$

93. $12T = -84$

94. $\frac{A}{-249} = 0$

95. $4L = 32$

96. What is the message?

Chapter TEST

In Exercises 1 and 2, write the integer that represents the phrase.

1. An altitude of 3000 feet

2. A loss of \$60

In Exercises 3 and 4, state the opposite and the absolute value of the integer.

3. 120

4. -54

5. Order the integers -4, 5, -3, 0, 2, and -1 from least to greatest.

6. Find the average of -5, 3, -4, -2, and -7.

In Exercises 7–12, simplify the expression.

7. $-4 + 8$

8. $-6 - (-3)$

9. $(-2)(-5)$

10. $(-3)(-4)(-5)$

11. $\frac{-12}{-6}$

12. $\frac{36}{-9}$

In Exercises 13–15, use the graph at the right.

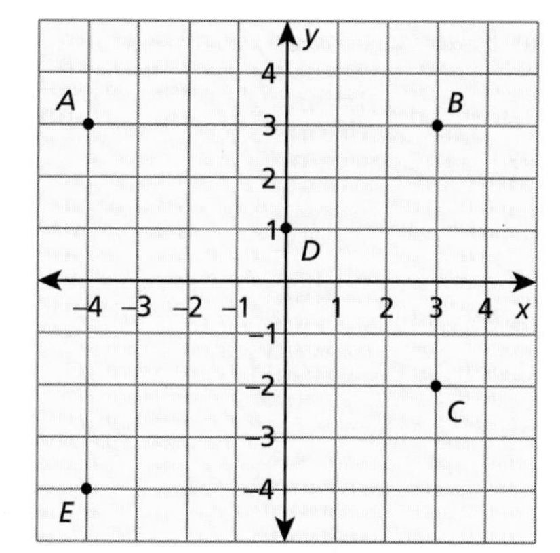

13. Write the coordinates of A, B, C, D, and E.

14. Which point lies in quadrant III?

15. The points A, B, and C form three vertices of a rectangle. Write the coordinates of the fourth vertex. Find the perimeter and area of the rectangle.

In Exercises 16–18, use mental math to solve the equation.

16. $x + 2 = -4$

17. $2x = -6$

18. $\frac{x}{9} = -2$

In Exercises 19 and 20, use the following information.

The low temperatures for 6 days in a row are $-7°$, $-3°$, $-6°$, $3°$, $0°$, and $-2°$.

19. If the average low temperature for the week was $-2°$, what was the low temperature on the 7th day?

20. If the average low temperature for the week was $-1°$, what was the low temperature on the 7th day?

21. Your bowling-league average is the sum of your scores divided by the number of games bowled. On your first night you bowled games of 128, 99, and 109. On the second night you bowled 117, 101, and 130. What was your average after the two nights?

In Exercises 1–6, describe the pattern. Then list the next 3 numbers or letters. (1.1)

1. 20, 18, 16, 14, ? ? ?

2. 1, 5, 9, 13, ? ? ?

3. $\frac{1}{2}, \frac{3}{4}, \frac{5}{6}, \frac{7}{8},$? ? ?

4. $\frac{14}{13}, \frac{12}{11}, \frac{10}{9}, \frac{8}{7},$? ? ?

5. Z, W, T, Q, ? ? ?

6. A, Z, C, X, ? ? ?

In Exercises 7–15, evaluate the expression using a calculator. Round to two decimal places when necessary. (1.3, 1.4)

7. 4^8

8. $\left(\frac{3}{4}\right)^5$

9. $\sqrt{48}$

10. $\sqrt{352}$

11. $(3.8)^7$

12. $\sqrt{26.19}$

13. $150 - 60 \div 3 \cdot 4$

14. $35 \div (19 - 12) + 4^5$

15. $3^3 + (14 + 8) \cdot 12 - 11$

In Exercises 16–24, evaluate the expression. (1.4, 3.1–3.6)

16. $|-3| - |5|$

17. $-|-4| + |-3|$

18. $-5 + 12 - 9 - 13$

19. $21 - 32 - 1 + 4$

20. $(10)(-11)$

21. $\frac{-144}{-2}$

22. $(-3)(-4)(6)$

23. $2^2 + (3 - 4)^2 \cdot 9$

24. $20 - (5 - 8)^3 \div 9$

In Exercises 25–28, decide whether the figure is a polygon. If it is, name it. If it is not, explain. (1.7)

25.

26.

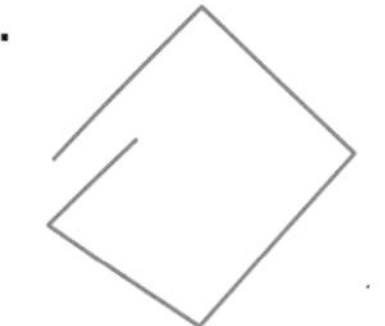

27.

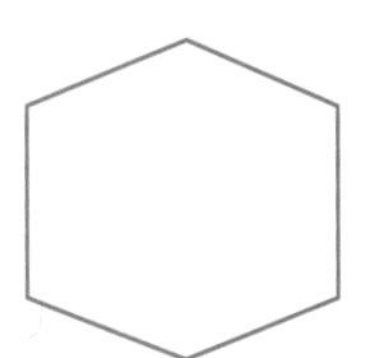

28.

In Exercises 29–40, write the expression without parentheses and combine like terms when possible. Then evaluate it when $x = 3$, $y = 4$, and $z = 6$. (1.5, 2.1, 2.2, 3.1–3.5)

29. $3(x + y + 3)$

30. $4(x + y + z + 3)$

31. $2(z + 3y)$

32. $5(z - 2x)$

33. $7x + x + z + y$

34. $9y - 2y - z - 2$

35. $4z + 6z - 9y - 25$

36. $-16 + 5x - 3x + y$

37. $6(5x - 3x + x)$

38. $3(2y + y) + 16$

39. $z(y + 2) + |-y|$

40. $|-z| + 3(y - 1)$

In Exercises 41–44, plot the number on a number line. (3.1)

41. -3

42. 5

43. -4

44. 0

In Exercises 45–48, plot the ordered pairs on one coordinate plane. (3.8)

45. $A(2, -3)$

46. $B(-1, 0)$

47. $C(-4, -5)$

48. $D(2, 4)$

In Exercises 49–64, solve the equation or inequality. (2.3–2.5, 2.9, 3.7)

49. $x + 7 = 16$

50. $y + 5 = -6$

51. $z - 8 = -16$

52. $a - 8 = 13$

53. $10b = 100$

54. $-9x = 36$

55. $\frac{y}{-12} = 8$

56. $\frac{m}{4} = 16$

57. $n + 6 < 7$

58. $p + 4 > 12$

59. $x - 2 \geq 5$

60. $y - 11 \leq 9$

61. $15c > 30$

62. $26 > 13t$

63. $\frac{q}{3} \leq 25$

64. $40 \leq \frac{s}{9}$

In Exercises 65–68, write an algebraic equation or inequality for the sentence. Then solve the equation or inequality. (2.7, 2.9, 3.7)

65. -12 is the sum of a number and 9.

66. The difference of a number and 16 is less than or equal to 20.

67. The product of 7 and a number is greater than 91.

68. Three is the quotient of a number and 4.

In Exercises 69 and 70, write the coordinates of the vertices of each numbered figure. Then find the perimeter of each figure, and find the area of the entire shaded region. (1.4, 3.8)

69.

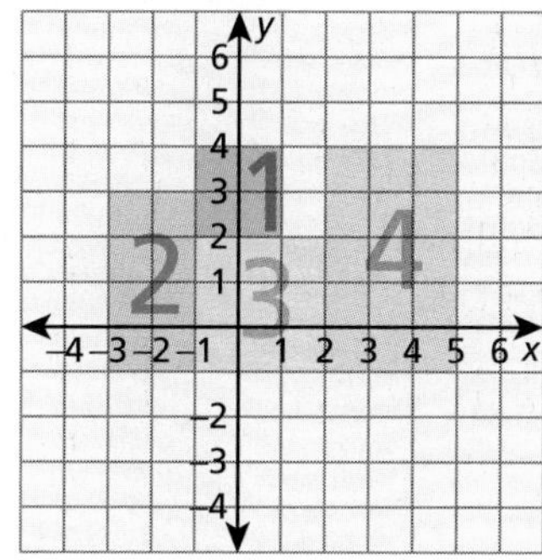

70.

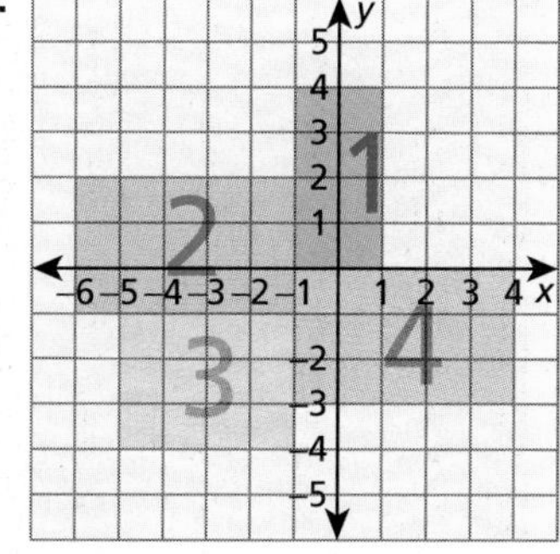

Roller Coaster **In Exercises 71 and 72, use the following information. (1.5)**

You and a friend are riding on a racing roller coaster that is 1.4 miles long and has dual tracks, with two sets of cars running side by side. You are in one set of cars and your friend is in the other. You finish the ride in 1.3 minutes and your friend finishes in 1.4 minutes.

71. How fast did your set of cars travel in miles per hour?

72. How fast did your friend's set of cars travel in miles per hour?

73. ***Amusement Park*** At an amusement park, you play x games that cost \$1 each and y games that cost \$2 each. Altogether, you spend \$12 on games. Construct a table that lists all possible values of x and y. **(3.8)**

CHAPTER 4

Exploring the Language of Algebra

LESSONS

The American bald eagle population has made a remarkable recovery over the past 30 years due in large part to the 1972 banning of the pesticide DDT and the 1973 passage of the Endangered Species Act. On July 4, 1994, the United States Fish and Wildlife Service announced that the bald eagle can now be upgraded to the threatened species list.

Equations can be used to model a wide variety of real-life situations. An example is the level of DDT residue found in eagle eggs. An approximate equation for the DDT residue levels R over time t is

$$R = 120 - 10(t - 1)$$

where 1967 is represented by $t = 0$.

In this chapter, you will learn how to use other multi-step equations to model real-world problems.

LESSON INVESTIGATION 4.1

Solving Two-Step Equations

Materials Needed: algebra tiles

In this investigation, you will use algebra tiles to model and solve two-step equations.

Example ***Using Algebra Tiles***

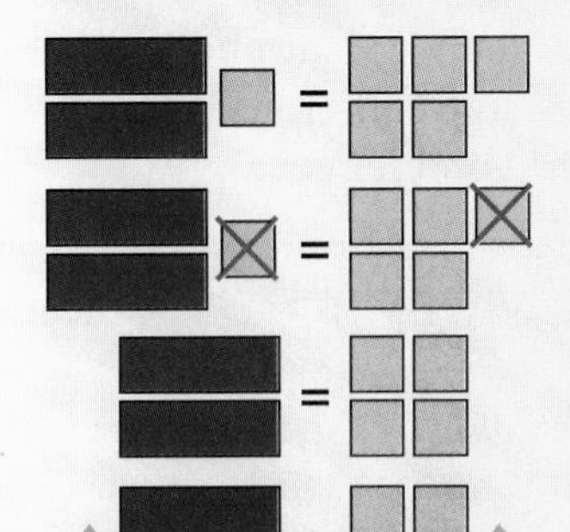

Original equation: $2x + 1 = 5$

To isolate the x-tiles, remove (subtract) a 1-tile from each side.

Transformed equation: $2x = 4$

To isolate one x-tile, divide each side into two groups and discard one group from each side.

Solution is $x = 2$. ■

Exercises

In Exercises 1 and 2, an equation has been modeled and solved with algebra tiles. Write the equation and its solution. Explain your steps.

1. 

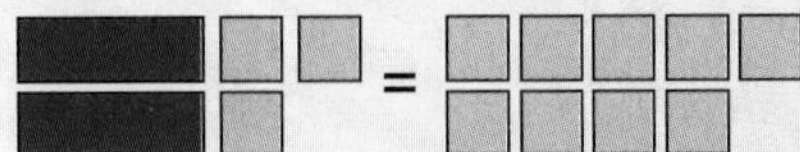

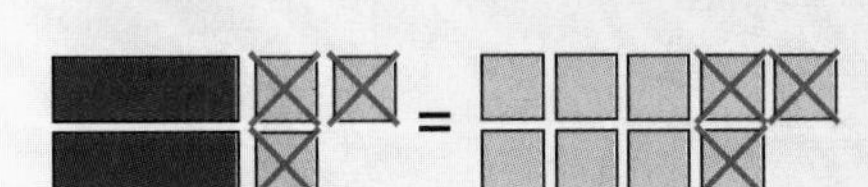

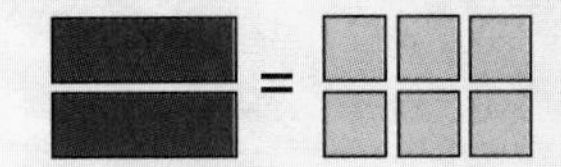

2. 

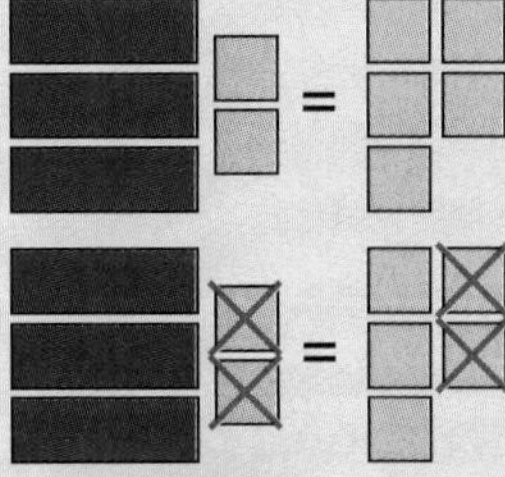

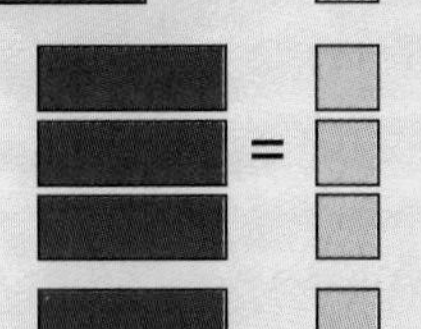

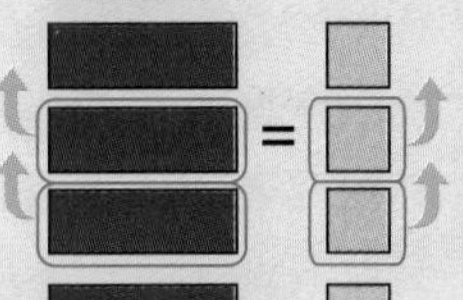

In Exercises 3–6, use algebra tiles to model and solve the equation.

3. $3x + 2 = 14$ **4.** $15 = 2n + 1$ **5.** $13 = 5 + 4y$ **6.** $6m + 3 = 15$

4.1 Solving Two-Step Equations

What you should learn:

How to use two transformations to solve a two-step equation

How to solve real-life problems using the work-backwards and make-a-table strategies

Why you should learn it:

You will need to use two or more transformations to solve most equations that model real-life situations, such as analyzing a tennis-club membership.

Study Tip...

Remember that addition and subtraction are inverse operations and that multiplication and division are inverse operations.

Goal 1 Using Two Transformations

Many equations require two or more transformations. Here are some guidelines that can help you decide how to start. Once you have found a solution, be sure to check it in the original equation.

1. Simplify both sides of the equation (if needed).
2. Use inverse operations (see Study Tip) to isolate the variable.

Example 1 *Solving an Equation*

Solve $3x + 8 = 2$.

Solution Remember that your goal is to isolate the variable.

$3x + 8 = 2$ *Rewrite original equation.*

$3x + 8 - 8 = 2 - 8$ *To isolate the x-term, subtract 8 from each side.*

$3x = -6$ *Simplify.*

$\frac{3x}{3} = \frac{-6}{3}$ *To isolate x, divide each side by 3.*

$x = -2$ *Simplify.*

The solution is -2. Because $3(-2) + 8 = -6 + 8 = 2$, the solution checks in the original equation. ■

Example 2 *Solving an Equation*

Solve $\frac{x}{-4} - 8 = 1$.

Solution

$\frac{x}{-4} - 8 = 1$ *Rewrite original equation.*

$\frac{x}{-4} - 8 + 8 = 1 + 8$ *Add 8 to each side.*

$\frac{x}{-4} = 9$ *Simplify.*

$-4 \cdot \frac{x}{-4} = -4 \cdot 9$ *Multiply each side by -4.*

$x = -36$ *Simplify.*

The solution is -36. Check this in the original equation.

Goal 2 Modeling Real-Life Situations

Example 3 *Problem Solving: Work Backwards*

Real Life
Monthly Dues

In 1993, there were about 220,000 tennis courts in the United States. Of these, about 14,000 were indoor courts. (Source: U.S. Tennis Association)

You are joining a community tennis club. The annual membership fee is \$50, and a tennis court rents for \$10 per hour. You plan to spend no more than \$190 playing tennis during the year. How many hours can you play?

Solution With some problems, you have to work backwards from the given facts to solve the problem.

Verbal Model

Total spent = Annual fee + Hourly rate • Hours of tennis

Labels

Total spent = 190 (dollars)
Annual fee = 50 (dollars)
Hourly rate = 10 (dollars per hour)
Number of hours played = n (hours)

Algebraic Model

$$190 = 50 + 10 \cdot n$$
$$190 - 50 = 50 + 10n - 50$$
$$140 = 10n$$
$$\frac{140}{10} = \frac{10n}{10}$$
$$14 = n$$

Answer and Check. You can play 14 hours. When you check this result, don't just check the numbers. You also need to check the units of measure.

$$(50 \text{ dollars}) + \frac{10 \text{ dollars}}{\text{hour}} \cdot (14 \text{ hours})$$
$$= 50 \text{ dollars} + 140 \text{ dollars} = 190 \text{ dollars}$$

■

Communicating about MATHEMATICS

SHARING IDEAS about the Lesson

Make a Table Another way to answer the question in Example 3 is to use a table. Copy and complete the table. Then use the result to answer the question.

Hours	1	2	3	4	5	6	7	8	9	10	11	12	13	14
Cost (\$)	?	?	?	?	?	?	?	?	?	?	?	?	?	?

EXERCISES

Guided Practice

CHECK for Understanding

In Exercises 1–3, describe the first step you would use to solve the equation.

1. $3x - 4 = 2$

2. $5x + 3 = 15$

3. $-2 = 2 - 4x$

4. *Error Analysis* Describe the error.

$$3x - 8 = 12$$
$$\frac{3x}{3} - 8 = \frac{12}{3}$$
$$x - 8 = 4$$
$$x = 12$$

In Exercises 5–8, match the equation with an equivalent equation.

a. $2x = -8$ **b.** $2x = 8$ **c.** $2x = -24$ **d.** $2x = 24$

5. $2x + 8 = 16$ **6.** $2x - 8 = 16$ **7.** $2x - 8 = -16$ **8.** $2x + 8 = -16$

Independent Practice

In Exercises 9–23, solve the equation. Then check your solution.

9. $3x + 15 = 24$

10. $4x + 11 = 31$

11. $6p + 8 = 2$

12. $5q + 14 = 4$

13. $-2r - 4 = 22$

14. $-3s - 5 = -20$

15. $\frac{t}{2} + 6 = 10$

16. $\frac{z}{3} + 17 = 21$

17. $\frac{x}{4} - 2 = -7$

18. $\frac{n}{3} - 5 = 20$

19. $\frac{x}{-5} + 1 = 10$

20. $\frac{m}{-2} + 2 = -1$

21. $7m - 105 = 350$

22. $14t - 280 = 84$

23. $3x + \frac{1}{2} = \frac{7}{2}$

24. *Modeling Equations* Sketch an algebra-tile solution for the equation $3x + 4 = 7$. Use the samples on page 148.

In Exercises 25–32, write the sentence as an equation. Then solve it.

25. 3 times a number, plus 7, is 34.

26. 8 times a number, plus 12, is 100.

27. One fourth of a number, minus 2, is 5.

28. Half of a number, plus 13, is 30.

29. The sum of 21 and $7x$ is -14.

30. The sum of 84 and $\frac{x}{2}$ is -36.

31. The difference of 57 and $9n$ is 129.

32. The quotient of $5y$ and 4 is 25.

Diagramming Equations **In Exercises 33 and 34, the upper line segment has the same length as the lower double line segment. Write the implied equation and solve for *x*.**

33.

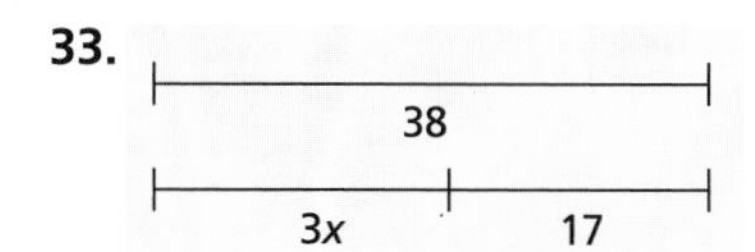

34.

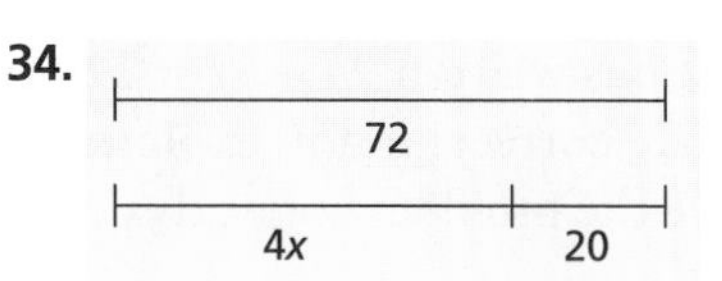

35. *A Folk Art Museum* You and a friend are going to visit an art museum. Your friend will meet you at your apartment. Then both of you will ride the subway uptown and walk 2 blocks east to the museum. Write the inverse of this plan.

36. *Telephone Rates* You are at a phone booth and need to make a long-distance call home. The call will cost 25 cents for the first minute and 15 cents for each additional minute or fraction of a minute. You have 95 cents in your pocket. How many minutes can you talk?

    a. Write a verbal model of the problem.

    b. Assign labels to each part of your verbal model.

    c. Write and solve an algebraic model of the problem.

    d. Answer the question. Then check your solution.

Panamanian Molas ***Folk museums of American art often contain molas that illustrate scenes of Kuna experiences of other cultures. This one, called*** **Masked Dancers,** ***shows African dancers and is from the Rio Sidra Island.***

37. *Automotive Repair* You are the service manager at an automotive repair shop. You charge \$22 per hour of labor plus the cost of any parts. One car needed \$156 of new parts and the final bill for the car was \$321. How long did it take to repair the car?

38. *Problem Solving* Describe a real-life situation that can be modeled by the equation $5x - 7 = 120$.

Integrated Review

Making Connections within Mathematics

Measurement Sense **In Exercises 39–42, simplify the expression. Include the appropriate unit of measure in your result.**

39. $\frac{22 \text{ miles}}{\text{hour}} \times (3 \text{ hours})$

40. $\frac{1.45 \text{ dollars}}{\text{pound}} \times (6 \text{ pounds})$

41. $\frac{18 \text{ meters}}{\text{second}} \times (12 \text{ seconds})$

42. $\frac{12 \text{ dollars}}{\text{square yard}} \times (3 \text{ yards}) \times (5 \text{ yards})$

Exploration and Extension

Writing Equations **In Exercises 43–46, write a two-step equation that has the given solution. (There are many correct answers.)**

43. $x = 1$

44. $x = 3$

45. $x = -3$

46. $x = -2$

47. *You Be the Teacher* A classmate asks you for help on Exercise 31. Your classmate has written $9n - 57 = 129$ and determined that $n \approx 20.67$. Is your classmate correct? If not, explain how you could help your classmate solve the problem correctly.

4.2 Solving Multi-Step Equations

What you should learn:

Goal 1 How to use three or more transformations to solve an equation

Goal 2 How to solve real-life problems using multi-step equations

Why you should learn it:

Many real-life problems can be modeled with equations whose solutions require three or more transformations.

Goal 1 Solving Multi-Step Equations

Before applying inverse operations to solve an equation, you should check to see whether one or both sides of the equation can be simplified by *combining like terms.*

Example 1 *Simplifying First*

Solve the equation $2x + 3x - 4 = 11$.

Solution

$2x + 3x - 4 = 11$	*Rewrite original equation.*
$5x - 4 = 11$	*Combine like terms:* $2x + 3x = 5x$.
$5x - 4 + 4 = 11 + 4$	*Add 4 to each side.*
$5x = 15$	*Simplify.*
$\frac{5x}{5} = \frac{15}{5}$	*Divide each side by 5.*
$x = 3$	*Simplify.*

The solution is 3. Because $2(3) + 3(3) - 4 = 6 + 9 - 4 = 11$, the solution checks in the original equation. ■

As you become more of an expert in solving equations, you may want to perform some of the solution steps mentally. If you do this, don't forget to check your solution.

Study Tip...
When some people use the "expert" solver format, they say they are skipping steps. That, however, isn't really what is happening. They aren't skipping steps—they are simply doing some of the steps mentally. For instance, in Example 2, which steps were performed mentally?

Example 2 *"Expert" Equation Solver Format*

Solve the equation $-13 = 3n + 3 + n$.

Solution

$-13 = 3n + 3 + n$	*Rewrite original equation.*
$-13 = 4n + 3$	*Combine like terms.*
$-16 = 4n$	*Subtract 3 from each side.*
$-4 = n$	*Divide each side by 4.*

The solution is -4. Because $3(-4) + 3 + (-4) = -12 + 3 + (-4) = -13$, the solution checks in the original equation. ■

Goal 2 Modeling Multi-Step Problems

Example 3 *Modeling a Real-Life Problem*

Real Life **Poster Sales**

Your wildlife club has made the poster shown above. To print the poster, a printer charges \$250, plus \$2 per poster. You plan to sell each poster for \$5. How many do you need to sell to make a profit of \$300?

Solution Here is a partial solution. You are asked to complete the solution below.

Verbal Model Profit = Income − Expenses

$$300 = 5 \cdot \text{Number of posters} - \left(250 + 2 \cdot \text{Number of posters}\right)$$

Labels Number of posters = n ■

Communicating about MATHEMATICS

SHARING IDEAS about the Lesson

Writing an Algebraic Model Write an equation that represents the verbal model in Example 3. Then solve the equation and answer the question given in Example 3.

EXERCISES

Guided Practice

CHECK for Understanding

Logical Reasoning **In Exercises 1–3, explain each step of the solution.**

1. $3x - x + 8 = -16$
$2x + 8 = -16$
$2x = -24$
$x = -12$

2. $4x + 7 - 5x = 9$
$-x + 7 = 9$
$-x = 2$
$(-1)x = 2$
$x = -2$

3. $-5 = -5x - 3x + 11$
$-5 = -8x + 11$
$-16 = -8x$
$2 = x$

Independent Practice

In Exercises 4–9, decide whether the given value is a solution of the equation. If not, find the solution.

4. $4x - x - 5 = -8;\ x = 1$

5. $9y - 6 - 7y = -12;\ y = -9$

6. $-13 = 3x + 2x - x + 7;\ x = -5$

7. $6y - y + 3y + 26 = -30;\ y = -2$

8. $4t - 2t - 8 = 1;\ t = 3$

9. $2 = 3a - 8a + 17;\ a = 3$

In Exercises 10–21, solve the equation. Check your solution.

10. $8x + 2x - 4 = 6$

11. $6y - 3y + 2 = -16$

12. $42 = 8a - 2a + 12$

13. $6m - 2m - 6 = -60$

14. $5x - 3x + 12 = 8$

15. $n - 3n + 8 = -8$

16. $2y - 7y - 4y = 81$

17. $9p - 6p - 6p = 24$

18. $\frac{3}{2}x - \frac{1}{2}x - 2 = 4$

19. $\frac{5}{2}x - \frac{1}{2}x - 3 = 5$

20. $4(x + 1) = 8$

21. $-3(x + 1) = 0$

In Exercises 22 and 23, write an equation that represents the sentence. Then solve.

22. The sum of $3x$ and $2x$ and $7x$ and 6 is 42.

23. 5 subtracted from the difference of $4y$ and y is -29.

Geometry **In Exercises 24 and 25, use the given information to write an equation. Then solve the equation.**

24. The sum of the measures of two **complementary angles** is 90°.

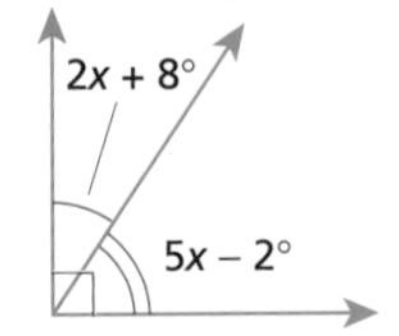

25. The sum of the measures of the angles of a triangle is 180°.

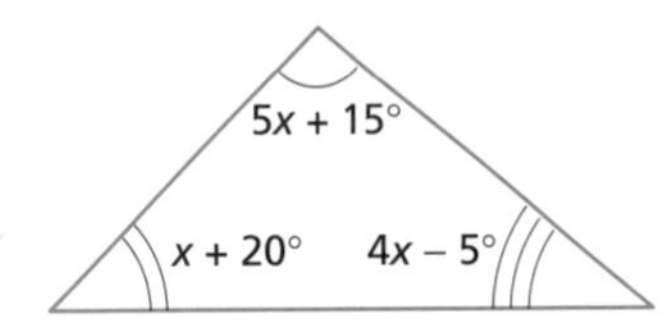

26. *Summer Job* You have a job mowing lawns Monday through Saturday. The table shows the number of lawns you mowed each day during a week. During that week, you earned $126. You earn x dollars per lawn.

M	T	W	Th	F	S
1	2	2	1	3	5

a. Write an equation that represents the amount you earn in a week.

b. Solve the equation to find the amount you get paid per lawn.

27. *Printing Posters* In Example 3 on page 154, suppose the printer raises the price to $275 and $3 per poster. How many posters do you now need to sell to make a profit of $300?

28. *Starting a Business* You start a business selling bottled fruit juices. You invest $10,000 for equipment. Each bottle costs you $0.30 to make. You sell each bottle for $0.75. How many bottles must you sell to earn a profit of $2,000?

Tom Scott (top left) and Tom First (top right) are shown with 2 of their employees. They began a fruit juice business called Nantucket Nectors. In 1993, the business sold $1.3 million worth of fruit juices. (Source: The Boston Sunday Globe)

Integrated Review

Describing Patterns **In Exercises 29–34, simplify the expression. Then complete the table for each expression and describe the pattern.**

x	-2	-1	0	1	2
Value of Expression	?	?	?	?	?

29. $9x - 4x + 2x$

30. $6x + 3x - x$

31. $5x + 11x - 4x + 5 - 8$

32. $8x - 7x + 2x - 6 + 12$

33. $2x - 5x - 8x$

34. $3x - 10x + 2x$

Exploration and Extension

Coordinate Geometry **In Exercises 35–37, for each equation complete the table showing several solutions of the equation. Write the solutions as ordered pairs. Then plot the ordered pairs in a coordinate plane. Describe the pattern. (For a sample, look at Lesson 3.8.)**

x	-2	-1	0	1	2
y	?	?	?	?	?

35. $x + y = 7$

36. $2x - y = 5$

37. $-2x + y = 1$

USING A CALCULATOR

Checking Solutions

When you check a solution that has been rounded, the solution will only make the original equation *approximately* true.

Example ***Checking a Rounded Solution***

Solve the equation $34x - 5 - 23x = 15$. Check your solution.

Solution

$34x - 5 - 23x = 15$	*Rewrite original equation.*
$11x - 5 = 15$	*Combine like terms.*
$11x - 5 + 5 = 15 + 5$	*Add 5 to each side.*
$11x = 20$	*Simplify.*
$\frac{11x}{11} = \frac{20}{11}$	*Divide each side by 11.*
$x \approx 1.82$	*Round to 2 decimal places.*

The solution is $\frac{20}{11}$ or approximately 1.82. You can check the solution in the original equation as follows.

$34x - 5 - 23x = 15$	*Original equation*
$34(1.82) - 5 - 23(1.82) \stackrel{?}{=} 15$	*Substitute 1.82 for x.*
$61.88 - 5 - 41.86 \stackrel{?}{=} 15$	*Use a calculator.*
$15.02 \approx 15$	*Approximate check* ✔

■

You could perform the check on a scientific calculator using the following keystrokes. (Some calculators will give you a wrong answer because they do not follow the order of operations.)

Calculator Keystrokes	Display
34 [×] 1.82 [−] 5 [−] 23 [×] 1.82 [=]	15.02

Exercises

In Exercises 1–8, solve the equation. Round your solution to 2 decimal places. Check your rounded solution in the original equation. Does it check exactly? If not, why?

1. $2x + 5x - 7 = 9$

2. $3y - 7 + 4y = 10$

3. $2n + 7n - 13 = -4$

4. $-5a + 17 + 2a = 8$

5. $12 = -2x + 5x - 9$

6. $-13 = 4y + 8y + 9$

7. $-4n - 14 - 5n = 11$

8. $5m + 8m - 14 = 23$

4.3 Two-Step Equations and Problem Solving

What you should learn:

Goal 1 How to solve an equation by multiplying by a reciprocal

How to use two-step equations to model real-life problems

Why you should learn it:

You can use two-step equations as algebraic models for analyzing real-life graphical data, such as populations of songbirds.

Need to Know

When you multiply a number by its reciprocal, you obtain 1. Here are some examples.

$5 \cdot \frac{1}{5} = 1$

$\left(-\frac{1}{3}\right) \cdot (-3) = 1$

$(-2) \cdot \left(-\frac{1}{2}\right) = 1$

$\frac{1}{4} \cdot 4 = 1$

Goal 1 Using Reciprocals

Often, there is not just one way to solve a problem. For instance, Example 1 shows two ways to solve $3x - 4 = 11$.

Example 1 *Comparing Two Solutions*

a.

$3x - 4 = 11$	*Original equation*
$3x = 15$	*Add 4 to each side.*
$\frac{3x}{3} = \frac{15}{3}$	*Divide each side by 3.*
$x = 5$	*Simplify.*

b.

$3x - 4 = 11$	*Original equation*
$3x = 15$	*Add 4 to each side.*
$\frac{1}{3} \cdot 3x = \frac{1}{3} \cdot 15$	*Multiply each side by $\frac{1}{3}$.*
$x = 5$	*Simplify.*

The number $\frac{1}{3}$ is the **reciprocal** of 3. Notice that multiplying by the reciprocal of a number produces the same result as dividing by the number. ■

The solution is 5. Because $3(5) - 4 = 15 - 4 = 11$, the solution checks in the original equation.

Example 2 *Multiplying by a Reciprocal*

a.

$\frac{1}{3}x = 12$	*Original equation*
$3 \cdot \frac{1}{3}x = 3 \cdot 12$	*Multiply each side by 3.*
$x = 36$	*Simplify.*

b.

$-\frac{1}{4}t + 2 = 6$	*Original equation*
$-\frac{1}{4}t + 2 - 2 = 6 - 2$	*Subtract 2 from each side.*
$-\frac{1}{4}t = 4$	*Simplify.*
$-4 \cdot \left(-\frac{1}{4}t\right) = -4 \cdot 4$	*Multiply each side by -4.*
$t = -16$	*Simplify.*

Check the solutions in the original equations. ■

Goal 2 Modeling with Two-Step Equations

Each year from 1940 through the present, the number of breeding pairs of songbirds in Rock Creek Park in Washington, D.C., have been counted. Many factors have contributed to the birds' decline. (Source: National Audubon Society)

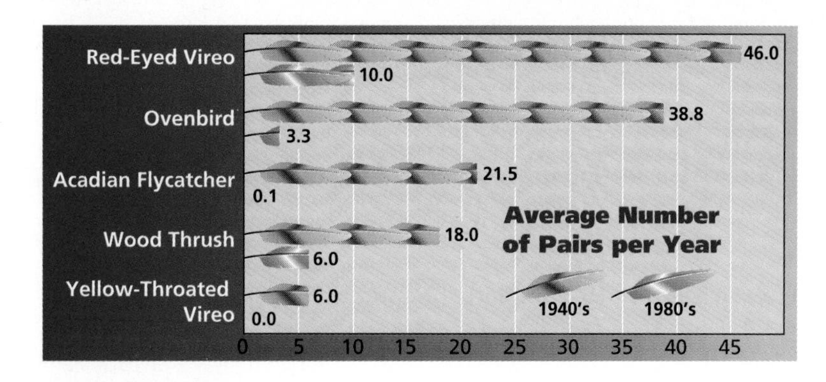

Real Life
Endangered Species

Example 3 *Modeling Real Life*

A biologist uses the data in the graph to model the number W of pairs of wood thrush in Rock Creek Park.

$$W = -0.3t + 18$$

In this model, $t = 0$ represents 1940. Use the model to predict the year in which no pairs of wood thrush $(W = 0)$ will be seen in Rock Creek Park.

Need to Know

In Example 3, the year 1940 is represented by $t = 0$. This means that $t = 10$ represents 1950, $t = 20$ represents 1960, $t = 30$ represents 1970, and so on. The notation $0 \leftrightarrow 1940$ indicates that 1940 is represented by 0.

Solution Substitute 0 for W in the equation and solve for t.

$W = -0.3t + 18$	*Model for number of pairs of wood thrush*
$0 = -0.3t + 18$	*Substitute 0 for W.*
$-18 = -0.3t$	*Subtract 18 from each side.*
$\frac{-18}{-0.3} = \frac{-0.3t}{-0.3}$	*Divide each side by −0.3.*
$60 = t$	*Use a calculator.*

Because $t = 0$ represents 1940, it follows that $t = 60$ represents 2000. Thus, you can predict that no pairs of wood thrush will be seen in the year 2000. ■

Communicating about MATHEMATICS

SHARING IDEAS about the Lesson

Making a Prediction The number R of pairs of red-eyed vireo in Rock Creek Park can be modeled by $R = -0.9t + 46$, where $t = 0$ represents 1940. How many pairs would you expect to see in 1955? During which year would you expect to see no pairs?

EXERCISES

Guided Practice

CHECK for Understanding

In Exercises 1–4, find the reciprocal of the number, if possible.

1. $\frac{1}{3}$ **2.** $-\frac{1}{4}$ **3.** -5 **4.** 0

In Exercises 5 and 6, complete the solution.

5.
$$7x = -28$$
$$\boxed{?} \cdot 7x = \boxed{?} \cdot (-28)$$
$$x = \boxed{?}$$

6.
$$-\frac{1}{4}x = 12$$
$$\boxed{?} \cdot \left(-\frac{1}{4}x\right) = \boxed{?} \cdot 12$$
$$x = \boxed{?}$$

In Exercises 7–10, let $t = 0$ represent 1950. What year is represented by the given value of t?

7. $t = 17$ **8.** $t = 45$ **9.** $t = -10$ **10.** $t = -25$

Independent Practice

In Exercises 11–14, describe two different ways to solve the equation.

11. $3x = -15$ **12.** $-2x = 14$ **13.** $27 = 9n$ **14.** $-48 = -4y$

In Exercises 15–29, solve the equation.

15. $2y - 12 = 4$ **16.** $5n - 21 = 24$ **17.** $-\frac{1}{8}x + 14 = 6$

18. $-\frac{1}{10}m - 11 = 1$ **19.** $-12t - 7 = -15$ **20.** $21z - 16 = 12$

21. $\frac{1}{5}x - 3 = -2$ **22.** $-\frac{1}{3}y + 27 = 39$ **23.** $5r + 15 = 10$

24. $-11t + 16 = -6$ **25.** $-\frac{1}{7}b + 2 = 1$ **26.** $\frac{1}{5}p - 3 = 0$

27. $\frac{2}{3}x - \frac{1}{3}x = 12$ **28.** $3(x + 7) = 27$ **29.** $-2(t + 2) = 6$

In Exercises 30 and 31, use a sketch to find the answer.

30. *Geometry* A rectangle has a perimeter of 34 inches. Its length is 5 inches more than its width, x.

a. Make a sketch of the rectangle and label its side lengths.

b. Find the rectangle's dimensions.

31. *Geometry* A triangle has a perimeter of 30 centimeters. Side a is 7 centimeters shorter than Side b, and Side c is 1 centimeter longer than Side b. Side b has a length of x centimeters.

a. Make a sketch of the triangle and label its side lengths.

b. Find the lengths of the triangle's sides.

32. *Population* In 1990, the population of Nebraska was about 71,700 less than 3 times the population of Alaska. The total population of the two states was about 2,128,400. Find the 1990 population of each state. *(Source: U.S. Bureau of the Census)*

Verbal Model Total Pop. = Pop. of Nebraska + Pop. of Alaska

↓

Labels Total Population = 2,128,400
Population of Alaska = A
Population of Nebraska = $3A - 71{,}700$

a. Write the algebraic model.

b. Solve the algebraic model.

c. Answer the question.

Although Alaska ranks 49th in population, it ranks first in area. Its area is about one-fifth that of the area of the lower 48 states.

33. *Population* In 1990, the population of Utah was about 79,300 more than half the population of Connecticut. The total population of the two states was about 5,010,000. Find the 1990 population of each state. *(Source: U.S. Bureau of the Census)*

Integrated Review

Making Connections within Mathematics

Analyzing Salary Data **In Exercises 34–37, use the data in the bar graph. The data represents the average starting salaries for public school teachers and accountants.** *(Source: Northwestern University Placement Center)*

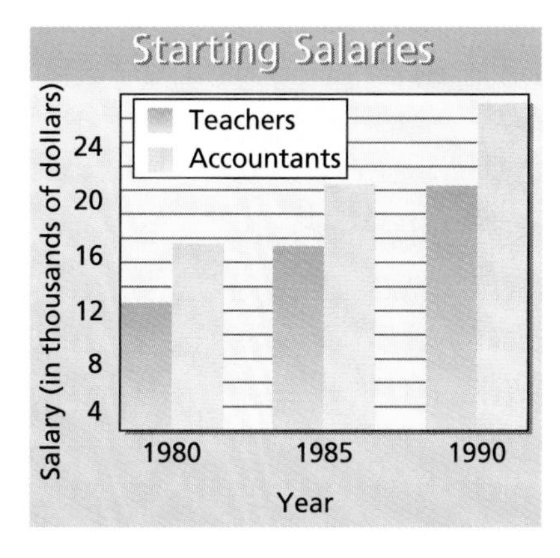

34. Estimate the starting salary for a teacher in 1980.

35. Estimate the starting salary for an accountant in 1985.

36. From 1980 to 1990, which average starting salary increased by the larger amount?

37. From the data, what would you predict the average starting salary for a teacher to be in 1995?

Exploration and Extension

In Exercises 38–45, use a calculator to find the reciprocal of the given number. Then verify that the product of the number and its reciprocal is 1.

Sample: To find the reciprocal of -0.5, enter 0.5 [+/-] [1/x].

38. 0.25	**39.** -0.125	**40.** -25	**41.** 32
42. $\frac{1}{2}$	**43.** $-\frac{5}{4}$	**44.** -100	**45.** 500

46. What happens if you try to use a calculator to find the reciprocal of 0? Explain.

Mixed REVIEW

In Exercises 1–3, evaluate the expression when $a = 4$ and $b = 5$. (3.4, 3.6)

1. $6b - 3$ **2.** $8b \div (-a)$ **3.** $a^2 - 2b - 3$

4. Describe the steps used to solve the equation $4x - 2 = 12$. **(4.2)**

5. Plot the points $(3, 2)$ and $(-2, 4)$ in a coordinate plane. **(3.8)**

In Exercises 6–8, write the expression without parentheses. (2.1)

6. $4(3 + 2y)$ **7.** $4(2x - 3)$ **8.** $0.25(4m + 8)$

In Exercises 9–14, solve the equation. (4.2, 4.3)

9. $2y - 14 = 0$ **10.** $16a + 14 = 110$ **11.** $2.06r + 1.14r = 8.32$

12. $\frac{5}{2}x - \frac{1}{2}x = -1$ **13.** $\frac{m}{5} + 1 = 7$ **14.** $2c + 4c - c = 10$

Milestones BOOLEAN ALGEBRA

1700 1750 1800 1850 1900 1950 2000

Euler born, 1707 · Rosetta stone, 1799 · Braille, 1829 · Morse code, 1838 · Venn diagrams, 1880 · BASIC, 1964 · Fractals, 1975

A self-taught elementary school teacher in England named George Boole (1815–1864) revolutionized mathematics in 1847. In his algebra, statements and their logical relationships could be represented by symbols in a new kind of equation. The rules that he invented, now known as Boolean Algebra, are currently used in topology, fractal mathematics, circuitry, probability, computer science, and truth-function logic.

George Boole

To avoid confusion with standard algebra, a new set of symbols was developed for Boolean Algebra, and John Venn (1834–1923) adapted Euler's logic circles as "Venn diagrams" to help people understand the relationships between the symbols.

Symbol	Meaning
A, B, C, etc.	Individual sets that contain elements
$A \cup B$	The union of two sets A and B; contains all the elements in either A or B, or in both A and B
$A \cap B$	The intersection of sets A and B; contains only the elements that are in both A and B

Venn Diagram

A C B

In symbols: $A \cap B = C$

- *What is the intersection of the set of multiples of 2 and the set of multiples of 3?*
- *Define two sets* A *and* B. *Determine* $A \cup B$ *and* $A \cap B$ *for your sets.*

4.4 Solving Equations: The Distributive Property

What you should learn:

How to use the Distributive Property to solve equations

How to use the Distributive Property to model and solve real-life problems

Why you should learn it:

You can use the Distributive Property and the problem-solving plan to model and solve real-life problems, such as finding the sales necessary to obtain a bonus.

Need to Know

Another way to solve the equation in Example 2 is to first multiply each side by the reciprocal of $\frac{1}{4}$.

$$24 = \frac{1}{4}(x - 8)$$
$$4 \cdot 24 = 4 \cdot \frac{1}{4}(x - 8)$$
$$96 = x - 8$$
$$96 + 8 = x - 8 + 8$$
$$104 = x$$

Which strategy do you think is easier?

Goal 1 Using the Distributive Property

In Lesson 4.1, you studied the two guidelines for solving equations. The first guideline often involves the Distributive Property.

1. Simplify both sides of the equation (if needed).
2. Use inverse operations to isolate the variable.

Example 1 Using the Distributive Property

Solve $5y + 2(y - 3) = 92$.

Solution

$5y + 2(y - 3) = 92$	*Rewrite original equation.*
$5y + 2y - 6 = 92$	*Distributive Property*
$7y - 6 = 92$	*Combine like terms.*
$7y - 6 + 6 = 92 + 6$	*Add 6 to each side.*
$7y = 98$	*Simplify.*
$\frac{7y}{7} = \frac{98}{7}$	*Divide each side by 7.*
$y = 14$	*Simplify.*

The solution is 14. Check this in the original equation. ■

Example 2 Using the Distributive Property

Solve $24 = \frac{1}{4}(x - 8)$.

Solution

$24 = \frac{1}{4}(x - 8)$	*Original equation*
$24 = \frac{1}{4}x - \frac{1}{4}(8)$	*Distributive Property*
$24 = \frac{1}{4}x - 2$	*Simplify.*
$24 + 2 = \frac{1}{4}x - 2 + 2$	*Add 2 to each side.*
$26 = \frac{1}{4}x$	*Simplify.*
$4 \cdot 26 = 4 \cdot \frac{1}{4}x$	*Multiply each side by 4.*
$104 = x$	*Simplify.*

The solution is 104. Check this in the original equation. ■

Goal 2 Real-Life Modeling

Real Life
Sales Bonus

Example 3 *Solving Real-Life Problems*

You are a sales representative for a company that sells medical equipment. Your annual salary is \$30,000, plus a bonus. Your bonus is $\frac{1}{30}$ of the amount by which your sales exceed \$500,000. Create a table showing the amounts you need to sell to earn a total salary of \$30,000, \$40,000, \$50,000, and \$60,000.

Solution

In 1993, over 9 million Americans worked in sales. Of these, more than half had salary plans that included commissions or bonuses based on sales. (Source: U.S. Bureau of Census)

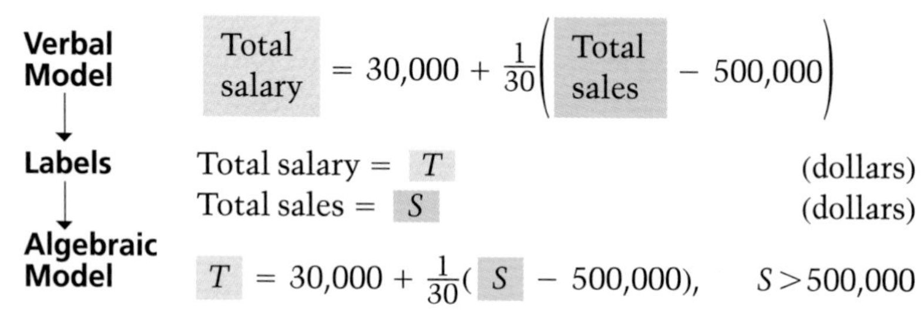

Verbal Model $\boxed{\text{Total salary}} = 30{,}000 + \frac{1}{30}\left(\boxed{\text{Total sales}} - 500{,}000\right)$

Labels Total salary = T (dollars)
Total sales = S (dollars)

Algebraic Model $T = 30{,}000 + \frac{1}{30}(S - 500{,}000), \quad S > 500{,}000$

To find the sales you need to earn a total salary of \$40,000, substitute 40,000 for T, and solve for S.

$$40{,}000 = 30{,}000 + \frac{1}{30}(S - 500{,}000)$$
$$10{,}000 = \frac{1}{30}(S - 500{,}000)$$
$$30 \cdot 10{,}000 = 30 \cdot \frac{1}{30}(S - 500{,}000)$$
$$300{,}000 = S - 500{,}000$$
$$800{,}000 = S$$

To complete the table, you can solve for the values of S that will produce total salaries of \$50,000 and \$60,000 in a similar way.

Salary, T	\$30,000	\$40,000	\$50,000	\$60,000
Sales, S	Up to \$500,000	\$800,000	\$1,100,000	\$1,400,000

■

Communicating about MATHEMATICS

Cooperative Learning

SHARING IDEAS about the Lesson

Finding a Pattern Describe the pattern given in the table in Example 3 to your partner. Then use the pattern to predict the amount you would have to sell to earn a salary of \$70,000. Finally, use the algebraic model to verify your answer.

EXERCISES

Guided Practice

CHECK for Understanding

1. *Writing* In your own words, describe a set of guidelines for solving an equation.
2. Explain how your guidelines can be used to solve the equation
   $\frac{1}{7}(x - 5) = 9.$
3. For the equation in Exercise 2, is it easier to apply the Distributive Property first or multiply both sides by the reciprocal of $\frac{1}{7}$ first? Explain.
4. *Logical Reasoning* Justify each step of the solved equation.

$$\begin{aligned} 6x + 4(x - 3) &= 8 \\ 6x + 4x - 12 &= 8 \\ 10x - 12 &= 8 \\ 10x - 12 + 12 &= 8 + 12 \\ 10x &= 20 \\ \frac{10x}{10} &= \frac{20}{10} \\ x &= 2 \end{aligned}$$

Independent Practice

Error Analysis **In Exercises 5–7, describe the error. Then solve the equation.**

5. $$\begin{aligned} 2(x + 2) &= 4 \\ 2x + 2 &= 4 \\ 2x &= 2 \\ x &= 1 \end{aligned}$$

6. $$\begin{aligned} 5x - 7x + 5 &= 3 \\ -2x + 5 &= 3 \\ -2x &= -2 \\ x &= -1 \end{aligned}$$

7. $$\begin{aligned} -2x - 4x + 6 &= 10 \\ -2x + 6 &= 10 \\ -2x &= 4 \\ x &= -2 \end{aligned}$$

In Exercises 8–22, solve the equation. Check your solution.

8. $x + 4(x + 6) = -1$
9. $1 = y + 3(y - 9)$
10. $3x + 2(x + 8) = 21$
11. $2z + 5(z - 2) = -31$
12. $4(p - 2) + 7p = 14$
13. $11 = 6(t - 4) - 13$
14. $3(4 - s) - 5s = 52$
15. $7(1 + r) - 5r = -5$
16. $5(2n + 3) = 65$
17. $\frac{3}{4}x - \frac{2}{4}x + 12 = -8$
18. $14 = \frac{1}{4}(q - 9)$
19. $-3(y + 4) = 18$
20. $2 = -2(n - 3)$
21. $8(4z - 7) = -56$
22. $6(2n - 5) = 42$

23. *It's Up to You* Solve the equation $\frac{1}{3}(x - 6) = 6$ in two ways. Which way do you prefer? Why?
    a. Use the Distributive Property first.
    b. Multiply by a reciprocal first.

Geometry **In Exercises 24–26, write an equation for the area of the rectangle. Then solve for *x*.**

24. Rectangle with width $x + 3$ and height 9. Area is 63 square units.

25. Rectangle with width $x - 5$ and height 12. Area is 48 square units.

26. Rectangle with width $2x - 1$ and height 5. Area is 35 square units.

In Exercises 27 and 28, write an equation and solve the equation for x.

27. *Geometry* The sum of the measures of the angles of a triangle is 180°.

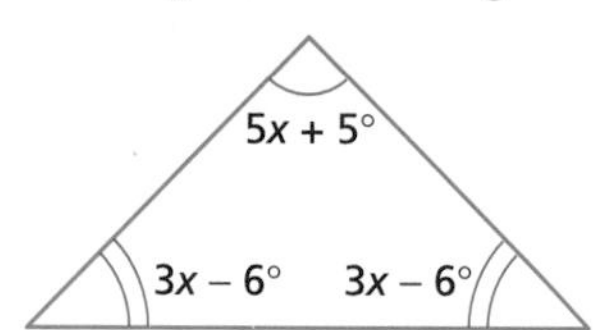

28. *Volleyball Court* The perimeter of a volleyball court is 180 feet.

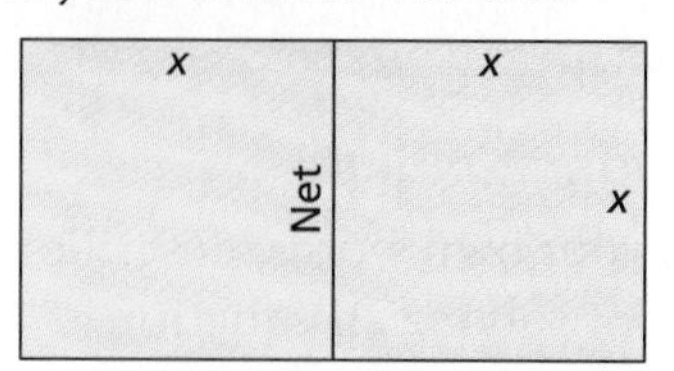

29. *Barbershop* Imagine that you own a barbershop. You provide haircuts, stylings, and manicures for x dollars each, and perms for $(x + 40)$ dollars. This week your income came from providing 96 haircuts, 80 stylings, 14 manicures, and 24 perms. In a week, you pay $800 in expenses.

a. Write an algebraic model that represents your income if this week's profit is $3,370.

b. How much does a haircut cost?

30. *Temperature* To convert a Celsius temperature to a Fahrenheit temperature, you can use the equation $C = \frac{5}{9}(F - 32)$.

a. What is 35°C on the Fahrenheit scale?

b. What is 40°C on the Fahrenheit scale?

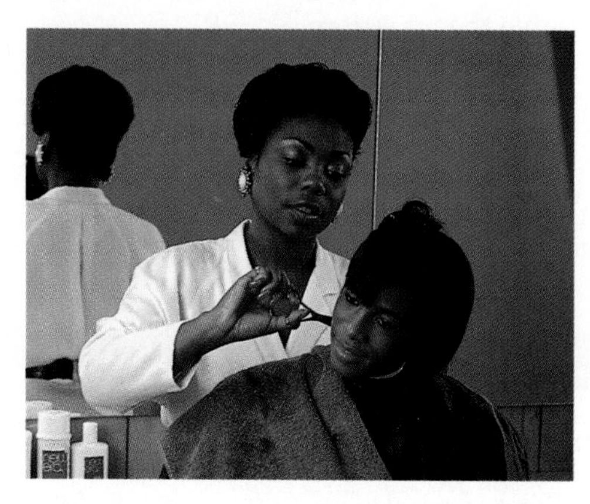

In 1987, approximately 390,700 Americans worked in barbershops and beauty shops.

Integrated Review

Making Connections within Mathematics

Mental Math **In Exercises 31–34, each equation has two solutions. Use mental math to find both solutions.**

31. $x^2 = 4$ **32.** $x^2 = 9$ **33.** $|x| = 3$ **34.** $|x| = 5$

35. *Estimation* Which of the following is the best estimate for the weight (in pounds) of this textbook?

a. $\frac{1}{2}$ **b.** 3 **c.** 6 **d.** 9

36. *Estimation* Which of the following is the best estimate for the area (in square inches) of this textbook's front cover?

a. 20 **b.** 40 **c.** 80 **d.** 160

Exploration and Extension

Number Puzzles **In Exercises 37 and 38, write an equation that represents the statements. Then solve the equation to find the number.**

37. A number is added to 10. The result is multiplied by 2. The original number is added to the product to obtain 71.

38. A number is decreased by 6. The result is multiplied by $\frac{1}{3}$ to obtain 15.

Mid-Chapter SELF-TEST

Take this test as you would take a test in class. The answers to the exercises are given in the back of the book.

1. A triangle has a perimeter of 16 centimeters. Side a is 3 centimeters shorter than Side c. Side b is 1 centimeter longer than Side c. Side c is p centimeters in length. Sketch the triangle and find its dimensions. **(4.3)**

In Exercises 2–4, find the reciprocal. (4.3)

2. $\frac{3}{2}$ **3.** $-\frac{8}{21}$ **4.** $\frac{16}{5}$

5. Choose any nonzero number. Multiply the number by its reciprocal. What is the result?

In Exercises 6–11, solve the equation. Then check your solution. (4.1)

6. $2y - 4 = 10$ **7.** $4t + 16 = 0$ **8.** $8 - 2b = 2$

9. $\frac{1}{2}r + 6 = 8$ **10.** $6m + 5 = -1$ **11.** $20p - 8 = 32$

In Exercises 12–15, solve the equation. Then check your solution. (4.2, 4.4)

12. $9s + 6s - 12s = 15$ **13.** $7t - 10t - t = 24$

14. $19 + 12p - 17p = -1$ **15.** $3(n + 4) + 1 = 28$

In Exercises 16–18, write an equation that represents the verbal sentence. Then solve the equation. (4.1, 4.2)

16. The sum of $2x$ and 3 is 21.

17. The difference of $16x$ and 30 is -2.

18. 17 is the sum of $2x$, x, x, and -4.

In Exercises 19 and 20, use the following information. (4.3)

You have just arrived in Little Rock, Arkansas. You see the road sign at the right. Two of the arrows are broken. Use the clues to find the distances to Washington, D.C. and New Orleans, Louisiana.

19. The distance from Little Rock to Dallas is equal to one-third the difference of the distance to Washington, D.C. and 57. What is the distance to Washington, D.C?

20. The distance from Little Rock to Phoenix is the sum of 74 and 3 times the distance to New Orleans. What is the distance to New Orleans?

LESSON INVESTIGATION 4.5
Equations with Variables on Both Sides

Materials Needed: algebra tiles

In this investigation, you will use algebra tiles to model and solve equations that have variables on both sides.

Example ***Using Algebra Tiles***

Solve the equation $2x + 2 = x + 4$.

Solution

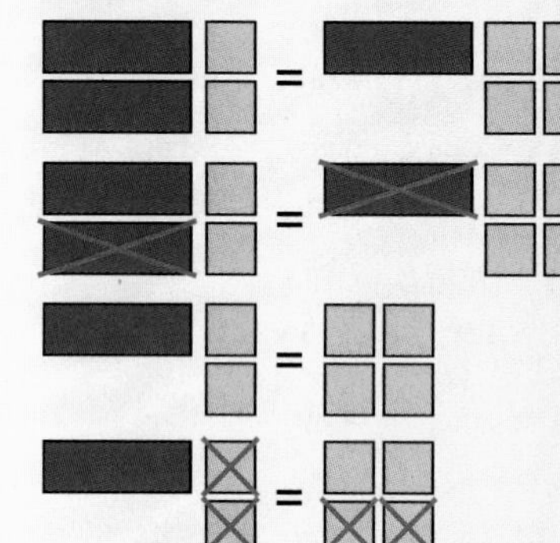

Original equation: $2x + 2 = x + 4$

Remove (subtract) an x-tile from each side so that you don't have x-tiles on each side.

Transformed equation: $x + 2 = 4$

To isolate x, remove (subtract) two 1-tiles from each side.

Solution is $x = 2$. ■

Exercises

In Exercises 1 and 2, an equation has been modeled and solved with algebra tiles. Write the equation and its solution.

1.

2.

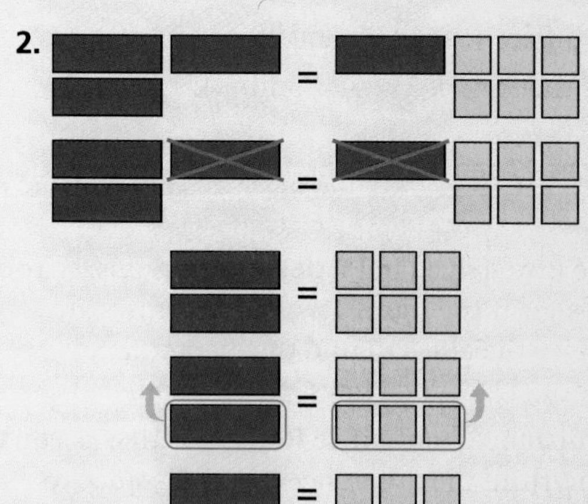

In Exercises 3–6, use algebra tiles to model and solve the equation.

3. $4x + 3 = 3x + 5$

4. $y + 9 = 3y + 3$

5. $3n + 5 = n + 11$

6. $5m + 2 = 2m + 14$

4.5 Solving Equations: Variables on Both Sides

What you should learn:

Goal 1 How to solve equations with variables on both sides

Goal 2 How use equations to model problems in geometry

Why you should learn it:

You can use algebra to solve problems in other branches of mathematics, such as finding the perimeter of a square or a triangle.

Goal 1 Collecting Variables on One Side

Some equations, such as $2x + 3 = 3x + 5$, have variables on both sides. The strategy for solving such equations is to *collect like variables* on the same side. We suggest that you collect variables on the side with the term that has the greater variable coefficient.

Example 1 *Collecting Like Variables*

Solve $2x + 3 = 3x + 5$.

Solution Remember that your goal is to isolate the variable.

$2x + 3 = 3x + 5$	*Rewrite original equation.*
$2x + 3 - 2x = 3x + 5 - 2x$	*Subtract 2x from each side.*
$3 = x + 5$	*Simplify.*
$3 - 5 = x + 5 - 5$	*Subtract 5 from each side.*
$-2 = x$	*Simplify.*

The solution is -2. Check this in the original equation. ■

Example 2 *Collecting Like Variables*

Solve $5x - 4 = 3(x - 8)$.

Solution

$5x - 4 = 3(x - 8)$	*Rewrite original equation.*
$5x - 4 = 3x - 24$	*Distributive Property*
$5x - 4 - 3x = 3x - 24 - 3x$	*Subtract 3x from each side.*
$2x - 4 = -24$	*Simplify.*
$2x - 4 + 4 = -24 + 4$	*Add 4 to each side.*
$2x = -20$	*Simplify.*
$\frac{2x}{2} = \frac{-20}{2}$	*Divide each side by 2.*
$x = -10$	*Simplify.*

The solution is -10. Check this in the original equation. ■

Study Tip...

Some people prefer to always collect like variables on the left side. Notice what happens in Example 1 when you collect variables on the left side.

$$2x + 3 = 3x + 5$$
$$2x + 3 - 3x = 3x + 5 - 3x$$
$$-x + 3 = 5$$
$$-x + 3 - 3 = 5 - 3$$
$$-x = 2$$
$$\frac{-x}{-1} = \frac{2}{-1}$$
$$x = -2$$

Which strategy do you prefer?

Goal 2 Modeling Problems in Geometry

Example 3 *Comparing Perimeters*

Find the value of x so that the rectangle and the triangle have the same perimeter.

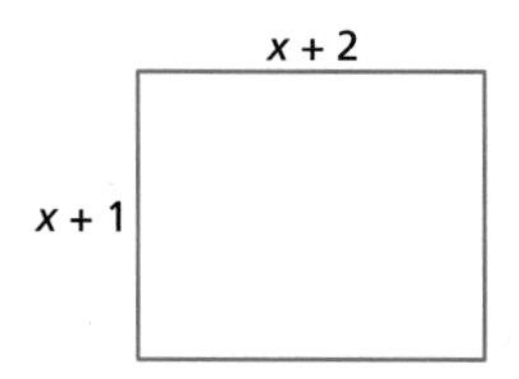

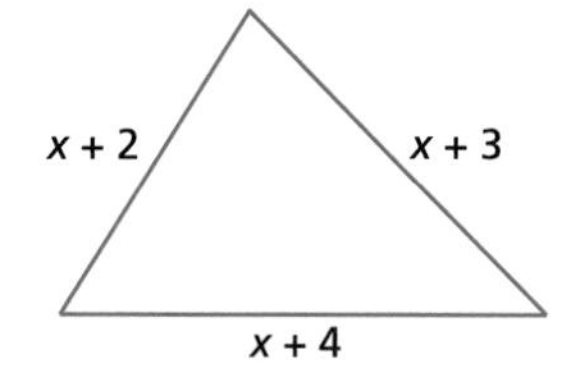

Solution

Rectangle's perimeter	=	Triangle's perimeter	*Write verbal model.*

$$2(x+1) + 2(x+2) = (x+2) + (x+3) + (x+4)$$
$$2x + 2 + 2x + 4 = 3x + 9$$
$$4x + 6 = 3x + 9$$
$$x + 6 = 9$$
$$x = 3$$

When $x = 3$, each figure has a perimeter of 18. ■

Example 4 *Creating an Equilateral Triangle*

A triangle is **equilateral** if its sides all have the same length. Find a value of x so that $\triangle ABC$ is equilateral.

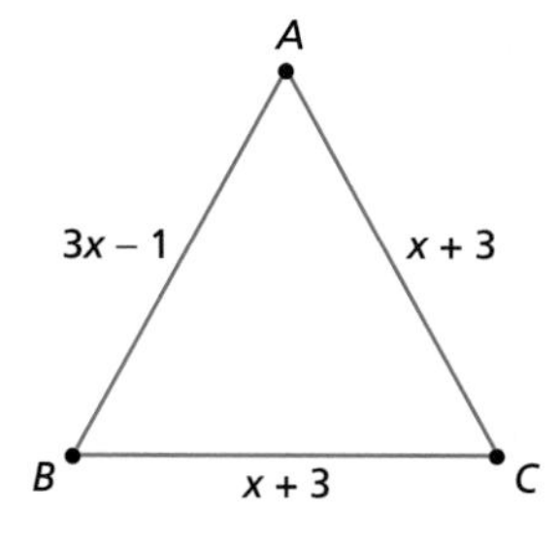

Solution

Length of side AB	=	Length of side AC or BC	*Write verbal model.*

$3x - 1 = x + 3$

$2x - 1 = 3$ *Subtract x from each side.*

$2x = 4$ *Add 1 to each side.*

$x = 2$ *Divide each side by 2.*

When $x = 2$, each side will have a length of 5. ■

Communicating about MATHEMATICS

SHARING IDEAS about the Lesson

Extending the Example In Example 3, which values of x make the rectangle's perimeter larger than the triangle's perimeter? Use a table to obtain your answer.

EXERCISES

Guided Practice

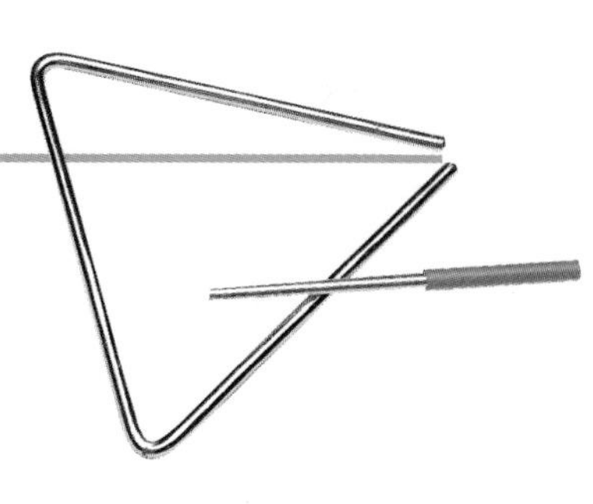

CHECK for Understanding

1. *Writing* In your own words, describe a strategy for solving an equation that has variables on both sides.
2. *Language Connections* The preface *equi* means "equal" and the world *lateral* means "side." Explain how these words are related to the term *equilateral triangle.*

In Exercises 3–6, on which side of the equation would you collect the variable terms? Why?

3. $2x - 4 = x$ **4.** $4y + 10 = 6y$ **5.** $x + 15 = 4x + 6$ **6.** $3x - 5 = x + 3$

Independent Practice

In Exercises 7–10, match the equation with its solution.

a. $n = 9$ **b.** $n = 5$ **c.** $n = 4$ **d.** $n = 3$

7. $4n + 1 = 2n + 7$ **8.** $4n - 1 = 2n + 7$ **9.** $2(n - 1) = n + 7$ **10.** $2(n + 1) = n + 7$

In Exercises 11–22, solve the equation. Then check your solution.

11. $7x + 12 = 13x$ **12.** $-2x + 6 = -x$ **13.** $10x + 17 = 4x - 1$

14. $-5x + 6 = x + 12$ **15.** $6(x - 3) = 4(x + 3)$ **16.** $2(x - 9) = 3(x - 6)$

17. $-6 - 9t = 5(2 - t)$ **18.** $4(7 + y) = 16 - 2y$ **19.** $10(2n + 10) = 120n$

20. $7y = 3(5y - 8)$ **21.** $\frac{7}{2}t + 12 = 6 + \frac{5}{2}t$ **22.** $-13 - \frac{1}{12}s = \frac{11}{12}s + 2$

Diagramming Equations **In Exercises 23 and 24, write the equation implied by the model. Then solve it.**

23.

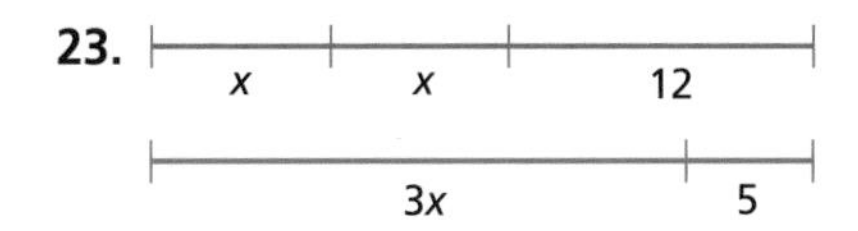

24.

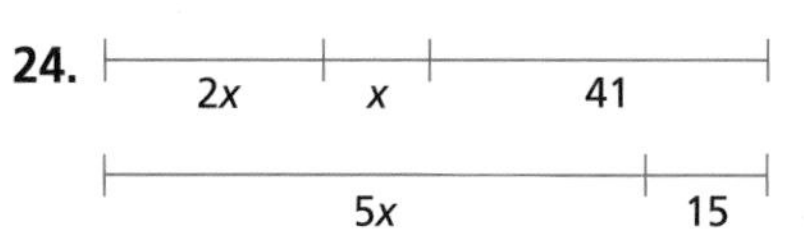

25. One less than three times a number is equal to the same number plus 19. What is the number?

26. Four times a number plus seventeen is equal to seven times the same number minus 7. What is the number?

27. *Geometry* Find the value of x so that the rectangle and the triangle have the same perimeter. What is the perimeter?

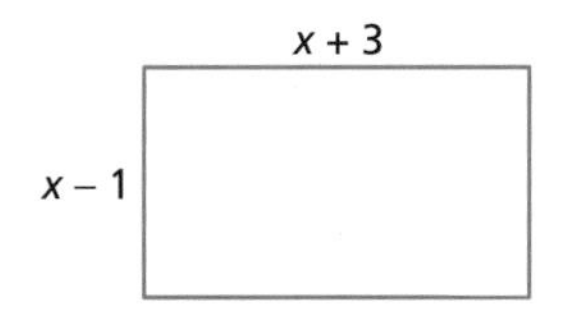

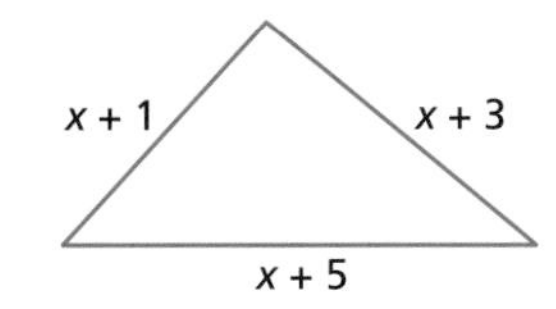

28. *Geometry* Find the value of x so that the triangle is equilateral.

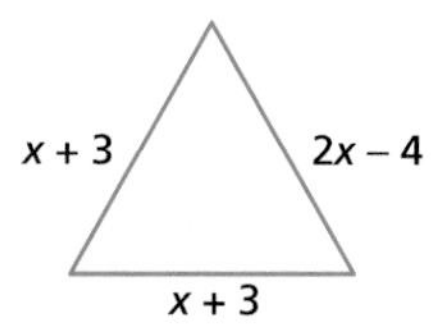

29. *Geometry* Find the value of x so that the figure is a square.

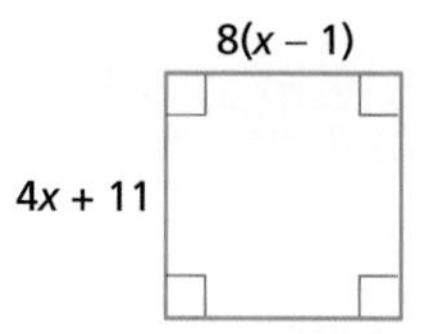

Rabbit versus Turtle **In Exercises 30–32, use the following information.**

A rabbit and a turtle are in a race. The rabbit is running 25 feet per second and the turtle is running 0.15 feet per second. The turtle was given a head start of 1000 feet. How long will it take the rabbit to catch up to the turtle? (Let t represent the time in seconds.)

30. Which equation is a correct model for this problem?

a. $0.15t = 25t + 1000$

b. $25t = 0.15t + 1000$

c. $0.15t - 1000 = 25t$

31. Solve the correct equation in Exercise 30 and interpret the result.

32. If the race is 1010 feet long, who will win?

The Lower Keys Rabbit is found in the Florida Keys. It is just one of 37 mammals in the United States that is on the endangered species list.

Integrated Review

Making Connections within Mathematics

Magic Squares **In Exercises 33 and 34, fill in the squares so that the sum of the integers in each row, column, and diagonal is the same.**

33.

1		
	2	
5		3

34.

–5		
	–4	
–6		–3

Exploration and Extension

Equation Sense **In Exercises 35–38, match the equation with the number of solutions. Explain your reasoning.**

a. No solution

b. Exactly one solution

c. Exactly two solutions

d. Many solutions

35. $10x + 35 = 5(7 + 2x)$

36. $3x + (1 - x) = 2x$

37. $2|x| + 1 = 5$

38. $4x + 7 = 2 - x$

4.6 Problem Solving Strategies

What you should learn:

Goal 1 How to use tables and graphs to solve real-life problems

Goal 2 How to use a general problem-solving plan

Why you should learn it:

To be an efficient problem solver, you should be able to solve real-life problems in more than one way.

Goal 1 Using Tables and Graphs

When solving a real-life problem, remember that there are usually several ways to solve the problem. Often, it helps to solve the problem in more than one way and then compare results.

Example 1 *Using Tables and Graphs*

In 1970, Americans drank an average of 31 gallons of milk and 24 gallons of soda pop. For the next 20 years, milk consumption dropped by 0.26 gallons a year, and soda pop consumption increased by 0.89 gallons a year. During which year did Americans drink the same amount of each?

Solution The table shows the consumptions for 3-year intervals. During each 3-year interval milk consumption decreased by 3(0.26) or 0.78 gallons, and soda pop consumption increased by 3 (0.89) or 2.67 gallons.

Year	1970	1973	1976	1979	1982	1985	1988	1991
Milk	31.00	30.22	29.44	28.66	27.88	27.10	26.32	25.54
Soda	24.00	26.67	29.34	32.01	34.68	37.35	40.02	42.69

From the table, you can see that the consumption of milk and soda pop was about the same in 1976. A graph of the data in the table helps you visualize the two consumption rates.

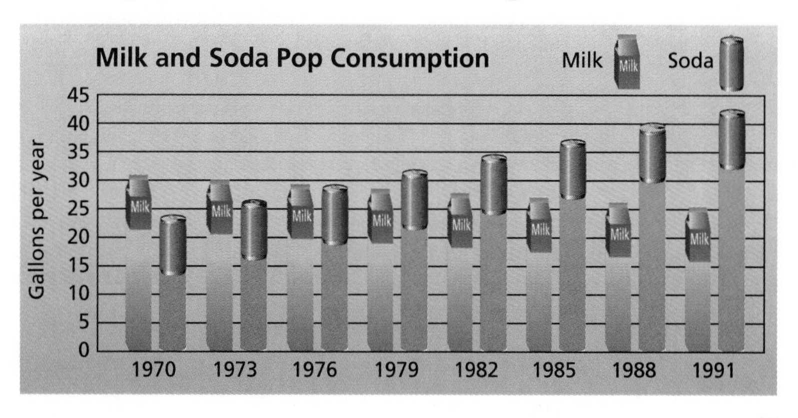

■

Goal 2 Using a Problem-Solving Plan

In Lesson 2.8, you studied a general problem-solving plan.

Write a verbal model. → Assign labels. → Write an algebraic model. → Solve algebraic model. → Answer the question. → Check

Example 2 shows how to use this plan to solve the problem given in Example 1.

Real Life
Nutrition

Example 2 *Using a Problem-Solving Plan*

Show how to use the general problem-solving plan to solve the "milk and soda pop problem" described in Example 1.

Solution

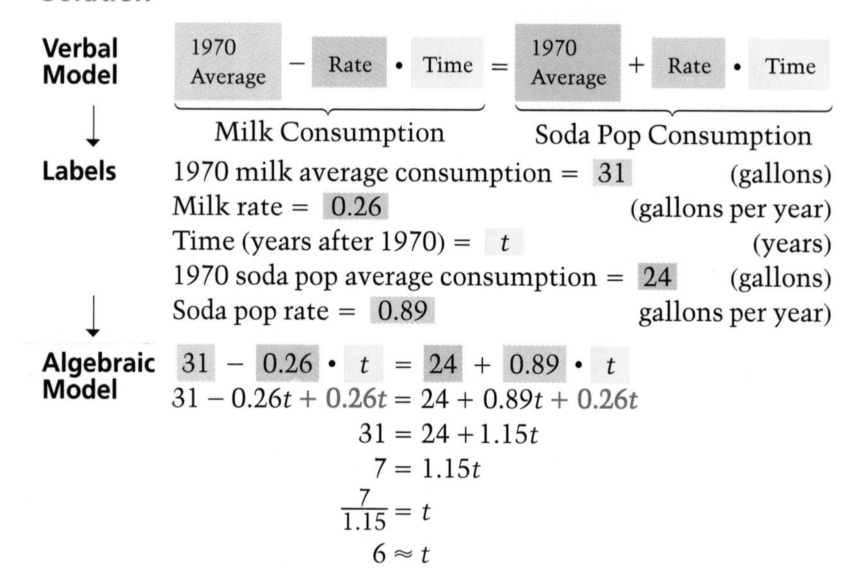

Verbal Model

$\underbrace{\boxed{\text{1970 Average}} - \boxed{\text{Rate}} \cdot \boxed{\text{Time}}}_{\text{Milk Consumption}} = \underbrace{\boxed{\text{1970 Average}} + \boxed{\text{Rate}} \cdot \boxed{\text{Time}}}_{\text{Soda Pop Consumption}}$

Labels

1970 milk average consumption = 31 (gallons)
Milk rate = 0.26 (gallons per year)
Time (years after 1970) = t (years)
1970 soda pop average consumption = 24 (gallons)
Soda pop rate = 0.89 gallons per year)

Algebraic Model

$$31 - 0.26 \cdot t = 24 + 0.89 \cdot t$$
$$31 - 0.26t + 0.26t = 24 + 0.89t + 0.26t$$
$$31 = 24 + 1.15t$$
$$7 = 1.15t$$
$$\frac{7}{1.15} = t$$
$$6 \approx t$$

Because $t \approx 6$, it follows that Americans drank the same amount of milk and soda pop in 1976. ■

Communicating about MATHEMATICS

Cooperative Learning

SHARING IDEAS about the Lesson

Estimating How Much With others in your group, estimate the number of gallons of milk and soda pop that you drink in a year.

EXERCISES

Guided Practice

CHECK for Understanding

1. *Writing* In your own words, describe the general problem-solving plan.

Using a Problem-Solving Plan **In Exercises 2–5, use the following.**

You are 60 inches tall and your cousin is 56 inches tall. Your cousin is growing at a rate of $\frac{4}{3}$ inch per year and you are growing at a rate of $\frac{1}{3}$ inch per year. When will you be the same height?

Your height now	+	Your rate of growth	•	Number of years	=	Cousin's height now	+	Cousin's rate of growth	•	Number of years

2. Assign labels to each part of the verbal model. Indicate the units of measure.
3. Write an algebraic model.
4. Solve the algebraic model.
5. When will you be the same height?

Independent Practice

Temperature **In Exercises 6–12, use the following.**

At 3:00 P.M., the temperature is 86°F in Santa Fe, New Mexico, and is decreasing at a rate of 3 degrees per hour. At the same time, the temperature is 56°F in Minot, North Dakota, and is increasing at a rate of 2 degrees per hour. When will the temperatures be the same?

6. Show how to use a table and graph to solve the problem.
7. Write a verbal model.
8. Assign labels to each part of the model.
9. Write an algebraic model.
10. Solve the algebraic model.
11. How long will it take for the temperatures to be the same?
12. At what time will the temperatures be the same?
13. *It's Up to You* You are considering joining one of two karate clubs in your area. At Club 1, there is no membership fee and lessons cost $6.00 per hour. At Club 2, there is an annual membership fee of $30, and lessons cost $4.50 per hour. Explain how you would decide which club to join.

The temperature in the desert can drop as much as 80°F at night.

14. *Interactive CDs* You want to join an interactive CD club to buy interactive books on compact discs. Your friend says that it is more economical to buy the CDs from a bookstore. You decide to consider your options. The CD club has a membership fee of \$50 and each interactive book costs \$25. The bookstore charges \$35 for each interactive book.

a. Complete the table and interpret the results.

Number of CDs, x	1	2	3	4	5	6	7	8	9	10
Bookstore Cost (\$)	?	?	?	?	?	?	?	?	?	?
CD Club Cost (\$)	?	?	?	?	?	?	?	?	?	?

b. How many CDs do you have to buy for the costs of both options to be the same?

c. Is your friend correct?

d. Show how the problem can be solved with a bar graph.

15. *Geometry* The perimeters of the regular hexagon and the square shown below are equal. Find the side lengths and perimeter of each figure.

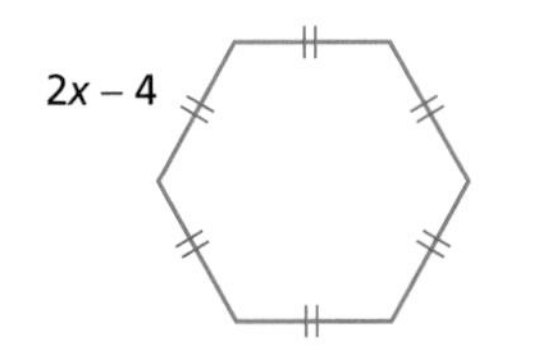

$x + 6$

16. *Breaking Even* You own a small business that produces skateboards. You want to determine how many skateboards must be sold to break even. Your costs are \$1500 plus \$15 in materials for each skateboard. You sell each skateboard for \$32. To break even, your total cost must be equal to your total income. Use the following model to find the number of skateboards you must sell to break even.

Total Cost = Total Income

\$1500 + 15 • Number sold = 32 • Number sold

Integrated Review

Making Connections within Mathematics

Equation Sense **In Exercises 17–22, solve the equation. Check your solution.**

17. $6x + 12 = 84$

18. $-3x - 25 = -4$

19. $-4y + 10 + 7y = 37$

20. $11y + 16 + 5y = -44$

21. $m + 2(m - 3) = 4m + 2$

22. $9 - 2(n + 1) = 10 - 3n$

Verbal Phrases **In Exercises 23 and 24, write an algebraic expression that represents the verbal phrase.**

23. The change in temperature in h hours when it decreases at a rate of 7° per hour.

24. The change in speed in s seconds when it increases at a rate of 6 feet per second.

Exploration and Extension

25. *Logic Puzzle* Bev, Kim, Lee, Ron, and Sue each have a favorite sport. No two of them have the same favorite sport. Copy the table below. Then use the clues to determine their favorite sports. (Hint: In each box, put a O for true and an X for false.)

a. Ron's favorite sport is not basketball.

b. Bev does not like basketball or soccer.

c. Sue's favorite sport is volleyball.

d. Kim does not like golf.

e. Lee's favorite sport is the sport that Kim does not like.

f. Three of the sports are basketball, soccer, and tennis.

	B	G	S	T	V
Bev					
Kim					
Lee					
Ron					
Sue					

B=basketball, G=golf, S=soccer, T=tennis, V=volleyball

Mixed REVIEW

In Exercises 1–8, solve the equation. (4.5)

1. $4a = 28$ **2.** $6 - 9x = -48$ **3.** $4.5r - 2 = 7$ **4.** $12x + 3x = 30$

5. $3a = 2 + 2a$ **6.** $14s - 6 = 12s$ **7.** $z + 2(4 - z) = 4$ **8.** $\frac{4}{3}p = \frac{12}{3} + \frac{2}{3}p$

In Exercises 9–12, write the verbal sentence as an algebraic equation and solve. (4.3)

9. Six times a number is 3.

10. The sum of $3x$ and 2 is $8x$.

11. $5x$ is the difference of 1 and x.

12. 2 times the sum of x and 2 is $-x$.

In Exercises 13 and 14, find the average of the numbers. (3.6)

13. 14, 24, 20, 12, 17, 22, 7, 20

14. $-2.4, 1.1, 2.7, -1.4, -0.5$

In Exercises 15–20, solve the inequality. (2.9)

15. $2x \geq 1$ **16.** $7y < 28$ **17.** $48 < 16y$

18. $3c < 21$ **19.** $\frac{r}{4} \geq 6$ **20.** $0.2p \leq 12$

4.7 Solving Equations Using Technology

What you should learn:

How to solve equations involving rounding with decimals

How to use a table to solve problems

Why you should learn it:

Many real-life problems involve decimal amounts, such as money, distances, and averages.

Technology
Using a Calculator

Goal 1 Equations Involving Decimals

LESSON INVESTIGATION

Investigating Round-Off Error

Group Activity In the following solutions, the red numbers are rounded to 2 decimal places. Which solution is more accurate? What does that tell you about rounding before the final step?

Rounding Early

$$0.12(3.45x - 2.80) = 9.45$$
$$0.414x - 0.336 = 9.45$$
$$0.41x - 0.34 \approx 9.45$$
$$0.41x \approx 9.79$$
$$x \approx \frac{9.79}{0.41}$$
$$x \approx 23.88$$

Rounding at Final Step

$$0.12(3.45x - 2.80) = 9.45$$
$$0.414x - 0.336 = 9.45$$
$$0.414x = 9.786$$
$$x = \frac{9.786}{0.414}$$
$$x \approx 23.64$$

Example 1 *Solving an Equation*

Solve $3.56x + 4.78 = 2.69(1.20x - 4.18)$. Round the solution to 2 decimal places.

Solution

$$3.56x + 4.78 = 2.69(1.20x - 4.18)$$
$$3.56x + 4.78 = 3.228x - 11.2442$$
$$3.56x + 4.78 - 3.228x = 3.228x - 11.2442 - 3.228x$$
$$0.332x + 4.78 = -11.2442$$
$$0.332x + 4.78 - 4.78 = -11.2442 - 4.78$$
$$0.332x = -16.0242$$
$$\frac{0.332x}{0.332} = \frac{-16.0242}{0.332}$$
$$x \approx -48.26 \quad \textit{Round to 2 decimal places.}$$

The solution is ≈ -48.26. Check this in the original equation. ■

Goal 2 Solving Problems with a Table

Example 2 *Using a Table*

Real Life
Telephone Costs

You are making a long-distance call to a friend. The call costs \$3.56 for the first minute and \$1.68 for each additional minute. You don't want to spend more than \$20.00 on the call. How long can you talk?

Solution

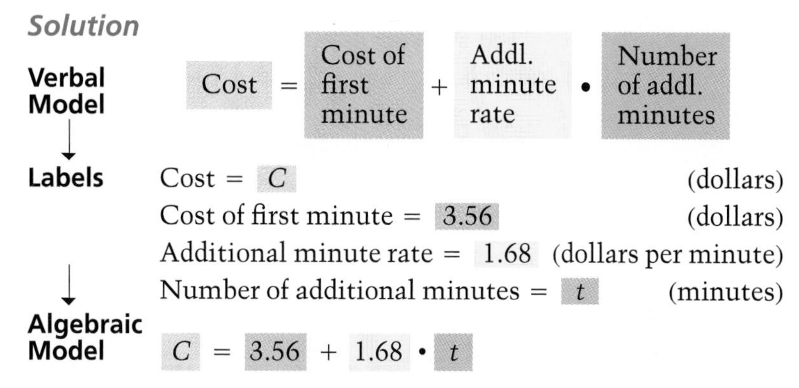

Verbal Model Cost = Cost of first minute + Addl. minute rate • Number of addl. minutes

Labels Cost = C (dollars)

Cost of first minute = 3.56 (dollars)

Additional minute rate = 1.68 (dollars per minute)

Number of additional minutes = t (minutes)

Algebraic Model $C = 3.56 + 1.68 \cdot t$

You want to find the values of t such that C is less than or equal to \$20.00. One way to do this is with a table. From the table, you can see that you can talk for up to 10 minutes (1 minute plus 9 additional minutes).

Almost all Americans live in homes with phones. Nearly half, however, choose to not have their phone number listed in the telephone book. (Source: Maritz Marketing Service, Inc.)

Additional minutes	1	2	3	4	5
Cost (dollars)	5.24	6.92	8.60	10.28	11.96

Additional minutes	6	7	8	9	10
Cost (dollars)	13.64	15.32	17.00	18.68	20.36

■

Communicating about MATHEMATICS

SHARING IDEAS about the Lesson

Extending the Example You often make telephone calls to your cousin who lives in another state. Which long-distance company would you choose? Explain your reasoning.

- **Company A** charges \$1.48 for the first minute and \$0.74 for each additional minute.
- **Company B** charges \$2.74 for the first minute and \$0.68 for each additional minute.

EXERCISES

Guided Practice

CHECK for Understanding

1. *Reasoning* Which of the following solutions of $9x = 1$ is more likely to have been obtained using a calculator or computer? Why?

   a. $x \approx 0.111$ b. $x = \frac{1}{9}$

2. Solve $0.25(3.2x + 4.1) = 7.2$. Round the result to two decimal places.

3. Explain the difference between $x = 2.5$ and $x \approx 2.5$.

4. *Measurement Sense* The total weight of 11 people is 1204 pounds. Which of the following is the better way to list their average weight? Why?

   a. 109.45455 pounds b. 109.5 pounds

Independent Practice

Error Analysis **In Exercises 5 and 6, describe the error. Write a correct solution.**

5.
$$\begin{aligned} 1.1(2.5x - 3.5) &= 11.2 \\ 2.8x - 3.9 &= 11.2 \\ 2.8x &= 15.1 \\ x &\approx 5.39 \end{aligned}$$

6.
$$\begin{aligned} 0.26(2.39x - 4.91) &= 10.64 \\ 0.62x - 1.28 &= 10.64 \\ 0.62x &= 11.92 \\ x &= 19.23 \end{aligned}$$

In Exercises 7–20, use a calculator to solve the equation. Round your result to two decimal places.

7. $3x + 12 = 17$
8. $13y + 22 = 16$
9. $29t - 17 = -86$
10. $15 - 11x = 108$
11. $6(4x - 12) = 8x + 9$
12. $13x - 22 = 2(9x + 10)$
13. $1.3y + 22.1 = 12.9$
14. $-7.4m + 36.4 = 9.5$
15. $0.15(9.85x + 3.70) = 4.65$
16. $2.16(3.47x - 8.60) = 17.59$
17. $0.19t - 1.57 = 0.46t$
18. $2.4x + 13.7 = 8.1x - 22.5$
19. $3.14x + 17.5 = 9.77x + 24.1$
20. $19.5(13.3 - 4.4x) = 7.2(0.8x - 11.6)$
21. *Sales Tax* You purchase an item. The sales tax rate is 0.06 and the total cost of the item is \$6.35. Let p represent the price of the item (not including sales tax). Solve the following equation to find the price of the item.
    $p + 0.06p = 6.35$
22. *Sales Tax* You purchase an item. The sales tax rate is 0.05 and the total cost of the item is \$2.62. Let p represent the price of the item (not including sales tax). Solve the following equation to find the price of the item.
    $p + 0.05p = 2.62$

Mailing a Letter **In Exercises 23 and 24, use the following information.**

You are mailing a letter with pictures inside to your friend. Postage costs \$.29 for the first ounce and \$.23 for each additional ounce. How heavy can your letter be if you have \$2.00 to spend?

23. Write a verbal model for the problem.

24. Create a table similar to the one in Example 2 to solve the problem.

Using a Label **In Exercises 25 and 26, use the following information.**

You are renting a car. The rental costs are \$35 plus 23 cents per mile. You don't want to spend more than \$140 on the car rental. How many miles can you drive the rental car?

25. Write a verbal model for the problem.

26. Create a table similar to the one in Example 2 to solve the problem.

Integrated Review

Making Connections within Mathematics

Computation Sense **In Exercises 27–30, use a calculator to evaluate the expression. Round the result to two decimal places.**

27. (4.21)(16.07)

28. (0.965)(19.68)

29. $31.02 \div 5.76$

30. $21.555 \div 4.148$

Verbal Phrases **In Exercises 31–34, use a calculator to evaluate the expression. Round the result to two decimal places.**

31. The square root of 4.52

32. The square of 4.52

33. 0.9 raised to the fifth power

34. The reciprocal of 0.3

Exploration and Extension

35. *The Daytona 500* The Daytona 500 is one of the most famous stock car races in the world. The race is 500 miles long. In 1994, the winner of the Daytona 500 finished the race in 3 hours, 11 minutes, and 10 seconds. What was the average speed in miles per hour of the winner? (*Hint:* Convert 11 minutes and 10 seconds to hours and use the distance formula, $d = rt$.)

36. To determine race position, all drivers must run two laps of the 2.5 mile track. The faster of the two laps is recorded. In 1994, #1 pole position was won by Loy Allen at 190.158 miles per hour. What was his lap time in seconds?

The 1994 Daytona 500 winner, Sterling Marlin, won \$253,575. It was his first victory in 279 races.

4.8 Formulas and Variables in Geometry

What you should learn:

Goal 1 How to use formulas from geometry to solve equations

Goal 2 How to use geometry formulas to solve real-life problems

Why you should learn it:

You can use formulas from geometry to model real-life problems, such as approximating the area of a city.

Goal 1 Using Formulas from Geometry

The following investigation shows how you can use the formula for the area of a rectangle (Area $=$ length $\times$ width) to find a formula for the area of a triangle.

LESSON INVESTIGATION

Investigating the Area of a Triangle

Group Activity Use dot paper to draw several triangles. (Two samples are shown below.) For each triangle, draw a rectangle whose area is twice the area of the triangle. Use the result to write a formula for the area of a triangle.

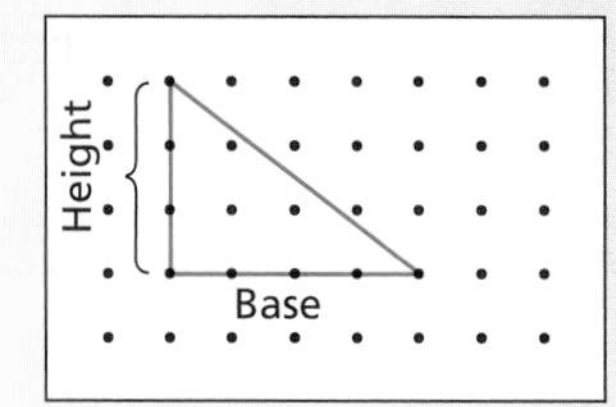

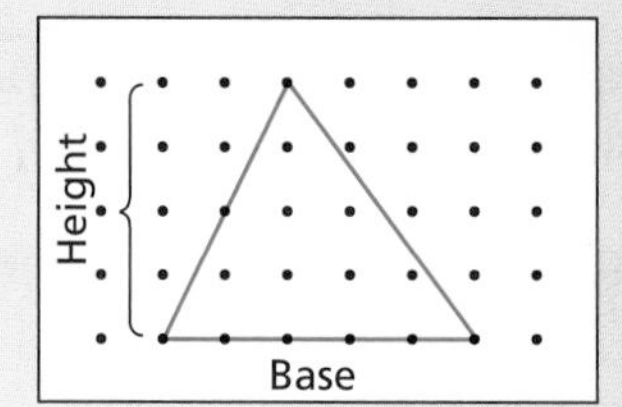

Example 1 *Finding the Dimensions of a Rectangle*

The length of a rectangle is 3 less than twice its width. The perimeter is 24 units. Find the dimensions of the rectangle.

Solution Let x represent the width of the rectangle. Then the length of the rectangle is $2x - 3$.

x

$2x - 3$

$$\text{Perimeter} = 2 \cdot \text{Width} + 2 \cdot \text{Length}$$

$$24 = 2 \cdot x + 2 \cdot (2x - 3)$$

$$24 = 2x + 4x - 6$$

$$30 = 6x$$

$$5 = x$$

The width is 5 units and the length is $2(5) - 3$ or 7 units. ■

Goal 2 Modeling Real Life with Formulas

Connections
Geometry

Example 2 *Estimating an Area*

Use the map to approximate the area of Detroit, Michigan.

Solution One way to answer this question is to sketch a triangle whose area appears to be about the same as Detroit's area.

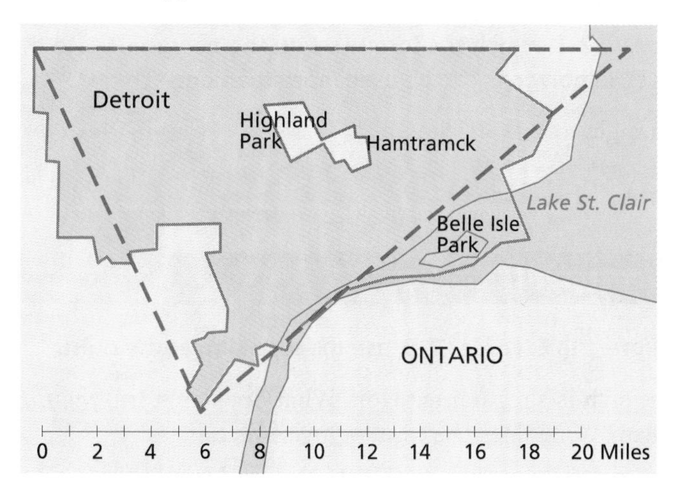

Detroit is the largest city in Michigan. It was founded in 1701 and is one of the world's greatest manufacturing centers.

From the scale on the map, you can see that the triangle has a base of about 22 miles and a height of about 13 miles.

Area of Detroit ≈ Area of triangle

$$\text{Area} = \frac{1}{2}(\text{base})(\text{height})$$
$$= \frac{1}{2}(22)(13)$$
$$= 143 \text{ square miles}$$

You can estimate Detroit's area to be about 143 square miles. ■

Communicating about MATHEMATICS

SHARING IDEAS about the Lesson

Comparing Perimeter and Area The triangle described in Example 2 gives a good approximation for the area of Detroit. (In fact, it is correct to the nearest square mile.) Do you think the triangle could be used to approximate the perimeter of Detroit? Explain your reasoning.

EXERCISES

Guided Practice

CHECK for Understanding

1. *Draw a Diagram* State a formula used in geometry. Sketch a figure that corresponds to the formula. Label the variables on the figure.

In Exercises 2–6, match the formula with the polygon. State the formula in words. (The polygons can be used more than once.)

a. Triangle **b.** Square **c.** Rectangle

2. $A = s^2$ **3.** $P = 2l + 2w$ **4.** $A = \frac{1}{2}bh$ **5.** $P = 4s$ **6.** $A = bh$

Independent Practice

Stop Sign **In Exercises 7–9, use the stop sign at the right.**

7. The sign is a regular octagon. What does the term *regular* mean? What does the term *octagon* mean?
8. The sign has a perimeter of 126 inches. How long is each side?
9. If each side were 16 inches long, what would the perimeter be?

In Exercises 10–15, solve for *x* and find the dimensions of the polygon.

10. *Rectangle*
Perimeter: 36 units
Width: x
Length: $4x - 2$

11. *Rectangle*
Perimeter: 58 units
Width: $6x - 4$
Length: $3x + 6$

12. *Square*
Perimeter: 16 units
Side: $x - 5$

13. *Regular Pentagon*
Perimeter: 90 units
Side: $5x - 7$

14. *Triangle*
Area: 155 square units
Height: 10 units
Base: $3x + 10$

15. *Rectangle*
Area: 225 square units
Width: 9 units
Length: $2x + 9$

Geometry **In Exercises 16–18, find the measure of each angle.**

16. The sum of the measures of $\angle 1$ and $\angle 2$ is 90°.

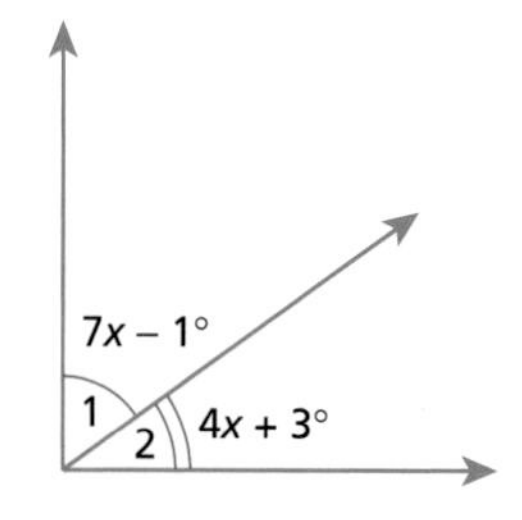

17. The sum of the measures of $\angle 1$ and $\angle 2$ is 180°.

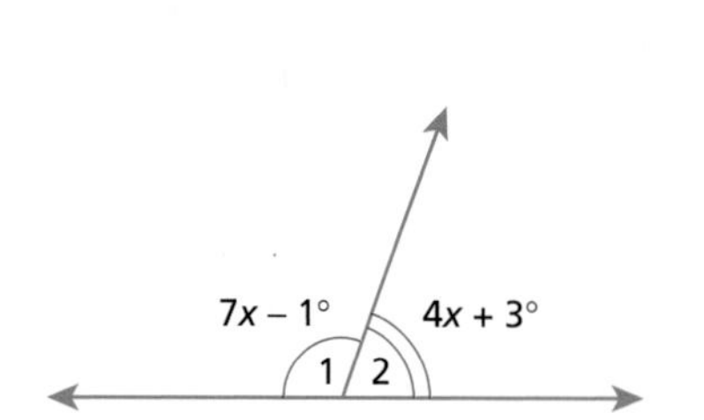

18. The sum of the measures of $\angle 1$, $\angle 2$, and $\angle 3$ is 180°.

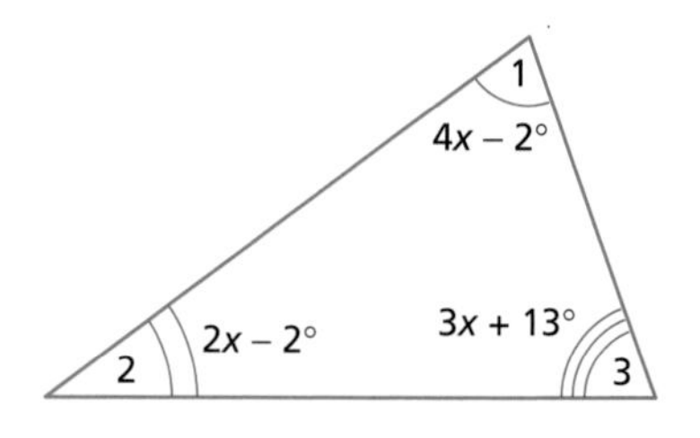

19. ***Television Screen*** The region shown below has a total area of 285 square inches. Find the area of the television screen.

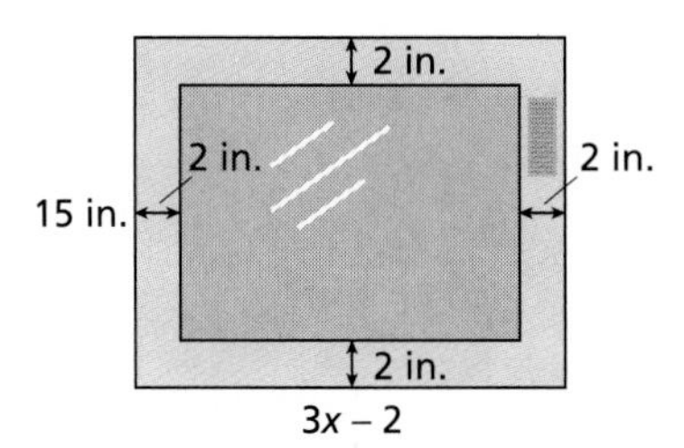

20. ***Bricklaying*** The fronts of the 7 bricks have a total area of 112 square inches. Find the dimensions of the front of each brick.

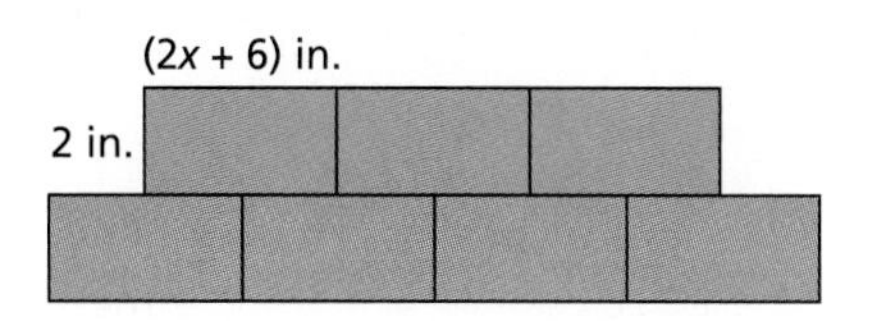

21. ***Baseball Diamond*** The perimeter of a baseball diamond is 360 feet. Find the distance between 1st base and 2nd base.

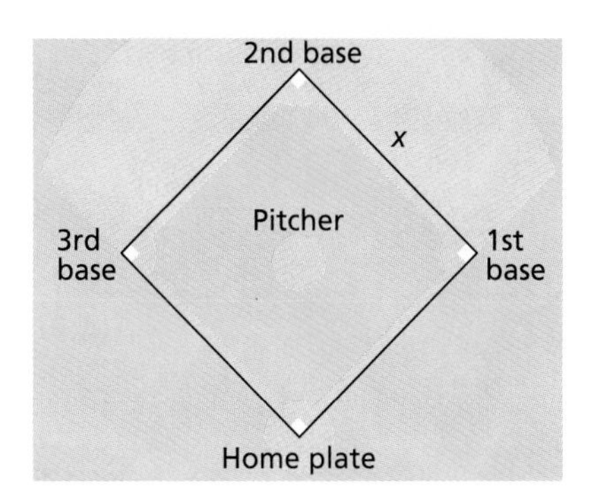

22. ***Nevada*** Use the map below to approximate the area of Nevada. Explain your method.

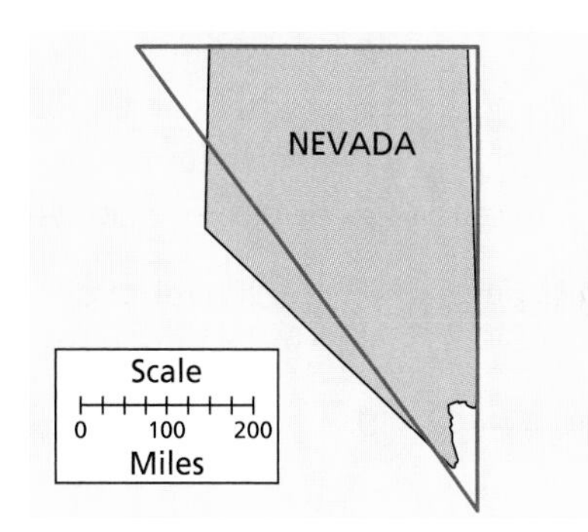

Integrated Review

Making Connections within Mathematics

Perimeter **In Exercises 23 and 24, find the perimeter of the square.**

23. The perimeter of a square that has the same area as a 2×8 rectangle.

24. The perimeter of a square that has the same area as a 4×16 rectangle.

Exploration and Extension

25. ***Investigating the Area of a Parallelogram*** A parallelogram is a quadrilateral with its opposite sides parallel, as shown. Cut a parallelogram out of paper. Cut the parallelogram as shown and rearrange the pieces to form a rectangle. Use the result to write a formula for the area of a parallelogram.

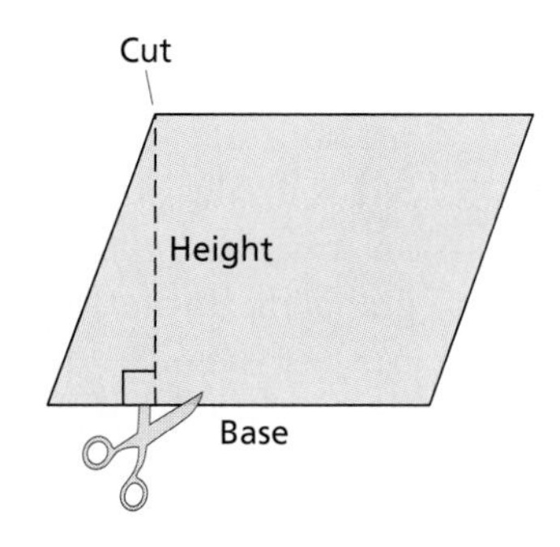

4

Chapter Summary

What did you learn?

Skills

1. Solve two-step equations. **(4.1)**
   - **General Guidelines** **(4.1)**
     - **a.** Simplify both sides of an equation (if needed).
     - **b.** Use inverse operations to isolate the variable.
2. Solve multi-step equations. **(4.2)**
   - Use an "expert" equation-solver format. **(4.2)**
3. Use reciprocals to solve equations. **(4.3)**
4. Use the Distributive Property to solve equations. **(4.4)**
5. Solve equations with variables on both sides. **(4.5)**
   - Collect variables on the side with the greater variable coefficient. **(4.5)**
6. Solve equations involving decimals. **(4.7)**

Problem-Solving Strategies

7. Model and solve real-life problems. **(4.1–4.8)**
8. Model and solve geometry problems. **(4.5, 4.8)**

Exploring Data

9. Use tables and graphs to solve problems. **(4.6, 4.7)**

Why did you learn it?

One of the most common ways to model a real-life situation is with an equation. Knowing how to solve equations helps you answer questions about these real-life situations. For instance, in this chapter you learned how to use equations to analyze a tennis club membership, plan the sale of a wildlife poster, predict the number of songbirds in a region, compute your sales bonus, and approximate the area of a city.

How does it fit into the bigger picture of mathematics?

Throughout this course, you will be given many opportunities to practice problem solving. One of the things you learned in this chapter is that there are many ways to solve problems. For instance, you learned that dividing both sides of an equation by a number produces the same result as multiplying both sides of the equation by the reciprocal of the number. You also studied examples of real-life problems that can be solved algebraically (by solving an equation), numerically (by using a table), and graphically (by using a graph). In future chapters and in future mathematics courses, remember that the best overall problem-solving strategy is to consider a variety of approaches. In fact, solving problems in more than one way and comparing the results is an excellent way to become an expert problem solver.

Chapter REVIEW

In Exercises 1–4, state the inverse of each. (4.1)

1. Subtracting 3 from a number

2. Adding 17 to a number

3. Multiplying a number by -4.2

4. Dividing a number by 6.5

In Exercises 5–10, solve the equation. Check your solution. (4.1)

5. $4x + 11 = 35$

6. $9y + 19 = -8$

7. $5t - 72 = 28$

8. $-7t - 25 = 38$

9. $15 - \frac{1}{2}n = 7$

10. $36 - \frac{1}{3}m = 52$

In Exercises 11–14, write an example of the given type of equation. Then solve the equation. (4.2–4.6)

11. One-step equation

12. Two-step equation

13. Three-step equation

14. Equation with variables on both sides

In Exercises 15–20, solve the equation. Check your solution. (4.2)

15. $7x - 5x + 9 = 21$

16. $5s + 11s - 6 = 74$

17. $3r - 7r + 22 = -14$

18. $21p - 73 - 46p = 27$

19. $\frac{2}{5}m + 9 - \frac{1}{5}m = 15$

20. $-31 = \frac{2}{3}n - 26 + \frac{1}{3}n$

Diagramming Equations **In Exercises 21 and 22, write the equation implied and solve. (4.1, 4.2)**

21.

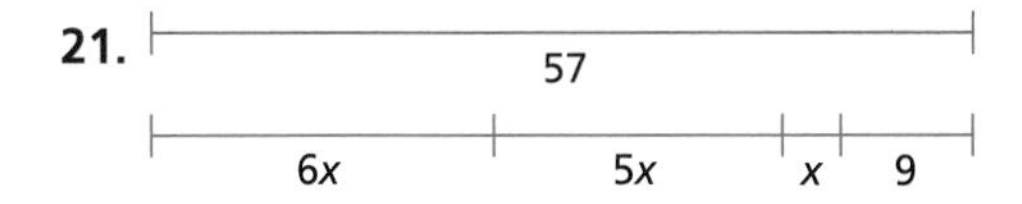

22.

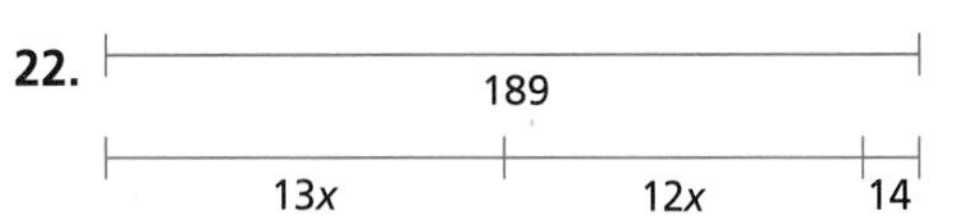

23. *Writing* In your own words describe the reciprocal of a number. **(4.3)**

24. *Number Sense* There are only two numbers that are their own reciprocals. What are they? **(4.3)**

In Exercises 25–28, find the reciprocal of the number. (4.3)

25. 7

26. -24

27. $-\frac{1}{2}$

28. $\frac{15}{8}$

In Exercises 29–34, solve the equation. (4.3, 4.4)

29. $\frac{1}{4}(z + 8) = 10$

30. $\frac{1}{3}(t - 9) = 4$

31. $11(x - 6) - 17 = 38$

32. $15(2m + 5) = -60$

33. $\frac{1}{2}n + \frac{1}{2}(n - 2) = -10$

34. $17 = 3n + \frac{1}{5}(15n - 5)$

In Exercises 35–40, solve the equation. Then check your solution. (4.5)

35. $15x - 56 = 7x$

36. $-18y + 175 = 7y$

37. $3(3 - t) = 5(2t + 7)$

38. $23(2n + 3) = 89n + 26$

39. $\frac{3}{4}m + 8 = 22 - \frac{1}{4}m$

40. $\frac{2}{3}t - 18 = \frac{1}{3}t - 20$

Analyzing Advertising Expenses **In Exercises 41–44, use the following information. (4.3)**

The graph at the right shows the amount A (in billions of dollars) spent on advertising in the United States. This amount can be modeled by the equation

$$A = 22.33 + 7.12t$$

where $t = 0$ represents 1975. ***(Source: McCann-Erickson, Inc.)***

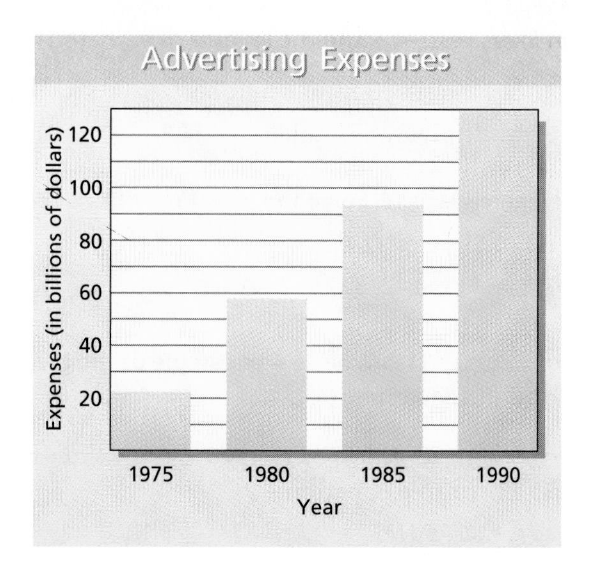

41. What year is represented by $t = 12$?

42. What value of t represents 1983?

43. Use the model to find the amount of money spent on advertising in 1990.

44. Use the model to predict the amount of money spent on advertising in the year 2000.

45. Explain how to solve the equation $12n - 39 = 6n + 3$. **(4.5)**

46. Explain how to solve the equation $3(n + 6) = 4n + 6 - 2n$. **(4.3–4.5)**

47. *Geometry* Find x so that the triangle is equilateral **(4.5)**

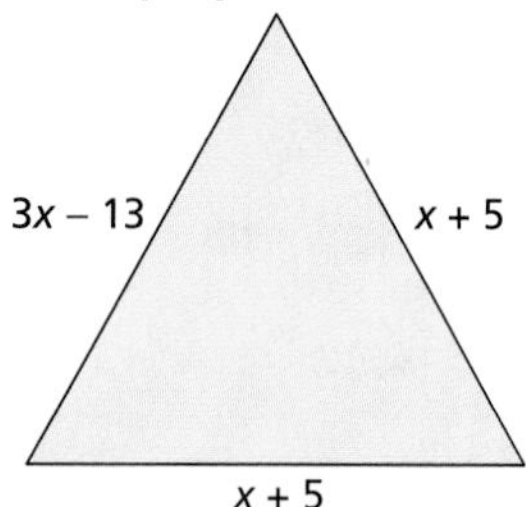

48. *Geometry* Find n so that the square and the triangle have the same perimeter. **(4.5)**

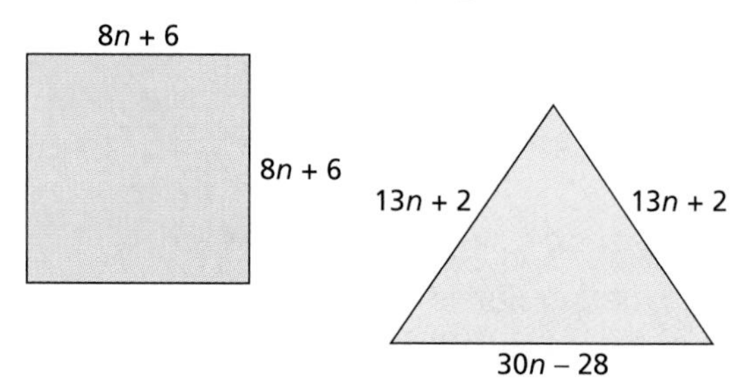

Canoeing **In Exercises 49 and 50, use the following information. (4.6)**

You and your friends go canoeing. The first canoe departs at 11:00 A.M. and travels at a rate of 4 miles per hour. Your canoe leaves an hour later and travels at a rate of 6 miles per hour.

49. Explain why the first canoe is 4 miles ahead of your canoe when you begin.

50. Use a table to find the time that your canoe will catch up to the first canoe.

The highest speed obtained in a canoe is 13.29 miles per hour. The Norwegian four-man team achieved that speed in the 1988 Olympic Games.

51. *Estimation* Round each number to two decimal places. **(4.7)**

a. 8.39154 **b.** −13.657 **c.** 25.9981

52. *Error Analysis* Describe the error. Write a correct solution. **(4.7)**

$$0.14(3.72x - 5.84) = 20.91$$
$$0.52x - 0.82 = 20.91$$
$$0.52x = 21.73$$
$$x \approx 41.79$$

In Exercises 53–56, use a calculator to solve the equation. Round your result to 2 decimal places. (4.7)

53. $4.7x - 9.5 = 13.2$

54. $-3.69y + 14.24 = 57.83$

55. $2.12(4.86t - 3.79) = 19.21$

56. $7.05(13.29n - 6.95) = -194.56$

Using a Table **In Exercises 57 and 58, use the following information.**

You want to purchase a new video game. You have saved $52.89. The sales tax rate is 0.055. The price of the game plus the sales tax cannot be more than the money you have saved. What is the maximum price the game can cost? **(4.6, 4.7)**

57. Write a verbal model for the problem.

58. Create a table to answer the question.

Geometry **In Exercises 59 and 60, find the area of the triangle. (4.8)**

59.

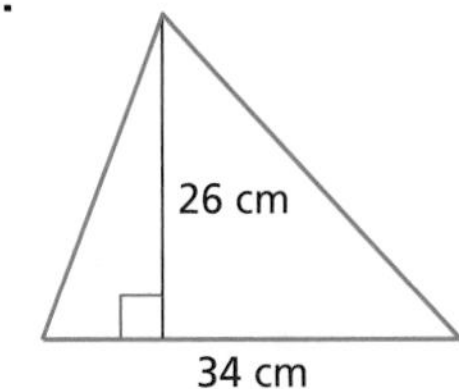

60.

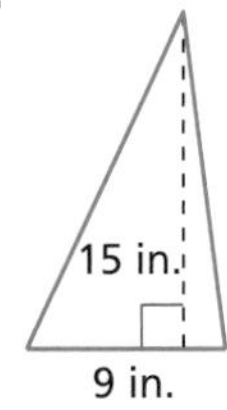

Patio Area **In Exercises 61–65, use the patio diagram shown at the right. (4.8)**

61. The area of the nonsquare rectangle is 240 square feet. Find the width of this rectangle.

62. Find the area of the square.

63. Find the base and height of the triangle.

64. Find the area of the triangle.

65. Find the total area of the patio.

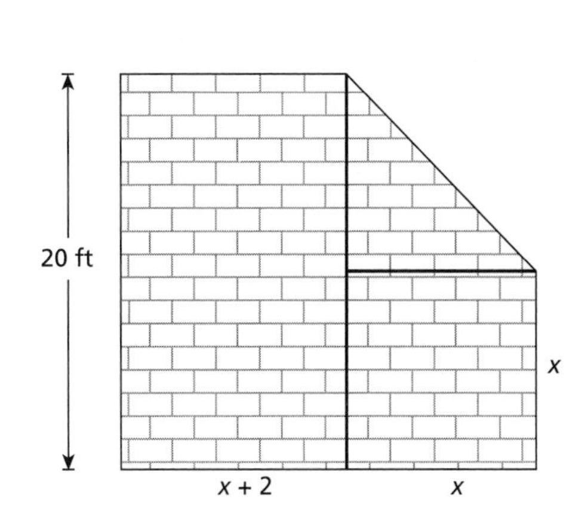

FOCUS ON *Endangered Species*

Real Life Connections

66. *California Condors* Solve the following equation to find the number of miles a California condor can fly without flapping its wings.

$$3.56x + 4.92x - 10.79x = 17.81x + 50.2 - 251.4$$

67. *The Duck and the Falcon* You are studying the flying speeds of two endangered species of birds: the Laysan duck and the Peregrine falcon. You observed a Laysan duck fly a distance of 2000 feet in the same time that a Peregrine falcon flew a distance of 5000 feet. From previous observations, you know that a Laysan duck can fly at a rate of about 80 feet per second. Use the following model to find how fast the Peregrine falcon was flying.

Verbal Model $\dfrac{\text{2000 feet}}{\text{Duck's speed}} = \dfrac{\text{5000 feet}}{\text{Falcon's speed}}$

Labels Duck's speed = 80 (feet per second)
Falcon's speed = s (feet per second)

The California condor is the largest flying land bird found in North America. In 1991, it was reported that about 32 survive, all in captivity.

Word Scramble **In Exercises 68–77, the name of an endangered species has been scrambled. Unscramble the word.**

Unscramble The Endangered Species

68. G E I T R
69. C A T E H E H
70. O A L I R L G
71. N Y E M O K
72. P R O L A E D
73. S E O N H R C R O I
74. L H A W E
75. A J U G R A
76. O C C I D E R L O
77. O U G A R N A N T

78. After unscrambling the words in Exercises 68–77, write the letters in the purple boxes on a piece of paper. Then unscramble the letters to decode the following message.

_ _ _ _ _ _ _ _ _ _ _ _ F _

79. *Problem Solving* After decoding the message, list some environmental situations that relate to the message.

In Exercises 1–8, solve the equation.

1. $4y - 2 = 18$
2. $3 - 3a = 21$
3. $12(r - 2) = 36$
4. $8x + 4 - 3x = 19$
5. $7s - 12 = s$
6. $\frac{1}{2}(x + 8) = 4$
7. $0.7x = 1.3x - 1.2$
8. $p + 2(p - 1) = 2p$

In Exercises 9 and 10, write the reciprocal of the number.

9. $-\frac{1}{2}$
10. 10

In Exercises 11–13, use a calculator to solve the equation. Round your answer to 2 decimal places.

11. $4x + 2 = 17$
12. $1.1(2.2x + 3.3) = 4.4$
13. $5(3 + x) - 2x = 17$

14. The sum of the two angles is 180°. Find the measure of each angle.

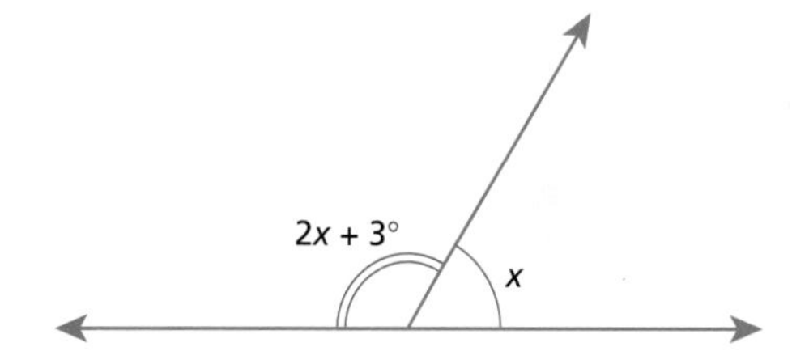

In Exercises 15–17, use the diagram of the swimming pool. The pool is surrounded by a fence whose length is 172 feet.

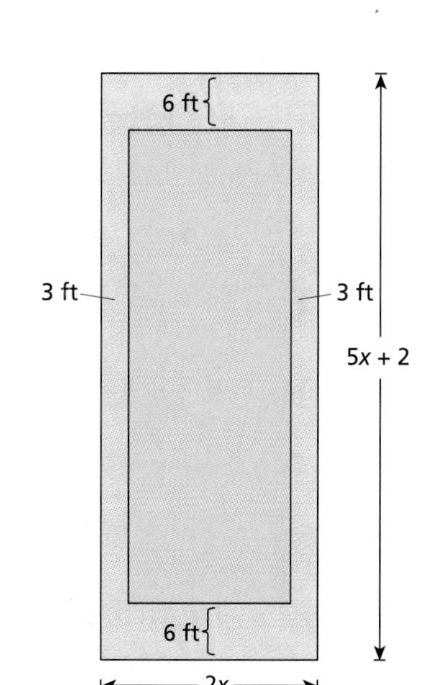

15. Write an equation for the length of the fence and solve for x.
16. What are the dimensions of the swimming pool?
17. What is the area of the swimming pool?

In Exercises 18–20, use the following information.

When you started your homework assignment, your friend already had 6 exercises done. You can do about 3 exercises per minute, whereas your friend can only do 2 exercises per minute.

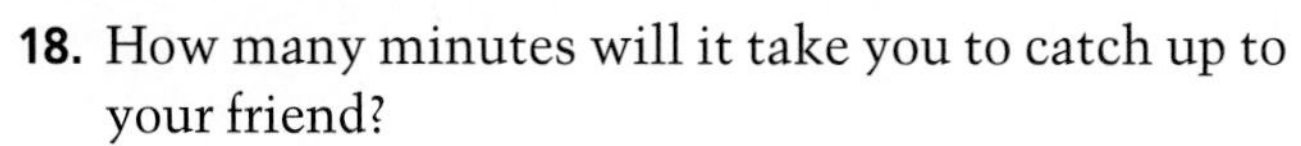

18. How many minutes will it take you to catch up to your friend?
19. When you catch up, how many exercises will you have done?
20. Copy and complete the table.

Minutes	0	1	2	3	4	5	6	7	8
Number of Exercises You Have Solved	0	3	?	?	?	?	?	?	?
Number of Exercises Your Friend Has Solved	6	8	?	?	?	?	?	?	?

CHAPTER 5

Exploring Data and Graphs

LESSONS

Because over 160,000,000 metric tons of garbage are generated annually in the United States, space to put it is filling up at an alarming rate. Almost half the states will fill up all their existing landfills within 10 years. To solve this problem, we need to use less packaging and recycle more of it.

Real Life
Recycling

In 1990, over 8 million pounds of packaging was produced in the United States. Because it is very versatile, plastic is one of the most popular materials used in packaging today. Plastic containers often have symbols like the one shown here to identify what type of material is in them.

In this chapter you will learn to interpret data and graphs in a variety of different forms. This will help you to determine the truth of advertising claims made about a product's environmental impact.

LESSON INVESTIGATION 5.1

Picture Graphs

Materials Needed: colored pencils, graph paper

In this investigation you will conduct a survey and organize the results graphically.

Example ***Conducting a Survey***

In the exercises below, you are asked to conduct a survey to find things that people are afraid of. Such a survey was conducted with 200 people by the magazine *Psychology Today.* The results of that survey are shown below.

Scary Thing	Number of People
Snakes	81
Public Speaking	51
High Places	37
Mice	32
Flying on a Plane	32
Spiders and Insects	22

These results can be shown graphically in several ways. You could use a standard bar graph, or you could modify the bar graph with symbols, as shown below.

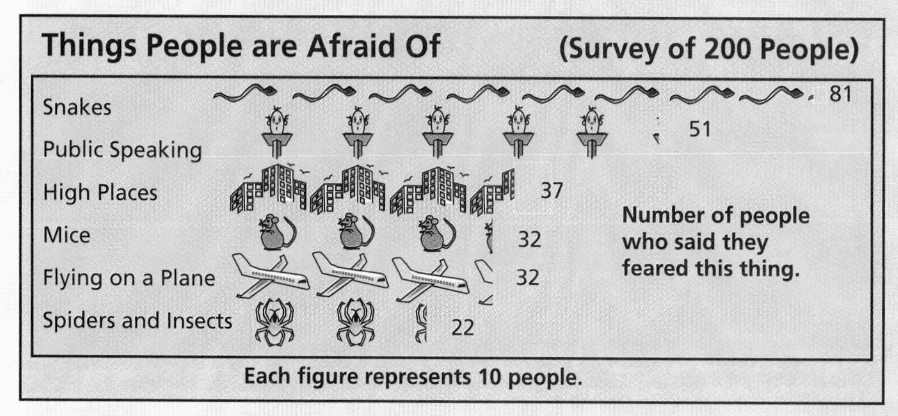

■

Exercises

1. *Group Project* With others in your group, decide how to conduct a survey to find things that people are afraid of. After conducting the survey, organize your results. Then represent your results graphically.
2. How do the results of your survey compare with the survey reported in *Psychology Today*?

5.1 Exploring Picture Graphs and Time Lines

What you should learn:

Goal 1 How to read and make picture graphs

Goal 2 How to read and make time lines

Why you should learn it:

Being able to draw and interpret picture graphs and time lines helps you understand the graphs and time lines you see in newspapers and magazines.

Goal 1 Using Picture Graphs

One of your goals in this course is to learn to use mathematics to communicate with others. In this chapter, you will learn many ways that graphs are used to communicate. For instance, Example 1 shows how a **picture graph** can be used to compare the prices of a movie in different cities.

Example 1 *Making a Picture Graph*

The average price of a movie in 1993 in each of several cities is shown below. Use a picture graph to represent this data. ***(Source: Runzheimer International)***

City	Price	City	Price
London, England	\$7.50	Los Angeles, U.S.A.	\$5.84
Mexico City, Mexico	\$2.57	Stockholm, Sweden	\$8.91
Tokyo, Japan	\$17.31	Toronto, Canada	\$6.07

Solution One way to represent the price of a movie ticket in the 6 cities is shown below.

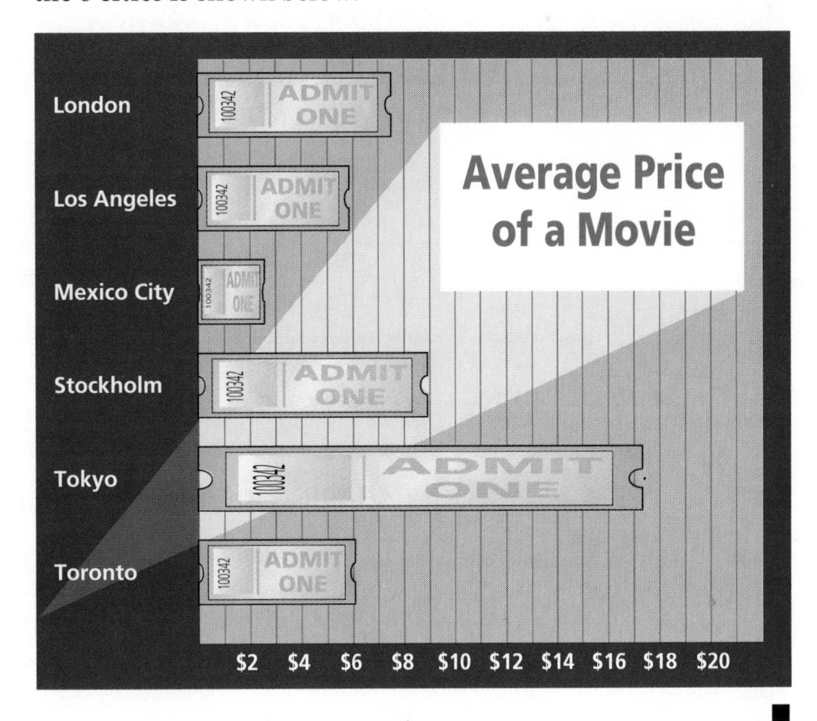

■

Goal 2 Using Time Lines

A **time line** is a graph that shows the dates of several occurrences. For instance, Example 2 shows a time line that gives information about the history of aviation.

Example 2 *Drawing a Time Line*

Real Life
Aviation History

The numbers of air passenger trips in America for several years is shown below. The most advanced passenger aircraft at the time is also listed. Draw a time line for this data. (**Source:** ***America by the Numbers***)

In the 1930's, United Airlines became the first American airline to employ stewardesses. All applicants had to be registered nurses.

Date	Passenger Trips	Aircraft
1783	2	Fire Balloon
1800	100	Gas Balloon
1906	1,000	Biplane
1913	20,000	Monoplane
1935	4,000,000	"Clipper"
1952	100,000,000	"Comet"
1960	200,000,000	Jet Prop
1970	400,000,000	Turbojet
1990	700,000,000	Supersonic Jet

Solution One way to draw the time line is shown below. Notice that the dates are placed on a number line.

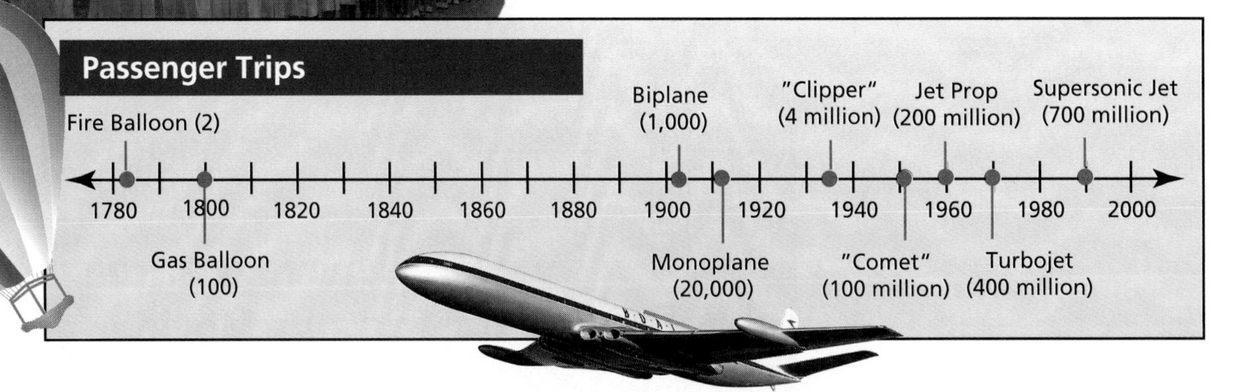

Communicating about MATHEMATICS

Connections
Language Arts

SHARING IDEAS about the Lesson

The Meaning of Words The prefix "bi" means *two* and the prefix "mono" means *one*. What do you think the words "biplane" and "monoplane" mean?

EXERCISES

Guided Practice

CHECK for Understanding

In Exercises 1 and 2, refer to Example 1.

1. Use the picture graph to rank the cities from highest to lowest according to average ticket price.
2. Redraw the picture graph using the symbol ADMIT ONE to represent \$2.00.

In Exercises 3 and 4, refer to Example 2.

3. How many years does each unit on the time line represent?
4. According to the time line, in which 20-year period were the most changes made in passenger aircraft?

Independent Practice

In Exercises 5–6, draw a time line that represents each set of data.

5. *Price of a U.S. Postage Stamp* Between 1975 and 1991, the cost of a first-class postage stamp increased several times.

Sep. 14, 1975	10¢	May 29, 1978	15¢	Nov. 1, 1981	20¢	Apr. 3, 1988	25¢
Dec. 31, 1975	13¢	Mar. 22, 1981	18¢	Feb. 17, 1985	22¢	Feb. 3, 1991	29¢

6. *Postmaster Generals* The years in which selected postmaster generals took office are listed below.

   Benjamin Franklin, 1775; Timothy Pickering, 1795;
   William Barry, 1829; Montgomery Blair, 1861;
   John Wanamaker, 1889; Frank Hitchcock, 1909;
   James Farley, 1933; Lawrence O'Brien, 1965

Air Travel **In Exercises 7–10, use the picture graph at the right. The graph shows the number of passengers that traveled on the five most heavily traveled airlines in 1991.** (*Source: Air Transportation Association of America*)

Top Five Airlines
Number of passengers in 1991:
Northwest
USAir
United
Delta
American
= 10 million passengers

7. How many passengers does one airplane represent?
8. Estimate the number of passengers that traveled on *United.*
9. How many more people traveled on *American* than on *USAir*?
10. If one airplane represented 20 million passengers, how would the picture graph change?
11. *International Breakfasts* The table at the right lists the 1993 annual consumption of breakfast cereal per person in the United States, Great Britain, Canada, and France. Use the data to create a picture graph. (*Source: Gale Book of Averages*)

Country	Consumption (in pounds)
Great Britain	7.34
Canada	6.02
United States	11.9
France	1.78

History of New Hampshire **In Exercises 12–15, use the time line. The time line gives a brief history of New Hampshire.** (*Source: PCUSA*)

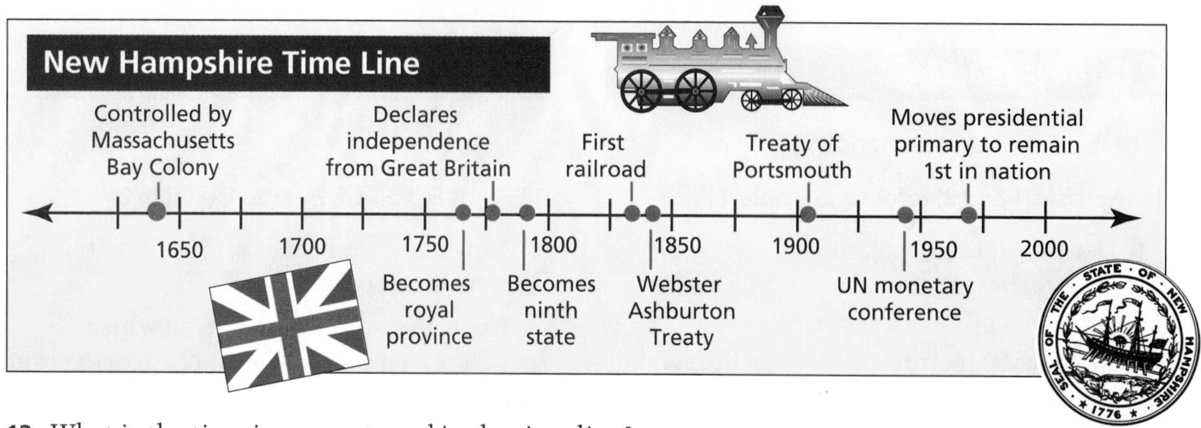

12. What is the time increment used in the time line?

13. Estimate the year that the first railroad appeared in New Hampshire.

14. What event in New Hampshire's history occurred in 1776?

15. Name 2 events that occurred in the 18th century.

16. ***History*** The data at the right lists several events and the years in which they occurred. Make a time line for this data.

Year	Event
1793	Eli Whitney invents cotton gin.
1816	David Brewster invents kaleidoscope.
1837	Samuel Morse patents successful telegraph.
1859	Edwin Drake drills first oil well.
1876	Alexander Bell invents telephone.
1898	Marie Curie discovers radium.
1934	Wallace Carothers invents nylon.
1948	Julian Percy invents synthetic cortisone.

Integrated Review

Making Connections within Mathematics

Mental Math **In Exercises 17–20, evaluate the expression.**

17. $200 \times 3\frac{1}{2}$ **18.** $300 \times 4\frac{1}{3}$ **19.** $(10{,}000)\left(4\frac{3}{4}\right)$ **20.** $(15{,}000)\left(6\frac{2}{5}\right)$

Average **In Exercises 21 and 22, find the average of the numbers.**

21. 18.4, 16.8, 17.5, 19.0, 14.7, 17.9, 16.1, 15.2, 14.9, 15.9

22. −5.61, −3.25, −4.50, −6.13, −4.75, −4.34, −5.07, −6.00, −4.21, −3.97

Exploration and Extension

Creating a Time Line **In Exercises 23 and 24, research the given topic. Create a time line that shows the important events that occurred.**

23. The history of your state, province, or region

24. The history of mathematics from A.D. 1000 to A.D. 2000

5.2 Exploring Bar Graphs and Histograms

What you should learn:

Goal 1 How to use bar graphs to represent data

Goal 2 How to use histograms to represent data

Why you should learn it:

Being able to draw and interpret bar graphs and histograms helps you communicate about real-life situations, such as comparing goals and achievements.

Real Life **American Dream**

Have you ever dreamed of owning your own business? That dream came true for Shiree Sanchez, who started a successful direct marketing business.

Goal 1 Using Bar Graphs

There are three basic types of **bar graphs:** simple bar graphs, double (or triple) bar graphs, and stacked bar graphs. When you use a bar graph to represent real-life data, you first need to decide which type of bar graph will best represent the data.

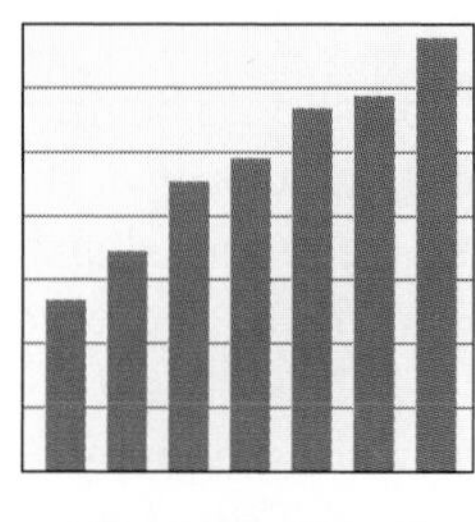

Simple Bar Graph

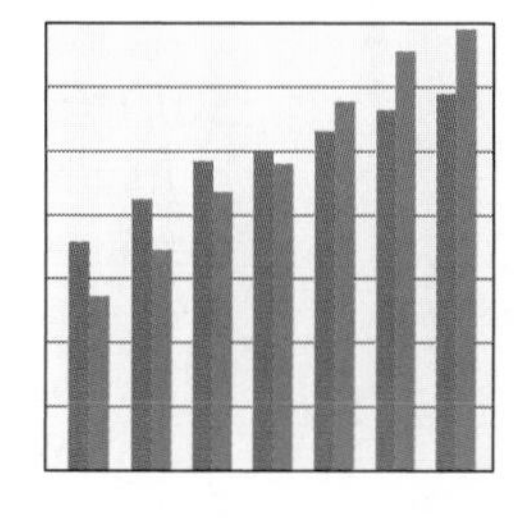

Double Bar Graph

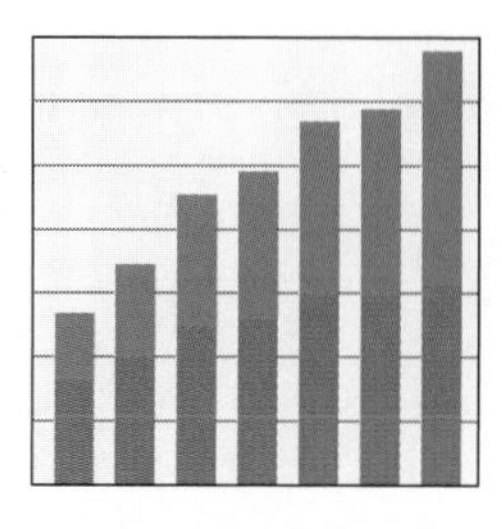

Stacked Bar Graph

Example 1 *Drawing a Bar Graph*

A survey asked 250 adults about their personal goals, and whether they had achieved their goals. Represent these results with a bar graph. ***(Source: Roper Organization)***

Goal	Have/Had Goal	Achieved Goal
Own a home	157	109
Happy marriage	140	100
Own a car	131	150
Have children	131	113
Become rich	113	7
Interesting job	111	60

Solution This data is best represented by a double bar graph.

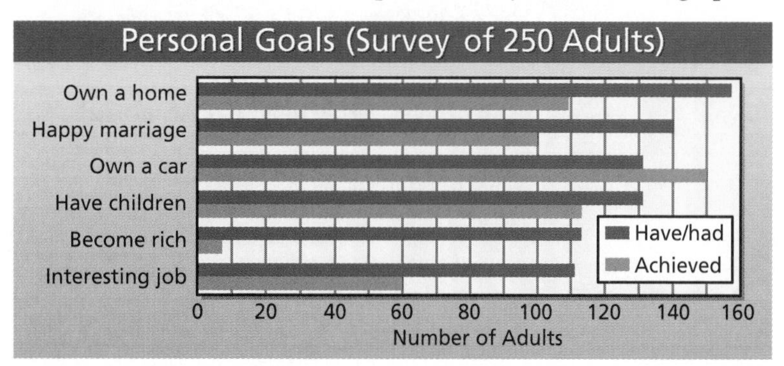

Goal 2 Using Histograms

A **histogram** is a bar graph in which the bars represent intervals of numbers.

Example 2 *Drawing a Histogram*

You have taken a survey of the heights (in inches) of 30 students. Show how this data can be organized by a histogram.

$58\frac{1}{2}$, 65, 60, $61\frac{1}{2}$, $58\frac{1}{2}$, 63, $64\frac{1}{2}$, $66\frac{1}{2}$, 62, 67,
59, $62\frac{1}{2}$, $55\frac{1}{2}$, 68, $56\frac{1}{2}$, 59, 60, 62, $63\frac{1}{2}$, 54,
56, 57, 64, 65, 67, 58, $60\frac{1}{2}$, 62, $64\frac{1}{2}$, 62

Frequency Distribution

Internal	Tally	Total
54–55.5	II	2
56–57.5	III	3
58–59.5	~~IIII~~	5
60–61.5	IIII	4
62–63.5	~~IIII~~ II	7
64–65.5	~~IIII~~	5
66–67.5	III	3
68–69.5	I	1

Solution You should begin by deciding which intervals will help you see the patterns of the heights. Then, use the intervals to construct a **frequency distribution** that shows the number of events or items in each interval. Notice that the frequency distribution shows how many heights are in each interval. Using the numbers in the frequency distribution, you can draw a histogram to represent the data.

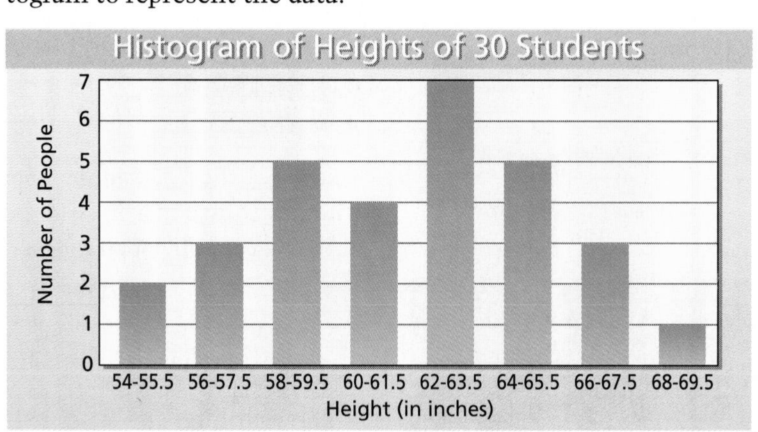

■

Communicating about MATHEMATICS

SHARING IDEAS about the Lesson

It's Up to You Use the intervals 54–57.5, 58–61.5, 62–65.5, and 66–69.5 to organize the data in Example 2. Then compare the resulting histogram to the one above. Which histogram do you think gives a fairer representation of the data? Explain.

EXERCISES

Guided Practice

CHECK for Understanding

1. How do you make a frequency distribution from given data?

2. How is a frequency distribution used?

In Exercises 3 and 4, refer to Example 1.

3. What was achieved by the most people? By the fewest?

4. What goal did most adults have? How many did not achieve that goal?

In Exercises 5 and 6, refer to Example 2.

5. What interval contains the greatest number of students? What interval contains the least?

6. What two intervals contain the same number of students?

Independent Practice

In Exercises 7 and 8, which type of bar graph do you think would best represent the data? Explain.

7. *An Apple a Day* In a survey, people were asked which type of apple they preferred: 39 preferred Red Delicious, 24 preferred Golden Delicious, 20 preferred Granny Smith, and 10 preferred McIntosh. (***Source: USA Today***)

8. *Food Preferences* The data below shows the average number of meals that were pizza, turkey, or pasta (consumed at home) out of 100, for different years. (***Source: NPD Group's National Eating Trends Service***)

1983	Pizza: 3	Turkey: 2	Pasta: 4
1986	Pizza: 4	Turkey: 1	Pasta: 4
1989	Pizza: 4	Turkey: 2	Pasta: 4
1992	Pizza: 5	Turkey: 2	Pasta: 5

9. *Interpreting Data* For the survey in Exercise 7, eight people preferred other types of apples and 17 people said they didn't like any kind of apples. Should these results be included in a bar graph? If so, how?

10. *Computers* A survey asked 100 parents of children ages 6 to 17 the skills that they believe their children develop from using computers. Represent the results with a bar graph. Then explain why you chose the type of bar graph you did. (***Source: Fuji Photo Film USA, Inc.***)

Skill	Number of Parents
Word processing/typing	32
Reading	26
Mathematics	26
Hand-eye coordination	22
Writing	18
Thinking/reasoning	18
Analytical problem solving	9
Speed	2

Youth Population **In Exercises 11–15, use the histogram that represents the projected population (in millions) of young people ages 3–17 for the years 1993, 1995, and 2005.** ***(Source: Census Bureau Projections)***

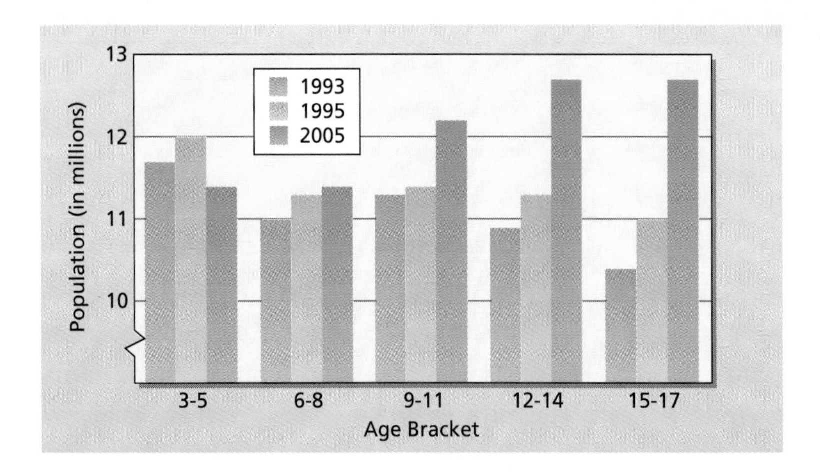

11. Which age brackets have about the same population in 2005?

12. Which age bracket shows a decrease in population from 1993 to 2005?

13. Which age bracket shows the largest increase through the years?

14. How much did the 9–11 age group increase from 1995 to 2005?

15. What conclusions can you make from the bar graph?

16. *Environment* You have taken a survey of the ages of 40 volunteers cleaning a local park. Show how this data can be organized by a histogram.

13	17	8	9	12	13	13	16	14	10
15	9	15	10	11	16	13	11	10	17
13	10	12	9	6	14	16	11	9	9
15	14	11	12	10	7	9	13	11	10

Integrated Review

Making Connections within Mathematics

Geoboard Shapes **In Exercises 17 and 18, use the geoboard.**

17. Count the number of "non-tilted" squares that can be formed on the geoboard. Organize the squares according to their perimeters and illustrate the results with a histogram.

18. Count the number of "non-tilted" rectangles that can be formed on the geoboard. Organize the rectangles according to their areas and illustrate the results with a histogram.

Exploration and Extension

19. *School Enrollment* The table shows the fall enrollment (in millions of students) in grades K–8 for public and private schools. Represent the results with a stacked bar graph. ***(Source: Bureau of Census)***

Type of School	1980	1985	1990	1995
Public	27.7	27.0	29.7	31.7
Private	4.0	4.2	4.1	4.3

5.3 Exploring Line Graphs

What you should learn:

Goal 1 How to use line graphs to represent data

Goal 2 How to use line graphs to explore patterns in geometry

Why you should learn it:

Line graphs can help you communicate about real-life situations, such as the numbers of female and minority members of congress.

Real Life
U.S. Congress

Student reporters for* National Geographic World, *Mark Fleming and Charis Willis, talk with Carrie Meek, who became a U.S. Congresswoman from Florida in 1993.

Goal 1 Using Line Graphs

Line graphs are often used to show trends over intervals of time. A *simple* line graph shows changes in one quantity. A *double* line graph shows changes in two quantities.

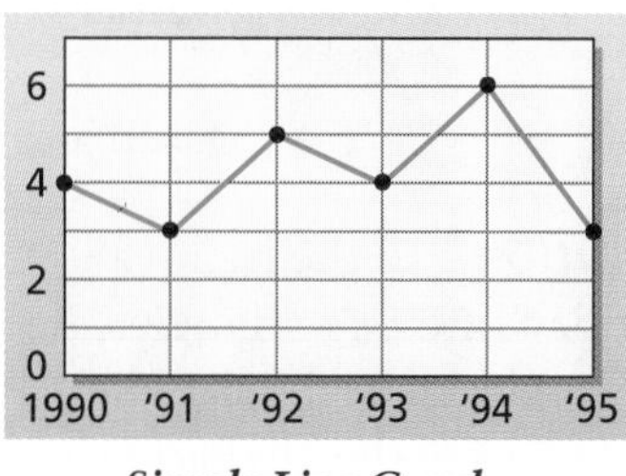

Simple Line Graph

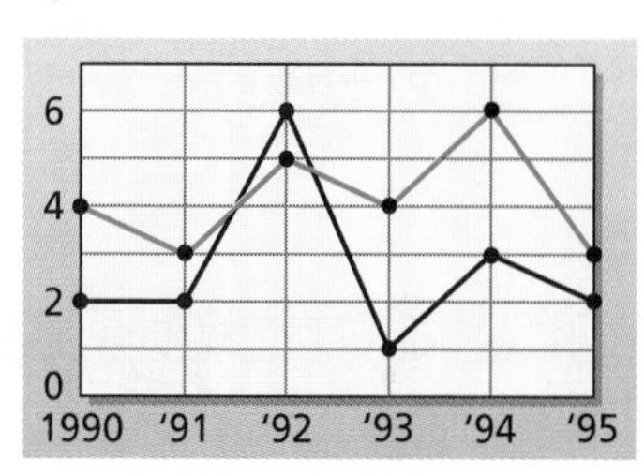

Double Line Graph

Example 1 *Interpreting a Line Graph*

The *triple* line graph below shows the numbers of female, African American, and Hispanic members of the United States House of Representatives from 1981 to 1993. (Representatives are elected for 2-year terms.) During which years did every group increase in members?

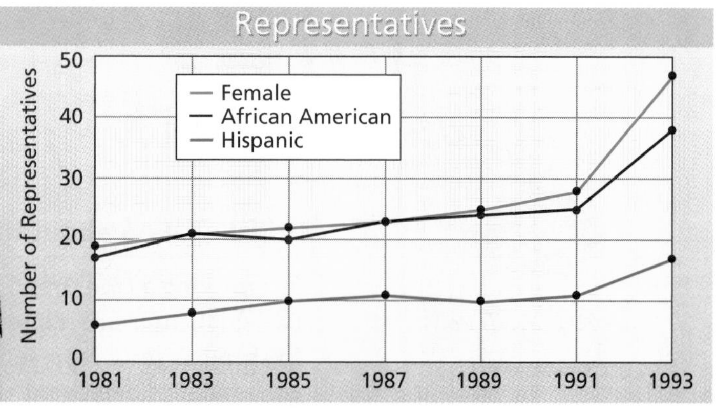

Solution The number of females increased every 2 years. The number of African Americans increased every 2 years except from 1983 to 1985. The number of Hispanics increased every 2 years except from 1987 to 1989. So every group increased in members in 1981–1983, 1985–1987, and 1989–1993. ■

The star in this Amish quilt is made up of triangular pieces of cloth.

Goal 2 Exploring Patterns in Geometry

Example 2 *Finding a Geometric Pattern*

The sum of the measures of the angles of a triangle is 180°. Use this result to find the sum of the measures of the angles of a quadrilateral, pentagon, hexagon, and heptagon. Use a line graph to represent your results.

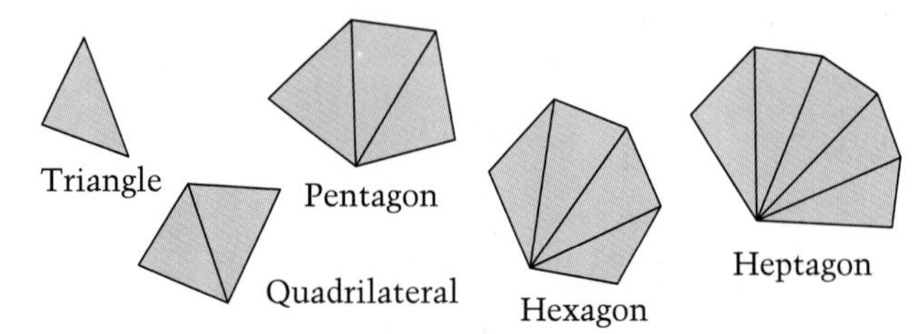

Solution In the figure above, notice that a quadrilateral can be divided into two triangles. Because the sum of the measures of the angles of each triangle is 180°, it follows that the sum of the measures of the angles of the quadrilateral is

$2(180°)$ or $360°$.

You can find the sum of the measures of the angles of the other polygons in a similar way.

$2(180°) = 360°$	*Sum of angle measures of quadrilateral*
$3(180°) = 540°$	*Sum of angle measures of pentagon*
$4(180°) = 720°$	*Sum of angle measures of hexagon*
$5(180°) = 900°$	*Sum of angle measures of heptagon*

These results are summarized in the line graph at the left. ■

Angle Measures of Polygons

Sum of Angle Measures (in degrees): 0, 200, 400, 600, 800, 1000

Number of Sides: 3, 4, 5, 6, 7

Communicating about MATHEMATICS

Cooperative Learning

SHARING IDEAS about the Lesson

Interpreting a Line Graph Graphs can help you see how two quantities are related to each other. For instance, in Example 2, let n represent the number of sides of the polygon, and let S represent the sum of the angle measures of the polygon. Use the line graph shown above to describe to your partner how n and S are related. Use the graph to predict the value of S when $n = 8$. Explain how you could verify your prediction.

EXERCISES

Guided Practice

CHECK for Understanding

Interpreting a Line Graph **In Exercises 1–4, use the line graph in Example 1.**

1. Estimate the number of female representatives in 1993.
2. Estimate the number of Hispanic representatives in 1993.
3. Estimate the increase in female representatives from 1991 to 1993.
4. Estimate the increase in African American representatives from 1991 to 1993.

Independent Practice

Handling Garbage **In Exercises 5–12, use the line graph.** ***(Source: Franklin Associates)***

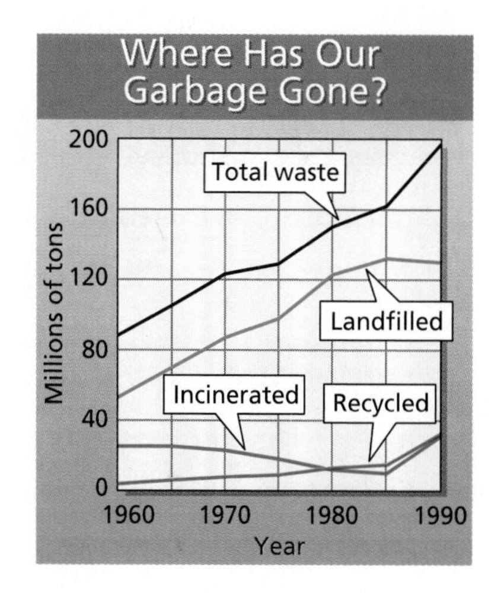

5. Name the units of the horizontal and vertical axes.
6. What are the four quantities shown in the line graph?
7. Estimate the total waste in 1980 and 1990.
8. Estimate the amount of incinerated garbage in 1970.
9. Which quantities increased every year?
10. During which time period did the amount of incinerated garbage decrease?
11. *Reasoning* What is the relationship between the four quantities in the line graph?
12. *Think about It* Why do you think landfill garbage is decreasing?

A Geometric Pattern **In Exercises 13 and 14, use the triangles shown below.**

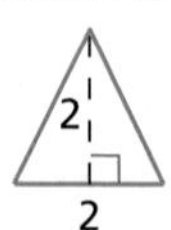

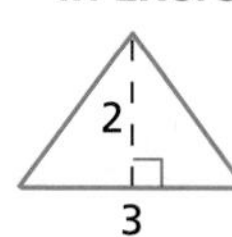

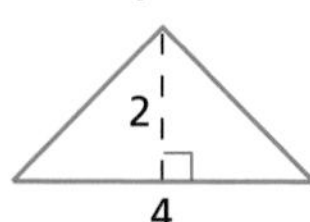

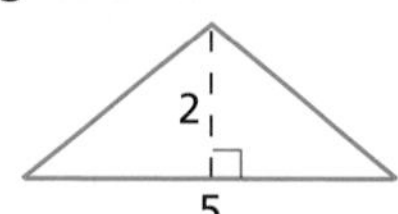

13. Create a table that lists the base, height, and area of each triangle.
14. Make a line graph showing the relationship between the base and the area.
15. *Airline Profits* The total profit or loss (in billions of dollars) for airlines in the United States is given in the table. Construct a line graph for this data. ***(Source: Air Transport Association)***

Year	1983	1984	1985	1986	1987	1988	1989	1990	1991	1992
Profit or Loss	−0.2	0.8	0.8	−0.2	0.7	1.7	0.1	−3.9	−1.9	−4.0

Year	Public	Private
1981	909	4113
1983	1148	5093
1985	1318	6121
1987	1537	7116
1989	1781	8446
1991	2137	10,017

College Costs **In Exercises 16 and 17, use the table, which lists the average annual cost (in dollars) of tuition and fees of four-year public and private colleges for several years.** *(Source: The College Board)*

16. Make a line graph of the data.

17. Are the costs for a public college increasing at a faster or slower rate than the costs for a private college? Explain your reasoning.

18. *Writing* Write a paragraph describing a real-life situation that could be represented by the line graph at the right. Include the units of measure for the data, and explain how the graph can be used to answer questions about the data.

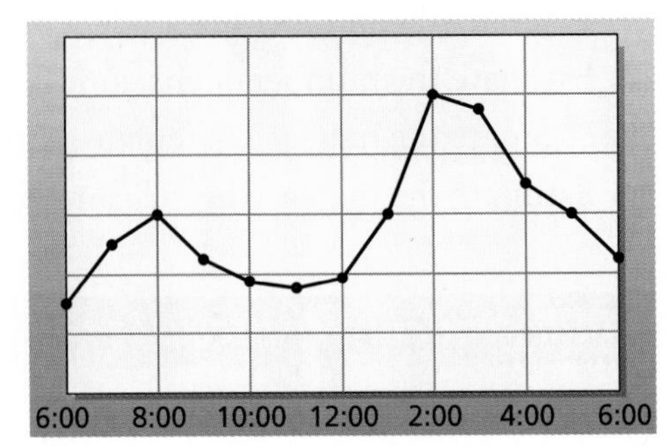

Integrated Review

Making Connections within Mathematics

Mental Math **In Exercises 19–22, evaluate the expression.**

19. $\frac{1}{2}(8)(4)$ **20.** $\frac{1}{2}(16)(7)$ **21.** $\sqrt{5^2+12^2}$ **22.** $\sqrt{6^2+8^2}$

23. *Estimation* Which is the best estimate for the distance (in miles) between Madison, Wisconsin, and Atlanta, Georgia?

a. 7 **b.** 70 **c.** 700 **d.** 7000

24. *Estimation* Which is the best estimate for the area (in square inches) of this page?

a. 20 **b.** 40 **c.** 60 **d.** 80

Exploration and Extension

Reading Graphs **In Exercises 25–30, use the line graph to find the missing coordinate of the ordered pair.** *(Source: U.S. Department of Energy)*

25. (1950, ?)

26. (?, 119.1)

27. (?, 56.7)

28. (?, 114.9)

29. (1955, ?)

30. (?, 31.2)

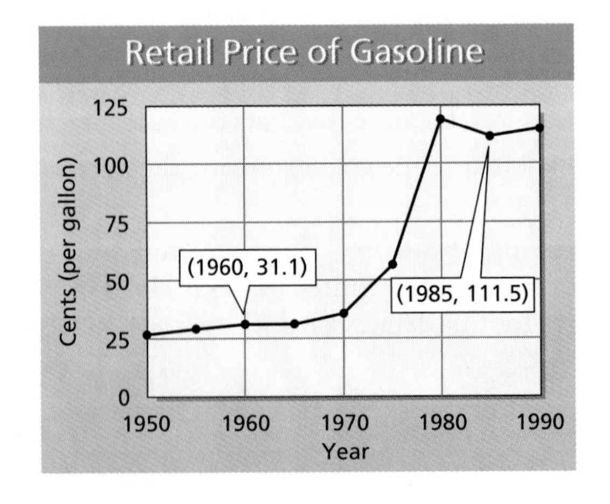

Mixed REVIEW

In Exercises 1–5, use the graph at the right. (5.2) ***(Source: American Veterinary Association)***

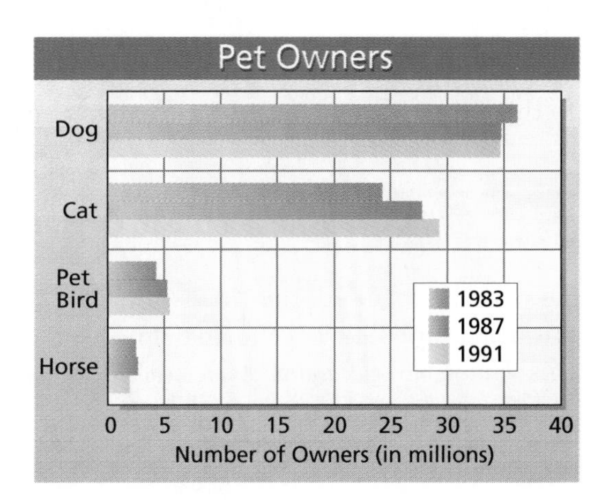

1. What type of graph is used?
2. Which types of pet owners increased?
3. Estimate the number of cat owners in 1983.
4. Estimate the number of dog owners in 1991.
5. Draw a line graph to represent the data. Use the horizontal axis for the year and the vertical axis for the number of owners. Which graph gives a better representation? Explain.

In Exercises 6–9, solve the equation. (4.3–4.5)

6. $3x + 4 = x$ **7.** $2(2x + 1) = 8$ **8.** $\frac{1}{3}(6x - 3) = 9x$ **9.** $4t + 6t = 8t - 13$

Career Interview

Recycling Engineer

Stephen Morgan designs, starts up and troubleshoots systems to recycle paper at paper mills. He travels all over the United States, and to foreign countries and helps to insure that the paper brought from the recycling bins is quickly and easily made into new paper products.

Q: ***What led you into this career?***
A: When I learned about it in college, I felt it was an interesting, new and exciting field.
Q: ***What is your favorite part of your job?***
A: I like to travel, and the different people I meet together with the variety of the work keeps it exciting and fun.
Q: ***Do you use high school mathematics in your work?***
A: Yes, I use algebra a great deal in my work, solving equations for unknown values. For example, I use it when I calculate mass balances as treated pulp is being moved from one place to another inside the mills. In addition, I also need to understand geometry when planning an installation.
Q: ***What would you like to tell students about math?***
A: Math is very important, it makes life easier to understand. Knowing math makes you aware of your surroundings.

LESSON INVESTIGATION 5.4

Organizing Data

Materials Needed: colored pencils, paper

In this investigation you will use a map to organize data.

0-5.9 6-11.9

12-17.9 18-23.9

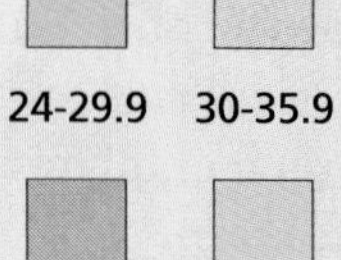

24-29.9 30-35.9

36-41.9 42-47.9

48-53.9 54-59.9

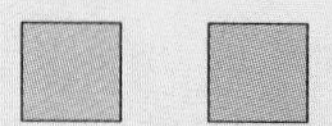

60-65.9 66-71.9

Example — *Using a Map to Organize Data*

The numbers (in millions) of visitors to state parks and recreation areas for 1991 are shown below. Use a map to organize this data. ***(Source: National Association of State Park Directors)***

AK	6.8	HI	19.1	ME	2.4	NJ	11.0	SD	5.9
AL	6.1	IA	12.1	MI	25.3	NM	4.3	TN	27.0
AR	6.9	ID	2.5	MN	8.0	NV	2.6	TX	24.0
AZ	2.2	IL	34.6	MO	15.0	NY	60.7	UT	4.9
CA	70.4	IN	10.5	MS	3.9	OH	67.2	VA	3.9
CO	8.7	KS	4.1	MT	1.7	OK	16.0	VT	1.0
CT	6.7	KY	27.3	NC	9.5	OR	39.5	WA	46.8
DE	3.2	LA	1.1	ND	1.0	PA	36.3	WI	12.3
FL	13.1	MA	12.0	NE	9.2	RI	5.1	WV	8.3
GA	16.3	MD	7.8	NH	2.8	SC	8.0	WY	2.0

Solution Begin with a state map. Then choose different colors to represent different intervals. Color each state accordingly.

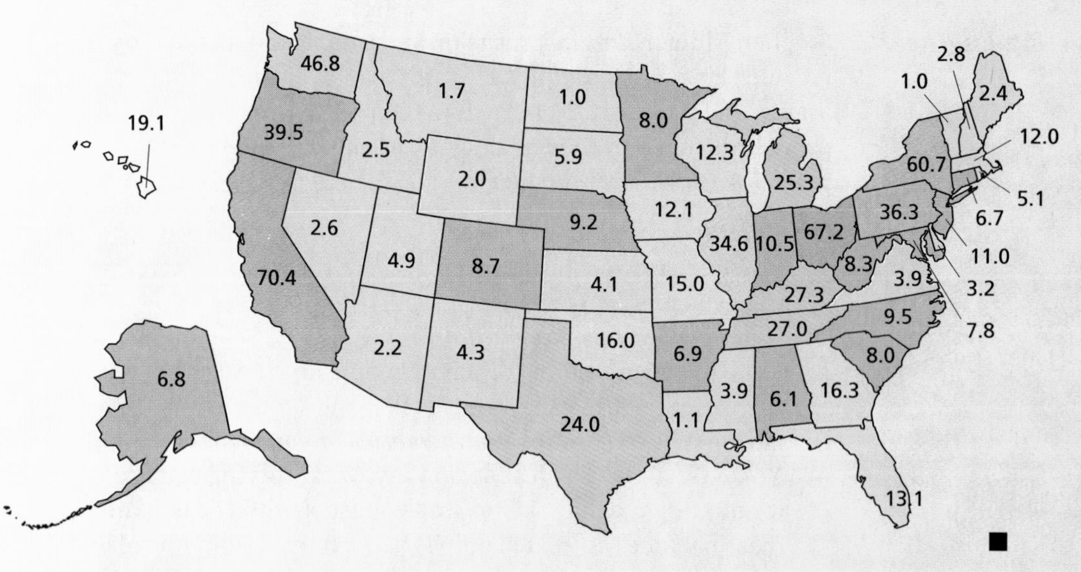

■

Exercises

1. Which 8 states had the most visitors?
2. Which 8 states had the least visitors?
3. Yellowstone Park, in Wyoming and Montana, is the most popular park in the United States. How does this information fit with that given above?

5.4 Problem Solving: Choosing an Appropriate Graph

What you should learn:

How to choose an appropriate graph to represent data

How to use graphs to make presentations

Why you should learn it:

Graphs can help you communicate about real-life situations, such as showing the numbers of miles walked by individuals in different occupations.

Real Life **Occupations**

Study Tip...

When deciding which type of graph to use, here are some guidelines.

1. *Use a bar graph when the data falls into distinct categories and you want to compare totals.*
2. *Use a line graph when you want to show the relationship between consecutive amounts or data over time.*
3. *Use a picture graph for informal presentations in which you want a high visual appeal.*

Goal 1 Choosing Appropriate Graphs

So far in the chapter, you have studied picture graphs, time lines, bar graphs, and line graphs. When you are using a graph to organize and present data, you must first decide which type of graph to use.

Example 1 *Organizing Data with a Graph*

The following data lists the daily average numbers of miles walked by people while working at their jobs. Organize the data graphically. ***(Source: American Podiatry Association)***

Occupation	Miles Walked per Day
Actor	3.2 miles
Mail Carrier	4.4 miles
Medical Doctor	3.5 miles
Nurse	3.9 miles
Police Officer	6.8 miles
Retail Salesperson	3.4 miles
Secretary	3.3 miles
Television Reporter	4.2 miles

Solution You can use either a picture graph or a bar graph. The bar graph below is horizontal. This makes it easier to label each bar. Also notice that the occupations are listed in order of the number of miles walked.

Average Number of Miles Walked in a Day

Police
Mail Carrier
Television Reporter
Nurse
Doctor
Retail Salesperson
Secretary
Actor

0 1 2 3 4 5 6 7

Goal 2 Making Graphical Presentations

Example 2 *Making a Presentation*

Real Life
Fashion Color Preferences

In 1993, the *Pantone Color Institute* took a survey asking adults which color of clothing they preferred for dressy occasions. The results are listed below. You work for a clothing designer and want to use this data in a presentation. How would you organize the data graphically? ***(Source: Pantone Color Institute of Carlstadt, New Jersey)***

Color	Number of Women	Number of Men
Black	727	702
Blue	551	473
Gray	223	649
Red	222	78
Brown	102	198

What do young people wear for dressy occasions? The data in this graph represents the color preferences of adults. ***If you were to survey your classmates, do you think your results would be similar?***

Solution One possibility is to use a picture graph, like that shown below.

Color Preference for Dressy Occasions
Each symbol represents 100 people who preferred the indicated color.

Men
Women
Black
Blue
Gray
Red
Brown

■

Communicating about MATHEMATICS

SHARING IDEAS about the Lesson

It's Up to You In Example 2, would it make sense to use a simple bar graph to display the data? A stacked bar graph? A double bar graph? A line graph? Explain your reasoning.

EXERCISES

Guided Practice

CHECK for Understanding

In Exercises 1 and 2, refer to Example 1.

1. On the average, how many more miles a day do police officers walk than mail carriers?
2. On the average, which occupation walks half as many miles in a day as police officers?

In Exercises 3 and 4, refer to Example 2.

3. How many more men than women prefer to wear gray?
4. Describe how this picture graph would be effective in giving a presentation.

Independent Practice

In Exercises 5–8, choose a type of graph that best represents the data. Explain why you chose that type, and then draw the graph.

5. *Hobbies* The number (in millions) of people who have a hobby ***(Source: Hobby Industries of America)***

Craft or Hobby	Number
Sewing/needlecrafts	46.4
Candy making/ cake decorating	14.9
Painting/drawing	13.9
Plastic model kits	11.1
Ceramics	9.3
Flowers	8.4
Model railroading	5.6

6. *Videocassette Recorders* The average number (out of 100) of people who own videocassette recorders ***(Source: The Roper Organization)***

Year	Number
1980	3
1983	10
1985	19
1987	50
1988	64
1990	65
1992	68

7. *Owning Cats* The average number (out of 100) of cat owners who state the reason for owning a cat ***(Source: Gallup Poll)***

Reason for Owning a Cat	Number
Someone to play with	93
Companionship	84
Help children learn responsibility	78
Someone to communicate with	62
Security	51

8. *Dog Tricks* The number (in millions) of dogs who can perform tricks ***(Source: Pet Food Institute)***

Trick	Number
Sit	5.3
Shake paw	3.8
Roll over	2.9
Speak	2.7
Stand on hind legs	1.9
Sing	0.8
Fetch newspaper	0.4

9. *Little League* The graph shows the number of kids playing Little League baseball around the world from 1940 through 1993.

   a. Estimate the number of kids who played in 1980, 1985, and 1990.

   b. Estimate the increase in players from 1940 to 1993.

   c. During which decade did the number of players remain about the same?

   d. During which decade did the greatest increase occur?

   e. Why is a line graph a good choice for presenting this information?

Kids Playing Little League Baseball around the World

2,500,000
2,000,000
1,500,000
1,000,000
500,000
0
2.4 million
112
'40 '50 '60 '70 '80 '90 '93
Year

10. *Organizing Data* You work for a telephone company and are asked to find the number of telephones that Americans have in their homes. You research the question and obtain the data shown at the right. (The number of people is shown in millions.) You are to present the data in a company meeting. How would you organize the data graphically?

Number of Phones in Home	Number of People
1	61.8
2	80.2
3	50.4
4	22.9
5 or more	13.7

Integrated Review

Making Connections within Mathematics

Modeling Numbers **In Exercises 11–14, draw the number line that is described.**

11. Low number: 0, High number: 15, 5 intervals
12. Low number: 0, High number: 16, 4 intervals
13. Low number: 0, High number: 64, 8 intervals
14. Low number: 0, High number: 150, 6 intervals

Exploration and Extension

15. *Puzzle* Copy the diagram at the right on a piece of paper. Then place the numbers from 1 to 14 in the small circles so that the sum of the numbers in each ring is 50. You can use each number only once. Six of the numbers have been placed in the small circles for you.

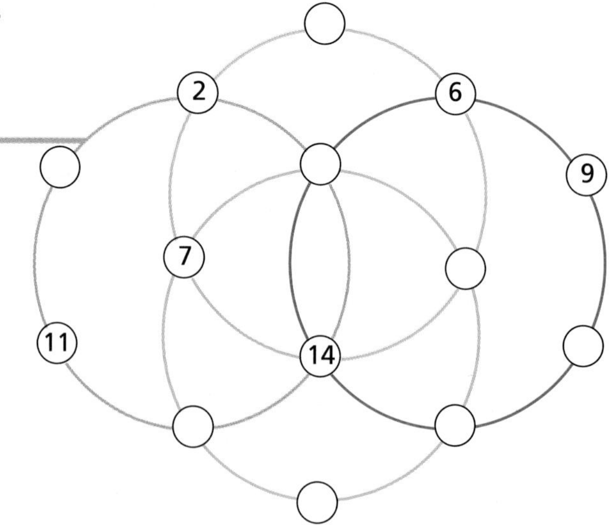

Mid-Chapter SELF-TEST

Take this test as you would take a test in class. The answers to the exercises are given in the back of the book.

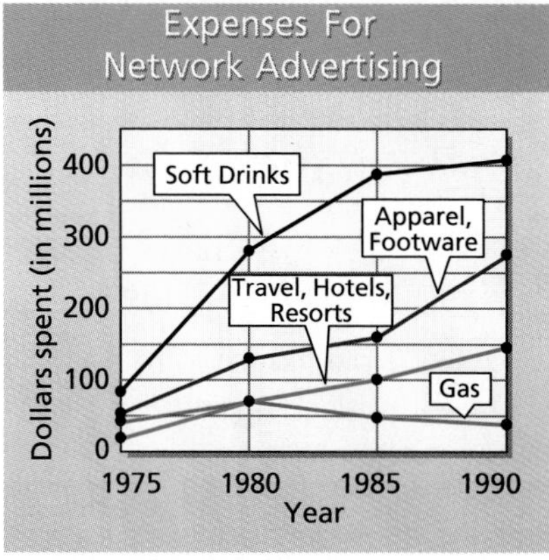

In Exercises 1–6, use the graph, which shows the expenses for network advertising. *(Source: Television Bureau of Advertising)* **(5.3)**

1. What type of graph is this?
2. Which group had the steadiest increase?
3. Name the only group whose expenses decreased.
4. Which two groups spent about the same amount in 1980?
5. Which group had the sharpest increase from 1985 to 1990?
6. Which group spent about $49 million in 1985?
7. ***Literature*** The table lists the year of birth of several famous authors. Create a time line using the given information. **(5.1)**

Author	Year of Birth
Emily Dickinson	1830
Mark Twain	1835
Jack London	1876
John Steinbeck	1902
Alice Walker	1944

8. The table below shows the price of a share of stock for each company in March of 1994. Draw a graph that best represents the data. Explain why you chose that type of graph. **(5.4)**

Company	Stock Price
Sears, Roebuck	$48
Wal-Mart Stores	$28
JC Penney	$55
K Mart Corp.	$19
The Gap, Inc.	$45

9. Draw a graph that best represents the net profit (in millions of dollars) of Blockbuster Entertainment for the years given. Explain why you chose that type of graph. **(5.4)**

   1993: $243.6 1992: $142.0 1991: $93.7
   1990: $68.7 1989: $44.2 1988: $15.5

In Exercises 10–13, use the data in the graph, which shows the number of people who doodle. *(Source: Faber-Castell)* **(5.2)**

10. What type of graph is this?
11. What seems to be the most common reason for doodling?
12. Estimate the number of people who doodle while they are thinking or solving problems.
13. What is the reason that about 82 million people doodle?

5.5 Problem Solving: Misleading Graphs

What you should learn:

Goal 1 How to recognize misleading bar graphs

Goal 2 How to recognize misleading line graphs

Why you should learn it:

Knowing when a graph is misleading helps you interpret real-life information that is presented graphically.

Goal 1 Misleading Bar Graphs

Investigating Bar Graphs

Group Activity Both bar graphs represent the same data. Which graph is misleading? Why?

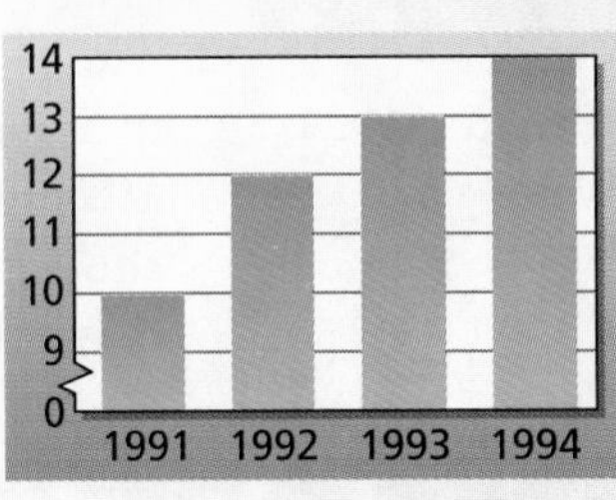

Broken Vertical Scale

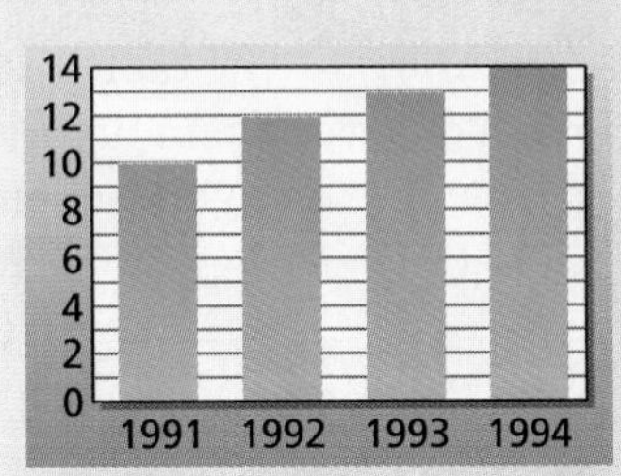

Unbroken Vertical Scale

Example 1 *Interpreting a Bar Graph*

Real Life **Pet Behavior**

From the graph, about how many more dogs sleep in or on their owners' beds as outside or in a garage? ***(Source: Gallup)***

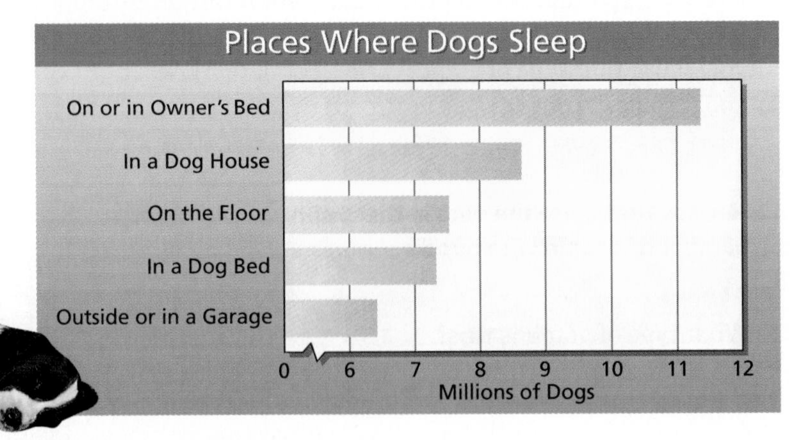

Solution The size of the bars makes it appear that about 4 times as many dogs sleep in or on their owners' beds as sleep outside or in a garage. From the scale, however, you can see that it is really only about twice as many. ■

Goal 2 Misleading Line Graphs

Example 2 *Interpreting Line Graphs*

Each of the line graphs shows the numbers of paperback books and hardback books sold in the United States from 1985 through 1991. Which graph is misleading? ***(Source: Book Industry Study Group)***

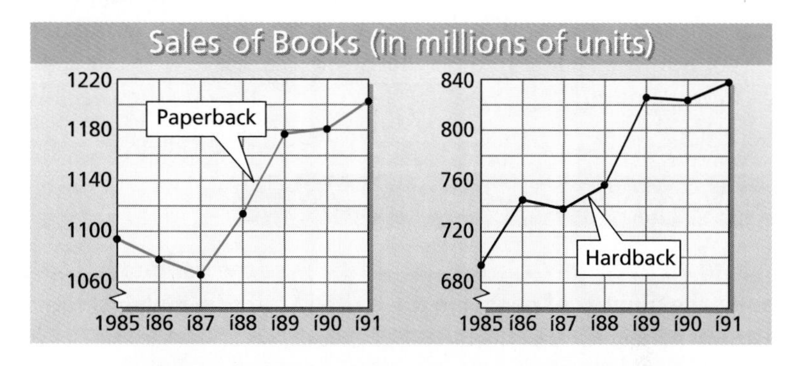

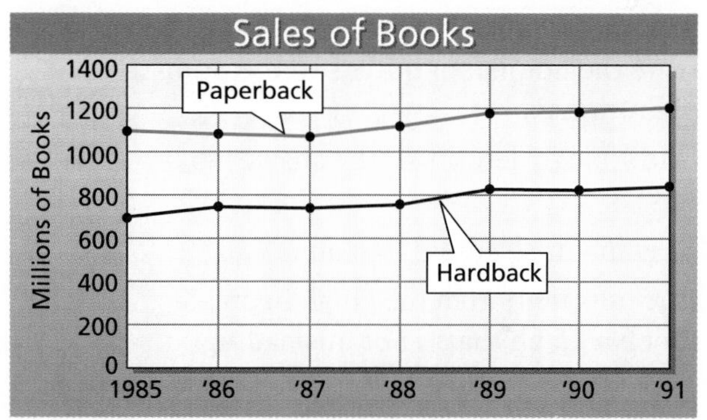

Solution The first graph is misleading; the second is not. The visual impression of the first graph makes it appear that paperbacks and hardbacks sold equally well. Moreover, the broken vertical scales accentuate the changes from one year to the next. ■

Communicating about MATHEMATICS

SHARING IDEAS about the Lesson

Research Look through newspapers or magazines to find examples of graphs that are misleading. Bring the graphs to class. Do you think the graphs were intended to be misleading? Explain your reasoning.

EXERCISES

Guided Practice

CHECK for Understanding

Business **In Exercises 1 and 2, use the graphs in the Lesson Investigation on page 214.**

1. Suppose that the graphs represent your company's profits. Which graph makes your profits appear to have increased more rapidly?

2. Suppose that the graphs represent your company's expenses. Which graph makes your expenses appear to have increased more slowly?

Independent Practice

Playing an Instrument **In Exercises 3–6, use the bar graph. The graph shows the number of people in the United States (in millions) that play each instrument.** *(Source: The* Unofficial *U.S. Census)*

3. Judging only from the length of the bars, compare the number of flutists to the number of drummers.
4. Use the scale to determine the answer to Exercise 3.
5. Is this graph misleading? Explain.
6. Use the information in the graph to create another bar graph that is *not* misleading.

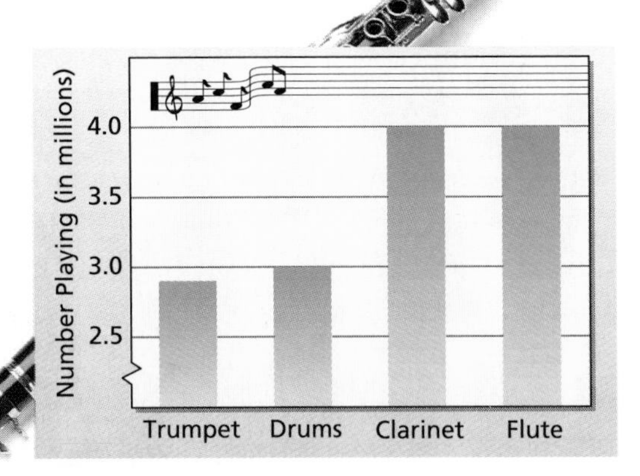

Weekly Earnings **In Exercises 7–10, use the line graph. This line graph is similar to the one that appeared in *The Wall Street Journal.* The graph shows the weekly earnings for factory workers from February, 1991 through January, 1994.**

7. Without looking at the vertical scale, compare the weekly pay in February, 1991 to the weekly pay in December, 1993.
8. Use the scale to determine the answer to Exercise 7.
9. Write a paragraph explaining why this graph is misleading.
10. Use the information in the line graph to create another line graph that is *not* misleading.

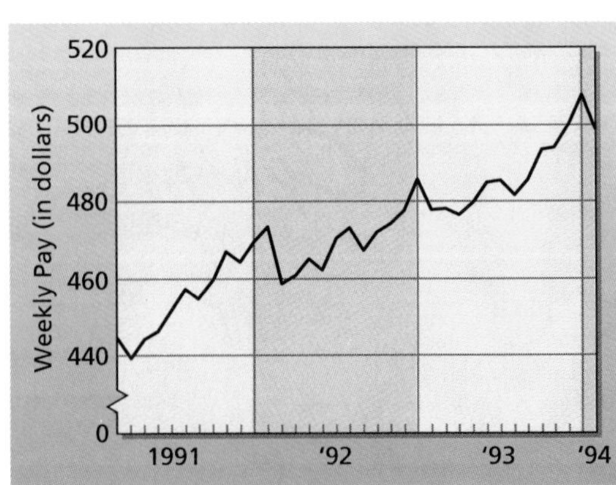

Comparing Graphs **In Exercises 11 and 12, use the two graphs below.**

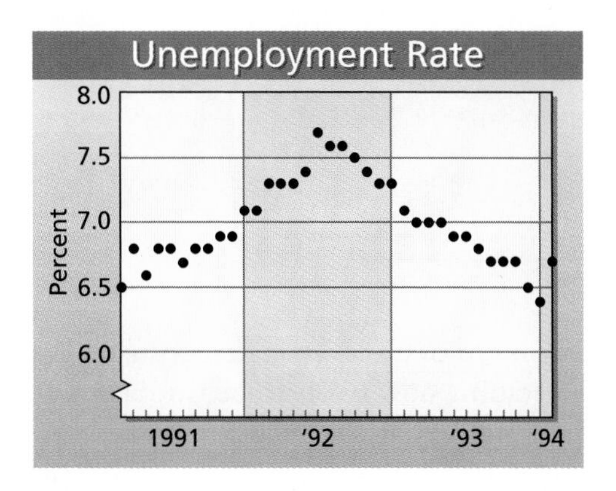

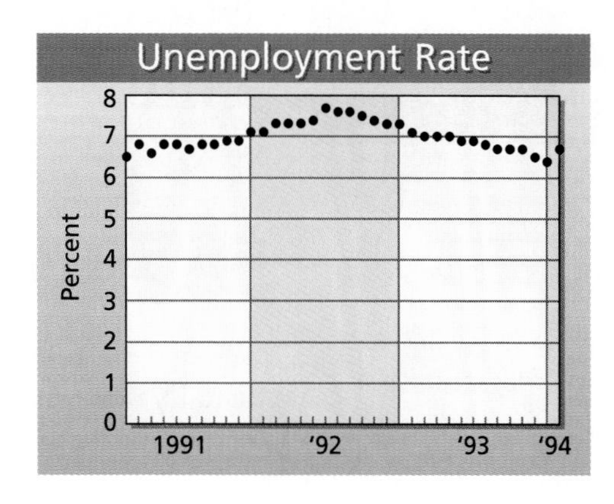

11. A candidate who has been in office is seeking reelection. Which graph might the candidate use? Why?

12. A candidate who has not been in office is seeking election. Which graph might the candidate use? Why?

13. *Research Project* Use a table or graph of information to create a misleading graph. Explain how the graph is misleading.

Integrated Review

Making Connections within Mathematics

Logical Thinking **In Exercises 14–17, decide whether the information in the graph allows you to conclude that the statement is true. Explain your reasoning.** ***(Source: Bureau of the Census)***

14. More new homes in 1990 have fireplaces than new homes in 1970.

15. In 1970, more than half of all new homes had only one bathroom.

16. The percent of new homes built with airconditioning in 1990 is triple that of 1970.

17. In 1970, more than 50% of all new homes were built without a garage.

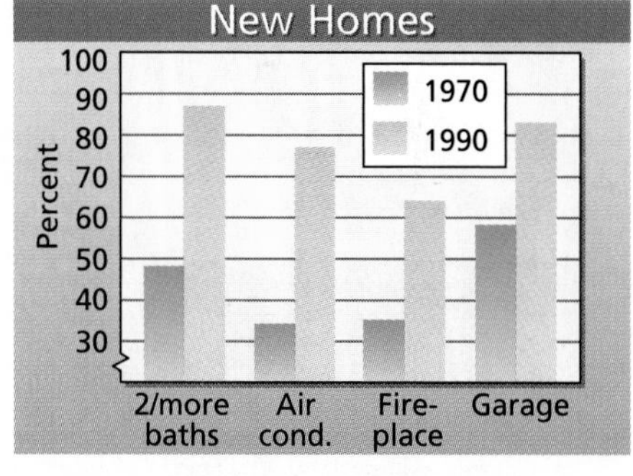

Exploration and Extension

18. *Average Home Prices* The picture graph shows the average sales price of a new one-family house in 1970 and in 1990. Explain why the graph is misleading.

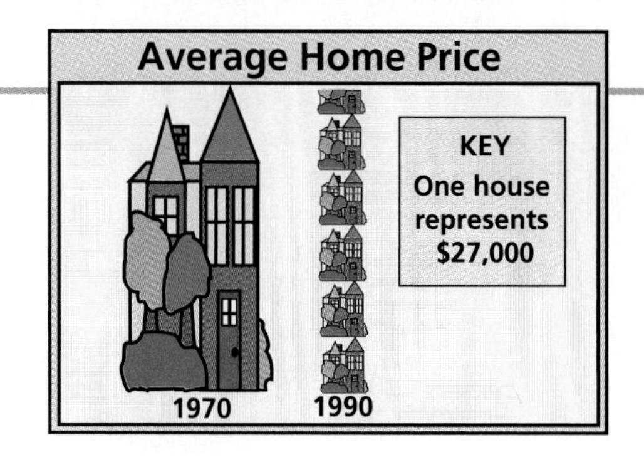

5.6 Statistics: Line Plots

What you should learn:

Goal 1 How to use line plots to organize data

Goal 2 How to use organized data to help make decisions

Why you should learn it:

Data is more meaningful and useful when it is organized.

Goal 1 Using Line Plots

When data from an experiment or a survey is first collected, it is usually not organized. Deciding how to organize the data is a critical part of a branch of mathematics called *statistics.*

LESSON INVESTIGATION

Investigating Data Organization

Group Activity The "raw data" (unorganized data) below lists the ages of a herd of deer at Presque Isle State Park. With other members of your group, discuss different ways that the data could be organized. As a group, choose one method and use it to organize the data.

3, 1, 1, 5, 11, 12, 5, 7, 7, 3, 2, 4, 1, 10, 11, 4, 4, 4,
1, 5, 6, 6, 7, 4, 1, 3, 2, 1, 9, 2, 12, 4, 4, 1, 8, 9,
1, 3, 3, 2, 5, 1, 3, 1, 2, 2, 1, 1, 2, 2, 1, 1, 3, 3,
1, 6, 2, 6, 1, 1, 8, 3, 2, 2, 2, 2, 5, 5, 1, 3, 8, 2

Real Life Wildlife

Example 1 *Using a Line Plot*

One way to organize the data shown above is with a line plot. To begin, draw a number line that includes all integers from 1 through 12. For each number in the list, place an × above the coordinate on the number line, as shown below.

Each × represents one deer of that age. For instance, there are 18 1-year-olds and 14 2-year-olds.

Age	1	2	3	4	5	6	7	8	9	10	11	12
Count	18	14	10	7	6	4	3	3	2	1	2	2

■

Goal 2 Using Data in Decision Making

Real Life
Recycling

Example 2 *Interpreting a Survey*

You are trying to determine the "environmental awareness" of a group of people. You ask each person in the group to complete a survey. One of the questions asks about the importance of recycling aluminum cans. The results are shown below. What would you conclude from the survey?

1. Very important. I always recycle aluminum cans.
2. Quite important. I usually recycle aluminum cans.
3. Somewhat important. I sometimes recycle aluminum cans.
4. Not very important. I recycle aluminum cans if it is convenient.
5. Not at all important. I never recycle aluminum cans.

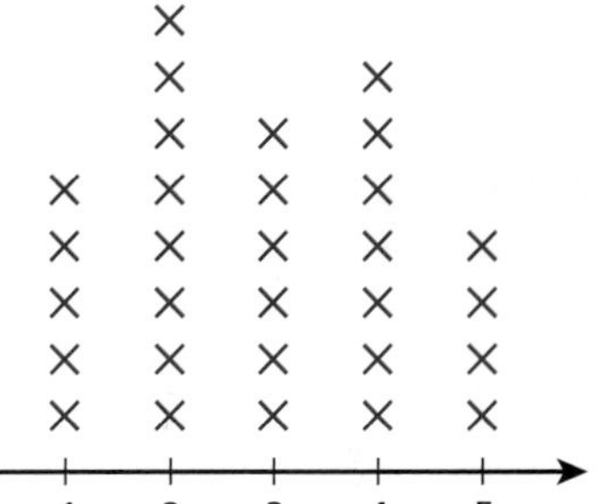

Each year in the United States, millions of tons of waste are buried in landfills.

Solution From the survey, it appears that many of the people in the group are not as aware of the importance of recycling as they should be. Four of the people said that they never recycle aluminum cans and another seven said that they only recycle if it is convenient (they might happen to be near a recycling container). ■

Communicating about MATHEMATICS

SHARING IDEAS about the Lesson

Connections
Art and Design

Poster Design You are designing a poster to increase the environmental awareness of a group of people. Which of the following do you think would be useful on the poster? How would you present the information? ***(Source: Environmental Protection Agency)***

- In 1993, over 200 million tons of nonindustrial waste was created in the United States.
- In 1993, the national average was 4.2 pounds of nonindustrial waste per person per day.
- In 1993, 3.6 billion pounds of aluminum cans were recycled in the United States.

EXERCISES

Guided Practice

CHECK for Understanding

1. *Writing* In your own words, describe the meaning of statistics.

In Exercises 2 and 3, refer to Example 1 on page 218.

2. How many 5-year-old deer are there?
3. What do you notice about the population of the deer as they increase in age?

Independent Practice

Organizing Data **In Exercises 4 and 5, decide whether the data could have been organized with a line plot. Explain your reasoning.**

4. The following data shows the heights (in inches) of 35 eighth graders.

Height (in inches)	Number of 8th graders
60	3
62	5
64	8
65	9
66	5
68	3
72	2

5. A survey was taken of the number of cars on the road of different ages. ***(Source: Motor Vehicle Manufacturers Association)***

Age of Car	Number on Road
Less than 1 year	7,812,000
1–5 years	47,569,500
5–10 years	42,687,000
10–15 years	27,760,500
15 years or older	13,671,000

6. *Homework* Sixty students in a class were asked to keep track of the number of hours each spent doing homework during a specific week. The results are shown below. Organize the data in a line plot.

3, 8, 1, 12, 6, 8, 3, 7, 11, 10, 8, 10, 15, 2, 1,
6, 14, 4, 15, 9, 10, 13, 14, 4, 11, 7, 6, 7, 10, 9,
5, 11, 10, 8, 15, 4, 12, 9, 5, 8, 11, 15, 10, 6, 5,
13, 5, 12, 10, 12, 5, 10, 11, 5, 10, 8, 12, 8, 12, 15

7. *Siblings* Thirty students in a class were asked the number of brothers and/or sisters each has. The results are shown below. Organize the data in a line plot.

4, 8, 2, 1, 3, 0, 2, 7, 0, 5, 4, 3, 1, 5, 3,
2, 2, 5, 3, 6, 3, 2, 4, 2, 6, 2, 3, 5, 3, 4

8. *Phone Numbers* You take a survey of the digits in the phone numbers of ten students in your class. You organize the data in a line plot, as shown at the right. Let × represent that digit occurring in a phone number.

a. Which digit occurs the most in the phone numbers? Which occurs the least?

b. All the phone numbers have the same exchange (first three digits). List all possible combinations of the exchange.

9. *Jogging* You conduct a survey among 30 joggers to find the average number of miles each jogs in a day. You organize the numbers in a line plot, as shown at the right. Let × represent a person who jogs that number of miles.

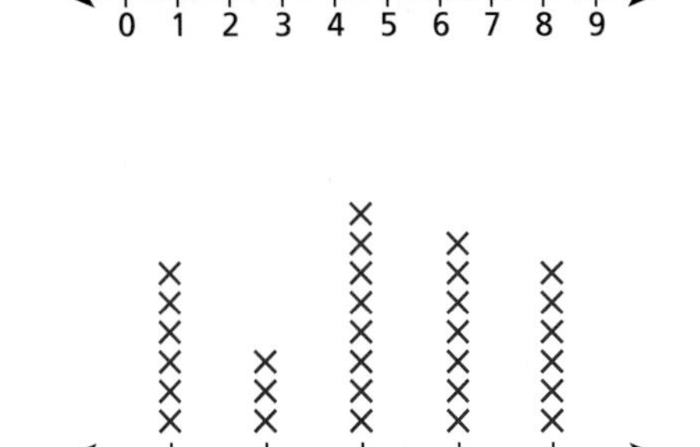

a. How many people jog 5 miles a day? What is the total number of miles jogged by all who jog 5 miles a day?

b. Find the average of the number of miles that all 30 people jogged.

10. *NFL Wins* The table at the right shows the number of wins of the teams in the 1993 NFL regular season. Organize the number of wins with a line plot. ***(Source: National Football League)***

Team Name	Wins	Team Name	Wins
Buffalo Bills	12	Washington Redskins	4
New York Jets	8	Dallas Cowboys	12
Miami Dolphins	9	Philadelphia Eagles	8
New England Patriots	5	New York Giants	11
Indianapolis Colts	4	Phoenix Cardinals	7
Houston Oilers	12	Detroit Lions	10
Pittsburgh Steelers	9	Chicago Bears	7
Cleveland Browns	7	Minnesota Vikings	9
Cincinnati Bengals	3	Green Bay Packers	9
Denver Broncos	9	Tampa Bay Buccaneers	5
Kansas City Chiefs	11	New Orleans Saints	8
Los Angeles Raiders	10	Atlanta Falcons	6
Seattle Seahawks	6	San Francisco 49ers	10
San Diego Chargers	8	Los Angeles Rams	5

11. *Geometry* The region at the right contains many different sizes of squares and rectangles.

a. Find the area of each figure in the region.

b. Organize the data in a line plot.

c. What is the total area of the region?

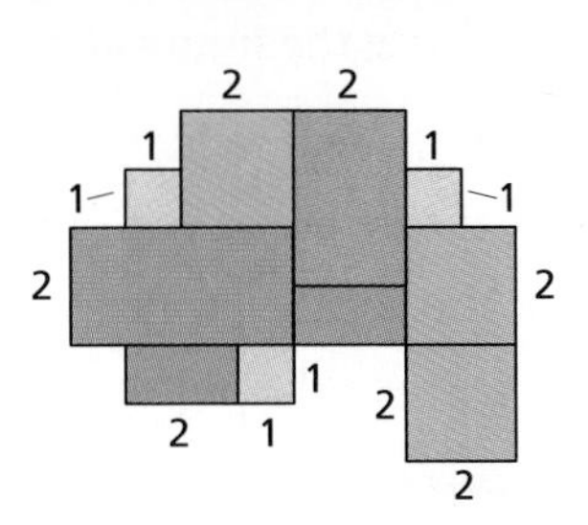

Integrated Review

Making Connections within Mathematics

Mental Math **In Exercises 12–17, solve the equation. Then check your solution.**

12. $7x + 3x + 12 = 32$

13. $16 = 3y - 8y - 9$

14. $-13 = \frac{p}{8} - 11$

15. $\frac{p}{12} + 14 = 17$

16. $5n - 20 = 3n$

17. $-9z = 18 - 3z$

Exploration and Extension

18. *Data Analysis* Conduct a survey of 20 people on a topic of your choice. Organize the data in a line plot. Write five questions that will require a classmate to interpret your line plot.

Mixed REVIEW

In Exercises 1–3, use the graphs, which compare auto sales in January, 1993 and January, 1994. *(Source: Auto Data)* **(5.5)**

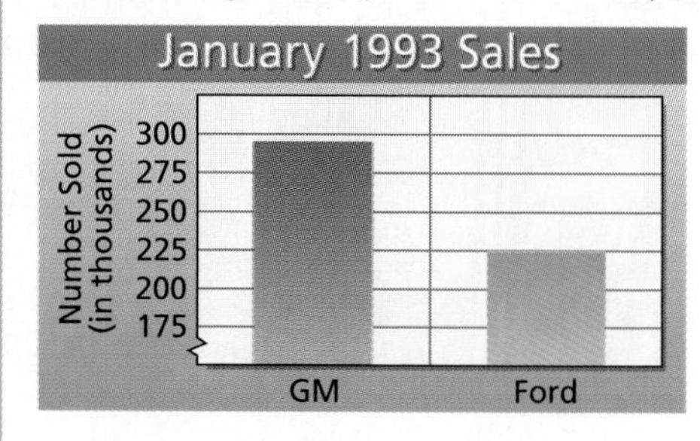

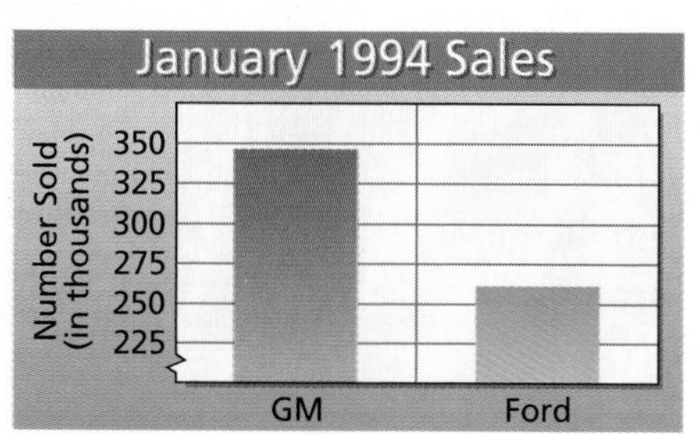

1. Both graphs are misleading, but which is more misleading? Why?

2. From the graph, it appears that Ford sold fewer cars in 1994 than in 1993. Why is this?

3. How could you represent this data with a single graph that was not misleading?

In Exercises 4–9, solve the equation. (4.1–4.7)

4. $5(n - 2) = 0$

5. $6r - 2 = 2r$

6. $2r + 6r = 4r - 28$

7. $2(p + 1) = -2(p + 1)$

8. $\frac{4}{5}s + 3.2 = -\frac{1}{5}s$

9. $\frac{1}{7}(x + 1) = 6$

In Exercises 10–13, use the figure at the right. (4.8)

10. What are the inside dimensions of the picture frame?

11. What is the inside perimeter of the picture frame?

12. What is the area of the picture?

13. What is the area of the picture frame?

2 in. 45 in. 2 in.
2 in.
30 in.
2 in.

USING A GRAPHING CALCULATOR

Make a Graph

There are many ways that technology can help you organize data. For instance, this bar graph was produced with a graphing calculator. The bar graph represents the data given in Example 1 on page 218.

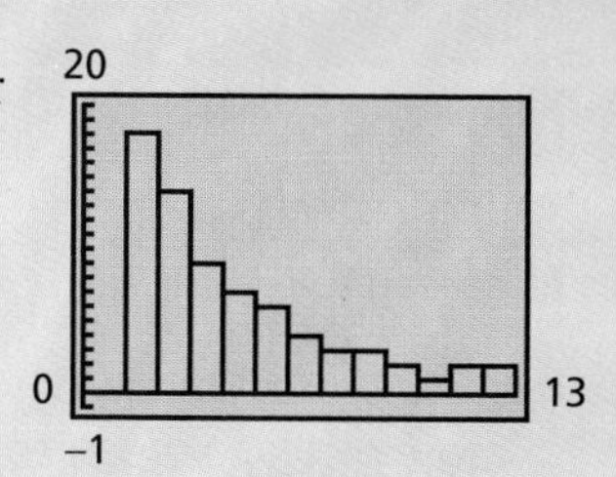

Age of Deer	1	2	3	4	5	6	7	8	9	10	11	12
Number of Deer	18	14	10	7	6	4	3	3	2	1	2	2

The following keystrokes show how to create the bar graph on several graphing calculators.*

TI-82

Window Xmin = 0 Ymin = −1
Xmax = 13 Ymax = 20
Xscl = 1 Yscl = 1

[STAT] [ENTER] (Edit)

L1(1) = 1 L2(1) = 18 L1(7) = 7 L2(7) = 3
L1(2) = 2 L2(2) = 14 L1(8) = 8 L2(8) = 3
L1(3) = 3 L2(3) = 10 L1(9) = 9 L2(9) = 2
L1(4) = 4 L2(4) = 7 L1(10) = 10 L2(10) = 1
L1(5) = 5 L2(5) = 6 L1(11) = 11 L2(11) = 2
L1(6) = 6 L2(6) = 4 L1(12) = 12 L2(12) = 2

[2nd] [STAT PLOT] [ENTER] (Plot1)

Choose the following:
On, Type: (bar icon), Xlist: L1, Freq: L2

[GRAPH]

Casio *fx-7700GE, fx-9700GE*

Range Xmin = 0 Ymin = −1
Xmax = 13 Ymax = 20
Xscl = 1 Yscl = 1

[MENU] 3 (SD)
[SHIFT] [SET UP]

Choose: GRAPH TYPE :RECT
DRAW TYPE :PLOT
STAT DATA :STO
STAT GRAPH :DRAW
M-DISP/COPY:M-DISP

[EXIT] [SHIFT] [Defm] 12 [EXE]

1 [F3] 18 [F1] 5 [F3] 6 [F1] 9 [F3] 2 [F1]
2 [F3] 14 [F1] 6 [F3] 4 [F1] 10 [F3] 1 [F1]
3 [F3] 10 [F1] 7 [F3] 3 [F1] 11 [F3] 2 [F1]
4 [F3] 7 [F1] 8 [F3] 3 [F1] 12 [F3] 2 [F1]

[GRAPH] [EXE]

Exercises

In Exercises 1 and 2, use a graphing calculator or a computer program to draw a bar graph of the data in the table.

1.

Age of Deer	1	2	3	4	5	6	7	8	9	10	11	12
Number of Deer	15	12	8	9	5	3	2	0	1	1	0	1

2.

Age of Deer	1	2	3	4	5	6	7	8	9	10	11	12
Number of Deer	24	22	18	15	12	8	9	5	4	6	2	1

* Keystrokes for other calculators are listed in *Technology—Keystrokes for Other Graphing Calculators* found at the end of this text.

5.7 Statistics: Scatter Plots

What you should learn:

Goal 1 How to use scatter plots to organize data

Goal 2 How to use scatter plots to help make decisions

Why you should learn it:

Scatter plots can help you understand how two real-life quantities are related.

Goal 1 Using Scatter Plots

A **scatter plot** is the graph of a collection of ordered pairs of numbers (x, y). If the y-coordinates tend to increase as the x-coordinates increase, then x and y have a **positive correlation.** If the y-coordinates tend to decrease as the x-coordinates increase, then x and y have a **negative correlation.** If no pattern exists between the coordinates, then x and y have **no correlation.**

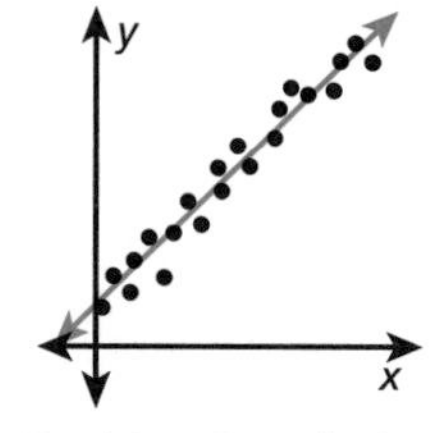

Positive Correlation

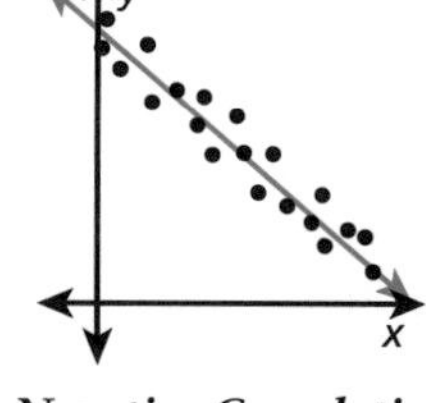

Negative Correlation

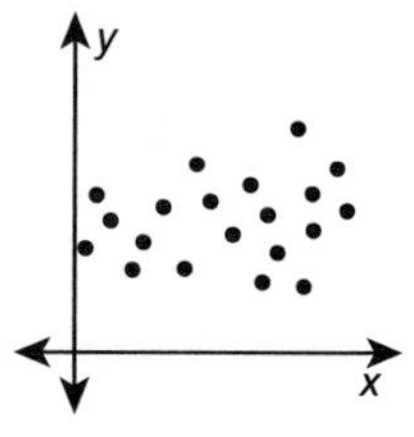

No Correlation

Connections **Biology**

Example 1 *Drawing a Scatter Plot*

The ordered pairs below show the wrist measurements and elbow-to-fingertip measurements for 21 students. Draw a scatter plot of this data. What can you conclude?

(12, 32), (16, 42), (15, 40), (15, 39), (15, 37), (16, 41), (17, 43), (14, 38), (15, 39), (16, 42), (17, 44), (14, 34), (15, 41), (13, 34), (13, 35), (14, 37), (16, 40), (17, 43), (13, 31), (15, 39), (14, 38)

Solution From the scatter plot, you can see that the two measurements have a positive correlation. So, people with larger wrists tend to have longer arms and vice versa.

Need to Know

The dashed line approximates the pattern of the data points. There should be as many points above the line as below the line.

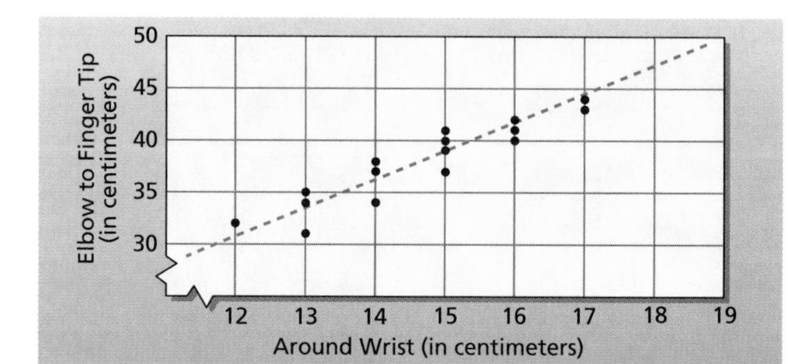

Goal 2 Decision Making with Scatter Plots

Real Life
Farming

In 1990, South Dakota produced about 223,000,000 bushels of grain corn, mostly in the area around Sioux Falls.

Example 2 *Interpreting a Scatter Plot*

The scatter plot below shows several temperatures per week (in degrees Celsius) taken from January through May in Sioux Falls, South Dakota. Each temperature represents the low temperature for that day. You live near Sioux Falls and are planning a garden. You should plant your corn when you are fairly certain that the low temperature will remain *above* freezing. From last year's temperatures, when do you think it would be safe to plant corn in your garden? ***(Source: PC USA)***

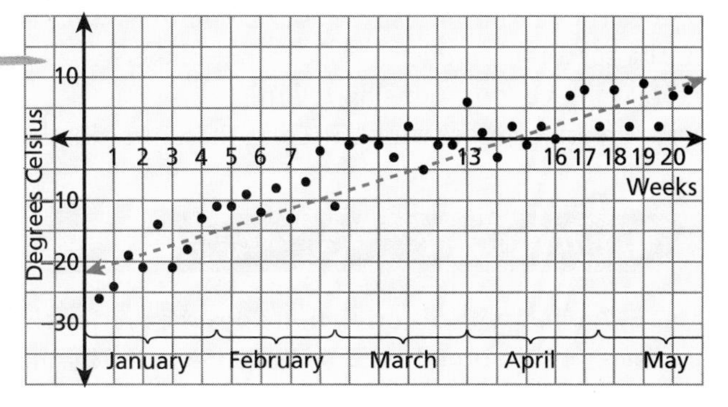

Solution Because freezing is 0° on the Celsius scale, it seems that it would be good to wait until the middle of April to plant corn in your garden. ■

Communicating about MATHEMATICS

Cooperative Learning

SHARING IDEAS about the Lesson

Determining Correlations Discuss with your partner the correlation of each of the following pairs. Do you think they have a positive correlation, a negative correlation, or no correlation? In each case, explain how you could collect and organize data that would test your hypothesis.

A. A person's height and shoe size

B. A baseball player's batting average and shoe size

C. The number of games a baseball team won during a season and the team's ranking during the season

D. A person's salary and the number of years that he or she went to school

EXERCISES

Guided Practice

CHECK for Understanding

Interpreting a Scatter Plot **In Exercises 1–4, use the scatter plot at the right. The scatter plot compares the number of hits x of 30 softball players during the first half of the season with the number of runs y batted in.**

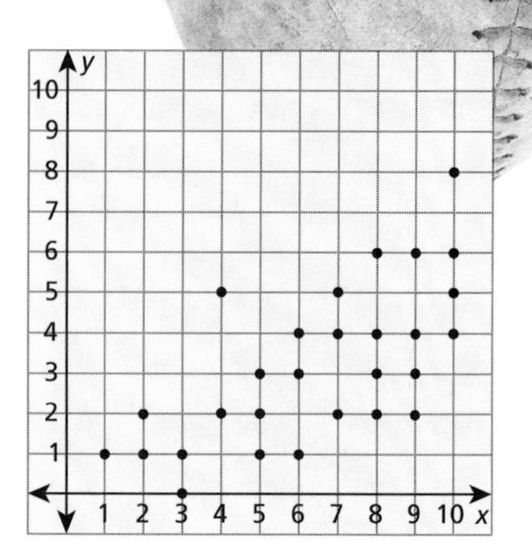

1. Do x and y have a positive correlation, a negative correlation, or no correlation?
2. Why does the scatter plot show only 28 points?
3. From the scatter plot, does it appear that players with more hits tend to have more runs batted in?
4. Can a player have more runs batted in than hits? Explain.

Independent Practice

In Exercises 5–7, what type of correlation does the scatter plot have? Describe a real-life situation that could be represented by the scatter plot.

5.

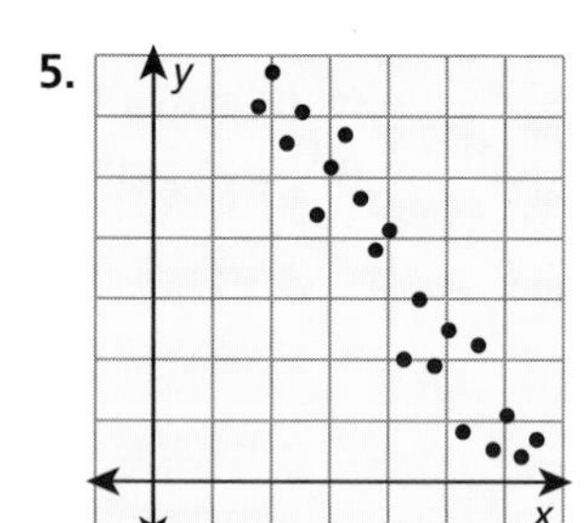

6.

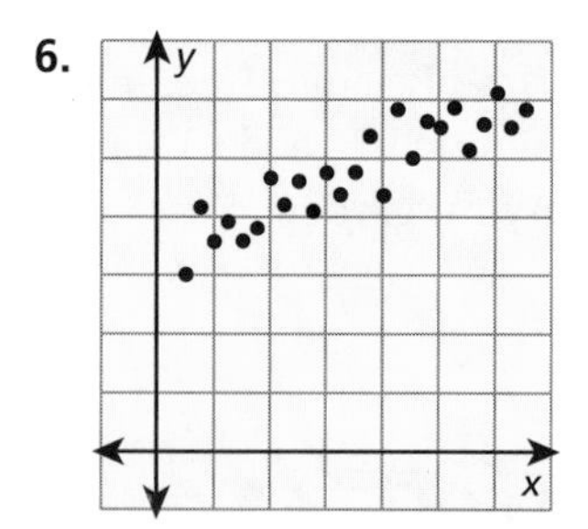

7.

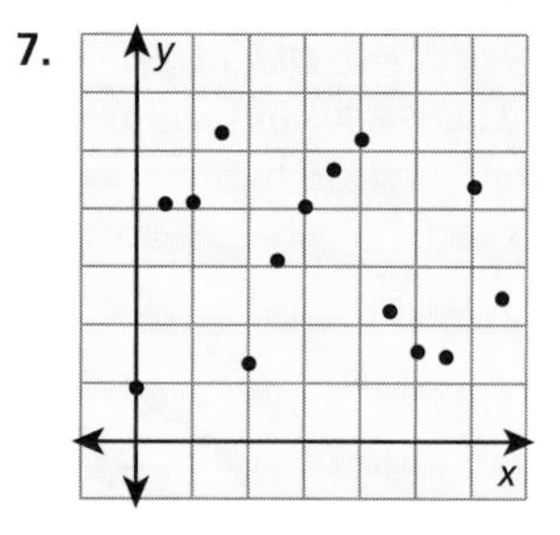

Think about It **In Exercises 8–11, decide whether a scatter plot relating the two quantities would tend to have a positive, negative, or no correlation. Explain.**

8. The age and value of a car
9. A student's study time and test scores
10. The height and age of a pine tree
11. A student's height and test scores

Making a Scatter Plot **In Exercises 12–15, use the data in the table. The table compares the altitude A (in thousands of feet) with the air pressure P (in pounds per square feet).**

Altitude	0	5	10	15	20	25	30	35	40	45	50
Pressure	14.7	12.3	10.2	8.4	6.8	5.4	4.5	3.5	2.8	2.1	1.8

12. Construct a scatter plot of the data.
13. How are A and P related?
14. Estimate the air pressure at 42,500 feet.
15. Estimate the altitude at which the air pressure is 5.0 pounds per square foot.

Interpreting a Scatter Plot **In Exercises 16–19, use the scatter plot at the right. The scatter plot shows the number of subscribers to basic cable TV from 1983 through 1992.** ***(Source: A. C. Nielsen)***

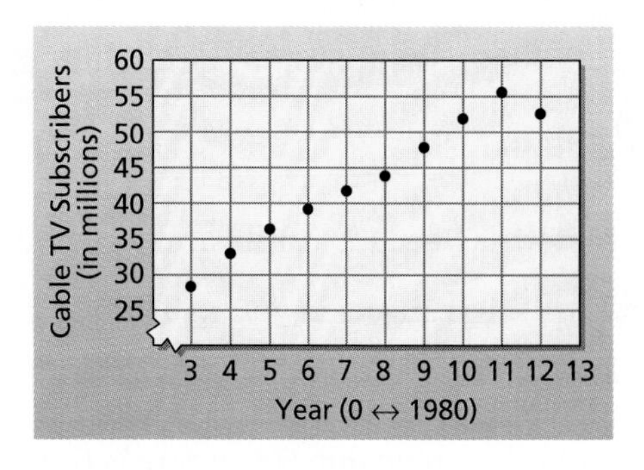

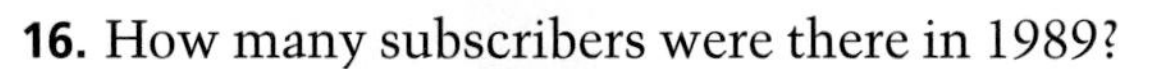

16. How many subscribers were there in 1989?

17. How are the years and subscribers related? Explain.

18. Copy the scatter plot on graph paper and draw a line that appears to best fit the points.

19. Use the results of Exercise 18 to estimate the number of subscribers in the year 2000.

Geometry **In Exercises 20 and 21, use the rectangle at the right.**

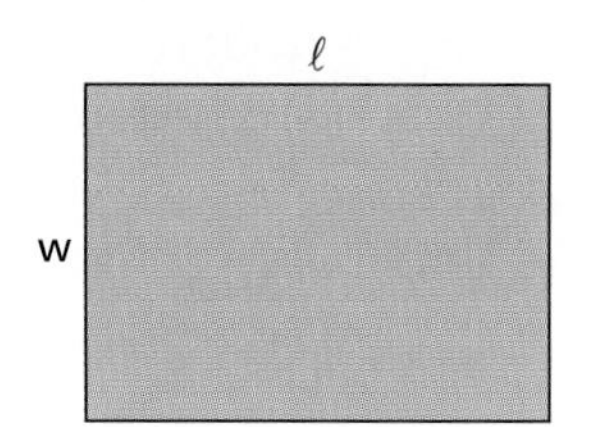

20. Complete the table.

Length, ℓ	1	2	3	4	5
Width, w	?	?	?	?	?
Perimeter	12	12	12	12	12

21. Construct a scatter plot using the lengths ℓ and widths w and discuss how they relate to each other.

Integrated Review

Making Connections within Mathematics

Coordinate Geometry **In Exercises 22–25, use the coordinate plane at the right.**

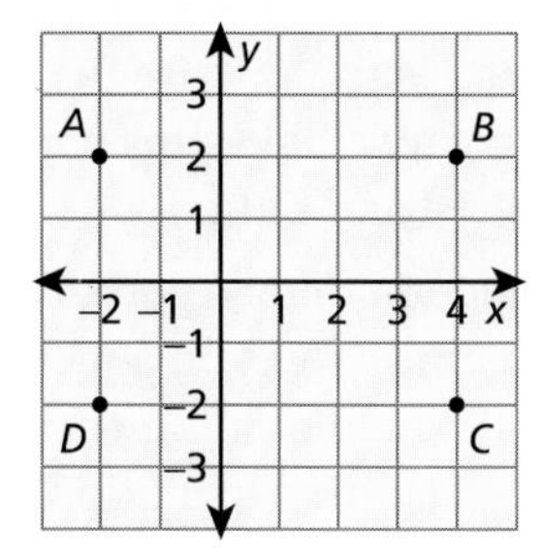

22. Which pairs of points have the same x-coordinate?

23. Which pairs of points have the same y-coordinate?

24. State the quadrant in which each point lies.

25. Find the perimeter and area of rectangle $ABCD$.

Exploration and Extension

In Exercises 26 and 27, draw a scatter plot and describe the result. A sample is shown at the right.

Technology To draw a scatter plot using a graphing calculator, enter the data as shown on page 223. Then, follow your calculator's procedure for drawing a scatter plot.

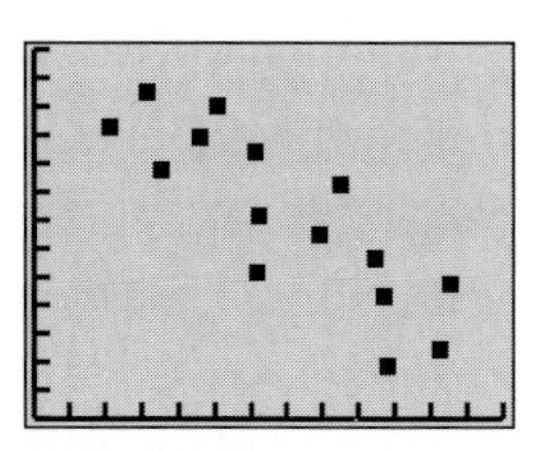

26. (1, 12), (2, 11), (3, 9), (3, 10), (4, 8), (5, 8), (6, 7), (7, 5), (8, 6), (9, 5), (10, 6)

27. (1, 3), (2, 3), (2, 4), (3, 6), (4, 5), (5, 5), (6, 7), (7, 7), (8, 9), (9, 11), (10, 10)

LESSON INVESTIGATION 5.8

Probability

Materials Needed: cardboard, paper clip, pencil

In this investigation you will investigate the probability of an event.

Description of the Problem Stevenson High School is playing Liberty High School for the state championship. The score is 65 to 66 in favor of Liberty. With one second left, a Liberty player fouls Laurie, a player from Stevenson. Laurie makes 6 out of every 10 free throws. Which of the following is more likely to happen?

a. Stevenson Wins Laurie makes two shots.

b. Stevenson Loses Laurie misses her first shot. (In this case, she doesn't shoot again.)

c. Overtime Play Laurie makes her first shot but misses her second shot. (In this case, the winner is determined in overtime play.)

There are three approaches to this problem. Try each approach in class. Compare the results of the three approaches.

1. **The Intuitive Approach** What outcome do you *think* is more likely? Each person in the class should write his or her guess on a slip of paper. The slips of paper should be collected and tallied.

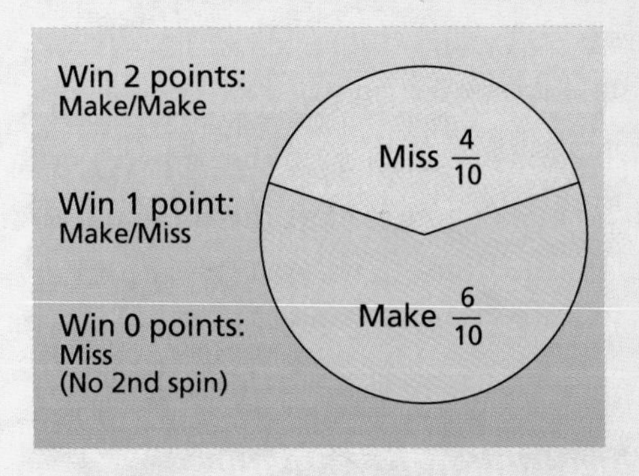

2. **The Experimental Approach** Copy the circle at the right on a piece of cardboard. Hold a pencil point at the center of the circle, and spin a paper clip that encircles the pencil point. Record the outcomes for several "shots." Which outcome occurred more often?

3. **The Theoretical Approach** The area model at the right can be used to determine the following probabilities.

   The probability of Stevenson winning in regulation time is [?].

   The probability of Stevenson losing in regulation time is [?].

   The probability of overtime play is [?].

   Which outcome has the greatest probability?

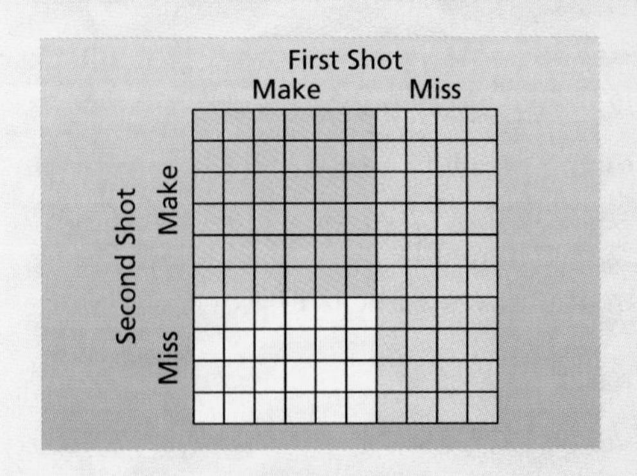

5.8 Exploring Probability

What you should learn:

Goal 1 How to compute the probability of an event

Goal 2 How to use concepts of probability to solve real-life problems

Why you should learn it:

Many events in real life are not certain. Probability can help you determine the likelihood that such events will actually occur.

Goal 1 Finding the Probability of an Event

The **probability of an event** is a measure of the likelihood that the event will occur. Probability is measured on a scale from 0 to 1.

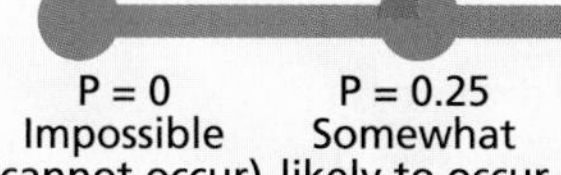

P = 0	P = 0.25	P = 0.5	P = 0.75	P = 1
Impossible (cannot occur)	Somewhat likely to occur	Equally likely to occur or not occur	Quite likely to occur	Certain to occur

To find the probability that an event will be "favorable," divide the number of ways that it can occur favorably by the total number of ways that the event can occur.

Probability of an Event

Let S be a set that has equally likely outcomes. IF E is a subset of S, then the probability that E will occur is

$$\text{Probability of } E = \frac{\text{Number of outcomes in } E}{\text{Number of outcomes in } S}.$$

Example 1 *Finding Probabilities*

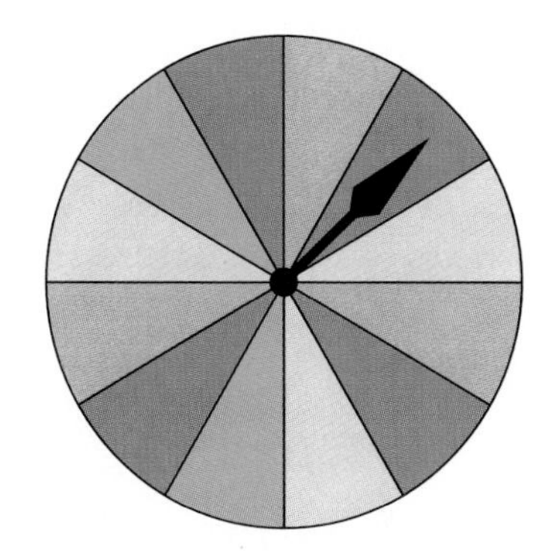

***The spinner can land on 12 different regions. This set of 12 possible outcomes is called the* sample space.**

a. The probability that the spinner at the left will land on red is

$$P = \frac{\text{Number of red regions}}{\text{Total number of regions}} = \frac{3}{12} = 0.25.$$

This means that if you spin the spinner 100 times, it should land on a red region about 25 times.

b. You have 15 pennies in your pocket. Two are Canadian and the rest are U.S. pennies. If you randomly choose one penny, the probability that it will be a U.S. penny is

$$P = \frac{\text{Number of U.S. pennies}}{\text{Total number of pennies}} = \frac{13}{15} \approx 0.87.$$

Goal 2 Using Probability in Real Life

To find characteristics of large groups, such as the population of the United States, pollsters sample a small group within the large group. If the small group is selected randomly, then its characteristics can be used to model those of the large group.

Real Life
Water Conservation

Example 2 *Conducting a Poll*

You are taking a poll to find how long adults take showers. You ask 600 adults and obtain the following data.

Length of Shower	Number Taking
1 minute or less	1
Between 1 and 5 minutes	111
Between 5 and 10 minutes	360
Between 10 and 15 minutes	109
Between 15 and 20 minutes	16
20 minutes or more	3

A typical shower uses 5 gallons of water per minute. If people limited their showers to 5 minutes, millions of gallons of water could be saved each day.

If you ask another adult how long he or she takes a shower, what is the probability that he or she will answer "between 5 and 10 minutes"? ***(Source: John O. Butler Company)***

Solution Of the 600 people sampled, 360 answered "between 5 and 10 minutes." Assuming that the sample is representative of the entire population, you can reason that the probability is

$$P = \frac{\text{Number of people answering "5 to 10 minutes"}}{\text{Number of people in survey}}$$

$$= \frac{360}{600} = 0.6.$$

The probability that a person will answer "between 5 and 10 minutes" is 0.6. ■

Communicating about MATHEMATICS

SHARING IDEAS about the Lesson

Extending the Example Use the data in Example 2 to compute the probabilities that an adult chosen at random will take showers that last for the other indicated times.

EXERCISES

Guided Practice

CHECK for Understanding

1. How do you find the probability of an event?
2. Describe what a "probability of 0.5" means.
3. If the probability of rain is 0.8, is it likely to rain? Explain.
4. *School Day* Think about events that may occur during a school day.
   a. Name three events whose probability of occurring is 1.
   b. Name three events whose probability of occurring is 0.
   c. Name three events whose probability of occurring is about 0.8.

Independent Practice

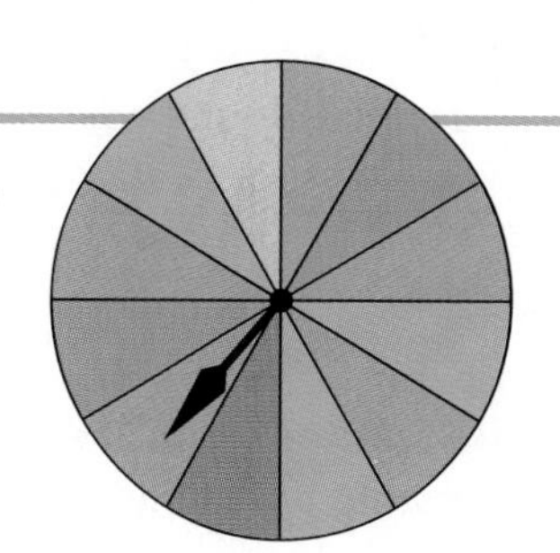

Spinning a Spinner **In Exercises 5–8, find the probability of the spinner landing on the color.**

5. Green 6. Red 7. Purple 8. Blue

Choosing Letters **In Exercises 9–12, consider the following.**

Write each letter in the word **ALABAMA** on a separate scrap of paper and put them in a bag. Without looking, choose one.

9. What is the probability of choosing an A?
10. What is the probability of choosing an M?
11. Choose a scrap of paper from the bag, record the letter, and replace the scrap. Do this 50 times. Which letter did you choose most often?
12. Divide the number of times you choose "A" by 50. This number is the experimental probability. Compare this to the probability obtained in Exercise 9.

Tossing a Coin **In Exercises 13–16, consider the probability of tossing a coin.**

13. What is the probability of tossing a head in one coin toss?
14. Toss a coin 30 times and record your results.
15. Divide the number of times you tossed a head by 30. This is the experimental probability of tossing a head.
16. Compare the experimental probability with the probability obtained in Exercise 13.
17. *Coloring a Spinner* Copy the spinner shown at the right. Then use the following statements to color the spinner.
    - Probability of red is 0.1
    - Probability of blue is 0.4
    - Probability of yellow is 0.3
    - Probability of green is 0.2

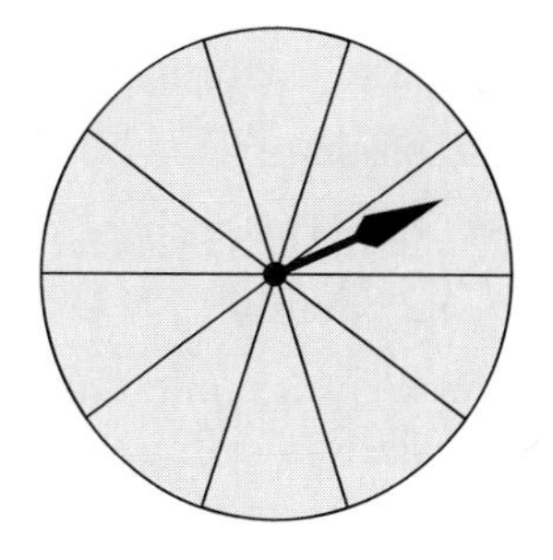

18. *Reasoning* Is there only one way to color the spinner in Exercise 17? If not, show how the spinner could be colored a different way.

Taking a Poll **In Exercises 19–21, you are taking a poll to find the blood types of 200 people. You obtain the results shown in the table at the right.** ***(Source: American Association of Blood Banks)***

Blood Type	Number
O^+	74
A^+	71
B^+	17
O^-	14
A^-	12
AB^+	6
B^-	4
AB^-	2

19. From the results of your survey, what is the probability that a randomly chosen person has the following blood type?

a. O^+ **b.** B^- **c.** A^+ or A^-

20. What is the probability that a person has a positive blood type?

21. Using only the result from Exercise 20, what is the probability that a person has a negative blood type? Explain your reasoning.

Integrated Review

Making Connections within Mathematics

Decimal Sense **In Exercises 22–29, use a calculator to write the fraction as a decimal. Round your result to 2 decimal places.**

22. $\frac{12}{15}$ **23.** $\frac{8}{15}$ **24.** $\frac{3}{7}$ **25.** $\frac{4}{7}$

26. $\frac{13}{24}$ **27.** $\frac{7}{24}$ **28.** $\frac{11}{20}$ **29.** $\frac{9}{20}$

More Spinners **In Exercises 30–32, all but one probability is given. Find the missing probability.**

30.

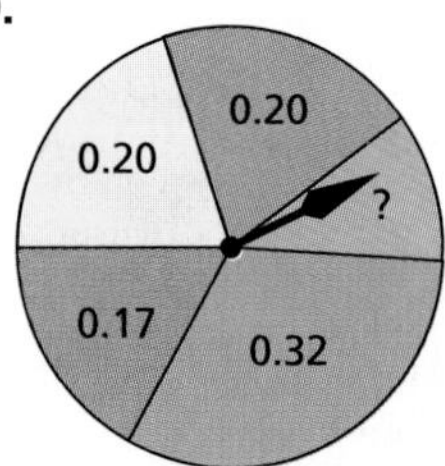

31.

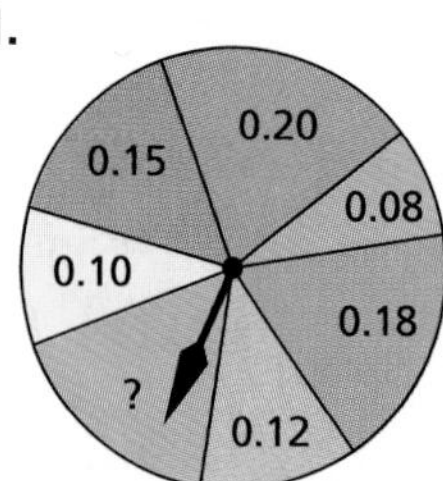

32.

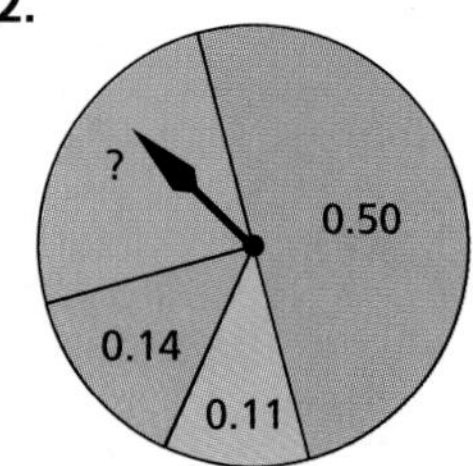

Exploration and Extension

Making a Conjecture **In Exercises 33–35, use the following information.**

A coin bank has 26 pennies, 15 nickels, 21 dimes, and 18 quarters. When you pick the bank up, a single coin falls out.

33. Find the probability that the coin is a

a. penny. **b.** nickel. **c.** dime. **d.** quarter.

34. Find the probability that the coin is *not* a

a. penny. **b.** nickel. **c.** dime. **d.** quarter.

35. Write a conjecture about the relationship between the probability that an event *will* occur and the probability that the event will *not* occur.

5

Chapter Summary

What did you learn?

Skills

1. Represent data using a
   - picture graph. **(5.1)**
   - time line. **(5.1)**
   - bar graph. **(5.2)**
   - line graph. **(5.3)**
   - scatter plot. **(5.7)**
2. Use a frequency distribution to organize data. **(5.2)**
3. Choose appropriate graphs to represent data. **(5.4)**
4. Recognize misleading bar graphs and line graphs. **(5.5)**
5. Use a line plot to organize data. **(5.6)**
6. Find the probability of an event. **(5.8)**

Problem-Solving Strategies

7. Model and solve real-life problems. **(5.1–5.8)**
8. Model and solve geometry problems. **(5.1–5.8)**

Exploring Data

9. Use tables and graphs to solve problems. **(5.1–5.8)**
10. Use probability concepts to solve problems.

Why did you learn it?

Almost every part of your life involves data. At school, on television, in magazines, and in newspapers, you are presented with data. Data is easier to understand when it is organized and presented visually with a graph. For instance, in this chapter you saw how a graph can be used to compare movie prices, trace the history of aircraft, compare people's goals and achievements, find patterns in geometry, analyze color preferences for clothing, compare the sales of hardback books and paperback books. From these examples, you can see that graphs are helpful in analyzing data in many walks of life.

How does it fit into the bigger picture of mathematics?

Organizing data is part of a branch of mathematics called *statistics.* In this chapter, you were introduced to some basic strategies for organizing, presenting, and interpreting data. One thing you learned is that there is seldom a "best" way to present data graphically. There are, however, some "bad" ways to present data graphically. For instance, you learned that breaks in the vertical scale can create false impressions with bar graphs and line graphs.

Throughout your study of mathematics, remember that a graph can help you recognize patterns. In fact, "drawing a graph" is an important problem-solving strategy in almost every branch of mathematics.

Chapter REVIEW

Sports Activities **In Exercises 1–3, use the picture graph at the right. The graph shows the number of young people ages 12–17 that participate in selected sports activities.** *(Source: National Sporting Goods Association)* **(5.1)**

Sports Activities

Bicycling
Camping
Volleyball
Bowling
Fishing (fresh water)

1 picture represents 1 million people

1. How many people does each picture represent?
2. Estimate the number of young people who participate in each activity.
3. About how many more young people camp than bowl?

Yogurt Consumption **In Exercises 4–6, use the graph at the right. The graph shows the average number of pounds of yogurt eaten in the United States in a year.** *(Source: The National Yogurt Association)* **(5.1, 5.2)**

4. How many pounds does each yogurt cup symbol represent?
5. From 1971 to 1991, how much did yogurt consumption increase?
6. Represent the results in a bar graph.

Aerobic Shoes **In Exercises 7–9, use the bar graph at the right. The graph shows the average number of aerobic shoe buyers (out of 100 buyers) who are different ages.** *(Source: National Sporting Goods Association)* **(5.2, 5.5)**

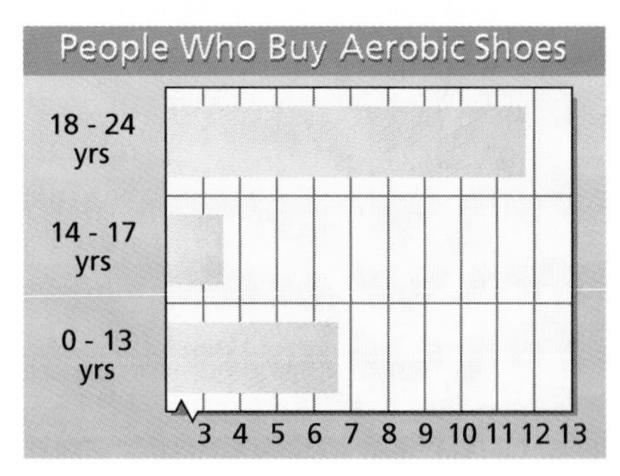

7. Without looking at the scale, compare the number of buyers who are under 14 to those who are in the 14 to 17 age group.
8. Why is this graph misleading?
9. Redraw the bar graph so that it is not misleading.

The History of Computers **In Exercises 10–13, use the time line.** *(Source: The 1993 Universal Almanac)* **(5.1)**

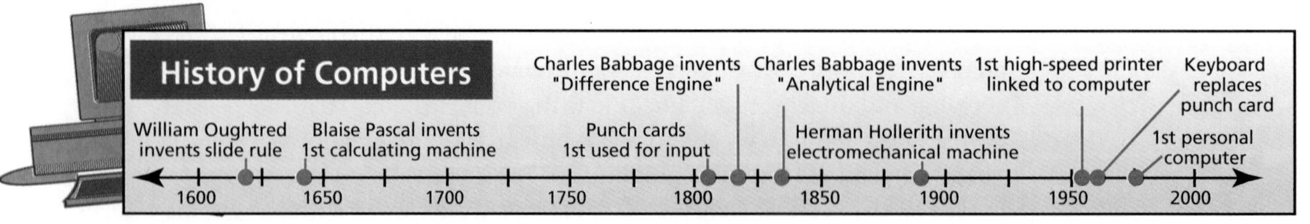

10. What is the time increment used in the time line?
11. Estimate the year that the first calculating machine was invented.
12. Estimate the year the Difference Engine was invented.
13. Name two computer developments of the 20th century.

In Exercises 14 and 15, organize the data and represent your results graphically. Explain why you used the type of graph you chose. (5.1–5.6)

14. *Mall Shoppers* A pollster in a mall is surveying opinions of shoppers who are from 13 to 20 years old. At the end of the day, the pollster had surveyed 50 people in this age group. The ages of the 50 people are listed below.

20, 18, 16, 19, 14, 17, 15, 20, 16, 13, 17, 15, 18,
16, 20, 15, 17, 17, 14, 15, 19, 18, 16, 18, 20, 20,
19, 15, 16, 17, 19, 16, 14, 16, 19, 16, 15, 17, 18,
16, 14, 19, 15, 18, 13, 16, 17, 15, 17, 18

15. *Test Grades* The following data shows the grades received by students on a math test.

84, 90, 63, 79, 83, 63, 84, 79, 60, 73, 76, 87,
99, 95, 75, 82, 85, 71, 95, 87, 76, 84, 100, 95,
90, 100, 87, 98, 93, 82, 94, 95, 81, 100, 96, 87

16. *Geometric Pattern* Each of the five triangles is an isosceles right triangle, which means that each has a right angle and two sides of the same length. Draw a scatter plot that relates the length of a shorter side with the length of the longer side. Describe the relationship between the two side lengths. **(5.7)**

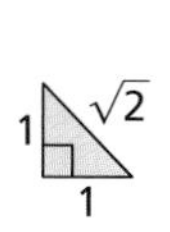

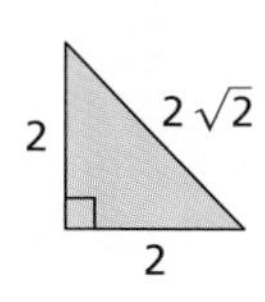

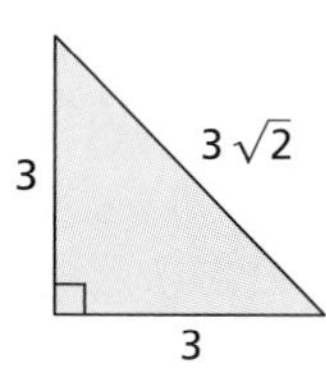

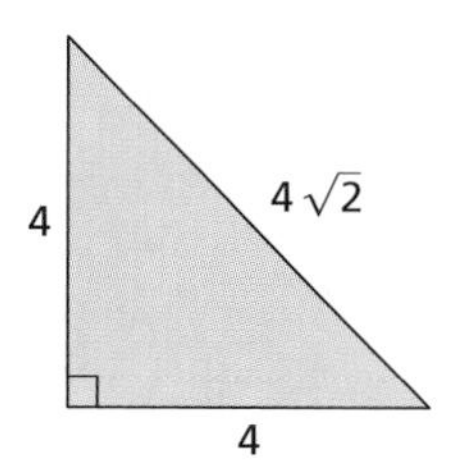

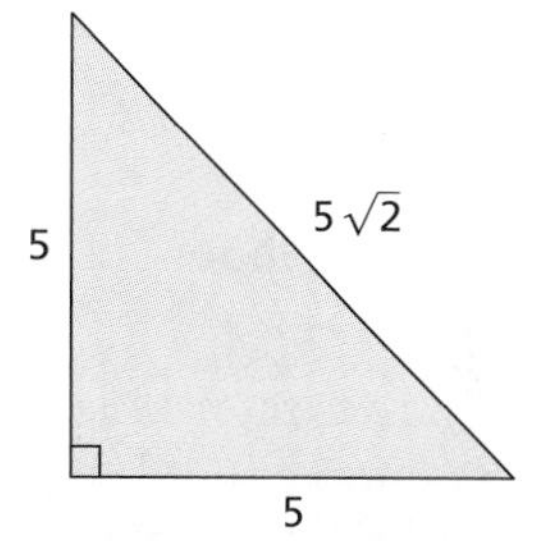

17. *U.S. Population* The ordered pairs below show the year and the United States population (in millions) for that year. Draw a scatter plot of this data. Describe the pattern. *(Source: U.S. Bureau of Census)* **(5.7)**

(1980, 227), (1981, 228), (1982, 231), (1983, 233),
(1984, 235), (1985, 237), (1986, 239), (1987, 241),
(1988, 243), (1989, 246), (1990, 248), (1991, 251)

18. *Probability* A bag contains 9 red marbles, 7 blue marbles, and 8 white marbles. Without looking at the color, you choose one marble from the bag. What is the probability that the marble is

a. red? **b.** blue? **c.** white?

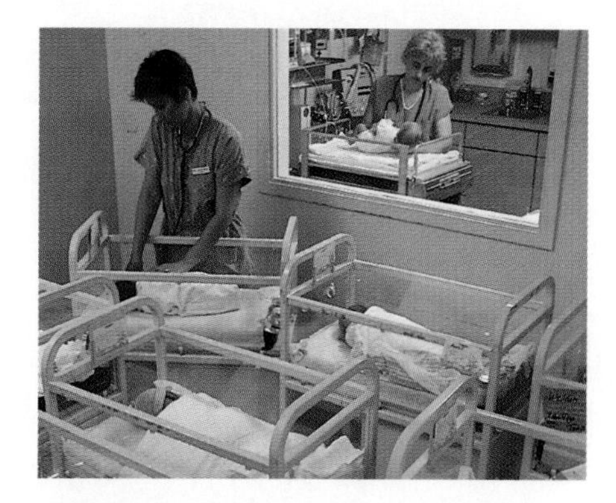

Since 1980, the population of the United States has been increasing due to the rise in the birth rate, increases in immigration, and the increased life expectancy of its citizens.

Analyzing a Company **In Exercises 19–22, use the table, which lists the gross income and profit (in millions of dollars) for *WMX Technologies, Inc.* for 1984 through 1993. In 1993, this company was the world's largest solid waste collection and disposal company.** *(Source: WMX Technologies, Inc.)*

Year	1984	1985	1986	1987	1988	1989	1990	1991	1992	1993
Gross Income	1,315	1,625	2,018	2,758	3,566	4,459	6,034	7,551	8,661	9,110
Profit	143	172	222	327	464	562	709	787	830	775

19. Create a double line graph for this data.

20. From the graph, which appears to be increasing more rapidly: gross income or profit? Explain.

21. In which year was the difference between gross income and profit the greatest?

22. When a company's gross income increases, does it necessarily follow that its profit increases? Explain.

What's in the Trash? **In Exercises 23–26, use the picture graph.** *(Source: U.S. Environmental Protection Agency)*

23. How much trash does each trash can symbol represent?

24. Estimate the amount of food in 100 pounds of trash.

25. Estimate the amount of paper products in 100 pounds of trash.

26. Based on the graph, where do you think the greatest effort should be placed on recycling?

Recycling of aluminum cans, plastic and glass containers, and newspapers can reduce the 100 pounds of trash by about 15 pounds.

Chapter TEST

In Exercises 1–3, use the graph at the right. *(Source: Leading National Advertisers, Inc.)*

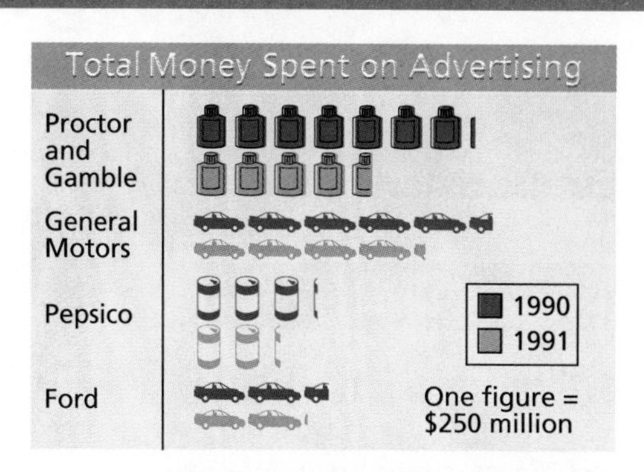

1. Name the type of graph.
2. Which company spent the most money on advertising in 1990?
3. Estimate the difference in the amount of money General Motors spent on advertising in 1990 and 1991.

In Exercises 4–6, use the graph at the right. *(Source: U.S. Bureau of Labor Statistics)*

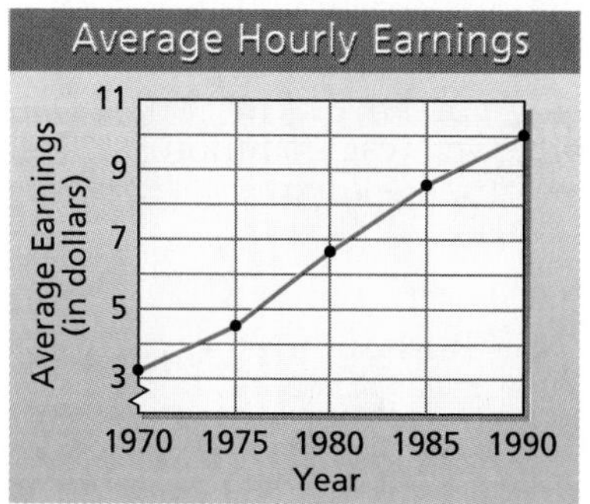

4. Name the type of graph.
5. *Without looking* at the scale, compare the 1970 average hourly wage to the 1990 average hourly wage.
6. Explain why this graph is misleading.
7. ***Sports Records*** The table lists the number of pitchers who were twenty-game winners in Major League Baseball by decade. Choose a graph that best represents the data. Then draw the graph.

Decade	1940's	1950's	1960's	1970's	1980's	1990's
Number of 20-game winners	55	60	73	96	37	45

Business Records **In Exercises 8 and 9, use the following information.**

The management of a company is recording the arrival times of its employees. They are keeping track of the number of minutes each employee is late for a specific day. The results are shown below.

0, 5, 7, 2, 1, 0, 0, 1, 0, 0, 2, 0, 7, 10,
0, 8, 4, 5, 0, 12, 0, 0, 3, 1, 2, 5, 6, 10

8. Organize the data in a line plot.
9. How many people arrived at work on time?

Scatter Plots **In Exercises 10 and 11, use the data at the right.**

(0, 1) (4, 6) (3, 3)
(2, 3) (7, 9) (5, 7)
(6, 6) (1, 2) (8, 11)

10. Create a scatter plot of the data (x, y) shown at the right.
11. Do x and y have a positive correlation, negative correlation, or no correlation?
12. ***Probability*** A box contains 12 Ping-Pong balls numbered from 1 through 12. One ball is chosen at random. What is the probability that the number on the ball is less than 4?

CHAPTER 6

Exploring Number Theory

LESSONS

Computers have changed the way music is produced and recorded. They ease and speed the process of both composition and communication with other artists. Using synthesizers, composers can now write, edit, and listen to their pieces all in one sitting, and without needing musicians.

Real Life
Making Music

The symbolic language of music is designed to tell the person reading it many things, including meter, tempo, melody, pitch, and key. The clef at the beginning of each staff tells the reader the name and pitch of each note, and the time signature indicates the music's meter.

Numbers can be used to describe musical notes, meter, and tempo. They can also be used to express a great many patterns in nature. These uses of numbers will be explored in this chapter.

6.1 Divisibility Tests

What you should learn:

Goal 1 How to use divisibility tests

Goal 2 How to factor natural numbers

Why you should learn it:

You can use divisibility tests to determine some possible dimensions of a rectangular region with a given area.

Goal 1 Using Divisibility Tests

If one natural number divides evenly into another natural number, then the second number is **divisible** by the first. For instance, 198 is divisible by 9 because $198 \div 9 = 22$, but 198 is not divisible by 4 because 4 does not divide evenly into 198.

Divisibility Tests

A natural number is divisible by

2 if the number is even.
3 if the sum of its digits is divisible by 3.
4 if the number formed by its last 2 digits is divisible by 4.
5 if its last digit is 0 or 5.
6 if the number is even and divisible by 3.
8 if the number formed by its last 3 digits is divisible by 8.
9 if the sum of its digits is divisible by 9.
10 if its last digit is 0.

Example 1 *Using a Divisibility Test*

Decide whether 534 is divisible by 2, 3, 4, 5, 6, 8, 9, and 10.

Solution

n	Is 534 divisible by n?	Reason
2	Yes	534 is even.
3	Yes	$5 + 3 + 4 = 12$, and 12 is divisible by 3.
4	No	34 is not divisible by 4.
5	No	The last digit of 534 is not 0 or 5.
6	Yes	534 is even and divisible by 3.
8	No	$534 \div 8 = 66.75$.
9	No	$5 + 3 + 4 = 12$, and 12 is not divisible by 9.
10	No	The last digit of 534 is not 0. ■

Study Tip
To decide whether a 3-digit number is divisible by 8, divide the number by 2, and test whether the result is divisible by 4. For instance, 534 is not divisible by 8 because $534 \div 2 = 267$, and 267 is not divisible by 4.

You can also use a calculator to decide whether one number is divisible by another. For instance, 3007 is divisible by 31 because $3007 \div 31 = 97$.

Goal 2 Factoring Natural Numbers

A natural number is **factored** when it is written as the product of two or more natural numbers. For instance, 28 can be factored as $4 \cdot 7$. The numbers 4 and 7 are **factors** of 28. Divisibility tests can be used to factor a number. For instance, because $39 \div 3 = 13$, you can conclude that $39 = 3 \cdot 13$.

Quotient: $39 \div 3 = 13$ (39 is the Dividend, 3 is the Divisor, 13 is the Quotient)

Product: $3 \cdot 13 = 39$ (3 is a Factor, 13 is a Factor, 39 is the Product)

Connections
Geometry

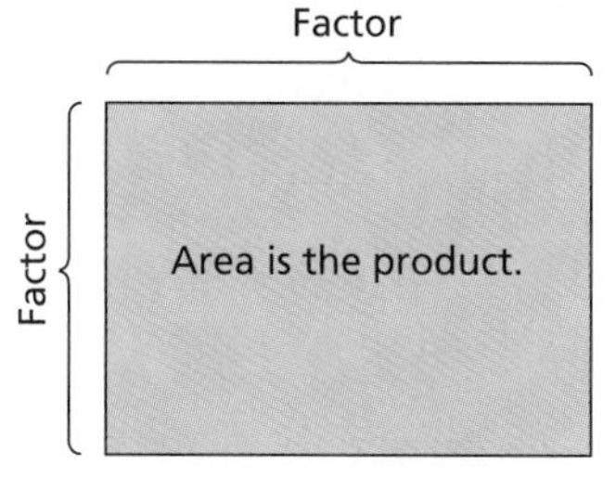

The area of a rectangle is the product of its length and width.

Example 2 *Finding Factors of a Number*

A rectangle has an area of 24 square units. The lengths of the sides are natural numbers. Name the possible dimensions. Find the factors of 24.

Solution You are hunting for pairs of numbers whose product is 24. That is, you are hunting for ways to factor 24 into the product of two numbers. From the rectangles below, you can see that the possible side lengths are 1 by 24, 2 by 12, 3 by 8, and 4 by 6. The factors of 24 are 1, 2, 3, 4, 6, 8, 12, and 24.

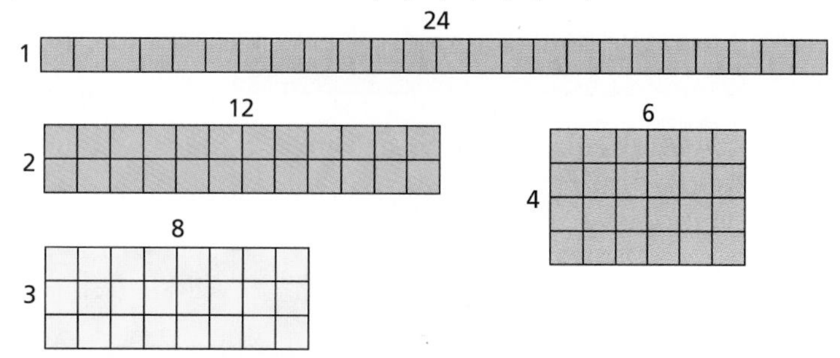

■

Real Life
Banking

Communicating about MATHEMATICS

Cooperative Learning

SHARING IDEAS about the Lesson

Cashing a Check You work as a bank teller. A person brings you a check for \$80. The person asks for bills that are all the same denomination. Explain to your partner how you can use divisibility tests to decide how many ways this can be done. Then explain how you can use factors to decide how many bills of each denomination you would give the person.

EXERCISES

Guided Practice

CHECK for Understanding

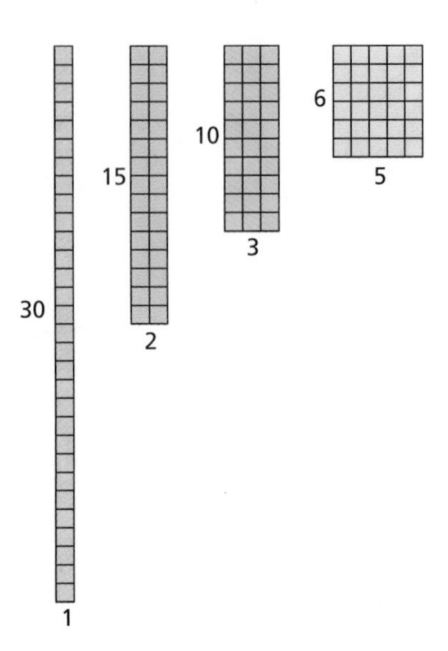

1. Explain what it means to say that one natural number is divisible by another natural number.
2. Decide whether 4485 is divisible by 2, 3, 4, 5, 6, 8, 9, and 10.
3. Name all the natural numbers that are factors of 36.
4. *Geometry* Each of the rectangles at the right has an area of 30 square units. Explain how to use these rectangles to find the factors of 30.

True or False? **In Exercises 5–8, decide whether the statement is true or false. Explain.**

5. A number divisible by 3 is also divisible by 9.
6. A number divisible by 10 is also divisible by 2.
7. The numbers 6 and 12 are factors of 72.
8. The number 18 is a factor of 9.

Independent Practice

In Exercises 9–16, use the divisibility tests to determine whether the number is divisible by 2, 3, 4, 5, 6, 8, 9, or 10.

9. 2160 **10.** 25,920 **11.** 192 **12.** 9756

13. 1234 **14.** 3725 **15.** 6859 **16.** 2401

17. *Reasoning* Which digits will make the number 34,?21 divisible by 3?

In Exercises 18–21, find the digit that makes the number divisible by 9.

18. 39,9?8 **19.** 5,43?,216,789 **20.** 12,?51 **21.** 2,546,?24

22. List all natural numbers less than 200 that are divisible by 3, 4, *and* 5.

23. *It's Up to You* Find the smallest natural number divisible by 2, 3, 4, 5, 6, 8, *and* 9. Explain how you found your answer.

24. *Logical Reasoning* If the numbers a and b are each divisible by 3, which of the following must also be divisible by 3? Explain your reasoning.

a. $a + b$ **b.** $a - b$ **c.** ab **d.** $\frac{a}{b}$

In Exercises 25–32, find all factors of the number.

25. 18 **26.** 36 **27.** 42 **28.** 45

29. 50 **30.** 100 **31.** 72 **32.** 96

In Exercises 33–36, write each number as the product of three factors. (Don't use 1 as a factor.)

33. 24 **34.** 36

35. 210 **36.** 144

37. *Geometry* The box at the right has a volume of 60 cubic units. The volume is the product of the box's length, width, and height. Use the box to write 60 as the product of three factors.

38. *Geometry* A rectangle has an area of 64 square units. The lengths of the sides are natural numbers. Find the possible dimensions.

Conservation Project **In Exercises 39–41, imagine you are part of a team that is planting 350 tree seedlings.**

39. Your team is instructed to plant the seedlings in straight rows with the same number of trees in each row. One team member wants to plant 15 rows. Is that possible? Explain why or why not.

40. Your team is instructed to plant the trees in a rectangular region. You can use any rectangle, so long as its area is 1600 square feet. If the side lengths (in feet) are natural numbers, what are the possible dimensions of the region?

41. Your team wants to construct a temporary fence around the seedlings. Of the dimensions found in Exercise 40, which have the smallest perimeter?

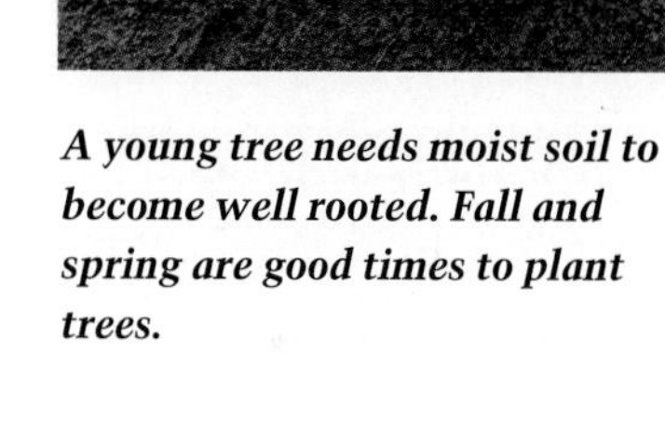

A young tree needs moist soil to become well rooted. Fall and spring are good times to plant trees.

42. Why can a number be divisible by 3 and not be divisible by 9?

43. Why is a number that is divisible by 8 also divisible by 4?

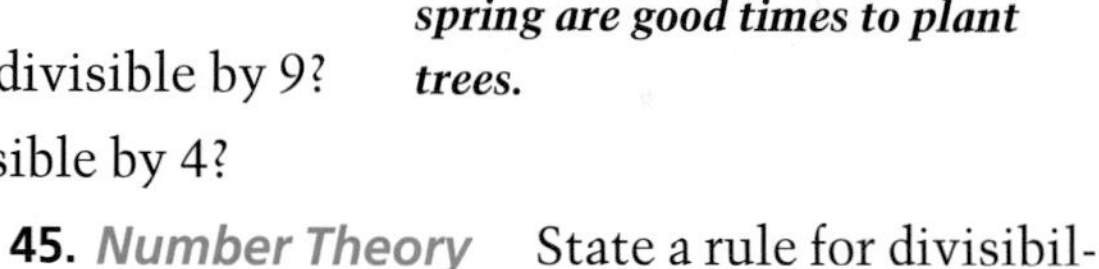

44. *Number Theory* State a rule for divisibility by 50.

45. *Number Theory* State a rule for divisibility by 20.

Integrated Review

Making Connections within Mathematics

Probability **In Exercises 46–49, one of the digits 2, 3, 5, 7, or 8 is randomly selected and used to form the indicated 3-digit number. Find the probability that the number is divisible by 4.**

46. 6?4 **47.** 87? **48.** ?32 **49.** ?94

Exploration and Extension

50. *A Division Pattern* Complete each statement. Then describe the pattern.

$56 \div 4 = \boxed{?}$, $156 \div 4 = \boxed{?}$, $256 \div 4 = \boxed{?}$, $356 \div 4 = \boxed{?}$

51. Using your pattern, predict $756 \div 4$. Check your answer.

52. *Writing* Write a paragraph explaining why the divisibility test for 4 works.

LESSON INVESTIGATION 6.2

An Area Model for Multiplication

Materials Needed: square tiles or graph paper

In this activity, you will investigate factors of natural numbers.

Example ***Finding Factors of a Number***

Find all possible factors of 16.

Solution As shown below, 16 tiles can be used to form 3 sizes of rectangles: a 1-by-16 rectangle, a 2-by-8 rectangle, and a 4-by-4 rectangle. This implies that 16 has five factors: 1, 2, 4, 8, and 16.

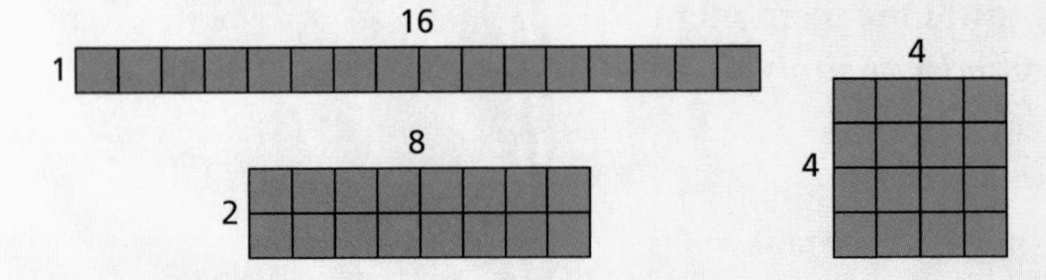

Exercises

Copy and complete the table.

Number, *n*	Number of Rectangles	Dimensions of Rectangles	Factors of *n*
1	?	?	?
2	?	?	?
3	?	?	?
4	?	?	?
5	?	?	?
6	?	?	?
7	?	?	?
8	?	?	?
9	?	?	?
10	?	?	?
11	?	?	?
12	?	?	?
13	?	?	?
14	?	?	?
15	?	?	?
16	3	1 by 16, 2 by 8, 4 by 4	1, 2, 4, 8, 16

6.2 Factors and Primes

What you should learn:

Goal 1 How to classify natural numbers as prime or composite

Goal 2 How to actor algebraic expressions and use factorizations to solve real-life problems.

Why you should learn it:

You can use the factors of a number to determine ways the number can be written as a product.

Goal 1 Classifying Primes and Composites

Natural numbers can be classified according to the number of factors they have.

Prime and Composite Numbers

1. A natural number is **prime** if it has exactly two factors, itself and 1. For instance, 2, 3, 5, and 7 are prime.
2. A natural number is **composite** if it has three or more factors. For instance, 4, 6, and 8 are composite.
3. The natural number 1 is neither prime nor composite.

LESSON INVESTIGATION

Investigating Prime Patterns

Group Activity Write the natural numbers from 1 through 96 as shown below. Circle all primes. Describe any pattern you observe.

1	7	13	19	25	31	37	43	49	55	61	67	73	79	85	91
2	8	14	20	26	32	38	44	50	56	62	68	74	80	86	92
3	9	15	21	27	33	39	45	51	57	63	69	75	81	87	93
4	10	16	22	28	34	40	46	52	58	64	70	76	82	88	94
5	11	17	23	29	35	41	47	53	59	65	71	77	83	89	95
6	12	18	24	30	36	42	48	54	60	66	72	78	84	90	96

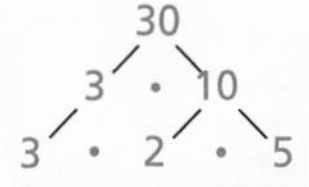

First factor 30 as 3 times 10. Then factor 10 as 2 times 5.

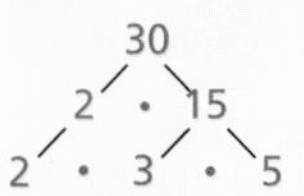

First factor 30 as 2 times 15. Then factor 15 as 3 times 5.

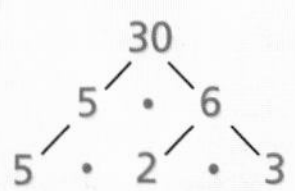

First factor 30 as 5 times 6. Then factor 6 as 2 times 3.

Example 1 *Prime Factorization*

The **prime factorization** of 24 is $2 \cdot 2 \cdot 2 \cdot 3$. Write the prime factorization of 30.

Solution Begin by factoring 30 as the product of two numbers other than 1 and itself. Continue factoring until all factors are prime. The **tree diagrams** at the left show three ways this can be done. In each case, you obtain the factors 2, 3, and 5. This implies that the prime factorization of 30 is

$30 = 2 \cdot 3 \cdot 5.$ *Prime factorization of 30.* ■

Goal 2 Factoring Algebraic Expressions

The factoring technique shown in Example 1 can be extended to negative integers and expressions involving variables. For instance, $-6ab^2$ can be written as $-6ab^2 = (-1) \cdot 2 \cdot 3 \cdot a \cdot b \cdot b$.

Example 2 *Factoring Algebraic Expressions*

	Expression	Expanded Form	Exponent Form
a.	-24	$(-1) \cdot 2 \cdot 2 \cdot 2 \cdot 3$	$(-1) \cdot 2^3 \cdot 3$
b.	$63a^3$	$3 \cdot 3 \cdot 7 \cdot a \cdot a \cdot a$	$3^2 \cdot 7 \cdot a^3$
c.	$18x^2y$	$2 \cdot 3 \cdot 3 \cdot x \cdot x \cdot y$	$2 \cdot 3^2 \cdot x^2 \cdot y$

■

Example 3 *Problem Solving: Consider All Causes*

$5 is to be divided evenly among the people in a group. How many people can be in the group?

Solution $5 is equal to 500 pennies. You must find all the factors of 500 since each factor is a possible group size.

Group Size	Amount Given to Each Person	Check
1	$5.00	1 • $5.00 = $5.00
2	$2.50	2 • $2.50 = $5.00
4	$1.25	4 • $1.25 = $5.00
5	$1.00	5 • $1.00 = $5.00
10	$0.50	10 • $0.50 = $5.00
20	$0.25	20 • $0.25 = $5.00
25	$0.20	25 • $0.20 = $5.00
50	$0.10	50 • $0.10 = $5.00
100	$0.05	100 • $0.05 = $5.00
125	$0.04	125 • $0.04 = $5.00
250	$0.02	250 • $0.02 = $5.00
500	$0.01	500 • $0.01 = $5.00

■

Japanese Money ***The Japanese yen is issued in denominations of 5, 10, 50, 100, and 500. Using only these denominations, how could you evenly divide 500 yen with each person in a group?***

Communicating about MATHEMATICS

SHARING IDEAS about the Lesson

Square Numbers Explain why every square number must have an *odd* number of factors. If a natural number has an odd number of factors, must it be square? Explain.

EXERCISES

Guided Practice

CHECK for Understanding

Number Sense **In Exercises 1–4, use the number sieve.**

2, 3, ~~4~~, 5, ~~6~~, 7, ~~8~~, 9, ~~10~~, 11, ~~12~~, 13, ~~14~~, 15, ~~16~~, 17, ~~18~~, 19, ~~20~~, 21, ~~22~~, 23, ~~24~~, 25, ~~26~~, 27, ~~28~~, 29, ~~30~~

2, 3, 4, 5, ~~6~~, 7, 8, ~~9~~, 10, 11, ~~12~~, 13, 14, ~~15~~, 16, 17, ~~18~~, 19, 20, ~~21~~, 22, 23, ~~24~~, 25, 26, ~~27~~, 28, 29, ~~30~~

2, 3, 4, 5, 6, 7, 8, 9, ~~10~~, 11, 12, 13, 14, ~~15~~, 16, 17, 18, 19, ~~20~~, 21, 22, 23, 24, ~~25~~, 26, 27, 28, 29, ~~30~~

1. What type of number is crossed out in the 1st row? The 2nd row? The 3rd row?
2. What type of number has not been crossed out in any row?
3. Extend each row to 40. Which of the numbers 31–40 would not be crossed out?
4. What natural number is neither prime nor composite?

Independent Practice

In Exercises 5–8, is the number prime or composite? Explain.

5. 17
6. 9
7. 35
8. 27

In Exercises 9–12, write the factorization represented by the tree diagram.

9. 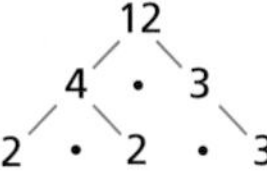

10. 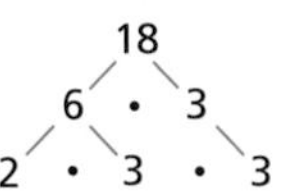

11. 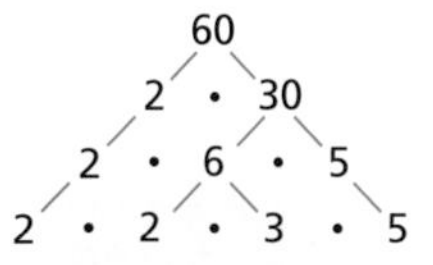

12. 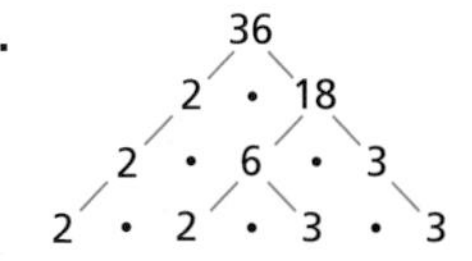

In Exercises 13–20, write the prime factorization of the number. Write your answer in exponent form.

13. 36
14. 63
15. 84
16. 100
17. 32
18. 64
19. 72
20. 90

In Exercises 21–28, write the expression in expanded form and exponent form.

21. -27
22. -28
23. $9x^3$
24. $125y^4$
25. $8a^3b^2$
26. $12p^4q$
27. $-45mn^3$
28. $-50s^2t^5$

In Exercises 29–32, evaluate the expression.

29. $2^3 \cdot 3 \cdot 5$
30. $3^2 \cdot 2 \cdot 13$
31. $-1 \cdot 3^2 \cdot 5 \cdot 13$
32. $-1 \cdot 2^3 \cdot 3 \cdot 7$

In Exercises 33–36, list all possible factors of the number.

33. 8
34. 16
35. 32
36. 64

37. *Number Sense* Consider the results of Exercises 33–36. If you double a number, does the list of all possible factors of the number double? Explain your reasoning.

38. *Geometry* If the lengths of the sides of a triangle are consecutive integers, could the perimeter be a prime number? Explain.

39. *Goldbach's Conjecture* A famous unproven conjecture by Christian Goldbach (1690–1764) states that every even natural number except 2 is the sum of two prime numbers. Write the even numbers from 20 to 40 as the sum of two primes.

40. *Twin Primes* Another famous unproven conjecture deals with primes whose difference is 2, such as 3 and 5, 5 and 7, and 11 and 13. Write the next five twin prime pairs.

41. *Dividing Money* $4.50 is divided evenly among the people in a group. How many people can be in the group?

42. *Dividing into Groups* A class is divided into more than two groups. If each group is the same size, could the number of students in the class be prime? Explain.

U.S. Flag History **In Exercises 43–46, design a rectangular pattern for the stars in the United States flag. Make one star for each state and make the rectangle's sides as close to the same lengths as possible.**

43. Flag of 1795: 15 states

44. Flag of 1818: 20 states

45. Flag of 1861: 34 states

46. Flag of 1994: 50 states

The flag from 1912 to 1959 served as the national flag longer than any other. The 13 stripes represent the 13 original states.

Integrated Review

Making Connections within Mathematics

Probability **In Exercises 47 and 48, consider the following. The natural numbers from 1 through 30 are written on slips of paper and placed in a box. One number is randomly selected from the box.**

47. What is the probability that the number is prime?

48. What is the probability that the number is composite?

Exploration and Extension

Crossnumber Puzzles **In Exercises *49* and *50*, copy and complete the puzzle using single digits. Each digit must be a factor of the number at the beginning of its row and column. You can use each digit only once.**

49.

╳	18	8	35
42			
60			
72			

50.

╳	36	30	56
35			
24			
18			

LESSON INVESTIGATION 6.3

Modeling Common Factors

Materials Needed: graph paper, pencil

In this activity, you will investigate common factors of numbers.

Example ***Finding Common Factors***

You are tiling a 12-by-16 floor with square tiles. The tiles cannot overlap, and you can't cut any of the tiles. Can the room be tiled using only 1-by-1 tiles? 2-by-2 tiles? 3-by-3 tiles? 4-by-4 tiles? 5-by-5 tiles? 6-by-6 tiles?

Solution The solutions are shown below. Notice that the 1-by-1, 2-by-2, and 4-by-4 tiles are the only ones that work. The reason is that 1, 2, and 4 are common factors of 12 and 16. The greatest common factor of 12 and 16 is 4.

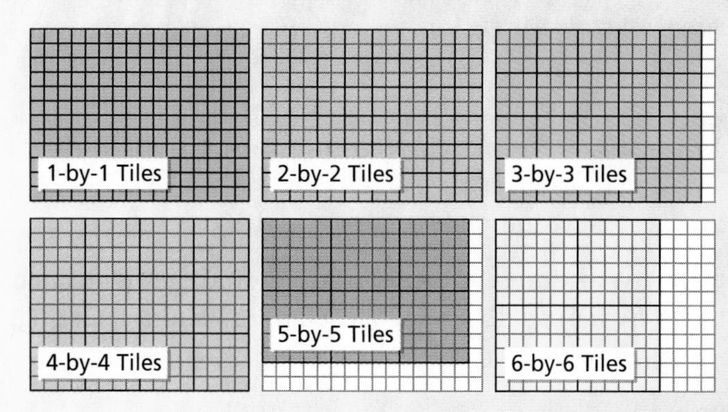

■

Exercises

In Exercises 1–4, decide which size tiles can be used to tile the room. Sketch your results. Which numbers are common factors of the width and length of the room? What is the greatest common factor?

1. 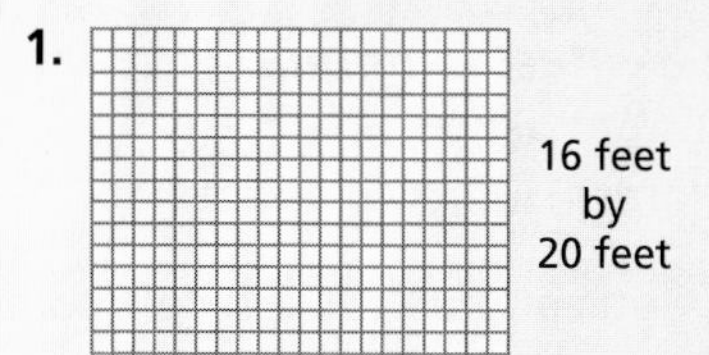

16 feet by 20 feet

2. 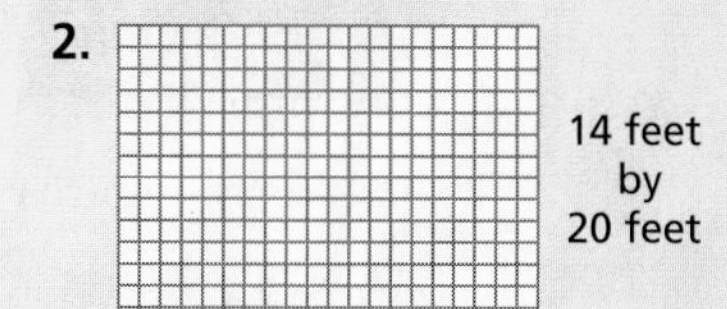

14 feet by 20 feet

3. 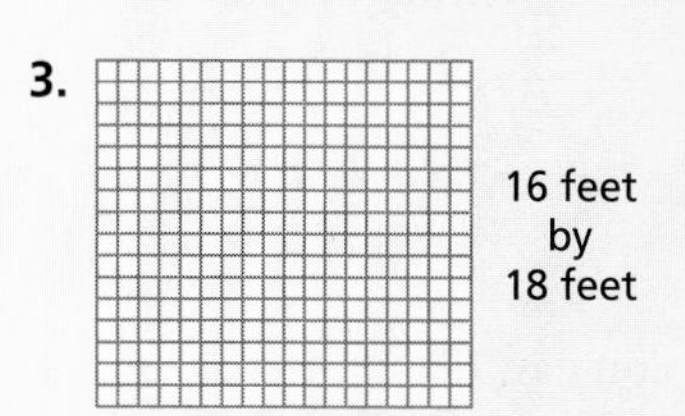

16 feet by 18 feet

4. 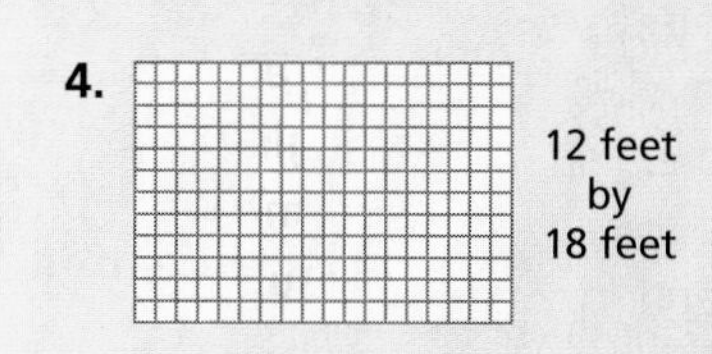

12 feet by 18 feet

6.3 Greatest Common Factor

What you should learn:

Goal 1 How to find the greatest common factor of two numbers or expressions

Goal 2 How to use the greatest common factor of numbers to solve real-life problems

Why you should learn it:

You can use the greatest common factor of two numbers to solve real-life problems, such as comparing frequencies of notes on a musical scale.

Goal 1 Finding Common Factors

In Lessons 6.1 and 6.2, you studied factors of numbers. In this lesson, you will study common factors of two numbers.

Common Factors and Greatest Common Factor

Let m and n be natural numbers.

1. A number that is a factor of both m and n is a **common factor** of m and n.
2. Of all common factors of m and n, the largest is called the **greatest common factor.**

Example 1 *Finding the Greatest Common Factor*

Find the greatest common factor of 16 and 20.

Solution With small numbers, you can find the greatest common factor by listing all factors of each number and selecting the largest common factor.

Number	16	20
Factors	1, 2, 4, 8, 16	1, 2, 4, 5, 10, 20

From the lists, you can see that 1, 2, and 4 are common factors of 16 and 20. Of these, 4 is the greatest common factor. ■

Example 2 *Finding the Greatest Common Factor*

Find the greatest common factor of 180 and 378.

Solution With larger numbers, you can find the greatest common factor by writing the prime factorization of each.

Number	180	378
Prime Factorization	$2 \cdot 2 \cdot 3 \cdot 3 \cdot 5$	$2 \cdot 3 \cdot 3 \cdot 3 \cdot 7$

From the prime factorizations, you can reason that the greatest common factor is $2 \cdot 3 \cdot 3$ or 18. ■

Greatest Common Factor of Algebraic Expressions

The concept of a greatest common factor also applies to algebraic expressions. For instance, the greatest common factor of $18x^2y$ and $12xy$ is $6xy$.

Goal 2 Using Greatest Common Factors

Example 3 *Using Greatest Common Factors*

Real Life
Music

Notes on a musical scale are characterized by their frequencies. The higher the frequency, the higher the note. On a piano, the A above middle C has a frequency of 440. What is the greatest common factor of the three A's shown on the following piano keyboard? What is the greatest common factor of the three G's?

Modern symphony orchestras have from 90 to 120 players. The violins, the highest pitched string instruments, usually carry the melody in orchestral music.

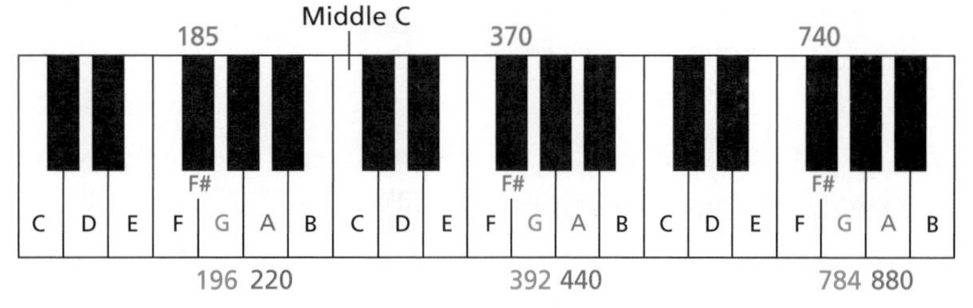

Solution The three A's have frequencies of 220, 440, and 880. Because 440 and 880 are each divisible by 220, it follows that 220 is the greatest common factor of the three numbers.

The three G's have frequencies of 196, 392, and 784. Because 392 and 784 are each divisible by 196, it follows that 196 is the greatest common factor of the three numbers. ■

In music, the notes A-220 and A-440 are an *octave* apart. The notes F#-185 and F#-370 are also an octave apart. (F# is read as F-sharp.) From Example 3, you can see that if two notes are an octave apart, then the higher note has twice the frequency of the lower note.

Communicating about MATHEMATICS

SHARING IDEAS about the Lesson

Relatively Prime Numbers Two natural numbers are **relatively prime** if their greatest common factor is 1. For instance, the numbers 8 and 21 are relatively prime. Find the greatest common factor of the following pairs. State whether the two numbers are relatively prime.

A. 135 and 224 **B.** 135 and 225 **C.** 134 and 224

Must two prime numbers be relatively prime? Explain.

EXERCISES

Guided Practice

CHECK for Understanding

In Exercises 1–4, find the common factors of the two numbers. What is the greatest common factor?

1. Factors of 12: 1, 2, 3, 4, 6, 12
Factors of 18: 1, 2, 3, 6, 9, 18

2. Factors of 24: 1, 2, 3, 4, 6, 8, 12, 24
Factors of 16: 1, 2, 4, 8, 16

3. Factors of 20: 1, 2, 4, 5, 10, 20
Factors of 35: 1, 5, 7, 35

4. Factors of 39: 1, 3, 13, 39
Factors of 25: 1, 5, 25

5. Two natural numbers are relatively prime if their greatest common factor is [?].

6. Decide whether 160 and 189 are relatively prime.

Independent Practice

In Exercises 7–14, find the greatest common factor of the numbers.

7. 20, 32 **8.** 36, 54 **9.** 90, 210 **10.** 126, 216
11. 1008, 1080 **12.** 546, 1995 **13.** 128, 256 **14.** 255, 256

In Exercises 15–18, find the greatest common factor of the expressions.

15. $2y^2z$, $8yz^2$ **16.** $3x^2y^2$, $15x^2y$ **17.** $9r^2z$, $21rz$ **18.** $42s^3t^4$, $70s^4t^3$

In Exercises 19–22, find two pairs of numbers that have the given greatest common factor. (There are many correct answers.)

19. 4 **20.** 6 **21.** 21 **22.** 18

In Exercises 23–26, decide whether the numbers are relatively prime.

23. 384, 945 **24.** 80, 189 **25.** 120, 336 **26.** 220, 315

Geometry **In Exercises 27–30, find the area and perimeter of the rectangle. Are the two measures relatively prime? Explain.**

27.
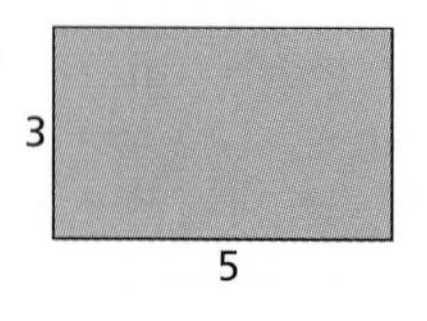

28.
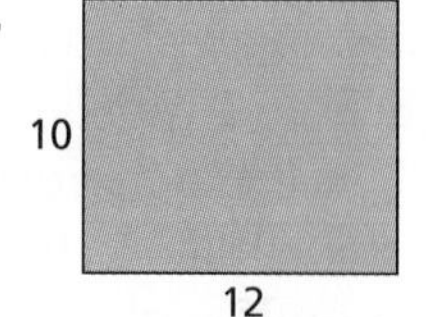

29.
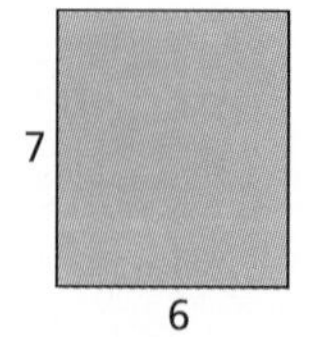

30.
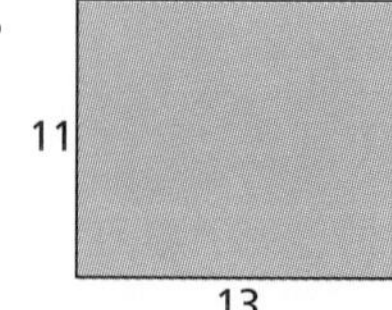

31. *Sequences* Find the greatest common factor of the terms in the following sequence: 6, 12, 18, 24, 30, ...

32. *Sequences* Find the greatest common factor of the terms in the following sequence: 8, 12, 16, 20, 24, ...

33. *Physical Fitness Equipment* You and some of your classmates have raised money to purchase a stair climber for $285 and a weight set for $418 for a community gym. Each person raised an equal share of the $285 and $418. How many people are in your fundraising group?

34. *Geometry* Three strings have lengths of 39 centimeters, 52 centimeters, and 65 centimeters. You want to cut the strings so that the resulting pieces are all the same length. How can you make the pieces as long as possible?

39 cm

52 cm

65 cm

35. *True or False?* The greatest common factor of $2^2 \cdot 3 \cdot 5 \cdot 19$ and $2 \cdot 3^2 \cdot 7 \cdot 19$ is 19.

36. *True or False?* If n and m are different primes, then they are relatively prime.

Number Theory **In Exercises 37–40, choose two different prime numbers *p* and *q*. Decide whether the statement about the greatest common factor (GCF) is true. Could your answer change by choosing different prime numbers? Explain.**

37. The GCF of p and pq is p.

38. The GCF of q and q^2 is q.

39. The GCF of p^2q and pq^2 is pq.

40. The GCF of p^2 and q^2 is 1.

Integrated Review

Making Connections within Mathematics

Factor Form **In Exercises 41–44, write the prime factorization of the number.**

41. 124

42. 196

43. 900

44. 3300

Exponent Form **In Exercises 45 and 46, write the expression in exponent form. Then simplify the expression.**

45. $2 \cdot 2 \cdot 3 \cdot 3 \cdot 3 \cdot 5 \cdot 7 \cdot 7$

46. $2 \cdot 3 \cdot 3 \cdot 5 \cdot 5 \cdot 5 \cdot 11 \cdot 11$

Exploration and Extension

Tiling **In Exercises 47 and 48, you are tiling two rooms with a single size of square tile. The tiles cannot overlap or be cut. Find the largest size tile that you can use to tile both rooms. Illustrate your solution with a sketch.**

47. Room 1: 8 feet by 10 feet
Room 2: 14 feet by 16 feet

48. Room 1: 8 feet by 16 feet
Room 2: 12 feet by 20 feet

Mixed REVIEW

1. Use the divisibility tests to decide whether 612 is divisible by 2, 3, 4, 5, 6, 8, 9, and 10. **(6.1)**
2. Use a calculator to decide whether 612 is divisible by 11, 12, 13, 14, 15, 16, 17, 18, 19, and 20. **(6.1)**

In Exercises 3–6, you randomly select one letter from the word DIVISIBILITY. What is the probability that the letter is as described? (5.8)

3. S **4.** I **5.** D or Y **6.** Not an I

In Exercises 7–10, write the prime factorization of the number. (6.2)

7. 87 **8.** 98 **9.** 76 **10.** 88

In Exercises 11–14, solve the equation. Show your work. (4.2, 4.4, 4.5)

11. $\frac{1}{2}(x + 2) = 4$ **12.** $\frac{1}{4}(r - 1) = 0$ **13.** $7(y + 2) = 5y$ **14.** $3.2 + 1.2s = -0.4s$

Milestones MUSIC OF THE GREEKS

600 B.C. 200 B.C. 200 A.D. 600 A.D. 1000 A.D. 1400 A.D. 1800 A.D.

Harp ? B.C. | Pythagoras, Greece 540 B.C. | Plainsong 400-800 | Bowed Instruments, China 900 | Sitar, India 1200 | Violin, Cello, Italy, 1500's | Banjo, Africa 1700's

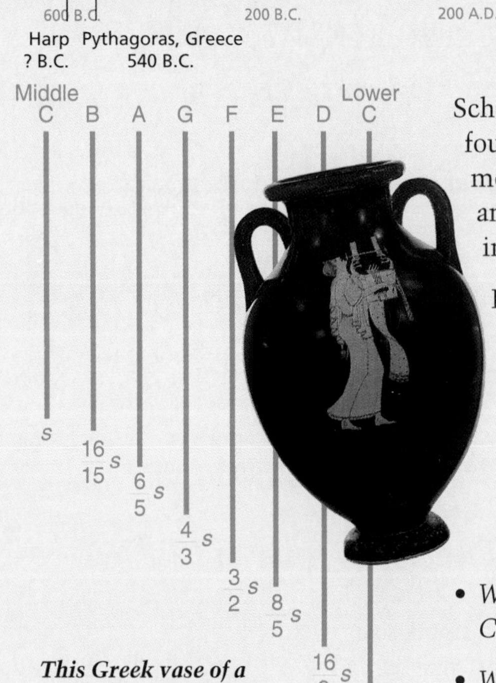

This Greek vase of a boy playing a Kithara is from about 490 B.C.

Scholars in ancient Greece (600 B.C.–500 B.C.) believed that four subjects ruled their universe—music, geometry, arithmetic, and astronomy. Pythagoras, a Greek mathematician and philosopher, and his followers looked for a numerical interpretation for all natural phenomena.

Pythagoreans found that musical sounds made by equally tight lyre strings were related to their lengths. So, starting with any note, you can go down the scale merely by changing the length of the string as shown in the chart. Strings whose lengths were in ratios made up of small integers, like 1 to 2, had a more "agreeable" sound than strings whose ratios were made up of larger integers, like 23 to 13.

- *Which notes do you think are "more agreeable" to Middle C?*
- *Which notes would be "less agreeable"?*

Try playing the notes together to check your answers.

6.4 Least Common Multiple

What you should learn:

Goal 1 How to find the least common multiple of two numbers

Goal 2 How to use a least common multiple to solve problems in geometry

Why you should learn it:

You can use the least common multiple of two numbers to solve real-life problems, such as analyzing the movements of gears.

Goal 1 Finding a Least Common Multiple

Least Common Multiple

Let m and n be natural numbers.

1. A number that is a multiple of both m and n is a **common multiple** of m and n.
2. Of all common multiples of m and n, the smallest is called the **least common multiple.**

Example 1 *Finding the Least Common Multiple*

Find the least common multiple of 6 and 9.

Solution With small numbers, you can find the least common multiple by listing multiples of each number. The smallest duplicate in the two lists is the least common multiple.

Number	Multiples
6	6, 12, 18, 24, 30, . . .
9	9, 18, 27, 36, . . .

From the lists, you can see that 18 is the least common multiple. ■

Example 2 *Finding the Least Common Multiple*

Find the least common multiple of 180 and 378.

Solution With larger numbers, you can find the least common multiple by writing the prime factorization of each.

Number	Prime Factorization
180	$2 \cdot 2 \cdot 3 \cdot 3 \cdot 5$
378	$2 \cdot 3 \cdot 3 \cdot 3 \cdot 7$

From the prime factorizations, you can reason that the least common multiple must contain the factors $2 \cdot 2$, $3 \cdot 3 \cdot 3$, 5, and 7. So the least common multiple is $2 \cdot 2 \cdot 3 \cdot 3 \cdot 3 \cdot 5 \cdot 7 = 3780$. Check to see that 3780 is divisible by 180 and by 378. ■

Least Common Multiple of Algebraic Expressions

The concept of a least common multiple also applies to algebraic expressions. For instance, the least common multiple of $6x$ and $4xy$ is $12xy$.

Goal 2 Using Least Common Multiples

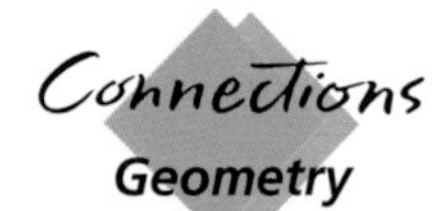

Example 3 Using a Least Common Multiple

You have a box of tiles, each of which is 3 inches by 5 inches. Without overlapping or cutting the tiles, what is the least number of tiles you must use to form a square region?

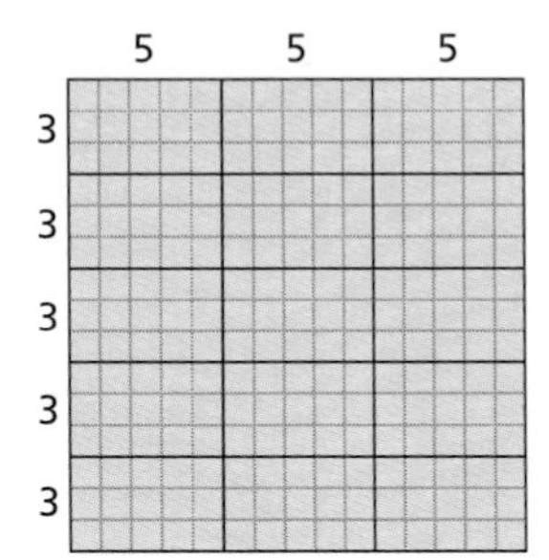

Solution As shown at the left, the least number of tiles is 15. This follows from the fact that 15 is the least common multiple of 3 and 5. ■

LESSON INVESTIGATION

Investigating Least Common Multiples

Group Activity Assume that you are given a box of tiles of each size shown below. For each size, find the least number of tiles you can use to form a square. How do the results relate to the least common multiple of the lengths of the sides of each tile?

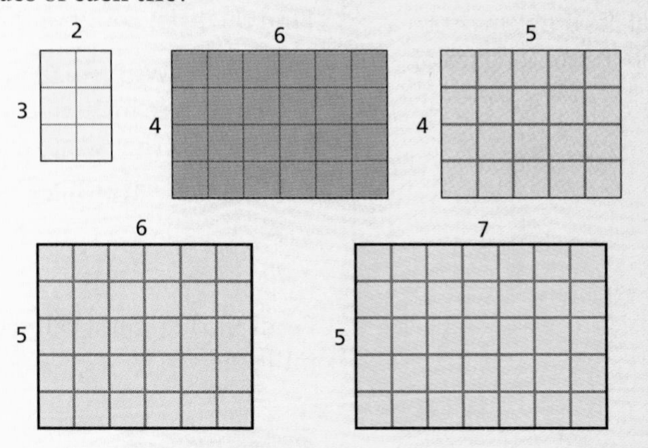

Communicating about MATHEMATICS

SHARING IDEAS about the Lesson

Least Common Multiple The least common multiple of 4, 5, and 6 is 60. Find the least common multiple of each of the following triples. Explain your reasoning.

A. 6, 8, 5 **B.** 3, 4, 6 **C.** 10, 6, 15

EXERCISES

Guided Practice

CHECK for Understanding

In Exercises 1 and 2, find the three missing multiples of the number.

1. Multiples of 4: [?], 8, [?], 16, 20, [?]
2. Multiples of 6: 6, [?],18, [?],30, [?]
3. What is the least common multiple (LCM) of 4 and 6?
4. *It's Up to You* In your own words, describe two ways to find the least common multiple of two numbers. Which way would you use to find the least common multiple of 10 and 16? Which way would you use to find the least common multiple of 112 and 204?

In Exercises 5–7, match the number with its prime factorization. Then find the least common multiple of the three numbers.

a. $2^2 \cdot 3 \cdot 5^2$ b. $2 \cdot 3^2 \cdot 5$ c. $2^3 \cdot 5^2$

5. 200
6. 90
7. 300
8. What is the least common multiple of $2a^2b^3$ and $4ab^4$?

Independent Practice

In Exercises 9–20, list the first several multiples of each number. Use the lists to find the least common multiple.

9. 3, 7
10. 7, 8
11. 6, 8
12. 3, 9
13. 8, 10
14. 10, 15
15. 10, 26
16. 4, 22
17. 3, 4, 18
18. 3, 6, 9
19. 5, 10, 20
20. 6, 9, 18

In Exercises 21–32, write the prime factorization of each expression. Use the results to find the least common multiple.

21. 90, 108
22. 7, 8
23. 125, 500
24. 160, 432
25. 135, 375
26. 225, 324
27. 144, 162
28. $16x$, $32x^4$
29. $7s^2t$, $49st^2$
30. $2x^3y$, $3xy^5$
31. $3m^4n^4$, $7m^6n^2$
32. $4a^6b^3$, $8a^7b^5$
33. *Reasoning* If two numbers are relatively prime, then what is their least common multiple? Give two examples.
34. *Reasoning* If one number is a multiple of another, then what is their least common multiple? Give two examples.

In Exercises 35–38, use the results of Exercises 33 and 34 to find the least common multiple.

35. 3, 8
36. 8, 9
37. 3, 6
38. 8, 24

Number Sense **In Exercises 39–42, find all pairs of numbers that satisfy the conditions.**

39. Two prime numbers whose LCM is 35
40. Two composite numbers whose LCM is 16
41. Two square numbers whose LCM is 36
42. Two even numbers whose LCM is 12
43. *Tiling* You have a box of tiles, each of which is 4 inches by 14 inches. Without overlapping or cutting the tiles, what is the least number of tiles you must use to form a square region? Draw a diagram and explain your answer.
44. *Stoplights* Consider three stoplights on a street. The first stoplight is red 3 minutes out of every 6 minutes. The second stoplight is red 4 minutes out of every 8 minutes. The third stoplight is red 5 minutes out of every 10 minutes. At 2:00 P.M., each stoplight turns red. When is the next time all three stoplights turn red?
45. *Gears* The gears at the right are rotating. Gear A has 14 teeth and Gear B has 12 teeth. How many complete revolutions must each gear make for the gears to align again as shown?

Some clocks operate with gears. These clocks are called analog clocks. As the gears revolve, the hands of the clock move.

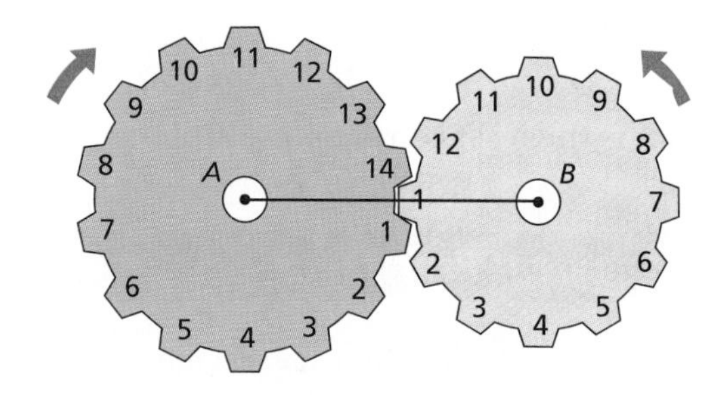

Integrated Review

Making Connections within Mathematics

Number Patterns **In Exercises 46 and 47, find the least common multiple of each pair. Then describe the pattern formed by the common multiples.**

46. 1 and 2, 2 and 3, 3 and 4, 4 and 5, 5 and 6, 6 and 7
47. 2 and 4, 4 and 6, 6 and 8, 8 and 10, 10 and 12, 12 and 14

Exploration and Extension

Age Riddles **In Exercises 48 and 49, find the age of each person.**

48. I have lived less than a half century. My age is a multiple of 8. Next year my age will be a multiple of 11.
49. I have lived more than 2 decades but less than 4 decades. Last year my age was a multiple of 3. This year my age is a multiple of 4. What is my age?

According to the 1990 census, the median age in the United States is 32.9 years, up 2.9 years from 1980.

6.5 Simplifying and Comparing Fractions

What you should learn:

How to simplify a fraction

How to compare two fractions

Why you should learn it:

Rewriting a fraction in a different form helps you decide whether one fraction is larger than another.

As fractions, $\frac{3}{4}$ and $\frac{6}{8}$ are equivalent. In music, however, three-quarter time is not equivalent to six-eights time. The first has 3 beats to a measure and the second has 6. African drum songs are often in six-eights time to build excitement or to wake people up.

Goal 1 Simplifying a Fraction

Two fractions are **equivalent** if they have the same decimal form. For instance, the fractions $\frac{1}{2}$ and $\frac{2}{4}$ are equivalent because each is equal to 0.5. The following hexagons show that the fractions $\frac{2}{3}$, $\frac{4}{6}$, and $\frac{8}{12}$ are also equivalent. The first of these, $\frac{2}{3}$, is in **simplest form**. Writing a fraction in reduced form is called **simplifying** the fraction.

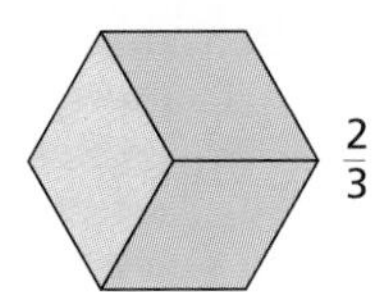

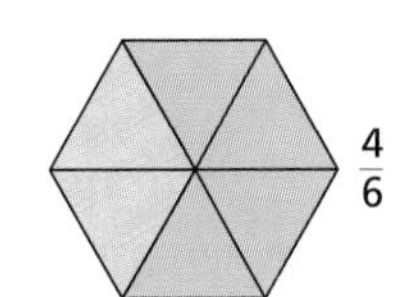

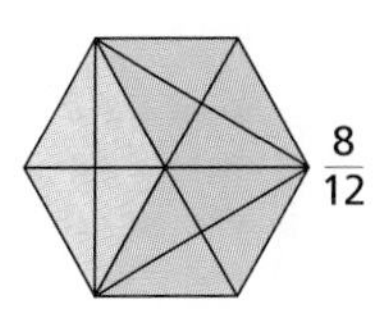

Example 1 *Simplifying a Fraction*

Simplify the fraction $\frac{12}{20}$.

Solution To simplify the fraction, factor its numerator and denominator. Then divide the numerator and denominator by any common factors.

$$\frac{12}{20} = \frac{\overset{1}{\cancel{2}} \cdot \overset{1}{\cancel{2}} \cdot 3}{\underset{1}{\cancel{2}} \cdot \underset{1}{\cancel{2}} \cdot 5} = \frac{3}{5}$$

Another way to simplify a fraction is to divide the numerator and denominator by their greatest common factor.

$$\frac{12}{20} = \frac{12 \div 4}{20 \div 4} = \frac{3}{5}$$

■

Example 2 *Simplifying a Fraction with Variables*

Simplify the fraction $\frac{4x^2}{6x}$.

Solution

$$\frac{4x^2}{6x} = \frac{\overset{1}{\cancel{2}} \cdot 2 \cdot \overset{1}{\cancel{x}} \cdot x}{\underset{1}{\cancel{2}} \cdot 3 \cdot \underset{1}{\cancel{x}}} = \frac{2x}{3}$$

■

Goal 2 Comparing Fractions

Example 3 *Comparing Fractions*

Which fraction is larger, $\frac{7}{12}$ or $\frac{9}{16}$?

Solution Begin by rewriting the fractions with a common denominator. The common denominator should be the least common multiple of the two original denominators. The least common multiple of 12 and 16 is 48. The fraction $\frac{7}{12}$ can be rewritten as

$$\frac{7}{12} \times \frac{4}{4} = \frac{28}{48}$$ *Multiply by $\frac{4}{4}$ to get a denominator of 48.*

The fraction $\frac{9}{16}$ can be written as

$$\frac{9}{16} \times \frac{3}{3} = \frac{27}{48}$$ *Multiply by $\frac{3}{3}$ to get a denominator of 48.*

By comparing the rewritten forms, you can see that $\frac{7}{12}$ is larger than $\frac{9}{16}$. ■

Study Tip
Another way to compare two fractions is to write them as decimals. From $\frac{7}{12} = 0.5833\ldots$ and $\frac{9}{16} = 0.5625$, you can see that $\frac{7}{12}$ is larger than $\frac{9}{16}$.

Example 4 *Comparing Fractions*

Real Life
Food Service

You ate 4 pieces of a small pizza that was cut into 6 equal pieces. Your friend ate 5 pieces of a small pizza that was cut into 8 equal pieces. Who ate more pizza?

Solution You ate $\frac{4}{6}$ of a pizza and your friend ate $\frac{5}{8}$ of a pizza. Because

$$\frac{4}{6} \times \frac{4}{4} = \frac{16}{24} \text{ and } \frac{5}{8} \times \frac{3}{3} = \frac{15}{24}$$

it follows that you ate more pizza. ■

Communicating about MATHEMATICS

SHARING IDEAS about the Lesson

Comparing Fractions In Example 3, the fraction $\frac{7}{12}$ was multiplied by $\frac{4}{4}$. Explain why this procedure produced an equivalent fraction.

EXERCISES

Guided Practice

CHECK for Understanding

Modeling Fractions **In Exercises 1–4, write a fraction that represents the portion of the region that is blue. Then simplify the fraction.**

1. 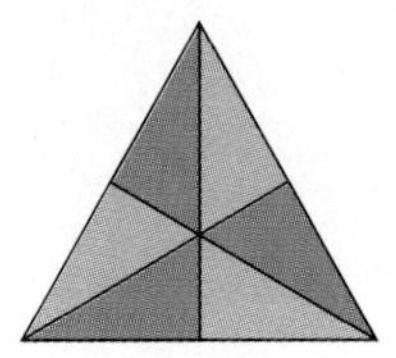

2. 

3. 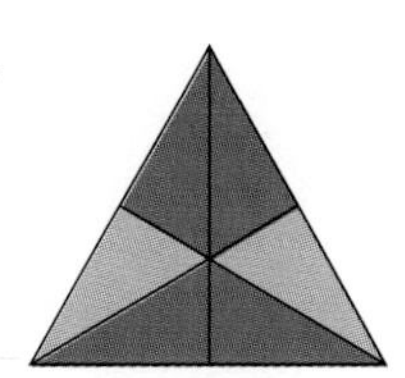

4. 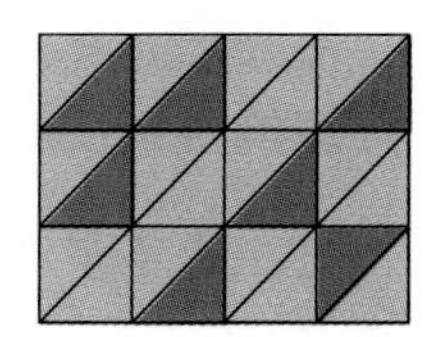

Fraction Strips **In Exercises 5 and 6, write the fraction that is represented by each fraction strip. Then rewrite the fractions with a common denominator and decide which is larger.**

5.

6. 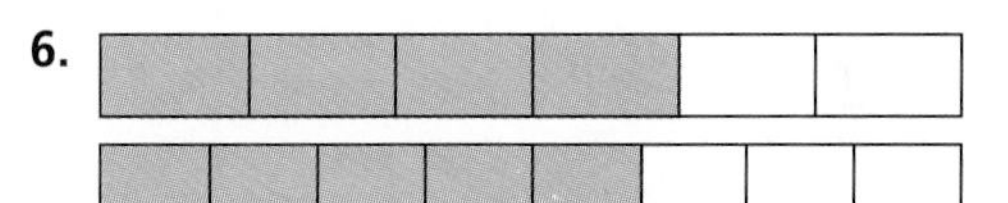

Independent Practice

In Exercises 7–14, what is the greatest common factor of the numerator and denominator? Use your answer to simplify the fraction.

7. $\frac{14}{20}$ **8.** $\frac{16}{36}$ **9.** $\frac{9}{42}$ **10.** $\frac{63}{105}$

11. $\frac{10}{75}$ **12.** $\frac{8}{28}$ **13.** $\frac{36}{54}$ **14.** $\frac{117}{143}$

In Exercises 15–22, simplify the variable expression.

15. $\frac{2ab}{8b^2}$ **16.** $\frac{3x^2y}{9y}$ **17.** $\frac{25z^2}{150z^3}$ **18.** $\frac{22s^3t}{55s^3t^2}$

19. $\frac{6yz}{8y}$ **20.** $\frac{15x}{21x^2}$ **21.** $\frac{28p^2q^2}{42p^3q^3}$ **22.** $\frac{34m^2}{68mn}$

23. ***Reasoning*** Explain how you can determine whether a fraction is in simplest form.

24. Which of the following is not equivalent to $\frac{7}{8}$: $\frac{14}{16}$, $\frac{17}{18}$, $\frac{21}{24}$?

In Exercises 25–28, write 3 fractions that are equivalent to the given fraction.

25. $\frac{1}{2}$ **26.** $\frac{2}{5}$ **27.** $\frac{10}{22}$ **28.** $\frac{8}{18}$

Comparing Fractions **In Exercises 29–36, complete the statement with $<$, $>$, or $=$.**

29. $\frac{1}{7}\ \boxed{?}\ \frac{1}{6}$ **30.** $\frac{18}{38}\ \boxed{?}\ \frac{27}{57}$ **31.** $\frac{1}{12}\ \boxed{?}\ \frac{1}{13}$ **32.** $\frac{8}{14}\ \boxed{?}\ \frac{6}{13}$

33. $\frac{7}{8}\ \boxed{?}\ \frac{8}{9}$ **34.** $\frac{0}{2}\ \boxed{?}\ \frac{0}{100}$ **35.** $\frac{15}{39}\ \boxed{?}\ \frac{5}{13}$ **36.** $\frac{26}{50}\ \boxed{?}\ \frac{27}{51}$

37. *Ordering Fractions* Order $\frac{2}{3}, \frac{6}{18}, \frac{8}{10}, \frac{1}{1}, \frac{3}{6}, \frac{5}{12}$ from least to greatest.

Photography **In Exercises 38 and 39, use the following information.**

Shutter speed is the length of time the shutter is open. Many cameras allow a photographer to adjust the shutter speed. A slow shutter speed allows more light to expose film than does a fast shutter speed.

This photograph appears blurry because the moving vehicles were photographed using a slow shutter speed.

38. The shutter speed on your camera is $\frac{1}{250}$ of a second. You want to decrease the shutter speed. Which of the following would be appropriate: $\frac{1}{500}$ of a second or $\frac{1}{125}$ of a second? Explain.

39. Your friend's camera is letting in too much light. You tell him that he needs to increase the shutter speed. The shutter speed setting is at $\frac{1}{250}$ of a second. Which of the following would be appropriate: $\frac{1}{500}$ of a second or $\frac{1}{125}$ of a second? Explain.

Simplifying Fractions **In Exercises 40–42, use the bar graph, which shows the average number (in millions) of *M & Ms* sold in a day.** *(Source: Mars Company)*

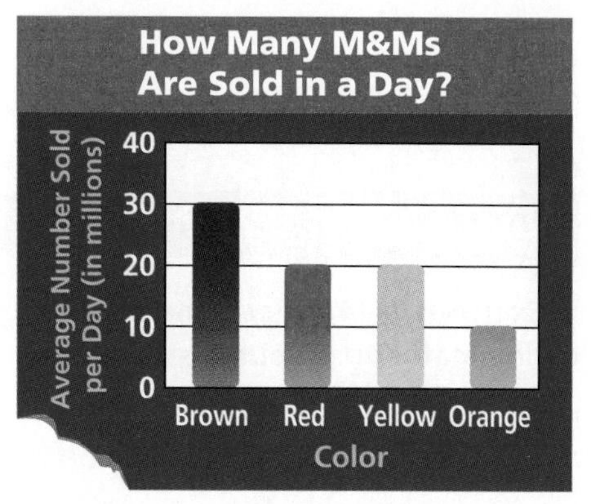

40. Express the number of brown *M & Ms* as a fraction of all brown, red, yellow, and orange *M & Ms*. Simplify.

41. Express the number of red *M & Ms* as a fraction of all brown, red, yellow, and orange *M & Ms*. Simplify.

42. How is the fraction $\frac{1}{3}$ related to the bar graph?

Integrated Review

Making Connections within Mathematics

GCF and LCM **In Exercises 43–46, find the GCF and LCM of the numbers.**

43. 10, 12 **44.** 15, 18 **45.** 75, 100 **46.** 36, 84

47. *Estimation* Which best estimates the height of the Washington Monument?

a. 5.55 feet **b.** 55.5 feet
c. 555 feet **d.** 5550 feet

48. *Estimation* Which best estimates the length of the Mississippi River?

a. 234 miles **b.** 2340 miles
c. 23,400 miles **d.** 234,000 miles

Exploration and Extension

49. *Comparing Pay Scales* Which pay scale would you prefer? Explain.

Pay scale 1: \$4 for each $\frac{1}{2}$ hour worked. **Pay scale 2:** \$5 for each $\frac{2}{3}$ hour worked.

50. Find the hourly rates for the pay scales in Exercise 49.

Mid-Chapter SELF-TEST

Take this test as you would take a test in class. The answers to the exercises are given in the back of the book.

In Exercises 1 and 2, use the divisibility tests to decide whether the number is divisible by 2, 3, 4, 5, 6, 8, 9, and 10. (6.1)

1. 510

2. 1360

3. Use a calculator to decide whether 816 is divisible by 11, 12, 13, 14, 15, 16, 17, 18, 19, or 20. **(6.1)**

4. List all the factors of the number 56.

In Exercises 5–7, write the prime factorization of the number. (6.2)

5. 80

6. 44

7. 105

In Exercises 8–10, find the greatest common factor. (6.3)

8. 12, 60

9. 36, 15

10. 135, 45, 25

In Exercises 11–13, find the least common multiple. (6.4)

11. 13, 5

12. 14, 21

13. $6x, 9x^2$

In Exercises 14–16, simplify the expression. (6.5)

14. $\frac{5}{25}$

15. $\frac{45}{306}$

16. $\frac{8y^2}{24y}$

In Exercises 17 and 18, find the least number of tiles that can be used to form a square given a box of tiles of the size shown. How is the length of the square's sides related to the lengths of the sides of the tile? (6.4)

17.

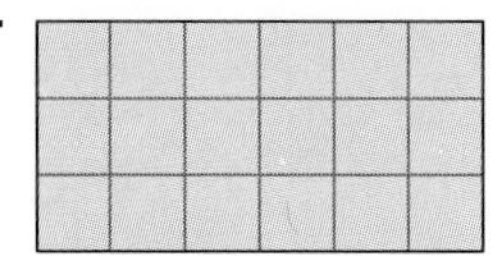

18.

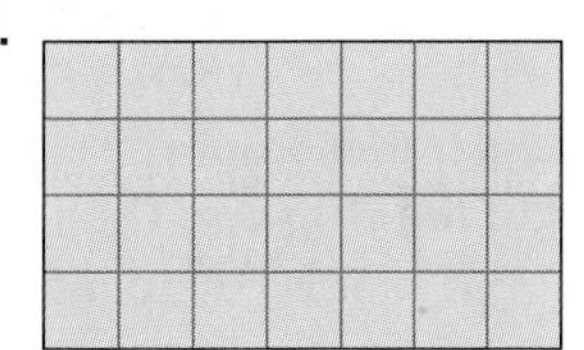

In Exercises 19 and 20, imagine that you are stacking 2 different-sized boxes as shown at the right. One box size is 6 inches high and the other is 14 inches high. (6.4)

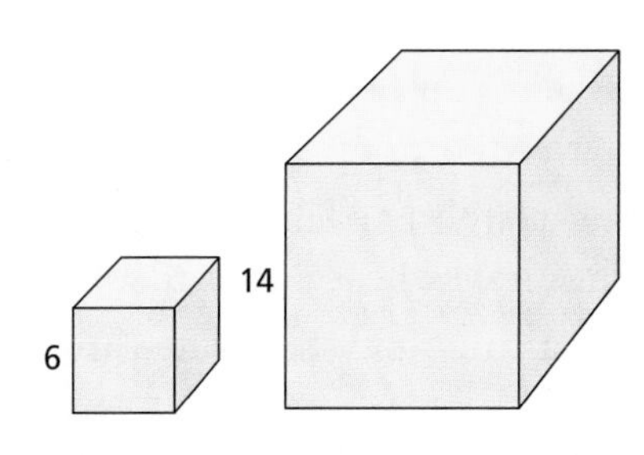

19. To obtain two stacks of the same height, what is the least number of boxes of each size that must be used?

20. In Exercise 19, is the height of each stack the least common multiple of 6 and 14 or the greatest common factor of 6 and 14?

LESSON INVESTIGATION 6.6

Folding Fraction Strips

Materials Needed: paper, scissors

In this activity, you will investigate properties of fractions by folding strips of paper to specified lengths.

Example — *Folding Fractions*

Each person in your group is given a strip of paper that is 3 inches long. Without using a ruler, each person is asked to fold the strip and mark a portion that is $\frac{3}{4}$ inch long. The following three methods were used. Is each method correct? Explain your reasoning.

a. 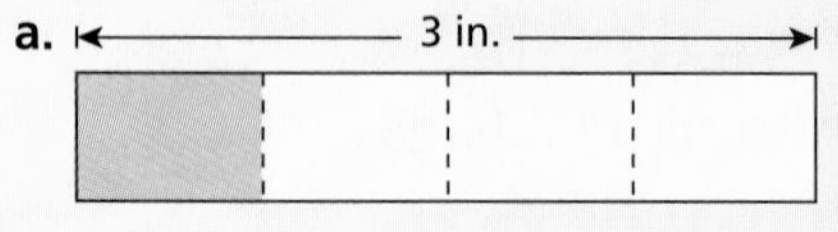

Fold the strip in half. Then fold it in half again. Each of four parts has a length of $\frac{3}{4}$ inch.

b. 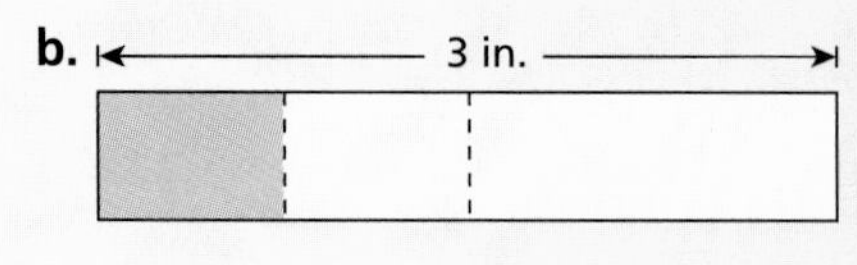

Fold the strip in half. Unfold it and fold one of the halves in half. Each of two smaller parts has a length of $\frac{3}{4}$ inch.

c. 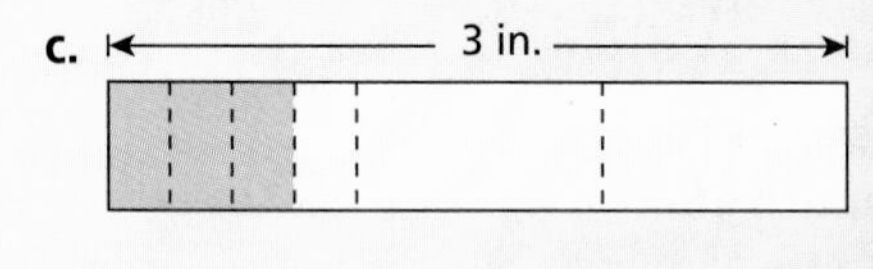

Fold the strip into 3 equal parts. Unfold it and fold one of the parts into 4 equal parts. Shade 3 of the smaller parts. ■

Exercises

In Exercises 1–4, use a ruler to measure a strip of paper that has the indicated length. Cut the strip out. Then, without using the ruler, fold the strip in a way that allows you to mark a portion of the strip that has the indicated length.

1. Length of paper strip: 4 inches
   Length of portion: $\frac{3}{4}$ inch
2. Length of paper strip: 4 inches
   Length of portion: $\frac{4}{3}$ inch
3. Length of paper strip: 5 inches
   Length of portion: $\frac{5}{6}$ inch
4. Length of paper strip: 3 inches
   Length of portion: $\frac{4}{3}$ inch
5. Draw a circle and label its area as 5 square units. Then shade a portion of the circle whose area is $\frac{5}{4}$ square units.
6. Draw a circle and shade one-thrd of the circle. Label the area of the shaded region as $\frac{4}{3}$ square units. What is the area of the unshaded region?

6.6 Rational Numbers and Decimals

What you should learn:

Goal 1 How to show that a number is rational

Goal 2 How to write a decimal as a fraction

Why you should learn it:

Knowing how to rewrite numbers in decimal form or as fractions helps you interpret results given by calculators and computers.

Goal 1 Identifying Rational Numbers

Throughout history, when people have studied numbers, they found that some numbers have special properties. For instance, you studied special properties of prime numbers in Lesson 6.2. In this lesson, you will study properties of **rational numbers.**

Rational Number

A number is **rational** if it can be written as the quotient of two integers. Numbers that cannot be written as the quotient of two integers are called **irrational.**

Rational Numbers	*Irrational Numbers*
$\frac{1}{2}, \frac{-3}{5}, \frac{9}{4}, \frac{5}{1}$	$\sqrt{2}, \sqrt{3}, \sqrt{5}$

Example 1 *Recognizing Rational Numbers*

Show that the following numbers are rational.

a. 4 **b.** 0.5 **c.** -3

Solution To show that a number is rational, you must show that it can be written as the quotient of two integers.

a. 4 is rational because it can be written as $4 = \frac{4}{1}$.

b. 0.5 is rational because it can be written as $0.5 = \frac{1}{2}$.

c. -3 is rational because it can be written as $-3 = \frac{-3}{1}$. ■

The Set of Rational Numbers

Integers

Whole Numbers

Natural Numbers

This Venn diagram shows that each natural number, whole number, and integer is a rational number.

Example 2 *Showing a Mixed Number Is Rational*

Show that the mixed number $1\frac{1}{4}$ is rational.

Solution $1\frac{1}{4}$ can be written as

$$1\frac{1}{4} = 1 + \frac{1}{4} = \frac{4}{4} + \frac{1}{4} = \frac{5}{4}.$$

Since $1\frac{1}{4}$ can be written as the quotient of the integers 5 and 4, it is rational. ■

Goal 2 Writing Decimals as Fractions

Decimals can be *terminating, repeating,* or *nonrepeating.* In decimal form, every rational number is either terminating or repeating, and every irrational number is nonrepeating.

Number	Decimal Form	Comment
$\frac{3}{8}$	0.375	Rational, terminating
$\frac{16}{11}$	$1.454545\ldots = 1.\overline{45}$	Rational, repeating
$\sqrt{2}$	$1.414213562\ldots$	Irrational, nonrepeating

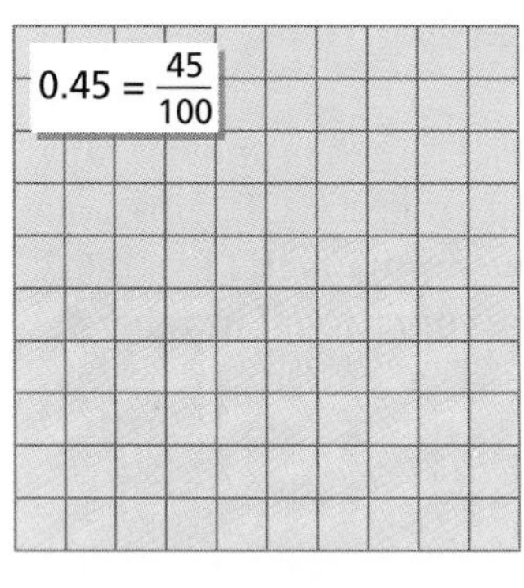

Each small square has an area of 0.01 or $\frac{1}{100}$.

Example 3 *Writing Decimals as Fractions*

Write the following decimals as fractions.

a. 0.45 **b.** $0.090909\ldots = 0.\overline{09}$

Solution

a. This terminating decimal represents 45 hundredths, as shown in the figure at the left. You can write it as a fraction as follows.

$$0.45 = \frac{45}{100} \quad \textit{Write as 45 hundredths.}$$
$$= \frac{5 \cdot 9}{5 \cdot 20} \quad \textit{Factor.}$$
$$= \frac{9}{20} \quad \textit{Simplify.}$$

b. To write a repeating decimal as a fraction, use the following strategy.

$$x = 0.090909\ldots \quad \textit{Let x represent the number.}$$
$$100x = 9.090909\ldots \quad \textit{Multiply each side by 100.}$$
$$99x = 9 \quad \textit{Subtract 1st equation from 2nd.}$$
$$x = \frac{9}{99} \quad \textit{Divide each side by 99.}$$
$$x = \frac{1}{11} \quad \textit{Simplify.}$$

■

Need to Know

In part **b,** each side was multiplied by 100 because 0.090909 . . . has *two* repeating digits. For one repeating digit, multiply by 10. For three repeating digits, multiply by 1000.

Communicating about MATHEMATICS

SHARING IDEAS about the Lesson

Writing Fractions Write each decimal number as a fraction. Explain your reasoning.

A. $0.6666\ldots = 0.\overline{6}$ **B.** $1.6666\ldots = 1.\overline{6}$

C. 2.25 **D.** $0.2222\ldots = 0.\overline{2}$

EXERCISES

Guided Practice

CHECK for Understanding

In Exercises 1–5, state whether the number is rational or irrational. Write the rational number as the quotient of two integers.

1. -3 **2.** $\sqrt{6}$ **3.** $2\frac{3}{5}$ **4.** 7 **5.** 0.4

Modeling Decimals **In Exercises 6–9, write the decimal that is represented by the blue portion of the grid. Then write the number as a fraction and simplify. (Each small square has an area of 0.01.)**

6.
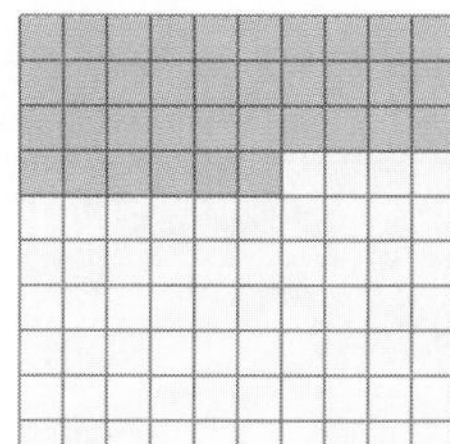

7.
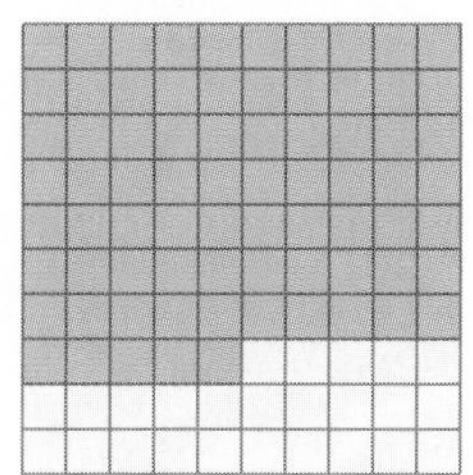

8.
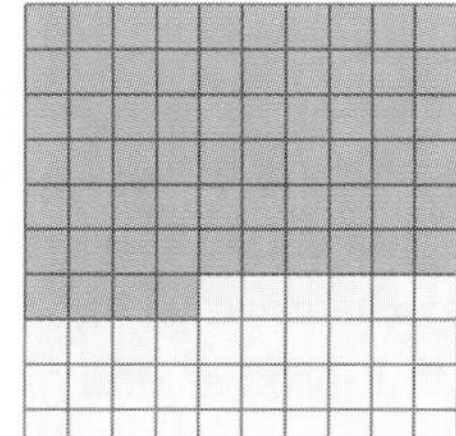

9.
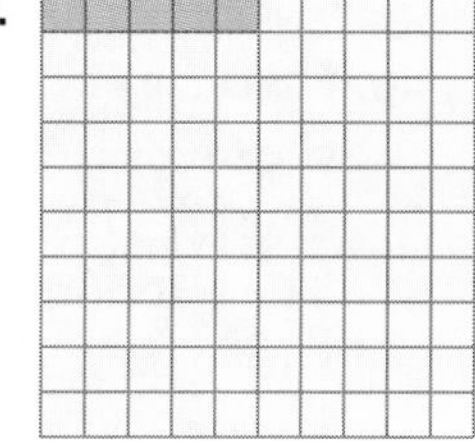

10. *Writing* Explain how to write a mixed number as a quotient of two integers.

Independent Practice

In Exercises 11–18, write the number as a fraction in simplest form.

11. 5 **12.** 0.75 **13.** 0.25 **14.** -9

15. $1\frac{1}{6}$ **16.** $2\frac{2}{9}$ **17.** $-1\frac{5}{8}$ **18.** $-2\frac{4}{5}$

In Exercises 19–26, decide whether the number is rational or irrational. Then write the decimal form of the number and state whether the decimal is terminating, repeating, or nonrepeating.

19. $\frac{3}{5}$ **20.** $\frac{9}{11}$ **21.** $\sqrt{8}$ **22.** $\sqrt{9}$

23. $\frac{8}{15}$ **24.** $\frac{7}{10}$ **25.** $\frac{7}{2}$ **26.** $\frac{13}{12}$

In Exercises 27–34, write the decimal as a fraction. Simplify the result.

27. 0.8 **28.** 0.35 **29.** 0.84 **30.** 0.64

31. $0.\overline{45}$ **32.** $0.\overline{86}$ **33.** $2.\overline{3}$ **34.** $1.\overline{135}$

In Exercises 35–40, match the rational number with its decimal form.

a. 0.4 **b.** 3.08 **c.** 2.12 **d.** $0.2\overline{7}$ **e.** $0.08\overline{3}$ **f.** $0.\overline{296}$

35. $\frac{10}{120}$ **36.** $\frac{6}{15}$ **37.** $2\frac{3}{25}$ **38.** $\frac{5}{18}$ **39.** $\frac{8}{27}$ **40.** $3\frac{6}{75}$

Fraction Patterns **In Exercises 41 and 42, write each rational number in decimal form. Then describe the pattern.**

41. $\frac{1}{11}, \frac{2}{11}, \frac{3}{11}, \frac{4}{11}, \frac{5}{11}, \frac{6}{11}$

42. $\frac{1}{2}, \frac{3}{4}, \frac{5}{6}, \frac{7}{8}, \frac{9}{10}, \frac{11}{12}$

Geometry **In Exercises 43–45, find the perimeter of the figure. Write the result in three ways: in fraction form, as a mixed number, and as a decimal.**

43.

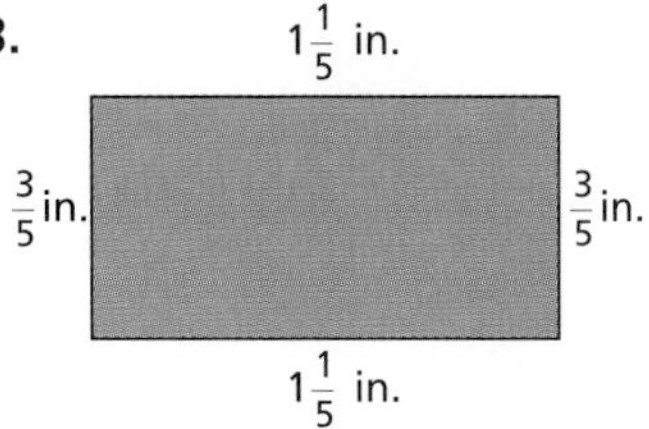

44.

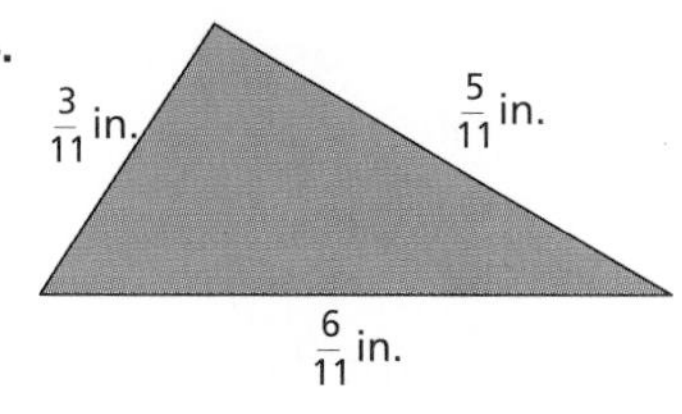

45.

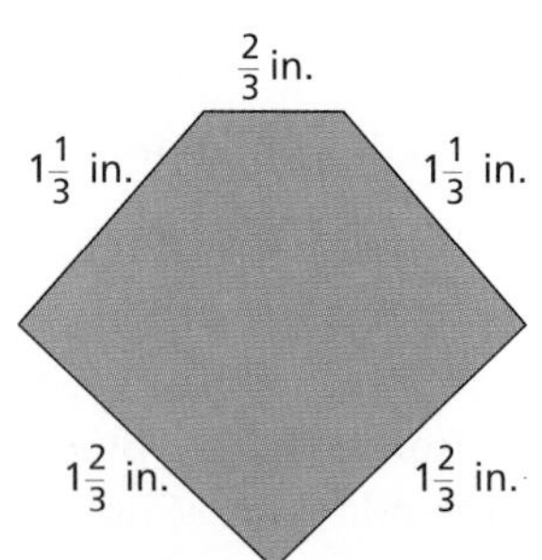

46. *Favorite Foods* A survey was taken to determine the favorite foods of students between the ages of 6 and 12. The top five choices are shown at the right. The fraction indicates the portion of those surveyed who chose that food as one of their favorites. Write each fraction in decimal form. Then order the top five foods from most favorite to least favorite.

Favorite Food	Number
Cheeseburgers	$\frac{21}{50}$
Chicken nuggets	$\frac{13}{25}$
Hot dogs	$\frac{9}{20}$
Macaroni and cheese	$\frac{41}{100}$
Pizza	$\frac{41}{50}$

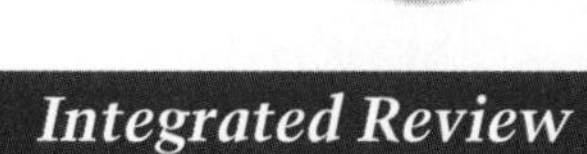

Integrated Review

Making Connections within Mathematics

Logical Reasoning **In Exercises 47–50, use the Venn diagram on page 265 to help you decide whether the statement is true or false. Explain.**

47. All integers are rational numbers.

48. All whole numbers are natural numbers.

49. All rational numbers are whole numbers.

50. All natural numbers are integers.

Exploration and Extension

Percents **In Exercises 51–54, write the percent as a fraction, then write the fraction in decimal form. What do you notice?**

Percent means "per hundred." A percent is a fraction whose denominator is 100. The symbol for percent is %. For instance, 25 percent can be written as 25% or $\frac{25}{100}$.

51. 24%

52. 83%

53. 56%

54. 12%

6.7 Powers and Exponents

What you should learn:

Goal 1 How to evaluate powers that have negative and zero exponents

Goal 2 How to multiply and divide powers

Why you should learn it:

Knowing how to evaluate numbers with negative and zero exponents helps you interpret results given by calculators and computers.

Sand is composed of tiny pieces of rocks that vary in width from 20^{-2} inch to 12^{-1} inch. Can you write these widths as fractions?

Goal 1 Using Negative and Zero Exponents

You already know how to evaluate powers that have positive integer exponents. For instance, $2^3 = 2 \cdot 2 \cdot 2 = 8$ and $(-3)^2 = (-3) \cdot (-3) = 9$. In this lesson, you will learn how to evaluate powers that have negative integer or zero exponents.

LESSON INVESTIGATION

Investigating Negative and Zero Exponents

Group Activity Use a calculator to write each power in decimal form.

$10^3 \quad 10^2 \quad 10^1 \quad 10^0 \quad 10^{-1} \quad 10^{-2} \quad 10^{-3}$

$2^3 \quad 2^2 \quad 2^1 \quad 2^0 \quad 2^{-1} \quad 2^{-2} \quad 2^{-3}$

What patterns can you discover? What does it mean to have a zero or negative exponent?

Sample Keystrokes for 10^{-2}: 10 [y^x] 2 [+/-] [=]

In this investigation, you may have discovered the following definitions.

Negative and Zero Exponents

Let n be a positive integer and let a be a nonzero number.

$$a^{-n} = \frac{1}{a^n} \qquad \text{and} \qquad a^0 = 1$$

Example 1 *Evaluating Powers*

a. $2^{-2} = \frac{1}{2^2} = \frac{1}{4}$

b. $-3^{-2} = -\frac{1}{3^2} = -\frac{1}{9}$

c. $(-4)^{-2} = \frac{1}{(-4)^2} = \frac{1}{16}$

d. $4^0 = 1$

e. $x^{-1} = \frac{1}{x^1} = \frac{1}{x}$

f. $2b^{-2} = 2(b^{-2}) = 2(\frac{1}{b^2}) = \frac{2}{b^2}$

■

Goal 2 Multiplying and Dividing Powers

To multiply two powers with the same base, add their exponents. To divide two powers with the same base, subtract the exponent of the denominator from the exponent of the numerator.

Multiplying and Dividing Powers

1. $a^m \cdot a^n = a^{m+n}$ **2.** $\frac{a^m}{a^n} = a^{m-n}$

Example 2 *Multiplying and Dividing Powers*

	Using Exponent Rules	Using Factors
a.	$4^2 \cdot 4^3 = 4^{2+3} = 4^5$	$4^2 \cdot 4^3 = \overbrace{4 \cdot 4}^{4^2} \cdot \overbrace{4 \cdot 4 \cdot 4}^{4^3} = 4^5$
b.	$\frac{3^3}{3^2} = 3^{3-2} = 3^1 = 3$	$\frac{3^3}{3^2} = \frac{\cancel{3} \cdot \cancel{3} \cdot 3}{\cancel{3} \cdot \cancel{3}} = 3$
c.	$2^4 \cdot 2^{-2} = 2^{4+(-2)} = 2^2$	$2^4 \cdot \frac{1}{2^2} = \frac{\cancel{2} \cdot \cancel{2} \cdot 2 \cdot 2}{\cancel{2} \cdot \cancel{2}} = 2^2$
d.	$\frac{5}{5^3} = 5^{1-3} = 5^{-2}$	$\frac{5}{5^3} = \frac{\cancel{5}}{\cancel{5} \cdot 5 \cdot 5} = \frac{1}{5^2}$

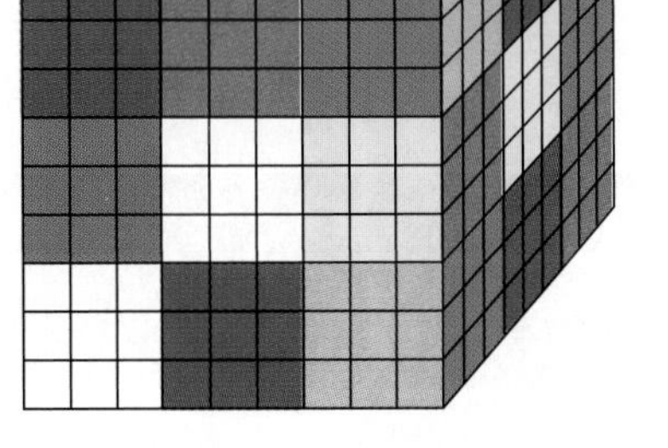

Example 3 *Rubik's Cubes*

You have a stack of Rubik's Cubes, as shown at the left. How many small cubes are in the stack?

Solution Each Rubik's Cube is composed of 3^3 small cubes, and there are 3^3 Rubik's Cubes in the stack. This means that the stack has $3^3 \cdot 3^3 = 3^{3+3} = 3^6 = 729$ small cubes.

You can check this result by reasoning that each Rubik's Cube is composed of 27 small cubes, which means that the stack has $27 \cdot 27 = 729$ small cubes. ■

Communicating about MATHEMATICS

Cooperative Learning

SHARING IDEAS about the Lesson

You Be the Teacher How would you explain to a partner that $3^4 \cdot 3^5$ is equal to 3^9?

EXERCISES

Guided Practice

CHECK for Understanding

1. *Writing* In your own words, state the definitions for negative integer and zero exponents.

In Exercises 2–5, rewrite the expression without using negative or zero exponents.

2. 4^{-1}
3. 5^{-2}
4. 100^0
5. x^{-3}

6. *Writing* In your own words, state how to multiply and divide two powers that have the same base.

In Exercises 7–10, simplify the expression, if possible.

7. $p^5 \cdot p^2$
8. $r^2 \cdot s^3$
9. $\frac{m^6}{n^4}$
10. $\frac{x^4}{x^2}$

Independent Practice

In Exercises 11–18, simplify the expression.

11. 3^{-2}
12. -10^{-3}
13. 16^0
14. $(-9)^2$
15. t^{-4}
16. $2x^{-3}$
17. $3s^{-2}$
18. r^0

In Exercises 19–26, simplify the expression.

19. $(-6)^{-3} \cdot (-6)^5$
20. $8^0 \cdot 8^4$
21. $x^{25} \cdot x^{-10}$
22. $y^{-6} \cdot y^4$
23. $\frac{7^5}{7^4}$
24. $\frac{-9^2}{-9^4}$
25. $\frac{a^{12}}{a^0}$
26. $\frac{b^7}{b^{10}}$

In Exercises 27–30, use a calculator to evaluate the expression. If necessary, round the result to 3 decimal places.

27. 2.5^{-4}
28. 5.5^{-2}
29. $5.5^3 \cdot 5.5^2$
30. $\frac{0.5^3}{0.5^6}$

Mental Math **In Exercises 31–34, solve the equation for *n*.**

31. $\frac{2^5}{2^2} = 2^n$
32. $\left(\frac{1}{2}\right)^n = 1$
33. $3^{-3} \cdot 3^n = 3^3$
34. $4^{-5} = \frac{1}{4^n}$

35. *Guess, Check, and Revise* Find the largest value of n such that $2^n < 100{,}000$.

36. *Guess, Check, and Revise* Find the largest value of n such that $3^{-n} > 0.00001$.

In Exercises 37 and 38, write an expression for the verbal phrase.

37. The product of seven raised to the tenth power and seven raised to the negative fourth power

38. The quotient of two raised to the negative fifth power and two raised to the third power

Comparing Powers **In Exercises 39–42, complete the statement with $<, >,$ or $=$.**

39. 3^{10} ? $3 \cdot 3^9$ **40.** 2^{-5} ? 5^{-2} **41.** $\frac{4^3}{4^2}$? $\frac{4^2}{4^3}$ **42.** $\frac{4^{14}}{4^3}$? 4^{10}

Sequences **In Exercises 43 and 44, rewrite the sequence as powers of 10. Describe the pattern and write the next three terms.**

43. 1, 10, 100, 1000, ?, ?, ? **44.** 1, 0.1, 0.01, 0.001, 0.0001, ?, ?, ?

45. *Geometry* The lengths of the sides of a rectangle are 2^3 meters and 2^4 meters. Write the area of the rectangle as a power of 2.

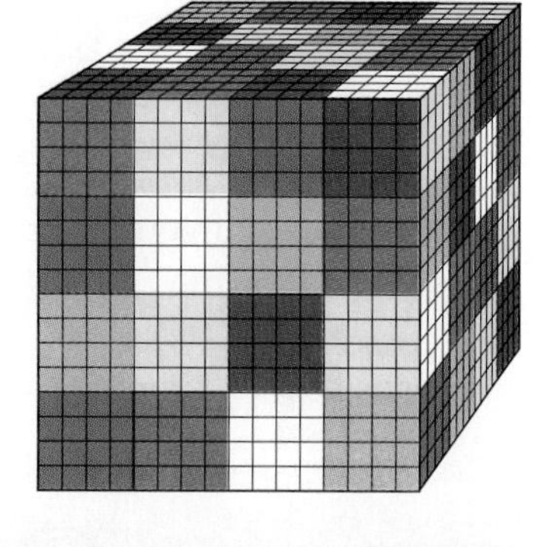

46. *Geometry* Suppose a Rubik's Cube contained 4^3 cubes. How many small cubes would be in the stack of Rubik's Cubes shown at the right?

47. *U.S. Population* In 1990, the United States had a population of about 250 million. Which of the following expressions represents this number? ***(Source: U.S. Bureau of Census)***

a. $2.5(10^6)$ **b.** $2.5(10^7)$ **c.** $2.5(10^8)$

48. *Beef Consumption* In 1990, the United States had a population of about 250 million.In that year, Americans ate about 16 billion pounds of beef. Find the average amount eaten by each American by simplifying the following expression. ***(Source: U.S. Department of Agriculture)***

$$\frac{1.6\,(10^{10})\text{ pounds}}{2.5\,(10^8)\text{ people}}$$

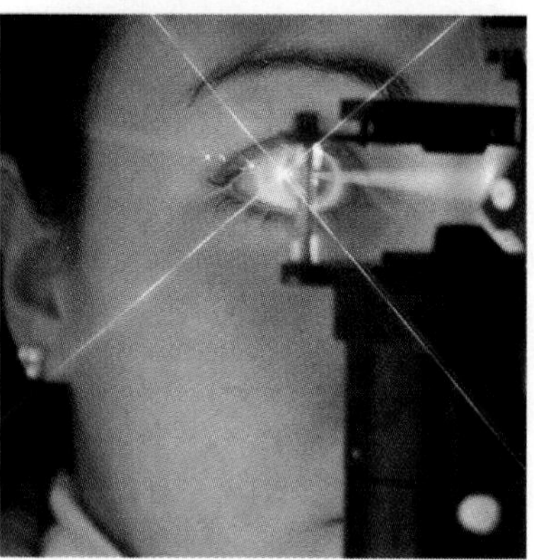

Laser is an acronym for Light Amplification by Stimulated Emission of Radiation.

49. *Biology* Most cells of living organisms are about 10 micrometers wide. One micrometer is 10^{-6} meters, which means that a typical cell is about $10 \cdot 10^{-6}$ meters wide. Write this measurement as a power of 10.

50. *Lasers* A laser is a device that amplifies light. A laser beam can be focused on a point that is just 0.05^2 millimeters wide. Write this measure as a fraction.

Integrated Review

Making Connections within Mathematics

Error Analysis **In Exercises 51–54, explain why the answer is incorrect. Then correct it.**

51. $3^{-2} = 3 \cdot -2 = -6$

52. $4^2 + 4^3 = 4^{2+3} = 4^5$

53. $7^2 \cdot 7^3 = 7^{2 \cdot 3} = 7^6$

54. $\frac{6^4}{6^8} = 6^{8-4} = 6^4$

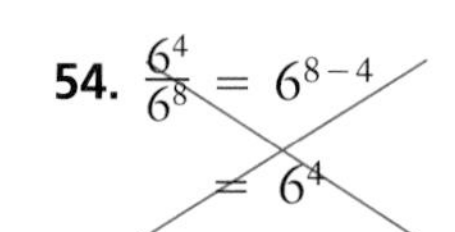

Reasoning **In Exercises 55–58, decide whether the answer is positive or negative. Explain.**

55. $(-1)^{16}$ **56.** $(-2)^{37}$ **57.** $(-3)^{42}$ **58.** $(-4)^{51}$

Exploration and Extension

59. *Think About It* Each cube below is made of smaller cubes. Each small yellow cube contains a \$20 bill. Each small red cube contains a \$10 bill. Each small blue cube contains a \$5 bill. If you were allowed to keep the money contained in one of the large cubes, which cube would you choose? Explain your reasoning.

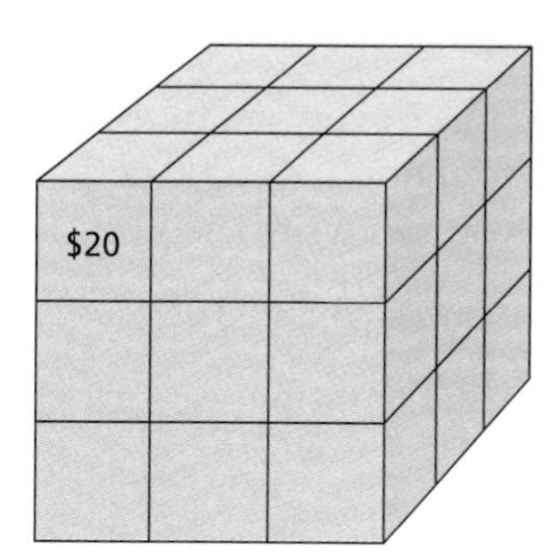

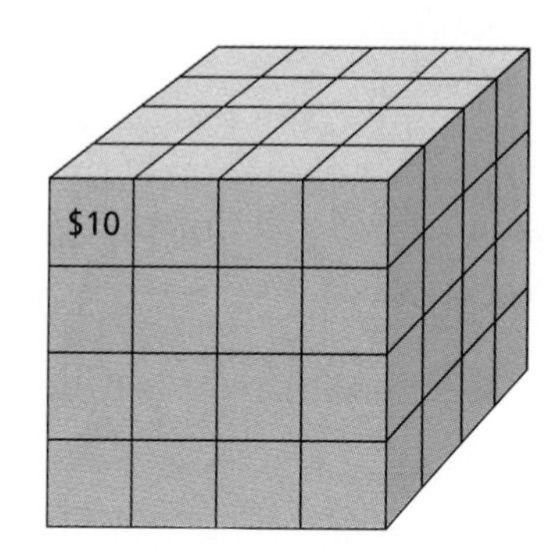

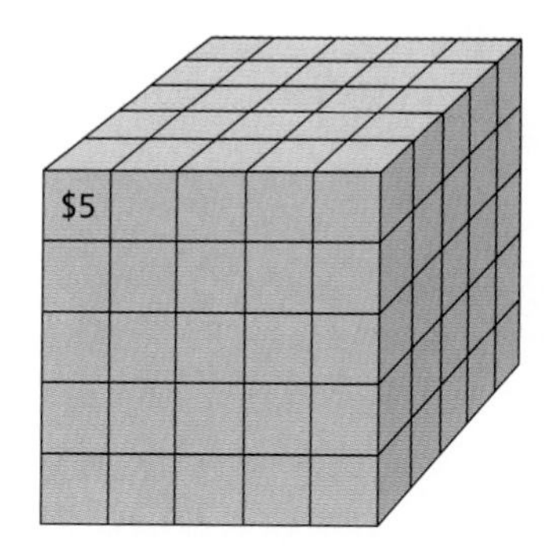

Mixed REVIEW

In Exercises 1–6, write the number in decimal form. Is the decimal terminating, repeating, or nonrepeating? (6.6)

1. $\frac{1}{27}$ **2.** $\frac{3}{8}$ **3.** $\frac{3}{9}$

4. $\frac{19}{18}$ **5.** $\frac{14}{9}$ **6.** $\frac{61}{111}$

In Exercises 7–15, solve the equation. Give the answer in decimal form. If necessary, round the result to 3 decimal places. (4.4, 6.6, 6.7)

7. $18x = 25$ **8.** $25y + 18 = 0$ **9.** $2(10a + 15) = 10$

10. $4x - \frac{1}{2} = \frac{5}{2}$ **11.** $0.5 - 50a = 200a$ **12.** $\frac{1}{10}(p + 2) = 1$

13. $32p - 10 = 10^{-2}$ **14.** $3(33x) = 313$ **15.** $3^2(111x) + 3142 = 0$

Coordinate Geometry **For Exercises 16–18, plot the points in the same coordinate plane. State in what quadrant the point is located. (3.8)**

16. $A(-2, 4)$ **17.** $B(0, 2)$ **18.** $C(3, -1)$

19. The points in Exercises 16–18 all lie on a line. Find another point that lies on the same line. **(3.8)**

20. Find the average of 0.1, −0.4, 0.2, −0.3, 0.1.

6.8 Scientific Notation

What you should learn:

How to use scientific notation to represent numbers

How to use scientific notation to solve real-life problems

Why you should learn it:

Scientific notation is a convenient way to represent numbers whose absolute values are very small or very large.

Goal 1 Using Scientific Notation

Many numbers in real life are very large or very small. For instance, the population of the world is about 5,500,000,000. Instead of writing this many zeros, you can write

$$5{,}500{,}000{,}000 = 5.5 \times 1{,}000{,}000{,}000 = 5.5 \times 10^9.$$

The form on the right is called **scientific notation.**

Scientific Notation

A number is written in **scientific notation** if it has the form

$c \times 10^n$

where c is greater than or equal to 1 and less than 10.

Example 1 *Writing Numbers in Scientific Notation*

	Decimal Form	Product Form	Scientific Notation
a.	3,400	3.4×1000	3.4×10^3
b.	56,000,000	$5.6 \times 10{,}000{,}000$	5.6×10^7
c.	0.00923	9.23×0.001	9.23×10^{-3}
d.	0.0000004	4×0.0000001	4×10^{-7}

■

Need to Know

When writing powers of 10, remember that positive exponents correspond to large numbers and negative exponents correspond to small numbers.

1,000,000	10^6
100,000	10^5
10,000	10^4
1,000	10^3
100	10^2
10	10^1
1	10^0
0.1	10^{-1}
0.01	10^{-2}
0.001	10^{-3}

In Example 1, notice that the exponent of 10 indicates the number of places the decimal point is moved. For instance, in part **a** the decimal point is moved 3 places and in part **d** the decimal point is moved 7 places.

Decimal Form	Scientific Notation
3,400.00	3.4×10^3

Move decimal point 3 places to the left.

Decimal Form	Scientific Notation
0.0000004	4×10^{-7}

Move decimal point 7 places to the right.

Goal 2 Using Scientific Notation in Real Life

To multiply two numbers that are written in scientific notation, you can use the rule for multiplying powers with like bases. Here is an example.

$(3.2 \times 10^5) \times (4 \times 10^6)$	
$= 3.2 \times 4 \times 10^5 \times 10^6$	*Reorder.*
$= (3.2 \times 4) \times (10^5 \times 10^6)$	*Regroup.*
$= 12.8 \times 10^{11}$	*Multiply.*
$= 1.28 \times 10^{12}$	*Scientific notation.*

This can also be written as 1,280,000,000,000 or 1.28 trillion.

Example 2 *Multiplying with Scientific Notation*

You work in a warehouse that stores paper. You are storing paper that has a thickness of 4.4×10^{-3} inch. The paper comes in packages of 500 sheets. Each carton of paper has a stack of 5 packages. How tall is a stack of 10 cartons?

In 1990, Americans used an average of about 600 pounds of paper and cardboard per person.

Solution

Verbal Model Height of stack = Number of sheets × Thickness of sheet

Labels Number of sheets = 10(5)(500) = 25,000 (sheets)
Thickness of sheet = 4.4×10^{-3} (inches per sheet)

Algebraic Model

$$\begin{aligned} \text{Height} &= 25{,}000 \times (4.4 \times 10^{-3}) \\ &= (2.5 \times 10^4) \times (4.4 \times 10^{-3}) \\ &= (2.5 \times 4.4) \times (10^4 \times 10^{-3}) \\ &= 11 \times 10^1 \\ &= 110 \end{aligned}$$

The stack is 110 inches high. ■

Communicating about MATHEMATICS

SHARING IDEAS about the Lesson

Scientific Notation State whether the number is in scientific notation. If it isn't, rewrite the number in scientific notation.

A. 12.4×10^{-3} **B.** 3.8×10^{-2} **C.** 0.5×10^4

EXERCISES

Guided Practice

CHECK for Understanding

1. Which of the following is written in scientific notation?

a. 12.3×10^3 **b.** 1.23×10^4 **c.** 0.123×10^5

In Exercises 2 and 3, find the power of 10.

2. $350{,}000 = 3.5 \times 10^{\boxed{?}}$

3. $0.00943 = 9.43 \times 10^{\boxed{?}}$

In Exercises 4 and 5, write the number in decimal form.

4. 6.25×10^5

5. 8.7×10^{-6}

6. *World Populations* One of the following is the approximate 1990 population of China. The other is the approximate 1990 population of Canada. Which is which? Explain your reasoning.

a. 2.6×10^7 **b.** 1.5×10^9

Independent Practice

In Exercises 7–14, write the number in scientific notation.

7. 5000 **8.** 643,000 **9.** 0.00041 **10.** 0.18

11. 32,610,000 **12.** 5,730,000,000 **13.** 0.000000012 **14.** 0.000008

In Exercises 15–22, write the number in decimal form.

15. 5.7×10^{-3} **16.** 3.41×10^{-6} **17.** 2.50×10^4 **18.** 2.4×10^9

19. 6.2×10^{10} **20.** 8.59×10^5 **21.** 3.63×10^{-7} **22.** 5.99×10^{-1}

In Exercises 23–28, decide whether the number is in scientific notation. If it is not, rewrite the number in scientific notation.

23. 5.3×10^{-5} **24.** 0.392×10^6 **25.** 25.6×10^8

26. 3.7×10^9 **27.** 791×10^{-4} **28.** 68.8×10^3

In Exercises 29–32, evaluate the product. Write the result in scientific notation and in decimal form.

29. $(6.2 \times 10^2)(8 \times 10^3)$

30. $(4.5 \times 10^{-3})(3.4 \times 10^5)$

31. $(0.3 \times 10^{-4})(0.6 \times 10^{-1})$

32. $(9.7 \times 10^4)(2.4 \times 10^2)$

Number Sense **In Exercises 33 and 34, decide which is larger. Explain.**

33. 1×10^9 or 9×10^8

34. 5×10^{-5} or 1×10^{-4}

In Exercises 35 and 36, write the number in scientific notation.

35. A thunderstorm cloud holds about 6,000,000,000,000 raindrops.

36. The adult human body contains about 100,000,000,000,000 cells.

37. *Musical Instruments* You are writing a report and have collected the information in the table, which shows the number of people who play the six most popular instruments. Rewrite the table so that the numbers are in decimal form. Which table do you think would be better to include in your report? Explain your reasoning. ***(Source: American Music Conference)***

Instrument	Piano	Guitar	Organ	Flute	Clarinet	Drums
Number	2.06×10^7	1.89×10^7	6.3×10^6	4×10^6	4×10^6	3×10^6

38. *Density* The density of an element is related to its weight. Light elements such as oxygen have a smaller density (in grams per cubic centimeter) than heavy elements such as iron. Write each of the following densities in scientific notation. Then order the elements from lightest to heaviest.

Element	**Density**
Chlorine	0.00295
Helium	0.0001664
Hydrogen	0.00008375
Nitrogen	0.001165
Oxygen	0.001332

The Milky Way galaxy contains the sun, Earth, and the rest of the solar system. Its diameter is about 100,000 light-years. Can you find its diameter in miles?

39. *Milky Way* Some stars in the Milky Way are 8×10^4 light-years from Earth. A light-year is 5.88×10^{12} miles. Write 8×10^4 light-years in miles.

Integrated Review

Making Connections within Mathematics

Geometry **In Exercises 40 and 41, find the area of the figure.**

40.

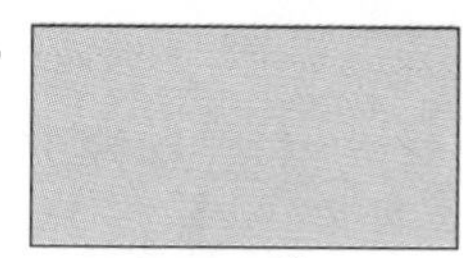

41. (Triangle with height 50×10^1 m and base 640×10^{-2} m)

Exploration and Extension

Division **In Exercises 42–44, divide. Explain your process.**

42. $\dfrac{3.6 \times 10^8}{1.2 \times 10^8}$

43. $\dfrac{3.6 \times 10^8}{1.2 \times 10^7}$

44. $\dfrac{3.6 \times 10^8}{1.2 \times 10^6}$

USING A CALCULATOR

Scientific Notation

With a scientific calculator, you may enter and display numbers whose absolute values are very small or very large using scientific notation.

Example 1 *Entering Numbers in Scientific Notation*

Number	Keystrokes	Display
8.75×10^{-15}	8.75 [EE] 15 [+/-]	8.75 −15
3.629×10^{12}	3.629 [EE] 12	3.629 12

With a scientific calculator, you enter the decimal portion of the number and the exponent. The base of 10 is not entered or displayed. Some scientific calculators have an [EXP] key instead of an [EE] key. ■

Example 2 *Reading the Calculator Display*

Perform the indicated operation.

a. $232{,}000 \times 1{,}500{,}000$ **b.** $0.003 \div 1{,}500{,}000$

Solution

a. The calculator displays 3.48 11. Because it is understood that the base of the exponent is 10, the display is read as 3.48×10^{11} or "three and forty-eight hundredths times ten to the eleventh power."

b. The calculator displays 2 −09. The display is read as 2×10^{-9} or "two times ten to the negative ninth power." ■

Exercises

In Exercises 1–8, write the result of the operation.

1. $(3.6 \times 10^4)(6.3 \times 10^2)$

2. $(9.83 \times 10^{10})(5.2 \times 10^8)$

3. $(1.35 \times 10^{-3})(8.2 \times 10^{-9})$

4. $(4.7 \times 10^{-7})(2.65 \times 10^{-5})$

5. $(422{,}000)(135{,}000)$

6. $(9{,}364{,}000)(2150)$

7. $(0.014) \div (560{,}000)$

8. $(9.12 \times 10^{-3}) \div (2.4 \times 10)$

9. Use the power key as shown to evaluate the expressions in Exercises 2 and 8.

Exercise 2: 9.83 [×] 10 [y^x] 10 [×] 5.2 [×] 10 [y^x] 8

Exercise 8: 9.12 [×] 10 [y^x] 3 [+/-] [÷] 2.4 [×] 10

What do you notice? Why is it a good idea to use the [EE] key, rather than the power key, when computing with scientific notation?

6.9 Exploring Patterns

What you should learn:

Goal 1 How to recognize number patterns

Goal 2 How to recognize patterns in a coordinate plane

Why you should learn it:

Being able to recognize patterns helps you find the patterns in real-life processes.

Goal 1 Recognizing Number Patterns

In earlier lessons, you studied several number patterns.

Name	Numbers	Pattern
Square **(1.3)**	1, 4, 9, 16, 25, ...	n^2
Cubic **(1.8)**	1, 8, 27, 64, 125, ...	n^3
Triangular **(1.8)**	1, 3, 6, 10, 15, ...	$\frac{1}{2}n(n+1)$
Prime **(6.2)**	2, 3, 5, 7, 11, ...	None is known.
Fibonacci **(1.8)**	1, 1, 2, 3, 5, ...	Each (after 1, 1) is the sum of two previous numbers.

For thousands of years, people have studied number patterns. Some patterns are simple, but some are very difficult. No one has been able to write a formula for the nth prime number.

Example 1 *Perfect Numbers*

A natural number is called **perfect** if it is equal to the sum of its factors except itself. It is **deficient** if the sum of its factors, except itself, is less than the number, and it is **abundant** if the sum of its factors, except itself, is greater than the number. Classify the natural numbers from 2 through 12.

Solution

Number	Factors	Sum of Factors	Type
2	1	1	Deficient
3	1	1	Deficient
4	1, 2	1 + 2 = 3	Deficient
5	1	1	Deficient
6	1, 2, 3	1 + 2 + 3 = 6	Perfect
7	1	1	Deficient
8	1, 2, 4	1 + 2 + 4 = 7	Deficient
9	1, 3	1 + 3 =4	Deficient
10	1, 2, 5	1 + 2 + 5 =8	Deficient
11	1	1	Deficient
12	1, 2, 3, 4, 6	1 + 2 + 3 + 4 + 6 =16	Abundant

Notice that the only perfect number in the table is 6. ■

African Wall Painting
Women from Ghana use earth colors and bold geometric patterns for wall decorations.

Goal 2 Patterns in a Coordinate Plane

Geometry can help you find patterns among numbers. For instance, on page 39, geometry was used to find a pattern for triangular numbers.

Example 2 *Graphical Model for Rational Numbers*

For each of the following rational numbers, $\frac{a}{b}$, plot the ordered pair (b, a). Then describe the pattern.

a. $\frac{1}{2}, \frac{2}{4}, \frac{3}{6}, \frac{4}{8}$ **b.** $\frac{3}{-2}, \frac{-3}{2}, \frac{6}{-4}, \frac{-6}{4}$

Solution

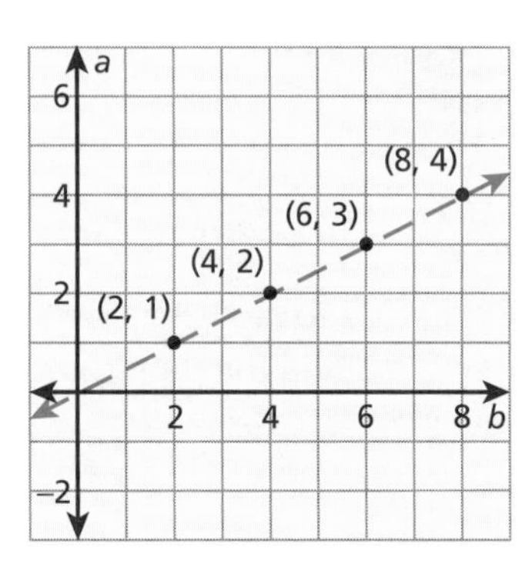

a. For $\frac{1}{2}$, plot the point $(2, 1)$.
For $\frac{2}{4}$, plot the point $(4, 2)$.
For $\frac{3}{6}$, plot the point $(6, 3)$.
For $\frac{4}{8}$, plot the point $(8, 4)$.
From the coordinate plane at the right, you can see that all four points lie on a line.

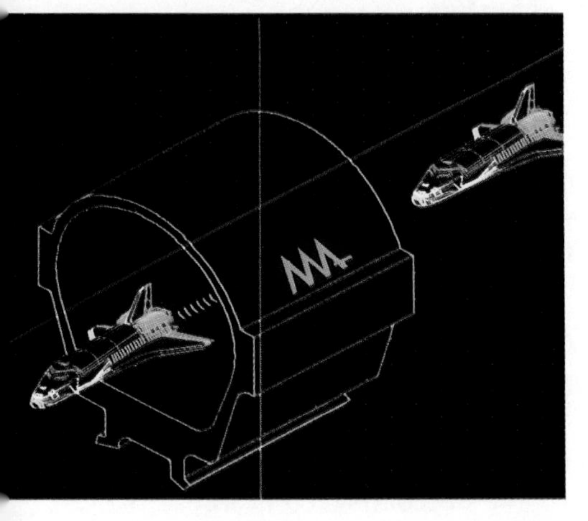

This art was produced on a computer. Computer programs that produce graphics such as this use coordinate systems to identify points on the computer screen.

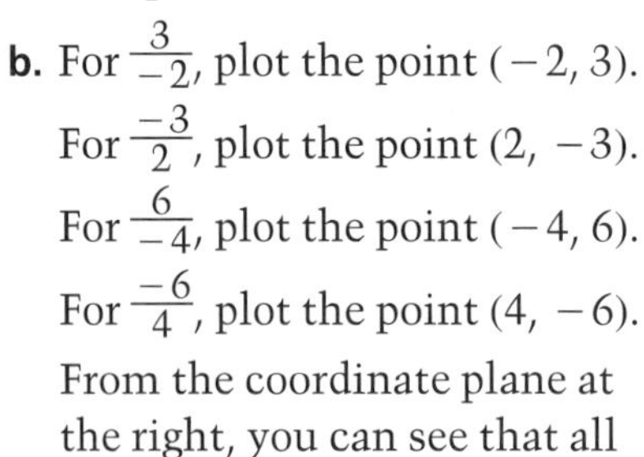

b. For $\frac{3}{-2}$, plot the point $(-2, 3)$.
For $\frac{-3}{2}$, plot the point $(2, -3)$.
For $\frac{6}{-4}$, plot the point $(-4, 6)$.
For $\frac{-6}{4}$, plot the point $(4, -6)$.
From the coordinate plane at the right, you can see that all four points lie on a line.

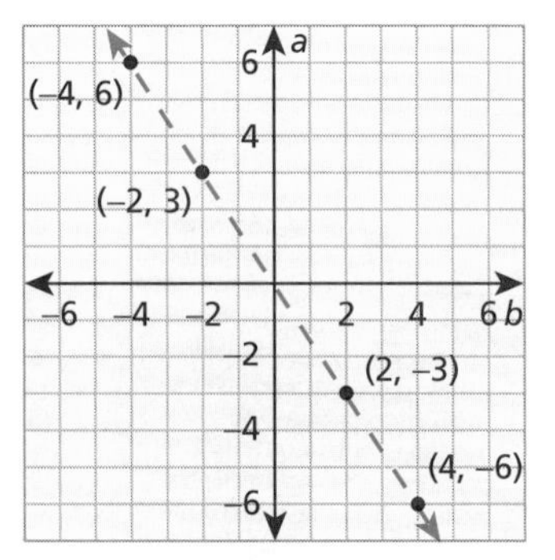

Communicating about MATHEMATICS

SHARING IDEAS about the Lesson

Making a Conjecture In Part **a** of Example 2, each of the four points lies on a line. Each of the four rational numbers is equivalent. Use these observations to write a conjecture. Test your conjecture by choosing other equivalent rational numbers and plotting the points that correspond to them.

EXERCISES

Guided Practice

CHECK for Understanding

1. List all the factors of 28. Then use the factors to show that 28 is a perfect number.

2. Complete the table.

n	1	2	3	4	5	6
$2n^2 + 1$	?	?	?	?	?	?

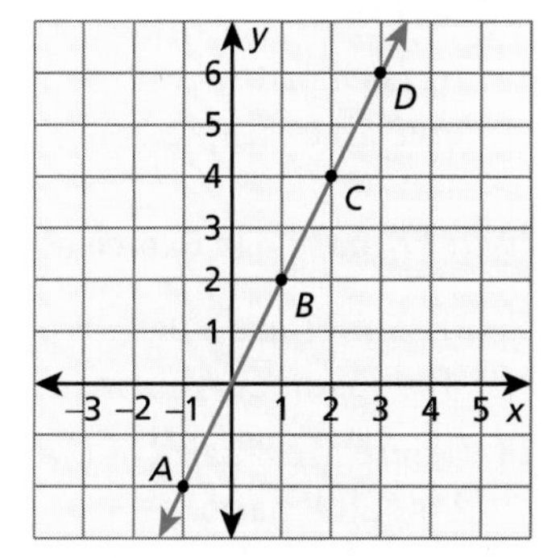

In Exercises 3 and 4, use the coordinate plane at the right.

3. Identify the coordinate of points A, B, C, and D.

4. *A Coordinate Pattern* Describe the pattern of the coordinates. If the x-coordinate is 11, what is the y-coordinate?

Independent Practice

In Exercises 5–8, use the formula to construct a table similar to that shown in Exercise 2.

5. $n^2 + 1$

6. $n^2 + n$

7. 2^{n-1}

8. 2^{1-n}

Sequences **In Exercises 9–12, describe the pattern. Then list the next three terms in the sequence.**

9. 0, 3, 8, 15, ?, ?, ?

10. $1, \frac{1}{3}, \frac{1}{9}, \frac{1}{27},$?, ?, ?

11. 1, 2, 2, 4, 8, ?, ?, ?

12. 2, 4, 12, 48, 240, ?, ?, ?

13. *Fibonacci Sequence* The sequence 1, 1, 2, 4, 7, 13, 24, . . . is similar to the Fibonacci sequence. Describe the pattern. Then list the next three terms.

14. *It's Up to You* Make up your own "Fibonacci-like" sequence.

Figurate Numbers **In Exercises 15–18, each figure represents a figurate number. Predict the next two numbers in the sequence. Then draw figures to check your predictions.**

15.

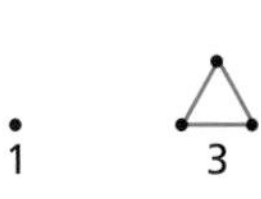

6

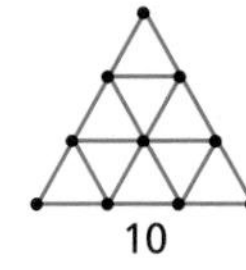

16.

1

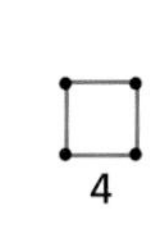

9

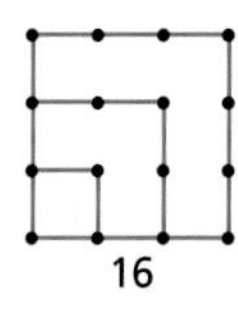

17.

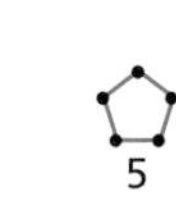

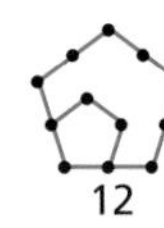

18.

1

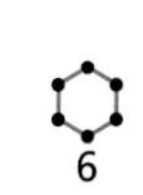

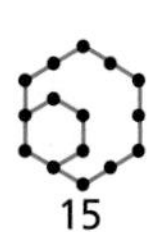

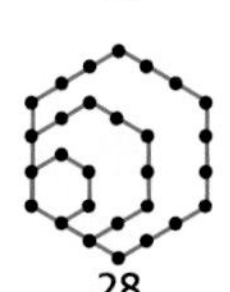

Symmetric Primes **In Exercises 19–22, use the following information.**

Let n be a natural number. Two primes are *symmetric primes of n* if their average is n. For instance, 7 and 13 are symmetric primes of 10 because the average of 7 and 13 is 10.

19. Find the other pair of symmetric primes of 10.

20. Explain why 2 and 18 are not symmetric primes of 10.

21. List the six pairs of symmetric primes of 50.

22. If n is a prime number, can it have a pair of symmetric primes? If so, give an example.

Twin Primes **In Exercises 23–26, use the following information.**

Twin primes are a pair of prime numbers whose difference is 2. For instance, the prime numbers 3 and 5 are twin primes.

23. There are 8 pairs of twin primes that are less than 100. List all 8 pairs.

24. Find the average of each pair in Exercise 23. For instance, the average of 5 and 7 is 6. Except for 3 and 5, do you think that the average of each pair of twin primes is divisible by 6? Explain your reasoning.

25. Consider the following pairs of twin primes: 11 and 13, 41 and 43, and 71 and 73. Describe the pattern and determine the next pair. Is it a pair of twin primes?

26. Is the number 211 and another number a twin prime? Explain how you obtained your answer.

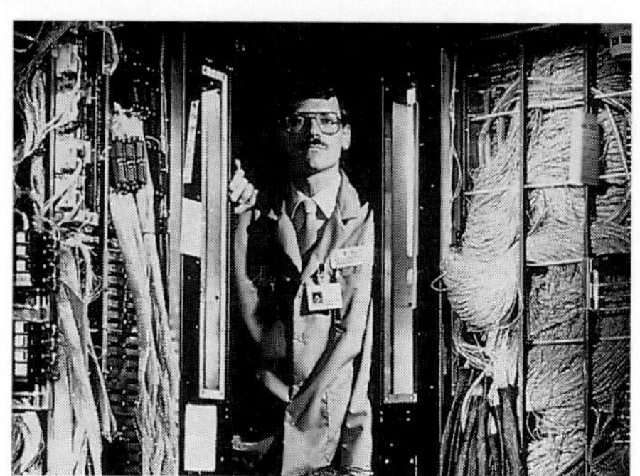

David Slowinski has designed software for Cray supercomputers to find prime numbers of the form $2^p - 1$, where p is prime. In 1994, his software produced the largest known prime number $2^{859,433} - 1$, a number that is 258,716 digits long.

Integrated Review

Making Connections within Mathematics

Factors **In Exercises 27–33, list all the factors of the number.**

27. 20 **28.** 42 **29.** 72 **30.** 90

Exploration and Extension

Perfect Numbers **In Exercises 31–33, use the following information.**

The ancient Greeks were the first to discover perfect numbers. Every even perfect number is of the form $2^{p-1}(2^p - 1)$ where p and $2^p - 1$ are prime.

31. Complete the columns in the table for 2^{p-1} and $2^p - 1$ when $p = 5$. Is $2^p - 1$ a prime number?

32. Complete the last column to find the third perfect number.

33. List all the factors of the third perfect number and show the sum of the factors equals the number.

p	2^{p-1}	$2^p - 1$	$2^{p-1}(2^p - 1)$
2	2	3	6
3	4	7	28
5	?	?	?

6 Chapter Summary

What did you learn?

Skills

1. Use divisibility tests. **(6.1)**
   - Factor a natural number. **(6.1)**
   - Factor an algebraic expression. **(6.2)**
2. Classify a natural number as prime or composite. **(6.2)**
   - Write the prime factorization of a natural number. **(6.2)**
3. Find the greatest common factor of two natural numbers. **(6.3)**
4. Find the least common multiple of two natural numbers. **(6.4)**
5. Simplify a fraction. **(6.5)**
   - Compare two fractions. **(6.5)**
   - Identify rational numbers. **(6.6)**
   - Write a rational number in decimal form. **(6.6)**
6. Evaluate powers that have negative and zero exponents. **(6.7)**
7. Use scientific notation to write decimal numbers. **(6.8)**
   - Multiply numbers in scientific notation. **(6.8)**

Problem-Solving Strategies

8. Recognize and describe number patterns. **(6.9)**
9. Model and solve real-life problems. **(6.1–6.9)**

Exploring Data

10. Use tables and graphs to solve problems. **(6.1–6.9)**

Why did you learn it?

Knowing properties of natural numbers helps you understand many different types of real-life situations. For instance, suppose you are an employer who is determining the annual salary of an employee. If you pay your employees each month, you would want the annual salary to be divisible by 12, but if you pay them each week, you would want the annual salary to be divisible by 52. Knowing how to use negative exponents, zero exponents, and scientific notation helps you solve problems that involve very large or very small numbers. For instance, if you become an astronomer, you will need to measure distances to other planets and other solar systems.

How does it fit into the bigger picture of mathematics?

Number theory is one of the oldest branches of mathematics. In this chapter, you learned the names of several types of numbers, such as prime numbers, composite numbers, and rational numbers. You also learned that each type of number has special properties. For instance, rational numbers are the only type of numbers whose decimal forms are either terminating or repeating.

Chapter REVIEW

In Exercises 1–8, use the Divisibility Tests to determine whether the number is divisible by 2, 3, 4, 5, 6, 8, 9, or 10. (6.1)

1. 2560 **2.** 16,480 **3.** 342 **4.** 4212

5. 245 **6.** 3845 **7.** 4968 **8.** 2721

In Exercises 9–16, decide whether the number is prime or composite. If it is composite, list all its factors. (6.1, 6.2)

9. 15 **10.** 9 **11.** 13 **12.** 38

13. 46 **14.** 50 **15.** 64 **16.** 29

In Exercises 17–28, write the prime factorization of the expression. (6.2)

17. 80 **18.** 96 **19.** 120 **20.** 136

21. -135 **22.** -252 **23.** $40x^4$ **24.** $108y^2$

25. $12a^6b^2$ **26.** $21p^3q$ **27.** $-81st^5$ **28.** $-48m^4n^7$

In Exercises 29–36, find the greatest common factor and the least common multiple. (6.3, 6.4)

29. 5, 15 **30.** 8, 18 **31.** 216, 240 **32.** 405, 450

33. $2x^4y^3, 4xy^8$ **34.** $9ab^5, 18a^2b^3$ **35.** $6y^5z^4, 14y^4z^5$ **36.** $9mn^4, 12mn$

In Exercises 37–44, simplify the expression. (6.5)

37. $\frac{4}{28}$ **38.** $\frac{5}{50}$ **39.** $\frac{8}{46}$ **40.** $\frac{12}{32}$

41. $\frac{2a^2b}{18a}$ **42.** $\frac{7rs}{63r^2}$ **43.** $\frac{36z}{54z^8}$ **44.** $\frac{44x^4y}{99x^4y^3}$

In Exercises 45–48, complete the statement with $<$, $>$, or $=$. (6.5)

45. $\frac{1}{3}$? $\frac{1}{4}$ **46.** $\frac{2}{5}$? $\frac{8}{20}$ **47.** $\frac{6}{14}$? $\frac{21}{49}$ **48.** $\frac{26}{34}$? $\frac{13}{17}$

In Exercises 49–56, decide whether the number is rational or irrational. Then write the number in decimal form and state whether the decimal is terminating, repeating, or nonrepeating. (6.6)

49. $\frac{3}{18}$ **50.** $\sqrt{11}$ **51.** $-\frac{5}{8}$ **52.** $\frac{15}{20}$

53. $\frac{18}{15}$ **54.** $-\frac{8}{9}$ **55.** $\sqrt{43}$ **56.** $\frac{29}{27}$

In Exercises 57–68, simplify the expression. (6.7)

57. 4^{-1} **58.** $(-5)^{-3}$ **59.** $(-7)^{-2}$ **60.** 12^0

61. x^{-4} **62.** $(-y)^{-2}$ **63.** $3^{-4} \cdot 3$ **64.** $\frac{8^5}{8^3}$

65. $-10a^5 \cdot 10a$ **66.** $2y^7 \cdot 4x^9y^2$ **67.** $\frac{6m^4n^5}{3m^6n}$ **68.** $\frac{9r^2st}{15r^2s^6}$

In Exercises 69–72, write the number in decimal form. (6.8)

69. 3.9×10^7 **70.** 6.8×10^{-6} **71.** 9.46×10^{-4} **72.** 7.52×10^5

In Exercises 73–76, write the number in scientific notation. (6.8)

73. 1,200,000,000 **74.** 456,000 **75.** 0.00045 **76.** 0.0000592

In Exercises 77–80, find the product. Write the result in scientific notation. (6.8)

77. $(1.5 \times 10^3)(2.5 \times 10^4)$

78. $(2.4 \times 10^{-1})(5.0 \times 10^8)$

79. $(1.0 \times 10^{-8})(1.0 \times 10^8)$

80. $(3.6 \times 10^{-4})(2.0 \times 10^{-2})$

Sequences **In Exercises 81–84, describe the pattern. Then write the next three terms in the sequence. (6.9)**

81. $\frac{1}{2}, \frac{1}{3}, \frac{1}{4}, \frac{1}{5}$, ?, ?, ?

82. $-\frac{1}{2}, -\frac{2}{3}, -\frac{3}{4}, -\frac{4}{5}$, ?, ?, ?

83. $-\frac{7}{1}, -\frac{6}{2}, -\frac{5}{3}, -\frac{4}{4}$, ?, ?, ?

84. $\frac{1}{2}, \frac{1}{4}, \frac{1}{6}, \frac{1}{8}$, ?, ?, ?

85. ***Number Sense*** Is it possible to find a number that is divisible by 2 and not divisible by 4? If so, give an example. **(6.1)**

86. ***Number Sense*** Is it possible to find a number that is divisible by 4 and not divisible by 2? If so, give an example. **(6.1)**

87. ***Bring on the Sun*** A survey was conducted to determine how important "a lot of sun" is to people when selecting a vacation spot. The table shows the fractions of the people who answered "very," "not very," and "not at all." Write each fraction as a decimal. What do you notice about the sum of the fractions? (***Source: USA Today***) **(6.5, 6.6)**

How Important?	Fraction
Very	$\frac{69}{100}$
Not very	$\frac{16}{100}$
Not at all	$\frac{15}{100}$

88. ***World Waters*** The table below shows the names of some oceans and seas of the world and their areas in square miles. Write each area in scientific notation. **(6.8)**

Ocean or Sea	Area (in square miles)
Pacific	64,186,000
Atlantic	33,420,000
Indian	28,350,000
Arctic	5,105,000
Caribbean	971,400

Earth contains about 140 million square miles of water, which is about $\frac{7}{10}$ of its total surface.

FOCUS ON *Composing Music*

Real Life Connections

Musical Notes **In Exercises 89–92, use the diagram of musical notes.**

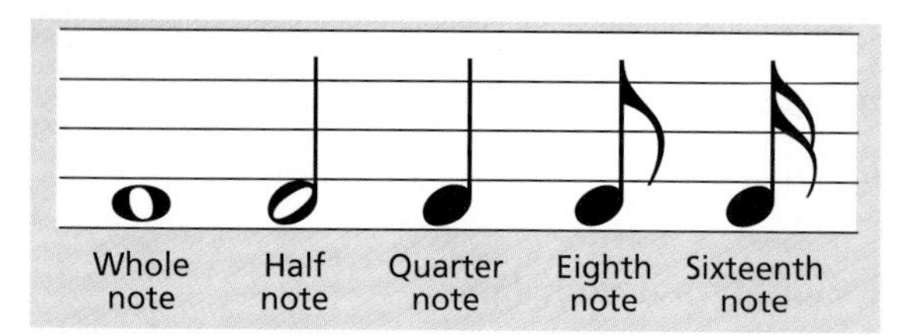

89. Write the names of the notes as fractions.

90. Describe the pattern of the fractions in Exercise 89.

91. If a whole note is held for four beats and a half note is held for two beats, how long would you hold a quarter note, an eighth note, and a sixteenth note?

92. Describe the pattern of the beats of the notes in Exercise 91.

Note	Number of Beats
Half	2
Dotted half	3
Quarter	1
Dotted quarter	$1\frac{1}{2}$
Eighth	$\frac{1}{2}$
Dotted eighth	$\frac{3}{4}$
Sixteenth	$\frac{1}{4}$
Dotted sixteenth	$\frac{3}{8}$

Use this table with Exercises 93 and 94.

Dotted Notes **In Exercises 93 and 94, use the following diagram and the table at the upper right.**

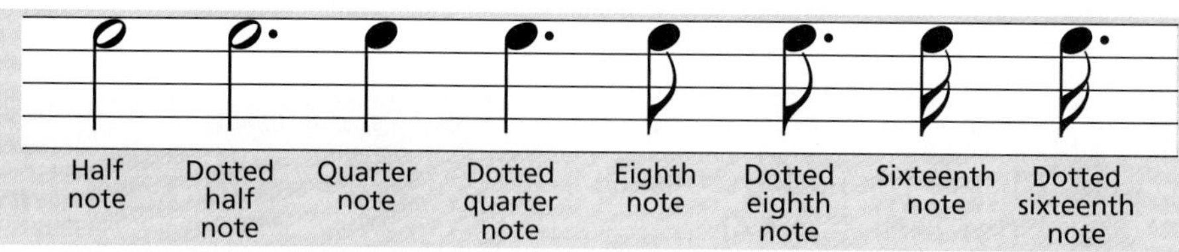

93. Dotting a note changes the number of beats that the note is held. The number of beats to hold a note and its related dotted note are given in the table above. How does the dot change the length of time a note is held?

94. Describe the pattern for the number of beats of the dotted-half note, dotted-quarter note, dotted-eighth note, and dotted-sixteenth note.

Reading Music **In Exercises 95 and 96, use the musical phrase below.**

95. Describe a pattern for the number of beats for the notes in this phrase.

96. Describe a pattern for the position of the notes on the musical staff.

97. If you can play an instrument or read music, create your own musical compositions. Describe any patterns used in creating your piece.

Chapter TEST

1. Use the divisibility tests to decide whether 1224 is divisible by 2, 3, 4, 5, 6, 8, 9, or 10.

2. Decide whether 1224 is divisible by 11, 12, 13, 14, 15, 16, 17, 18, 19, or 20.

In Exercises 3 and 4, write the prime factorization of the number.

3. 120
4. 125

In Exercises 5 and 6, write the prime factorization of the expression.

5. $99ab$
6. $121x^2$

In Exercises 7 and 8, find the greatest common factor.

7. 48, 36
8. 56, 98

In Exercises 9–11, find the least common multiple.

9. 10, 35
10. 5, 18
11. 7, 10, 14

12. Simplify the fraction $\frac{20}{800}$.
13. Decide which fraction is larger: $\frac{3}{11}$ or $\frac{5}{22}$.
14. Write 0.65 as a fraction and simplify.
15. Simplify the expression $x^9 \cdot x^{-4}$.

In Exercises 16–19, match the scientific notation with its decimal form.

a. 0.00016
b. 0.016
c. 160
d. 16,000

16. 1.6×10^4
17. 1.6×10^{-4}
18. 1.6×10^{-2}
19. 1.6×10^2

20. The first 3 square numbers are represented below. Draw the next 2 square numbers.

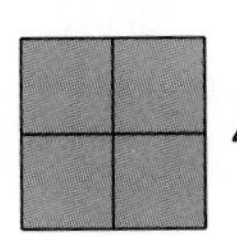

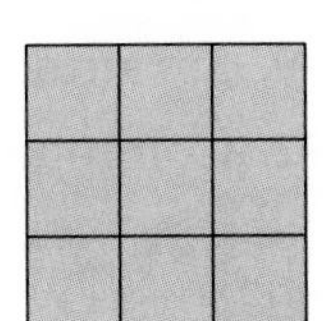

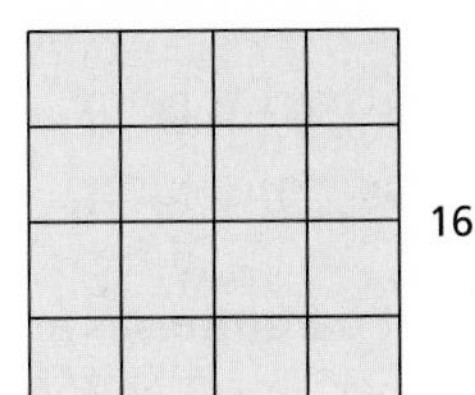

In Exercises 21–23, use the following information.

Scientists have measured the speed of light to be about 300,000 kilometers per second. It takes light about 500 seconds to travel from the sun to Earth.

21. Write the speed of light in scientific notation.
22. Write the time (in seconds) that it takes light to travel from the sun to Earth in scientific notation.
23. Approximate the distance between the sun and Earth by simplifying the following expression.

$$\text{Distance} = \left(300{,}000\ \frac{\text{kilometers}}{\text{second}}\right)(500\ \text{seconds}).$$

It takes light about 8 minutes and 20 seconds to travel from the sun to Earth. Direct sunlight can be harmful to the eyes.

Cumulative REVIEW ▪ Chapters 1–6

Geometric Patterns **In Exercises 1 and 2, describe the pattern and draw the next figure. Then write an expression for the perimeter of the regular polygon and evaluate the expression when $x = 2$. (1.7, 2.2)**

1.

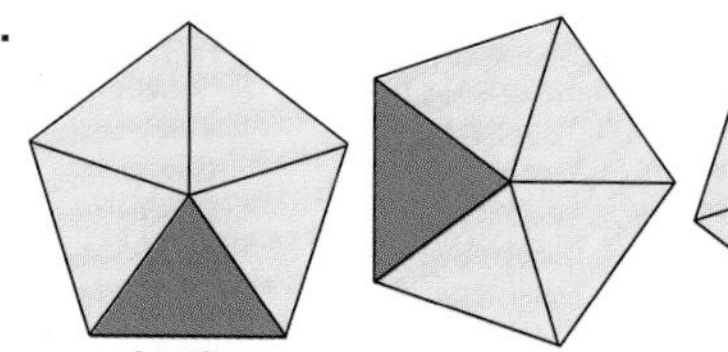

2.

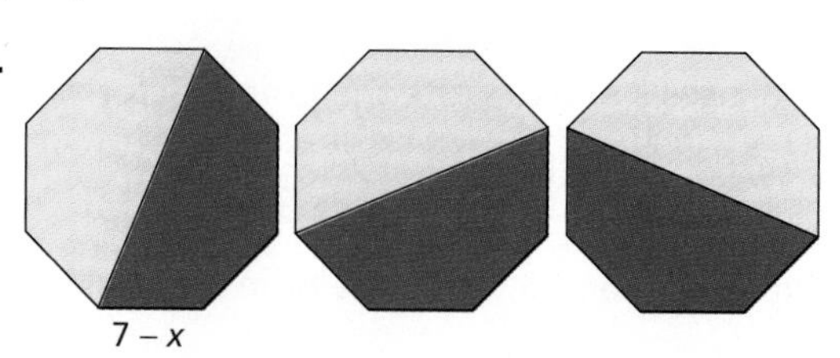

In Exercises 3–10, use a calculator to evaluate the expression. Round to two decimal places if necessary. (1.3, 1.4)

3. 5^7
4. $\left(\frac{4}{9}\right)^4$
5. $\sqrt{84}$
6. $(5^4 - 20) \div 11 + 9$
7. $\sqrt{512}$
8. $(6.3)^6$
9. $\sqrt{17.82}$
10. $2^5 + (24 - 6) \cdot 3$

In Exercises 11–16, evaluate the expression. (1.4, 3.1–3.6)

11. $-|-8| + 17 - 13 - 5$
12. $|-6| - 9 - 4 + 21$
13. $(-5)(3)(-6)(-2)$
14. $\frac{-625}{-5}$
15. $4^3 + (2 - 7)^2 \div 5$
16. $32 - (4 - 7)^3 \cdot 3$

In Exercises 17–22, match the term or property with its definition. (1.5, 2.1, 2.4, 2.5)

a. A collection of numbers, variables, operations, and grouping symbols
b. A letter used to represent one or more numbers
c. $ab + ac = a(b + c)$
d. $a(bc) = (ab)c$
e. $a + b = b + a$
f. Replacing a variable by a number

17. Variable
18. Distributive Property
19. Substitution
20. Associative Property
21. Algebraic expression
22. Commutative Property

In Exercises 23–28, write the expression without parentheses and combine like terms when possible. Then evaluate it when $x = -2$, $y = 4$, and $z = 5$. (3.1–3.5)

23. $5(x + y + z)$
24. $-4(x - y + 2z)$
25. $3y - 6x + 3z - 5y$
26. $-2(4x + 3x + y)$
27. $-7 + x(z + y)$
28. $2(3z - z) - |-y|$

In Exercises 29–36, write an algebraic equation or inequality for the sentence. Then solve. (2.3–2.5, 2.7, 2.9, 3.7)

29. -18 is the difference of x and 9.
30. The sum of n and 16 is 3.
31. The product of y and -8 is -104.
32. 12 is the quotient of y and 4.
33. 5 is greater than or equal to the quotient of a and 13.
34. The difference of b and 8 is greater than or equal to -15.
35. -5 is less than the sum of z and 9.
36. 147 is greater than the product of m and 7.

Coordinate Geometry **In Exercises 37–40, plot the points in a coordinate plane. Connect the points to form a rectangle. Then find the perimeter and area of the rectangle. (3.8)**

37. $(-2, 2), (-2, 4), (-4, 2), (-4, 4)$

38. $(1, 1), (6, 1), (1, -2), (6, -2)$

39. $(0, -1), (0, -4), (-3, -1), (-3, -4)$

40. $(-2, 3), (-6, 3), (-2, 0), (-6, 0)$

41. ***Problem-Solving Plan*** Order the steps (from 1 through 6) of the problem-solving plan. **(2.8)**

- Solve.
- Check your answer.
- Answer the question.
- Write a verbal model.
- Assign labels.
- Write an algebraic model.

UPC Code **In Exercises 42–45, use the following information to decide whether the UPC code checks. Explain your reasoning.**

Most retail products contain a bar code called the Universal Product Code (UPC), as shown at the right. When a computer scanner reads a bar code, the computer checks the code using the following steps.

- Add the digits in the odd-numbered positions together. Multiply by 3.
- Add the digits in the even-numbered positions together.
- Add the results of steps 1 and 2.
- Subtract the result of step 3 from the next highest multiple of 10.

$3(0 + 8 + 0 + 0 + 8 + 9) +$
$(2 + 4 + 0 + 1 + 6) = 88$
The next highest multiple of 10 is 90. Therefore, $90 - 88 = 2$, ***which is the check digit.***

42. 0 43699 20450 2

43. 0 35902 11234 9

44. 0 25401 42232 1

45. 0 38322 56613 3

46. ***Operation Sense*** Identify the inverse of each operation. **(4.1)**

a. Addition **b.** Subtraction **c.** Multiplication **d.** Division

Mental Math **In Exercises 47 and 48, what operation would you use to solve the equation? (4.1)**

47. $x - 4 = 18$

48. $3x = 27$

In Exercises 49–52, find the reciprocal of the number. (4.3)

49. 5

50. -1

51. $-\frac{1}{4}$

52. $\frac{2}{9}$

In Exercises 53–58, solve the equation. Then check your solution. (4.1, 4.2)

53. $6x - 17 = 7$

54. $4y + 13 = -19$

55. $-\frac{n}{7} + 9 = 5$

56. $8m + 3m - 2 = 9$

57. $\frac{3}{4}z - \frac{1}{4}z + 6 = 12$

58. $15t + 14 - 7t = 30$

Geometry **In Exercises 59 and 60, write an equation that equates the sum of the measures of the angles of the triangle to 180°. Then solve for x and find the measures of each angle. (4.4)**

59.

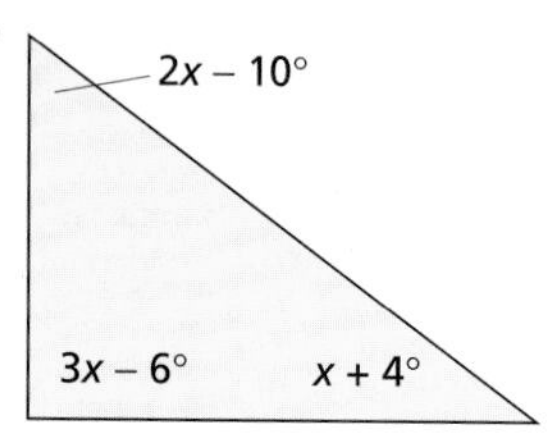

60.

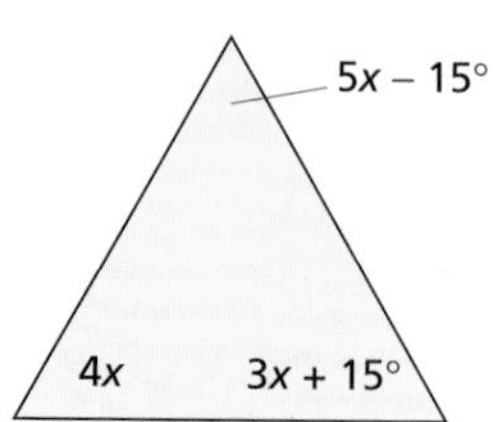

Distributive Property **In Exercises 61–64, simplify the expression. (4.4)**

61. $3x + 2(x + 1)$

62. $4(y - 1) - 2y$

63. $-3n + 2(4n - 5)$

64. $-5(6s - 3) + 7s$

In Exercises 65–72, solve the equation. Then check your solution. (4.4, 4.5)

65. $10y - 27 = y$

66. $13x - 80 = 60 - 7x$

67. $18x + 53 = 4x - 31$

68. $5t + 3t + 15 = 39$

69. $3(4t + 3) = 2t$

70. $b + 6 = 2(b - 2)$

71. $2(4n + \frac{1}{2}) = 10n$

72. $3(y - 2) + 2 = -y$

In Exercises 73 and 74, solve the equation. Round the result to 2 decimal places. (4.7)

73. $15x - 21 = -42x + 89$

74. $14.1(2.37y + 5.6) = 0.71y - 29.3$

75. *Geometry* Find the area of a triangle with height 4 inches and base 3 inches. **(4.8)**

76. *Geometry* Find the area of a square that has a perimeter of 36 centimeters. **(4.8)**

Population **In Exercises 77–80, use the graph at the right, which shows the populations (in millions) of males and females in the United States for 1920 through 1990. In the graph, t = 0 represents 1920.** ***(Source: U.S. Bureau of Census)*** **(5.3, 5.5)**

77. Does each graph increase during each decade?

78. Estimate the female population in 1960.

79. During which decade did the female population first exceed the male population?

80. Is this graph misleading? Explain.

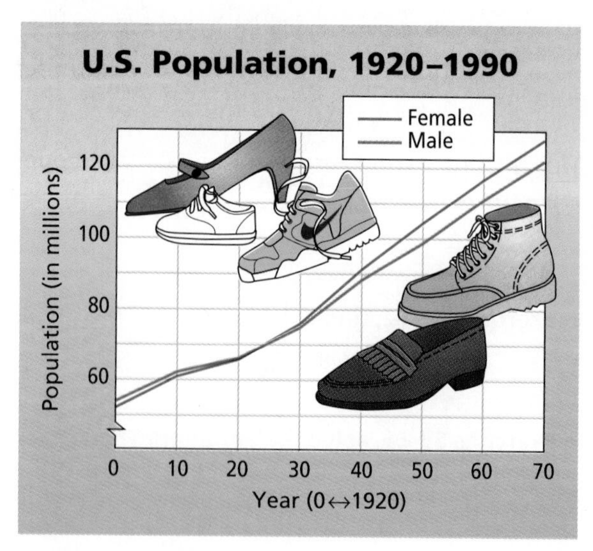

Gas Mileage **In Exercises 81–84, use the scatter plot at the right, which compares the speed and gas mileage of a typical automobile. (5.7)**

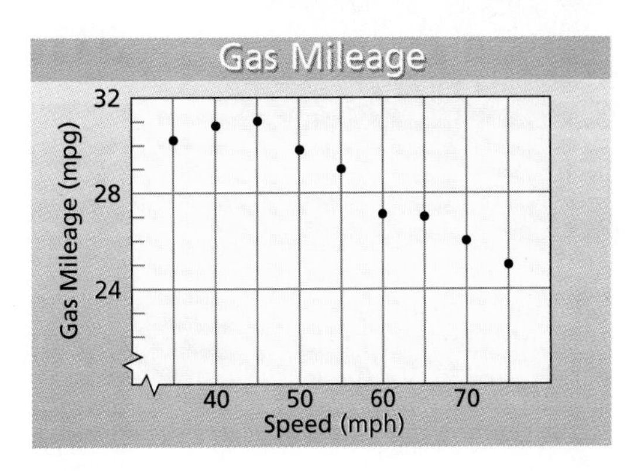

81. Do the speed and gas mileage have a positive correlation, a negative correlation, or no correlation? Explain your reasoning.

82. Write a sentence that describes the relationship between speed and gas mileage.

83. Estimate the gas mileage for a speed of 60 miles per hour.

84. Estimate the speed that corresponds to the maximum gas mileage.

85. *Radio Stations* The table lists the number of types of radio stations in the United States in 1991. Represent this data graphically. ***(Source: Radio Information Center)*** **(5.4)**

Type of Music	Number of Stations
Country	2314
Adult Contemporary	1898
Golden Oldies	729
Contemporary Hits	705
Easy Listening	268
Soft Contemporary	182
Classic Rock	127

Probability **In Exercises 86–88, find the probability that the spinner will land on the given color. (5.8)**

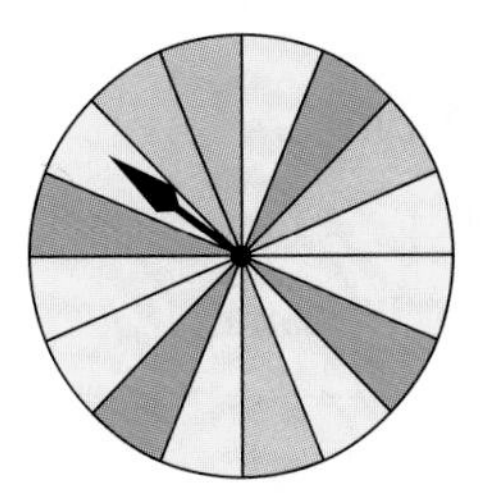

86. Blue **87.** Red **88.** Orange

89. What is the probability that the spinner will not land on green?

In Exercises 90–95, find the greatest common factor and least common multiple. (6.2–6.4)

90. 7, 49 **91.** 4, 18 **92.** 270, 450

93. 864, 972 **94.** $6x^2y$, $8xy^3$ **95.** $9a^2b$, $12ab^4$

In Exercises 96–99, simplify the fraction. Then write in decimal form. Round your answer to 3 decimal places, if necessary. (6.5, 6.6)

96. $\frac{6}{48}$ **97.** $\frac{25}{45}$ **98.** $\frac{10}{15}$ **99.** $\frac{52}{54}$

In Exercises 100–107, simplify the expression. (6.7)

100. 5^0 **101.** 3^{-2} **102.** a^{-4} **103.** $\frac{x}{x^3}$

104. $\frac{8a}{10a^2}$ **105.** $4^2 \cdot 4^{-2}$ **106.** $\frac{3^4}{3^5}$ **107.** $\frac{10^0}{10^4}$

108. Write 1.45×10^4 in decimal form.

109. Write 0.000052 in scientific notation.

CHAPTER

7

Rational Numbers and Percents

LESSONS

According to scientists, the world's water supply neither grows nor diminishes, it recycles itself by changing form and location. Three quarters of the earth's surface is covered by about 1350 million cubic kilometers of water in the form of oceans, rivers, lakes, snowcaps, and ice fields.

Real Life Water Resources

About 97% of the earth's water resources are in the form of salt water as oceans and seas. Over 2% of the remaining fresh-water resources are frozen in the Arctic and Antarctic ice-caps or buried deep beneath the earth's surface. Less than 0.5% of the total water resources is available to maintain all life on earth.

Much of the real-life data you will encounter is given in the form of rational numbers as decimals or percents. In this chapter, you will explore how to use rational numbers in describing real-life situations.

7.1 Addition and Subtraction of Like Fractions

What you should learn:

How to add like fractions

How to subtract like fractions

Why you should learn it:

You can use addition and subtraction of like fractions to solve real-life problems, such as comparing the times of two runners.

Goal 1 Adding Like Fractions

Like fractions are fractions that have the same denominator.

Like Fractions	Unlike Fractions
$\frac{1}{5}$ and $\frac{3}{5}$	$\frac{1}{2}$ and $\frac{1}{3}$
$\frac{a}{c}$ and $\frac{b}{c}$	$\frac{a}{c}$ and $\frac{a}{d}$

Adding Like Fractions

To add like fractions, add the numerators and write the sum over the denominator.

Numerical Example

$$\frac{1}{5} + \frac{3}{5} = \frac{1+3}{5} = \frac{4}{5}$$

Variable Example

$$\frac{a}{c} + \frac{b}{c} = \frac{a+b}{c}$$

Example 1 *Adding Like Fractions*

a. $\frac{5}{8} + \frac{7}{8} = \frac{5+7}{8}$ *Add numerators.*

$= \frac{12}{8}$ *Simplify numerator.*

$= \frac{\not{4} \cdot 3}{\not{4} \cdot 2}$ *Factor numerator and denominator.*

$= \frac{3}{2}$ *Simplify fraction.*

b. $\frac{-3}{10} + \frac{-5}{10} = \frac{-3+(-5)}{10}$ *Add numerators.*

$= \frac{-8}{10}$ *Simplify numerator.*

$= \frac{-4 \cdot \not{2}}{5 \cdot \not{2}}$ *Factor numerator and denominator.*

$= -\frac{4}{5}$ *Simplify fraction.*

c. $1\frac{2}{6} + 1\frac{3}{6} = \frac{8}{6} + \frac{9}{6}$ *Rewrite as improper fractions.*

$= \frac{8+9}{6}$ *Add numerators.*

$= \frac{17}{6}$ *Simplify numerator.*

d. $\frac{6x}{5} + \frac{3x}{5} = \frac{6x+3x}{5}$ *Add numerators.*

$= \frac{9x}{5}$ *Simplify numerator.* ■

Model for $\frac{1}{5}$

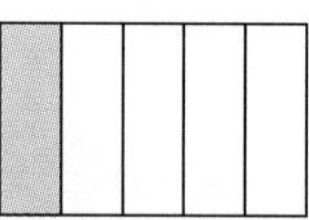

Model for $\frac{3}{5}$

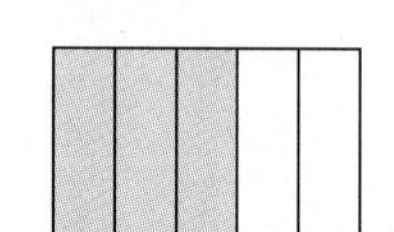

Model for Sum

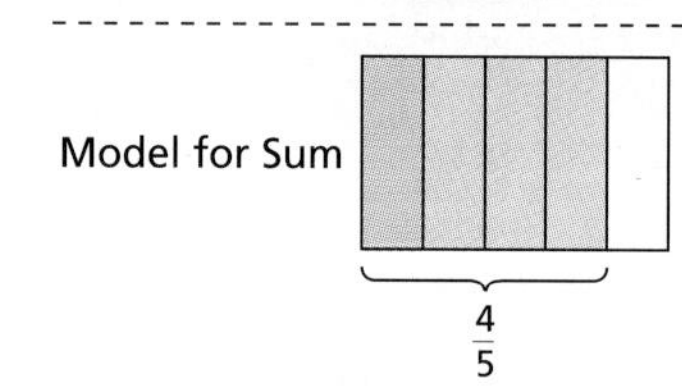

This geometric model shows that the sum of $\frac{1}{5}$ and $\frac{3}{5}$ is $\frac{4}{5}$.

Goal 2 Subtracting Like Fractions

Subtracting Like Fractions

To subtract like fractions, subtract the numerators and write the difference over the denominator.

Numerical Example

$\frac{3}{5} - \frac{1}{5} = \frac{3-1}{5} = \frac{2}{5}$

Variable Example

$\frac{a}{c} - \frac{b}{c} = \frac{a-b}{c}$

Example 2 *Subtracting Like Fractions*

You are helping two of your friends train for a track meet. You time your friends on a 100-meter sprint. One friend's time is $71\frac{3}{4}$ seconds and the other's time is $73\frac{1}{4}$ seconds. How much longer did your second friend take to complete the sprint?

Solution Subtract the first friend's time from the second friend's time.

Difference in times	=	Second friend's time	−	First friend's time

$\text{Difference} = 73\frac{1}{4} - 71\frac{3}{4}$ *Substitute for times.*

$= \frac{293}{4} - \frac{287}{4}$ *Write mixed numbers as fractions.*

$= \frac{293-287}{4}$ *Subtract numerators.*

$= \frac{6}{4}$ *Simplify.*

$= \frac{3}{2}$ *Simplify.*

Your second friend took $1\frac{1}{2}$ seconds longer to complete the sprint. ■

By the age of 14, Angela T. Williams, from Ontario, California, had won 15 national sprinting championships and set 6 national sprinting records.

Communicating about MATHEMATICS

SHARING IDEAS about the Lesson

Solving Equations Use the rules for adding and subtracting fractions to solve the equations.

A. $x - \frac{6}{7} = \frac{2}{7}$ **B.** $y - \frac{4}{3} = \frac{5}{3}$

C. $m + \frac{5}{4} = \frac{7}{4}$ **D.** $n + \frac{1}{8} = \frac{5}{8}$

EXERCISES

Guided Practice

CHECK for Understanding

In Exercises 1 and 2, write the indicated sum or difference.

1. 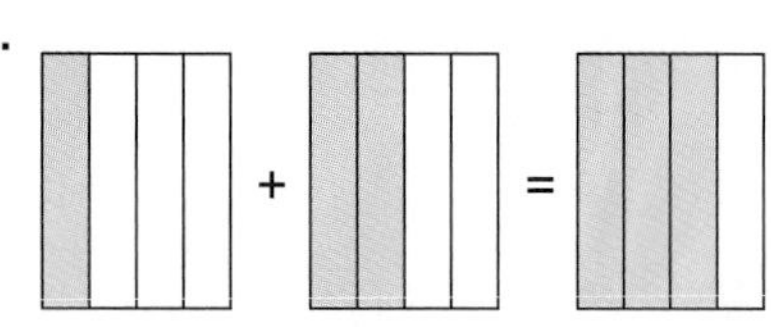

2. 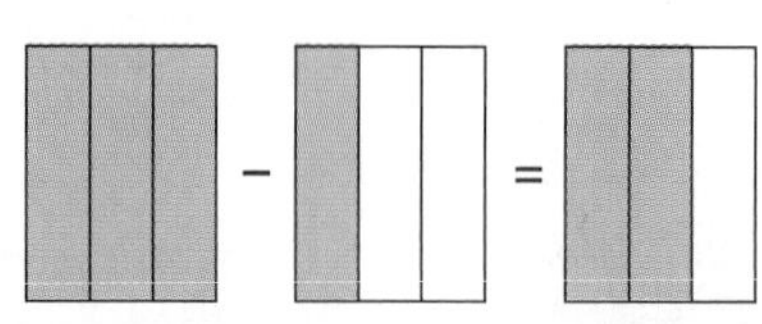

3. *Writing* In your own words, describe how to add like fractions. Give an example.

4. *Writing* In your own words, describe how to subtract like fractions. Give an example.

Independent Practice

In Exercises 5–12, add or subtract. Then simplify, if possible.

5. $\frac{2}{6} + \frac{3}{6}$
6. $\frac{8}{12} - \frac{4}{12}$
7. $\frac{-8}{15} - \frac{7}{15}$
8. $\frac{-3}{8} + \frac{-7}{8}$
9. $\frac{4}{11} - \frac{10}{11}$
10. $\frac{-8}{5} + \frac{2}{5}$
11. $3\frac{1}{2} + 1\frac{1}{2}$
12. $3\frac{2}{3} - 4\frac{1}{3}$

In Exercises 13–20, add or subtract. Then simplify, if possible.

13. $\frac{x}{2} + \frac{4x}{2}$
14. $\frac{12y}{10} - \frac{4y}{10}$
15. $\frac{-a}{5} - \frac{4a}{5}$
16. $\frac{-3b}{4} + \frac{6b}{4}$
17. $\frac{1}{z} + \frac{6}{z}$
18. $\frac{2}{4t} - \frac{9}{4t}$
19. $\frac{4}{5b} - \frac{1}{5b}$
20. $\frac{1}{8x} + \frac{3}{8x} - \frac{7}{8x}$

In Exercises 21–28, solve the equation. Then simplify, if possible.

21. $x + \frac{2}{3} = \frac{4}{3}$
22. $y - \frac{6}{8} = \frac{5}{8}$
23. $m + \frac{19}{5} = \frac{4}{5}$
24. $n - \frac{1}{6} = \frac{-9}{6}$
25. $s + \frac{5}{4} = \frac{-9}{4}$
26. $t - \frac{8}{11} = \frac{-6}{11}$
27. $3x + \frac{1}{2} = \frac{5}{2}$
28. $2z - \frac{8}{7} = \frac{6}{7}$

In Exercises 29–32, use a calculator to evaluate the expression as a decimal rounded to two decimal places.

29. $\frac{1}{7} + \frac{3}{7}$
30. $\frac{7}{6} - \frac{3}{6}$
31. $\frac{5}{9} - \frac{8}{9}$
32. $\frac{5}{16} - \frac{11}{16}$

Patterns **In Exercises 33 and 34, add or subtract. Then describe the pattern and write the next three numbers in the pattern.**

33. $\frac{1}{8} + \frac{2}{8} = \boxed{?}$
$\frac{3}{8} + \frac{4}{8} = \boxed{?}$
$\frac{5}{8} + \frac{6}{8} = \boxed{?}$
$\frac{7}{8} + \frac{8}{8} = \boxed{?}$

34. $\frac{10}{2} - \frac{1}{2} = \boxed{?}$
$\frac{-9}{2} - \frac{2}{2} = \boxed{?}$
$\frac{8}{2} - \frac{3}{2} = \boxed{?}$
$\frac{-7}{2} - \frac{4}{2} = \boxed{?}$

Correct or Incorrect? **In Exercises 35 and 36, decide whether the addition is correct or incorrect. If it is incorrect, write a correct version.**

35. $\frac{3}{5} + \frac{1}{5} = \frac{3+1}{5+5}$
$= \frac{4}{10}$
$= \frac{2}{5}$

36. $1\frac{1}{3} + 2\frac{1}{3} = 1 + \frac{1}{3} + 2 + \frac{1}{3}$
$= 3 + \frac{2}{3}$
$= 3\frac{2}{3}$

Modeling Fractions **In Exercises 37 and 38, write the indicated sum or difference.**

37.

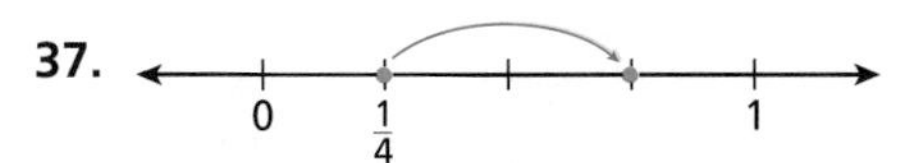

38.

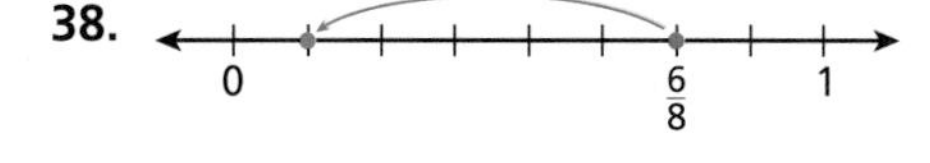

39. *Baking* The recipe below will make a single batch of banana bread. Rewrite the recipe for a double batch.

Recipe for Banana Bread
From Grandma
Ingredients
1 3/4 cups all- purpose flour
2/3 cup of sugar
2 teaspoons baking powder
1/2 teaspoon baking soda
1/4 teaspoon salt
1 cup mashed ripe banana
1/3 cup shortening, margerine, or butter
2 tablespoons milk
2 eggs
1/4 cup chopped nuts

Mammals **In Exercises 40–42, use the table, which shows the nose-to-tail lengths of some mammals.**

40. How much longer is the Siberian tiger than the African lion?

41. How much shorter is the polar bear than the Siberian tiger?

42. How much shorter is the polar bear than the African lion?

Mammal	Length in feet
Siberian tiger	$10\frac{4}{12}$
African lion	$9\frac{1}{12}$
Polar bear	$7\frac{9}{12}$

Integrated Review

Making Connections within Mathematics

Geometry **In Exercises 43 and 44, find the perimeter of the figure.**

43.

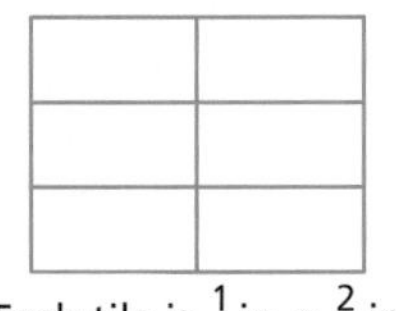

Each tile is $\frac{1}{3}$ in. × $\frac{2}{3}$ in.

44.

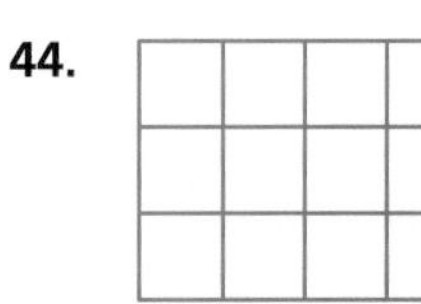

Each tile is $\frac{1}{4}$ in. × $\frac{1}{4}$ in.

Exploration and Extension

Fraction Riddles **In Exercises 45 and 46, find the fraction that is described by the clues given.**

45. The fraction is between 0 and 1. It is more than $\frac{1}{4}$. If you add $\frac{2}{8}$ to it, the fraction will be equivalent to $\frac{10}{16}$.

46. The fraction is between -1 and 0. It is less than $-\frac{3}{8}$. If you subtract $\frac{2}{4}$ from it, the fraction will be equal to $-1\frac{1}{4}$.

LESSON INVESTIGATION 7.2

A Model for Adding and Subtracting Fractions

Materials Needed: fraction strips

In this activity, you will use fraction strips to investigate the sum and difference of fractions with unlike denominators.

Example ***Adding and Subtracting Unlike Fractions***

Use fraction strips to model the following.

a. $\frac{2}{3} + \frac{1}{6}$ **b.** $\frac{2}{3} - \frac{1}{4}$

Solution Begin by choosing fraction strips to represent $\frac{2}{3}$, $\frac{1}{6}$, and $\frac{1}{4}$.

a. To add the fractions, place the shaded parts together and find another fraction strip that has this length. From this, you can conclude that the sum is $\frac{5}{6}$.

$\frac{1}{3}$ $\frac{1}{3}$ $\frac{1}{6}$

$\frac{1}{6}$ $\frac{1}{6}$ $\frac{1}{6}$ $\frac{1}{6}$ $\frac{1}{6}$

b. To subtract the fractions, overlap the shaded parts and find another fraction strip that has the same length as the difference. From this, you can conclude that the difference is $\frac{5}{12}$.

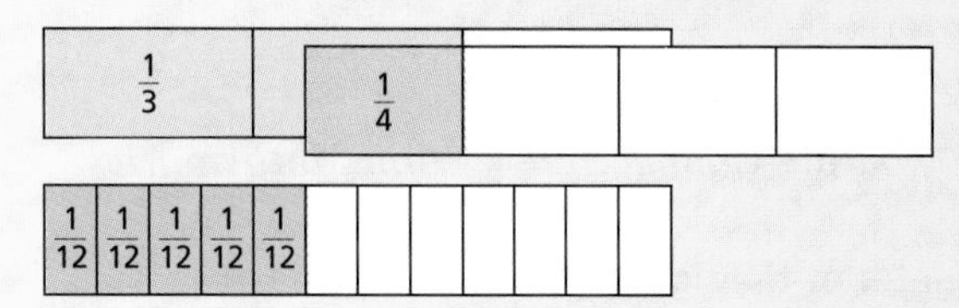

■

Exercises

In Exercises 1, and 2, write the sum or difference modeled by the fraction strips.

1.

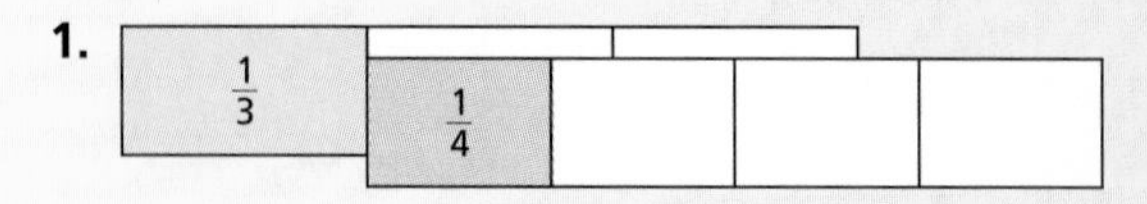

2.

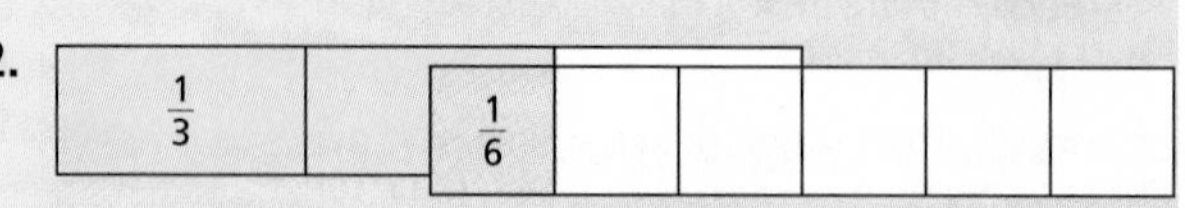

In Exercises 3–10, use fraction strips to find the sum or difference.

3. $\frac{2}{5} + \frac{1}{10}$ **4.** $\frac{2}{3} + \frac{1}{4}$ **5.** $\frac{1}{4} + \frac{1}{6}$ **6.** $\frac{1}{6} + \frac{3}{4}$

7. $\frac{2}{5} - \frac{1}{10}$ **8.** $\frac{2}{3} - \frac{1}{4}$ **9.** $\frac{5}{6} - \frac{1}{4}$ **10.** $\frac{3}{4} - \frac{1}{6}$

7.2 Addition and Subtraction of Unlike Fractions

What you should learn:

Goal 1 How to add and subtract unlike fractions

Goal 2 How to use addition and subtraction of fractions to solve real-life problems

What you should learn:

You can use addition and subtraction of unlike fractions to solve real-life problems, such as comparing the ways electric power is produced.

Goal 1 Add and Subtract Unlike Fractions

The **least common denominator** of two fractions is the least common multiple of their denominators. For instance, the least common denominator of $\frac{1}{2}$ and $\frac{1}{3}$ is 6. In this lesson, you will learn how to use least common denominators to add and subtract unlike fractions.

Addition and Subtraction of Unlike Fractions

To add or subtract unlike fractions, rewrite the fractions so that they have a common denominator. Then add or subtract the resulting like fractions.

Numerical Example

$\frac{1}{2} + \frac{1}{3} = \frac{3}{6} + \frac{2}{6} = \frac{5}{6}$

Variable Example

$\frac{3}{x} - \frac{1}{2x} = \frac{6}{2x} - \frac{1}{2x} = \frac{5}{2x}$

Example 1 *Add and Subtract Unlike Fractions*

a. $\frac{5}{6} + \frac{3}{8} = \frac{5}{6} \cdot \frac{4}{4} + \frac{3}{8} \cdot \frac{3}{3}$ *Least common denominator is 24.*

$= \frac{20}{24} + \frac{9}{24}$ *Rewrite as like fractions.*

$= \frac{29}{24}$ *Add like fractions.*

b. $\frac{7}{12} - \frac{3}{4} = \frac{7}{12} - \frac{3}{4} \cdot \frac{3}{3}$ *Least common denominator is 12.*

$= \frac{7}{12} - \frac{9}{12}$ *Rewrite as like fractions.*

$= \frac{-2}{12}$ *Subtract like fractions.*

$= -\frac{1}{6}$ *Simplify fraction.*

c. $\frac{2}{a} + \frac{3}{2} = \frac{2}{a} \cdot \frac{2}{2} + \frac{3}{2} \cdot \frac{a}{a}$ *Least common denominator is 2a.*

$= \frac{4}{2a} + \frac{3a}{2a}$ *Rewrite as like fractions.*

$= \frac{4 + 3a}{2a}$ *Add like fractions.*

d. $\frac{x}{4} - \frac{x}{5} = \frac{x}{4} \cdot \frac{5}{5} - \frac{x}{5} \cdot \frac{4}{4}$ *Least common denominator is 20.*

$= \frac{5x}{20} - \frac{4x}{20}$ *Rewrite as like fractions.*

$= \frac{x}{20}$ *Subtract like fractions.* ■

Need to Know

The rule for adding and subtracting unlike fractions can be thought of as having two steps.

1. Rewrite fractions as like fractions.
2. Add or subtract the like fractions.

Goal 2 Solving Real-Life Problems

Example 2 Adding Unlike Fractions

Real Life
Electric Power

The circle graph below compares the different ways that electricity is produced in the United States. Show that the sum of the five fractions is 1. ***(Source: Electrical Power Annual)***

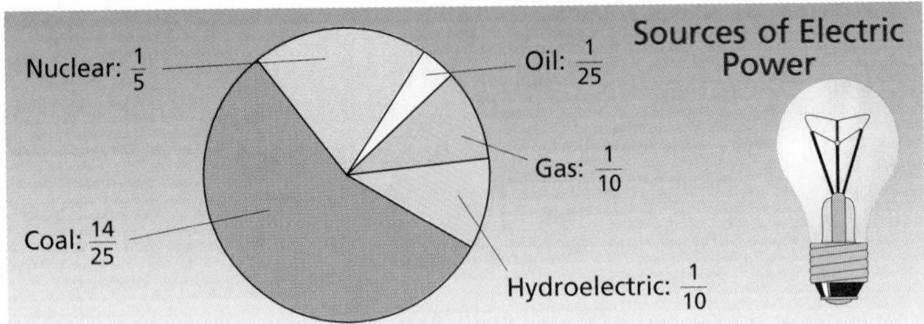

The Grand Coulee Dam in Washington is the largest dam in the United States. It contains 12 million cubic yards of concrete.

Solution The least common denominator of the five fractions is 50.

$$\frac{14}{25} + \frac{1}{5} + \frac{1}{25} + \frac{1}{10} + \frac{1}{10}$$
$$= \frac{14}{25} \cdot \frac{2}{2} + \frac{1}{5} \cdot \frac{10}{10} + \frac{1}{25} \cdot \frac{2}{2} + \frac{1}{10} \cdot \frac{5}{5} + \frac{1}{10} \cdot \frac{5}{5}$$
$$= \frac{28}{50} + \frac{10}{50} + \frac{2}{50} + \frac{5}{50} + \frac{5}{50}$$
$$= \frac{28 + 10 + 2 + 5 + 5}{50}$$
$$= \frac{50}{50}$$
$$= 1$$

Thus, the sum of the five fractions is 1. ■

Communicating about MATHEMATICS

SHARING IDEAS about the Lesson

Geoboard Fractions Assume that the area of the region inside the pegs of each geoboard is 1. Write the fraction that represents the area of each shaded region. Then find the total area of the two shaded regions.

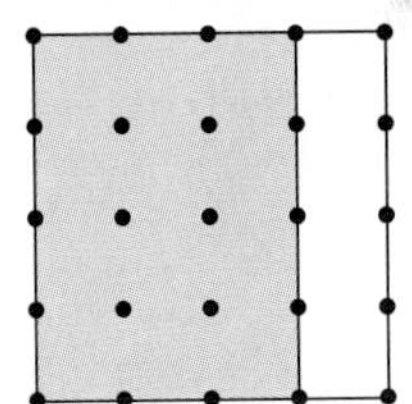

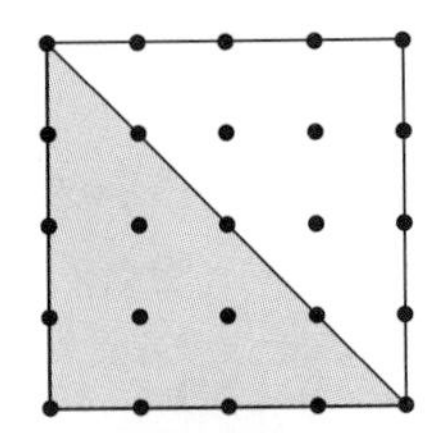

EXERCISES

Guided Practice

CHECK for Understanding

In Exercises 1 and 2, the area of each region is 1. Write the fraction that represents the area of each blue region. Then add the fractions.

1. 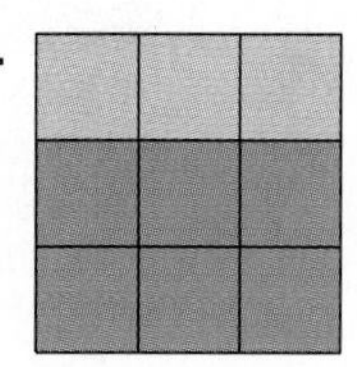

2. 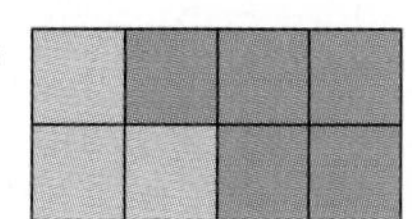 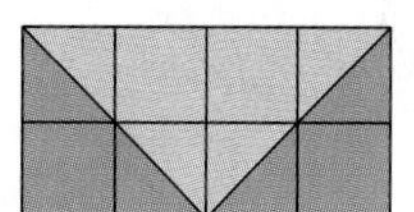

In Exercises 3–6, find the sum or difference. Then simplify, if possible. Explain your steps and identify the least common denominator of the fractions.

3. $\frac{2}{5} + \frac{1}{3}$ **4.** $\frac{4}{5} - \frac{3}{10}$ **5.** $\frac{a}{2} - \frac{a}{3}$ **6.** $\frac{4}{t} + \frac{1}{2t}$

Independent Practice

In Exercises 7–14, find the sum or difference. Then simplify, if possible.

7. $\frac{1}{6} + \frac{7}{12}$ **8.** $\frac{2}{3} - \frac{3}{8}$ **9.** $\frac{-1}{2} + \frac{-7}{12}$ **10.** $\frac{7}{9} - \frac{4}{5}$

11. $\frac{-11}{15} + \frac{2}{5}$ **12.** $\frac{-3}{7} - \frac{1}{3}$ **13.** $\frac{-3}{10} + \frac{7}{8}$ **14.** $\frac{1}{2} + \frac{5}{6} - \frac{7}{9}$

In Exercises 15–22, find the sum or difference. Then simplify, if possible.

15. $\frac{x}{3} + \frac{x}{6}$ **16.** $\frac{a}{8} - \frac{a}{12}$ **17.** $\frac{2}{x} + \frac{9}{10}$ **18.** $\frac{4}{a} - \frac{11}{b}$

19. $\frac{-2}{3t} - \frac{4}{9t}$ **20.** $\frac{-7}{rs} + \frac{4}{s}$ **21.** $\frac{2}{mn} - \frac{1}{3mn}$ **22.** $1\frac{2}{3} + 1\frac{3}{4}$

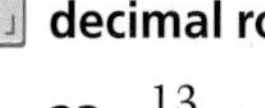

In Exercises 23–26, use a calculator to find the sum or difference as a decimal rounded to two decimal places.

23. $\frac{13}{18} + \frac{9}{10}$ **24.** $\frac{5}{28} + \frac{4}{7}$ **25.** $\frac{8}{9} - \frac{23}{135}$ **26.** $\frac{16}{27} - \frac{63}{108}$

Geometry **In Exercises 27 and 28, find the perimeter of the figure.**

27. 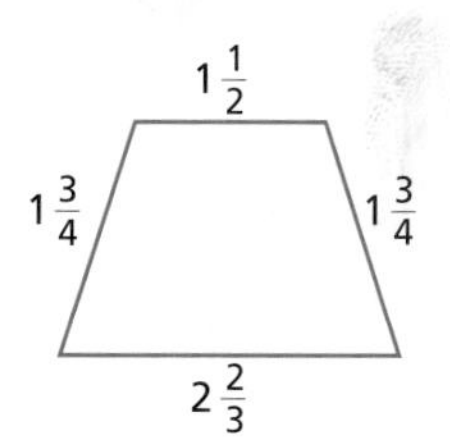

28. 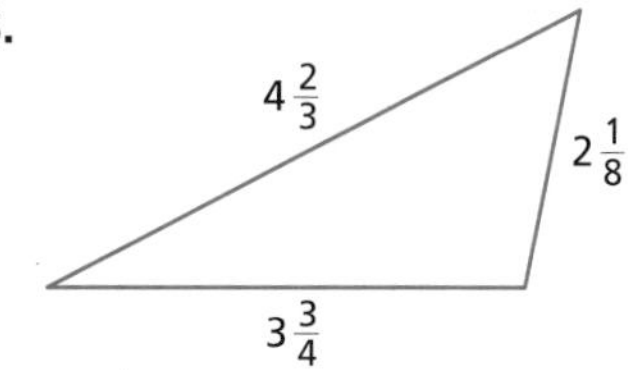

Sequence **In Exercises 29–32, consider the following sequence.**

$\frac{1}{2}, -\frac{2}{3}, \frac{3}{4}, -\frac{4}{5}, \ldots$

29. Describe the pattern.

30. Write the next 2 terms of the sequence.

31. Find the sum of the first 3 terms.

32. Find the sum of the first 5 terms.

Science **In Exercises 33–35, use the circle graph at the right, which shows what most influences students in grades 3–12 in their interest in science. The fractions represent portions of the student population.** *(Source: USA Today)*

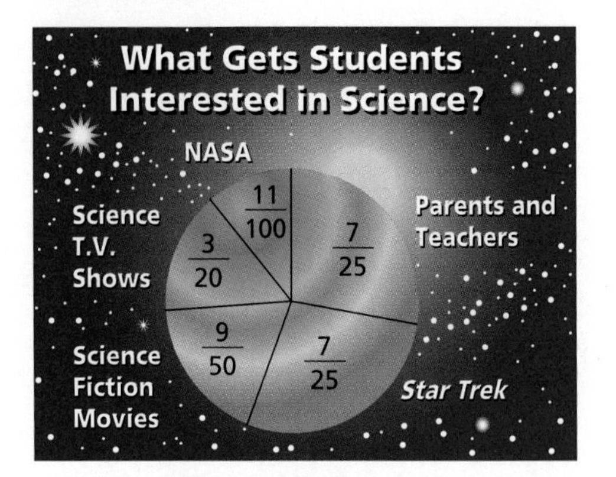

33. What portion of students are influenced most by Star Trek and science fiction movies?

34. Find the difference in the portions influenced by science television shows and by NASA.

35. Show that the sum of the five fractions is 1.

36. *Comics* You are making a comic book that is $4\frac{1}{2}$ inches by $6\frac{1}{3}$ inches. Each page of the comic book has a bottom and a top margin of $\frac{2}{3}$ inch and a left and right margin of $\frac{1}{4}$ inch, as shown at the right.

a. What is the perimeter of the page?

b. What are the dimensions of the printed portion of the page?

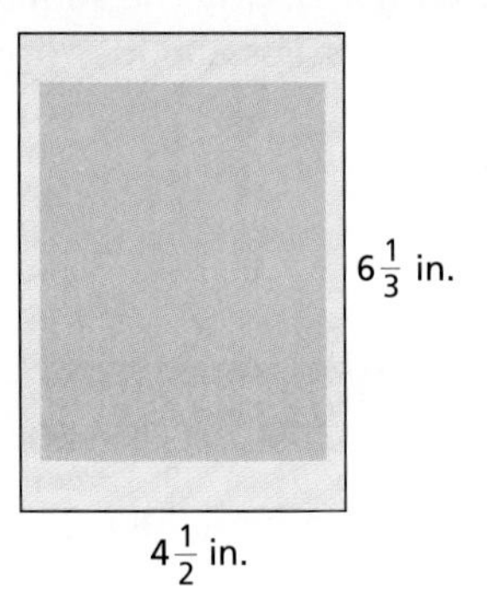

Integrated Review

Making Connections within Mathematics

Estimating Sums **In Exercises 37–42, estimate the sum by rounding each fraction to the nearest $\frac{1}{2}$ and adding.**

37. $\frac{5}{6} + \frac{3}{8}$

38. $\frac{2}{3} + \frac{5}{16}$

39. $\frac{1}{10} + \frac{3}{4} + \frac{7}{8}$

40. $\frac{7}{24} + \frac{3}{7} + \frac{13}{16}$

41. $1\frac{3}{11} + 2\frac{9}{13}$

42. $4\frac{4}{9} + 2\frac{1}{20}$

Exploration and Extension

Modeling Least Common Denominators **In Exercises 43–46, use the sample at the right to represent the sum with graph paper.**

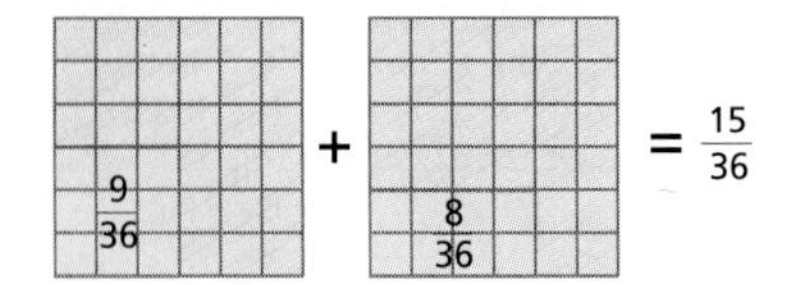

43. $\frac{3}{4} + \frac{5}{9}$

44. $\frac{1}{6} + \frac{1}{5}$

45. $\frac{2}{5} + \frac{3}{10}$

46. $\frac{1}{6} + \frac{1}{8}$

7.3 Exploring Fractions and Decimals

What you should learn:

Goal 1 How to add and subtract fractions by writing the fractions as decimals

Goal 2 How to use addition and subtraction of decimals to solve real-life problems

Why you should learn it:

You can use addition and subtraction of decimals to solve real-life problems, such as interpreting the results of a survey.

Goal 1 Adding and Subtracting Decimals

In Lessons 7.1 and 7.2, you studied ways to add and subtract fractions. Another way to add and subtract fractions is to rewrite the fractions as decimals or use geometric models. Which of the following ways of adding $\frac{1}{2}$ and $\frac{1}{4}$ do you prefer?

Adding as Fractions

$$\frac{1}{2} + \frac{1}{4} = \frac{1}{2} \cdot \frac{2}{2} + \frac{1}{4} = \frac{2}{4} + \frac{1}{4} = \frac{3}{4}$$

Adding as Decimals

$\frac{1}{2} = 0.5$ and $\frac{1}{4} = 0.25$

$$\begin{array}{r} 0.5 \\ +\ 0.25 \\ \hline 0.75 \end{array}$$

Adding with Geometric Models

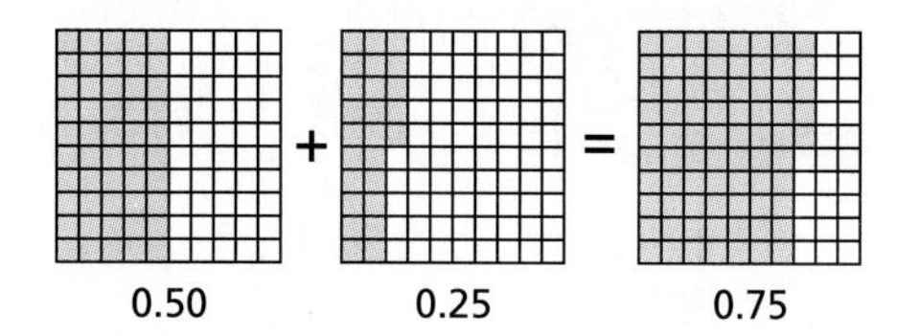

When you add or subtract fractions by first rewriting the fractions as decimals, remember that the resulting sums or differences may be only approximate.

Study Tip...

Example 1 asks you to round the result to 2 decimal places. To help avoid a round-off error, you should begin by rounding the numbers to 3 decimal places—one place more than is required in the final result. What would happen if you began by rounding the fractions to 0.31 and 0.63?

Example 1 *Adding and Subtracting Decimals*

Evaluate the expression by first rewriting in decimal form. Round the result to 2 decimal places.

a. $\frac{4}{13} + \frac{5}{8}$ **b.** $\frac{8}{11}x - \frac{3}{7}x$

Solution

a. $\frac{4}{13} + \frac{5}{8} \approx 0.308 + 0.625$ *Write as rounded decimals.*

$= 0.933$ *Add decimals.*

≈ 0.93 *Round to 2 decimal places.*

b. $\frac{8}{11}x - \frac{3}{7}x \approx 0.727x - 0.429x$ *Write as rounded decimals.*

$= 0.298x$ *Subtract decimals.*

$\approx 0.30x$ *Round to 2 decimal places.*

■

Goal 2 Solving Real-Life Problems

Real Life **Psychology**

Psychologists study the ways people and animals think and relate to one another. An important part of psychology is studying the way people think about themselves.

Example 2 *Adding and Subtracting Decimals*

In a poll, adults were asked how they would describe their own physical appearance. The results are shown in the circle graph. What portion of the adults described themselves as "pretty or handsome"? ***(Source: Cosmopolitan Magazine)***

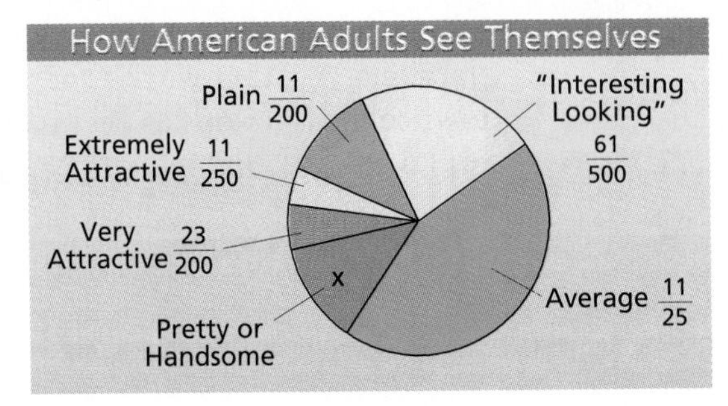

Solution The fractions for the six parts have a sum of 1. To find the portion that answered "pretty or handsome," add the other five fractions and subtract the result from 1.

$$\begin{aligned} x &= 1 - \left(\frac{11}{250} + \frac{23}{200} + \frac{11}{25} + \frac{61}{500} + \frac{11}{200}\right) \\ &= 1 - (0.044 + 0.115 + 0.44 + 0.122 + 0.055) \\ &= 1 - 0.776 \\ &= 0.224 \end{aligned}$$

This means that about 224 out of every 1000 adults consider themselves to be pretty or handsome. ■

Communicating about MATHEMATICS

Cooperative Learning

SHARING IDEAS about the Lesson

Modeling Fractions Work with a partner. Explain how each model represents $\frac{7}{10}$. Describe other geometric models for this fraction.

A.

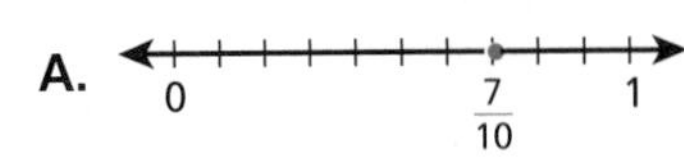

B.

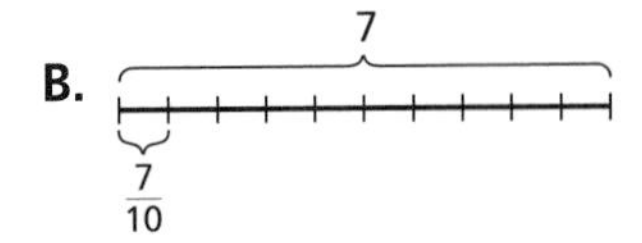

C.

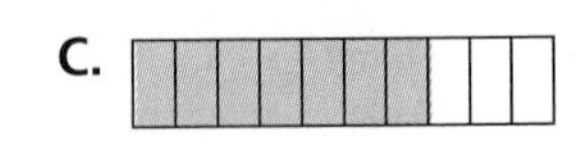

EXERCISES

Guided Practice

CHECK for Understanding

1. Explain each of the following steps.

$$\frac{7}{9} + \frac{11}{19} \approx 0.778 + 0.579$$
$$= 1.357$$
$$\approx 1.36$$

2. Which is greater: $\frac{2}{3}$ or 0.67? Explain your reasoning.

3. Use a calculator to evaluate $\frac{2}{5} - \frac{10}{13}$. Round your results to two decimal places.

4. The sum of the fractions in the circle graph is 1. Use a calculator to solve for x. Round your result to two decimal places.

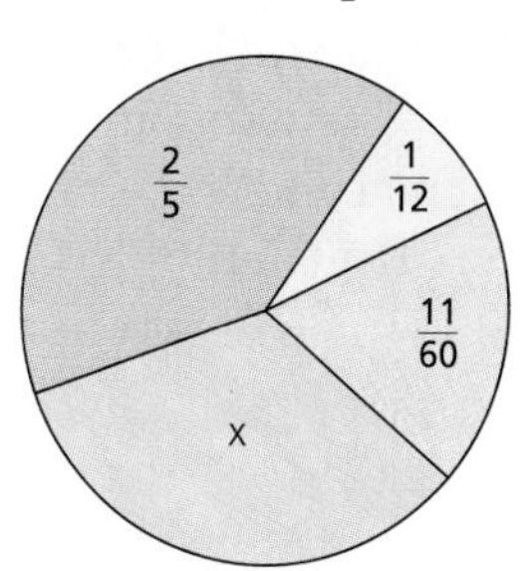

Independent Practice

In Exercises 5–8, evaluate the expression.

5. $0.31 + 0.55$
6. $1.823 + 0.021$
7. $3.73 - 2.09$
8. $2.009 - 1.793$

In Exercises 9 and 10, write the expression represented by the model. Then evaluate the expression by first converting to decimals. Round your result to two decimal places.

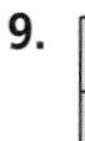

9.

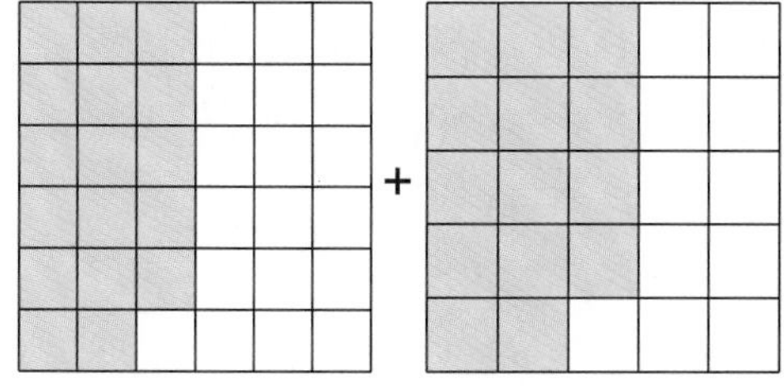

10.

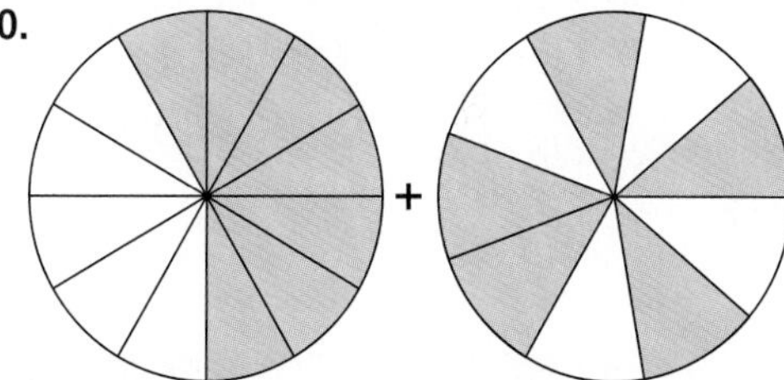

Modeling Fractions **In Exercises 11–14, sketch a geometric model of the fraction. (Use a different model for each exercise.)**

11. $\frac{11}{16}$
12. $\frac{24}{50}$
13. 0.45
14. 0.625

In Exercises 15–22, evaluate the expression by first rewriting in decimal form. Round your result to two decimal places.

15. $\frac{73}{111} + \frac{54}{109}$
16. $\frac{82}{89} - \frac{76}{127}$
17. $\frac{17}{35}y - \frac{14}{41}y$
18. $\frac{24}{31}n + \frac{7}{15}n$
19. $1 - \left(\frac{21}{56} + \frac{32}{99} + \frac{3}{25}\right)$
20. $2x - \left(\frac{4}{9}x + \frac{3}{7}x + \frac{11}{20}x\right)$
21. $2\frac{3}{4} + 3\frac{1}{8} - 1\frac{9}{10} + 4\frac{7}{10}$
22. $1\frac{5}{6} - 2\frac{3}{5} - \frac{17}{15} + 3\frac{5}{11}$

Government Spending **In Exercises 23–25, use the circle graph, which shows how the United States federal government spent its money in 1992.** ***(Source: Office of Management and Budget)***

National Defense: x
Deposit Insurance Payments: $\frac{3}{50}$
Other Federal Operations: $\frac{7}{100}$
State and Local Grants: $\frac{3}{25}$
Federal Medical and Retirement Programs: $\frac{41}{100}$
Net Interest: $\frac{7}{50}$

U.S. Spending in 1992

23. Find the sum of the portions for all categories other than national defense.

24. What portion was spent on the national defense?

25. Find the sum of the portions spent on state and local grants and "other federal operations."

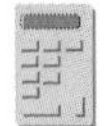

26. *Error Analysis* You are asked to find the sum of $\frac{15}{31}$ and $\frac{18}{39}$ and to round the result to *two* decimal places. Which of the two solutions is better? Explain.

a. $\frac{15}{31} + \frac{18}{39} \approx 0.48 + 0.46$
$= 0.94$

b. $\frac{15}{31} + \frac{18}{39} \approx 0.484 + 0.462$
$= 0.946$
≈ 0.95

Integrated Review

Making Connections within Mathematics

Decimal Sense **In Exercises 27–30, write each fraction as a decimal rounded to three decimal places.**

27. $\frac{47}{99}$ **28.** $\frac{63}{200}$ **29.** $-\frac{12}{43}$ **30.** $-\frac{79}{145}$

Number Sense **In Exercises 31–36, complete each statement with $<$, $>$, or $=$.**

31. 0.9876 ? 0.9853 **32.** -0.4201 ? -0.4199 **33.** 1.0001 ? 1.1

34. $\frac{13}{21}$? 0.62 **35.** $-\frac{5}{13}$? -0.38 **36.** $\frac{6}{17}$? 0.353

Exploration and Extension

Area Models **In Exercises 37–39, each of the geometric models is an area model. Each circle has an area of 1 square unit. Write the fraction and decimal that are represented by the shaded part of the model.**

37.

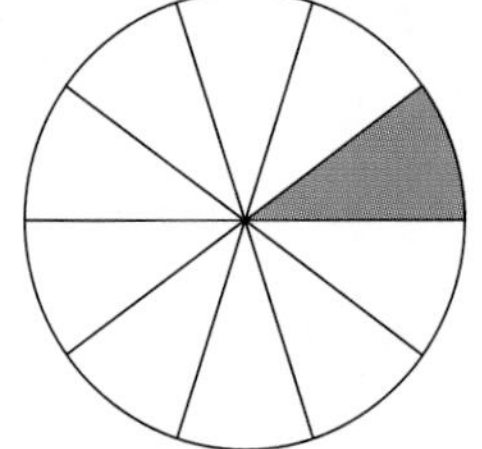

38.

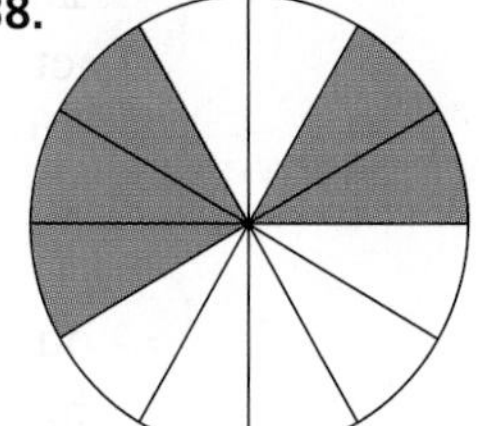

39.

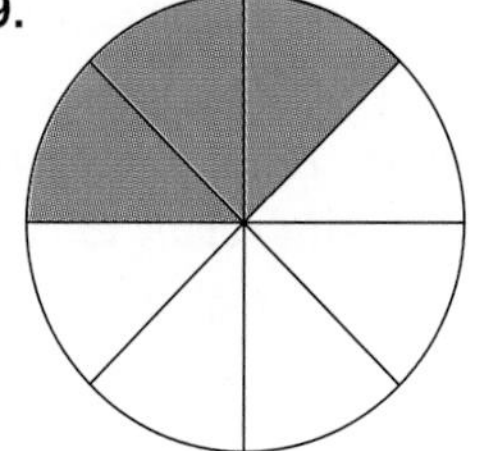

40. Find the total area of the shaded parts in Exercises 37–39.

Mixed REVIEW

In Exercises 1–4, solve the equation. (7.2)

1. $a + \frac{2}{3} = \frac{4}{3}$ **2.** $y + \frac{1}{4} = \frac{3}{4}$ **3.** $\frac{1}{4}(4y + 8) = 16$ **4.** $\frac{14}{9} = b + \frac{13}{9}$

In Exercises 5–8, write the fractions as decimals and solve the equation. Round your answer to two decimal places. (7.3)

5. $\frac{1}{4} + x = \frac{1}{8}$ **6.** $\frac{1}{5}a = \frac{14}{5}$ **7.** $y + \frac{1}{100} = \frac{1}{10}$ **8.** $\frac{1}{20} = c + \frac{3}{16}$

In Exercises 9–12, find the least common multiple. (6.4)

9. 2 and 7 **10.** 10 and 16 **11.** 3, 4, and 5 **12.** 4, 5, and 6

In Exercises 13–16, evaluate the expression. Then simplify, if possible. (7.2)

13. $\frac{1}{20} + \frac{1}{5} + \frac{1}{4}$ **14.** $\frac{2}{3} + \frac{4}{5} - \frac{1}{5}$ **15.** $\frac{99}{100} - \frac{1}{2} - \frac{6}{25}$ **16.** $\frac{4}{5} - \frac{1}{25} + \frac{6}{25}$

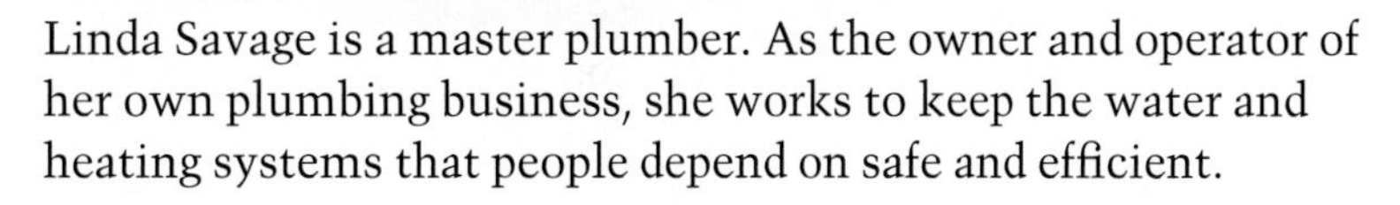

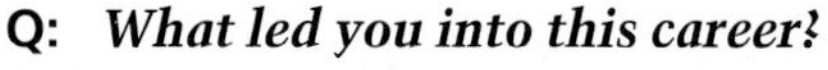

Plumber

Linda Savage is a master plumber. As the owner and operator of her own plumbing business, she works to keep the water and heating systems that people depend on safe and efficient.

Q: ***What led you into this career?***

A: I was working as a tool person at a construction site, and the master plumber there convinced me to try plumbing.

Q: ***What special training did you have to take after high school?***

A: I spent four years in apprenticeship and going to night school, and then after I passed the test to become a journeyman, I spent one more year working and going to night school in order to become a master plumber.

Q: ***What math did you take in school?***

A: Algebra 1 and 2, geometry, and some trigonometry. Since then, I have had to learn lots of other math to get licensed.

Q: ***What would you like to tell kids who are in school about math?***

A: Take as many math courses as you can, because most careers have lots of math, and many people are failing licensing tests because they didn't take enough.

7.4 Multiplication of Rational Numbers

What you should learn:

Goal 1 How to multiply rational numbers

Goal 2 How to use multiplication of rational numbers to solve real-life problems

Why you should learn it:

You can use multiplication of rational numbers to solve real-life problems, such as finding the floor area of a greenhouse.

Goal 1 Multiplying Rational Numbers

When adding or subtracting fractions, you have to consider whether they have like denominators or unlike denominators. In this lesson, you will learn that the rule for multiplying fractions applies whether the denominators are like or unlike.

Multiplying Rational Numbers

To multiply two rational numbers, $\frac{a}{b}$ and $\frac{c}{d}$, multiply the numerators and multiply the denominators.

Numerical Example $\frac{1}{5} \cdot \frac{3}{4} = \frac{1 \cdot 3}{5 \cdot 4} = \frac{3}{20}$

Variable Example $\frac{m}{n} \cdot \frac{r}{s} = \frac{m \cdot r}{n \cdot s} = \frac{mr}{ns}$

Example 1 *Multiplying Rational Numbers.*

a. $\frac{5}{8} \cdot \frac{-2}{3} = \frac{5 \cdot (-2)}{8 \cdot 3}$ *Multiply numerators and multiply denominators.*

$= \frac{-10}{24}$ *Simplify.*

$= \frac{\cancel{2} \cdot (-5)}{\cancel{2} \cdot 12}$ *Factor numerator and denominator.*

$= \frac{-5}{12}$ *Simplify fraction.*

b. $1\frac{2}{3} \cdot 3\frac{4}{5} = \frac{5}{3} \cdot \frac{19}{5}$ *Rewrite as improper fractions.*

$= \frac{\cancel{5} \cdot 19}{3 \cdot \cancel{5}}$ *Multiply.*

$= \frac{19}{3}$ *Simplify fraction.*

c. $\frac{6x}{5} \cdot 10 = \frac{6x}{5} \cdot \frac{10}{1}$ *Rewrite 10 as $\frac{10}{1}$.*

$= \frac{6x \cdot 10}{5 \cdot 1}$ *Multiply.*

$= \frac{6x \cdot \cancel{5} \cdot 2}{\cancel{5}}$ *Factor numerator.*

$= 12x$ *Simplify fraction.* ■

$\frac{1}{12}$	$\frac{1}{12}$	$\frac{1}{12}$	
$\frac{1}{12}$	$\frac{1}{12}$	$\frac{1}{12}$	

This area model for multiplication shows that the product of $\frac{3}{4}$ and $\frac{2}{3}$ is $\frac{6}{12}$ or $\frac{1}{2}$.

Goal 2 Solving Real-Life Problems

Real Life
Construction

This lean-to greenhouse is found in one corner of a rectangular patio whose outside dimensions are $20\frac{1}{2}$ feet by 24 feet.

Example 2 *Finding the Area of a Region*

The greenhouse shown at the left was built from prefabricated glass panels, each of which is $3\frac{1}{6}$ feet wide. What is the floor area of the greenhouse?

$3\frac{1}{6}$ ft $3\frac{1}{6}$ ft $3\frac{1}{6}$ ft $3\frac{1}{6}$ ft
$3\frac{1}{6}$ ft $3\frac{1}{6}$ ft $3\frac{1}{6}$ ft

Solution The two sides of the greenhouse are each 3 panels wide. The front of the greenhouse is 4 panels wide.

$$\text{Width} = 3\left(3\tfrac{1}{6}\right) = 3\left(\tfrac{19}{6}\right) = \tfrac{19}{2}\text{ feet} \qquad \text{Length} = 4\left(3\tfrac{1}{6}\right) = 4\left(\tfrac{19}{6}\right) = \tfrac{38}{3}\text{ feet}$$

To find the area, multiply the width by the length.

$$\begin{aligned}\text{Area} &= (\text{Width}) \times (\text{Length})\\ &= \frac{19}{2} \times \frac{38}{3}\\ &= \frac{19 \cdot 38}{2 \cdot 3}\\ &= \frac{19 \cdot 2 \cdot 19}{2 \cdot 3}\\ &= \frac{19^2}{3}\\ &= \frac{361}{3}\text{ square feet}\end{aligned}$$

The greenhouse has an area of $\frac{361}{3}$ or about 120.3 square feet. You can check this by reworking the problem with decimals. Using a width of 9.5 feet and a length of 12.667 feet, you obtain an area of (9.5)(12.667) or about 120.3 square feet. ■

Communicating about MATHEMATICS

SHARING IDEAS about the Lesson

Problem Solving In Example 2, find the area of the patio that is not covered by the greenhouse. Explain how you did it.

EXERCISES

Guided Practice

CHECK for Understanding

1. *Writing* Explain in your own words how to multiply fractions.

In Exercises 2–5, multiply. Then simplify, if possible.

2. $\frac{4}{7} \cdot \frac{3}{5}$ 3. $\frac{4}{7} \cdot \frac{7}{4}$ 4. $\frac{5x}{9} \cdot \frac{2}{4x}$ 5. $1\frac{3}{5} \cdot 2\frac{1}{2}$

Area Models **In Exercises 6–9, the large square is 1 unit by 1 unit. Find the dimensions and area of the green region.**

6. 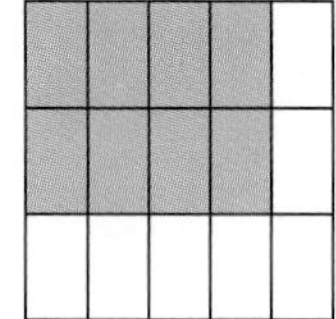

7. 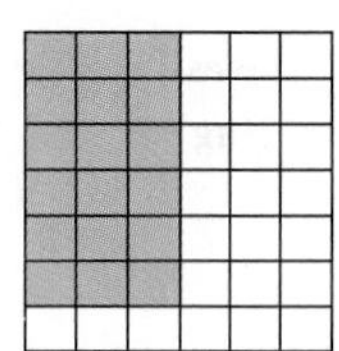

8. 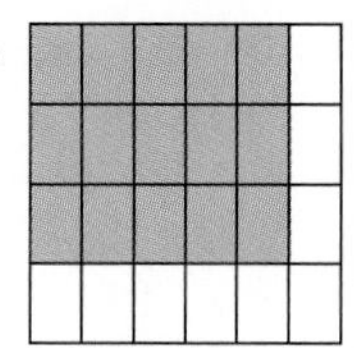

9. 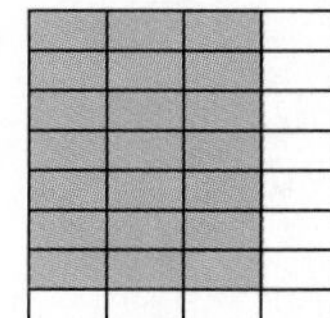

Independent Practice

In Exercises 10–17, multiply. Then simplify, if possible.

10. $\frac{1}{4} \cdot \frac{4}{5}$ 11. $\frac{-2}{3} \cdot \frac{8}{9}$ 12. $\frac{-5}{6} \cdot \frac{-3}{4}$ 13. $1\frac{2}{5} \cdot 2\frac{2}{7}$

14. $1\frac{1}{5} \cdot \left(-6\frac{2}{3}\right)$ 15. $-4\frac{1}{2} \cdot \left(-2\frac{5}{9}\right)$ 16. $\frac{2}{3} \cdot \frac{-4}{7} \cdot \frac{4}{5}$ 17. $\frac{-8}{18} \cdot \frac{2}{3} \cdot \frac{-3}{8}$

In Exercises 18–25, multiply. Then simplify, if possible.

18. $\frac{5x}{6} \cdot 12$ 19. $7 \cdot \frac{8y}{3}$ 20. $\frac{-2x}{9} \cdot \frac{7}{4x}$ 21. $\frac{16z}{11} \cdot \frac{-11}{10z}$

22. $\frac{-13t}{20} \cdot \frac{-1}{2}$ 23. $\frac{-5}{6} \cdot \frac{-6a}{15}$ 24. $\frac{-16x^2}{9} \cdot \frac{9}{4x}$ 25. $\frac{-6y^5}{7} \cdot \frac{-3}{14y^2}$

Geometry **In Exercises 26–28, find the area of the figure.**

26. 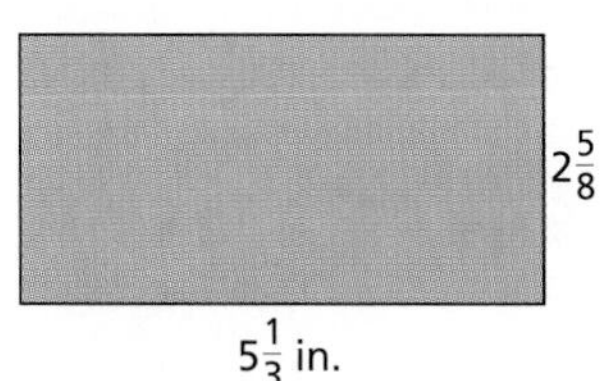

27. 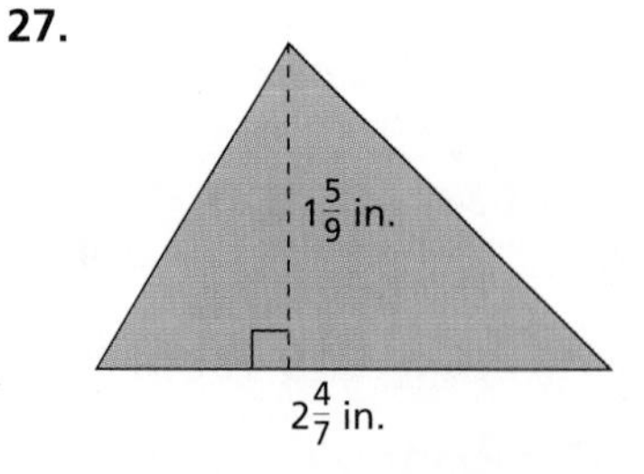

28. 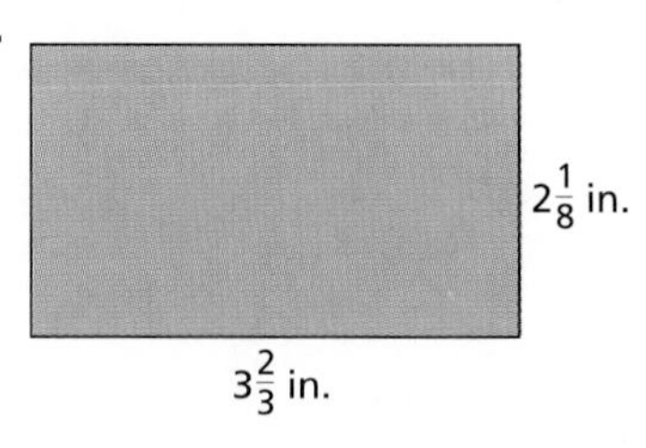

In Exercises 29–32, use a calculator to multiply. Round your result to three decimal places.

29. $\frac{9}{16} \cdot \frac{6}{13}$ 30. $\frac{17}{25} \cdot 2\frac{3}{4}$ 31. $\frac{23}{48} \cdot (-7)$ 32. $\frac{-21}{32} \cdot \frac{2}{5}$

In Exercises 33–36, write each decimal as a fraction. Then multiply.

33. $0.25; 0.\overline{6}$ **34.** $0.\overline{3}; 0.75$ **35.** $0.2; 0.625$ **36.** $0.375; 0.7$

Dinosaurs **In Exercises 37–42, use the information given with the photograph to find the length of the dinosaur.**

37. Torosaurus: Length is $\frac{1}{3}$ of Diplodocus

38. Stegosaurus: Length is $\frac{2}{3}$ of Torosaurus

39. Ankylosaurus: Length is $\frac{3}{4}$ of Stegosaurus

40. Tyrannosaurus Rex: Length is $2\frac{2}{3}$ of Ankylosaurus

41. Ornitholestes: Length is $\frac{3}{20}$ of Tyrannosaurus Rex

42. Brachiosaurus: Length is $13\frac{1}{3}$ of Ornitholestes.

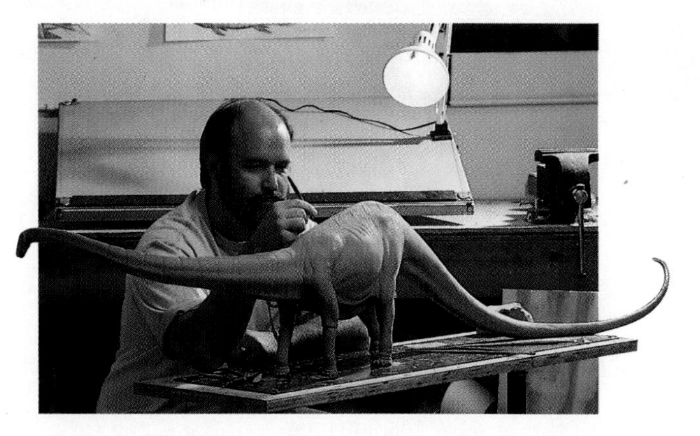

The longest known dinosaur was the Diplodocus measuring 90 feet.

Coins **In Exercises 43–46, represent the money shown as a fraction of a dollar.**

43.

44.

45.

46.

Integrated Review

Making Connections within Mathematics

Mental Math **In Exercises 47–52, solve the equation.**

47. $4x = \frac{5}{6}$ **48.** $-9x = \frac{3}{7}$ **49.** $-12x = \frac{-12}{19}$

50. $3x + 5x = \frac{4}{3}$ **51.** $13x = \frac{6}{11} + 8x$ **52.** $4x = 10x - \frac{12}{13}$

Exploration and Extension

The Number Line **In Exercises 53–55, use the number line below.**

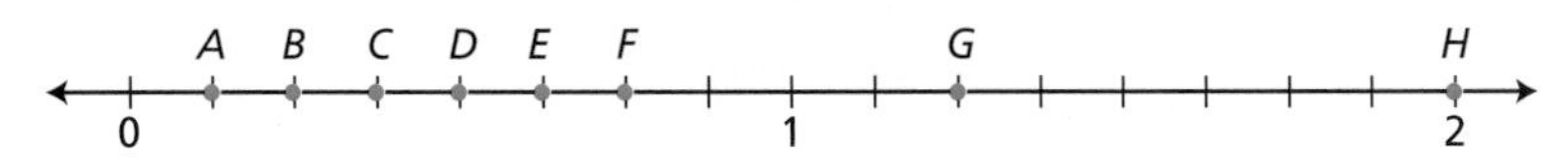

53. What point on the number line represents the product of D and F?

54. What point on the number line represents the product of D and G?

55. What point on the number line represents the product of H and A?

7.5 Division of Rational Numbers

What you should learn:

Goal 1 How to divide rational numbers

Goal 2 How to use division of rational numbers to solve real-life measurement problems

Why you should learn it:

You can use use division of rational numbers to solve real-life problems, such as finding the number of horses that can be grazed on a pasture.

Goal 1 Dividing Rational Numbers

Suppose you took 7 rides on a water slide in a half hour. How many rides could you take in an hour?

$$\frac{7 \text{ rides}}{\frac{1}{2} \text{ hour}} = \frac{14 \text{ rides}}{\text{hour}}$$

Dividing by $\frac{1}{2}$ produces the same result as multiplying by 2.

Dividing Rational Numbers

To divide by a fraction, multiply by its reciprocal.

Numerical Example

$\frac{1}{5} \div \frac{3}{4} = \frac{1}{5} \cdot \frac{4}{3} = \frac{4}{15}$

Variable Example

$\frac{a}{b} \div \frac{c}{d} = \frac{a}{b} \cdot \frac{d}{c} = \frac{ad}{bc}$

Example 1 Dividing Rational Numbers

a. $\frac{3}{4} \div 3 = \frac{3}{4} \cdot \frac{1}{3}$ *Reciprocal of 3 is $\frac{1}{3}$.*

$= \frac{\cancel{3} \cdot 1}{4 \cdot \cancel{3}}$ *Multiply fractions.*

$= \frac{1}{4}$ *Simplify fraction.*

b. $\frac{-2}{3} \div \frac{-4}{5} = \frac{-2}{3} \cdot \frac{5}{-4}$ *Reciprocal of $\frac{-4}{5}$ is $\frac{5}{-4}$.*

$= \frac{-10}{-12}$ *Multiply fractions.*

$= \frac{5}{6}$ *Simplify fraction.*

c. $3 \div 4\frac{1}{2} = 3 \div \frac{9}{2}$ *Write mixed number as fraction.*

$= 3 \cdot \frac{2}{9}$ *Reciprocal of $\frac{9}{2}$ is $\frac{2}{9}$.*

$= \frac{6}{9}$ *Multiply.*

$= \frac{2}{3}$ *Simplify fraction.*

d. $\frac{x}{2} \div 3 = \frac{x}{2} \cdot \frac{1}{3}$ *Reciprocal of 3 is $\frac{1}{3}$.*

$= \frac{x}{6}$ *Multiply fractions.* ■

Study Tip...

Note that negative fractions can be written in several ways. For instance,

$-\frac{4}{5}$, $\frac{-4}{5}$, and $\frac{4}{-5}$

are all equivalent.

Goal 2 Solving Real-Life Problems

Example 2 *Multiplying and Dividing Fractions*

Real Life
Ranching

In 1991, Americans owned about 4.9 million horses. Of the families who owned horses, the average number owned was 2.5.

You own a horse ranch. Your pasture is rectangular, with a width of $\frac{1}{2}$ mile and a length of $\frac{3}{4}$ mile. The recommended grazing area for each horse is $\frac{3}{2}$ acres. There are 640 acres in a square mile. What is the maximum number of horses you should have in your pasture?

Solution Your pasture has an area of $\frac{1}{2} \cdot \frac{3}{4}$, or $\frac{3}{8}$ square mile. To find the number of acres in your pasture, multiply by 640.

$$\frac{640 \text{ acres}}{1 \text{ square mile}} \cdot \left(\frac{3}{8} \text{ square mile}\right) = 240 \text{ acres}$$

To find the maximum number of horses, divide 240 acres by $\frac{3}{2}$ (acres per horse).

$$\begin{aligned} 240 \text{ acres} \div \left(\frac{3}{2}\,\frac{\text{acres}}{\text{horses}}\right) &= 240 \text{ acres} \cdot \frac{2}{3}\,\frac{\text{horses}}{\text{acres}} \\ &= 240 \cdot \frac{2}{3} \text{ horses} \\ &= 160 \text{ horses} \end{aligned}$$

You can have up to 160 horses in your pasture. ■

Communicating about MATHEMATICS

Cooperative Learning

SHARING IDEAS about the Lesson

Division Model Work with a partner. Divide the rectangle into squares, each of which has an area of $\frac{4}{25}$ square unit. How many small squares did you form? Show how to solve the problem both geometrically and algebraically.

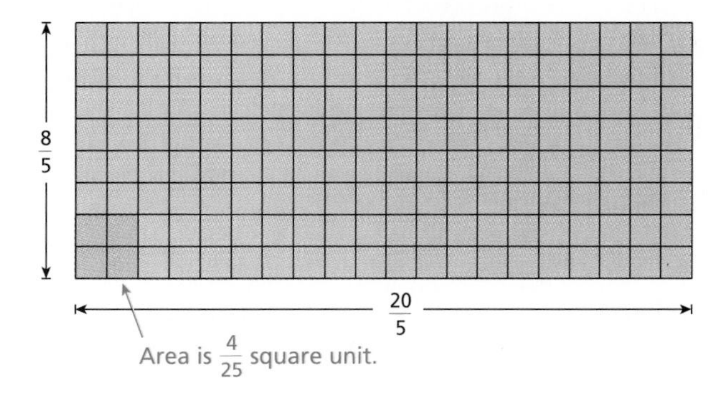

EXERCISES

Guided Practice

CHECK for Understanding

In Exercises 1–4, write the reciprocal.

1. $\frac{1}{5}$ **2.** $-\frac{2}{3}$ **3.** 7 **4.** $\frac{4}{t}$

Error Analysis **In Exercises 5 and 6, describe the error. Then correct it.**

5. $\frac{5}{6} \div 3 = \frac{5}{6} \cdot \frac{3}{1} = \frac{5 \cdot 3}{2 \cdot 3} = \frac{5}{2}$

6. $\frac{-4}{3} \div \frac{3}{2} = \frac{-4}{3} \cdot \frac{2}{3} = \frac{-4 \cdot 2}{3} = -\frac{8}{3}$

In Exercises 7–10, simplify the expression.

7. $\frac{1}{2} \div \frac{5}{6}$ **8.** $6 \div \frac{4}{9}$ **9.** $\frac{n}{3} \div \frac{3}{2}$ **10.** $3\frac{1}{2} \div \frac{4}{x}$

Independent Practice

In Exercises 11–14, write the reciprocal.

11. $\frac{1}{4}$ **12.** $\frac{3}{x}$ **13.** $\frac{7a}{5}$ **14.** $-2\frac{2}{3}$

Error Analysis **In Exercises 15–17, describe the error. Then correct it.**

15. $-\frac{3}{2} \div \frac{3}{5} = -\frac{2}{3} \cdot \frac{3}{5} = \frac{-2 \cdot 3}{3 \cdot 5} = -\frac{2}{5}$

16. $8 \div 2\frac{1}{2} = 8 \cdot \frac{5}{2} = \frac{40}{2} = 20$

17. 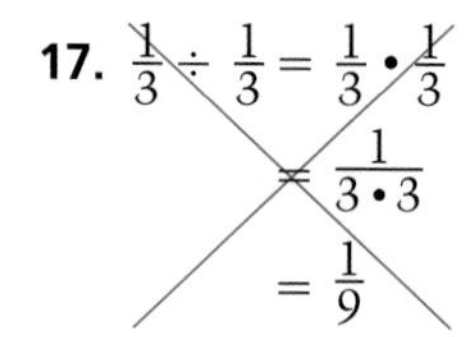
$\frac{1}{3} \div \frac{1}{3} = \frac{1}{3} \cdot \frac{1}{3} = \frac{1}{3 \cdot 3} = \frac{1}{9}$

Mental Math **In Exercises 18–21, simplify the expression.**

18. $\frac{1}{4} \div 2$ **19.** $\frac{1}{4} \div 3$ **20.** $\frac{1}{4} \div 4$ **21.** $\frac{1}{4} \div 5$

22. *Finding a Pattern* In Exercises 18–21, write each result as a decimal, rounded to two decimal places. Describe the pattern. What happens when you divide a number by larger and larger numbers?

In Exercises 23–26, simplify the expression.

23. $\frac{3}{2} \div \frac{1}{2}$ **24.** $\frac{3}{2} \div \frac{1}{3}$ **25.** $\frac{3}{2} \div \frac{1}{4}$ **26.** $\frac{3}{2} \div \frac{1}{5}$

27. *Finding a Pattern* In Exercises 23–26, write each result as a decimal. Describe the pattern. What happens when you divide a number by smaller and smaller numbers?

In Exercises 28–39, simplify the expression.

28. $\frac{3}{4} \div 2$

29. $3 \div \frac{-5}{6}$

30. $\frac{-1}{2} \div \frac{1}{3}$

31. $\frac{7}{4} \div \frac{1}{-4}$

32. $3\frac{1}{2} \div \frac{3}{4}$

33. $\frac{4}{5} \div 1\frac{1}{2}$

34. $\frac{x}{2} \div (-4)$

35. $\frac{-3}{5} \div \frac{9}{x}$

36. $6\frac{2}{3} \div a$

37. $n \div 1\frac{1}{4}$

38. $\frac{1}{y} \div \frac{4}{y}$

39. $\frac{3b}{2} \div \frac{9b}{5}$

40. *Baby-sitting* You baby-sat your three-year old sister and received $4.00 for working $1\frac{2}{3}$ hours. What was your hourly wage? Explain how to check your answer.

41. *Baby-sitting* You baby-sat your neighbor's son for $3\frac{3}{4}$ hours. You earned $12.00. What was your hourly wage? Explain how to check your answer.

42. *Pizza* You ordered 3 pizzas to be delivered when your friends come over. You think each person will eat $\frac{3}{8}$ of a pizza. How many people can you feed?

43. *Water Pitcher* A pitcher holds $\frac{3}{4}$ gallon of water. If each water glass holds $\frac{1}{16}$ gallon, how many glasses can be filled?

Geometry **In Exercises 44–47, write an equation that allows you to solve for *x*. Then solve the equation.**

44.

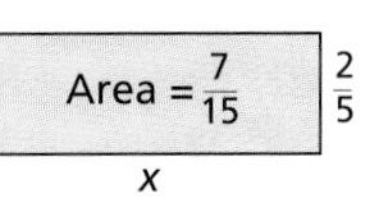

45.

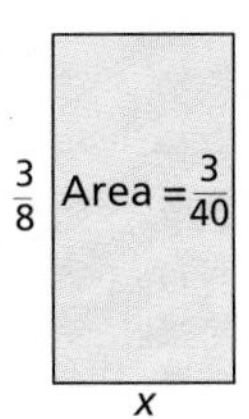

46.

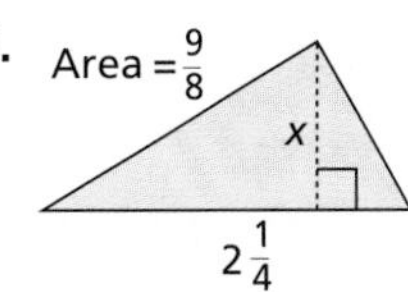

47.

Area = $\frac{63}{4}$ $3\frac{1}{2}$

x

Integrated Review

Making Connections within Mathematics

Unit Analysis **In Exercises 48–51, determine the unit of measure of the quotient.**

48. $\dfrac{\text{miles}}{\left(\dfrac{\text{miles}}{\text{hour}}\right)}$

49. $\left(\dfrac{\text{miles}}{\text{hour}}\right)(\text{hours})$

50. $\left(\dfrac{\text{dollars}}{\text{pound}}\right)(\text{pounds})$

51. $\dfrac{\text{liters}}{\left(\dfrac{\text{liters}}{\text{minute}}\right)}$

Exploration and Extension

Number Sense **In Exercises 52–55, use the restrictions for the numbers *a*, *b*, and *c* at the right. Complete the statement with $<$, $>$, $=$, or CBD (can't be determined). Give examples to illustrate your answer.**

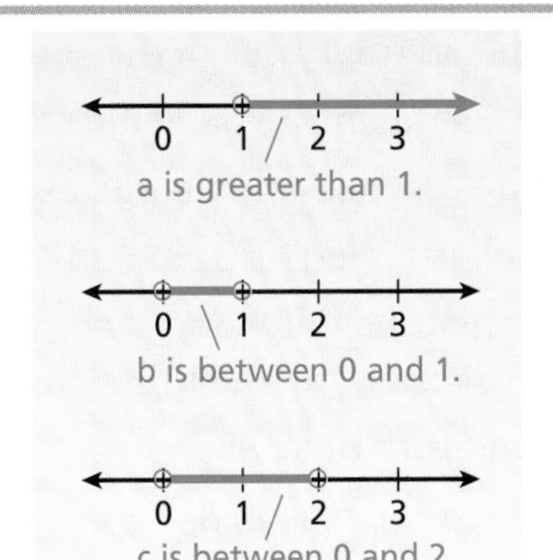

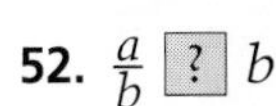

52. $\frac{a}{b}$ [?] b

53. $b \cdot c$ [?] c

54. $c \div a$ [?] b

55. $a \div a$ [?] $\frac{b}{b}$

Mid-Chapter SELF-TEST

Take this test as you would take a test in class. The answers to the exercises are given in the back of the book.

In Exercises 1–4, find the sum or difference and simplify, if possible. (7.1, 7.2)

1. $\frac{1}{11} + \frac{3}{11}$ **2.** $\frac{5}{6} - \frac{1}{6}$ **3.** $\frac{7}{10} + \frac{4}{25}$ **4.** $\frac{9}{10} - \frac{1}{2}$

In Exercises 5–8, find the product or quotient and simplify, if possible. (7.4, 7.5)

5. $-\frac{4}{7} \cdot \frac{7}{8}$ **6.** $\frac{2}{3} \cdot \frac{3}{4} \cdot \frac{4}{5}$ **7.** $\frac{7}{10} \div 2$ **8.** $\frac{2}{5} \div \left(-\frac{6}{5}\right)$

In Exercises 9–12, solve the equation. (7.1–7.5)

9. $\frac{4}{5} + a = \frac{1}{4}$ **10.** $x - \frac{1}{5} = \frac{4}{5}$ **11.** $\frac{1}{3}x = \frac{5}{6}$ **12.** $-4x = \frac{4}{5}$

13. Find the perimeter and area of the rectangle. **(7.2, 7.4)**

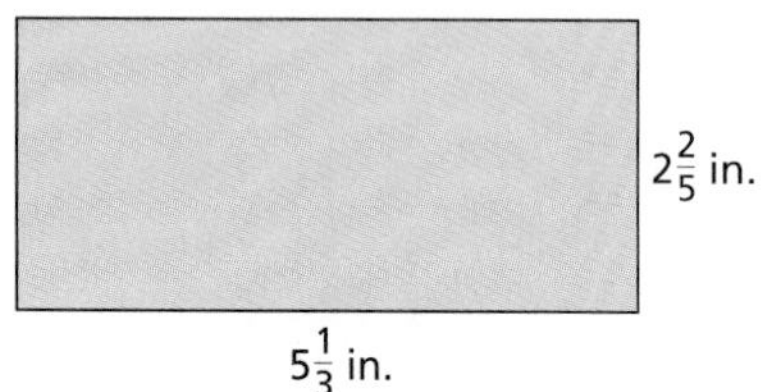

14. The large square has an area of 1. Find the area of the blue region. **(7.4)**

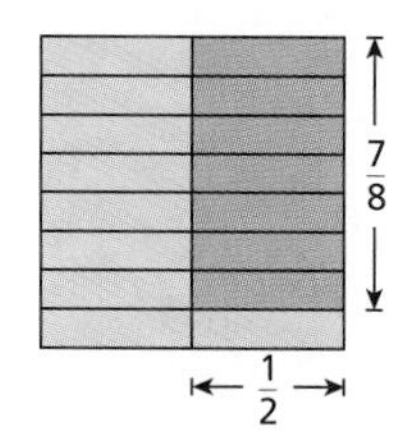

15. A pair of unwashed jeans is 32 inches long. Washing shrinks the jeans to $\frac{7}{8}$ of their original length. How long are they after washing? **(7.4)**

16. A piece of lumber is $8\frac{3}{4}$ feet long. The lumber is cut into 5 pieces of equal lengths. How long is each piece? **(7.5)**

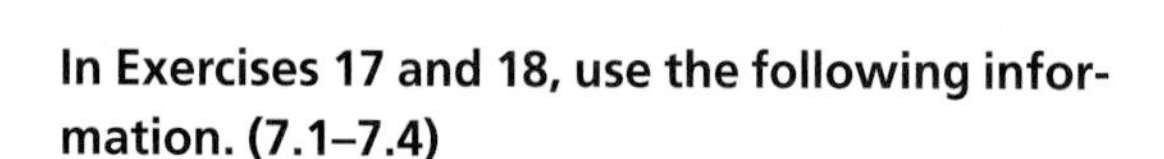

In Exercises 17 and 18, use the following information. (7.1–7.4)

You are shopping for shoes and find a clearance rack with shoes that are $\frac{1}{3}$ to $\frac{1}{2}$ off the original price. The original price of the shoes is $30.

17. What is the most you could expect to pay?

18. What is the least you could expect to pay?

19. Which is larger: 0.72 or $\frac{8}{11}$? **(7.3)**

20. Which of the following cannot be represented exactly with United States coins?

a. One-fifth of $2.85

b. One-fourth of $2.85

c. One-third of $2.85

In 1990, about one fifth of all apparel stores in the United States were shoe stores (Source: U.S. Bureau of the Census)

LESSON INVESTIGATION 7.6

Modeling Portions of Regions

Materials Needed: graph paper or grid paper

In this activity, you will use graph paper to investigate the concept of a portion of a region.

Example — *Finding Portions of Regions*

On a piece of graph paper, outline a square that is 5 units by 5 units. Then shade $\frac{1}{4}$ of the square.

Solution There are many ways to shade one-fourth of the square. Three ways are shown below. Notice that the entire region has an area of 25 square units, and the shaded region has an area of $6\frac{1}{4}$ square units.

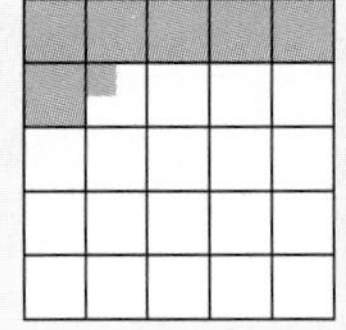

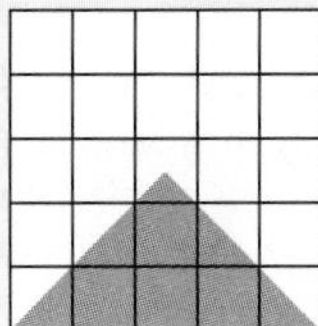

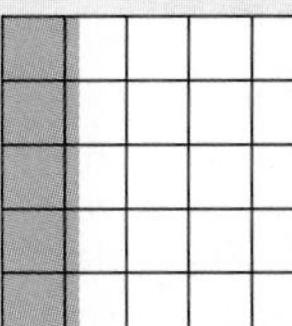

■

Exercises

In Exercises 1–3, what portion of the 10-by-10 rectangle is shaded blue?

1.

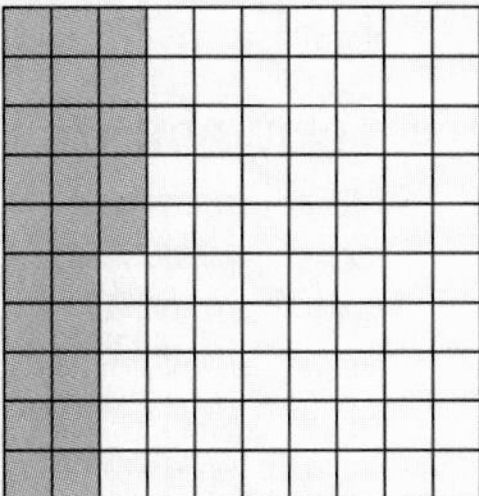

2.

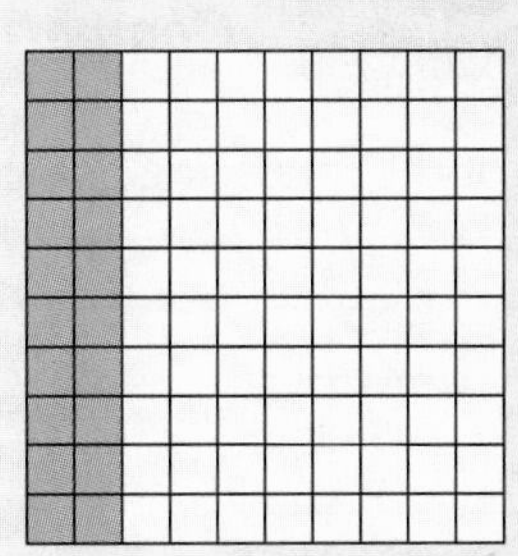

3.

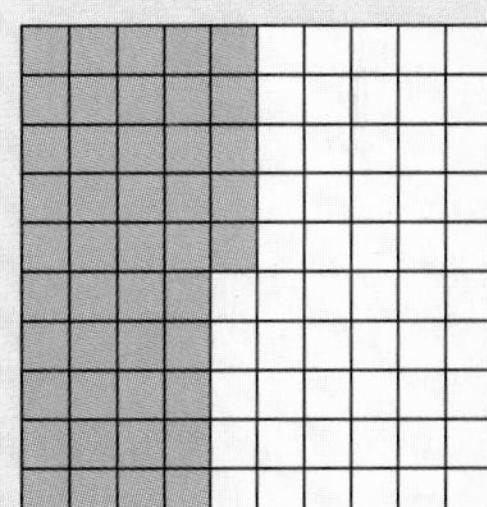

In Exercises 4–6, outline the indicated region on graph paper. Then shade the indicated number of unit squares.

4. 4×10 Rectangle
Shade 8 unit squares.

5. 5×5 Square
Shade 10 unit squares.

6. 5×9 Rectangle
Shade 9 unit squares.

7. In Exercises 4–6, what portion of each rectangle is shaded?

8. Shade 8 unit squares on a piece of graph paper. Then draw a rectangle so that the shaded unit squares form one-third of the rectangle.

9. Shade 12 unit squares on a piece of graph paper. Then draw a rectangle so that the shaded unit squares form one third of the rectangle.

7.6 Exploring Percents

What you should learn:

Goal 1 How to write portions as percents

Goal 2 How to use percents to solve real-life problems

Why you should learn it:

You can use percents to solve real-life problems, such as finding patterns in ancient writings.

Goal 1 Writing Percents

A **portion** is a fraction that compares the measure of part of a quantity to the measure of the whole quantity. For instance, if you own 12 T-shirts, 3 of which are black, then you can say that $\frac{3}{12}$ or $\frac{1}{4}$ of your T-shirts are black. For these portions, the denominators are 12 and 4. When the denominator is 100, the portion is called a **percent.**

Percent

A **percent** is a portion whose denominator is 100. The symbol % means *percent.* Here is an example.

Fraction Form	Percent Symbol Form	Verbal Form
$\frac{65}{100}$	65%	65 percent

Example 1 *Comparing Percents*

Which of the following has the greatest percent of its area shaded?

a. 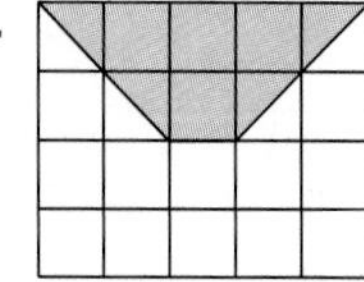

b. 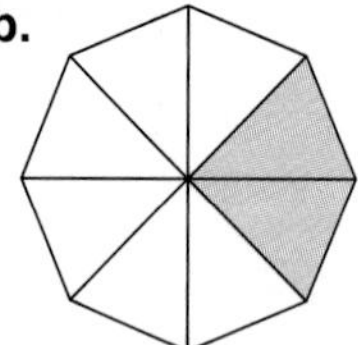

c. 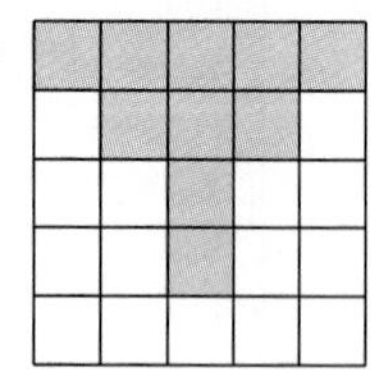

Study Tip...
In Example 1, notice that rewriting each portion as a percent makes it easier to compare the portions. This is one of the main reasons for using percents.

Solution

a. The portion of the region that is shaded is

$\frac{6}{20} = \frac{6}{20} \cdot \frac{5}{5} = \frac{30}{100} = 30\%.$

b. The portion of the region that is shaded is

$\frac{2}{8} = \frac{1}{4} = \frac{1}{4} \cdot \frac{25}{25} = \frac{25}{100} = 25\%.$

c. The portion of the region that is shaded is

$\frac{10}{25} = \frac{10}{25} \cdot \frac{4}{4} = \frac{40}{100} = 40\%.$

The third figure has the greatest percent of its area shaded. ■

Goal 2 Solving Real-Life Problems

Real Life
Archaeology

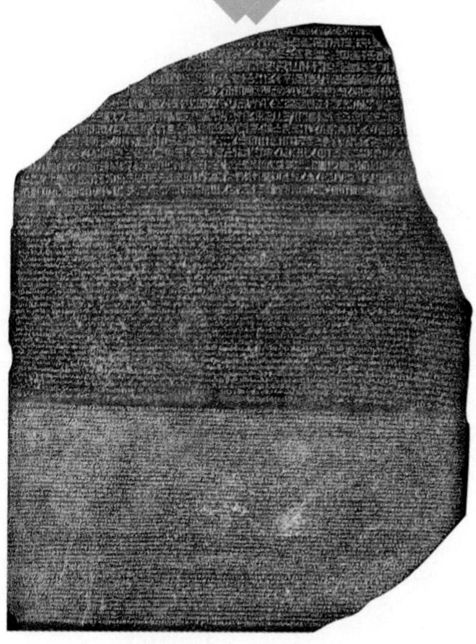

The Rosetta Stone ***For hundreds of years, no one could decode Egyptian hieroglyphics. This stone, called the Rosetta Stone, helped solve the mystery. It contains the same message in three languages.***

Example 2 *Writing Percents*

In the ancient Egyptian hieroglyph shown below, what percent of the symbols is the symbol for water, ~~~~ ?

Solution The hieroglyph contains 20 symbols, five of which are the symbol for water.

$$\text{Portion} = \frac{\text{Water symbols}}{\text{Total symbols}}$$
$$= \frac{5}{20}$$

To rewrite this portion as a percent, multiply the numerator and denominator by 5.

$$\frac{5}{20} = \frac{5}{20} \cdot \frac{5}{5}$$ *Multiply by $\frac{5}{5}$.*

$$= \frac{25}{100}$$ *Fraction form*

$$= 25\%$$ *Percent form*

Thus, 25% of the symbols are symbols for water. ■

Communicating about MATHEMATICS

SHARING IDEAS about the Lesson

Writing Percents What percent of each square is shaded blue? For which square is it easiest to find the percent? Why?

A.

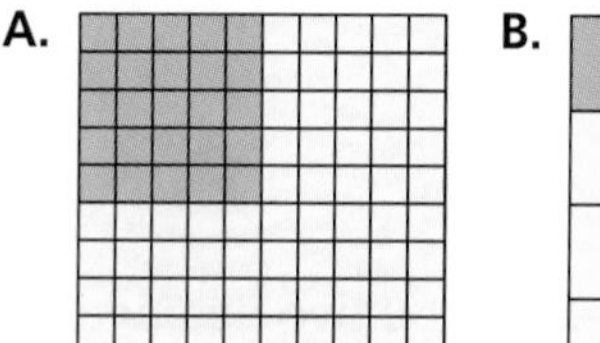

B.

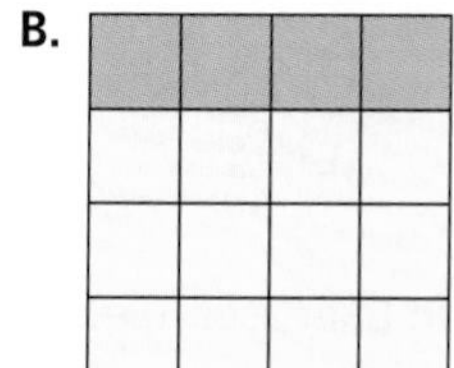

C.

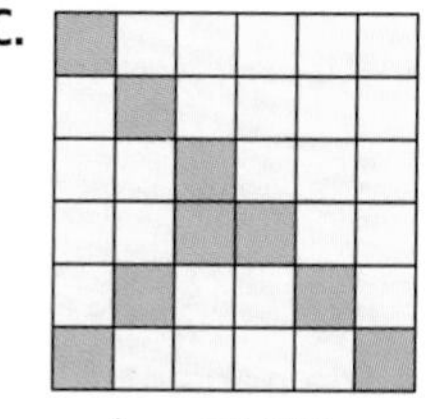

From these results, what common fraction is equal to 25%?

EXERCISES

Guided Practice

CHECK for Understanding

1. Copy and complete the table.

Fraction Form	Percent Symbol Form	Verbal Form
$\frac{24}{100}$		
	83%	
		49 percent

2. What percent of the figure is shaded?

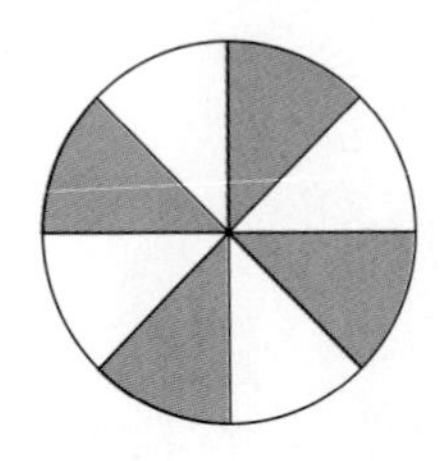

3. Write $\frac{9}{20}$ as a percent.

4. ***Modeling Percents*** Draw two geometric models for 60%.

Independent Practice

In Exercises 5–8, determine the percent of the figure that is shaded blue.

5.

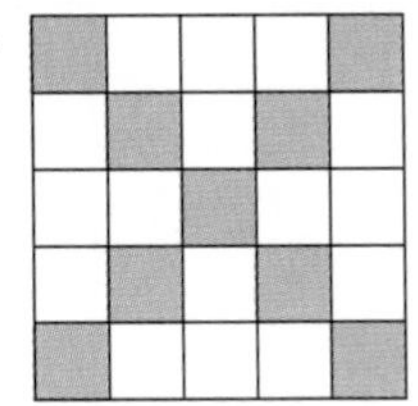

6.

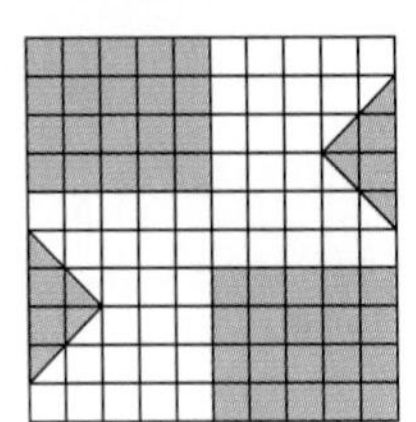

7.

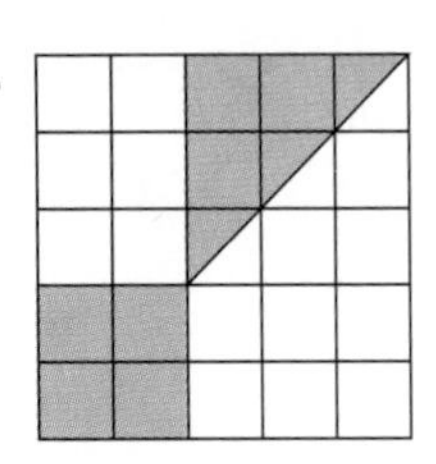

8.

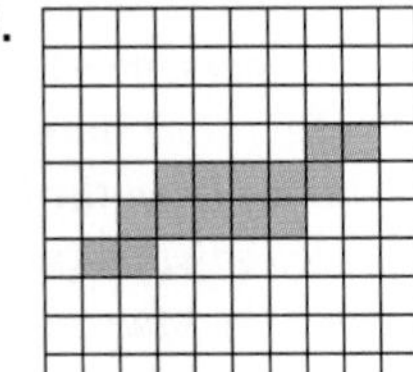

9. Which of the following has the least percent of its area shaded blue? Which has the greatest?

a.

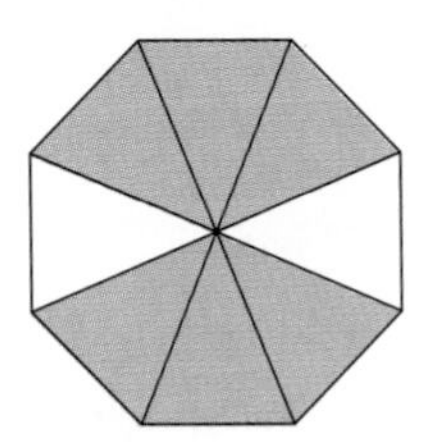

b.

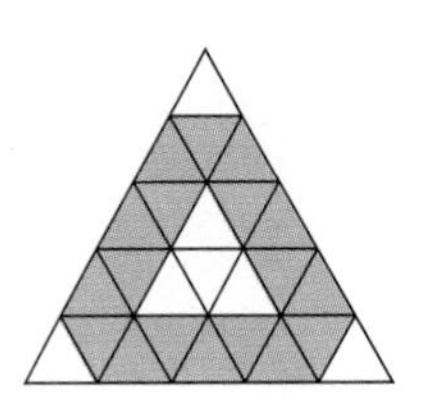

c.

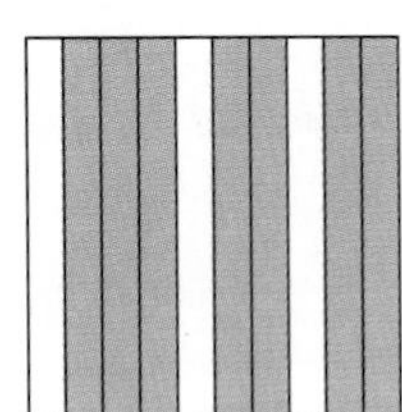

d.

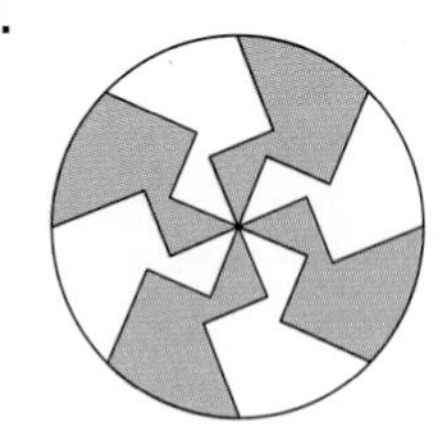

In Exercises 10–17, write each portion as a percent.

10. $\frac{1}{10}$ **11.** $\frac{1}{20}$ **12.** $\frac{31}{50}$ **13.** $\frac{7}{25}$

14. $\frac{24}{32}$ **15.** $\frac{18}{40}$ **16.** $\frac{45}{150}$ **17.** $\frac{180}{300}$

Modeling Percents **In Exercises 18–21, draw *two* geometric models for the percent.**

18. 25% **19.** 20% **20.** 80% **21.** 100%

Estimation **In Exercises 22 and 23, use the map at the right. The blue regions on the map indicate water.**

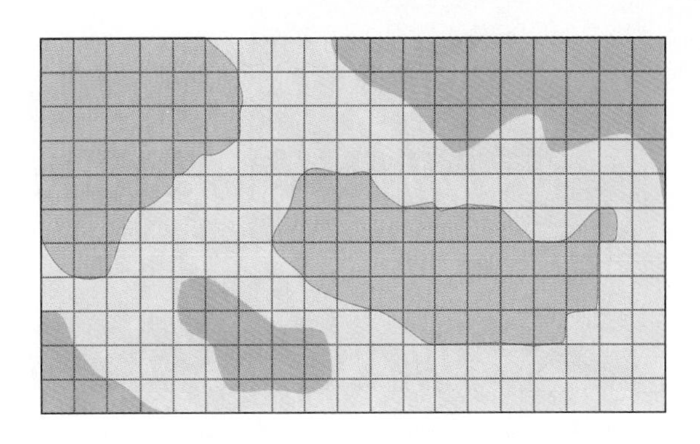

22. About what percent of the region shown on the map is water?

23. Explain how to use your answer to Exercise 22 to estimate the percent of the region that is *not* water.

In Exercises 24 and 25, determine which figure does not fit the pattern. Explain, using percents.

24. a.

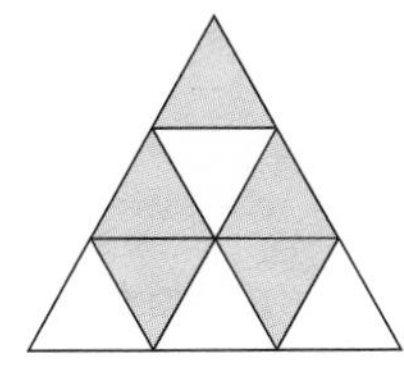

b.

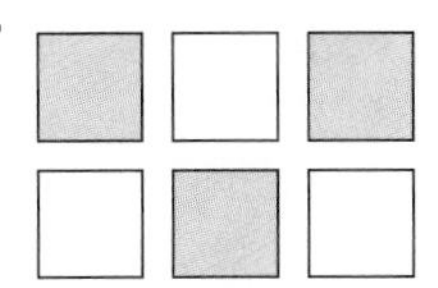

c.

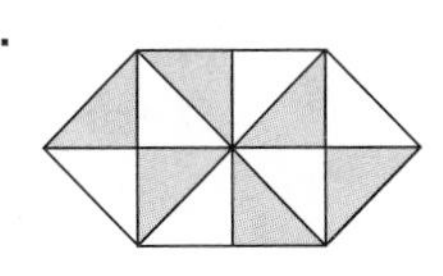

d.

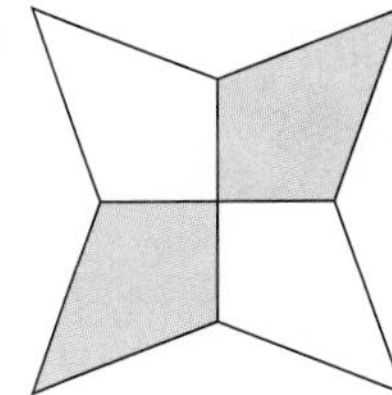

25. a.

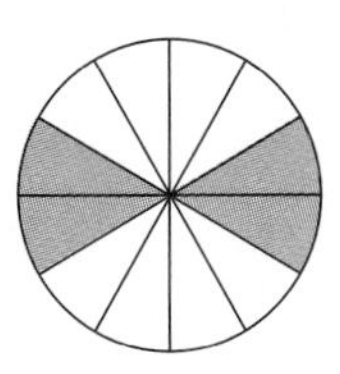

b.

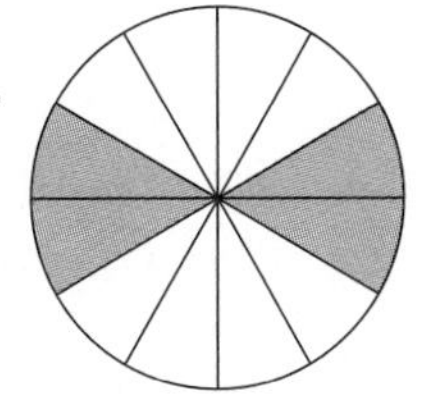

c.

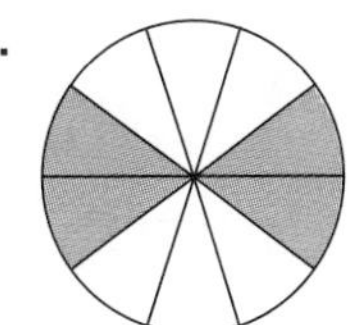

d.

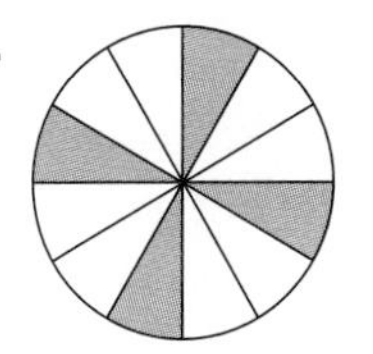

26. *Research Project* Find an advertisement or an article in a magazine or newspaper that uses percents. Then write a paragraph about how percent is used in the article or advertisement.

Integrated Review

Making Connections within Mathematics

Word Meanings **In Exercises 27–32, match the "cent word" with its meaning.**

a. Century **b.** Centennial **c.** Centavo

d. Centimeter **e.** Centigrade **f.** Centipede

27. 1/100th of a peso

28. 100-base temperature scale

29. 100 years

30. 1/100th of a meter

31. 100th anniverary

32. Bug with "100 legs"

Exploration and Extension

Estimating Percents **In Exercises 33–36, estimate the percent of a typical school day that you perform each activity.**

33. Eat **34.** Sleep **35.** Attend school **36.** Do homework

7.7 Percents, Decimals, and Fractions

What you should learn:

Goal 1 How to write percents as decimals and how to write decimals as percents

Goal 2 How to write fractions as percents and how to write percents as fractions

Why you should learn it:

Knowing how to interpret percents helps you understand data that is presented with circle graphs, such as a circle graph for the age distribution in the United States.

Goal 1 Writing Percents as Decimals

In real life, percents are written in several different forms. For instance, a 35% discount at a store can be written in the following ways.

Percent Form	Verbal Form	Fraction Form	Decimal Form
35%	35 percent	$\frac{35}{100}$ or $\frac{7}{20}$	0.35

The first two forms are used for communication and the second two forms are used for computation.

Percent Form and Decimal Form

1. To write a percent as a decimal, remove the percent sign and divide by 100.
2. To write a decimal as a percent, multiply the decimal by 100%.

Example 1 *Rewriting Percents as Decimals*

When you are rewriting a percent in decimal form, remember that *percent* means *per hundred.*

a. $14\% = \frac{14}{100} = 0.14$

b. $0.5\% = \frac{0.5}{100} = 0.005$

c. $100\% = \frac{100}{100} = 1$

d. $125\% = \frac{125}{100} = 1.25$

e. $12.5\% = \frac{12.5}{100} = 0.125$

f. $33\frac{1}{3}\% \approx \frac{33.3}{100} = 0.333$ ■

Need to Know

In Examples 1 and 2, notice that some percents are greater than 100%. Also notice that some percents are less than 1%.

Example 2 *Rewriting Decimals as Percents*

To rewrite a decimal as a percent, multiply the decimal by 100%.

a. $0.28 = 0.28(100\%) = 28\%$

b. $0.346 = 0.346(100\%) = 34.6\%$

c. $0.9 = 0.9(100\%) = 90\%$

d. $2.3 = 2.3(100\%) = 230\%$

e. $1.045 = 1.045(100\%) = 104.5\%$

f. $0.001 = 0.001(100\%) = 0.1\%$ ■

Age Distribution in the U.S.

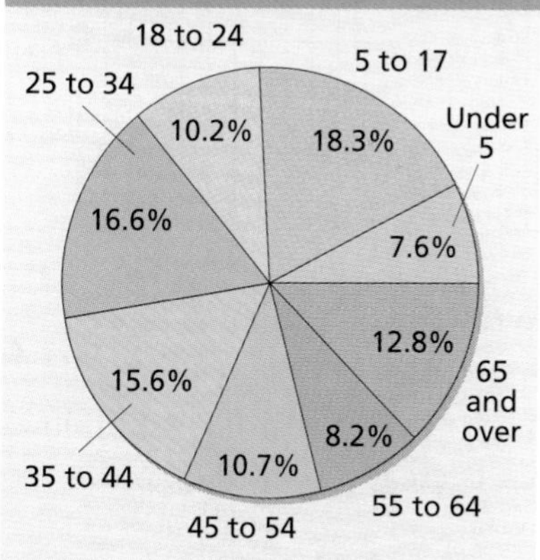

The parts of a circle graph are often labeled with percents. This circle graph shows the age distribution in the United States in 1992. The sum of the percents in a circle graph is 100%. Check that the sum of the percents in this circle graph is 100%.

Goal 2 Writing Fractions as Percents

To rewrite a fraction as a percent, first rewrite the fraction in decimal form. Then multiply by 100%. To rewrite a percent in fraction form, divide by 100%. Then simplify, if possible.

Example 3 *Rewriting Fractions as Percents*

a. $\frac{7}{8} = 0.875 = 0.875(100\%) = 87.5\%$

b. $\frac{1}{3} \approx 0.333 = 0.333(100\%) = 33.3\%$

c. $\frac{12}{5} = 2.4 = 2.4(100\%) = 240\%$ ■

Example 4 *Rewriting Percents as Fractions*

a. $72\% = \frac{72}{100} = \frac{18}{25}$ **b.** $125\% = \frac{125}{100} = \frac{5}{4}$ ■

Commonly Used Percents

$0 = 0\%$	$\frac{1}{10} = 0.1 = 10\%$	$\frac{1}{8} = 0.125 = 12.5\%$
$\frac{1}{6} \approx 0.167 = 16.7\%$	$\frac{1}{5} = 0.2 = 20\%$	$\frac{1}{4} = 0.25 = 25\%$
$\frac{3}{10} = 0.3 = 30\%$	$\frac{1}{3} \approx 0.333 = 33.3\%$	$\frac{3}{8} = 0.375 = 37.5\%$
$\frac{2}{5} = 0.4 = 40\%$	$\frac{1}{2} = 0.5 = 50\%$	$\frac{3}{5} = 0.6 = 60\%$
$\frac{5}{8} = 0.625 = 62.5\%$	$\frac{2}{3} \approx 0.667 = 66.7\%$	$\frac{7}{10} = 0.7 = 70\%$
$\frac{3}{4} = 0.75 = 75\%$	$\frac{4}{5} = 0.8 = 80\%$	$\frac{5}{6} \approx 0.883 = 83.3\%$
$\frac{7}{8} = 0.875 = 87.5\%$	$\frac{9}{10} = 0.9 = 90\%$	$1 = 100\%$

Communicating about MATHEMATICS

SHARING IDEAS about the Lesson

Circle Graphs What percent of the circle graph is blue?

A.

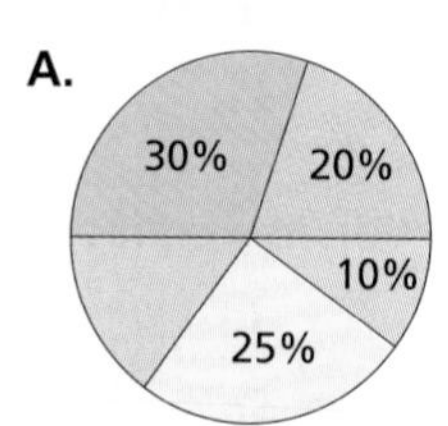

B.

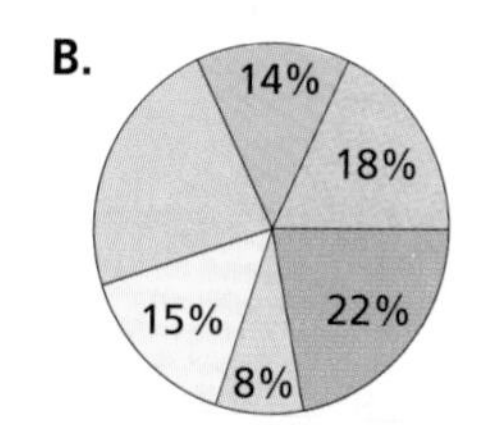

C.

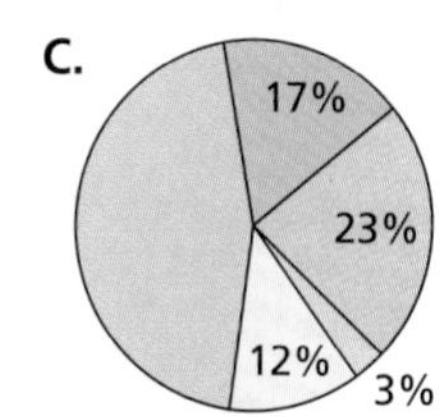

EXERCISES

Guided Practice

CHECK for Understanding

In Exercises 1–4, what portion of the figure is blue? Express your answer as a percent and as a decimal.

1. 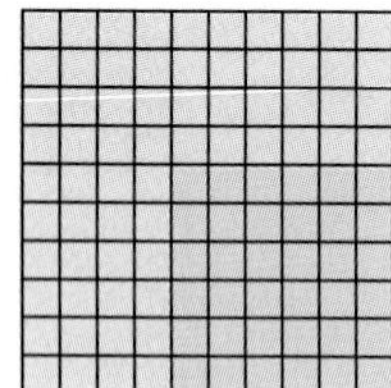

2. 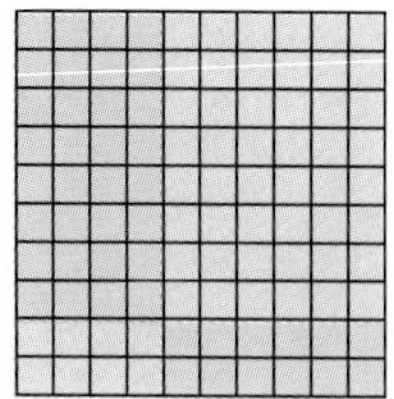

3. 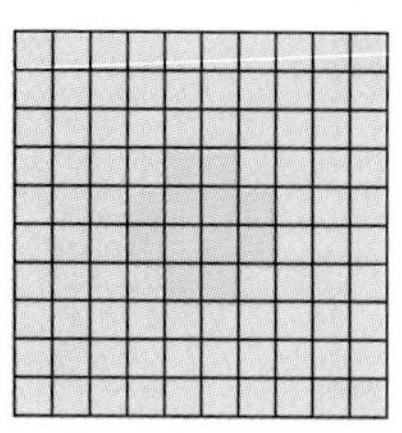

4. 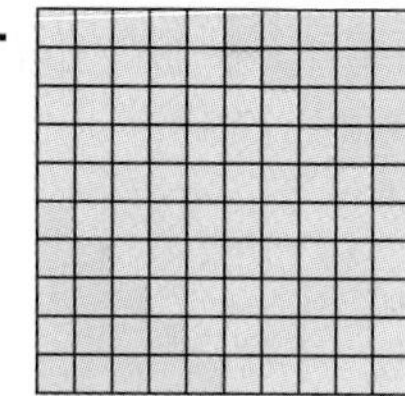

5. Explain how to rewrite a percent as a decimal. Then rewrite 48% as a decimal.

6. Explain how to rewrite a decimal as a percent. Then rewrite 0.045 as a percent.

7. Explain how to rewrite a fraction as a percent. Then rewrite $\frac{5}{16}$ as a percent.

8. *Problem Solving* Give a real-life example of a percent that is greater than 100%.

Independent Practice

In Exercises 9–16, rewrite the percent as a decimal.

9. 36% **10.** 47% **11.** 115% **12.** 265%

13. 89.7% **14.** 1.44% **15.** $14\frac{2}{3}\%$ **16.** $36\frac{3}{4}\%$

In Exercises 17–24, rewrite the decimal as a percent.

17. 0.25 **18.** 0.826 **19.** 0.7 **20.** 0.9453

21. 1.4 **22.** 3.083 **23.** 0.009 **24.** 0.055

In Exercises 25–28, complete the statement with $<$, $>$, or $=$.

25. $\frac{3}{8}$? 3.75% **26.** $\frac{7}{16}$? 43.75% **27.** $\frac{1}{25}$? 4% **28.** $\frac{3}{50}$? 0.6%

Estimation **In Exercises 29–32, match the percent with the portion of the American population that you think has the indicated characteristic.**

a. Over 85 years old **b.** Female **c.** Watch television **d.** Under 10 years old

29. 51% **30.** 93% **31.** 15% **32.** 1%

In Exercises 33–40, rewrite the percent as a fraction in simplest form.

33. 52% **34.** 75% **35.** 6% **36.** 8%

37. 160% **38.** 248% **39.** 102% **40.** 95%

In Exercises 41–48, rewrite the fraction as a percent.

41. $\frac{13}{208}$ **42.** $\frac{52}{650}$ **43.** $\frac{78}{99}$ **44.** $\frac{375}{450}$

45. $\frac{104}{64}$ **46.** $\frac{429}{286}$ **47.** $\frac{180}{54}$ **48.** $\frac{357}{252}$

In Exercises 49–52, what percent of the entire region is blue?

49.

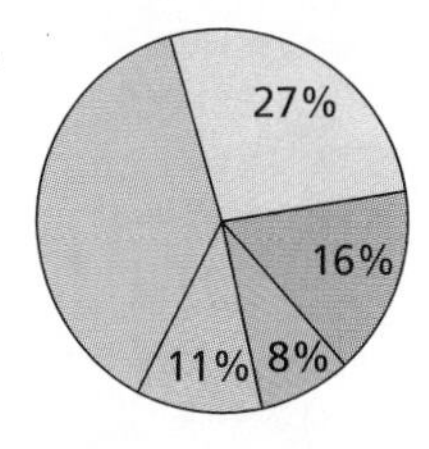

50.

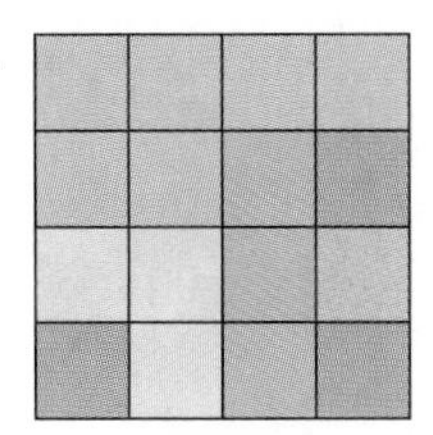

51.

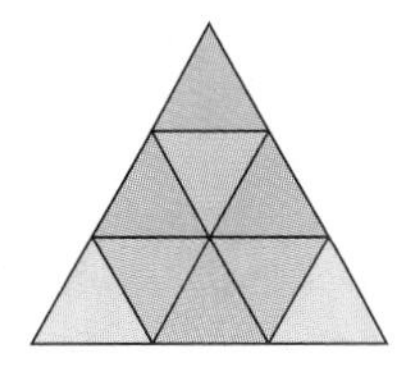

52.

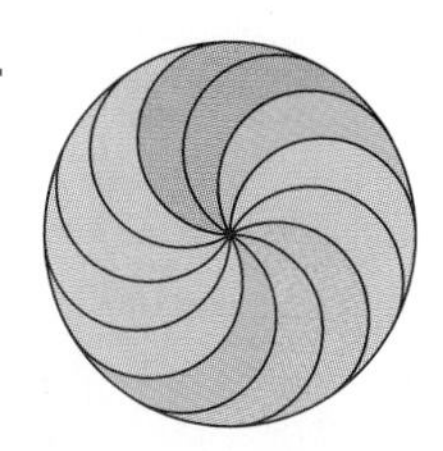

Cameras In Exercises 53–56, use the circle graph at the right. *(Source: Photo Marketing Association International)*

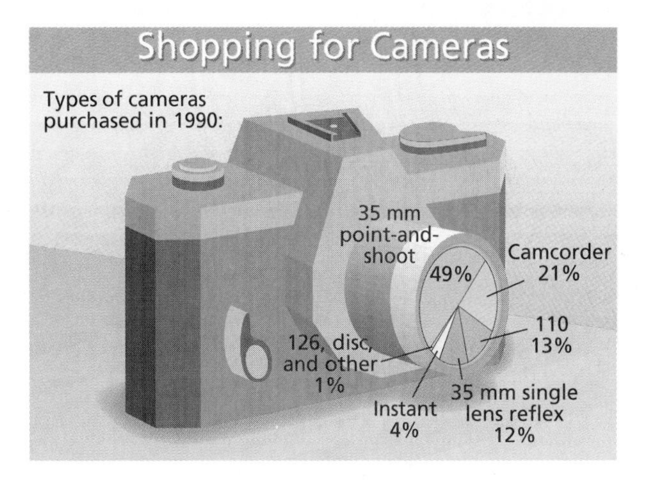

53. More than 3 out of every 20 American households purchased at least one new camera in 1990. What percent is this?

54. Create a table showing each percent rewritten as a decimal and a fraction in simplest form.

55. Show that the sum of the percents is 100%.

56. Which of the percents is three times as large as another?

Test Grades In Exercises 57–60, use the table at the right that shows the grades on a math test.

57. Create a table showing the fraction, decimal, and percent of the number of students receiving each letter grade.

58. What percentage of students passed the test (got a D or better)?

59. What percentage of students received a C or better?

60. Create a bar graph showing the grade distribution.

Grade	Number of Students
A	ЖЖЖ I
B	ЖЖ IIII
C	Ж
D	III
F	II

Integrated Review

Making Connections within Mathematics

Estimation In Exercises 61–66, estimate the amount and tell how you estimated.

61. $\frac{1}{5}$ of 895 **62.** $\frac{4}{9}$ of 365 **63.** $\frac{2}{3}$ of \$14.95

64. 34% of \$27.65 **65.** 83% of 320 **66.** 37% of \$49.29

Exploration and Extension

Estimation **In Exercises 67–69, estimate the percent that is represented by each part of the circle graph. Which parts are easiest to estimate? Why?**

67.

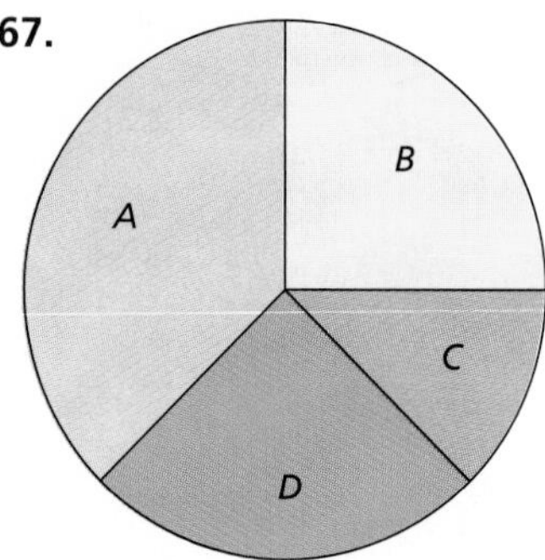

68.

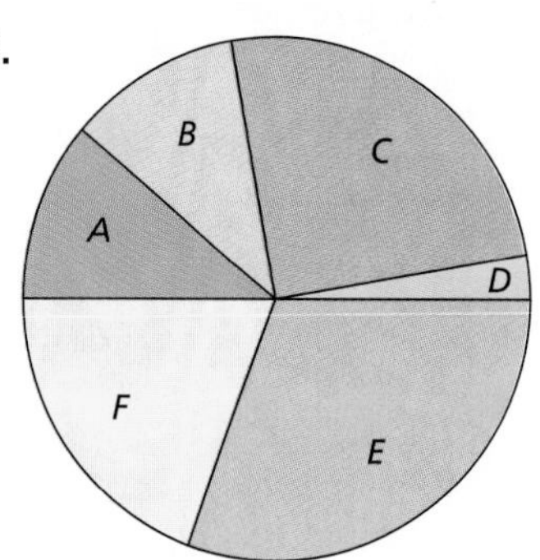

69.

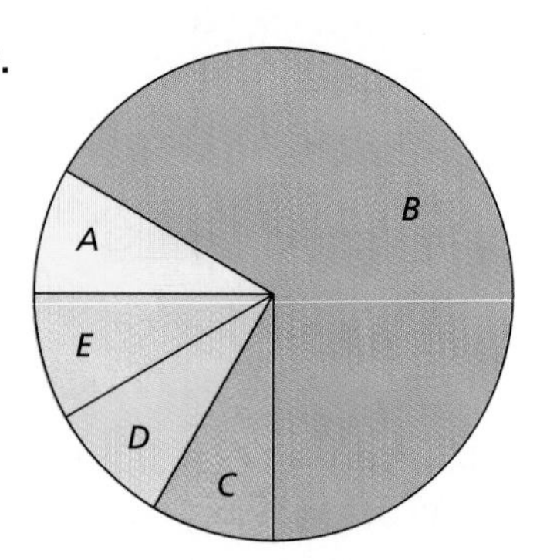

Mixed REVIEW

In Exercises 1–4, solve the equation. (7.2)

1. $\frac{1}{2} = \frac{1}{4} + r$

2. $1 = \frac{1}{6} + a$

3. $\frac{2}{5}z - 2 = \frac{1}{5}z$

4. $-\frac{6}{7} + 3p = -\frac{1}{7}$

In Exercises 5–10, use mental math to solve the equation. (7.7)

5. $58\% + x = 67\%$

6. $32\% = b - 18\%$

7. $50\% - r = 25\%$

8. $84\% = 12\% + x$

9. $112\% = p - 24\%$

10. $42\% = 100\% - f$

In Exercises 11–16, solve the equation. (7.4, 7.5)

11. $\frac{1}{3}x = \frac{1}{2}$

12. $\frac{1}{4} = \frac{1}{2}y$

13. $3b - \frac{1}{2} = \frac{1}{3}$

14. $\frac{2}{3} = 2d + \frac{1}{6}$

15. $\frac{10}{3} + 2x = \frac{1}{3}$

16. $\frac{5}{6} + 5y = \frac{5}{2}$

17. What does it mean when the weatherperson says, "There is a 50% chance of rain"?

In Exercises 18–20, write the verbal sentence as an equation and solve. (7.2, 7.4, 7.7)

18. x times $\frac{3}{4}$ is 18.75.

19. The product of $\frac{1}{4}$ and a is $\frac{2}{3}$.

20. The sum of x and $\frac{1}{3}$ is $\frac{1}{2}$.

7.8 Finding a Percent of a Number

What you should learn:

Goal 1 How to find a percent of a number

Goal 2 How to use percents to solve real-life problems

Why you should learn it:

Knowing how to find a percent of a number helps you solve real-life problems, such as finding how many people voted for a presidential candidate.

Goal 1 Finding a Percent of a Number

One way to find a percent of a number is to multiply the *decimal form* of the percent by the number.

Example 1 *Finding Percents of Numbers*

a. Find 36% of 825. **b.** Find 150% of 38.

Solution

a. Begin by writing 36% as 0.36. Then multiply by 825.

$0.36 \times 825 = 297$

Thus, 36% of 825 is 297.

b. Begin by writing 150% as 1.5. Then multiply by 38.

$1.5 \times 38 = 57$

Thus, 150% of 38 is 57. ■

Example 2 *Finding a Percent of an Area*

40% of 60 squares is 24 squares.

The rectangle at the left has 60 small squares. How many small squares are in 40% of the rectangle?

Solution One way to answer this question is to shade 40% or $\frac{4}{10}$ of the rectangle and count the number of squares. Doing this, you can see that 40% of the rectangle consists of 24 squares. Another way to answer the question is to rewrite 40% as 0.4, then multiply by 60.

$0.4 \times 60 = 24$

Thus, 40% of 60 squares is 24 squares. ■

With some common percents, you can use mental math to find the percent of a number. Here are two examples.

Problem	Mental Math	Answer
50% of 30	One half of 30	15
$33\frac{1}{3}$% of 36	One third of 36	12

Study Tip...
$\frac{1}{3}$ of 100 is $33\frac{1}{3}$. So, $\frac{1}{3}$ of 100% is $33\frac{1}{3}$%.

Goal 2 Solving Real-Life Problems

Real Life Politics

Example 3 Finding Percents of Numbers

In the 1992 election for the president of the United States, about 104,425,000 Americans voted. The total number of people who voted is called the *popular vote*. The percent of the popular vote received by the three main candidates is shown below. How many people voted for each of the three candidates?

	Candidate	Political Party	Percent of Popular Vote
a.	George Bush	Republican	37.4%
b.	Bill Clinton	Democrat	43.0%
c.	Ross Perot	Independent	18.9%

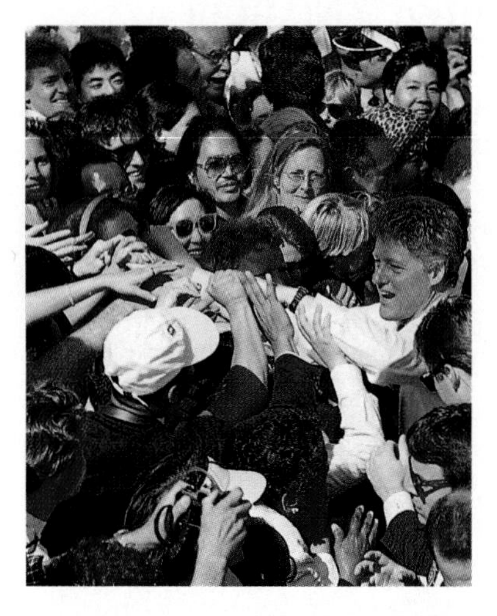

In the United States, presidents are elected by electoral votes, not popular votes. For instance, in 1992, Bill Clinton received only 43% of the popular vote, but he received 68.8% of the electoral votes.

Solution

a. To find how many people voted for George Bush, rewrite 37.4% as 0.374, and multiply by the total popular vote.

$0.374 \times 104{,}425{,}000 = 39{,}054{,}950$ *Bush*

b. To find how many people voted for Bill Clinton, rewrite 43.0% as 0.43, and multiply by the total popular vote.

$0.43 \times 104{,}425{,}000 = 44{,}902{,}750$ *Clinton*

c. To find how many people voted for Ross Perot, rewrite 18.9% as 0.189, and multiply by the total popular vote.

$0.189 \times 104{,}425{,}000 = 19{,}736{,}325$ *Perot*

From these results, how can you tell that some Americans voted for candidates other than Bush, Clinton, or Perot? ■

Communicating about MATHEMATICS

SHARING IDEAS about the Lesson

Using Area Models Copy the rectangles on graph paper. Then shade 60% of each rectangle. How many squares did you shade? Explain how to check your results.

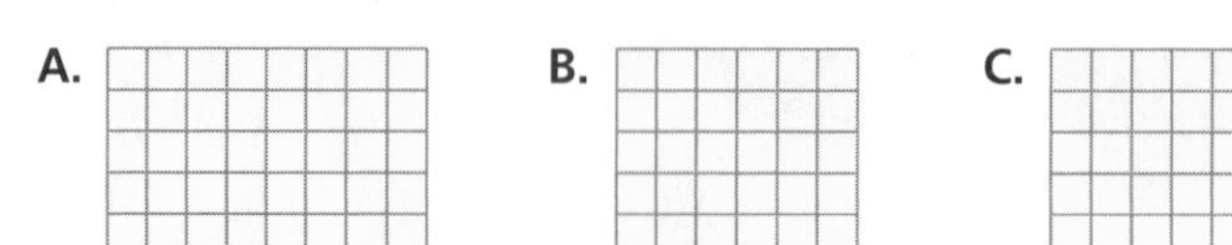

EXERCISES

Guided Practice

CHECK for Understanding

1. *Writing* Explain in your own words how to find the percent of a number.

2. Find 45% of 380.

3. *Mental Math* Use mental math to evaluate the expression.

   a. 10% of 48 b. $33\frac{1}{3}\%$ of 96 c. 50% of 64 d. 200% of 23

Independent Practice

In Exercises 4–11, write the percent as a decimal. Then multiply to find the percent of the number.

4. 16% of 50
5. 80% of 285
6. 76% of 375
7. 340% of 5
8. 120% of 35
9. 250% of 46
10. 0.8% of 500
11. 6.5% of 800

In Exercises 12–15, match the percent phrase with the fraction phrase. Then find the percent of the number.

a. $\frac{1}{8}$ of 120 b. $\frac{1}{3}$ of 120 c. $\frac{3}{5}$ of 120 d. $\frac{1}{4}$ of 120

12. 25% of 120
13. 60% of 120
14. 12.5% of 120
15. $33\frac{1}{3}\%$ of 120

In Exercises 16–18, copy the rectangle and shade the indicated number of squares.

16. 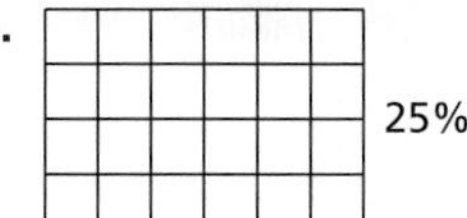

17. 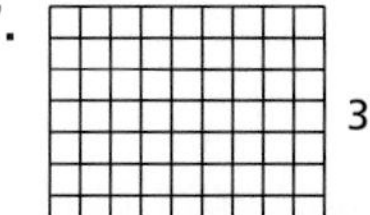

18. 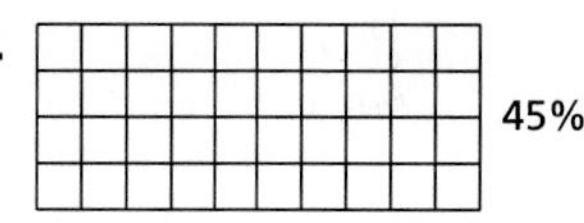

In Exercises 19–22, use the percent key on a calculator to find the percent of the number. Round the result to 2 decimal places.

Sample (50% of 62): 50  **62**  **50** 

19. 96% of 25
20. 185% of 40
21. 624% of 50
22. 1.5% of 800

Geometry **In Exercises 23–26, use the rectangle at the right.**

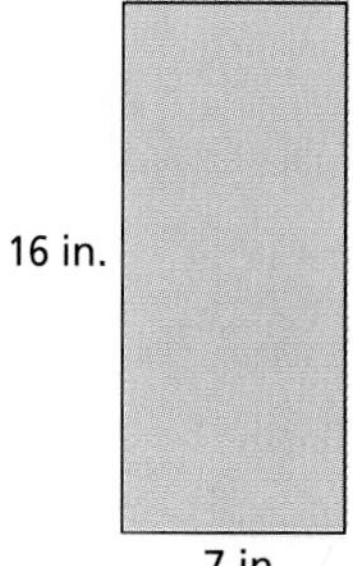

23. Find the perimeter and area of the rectangle.
24. Draw a new rectangle whose dimensions are 150% of the given rectangle. Find the perimeter and area of the new rectangle.
25. Find 150% of the perimeter of the given rectangle. Is the result equal to the perimeter of the new rectangle? Why or why not?
26. Find 150% of the area of the given rectangle. Is the result equal to the area of the new rectangle? Why or why not?

Geometry **In Exercises 27–29, compare the width, length, and perimeter of the blue rectangle with the large rectangle. For instance, in the sample, the width, length, and perimeter of the blue rectangle are each 50% of the width, length, and perimeter of the large rectangle.**

Sample

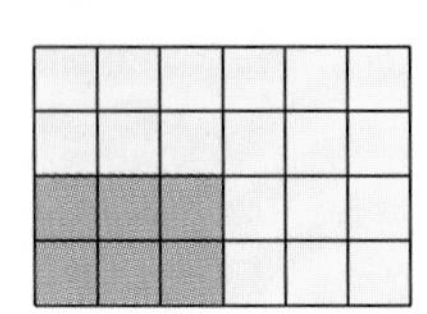

27.

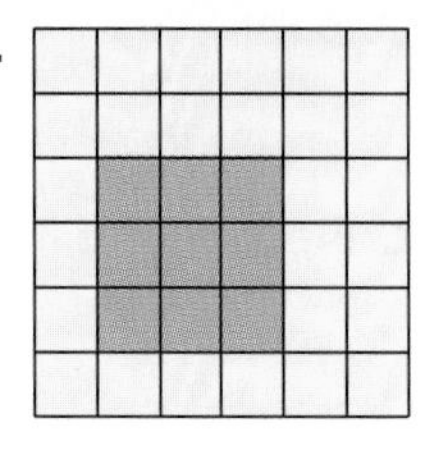

28.

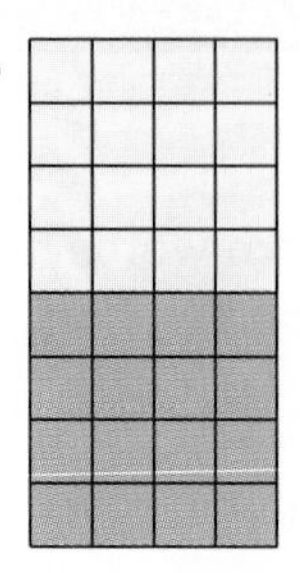

29.

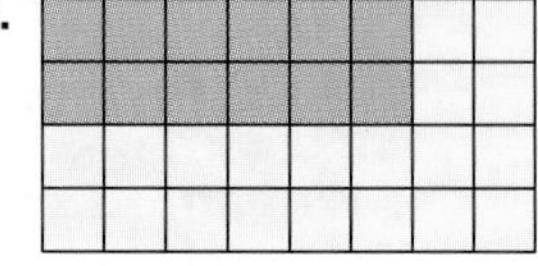

30. *Shopping* You go to the mall to buy a sweater that has a discount of 35%. The original price of the sweater is $40.

a. How much is the discount? That is, what is 35% of $40?
b. How much do you have to pay for the sweater?

Water Resources **In Exercises 31–33, use the following information. Earth has about 326 million cubic miles of water. Of that, 3% is freshwater.**

31. How many cubic miles of freshwater are on Earth?
32. How many cubic miles of salt water are on Earth?
33. A cubic mile of water contains about 9.5×10^{11} gallons. How many gallons of freshwater are on Earth?

Integrated Review

Number Sense **In Exercises 34–37, which item does not fit in the list? Explain.**

34.	$\frac{3}{8}$	$1\frac{2}{8}$	37.5%	37.5 percent	$\frac{75}{200}$
35.	0.125	125%	$1\frac{1}{4}$	$\frac{5}{4}$	$\frac{125}{100}$
36.	4	16% of 25	10% of 40	40%	4.0×10^0
37.	4% of 150	0.06	75% of 8	6	$\frac{1}{2}$ of 12

Exploration and Extension

Patterns **In Exercises 38 and 39, evaluate each expression. Then describe the pattern and write the next three numbers in the sequence.**

38. 1% of 500, 2% of 600, 3% of 700, 4% of 800, 5% of 900, 6% of 1000
39. 25% of 4, 50% of 8, 75% of 12, 100% of 16, 125% of 20, 150% of 24

USING A SPREADSHEET OR A GRAPHING CALCULATOR

Interest in a Savings Account

The interest paid in a savings account is a percent of the amount you deposit. For instance, if you deposit \$50 in an account that pays 6% annual interest, then at the end of one year you will earn 6% of \$50 or \$3 in interest. Your new balance will then be \$53. If you leave the money in the bank for a second year, then you will earn 6% of \$53 or \$3.18 in interest. The table shows the balances in your account during a 10-year period.

Year	Interest	Balance
1	$0.06 \cdot 50.00 = \$3.00$	$50.00 + 3.00 = \$53.00$
2	$0.06 \cdot 53.00 = \$3.18$	$53.00 + 3.18 = \$56.18$
3	$0.06 \cdot 56.18 = \$3.37$	$56.18 + 3.37 = \$59.55$
4	$0.06 \cdot 59.55 = \$3.57$	$59.55 + 3.57 = \$63.12$
5	$0.06 \cdot 63.12 = \$3.79$	$63.12 + 3.79 = \$66.91$
6	$0.06 \cdot 66.91 = \$4.01$	$66.91 + 4.01 = \$70.92$
7	$0.06 \cdot 70.92 = \$4.26$	$70.92 + 4.26 = \$75.18$
8	$0.06 \cdot 75.18 = \$4.51$	$75.18 + 4.51 = \$79.69$
9	$0.06 \cdot 79.69 = \$4.78$	$79.69 + 4.78 = \$84.47$
10	$0.06 \cdot 84.47 = \$5.07$	$84.47 + 5.07 = \$89.54$

X	Y1
1.00	53.00
2.00	56.18
3.00	59.55
4.00	63.12
5.00	66.91
6.00	70.93
7.00	75.18
8.00	79.69
9.00	84.47
10.00	89.54

The different balance in the 6th year occurs because of the way the balance is rounded.

A formula for finding the balance A after n years is

$$A = P(1 + r)^n$$

where P is the original deposit and r is the annual interest rate *in decimal form.* You can use a spreadsheet on a computer or a calculator to construct tables to find the balances listed above. For instance, the table at the right was made with a *TI-82* calculator.

Exercises

1. You have deposited \$120 in a savings account that pays an annual interest rate of 7%. Construct a table that shows your balances after 1, 2, 3, 4, and 5 years.

In Exercises 2–5, use the formula given above to find the balance in your account after the indicated number of years.

	Deposit	Interest Rate	Number of Years
2.	\$80.00	7%	8 years
3.	\$800.00	6%	20 years
4.	\$100.00	8%	100 years
5.	\$1000.00	8%	50 years

7.9 Problem Solving with Percents

What you should learn:

Goal 1 How to use percents to solve real-life consumer problems

Goal 2 How to use percents to help organize data

Why you should learn it:

You can use percents to solve real-life problems, such as calculating the sales tax on a purchase.

Goal 1 Using Percents in Real life

Most states charge a sales tax when you purchase certain types of items. To find the amount of the sales tax, change the percent to a decimal and multiply by the amount of the purchase. For instance, in a state that has a 6% sales tax, the tax on a $54.80 purchase is

$0.06 \times 54.80 \approx \$3.29.$

Here are some sales tax rates for different states.

State	Percent	State	Percent
California	6%	Ohio	5%
Connecticut	8%	Oregon	0%
Missouri	4.225%	Texas	6.25%
New Jersey	7%	Wyoming	3%

Example 1 Finding Sales Tax Amounts

Real Life **Retail Business**

You have $81.00 and want to buy a necklace that costs $76.95. In which of the states listed above would you have enough money to buy the necklace?

Solution

State	Sales Tax	Total Cost	Enough?
California	$0.06 \times 76.95 = \$4.62$	$81.57	No
Connecticut	$0.08 \times 76.95 = \$6.16$	$83.11	No
Missouri	$0.04225 \times 76.95 = \$3.25$	$80.20	Yes
New Jersey	$0.07 \times 76.95 = \$5.39$	$82.34	No
Ohio	$0.05 \times 76.95 = \$3.85$	$80.80	Yes
Oregon	None	$76.95	Yes
Texas	$0.0625 \times 76.95 = \$4.81$	$81.76	No
Wyoming	$0.03 \times 76.95 = \$2.31$	$79.26	Yes

You would have enough money to buy the necklace in Missouri, Ohio, Oregon, and Wyoming. ■

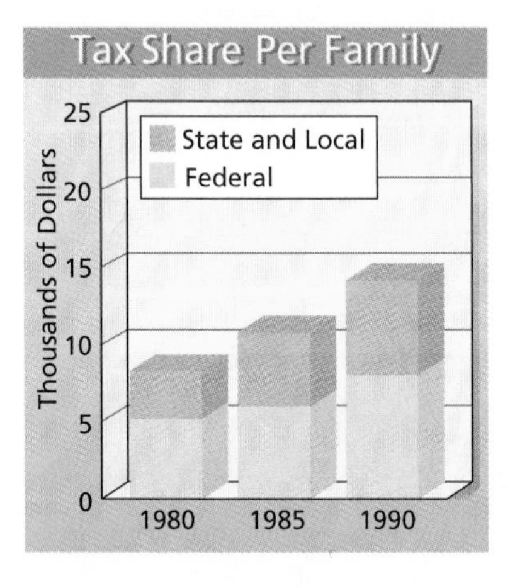

This stacked bar graph shows the average share per family of all taxes paid to federal, state, and local governments. (Source: U.S. Bureau of Census)

Americans pay many types of taxes to the federal government, state governments, and local governments. Governments use taxes to provide services, such as roads and health care.

Goal 2 Using Percents to Organize Data

Real Life Taking a Poll

Isabel Valdés owns a market research company called Hispanic Market Connections, Inc. Market research companies use surveys to find people's opinions.

Example 2 *Organizing Data*

You are surveying 250 people. You ask them to read each common saying and state whether they believe it is true. Show how you could use percents to organize the results. ***(Source: R.H. Bruskin)***

	Saying	Number Answering True
a.	*Beauty is only skin deep.*	205
b.	*Don't put all your eggs in one basket.*	218
c.	*Look before you leap.*	240
d.	*The early bird catches the worm.*	188
e.	*The grass is always greener on the other side of the fence.*	95
f.	*What's good for the goose is good for the gander.*	143

Solution To find the percent who answered true, divide the number who answered true by the total number surveyed.

a. $\frac{205}{250} = 0.82 = 82\%$ **b.** $\frac{218}{250} = 0.872 = 87.2\%$

c. $\frac{240}{250} = 0.96 = 96\%$ **d.** $\frac{188}{250} = 0.752 = 75.2\%$

e. $\frac{95}{250} = 0.38 = 38\%$ **f.** $\frac{143}{250} = 0.572 = 57.2\%$

After calculating the percents, you could organize the results with a bar graph like the one shown below.

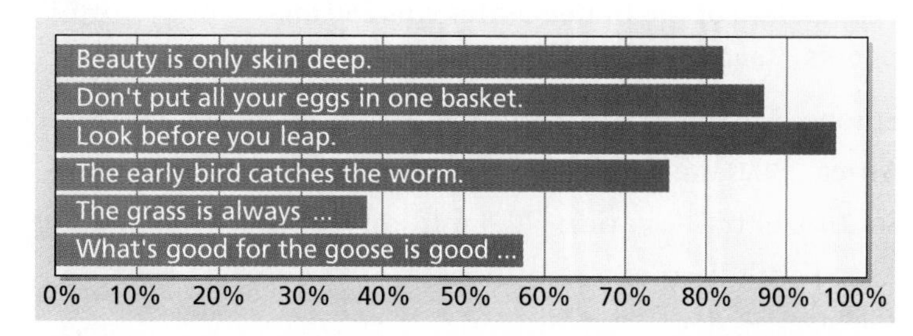

■

Communicating about MATHEMATICS

SHARING IDEAS about the Lesson

Interpreting a Survey If the survey in Example 2 had been taken with 400 people, how many do you think would have answered true to each saying? Explain.

EXERCISES

Guided Practice

CHECK for Understanding

1. *Sales Tax* The sales tax rate on a $19.79 item is 5.75%. What is the total cost of the item?

Restaurant Bill **In Exercises 2 and 3, use the following information.**

At a restaurant, the cost of the meal is $27.53. The sales tax is 5.75%.

2. What is your total cost, including the sales tax?
3. If you leave a 15% tip (of your total cost), how much will you leave for the tip?
4. In a survey of 175 adults, 104 of them said that they attend at least one movie per year. What percent does this represent?

Independent Practice

In Exercises 5–8, the price of an item is given. Find the total cost of the item, including a 4% sales tax.

5. $4.65
6. $10.39
7. $50
8. $463.87

National Origins **In Exercises 9–14, find the percent of the 1990 United States population that traced its origin to the given continent. In 1990, the total population of the United States was 250 million people.** ***(Source: Bureau of the Census)***

In 1990, about 11 million Americans traced their origins by saying "I'm just American—a little bit of everything."

9. Europe, Australia: 177.3 million
10. Africa: 30.0 million
11. South/Central America: 22.4 million
12. Asia, South Pacific: 7.3 million
13. Native American: 2.0 million
14. "American": 11.0 million
15. *Starting a Savings Account* You deposit $350 into a savings account that pays 5.75% in simple interest. (This means that at the end of one year your savings account will earn 5.75% in interest.) If you make no other deposits or withdrawals during the year, how much money will be in the account after one year?
16. *Developing a Budget* Members of a small town council determine that the town must decrease its operating cost by 17%. The present operating cost is $476,000. The council reduces the operating cost to $397,000. Is that enough? Explain.

Payroll Taxes **In Exercises 17–22, use the following.**

Your gross pay for one week is \$650. To find your take-home pay, you must subtract each of the following taxes from your gross pay: Federal Income Tax: 18.1%, State Income Tax: 2.8%, Social Security Tax: 6.2%, Local Income Tax: 1%, and Medicare Tax: 1.5%.

17. What is your Federal income tax?

18. What is your state income tax?

19. What is your Social Security tax?

20. What is your local income tax?

21. What is your Medicare tax?

22. What is your take-home pay?

Newspaper Subscribers **In Exercises 23–26, use the following.**

You work for a newspaper that has 60,200 subscriptions: 31,500 for the morning paper, 10,800 for the evening paper, and 17,900 for the Sunday paper.

23. What percent of your subscriptions are for the morning paper?

24. What percent of your subscriptions are for the evening paper?

25. What percent of your subscriptions are for the Sunday paper?

26. *Estimation* If the number of subscriptions increased to 72,000, how many do you think would be for the morning paper? Explain.

Integrated Review

Making Connections within Mathematics

Estimation **In Exercises 27–30, use the following.**

In 1993, a survey was taken of 240 working mothers. The percent who said that the indicated condition was "very important" to them is given in the graph. ***(Source: Kudos/Gallop poll)***

What Working Moms Want

Condition	Percent
Parental leave (job guarantee)	77%
Paid maternity leave	72%
Flex time	63%
Company-provided or -sponsored day care	58%

27. How many said that a guaranteed job was very important after returning from parental leave?

28. How many said that paid maternity leave was very important?

29. How many said that flexible working hours were very important?

30. How many said that a company that provided day care was very important?

Exploration and Extension

31. *Conducting a Survey* Use the questions below to conduct a survey of at least 20 people. Organize your data using percents and a graph.

Almost every day:

a. do you eat breakfast?

b. do you talk on the phone?

c. do you watch television?

d. do you listen to music?

e. do you read part of a newspaper, magazine, or book?

7

Chapter Summary

What did you learn?

Skills

1. Add and subtract fractions
   - with like denominators. **(7.1)**
   - with unlike denominators. **(7.2)**
   - by writing decimals. **(7.3)**
2. Multiply fractions. **(7.4)**
3. Divide fractions. **(7.5)**
4. Perform operations with rational expressions that contain variables. **(7.1–7.5)**
5. Write a percent as a portion with a denominator of 100. **(7.6)**
6. Write percents in decimal form. **(7.7)**
   - Write decimals as percents. **(7.7)**
   - Write fractions as percents. **(7.7)**
   - Write percents as fractions. **(7.7)**
7. Find a percent of a number. **(7.8)**

Problem-Solving Strategies

8. Model and solve real-life problems. **(7.1–7.9)**
   - Use geometry to model operations with fractions, decimals, and percents **(7.1–7.9)**

Exploring Data

9. Use tables and graphs to solve problems. **(7.1–7.9)**
   - Interpret circle graphs. **(7.2, 7.3, 7.7)**

Why did you learn it?

You can use operations with fractions and percents to answer questions about real-life situations. For instance, in this chapter, you saw how subtraction of fractions can be used to compare running times in a 100-meter sprint, and how percents can be used to organize the results of a survey. You also saw how circle graphs can be used to represent parts of a whole, for instance, the different ways that electric power is produced.

How does it fit into the bigger picture of mathematics?

Numbers can be written in several different ways. For instance, the number $\frac{1}{4}$ can be written as $\frac{2}{8}$, or 0.25, or $\frac{25}{100}$, or as 25%. In this chapter, you learned that the "best" way to write a number depends on the way the number is being used. For instance, if you are advertising a clothing sale, you might choose 25% as the best way to describe your store's discount. However, when you are calculating the discount on a purchase, it is better to use the decimal form, 0.25, or the fraction form, $\frac{1}{4}$.

Chapter REVIEW

In Exercises 1–16, evaluate the expression. Then simplify, if possible. (7.1, 7.2, 7.4, 7.5)

1. $\frac{4}{9}+\frac{2}{9}$ **2.** $\frac{3}{10}-\frac{7}{10}$ **3.** $\frac{12}{13}-\frac{1}{2}$ **4.** $\frac{-5}{14}+\frac{-3}{4}$

5. $-1\frac{3}{4}\cdot 2\frac{1}{4}$ **6.** $\frac{-6}{7}\cdot\frac{-2}{5}$ **7.** $-8\div\frac{-4}{5}$ **8.** $3\frac{3}{4}\div\left(-2\frac{1}{2}\right)$

9. $\frac{-5}{8t}+\frac{3}{8t}$ **10.** $\frac{-x}{6}+\frac{-5x}{6}$ **11.** $\frac{-9}{xy}-\frac{4}{x}$ **12.** $\frac{a}{3}+\frac{2a}{5}$

13. $15\cdot\frac{3m}{4}$ **14.** $\frac{-36n^2}{7}\cdot\frac{7}{6n}$ **15.** $\frac{-8}{y}\div\frac{10}{y}$ **16.** $\frac{-b^3}{6}\div\frac{-b}{3}$

In Exercises 17–28, solve the equation. (7.1, 7.2, 7.4, 7.5)

17. $x+\frac{3}{14}=\frac{9}{14}$ **18.** $y+\frac{5}{6}=\frac{1}{6}$ **19.** $10z-\frac{11}{2}=\frac{1}{2}$ **20.** $a+\frac{4}{7}=\frac{-5}{14}$

21. $b+\frac{3}{4}=\frac{2}{5}$ **22.** $18t-\frac{9}{2}=\frac{-12}{8}$ **23.** $-7s=\frac{2}{5}$ **24.** $2x=6x-\frac{6}{10}$

25. $\frac{1}{14}y=\frac{5}{6}$ **26.** $25z+5=-8$ **27.** $8y=19+14y$ **28.** $\frac{15}{9}=\frac{-1}{3}t-\frac{7}{2}$

Geometry **In Exercises 29–31, find the perimeter of the figure. (7.1, 7.2)**

29.

Each tile is $\frac{1}{4}$ in. × $\frac{3}{4}$ in.

30.

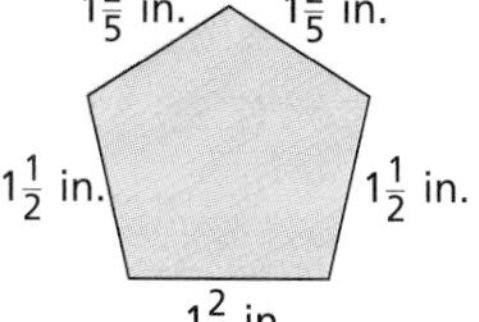

31.

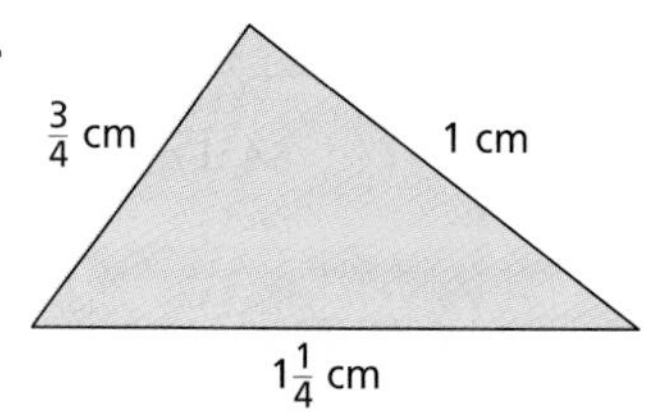

Geometry **In Exercises 32–34, find the area of the figure. (7.4)**

32.

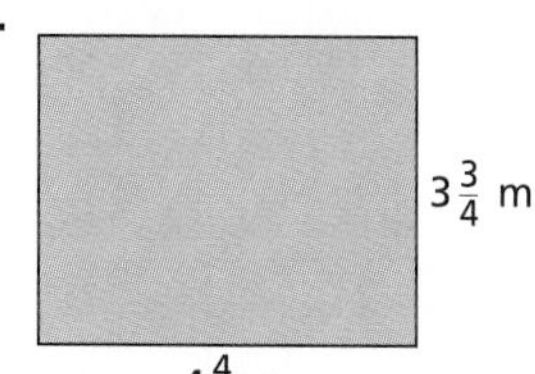

33.

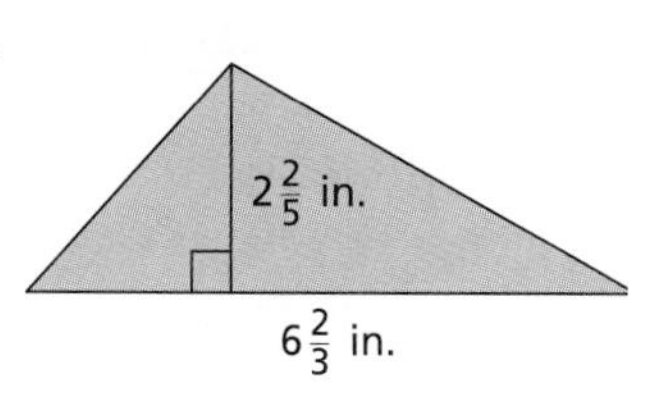

34.

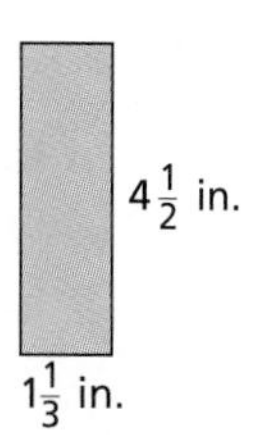

In Exercises 35–46, use a calculator to evaluate the expression. Round your result to two decimal places. (7.3)

35. $\frac{3}{14}-\frac{9}{14}$ **36.** $\frac{3}{7}+\frac{5}{7}$ **37.** $-\frac{5}{9}+\frac{35}{89}$ **38.** $\frac{7}{25}+\frac{13}{25}$

39. $\frac{-8}{15}\cdot\frac{-7}{12}$ **40.** $\frac{13}{19}\cdot\frac{-4}{5}$ **41.** $\frac{16}{19}\div 5$ **42.** $-\frac{48}{55}\div\frac{6}{11}$

43. $\frac{65}{121}+\frac{99}{148}$ **44.** $\frac{59}{252}-\frac{86}{133}$ **45.** $\frac{12}{47}x-\frac{14}{15}x$ **46.** $\frac{43}{56}y+\frac{9}{13}y$

47. Translate the following sentence to an equation. Then solve the equation. *The sum of one fourth and twice a number is equal to one half.* **(7.2, 7.5)**

48. Translate the following sentence to an equation. Then solve the equation. *The product of three eighths and a number is equal to one half.* **(7.4, 7.5)**

In Exercises 49–52, write the fraction that represents the portion of the figure's area that is blue. Then write the fraction as a percent. (7.1, 7.2, 7.4, 7.6, 7.7)

49. 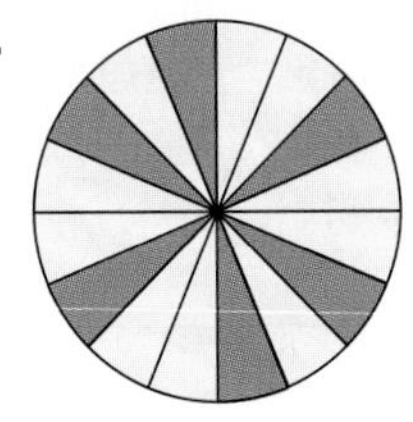

50. 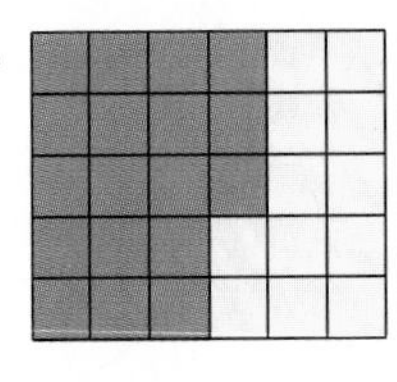

51. 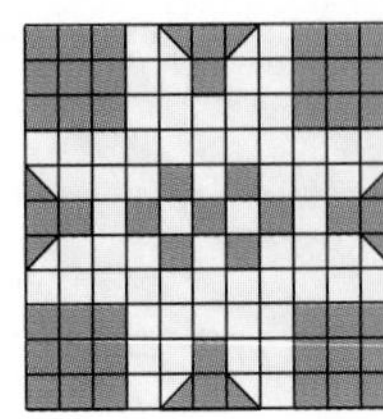

52. 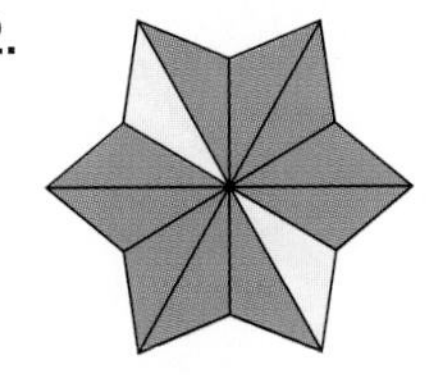

In Exercises 53–56, rewrite as a percent. (7.6, 7.7)

53. 0.745 **54.** 0.029 **55.** $\frac{7}{8}$ **56.** $\frac{27}{5}$

In Exercises 57–60, rewrite the percent as a fraction. Then simplify. (7.7)

57. 46% **58.** 88% **59.** 180% **60.** 104%

In Exercises 61–64, find the percent of the number. (7.8)

61. 26% of 50 **62.** 78% of 150 **63.** 350% of 80 **64.** 520% of 105

Computation Sense **In Exercises 65–68, match the operations. (7.7)**

a. Divide by 20. **b.** Divide by 4. **c.** Divide by 8. **d.** Divide by 5.

65. Multiply by 25%. **66.** Multiply by 5%. **67.** Multiply by 20%. **68.** Multiply by $12\frac{1}{2}\%$

Geometry **In Exercises 69 and 70, find the total area of the figure. Then find the percent of the area that is red, blue, and yellow. (7.1–7.8)**

69. 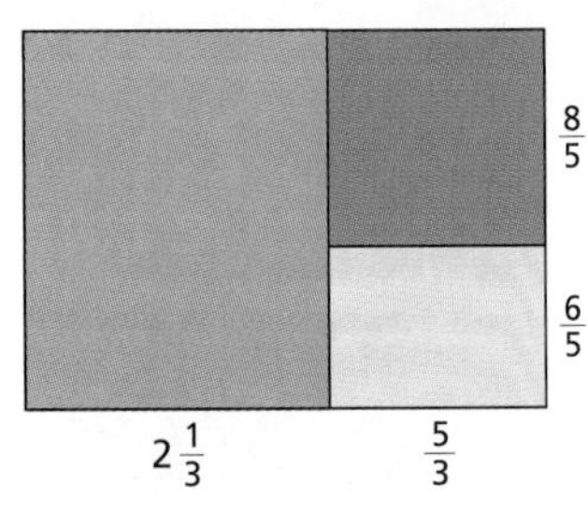

70. 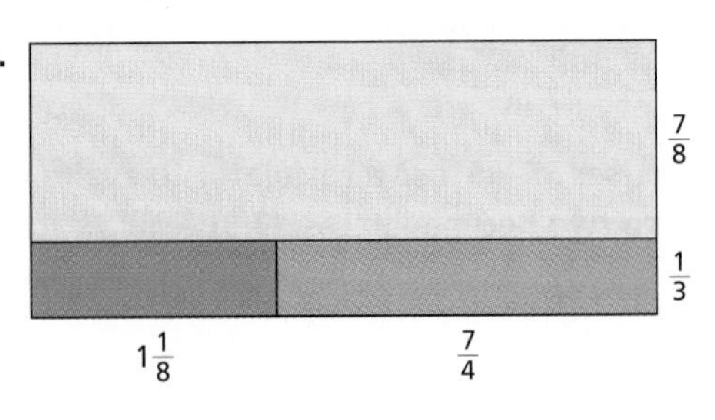

Error Analysis **In Exercises 71–73, describe the error. Then correct it. (7.2, 7.5, 7.7)**

71. $\frac{7}{20} - \frac{2}{10} = \frac{7-2}{20-10} = \frac{5}{10} = \frac{1}{2}$

72. 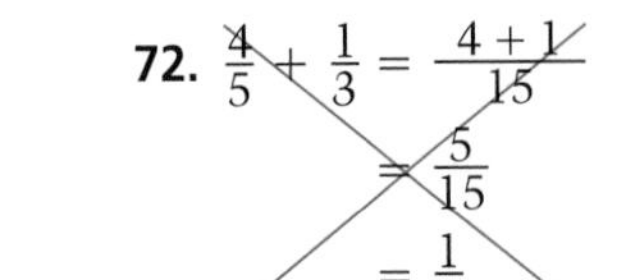

73. $\frac{-2}{9} \div \frac{9}{5} = \frac{-2}{9} \cdot \frac{5}{9} = \frac{-2 \cdot 5}{9} = \frac{-10}{9}$

The Mail **In Exercises 74–77, use the circle graph at the right, which shows the makeup of mail in the United States.** *(Source: National Postal Museum)* **(7.1–7.2)**

What Makes Up Our Mail?

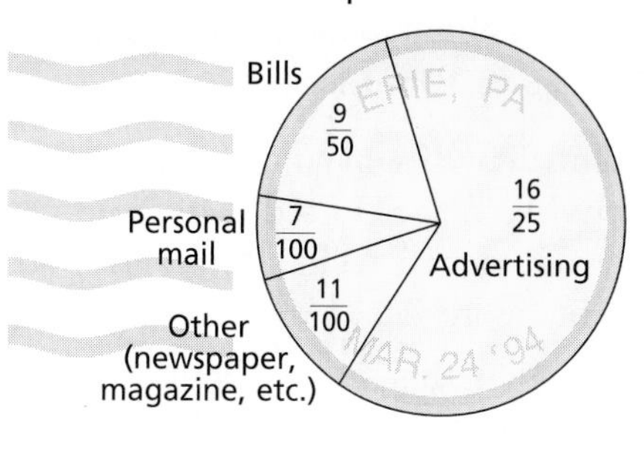

74. Find the sum of the portions representing personal mail and newspapers and magazines.

75. Find the difference in the portions of mail that are advertising and bills.

76. Which portions have a sum of $\frac{3}{4}$?

77. What is the sum of the four portions?

78. *Sales Tax* You are buying a Navaho bracelet in New Mexico, where the sales tax is 4.75%. The price of the bracelet is $31.99. What is the total cost, including sales tax? **(7.9)**

79. *Loan Payment* You are borrowing $1200 from a bank. At the end of a year you will pay back all of the loan, plus an interest of 11.25%. How much will you owe at the end of the year? **(7.9)**

80. *Test Scores* Your scores on five 100-point tests are 73%, 82%, 79%, 94%, and 98%. What is your average percent? **(7.9)**

In Exercises 81 and 82, which number does not fit the list? Explain. (7.5)

81. $-\frac{1}{2}, \frac{-1}{2}, \frac{-1}{-2}, -\frac{-1}{-2}, \frac{1}{-2}$

82. $\frac{-3}{8}, -\frac{3}{8}, \frac{3}{-8}, -\frac{-3}{8}, -\frac{-3}{-8}$

83. *National Forests and Parks* In a survey, 1500 people were asked whether national forests and parks should collect fees for various uses. Use percents and a graph to organize the results of the survey. (Some of those polled had no opinion.) **(7.9)** *(Source: Roper Organization)*

	Should be free	Should pay a fee
Logging	210	1155
Hard-rock mining	210	1140
Oil drilling	225	1095
Ski area development	360	1005
Livestock grazing	540	810
Commercial fishing	585	765
Game hunting	600	765
Recreational fishing	1005	405
Boating and yachting	1050	360
Hiking	1275	180

Hiking fees would be used to offset the expenses related to establishing, marking, and maintaining trails.

Water Resources **In Exercises 84 and 85, answer the questions about water resources in the United States. (7.9)** *(Source: World Book)*

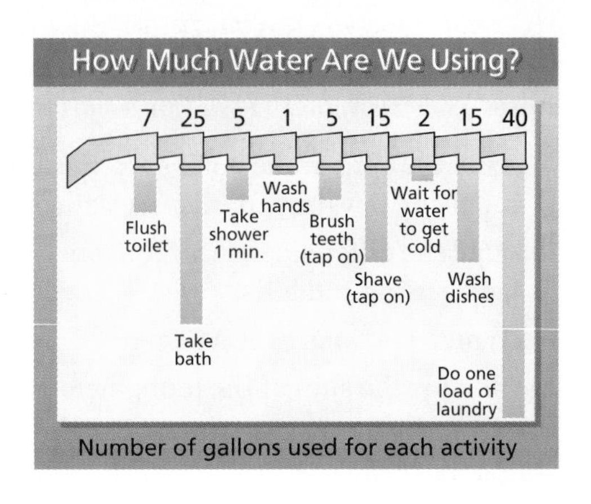

84. In 1990, fifteen percent of the 250 million people living in the United States obtained their water from private wells. What number of people does this represent?

85. In 1990, there were about 70,000 public water-supply systems in the United States. Of these about 800 served 50,000 people or more. What percent of the public water-supply systems served 50,000 people or more? **(7.9)**

Water Use **In Exercises 86–88, use the following information. (7.9)**

Approximately 340 billion gallons of water are used in the United States each day. Twenty-two percent comes from groundwater (wells) and the rest comes from surface water (rivers and lakes). *(Source: U.S. Geological Survey)*

86. In a typical day, how many gallons of groundwater are used?

87. What percent of water used in the United States comes from rivers and lakes?

88. Two-thirds of the groundwater used in the United States is used to irrigate crops. Estimate this amount.

North American Waterfalls **In Exercises 89–92, use the following information. (7.9)**

Bridal Veil Falls in California is 189 meters high.
Ribbon Falls in California is 491 meters high.
Nevada Falls in California is 181 meters high.
Multnomah Falls in Oregon is 165 meters high.

89. Niagara Falls, between New York and Ontario, is $\frac{2}{7}$ the height of Bridal Veil Falls. How high is Niagara Falls?

90. Yosemite Lower Falls in California is 20% of the height of Ribbon Falls. How high is Yosemite Lower Falls?

91. Multnomah Falls is part of a series of waterfalls. Multnomah Falls represents $\frac{7}{8}$ of the total height of the series. What is the total height of the series?

92. Nevada Falls in California is about 50% as high as Takakkaw Falls in British Columbia. How high is Takakkaw Falls?

Chapter TEST

In Exercises 1–4, find the sum or difference and simplify.

1. $\frac{1}{5} + \frac{3}{5}$
2. $\frac{11}{12}x - \frac{7}{12}x$
3. $\frac{4}{5} + \frac{1}{10}$
4. $\frac{11}{18} - \frac{2}{3}$

In Exercises 5–8, find the product or quotient and simplify.

5. $\frac{4}{5} \cdot \frac{1}{2}$
6. $\frac{4}{5} \div \frac{1}{2}$
7. $1\frac{1}{3} \times 1\frac{1}{2}$
8. $-\frac{5s}{6} \div \frac{3s}{5}$
9. Write $\frac{10}{25}$ as a percent.
10. Write 0.365 as a percent.
11. Write 24.5% as a decimal.
12. Write 32% as a fraction and simplify.
13. Find 85% of 16.
14. 12 is what percent of 48?
15. Find the total area of the figure below. Then find the percent of the area that is red, blue, and yellow.

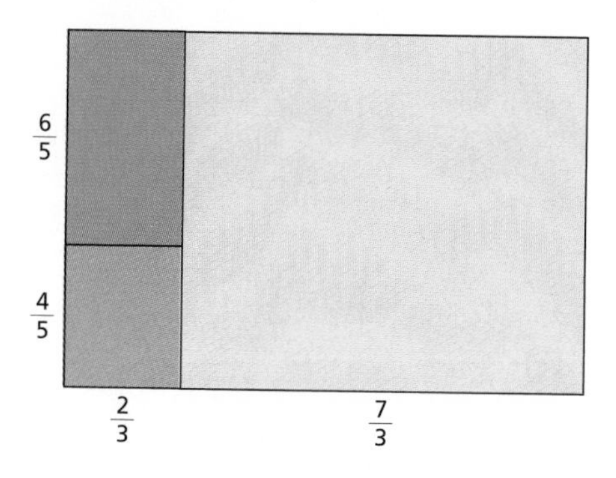

16. The circle graph shows the percents of types of books sold in the United States. If you owned a bookstore that sold 2546 books in a week, how many would you expect to be fiction?

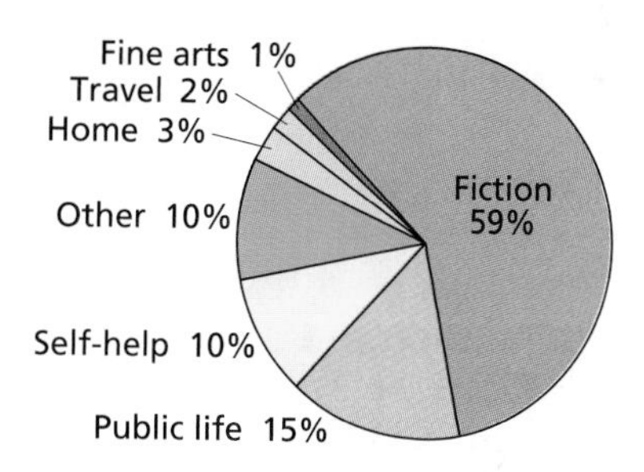

In Exercises 17 and 18, use the following information.

You are the manager of a clothing store. On one of your clearance racks, you have 65 shirts. Of these, 18 are T-shirts. Of the 18 T-shirts, 5 are blue and 8 are white.

17. What percent of the total number of shirts are T-shirts?
18. What percent of the total number of T-shirts are blue?

In Exercises 19 and 20, use the following information.

In 1990, a survey was taken to determine the amount of television that American adults watch in a day. The results were: 1 hour or less (18.4%), between 1 and 5 hours (68.8%), between 5 and 10 hours (10.2%), between 10 and 15 hours (1.9%), and over 15 hours (0.7%). ***(Source: Impact Resources, Inc.)***

19. Of 415 American adults, how many would you expect to watch television for 1 hour or less each day?
20. Of 824 American adults, how many would you expect to watch between 1 and 5 hours of television a day?

Sporting events, like the 1992 Barcelona Olympics, attract large audiences worldwide.

CHAPTER

8

Proportion, Percent, and Probability

LESSONS

The information superhighway of computer mail, cellular phones, and modems has been made possible by advances in fiber optics. These transparent glass fibers have revolutionized the fields of telecommunication and surgical medicine. A single glass fiber, whose diameter is 1.5×10^{-4} millimeters, can carry 2.5 trillion pages of data per second. In medicine, bundles of these flexible rods are used in laser surgery and in fiberscopes.

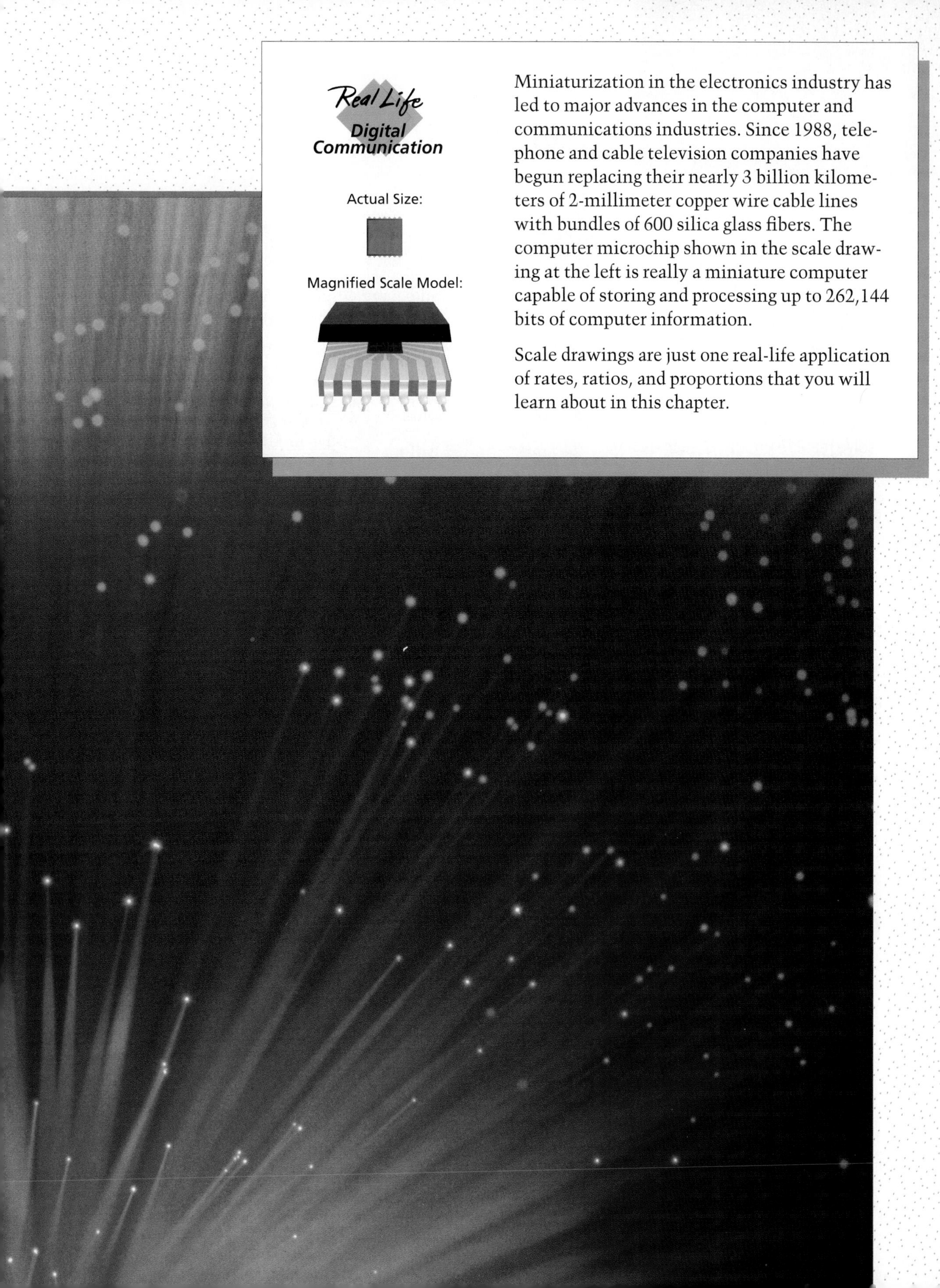

Miniaturization in the electronics industry has led to major advances in the computer and communications industries. Since 1988, telephone and cable television companies have begun replacing their nearly 3 billion kilometers of 2-millimeter copper wire cable lines with bundles of 600 silica glass fibers. The computer microchip shown in the scale drawing at the left is really a miniature computer capable of storing and processing up to 262,144 bits of computer information.

Scale drawings are just one real-life application of rates, ratios, and proportions that you will learn about in this chapter.

8.1 Exploring Rates and Ratios

What you should learn:

Goal 1 How to find rates

Goal 2 How to find ratios

Why you should learn it:

You can use rates and ratios to solve real-life problems, such as finding unit prices.

Goal 1 Finding Rates

If two quantities a and b have different units of measure, then the **rate** of a *per* b is $\frac{a}{b}$. Rates are used in almost all parts of real life. Here are two examples.

a	b	Rate
\$21.00	3 hours	$\frac{21 \text{ dollars}}{3 \text{ hours}} = 7$ dollars per hour
240 miles	12 gallons	$\frac{240 \text{ miles}}{12 \text{ gallons}} = 20$ miles per gallon

Real Life
Average Speed

Example 1 *Finding a Rate*

You drove 200 miles in 4 hours. What was your average speed?

Solution To find the rate, divide the distance by the time.

$$\text{Rate} = \frac{\text{Distance}}{\text{Time}}$$ *Verbal model*

$$\text{Rate} = \frac{200 \text{ miles}}{4 \text{ hours}}$$ *Substitute for distance and time.*

$$= 50 \text{ miles per hour}$$ *Simplify.*

Your rate (or average speed) was 50 miles per hour. ■

Real Life
Unit Price

Example 2 *Finding a Rate*

A 16-ounce box of breakfast cereal costs \$2.89, and a 20-ounce box costs \$3.49. Which is the better buy?

Solution

Price	Weight	Unit Price
\$2.89	16 ounces	$\frac{\$2.89}{16 \text{ ounces}} \approx 0.181$ dollars per ounce
\$3.49	20 ounces	$\frac{\$3.49}{20 \text{ ounces}} \approx 0.175$ dollars per ounce

Because the larger box has the smaller unit price, it follows that it is the better buy. ■

The mechanical praying mantis at the right is 17 feet 4 inches long. It was built by Creative Presentations. This company built several huge insect robots for a traveling museum show called Backyard Monsters. *The company started by studying live insects.*

Real Life
Robotics

Goal 2 Finding Ratios

If two quantities a and b have the same units of measure, then the **ratio** of *a to b* is $\frac{a}{b}$.

Example 3 *Finding a Ratio*

To build the robot mantis shown above, Creative Presentations used a real praying mantis that was 4 inches long. What is the ratio of the robot mantis's length to the real mantis's length?

Solution To find a ratio, both quantities must have the same unit of measure. In inches, the length of the robot mantis is

$$17 \text{ feet} + 4 \text{ inches} = 17(12 \text{ inches}) + 4 \text{ inches} = 208 \text{ inches}.$$

The ratio of the robot's length to the real mantis's length is

$$\frac{\text{Robot's length}}{\text{Mantis's length}} = \frac{208 \text{ inches}}{4 \text{ inches}} = \frac{52}{1} \quad \text{or} \quad 52 \text{ to } 1.$$

The ratio of the robot mantis's length to the real mantis's length is 52 to 1. In other words, the robot is 52 times as long as the real mantis. ■

Study Tip...
In Example 3, notice that both quantities were written in the same units of measure before dividing to find the ratio.

Communicating about MATHEMATICS

SHARING IDEAS about the Lesson

Rate or Ratio? Decide whether the quotient is a rate or a ratio. Then simplify the quotient.

A. $\frac{10{,}000 \text{ people}}{4 \text{ years}}$ **B.** $\frac{10{,}000 \text{ people}}{4{,}000 \text{ people}}$ **C.** $\frac{\$6{,}000}{4 \text{ years}}$

EXERCISES

Guided Practice

CHECK for Understanding

1. ***Problem Solving*** Explain the difference between a rate and a ratio. Give an example of each.

In Exercises 2–5, state whether the quotient is a rate or a ratio. Then simplify.

2. $\frac{100 \text{ miles}}{4 \text{ hours}}$
3. $\frac{10 \text{ inches}}{4 \text{ inches}}$
4. $\frac{8 \text{ balloons}}{24 \text{ balloons}}$
5. $\frac{\$27.99}{3 \text{ pounds}}$

6. You work for 12 hours and get paid \$60. What is your rate of pay? Include units of measure in your answer.
7. Which of the following measures the rate at which an automobile uses gasoline?
   a. Miles per hour
   b. Gallons per mile
   c. Miles per gallon
8. Describe six ratios that compare the colors of the nine triangles below.

Independent Practice

In Exercises 9–12, determine whether the quotient is a rate or a ratio. Then simplify.

9. $\frac{16 \text{ students}}{18 \text{ students}}$
10. $\frac{120 \text{ meters}}{15 \text{ seconds}}$
11. $\frac{88 \text{ points}}{4 \text{ games}}$
12. $\frac{3 \text{ cars}}{5 \text{ cars}}$

In Exercises 13–16, write the verbal phrase as a rate or a ratio. Explain why the phrase is a rate or a ratio.

13. Recommended by 4 out of 5 doctors
14. Approved by 9 out of 10 inspectors
15. Traveled 1600 miles in 3 days
16. Answered 40 questions in 60 minutes

In Exercises 17–24, write each quotient as a ratio and simplify.

17. $\frac{2 \text{ feet}}{18 \text{ inches}}$
18. $\frac{2640 \text{ feet}}{1 \text{ mile}}$
19. $\frac{2 \text{ minutes}}{300 \text{ seconds}}$
20. $\frac{1 \text{ hour}}{3600 \text{ seconds}}$
21. $\frac{640¢}{\$4}$
22. $\frac{2 \text{ gallons}}{10 \text{ quarts}}$
23. $\frac{200 \text{ centimeters}}{3 \text{ meters}}$
24. $\frac{2 \text{ liters}}{50 \text{ milliliters}}$

25. ***Concert Tickets*** A concert sold out in 6 hours. 9000 tickets were sold for the concert. At what rate did the tickets sell?
26. ***Snowfall*** The record for snowfall in a 24-hour period is 76 inches at Silver Lake, Colorado, on April 14–15, 1921. At what rate did the snow fall on that day?

Unit Pricing **In Exercises 27–30, decide which is the better buy. Explain your reasoning.**

27. a. 6 apples for \$1.18

b. 10 apples for \$1.79

28. a. 12-ounce box for \$2.69

b. 18-ounce box for \$3.99

29. a. 2 pounds, 4 ounces for \$7.89

b. 5 pounds, 2 ounces for \$18.89

30. a. six 12-ounce cans for \$2.69

b. one half-gallon bottle for \$4.59

31. *Ballooning* It took 86 hours for Joe Kittinger to cross the Atlantic in his balloon, *Rosie O'Grady*. What was the average speed of the balloon?

Ex-USAF Colonel Joe Kittinger is the first man to complete a solo transatlantic crossing by balloon. Between September 14–18, 1984, he flew from Caribou, Maine, to Montenotte, Italy, approximately 3543 miles.

32. *Building a Model Railroad* On an N-gauge model train set, a tank car is 3.75 inches long. An actual tank car is 50 feet long. What is the ratio of the length of the actual tank car to the length of the model tank car?

Geometry **In Exercises 33–36, find the ratio of the blue figure's perimeter to the yellow figure's perimeter. Then find the ratio of the blue figure's area to the yellow figure's area. Which ratio is greater?**

33. 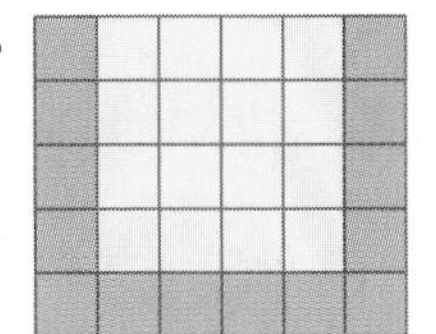

34. 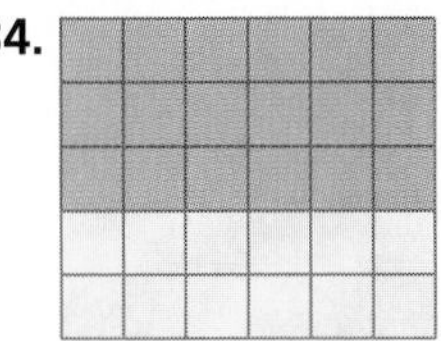

35. 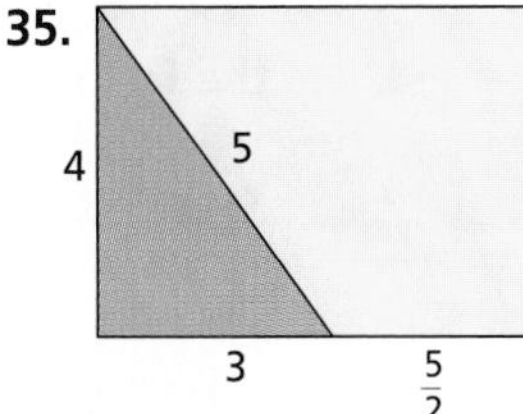

36. 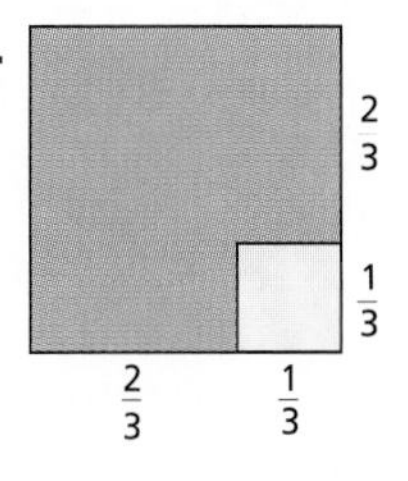

Integrated Review

Making Connections within Mathematics

Measurement Sense **In Exercises 37–42, convert the measure as indicated.**

37. 384 inches to feet

38. 115 meters to centimeters

39. 10,800 seconds to hours

40. 584 cents to dollars

41. 10 pints to quarts

42. 3 gallons to quarts

Exploration and Extension

Geometry **In Exercises 43 and 44, refer to $\triangle ABC$.**

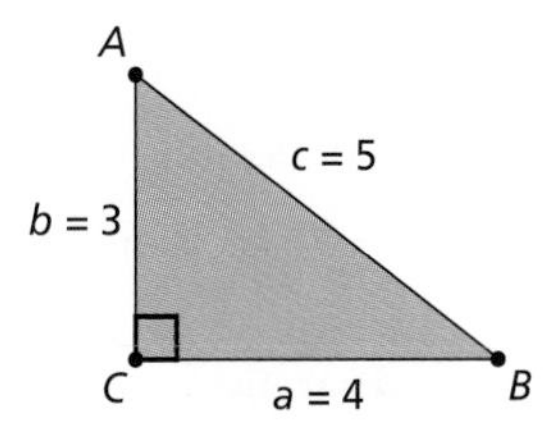

43. Find the following ratios.

a. c to a **b.** c to b **c.** b to a

44. Draw a second triangle, $\triangle DEF$, that is larger than $\triangle ABC$ but whose sides have the same ratios.

LESSON INVESTIGATION 8.2

Ratios

In this activity, you will find the length-to-width ratios of several rectangles.

Example — *Finding Ratios*

For each rectangle, find the ratio of the length to the width. Which rectangle do you think is most artistically pleasing to look at?

a. 

b. 

c. 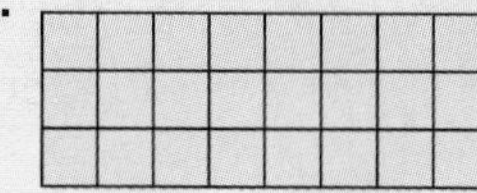

Solution

a. $\frac{6}{5} = 1.2$ **b.** $\frac{8}{5} = 1.6$ **c.** $\frac{8}{3} \approx 2.67$

For most people, the second rectangle is most pleasing. They find the first rectangle too square and the third too skinny. ■

Exercises

Golden Ratio **In Exercises 1–4, use the following rectangles.**

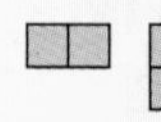 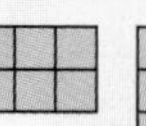 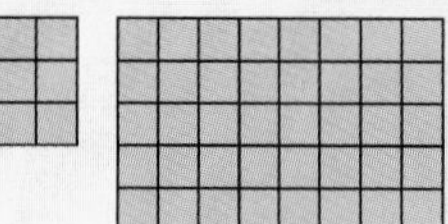 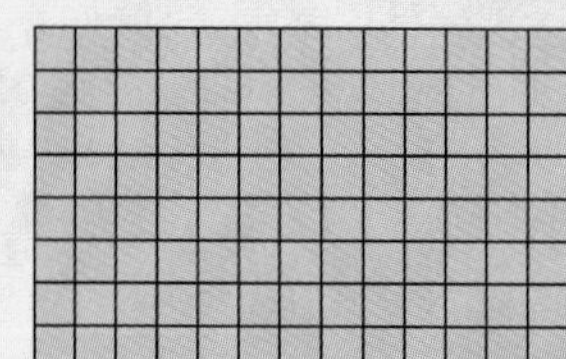

1. For each rectangle, find the ratio of its length to its width.
2. Describe the patterns of the lengths and widths. What are the lengths and widths of the next three rectangles in the pattern?
3. Use the results of Exercises 1 and 2 to complete the table.

Length	2	3	5	8	13	?	?	?
Width	1	2	3	5	8	?	?	?
Ratio	?	?	?	?	?	?	?	?

4. The **golden ratio** is the number $\frac{\sqrt{5}+1}{2}$. Use a calculator to round this number to 3 decimal places. How does the result relate to the table in Exercise 3?

8.2 Solving Proportions

What you should learn:

How to solve proportions

How to write proportions for similar triangles

Why you should learn it:

You can use proportions to solve problems in geometry, such as finding the lengths of the sides of similar triangles.

Goal 1 Solving Proportions

An equation that equates two ratios is a **proportion.** For instance, if the ratio $\frac{a}{b}$ is equal to the ratio $\frac{c}{d}$, then the proportion is

$\frac{a}{b} = \frac{c}{d}$. *Proportion*

This is read as "a is to b as c is to d."

Example 1 *Solving a Proportion*

Solve the proportion for x: $\frac{x}{2} = \frac{3}{4}$.

Solution

$\frac{x}{2} = \frac{3}{4}$	*Rewrite original proportion.*
$2 \cdot \frac{x}{2} = 2 \cdot \frac{3}{4}$	*Multiply each side by 2.*
$x = \frac{6}{4}$	*Simplify.*
$x = \frac{3}{2}$	*Simplify.*

The solution is $x = \frac{3}{2}$. Check this in the original proportion. ■

Example 2 *Solving a Proportion*

Solve the proportion for m: $\frac{3}{m} = \frac{5}{8}$.

Solution To solve this proportion, you can use the **Reciprocal Property,** which states that if $\frac{a}{b} = \frac{c}{d}$, then $\frac{b}{a} = \frac{d}{c}$.

$\frac{3}{m} = \frac{5}{8}$	*Rewrite original proportion.*
$\frac{m}{3} = \frac{8}{5}$	*Reciprocal Property*
$3 \cdot \frac{m}{3} = 3 \cdot \frac{8}{5}$	*Multiply each side by 3.*
$m = \frac{24}{5}$	*Simplify.*

The solution is $m = \frac{24}{5}$. Check this in the original proportion. Try using the Cross Product Property to solve this proportion. Which method do you prefer? ■

Cross Product Property

You can solve or check a proportion by using the **Cross Product Property,** which states that if $\frac{a}{b} = \frac{c}{d}$, then $ad = bc$. For instance, you can check the solution of Example 1 as follows.

$$\frac{x}{2} \stackrel{?}{=} \frac{3}{4}$$

$$\frac{\frac{3}{2}}{2} \stackrel{?}{=} \frac{3}{4}$$

$$\frac{3}{2} \cdot 4 \stackrel{?}{=} 2 \cdot 3$$

$$6 = 6 \checkmark$$

Goal 2 Writing Proportions

> **Need to Know**
> **Similar triangles** have the same shape but not necessarily the same size. When two angles have the same measure, they are marked with the same symbols.

Two triangles are **similar** if they have the same angle measures. If two triangles are similar, then the ratios of **corresponding sides** are equal. For instance, the following triangles are similar because each has a 30° angle, a 60° angle, and a 90° angle. Because the ratios of corresponding sides are equal, it follows that

$$\frac{a}{d} = \frac{b}{e} = \frac{c}{f}.$$

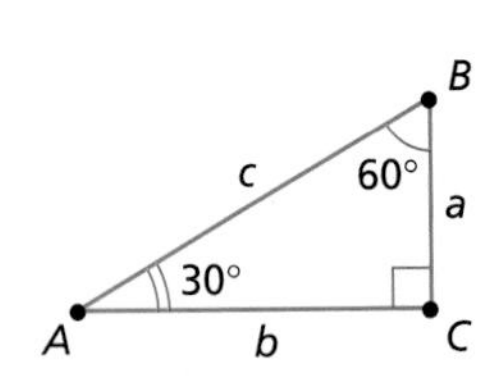

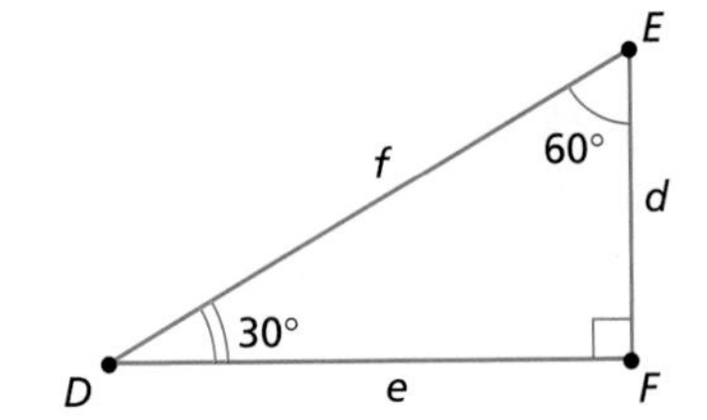

Example 3 *Writing a Proportion*

Connections Geometry

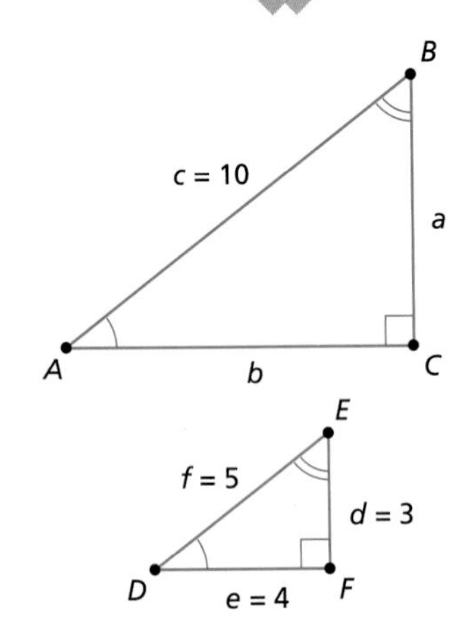

The triangles at the left are *similar*. Find a.

Solution Begin by writing a proportion that involves a. Then solve the proportion.

$\frac{a}{d} = \frac{c}{f}$	*Ratios of corresponding sides are equal.*
$\frac{a}{3} = \frac{10}{5}$	*Substitute for c, d, and f.*
$3 \cdot \frac{a}{3} = 3 \cdot \frac{10}{5}$	*Multiply each side by 3.*
$a = 6$	*Simplify.*

Try using this approach to find b. Note that each side of $\triangle ABC$ is twice as long as the corresponding side of $\triangle DEF$. ■

Communicating about MATHEMATICS

SHARING IDEAS about the Lesson

Writing Proportions Write a proportion for each statement. Then solve the proportion.

A. x is to 5 as 3 is to 30.
B. 4 is to n as 12 is to 30.
C. 5 is to 3 as m is to 5.
D. 4 is to 7 as 6 is to y.

EXERCISES

Guided Practice

CHECK for Understanding

1. *Think about It* A person paid \$2000 in property tax on an \$80,000 house. In the same neighborhood, another person paid \$3000 tax on a \$100,000 house. Is that fair? Explain how a proportion could be used to answer the question.
2. *Problem Solving* Proportions are often used to measure fairness or equity in real life. Exercise 1 gives one example. Describe some others.

In Exercises 3–6, solve the proportion. Explain how to check your answers.

3. $\frac{b}{3} = \frac{4}{12}$ **4.** $\frac{9}{x} = \frac{3}{5}$ **5.** $\frac{2}{3} = \frac{m}{36}$ **6.** $\frac{7}{18} = \frac{21}{y}$

In Exercises 7 and 8, state whether the triangles are similar. If they are, write three of the proportions that compare their sides.

7.

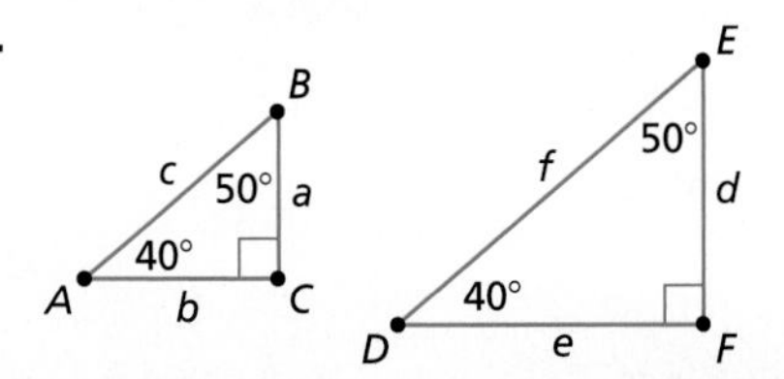

8.

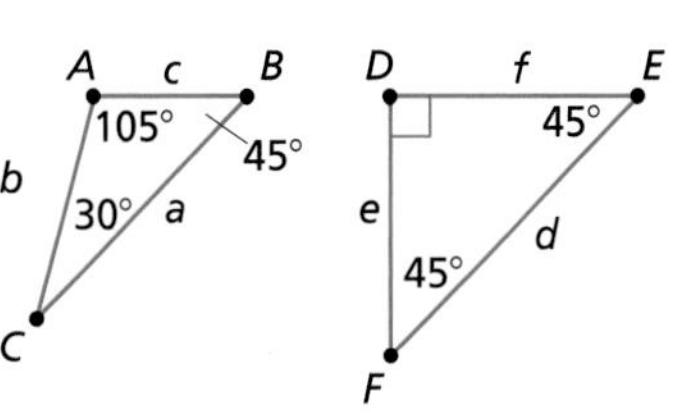

Independent Practice

In Exercises 9–12, decide whether the proportion is true. Explain.

9. $\frac{1}{4} \stackrel{?}{=} \frac{3}{12}$ **10.** $\frac{6}{16} \stackrel{?}{=} \frac{3}{7}$ **11.** $\frac{4}{9} \stackrel{?}{=} \frac{16}{36}$ **12.** $\frac{9}{7} = \frac{18}{15}$

In Exercises 13–20, solve the proportion. Check your solution.

13. $\frac{x}{3} = \frac{4}{9}$ **14.** $\frac{y}{5} = \frac{8}{5}$ **15.** $\frac{5}{7} = \frac{z}{2}$ **16.** $\frac{5}{12} = \frac{t}{2}$

17. $\frac{8}{m} = \frac{2}{5}$ **18.** $\frac{9}{x} = \frac{15}{2}$ **19.** $\frac{2}{3} = \frac{12}{b}$ **20.** $\frac{2.8}{y} = \frac{11}{2.5}$

In Exercises 21–26, write the sentence as a proportion. Then solve.

21. x is to 6 as 8 is to 9. **22.** y is to 5 as 6 is to 17. **23.** 3 is to 8 as m is to 24.

24. 2 is to 5 as 10 is to n. **25.** 5 is to 6 as 12 is to s. **26.** 2 is to t as 4 is to 13.

In Exercises 27–30, use a calculator to solve the proportion. Round your result to 2 decimal places.

Sample $\frac{3}{4} = \frac{x}{8}$: 3 [×] 8 [÷] 4 [=]

27. $\frac{14}{15} = \frac{x}{25}$ **28.** $\frac{y}{32} = \frac{16}{27}$ **29.** $\frac{p}{21} = \frac{8}{42}$ **30.** $\frac{14}{15} = \frac{x}{36}$

Geometry **In Exercises 31–34, find the missing lengths of the sides of the similar triangles.**

31.

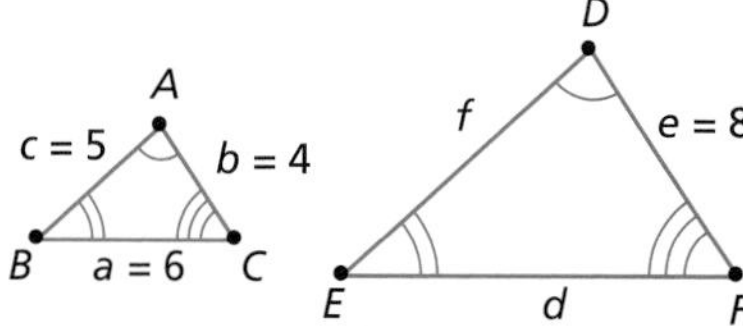

32.

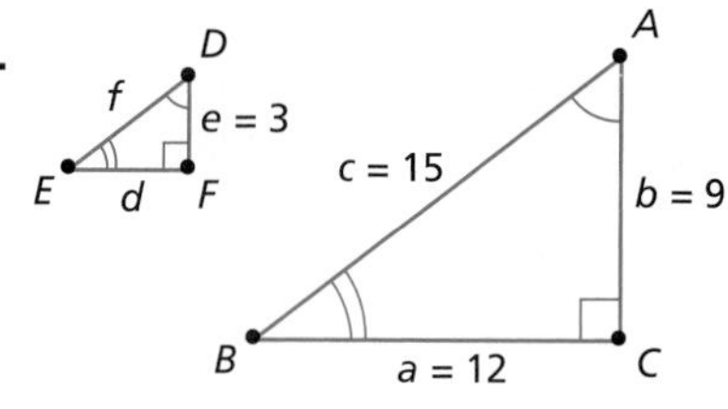

33.

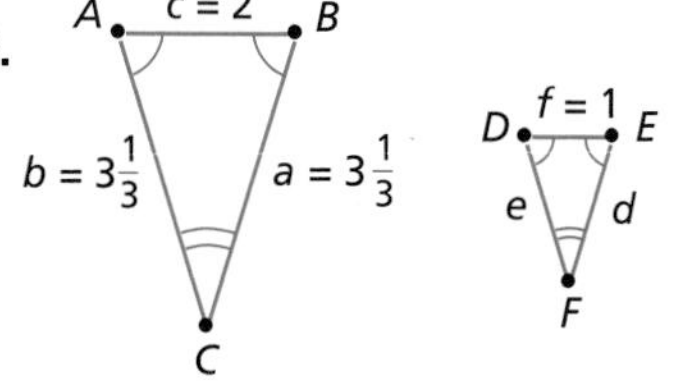

34.

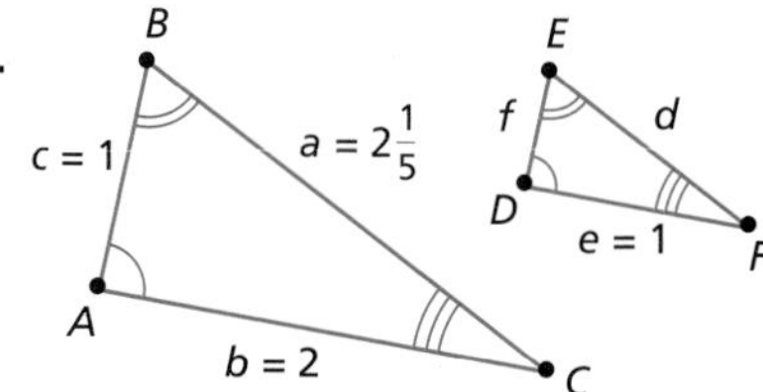

35. *Yogurt Shop* You work for a frozen yogurt shop. You sell yogurt cones at an average rate of 170 every 3 hours. The shop is open for 12 hours. About how many will be sold on a given day?

36. *Canoe Racing* Two canoes are competing in a 20-mile race. Canoe 1 is traveling at a rate of 4 miles every 25 minutes. Canoe 2 is traveling at a rate of 3 miles every 20 minutes.

a. How many minutes did it take each canoe to complete the race?

b. How many hours did it take for each canoe to complete the race?

c. How fast (in miles per hour) were the canoes traveling?

d. Which canoe won the race?

The only American canoeist to have won two Olympic gold medals is Greg Barton, who won the singles and doubles 1000-meter kayak races in 1988.

Integrated Review

Making Connections within Mathematics

Sequence **In Exercises 37 and 38, solve each proportion. Describe the pattern and find the next 2 numbers of the sequence.**

37. $\frac{1}{x} = \frac{2}{3}, \frac{2}{x} = \frac{3}{4}, \frac{3}{x} = \frac{4}{5}, \frac{4}{x} = \frac{5}{6}$

38. $\frac{x}{7} = \frac{1}{2}, \frac{x}{6} = \frac{1}{3}, \frac{x}{5} = \frac{1}{4}, \frac{x}{4} = \frac{1}{5}$

Exploration and Extension

39. *Photography Enlargements* You have a 3-inch by 5-inch post card that contains a photograph of a triangle. You enlarge the post card so that it is 6 inches by 10 inches. Is the enlarged triangle similar to the original? Explain your reasoning and illustrate your answer with a drawing.

8.3 Problem Solving Using Proportions

What you should learn:

How to use proportions to solve real-life problems

How to use similar triangles to measure objects indirectly

Why you should learn it:

You can use proportions to solve real-life problems, such as finding the dimensions of an airplane from a scale model.

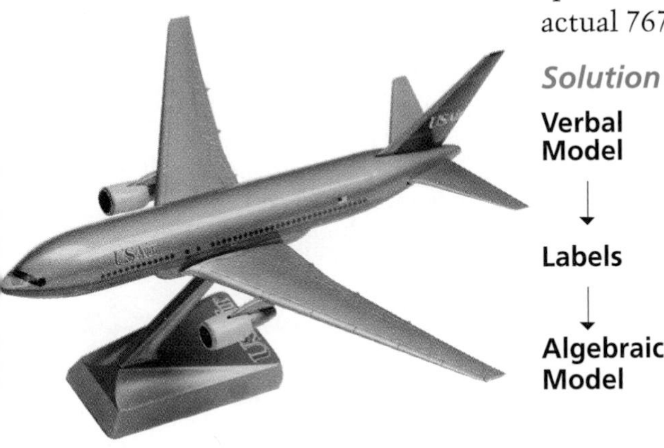

Goal 1 Solving Real-Life Problems

Proportions are used in architecture and manufacturing to construct scale models. For instance, if you are drawing house plans that have a 1-foot-to-$\frac{1}{8}$-inch scale, then the ratio of the actual house dimensions to the dimensions on the plans is

$$\frac{1 \text{ foot}}{\frac{1}{8} \text{ inch}} = \frac{12 \text{ inches}}{\frac{1}{8} \text{ inch}} \quad \textit{Rewrite 1 foot as 12 inches.}$$

$$= 12 \cdot \frac{8}{1} \quad \textit{Multiply by reciprocal of denominator.}$$

$$= 96. \quad \textit{Simplify.}$$

This means that each dimension of the actual house is 96 times as large as the corresponding dimension in the house plans.

Example 1 *Finding Dimensions*

You have purchased a scale model of a Boeing 767, as shown at the left. The model is constructed on a 1-to-250 scale. The wing span on the model is $7\frac{1}{2}$ inches. What is the wing span on an actual 767?

Solution

Verbal Model

$$\frac{\text{Model wing span}}{\text{767 wing span}} = \frac{1}{250}$$

Labels

Model wing span = 7.5 (inches)
767 wing span = x (inches)

Algebraic Model

$$\frac{7.5}{x} = \frac{1}{250} \quad \textit{Proportion}$$

$$\frac{x}{7.5} = 250 \quad \textit{Reciprocal Property}$$

$$7.5 \cdot \frac{x}{7.5} = 7.5 \cdot 250 \quad \textit{Multiply each side by 7.5.}$$

$$x = 1875 \quad \textit{Simplify.}$$

The wing span of the 767 is 1875 inches. To convert this measurement to feet, divide by 12 to obtain a wing span of $156\frac{1}{4}$ feet. ■

Goal 2 Indirect Measurement

In Lesson 8.2, you learned that the ratios of corresponding sides of similar triangles are equal. This property of similar triangles can be used to indirectly measure the heights of trees or buildings. For instance, in the drawing below, the ratio of the post's height to the tree's height is equal to the ratio of the post's shadow to the tree's shadow. Knowing the post's height and the lengths of the two shadows (at the same time of day) allows you to find the tree's height.

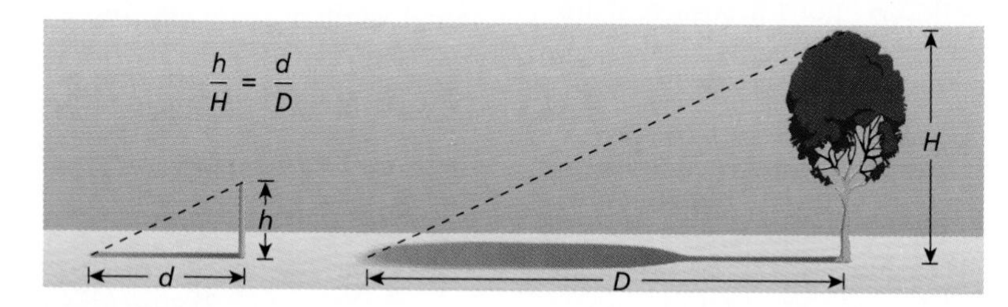

Real Life
Architecture

Example 2 *Finding the Height of a Building*

To estimate the height of the Transco Tower in Houston, Texas, you measure its shadow to be 55 meters. The shadow of a 5-meter post is 1 meter. How tall is the Transco Tower?

Solution

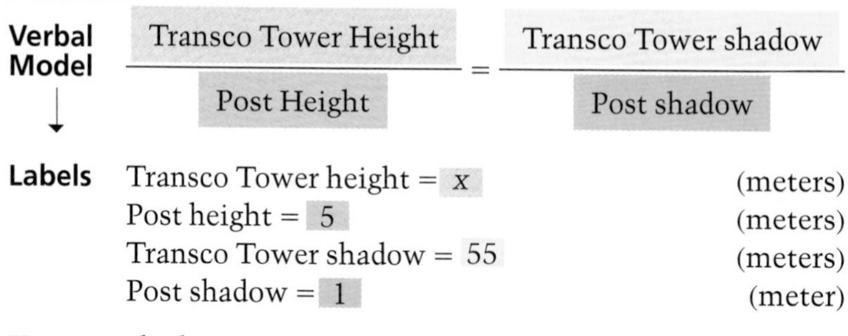

Verbal Model

$$\frac{\text{Transco Tower Height}}{\text{Post Height}} = \frac{\text{Transco Tower shadow}}{\text{Post shadow}}$$

Labels

Transco Tower height = x (meters)
Post height = 5 (meters)
Transco Tower shadow = 55 (meters)
Post shadow = 1 (meter)

You are asked to complete the solution below. ■

Communicating about MATHEMATICS

SHARING IDEAS about the Lesson

Extending the Example Example 2 contains only a partial solution. Use the verbal model and labels to write the algebraic model. Then solve the algebraic model to find the height of the Transco Tower in Houston.

EXERCISES

Guided Practice

CHECK for Understanding

Flagpole Height **In Exercises 1–4, use the following information.**

A flagpole is casting a 15-foot shadow. You are five feet tall and cast a 3-foot shadow. What is the height of the flagpole?

1. Sketch a diagram of the given information.
2. Which of the following verbal models is correct?

   a. $\dfrac{\text{Flagpole height}}{\text{Your height}} = \dfrac{\text{Your shadow}}{\text{Flagpole shadow}}$ b. $\dfrac{\text{Flagpole height}}{\text{Your height}} = \dfrac{\text{Flagpole shadow}}{\text{Your shadow}}$

3. Write an algebraic model for the problem. Let h represent the height of the flagpole.
4. Solve the algebraic model to find the height of the flagpole.

Independent Practice

Heartbeat Rate **In Exercises 5 and 6, use the following information.**

A typical heartbeat rate (or pulse) is 72 beats per minute. When exercising, this rate increases.

5. Assume you have the typical heartbeat rate. If you take your pulse for 10 seconds, how many beats would you feel?
6. After jogging one-half hour, you take your pulse for 6 seconds. You feel 11 beats. What is your heartbeat rate in beats per minute?

Estimation **In Exercises 7 and 8, use the following information.**

In 1993, an average of about 8800 Americans per day became teenagers.

7. About how many Americans became teenagers in April?
8. About how many Americans became teenagers in 1993?
9. *Estimation* If you sleep 220,000 hours by age 70, about how many hours will you have slept by age 40?

The President's Council on Physical Fitness and Sports recommends 30 minutes of continuous exercise at least 3 times a week.

10. *Estimation* If you pay \$1500 in property tax for a \$90,000 house, about how much would you pay for a \$110,000 house?

Computer Screens **In Exercises 11–13, use the following information.**

Computer screens are partitioned into pixels. Consider a rectangular computer screen that is $8\frac{1}{2}$ inches by 10 inches and has 5184 pixels per square inch. How many pixels are on the entire screen?

11. Which of the following verbal models is correct?

a. $\dfrac{\text{Pixels on screen}}{\text{Pixels in 1 in.}^2} = \dfrac{\text{Area of screen}}{\text{1 in.}^2}$ **b.** $\dfrac{\text{Pixels on screen}}{\text{Pixels in 1 in.}^2} = \dfrac{\text{10 inches}}{8\frac{1}{2}\text{ inches}}$

12. Assign labels to the correct verbal model and write an algebraic model.

13. Solve the algebraic model to find the number of pixels on the computer screen.

Magnification **In Exercises 14–16, use the following information.**

An amoeba, a one-celled organism, has been magnified 250 times by a microscope and photographed. In the photo, the amoeba is 25 millimeters wide. What is the actual width of the amoeba?

14. Which of the following verbal models is correct?

a. $\dfrac{\text{Photo amoeba width}}{\text{Actual amoeba width}} = \dfrac{250}{1}$ **b.** $\dfrac{\text{Photo amoeba width}}{\text{1 millimeter}} = \dfrac{\text{Actual amoeba width}}{\text{250 millimeters}}$

15. Assign labels to the verbal model and write an algebraic model.

16. Solve the algebraic model to find the width of the amoeba.

Integrated Review

Making Connections within Mathematics

Mental Math **In Exercises 17–20, use mental math to solve the proportion.**

17. $\frac{1}{2} = \frac{x}{8}$ **18.** $\frac{m}{6} = 40$ **19.** $\frac{5}{n} = \frac{10}{25}$ **20.** $\frac{2}{3} = \frac{8}{b}$

Exploration and Extension

Geometry **In Exercises 21–24, draw each figure. Then divide the figure into four congruent figures, each of which is similar to the larger figure. A sample is shown at the right.**

Two geometric figures are **congruent** if they have exactly the same size and shape. Two figures are **similar** if they have the same shape.

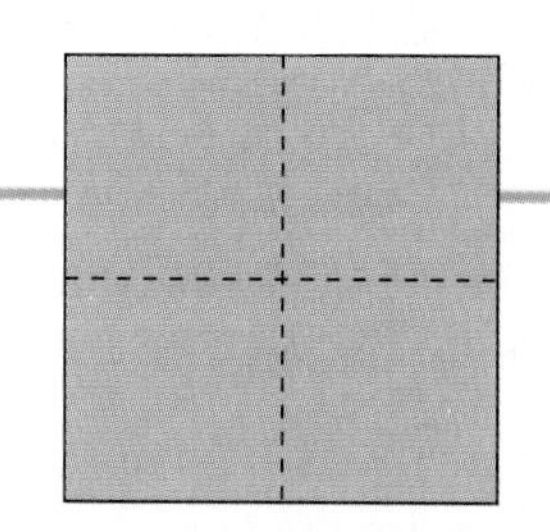

This square is divided into 4 congruent squares.

21.

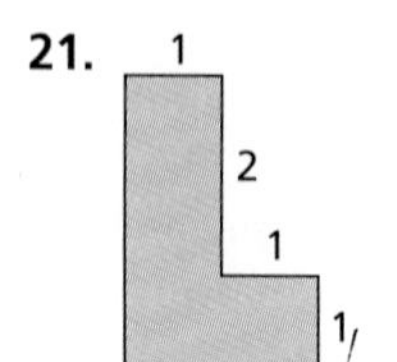

22.

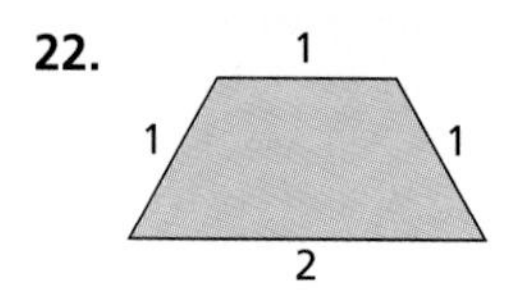

23.

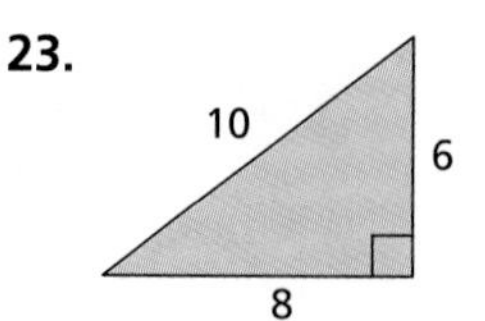

24.

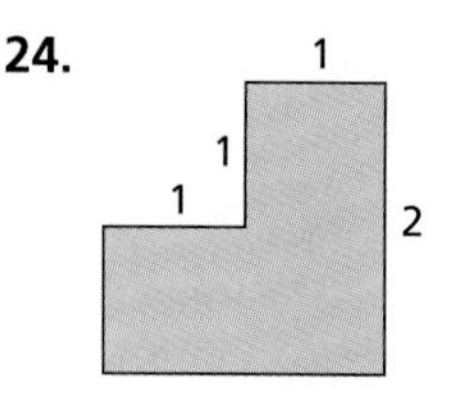

Mixed REVIEW

In Exercises 1–9, solve the equation. Check your solution. (4.2, 4.4, 4.5)

1. $4(x + 2) = 12$

2. $3y = 4y - 7$

3. $a + \frac{5}{3} = 2a - \frac{1}{3}$

4. $\frac{1}{12}p - \frac{1}{12} = \frac{5}{12}$

5. $5 + 2(q - 3) = 0$

6. $0.2s - 5.4 = 0$

7. $1 = \frac{1}{2}(p - 1)$

8. $\frac{1}{4}x + \frac{1}{4}x = 3$

9. $\frac{1}{3}p - 2 = 2$

In Exercises 10–13, write the percent in decimal form. (7.6–7.8)

10. 52%

11. 83%

12. 146%

13. 206%

In Exercises 14–17, determine whether the quotient is a rate or a ratio. Then simplify. (8.1)

14. $\frac{16 \text{ yards}}{2 \text{ jumps}}$

15. $\frac{28 \text{ points}}{4 \text{ quarters}}$

16. $\frac{2 \text{ animals}}{20 \text{ animals}}$

17. $\frac{4 \text{ feet}}{10 \text{ seconds}}$

Milestones THE ARABIAN LEGACY

400 · 600 · 800 · 1000 · 1200 · 1400 · 1600 · 1800 · 2000

Fall of Rome 476 · Mohammed born 570 · First printed book, China 868 · Printing Press 1450 · Electric Battery 1786 · Satellite Intelsat I 1965

This painting from 1494 shows Laila and Majnun, a Persian heroine and hero, at school.

By the ninth century, years of conflict had destroyed many European centers of learning. By contrast, from about 800 through 1100 the Arabic influence extended from Baghdad, Iraq, its center, westward to Europe and Africa, northward to Turkey and Russia, and eastward to India and China. Arabic scholars translated thousands of ancient Latin, Greek, Hebrew, and Hindu texts into Arabic, thus preserving them from total destruction.

The most famous Arabic author, Mohammed ibn-Musa al-Khowarizmi (?825), wrote a treatise on equations called *Hisab Al-jabr* meaning the "science of restoration or reunion." (In Spain, barbers, who were the bone setters of the day, called themselves "algebristi.") The person responsible for the mathematical term *algebra* was Leonardo of Pisa, also known as Fibonacci (1180–1250). While a child, he learned mathematics through Arabic texts. When he translated *Hisab Al-jabr* into Latin, it became translated as the "science of equations."

- *Fibonacci is most famous for his sequence 1, 1, 2, 3, 5, 8, . . . (see page 41). Find the quotient, or ratio, of successive terms for the first 10 terms. That is, find 1 ÷ 1, 2 ÷ 1, 3 ÷ 2, etc.*
- *What pattern do you notice in these ratios? (See page 348.)*

8.4 Solving Percent Equations

What you should learn:

Goal 1 How to find what percent one number is of another

Goal 2 How to solve a percent equation

Why you should learn it:

You can use percents to describe real-life situations, such as finding your correct percent on a quiz.

Goal 1 Finding Percents

In Chapter 7, you learned that percents can be written in different forms. For instance, 25% can be written as

$$25\%, \quad \frac{25}{100}, \quad 0.25, \quad \text{or} \quad 25 \text{ percent.}$$

The most common use of percents is to compare one quantity to another. For instance, because 3 is one-fourth of 12, you can say that 3 is 25% of 12.

The Percent Equation

The statement "a is p percent of b" is equivalent to the equation

$$\frac{a}{b} = \frac{p}{100}. \qquad \textit{Percent equation}$$

In this equation, b is the **base** and a is the number that is compared to the base.

Example 1 *Finding Percents*

a. You got 17 points on a 20-point quiz. You can find the percent p that you got correct as follows.

$\frac{17}{20} = \frac{p}{100}$ *Write percent equation.*

$0.85 = \frac{p}{100}$ *Divide to obtain decimal form.*

$85 = p$ *Multiply each side by 100.*

You got 85% correct. One way to describe this is to say that *17 is 85% of 20.*

b. On a 50-point quiz, you got 48 points plus 5 bonus points. You can find the percent p that you got correct as follows.

$\frac{53}{50} = \frac{p}{100}$ *Write percent equation.*

$1.06 = \frac{p}{100}$ *Divide to obtain decimal form.*

$106 = p$ *Multiply each side by 100.*

You got 106% correct. One way to describe this is to say that *53 is 106% of 50.* ■

One way to see that 17 is 85% of 20 is to use a grid with 100 squares. Divide the grid into twenty 1-by-5 rectangles. Then shade 17 of the rectangles and count the number of small squares that have been shaded.

Goal 2 Solving Percent Equations

There are three basic types of percent problems. Each can be solved by substituting the two given quantities into the percent equation and solving for the third quantity.

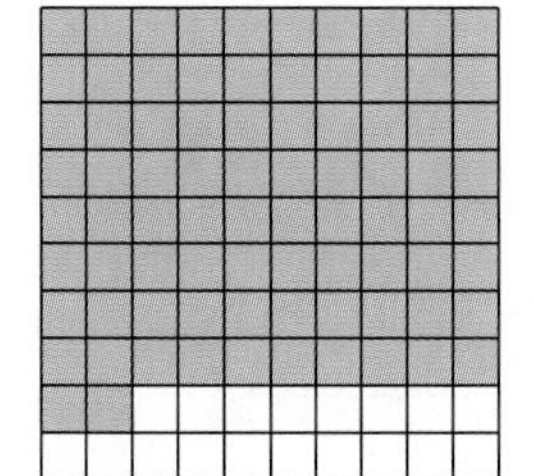

If each small square represents 2.5 people, then 82 squares must represent 82(2.5) or 205 people.

Question	Given	Percent equation
a is what percent of b?	a and b	Solve for p.
What is p percent of b?	p and b	Solve for a.
a is p percent of what?	a and p	Solve for b.

Example 2 *Solving Percent Equations*

a. You ask 250 people whether they prefer blue jeans or another color of jeans and 82% say they prefer blue jeans. You can find the number a of people who prefer blue jeans as follows.

$$\frac{a}{250} = \frac{82}{100}$$ *Write in fraction form.*

$$\frac{a}{250} = 0.82$$ *Write in decimal form.*

$$a = 250 \cdot 0.82$$ *Multiply each side by 250.*

$$a = 205$$ *Simplify.*

Thus, 205 people said they prefer blue jeans.

25%	25%	25%	25%
\$8.83	\$8.83	\$8.83	\$8.83

3(8.83) = \$26.49

4(8.83) = \$35.32

Geometric models like these can help you solve percent problems.

b. You paid \$26.49, which is 75% of the full price, for a sweater. You can find the full price b as follows.

$$\frac{26.49}{b} = \frac{75}{100}$$ *Write in fraction form.*

$$\frac{b}{26.49} = \frac{100}{75}$$ *Write reciprocal of each side.*

$$b = 26.49 \cdot \frac{100}{75}$$ *Multiply each side by 26.49.*

$$b = 35.32$$ *Simplify.*

The full price of the sweater is \$35.32. ■

Communicating about MATHEMATICS

Cooperative Learning

SHARING IDEAS about the Lesson

Using Percents Your aunt works in a clothing store. She tells you that a quick way to find $p\%$ of b is to change $p\%$ to decimal form and multiply by b. Is she correct? Explain your reasoning to your partner.

EXERCISES

Guided Practice

CHECK for Understanding

In Exercises 1 and 2, describe the strategy that was used to solve the problem.

1. **Problem:** What is 65% of 160?

   Solution: $\frac{a}{160} = \frac{65}{100}$

   $\frac{a}{160} = 0.65$

   $a = 104$

2. **Problem:** 50 is what percent of 40?

   Solution: $\frac{50}{40} = \frac{p}{100}$

   $1.25 = \frac{p}{100}$

   $125 = p$

In Exercises 3–5, identify the base. Then write and solve the percent equation.

3. 20 is what percent of 25?
4. What is 16% of 50?
5. 90 is 75% of what?
6. Is fifteen 250% of six? Explain your reasoning.

Independent Practice

In Exercises 7–16, solve the percent equation. Round your answer to 2 decimal places.

7. 22 is what percent of 30?
8. What is 33 percent of 165?
9. What is 2% of 360?
10. 66 is 120 percent of what number?
11. 34 is 50% of what number?
12. 45 is what percent of 20?
13. What is 110 percent of 110?
14. 6.06 is 20.2% of what number?
15. 71.5 is what percent of 90?
16. What is 25.5% of 270?

Mental Math **In Exercises 17–20, use mental math to solve the percent equation.**

17. 100 is 200 percent of what number?
18. What is 50% of 200?
19. 100 is 100% of what number?
20. 5 is what percent of 15?

Error Analysis **In Exercises 21 and 22, find and correct the error.**

21. **Problem:** 45 is what percent of 150?

    Percent Equation: ~~$\frac{150}{45} = \frac{p}{100}$~~

22. **Problem:** What is 24 percent of 50?

    Percent Equation: ~~$\frac{24}{50} = \frac{p}{100}$~~

Geometric Modeling **In Exercises 23–25, solve the percent equation. Then sketch a geometric model that illustrates your solution.**

23. $\frac{13}{25} = \frac{p}{100}$
24. $\frac{a}{20} = \frac{85}{100}$
25. $\frac{3}{b} = \frac{60}{100}$

26. *Test Scores* On a 35-question exam, you needed to get 90% of the questions correct to earn an *A*. You answered 31 questions correctly. Did you earn an *A*? Explain.

27. *Sales Tax* You buy a new Sega Genesis game. You pay \$54.99 for the game plus \$2.75 in sales tax. What is the sales tax percent?

Linda Ronstadt is a multiple-category grammy winner: country vocal (1975), pop vocal (1976), country duo or group (1987), best Mexican/American (1988 and 1992), pop duo or group (1989 and 1990), and best tropical Latin (1992).

28. *Kicking Field Goals* During the regular season, a field goal kicker made 75% of the field goals he attempted. He made 24 field goals. How many did he attempt?

Winning a Grammy **In Exercises 29 and 30, use the following information.**

When the Grammy Awards began in 1958, there were 28 categories in which to win an award. In 1993, there were 81 categories. *(Source: National Academy of Recording Arts and Sciences)*

29. The number of awards given in 1958 is what percent of the number of awards given in 1993?

30. The number of awards given in 1993 is what percent of the number of awards given in 1958?

Integrated Review

Making Connections within Mathematics

Decimal Sense **In Exercises 31–34, write the percent as a decimal.**

31. 62%
32. 21%
33. 18.24%
34. 98.04%

Percents **In Exercises 35–38, write the decimal as a percent.**

35. 0.1234
36. 0.1001
37. 0.9432
38. 0.7693

Exploration and Extension

Geometric Probability **In Exercises 39–41, a pebble is tossed and lands on the region. The pebble is equally likely to land anywhere on the region. What is the probability that the pebble will land on the green region? Give your answer in percent form.**

39.

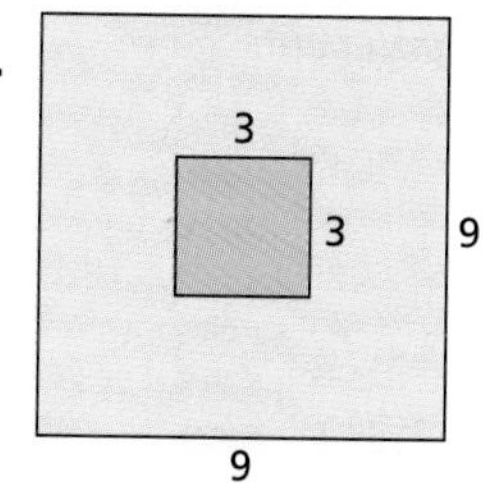

40.

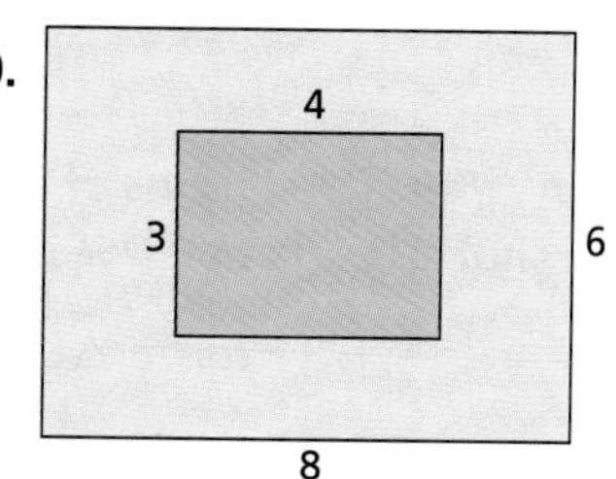

41.

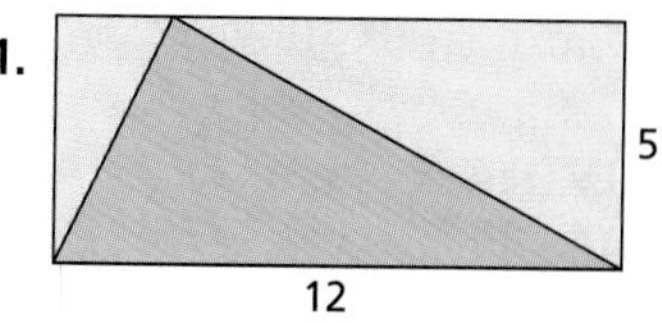

Mid-Chapter SELF-TEST

Take this test as you would take a test in class. The answers to the exercises are given in the back of the book.

In Exercises 1–4, is the quotient a rate or ratio? Simplify. (8.1)

1. $\frac{100 \text{ meters}}{18 \text{ seconds}}$ **2.** $\frac{18 \text{ lures}}{6 \text{ fishermen}}$ **3.** $\frac{42 \text{ pounds}}{3 \text{ pounds}}$ **4.** $\frac{5 \text{ days}}{1 \text{ week}}$

In Exercises 5 and 6, state whether the two triangles are similar. If they are, write three different proportions that compare the side lengths. (8.2)

5.

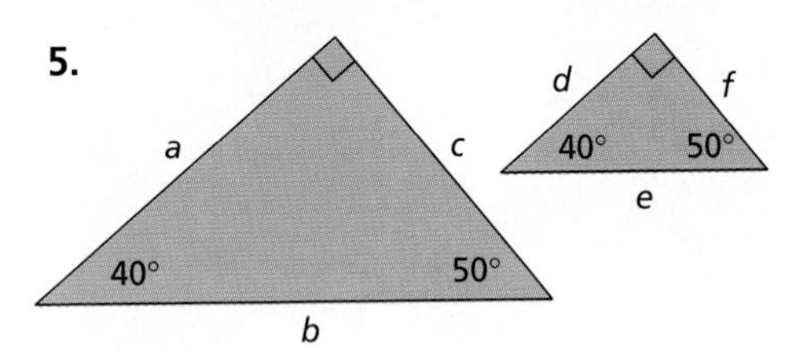

6.

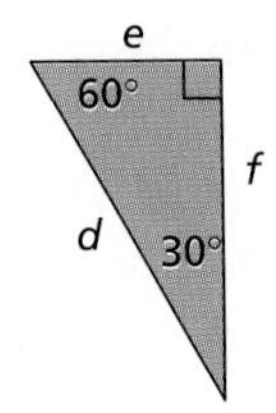

In Exercises 7–9, decide whether the equation is true. (8.2)

7. $\frac{2}{11} \stackrel{?}{=} \frac{10}{55}$ **8.** $\frac{250}{350} \stackrel{?}{=} \frac{2}{3}$ **9.** $\frac{5}{7} \stackrel{?}{=} \frac{25}{35}$

In Exercises 10–12, solve the percent equation. (8.4)

10. What is 35% of 18? **11.** 36 is what percent of 150? **12.** 18 is 150% of what?

In Exercises 13–16, solve the proportion. Check your solution. (8.2)

13. $\frac{x}{4} = \frac{8}{2}$ **14.** $\frac{3}{7} = \frac{a}{9}$ **15.** $\frac{18}{b} = \frac{25}{36}$ **16.** $\frac{y}{3} = \frac{10}{2}$

17. You are visiting the Gateway Arch in St. Louis, Missouri. To find out how tall the monument is, you measure the length of the shadow as 60 feet. At the same time, your own shadow is $\frac{1}{2}$ foot long. If you are $5\frac{1}{4}$-feet tall, how tall is the Gateway Arch? **(8.2)**

18. While in St. Louis, you decide to take a tour through an automobile manufacturing plant. Your tour guide tells you that 29% of the cars are manufactured with a driver's side airbag. If 47,000 cars are manufactured, how many have an airbag? **(8.4)**

The Gateway Arch in St. Louis, Missouri, is the nation's tallest monument.

In Exercises 19 and 20, $\triangle ABC$ is similar to $\triangle DEF$. The lengths of the sides of $\triangle ABC$ are $a = 8$, $b = 10$, and $c = 12$. The shortest side of $\triangle DEF$ is $d = 9$. (8.2)

19. In $\triangle DEF$, find the longest side. **20.** In $\triangle DEF$, find the third side.

8.5 Problem Solving Using Percents

What you should learn:

How to use percents to solve real-life problems

How to use percents to find discounts

Why you should learn it:

You can use percents to solve real-life problems, such as finding a discount.

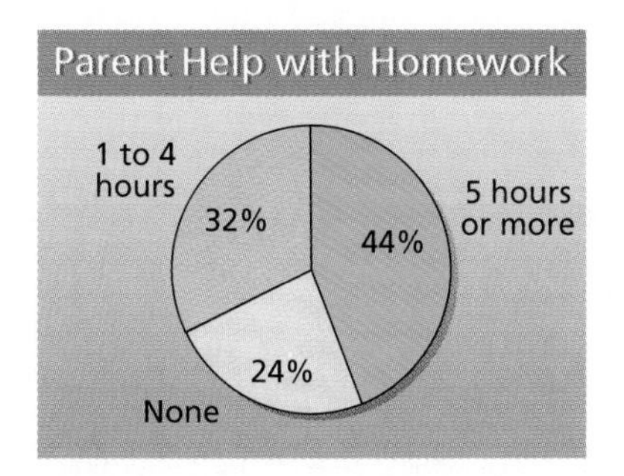

Goal 1 Solving Real-Life Problems

Knowing how to use percents is important in almost every part of real life. In this lesson, you will see how the problem-solving plan can be used to solve real-life problems involving percents.

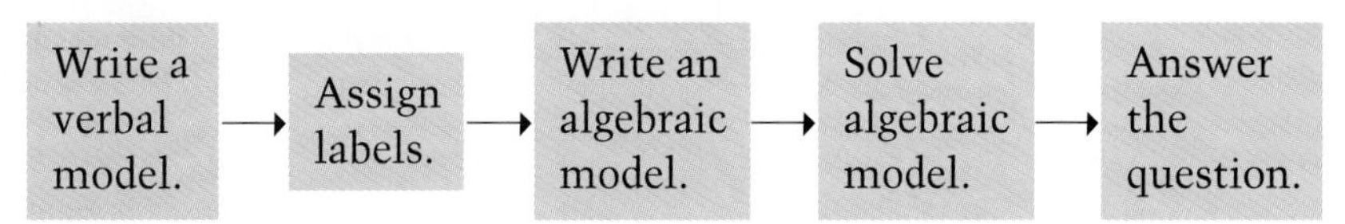

Example 1 *Using a Problem-Solving Plan*

In a survey, parents of elementary school children were asked how many hours per week they spent helping their children with homework. The results are shown in the circle graph at the left. Eighty-eight said they spent 1 to 4 hours. How many parents were surveyed? How many said "none"? How many said "5 hours or more"? ***(Source: 20/20 Research)***

Solution

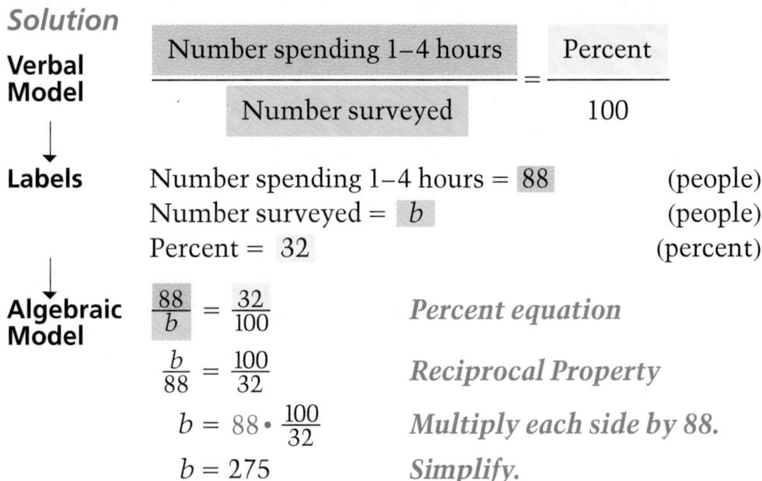

Verbal Model

$$\frac{\text{Number spending 1–4 hours}}{\text{Number surveyed}} = \frac{\text{Percent}}{100}$$

Labels

Number spending 1–4 hours = 88 (people)
Number surveyed = b (people)
Percent = 32 (percent)

Algebraic Model

$\frac{88}{b} = \frac{32}{100}$ ***Percent equation***

$\frac{b}{88} = \frac{100}{32}$ ***Reciprocal Property***

$b = 88 \cdot \frac{100}{32}$ ***Multiply each side by 88.***

$b = 275$ ***Simplify.***

There were 275 parents surveyed. To find the number who said "none," find 24% of 275. To find the number who said "5 hours or more," find 44% of 275.

$0.24 \cdot 275 = 66$ ***"None"***

$0.44 \cdot 275 = 121$ ***"5 hours or more"*** ■

Goal 2 Finding Discounts

When an item is "on sale" the difference between the regular price and the sale price is called the **discount.**

$$\text{Discount} = \text{Regular Price} - \text{Sale Price}$$

To find the **discount percent,** use the regular price as the base. For instance, if a $36 book is on sale for $27, then the discount is $9, and the discount percent is

$$\frac{\text{Discount}}{\text{Regular Price}} = \frac{9}{36}$$ *Divide discount by regular price.*

$$= 0.25$$ *Rewrite in decimal form.*

$$= 25\%.$$ *Rewrite in percent form.*

Real Life
Consumers

Example 2 *Finding a Discount*

Last week you bought a sweatshirt for $16.80. The store had a sign that stated that the price had been discounted 35%. This week you visit the store again and discover that the regular price is $22.68. Did you really get a 35% discount?

Solution The discount is

$$22.68 - 16.80 = \$5.88.$$ *Discount*

Then use the formula to find the discount percent.

$$\frac{5.88}{22.68} \approx 0.259$$ *Divide discount by regular price.*

$$= 25.9\%$$ *Write in percent form.*

The discount was only 25.9%, not 35%. ■

Communicating about MATHEMATICS

SHARING IDEAS about the Lesson

What Is Wrong? After seeing the regular price, you asked the store manager why the sweatshirt was listed as having a 35% discount. The manager tells you that 35% of $16.80 is $5.88, and the regular price is

$$\$16.80 + \$5.88 = \$22.68.$$

What is wrong with the manager's mathematics?

EXERCISES

Guided Practice

CHECK for Understanding

1. *Shoe Sale* You pay $38.15 for a pair of tennis shoes. The discount was $16.35.
   a. What is the original price?
   b. Based on the *original price,* what is the discount percent?
2. *Shoe Sale* You are shopping for shoes and see the sign at the right. Explain how you could find the original price of the shoes.

Independent Practice

Music **In Exercises 3 and 4, use the following.**

You take a survey in your classroom about favorite types of music. The circle graph shows the percent of students in each type. Nine students said that country was their favorite.

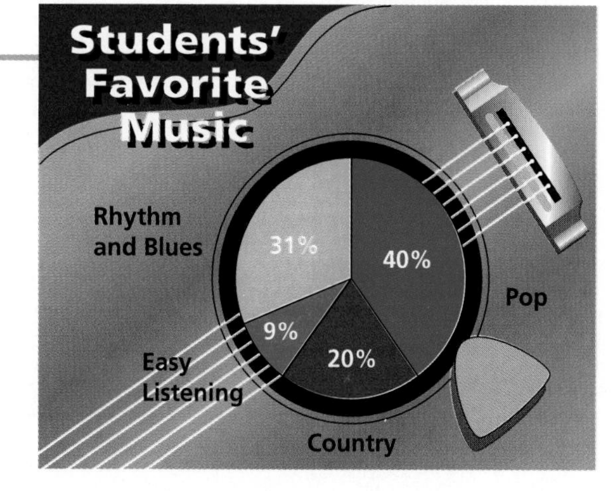

3. How many students were surveyed?
4. How many students said
   a. pop?
   b. rhythm and blues?
   c. easy listening?
5. *Spanish* In 1992, about 410,000 Americans took a course in Spanish. This represented 41% of those who took a language course. How many took a language course? ***(Source: Modern Language Association)***
6. *French* In 1992, 27.4% of the Americans who took a language course took a course in French. Use the result of Exercise 5 to find the number of Americans who took a course in French.

Pay Phones **In Exercises 7–9, use the following.**

The circle graph shows the categories and the amounts of money that make up the yearly income for the average pay phone. Coin revenue makes up 35% of the yearly income. ***(Source: USA Today)***

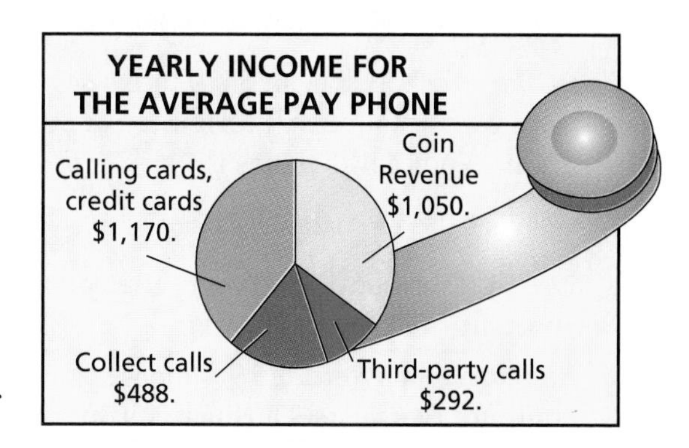

7. What is the yearly income for the average pay phone?
8. Describe another way to find the yearly income.
9. Find the percent of yearly income of
   a. calling and credit cards. b. collect calls. c. third-party calls.

Where Is Your Radio? **In Exercises 10–13, use the following information.**

The table shows the numbers, in millions, of radios in specific locations in American households. The percent of radios in bathrooms is about 4.3%.

Location	Number of Radios
Bedrooms	172.3
Living rooms	63.3
Kitchens	46.2
Bathrooms	14.7
Dining rooms	13.3
Other	33.2

10. How many radios do Americans own?

11. Find the percent of radios in bedrooms.

12. Find the percent of radios in living rooms.

13. Find the percent of radios in kitchens.

14. *Discount* You buy a pair of rollerblades. The sign at the sporting goods store stated that the price of rollerblades had been discounted 25%. The discount was $21.40. What was the regular price of the rollerblades?

Geometry **In Exercises 15 and 16, the blue region's percent of the total area is given.**

15. The blue region has an area of 16 square units. What is the area of the entire region?

68%

16. The yellow region has an area of 16 square units. What is the area of the entire region?

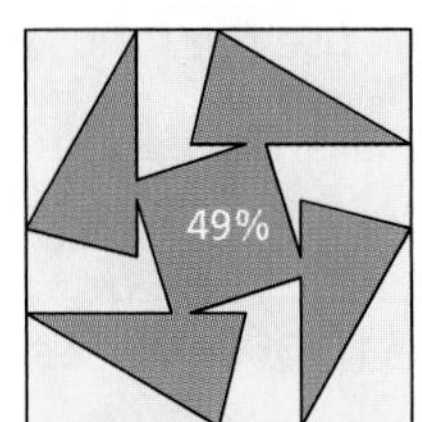

Integrated Review

Making Connections within Mathematics

Proportions **In Exercises 17–24, solve the proportion.**

17. $\frac{95}{x} = \frac{20}{100}$ **18.** $\frac{50}{100} = \frac{84}{y}$ **19.** $\frac{n}{90} = \frac{60}{100}$ **20.** $\frac{m}{40} = \frac{80}{100}$

21. $\frac{9}{60} = \frac{x}{100}$ **22.** $\frac{y}{100} = \frac{3}{5}$ **23.** $\frac{45}{50} = \frac{s}{100}$ **24.** $\frac{36}{90} = \frac{t}{100}$

Exploration and Extension

25. *Banking* You deposit $470 into a savings account. At the end of one year, your account will earn interest at the rate of 3.5%. A shortcut for finding your new balance is

$$470(1.035) = \text{balance}.$$

Use the Distributive Property to explain how you could obtain this shortcut.

26. *Banking* In Exercise 25, suppose that you leave the money in the account for two years. During that time, you make no additional deposits or withdrawals. What is your balance at the end of two years?

8.6 Exploring Percent of Increase or Decrease

What you should learn:

Goal 1 How to find a percent of increase

Goal 2 How to find a percent of decrease

Why you should learn it:

You can use percents to show how real-life quantities increase or decrease over time.

Goal 1 Finding a Percent of Increase

A **percent of increase** or **percent of decrease** tells how much a quantity has changed. For instance, the enrollments at Roosevelt Middle School for 1992 and 1994 are shown below.

1992 Enrollment: 400 **1994 Enrollment:** 420

From 1992 to 1994, the enrollment *increased* by 20 students. The percent of increase is

$$\text{Percent increase} = \frac{20}{400} = 0.05 = 5\%.$$

Percent of Increase or Decrease

The percent of change of a quantity is given by

$$\frac{\text{Actual change}}{\text{Original amount}}.$$

This percent is a **percent of increase** if the quantity increased and it is a **percent of decrease** if the quantity decreased.

Real Life Technology

Example 1 *Finding a Percent of Increase*

The bar graph shows the number of computers that were in use in public schools from 1985 through 1991. Find the percent of increase from 1990 to 1991. ***(Source: Market Data Retrieval)***

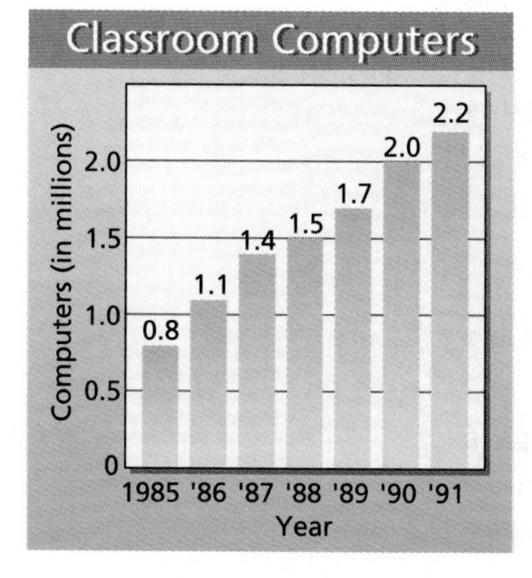

Solution There were 2.2 million in 1991 and 2 million in 1990. This means that the actual increase was 0.2 million computers. To find the percent of increase, find the ratio of 0.2 million (the increase) to 2 million.

$$\frac{\text{Number in 1991} - \text{Number in 1990}}{\text{Number in 1990}} = \frac{2.2\text{ million} - 2\text{ million}}{2\text{ million}}$$

$$= \frac{0.2\text{ million}}{2\text{ million}} = 0.1 = 10\%$$

The percent of increase is 10%. ■

Goal 2 Finding a Percent of Decrease

Real Life **Communication**

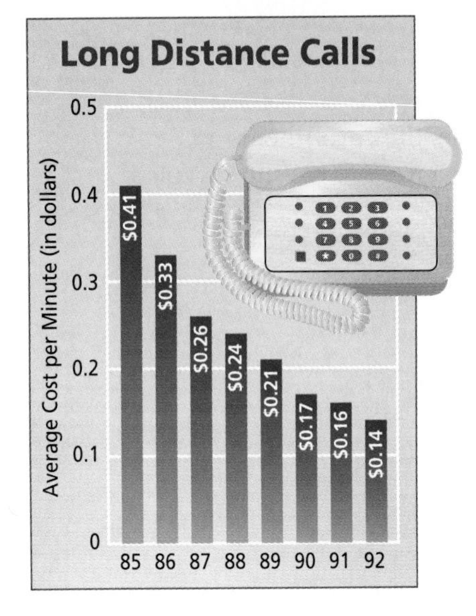

Example 2 *Finding a Percent of Decrease*

The bar graph at the left shows the averge cost per minute for long distance calls from 1985 through 1992. Find the percent of decrease from 1991 to 1992. ***(Source: Federal Communications Commission)***

Solution The average cost in 1991 was \$0.16, and the average cost in 1992 was \$0.14. This means that the actual decrease was \$0.02. To find the percent of decrease, find the ratio of \$0.02 to \$0.16.

$$\frac{\text{Average Cost in 1991} - \text{Average Cost in 1992}}{\text{Average Cost in 1991}} = \frac{\$0.16 - \$0.14}{\$0.16}$$

$$= \frac{\$0.02}{\$0.16}$$

$$= 0.125$$

$$= 12.5\%$$

The percent of decrease was 12.5%. ■

When you are finding percents of increase or decrease over time, notice that the earlier number is always the base. For instance, in Example 1, the 1990 number is the base because it is earlier than the 1991 number. In Example 2, the 1991 number is the base because it is earlier than the 1992 number.

Communicating about MATHEMATICS

SHARING IDEAS about the Lesson

Finding Percents of Decrease Use the data in the graph in Example 2 to find the percents of decrease over pairs of consecutive years. Then copy and complete the table. When did the greatest decrease occur? When did the greatest percent of decrease occur?

Years	85–86	86–87	87–88	88–89	89–90	90–91	91–92
Decrease	?	?	?	?	?	?	?
Percent	?	?	?	?	?	?	?

EXERCISES

Guided Practice

CHECK for Understanding

1. *Computers* In Example 1 on page 367, find the percent of increase in the number of computers from 1985 to 1991.

2. *Long-Distance Rates* In Example 2 on page 368, find the percent of decrease in the average cost per minute for long-distance telephone calls from 1985 to 1992.

In Exercises 3 and 4, state whether the quantities represent a percent of increase or a percent of decrease. Then find the percent.

3. Yesterday: $16.35
   Today: $18.21

4. May: 1056 units
   June: 972 units

Independent Practice

In Exercises 5–10, decide whether the change is an increase or a decrease and find the percent.

5. Before: 10, After: 12
6. Before: 15, After: 12
7. Before: 75, After: 60
8. Before: 110, After: 143
9. Before: 90, After: 200
10. Before: 260, After: 160

In Exercises 11–14, decide whether the change is an increase or a decrease and find the percent.

11. 1994: $171.33
    1995: $201.59

12. Regular Price: $31.99
    Sale Price: $22.39

13. Beginning Balance: $521.43
    End Balance: $413.68

14. Opening Price: $18.77
    Closing Price: $19.17

Finding a Pattern **In Exercises 15–18, use percents to describe the sequence's pattern. Then list the next three terms.**

15. 1, 2, 4, 8, ?, ?, ?
16. 4096, 1024, 256, 64, ?, ?, ?
17. 15625, 3125, 625, 125, ?, ?, ?
18. 1, 3, 9, 27, ?, ?, ?

Reasoning **In Exercises 19–22, decide whether the statement is true or false. Explain.**

19. Two times a number is a 100% increase of the number.
20. Half a number is a 50% decrease of the number.
21. A 20% decrease of 80 is 60.
22. A 25% increase of 100 is 125.

Problem Solving **In Exercises 23 and 24, describe a real-life situation that involves the given increase or decrease.**

23. An increase of 15%

24. A decrease of 30%

25. Copy and complete the table.

Original Number	New Number	Percent Change
45	72	?
45	18	?
400	?	25% Increase
400	?	25% Decrease

26. *Production* In August, a small company produced 32,562 units. In September, the company produced 28,894 units. Find the percent decrease from August to September.

27. *County Population* Riverside County, California, had a large percent increase in population between 1980 and 1990. The 1980 population was 663,199 and the 1990 population was 1,170,413. Find the percent increase of population. ***(Source: U.S. Bureau of Census)***

28. *Shoe Profits* The graph at the right shows the net profits (in millions of dollars) of Reebok for 1988 to 1994. Approximate the percent of change of net profit for each year. ***(Source: Reebok International, Ltd.)***

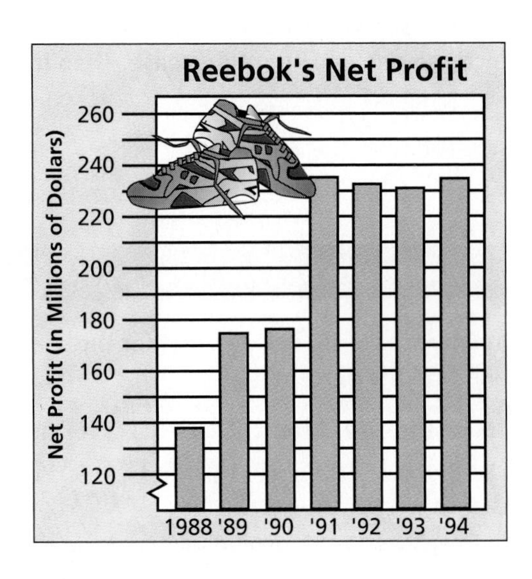

Salary Increase **In Exercises 29 and 30, use the following information.**

Your older sister tells you that she and a co-worker each received an annual salary increase. Your sister's salary was increased from $24,000 to $25,800 and her co-worker's salary was increased from $27,000 to $28,944.

29. Find the percent of increase in your sister's salary and her co-worker's salary.

30. Who received the larger raise? Who received the larger percent raise?

Integrated Review

Making Connections within Mathematics

31. *Estimation* Which percent best approximates the increase in the United States population each year?

a. 1% **b.** 10% **c.** 25% **d.** 50%

32. *Estimation* Which percent best approximates the decrease in the value of a dollar each year?

a. 6% **b.** 26% **c.** 46% **d.** 66%

33. What is 75% of 240?

34. 72 is what percent of 180?

35. 2.8 is 50% of what number?

36. 234.5 is 35% of what number?

Exploration and Extension

Geometry **In Exercises 37–40, use graph paper or dot paper to draw the indicated figure. Each small square in the figure has an area of one square unit.**

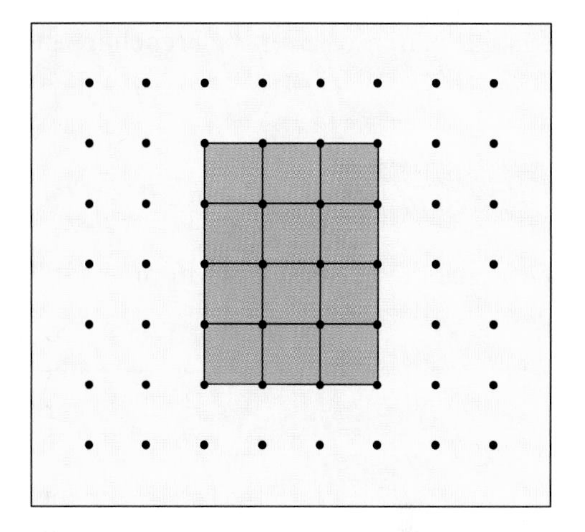

37. Draw a figure whose area is 25% less than the area of the figure at the right.

38. Draw a figure whose area is $66\frac{2}{3}\%$ less than the area of the figure at the right.

39. Draw a figure whose area is 75% greater than the area of the figure at the right.

40. Draw a figure whose area is $33\frac{1}{3}\%$ greater than the area of the figure at the right.

Mixed REVIEW

In Exercises 1–4, use the spinner at the right. All the divisions are the same size. (5.8)

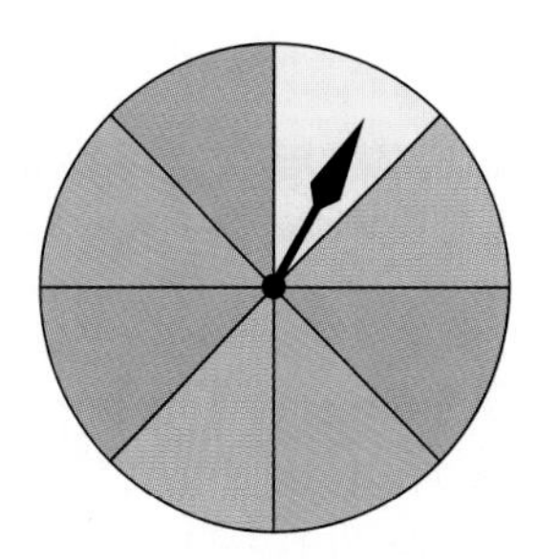

1. Find the probability that the spinner will stop on red.

2. Find the probability that the spinner will stop on blue.

3. Find the probability that the spinner will stop on green.

4. Find the probability that the spinner will stop on yellow.

In Exercises 5–8, evaluate the expression when $a = -8$ and $b = 2$. (3.1)

5. $|ab|$ **6.** $|a^2|$ **7.** $|a| - |b|$ **8.** $|a - b|$

9. On a number line, plot the integers 4, -3, -2, and 0. **(3.1)**

10. Order the numbers $\frac{1}{2}$, $\frac{1}{3}$, $\frac{5}{8}$, $\frac{3}{7}$, and $\frac{4}{9}$ from least to greatest. **(6.5, 7.7)**

11. What is the average of -1.02, -0.98, -1.01, -1, and -1.04? **(3.6)**

In Exercises 12–15, simplify the expression, if possible. Then evaluate the expression when $x = 5$ and $y = -4$. (3.3)

12. $3x + 2y + x$ **13.** $4x - 5y$ **14.** $xy - 2y$ **15.** $10x + 20y + 2x$

In Exercises 16–19, solve the percent problem. (8.4)

16. What is 35% of 80? **17.** 32 is 45% of what number?

18. 190 is what percent of 310? **19.** What is 116% of 92?

LESSON INVESTIGATION 8.7

Pascal's Triangle

In this activity, you will explore the patterns in Pascal's Triangle. This triangle is named after the French mathematician Blaise Pascal (1623–1662).

Example ***Exploring Pascal's Triangle***

Describe the pattern of the numbers in Pascal's Triangle. Then use the pattern to write the 8th row.

0th Row — 1

1st Row — 1 1

2nd Row — 1 2 1

3rd Row — 1 3 3 1

4th Row — 1 4 6 4 1

5th Row — 1 5 10 10 5 1

6th Row — 1 6 15 20 15 6 1

7th Row — 1 7 21 35 35 21 7 1

Solution The pattern of the numbers is as follows: The first and last numbers in each row are always 1, and every other number is the sum of the two numbers that lie directly above it. (For instance, 7 is the sum of 1 and 6. Similarly, 21 is the sum of 6 and 15.) Using this pattern, you can write the 8th row as follows.

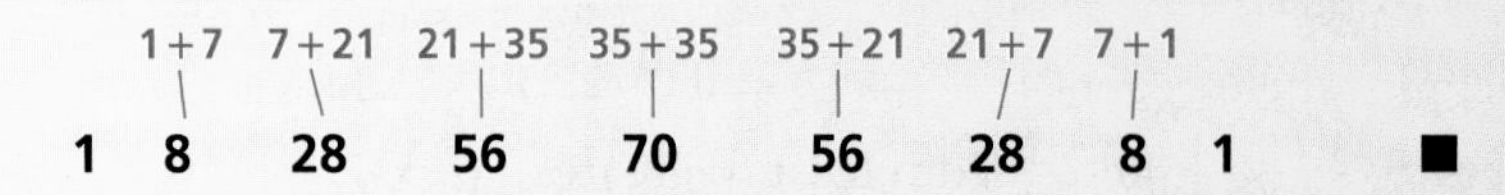

■

Exercises

1. Write the 9th row of Pascal's Triangle.
2. Write the 10th row of Pascal's Triangle.
3. Find the sum of the numbers in each of the first 8 rows of Pascal's Triangle. Then copy and complete the table.

Row	0	1	2	3	4	5	6	7
Sum	?	?	?	?	?	?	?	?

4. Describe the pattern in the table. Use your pattern to predict the sum of the 8th row. Then find the sum to confirm your prediction.

8.7 The Counting Principle and Probability

What you should learn:

Goal 1 How to the Counting Principle

Goal 2 How touse Pascal's Triangle to count the number of ways an event can happen

Why you should learn it:

You can use the counting principle to solve real-life problems, such as finding the number of different committees that can be chosen from a group.

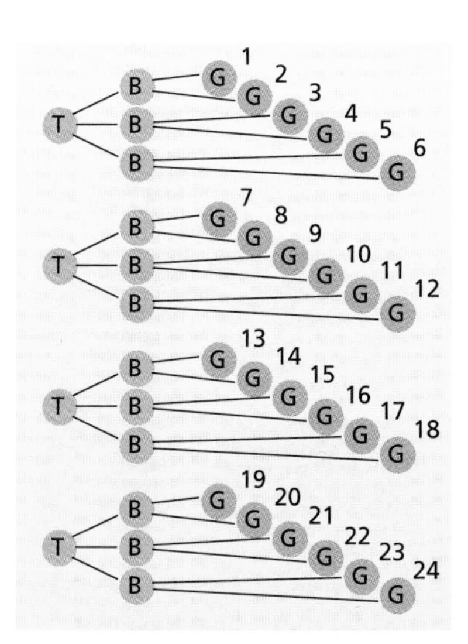

This tree diagram shows that 1 teacher, 1 boy, and 1 girl can be chosen from 4 teachers, 3 boys, and 2 girls in 24 different ways.

Goal 1 Using the Counting Principle

Suppose there are 30 people in your class, and each person has exactly 5 books. Then the total number of books is $30 \cdot 5$ or 150 books. This is an example of the **Counting Principle.**

The Counting Principle

If one event can occur in m ways and another event can occur in n ways, then the two events can occur in mn ways.

Example 1 *Forming a Committee*

The math club in your school has 5 officers: 3 boys and 2 girls. You are forming a committee of two officers to visit the math club at another school. You want the committee to have 1 boy and 1 girl. How many different committees are possible? Donna is one of the officers. What is the probability that she will be on the committee?

Solution With small numbers like this, you can find the sample space by listing the different committees.

Six different committees are possible. You can check this with the Counting Principle. Because you can choose a boy in 3 ways and a girl in 2 ways, it follows that you can choose 1 boy and 1 girl in $3 \cdot 2$ or 6 ways. If each committee is equally likely, then the probability that Donna will be chosen is

$$\frac{\text{Number of committees Donna is on}}{\text{Total number of committees}} = \frac{3}{6} = \frac{1}{2}.$$ ■

The Counting Principle can be applied to three or more events. For instance, suppose your school has 4 math teachers, and the committee is to consist of 1 math teacher, 1 boy, and 1 girl. Then the number of possible committees is $4 \cdot 3 \cdot 2$ or 24, as shown by the tree diagram at the left.

Goal 2 Using Pascal's Triangle

Real Life **Social Studies**

Almost all organizations such as schools, corporations, and governments use committees to help make decisions for the organization.

Example 2 *Forming a Committee*

After finding the number of ways that 1 boy and 1 girl can be chosen to form the committee in Example 1, someone in the math club objects. The person thinks that *any* two of the officers should be able to be on the committee—even if the two are both boys or both girls. How many ways can you choose 2 committee members out of the 5 officers?

Solution As in Example 1, you can find the sample space by listing the different committees.

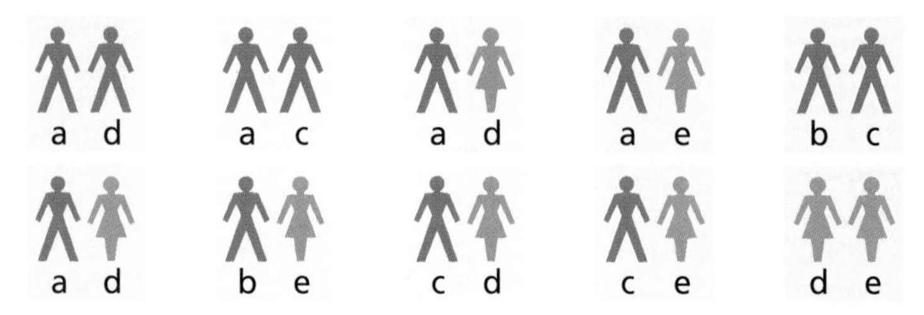

Ten committees are possible. ■

You can use the 5th row of Pascal's Triangle (see page 372) to find the number of ways you can choose 2 people from a group of 5 people.

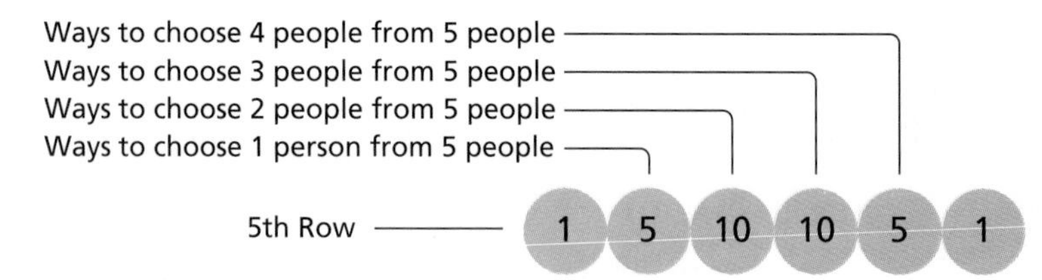

Communicating about MATHEMATICS

Cooperative Learning

SHARING IDEAS about the Lesson

Using Pascal's Triangle Work with a partner. Use Pascal's Triangle to find the number of committees that can be formed. Then verify your answer by actually listing the different committees.

A. Choose a committee of 2 from 6 people.

B. Choose a committee of 3 from 6 people.

C. Choose a committee of 2 from 7 people.

EXERCISES

Guided Practice

CHECK for Understanding

1. *Counting Principle* You have two pair of cutoffs and 4 T-shirts. Use the Counting Principle to determine how many different outfits you can wear by choosing 1 pair of cutoffs and 1 T-shirt.

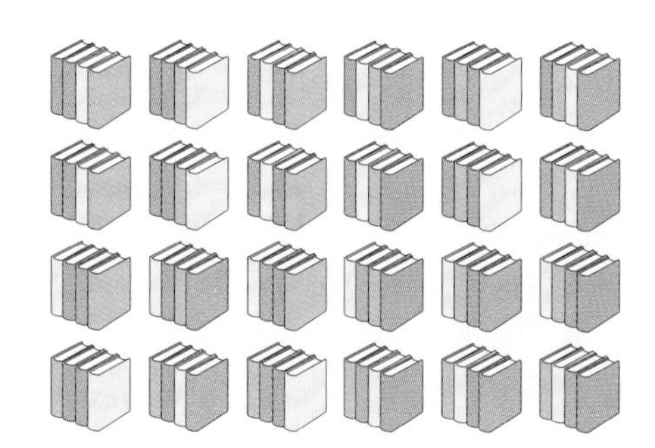

2. *Counting Principle* The diagram at the right shows how many ways you can order four books on a shelf. Explain how to use the Counting Principle to find the number of ways you can order the books.

3. *Using a Tree Diagram* You are writing a three-digit number. The first digit must be 1 or 7, the second digit must be 3, 6, or 9, and the number must be divisible by 5. Use a tree diagram to find how many numbers are possible.

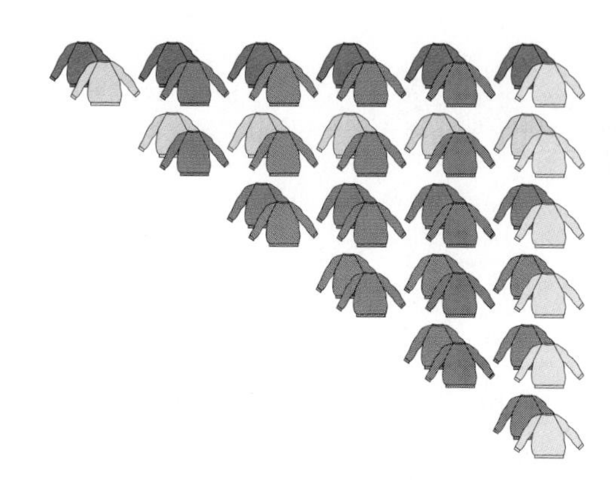

4. *Using Pascal's Triangle* The diagram at the right shows how many ways you can choose two sweaters from a group of 7 sweaters. Explain how to use Pascal's Triangle to find how many ways you can choose two sweaters from 7 sweaters.

Independent Practice

5. *Yogurt Sundae* A yogurt shop has 6 flavors of yogurt and 5 toppings. Each sundae consists of 1 yogurt flavor and 1 topping. Use the Counting Principle to find how many different sundaes are possible. Then confirm your answer by listing the different sundaes.

6. *Menu Choices* The menu at the right shows the choices of side dishes that come with each dinner. You are to choose 1 vegetable and 1 rice or potato. Use the Counting Principle to find how many different choices you have. Then confirm your answer by listing the different choices.

7. *Probability* In Exercise 6, suppose you are asked to make the choices for someone else's dinner. Without knowing their preferences, what is the probability that you will chose exactly the items they want?

8. *Taking a Test* You are taking a test with 8 questions. Each question must be answered true or false. Which of the following represents the number of ways you can answer the 8 questions? Explain.

a. $2 \cdot 2 \cdot 2 \cdot 2 \cdot 2 \cdot 2 \cdot 2 \cdot 2 = 2^8$ **b.** $2 + 2 + 2 + 2 + 2 + 2 + 2 + 2 = 8(2)$

9. *Standing in Line* In how many different orders can 2 people stand in line? In how many different orders can 3 people stand in line? In how many different orders can 4 people stand in line? Describe the pattern. Find the number of orders that 7 people can stand in line.

10. *Shopping* You are buying a sweatshirt. You have a choice of a pullover, button-down, or hooded sweatshirt. Each style comes in red, white, blue, gray, green, or purple.

a. How many different sweatshirts can you choose from?

b. Verify your answer with a tree diagram.

11. *Board Game* You are playing a board game and must spin each of four spinners. How many different arrangements can the spinners show?

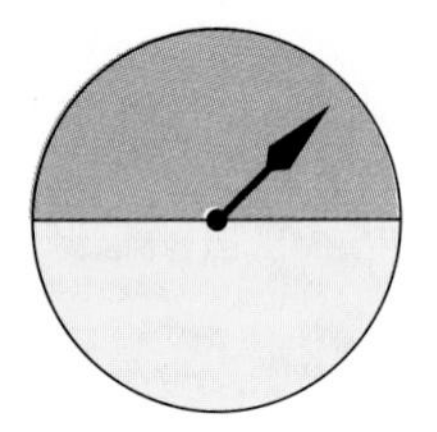
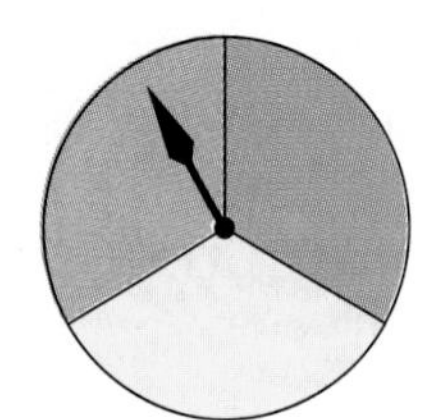
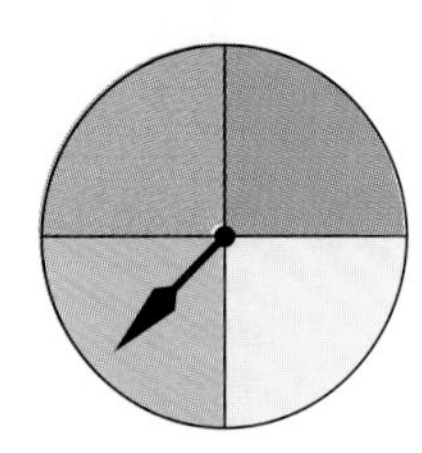
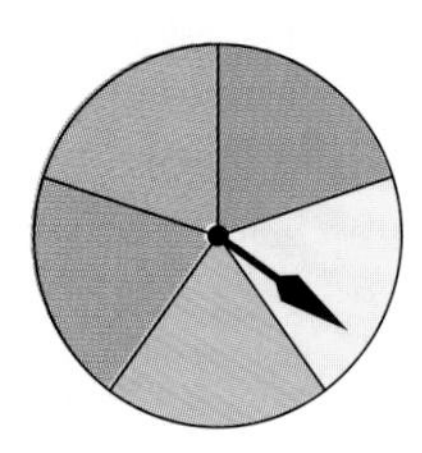

Pascal's Triangle **In Exercises 12–15, use Pascal's Triangle to find the number of ways to choose the following.**

12. Choose 3 books from 7 books.

13. Choose 4 pencils from 8 pencils.

14. Choose 5 CDs from 6 CDs.

15. Choose 3 photos from 9 photos.

Integrated Review

Making Connections within Mathematics

Probability **In Exercises 16–19, consider a school bus that has 3 seventh-grade students, 5 eighth-grade students, 4 ninth-grade students, and 6 tenth-grade students. One of the students is chosen. What is the probability that the student is in the indicated grade?**

16. Seventh grade **17.** Eighth grade **18.** Ninth grade **19.** Tenth grade

Exploration and Extension

Factorials **In Exercises 20–23, evaluate the factorial.**

The product of the whole numbers from 1 to n is called ***n*-factorial** and is denoted by the symbol $n!$. For instance, $3! = 3 \cdot 2 \cdot 1 = 6$.

20. $4!$ **21.** $5!$ **22.** $6!$ **23.** $7!$

24. In Exercise 11, express the number of arrangements as a factorial. Explain.

USING A COMPUTER

Modeling Probability with a Computer

There are two basic ways to compute the probability of an event: experimental and theoretical. The experimental approach often involves a **simulation,** which is an experiment that models a real-life situation.

Example *Using a Computer Simulation*

You are making a long distance telephone call to a friend. You remember your friend's number, but all you remember about the 3-digit area code is that the first digit is 2. If you randomly choose the second and third digits, what is the probability that the area code is correct? (The actual area code is 206.)

Solution One way to answer this question is to use a computer simulation, as shown below. Each time the program is run, it selects 1000 random numbers, each of the form 2??. The program then counts the number of times the area code 206 was chosen. To test the simulation, we ran the program 100 times. The results are shown in the line plot. The exercises below ask you to analyze these results.

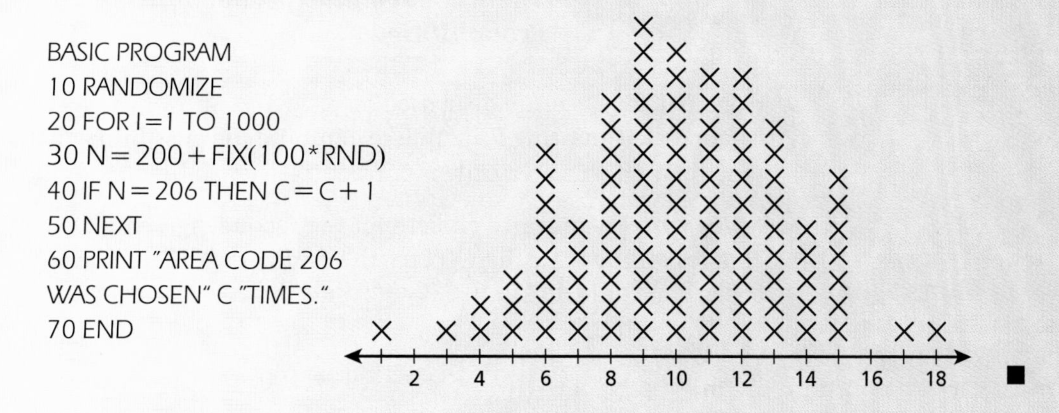

■

Exercises

1. For each of the 100 times the program was run, it selected 1000 numbers. How many numbers did it select all together?
2. Use the line plot to determine how many times the program selected the correct area code.
3. Use the results of Exercises 1 and 2 to find the experimental probability that the correct area code is selected.

8.8 Probability and Simulations

Goal 1 Compare theoretical and experimental probabilities

Goal 2 How to find the probability of a multistage event

You can use probability to solve real-life problems, such as finding the likelihood of dialing a correct telephone number.

Real Life
Telephone Area Codes

Goal 1 Comparing Probabilities

In the *Lesson Investigation* on page 377, you used a computer to *simulate* choosing an area code.

Using a computer is only one way to simulate an event. There are many other ways. For instance, you could use pieces of paper. Can you see how?

Describe a simulation for Example 1 that uses 10 pieces of paper

Example 1 *Theoretical and Experimental Probabilities*

Consider the question discussed on page 377. Show how you can use the Counting Principle to compute the possibility that a randomly chosen area code is correct.

Solution You have 10 choices for the second digit and 10 choices for the third digit. Thus, the number of different area codes you could dial is

$$\text{Choices for 2nd digit} \cdot \text{Choices for 3rd digit} = 10 \cdot 10 = 100.$$

Because only one of these area codes is correct, the probability that you will dial the correct area code is

$$\text{Probability} = \frac{1}{100} = 0.01.$$

This theoretical probability is close to the result that was obtained with the computer simulation on page 377. ■

Suppose that you remembered that the second and third digits are both even numbers. Would you be twice as likely to dial the correct area code or four times as likely?

200	201	202	203	204	205	206	207
208	209	210	211	212	213	214	215
216	217	218	219	220	221	222	223
224	225	226	227	228	229	230	231
232	233	234	235	236	237	238	239
240	241	242	243	244	245	246	247
248	249	250	251	252	253	254	255
256	257	258	259	260	261	262	263
264	265	266	267	268	269	270	271
272	273	274	275	276	277	278	279
280	281	282	283	284	285	286	287
288	289	290	291	292	293	294	295
296	297	298	299				

There are 100 3-digit area codes that begin with 2. The ones shown in blue have even second and third digits.

Goal 2 The Probability of a Multistage Event

In 1990, there were about 143 million automobiles in the United States. Of these, 17 million were in California. (Source: Federal Highway Commission)

Example 2 *Finding a Probability*

You live in a state whose automobile license plates have 3 letters followed by 3 digits.

a. How many different license plate "numbers" are possible?

b. Your license number is DBX-281. If one of the possible license-plate numbers is chosen at random, what is the probability that the number is yours?

Solution

a. For each of the three letters, there are 26 choices. For each of the three digits, there are 10 choices. Thus, the number of different license plate numbers is

$$26 \cdot 26 \cdot 26 \cdot 10 \cdot 10 \cdot 10 = 26^3 \cdot 10^3 = 17{,}576{,}000.$$

b. Because there are 17,576,000 different license numbers, the probability that yours is chosen is

$$\text{Probability} = \frac{1}{17{,}576{,}000}.$$

In other words, there is only 1 chance in over 17 million that your number will be chosen! ■

Communicating about MATHEMATICS

SHARING IDEAS about the Lesson

Finding Probabilities Your automobile license number has three letters followed by three digits. In each of the following cases, you have forgotten one or more of the letters or digits. If you choose the letters and numbers at random, what is the probability that you will be correct?

A. Welcome to America's — A?Q-568 — Winter Wonderland

B. M??-420 — Home of the Rockies

C. Settle Down in — BLT-30? — The Sunset State

D. A Tropical Paradise — S?N-4?3 — in Your Own Back Yard

EXERCISES

Guided Practice

CHECK for Understanding

Choosing a Committee **In Exercises 1–4, use the following information.**

The math class in your school has five officers: 3 boys and 2 girls. To form a committee of two officers, you select two names at random. (The possible committees are listed in Example 2 on page 374.)

1. How many committees are possible?
2. What is the probability the committee will consist of one boy and one girl?
3. What is the probability the committee will consist of two boys?
4. What is the probability the committee will consist of two girls?

Independent Practice

Spinning Spinners **In Exercises 5 and 6, use the following information.**

Each of the spinners at the right is spun once. On the first, the arrow is equally likely to land on any letter and on the second, the arrow is equally likely to land on any number.

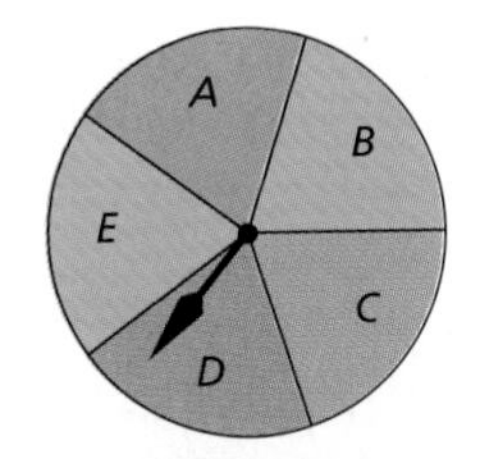

5. How many outcomes are possible?
6. Explain how to find the probability that the first spinner will land on A *and* the second will land on 1.

In Exercises 7–12, find the probability of each outcome.

7. C and 6
8. E and 5
9. A vowel and 4
10. A vowel and an even number
11. A consonant and 2
12. A consonant and an odd number

Tossing a Coin and Rolling a Die **In Exercises 13–17, a coin is tossed, then a six-sided die is rolled. Find the probability of the indicated outcome.**

13. Heads and 5
14. Tails and 1
15. Tails and an even number
16. Heads and a prime number
17. Heads and a number less than 3
18. *Simulation* Toss a coin and roll a six-sided die 40 times. Use the results to find the experimental probabilities for each of the outcomes in Exercises 13–17.

Rolling a Die and Choosing a Card **In Exercises 19–24, a die is rolled, then a card is chosen from a deck of standard playing cards. Find the probability of each outcome.**

19. A six and an ace of hearts

20. A three and a ten of diamonds

21. A two and a king

22. A five and a club

23. An odd number and a queen

24. An even number and a spade

Standard 52-Card Deck

♠ A	♠ K	♠ Q	♠ J	♠ 10	♠ 9	♠ 8	♠ 7	♠ 6	♠ 5	♠ 4	♠ 3	♠ 2
♥ A	♥ K	♥ Q	♥ J	♥ 10	♥ 9	♥ 8	♥ 7	♥ 6	♥ 5	♥ 4	♥ 3	♥ 2
♣ A	♣ K	♣ Q	♣ J	♣ 10	♣ 9	♣ 8	♣ 7	♣ 6	♣ 5	♣ 4	♣ 3	♣ 2
♦ A	♦ K	♦ Q	♦ J	♦ 10	♦ 9	♦ 8	♦ 7	♦ 6	♦ 5	♦ 4	♦ 3	♦ 2

Social Security Number **In Exercises 25–28, digits are chosen at random to complete the Social Security Number, as indicated. Find the probability that the number will be 256-18-9342.**

25. 256-1?-9342

26. 256-??-9342

27. ???-18-9342

28. 256-18-????

Integrated Review

Making Connections within Mathematics

Estimation **In Exercises 29–32, use the graph at the right to estimate the probability that a grandparent, chosen at random, is of the indicated age. The graph shows the ages of grandparents in the United States in 1993.** ***(Source: Doublebase Mediamark Research, Inc.)***

29. 44 or under

30. Between 45 and 54

31. Between 55 and 64

32. 65 or over

Exploration and Extension

Spinning a Spinner **In Exercises 33–38, each spinner is spun once. Find the probability of the indicated result.**

33. Blue, A, 4

34. Green, a vowel, 3

35. Yellow, C, an odd number

36. Orange, a consonant, 1

37. Green, a consonant, an even number

38. Yellow, a vowel, a number less than 5

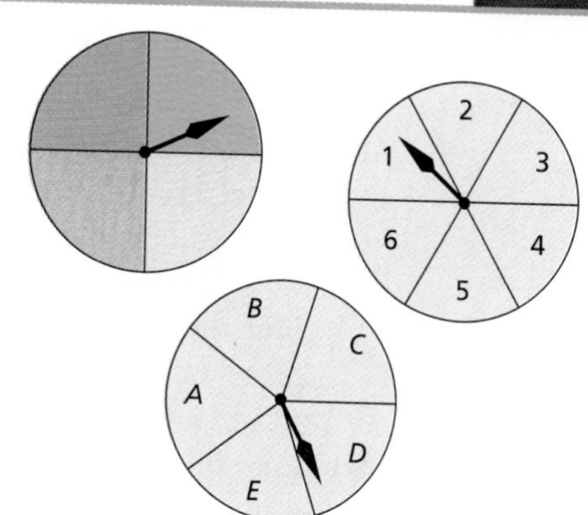

8

Chapter Summary

What did you learn?

Skills

1. Find a rate of one quantity per another quantity. **(8.1)**
2. Find a ratio of one quantity to another. **(8.1)**
3. Write and solve a proportion. **(8.2)**
   - Use similar triangles to write a proportion. **(8.2, 8.3)**
4. Solve a percent equation.
   - a is what percent of b? **(8.4)**
   - What is p percent of b? **(8.4)**
   - a is p percent of what? **(8.4)**
5. Use percents to solve discount problems. **(8.5)**
6. Find a percent of increase or decrease. **(8.6)**
7. Find the theoretical probability of an event
   - by using the Counting Principle. **(8.7)**
   - by using simulations as experimental models. **(8.8)**
8. Use Pascal's Triangle to find the number of ways an event can occur. **(8.7)**

Problem-Solving Strategies

9. Model and solve real-life problems. **(8.1–8.8)**

Exploring Data

10. Use tables and graphs to solve problems. **(8.1–8.8)**

Why did you learn it?

You can use rates, ratios, proportions, percents, and probabilities to answer questions about real-life situations. For instance, in this chapter, you saw how rates can be used to compare unit prices and determine which of two products is a better bargain. You also saw how ratios and proportions can be used to estimate the height of a building, how percents can be used to find the price of an item that is on sale, and how probability can be used to find the likelihood of dialing a forgotten area code.

How does it fit into the bigger picture of mathematics?

In this chapter, you learned that when you are using rational numbers to model a real-life situation, it is important to know the units of measure. For instance, if a is measured in miles and b is measured in hours, then the rate $\frac{a}{b}$ is measured in miles per hour. You also learned that when a and b have the same units of measure, then the ratio $\frac{a}{b}$ has no units of measure. Percents and probabilities are examples of ratios. Knowing this can help you check the units in your answers. For instance, 25% of 8 *dollars* is $0.25 \cdot 8$ or 2 *dollars*. Multiplying 8 dollars by 0.25 doesn't change the units of measure because 0.25 has no units of measure.

Chapter REVIEW

In Exercises 1–6, match the term with its description. (8.1–8.7)

a. Ratio
b. Counting Principle
c. Rate
d. Reciprocal Property
e. Percent Equation
f. Proportion

1. An equation that relates two ratios

2. If one event can occur in m ways and another can occur in n ways, then the two can occur in mn ways.

3. $\frac{a}{b} = \frac{p}{100}$

4. If $\frac{a}{b} = \frac{c}{d}$, then $\frac{b}{a} = \frac{d}{c}$.

5. $\frac{a}{b}$ where a and b have the same units of measure

6. $\frac{a}{b}$ where a and b have different units of measure

In Exercises 7–10, decide whether the quotient is a rate or a ratio. Then simplify. (8.1)

7. $\frac{51 \text{ shoes}}{9 \text{ shoes}}$
8. $\frac{25 \text{ boys}}{30 \text{ girls}}$
9. $\frac{184 \text{ miles}}{4 \text{ hours}}$
10. $\frac{14 \text{ bikes}}{70 \text{ bikes}}$

In Exercises 11–14, write the verbal phrase as a rate or a ratio. State whether the result is a rate or a ratio. (8.1)

11. 27 out of 50 people surveyed
12. 5 out of 6 students participating
13. Jogged 13 miles in 2 days
14. Solved 75 problems in 2 days

In Exercises 15–22, solve the proportion. Check your solution. (8.2)

15. $\frac{x}{4} = \frac{5}{8}$
16. $\frac{7}{9} = \frac{y}{3}$
17. $\frac{2}{z} = \frac{11}{6}$
18. $\frac{5}{2} = \frac{13}{a}$
19. $\frac{b}{8} = \frac{7}{11}$
20. $\frac{4}{15} = \frac{s}{2}$
21. $\frac{10}{3} = \frac{12}{t}$
22. $\frac{6}{7} = \frac{9}{x}$

In Exercises 23 and 24, the triangles are similar. Solve for x. (8.2)

23.

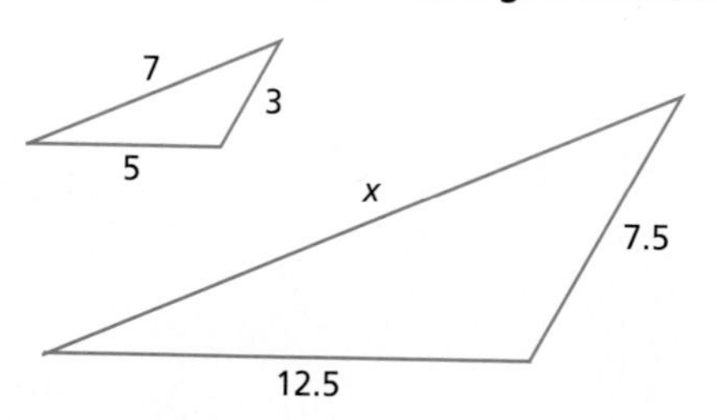

24.

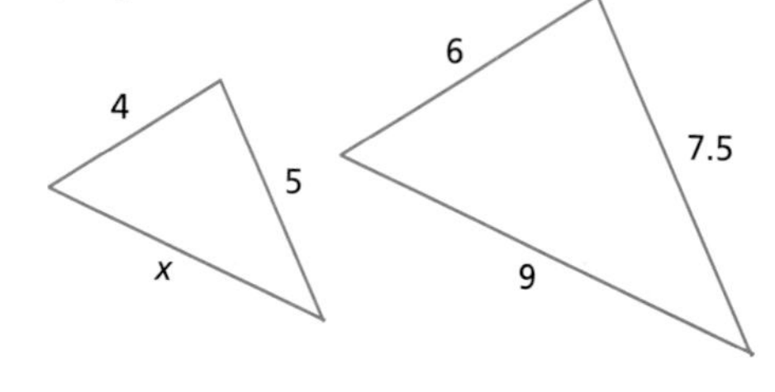

In Exercises 25–30, solve the percent equation. (8.4)

25. 35 is what percent of 40?
26. What is 68 percent of 22?
27. 44 is 25% of what number?
28. 55 is what percent of 25?
29. What is 86 percent of 7.5?
30. 86.4 is 64% of what number?

In Exercises 31–33, decide whether the change represents a percent increase or a percent decrease. Then find the percent. (8.6)

31. Before: 15, After: 21
32. Before: 112, After: 56
33. Before: 250, After: 325

In Exercises 34–36, decide whether the two quantities represent a percent of increase or a percent of decrease and find the percent. (8.6)

34. Monday: 845 people Next day: 169 people

35. Beginning Balance: $740.20 Ending Balance: $777.21

36. Regular Price: $65.50 Sale Price: $45.85

Error Analysis **In Exercises 37–39, find and correct the error. (8.2, 8.4, 8.5)**

37.

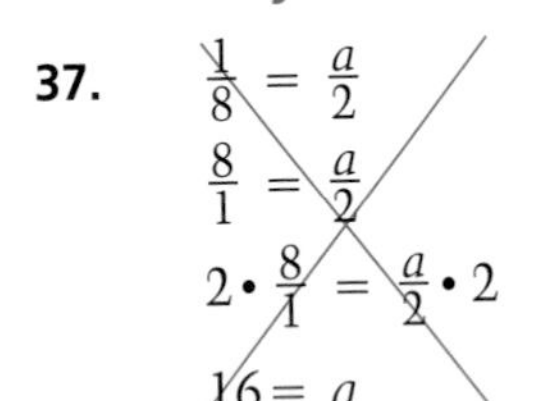

38. Sale price: $35

Discount: 25%

Regular price:

~~$35 + 35(0.25) = \$43.75$~~

39. Problem:

55 is 44% of what?

Percent equation:

~~$\frac{a}{55} = \frac{44}{100}$~~

40. What percent of 120 is 25% of 25%? **(8.6)**

41. What percent of 120 is 50% of 50%? **(8.6)**

Tossing Two Dice **In Exercises 42–53, use the following information.**

When two 6-sided dice are tossed, 36 different outcomes can occur. These outcomes are shown at the right. Use the table to find the probability that the indicated total is tossed.

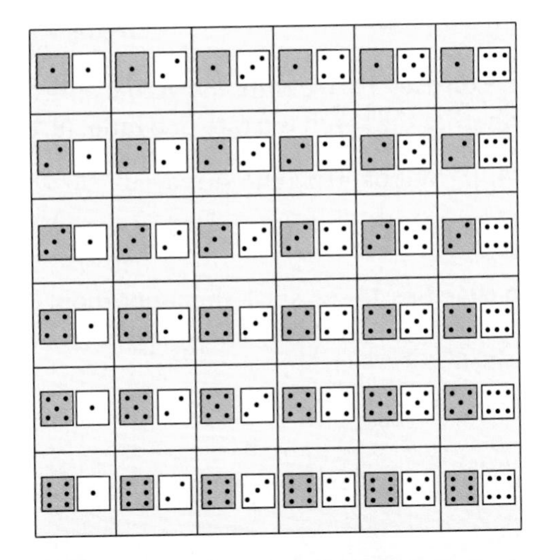

42. A total of 2

43. A total of 3

44. A total of 4

45. A total of 5

46. A total of 6

47. A total of 7

48. A total of 8

49. A total of 9

50. A total of 10

51. A total of 11

52. A total of 12

53. Graph your results.

Computer Passwords **In Exercises 54–57, use the following information.**

Each of the computer passwords below consists of 5 letters followed by 3 digits. You randomly choose letters or digits to complete each password. What is the probability that you will obtain the correct password? **(8.8)**

54. M L K ? P 1 2 ?

55. N ? A C ? 4 8 7

56. ? Z X W A ? ? 7

57. B ? F ? H ? 3 ?

The horizontal and vertical decoder bands on computer microchips act like coordinate axes dividing the memory cells into 4 quadrants.

Playing Basketball **In Exercises 58 and 59, use the diagram at the right. In the diagram, the sun is creating two similar triangles. (8.2, 8.3)**

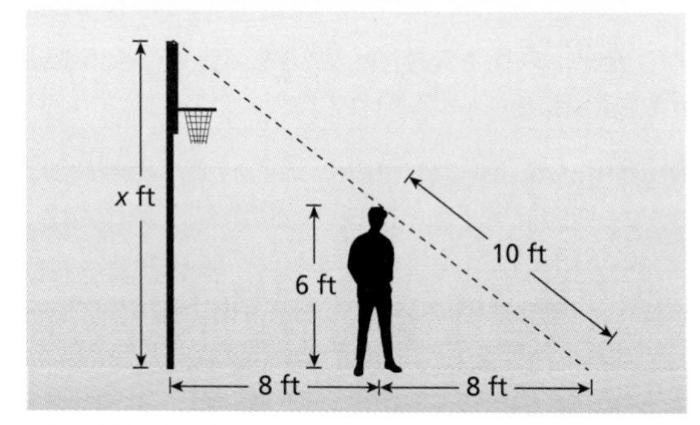

58. Explain why the following proportion is valid.

$$\frac{x}{8+8} = \frac{6}{8}$$

59. What is the height of the backboard?

In Exercises 60–62, rewrite the quotient as a ratio. Then simplify. (8.1)

60. $\frac{\text{2 feet}}{\text{10 inches}}$

61. $\frac{\text{8 ounces}}{\text{2 pounds}}$

62. 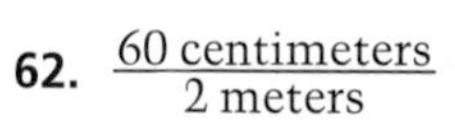 $\frac{\text{60 centimeters}}{\text{2 meters}}$

63. *Discount* A sign in a store states that all baseball hats are 45% off. The regular price is \$15.00. What is the sale price of the hat? **(8.5)**

64. *Discount* A sign in a store states that all baseball hats are 45% off. The discount is \$9.36. What was the regular price of the hat? **(8.5)**

65. *Beach Volleyball* Using the caption at the right, find the percent increase in the number of volleyball players from 1989 to 1992. **(8.6)**

In 1989, 10.3 million Americans played beach volleyball. In 1992, 13.2 million Americans played beach volleyball. (Source: Sporting Goods Manufacturers Association)

Pascal's Triangle **In Exercises 66 and 67, use Pascal's Triangle to find the number of teams that can be formed. Then verify your answer by actually listing the different teams. (8.7)**

66. Choose a team of 3 from 5 people.

67. Choose a team of 5 from 6 people.

Estimation **In Exercises 68–71, use the graph at the right, which shows by percents what happened to the 242 million automobile tires that were discarded in 1990 in the United States.** *(Source: National Solid Wastes Management Associates)*

68. About how many tires were exported?

69. About how many tires were recycled?

70. About how many tires were burned for energy?

71. If you randomly selected a discarded tire in 1990, what is the probability that it was dumped in a landfill or dumped illegally?

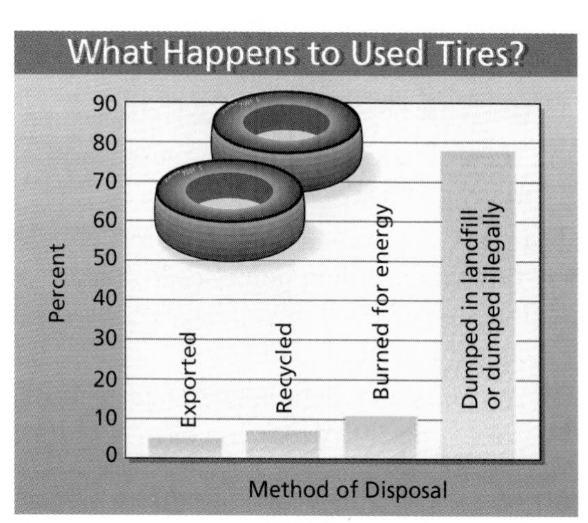

Modems **In Exercises 72–76, use the following information.**

Information in a computer is stored in units called bytes. A byte is made up of 8 bits. A bit is the smallest unit of information that can be stored. A modem allows two computers to communicate over a telephone line.

72. Some modems can transfer data at 14,400 bits per second. Is that a rate or a ratio? Explain.

73. How many bits of information are there in 36 bytes?

74. How many bytes are needed to store 776 bits?

75. A 10,000-byte file, which could be four pages of text, is to be transferred over a 1200-bit-per-second modem. How many seconds will the transfer take?

76. A 10,000-byte file is transferred over a 9600-bit-per-second modem. How many seconds will the transfer take?

A modem converts digital signals from a sending computer into analog signals that can travel over telephone lines. A modem attached to a receiving computer converts the analog signals back to digital signals.

Binary Codes **In Exercises 77–82, use the following information.**

Binary codes use the numbers zero and one in different combinations to represent symbols such as letters, numbers, and punctuation. The diagram at the right shows how many different codes are possible with one, two, three, or four digits.

77. The *Baudot binary system* uses 5 digits. How many codes are possible in this system?

78. The *BCD binary system* uses 6 digits. How many codes are possible in this system?

79. The *ASCII binary system* uses 7 digits. How many codes are possible in this system?

80. The *EBCDIC binary system* uses 8 digits. How many codes are possible in this system?

81. In ASCII, the code for the letter A is "1000001." You enter a 7-digit binary code at random. What is the probability that you have entered the ASCII code for the letter A?

82. You enter an 8-digit binary code at random. What is the probability that you have entered the EBCDIC code for the letter A?

Possible 1-Digit Codes
0, 1

Possible 2-Digit Codes
00, 01, 10, 11

Possible 3-Digit Codes
000, 001, 010, 011,
100, 101, 110, 111

Possible 4-Digit Codes
0000, 0001, 0010, 0011, 0100, 0101
0110, 0111, 1000, 1001, 1010, 1011
1100, 1101, 1110, 1111

In Exercises 1–4, use the similar triangles at the right.

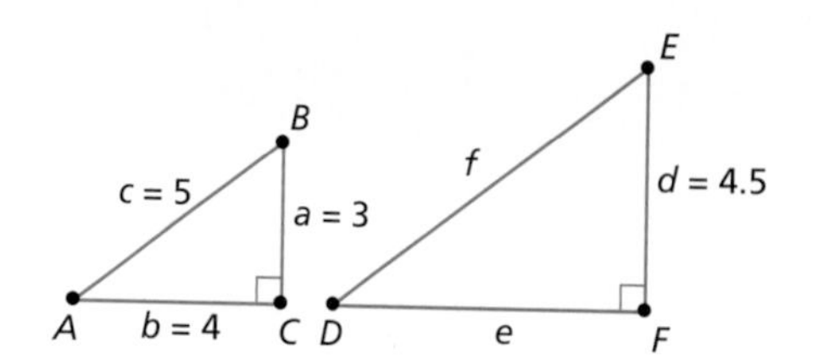

1. Find the ratio of a to d.
2. Solve for f.
3. Solve for e.
4. Find the ratio of b to e.

In Exercises 5 and 6, find the percent increase or decrease.

5. Before: 76 After: 95
6. Before: \$15.00 After: \$9.00
7. You rent a car for 5 days for \$195. What is the daily rental rate?
8. What is 82% of 115?
9. 56 is what percent of 70?

In Exercises 10–13, use the rectangle at the right. Each small rectangle is the same size.

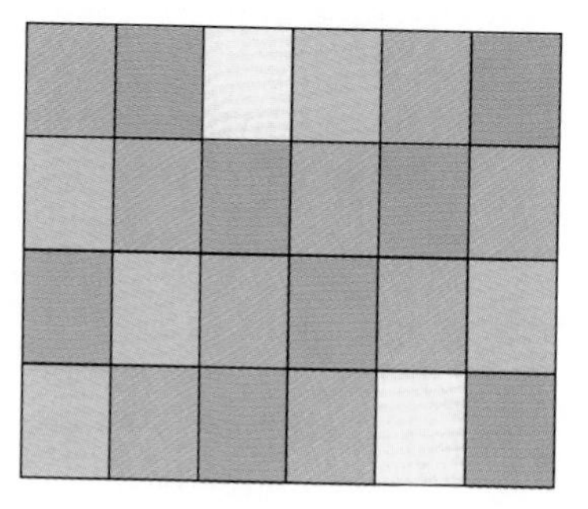

10. What percent of the large rectangle is red?
11. What percent of the large rectangle is yellow?
12. If one of the small rectangles is chosen at random, what is the probability that it will be green?
13. If one of the small rectangles is chosen at random, what is the probability that it will not be red?

In Exercises 14–20, use the given information.

Shadows A building is casting a 100-foot shadow at the same time that a 5-foot post is casting a 1-foot shadow.

14. Draw a diagram that shows the building, the post, and the two shadows.
15. Write a proportion that involves the height of the building, the height of the post, and the lengths of the two shadows.
16. Solve the proportion to find the height of the building.

Committees A two-person committee is to be chosen from 3 seventh-grade students and 3 eighth-grade students. The committee must have one student from each grade.

17. List the different committees that are possible.
18. Use the Counting Principle to confirm your list.

Birth Rates In 1990, about 10,500 women gave birth in the United States. Of these, 217 women gave birth to twins and 5 women gave birth to triplets.

19. Estimate the chance of having twins.
20. Estimate the chance of having triplets.

CHAPTER

9

Real Numbers and Inequalities

LESSONS

Solar photovoltaic generators of electricity, like the one shown here in California, have an efficiency rating of about 15% and currently cost between 25¢ and 35¢ per electrical unit. This is double the cost of nuclear power plants and 5 to 10 times more costly than traditional power plants.

Real Life
Solar Energy

December 21 or 22

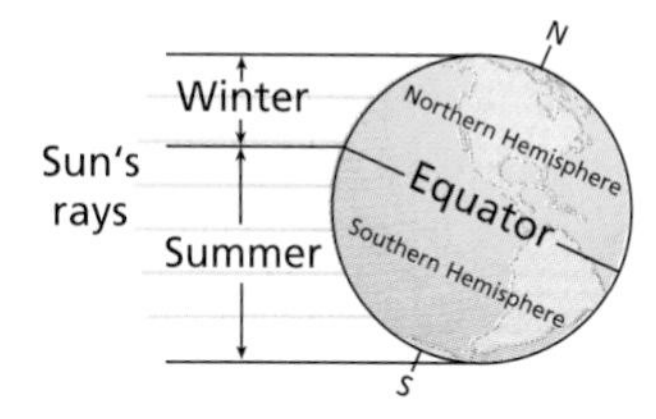

June 20 or 21

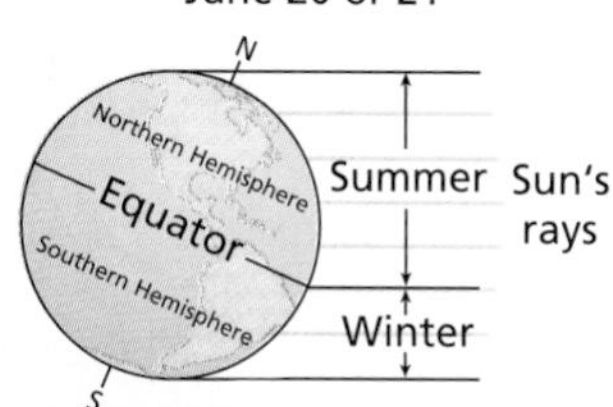

Seasonal changes affect the efficiency of solar panels. The closer to the vertical that the sun's rays strike the panel, the more solar energy they can produce. Since the Earth's axis is tilted 23.4° from the vertical, countries near the equator have the best potential for developing solar energy year round. As you move north or south from the equator, the sun's intensity decreases during the winter. In the United States, the only year-round solar energy plants are located in the Southwest.

The distance from the sun to Earth changes seasonally. (It is actually the least during our winter.) A verbal model for this information is 1.47×10^8 km is less than the distance to the sun is less than 1.52×10^8 km.

In this chapter you will learn to write inequalities and find all their real-number solutions.

9.1 Exploring Square Roots

What you should learn:

How to solve equations whose solutions are square roots

How to use square roots to solve real-life problems

Why you should learn it:

You can use square roots to solve real-life problems, such as finding the dimensions of a painting.

Goal 1 Using the Square Root Property

The square at the right has an area of 16 square units. Because each side has a length of x units, it follows that

$$x^2 = 16.$$

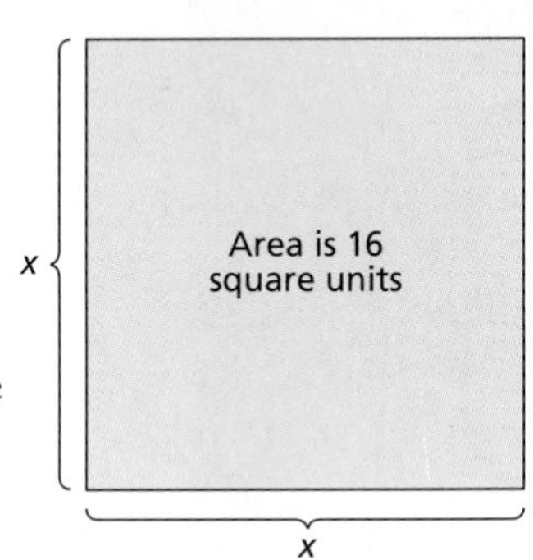

There are two numbers whose square is 16: -4 and 4. Because length must be positive, each side of the square has a length of 4 units.

Square Root Property

If a is a positive number, then $x^2 = a$ has two solutions.

1. $x = -\sqrt{a}$ is a solution because $(-\sqrt{a})^2 = a$.
2. $x = \sqrt{a}$ is a solution because $(\sqrt{a})^2 = a$.

Need to Know

If a is a perfect square, then the two solutions of $x^2 = a$ can be written without using a square root symbol. For instance, the two solutions of $x^2 = 1.44$ are -1.2 and 1.2.

If a is not a perfect square, then the two solutions of $x^2 = a$ should be written with square root symbols. For instance, the two solutions of $x^2 = 10$ are $-\sqrt{10}$ and $\sqrt{10}$.

Example 1 *Using Square Roots to Solve Equations*

a. $t^2 = 25$ — *Original equation*

$t = -\sqrt{25}$ — *$-\sqrt{25}$ or -5 is one solution.*

$t = \sqrt{25}$ — *$\sqrt{25}$ or 5 is the other solution.*

There are two solutions: -5 and 5. You can check these by observing that $(-5)^2 = 25$ and $5^2 = 25$.

b. $x^2 + 2 = 11$ — *Original equation*

$x^2 + 2 - 2 = 11 - 2$ — *Subtract 2 from each side.*

$x^2 = 9$ — *Simplify.*

$t = -\sqrt{9}$ — *$-\sqrt{9}$ or -3 is one solution.*

$t = \sqrt{9}$ — *$\sqrt{9}$ or 3 is the other solution.*

There are two solutions: -3 and 3. You can check these by observing that $(-3)^2 + 2 = 11$ and $3^2 + 2 = 11$. ■

Goal 2 Modeling Real-Life Problems

Example 2 *Finding the Dimensions of a Painting*

Real Life
Art

The square painting, "Falling Star," has an area of 3600 square inches. What are the dimensions of the painting?

Solution

Verbal Model Area of Square = (length of side)^2

↓

Labels Area of square = 3600 (square inches)
Length of side = s (inches)

↓

Algebraic Model

$3600 = s^2$ *Write algebraic model.*

$60 = s$ *Choose positive square root.*

The length of each side of the painting is 60 inches. You can check this by squaring 60 to obtain 60^2 or 3600. ■

"Falling Star" was painted by Robert Orduño, a native American artist. It depicts the story from Sioux mythology of Wohṗe, the only daughter of the Great Spirit WakanTanka, falling from the sky.

In Example 2, notice that s^2 is in *square* inches, which means that s is measured in inches. Here are two other examples.

1. If a square parking lot has an area of 120 *square meters,* then the length of each side of the parking lot is $\sqrt{120}$ *meters* or about 10.95 meters.
2. If a square book cover has an area of 400 *square centimeters,* then the length of each side of the cover is 20 *centimeters.*

Communicating about MATHEMATICS

SHARING IDEAS about the Lesson

Estimation The small squares of the graph paper are each 1 square unit. Estimate the side lengths of each larger square. Explain your reasoning. Check your estimate with a calculator.

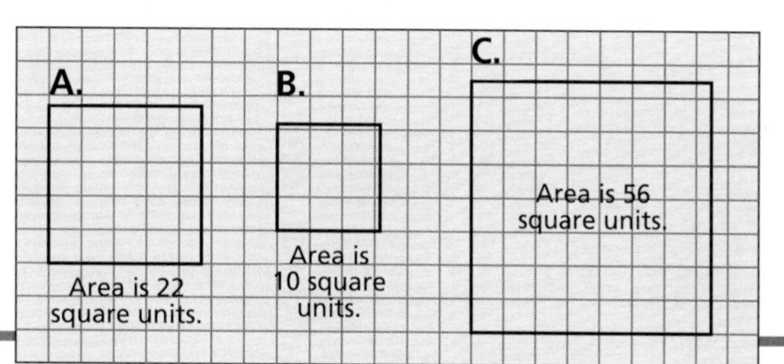

EXERCISES

Guided Practice

CHECK for Understanding

1. Explain why $r^2 = 4$ has two solutions.

In Exercises 2–5, write both square roots of the number. State whether the number is a perfect square.

2. 49 **3.** 5 **4.** 1.21 **5.** $\frac{25}{4}$

6. Solve the equation $x^2 = 14$. Use a calculator to approximate the solutions to three decimal places.

Independent Practice

In Exercises 7–14, write both square roots of the number.

7. 14 **8.** 27 **9.** 64 **10.** 169

11. 0.36 **12.** 1600 **13.** $\frac{64}{9}$ **14.** $\frac{49}{81}$

In Exercises 15–18, sketch the largest possible square that can be formed with the tiles. (You won't be able to use all the tiles.) How does the result help estimate the square root of a number?

15. 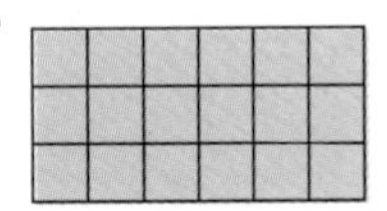

16. 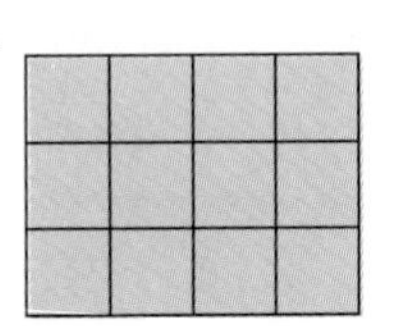

17. 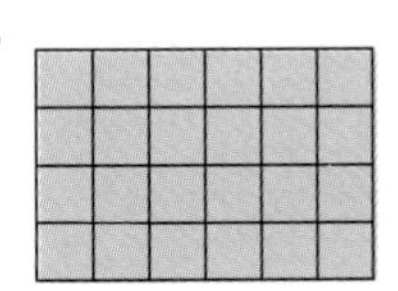

18. 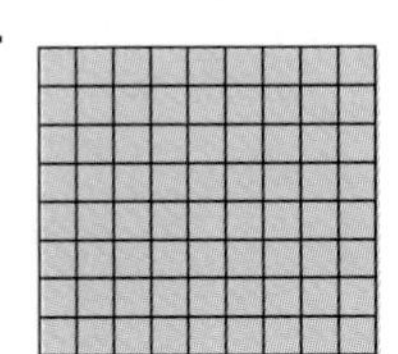

In Exercises 19–26, write both solutions of the equation. Round each solution to three decimal places, if necessary.

19. $t^2 = 9$ **20.** $x^2 = 100$ **21.** $p^2 = 22$ **22.** $r^2 = 17$

23. $b^2 + 2 = 27$ **24.** $y^2 - 6 = 30$ **25.** $3a^2 = 243$ **26.** $4s^2 = 49$

Estimation **In Exercises 27–30, the small squares of the graph paper are each 1 square unit. Estimate the side lengths of the blue square. Use a calculator to confirm your estimate.**

27. 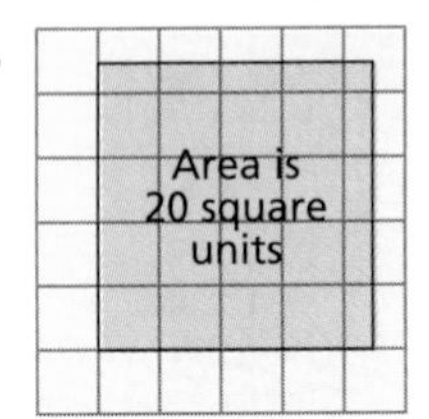

28. 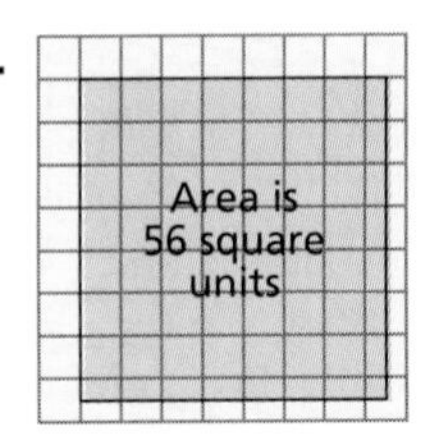

29. 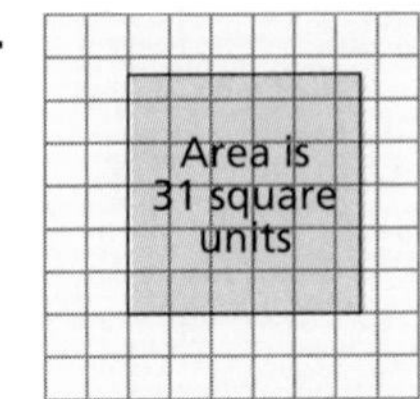

30.
Area is
39 square
units

In Exercises 31–34, write an algebraic equation for the sentence. Then solve the equation.

31. The positive square root of 25 is x.
32. y squared is 47.
33. The sum of a squared and 6 is 14.
34. The product of 3 and r squared is 27.

On a Clear Day **In Exercises 35 and 36, use the following information.**

On a clear day, the distance, d, in miles that you can see out to the ocean is approximated by the equation $d^2 = \frac{3}{2}h$, where h is the height in feet of your eyes above ground level.

35. You are standing on an observation deck overlooking the Atlantic Ocean. Your eyes are 10 feet above sea level. How far can you see on a clear day?
36. You climb to the top of the observation tower where your eyes are 30 feet above sea level. How far can you see on a clear day?

The apparent boundary of the earth and sky is called the horizon. The horizon seems nearer to an observer at ground level.

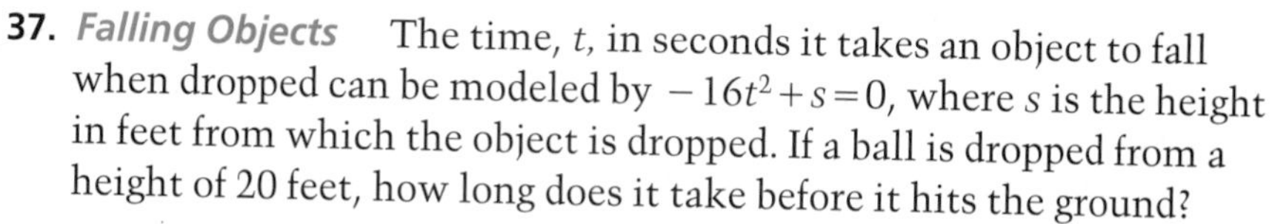

37. *Falling Objects* The time, t, in seconds it takes an object to fall when dropped can be modeled by $-16t^2 + s = 0$, where s is the height in feet from which the object is dropped. If a ball is dropped from a height of 20 feet, how long does it take before it hits the ground?
38. *Surface Area* The surface area of a cube is the sum of the areas of its faces. How long is each edge if the surface area is 216 square centimeters?

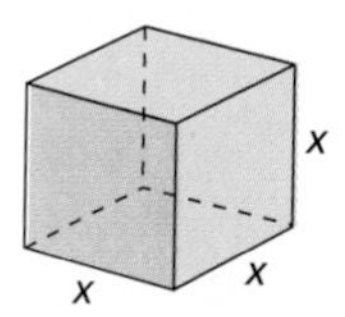

Integrated Review

Making Connections within Mathematics

39. *Estimation* Which number best estimates 4.5^2?

    a. 9 b. 16 c. 20 d. 25

40. *Estimation* Which number best estimates $\left(\frac{7}{16}\right)^2$?

    a. $\frac{1}{4}$ b. $\frac{3}{4}$ c. $\frac{7}{8}$ d. $\frac{49}{16}$

Exploration and Extension

41. *Coordinate Patterns* Copy and complete the table.

x	0	1	2	3	4	5	6	7	8	9	10
$y = \sqrt{x}$	0	1	1.4	?	?	?	?	?	?	?	?

42. Plot the data in the table in a coordinate plane. Describe the pattern.
43. Create another table for $y = -\sqrt{x}$. Plot the results in a coordinate plane and describe the pattern.

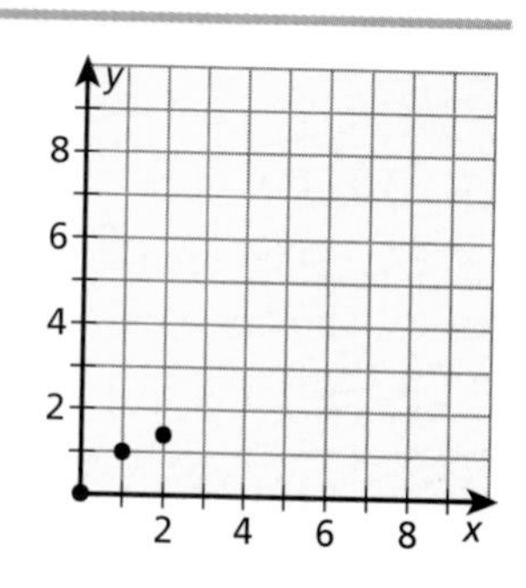

USING A CALCULATOR

Approximating Square Roots

By its definition, you know that when you square a square root of a number, you obtain the original number. For instance, 7 is a square root of 49 because $7^2 = 49$.

The number $\sqrt{49}$ can be written exactly as a decimal because 49 is a perfect square. However, many square roots cannot be written as exact decimals. For instance, no matter how many decimal places you list for $\sqrt{50}$, you will not obtain a number whose square is *exactly* 50.

Number of Decimal Places	Decimal Approximation	Check by Squaring
1	$\sqrt{50} \approx 7.1$	$(7.1)^2 = 50.41$
2	$\sqrt{50} \approx 7.07$	$(7.07)^2 = 49.9849$
3	$\sqrt{50} \approx 7.071$	$(7.071)^2 = 49.999041$
4	$\sqrt{50} \approx 7.0711$	$(7.0711)^2 = 50.00045521$
5	$\sqrt{50} \approx 7.07107$	$(7.07107)^2 = 50.0000309449$

Example ***Approximating Square Roots***

Use a calculator to evaluate the square root. Round your result to three decimal places, if necessary.

a. $\sqrt{2.25}$ **b.** $\sqrt{2.2}$

Solution

a. Because 2.25 is a perfect square, you can write

$\sqrt{2.25} = 1.5.$ *Check:* $1.5^2 = 2.25$

b. The number 2.2 is not a perfect square, so you must approximate its square root.

$\sqrt{2.2} \approx 1.483$ *Check:* $1.483^2 = 2.199289 \approx 2.2$ ■

Exercises

In Exercises 1–8, use a calculator to decide whether the square root can be written exactly as a decimal. If it can, write the exact decimal. If it can't, round the decimal to three places.

1. $\sqrt{1.21}$ **2.** $\sqrt{1.25}$ **3.** $\sqrt{17}$ **4.** $\sqrt{81}$

5. $\sqrt{10{,}000}$ **6.** $\sqrt{500}$ **7.** $\sqrt{12.96}$ **8.** $\sqrt{12.95}$

9. Make a table like that above that shows how you can obtain better and better approximations of $\sqrt{30}$ by listing more and more decimal places.

9.2 The Real Number System

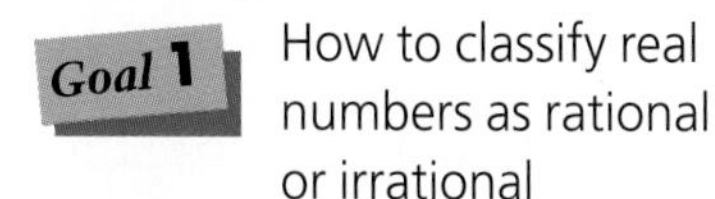

Goal 1 How to classify real numbers as rational or irrational

Goal 2 How to represent real numbers with a number line

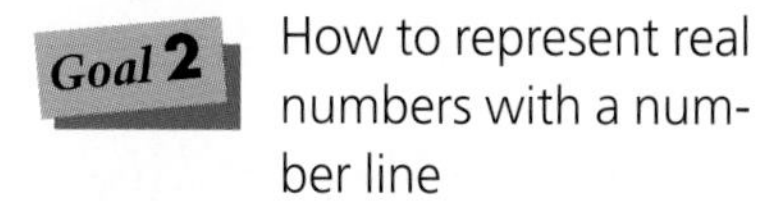

Many numbers that occur in real-life problems are not rational. Some real-life problems can be modeled only with irrational numbers.

Goal 1 Classifying Real Numbers

In Lesson 6.6, you learned that a *rational number* is a number that can be written as the quotient (or ratio) of two integers. *Irrational numbers* are numbers that cannot be written as the quotient of two integers. Together, the sets of all rational numbers and irrational numbers make up the set of **real numbers.** The decimal form of a rational number either terminates or repeats. The decimal form of an irrational number does not terminate and does not repeat.

Number	Type	Decimal Form	Decimal Type
$\frac{3}{4}$	Rational	$\frac{3}{4} = 0.75$	Terminating
$\frac{1}{11}$	Rational	$\frac{1}{11} = 0.0909\ldots = 0.\overline{09}$	Repeating
$\sqrt{3}$	Irrational	$\sqrt{3} = 1.7320508\ldots$	Nonrepeating

Example 1 describes a real-life length that is irrational.

Real Life
Floor Covering

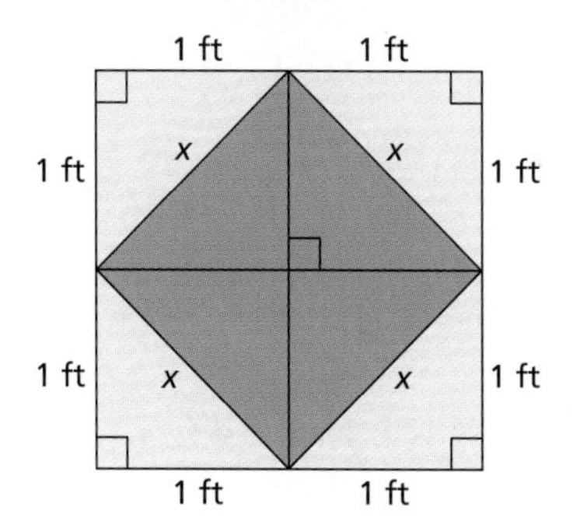

Example 1 *Measuring Sides of Triangles*

You are designing floor tiles that are right triangles. Eight of the tiles can be used to form a square that is 2 feet by 2 feet, as shown at the left. What are the lengths of the sides of each tile? Can these lengths be written exactly as decimals?

Solution The blue square is formed by 4 tiles. The large square is formed by 8 tiles and has an area of 4 square feet. This means that the blue square must have an area of 2 square feet.

Area of blue square $= x^2$

$2 = x^2$

By solving this equation, you can conclude that x is equal to $\sqrt{2}$. Thus, each tile has side lengths of 1 foot, 1 foot, and $\sqrt{2}$ feet. A length of 1 foot can be represented exactly as a decimal, but a length of $\sqrt{2}$ feet cannot. Rounding to two decimal places, you can approximate $\sqrt{2}$ feet as

$\sqrt{2} \approx 1.41$ feet.

Goal 2 Using a Number Line

Real numbers can be plotted on a number line. For example, the number line below shows the points that correspond to the real numbers $-\frac{3}{2}$, $\sqrt{2}$, -0.5, $\sqrt{3}$, and $\frac{8}{3}$.

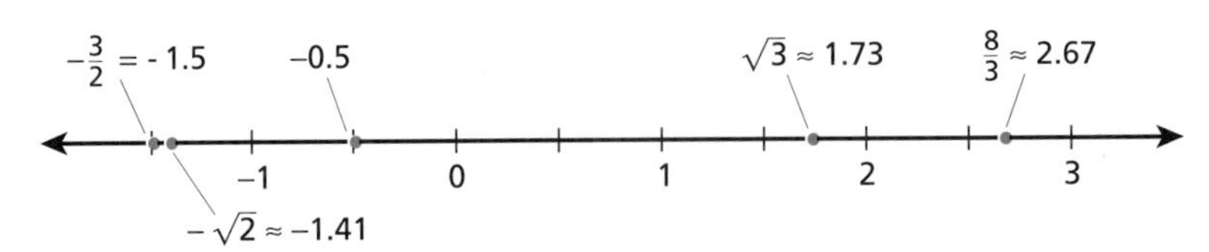

To plot a real number on a number line, you could first write the number in decimal form.

Example 2 *Comparing Numbers on a Number Line*

Plot each pair of numbers. Then complete the statement with $<$, $>$, or $=$.

a. $\sqrt{5}$ [?] $\frac{9}{4}$ **b.** $-\sqrt{7}$ [?] $-\frac{8}{3}$ **c.** $\sqrt{\frac{9}{4}}$ [?] $\frac{3}{2}$

Solution

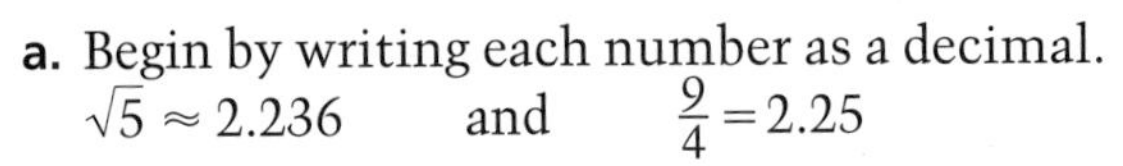

a. Begin by writing each number as a decimal.

$\sqrt{5} \approx 2.236$ and $\frac{9}{4} = 2.25$

Then, plot the numbers, as shown at the left. Because $\sqrt{5}$ is to the left of $\frac{9}{4}$, it follows that $\sqrt{5} < \frac{9}{4}$.

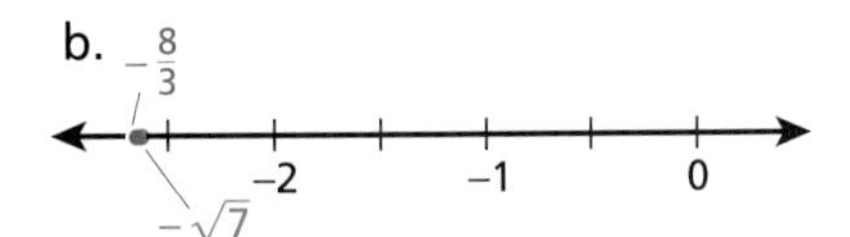

b. Begin by writing each number as a decimal: $-\sqrt{7} \approx -2.646$ and $-\frac{8}{3} \approx -2.667$. After plotting the numbers on a number line, you can conclude that $-\sqrt{7} > -\frac{8}{3}$.

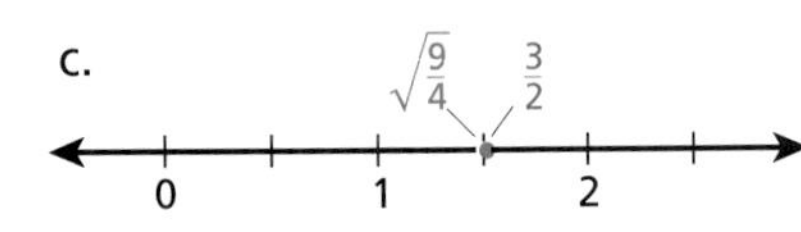

c. You could begin by writing each number as a decimal. In this case, however, you may notice that $\frac{9}{4}$ is a perfect square.

Because $\left(\frac{3}{2}\right)^2 = \frac{9}{4}$, it follows that $\sqrt{\frac{9}{4}} = \frac{3}{2}$. ■

Communicating about MATHEMATICS

SHARING IDEAS about the Lesson

Comparing Numbers Use a calculator to plot the set of numbers on a number line.

$\left\{\sqrt{\frac{3}{2}}, 0.83, -1, \sqrt{2}, -\frac{5}{4}, -\sqrt{3}, -\frac{6}{11}\right\}$

EXERCISES

Guided Practice

CHECK for Understanding

1. *Writing* In your own words, state the definition of a rational number.
2. *Writing* In your own words, explain how the decimal form of a rational number differs from the decimal form of an irrational number.
3. State whether the number is rational. Explain your reasoning.
   a. 0.123 **b.** $0.\overline{123}$ **c.** $0.123714356\ldots$

In Exercises 4–7, state whether the number is rational or irrational.

4. $\frac{8}{2}$ **5.** $\sqrt{5}$ **6.** $\sqrt{9}$ **7.** $-\sqrt{\frac{16}{9}}$

8. Plot the set of numbers on a number line. $\left\{0, \frac{1}{2}, -\frac{5}{3}, -2.1, -\sqrt{4}, \sqrt{6}, -\sqrt{8}, 3\right\}$

Independent Practice

Number Sense **In Exercises 9–16, determine whether the number is rational or irrational. Explain your reasoning.**

9. $\frac{11}{5}$ **10.** $-\frac{21}{16}$ **11.** $\sqrt{10}$ **12.** $-\sqrt{15}$

13. $\sqrt{1.44}$ **14.** $-\sqrt{\frac{100}{36}}$ **15.** $-\sqrt{\frac{3}{2}}$ **16.** $\sqrt{\frac{9}{6}}$

Logical Reasoning **In Exercises 17–20, complete the statement using *always, sometimes,* or *never.* Explain.**

17. A real number is [?] a rational number.
18. An irrational number is [?] a real number.
19. A negative integer is [?] an irrational number.
20. The square root of a number is [?] an irrational number.

In Exercises 21–24, evaluate the expression for $a = 2$, $b = 4$, and $c = 9$. Is the result rational? Explain.

21. $\sqrt{a} + \sqrt{b}$ **22.** $\sqrt{b} - \sqrt{c}$ **23.** $\sqrt{c} \cdot \sqrt{b}$ **24.** $\sqrt{b} \div \sqrt{a}$

In Exercises 25–30, match the number with its graph.

a b c d e f
−4 −3 −2 −1 0 1 2 3 4

25. $\sqrt{8}$ **26.** $-\sqrt{15}$ **27.** $\frac{-\sqrt{144}}{5}$ **28.** 0.49 **29.** $\sqrt{3.8}$ **30.** $-\sqrt{\frac{25}{16}}$

In Exercises 31–36, plot the number on a number line.

31. $-\frac{7}{2}$ **32.** $\frac{11}{3}$ **33.** $\sqrt{12}$

34. $-\sqrt{0.81}$ **35.** $\frac{\sqrt{3}}{2}$ **36.** $-\frac{\sqrt{30}}{3}$

In Exercises 37–42, complete the statement with <, >, or =.

37. $\sqrt{3}$? $\frac{23}{13}$ **38.** $-\sqrt{0.16}$? $-\frac{16}{41}$ **39.** $\sqrt{2.25}$? $\frac{3.6}{2.4}$

40. $\sqrt{\frac{49}{64}}$? $\frac{98}{112}$ **41.** $-\sqrt{2}$? $-\frac{\sqrt{19}}{3}$ **42.** $-\frac{25}{3}$? $-\frac{\sqrt{275}}{2}$

43. *Geometry* Decide whether the side lengths of the shaded square can be written exactly as decimals. Explain.

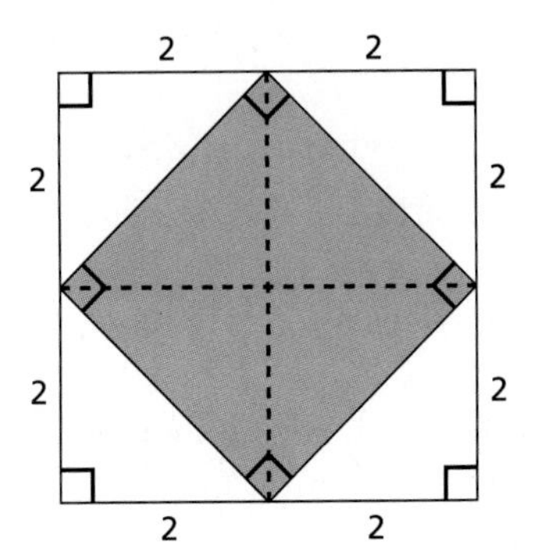

44. *Carpentry* You are helping rebuild a stairway in an apartment building. One of the boards needs to be $\sqrt{56}$ feet long. How accurately do you think the board should be measured? Explain your reasoning.

45. *Think About It* There are an *infinite* number of rational numbers. Do you think there are an infinite number of irrational numbers? Write a paragraph explaining your answer.

Integrated Review

Making Connections within Mathematics

46. *Venn Diagram* Copy the table. In your copy of the table, use check marks to indicate all the labels that describe the numbers. The Venn diagram can help you make your decisions.

	Natural	Whole	Integers	Rational	Irrational	Real
-5						
$\frac{15}{12}$						
$\sqrt{9}$						
$-\sqrt{\frac{6}{5}}$						
$\sqrt{11}$						
0						

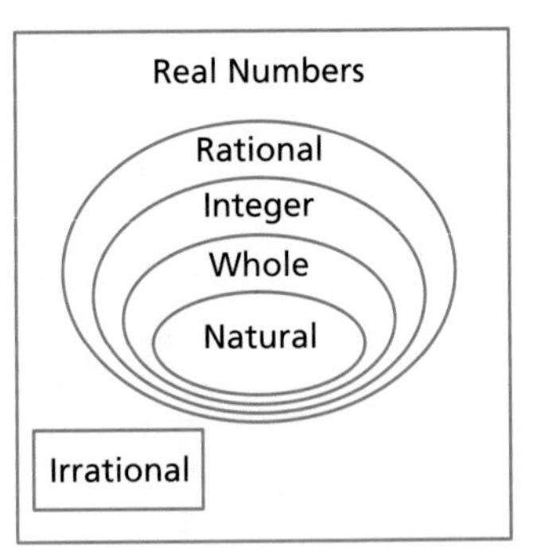

Exploration and Extension

Estimating Square Roots **In Exercises 47–52, estimate the square root. For example, $\sqrt{10}$ is greater than 3.1 (because $3.1^2 = 9.61$), but less than 3.2 (because $3.2^2 = 10.24$). Use a calculator to check your estimate.**

47. $\sqrt{7}$ **48.** $\sqrt{30}$ **49.** $\sqrt{50}$

50. $\sqrt{70}$ **51.** $\sqrt{110}$ **52.** $\sqrt{150}$

LESSON INVESTIGATION 9.3

Exploring Right Triangles

Materials Needed: metric dot paper, metric ruler

Example — *Exploring Right Triangles*

Use metric dot paper to draw a right triangle. Label the sides a, b, and c. Draw a square along each of the sides. Compare the areas of the squares. What can you conclude?

Solution Draw a right triangle, as shown below at the left. Next, draw a square along each side. From the dot pattern, you know that $a = 2$ centimeters and $b = 3$ centimeters. Using a metric ruler, you can approximate the length of the third side to be $c \approx 3.6$ centimeters. The areas of the squares are

$$a^2 = 2^2 = 4, \quad b^2 = 3^2 = 9, \quad \text{and} \quad c^2 \approx 3.6^2 = 12.96.$$

For this triangle, $a^2 + b^2$ is close to the approximation of c^2.

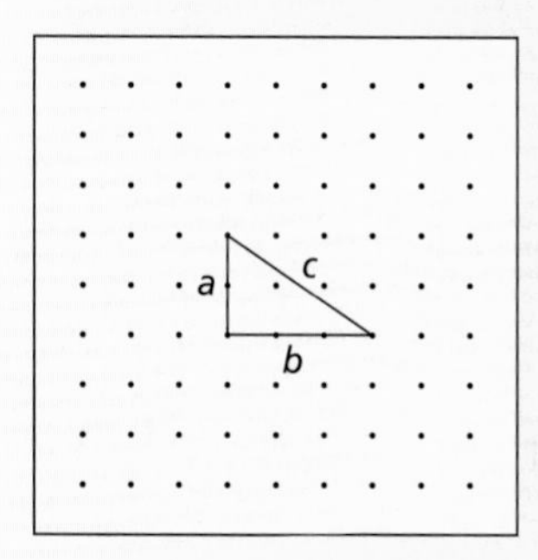

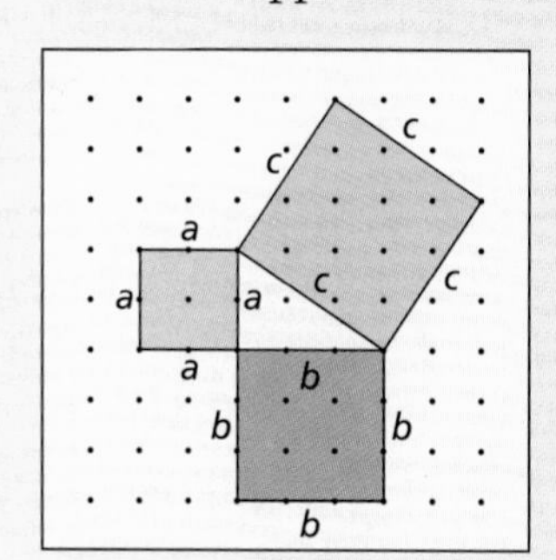

■

Exercises

In Exercises 1 and 2, copy the triangle and squares on metric dot paper. Compare the values of a^2, b^2, and c^2.

1.

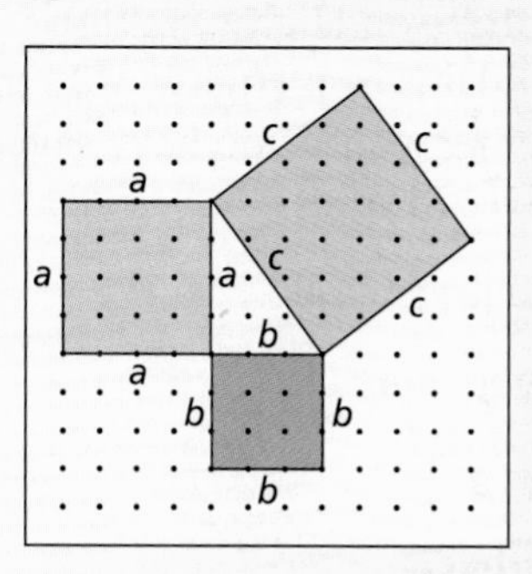

2.

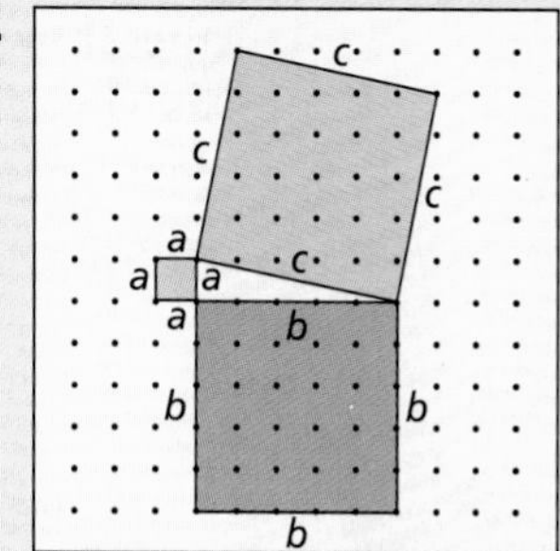

3. Repeat the procedure described in the example for several other triangles. What can you conclude?

9.3 The Pythagorean Theorem

What you should learn:

Goal 1 How to use the Pythagorean Theorem

Goal 2 How to solve a right triangle

Why you should learn it:

You can use the Pythagorean Theorem to solve real-life problems, such as finding a driving distance.

Goal 1 Using the Pythagorean Theorem

A right triangle is a triangle that has a right angle (one whose measure is 90°). The sides that form the right angle are the **legs** of the triangle, and the other side is the **hypotenuse.** The small square in the corner indicates which angle is the right angle.

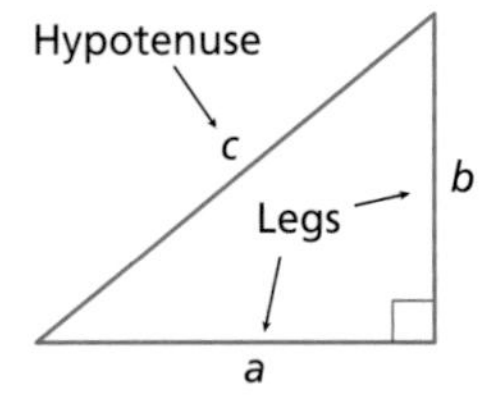

In the Lesson Investigation on page 399, you may have discovered a rule that is true of all right triangles. The rule is called the **Pythagorean Theorem,** and it is named after the Greek mathematician Pythagoras (about 585–500 B.C.).

> **Pythagorean Theorem**
>
> For any right triangle, the sum of the squares of the lengths of the legs, a and b, equals the square of the length of the hypotenuse, c.
>
> $a^2 + b^2 = c^2$ *Pythagorean Theorem*

Example 1 *Finding a Hypotenuse Length*

Real Life **Distance**

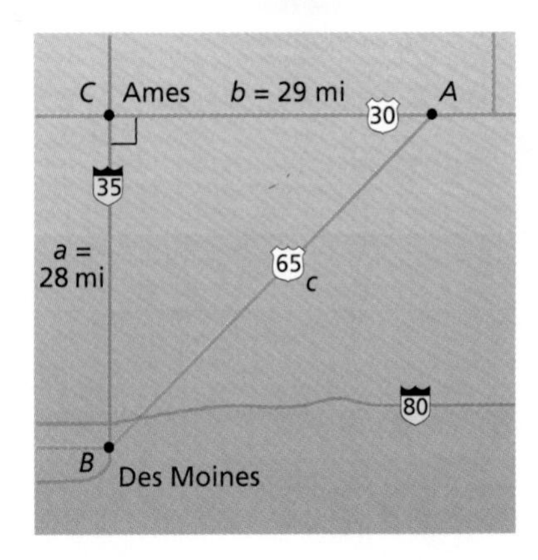

From the Highway 65 junction, you drive on Highway 30 to Ames, Iowa. Then, you turn south and drive on Highway 35 to Des Moines. How many miles of driving would you have saved if you had driven on Highway 65 to Des Moines?

Solution On the map, $\triangle ABC$ is a right triangle.

$a^2 + b^2 = c^2$	*Pythagorean Theorem*
$28^2 + 29^2 = c^2$	*Substitute for a and b.*
$1625 = c^2$	*Simplify.*
$\sqrt{1625} = c$	*Square Root Property*
$40.3 \approx c$	*Use a calculator.*

The distance along Highway 65 is about 40.3 miles. You would have saved $(28 + 29) - 40.3$ or about 16.7 miles. ■

Goal 2 Solving a Right Triangle

Using the lengths of two sides of a right triangle to find the length of the third side is called **solving a right triangle.** For instance, in Example 1 you used the lengths of the legs to find the length of the hypotenuse. Example 2 shows how to use the lengths of the hypotenuse and one of the legs to find the length of the other leg.

Connections
Geometry

Example 2 *Solving a Right Triangle*

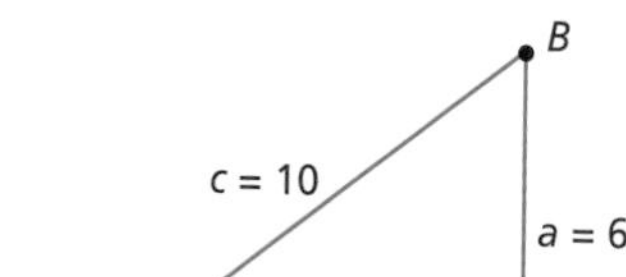

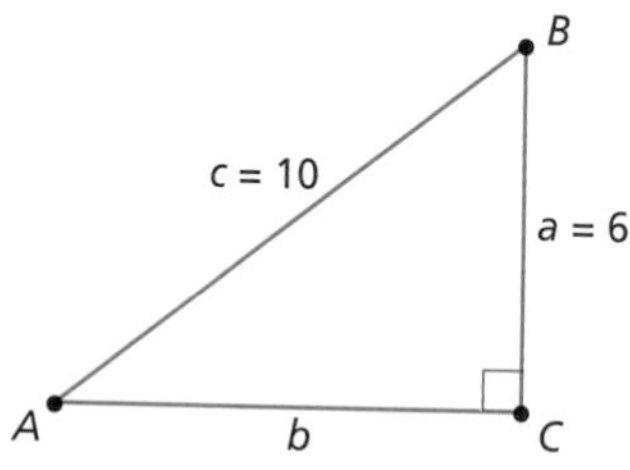

a. In $\triangle ABC$, $c = 10$ and $a = 6$. You can use the Pythagorean Theorem to find the length of the other leg.

$a^2 + b^2 = c^2$	*Pythagorean Theorem*
$6^2 + b^2 = 10^2$	*Substitute for a and c.*
$36 + b^2 = 100$	*Simplify.*
$b^2 = 64$	*Subtract 36 from each side.*
$b = 8$	*Square Root Property*

The length of the other leg is 8.

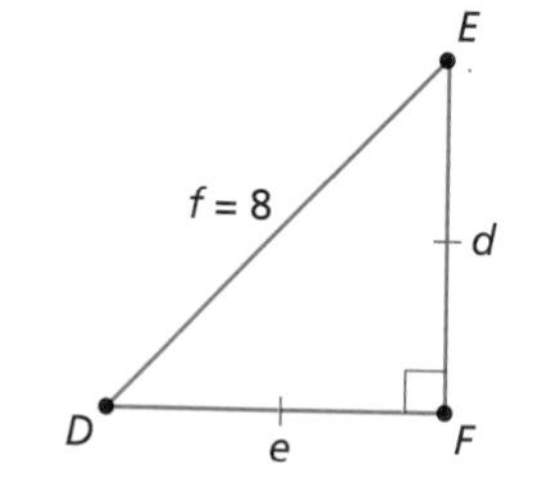

b. $\triangle DEF$, is **isosceles,** which means that its legs have the same length. The hypotenuse has a length of 8. You can use the Pythagorean Theorem to find the length of each leg.

$d^2 + e^2 = f^2$	*Pythagorean Theorem*
$d^2 + d^2 = 8^2$	*Substitute d for e and 8 for f.*
$2d^2 = 64$	*Simplify.*
$d^2 = 32$	*Divide each side by 2.*
$d = \sqrt{32}$	*Square Root Property*
$d \approx 5.66$	*Use a calculator.*

Each leg has a length of about 5.66 units. ■

Communicating about MATHEMATICS

Cooperative Learning

SHARING IDEAS about the Lesson

Pythagorean Triples A set of three natural numbers that represent the lengths of the sides of a right triangle is called a **Pythagorean Triple**. For instance, in Example 2a, the natural numbers 6, 8, and 10 form a Pythagorean Triple. With your partner, find other examples of Pythagorean Triples.

EXERCISES

Guided Practice

CHECK for Understanding

1. Draw a right triangle. Label its legs m and n and its hypotenuse t. How are the legs and hypotenuse related by the Pythagorean Theorem?
2. Can a right triangle have an obtuse angle? Explain.

In Exercises 3 and 4, a and b are the lengths of the legs of a right triangle and c is the length of the hypotenuse. Draw the right triangle on graph paper and estimate the missing length. Then use the Pythagorean Theorem to confirm your estimate.

3. $a = 3, b = \boxed{?}, c = 5$
4. $a = 6, b = 7, c = \boxed{?}$

Independent Practice

Logical Reasoning **In Exercises 5–8, decide whether the statement is *sometimes, always,* or *never* true.**

5. The Pythagorean Theorem can be applied to a triangle that is not a right triangle.
6. The hypotenuse is the longest side of a right triangle.
7. The legs of a right triangle are the same length.
8. In a right triangle, if a and b are integers, then c is an integer.

In Exercises 9–14, a and b are the lengths of the legs of a right triangle, and c is the length of the hypotenuse. Find the missing length.

9. $a = 7, b = 11$
10. $a = 16, c = 34$
11. $b = 12, c = 15$
12. $a = 6, c = 16.16$
13. $b = 42, c = 43.17$
14. $a = 18, b = 28$

In Exercises 15–22, if possible, draw a right triangle whose sides have the given lengths.

15. 8, 15, 17
16. 5, 12, 13
17. 9, 38, 41
18. 13, 36, 40
19. 7, 24, 26
20. 20, 21, 29
21. 30, 180, 181
22. 12, 35, 38

In Exercises 23–26, find the length of the third side.

23. 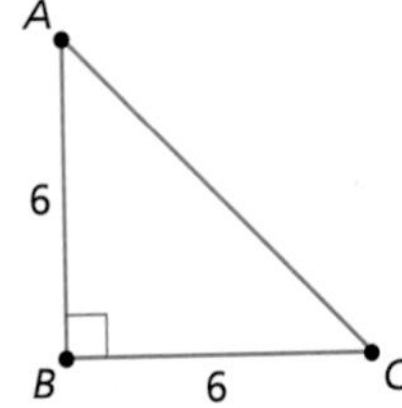

24. 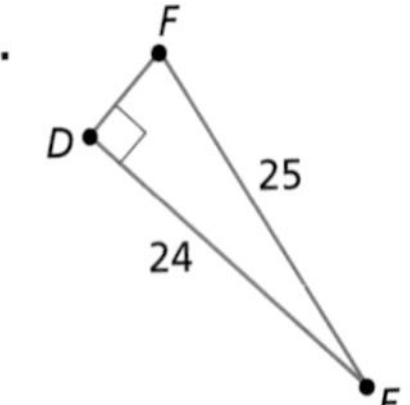

25. 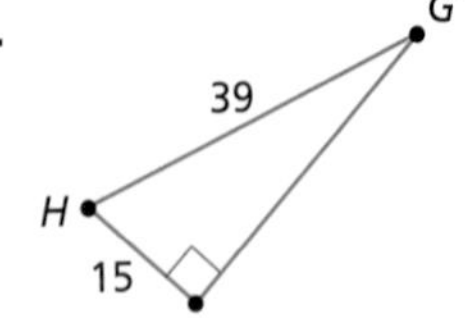

26. 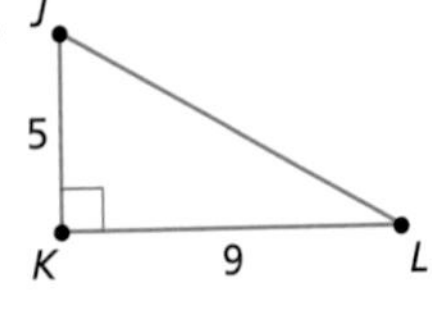

Volleyball **In Exercises 27–29, use the following information.**

You are setting up a volleyball net. To stay each pole, you use two ropes and two stakes as shown at the right.

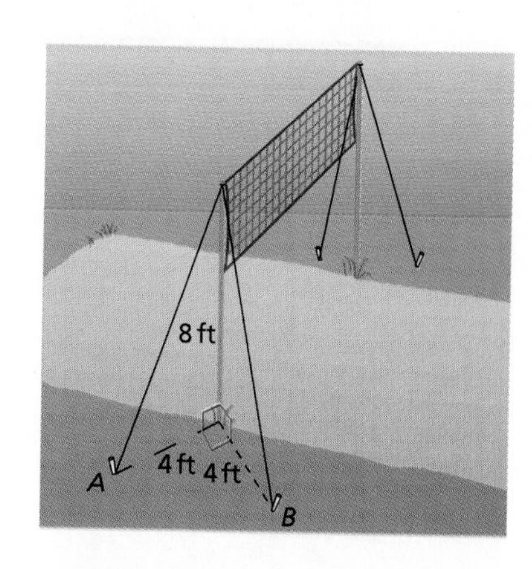

27. How long is each piece of rope?

28. Estimate the total length of the rope needed to stay both poles.

29. What is the distance between the stakes marked A and B?

30. *Visiting Friends* You ride your bike to your friend's house. You take the back roads as shown on the map at the right. How many miles would you have saved if you had traveled through town?

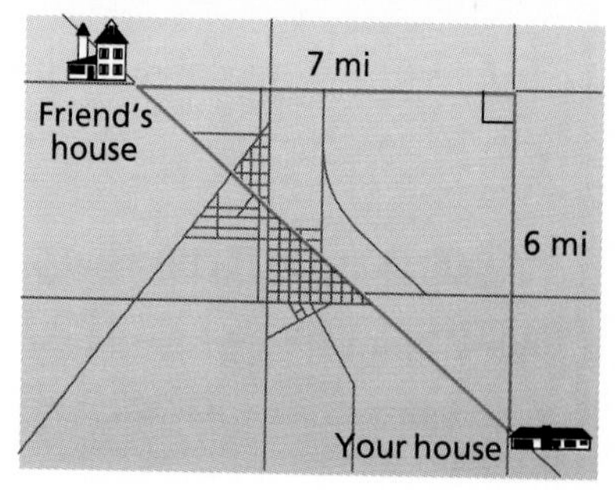

Drawing Right Triangles **In Exercises 31 and 32, you are given the lengths of two sides of a right triangle. Draw two *different* sizes of right triangles that have these side lengths and find the length of the third side.**

31. 8, 10

32. 5, 6

Integrated Review

Making Connections within Mathematics

Coordinate Graphing **In Exercises 33–38, plot the points in a coordinate plane. Decide whether the points can be connected to form a right triangle. If they can, find the lengths of the sides of the right triangle.**

33. $(1, 2), (-3, 2), (-3, -5)$

34. $(0, 1), (4, 1), (2, 2)$

35. $(2, 0), (-2, 0), (0, 1)$

36. $(-3, 1), (-3, -3), (3, -3)$

37. $(5, 1), (5, 4), (2, 3)$

38. $(4, 0), (-5, 0), (-5, -5)$

Exploration and Extension

Area of a Trapezoid **In Exercises 39–42, use the diagram at the right and the following information.**

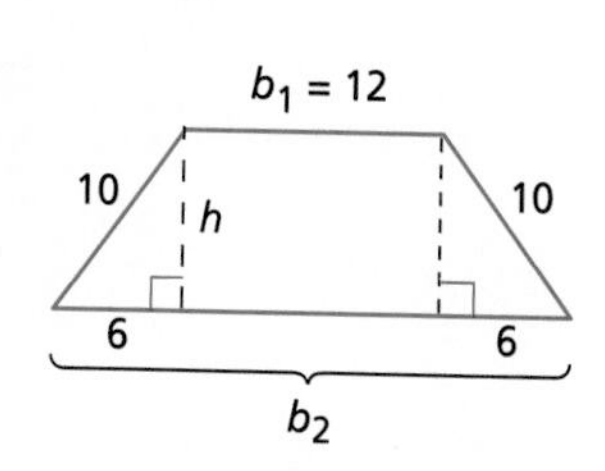

The area of a trapezoid is $A = \frac{1}{2}(b_1 + b_2) \cdot h$.

39. Find the base length, b_2.

40. Find the height, h.

41. Use the formula given above to find the area of the trapezoid.

42. Find the area of the trapezoid in another way and compare your result to the area found in Exercise 41.

Mixed REVIEW

In Exercises 1–4, find the side length of the square with the given area. (9.1)

1. 68 square units **2.** 121 square units **3.** 31.36 square units **4.** 53.29 square units

In Exercises 5–8, solve the right triangle. (9.3)

5.

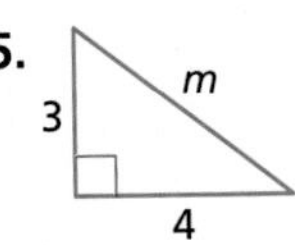

6.

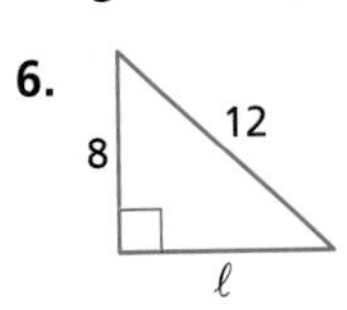

7.

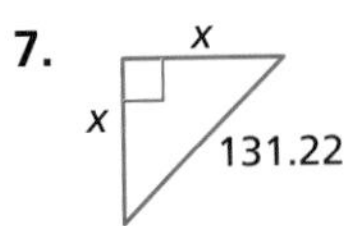

8. p 9 15

In Exercises 9–12, solve the inequality. (2.9)

9. $x + 2 \geq 9$ **10.** $2p < 14$ **11.** $5n > 8$ **12.** $y - 7 \leq 4$

In Exercises 13–16, find the greatest common factor. (6.3)

13. 24 and 39 **14.** 88 and 60 **15.** $100x$ and $222y$ **16.** 9, 12, 15

Career Interview

Solar Consultant and Construction Contractor

Michael A. Coca owns his own company, San Miguel Sun Dwellings, where he teaches others how to improve their homes with solar energy.

Q: ***What types of math do you use on your job?***

A: In order to determine the best way to improve the heating system of a house or business, heat-load analysis must be performed. This requires estimating how many square feet of different materials will be needed and drawing a one-dimensional scaled model of the construction.

Q: ***What led you to this career?***

A: It was exciting to be at the beginning of a new technology. I wanted to use what I learned to help people with the basic necessities of life, particularly shelter.

Q: ***What would you like to tell kids who are in school?***

A: Education is not just about learning math, English, science, and social studies. It is about learning how to live. Many of you have a vision of how you want to live when you grow up—you want to own a home, have a family, own a car, etc. In order to make that vision a reality, you must learn what it takes to get there.

Problem Solving Using the Pythagorean Theorem

What you should learn:

How to use properties of triangles to solve real-life problems

How to use the Pythagorean Theorem to measure indirectly

Why you should learn it:

You can use properties of triangles to solve real-life problems, such as finding the area of a park.

Goal 1 Modeling Real-Life Problems

You have now studied four properties of triangles.

- *Perimeter* The perimeter of a triangle is the sum of the lengths of its sides.
- *Area* The area of a triangle is one-half the product of its base and its height.
- *Similar Triangles* The ratios of corresponding sides of similar triangles are equal.
- *Pythagorean Theorem* In a right triangle, the sum of the squares of the lengths of the legs is equal to the square of the length of the hypotenuse.

In many real-life problems, you may need to use two or more of these properties to solve the problem.

Real Life
City Planning

Example 1 *Finding Area and Perimeter*

You are planning a city park as shown at the left. The park has the shape of a right triangle. You have measured one of the legs to be 50 meters and have measured the hypotenuse to be 86 meters. Do you need to take additional measurements to determine the perimeter and area of the park?

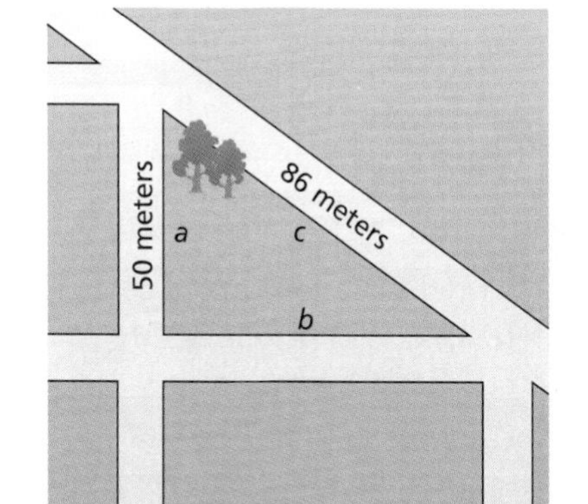

Solution Using the two measurements you already have, you can use the Pythagorean Theorem to find the length of the other leg.

$a^2 + b^2 = c^2$	*Pythagorean Theorem*
$50^2 + b^2 = 86^2$	*Substitute for a and c.*
$2500 + b^2 = 7396$	*Simplify.*
$b^2 = 4896$	*Subtract 2500 from each side.*
$b = \sqrt{4896}$	*Square Root Property*
$b \approx 70$	*Use a calculator.*

Thus the perimeter of the park is about 50 + 86 + 70 or 206 meters, and the approximate area is

$$\text{Area} = \frac{1}{2}ab = \frac{1}{2}(50)(70) = 1750 \text{ square meters.}$$

■

Goal 2 Indirect Measurement

In Lesson 8.3, you learned how to use ratios of corresponding sides of similar triangles to indirectly measure the height of a tree or a building. The Pythagorean Theorem can also be used to measure objects indirectly.

Example 2 *Using the Pythagorean Theorem*

Real Life
Street Maintenance

This huge sinkhole appeared in Atlanta, Georgia in June 1993. The sinkhole was over a hundred feet wide.

You work on a street maintenance crew for a city. Part of the street has caved in, as shown at the right. You don't have any ropes or tape measures that are long enough to stretch across the hole. Explain how you could measure the distance across the hole indirectly.

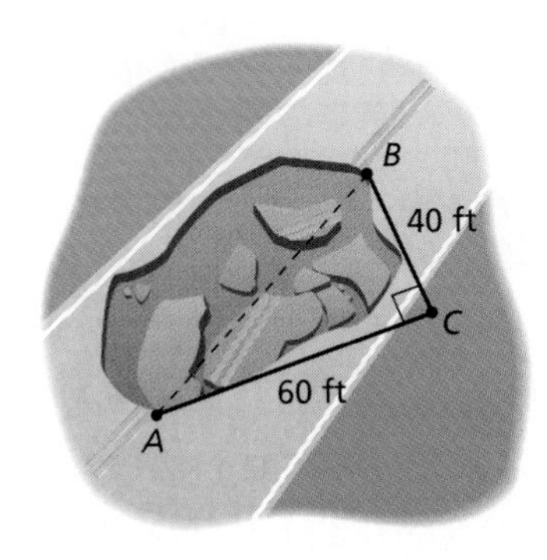

Solution Here is one way to solve the problem. Mark the vertices of a right triangle whose hypotenuse is the distance across the hole. Measure the legs of the right triangle, as shown in the figure. Then, use the Pythagorean Theorem to find the length of the hypotenuse.

$a^2 + b^2 = c^2$	*Pythagorean Theorem*
$40^2 + 60^2 = c^2$	*Substitute for a and b.*
$1600 + 3600 = c^2$	*Simplify.*
$5200 = c^2$	*Simplify.*
$\sqrt{5200} = c$	*Square Root Property*
$72.1 \approx c$	*Use a calculator.*

The distance across the hole is about 72.1 feet. ■

Communicating about MATHEMATICS

SHARING IDEAS about the Lesson

Indirect Measurement The three strawberries on this page form the vertices of a right triangle. Use a ruler to measure the two legs. Then *indirectly* find the length of the hypotenuse. Then measure the hypotenuse *directly* to check your result.

EXERCISES

Guided Practice

CHECK for Understanding

1. *Drawing Triangles* Draw sketches to illustrate the four properties of triangles that you have studied in this course.

2. *Playground* You are designing a neighborhood playground that has the shape of a right triangle, as shown at the right. You have measured one of the legs to be 45 meters and the hypotenuse to be 75 meters. Do you need to take additional measurements to determine the perimeter and the area of the playground? Explain.

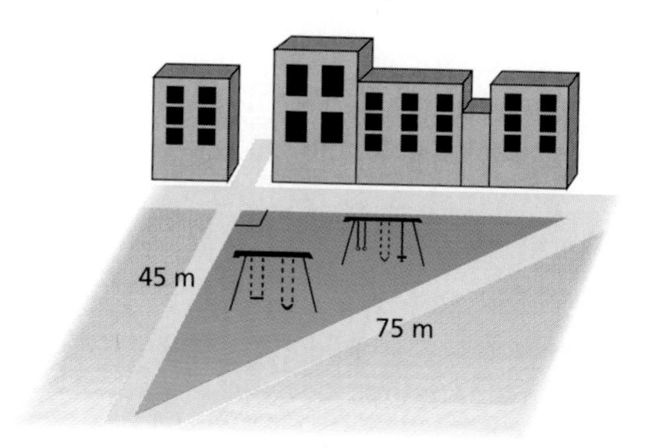

Independent Practice

In Exercises 3–6, find the perimeter and area of the figure.

3. A, 41, B 9 C

4.

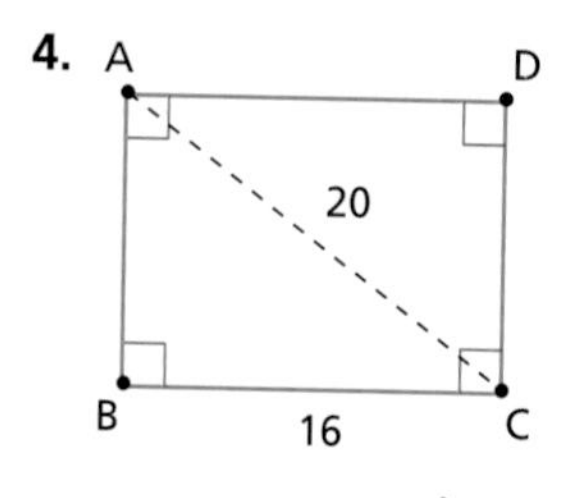

5.

6.

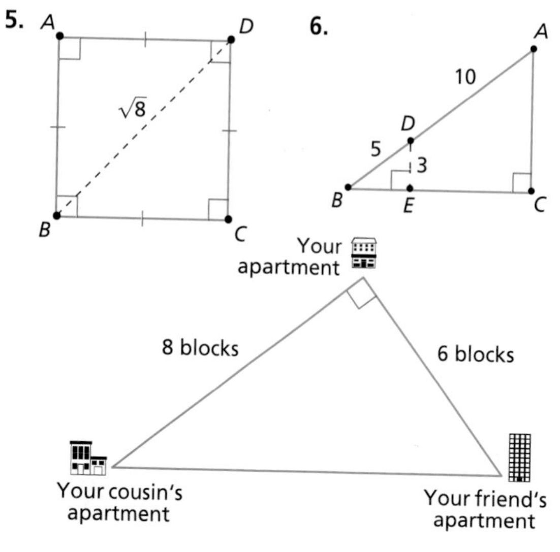

7. *Walking in the City* Your apartment is 6 blocks from your friend's apartment and 8 blocks from your cousin's apartment, as shown at the right. Each block is 500 feet long. You are walking from your friend's apartment to your cousin's apartment. Is the walk at least one mile? Explain your reasoning.

8. *Camping* You are setting up a camping tent, as shown below. What is the tallest that a person could be to stand in the tent? Explain your reasoning.

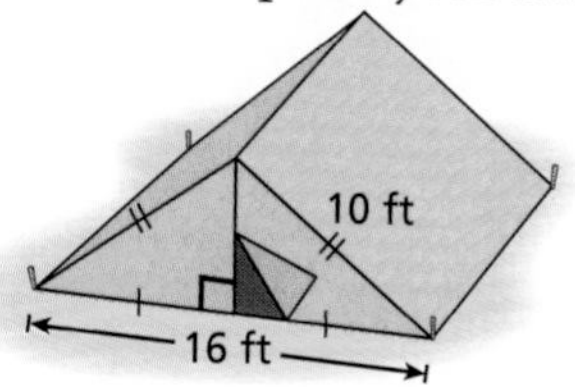

9. *Tree Height* How tall is the tree shown below? Explain your reasoning.

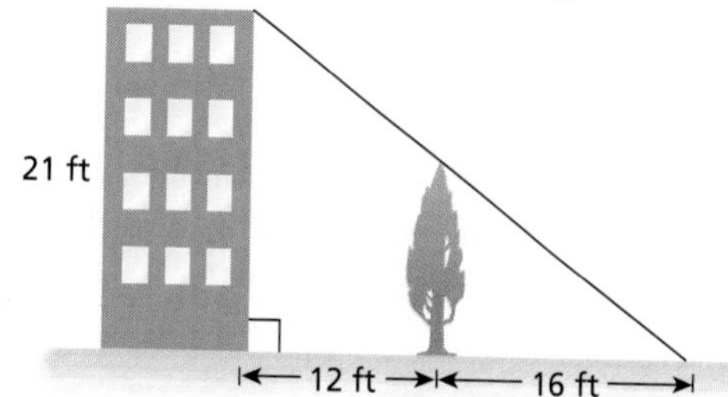

10. *Flying a Kite* You are flying a kite. You have let out 2000 feet of string. The sun is directly overhead and is casting a shadow that is 1325 feet from you. How high is the kite?

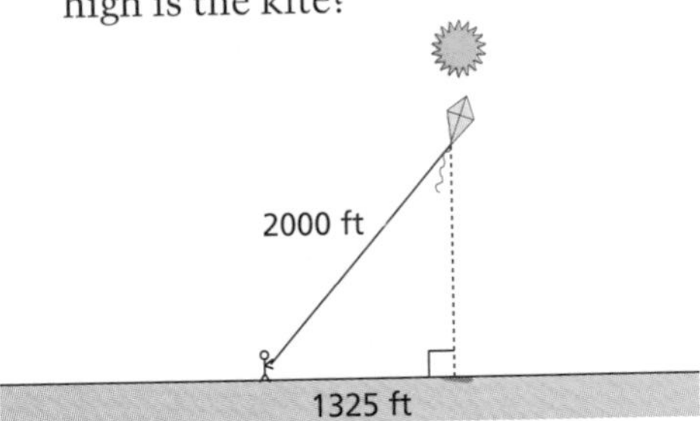

11. *Swimming* It takes you 10 seconds to swim the length of the pond shown below. How fast (in feet per second) did you swim?

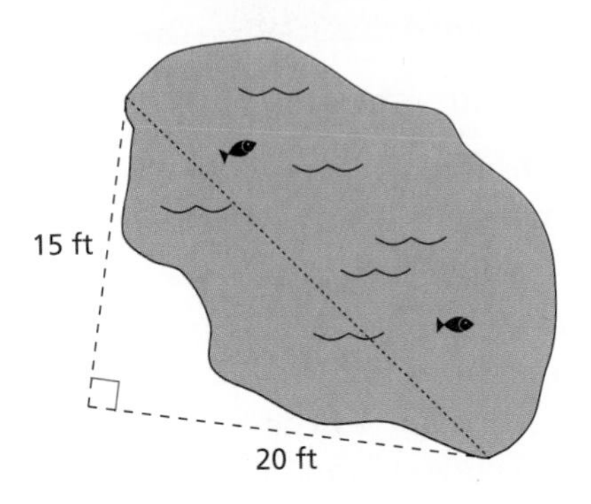

Integrated Review

Making Connections within Mathematics

Pythagorean Triples **In Exercises 12–14, use the table at the right that shows some Pythagorean Triples.**

a	b	c	$a^2 + b^2 = c^2$
3	4	5	$3^2 + 4^2 = 5^2$
6	8	10	$6^2 + 8^2 = 10^2$
9	12	15	$9^2 + 12^2 = 15^2$
12	16	20	$12^2 + 16^2 = 20^2$

12. Describe the pattern.

13. Find the next 3 Pythagorean Triples of the table.

14. *Think about It* If the table were extended indefinitely, would it contain all the possible Pythagorean Triples? If not, list some triples that would not appear in the table.

Exploration and Extension

15. *The Converse of the Pythagorean Theorem* The ancient Egyptians used a rope with equally spaced knots, as shown at the right. When the rope was held tight as shown, they concluded that the triangle was a right triangle. Which of the following results were they using? Explain your reasoning.

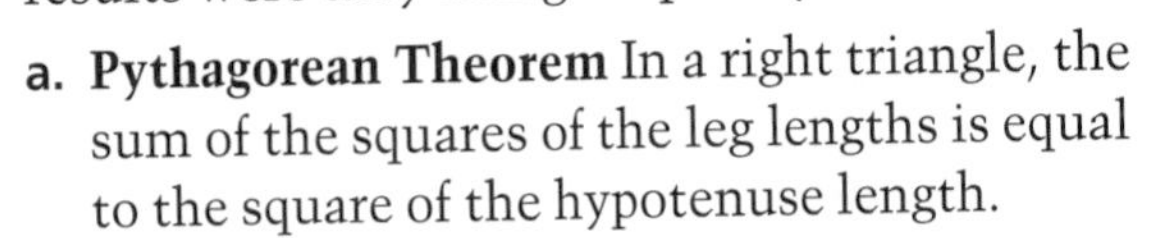

a. **Pythagorean Theorem** In a right triangle, the sum of the squares of the leg lengths is equal to the square of the hypotenuse length.

b. **Converse of the Pythagorean Theorem** In a triangle, if the sum of the squares of two side lengths is equal to the square of the third side length, then the triangle is a right triangle.

Converse of the Pythagorean Theorem **In Exercises 16–19, you are given the lengths of the sides of a triangle. From the given information, can you find the area of the triangle? Explain your reasoning.**

16. 4, 5, 6 **17.** 5, 12, 13 **18.** 6, 8, 10 **19.** 5, 7, 9

Mid-Chapter SELF-TEST

Take this test as you would take a test in class. The answers to the exercises are given in the back of the book.

In Exercises 1–4, find both square roots of the number. (9.1)

1. 16 **2.** 121 **3.** 0.49 **4.** 0.36

In Exercises 5 and 6, find the side lengths of the squares. (9.1)

5. 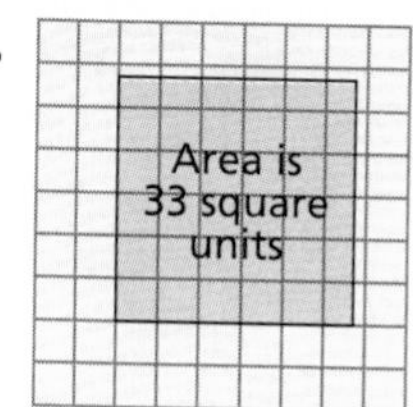

6. 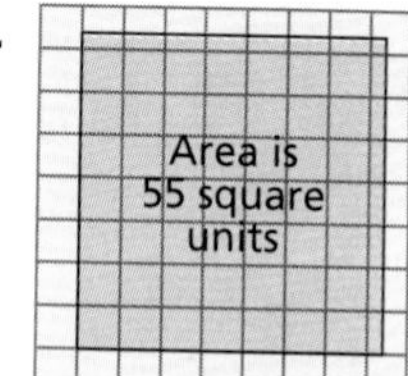

In Exercises 7 and 8, the small squares are each 1 square unit. Estimate the side lengths of the larger square. (9.1)

7. Area is 33 square units

8. Area is 55 square units

In Exercises 9–12, explain whether the number is rational or irrational. (9.2)

9. $\sqrt{250}$ **10.** $\sqrt{25}$ **11.** $\sqrt{2.5}$ **12.** $\sqrt{0.25}$

In Exercises 13–16, solve the triangle. Round your result to 2 decimal places. (9.3)

13. 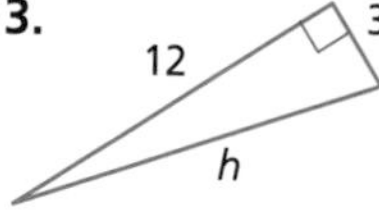

14. 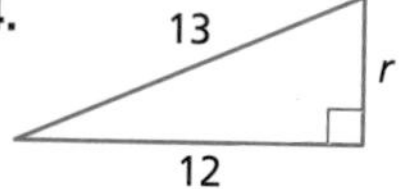

15. 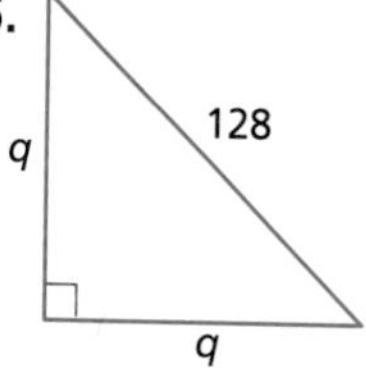

16. 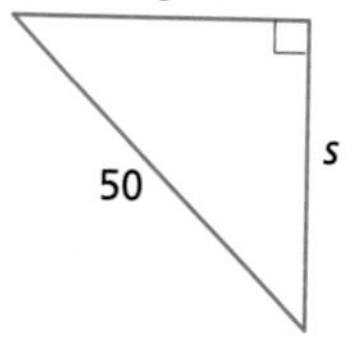

In Exercises 17–19, use the following information. (9.4)

You are building a bookshelf in the corner of your room, as shown to the right. The walls intersect at a right angle.

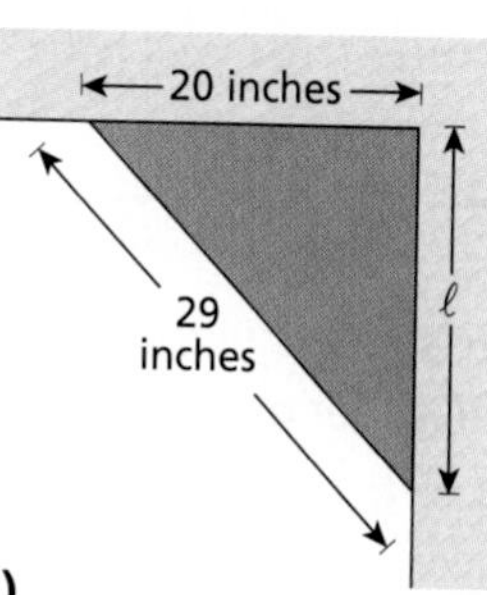

17. What is the length of ℓ?

18. What is the perimeter of the shelf?

19. What is the area of the shelf?

20. ***True or False?*** $\sqrt{5}$ is rational because it can be written as $\frac{\sqrt{5}}{1}$. **(9.2)**

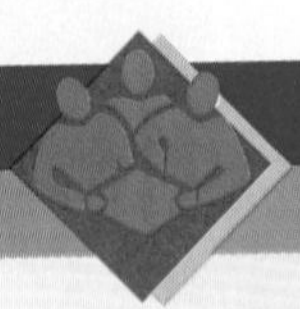

LESSON INVESTIGATION 9.5

Plotting Irrational Numbers

Materials Needed: compass, straightedge, graph paper

Example — *Plotting Irrational Numbers*

Plot the irrational number $\sqrt{20}$ on a number line.

Solution One way to plot the number is to use a calculator to approximate $\sqrt{20}$ as 4.5. Another way to plot the number is to use a compass, straightedge, and graph paper. Begin by drawing a right triangle whose legs have lengths of 2 and 4. By the Pythagorean Theorem, you can conclude that the hypotenuse has a length of $\sqrt{20}$. As shown in the figure below, draw a number line along the leg whose length is 4. Then use a compass to copy the hypotenuse's length onto the number line. Because the hypotenuse has a length of $\sqrt{20}$, you can conclude that the copied segment also has a length of $\sqrt{20}$.

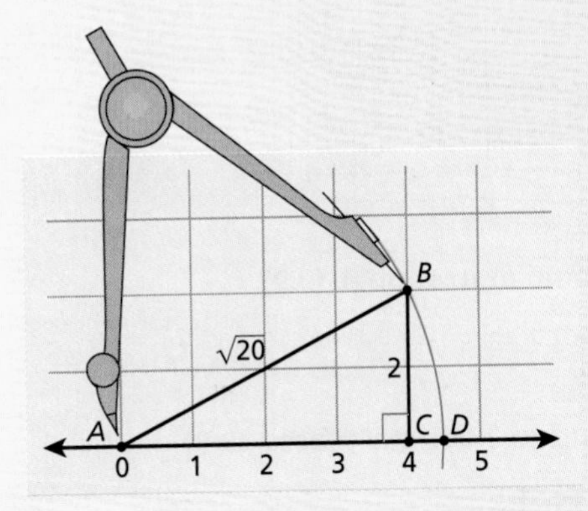

■

Exercises

In Exercises 1 and 2, state the irrational number that has been plotted on the number line.

1.

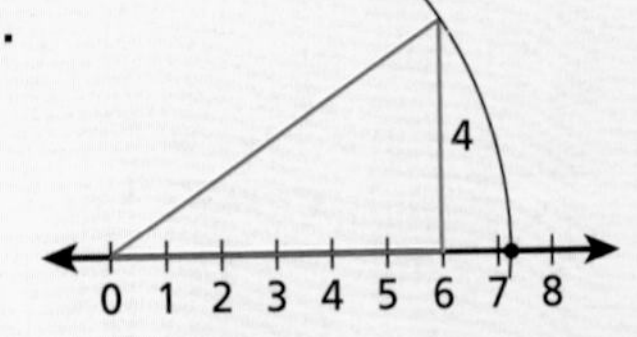

2.

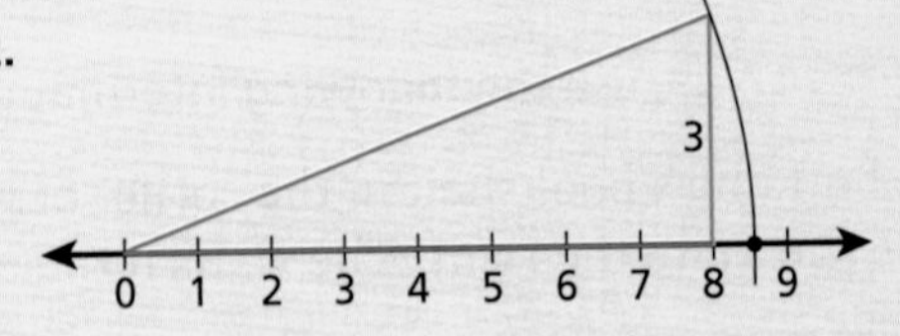

In Exercises 3–6, use a compass and straightedge to plot the irrational number on a number line.

3. $\sqrt{2}$ **4.** $\sqrt{5}$ **5.** $\sqrt{8}$ **6.** $\sqrt{10}$

9.5 Graphing Inequalities

What you should learn:

Goal 1 How to graph an inequality

Goal 2 How to write equivalent inequalities

Why you should learn it:

You can use inequalities to model and solve real-life problems, such as describing the size of an oil spill.

David Usher (second from left) is shown with his crew. His company, Marine Pollution Control, travels around the world helping to clean up after environmental accidents, such as oil spills.

Goal 1 Graphing Inequalities

In Lesson 2.9, you learned how to solve simple inequalities. Here is an example.

$x + 3 > 1$	*Original inequality*
$x + 3 - 3 > 1 - 3$	*Subtract 3 from each side.*
$x > -2$	*Simplify.*

The solution of this inequality is the set of all real numbers that are greater than -2. To graph this inequality on a number line, plot the number -2 with an open dot to show that -2 is not included. Then shade the part of the number line that is to the right of -2, as shown below.

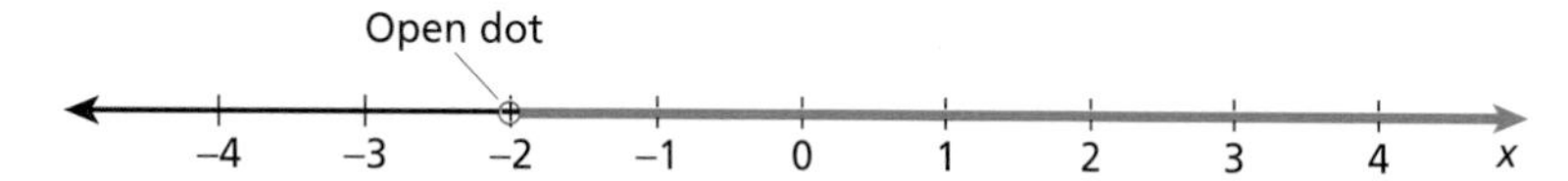

Inequalities are used to describe many types of real-life situations. For instance, the number of gallons in an ocean oil spill would have to be greater than zero.

There are four basic types of simple inequalities. Example 1 shows a sample of each type.

Example 1 *Graphing Inequalities on a Number Line*

Verbal Phrase	Inequality	Graph
a. All real numbers less than 2	$x < 2$	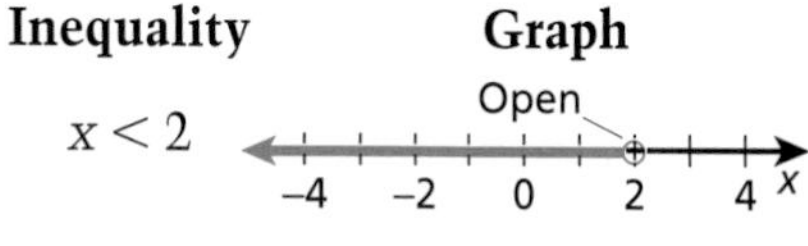
b. All real numbers greater than -3	$x > -3$	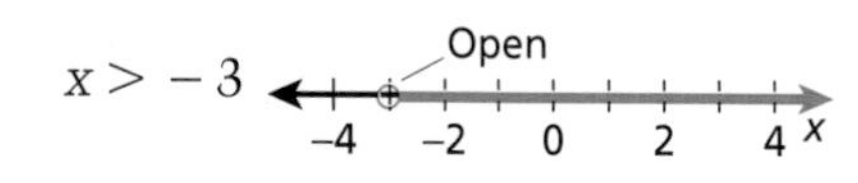
c. All real numbers less than or equal to -1	$x \leq -1$	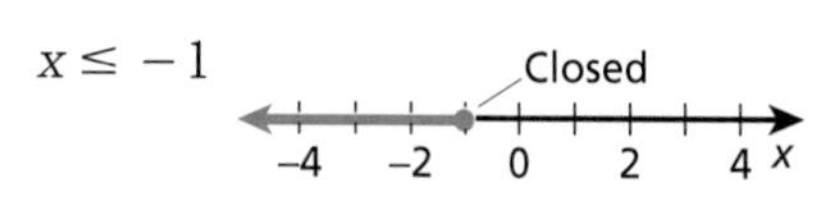
d. All real numbers greater than or equal to 0	$x \geq 0$	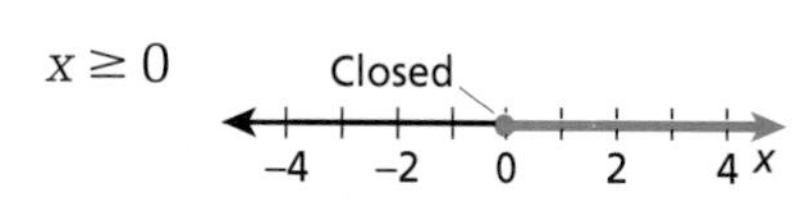

■

Goal 2 Writing Equivalent Inequalities

Each of the inequalities in Example 1 is written with the variable on the left. They can also be written with the variable on the right. For instance, $x < 2$ is equivalent to $2 > x$.

The graph of the inequality $x < 2$ is all real numbers that are less than 2.

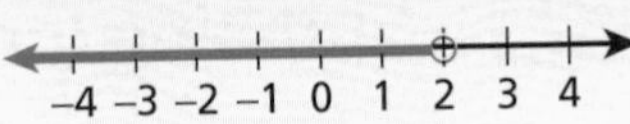

The graph of the inequality $2 > x$ is all real numbers that 2 is greater than.

–4 –3 –2 –1 0 1 2 3 4

To write an inequality that is equivalent to $x < 2$, move each number and letter to the other side, and reverse the inequality.

$x < 2 \Rightarrow 2 > x$

Example 2 *Writing Equivalent Inequalities*

For each of the following, write an equivalent inequality. State the inequality verbally.

a. $y > -3$ **b.** $0 \le m$ **c.** $4 \ge t$

Solution

a. The inequality $y > -3$ is equivalent to

$-3 < y.$

Either inequality can be written verbally as "the set of all real numbers that are greater than -3."

b. The inequality $0 \le m$ is equivalent to

$m \ge 0.$

Either inequality can be written verbally as "the set of all real numbers that are greater than or equal to 0."

c. The inequality $4 \ge t$ is equivalent to

$t \le 4.$

Either inequality can be written verbally as "the set of all real numbers that are less than or equal to 4." ■

Study Tip

One way to check that two inequalities are equivalent is to be sure that the inequality symbols "point" toward the same number or variable. For instance, in the inequalities $x < 2$ and $2 > x$, the inequality symbols point to x.

Similarly, in inequalities $y > -3$ and $-3 < y$, the inequality symbols point to -3.

Communicating about MATHEMATICS

SHARING IDEAS about the Lesson

Solving Inequalities Solve each inequality. Then graph the solutions.

A. $3 + b \le 5$ **B.** $-3 \ge x + 2$ **C.** $y - 4 < -8$

EXERCISES

Guided Practice

CHECK for Understanding

In Exercises 1–4, match the inequality with its graph.

a. 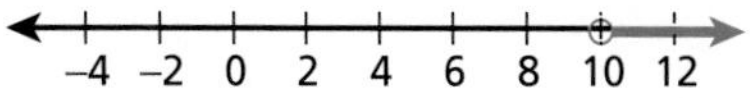

b. 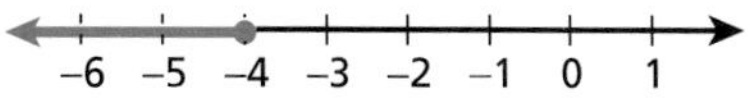

c. 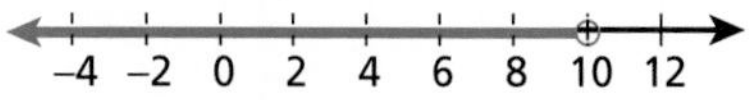

d. 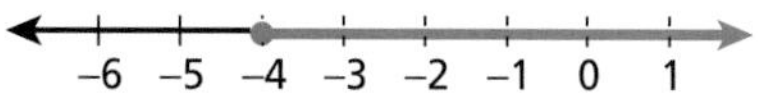

1. $x < 10$ **2.** $x \geq -4$ **3.** $x \leq -4$ **4.** $x > 10$

In Exercises 5–8, write two equivalent inequalities for the phrase.

5. All real numbers less than 15

6. All real numbers greater than or equal to 0

7. All real numbers greater than -3

8. All real numbers less than or equal to -11

In Exercises 9 and 10, solve the inequality. Then graph the solution.

9. $x + 5 < -2$ **10.** $x - 5 \geq -2$

Independent Practice

In Exercises 11–14, graph the inequality.

11. $x \geq 1$ **12.** $x < 0$ **13.** $x > 7$ **14.** $x \leq -2$

In Exercises 15–18, write the inequality represented by the graph.

15. 

16. 

17. 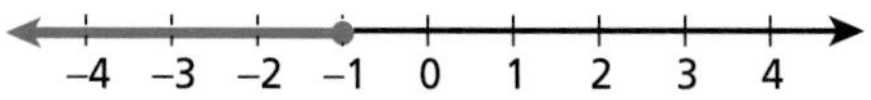

18. 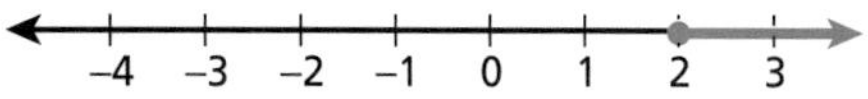

In Exercises 19–22, write the inequality given by the verbal phrase. Then graph the inequality.

19. All real numbers greater than $\sqrt{2}$

20. All real numbers less than or equal to $\sqrt{5}$

21. All real numbers less than $-\sqrt{3}$

22. All real numbers greater than or equal to $-\sqrt{6}$

In Exercises 23–28, solve the inequality. Then graph the solution.

23. $x + 3 \geq 2$ **24.** $5 > y + 2$ **25.** $-3 < n - 4$

26. $t - 1 \leq 7$ **27.** $z + 7 > -2$ **28.** $-5 \geq w - 4$

In Exercises 29–32, write an equivalent inequality. Then write the inequality verbally.

29. $x \le -20$ **30.** $y > -3$ **31.** $s < 17$ **32.** $m \ge 13$

In Exercises 33–36, write an algebraic model for the verbal phrase. Then solve.

33. t plus 17 is greater than 24.

34. x minus 5 is less than -19.

35. The sum of n and 5 is less than or equal to -9.

36. The difference of m and 10 is greater than or equal to 12.

37. *Solar Energy* The fastest speed attained by the *Sunraycer,* a solely solar-powered vehicle, is 48.71 miles per hour. Let S represent the speed of the *Sunraycer.* Which of the following best describes the *Sunraycer's* speed? Explain your reasoning.

a. $0 < S$ and $S < 48.71$
b. $0 < S$ and $S \le 48.71$
c. $0 \le S$ and $S < 48.71$
d. $0 \le S$ and $S \le 48.71$

On June 24, 1988, at Mesa, Arizona, Molly Brennan, shown here outside the Lincoln Memorial, drove the Sunraycer *at a record speed.*

38. The sun is about 93 million miles from Earth. In 1977, *Voyager I* was launched from Earth and moved out in the solar system (away from the sun). Let d represent the distance between *Voyager I* and the sun. Write an inequality that describes the values of d.

39. *Temperatures* The lowest possible temperature is -453°F. Let T represent the temperature of an object. Write an inequality that describes the possible values of T.

40. *Mammal Weights* The largest mammals are blue whales that can weigh up to 450,000 pounds. Let w be the weight of a mammal. Write two inequalities that describe the possible values of w.

Choosing the Better Graph **In Exercises 41 and 42, match the statement with the graph that you think is a better representation. Explain.**

a.

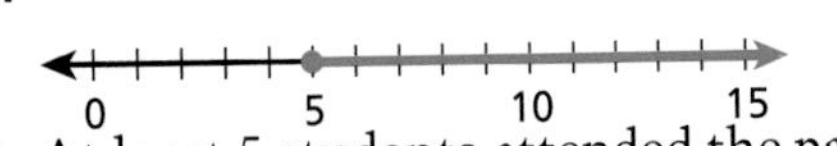

b.

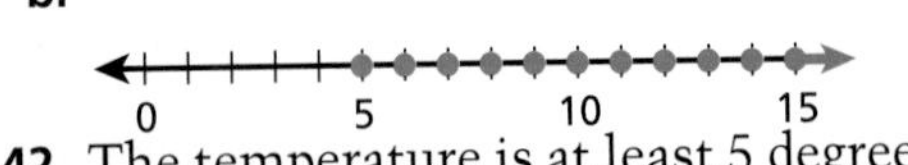

41. At least 5 students attended the party.

42. The temperature is at least 5 degrees.

Integrated Review

Making Connections within Mathematics

Mental Math **In Exercises 43–48, use mental math to solve the equation.**

43. $x - 19 = 40$ **44.** $y - 12 = -3$ **45.** $t + 9 = 21$

46. $s + 31 = 66$ **47.** $r - 19 = -1$ **48.** $n + 16 = -4$

Exploration and Extension

Compound Inequalities **In Exercises 49–52, match the compound inequality with its graph. State the inequality verbally.**

Sample: The statement *All real numbers greater than 0 and less than 3* is a **compound inequality**. It can be written as $0 < x < 3$ and its graph is

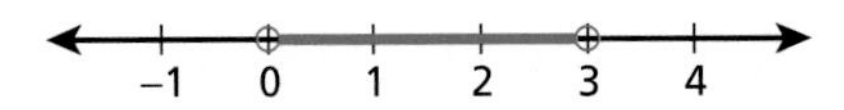

a.

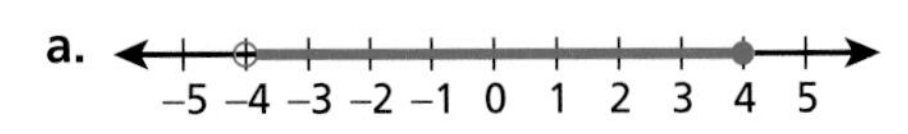

b. –5 –4 –3 –2 –1 0 1 2 3 4 5

c. –5 –4 –3 –2 –1 0 1 2 3 4 5

d. –5 –4 –3 –2 –1 0 1 2 3 4 5

49. $-4 < x < 4$ **50.** $-4 < x \leq 4$ **51.** $-4 \leq x < 4$ **52.** $-4 \leq x \leq 4$

Mixed REVIEW

In Exercises 1–5, use the diagram at the right. (8.3, 9.3, 9.4)

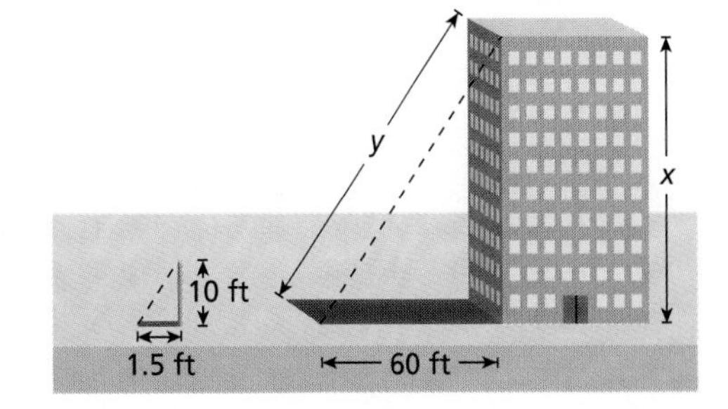

1. Find x, the height of the building.
2. Find y, the distance from the top of the building to the end of the building's shadow.
3. Find the distance from the top of the pole to the end of its shadow.
4. Write a proportion involving x, 60, and 10.
5. Write a proportion involving y, 60, and 1.5.

In Exercises 6–9, solve the inequality and graph the solution on a number line. (2.9, 9.5)

6. $x + 2 > 12$ **7.** $0 > x - 2$ **8.** $2x \leq 10$ **9.** $4x \geq x - 9$

In Exercises 10–15, write the prime factorization of the number. (6.2)

10. 70 **11.** 360 **12.** 270

13. 189 **14.** 369 **15.** 368

In Exercises 16–20, take the word REARRANGE and write each letter on a separate piece of paper. Put the pieces in a bag. What is the percent probability that you choose the indicated letter? (8.8)

16. R **17.** N **18.** G

19. A or E **20.** a letter other than R

9.6 Solving Inequalities: Multiplying and Dividing

What you should learn:

Goal 1 How to use properties of inequalities

Goal 2 How to use multiplication and division to solve an inequality

Why you should learn it:

You can use inequalities to model and solve real-life problems, such as finding real-estate commissions.

Goal 1 Using Properties of Inequalities

In this lesson, you will learn that there is an important difference between solving an equation and solving an inequality. You can discover this difference in the following investigation.

LESSON INVESTIGATION

■ Investigating Solutions of Inequalities

Group Activity Consider the inequality $-2x < 4$. You can check whether a number is a solution of the inequality by substituting the number for x. For instance, $x = 1$ is a solution because $-2(1)$ is equal to -2, which is less than 4. On the other hand, $x = -3$ is *not* a solution because $-2(-3)$ is equal to 6, which is not less than 4. Use a guess, check, and revise strategy to discover the solution of this inequality. Can you obtain your solution by dividing each side of the inequality by -2?

In the above investigation, you may have discovered that the solution of $-2x < 4$ is $x > -2$. To obtain this solution, you can divide each side of the original equation by -2, *provided* you reverse the direction of the inequality symbol.

$-2x < 4$	*Original inequality*
$\frac{-2x}{-2} > \frac{4}{-2}$	*Divide each side by -2 and reverse inequality.*
$x > -2$	*Simplify.*

Properties of Inequalities

1. Adding or subtracting the same number on each side of an inequality produces an equivalent inequality.
2. Multiplying or dividing each side of an inequality by the same *positive* number produces an equivalent inequality.
3. Multiplying or dividing each side of an inequality by the same *negative* number and *reversing the direction of the inequality symbol* produces an equivalent inequality.

Goal 2 Solving Inequalities

Example 1 *Solving Inequalities*

a. $-\frac{1}{2}x \geq 6$ *Original inequality*

$-2 \cdot \left(-\frac{1}{2}\right)x \leq -2 \cdot 6$ *Multiply each side by -2 and reverse inequality.*

$x \leq -12$ *Simplify.*

The solution is $x \leq -12$, which is the set of all real numbers that are less than or equal to -12.

b. $12 < 3m$ *Original inequality*

$\frac{12}{3} < \frac{3m}{3}$ *Divide each side by 3.*

$4 < m$ *Simplify.*

The solution is $4 < m$ or $m > 4$, which is the set of all real numbers that are greater than 4. ■

Example 2 *Solving an Inequality*

You are a real-estate agent and earn a 5% commission for each house you sell. What range of house prices will earn you a commission of at least \$4000?

Solution Let H represent the price of the house. Then your commission is 5% of H, or $0.05H$.

$0.05H \geq 4000$ *Commission is at least \$4000.*

$\frac{0.05H}{0.05} \geq \frac{4000}{0.05}$ *Divide each side by 0.05.*

$H \geq 80{,}000$ *Simplify.*

The price of the house you sell must be at least \$80,000. ■

Communicating about MATHEMATICS

SHARING IDEAS about the Lesson

It's Up to You Compare the following strategies used to solve the inequality $2 > -x$. Which do you prefer? Why?

A. Divide each side by -1 and reverse inequality.

B. Add x to each side, then subtract 2 from each side.

EXERCISES

Guided Practice

CHECK for Understanding

In Exercises 1–4, use $<$, $>$, $\le$, or $\ge$ to complete the inequality.

1. $4 > -x$
   -4 ? x
2. $-3 \ge -t$
   3 ? t
3. $3y > 15$
   y ? 5
4. $-2 \ge -\frac{1}{2}a$
   4 ? a

In Exercises 5–8, will the strategy require reversing the direction of the inequality?

5. Multiply both sides by -1.
6. Divide both sides by 4.
7. Multiply both sides by $\frac{1}{4}$.
8. Divide both sides by -5.

In Exercises 9–12, solve the inequality.

9. $4b > 24$
10. $-\frac{1}{4}x \le 14$
11. $-5 < 0.2h$
12. $36 \ge -\frac{1}{2}f$

Independent Practice

Error Analysis **In Exercises 13–15, describe the error.**

13. $-4x > 26$
$-\frac{1}{4} \cdot (-4)x > -\frac{1}{4} \cdot 26$
$x > -\frac{26}{4}$
$x > -\frac{13}{2}$

14. 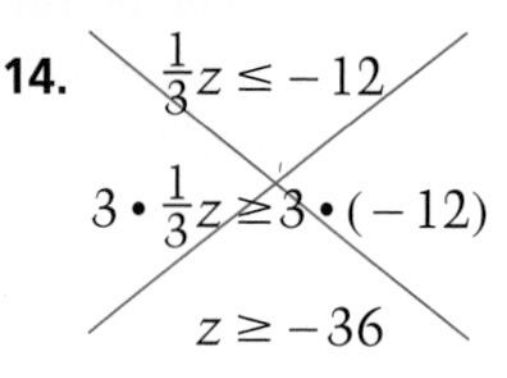

15. 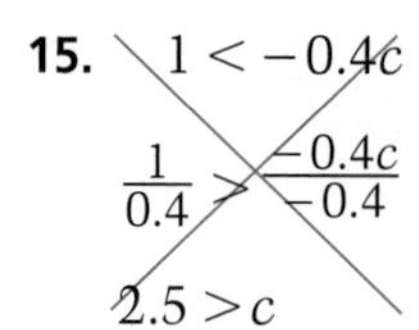

In Exercises 16–19, match the solution of the inequality to its graph.

a. 

b. 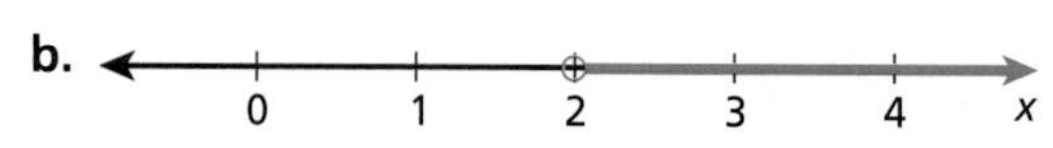

c. 

d. 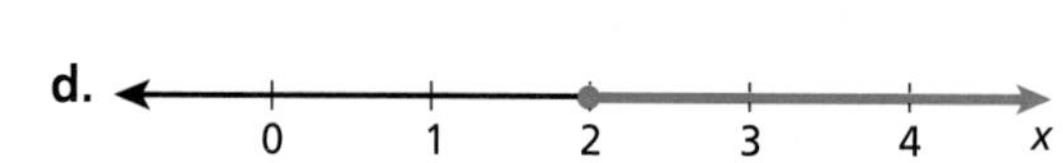

16. $0.7x \le 1.4$
17. $-1 > -\frac{1}{2}x$
18. $\frac{1}{8} \le \frac{1}{16}x$
19. $-3x > -6$

In Exercises 20–35, solve the inequality. Then graph the solution.

20. $3m < 4$
21. $2n \ge 5$
22. $\frac{x}{2} \le 8$
23. $\frac{y}{9} > 4$
24. $35 \ge -5b$
25. $\frac{1}{2} < -2a$
26. $-\frac{1}{2}z > 5$
27. $-\frac{1}{5}p \le 2$
28. $-\frac{3}{4}a \le -6$
29. $14 > -\frac{1}{3}n$
30. $4 < 0.8r$
31. $1.2x \ge 3.6$
32. $\frac{a}{6} > -2$
33. $\frac{3}{5}y \le -6$
34. $-1.4m \ge -5.6$
35. $-\frac{3}{2} < -\frac{1}{4}x$

36. *Fundraiser* You are selling sandwiches as a fundraiser for the softball team. You make a profit of 75¢ for each sandwich sold and the team needs to raise at least \$300.00. How many sandwiches must be sold?

37. *Buying a Video* The movie video you want to buy costs \$26.95. You earn \$3 an hour baby-sitting. What is the least number of hours you need to baby-sit to earn enough money to buy the video?

38. *Walking Speed* You want to walk 2 miles in less than 40 minutes. What must your speed be in miles per hour?

39. *Riding Speed* Your top bicycling speed is 40 kilometers per hour. What is the least amount of time it would take you to ride 60 kilometers?

Integrated Review

Making Connections within Mathematics

Logical Reasoning **In Exercises 40–43, decide whether the statement can be determined from the bar graph. Explain.** ***(Source: American Hotel and Motel Association)***

40. Some visitors come for more than one reason.

41. *About* the same number of people come for business reasons as come to visit friends.

42. *Exactly* the same number of people come for business reasons as come to visit friends.

43. Some people who come for holiday travel also come for business or to visit friends.

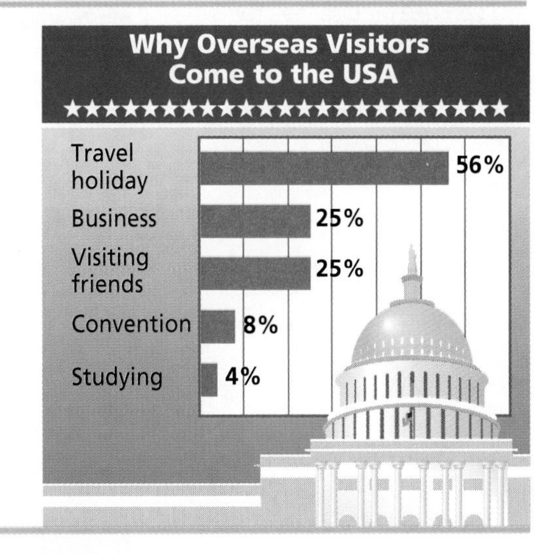

Exploration and Extension

44. *Game Strategy* With a partner, set up and play the following game. After playing the game, describe a playing strategy.

Setup: Write the numbers 1 to 9 on pieces of paper. Lay the pieces faceup on a table.

Object: Obtain three numbers whose sum is 15.

Rules: Players take turns choosing numbers, one number at a time.

End: The game ends when one player has chosen three numbers whose sum is 15.

45. Complete the magic square so that the sum of 3 horizontal, vertical, or diagonal numbers is 15.

		6
	5	
4		

46. Does the magic square in Exercise 45 influence your strategy for the game in Exercise 44? Explain.

9.7 Solving Multi-Step Inequalities

What you should learn:

Goal 1 How to solve multi-step inequalities

Goal 2 How to use multi-step inequalities to solve real-life problems

Why you should learn it:

You can use inequalities to model and solve real-life problems, such as problems that deal with nutrition.

Goal 1 Solving Multi-step Inequalities

In Lessons 9.5 and 9.6, you used properties of inequalities to solve inequalities that can be solved with one step. You can use these same properties to solve inequalities that require two or more steps. Remember that if you multiply or divide by a negative number, you must reverse the direction of the inequality symbol.

Example 1 Solving a Multi-step Inequality

Solve $2x + 1 \le 4$.

Solution

$2x + 1 \le 4$	*Rewrite original inequality.*
$2x + 1 - 1 \le 4 - 1$	*Subtract 1 from each side.*
$2x \le 3$	*Simplify.*
$\frac{2x}{2} \le \frac{3}{2}$	*Divide each side by 2.*
$x \le \frac{3}{2}$	*Simplify.*

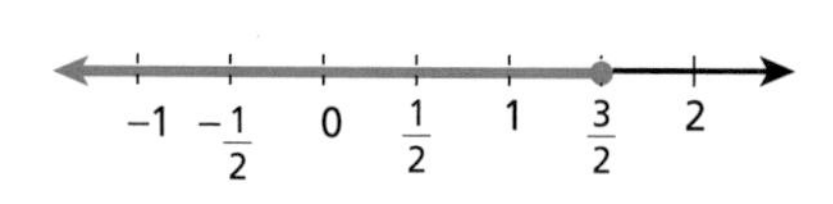

The solution is all real numbers that are less than or equal to $\frac{3}{2}$. A graph of the solution is shown at the left. ■

Example 2 Solving a Multi-step Inequality

Solve $-\frac{1}{3}m - 5 > 2$.

Solution

$-\frac{1}{3}m - 5 > 2$	*Rewrite original inequality.*
$-\frac{1}{3}m - 5 + 5 > 2 + 5$	*Add 5 to each side.*
$-\frac{1}{3}m > 7$	*Simplify.*
$(-3)\left(-\frac{1}{3}\right)m < (-3)(7)$	*Multiply each side by −3 and reverse the inequality.*
$m < -21$	*Simplify.*

The solution is all real numbers that are less than −21. ■

Study Tip...
You can check a solution of an inequality by substituting several numbers into the original inequality. For instance, when checking the solution in Example 1, numbers that are less than $\frac{3}{2}$ should yield true statements and numbers that are greater than $\frac{3}{2}$ should yield false statements.

Goal 2 Solving Real-Life Problems

Example 3 *Solving an Inequality*

Real Life **Nutrition**

You are baking a batch of 36 oatmeal chip cookies. Without the chips, the recipe has 4500 calories. You want each cookie to have less than 150 calories. Each chocolate chip has 3 calories. How many chips can you have in each cookie?

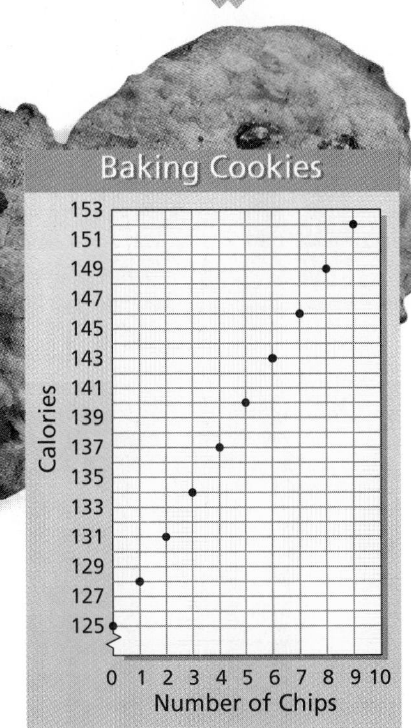

Solution Without any chocolate chips, each cookie would have $\frac{1}{36}(4500)$ or 125 calories.

Verbal Model $125 + 3 \cdot$ Chips per cookie < 150

Labels Number of chocolate chips per cookie = n (chips)

Algebraic Model

$$125 + 3 \cdot n < 150$$
$$3n < 25$$
$$n < 8\tfrac{1}{3}$$

You can use up to 8 chocolate chips per cookie. ■

You can also solve Example 3 with a table or with a graph. With either the table or the graph, notice that the number of calories exceeds 150 when the number of chocolate chips is greater than 8.

Number of Chips	0	1	2	3	4
Calories	125	128	131	134	137

Number of Chips	5	6	7	8	9
Calories	140	143	146	149	152

Communicating about MATHEMATICS

Cooperative Learning

SHARING IDEAS about the Lesson

Nutrition Work with a partner to plan a "macaroni and cheese" school lunch that is to contain at least 20 grams of protein. Without the macaroni and cheese, the lunch has 6 grams of protein. The macaroni and cheese has 2 grams of protein per ounce. What size servings of macaroni and cheese should you plan?

EXERCISES

Guided Practice

CHECK for Understanding

Reasoning **In Exercises 1 and 2, solve the inequality. Explain your steps.**

1. $3x - 2 \leq 13$

2. $4 < -\frac{1}{5}y + 2$

3. *Writing* In your own words, explain how to check a solution of an inequality. How is it different from checking a solution of an equation?

In Exercises 4 and 5, solve the inequality. Then graph the solution on a number line.

4. $-18 + 4y \geq -6y + 12$

5. $2(z + 1) < -3z + 2$

Independent Practice

Error Analysis **In Exercises 6–8, describe the error. Then correct it.**

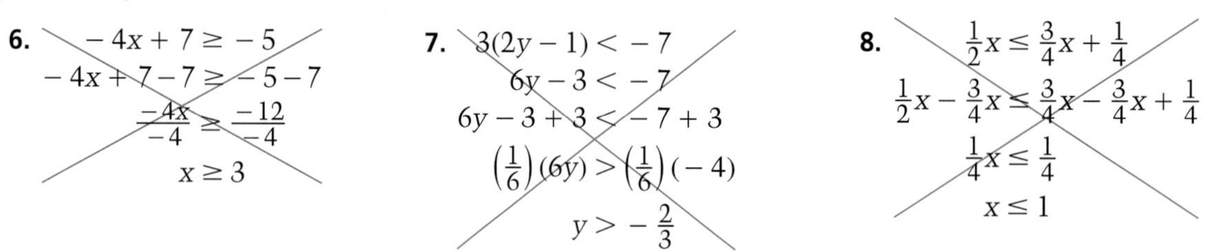

Logical Reasoning **In Exercises 9–12, decide whether the statement is sometimes, always, or never true.**

9. If $-5x + 9 \leq -11$, then $x = 3$.

10. If $7y - 20 \geq 2y + 15$, then $y = 7$.

11. If $4(2a - 1) < 8$, then $a < 1$.

12. If $2(-3b - 6) > 9b - 3$, then $b < \frac{3}{5}$.

In Exercises 13–16, match the inequality with its solution.

a. $x < -2$ **b.** $x < 2$ **c.** $x > -2$ **d.** $x > 2$

13. $2x + 13 > 9$

14. $-7x - 8 > 5x + 16$

15. $6 < 6(3 - x)$

16. $2(8 - 5x) < 4 - 4x$

In Exercises 17–25, solve the inequality.

17. $-11x + 3 < -30$

18. $\frac{1}{5}y + 12 \leq 8$

19. $5a + 6 \geq 14a - 9$

20. $8b - 9 < 2b - 13$

21. $\frac{3}{4}m \leq \frac{1}{4}m + 2$

22. $-\frac{1}{5}x > \frac{4}{5}x + 3$

23. $2(x + 1) \geq 3x - 2$

24. $4x + 1 \leq 2(x + 2)$

25. $-4x + 3 \geq -5x$

Consecutive Integers **In Exercises 26–28, let n, $n + 1$, and $n + 2$ be consecutive integers. Write the inequality that represents the verbal sentence. Then solve the inequality.**

26. The sum of 2 consecutive integers is less than or equal to 7.

27. The sum of 3 consecutive integers is more than 18.

28. The sum of 3 consecutive integers is less than 20.

Geometry **In Exercises 29 and 30, describe the possible values of x.**

29. The area of the rectangle is at least 28 square centimeters.

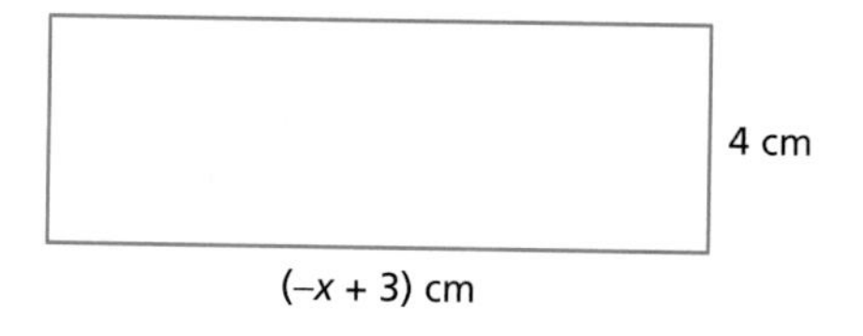

30. The perimeter of the triangle is less than or equal to 36 feet.

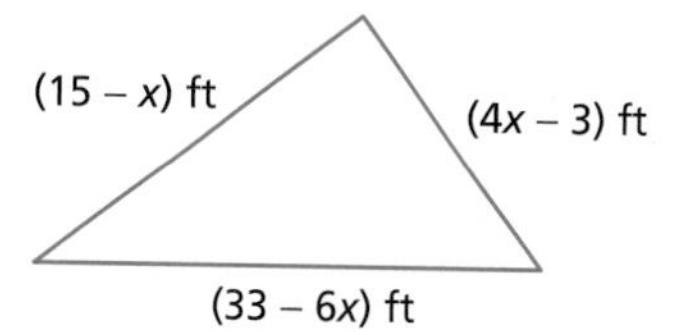

31. *Night Work* You are a food server at a restaurant. You earn \$4 an hour, plus tips. One night, you earned \$15 in tips and your total earnings were less than \$43. Describe the number of hours you could have worked.

32. *Carnival* You are going to a carnival. It costs \$10 to enter and \$0.25 each for tickets for games and rides. You don't want to spend more than \$20. Write an inequality that describes the number of tickets you can buy.

33. Create a table showing the results of Exercise 32.

34. *Riding a Bike* You rode your bike on a trail that is over 12 miles long. Your average speed was 15 miles per hour. Write an inequality that represents the time (in hours) that you rode your bike. Solve the inequality.

In a racing derby, the four horses in a row compete against each other. One of these very rare carousels is in Sandusky, Ohio; the other is in Rye, New York.

Integrated Review

Making Connections within Mathematics

Ordered Pairs **In Exercises 35–38, decide whether the ordered pair is a solution to the inequality. If not, find an ordered pair that *is* a solution.**

35. $x + 3y \le 6;\quad (0, 2)$

36. $2x - y > -4;\quad (-2, 1)$

37. $4x - 5y \le 21;\quad (3, -2)$

38. $-3x + 9y < 1;\quad (-3, -1)$

Exploration and Extension

39. *Equation Sense* In a coordinate plane, graph all ordered pairs, (x, y), for which the following is true: x and y are positive integers and $xy \le 8$.

40. *Equation Sense* In a coordinate plane, graph all ordered pairs, (x, y), for which the following is true: x and y are positive integers and $x + y \le 8$.

LESSON INVESTIGATION 9.8

Exploring Triangles

Materials Needed: paper, ruler

Example ***Exploring Triangles***

Draw a large triangle on a piece of paper. Then use a ruler to measure the lengths of the sides of the triangle. Compare the sum of the lengths of any two of the sides with the length of the third side.

Solution Begin by drawing a triangle. Then, use the ruler to measure the lengths of the sides of the triangle. In $\triangle ABC$ below, the lengths of the sides are about 6 inches, 4.75 inches, and 5.25 inches. You can compare the sum of the lengths of any two sides with the length of the third side, as follows.

Two Sides	Sum of Lengths of Two Sides	Length of 3rd Side
$\overline{AB}, \overline{AC}$	$4.75 + 6 = 10.75$ inches	$a = 5.25$ inches
$\overline{AB}, \overline{BC}$	$4.75 + 5.25 = 10$ inches	$b = 6$ inches
$\overline{AC}, \overline{BC}$	$6 + 5.25 = 11.25$ inches	$c = 4.75$ inches

In the table, notice that the sides of $\triangle ABC$ are denoted by $\overline{AB}$, $\overline{AC}$, and $\overline{BC}$, and the lengths of the sides are denoted by c, b, and a.

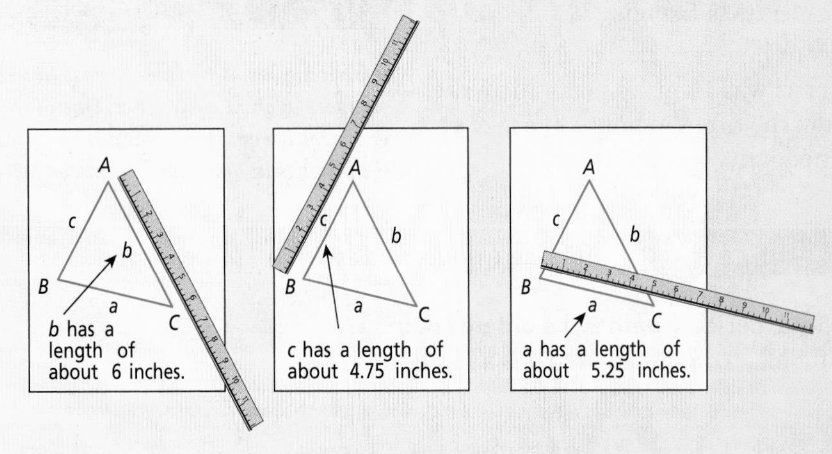

■

Exercises

1. Repeat the procedure shown in the example with two other triangles.
2. Compare your results with others in your class. Did anyone in the class find a triangle that has one side that is longer than the sum of the lengths of the other two sides? What can you conclude?
3. Let a, b, and c be the lengths of the sides of a triangle. Write three inequalities that compare a, b, and c.

9.8 The Triangle Inequality

What you should learn:

Goal 1 How to use the Triangle Inequality

Goal 2 How to use the Triangle Inequality to solve real-life problems

Why you should learn it:

You can use the Triangle Inequality to solve real-life problems, such as finding bounds on the distance between two locations.

Goal 1 Using the Triangle Inequality

In the *Lesson Investigation* on page 424, you may have discovered that the sum of the lengths of two sides of a triangle is always greater than the length of the third side. This result is called the **Triangle Inequality.**

> **The Triangle Inequality**
> The sum of the lengths of any two sides of a triangle is greater than the length of the third side.

In any triangle, the Triangle Inequality produces three inequalities. For instance, in $\triangle DEF$ at the right, you can write the following inequalities.

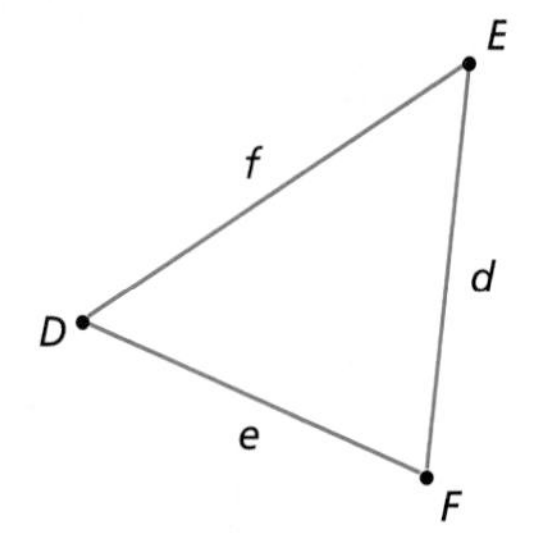

$$d + e > f$$
$$d + f > e$$
$$e + f > d$$

Example 1 *Using the Triangle Inequality*

You have moved to a new city, and are told that your apartment is 5 miles from your school and 6 miles from the restaurant where you have a part-time job. You are also told that your apartment, school, and restaurant do not lie on a straight line. Without knowing any other information, what can you say about the distance between your school and the restaurant?

Solution Begin by drawing a diagram, as shown at the left. Because your apartment, school, and restaurant do not lie on a line, they must form the vertices of a triangle. From the triangle, you can write the following inequalities:

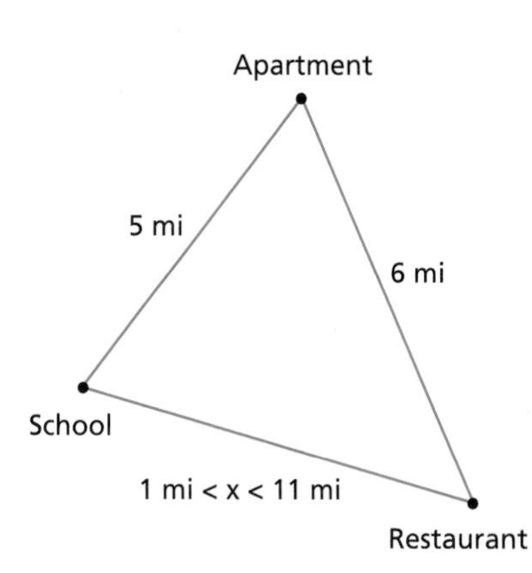

$$5 + x > 6 \quad \text{and} \quad 5 + 6 > x.$$

By solving these two inequalities, you can determine that the distance between your school and restaurant is *more than 1 mile* ($x > 1$) and *less than 11 miles* ($x < 11$). ■

Goal 2 Solving Real-Life Problems

Real Life
Art

Stained glass windows are made up of pieces of colored glass held together by strips of lead. Sunlight enhances the design effects.

Example 2 *Using the Triangle Inequality*

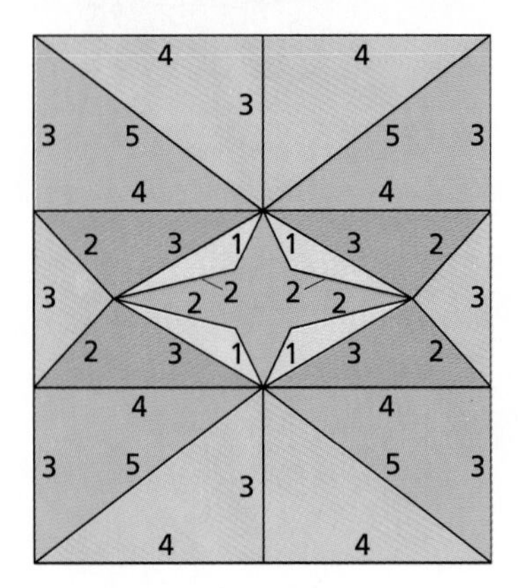

You are creating a design for a large stained-glass window. To begin, you draw a pattern and label the lengths (in feet) of each piece of glass. When a friend of yours looks at the plans, she says that there is something wrong with the measurements. How can she tell?

Solution There are four different sizes of triangles in the pattern. The smallest one must be mislabeled.

From the Triangle Inequality, it is possible to have a triangle whose sides are 3, 4, and 5. It is also possible to have a triangle whose sides are 2, 3, and 4, and one whose sides are 2, 2, and 3. It is *not,* however, possible to have a triangle whose sides have lengths of 1, 2, and 3 because the sum of the lengths of two sides would be equal to the length of the third side. This is not possible because the Triangle Inequality states that the sum of the lengths of any two sides must be *greater than* the length of the third side. ■

Communicating about MATHEMATICS

SHARING IDEAS about the Lesson

Using the Triangle Inequality Which of the following triangles must have at least one of its lengths listed incorrectly? Explain your reasoning.

A.

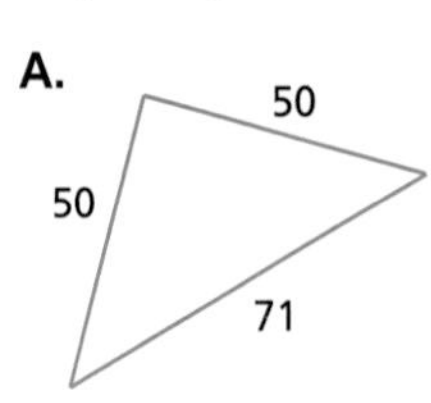

B.

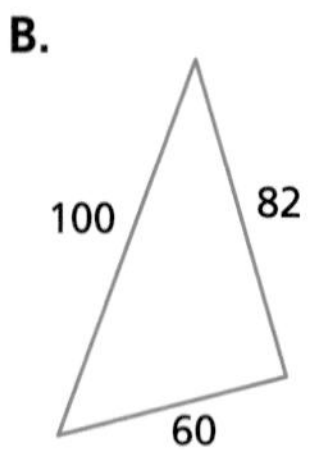

C.

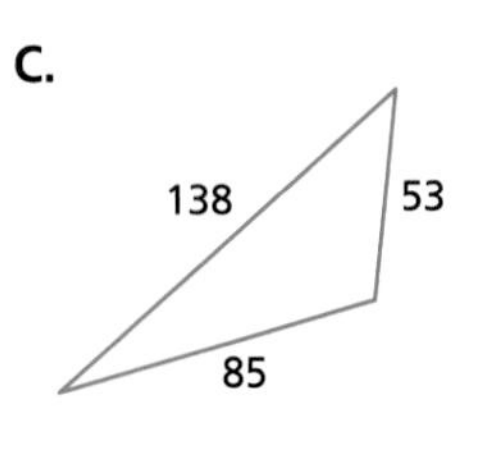

EXERCISES

Guided Practice

CHECK for Understanding

1. Can lengths of 5 cm, 6 cm, and 11 cm be sides of a triangle? Explain your answer with a sketch.
2. Which of the following are true statements about the lengths of the sides of $\triangle RST$? Explain.

   a. $r + s \geq t$　　b. $r > s + t$　　c. $r + s > t$

3. Which of the following cannot be the side lengths of a triangle? Explain.

   a. 3, 4, 6　　b. 4, 6, 10　　c. 5, 7, 11

In Exercises 4–6, the measures of two sides of a triangle are given. What can you say about the measure of the third side?

4. 3 and 5　　5. 9 and 11　　6. 16 and 20

Independent Practice

In Exercises 7–10, can the side lengths be correct? Explain.

7\. 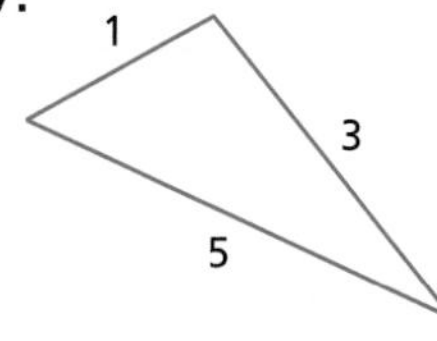

8\. 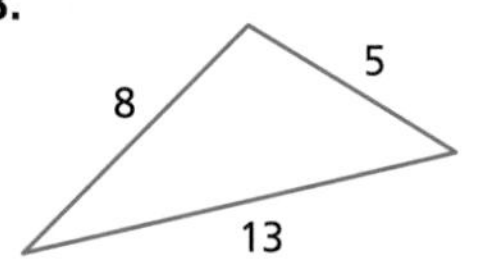

9\. 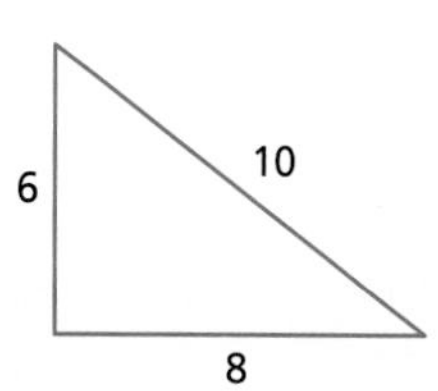

10\. 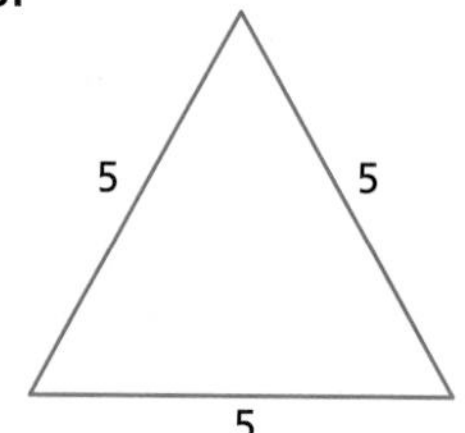

11. Copy and complete the table.

Measure of Side 1	Measure of Side 2	Measure of Side 3 is greater than	Measure of Side 3 is less than
3 cm	8 cm	?	?
9 in.	16 in.	?	?
10 ft	21 ft	?	?
30 m	45 m	?	?
100 cm	225 cm	?	?

In Exercises 12–17, can the numbers be side lengths of a triangle?

12. $\frac{5}{2}, \frac{7}{2}, \frac{9}{2}$　　13. $\sqrt{2}, \sqrt{3}, \sqrt{10}$　　14. 3.25, 6.79, 10.1

15. 4.06, 13.58, 17.21　　16. $\sqrt{15}, \sqrt{20}, \sqrt{65}$　　17. $\frac{1}{8}, \frac{1}{4}, \frac{1}{2}$

In Exercises 18–21, use the figure at the right to complete the statement.

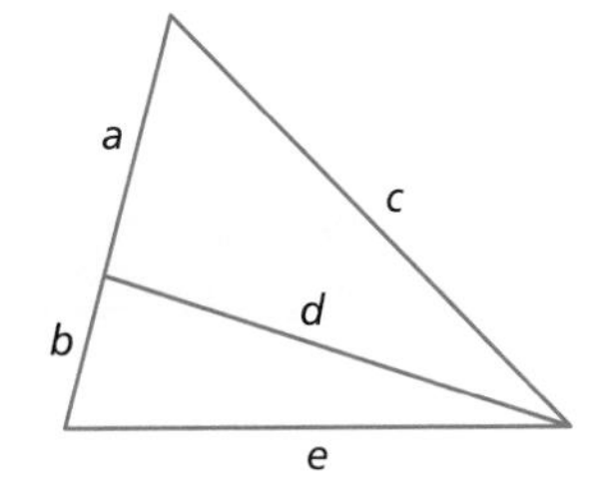

18. $b + d > \boxed{?}$

19. $a + b + \boxed{?} > e$

20. $a < d + \boxed{?}$

21. $b + \boxed{?} + e > c$

In Exercises 22 and 23, decide whether the string can be folded at the indicated points to form a triangle. Explain.

22. Fold Fold

23. Fold Fold

24. *Using Scissors* The blades on your scissors are both 3.75 inches long. The two tips of the blades and the pin that holds the scissors together can form a triangle. When a triangle is formed, how far apart can the tips of the blades be? Explain your reasoning.

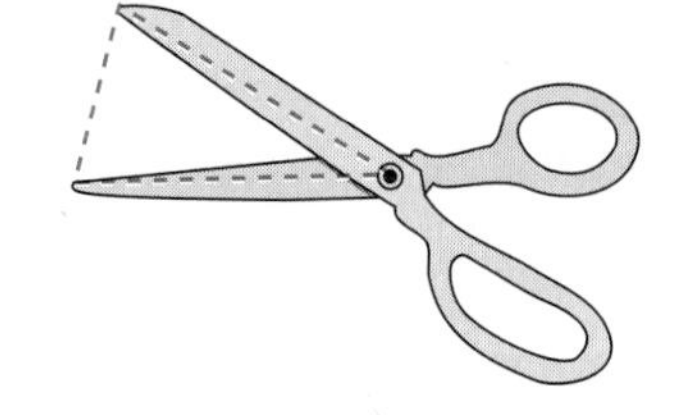

25. *Engineering* Engineers often use triangles to support structures that they build. After receiving the following partial blueprint design for a project, you discover that a mistake has been made. What is it?

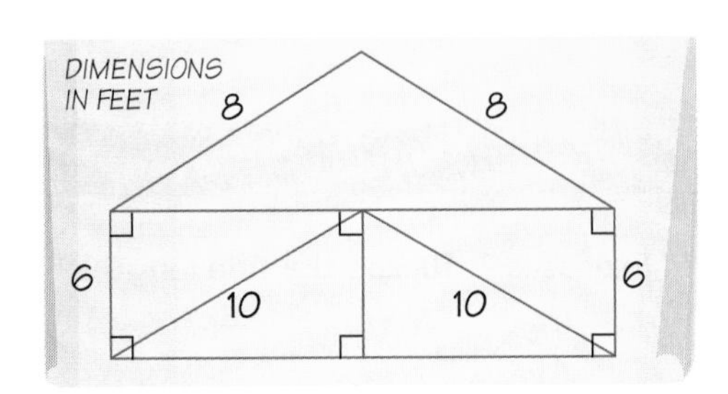

26. *Finding Perimeters* A triangle has side lengths of 11 centimeters and 14 centimeters. The perimeter of the triangle must be between what two numbers?

Integrated Review

Making Connections within Mathematics

Inequalities **In Exercises 27–30, solve the inequality.**

27. $2x + 17 > 33$

28. $21 - 3x \leq 36$

29. $4x + 2 < 3x + 7$

30. $6x - 3 \geq 2x + 25$

31. *Geometry* The triangle at the right is a right triangle. Find the perimeter of the triangle.

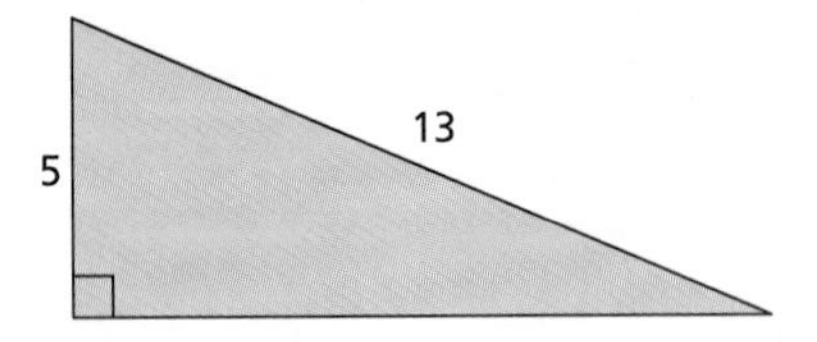

Exploration and Extension

Creating Triangles **In Exercises 32–34, you have 5 pencils whose lengths are 3 inches, 4 inches, 5 inches, 8 inches, and 10 inches.**

32. How many different sizes of triangles can you make with the pencils?

33. How many of the triangles are right triangles?

34. How many of the triangles have a perimeter that is divisible by 3?

Chapter Summary

What did you learn?

Skills

1. Use the Square Root Property to solve equations. **(9.1)**
2. Decide whether a real number is rational or irrational. **(9.2)**
   - Use the decimal representation of a real number to decide whether it is rational or irrational. **(9.2)**
   - Graph real numbers on a number line. **(9.2)**
3. Use the Pythagorean Theorem. **(9.3)**
   - Solve a right triangle. **(9.3)**
4. Graph an inequality on a number line. **(9.5)**
   - Decide whether two inequalities are equivalent. **(9.5)**
5. Solve an inequality.
   - Use addition or subtraction. **(9.5)**
   - Use multiplication or division. **(9.6)**
   - Use multiple steps. **(9.7)**
6. Use the Triangle Inequality. **(9.8)**

Problem-Solving Strategies

7. Model and solve real-life problems with equations. **(9.1–9.8)**
8. Model and solve real-life problems with inequalities. **(9.5–9.8)**

Exploring Data

9. Use tables and graphs to solve problems. **(9.1–9.8)**

Why did you learn it?

Not all of the numbers that occur in real-life situations are rational. For instance, you learned that if a right triangle has legs that are each 1-foot long, then the hypotenuse has a length of $\sqrt{2}$ feet or about 1.41 feet. In this chapter, you also learned that many real-life situations are better modeled by inequalities rather than equations. For instance, if you earn a commission of 5%, then the range of house prices H that will earn a commission of at least \$4000 is modeled by the inequality $0.05H \geq 4000$ or $H \geq 80{,}000$.

How does it fit into the bigger picture of mathematics?

Throughout history, people have developed different types of numbers to model different types of real-life situations. Thousands of years ago, primitive people needed only natural numbers such as 1, 2, 3, and 4. As civilization became more and more complicated, people needed rational numbers to measure half of a field or two-thirds of a sack of grain. Later, negative numbers and irrational numbers were needed to measure things such as temperatures below zero and lengths of hypotenuses of triangles. The study of different types of *real numbers* will help you prepare for future classes and occupations.

Chapter REVIEW

In Exercises 1–8, solve the equation. (9.1)

1. $x^2 = 49$ **2.** $y^2 = 144$ **3.** $35 = s^2$ **4.** $t^2 = 19$

5. $a^2 + 4 = 20$ **6.** $46 = b^2 - 18$ **7.** $5m^2 = 605$ **8.** $5n^2 = 25$

In Exercises 9–16, state whether the number is rational or irrational. Then match the number with its location on the number line. (9.2)

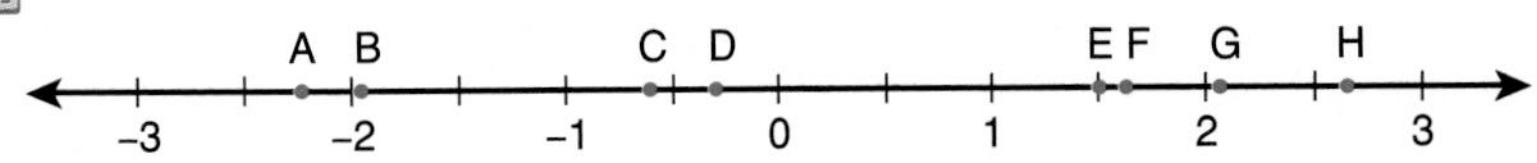

9. $\frac{13}{8}$ **10.** $-\frac{49}{25}$ **11.** $-\sqrt{5}$ **12.** $\sqrt{7}$

13. $\sqrt{2.25}$ **14.** $-\sqrt{0.09}$ **15.** $-\sqrt{\frac{3}{8}}$ **16.** $\sqrt{\frac{17}{4}}$

In Exercises 17–20, solve the right triangle. (9.3)

17.

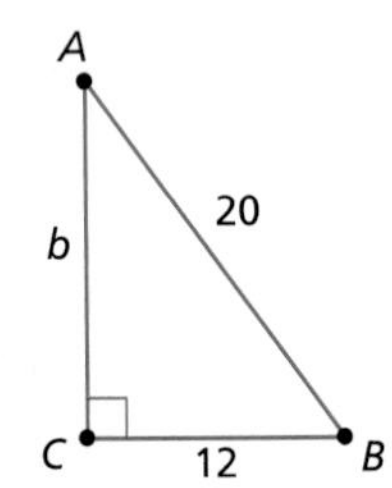

18.

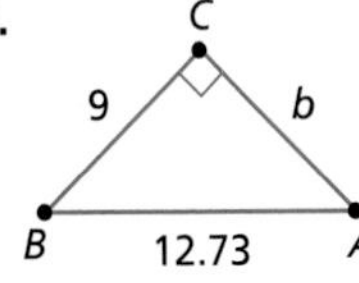

19.

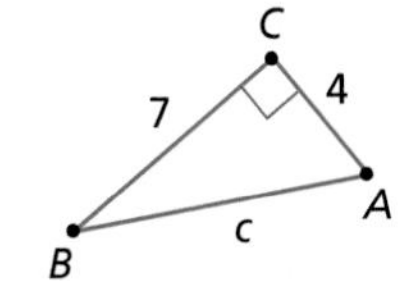

20.

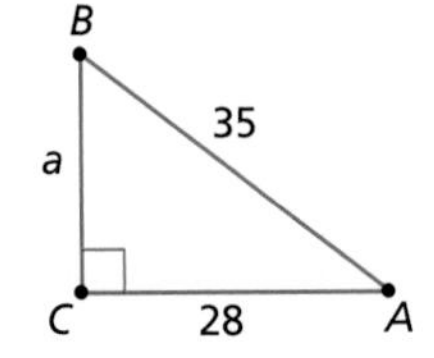

In Exercises 21–28, solve the inequality. Then write the solution in verbal form. (9.5, 9.6)

21. $x + 12 \le 7$ **22.** $-3 < y - 16$ **23.** $4 > m - 9$ **24.** $n + 8 \ge -19$

25. $5z \ge 40$ **26.** $\frac{x}{-2} < -10$ **27.** $-6p > \frac{2}{3}$ **28.** $-\frac{1}{4} \ge \frac{1}{10}q$

In Exercises 29–37, solve the inequality. Then graph the solution on a number line. (9.7)

29. $4x + 8 \le 13$ **30.** $-\frac{1}{7}y + 5 > 9$ **31.** $5x + 18 < -x$

32. $-6x + 12 \ge 3x - 3$ **33.** $-45 \le 4(-x - 9)$ **34.** $-\frac{1}{4}(2x + 8) < 5$

35. $3(-7 - x) > 19 + 5x$ **36.** $6(8 - 2x) \le 3(x - 14)$ **37.** $-3y + 4 \ge 2y - 6$

In Exercises 38–41, decide whether the triangle can have the given side lengths. Explain. (9.3, 9.8)

38.

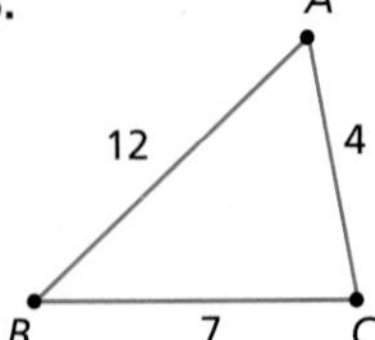

39.

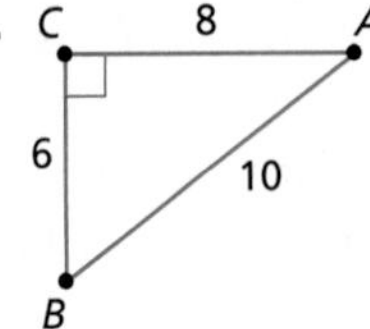

40.

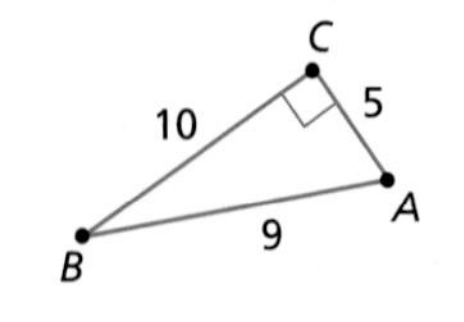

41.

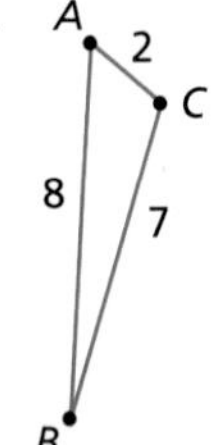

In Exercises 42–49, write an algebraic model for the sentence. Then solve. (9.1, 9.5–9.7)

42. The sum of b^2 and 11 is 75.

43. The product of 5 and t^2 is 125.

44. a plus 16 is less than 3.

45. x minus 7 is greater than 11.

46. y divided by -6 is greater than or equal to -8.

47. -4 times m is less than or equal to 52.

48. The difference of $9z$ and 14 is greater than 49.

49. The product of 2 and $(5 - x)$ is less than $(x + 11)$.

50. Explain why there are two solutions to $x^2 = 9$. **(9.1)**

51. *Monopoly* The board for a Monopoly game is square and has an area of 380.25 square inches. Write an equation that models the area of the board. Then solve to find its side length. **(9.1)**

52. *Geometry* The rectangle at the right has a perimeter that is at most 32 feet. Write an inequality that models this condition. Then solve the inequality. **(9.7)**

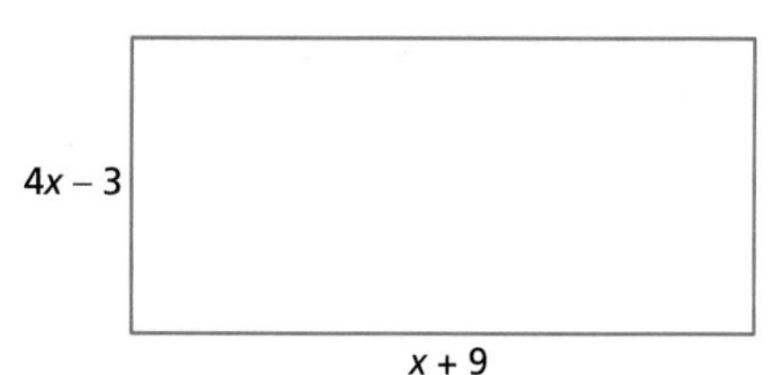

53. *Geometry* A triangle has side lengths 9 and 16. The perimeter of the triangle must be between what 2 numbers? **(9.8)**

In Exercises 54–56, find the length of the unlabeled side of the triangle. (9.3)

54.

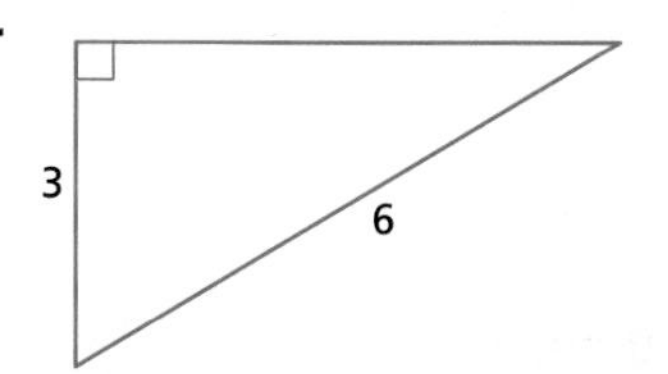

55.

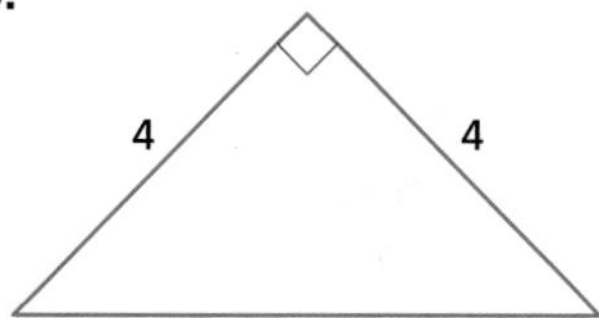

56.

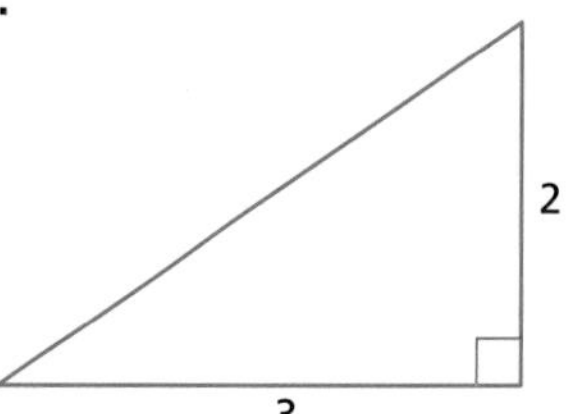

Bowling **In Exercises 57 and 58, use the following information.**

You and a friend go bowling. Your friend brings her own bowling shoes. You need to rent shoes, which costs \$1.25. Each game costs \$1.75. Your friend says she can spend at most \$7.00. You can spend at most \$10.00. **(9.5–9.7)**

57. Which of the following is the correct model for your friend's expenses? Solve the correct model and interpret the result.

a. $1.25x + 1.75 \leq 7$ **b.** $1.75x + 1.25 \leq 7$ **c.** $1.25x \leq 7$ **d.** $1.75x \leq 7$

58. Which of the following is the correct model for your expenses? Solve the correct model and interpret the result.

a. $1.25x + 1.75 \leq 10$ **b.** $1.75x + 1.25 \leq 10$ **c.** $1.25x \leq 10$ **d.** $1.75x \leq 10$

Solar Energy Facts **In Exercises 59–61, solve for *a, b,* or *c* to obtain the solar energy fact.**

a.

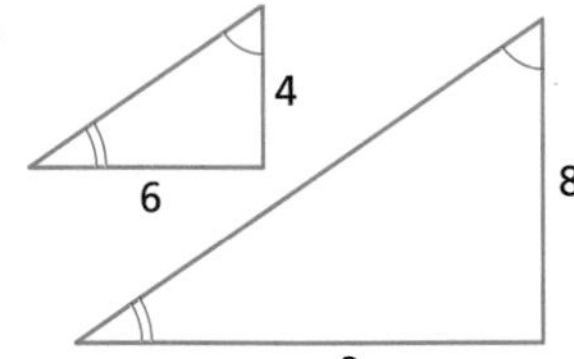

b.

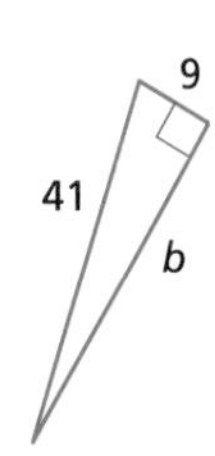

c.

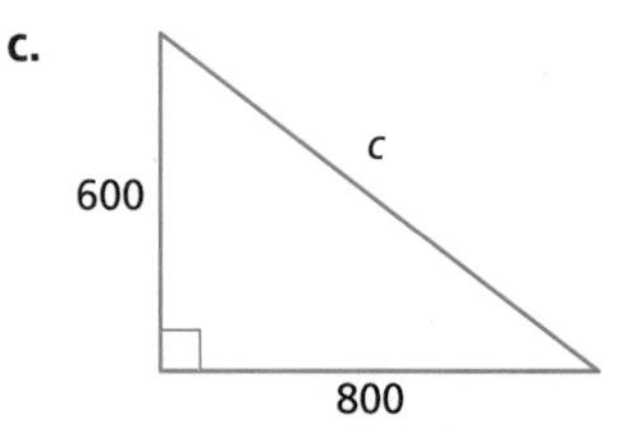

59. Every b minutes, the sun delivers as much energy to the earth's surface as all the people on earth use in one year.

60. The amount of solar energy falling on an area of 1 square yard on a sunny day is approximately c watts.

61. The solar energy that falls on the earth in a weeks is the equivalent of the world's entire initial reserves of coal and gas.

62. *The Odeillo Solar Oven* The Odeillo Solar Oven's mirror reflects the sun's rays into a solar furnace. Temperatures inside the furnace can reach up to 6870°F. Let T represent the temperature inside the furnace. Write an inequality for T.

63. *The Odeillo Solar Oven* The huge mirror actually consists of 9500 smaller square mirrors. The side length of each small mirror is about 18 inches. Approximate the total area of the mirror. Write your answer in square inches *and* square feet.

The Odeillo Solar Oven, built in France in the 1960s, is one of the most successful solar furnaces. It is also the largest solar-powered electric plant in Europe.

Types of Solar Energy **In Exercises 64 and 65, use the following information.**

There are two basic types of solar energy systems: photovoltaic systems and solar thermal systems. Photovoltaic systems convert sunlight directly into electricity by means of solar cells. These systems are quiet, require no fuel, and generate no pollution. Solar thermal systems do not generate electricity directly. Instead, they use the heat from the sun's rays to run generators that produce electricity.

64. What type of solar energy system do you think solar-powered calculators use? Explain.

65. What type of solar energy system does the Odeillo Solar Oven use? Explain.

Solar collectors are painted black since black absorbs sunlight more effectively. The glass covers protect the panels from the weather.

Chapter TEST

1. List both square roots of 225.
2. Solve the equation $a^2 + 3 = 39$.
3. A square has an area of 65.61 square units. What is the length of each side?
4. A right triangle has sides of 5 inches and 12 inches. Name two possibilities for the length of the third side.

In Exercises 5–8, match each number with its location on the number line.

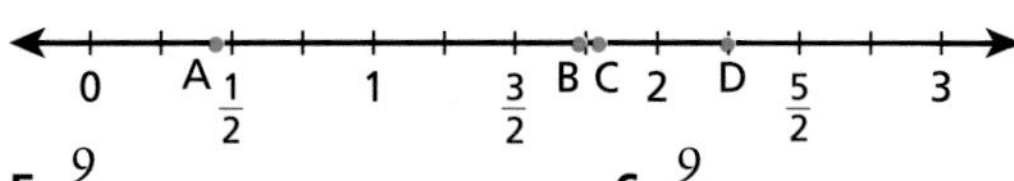

5. $\frac{9}{5}$
6. $\frac{9}{4}$
7. $\sqrt{3}$
8. $\sqrt{0.2}$

In Exercises 9–11, solve the triangle.

9. 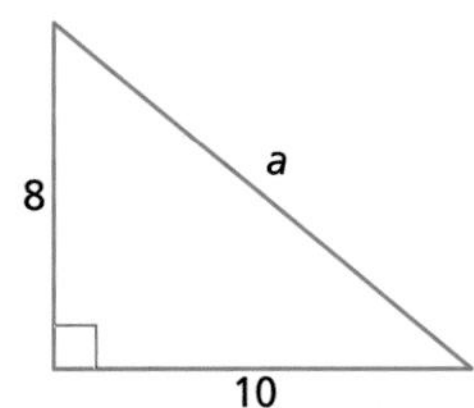

10. 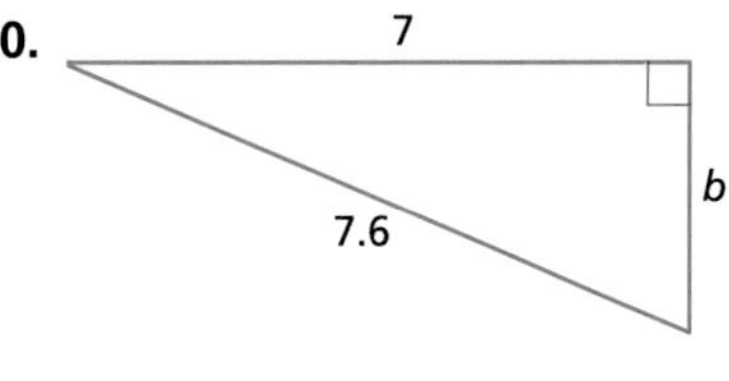

11. 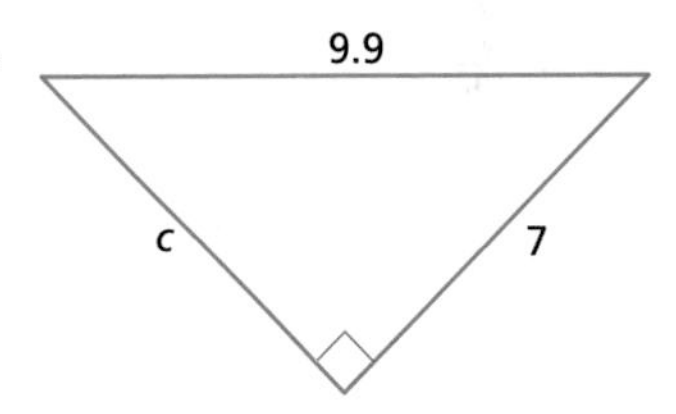

In Exercises12–14, graph the inequality on a number line.

12. $x \geq -4$
13. $x < 12$
14. $x > 5$

In Exercises 15–17, solve the inequality.

15. $-x < 2$
16. $-8r + 16 \leq 8$
17. $-2p \geq 5$
18. The measures of two sides of a triangle are 4 cm and 9 cm. What can you say about the measure of the third side?

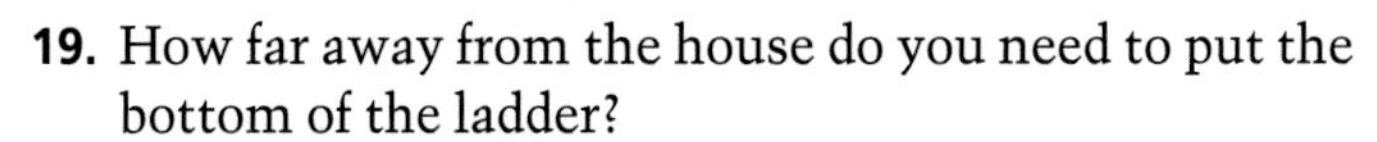

In Exercises 19–21, use the following.

You have a summer job of painting houses. You want to place a 20-foot ladder to the bottom of a window, as shown in the diagram at the right.

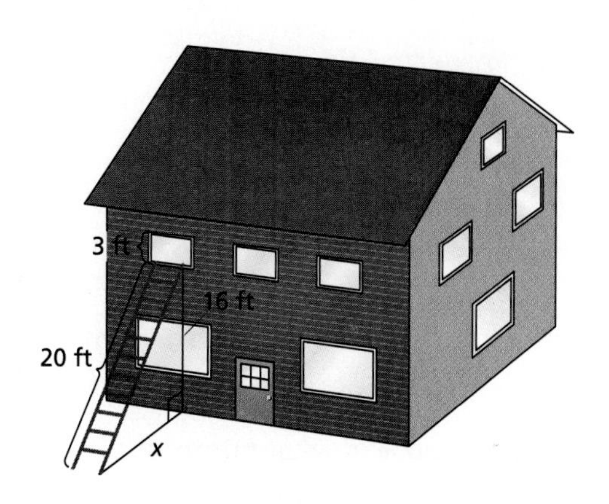

19. How far away from the house do you need to put the bottom of the ladder?
20. If you wanted to move the ladder just above the top of the window, how far away from the house would you have to put the bottom of the ladder?
21. To maintain balance, you don't want the bottom of the ladder to be closer than 5 feet from the house. What is the highest that the top of the ladder can be?

In Exercises 1–4, simplify the expression. (7.1–7.2)

1. $\frac{2}{7} + \frac{4}{7}$ **2.** $\frac{16x}{4} - \frac{14x}{4}$ **3.** $\frac{1}{4} + \frac{1}{6}$ **4.** $\frac{5}{8} - \frac{17}{32}$

In Exercises 5–8, use a calculator to evaluate the expression. Round the result to 2 decimal places. (7.3)

5. $\frac{64}{71} + \frac{57}{90}$ **6.** $\frac{104}{115}x - \frac{21}{37}x$ **7.** $1 - \left(\frac{3}{20} + \frac{7}{31}\right)$ **8.** $\frac{26}{9}t + \frac{85}{200}t$

In Exercises 9–12, simplify the expression. (7.4–7.5)

9. $\frac{5}{12} \cdot \frac{10}{3}$ **10.** $\frac{5}{2} \div \frac{1}{5}$ **11.** $\frac{7n}{4} \cdot 16$ **12.** $-\frac{6}{10} \div \frac{z}{5}$

In Exercises 13 and 14, find the perimeter and area of the figure. (7.1–7.5)

13. $5\frac{5}{6}$ in. $4\frac{1}{4}$ in.

14. $6\frac{7}{8}$ in. $6\frac{7}{8}$ in.

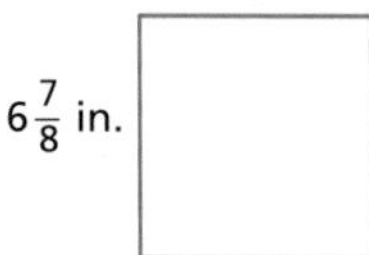

In Exercises 15–18, write the portion as a percent. (7.6)

15. $\frac{3}{10}$ **16.** $\frac{18}{25}$ **17.** $\frac{140}{175}$ **18.** $\frac{100}{250}$

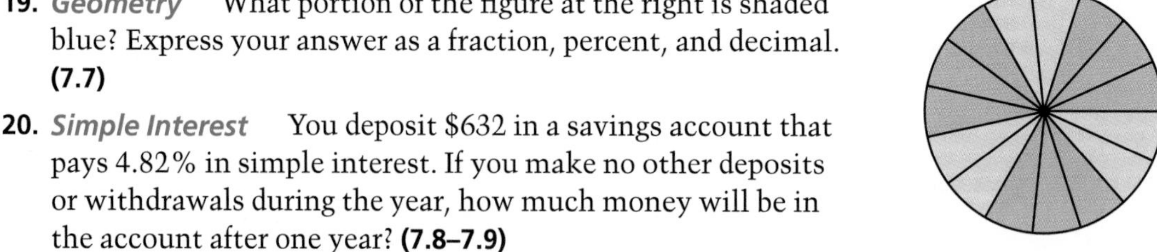

19. ***Geometry*** What portion of the figure at the right is shaded blue? Express your answer as a fraction, percent, and decimal. **(7.7)**

20. ***Simple Interest*** You deposit \$632 in a savings account that pays 4.82% in simple interest. If you make no other deposits or withdrawals during the year, how much money will be in the account after one year? **(7.8–7.9)**

In Exercises 21–24, decide whether the quotient is a rate or a ratio. Then simplify. (8.1)

21. $\frac{84 \text{ gallons}}{3 \text{ minutes}}$ **22.** $\frac{5 \text{ houses}}{3 \text{ houses}}$ **23.** $\frac{16 \text{ feet}}{24 \text{ inches}}$ **24.** $\frac{63 \text{ meters}}{1.5 \text{ seconds}}$

25. ***Property Tax*** If you pay \$2100 in property tax for a \$105,000 house, how much property tax would you pay for a \$140,000 house? **(8.3)**

In Exercises 26–29, solve the proportion. (8.2)

26. $\frac{3}{18} = \frac{t}{30}$ **27.** $\frac{12}{16} = \frac{27}{n}$ **28.** $\frac{1}{m} = \frac{1.5}{6}$ **29.** $\frac{x}{55} = \frac{5}{8}$

In Exercises 30–35, solve the percent equation. (8.4)

30. 63 is what percent of 90?

31. What is 85% of 40?

32. 80 is 50% of what number?

33. 95 is what percent of 125?

34. What is 52.5% of 230?

35. 105 is 150% of what number?

Favorite Movies **In Exercises 36–40, use the following. (8.5)**

You take a survey about favorite types of movies. The circle graph shows the results of your survey. Eighteen people said that science fiction movies were their favorite.

36. How many people were surveyed?

37. How many people said drama?

38. How many people said comedy?

39. How many people said western?

40. How many people said horror?

In Exercises 41 and 42, decide whether the change is an increase or decrease. Then find the percent. (8.6)

41. 1995: 207,100 units
1996: 215,025 units

42. Regular Price: $39.99
Sale Price: $25.99

Canadian Zip Codes **In Exercises 43–45, use the following. (8.7–8.8)**

In Canada, each zip code begins with a letter, then alternates between numbers and letters. How many zip codes have the indicated form?

43. L2R 1?0

44. H?C 4?9

45. L2R ???

46. Plot the set of numbers on a number line. **(9.2)**
$\left\{\frac{12}{15}, -\sqrt{3}, -\sqrt{\frac{81}{121}}, \frac{\sqrt{68}}{4}, -\frac{14}{29}, \sqrt{5.76}\right\}$

In Exercises 47–49, solve the right triangle. (9.3)

47.
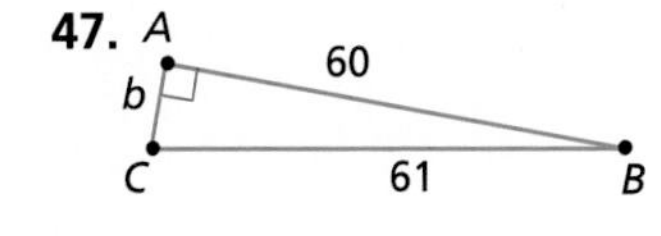

48.
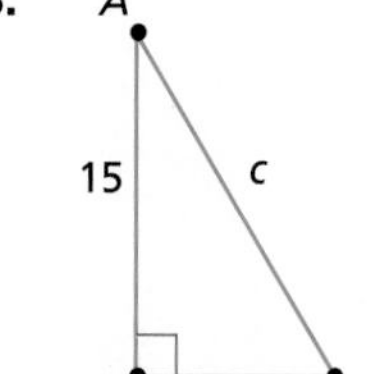

49.
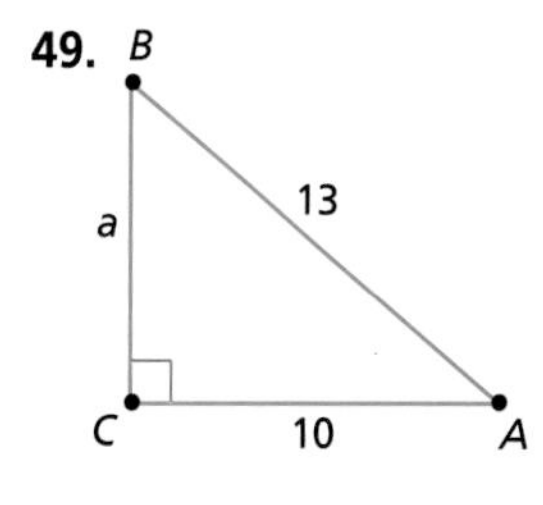

In Exercises 50–53, solve the inequality. Then graph the solution. (9.4–9.8)

50. $a + 12 < 7$

51. $b - 19 \geq -5$

52. $-4 \geq -6m$

53. $\frac{-n}{8} > \frac{2}{3}$

In Exercises 54–56, decide whether the numbers can be the lengths of three sides of a triangle. (9.8)

54. 5, 9, 13

55. $\frac{6}{17}, \frac{1}{4}, \frac{3}{8}$

56. $4, 7, \sqrt{128}$

CHAPTER

10 Geometry Concepts and Spatial Thinking

LESSONS

Breath-held divers have been recorded at depths of about −278 feet (feet below sea level). With the aid of SCUBA (Self Contained Underwater Breathing Apparatus) gear, divers have reached depths of −437 feet. By contrast, mammals like seals have been found at depths of about −1970 feet and sperm whales at about −3270 feet.

Real Life
Deep Sea Missions

Cathode Ray Display

Ships use sonar (SOund NAvigation and Ranging) to track icebergs, whales, and schools of fish. Sound pulses are sent out in all directions. When they encounter a large object, the sound wave bounces back as an echo. A cathode ray screen displays the echo as a blip of light.

The cathode ray display is a coordinate system. Angle measures are read along the outer rim. Distances from the center (the location of the ship) are read from the concentric circles. The measuring of angles and distances is just one of the real-life spatial relationships that you will encounter in this chapter.

LESSON INVESTIGATION 10.1

Exploring Three-Dimensional Objects

Materials Needed: toothpicks, small marshmallows

Example *Exploring Three-Dimensional Objects*

Use toothpicks and small marshmallows to build some three-dimensional objects. Each toothpick must have a marshmallow at each end, and each marshmallow must have at least 3 toothpicks stuck into it. After you have created the objects, count the number of marshmallows and toothpicks used. Then count the number of positions that the object can be turned so that the object can be set on a table.

Solution Two objects are shown below. The first uses 8 marshmallows and 12 toothpicks. This object can be placed on a table in 6 different positions. With each position, 4 marshmallows are on the table. The second object uses 5 marshmallows and 8 toothpicks. This object can be placed on a table in 5 different positions. With four of the positions, 3 marshmallows are on the table. With the fifth position, 4 marshmallows are on the table.

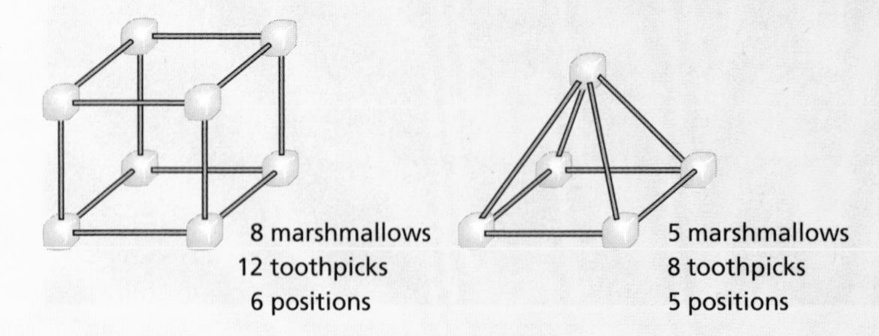

■

Exercises

Answer the following questions in groups. Before answering the questions, each member of your group should build at least two objects that are different from those shown above.

1. What is the fewest number of marshmallows used by a person in your group to form another object? Is this the fewest possible?

2. What is the fewest number of toothpicks used by a person in your group to form an object? Is this the fewest possible?

3. Using exactly 6 marshmallows, can you build two objects that do not look alike?

4. Name the shapes that are formed by the bottom of each object you have made as it rests on a table.

10.1 Exploring Points, Lines, and Planes

What you should learn:

How to identify points, lines, and planes in real-life situations

How to use geometry figures to solve real-life problems

Why you should learn it:

You can use geometry to solve real-life problems, such as planning a trip through several cities.

Goal 1 Identifying Points, Lines, and Planes

In Chapters 10, 11, and 12, you will learn how geometry can be used to model real-life situations. To become skilled in geometry, you must learn the meaning of the *words* of geometry and you must learn how to use the *properties* of geometry.

Pictured at the right are **points,** a **line,** a **ray,** a **line segment,** and a **plane.** You can't actually draw a line, ray, or plane because each extends forever in at least one direction. A line extends forever in two directions, a ray extends forever in only one direction, and a plane extends forever in many directions.

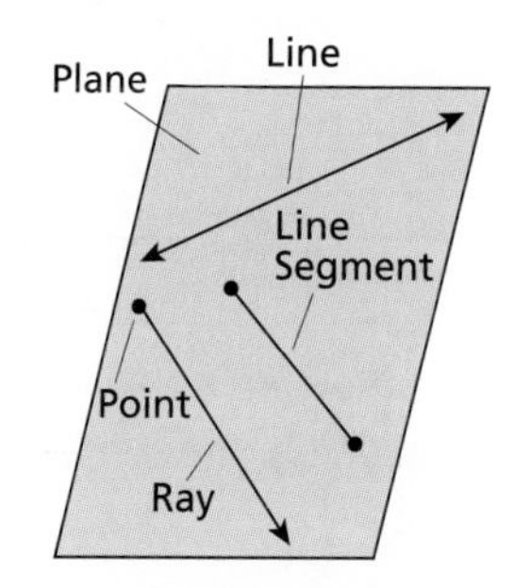

The **length** of a line segment $\overline{AB}$ is denoted by AB. Notice the difference between the symbols used to denote lines, rays, line segments, and lengths of line segments.

Line: R, T, S

Ray: P, O, Q

Line Segment: M, N

Name	Symbol
Line	$\overleftrightarrow{RS}$, $\overleftrightarrow{SR}$, $\overleftrightarrow{RT}$
Ray	$\overrightarrow{PQ}$, $\overrightarrow{PO}$
Line segment	$\overline{MN}$, $\overline{NM}$
Line segment length	MN, NM

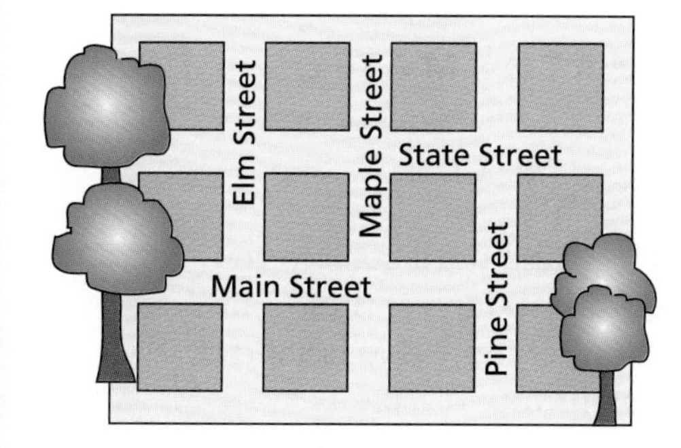

Exploring Parallel Lines

Two lines are **parallel** if they lie in the same plane and do not **intersect.** Which streets on the map at the left are parallel?

Solution Elm Street, Maple Street, and Pine Street are parallel. Main Street and State Street are also parallel. No other pair of streets are parallel because they intersect. In fact, the word *intersection* is used in everyday language to indicate the place where two streets or roads meet. ■

Goal 2 Solving Real-Life Problems

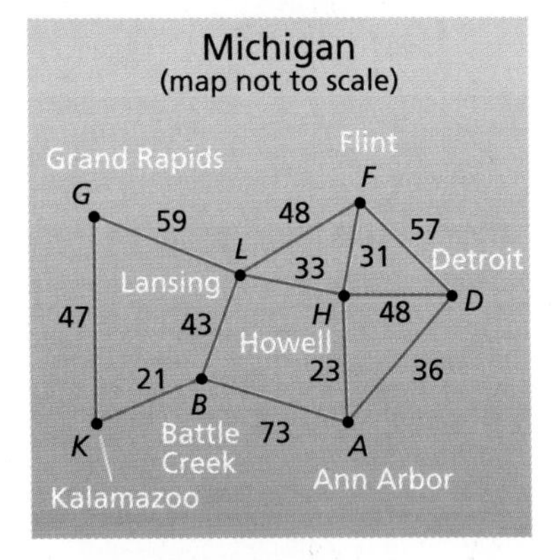

Example 2 *Using Line Segments*

You are planning a trip through eight cities in Michigan: Kalamazoo, Grand Rapids, Lansing, Howell, Flint, Ann Arbor, Battle Creek, and Detroit. You want to travel on the roads shown on the map at the left. Beginning and ending in Detroit, what is the least number of miles you can drive?

Solution Four possible routes are shown below. To find the distance along each route, add the lengths of the line segments that make up the route. For instance, the distance along Route 1 is

$$48 + 23 + 73 + 21 + 47 + 59 + 48 + 57 = 376 \text{ miles.}$$

In the *Communicating* feature below, you are asked to find the distances along the other routes.

The University of Michigan, located in Ann Arbor, was established in 1817.

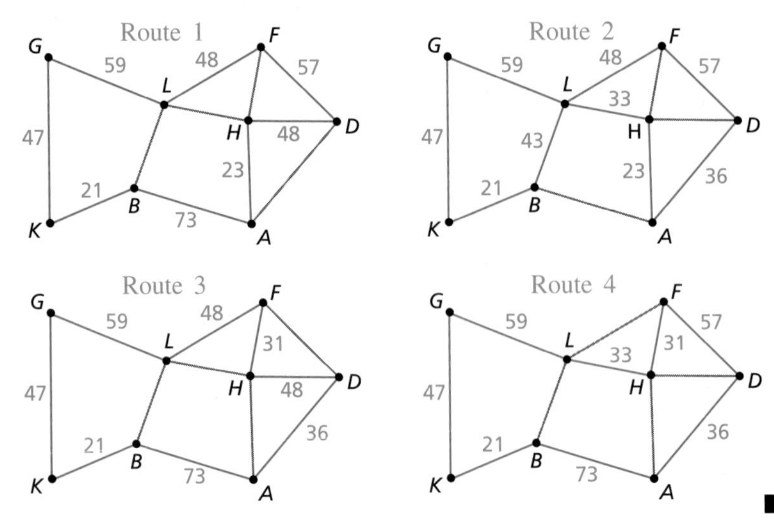

■

Communicating about MATHEMATICS

SHARING IDEAS about the Lesson

Comparing Distances In Example 2, find the distances you would travel along Routes 2, 3, and 4. Which of these four routes is the shortest? Is there another route that is shorter than the four routes shown? Explain your reasoning.

EXERCISES

Guided Practice

CHECK for Understanding

In Exercises 1–4, match the figure with its term.

a. A B　　b. S R　　c. T　　d. P Q

1. Ray　　**2.** Line Segment　　**3.** Line　　**4.** Plane

5. Write the symbol for each figure in Exercises 1–3.

6. *The Shape of Things* Describe examples of parallel lines and intersecting lines in your classroom.

Independent Practice

In Exercises 7–14, use the diagram at the right.

7. Write four other names for the line $\overleftrightarrow{CF}$.

8. Name 3 different line segments that lie on $\overleftrightarrow{AG}$.

9. Name 5 rays that have the same beginning point.

10. Name 2 lines that appear parallel.

11. Name 2 pairs of lines that intersect.

12. Are $\overrightarrow{EB}$ and $\overrightarrow{BE}$ the same ray? Explain.

13. Are $\overline{EI}$ and $\overline{IE}$ the same line segment? Explain.

14. Do DC and CD represent the same length? Explain.

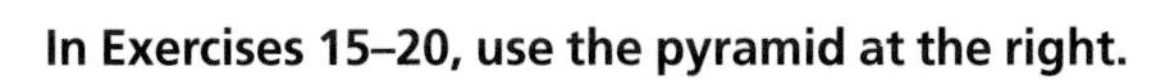

In Exercises 15–20, use the pyramid at the right.

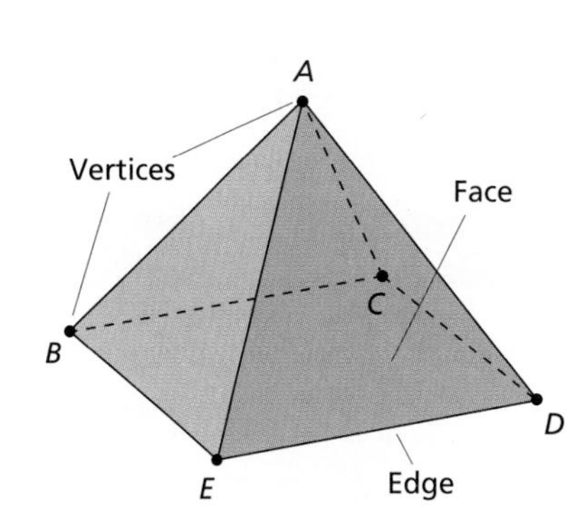

A **pyramid** is a space figure whose base is a polygon and whose other faces are triangles that share a common vertex.

15. How many planes form the pyramid's faces?

16. Name the line segments that form the pyramid's edges.

17. Name the points that form the pyramid's vertices.

18. Name four points that lie in the same plane.

19. Which line is parallel to $\overleftrightarrow{DE}$? to $\overleftrightarrow{CD}$?

20. Name four rays that have the same beginning point.

In Exercises 21–24, draw the indicated figure.

21. 3 lines that do not intersect

22. 3 lines that intersect in 1 point

23. 3 lines that intersect in 2 points

24. 3 lines that intersect in 3 points

25. *Visualizing Planes* Each figure shows three planes. Explain how the planes differ.

a. 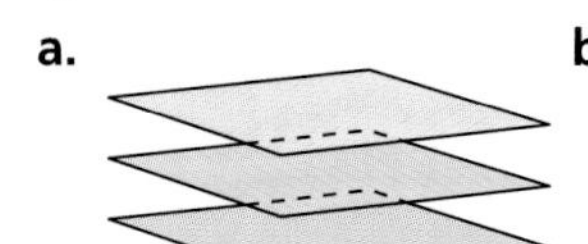

b. 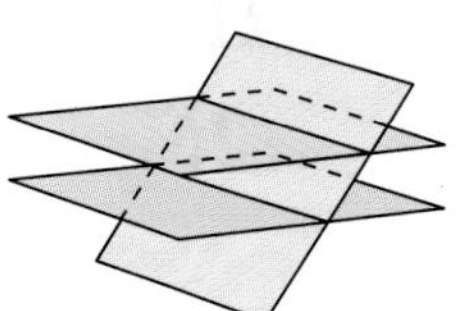

c. 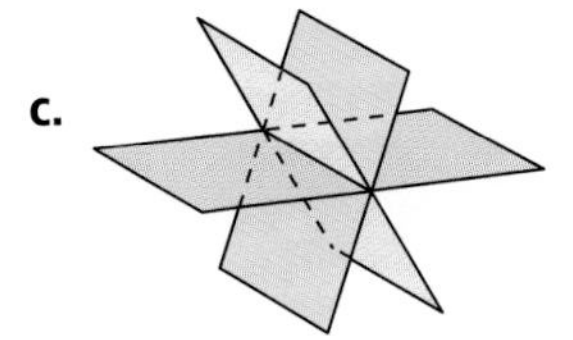

d. 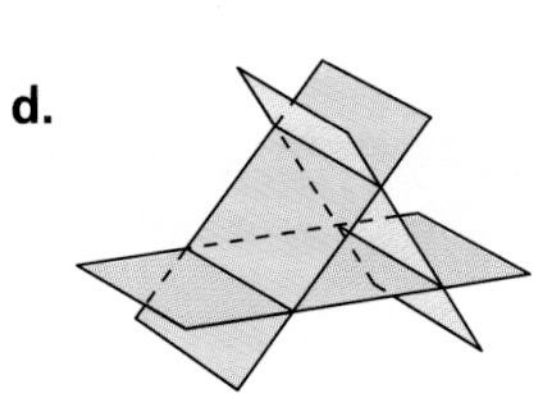

Architecture **In Exercises 26–30, use the building at the right.**

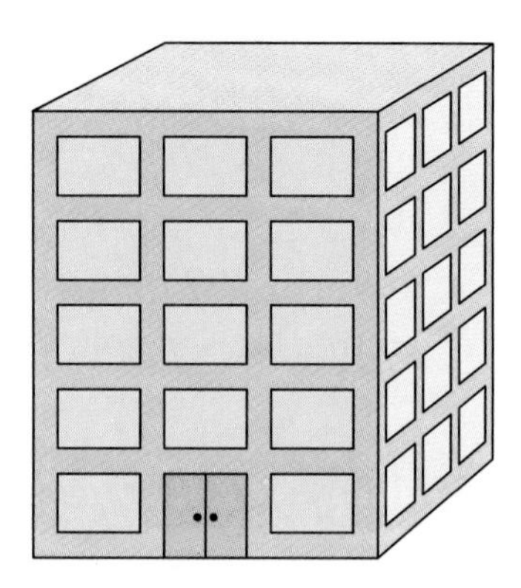

Each window in the building is 10 feet wide and 7 feet high. The front doors of the building are 10 feet wide and 10 feet high. The space between the windows is 3 feet, and the space between the windows and the edges of the building is 3 feet.

26. Describe some windows that lie in the same plane.

27. Is the base of every window parallel to the base of every other window? Explain.

28. What are the dimensions of the building's base?

29. How tall is the building?

30. What percent of the front of the building is glass?

Integrated Review

Making Connections within Mathematics

Drawing Lines **In Exercises 31–34, copy the points on a piece of paper. Then draw and name all the lines that pass through pairs of points.**

31. A B

32. A C B

33. A B D C

34. B A C E D

35. *Patterns* Describe how the pattern for the number of lines found in Exercises 31–34 are related to Pascal's triangle (page 372).

Exploration and Extension

Folding Cubes **In Exercises 36 and 37, use the nets (patterns) shown below.**

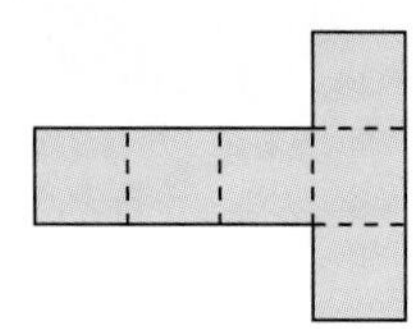

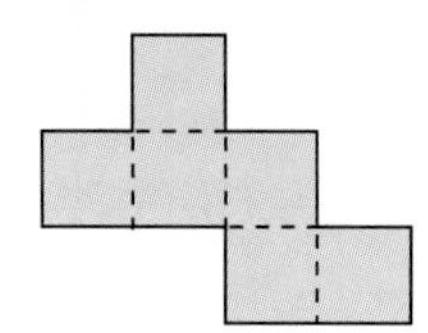

36. Use graph paper to draw, cut, and fold each net to form a cube.

37. Find a different net that you can cut and fold to form a cube.

10.2 Angles: Naming, Measuring, Drawing

What you should learn:

How to identify angles as acute, right, obtuse, or straight

How to use angle measures to analyze real-life situations

Why you should learn it:

You can use angles to solve real-life problems, such as finding angles that produce certain types of reflections in mirrors.

Babylonian Geometry ***The concept of degree measure began with the ancient Babylonians in northern Africa. They divided a full circle into 360 degrees.***

Goal 1 Identifying Angles

An **angle** consists of two rays that begin at the same point. The rays are the **sides** of the angle, and the point is the **vertex** of the angle. The angle at the right can be denoted by

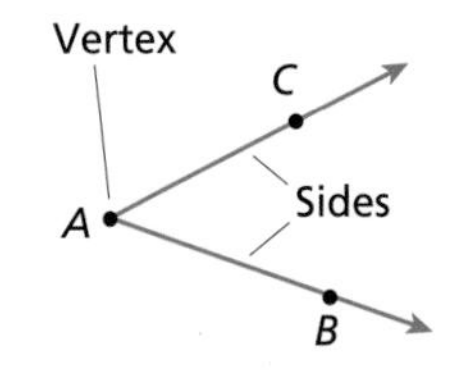

$\angle BAC$, $\angle CAB$, or $\angle A$.

The **measure** of $\angle A$ is denoted by $m\angle A$. A **protractor** can be used to approximate the measure of an angle. An **acute angle** measures between 0° and 90°. A **right angle** measures 90°. An **obtuse angle** measures between 90° and 180°. A **straight angle** measures 180°. Two lines or line segments that meet at a right angle are **perpendicular.**

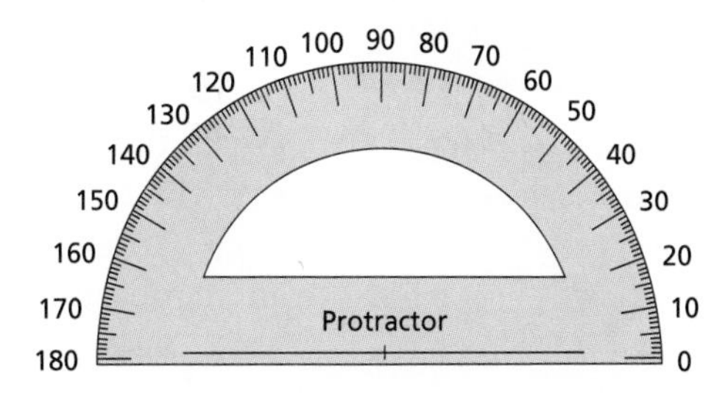

LESSON INVESTIGATION

Investigating Angle Measures

Group Activity Use a scrap of paper to fold an obtuse angle and an acute angle, as shown below. Use a protractor to measure each of your angles.

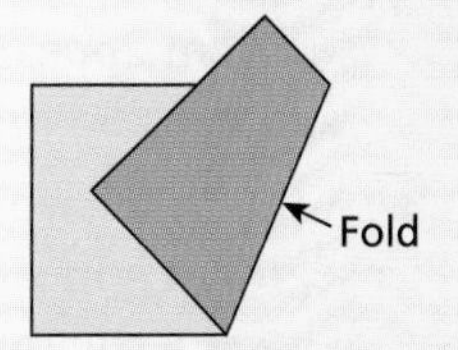

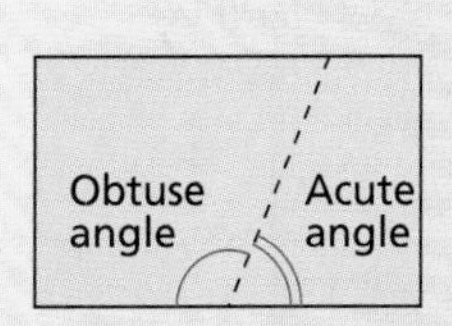

Two angles are **congruent** if they have the same measure. Are any of the angles folded by your group congruent?

Goal 2 Measuring Angles in Real Life

Example 1 *Measuring Angles*

Draw a line on a piece of paper. Then tape two mirrors together and place the mirrors on the paper so that a segment appears to form a hexagon and an octagon. Measure the angles between the two mirrors.

The property of mirrors that is illustrated in Example 1 is used to build kaleidoscopes.

Solution

As shown at the right, there is an acute angle that will make a segment appear to form a hexagon. Once you can see a hexagon, trace the angle between the two mirrors and use a protractor to measure the angle. You will find that the angle measures 60°.

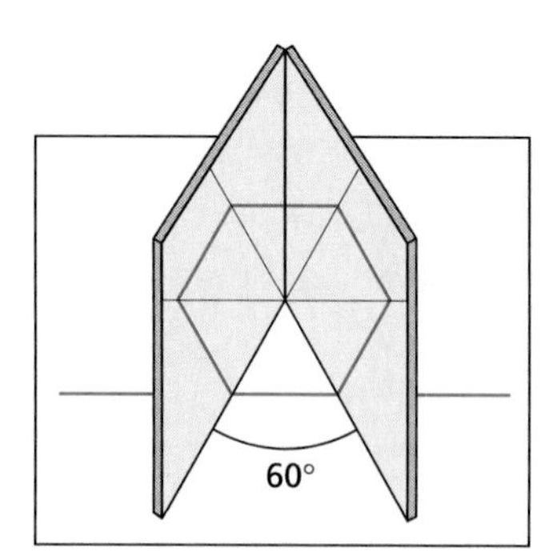

As shown at the right, there is an acute angle that will make a segment appear to form an octagon. Once you can see an octagon, trace the angle between the two mirrors and use a protractor to measure the angle. You will find that the angle measures 45°.

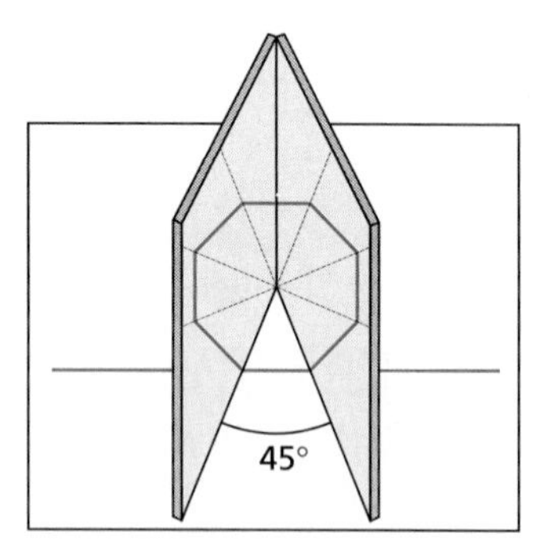

Communicating about MATHEMATICS

Cooperative Learning

SHARING IDEAS about the Lesson

Measuring Angles Work with a partner.

A. Adjust the two taped mirrors in Example 1 so that a segment appears to form a pentagon. What is the angle between the mirrors?

B. Adjust the two taped mirrors in Example 1 so that a segment appears to form a square. What is the angle between the mirrors?

C. Can you form other polygons with the mirrors? If so, what angle is associated with each polygon?

EXERCISES

Guided Practice

CHECK for Understanding

In Exercises 1–4, use the figure at the right.

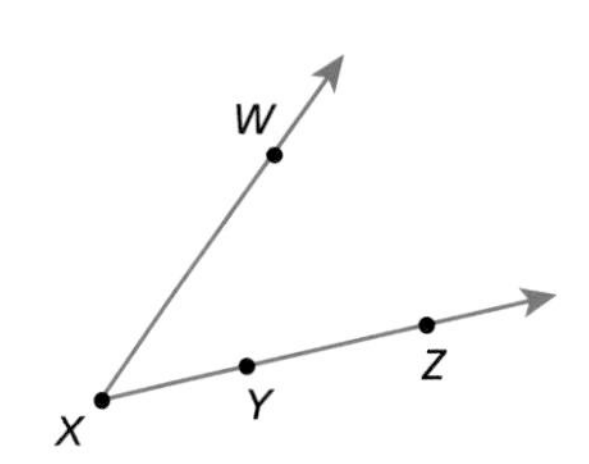

1. Name the vertex of $\angle WXZ$.

2. Name the sides of $\angle WXZ$.

3. State other names for $\angle WXZ$.

4. Which appears to have the greater measure: $\angle WXZ$ or $\angle WYZ$?

In Exercises 5–8, use a protractor to measure the angle. Is the angle acute, obtuse, right, or straight?

5.

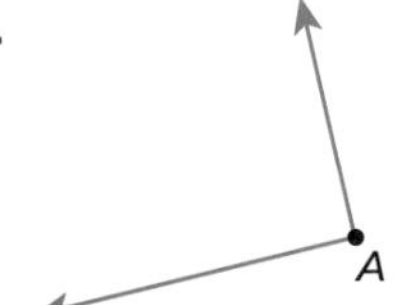

6.

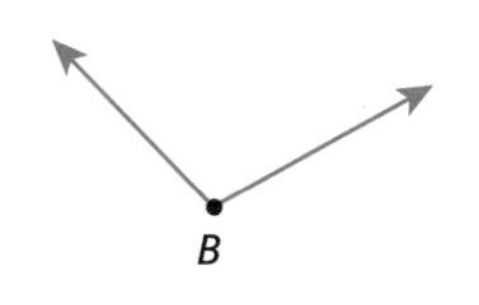

7.

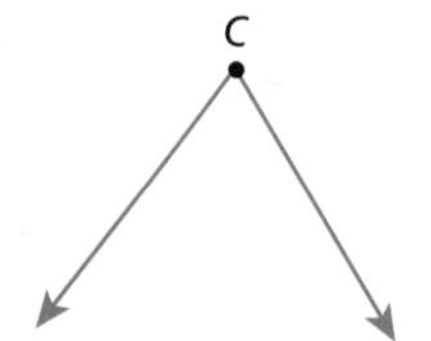

8.

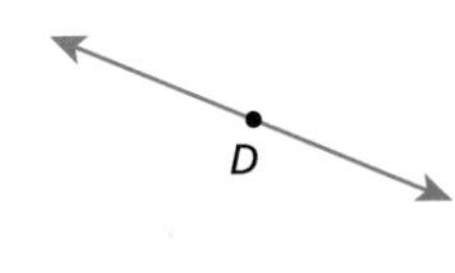

Independent Practice

In Exercises 9–14, use the figure at the right.

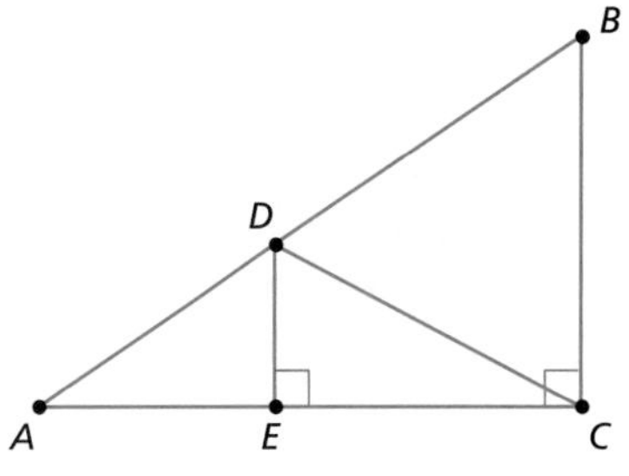

9. List four other names for $\angle B$.

10. How many acute angles are in the figure? List them.

11. How many obtuse angles are in the figure? List them.

12. How many right angles are in the figure? List them.

13. How many straight angles are in the figure? List them.

14. Identify the vertex and sides of $\angle CDE$. Explain why you can't simply name the angle as $\angle D$.

In Exercises 15–18, without using a protractor, match the angle with its measure.

a. $45°$ **b.** $90°$ **c.** $180°$ **d.** $135°$

15. **16.** **17.** **18.**

In Exercises 19–22, use a protractor to draw an angle with the indicated measure.

19. $30°$ **20.** $90°$ **21.** $125°$ **22.** $135°$

In Exercises 23–26, use a protractor to measure the angle.

23. 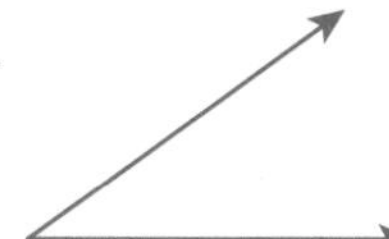

24. 

25. 

26. 

27. *Writing* In your own words, explain what it means for two angles to be congruent.

28. Which of the following angles are congruent?

a. 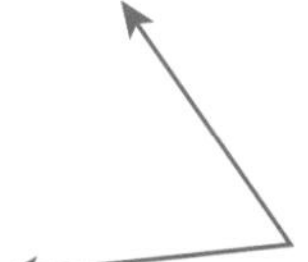

b. 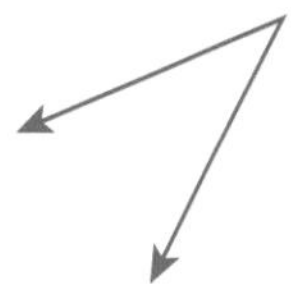

c. 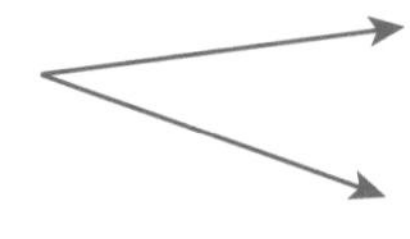

d. 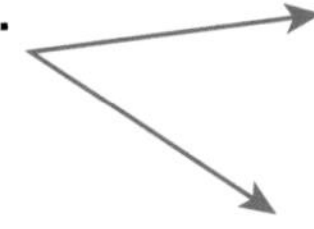

Clock Faces **In Exercises 29–32, determine what type of angle (acute, right, obtuse, or straight) the hands of a clock make at the given time.**

29. 6:00 P.M. **30.** 9:00 A.M. **31.** 4:00 A.M. **32.** 10:00 P.M.

Birds in Flight **In Exercises 33–35, measure the angle from the wingtip to the beak to the other wingtip. Name the type of angle formed.**

33. 

34. 

35. 

Integrated Review

Making Connections within Mathematics

Algebra **In Exercises 36–38, use the fact that the two angles are congruent to solve for *x*. Then find the measures of the angles.**

36. 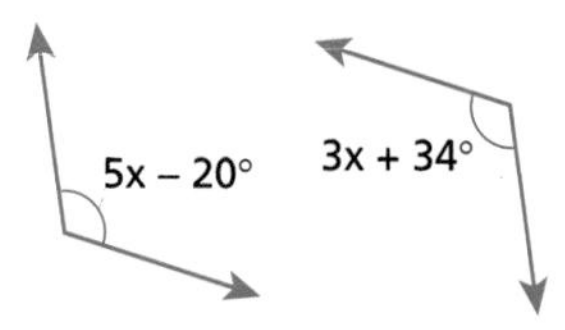

37. 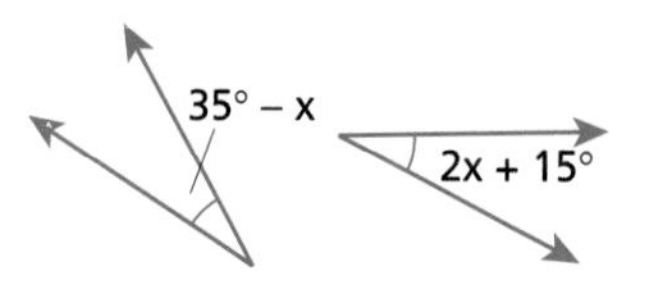

38. 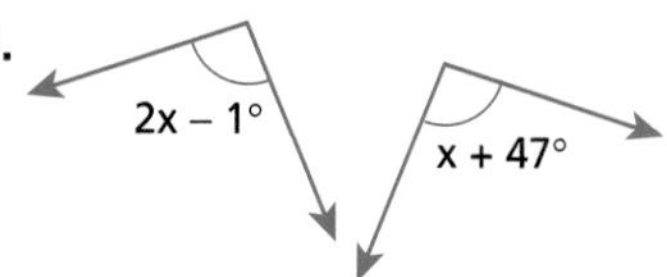

39. *Estimation* Which of the following is the best estimate for the angle of the tip of a star on a United States flag?

a. 5° **b.** 36° **c.** 90° **d.** 120°

Exploration and Extension

40. *Intersecting Lines* Use a straightedge to draw two lines that intersect to form two acute angles and two obtuse angles. Are the acute angles congruent? Are the obtuse angles congruent? Explain.

LESSON INVESTIGATION 10.3

Exploring Parallel Lines

Materials Needed: ruled paper, straightedge, colored pencils, scissors

Example ***Exploring Parallel Lines***

Use a pencil and straightedge to darken three lines on a piece of ruled paper. Then, use the pencil and straightedge to draw two other parallel lines, one on each side of the straightedge, as shown below. Draw these lines at a slant—not at right angles to rules on the paper. On another piece of paper, carefully trace one of the quadrilaterals formed by the parallel lines. Cut out the quadrilateral and use it to compare the angles formed by the parallel lines. What can you conclude?

Solution By moving the quadrilateral around on the paper, you can determine that $\angle 1$, $\angle 2$, and $\angle 3$ have the same measure, which means they are congruent. You can also determine that $\angle 4$, $\angle 5$, and $\angle 6$ have the same measure, which means that they are also congruent. ■

Exercises

1. Try the investigation described in the example. Make the slanted parallel lines meet the horizontal lines at any angle except a right angle. The five lines should form 24 angles. Use colored pencils to indicate angles that are congruent.
2. In the quadrilateral that you cut out, are any of the four angles congruent? How can you tell?
3. Draw two parallel lines. Then draw a third line that intersects each of the first two lines at a slant. How many angles are formed by the three lines? Use colored pencils to classify the angles into congruent groups.

10.3 Exploring Parallel Lines

What you should learn:

Goal 1 How to identify angles formed when two parallel lines intersect a third line

Goal 2 How to use a property of parallel lines to solve real-life problems

Why you should learn it:

You can use a property of parallel lines to solve real-life problems, such as measuring angles in construction projects.

Goal 1 Using a Property of Parallel Lines

In the *Investigation* on page 447, you may have discovered that when two parallel lines are intersected by a third line, several pairs of congruent angles are formed. To describe the results, it helps to classify pairs of angles as **vertical angles** and **corresponding angles.**

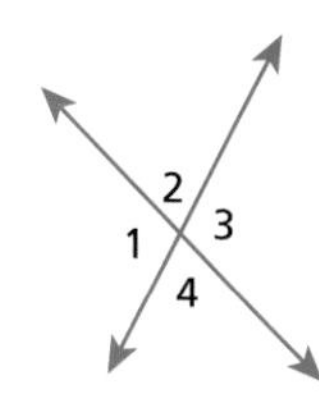

Vertical angles: $\angle 1$ and $\angle 3$, $\angle 2$ and $\angle 4$

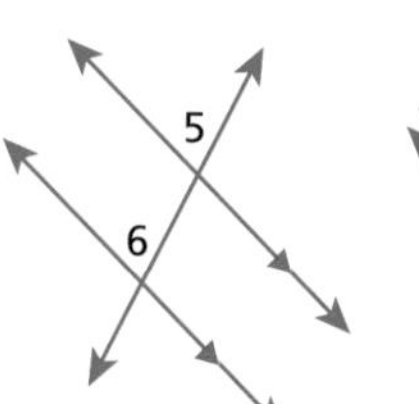

Corresponding angles: $\angle 5$ and $\angle 6$, $\angle 7$ and $\angle 8$

Vertical Angles and a Property of Parallel Lines

1. Vertical angles are congruent.
2. When two *parallel* lines are intersected by a third line, the corresponding angles are congruent. (If the lines are not parallel, then the corresponding angles are not congruent.)

Example 1 *Identifying Congruent Angles*

In the diagram at the left, identify all congruent vertical angles and all congruent corresponding angles.

Solution The three lines form four sets of vertical angles.

$\angle 1 \cong \angle 4$, $\angle 2 \cong \angle 3$ *Vertical angles*
$\angle 5 \cong \angle 8$, $\angle 6 \cong \angle 7$ *Vertical angles*

The symbol $\cong$ means "is congruent to." Because m and n are parallel, the three lines also form four sets of congruent corresponding angles.

$\angle 1 \cong \angle 5$, $\angle 2 \cong \angle 6$ *Corresponding angles*
$\angle 3 \cong \angle 7$, $\angle 4 \cong \angle 8$ *Corresponding angles* ■

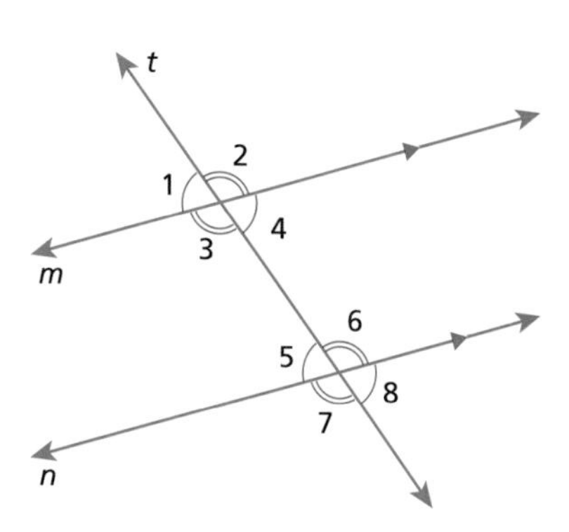

The red arrowheads indicate that the lines m and n are parallel.

Goal 2 Solving Real-Life Problems

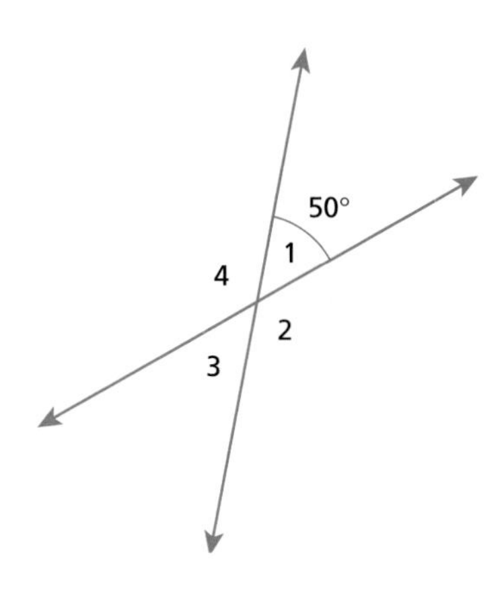

Example 2 *Finding Angle Measures*

In the diagram at the left, $m\angle 1 = 50°$. Find the measures of the other three angles.

Solution Because $\angle 1$ and $\angle 3$ are vertical angles, they are congruent and must have the same measure. Thus, $m\angle 3 = 50°$.

Because $\angle 1$ and $\angle 2$ combined form a straight angle, you know that the sum of their measures is 180°. Thus,

$$m\angle 2 = 180° - m\angle 1 = 180° - 50° = 130°.$$

Because $\angle 2$ and $\angle 4$ are vertical angles, they are congruent and you can conclude that $m\angle 4 = 130°$.

Real Life
Construction

Example 3 *Identifying Congruent Angles*

In the photo at the left, the lower man is sitting on a pair of cross beams. Draw a diagram of the beams and label the congruent angles.

Solution The beams form two pairs of parallel lines, as shown at the right. By using vertical angles and a property of parallel lines, you can conclude that

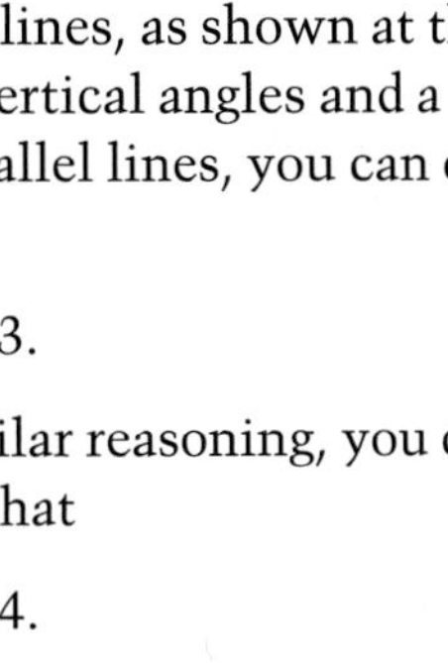

$$\angle 1 \cong \angle 3.$$

Using similar reasoning, you can also conclude that

$$\angle 2 \cong \angle 4.$$

In the *Communicating* feature below, you are asked to explain how you can use vertical angles and a property of parallel lines to make these conclusions. ■

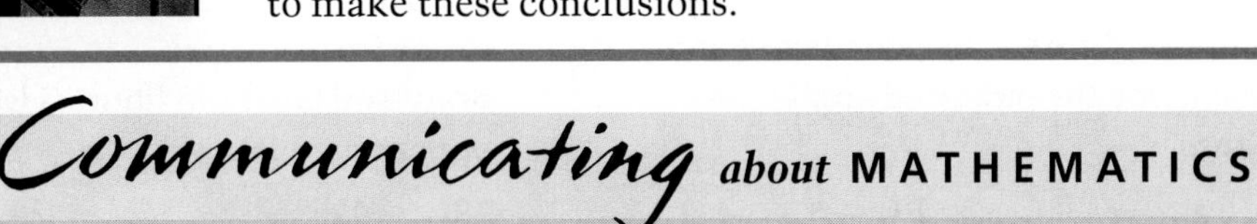

Communicating about MATHEMATICS

SHARING IDEAS about the Lesson

Extending the Example In Example 3, explain why $\angle 1 \cong \angle 3$ and $\angle 2 \cong \angle 4$. If $m\angle 1 = 84°$, what are the measures of the other three angles?

EXERCISES

Guided Practice

CHECK for Understanding

In Exercises 1–6 , use the figure at the right.

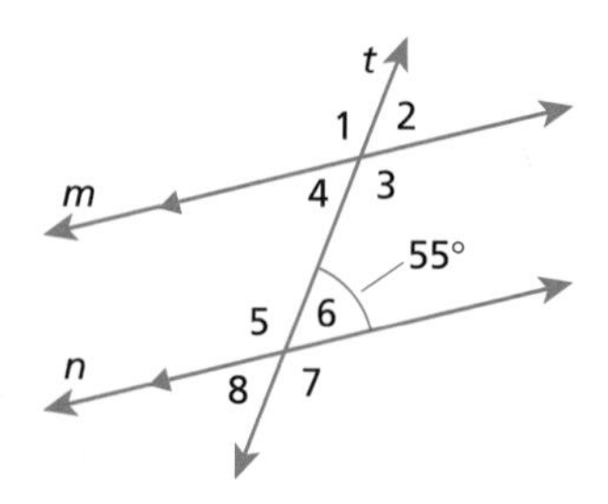

1. Which two lines are parallel?
2. Name four pairs of vertical angles.
3. Name four pairs of corresponding angles.
4. What is the measure of $\angle 2$? Explain your reasoning.
5. What is the measure of $\angle 4$? Explain your reasoning.
6. What is the measure of $\angle 8$? Explain your reasoning.
7. Two lines intersect to form four angles. One of the angles measures 45°. What do the other three angles measure? Explain your reasoning.
8. Draw two parallel lines. Then draw a line that intersects one of the lines to form a 60° angle. Of the eight angles that are formed, how many measure 60°?

Independent Practice

In Exercises 9–16, use the figure at the right.

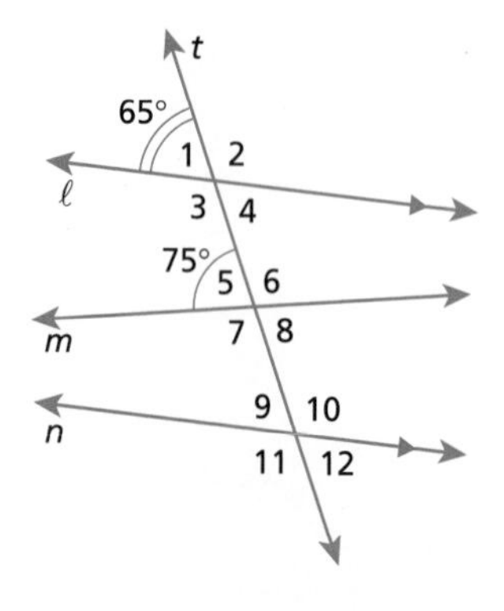

9. Which two lines are parallel?
10. Explain why $\angle 4$ is not congruent to $\angle 8$.
11. List all angles whose measure is 65°.
12. List all angles whose measure is 75°.
13. List all angles whose measure is 115°.
14. List all angles whose measure is 105°.
15. Name two corresponding angles that have different measures.
16. Name two corresponding angles that have the same measure.

Parallelogram Grids **In Exercises 17 and 18, explain why the indicated angles are congruent.**

17.

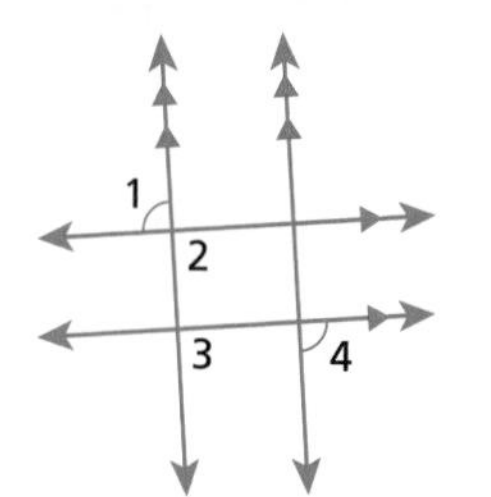

18.

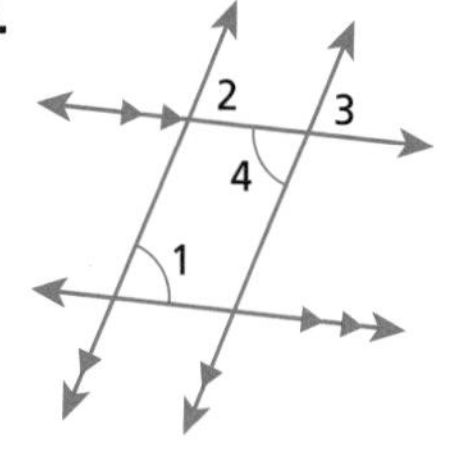

What Do You See? **In Exercises 19 and 20, draw and label the figure. List the lines that appear parallel.**

19.

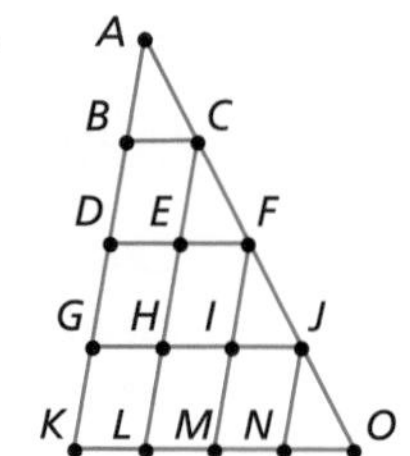

20.

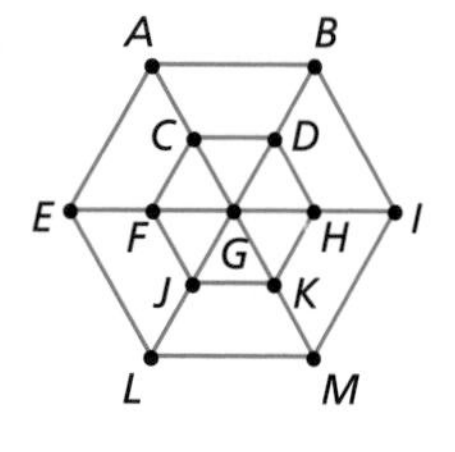

21. *Systematic Counting* How many triangles are in the figure in Exercise 19?

22. *Systematic Counting* How many trapezoids are in the figure in Exercise 19?

Walking **In Exercises 23–25, use the map at the right and the information below.**

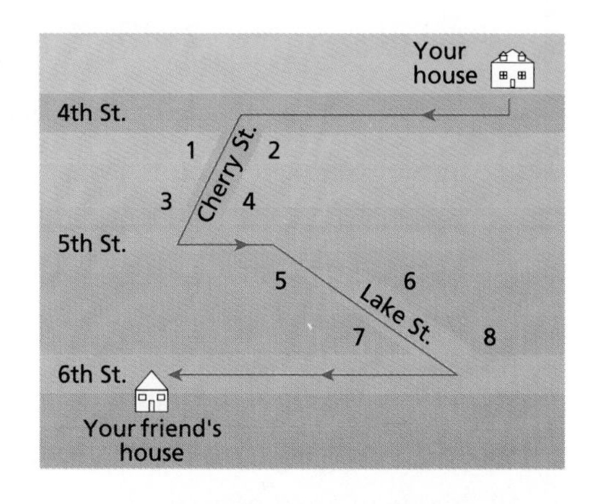

You live on 4th Street, which runs parallel to 5th and 6th Streets. The route you follow from your house to your friend's house is shown on the map. The turn from 4th Street onto Cherry Street is 117° and from Lake Street onto 6th Street is 34°.

23. Draw and label a diagram of the streets.

24. Identify all the congruent angles.

25. Find the measures of all the angles.

26. *Great Britain* Draw a diagram of the flag of Great Britain shown at the right. Is it true that every line segment in the flag is parallel to the top of the flag, the right side of the flag, or one of its two diagonals?

27. Use a protractor to measure several of the angles in Great Britain's flag. How many different angle measures are there? Explain your reasoning.

World Flags ***Great Britain's flag, known as the British Union Jack, was adopted in 1801.***

Integrated Review

Making Connections within Mathematics

Triangular Grid **In Exercises 28–32, draw the grid on triangular dot paper. Then outline the indicated figure.**

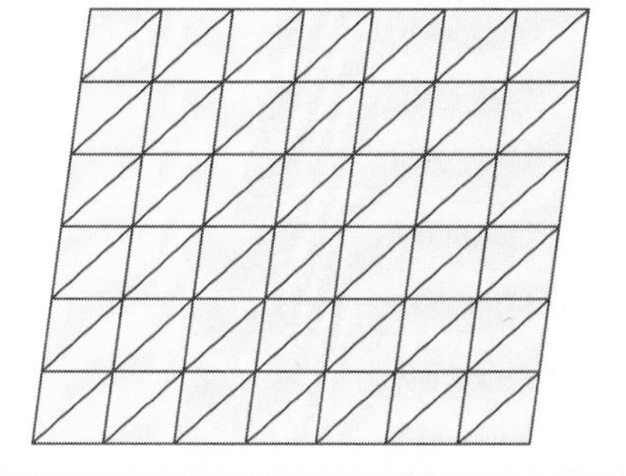

28. Trapezoid

29. Hexagon

30. Pentagon

31. Two similar triangles

32. Two parallelograms that have different shapes

Exploration and Extension

Alternate Interior Angles **In Exercises 33 and 34, use the figure at the right. In the figure, $\angle 3$ and $\angle 6$ are called alternate interior angles.**

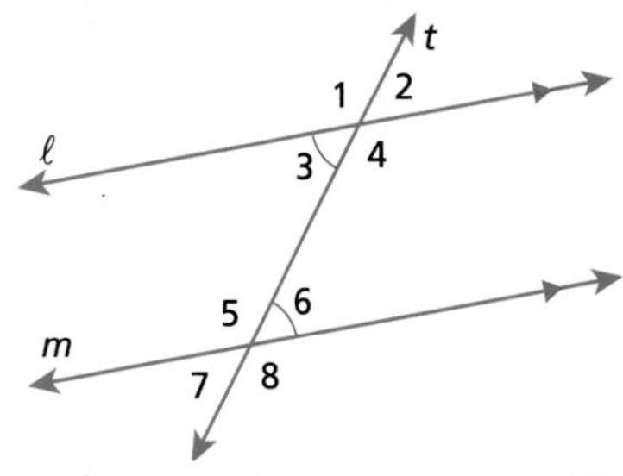

33. Explain why $\angle 3$ is congruent to $\angle 6$.

34. Name another set of alternate interior angles.

Mixed REVIEW

In Exercises 1–4, plot and label the points in the same coordinate plane. (3.8)

1. $A(3, 3)$ **2.** $B(3, -1)$ **3.** $C(-1, -1)$ **4.** $D(-1, 3)$

In Exercises 5–8, use the points in Exercises 1–4. (10.1, 10.2)

5. By connecting the points, what types of polygons can you form?

6. Use the points to identify parallel lines.

7. Use the points to identify perpendicular lines.

8. What is the length of AC? (Round your answer to 2 decimal places.)

In Exercises 9–13, use the figure at the right. (10.3)

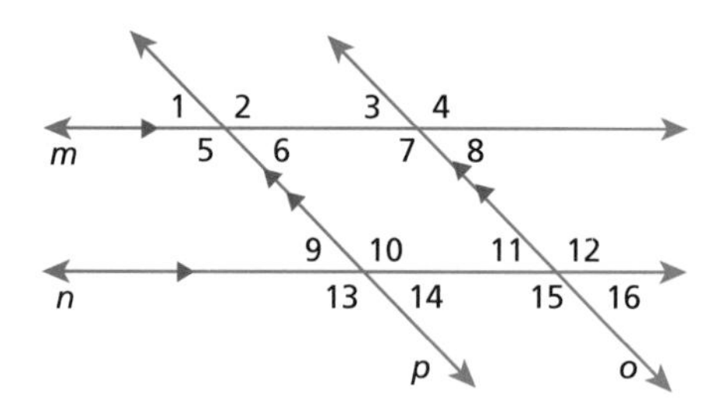

9. Name two angles that are corresponding angles to $\angle 6$.

10. Name two sets of vertical angles.

11. If $m\angle 1 = 45°$, what is $m\angle 2$?

12. If $m\angle 13 = 135°$, what is $m\angle 4$?

13. What is the relationship between $\angle 1$ and $\angle 6$?

Milestones NAVIGATION

200 400 600 800 1000 1200 1400 1600 1800

Hawaii settled 300-600 · Mayan's 500-700 · Viking 982-1000 · Aztecs, Incas 1200's-1500's · Cortez, 1519 · Jamestown 1607 · *Mayflower* 1620

Hawaii, the Aloha State, entered the union as the 50th state on August 21, 1959.

Before oceanography became a science in the 1700's, navigators had used daytime observations of the sun to determine their east-west locations (longitude) and the nighttime observations of the stars and planets to determine their north-south locations (latitude).

When Captain James Cook set out in the 1760's and 1770's to chart the Pacific, imagine his surprise at finding a vast triangular empire bordered by Hawaii, New Zealand, and Easter Island, off Chile. These Pacific Islands had all been settled by descendants of stone-age Melanesians and Micronesians who dominated the Pacific.

- *If it takes 24 hours for the Earth to rotate 360° from east to west, then how many minutes is 1° of longitude?*
- *The world is divided into 24 time zones, 1 hour apart. How many degrees of longitude wide is one time zone?*

10.4 Symmetry

Goal 1 How to identify line symmetry

Goal 2 How to identify rotational symmetry

You can use symmetry to describe objects that occur in nature, such as the radiance of the sun.

Goal 1 Identifying Line Symmetry

A figure has **line symmetry** if it can be divided by a line into two parts, each of which is the mirror image of the other.

Example 1 *Identifying Line Symmetry*

Identify the lines of symmetry in the figures.

a.

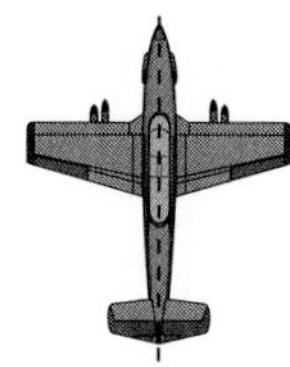

b.

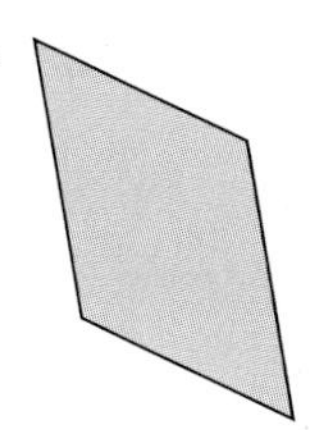

c.

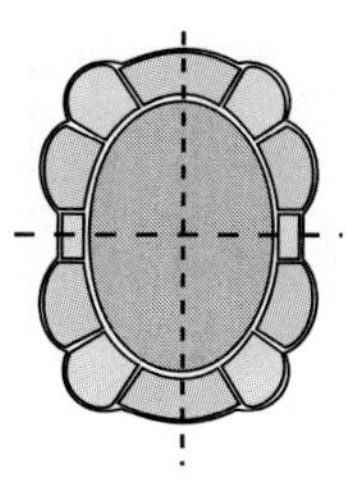

Solution

a. This figure has a vertical line of symmetry.

b. This figure has no line of symmetry.

c. This figure has a vertical line of symmetry *and* a horizontal line of symmetry. ■

LESSON INVESTIGATION

Investigating Line Symmetry

Group Activity With a partner, fold a sheet of paper, as shown in Diagram 1 or 2 below. Cut a design out of one or more edges of the folded paper. Predict the pattern when the paper is unfolded. Did symmetry help you make a prediction? Explain.

Diagram 1

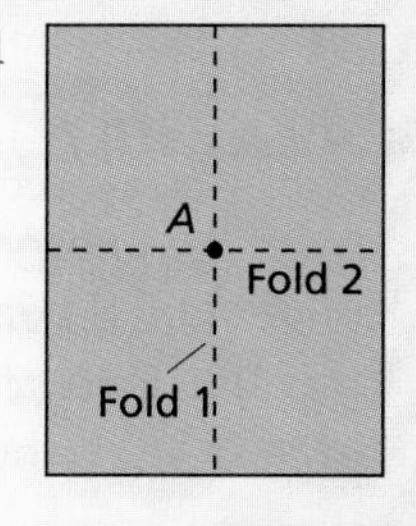

Diagram 2

Fold 2
B
Fold 1
Fold 2

A
Cut out

B
Cut out

Cut designs in one or more edges of the folded paper.

Goal 2 Identifying Rotational Symmetry

A figure has **rotational symmetry** if it coincides with itself after rotating 180° or less, either clockwise or counterclockwise, about a point.

Example 2 *Identifying Rotational Symmetry*

Identify any rotational symmetry in the figures.

a. 

b. 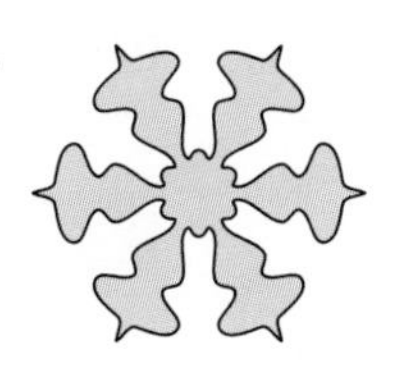

c. 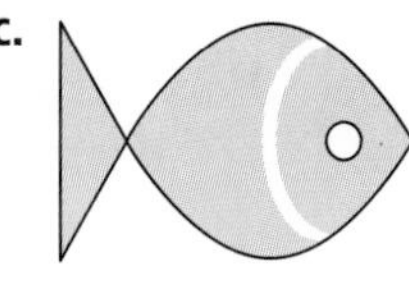

Solution

a. This figure has rotational symmetry. It will coincide with itself after being rotated 90° or 180° in either direction.

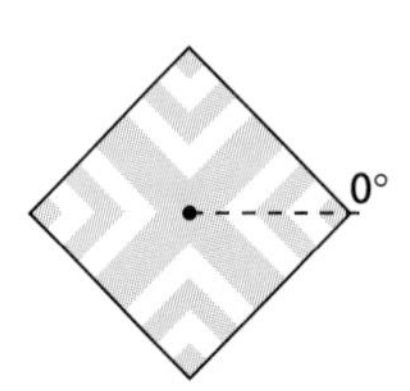

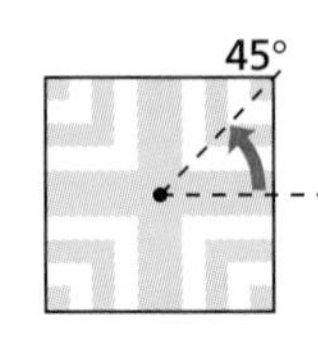

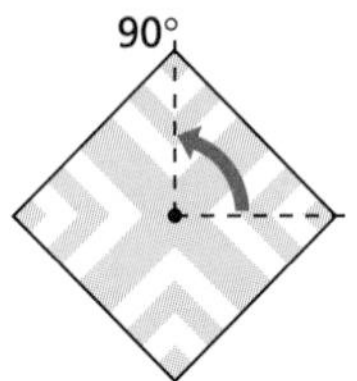

b. This figure has rotational symmetry. It will coincide with itself after being rotated 60°, 120°, or 180° in either direction.

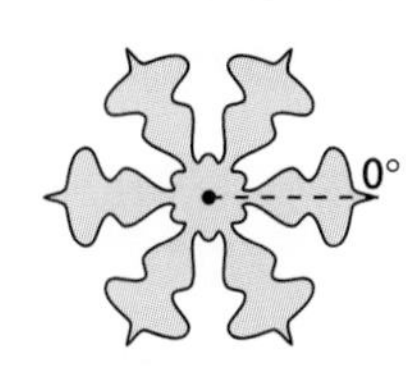

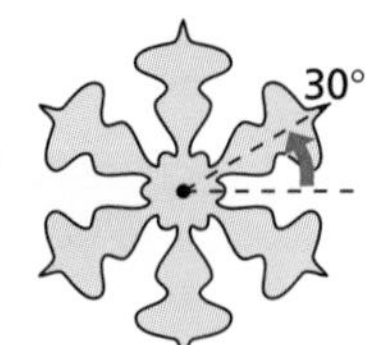

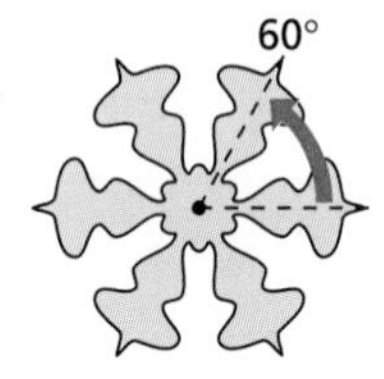

c. This figure has no rotational symmetry. (It does have line symmetry about a horizontal line.) ■

Many objects in nature have line symmetry or rotational symmetry. What types of symmetry does this photograph of the sun have?

Communicating about MATHEMATICS

SHARING IDEAS about the Lesson

Point Symmetry Trace the object in the photograph at the upper left. Then rotate the tracing and place it on the photograph. Can you find a rotation for which the tracing coincides with the original photo? What can you conclude?

EXERCISES

Guided Practice

CHECK for Understanding

1. *The Shape of Things* Name several objects in your classroom that have line symmetry.

2. *The Shape of Things* Name several objects in your classroom that have rotational symmetry.

In Exercises 3–6, identify any symmetry of the figure.

3. 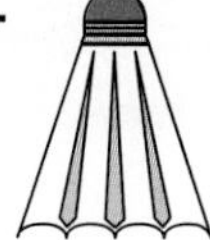

4. 

5. 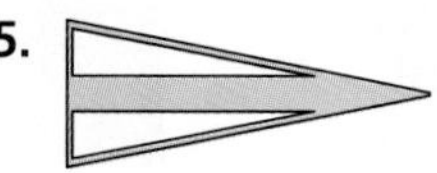

6. 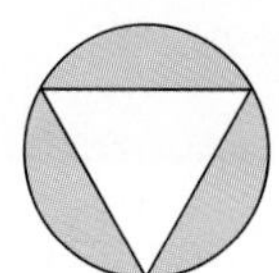

Independent Practice

In Exercises 7–10, identify any symmetry of the figure.

7. 

8. 

9. 

10. 

In Exercises 11–14, draw, if possible, a figure that has the given characteristics.

11. Exactly one line of symmetry

12. Exactly two lines of symmetry

13. Line, but not rotational, symmetry

14. Rotational, but not line, symmetry

In Exercises 15–18, copy the figure. Then shade one square so that the figure has the indicated symmetry.

15. Line

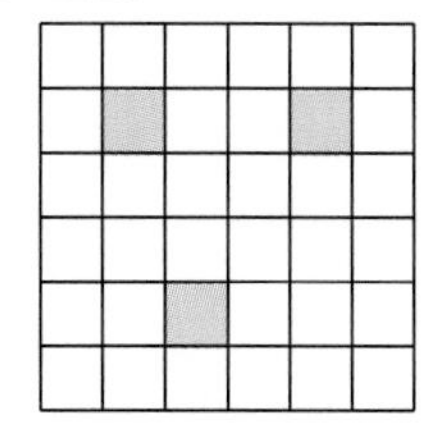

16. Rotational

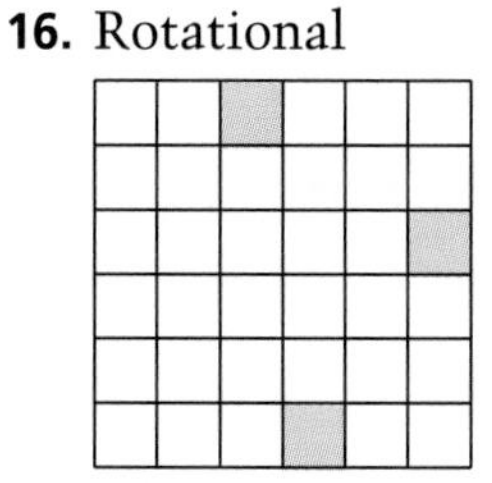

17. Rotational

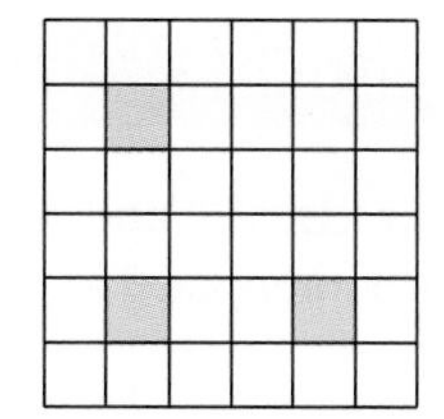

18. Line

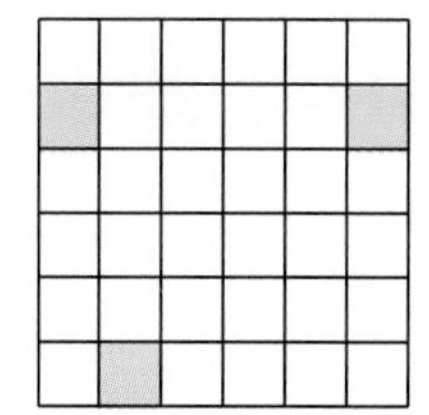

19. *Experimenting with Symmetry* Can you draw a figure that has a vertical and horizontal line of symmetry but does not have rotational symmetry?

20. *Experimenting with Symmetry* Can you draw a figure that has a vertical line of symmetry and rotational symmetry but does not have a horizontal line of symmetry?

Symmetry in Nature **In Exercises 21–23, identify any type of symmetry.**

21.

22.

23.

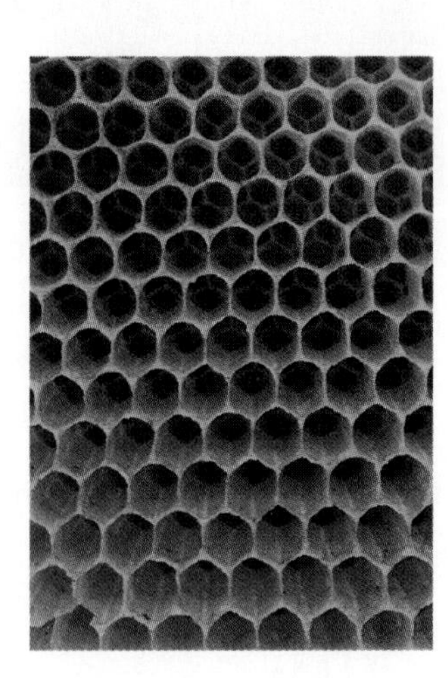

Letter Symmetry **In Exercises 24–26, trace the figure and find out what word is spelled using the indicated line of symmetry. (Hint: You can also use a mirror to solve the problems.)**

24. BOX **25.** DECK **26.** CHOICE

27. *Try It Yourself* List as many words as you can that have the type of symmetry used in Exercises 24–26. Try to write an entire sentence using such words.

Integrated Review

Making Connections within Mathematics

28. *Coordinate Geometry* Plot the point $A(4, 3)$ in a coordinate plane. Locate the points B, C, and D so that $ABCD$ is a rectangle that has the x-axis and y-axis as lines of symmetry.

29. *Coordinate Geometry* Plot the points $A(2, 1)$ and $B(-2, 3)$ in a coordinate plane. Locate the points C and D so that quadrilateral $ABCD$ is a parallelogram that has rotational symmetry about the origin, but no line symmetry.

Exploration and Extension

Symmetric Polygons **In Exercises 30 and 31, cut and label a 3-inch by 5-inch index card as shown.**

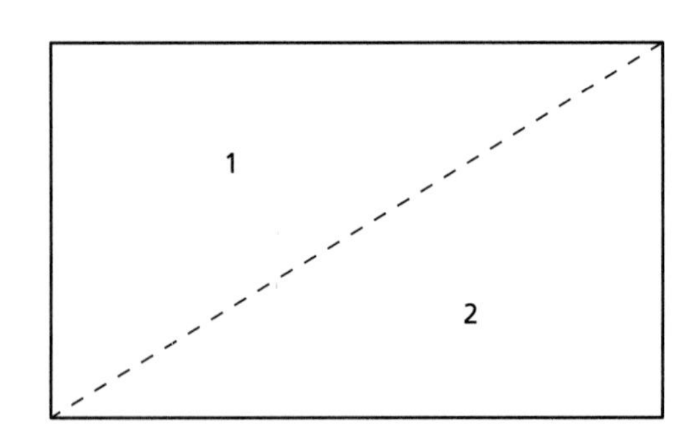

30. How many different polygons can be formed by placing the two triangles side by side so that their sides coincide?

31. Identify the type of symmetry of each polygon formed in Exercise 30.

10.5 Exploring Triangles

What you should learn:

How to identify triangles by their sides

How to identify triangles by their angles

Why you should learn it:

Triangles occur in a wide variety of real-life situations. Being able to identify types of triangles helps you communicate ideas about real-life situations.

The sails of these windsurfboards are scalene triangles—all three sides of each sail have different lengths.

Goal 1 Identifying Triangles by Their Sides

LESSON INVESTIGATION

Investigating Types of Triangles

Partner Activity Use the sheet of triangles provided by your teacher. Cut the triangles out. Then sort the triangles into two piles so that every triangle in one pile has a certain property and every triangle in the other pile does not have the property. Ask your partner to describe the property. Then, reverse roles and ask your partner to sort the triangles and see whether you can describe your partner's property.

Can you and your partner discover other properties that allow you to sort the triangles into two piles?

Triangles are classified by their sides into three categories. For a **scalene** triangle, all sides have different lengths. For an **isosceles** triangle, at least two sides have the same length. For an **equilateral** triangle, all three sides have the same length.

Example 1 *Classifying Triangles*

Classify each triangle according to its sides.

a.

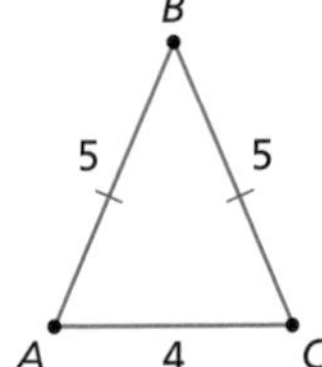

b.

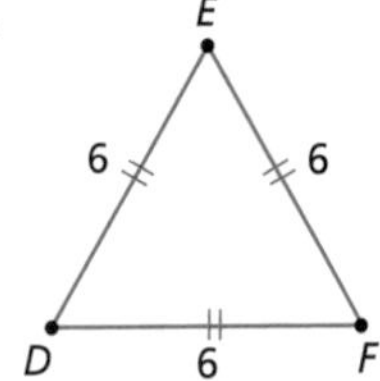

c.

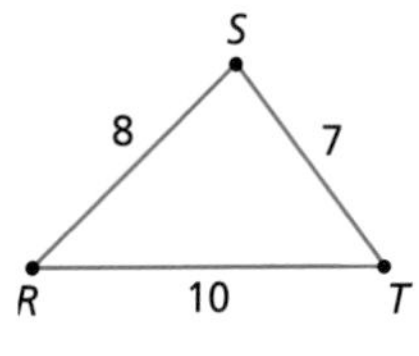

Solution

a. $\triangle ABC$ is isosceles because it has two sides of length 5.

b. $\triangle DEF$ is equilateral because each side has a length of 6.

c. $\triangle RST$ is scalene—all three sides have different lengths. ■

Goal 2 Identifying Triangles by Their Angles

Triangles are classified by their angles into four categories. A triangle is **acute** if all three angles are acute. An acute triangle is **equiangular** if all three angles have the same measure. A triangle is **obtuse** if one of its angles is obtuse. A triangle is **right** if one of its angles is a right angle.

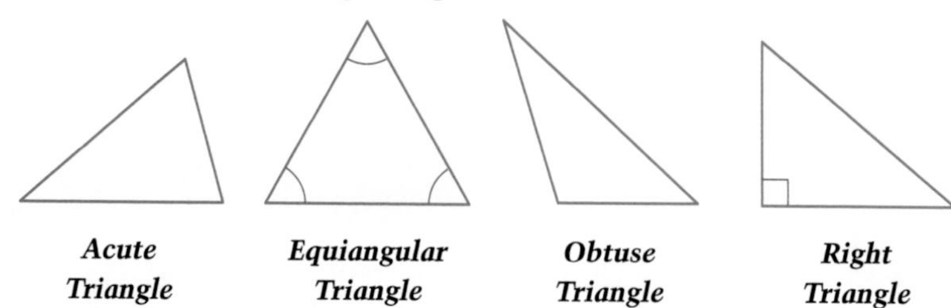

Comparing Sides and Angles

The diagram at the left shows part of an engine. Point C is the center of a circle. Point B stays in the same position on the circle, and point A moves clockwise around the circle. How does $\triangle ABC$ change as point A moves around the circle?

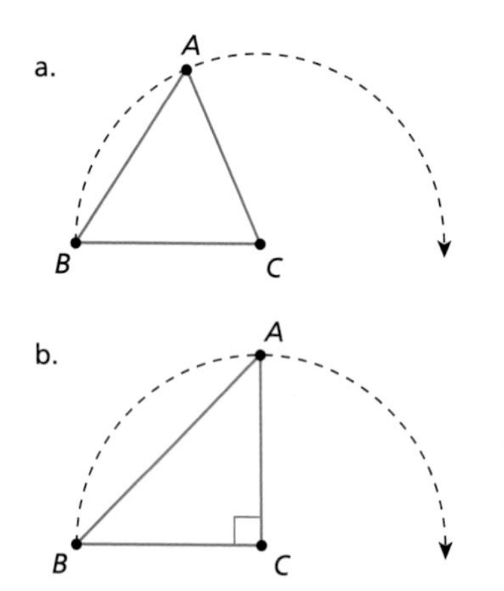

Solution

a. In this position, each angle of $\triangle ABC$ is acute, so $\triangle ABC$ is acute.

b. In this position, $\angle C$ is a right angle, so $\triangle ABC$ is a right triangle.

c. In this position, $\angle C$ is obtuse, so $\triangle ABC$ is obtuse. ■

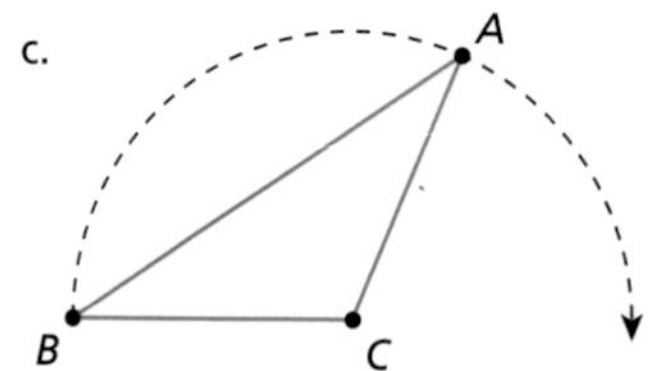

Communicating about MATHEMATICS

SHARING IDEAS about the Lesson

Classifying Triangles On a sheet of square dot paper, outline several 9-dot squares. How many *different* (noncongruent) triangles can you draw? Classify each triangle by its sides and angles.

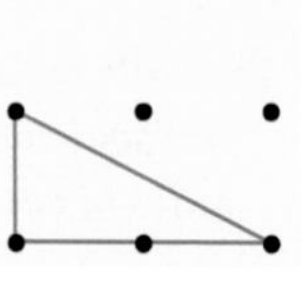

EXERCISES

Guided Practice

CHECK for Understanding

In Exercises 1–8, match each triangle with all words that describe it. Explain.

a. Isosceles **b.** Equiangular **c.** Scalene **d.** Obtuse
e. Equilateral **f.** Right **g.** Acute

1. 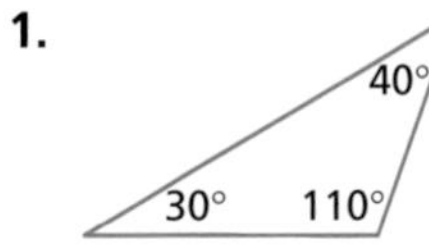

2. 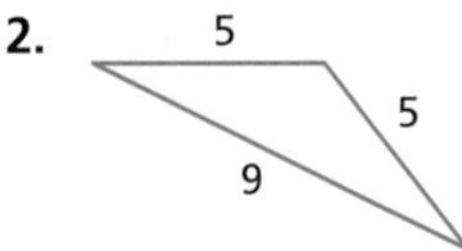

3. 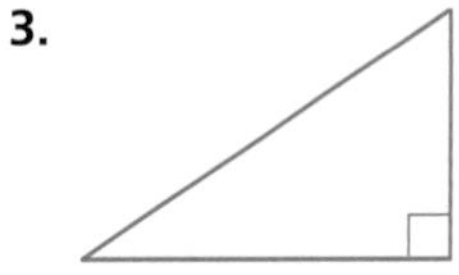

4. 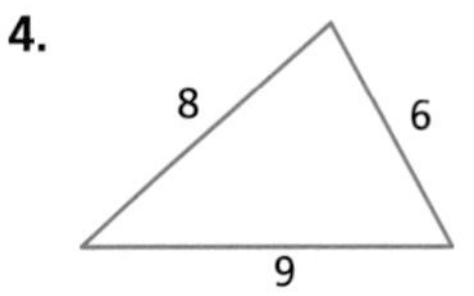

5. 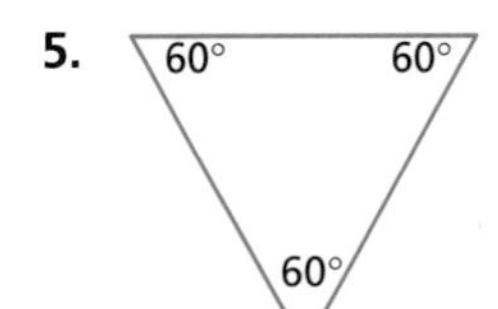

6. 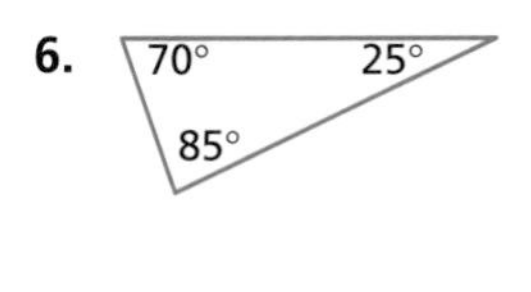

7. 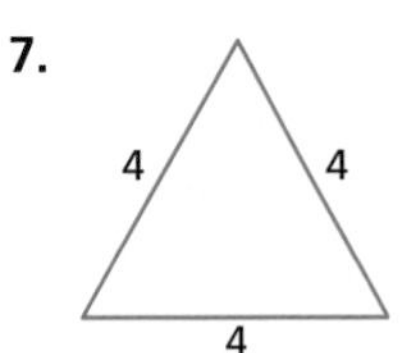

8. 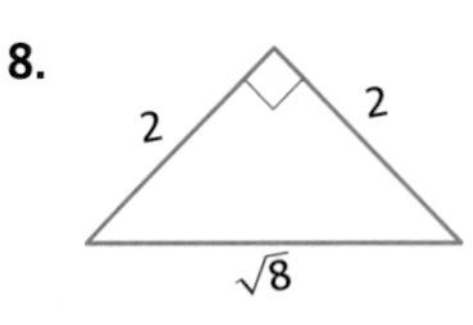

Independent Practice

Sketching Triangles **In Exercises 9–14, sketch the indicated type of triangle. Then label it with appropriate side or angle measures.**

9. Obtuse **10.** Acute **11.** Right scalene
12. Right isosceles **13.** Acute isosceles **14.** Obtuse scalene

In Exercises 15–20, classify the triangle according to its sides and angles.

15. 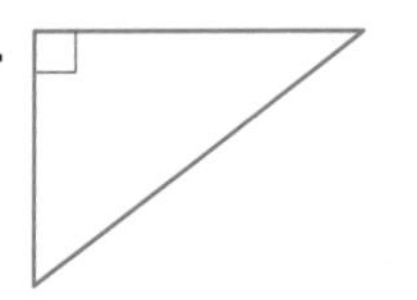

16. 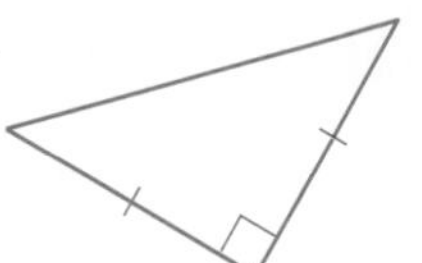

17. 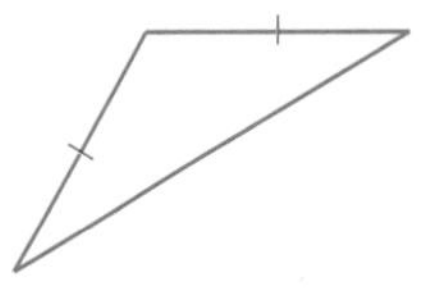

18. 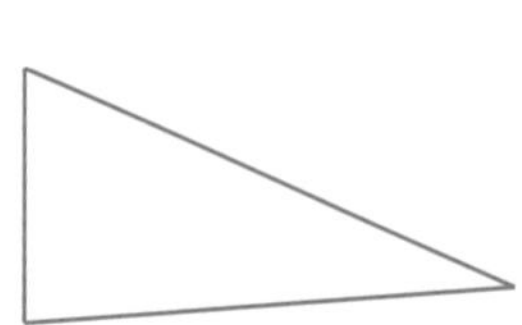

19. 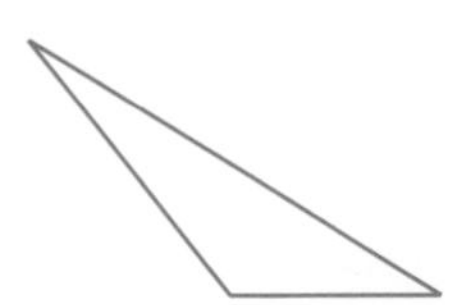

20. 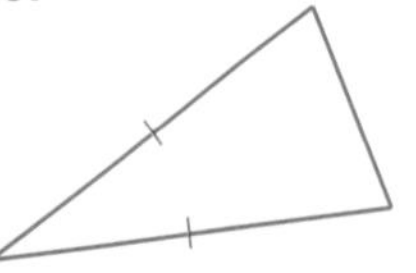

In Exercises 21–26, use a protractor to draw $\triangle ABC$ with the given angle measures. Then classify the triangle according to its sides and angles.

21. $m\angle A = 60°, m\angle B = 60°, m\angle C = 60°$
22. $m\angle A = 70°, m\angle B = 70°, m\angle C = 40°$
23. $m\angle A = 50°, m\angle B = 60°, m\angle C = 70°$
24. $m\angle A = 30°, m\angle B = 60°, m\angle C = 90°$
25. $m\angle A = 45°, m\angle B = 45°, m\angle C = 90°$
26. $m\angle A = 120°, m\angle B = 40°, m\angle C = 20°$

In Exercises 27–30, find the triangle's perimeter and area. Then classify the triangle according to its sides and angles. (The area of each small square on the grid is 1 square unit.)

27. 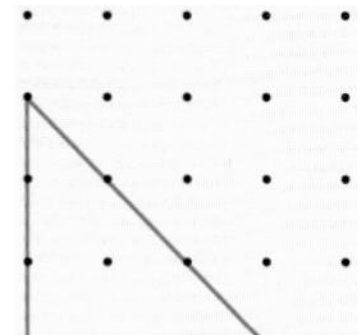

28. 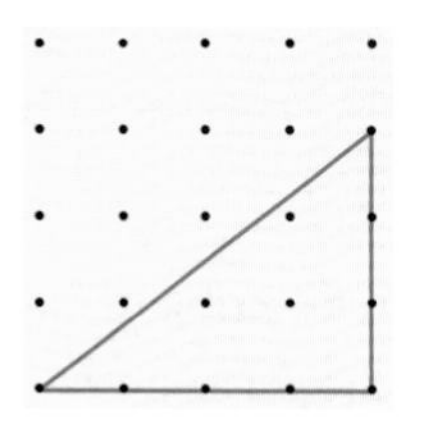

29. 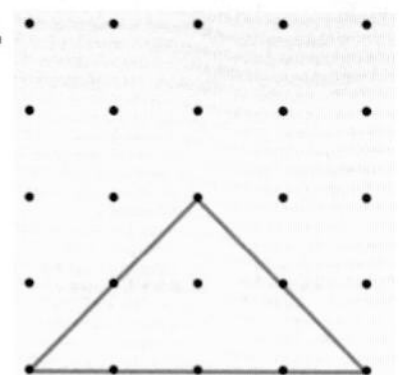

30. 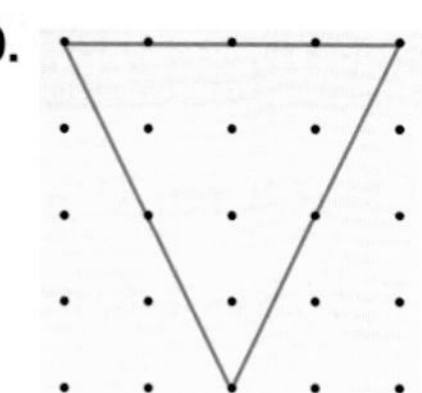

31. Explain the relationship between an equilateral triangle and an equiangular triangle.

32. *Think about It* Can a right triangle be equilateral? Explain your reasoning.

33. *The Shape of Things* The figures at the right are made with popsicle sticks and brads. Is the triangle rigid, or can you adjust its sticks to form a different shape of triangle? Is the rectangle rigid, or can you adjust its sticks to form a different shape of quadrilateral?

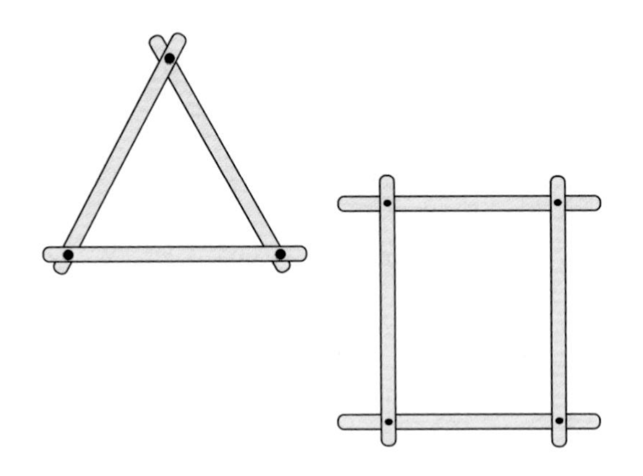

Bicycling **In Exercises 34 and 35, use the drawing of the bicycle at the right.**

34. Sketch the frame of the bicycle. Then label the frame with points and identify all of the triangles formed by the bicycle's frame.

35. *The Shape of Things* Based on your answers to Exercise 33, why do you think triangles are used in bicycle construction instead of quadrilaterals?

Integrated Review

Making Connections within Mathematics

Coordinate Geometry **In Exercises 36–39, plot the points on a coordinate plane. Then classify $\triangle ABC$ by its sides and angles.**

36. $A(3, 0)$, $B(3, -3)$, $C(0, -3)$

37. $A(0, 1)$, $B(4, -1)$, $C(-4, -1)$

38. $A(-1, -5)$, $B(-5, -5)$, $C(-4, 2)$

39. $A(-1, 3)$, $B(2, 3)$, $C(2, -3)$

Exploration and Extension

Intersections **In Exercises 40–43, sketch all possible intersections of the given figures.**

40. Triangle and line segment

41. Two triangles

42. Triangle and rectangle

43. Triangle and hexagon

Mid-Chapter SELF-TEST

Take this test as you would take a test in class. The answers to the exercises are given in the back of the book.

In Exercises 1–4, use the cube at the right. (10.1)

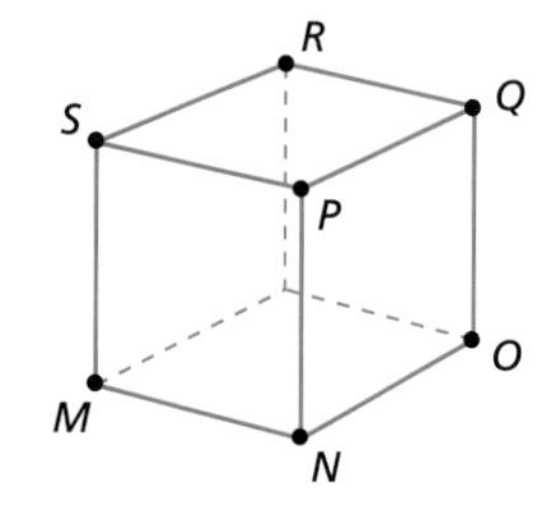

1. Name another point that lies in the same plane as M, N, and P.
2. Name two lines that are parallel to $\overleftrightarrow{SR}$.
3. Name the point of intersection of $\overline{SP}$ and $\overline{PQ}$.
4. Does the ray $\overrightarrow{NP}$ point up or down?

In Exercises 5–8, use the figure at the right. (10.2)

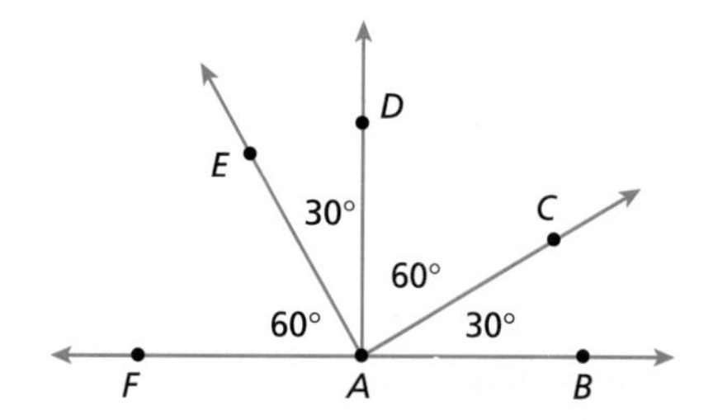

5. Name the right angles.
6. Name the acute angles.
7. Name the obtuse angles.
8. Name the straight angle.

In Exercises 9–12, use the figure at the right. (10.3)

In Exercises 9–11, use the words *vertical* or *corresponding.*

9. $\angle 1$ and $\angle 7$ are [?] angles.
10. $\angle 1$ and $\angle 3$ are [?] angles.
11. $\angle 2$ and $\angle 4$ are [?] angles.
12. Find the measure of each angle.

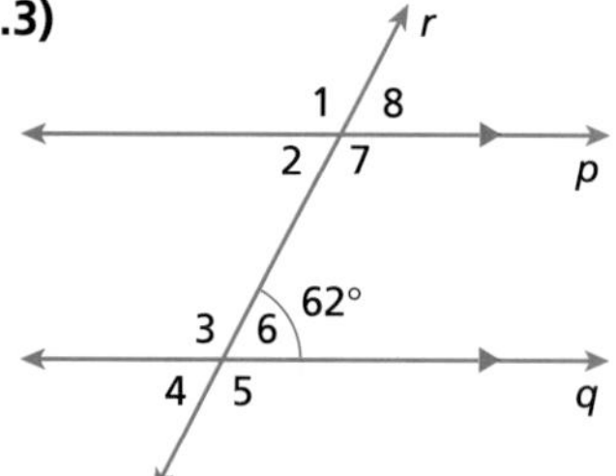

Line p and line q are parallel.

In Exercises 13–15, identify any symmetry of the figure. (10.4)

13.

14.

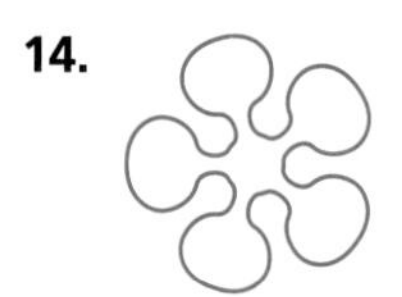

15.

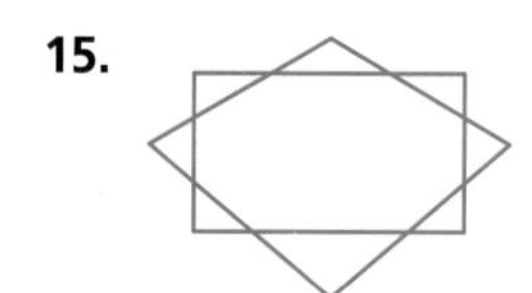

This microscopic diatom has both line and rotational symmetries.

In Exercises 16–19, classify the triangle by its sides and by its angles. (10.5)

16.

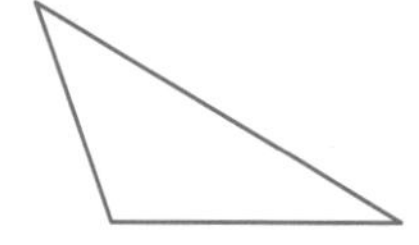

17.

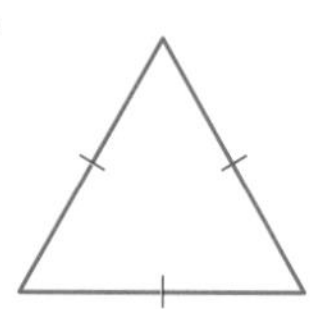

18.

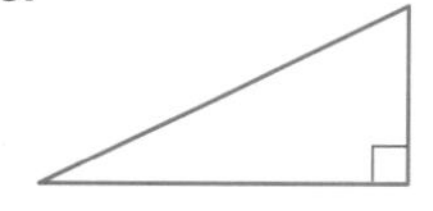

19.

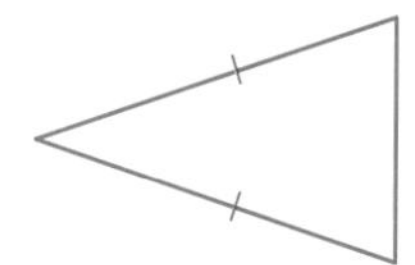

20. If a triangle is obtuse, does it have to be scalene? Illustrate your answer with a sketch.

10.6 Exploring Quadrilaterals

What you should learn:

How to identify quadrilaterals

How to identify quadrilaterals in real-life situations

Why you should learn it:

Being able to identify different types of quadrilaterals helps you communicate ideas about real-life situations.

Identifying Quadrilaterals

LESSON INVESTIGATION

Investigating Types of Quadrilaterals

Group Activity How many different sizes of quadrilaterals can you draw on a 9-dot square? (The two quadrilaterals at the right are not considered different.) Discuss your results with other members of your group. What types of quadrilaterals were you able to draw?

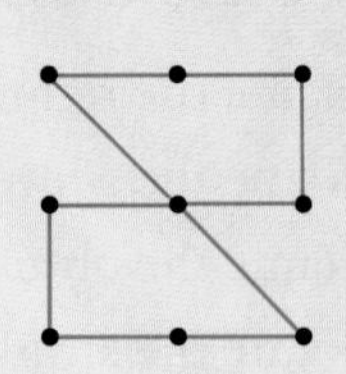

In the above investigation, you may have used some of the following terms to classify quadrilaterals.

Parallelogram: A quadrilateral with opposite sides parallel
Rectangle: A parallelogram with 4 right angles
Square: A rectangle with sides of equal length
Rhombus: A parallelogram with sides of equal length
Trapezoid: A quadrilateral with only one pair of parallel sides
Kite: A quadrilateral that is not a parallelogram, but has two pair of sides of equal length
Scalene quadrilateral: All sides have different lengths.

A trapezoid is **isosceles** if its nonparallel sides have the same length. The figures below show several types of quadrilaterals.

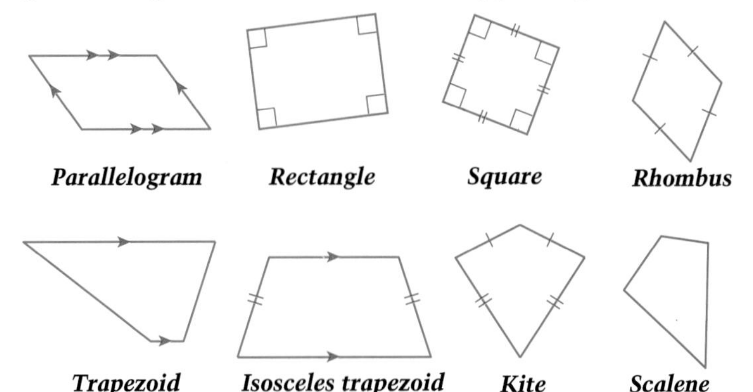

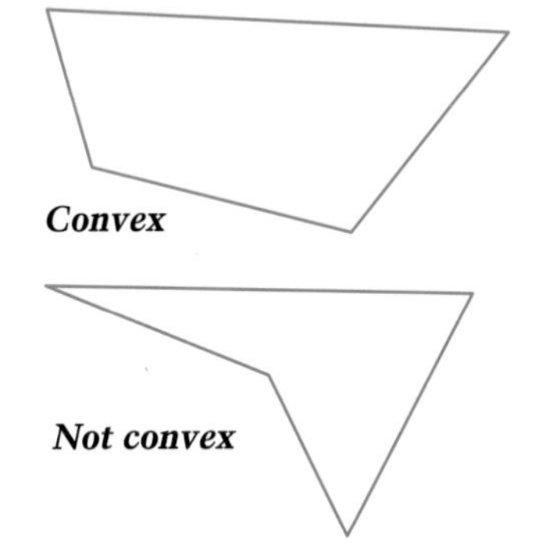

Quadrilaterals can be convex or not convex. (See page 37 for a definition of convex.)

Goal 2 Identifying Quadrilaterals in Real Life

The photo at the left was scanned into a computer. Then a computer design program was used to alter the shape of the photo.

Real Life
Photography

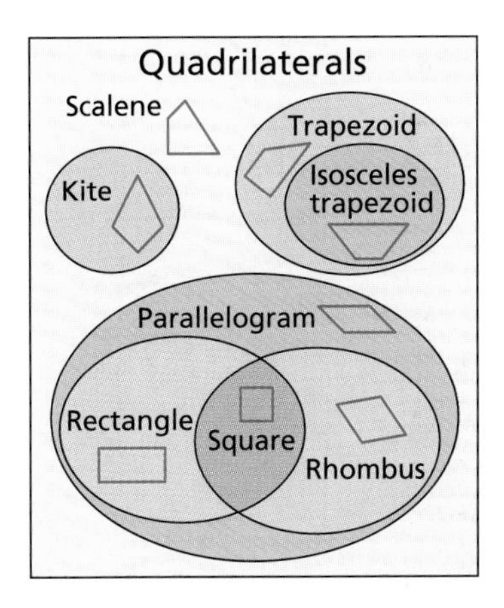

Example 1 *Classifying Quadrilaterals*

Each of the photographs shown above is a quadrilateral. Name *all* types of quadrilaterals that each photograph appears to be.

Solution The first photograph appears to have opposite sides that are parallel, which would make it a parallelogram. Moreover, it appears to have 4 right angles, which makes it a rectangle.

The second photograph also appears to have opposite sides that are parallel, which would make it a parallelogram. Moreover, it appears to have 4 right angles, which would make it a rectangle, *and* sides of equal length, which would make it a square and a rhombus.

The third photograph appears to have opposite sides that are parallel, which would make it a parallelogram. ■

Communicating about MATHEMATICS

Cooperative Learning

SHARING IDEAS about the Lesson

Interpreting Venn Diagrams The Venn diagram above shows that every square is a rectangle. Work with a partner. Use the diagram to write several other statements of the form "Every [?] is a [?]."

EXERCISES

Guided Practice

CHECK for Understanding

1. Name all quadrilaterals that:

a. have 2 pair of parallel sides.

b. have no parallel sides.

c. have exactly one pair of parallel sides.

d. have 4 congruent sides.

e. have no congruent sides.

f. have at least one pair of congruent sides.

2. *Writing* In your own words, state how to determine whether a quadrilateral is convex or not convex. Draw examples of convex and nonconvex quadrilaterals.

3. Explain the difference between a kite and a rhombus.

4. Explain the difference between a square and a rectangle.

Independent Practice

In Exercises 5–12, identify the quadrilateral from its appearance. Use the name that *best* describes the quadrilateral.

5.

6.

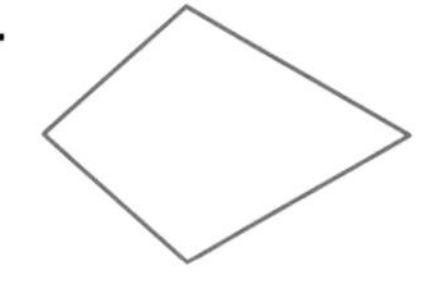

7.

8.

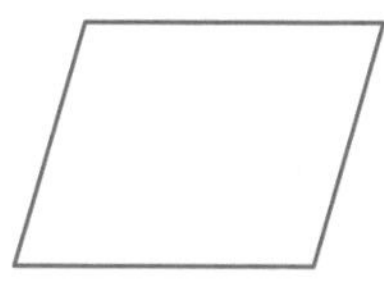

9.

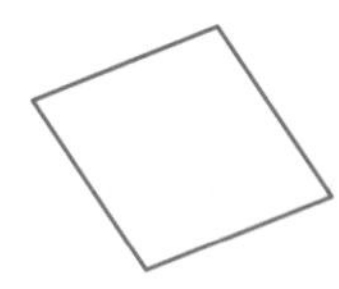

10.

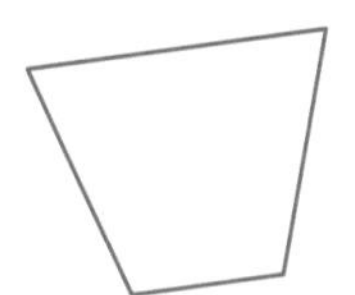

11.

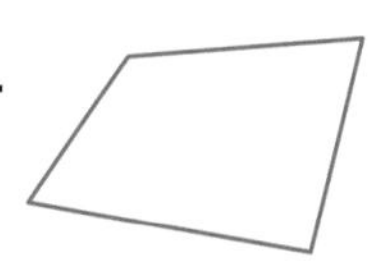

12.

Logical Reasoning **In Exercises 13–16, complete the statement with *always, sometimes,* or *never.* Explain.**

13. A quadrilateral is [?] a parallelogram.

14. A rectangle is [?] a rhombus.

15. A trapezoid is [?] a convex quadrilateral.

16. A rhombus is [?] a square.

In Exercises 17–20, find the values of x and y.

17. Isosceles trapezoid

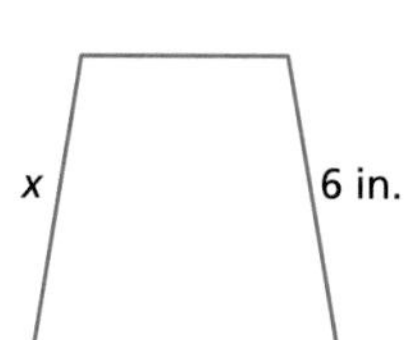

18. Square

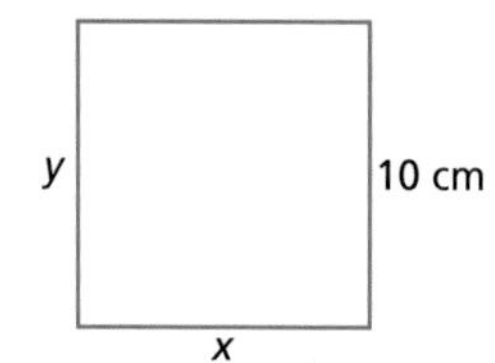

19. Parallelogram

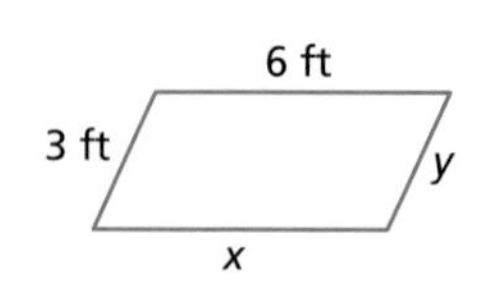

20. Kite

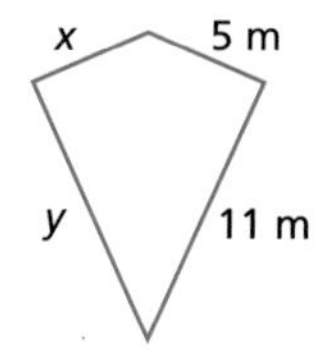

Drawing Figures **In Exercises 21–24, use the description to sketch the figure. If it is not possible, write *not possible*.**

21. A parallelogram with 2 pair of congruent sides

22. A quadrilateral with one pair of congruent sides and one pair of parallel sides

23. A parallelogram with no congruent sides

24. A quadrilateral with no congruent sides

Symmetry **In Exercises 25–32, match the quadrilateral with the description of its symmetry. Use each description that applies. Make a sketch to support your answers.**

a. No symmetry
b. Exactly 1 line of symmetry
c. Exactly 2 lines of symmetry
d. Exactly 3 lines of symmetry
e. Exactly 4 lines of symmetry
f. Rotational symmetry

25. Parallelogram **26.** Rectangle **27.** Square **28.** Rhombus
29. Trapezoid **30.** Isosceles trapezoid **31.** Kite **32.** Scalene quadrilateral

Making a Conjecture **In Exercises 33–36, use a straightedge and protractor.**

33. Draw any convex quadrilateral and measure its angles.

34. Repeat Exercise 33 using three different quadrilaterals. Record the sum of the measures of the angles for each quadrilateral.

35. Use the results of Exercises 33 and 34 to write a conjecture about the sum of the measures of the angles of any convex quadrilateral.

36. Use your conjecture to find $m\angle A$ for the quadrilateral at the right.

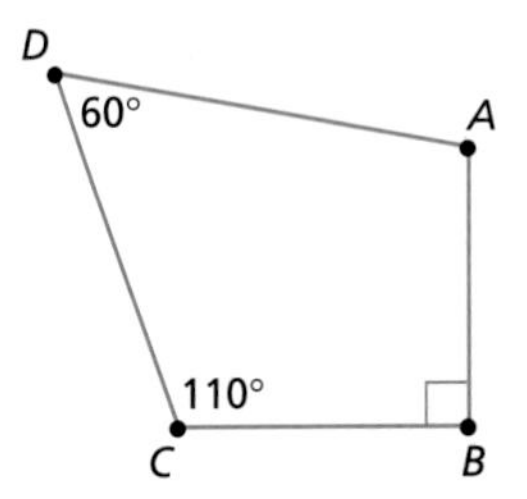

Bird Feeder **In Exercises 37–40, use the drawing of a bird feeder at the right.**

37. Draw each piece of wood that is used to build the bird feeder.

38. Identify each piece of wood that you sketched in Exercise 37.

39. Identify all pieces of wood that are congruent.

40. Design your own bird feeder. Sketch and identify the pieces of wood that you use.

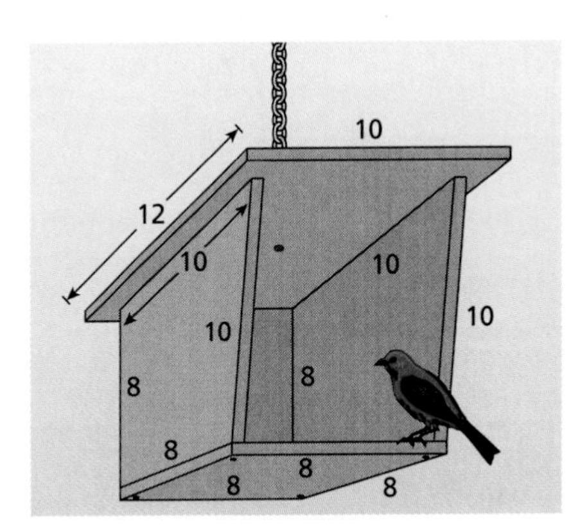

Integrated Review

Making Connections within Mathematics

Coordinate Geometry **In Exercises 41–44, plot the points in a coordinate plane. Then identify the quadrilateral *ABCD*.**

41. $A(0, 0)$, $B(2, 3)$, $C(8, 0)$, $D(2, -3)$

42. $A(-2, 3)$, $B(2, 3)$, $C(5, -2)$, $D(-5, -2)$

43. $A(0, 0)$, $B(3, 3)$, $C(9, 3)$, $D(6, 0)$

44. $A(2, 2)$, $B(2, -2)$, $C(-2, -2)$, $D(-2, 2)$

Exploration and Extension

Tangrams **In Exercises 45–52, use the tangram pieces at the right to construct the quadrilateral.**

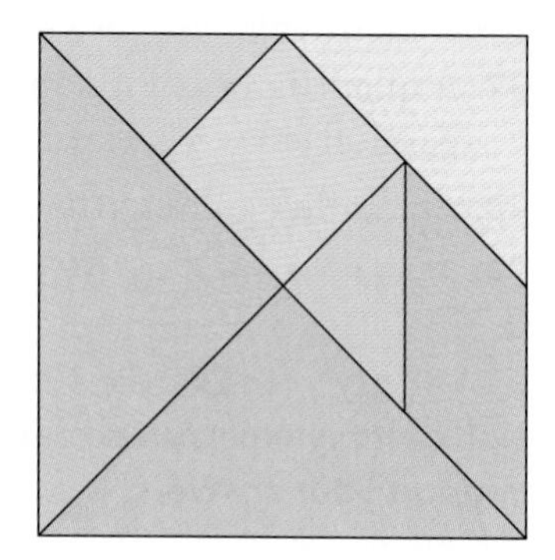

45. Use 3 tangram pieces to make a square.

46. Use 3 tangram pieces to make an isosceles trapezoid.

47. Use 4 tangram pieces to make a nonsquare rectangle.

48. Use 4 tangram pieces to make a nonrectangular parallelogram.

49. Use 4 tangram pieces to make a nonisosceles trapezoid.

50. Use 5 tangram pieces to make a square.

51. Use 5 tangram pieces to make an isosceles trapezoid.

52. Use 7 tangram pieces to make the figure at the right.

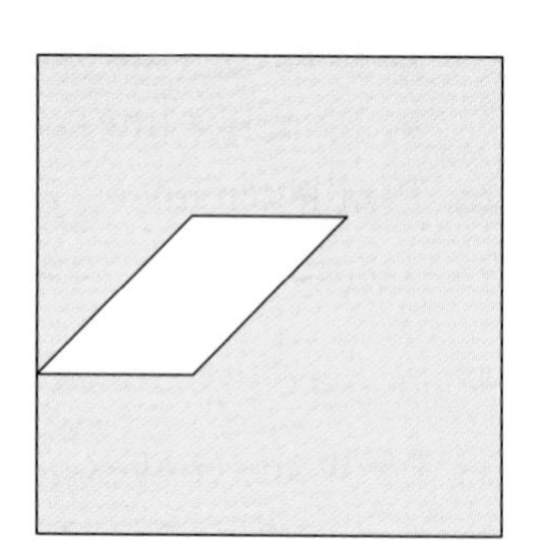

Chinese Tangrams ***A tangram is a square that has been cut into a square, a parallelogram, and 5 triangles. Tangram puzzles were invented by the Chinese more than 1000 years ago.***

Mixed REVIEW

In Exersises 1–3, plot the points and identify quadrilateral *BDCA*. (10.6)

1. $A(0, 2), B(-2, 0)$
$C(1, 1), D(0, -3)$

2. $A(-2, -1), B(-1, 1)$
$C(4, -1), D(3, 1)$

3. $A(0, 5), B(2, 4)$
$C(-2, 4), D(0, 0)$

In Exercises 4–9, solve the equation. (4.2, 7.2)

4. $2b - 2 = -2b$

5. $62p - 203 = 111 - 38p$

6. $2a + 4.04 = 16.08$

7. $\frac{1}{4}(3r - 1) = \frac{1}{4}$

8. $p^2 + 4 = 40$

9. $q - \frac{1}{2} = \frac{1}{3}$

In Exercises 10–14, rewrite the number in scientific notation. (6.8)

10. 2100 **11.** 0.00092 **12.** 16,000,000 **13.** 0.00000046 **14.** 92.4×10^{18}

In Exercises 15–20, simplify the expression. (7.2, 7.4, 7.5)

15. $\frac{1}{5} + \frac{2}{5}$

16. $\frac{4}{9} - \frac{2}{9}$

17. $\frac{4}{9} + \frac{1}{3}$

18. $\frac{4}{5} - \frac{3}{4}$

19. $\frac{3}{8} \times \frac{1}{2}$

20. $\frac{3}{10} \div \frac{9}{2}$

10.7 Polygons and Congruence

What you should learn:

How to recognize congruent polygons

How to identify regular polygons

Why you should learn it:

Congruent polygons can be used to create real-life floor tiles.

Islamic Art **This Morroccan mosaic is a prime example of the geometric tiling patterns found in Islamic art and architecture.**

Goal 1 Recognizing Congruent Polygons

Two polygons are **congruent** if they are exactly the same size and shape. To decide whether two polygons are congruent, you can trace each on paper, cut one out, and try to move the cut polygon so that it lies exactly on top of the other polygon.

LESSON INVESTIGATION

Investigating Congruent Polygons

Group Activity Use dot paper to draw a hexagon or octagon whose sides have the same lengths and whose angles have the same measures. In how many ways can you divide the hexagon or octagon into congruent polygons? The sample below shows an equilateral triangle divided into congruent polygons in four different ways.

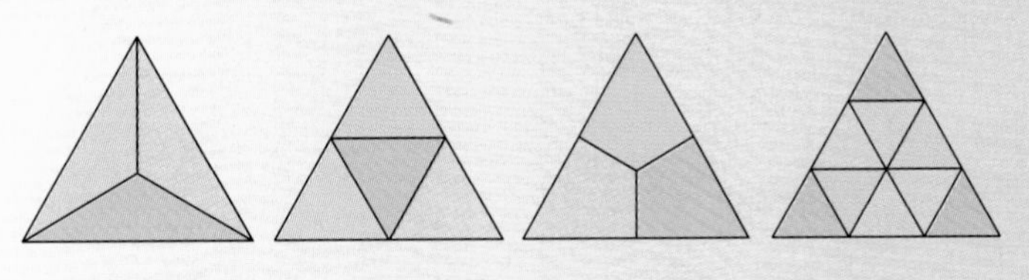

Example 1 *Identifying Congruent Polygons*

Which of the quadrilaterals are congruent?

a.

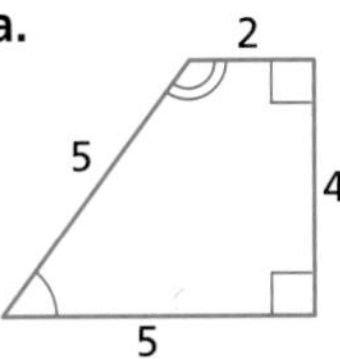

b.

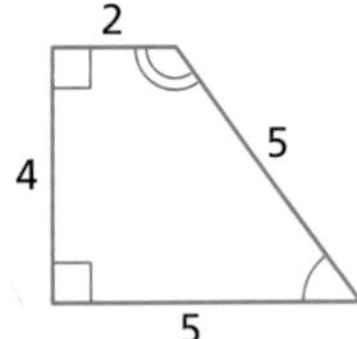

c.

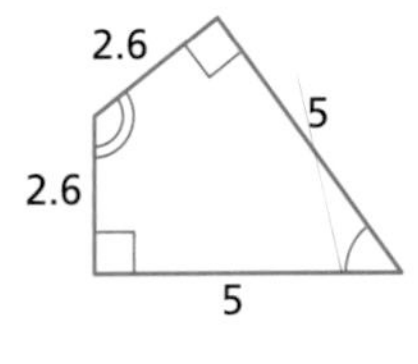

Solution The first two quadrilaterals are congruent. Try confirming this with tracing paper.

The third quadrilateral is not congruent to either of the first two. One reason for this is that its sides don't have the same lengths as the sides of either of the first two quadrilaterals. ■

Goal 2 Identifying Regular Polygons

A polygon is **regular** if each of its sides has the same length *and* each of its angles has the same measure. Four examples are shown below.

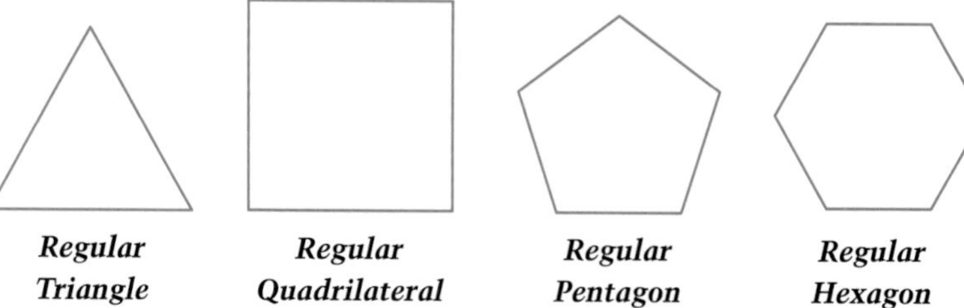
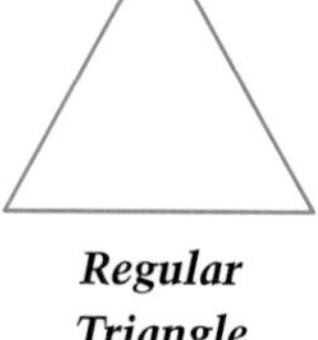
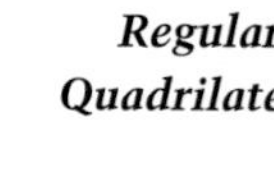

Tiling Design

Color and shading can be used to create illusions of space and form. In this design, the tilted tiles appear to radiate from a center point.

Example 2 *Tiling with Regular Polygons*

Which of the above regular polygons could be used to tile a floor with no gaps or overlapping tiles?

Solution You can solve this problem experimentally by tracing several polygons of each shape and trying to fit the polygons together. After doing this, you can discover that regular triangles, quadrilaterals, and hexagons can be used as tiles, but the regular pentagons cannot be used.

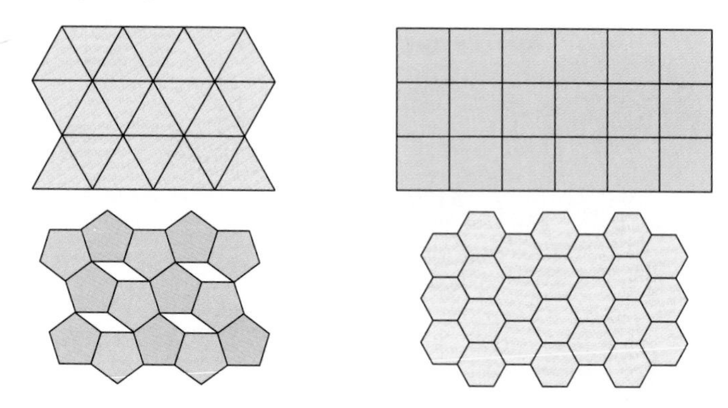

■

Communicating about MATHEMATICS

SHARING IDEAS about the Lesson

Extending the Example In Example 2, suppose you could mix different types of regular polygons to tile the floor. Draw some possible patterns.

EXERCISES

Guided Practice

CHECK for Understanding

1. Explain how you can tell whether two polygons are congruent.

In Exercises 2 and 3, two polygons are congruent. Which are they?

2. a. 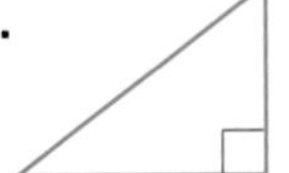

b. 

c. 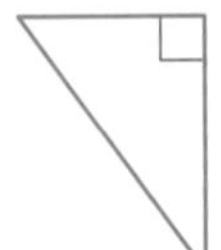

3. a. 

b. 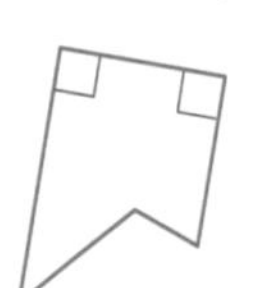

c. 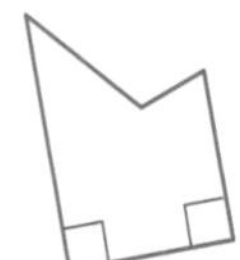

4. Name each polygon. Is the polygon regular?

a. 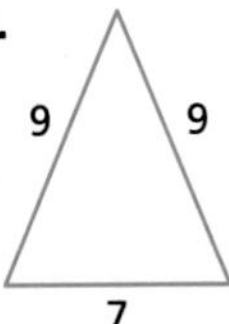

b. 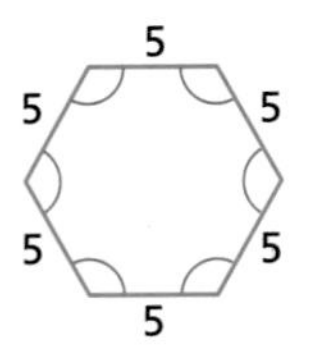

c. 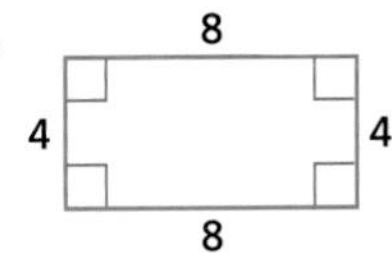

d. 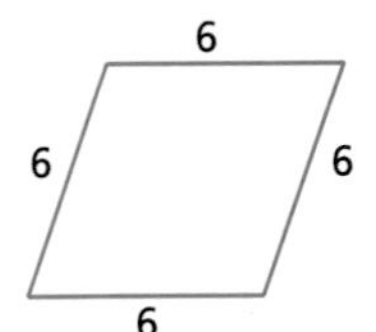

Independent Practice

In Exercises 5–8, match the quadrilateral with a congruent quadrilateral.

a. 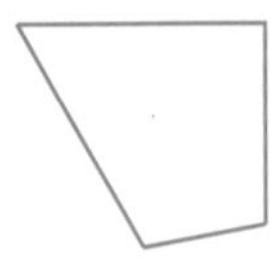

b. 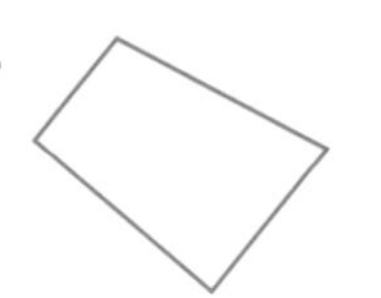

c. 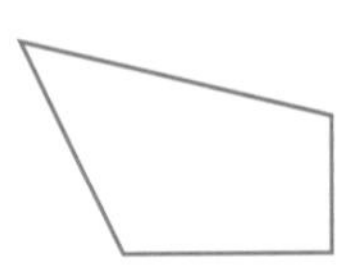

d. 

5. 

6. 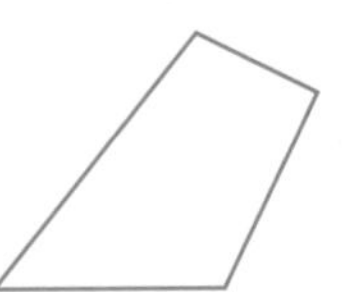

7. 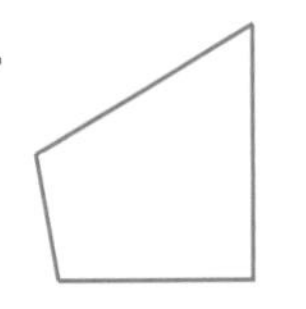

8. 

In Exercises 9 and 10, use the words congruent, equilateral, equiangular, and regular to describe the polygons.

9. a. 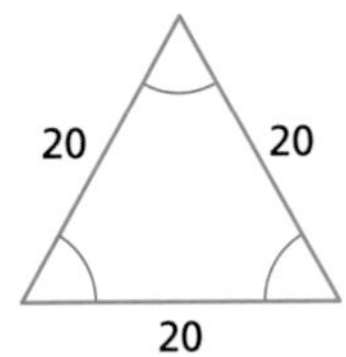

b. 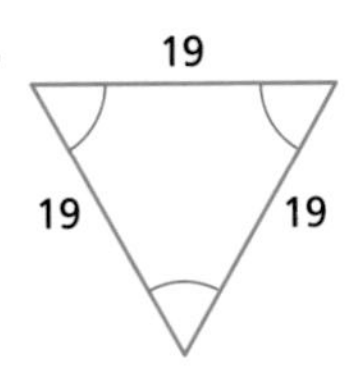

c. 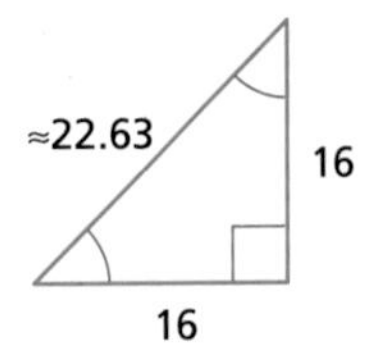

d. 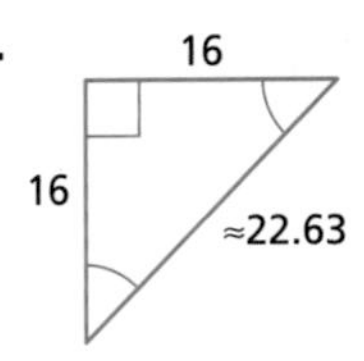

10. a. 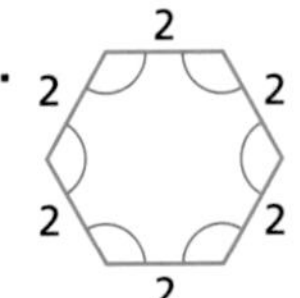

b. 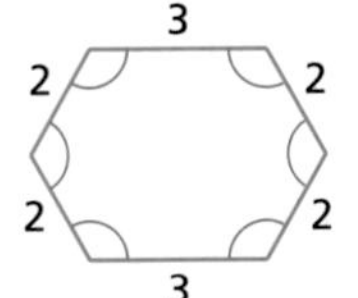

c. 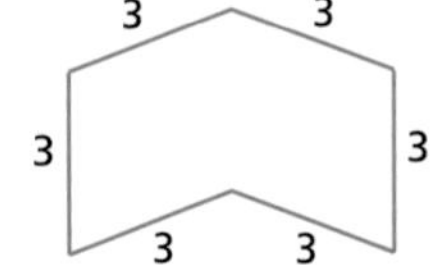

d. 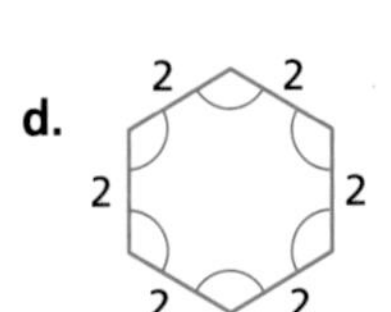

In Exercises 11–14, trace the figure. Then divide the figure into congruent regular triangles. Use as few triangles as possible.

11. 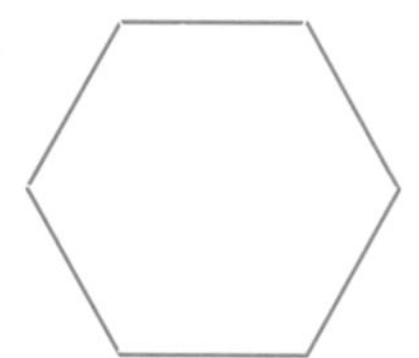

12. 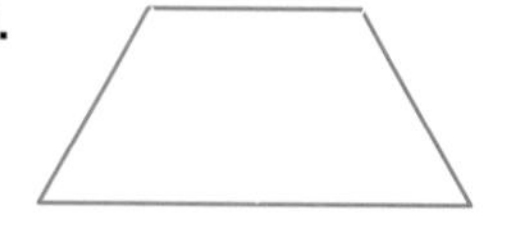

13. 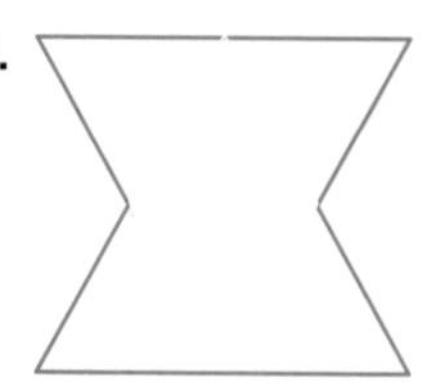

14. 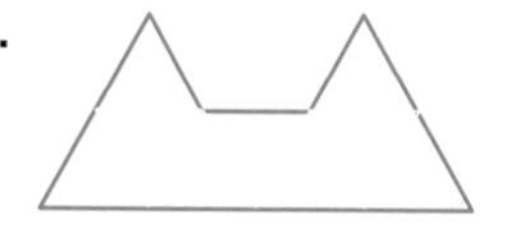

15. *Think about It* Discuss the relationship between equilateral, equiangular, and regular. If a polygon is equilateral, must it be equiangular? Explain.

Polygon Puzzle **In Exercises 16–18, use the figure at the right.**

16. The eight polygons at the right can be rearranged to form a square. What is the area of the square? What are the dimensions of the square?

17. Which of the polygons are equilateral? Which are equiangular?

18. Trace the eight figures on dot paper. Then rearrange the figures to form a square. (*Hint:* The result of Exercise 16 can help you.)

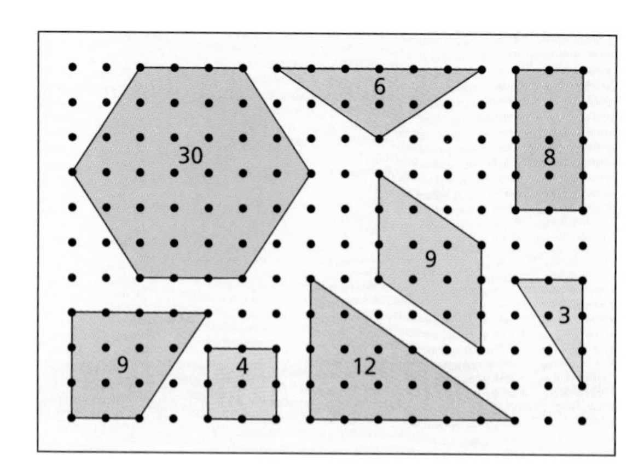

Integrated Review

Making Connections within Mathematics

Perimeter **In Exercises 19–21, solve for x and find the length of each side.**

19. The perimeter is 90.

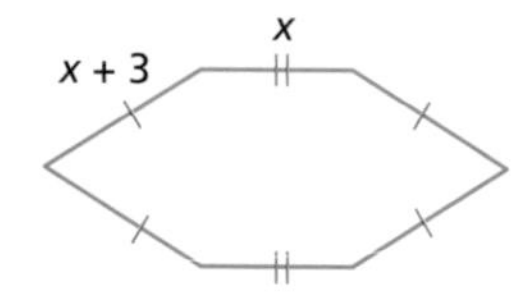

20. The perimeter is 42.

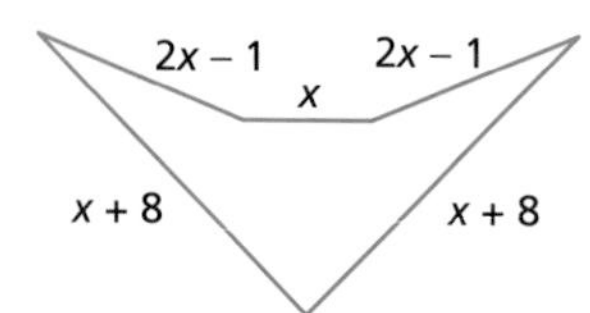

21. The perimeter is 62.

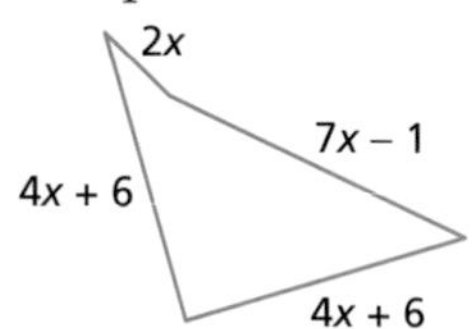

Exploration and Extension

Forming Squares **In Exercises 22–24, copy the figure. Then cut the figure into two parts and rearrange it to form a square. You can use only one straight cut.**

22. 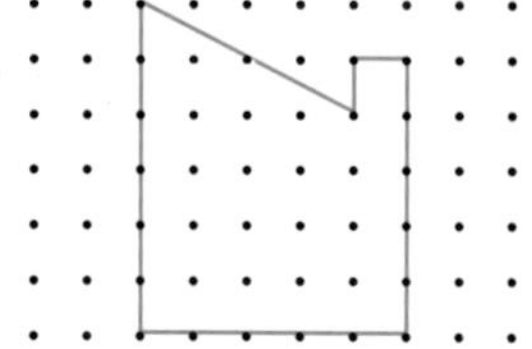

23. 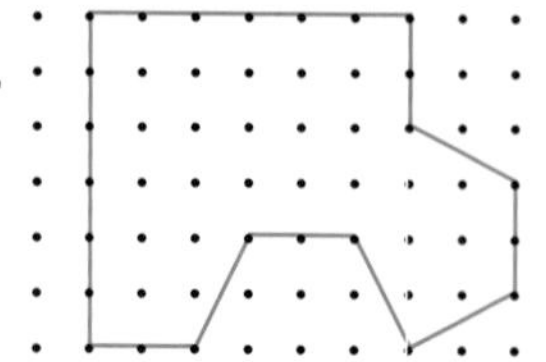

24. 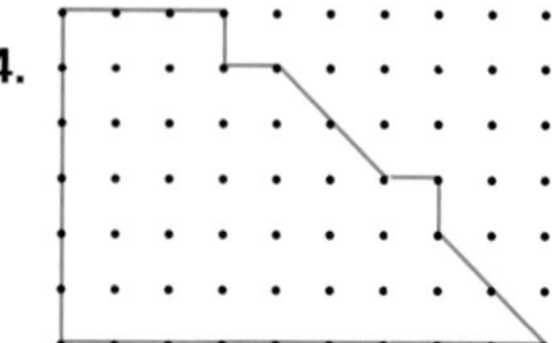

LESSON INVESTIGATION 10.8

Exploring Angles of Polygons

Materials Needed: paper, straightedge, scissors

Example ***Exploring Angles of Triangles***

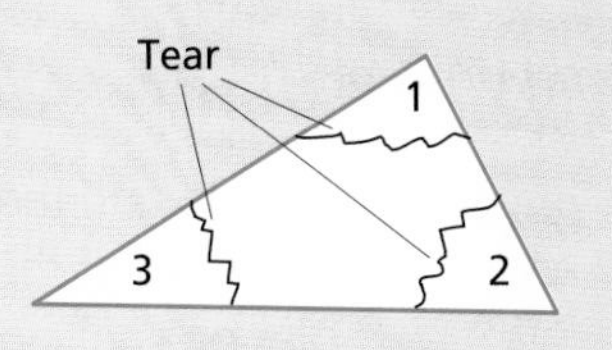

Draw a large triangle on a piece of paper. Cut the triangle out and tear off its three corners. Explain how you can use the three corners to confirm that the sum of the three angle measures is 180°.

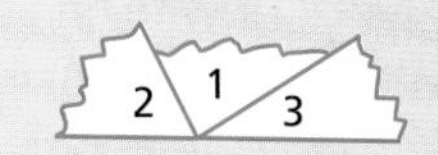

Solution You can arrange the three corners as shown at the right. If you place the corners carefully, you will see that they form a straight angle. This implies that the sum of the three angle measures is 180°. ■

Exercises

1. Work in a group to complete the table. Use the procedure described in Example 2 on page 204.

Polygon	Number of Sides	Number of Angles	Number of Triangles	Sum of Angle Measures	Measure of Each Angle in a Regular Polygon
Triangle	3	3	1	1(180°) = 180°	60°
Quadrilateral	4	4	2	2(180°) = 360°	90°
Pentagon	5	?	?	?	?
Hexagon	6	?	?	?	?
Heptagon	7	?	?	?	?
Octagon	8	?	?	?	?
Nonagon	9	?	?	?	?
Decagon	10	?	?	?	?
n-gon	n	?	?	?	?

2. Describe the pattern for the sum of the measures of the angles of a polygon. Use your result to find the sum of the measures of the angles of a 12-sided polygon.
3. Describe the pattern for the measure of each angle of a regular polygon. Use your result to find the measure of each angle of a regular 12-sided polygon.

10.8 Angles of Polygons

What you should learn:

Goal 1 How to find the measures of the angles of a polygon

Goal 2 How to find the measure of each angle of a regular polygon

Why you should learn it:

You can use the measures of the angles of a polygon to solve real-life problems, such as analyzing the design of a star.

Goal 1 Measuring the Angles of a Polygon

The angles of a polygon are called **interior angles.** Polygons also have **exterior angles,** as shown below.

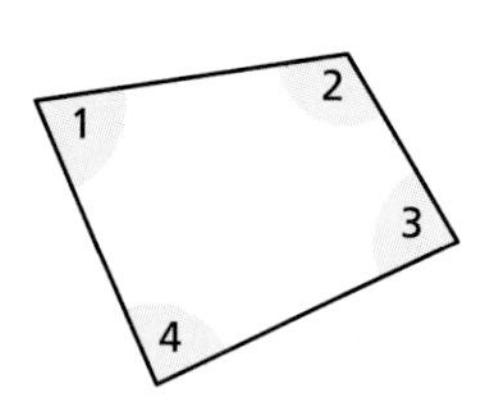

∠1, ∠2, ∠3, and ∠4 are interior angles of this polygon.

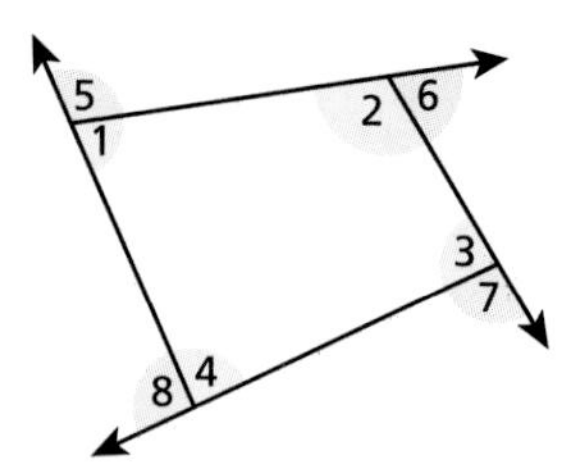

∠5, ∠6, ∠7, and ∠8 are exterior angles of this polygon.

LESSON INVESTIGATION

Investigating Exterior Angles of a Polygon

Group Activity Draw a polygon on a piece of paper and extend the sides to form one exterior angle at each vertex. Then cut out the exterior angles and tape them together. Discuss your result. Do you think the result is true of *any* polygon? Explain your reasoning.

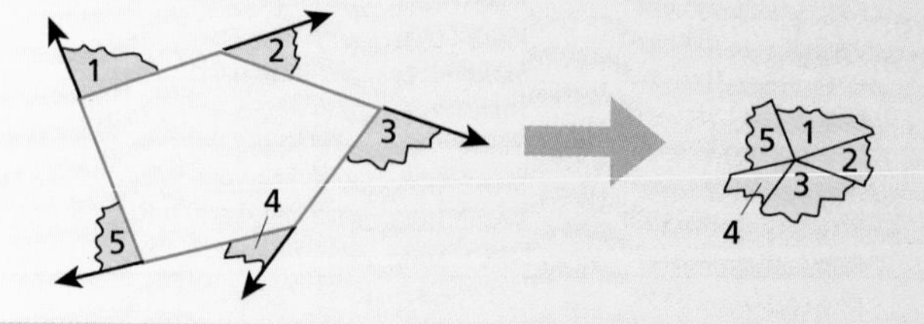

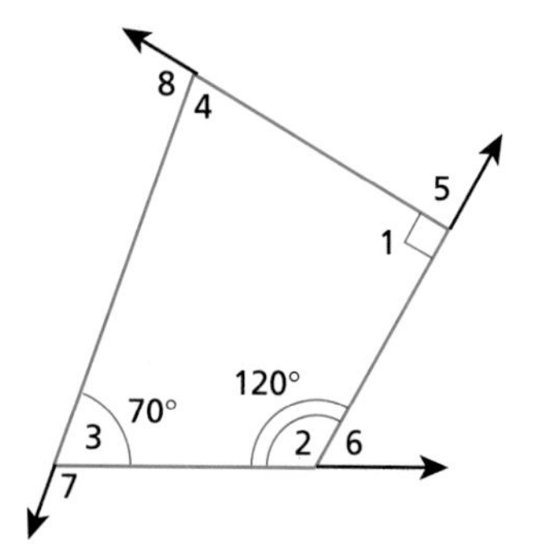

Example 1 *Measuring the Angles of a Polygon*

Find the measures of the angles of the polygon at the left.

Solution You are given that $m\angle 1 = 90°$, $m\angle 2 = 120°$, and $m\angle 3 = 70°$. Because the sum of the interior angles is 360°, it follows that $m\angle 4 = 360° - (90° + 120° + 70°)$ or 80°. Explain how you can find the measure of each exterior angle. ■

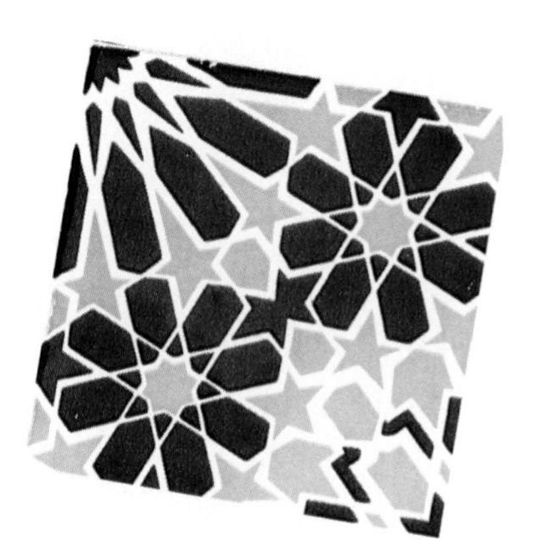

Mosaic star designs often contain both "regular" and "irregular" stars.

Goal 2 Measuring Angles in a Regular Polygon

Angle Measures of a Polygon

Consider a polygon with n sides.

1. The sum of the interior angle measures is $(n-2)(180°)$.
2. For a regular polygon, each interior angle measures $\frac{(n-2)(180°)}{n}$.
3. The sum of the exterior angle measures is 360°.
4. For a regular polygon, each exterior angle measures $\frac{360°}{n}$.

Example 2 *Measuring Angles in a Regular Polygon*

You are designing a flag that uses "regular" five-pointed stars. Sketch a five-pointed star. Then explain how to determine the measures of its angles.

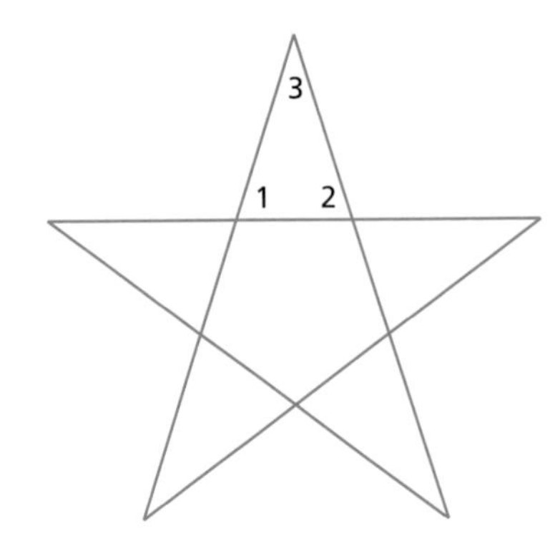

Solution A sketch of a five-pointed star is shown at the left. Notice that the star is built around a regular pentagon.

Each exterior angle of the regular pentagon has a measure of

$$\frac{360°}{5} = 72°. \qquad m\angle 1, m\angle 2$$

Because the measures of $\angle 1$ and $\angle 2$ are the same, it follows that the measure of $\angle 3$ is

$$180° - (m\angle 1 + m\angle 2) = 180° - 144° = 36°. \qquad m\angle 3$$

The measures of the other angles of the star can be found in a similar way. ■

Communicating about MATHEMATICS

SHARING IDEAS about the Lesson

Designing Stars

A. Draw a six-pointed "regular" star and label its angles. Explain how to find the measures of its angles.

B. Draw a seven-pointed "regular" star and label its angles. Explain how to find the measures of its angles.

C. Draw an eight-pointed "regular" star and label its angles. Explain how to find the measures of its angles.

EXERCISES

Guided Practice

CHECK for Understanding

1. Draw a triangle with a 60° exterior angle.

2. Draw a quadrilateral with a 60° interior angle.

In Exercises 3 and 4, find the measure of each labeled angle.

3. Hexagon (vertical line of symmetry)

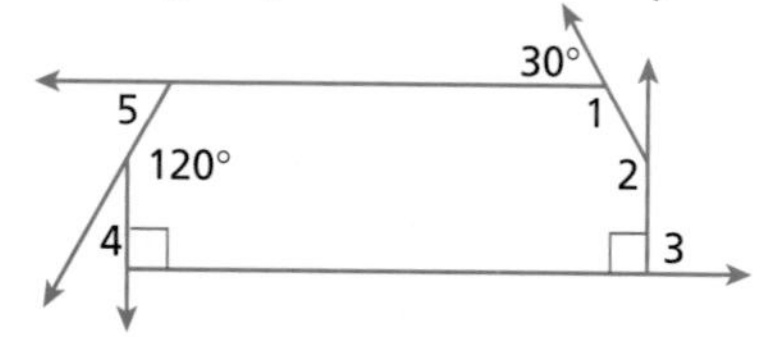

4. Regular hexagon.

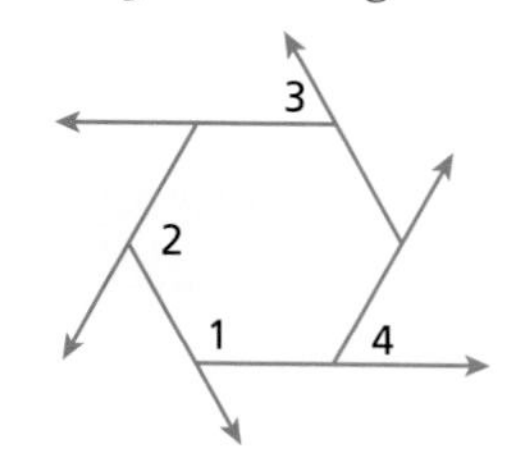

Independent Practice

In Exercises 5–8, use the pentagon at the right.

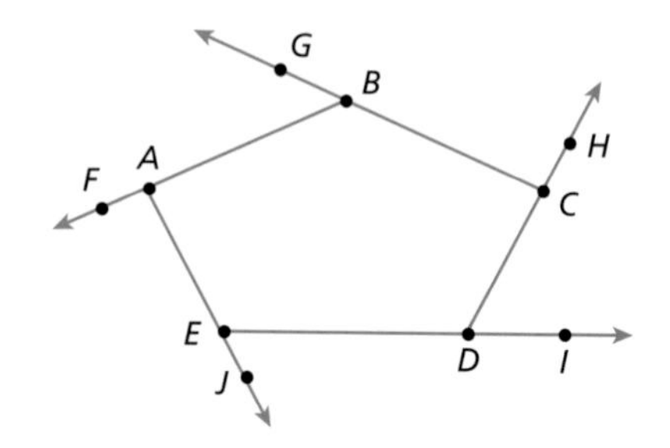

5. Name the interior angles of the pentagon. What is the sum of their measures?

6. Name the indicated exterior angles of the pentagon. What is the sum of their measures?

7. Does the pentagon appear to be regular?

8. If $m\angle ABG = 35°$, what is $m\angle ABC$?

In Exercises 9–11, find the measures of the labeled angles.

9.

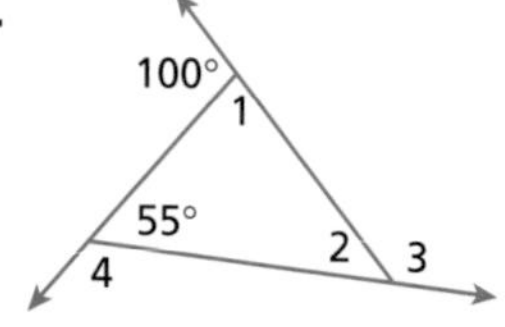

10.

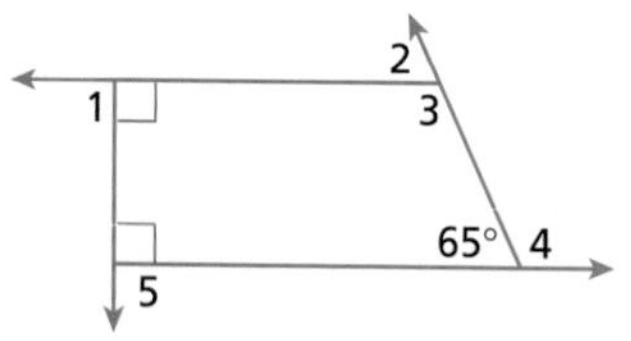

11.

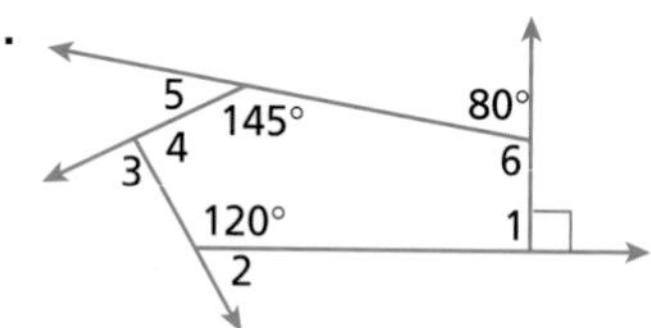

In Exercises 12–14, find the measure of each interior and exterior angle of the regular polygon. Illustrate your results with a sketch.

12. Regular octagon

13. Regular decagon

14. Regular 12-gon

Algebra **In Exercises 15–18, find the measure of each interior angle.**

15.

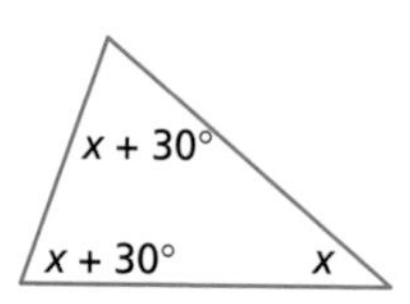

16.

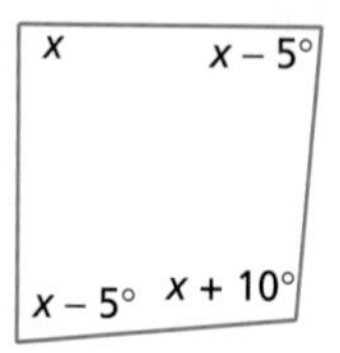

17.

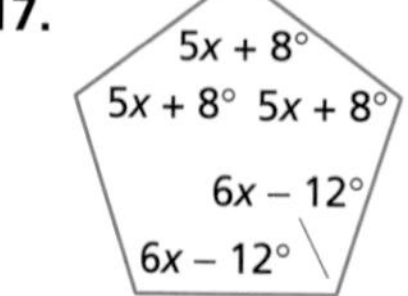

18.

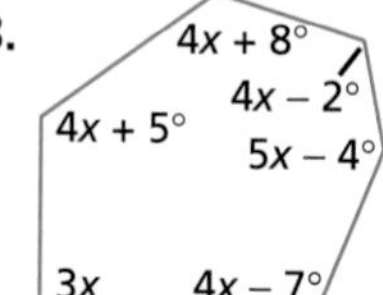

Sunrooms **In Exercises 19–22, use the photo at the right.**

19. Sketch the glass panes and describe their shapes.
20. Which panes appear congruent?
21. Find the angle measures of the glass triangle.
22. The ceiling meets the left wall of the sunroom at an angle of 117°. Find the angle measures of the glass trapezoid.
23. *Error Analysis* The polygon below has a vertical line of symmetry and exactly one of the angle measures is incorrect. Which measure is incorrect? What should it be?

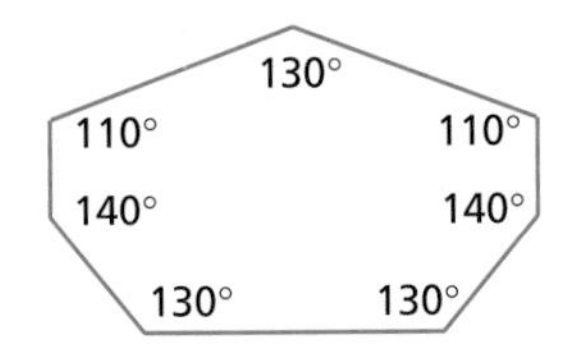

24. The square at the right is composed of 1 square, 4 equilateral triangles, and 4 isosceles triangles. What is the measure of the angles of each isosceles triangle?

Integrated Review

Making Connections within Mathematics

25. *Making a Table* Copy and complete the table.

Regular Polygon	Triangle	Square	Pentagon	Hexagon	Heptagon	Octagon
Interior Angle Measure	?	?	?	?	?	?

26. *Making a Scatter Plot* Use the data in Exercise 25 to make a scatter plot.
27. *Interpreting a Scatter Plot* Use the scatter plot to estimate the interior angle measure of a regular nonagon (9 sides). Confirm your result.

Exploration and Extension

28. *Estimation* Copy the hexagon at the right onto graph paper. Each small square has an area of 1 square unit. Estimate the area of the hexagon.

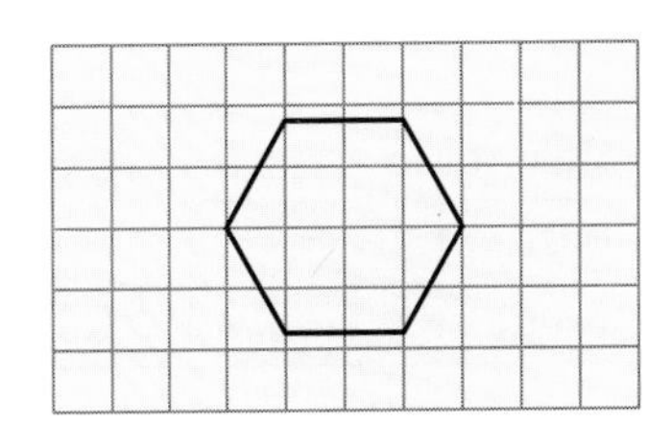

29. Show how to divide the hexagon into a rectangle and 2 triangles. Use these polygons to estimate the area of the hexagon. Compare the result to that obtained in Exercise 28.

USING A COMPUTER DRAWING PROGRAM

Measuring Angles

Some computer programs are able to measure lengths and angles. If you have access to such a program, try using it to duplicate the activity described in the following example.

Example ***Measuring Angles of Triangles***

Use a computer drawing program to draw the nine right triangles shown below. For each triangle, measure the upper angle. Describe the pattern.

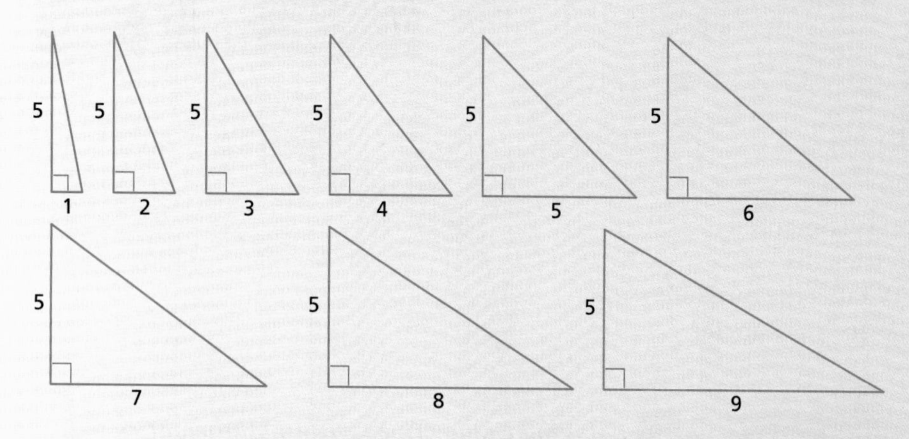

Solution Using a computer drawing program, you can obtain the measures shown in the table.

Height	5	5	5	5	5	5	5	5	5
Base	1	2	3	4	5	6	7	8	9
Angle	11.2°	21.8°	31.0°	38.7°	45.0°	50.2°	54.5°	58.0°	60.9°

From the table, you can see that as the base increases, the measure of the upper angle increases. ■

Exercise

Use a computer drawing program to draw nine right triangles. Let the height of each triangle be 6, and let the bases vary from 1 to 9. Measure the upper angle of each triangle and complete the table. How does this table compare to that given in the example.

Height	6	6	6	6	6	6	6	6	6
Base	1	2	3	4	5	6	7	8	9
Angle	?	?	?	?	?	?	?	?	?

10.9 Angle and Side Relationships

What you should learn:

Goal 1 How to compare side lengths and angle measures of a triangle

Goal 2 How to find the angle measures of an isosceles triangle

Why you should learn it:

You can use the measures of the angles and sides of a triangle to solve real-life problems, such as analyzing the angles in a house of cards.

Goal 1 Angles and Sides of Triangles

LESSON INVESTIGATION

Investigating Lengths and Angle Measures

Group Activity Use a ruler and compass to draw triangles with the following side lengths.

5 cm, 7 cm, 11 cm; 8 cm, 8 cm, 14 cm

Use a protractor to measure each angle.
Next, use a protractor and straightedge to draw triangles with the following angle measures.

40°, 60°, 80°; 20°, 60°, 100°; 35°, 35°, 110°

Use a ruler to measure the length of each side. What conclusions can you make about the relationships between side lengths and angle measures?

You may have discovered the following relationships between the side lengths and the angle measures of a triangle.

Angle and Side Relationships

1. In a triangle, the longest side is opposite the largest angle and the shortest side is opposite the smallest angle.
2. In an isosceles triangle, the angles opposite the sides of the same lengths have equal measures.

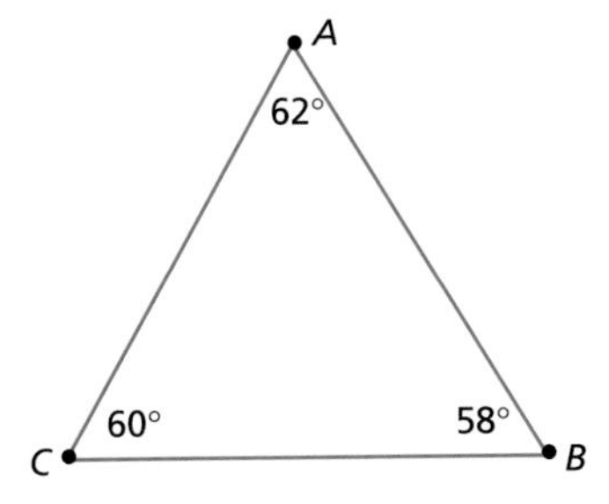

Example 1 *Comparing Sides and Angles*

In the triangle at the left, without using a ruler, state which side is longest and which is shortest.

Solution The largest angle is $\angle A$. This implies that the longest side is $\overline{BC}$. The smallest angle is $\angle B$. This implies that the shortest side is $\overline{AC}$. ■

Goal 2 Comparing Angles in Isosceles Triangles

Example 2 *Analyzing an Isosceles Right Triangle*

In an isosceles right triangle, what is the measure of each angle?

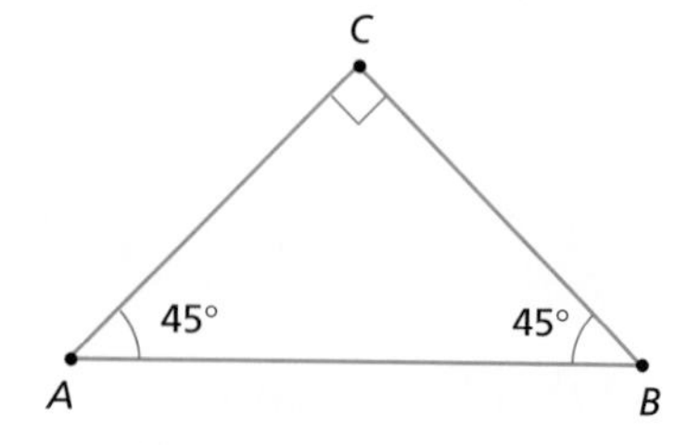

Solution Begin by drawing an isosceles right triangle, as shown at the left. Label the right angle $\angle C$, and the other two angles $\angle A$ and $\angle B$.

Because $\angle A$ and $\angle B$ are opposite the sides that are the same length, they must have equal measures. That is, $m\angle A = m\angle B$. Because $m\angle A + m\angle B = 90°$, you can conclude that

$$m\angle A = 45° \quad \text{and} \quad m\angle B = 45°.$$ ■

Example 3 *Measuring Angles*

You are building a house of cards, as shown at the left. Each triangle formed by the cards has exactly one angle that measures 30°. What are the measures of the other two angles?

Solution Because each triangle has two sides formed by cards of the same size, it follows that each triangle is isosceles. Thus, each triangle has two angles that have the same measure. The sum of these two angles is

$$180° - 30° = 150°.$$

Thus, each of the two "same-size" angles must measure $\frac{1}{2}(150°)$ or 75°. ■

Communicating about MATHEMATICS

SHARING IDEAS about the Lesson

Matching Triangles The angle measures and side lengths of three triangles are given below. Match the angle measures with the side lengths and explain your reasoning.

A. 30°, 60°, 90° **B.** 50°, 50°, 80° **C.** 30°, 30°, 120°

1. 10, 10, 17.3 **2.** 5, 8.66, 10 **3.** 10, 10, 12.9

EXERCISES

Guided Practice

CHECK for Understanding

1. Discuss the angle and side relationships of
   a. a scalene triangle. **b.** an isosceles triangle. **c.** a regular triangle.
2. *Writing* In your own words, explain why an isosceles right triangle must have two 45° angles.

In Exercises 3–6, identify the smallest angle, largest angle, shortest side, and longest side.

3.

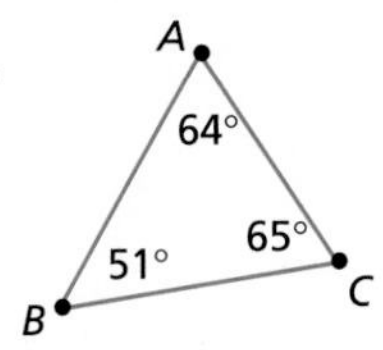

4.

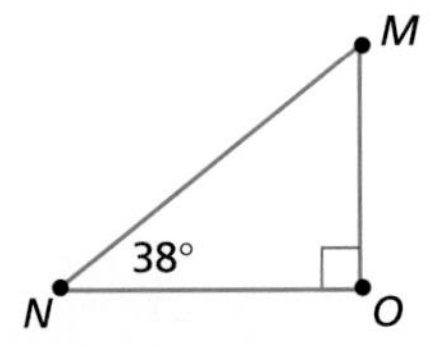

5.

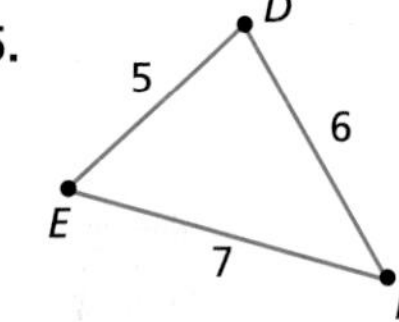

6.

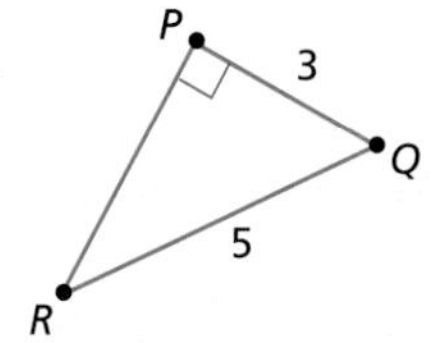

Independent Practice

In Exercises 7–12, use the figure at the right. List the angle measures of the triangle. Then name its shortest and longest sides.

7. $\triangle DEF$
8. $\triangle FGH$
9. $\triangle ADG$
10. $\triangle DFG$
11. $\triangle CGH$
12. $\triangle DGH$
13. What is the longest side of a right triangle? Explain.

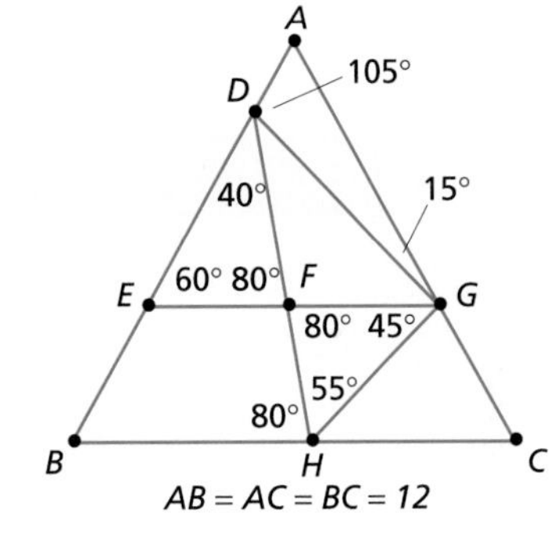

In Exercises 14–17, name the smallest angle of the right triangle. Explain.

14.

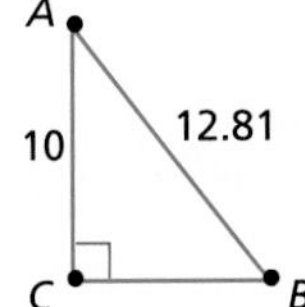

15.

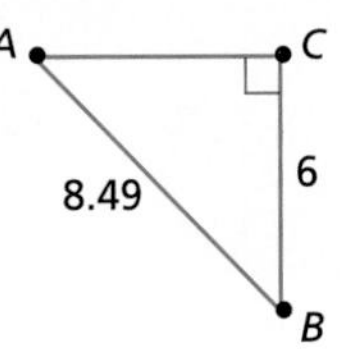

16.

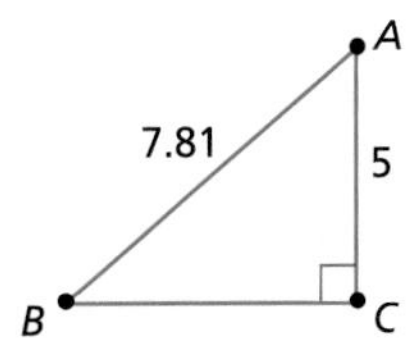

17.

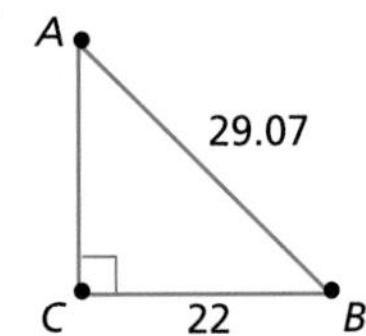

In Exercises 18–21, name the shortest and longest sides of the triangle. Explain.

18.

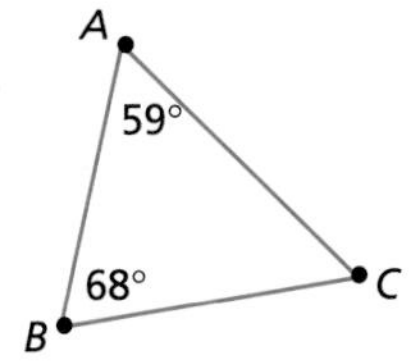

19.

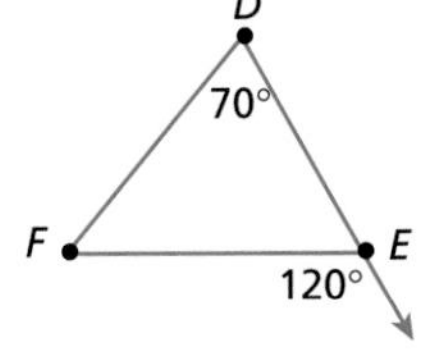

20.

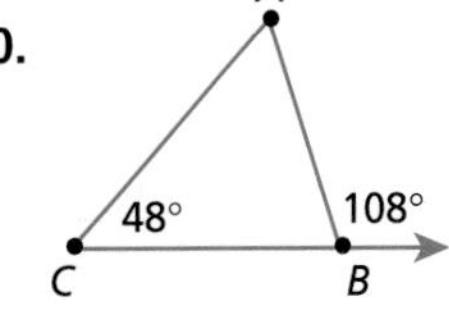

21.

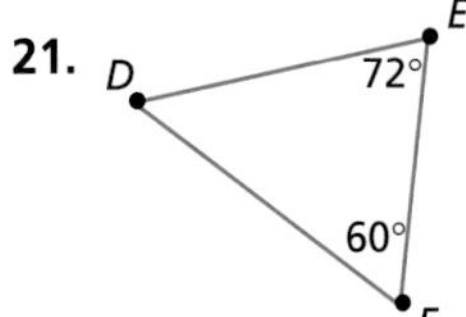

Algebra **In Exercises 22–25, solve for x. Then name the smallest angle, the largest angle, the shortest side, and the longest side of the triangle.**

22.

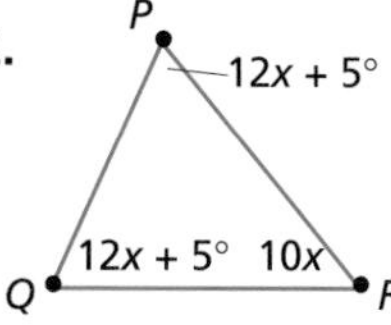

23.

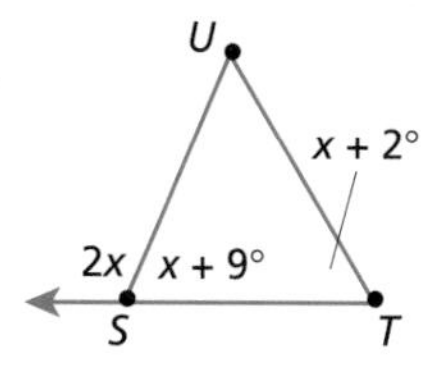

24.

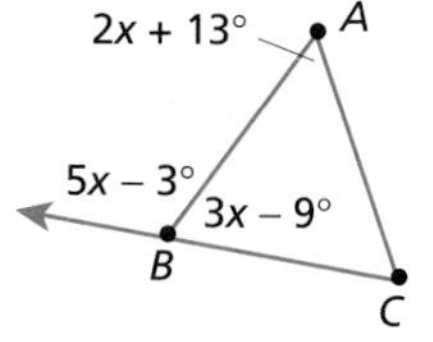

25.

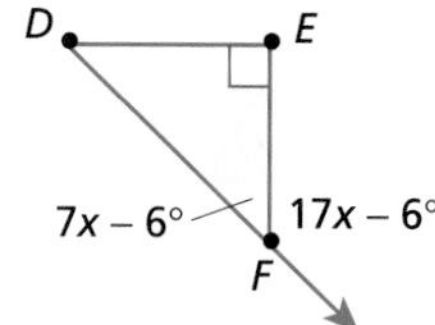

In Exercises 26–30, match the angle measures with the approximate side lengths. Explain your reasoning.

a. 40°, 70°, 70° **b.** 30°, 60°, 90° **c.** 60°, 60°, 60° **d.** 45°, 45°, 90° **e.** 25°, 25°, 130°

26. 5, 5, 7.1 **27.** 5, 5, 5 **28.** 5, 5, 9.1 **29.** 3, 5.2, 6 **30.** 5, 5, 3.4

Walking to the Library **In Exercises 31 and 32, use the following information.**

You and a friend are on a walk, as shown at the right. From your point of view, the angle between your friend and the library is 55°. From your friend's point of view, the angle between you and the library is 62°.

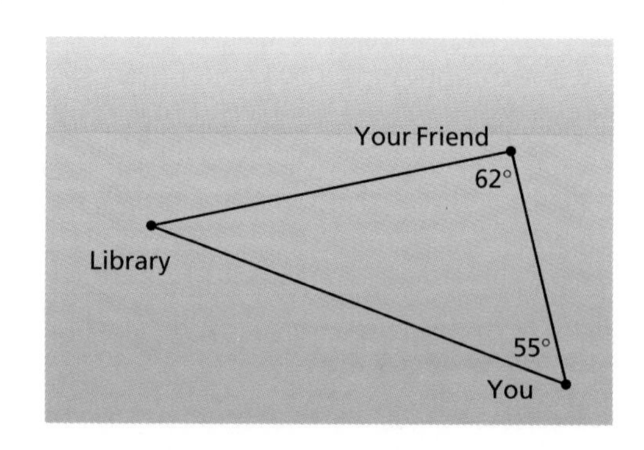

31. What is the third angle of the triangle?

32. Who has farther to walk to the library, you or your friend? Explain your reasoning.

Integrated Review

Making Connections within Mathematics

Movie Attendance **In Exercises 33–36, use the line graph at the right, which shows the percent of Americans who went to the movies every week.** *(Source: Motion Picture Association of America)*

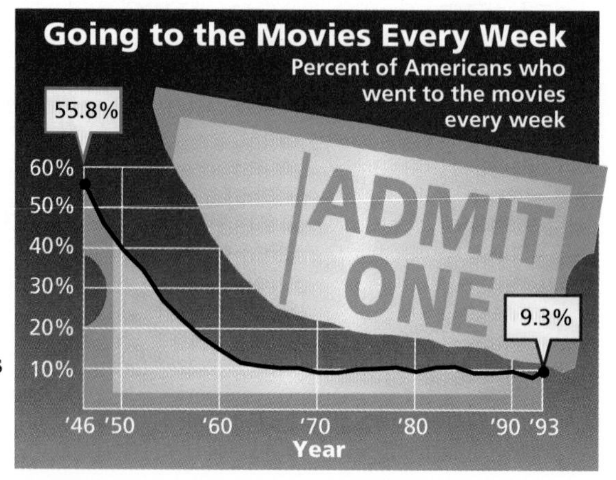

33. Approximate the percent in 1955.

34. Approximate the change from 1946 to 1993.

35. During which decade (50's, 60's, 70's, or 80's) did the greatest decrease occur? Explain.

36. During which decade (50's, 60's, 70's, or 80's) was the percent most stable? Explain.

Exploration and Extension

Coordinate Geometry **In Exercises 37–40, draw $\triangle ABC$ in a coordinate plane. Find the lengths of its sides. Then name its smallest and largest angles.**

37. $A(4, 0), B(0, 0), C(0, -4)$

38. $A(5, 4), B(5, -2), C(3, -2)$

39. $A(-2, -1), B(-2, -3), C(4, -3)$

40. $A(0, 2), B(3, -4), C(-3, -4)$

10

Chapter Summary

What did you learn?

Skills

1. Identify points, lines, and planes. **(10.1)**
   - Find the length of a line segment. **(10.1)**
   - Decide whether two lines are parallel. **(10.1)**
2. Classify angles as acute, right, obtuse, or straight. **(10.2)**
   - Use a protractor to measure an angle. **(10.2)**
3. Use a property of parallel lines. **(10.3)**
   - Identify vertical angles and corresponding angles. **(10.3)**
4. Identify line symmetry and rotational symmetry. **(10.4)**
5. Classify triangles by their sides and their angles. **(10.5)**
6. Classify quadrilaterals. **(10.6)**
7. Recognize congruent polygons and regular polygons. **(10.7)**
8. Measure the interior and exterior angles of a polygon. **(10.8)**
9. Compare side lengths and angle measures of a triangle. **(10.9)**
   - Use a property of isosceles triangles. **(10.9)**

Problem-Solving Strategies

10. Use geometric figures such as lines, segments, angles, and polygons to model and solve real-life problems. **(10.1–10.9)**

Exploring Data

11. Use tables and graphs to solve problems. **(10.1–10.9)**

Why did you learn it?

Geometry is used in almost all parts of real life to measure, compare, and explain the shapes of things. For instance, you learned that a kaleidoscope contains images that are pentagons, hexagons, octagons, or other polygons. Geometry is also used to measure angles in the construction industry, to manufacture sails for windsurfboards, and to analyze the motion of engine parts. No matter which profession you choose, geometry can help you be more successful.

How does it fit into the bigger picture of mathematics?

The word geo-metry means "earth-measuring," and that describes geometry very well. In this chapter, you learned how to use geometry to measure line segments and angles. In this chapter, you were introduced to dozens of geometric terms. Many of these are familiar (triangle, rectangle, line, plane, and so on), but many are probably new (scalene, isosceles trapezoid, and exterior angle). It is important to continue to build your mathematical vocabulary. Doing this will help you communicate mathematics to others. Remember, however, that memorizing the meaning of mathematical terms is not nearly as important a skill as being able to understand mathematical properties.

Chapter REVIEW

In Exercises 1–4, use the figure at the right. (10.1–10.2)

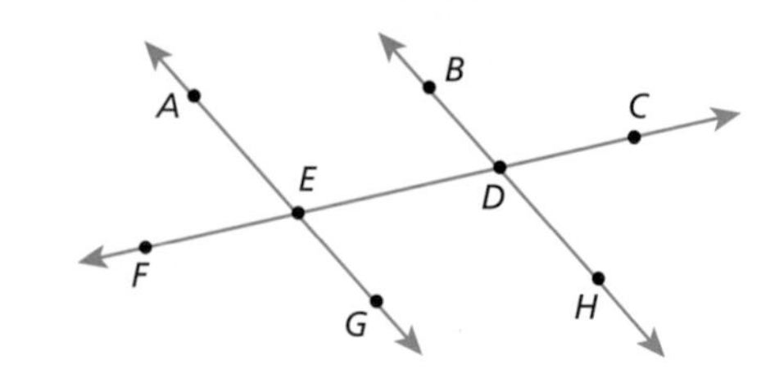

1. Write two other names for the line $\overleftrightarrow{EG}$.
2. Name five line segments that have D as an endpoint.
3. Write another name for the ray $\overrightarrow{HD}$.
4. Write three other names for $\angle AEC$.

In Exercises 5–8, use the drawing of an anvil at the right. (10.1)

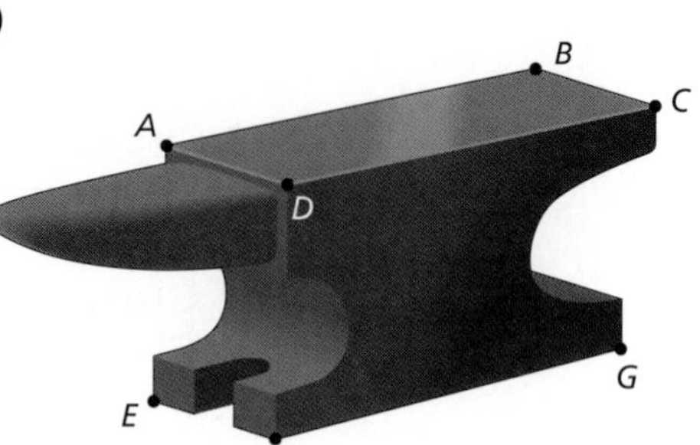

5. Name another point that lies in the same plane as C, D, and G.
6. Name four line segments that lie in the same plane as A, B, and C.
7. From the pattern, where do you think point F lies?
8. Name 3 lines that are parallel to $\overleftrightarrow{BC}$.

In Exercises 9–12, use a protractor to measure the angle. Is it acute, obtuse, right, or straight? (10.2)

9.

10. 

11.

12. 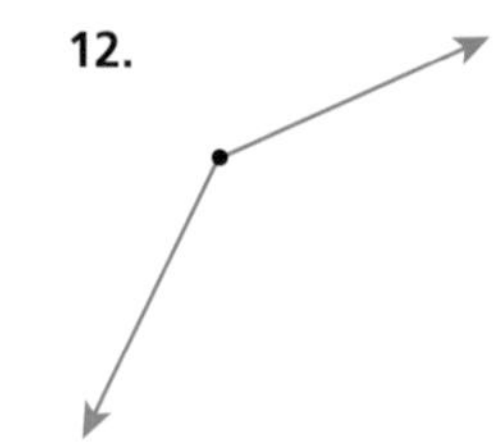

In Exercises 13–18, use the figure at the right. (10.3)

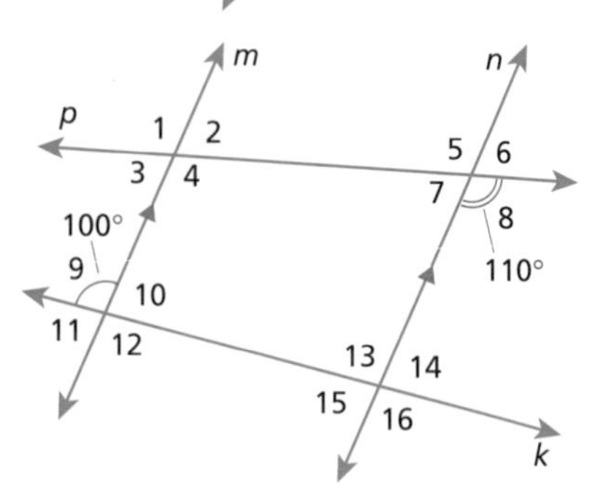

13. Which lines are parallel?
14. Name 2 pairs of vertical angles.
15. List all angles whose measure is 110°.
16. List all angles whose measure is 100°.
17. Which is greater: $m\angle 10$ or $m\angle 2$? Explain.
18. List 2 pairs of corresponding angles that have different measures.

In Exercises 19–22, identify any symmetry of the figure. (10.4)

19. 

20. 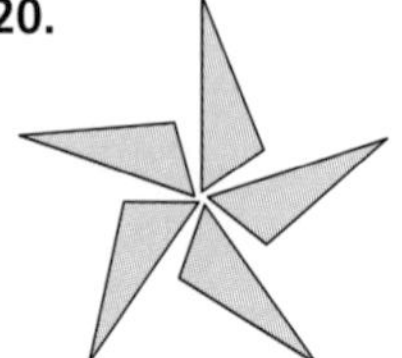

21. 

22. 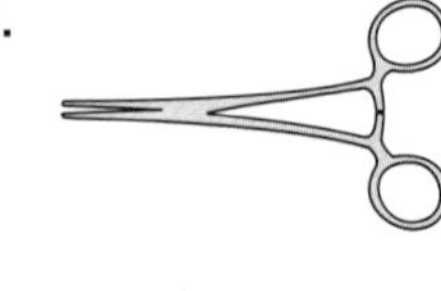

23. *Research Project* Use magazines or newspapers to find examples of company logos that have symmetry. Trace the logos and describe the symmetry. **(10.4)**

24. *Try It Yourself* Design a company logo that has at least one type of symmetry. Describe the company and explain how the logo relates to the company. **(10.4)**

In Exercises 25–27, draw $\triangle ABC$ with the given side lengths. Then classify the triangle according to its sides and angles. (10.5)

25. $a = 2, b = 3, c = 4$

26. $a = 5, b = 12, c = 13$

27. $a = 6, b = 6, c = 8$

In Exercises 28–41, match the polygon with a region at the right. Use each region exactly once. (10.5, 10.6)

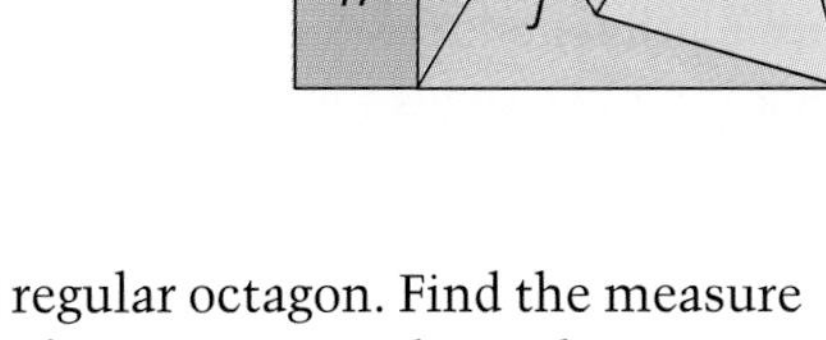

28. Rhombus
29. Isosceles trapezoid
30. Square
31. Isosceles triangle
32. Right triangle
33. Nonconvex quadrilateral
34. Rectangle
35. Parallelogram
36. Kite
37. Acute triangle
38. Obtuse triangle
39. Scalene quadrilateral
40. Trapezoid
41. Equilateral triangle

42. Draw a regular pentagon. Find the measure of each of its interior angles and exterior angles. **(10.8)**

43. Draw a regular octagon. Find the measure of each of its interior angles and exterior angles. **(10.8)**

44. Draw a regular hexagon. Then divide it into four congruent quadrilaterals. **(10.7)**

45. Draw a regular octagon. Then divide it into two congruent trapezoids and a rectangle. **(10.7)**

In Exercises 46 and 47, order the sides from the shortest to the longest. (10.9)

In Exercises 48 and 49, order the angles from the smallest to the largest. (10.9)

46.

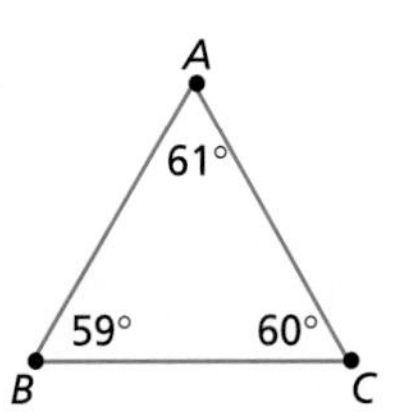

47.

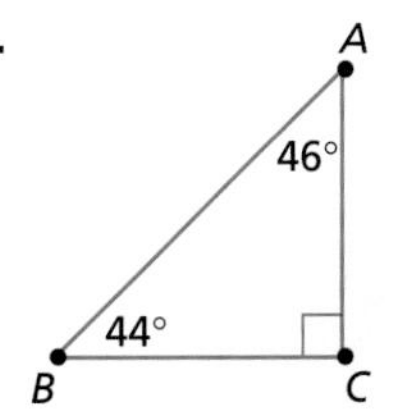

48.

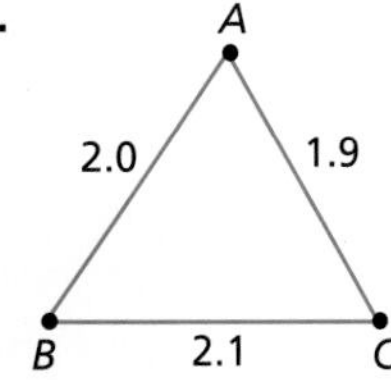

49.

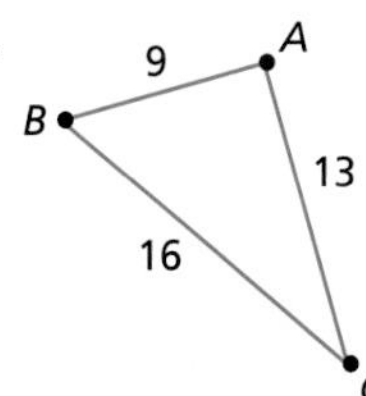

In Exercises 50–53, find the measures of the angles of the triangle. (10.9)

50.

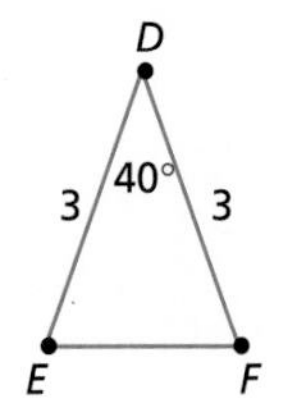

51.

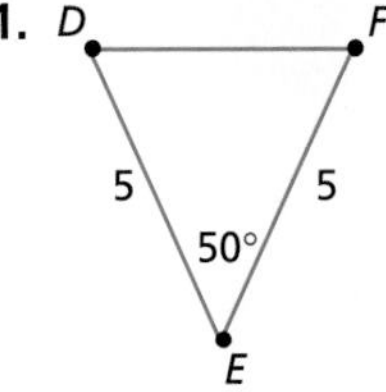

52.

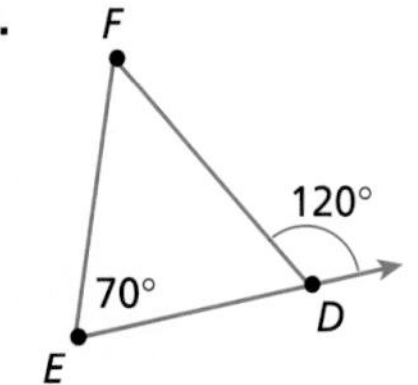

53.

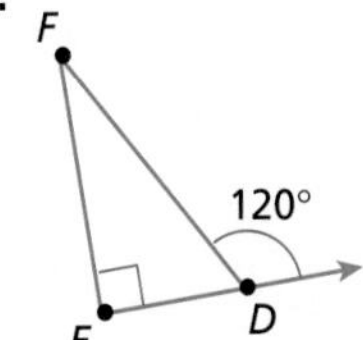

FOCUS ON *Deep Sea Missions*

In Exercises 54–57, use the diagram of the Remotely Operated Vehicle (ROV) traveling on the flat ocean floor.

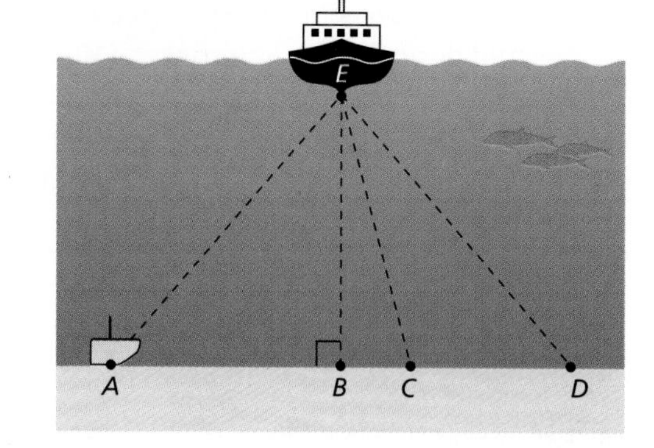

54. The ROV moves from point A to point B. What type of triangle is $\triangle ABE$?

55. Triangle AED is isosceles. If $\overline{AE}$ and $\overline{DE}$ are the same length, which two angles of $\triangle AED$ will have the same measure?

56. In $\triangle AEC$, $m\angle C$ is greater than $m\angle A$. Which side is longer, $\overline{AE}$ or $\overline{EC}$?

57. If $m\angle BCE = 75°$, what is the measure of $\angle BEC$?

In Exercises 58 and 59, use the following information.

In 1994 the MIT Sea Grant College Program in Cambridge, Massachusetts, launched Odyssey II into the ocean. This unmanned sub relays data to a communication node suspended from a buoy. The communication node sends the data to the base station.

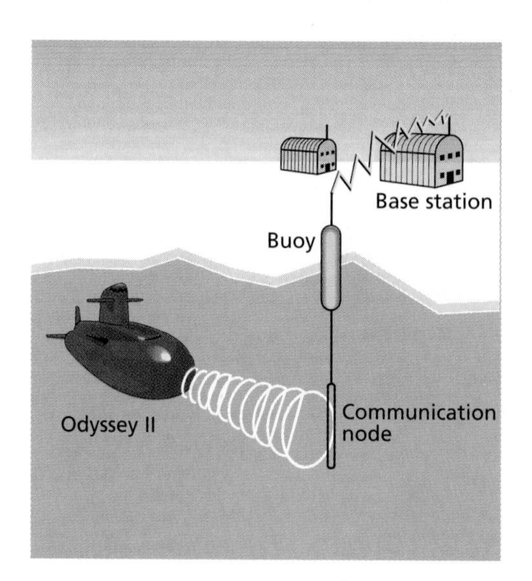

58. The unmanned sub leaves the communication node, travels a straight path for 1000 meters, and turns 90° to the left. The sub repeats this procedure until it returns to the communication node. If the sub remains at a constant depth and travels 4000 meters, what quadrilateral could describe the path of the sub?

59. If the unmanned sub travels a 5-sided path before returning to the communication node, what is the sum of the interior angles of the pentagon?

In Exercises 60–62, identify the (approximate) type of symmetry.

60.

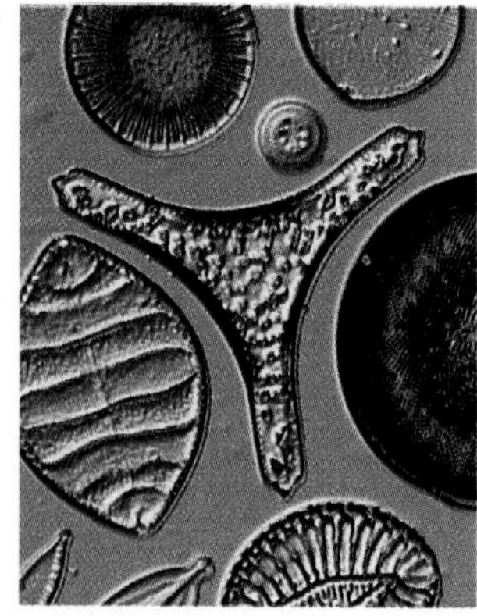

61.

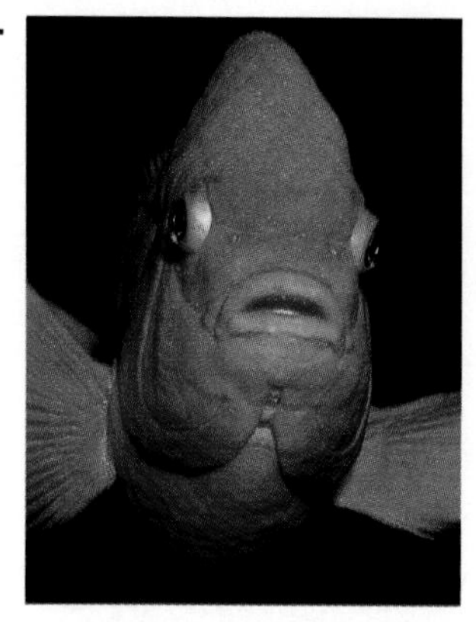

62.

Chapter TEST

In Exercises 1–3, classify the triangle by its sides and by its angles.

1. 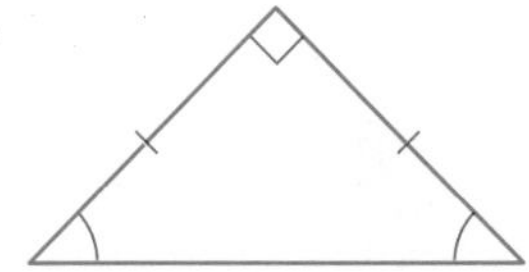

2. 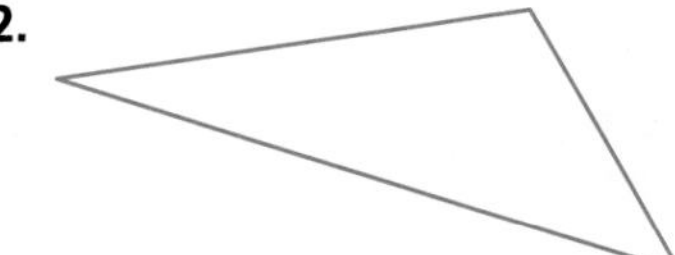

3. 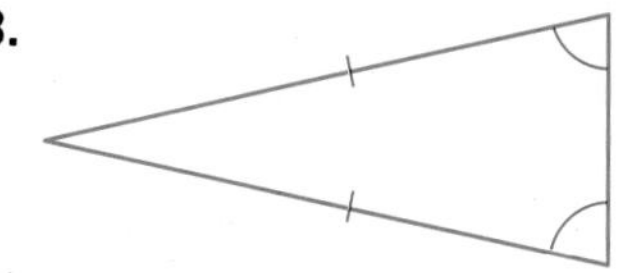

True or False? **In Exercises 4 and 5, decide whether the statement is true or false. Explain your reasoning.**

4. A regular quadrilateral is a square.

5. The opposite sides of a kite are parallel.

In Exercises 6–8, match the angle with its measure.

a. 155°

b. 135°

c. 42°

6. 

7. **8.** 

In Exercises 9–13, use the figure at the right.

9. Give another name for $\overleftrightarrow{MP}$.

10. List the rays that have P as a beginning point.

11. List 2 pairs of vertical angles.

12. Is $\angle KML \cong \angle PSR$? Explain.

13. Is $\angle PMN \cong \angle PRS$? Explain.

14. Draw a five-pointed star that has a regular pentagon as its center. Discuss the symmetry of the star.

In Exercises 15–17, use the figure at the right.

15. What is the measure of $\angle 1$?

16. What is the measure of $\angle 2$?

17. Which is the longest side?

In Exercises 18–20, use the diagram at the right, which shows a design for a laser television set.

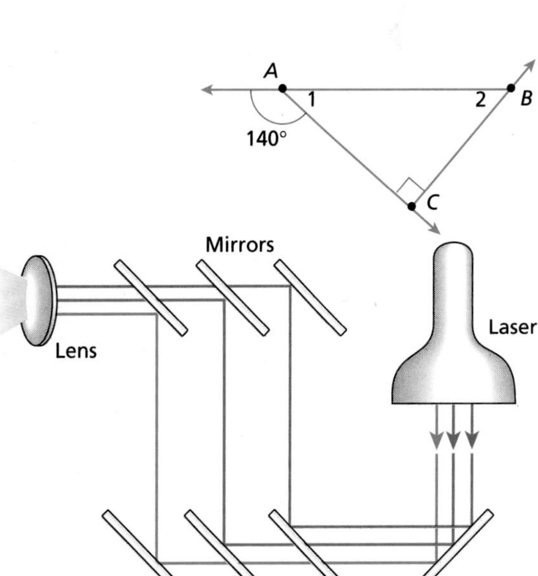

18. Describe the relationship between the 3 beams as they leave the laser unit.

19. How many right angles are formed by each color of laser beam?

20. How many of the 7 mirrors are parallel?

CHAPTER 11

Congruence, Similarity, and Transformations

LESSONS

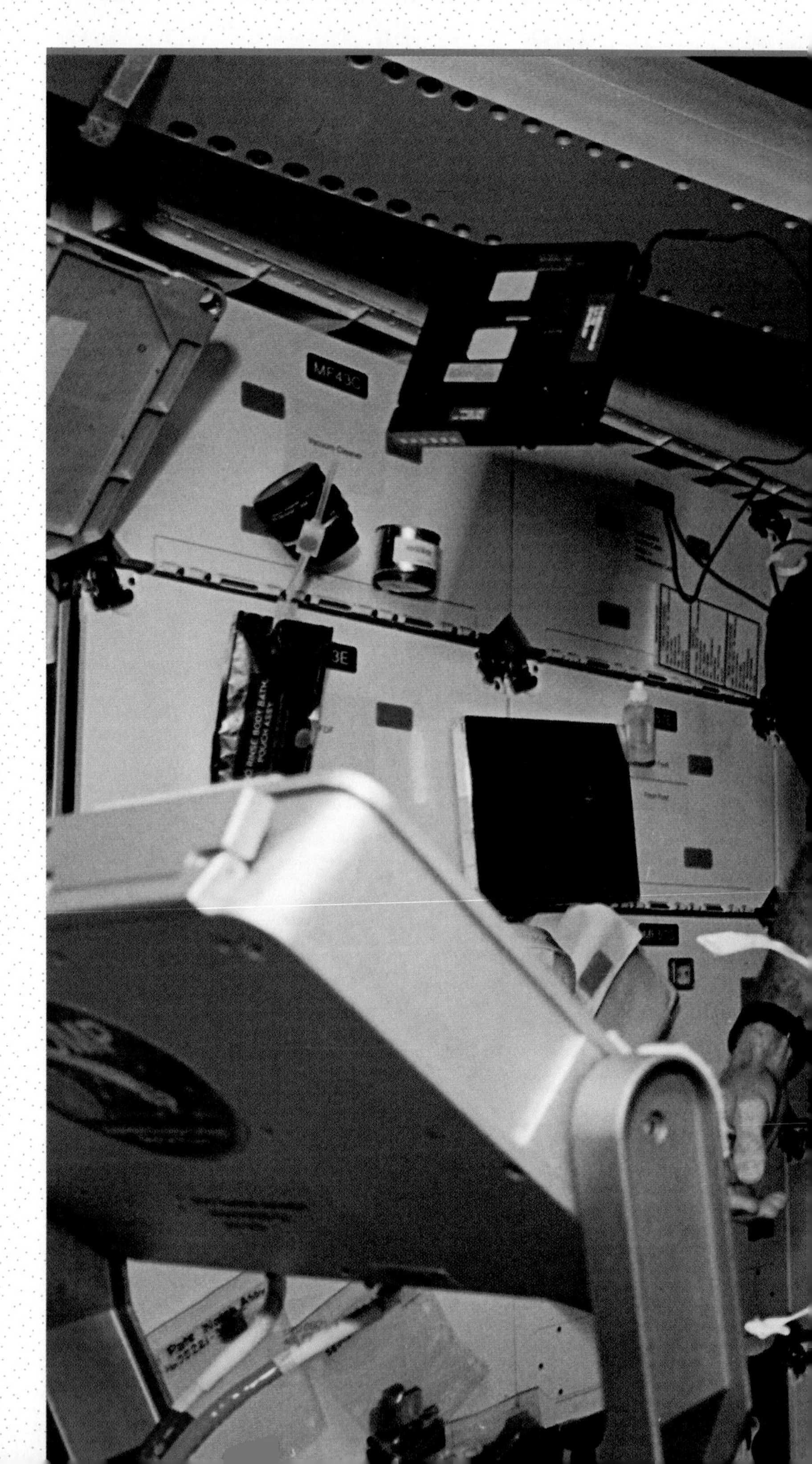

Daily exercise is an important factor in overcoming the adverse effects of weightlessness on the human body. So, NASA has installed exercise gear on the Space Shuttle* Discovery*'s mid-deck. Mission specialist Jerry M. Linenger reported that running on the treadmill was a pleasure—just like being on Earth.

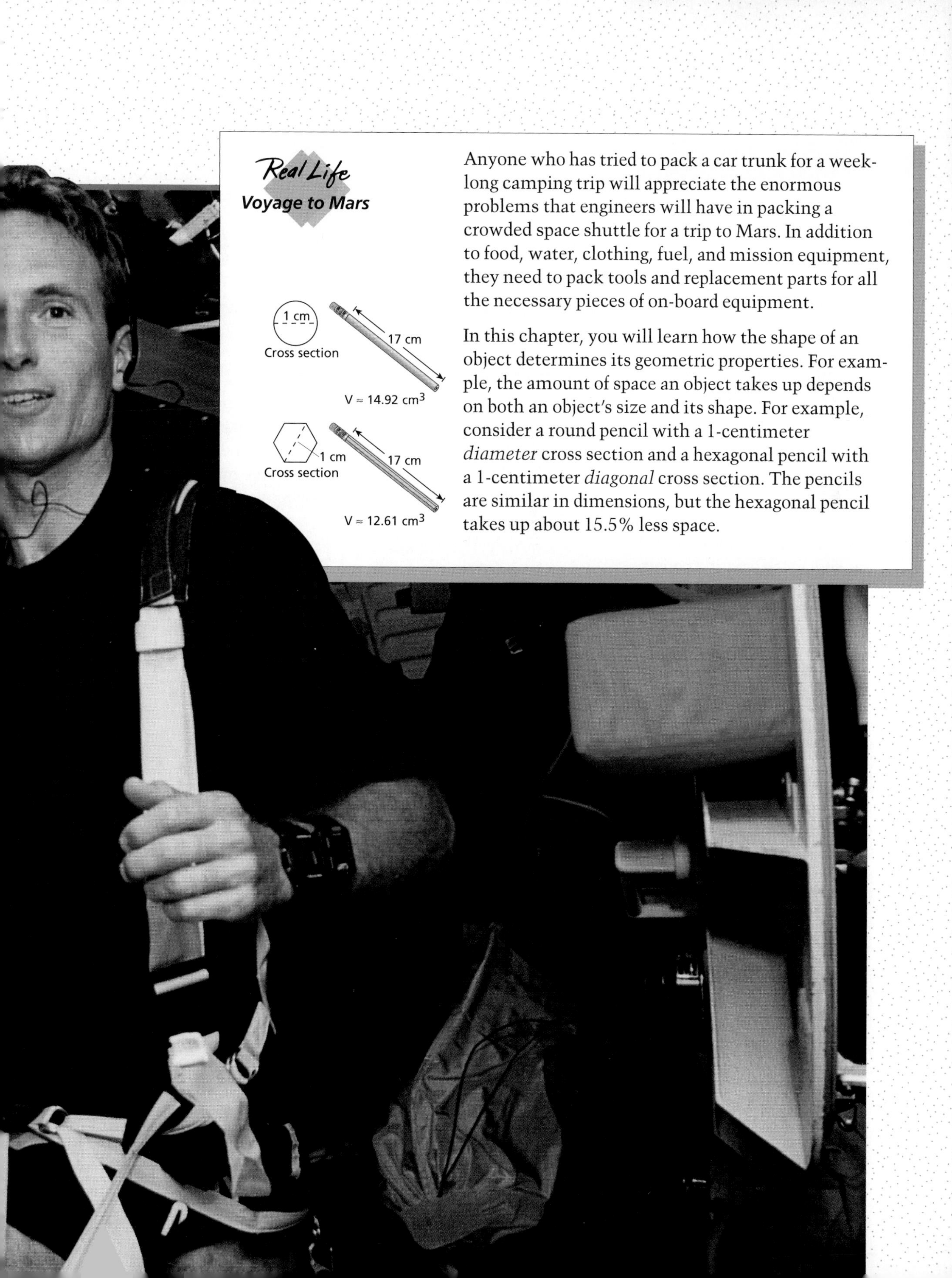

Anyone who has tried to pack a car trunk for a week-long camping trip will appreciate the enormous problems that engineers will have in packing a crowded space shuttle for a trip to Mars. In addition to food, water, clothing, fuel, and mission equipment, they need to pack tools and replacement parts for all the necessary pieces of on-board equipment.

In this chapter, you will learn how the shape of an object determines its geometric properties. For example, the amount of space an object takes up depends on both an object's size and its shape. For example, consider a round pencil with a 1-centimeter *diameter* cross section and a hexagonal pencil with a 1-centimeter *diagonal* cross section. The pencils are similar in dimensions, but the hexagonal pencil takes up about 15.5% less space.

11.1 Area and Perimeter

What you should learn:

Goal 1 How to find the area and perimeter of polygons

Goal 2 How to use area and perimeter to solve real-life problems

Why you should learn it:

You can use area and perimeter to solve real-life problems, such as finding the area and perimeter of a miniature golf course.

Goal 1 Finding Area and Perimeter

LESSON INVESTIGATION

Investigating Area and Perimeter

Group Activity Use a 4-inch by 6-inch index card. Draw lines on each side to form 24 one-inch squares. Draw a diagonal and cut to form two triangles. Label the length of each side. Then place the triangles side by side to form different polygons. Name each polygon and find its area and perimeter. By rearranging the triangles, can you change the area? Can you change the perimeter?

6 in.
≈ 7.2 in.
4 in.
≈ 7.2 in.
4 in.
6 in.

Area of Parallelograms and Trapezoids

Area of a Parallelogram: Area = (base) × (height)

Area of a Trapezoid: Area = $\frac{1}{2}$ (base 1 + base 2) × (height)

Example 1 *Finding Areas of Polygons*

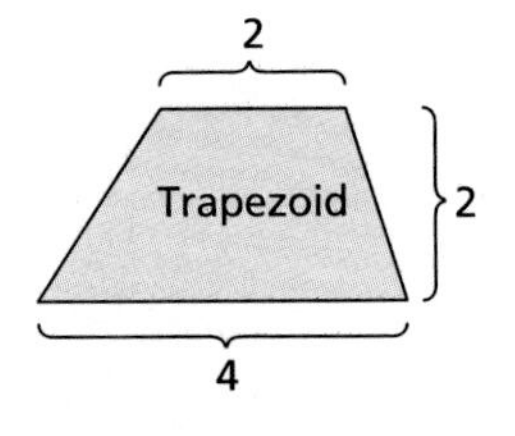

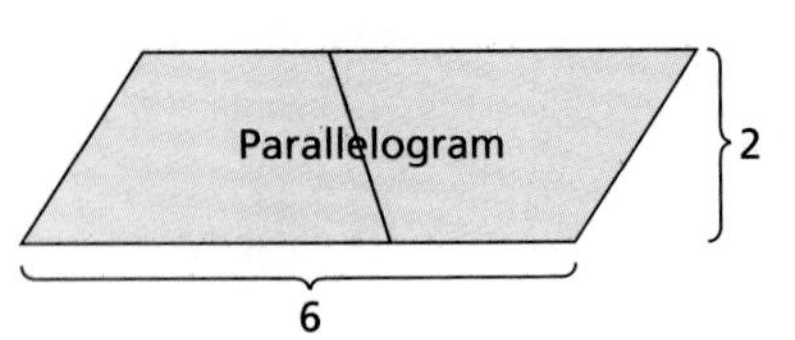

The trapezoid at the left has been duplicated. Then the trapezoid and its duplicate have been arranged to form a parallelogram. Find the area of each.

Solution For the trapezoid, the height is $h = 2$ and the lengths of the bases are $b_1 = 2$ and $b_2 = 4$.

$$\text{Area} = \frac{1}{2}(b_1 + b_2)h = \frac{1}{2}(2 + 4)(2) = 6$$

For the parallelogram, the height is $h = 2$ and the length of the base is $b = 6$.

$$\text{Area} = bh = (6)(2) = 12$$

The parallelogram has twice the area of the trapezoid. ■

Goal 2 Solving Real-Life Problems

Example 2 *Finding Areas*

You are designing a miniature golf course. The plan for one hole is shown at the left. Each square represents 1 square foot. What is the total area of the green portion of the hole?

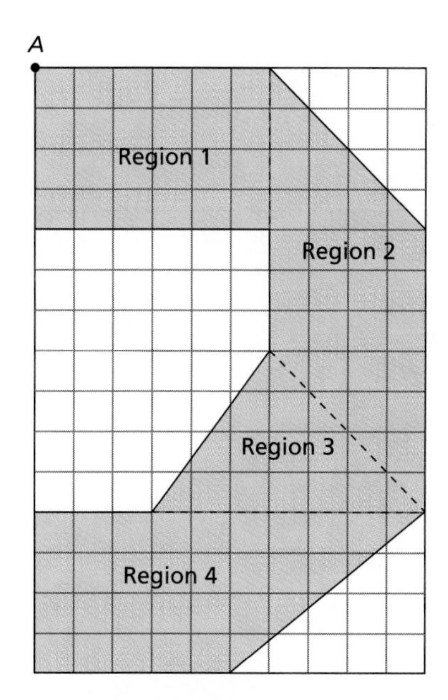

Each small square is 1 foot by 1 foot. The lengths of slanted sides can be found with the Pythagorean Theorem.

Solution One way to answer the question is to divide the green region into four smaller regions and find the area of each.

Region 1 (Rectangle):

Area $=$ (Length)(Width) $= 6 \times 4 = 24$ square feet

Region 2 (Parallelogram):

Area $=$ (Base)(Height) $= 7 \times 4 = 28$ square feet

Region 3 (Triangle):

Area $= \frac{1}{2}$(Base)(Height) $= \frac{1}{2}(7)(4) = 14$ square feet

Region 4 (Trapezoid):

$$\text{Area} = \tfrac{1}{2}(\text{Base 1} + \text{Base 2})(\text{Height})$$
$$= \tfrac{1}{2}(5 + 10)(4) = 30 \text{ square feet}$$

The total area of the green region is

$$\text{Total Area} = 24 + 28 + 14 + 30$$
$$= 96 \text{ square feet.}$$

■

In Example 2, the perimeter of the green region (starting at A) is

$$\text{Perimeter} = 6 + \sqrt{32} + 7 + \sqrt{41}$$
$$+ 5 + 4 + 3 + 5 + 3 + 6 + 4$$
$$= 43 + \sqrt{32} + \sqrt{41}$$
$$\approx 55.1 \text{ feet.}$$

Communicating about MATHEMATICS

SHARING IDEAS about the Lesson

Exploring Area Use a piece of graph paper to copy the outline of the green region shown in Example 2. Show other ways that the region can be divided into smaller regions. For each way, find the total area of the region.

EXERCISES

Guided Practice

CHECK for Understanding

In Exercises 1–4, find the area and perimeter of each figure.

1.

2.

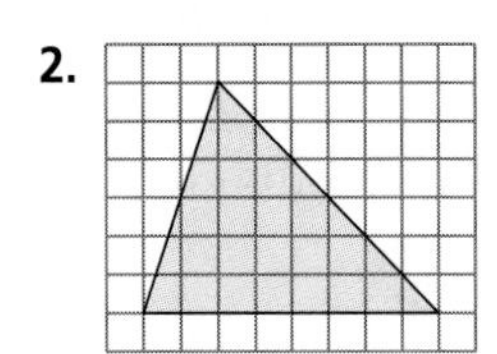

3.

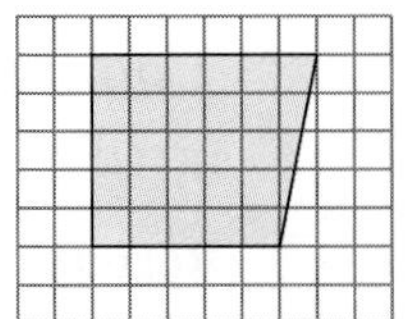

4.

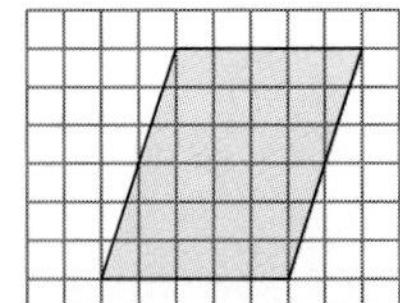

5. The parallelogram at the right has been cut and rearranged to form a rectangle. Find the area and perimeter of each. Does rearranging change the area? Does it change the perimeter?

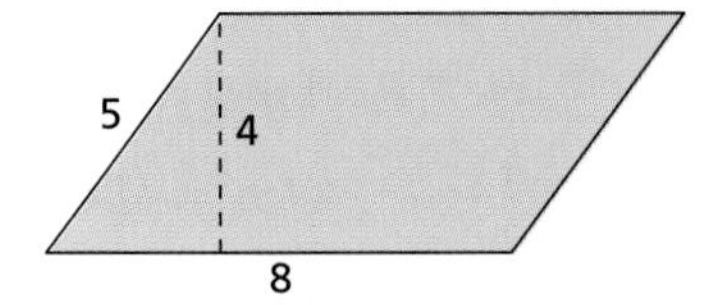

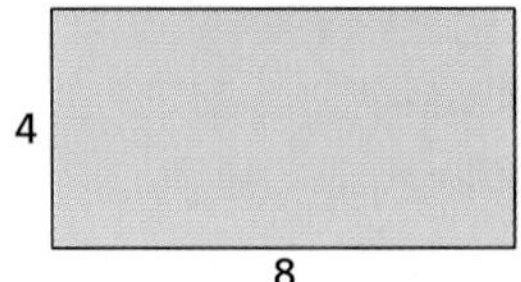

Independent Practice

In Exercises 6 and 7, find the area of each figure. Describe two ways to find the area of the second figure.

6.

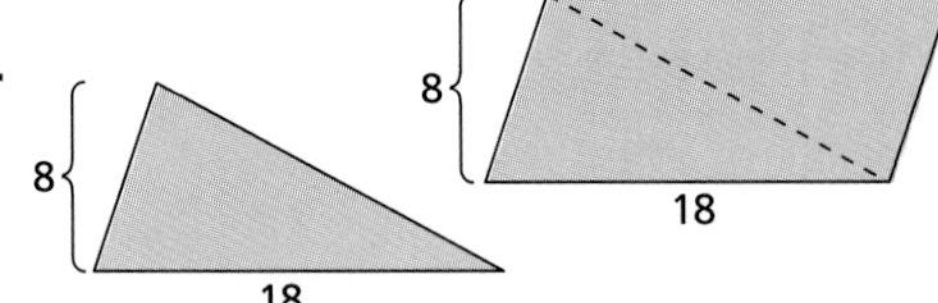

7.

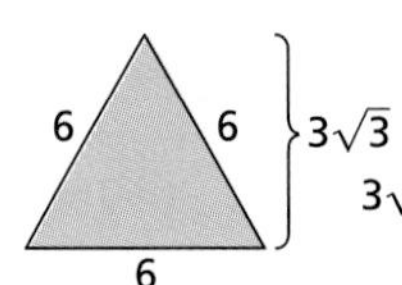

$3\sqrt{3}$ 6 12

In Exercises 8 and 9, show how the second figure can be used to find the area of the first. Then find the perimeter of each figure.

8.

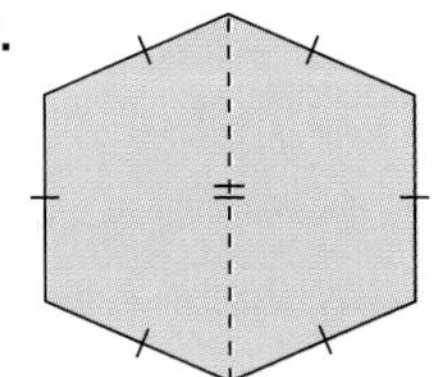

13 12 13 23

9.

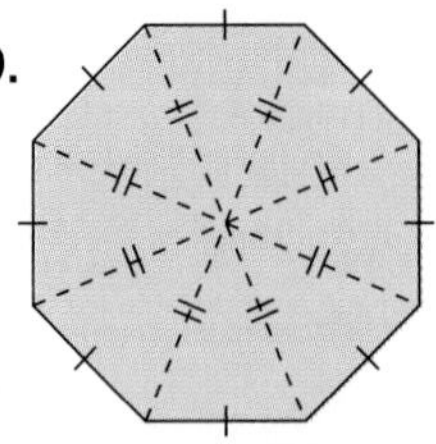

14 ≈13.08 10

Drawing "Equations" **In Exercises 10–14, make a sketch on dot paper to represent the "equation." Then find the area and perimeter of the final figure.**

10. (2 Congruent Right Triangles) + (1 Square) = (1 Parallelogram)

11. (1 Trapezoid) + (1 Triangle) = (1 Parallelogram)

12. (1 Isosceles Triangle) + (1 Isosceles Trapezoid) = (1 Pentagon)

13. (2 Isosceles Trapezoids) + (1 Rectangle) = (1 Octagon)

14. (2 Right Scalene Triangles) = (1 Kite)

Polygon Puzzle **In Exercises 15–18, use the figure at the right. Each small square of the grid has an area of 1 square unit.**

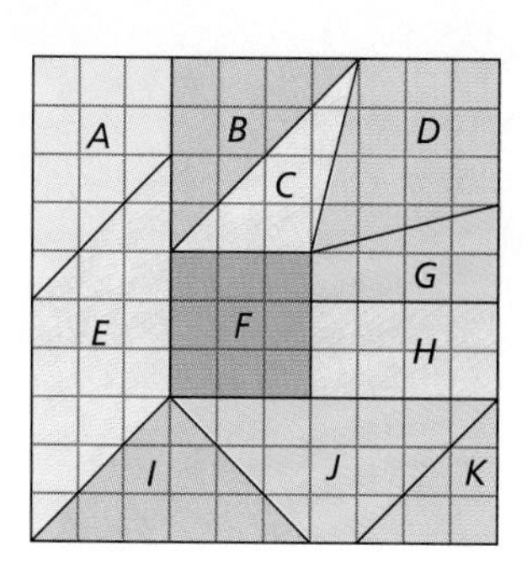

15. Name each type of polygon, A through K.
16. Find the area of each polygon.
17. Find the area of the entire figure.
18. Which polygons have the same area?

19. *Picture Frame* The polygons at the right can be put together to form a picture frame.
    a. Find the area of each polygon.
    b. Sketch the picture frame.
    c. Find the area of the picture frame.

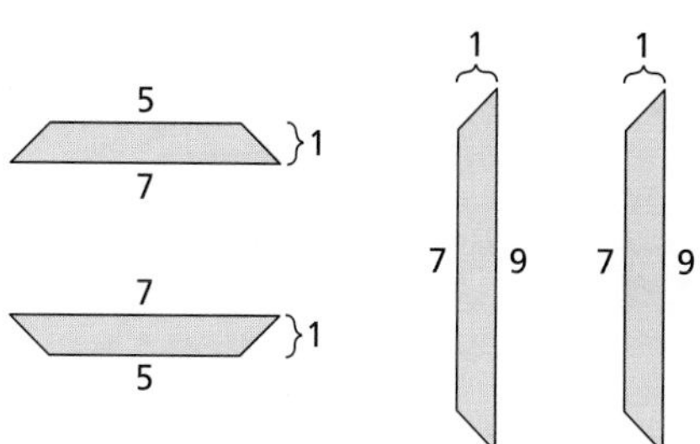

Geography **In Exercises 20–22, name the state. Then estimate its area.**

20.

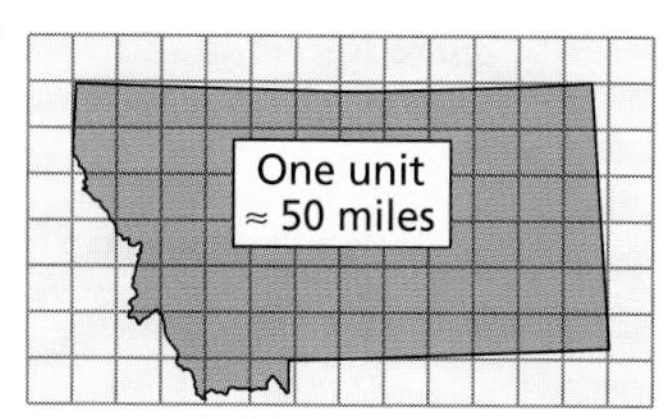

21.

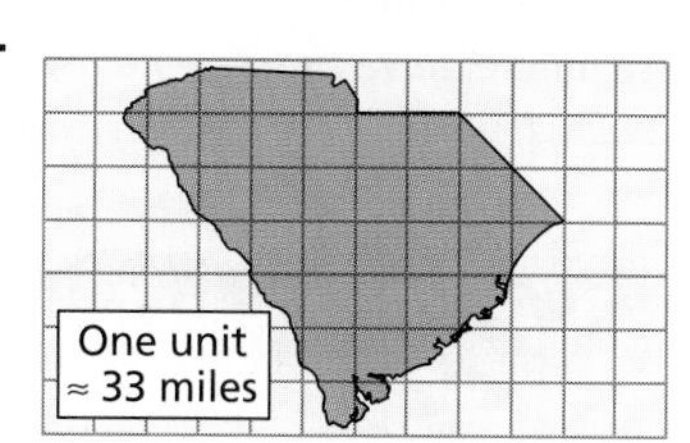

22.

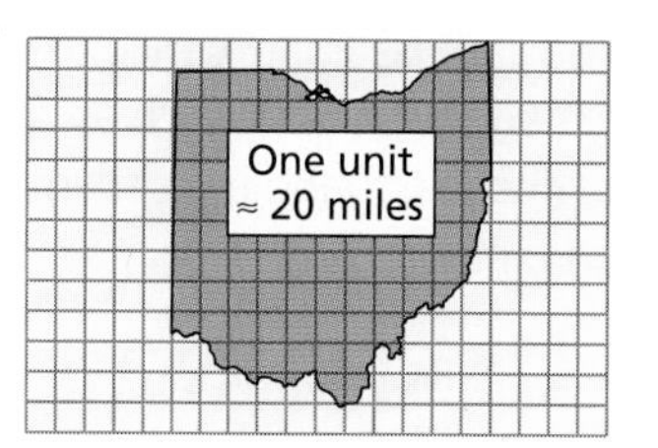

Integrated Review

Making Connections within Mathematics

Describing Patterns **In Exercises 23 and 24, find the area of each figure. Use a table and a graph to organize your results. Then describe the pattern.**

23.

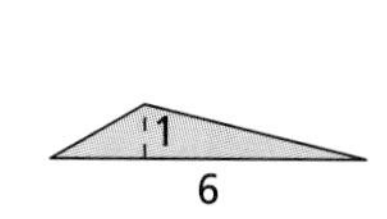

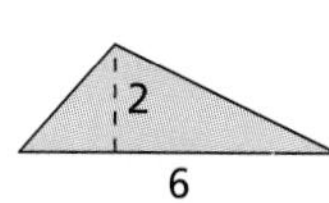

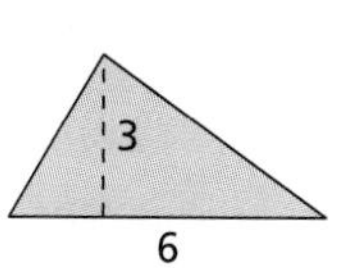

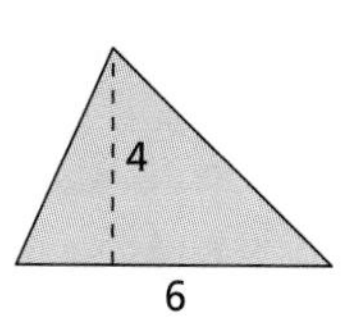

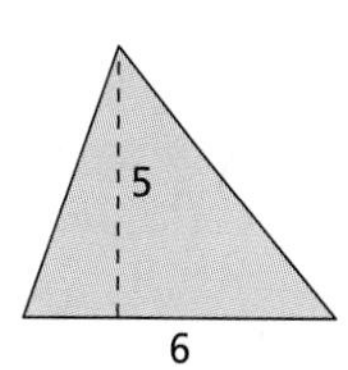

24.

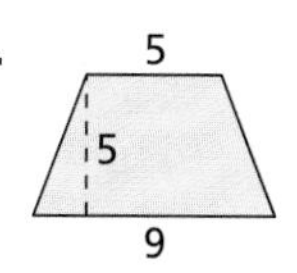

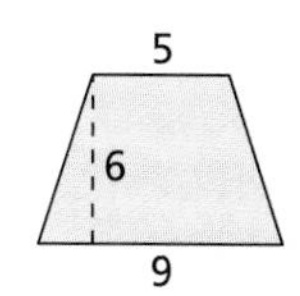

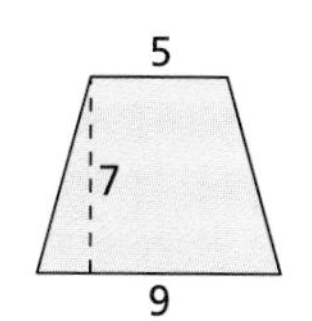

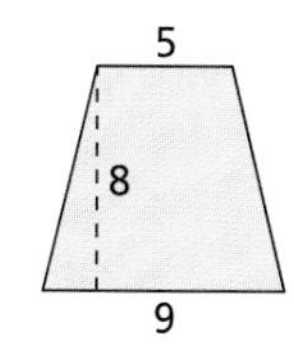

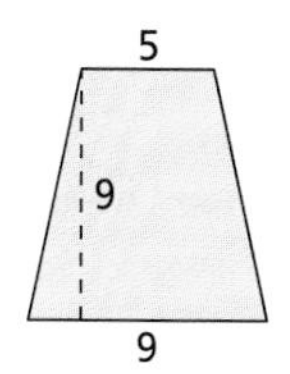

Exploration and Extension

25. *Nine Square Puzzle* Use 9 squares with sides of 1, 4, 7, 8, 9, 10, 14, 15, and 18 units to form a rectangle. Find the dimensions of the rectangle.

11.2 Exploring Congruence

What you should learn:

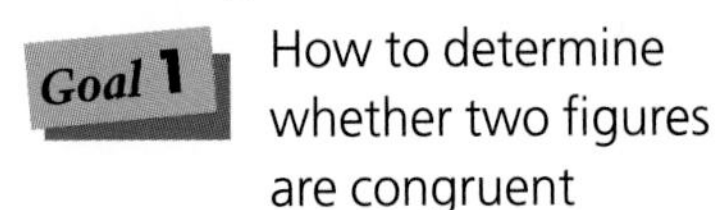

Goal 1 How to determine whether two figures are congruent

Goal 2 How to use congruence to solve real-life problems

Why you should learn it:

You can use congruence to solve real-life problems, such as comparing the triangles used in an art piece.

Goal 1 Congruence and Measure

In Chapter 10, you were introduced to the idea of *congruence.*

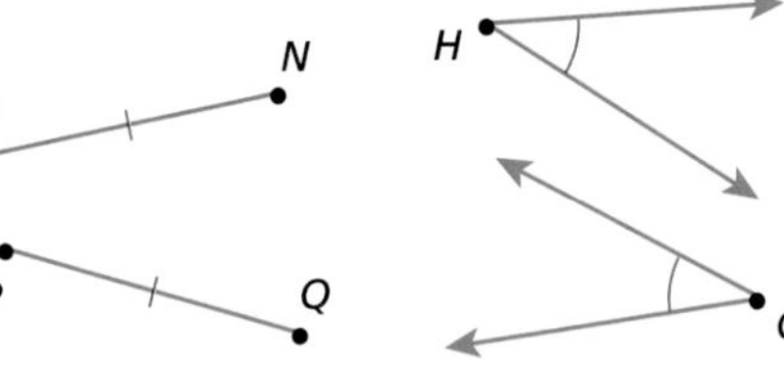

$\overline{MN} \cong \overline{PQ}$

Congruent line segments have the same length.

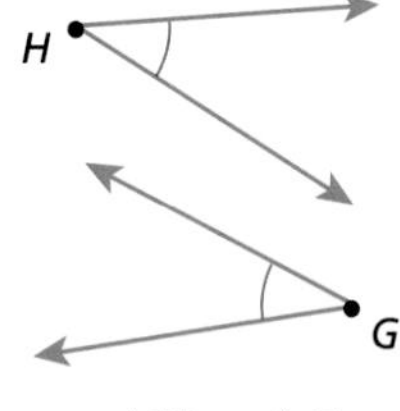

$\angle H \cong \angle G$

Congruent angles have the same measure.

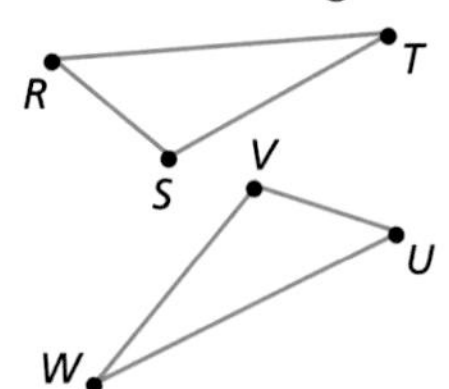

$\triangle RST \cong \triangle UVW$

Congruent triangles have the same size and shape.

LESSON INVESTIGATION

Investigating Congruence

Group Activity Plot the following points on graph paper: $A(-2, 7)$, $B(-4, 4)$, $C(3, 4)$, $D(-2, -3)$, $E(1, -5)$, $F(1, 2)$. Draw $\triangle ABC$ amd $\triangle DEF$. Do the triangles seem to be congruent? Find the length of each side. Then use a protractor to approximate the measure of each angle. What can you conclude? What can you say about the sides and angles of congruent triangles?

In the investigation, you may have discovered the following result about congruent triangles.

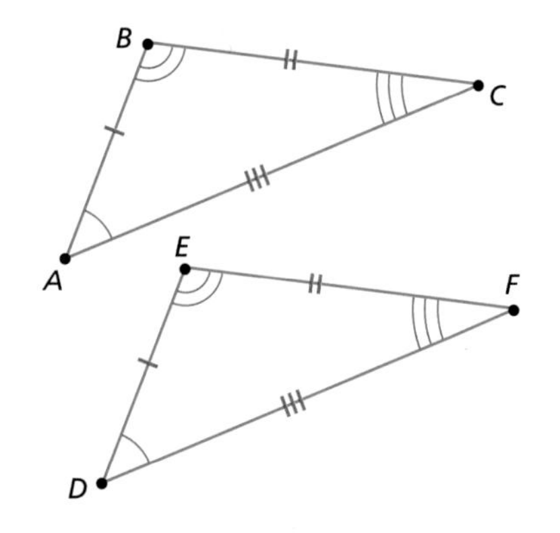

Congruent Triangles

$\triangle ABC \cong \triangle DEF$ if and only if the corresponding sides are congruent and the corresponding angles are congruent.

Corresponding Sides	Corresponding Angles
$\overline{AB} \cong \overline{DE}$	$\angle A \cong \angle D$
$\overline{BC} \cong \overline{EF}$	$\angle B \cong \angle E$
$\overline{AC} \cong \overline{DF}$	$\angle C \cong \angle F$

Goal 2 Using Congruence in Real Life

Real Life
Art

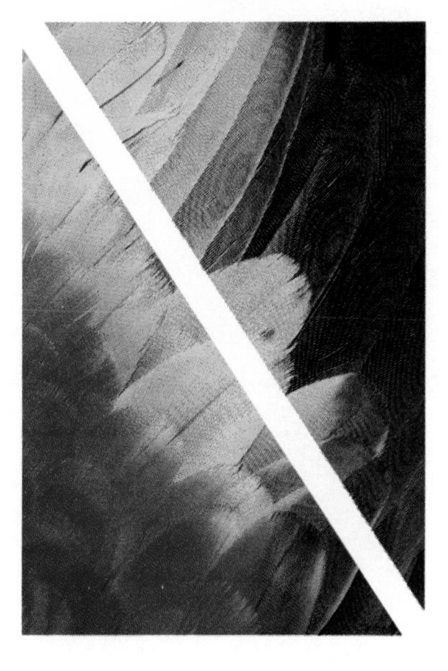

Example 1 *Showing Two Triangles Are Congruent*

You are using parts of photographs to create a collage. As part of the collage, you cut the photograph at the left along one of its diagonals. Are the resulting triangles congruent? Explain your reasoning.

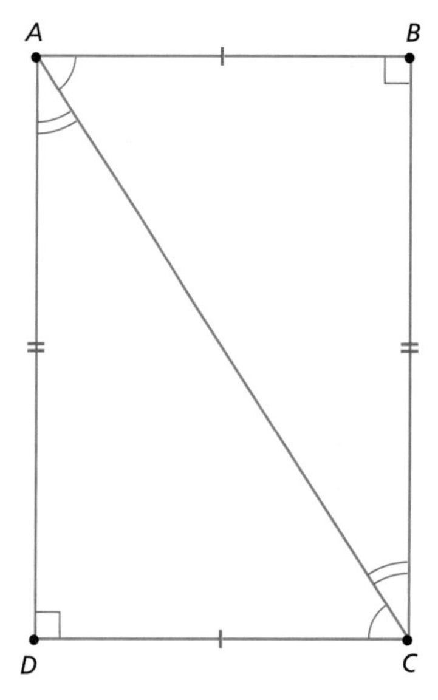

Solution The resulting triangles are congruent. To show this, draw a rectangle and one of its diagonals, as shown at the right. You can conclude that $\triangle ABC$ is congruent to $\triangle CDA$ because the corresponding sides and angles are congruent.

- $\overline{BC} \cong \overline{AD}$
- $\overline{AB} \cong \overline{CD}$
- $\overline{AC} \cong \overline{AC}$
- $\angle B \cong \angle D$
- $\angle CAB \cong \angle DCA$
- $\angle DAC \cong \angle ACB$

Explain why each of these statements is true. ■

Communicating about MATHEMATICS

Cooperative Learning

SHARING IDEAS about the Lesson

Area, Perimeter, and Congruence Discuss the following questions with a partner. Use graph paper to help illustrate your answers.

A. If two figures are congruent, must they have the same area and perimeter?

B. If two figures have exactly the same area, must they be congruent?

C. If two figures have the same perimeter, must they be congruent?

D. If two figures have the same area and perimeter, must they be congruent?

EXERCISES

Guided Practice

CHECK for Understanding

1. ***Writing*** In your own words, explain what it means for two figures to be congruent. Give examples of congruent figures in your classroom.

In Exercises 2–5, use the figure at the right. Write the congruence statement in words.

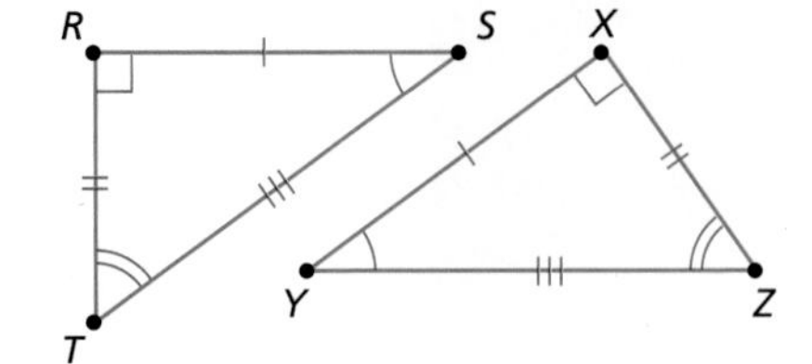

2. $\triangle RST \cong \triangle XYZ$
3. $\overline{RS} \cong \overline{XY}$
4. $\angle T \cong \angle Z$
5. What other congruence statements are indicated in the figure?

In Exercises 6–9, decide whether the figures are congruent. Explain.

6. 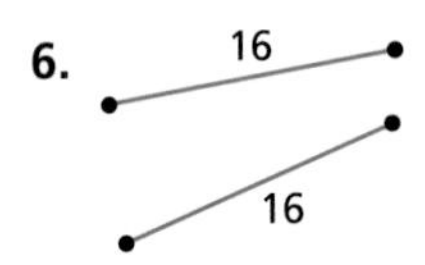

7. 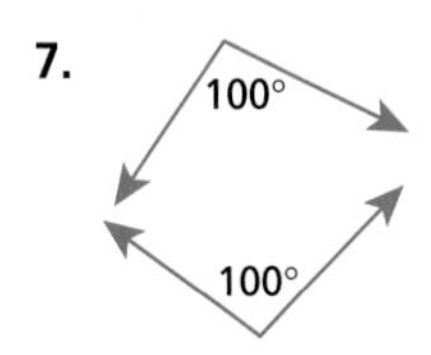

8. 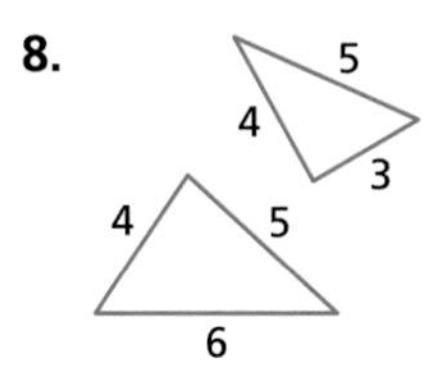

9. 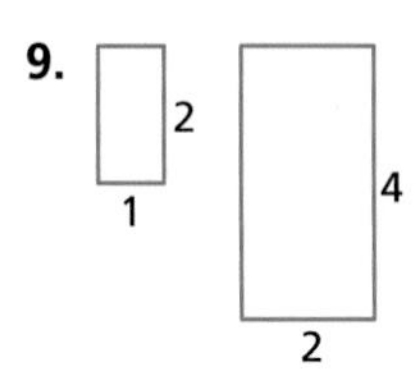

Independent Practice

In Exercises 10–15, use the fact that $\triangle HIJ \cong \triangle KLM$ to complete the statement.

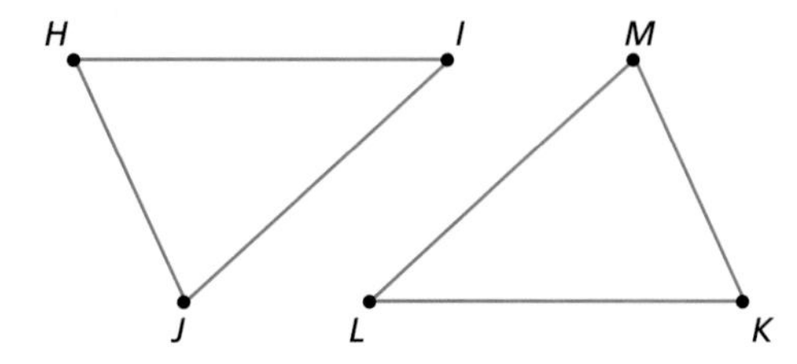

10. $\angle H \cong$ [?]
11. $\overline{JH} \cong$ [?]
12. $\overline{ML} \cong$ [?]
13. $\angle J \cong$ [?]
14. $\angle L \cong$ [?]
15. $\overline{LK} \cong$ [?]

16. ***Drawing*** Draw an obtuse triangle on cardboard. Cut the triangle out and trace two copies. Label the copies as $\triangle FGH$ and $\triangle MNO$. Write all possible congruence statements about the two triangles.
17. ***Drawing*** Draw an acute triangle on cardboard. Cut the triangle out and trace two copies. Label the copies as $\triangle FGH$ and $\triangle MNO$. Write all possible congruence statements about the two triangles.
18. ***Congruent Polygons*** List each pair of congruent polygons.

a. 

b. 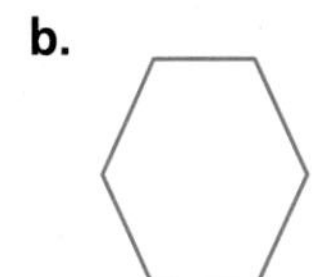

c. 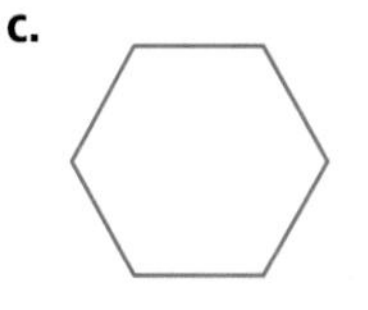

d. 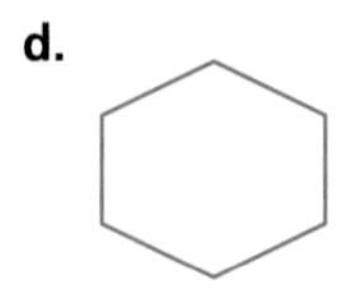

e. 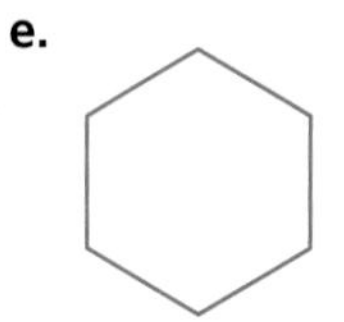

In Exercises 19–22, copy the figure on dot paper. Then divide it into 2 congruent parts. Give more than one answer, if possible.

19. 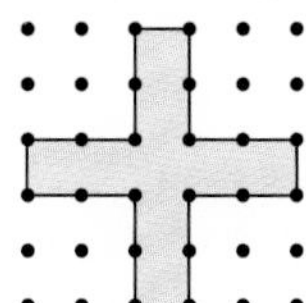

20. 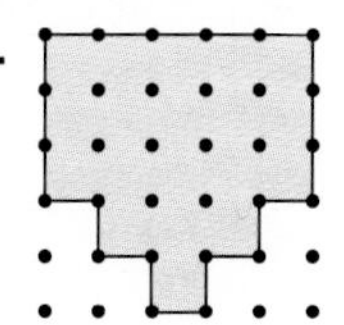

21. 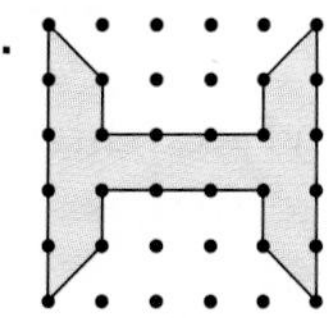

22. 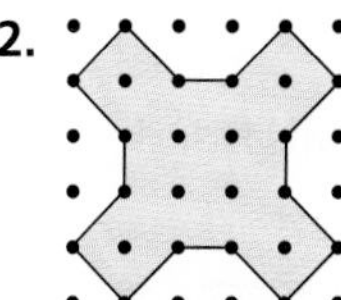

23. *Photography* Which of the photographs below are congruent? Explain how the photos could have been created.

a. 

b. 

c. 

24. *Think about It* Is there a type of triangle for which the following statements are both true? If so, sketch and describe such a triangle.

$\triangle ABC \cong \triangle ACB \qquad \triangle ABC \cong \triangle BAC$

Integrated Review

Making Connections within Mathematics

Coordinate Geometry **In Exercises 25–27, find the coordinates of the missing point so that the figures are congruent.**

25. 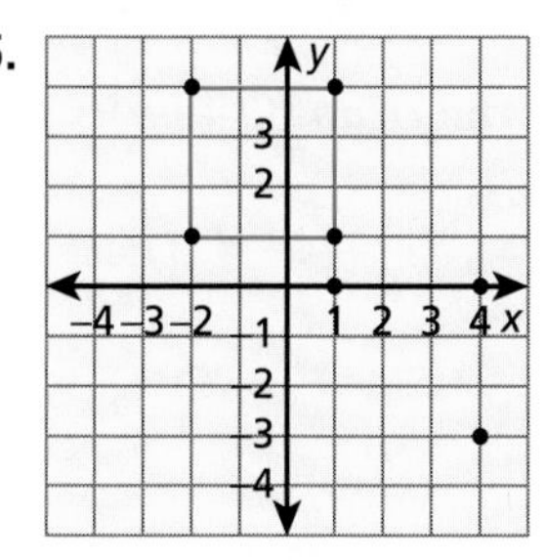

26. 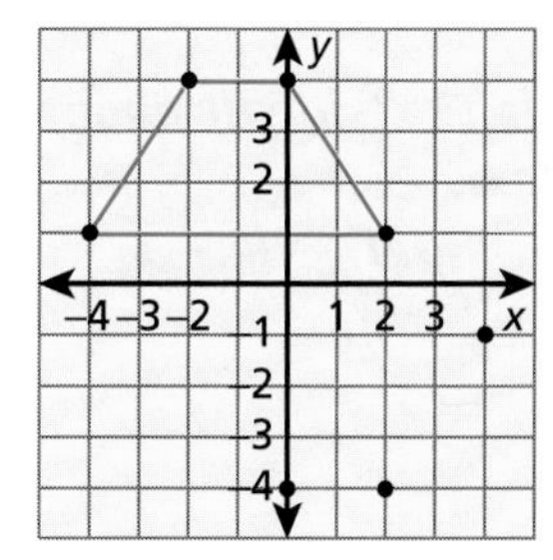

27. 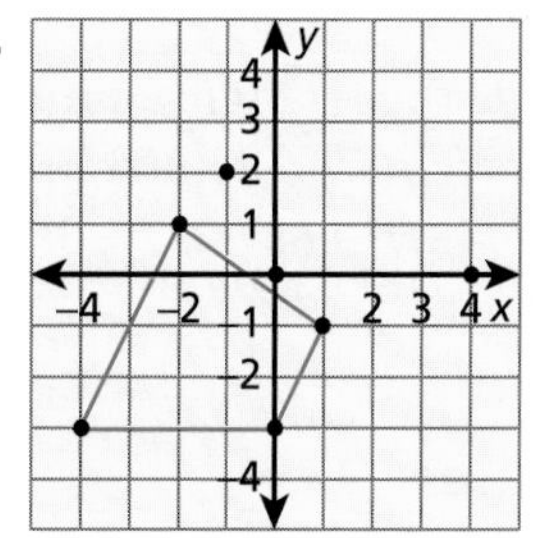

Exploration and Extension

Counting Squares **In Exercises 28–32, use the square grid at the right.**

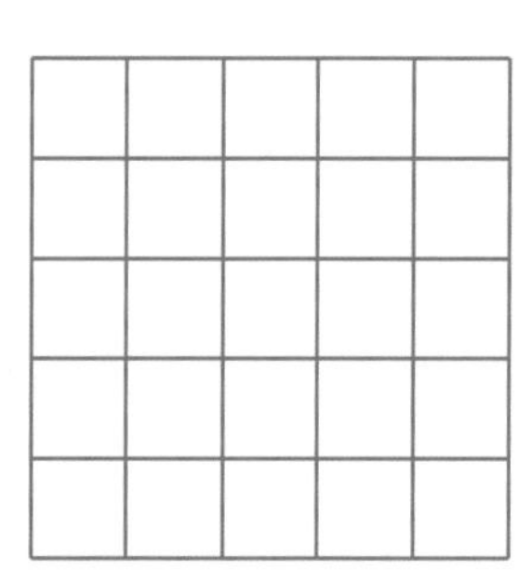

28. How many 1-by-1 squares are in the grid?

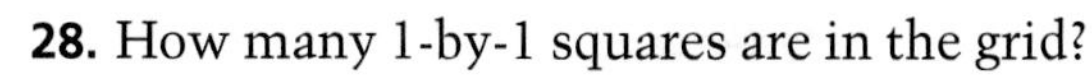

29. How many 2-by-2 squares are in the grid?

30. How many 3-by-3 squares are in the grid?

31. How many 4-by-4 squares are in the grid?

32. How many squares are in the grid?

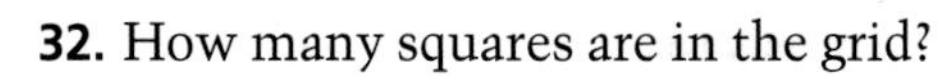

LESSON INVESTIGATION 11.3

Exploring Motion in a Plane

Materials Needed: graph paper and straightedge

Example — *Moving Objects in a Coordinate Plane*

Consider the triangle whose vertices are $A(0, 0)$, $B(3, 4)$ and $C(4, 1)$. Move each vertex according to the given motion rule. How is the new triangle related to the original triangle?

a. *Motion Rule:* Add 2 to x, add 3 to y.

b. *Motion Rule:* Keep x the same, replace y by $-y$.

c. *Motion Rule:* Subtract 4 from x, keep y the same.

Solution The results are shown below.

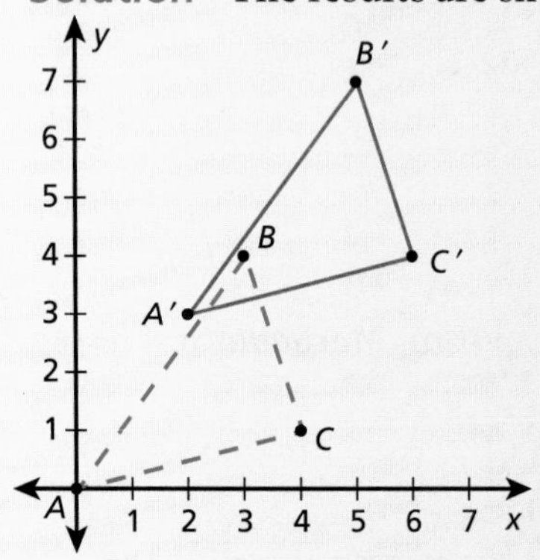

a. ***Slide the triangle 2 units to the right and 3 units up.***
$(4, 1) \to (4+2, 1+3)$

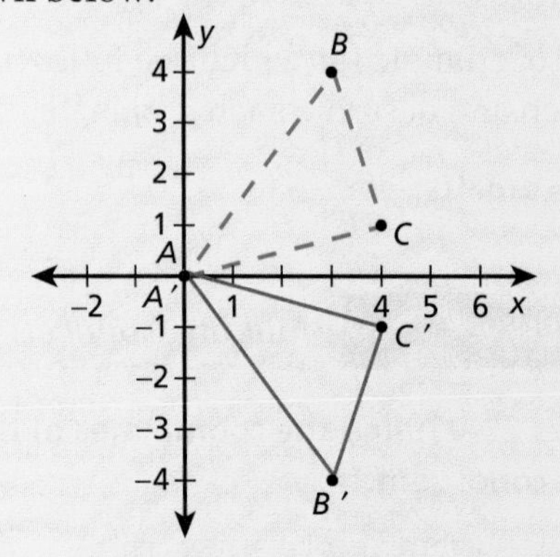

b. ***Flip the triangle about the x-axis.***
$(4, 1) \to (4, -1)$

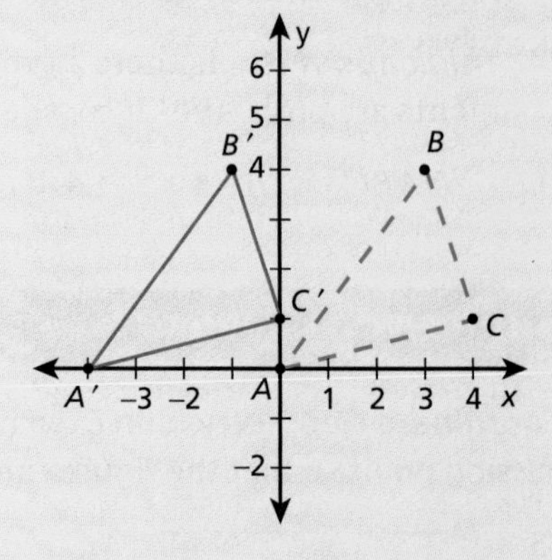

c. ***Slide the triangle 4 units to the left.***
$(4, 1) \to (4-4, 1)$

■

Exercises

In Exercises 1–8, sketch $\triangle ABC$ with $A(-4, -2)$, $B(-3, -4)$, and $C(-1, -1)$. Apply the motion rule and sketch the new triangle. Use tracing paper to decide whether the two triangles are congruent.

1. Add 5 to x, subtract 4 from y.
2. Keep x the same, replace y by $-y$.
3. Replace x by $-x$, add 2 to y.
4. Add 3 to x, add 4 to y.
5. Subtract 3 from x, add 2 to y.
6. Subtract 5 from x, replace y by $-y$.
7. Replace x by $-x$, keep y the same.
8. Double x, keep y the same.
9. In Exercises 1–8, is the new triangle congruent to the original triangle? If so, explain how the original can be moved to form the new triangle.

11.3 Line Reflections

What you should learn:

Goal 1 How to reflect a figure about a line

Goal 2 How to use properties of reflections to answer questions about real-life situations

Why you should learn it:

You can use properties of reflections to answer questions about real life, such as analyzing the symmetry of fabric pieces used to make clothing.

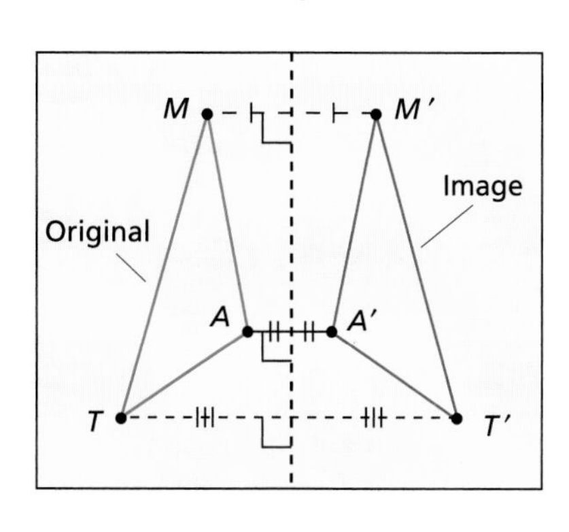

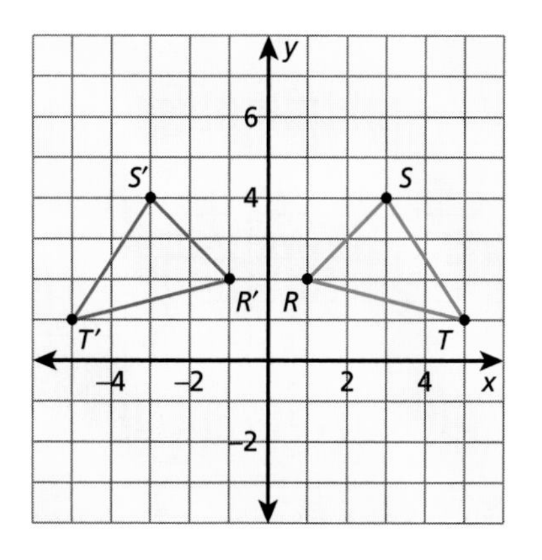

Goal 1 Reflecting Figures about Lines

LESSON INVESTIGATION

Investigating Line Reflections

Group Activity Draw $\triangle MAT$ on the left side of a piece of paper. Fold the paper in half. Hold the paper up to the light and mark the images of M, A, and T on the right side of the paper. Unfold the paper, label the images as M', A', and T'. Draw $\triangle M'A'T'$. What can you conclude? Draw the line segments $\overline{MM'}$, $\overline{AA'}$, and $\overline{TT'}$. How are these segments related to the fold in the paper?

In the investigation, you may have discovered the following properties of **reflections.**

Properties of Line Reflections

1. When a figure is reflected about a line, the image is congruent to the original figure.
2. In a reflection, the **reflection line** is perpendicular to and bisects each segment that joins an original point to its image.

Example 1 *Reflecting in a Coordinate Plane*

Consider the points $R(1, 2)$, $S(3, 4)$, and $T(5, 1)$ in a coordinate plane. Reflect each point about the y-axis. Then compare $\triangle RST$ with $\triangle R'S'T'$.

Solution From the figure at the left, you can see that the two triangles are congruent. Their orientation, however, is not the same. In $\triangle RST$, the vertices R, S, and T are in clockwise order; but in $\triangle R'S'T'$, the vertices are in counterclockwise order. This is comparable to the different orientations of your left and right hands. ■

Goal 2 Reflections and Line Symmetry

In Lesson 10.4, you studied line symmetry. The next example illustrates a connection between line symmetry and reflections.

Example 2 *Reflections and Line Symmetry*

Real Life
Clothing

The pattern pieces in Example 2 can be used to make the shirt above.

You work for a company that creates patterns for clothing. The pattern shown at the left is marked on a folded piece of cloth. When the cloth is cut and unfolded, you obtain the pieces shown at the right. Which pairs are reflections? Which pieces have line symmetry?

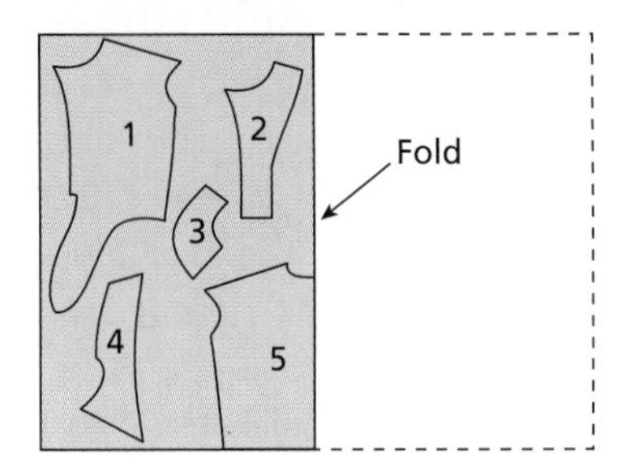

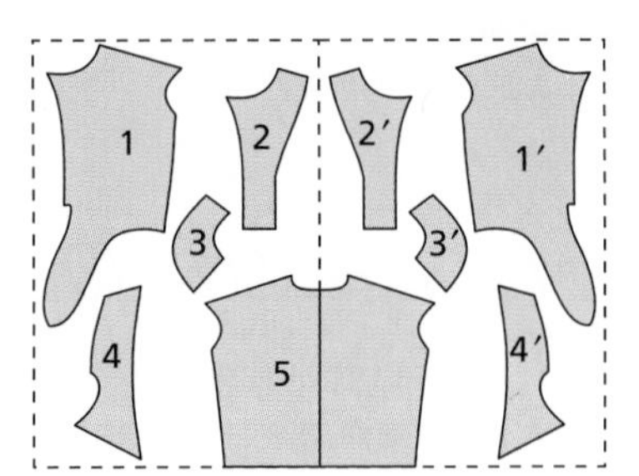

Solution Piece 1′ is a reflection of Piece 1. Piece 2′ is a reflection of Piece 2. Piece 3′ is a reflection of Piece 3. Piece 4′ is a reflection of Piece 4. Piece 5 has line symmetry, because it had one edge lying along the fold. ■

Communicating about MATHEMATICS

SHARING IDEAS about the Lesson

Area and Perimeter A trapezoid is reflected about the indicated lines. For each line, sketch the figure that consists of the original trapezoid and its image. Compare the area and perimeter of this new figure with the area and perimeter of the original trapezoid.

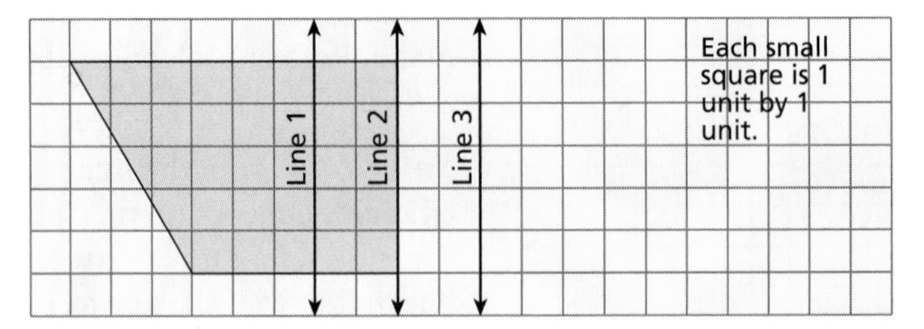

EXERCISES

Guided Practice

CHECK for Understanding

In Exercises 1–4, draw the figure at the right in a coordinate plane. Then draw the indicated reflection.

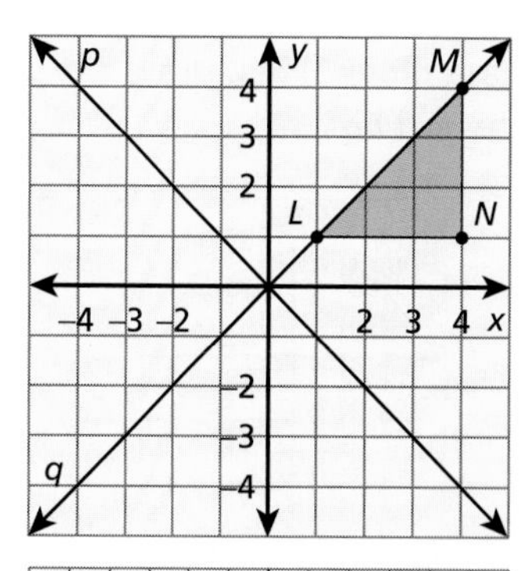

1. Reflect the triangle about the x-axis.
2. Reflect the triangle about the y-axis.
3. Reflect the triangle about Line p.
4. Reflect the triangle about Line q.

Independent Practice

In Exercises 5–8, name the image of $\triangle ABC$ after the indicated reflection(s). (The image is a triangle.)

5. Reflect $\triangle ABC$ about the x-axis.
6. Reflect $\triangle ABC$ about the y-axis.
7. Reflect $\triangle ABC$ about the x-axis, then about the y-axis.
8. Reflect $\triangle ABC$ about the y-axis, then reflect it again about the y-axis.
9. *What Do You See?* When you look at yourself in a mirror, your right side appears to be on your left side. Is this also true when you look at a photo of yourself? Explain.
10. *What Do You See?* You are walking by a truck delivery door at the back of a grocery store. A sign by the door reads as follows. Why would the sign be printed this way?

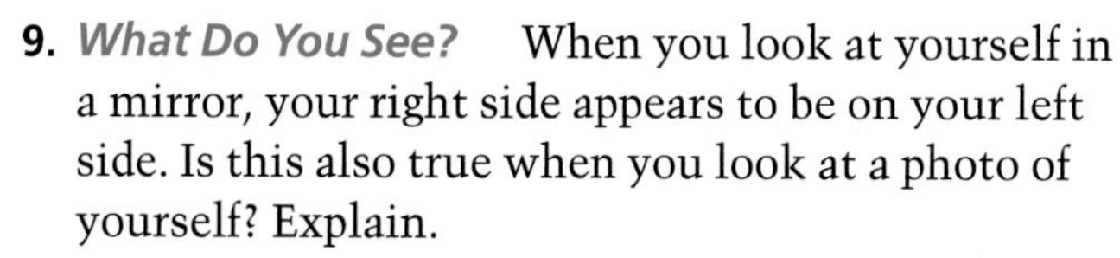

In Exercises 11–14, decide whether the red figure is a reflection of the blue figure in Line ℓ. If not, sketch the reflection of the blue figure.

11.

12.

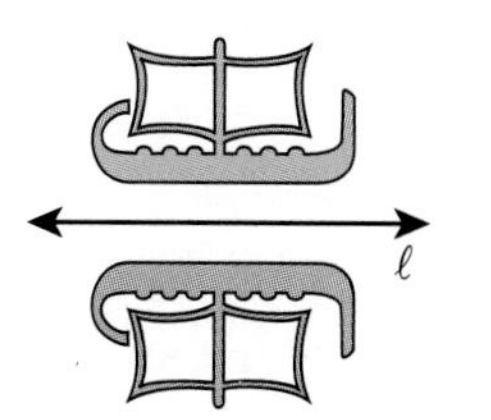

13.

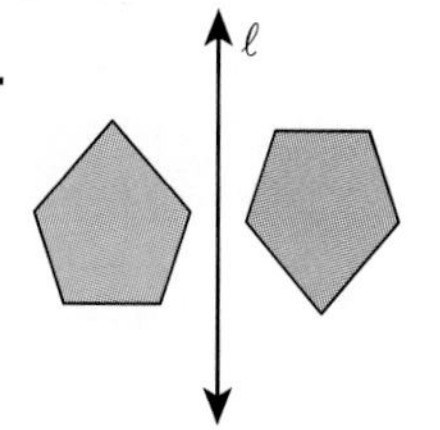

14.

Coordinate Geometry **In Exercises 15 and 16, sketch $\triangle ABC$ in a coordinate plane. Then sketch the images when the triangle is reflected about the *x*-axis and the *y*-axis.**

15. $A(1, 3), B(4, 1), C(2, -1)$

16. $A(0, 1), B(-2, -2), C(2, -2)$

Coordinate Geometry **In Exercises 17–20, draw the figures in a coordinate plane. Then draw the line that can be used to reflect the blue figure to the red figure.**

17.

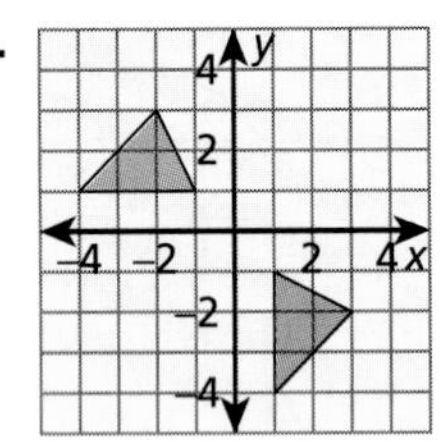

18.

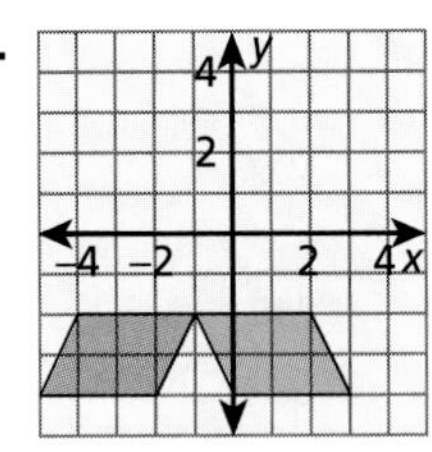

19.

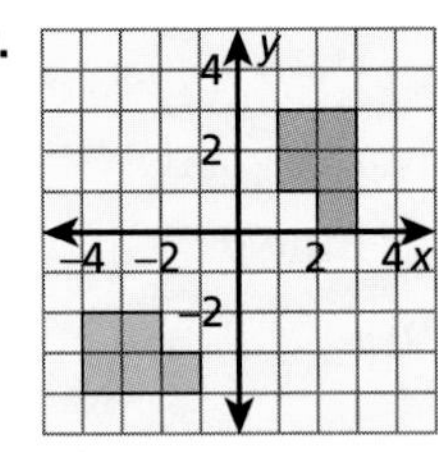

20.

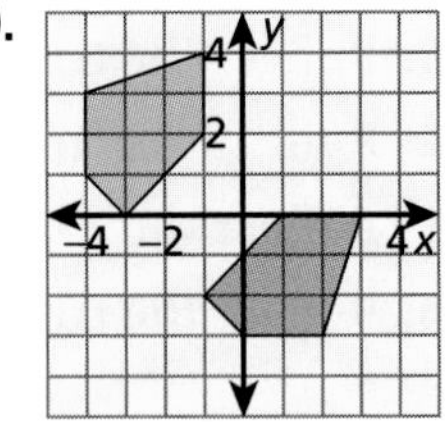

21. *Repeated Reflections* In a coordinate plane, reflect a rectangle four times: first about the *x*-axis, then the *y*-axis, then the *x*-axis, and then the *y*-axis. Compare the final image with the original rectangle.

22. *Alphabet Reflection* Which capital letters of the alphabet remain the same when reflected about a vertical line? Which letters remain the same when reflected about a horizontal line?

Playing Pool **In Exercises 23 and 24, use the diagram at the right.**

23. The ball is shot from point (0, 3). Describe the line of reflection of its path.

24. Suppose you shoot the ball in the same direction from point (0, 2). Sketch the path of the ball. Does the line of reflection change?

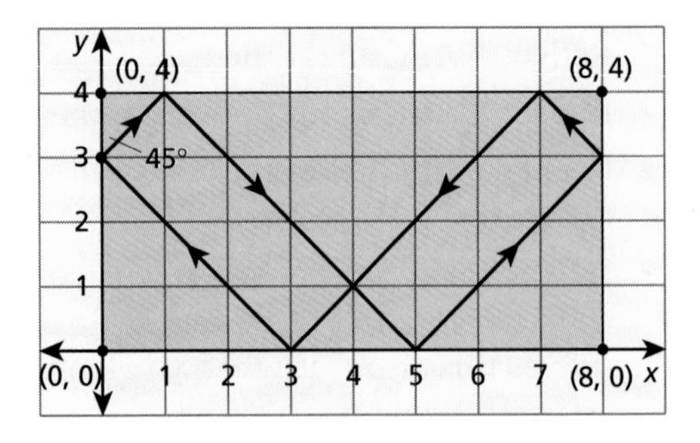

Integrated Review

Making Connections within Mathematics

Symmetrical Faces **In Exercises 25–28, state whether the face has line symmetry. Which faces are reflections of each other?**

25.

26.

27.

28.

Exploration and Extension

29. *Symmetrical Faces* Find photographs in magazines or newspapers of faces that appear to have line symmetry and of faces that appear not to have line symmetry. Which type of photo is easier to find?

Mixed REVIEW

In Exercises 1–3, sketch two *noncongruent* trapezoids that have the given measures. Find the area of each. (11.1)

1. $b_1 = 6, b_2 = 12, h = 2$ **2.** $b_1 = 7, b_2 = 8, h = 4$ **3.** $b_1 = 1, b_2 = 9, h = 1$

In Exercises 4–7, rewrite the decimal as a simplified fraction. (6.6)

4. $0.\overline{1}$ **5.** $0.\overline{2}$ **6.** $0.\overline{3}$ **7.** $0.\overline{4}$

8. Use the result of Exercises 4–7 to describe a quick way to rewrite $\frac{n}{9}$ as a decimal.

In Exercises 9–12, rewrite as a decimal. Round to two places. (6.7)

9. $\frac{8}{15}$ **10.** $3\frac{2}{7}$ **11.** $-\frac{8}{7}$ **12.** $-4\frac{5}{6}$

In Exercises 13–16, find the percent increase or decrease. (8.6)

13. Before: \$42.12 **After:** \$40.18

14. Before: 3.6 lb **After:** 7.6 lb

15. Before: 77.1° **After:** 66.1°

16. Before: 4.4 ft **After:** 6.3 ft

Career Interview

Research Scientist

Carol Stoker is a research scientist at NASA Ames Research Center in Moffett Field, California. She has spent the past two years working in Antarctica because its environment most closely mirrors the conditions astronauts would face on the Moon and Mars.

Q: ***What led you into this career?***

A: My primary interest has always been anything to do with the human exploration of Mars.

Q: ***What math do you use in your work?***

A: I use lots of algebra, geometry, and calculus. Although I am always using computers, I must do the thinking. I need to understand the logic behind why things work.

Q: ***What would you like to tell kids in school about math?***

A: Math is important for coping with everyday life decisions, not just for high-tech careers. Living in Antarctica where Radio Shack and K-Mart are not around the corner, I must estimate and order what supplies I will need for my entire trip. If a piece of equipment breaks, I can't just call out for repairs. I must use my problem-solving skills to figure out the solution on my own.

11.4 Rotations

What you should learn:

How to rotate a figure about a point

How to use properties of rotations to answer questions about real-life situations

Why you should learn it:

You can use properties of rotations to answer questions about real life, such as finding the angle of rotation of the minute hand of a clock.

Goal 1 Rotating Figures about Points

In Lesson 11.3, you studied one type of **transformation**—a reflection about a line. In this lesson, you will study another type of transformation—a **rotation** about a point. Three examples are shown below.

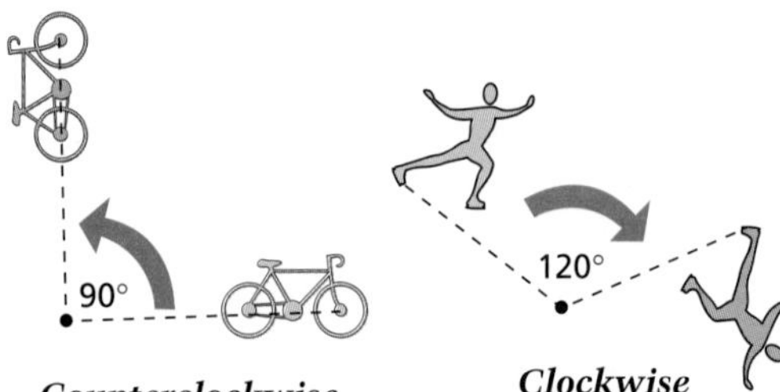

Clockwise rotation of 60° ***Counterclockwise rotation of 90°*** ***Clockwise rotation of 120°***

Example 1 *Finding an Angle of Rotation*

The blue figure is rotated about the origin to become the red figure. Find the angle of rotation. Find the image's vertices.

a. 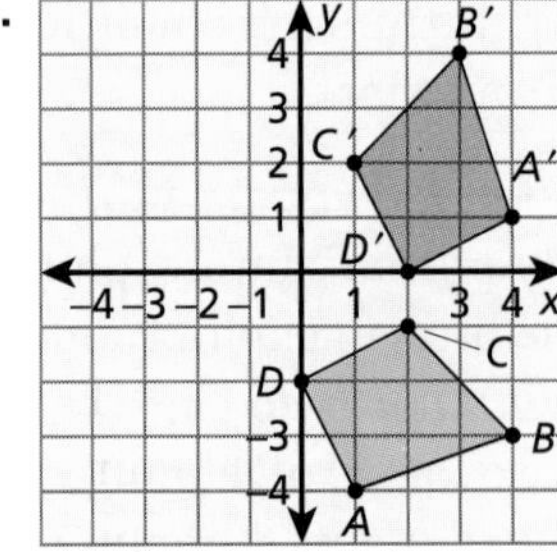

b. 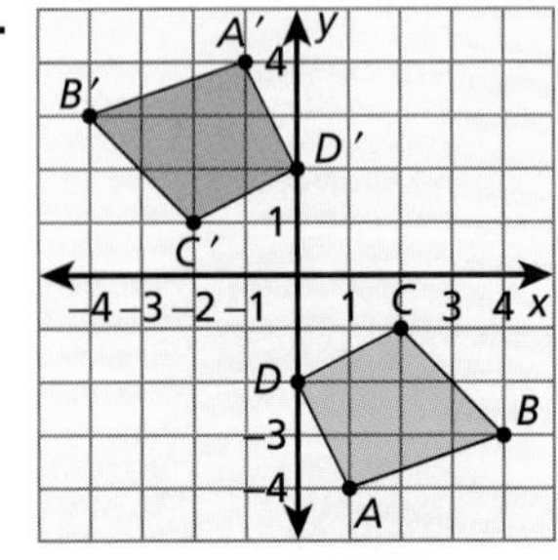

Study Tip...

In Example 1, notice the patterns when an object is rotated 90° or 180° counterclockwise about the origin. For 90°, the point (x, y) is rotated to the point (−y, x). What is the pattern for 180°?

Solution

a. The blue figure is rotated counterclockwise 90°. The vertices of the original figure and the image are as follows.

$A(1, -4)$, $B(4, -3)$, $C(2, -1)$, $D(0, -2)$
$A'(4, 1)$, $B'(3, 4)$, $C'(1, 2)$, $D'(2, 0)$

b. The blue figure is rotated clockwise 180°. The vertices of the original figure and the image are as follows.

$A(1, -4)$, $B(4, -3)$, $C(2, -1)$, $D(0, -2)$
$A'(-1, 4)$, $B'(-4, 3)$, $C'(-2, 1)$, $D'(0, 2)$ ■

Goal 2 Properties of Rotations

Rotational symmetry is a special case of a rotation in which the figure fits back on itself.

LESSON INVESTIGATION

Investigating Rotations about a Point

Group Activity Copy each figure onto dot paper. Then copy the figure onto a piece of tracing paper. Rotate the figure the indicated number of degrees about Point *P*. Is the image congruent to the original figure?

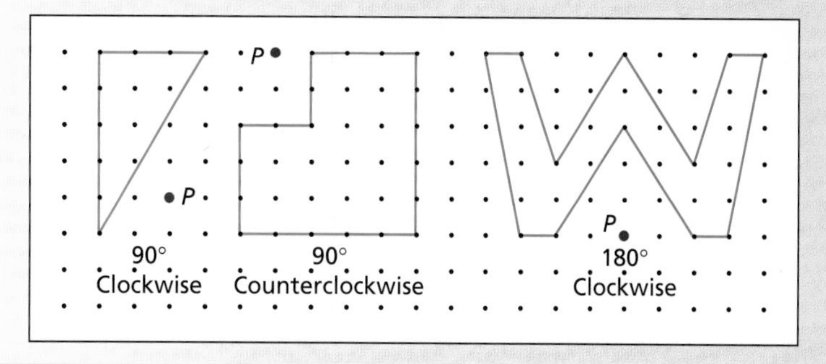

In this investigation, you may have discovered that rotating a figure produces a congruent figure. Lengths, perimeters, and areas are not changed by rotations. Orientations of figures and locations of points *are* changed. For instance, when the number 6 is rotated 180°, it becomes the number 9.

Communicating about MATHEMATICS

SHARING IDEAS about the Lesson

Rotating Clock Hands Each pair of clocks shows a beginning time and an ending time. Find the angle of rotation of the minute hand. Explain how you obtained the angle.

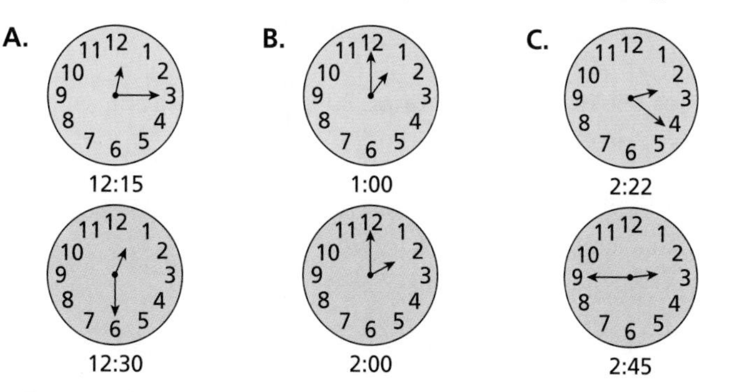

EXERCISES

Guided Practice

CHECK for Understanding

In Exercises 1–4, the blue figure has been rotated to produce the red figure. Estimate the angle and direction of the rotation.

1.  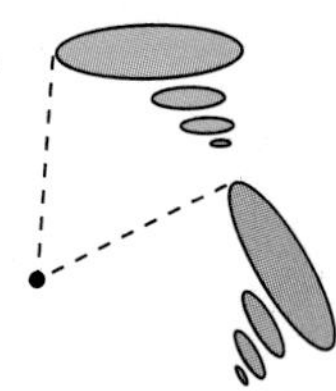

2.

3. 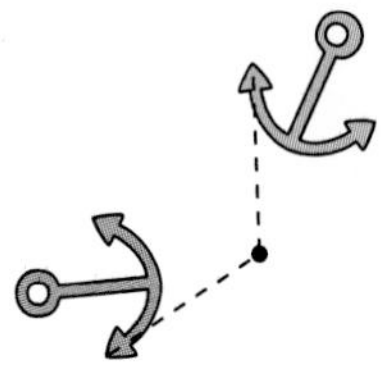

4. 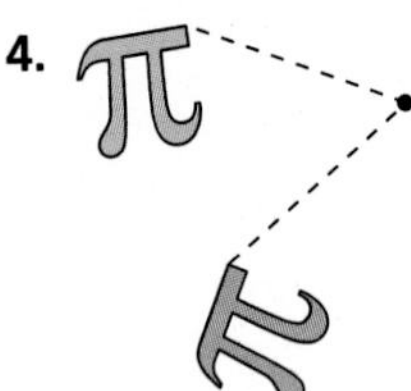

5. Describe the relationship between rotations and rotational symmetry.

6. *The Shape of Things* Describe two objects in your classroom that are rotations of each other.

Independent Practice

In Exercises 7–10, the blue figure is rotated clockwise about the origin to produce the red figure. Find the angle of rotation.

7. 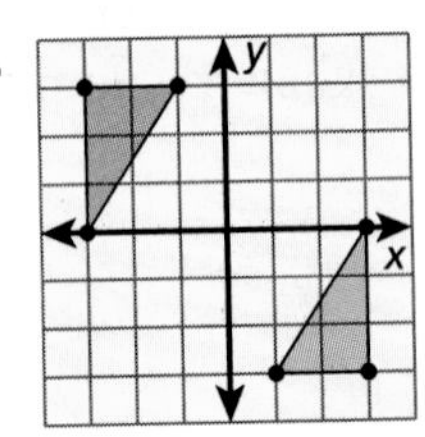

8. 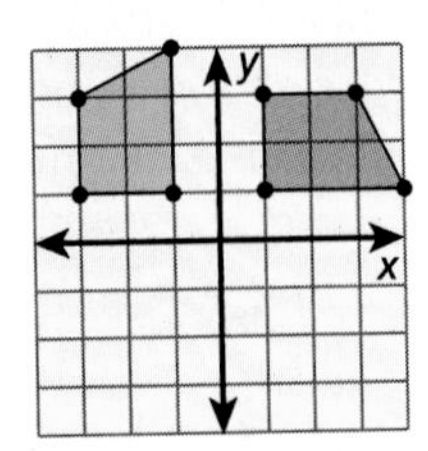

9. 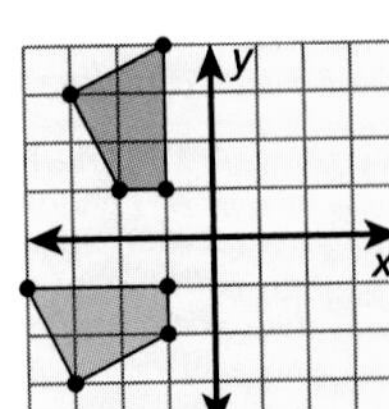

10. 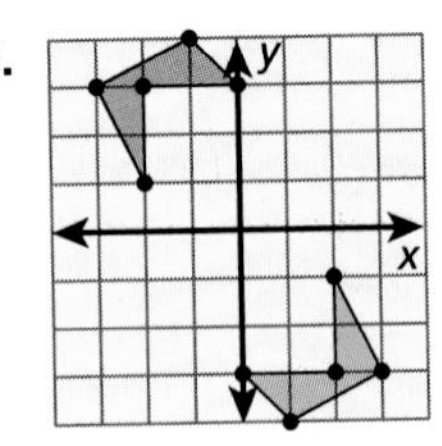

In Exercises 11–14, estimate the angle and direction of rotation.

11. 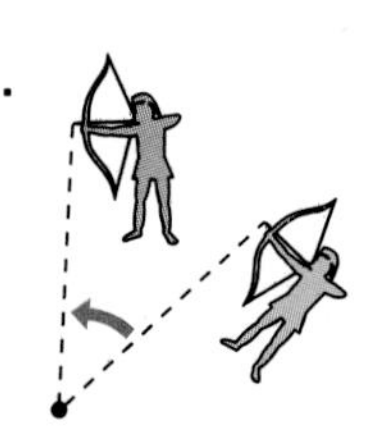

12. 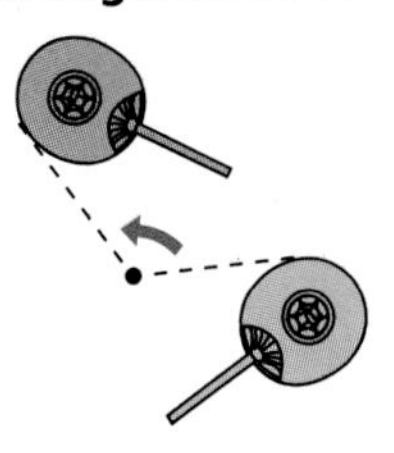

13. 

14. 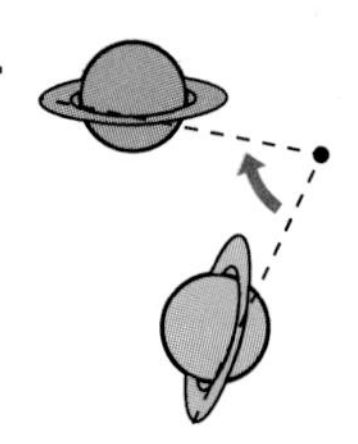

In Exercises 15–18, $\triangle ABC$ is rotated 90° clockwise about the origin to produce $\triangle A'B'C'$.

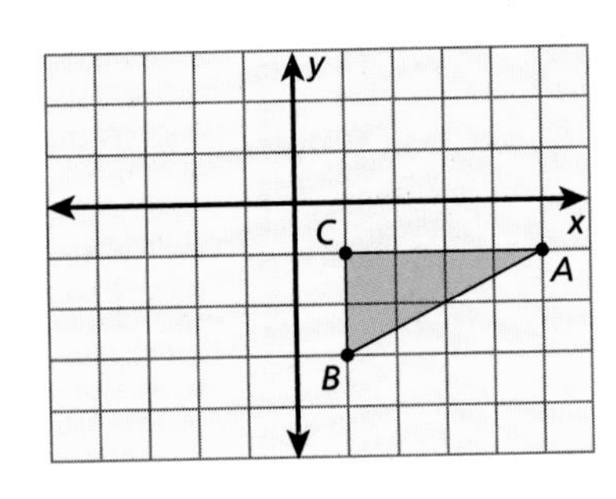

15. Find the measure of each.
    a. $\overline{A'C'}$ b. $\overline{B'C'}$ c. $\overline{A'B'}$

16. What type of angle is $\angle C'$?

17. Find the perimeter of $\triangle A'B'C'$. Explain.

18. Find the area of $\triangle A'B'C'$. Explain.

In Exercises 19–21, draw the figure on tracing paper. Rotate the tracing paper 90° clockwise about point *P*. Then trace the figure again.

19. 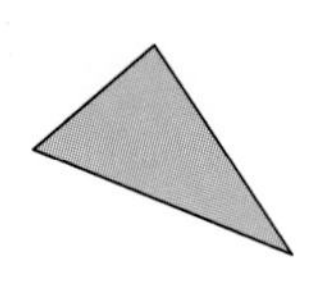

• *P*

20. 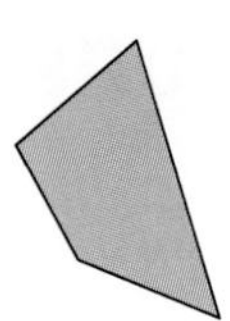

P •

21. 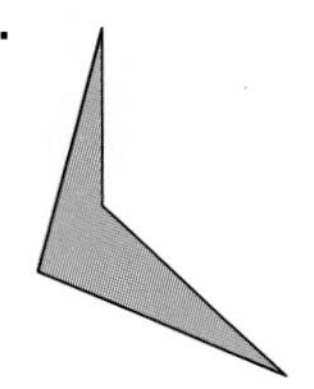

P •

In Exercises 22–26, find the image of the point, segment, or triangle.

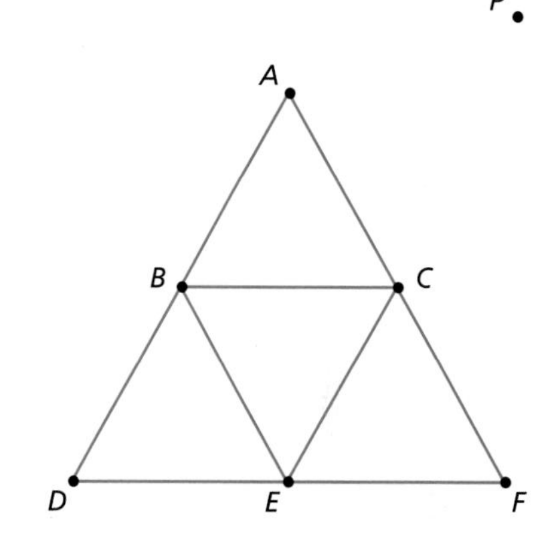

22. 60° clockwise rotation of D about E

23. 60° clockwise rotation of $\overline{BC}$ about E

24. 60° counterclockwise rotation of $\triangle ABC$ about C

25. 180° rotation of $\overline{AC}$ about C

26. 120° clockwise rotation of $\overline{CF}$ about C

Integrated Review

Making Connections within Mathematics

Rotational Symmetry **In Exercises 27–30, find the smallest angle that the figure can be rotated to coincide with itself.**

27. 

28. 

29. 

30. 

Exploration and Extension

Reflections and Rotations **In Exercises 31–34, use the figure at the right.**

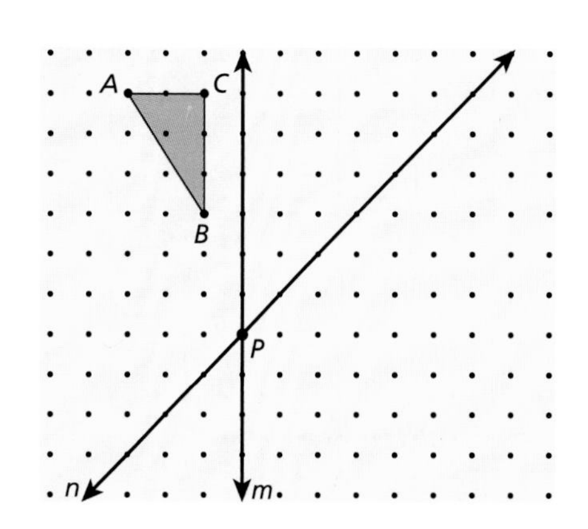

31. Copy the figure onto dot paper. Include the labels.

32. Reflect $\triangle ABC$ in Line m to produce $\triangle DEF$.

33. Reflect $\triangle DEF$ in Line n to produce $\triangle GHI$.

34. What single transformation relates $\triangle ABC$ to $\triangle GHI$?

11.5 Translations

What you should learn:

How to translate a figure in a plane

How to represent translations in a coordinate plane

Why you should learn it:

You can use translations to perform real-life actions, such as creating a computer graphic.

Goal 1 Translating Figures in a Plane

The third type of transformation you will study is a **translation** (or **slide**). When a figure is translated, each point of the figure is moved the same distance. The image of a translation is congruent to the original *and* has the same orientation as the original. The only difference between the original and the image is their locations in the plane.

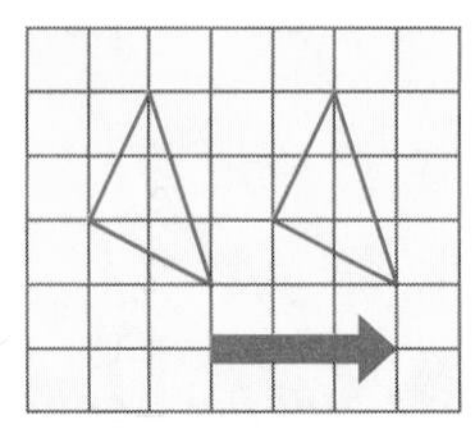

Slide 3 units right

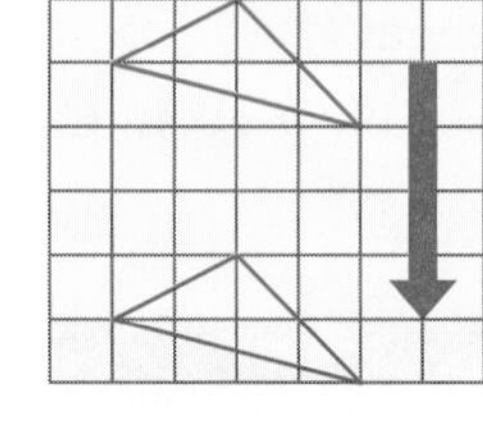

Slide 4 units down

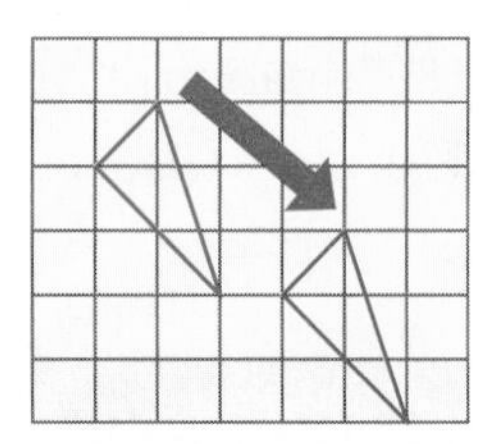

Slide 3 units right and 2 units down

Example 1 *Drawing Translations*

Draw each figure on dot paper. Then translate the figure the direction and distance indicated by the arrow. Write a verbal description of the translation.

a.

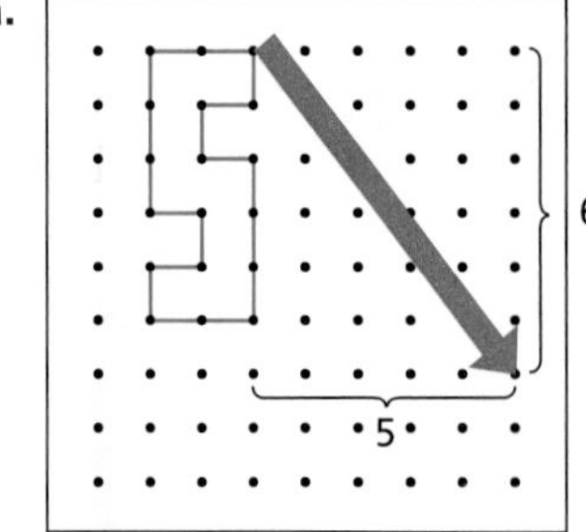

b.

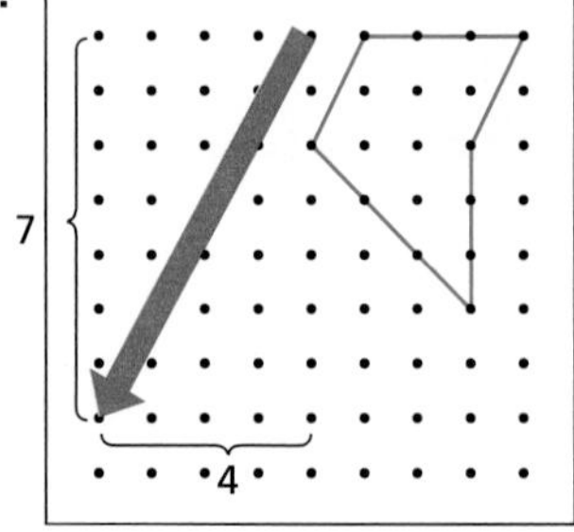

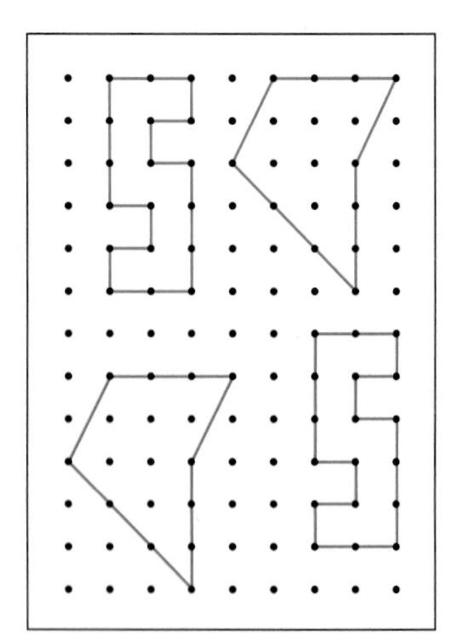

Solution The translated figures are shown at the left.

a. This figure has been translated 5 units to the right and 6 units down.

b. This figure has been translated 4 units to the left and 7 units down. ■

Goal 2 Translating in a Coordinate Plane

Transformations (reflections, rotations, and translations) are used in computer graphics to produce animated motion.

Example 2 *Translating in a Coordinate Plane*

Draw the parallelogram whose vertices are $A(-4, 3)$, $B(-1, 4)$, $C(3, 3)$, $D(0, 2)$. Then use the following motion rule to translate each vertex.

Original Figure		**Image**
(x, y)	$\Rightarrow$	$(x + 2, y - 5)$

Finally, write a verbal description of the transformation.

Solution The motion rule tells you to add 2 to each x-coordinate and subtract 5 from each y-coordinate.

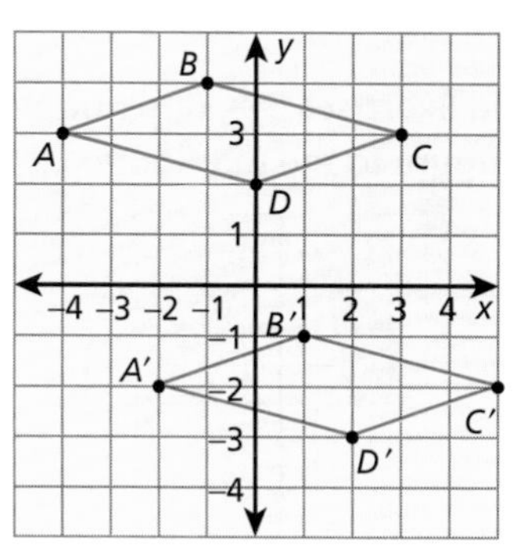

Original		**Image**
$A(-4, 3)$	$\Rightarrow$	$A'(-2, -2)$
$B(-1, 4)$	$\Rightarrow$	$B'(1, -1)$
$C(3, 3)$	$\Rightarrow$	$C'(5, -2)$
$D(0, 2)$	$\Rightarrow$	$D'(2, -3)$

You can describe this translation verbally as *each point is slid 2 units to the right and 5 units down.* ■

Communicating about MATHEMATICS

SHARING IDEAS about the Lesson

Computer Graphics You are working as a computer artist for a company that produces animations. What type of transformation is used to move the penguin from each screen to the next?

EXERCISES

Guided Practice

CHECK for Understanding

In Exercises 1–4, describe the translation verbally.

1. 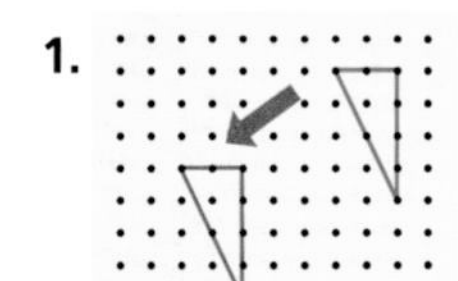

2. 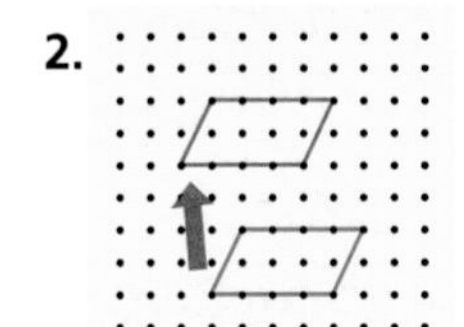

3. 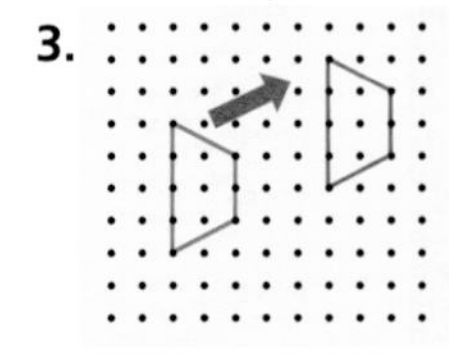

4. 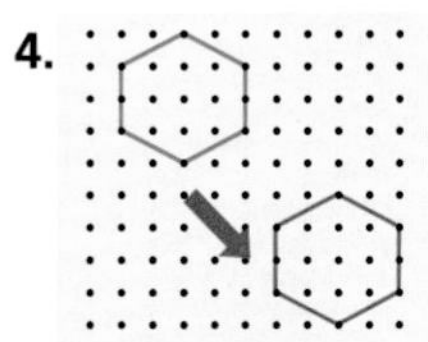

In Exercises 5–8, describe the transformation that maps the blue figure to the red figure.

5. 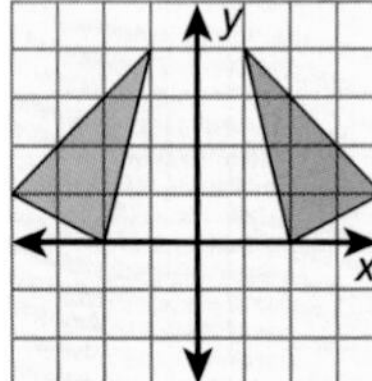

6. 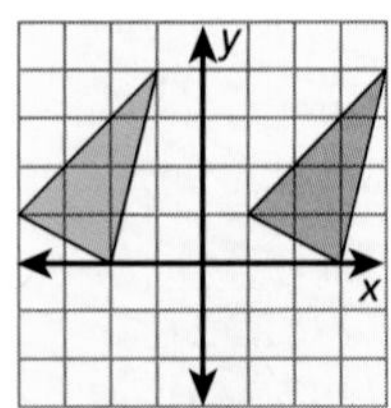

7. 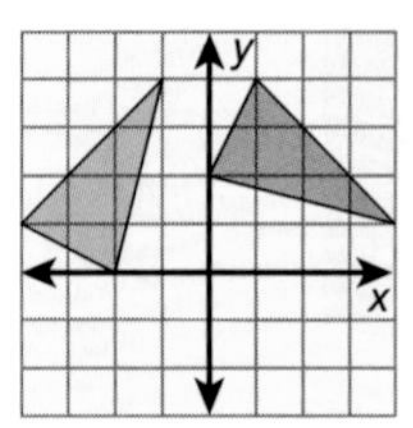

8. 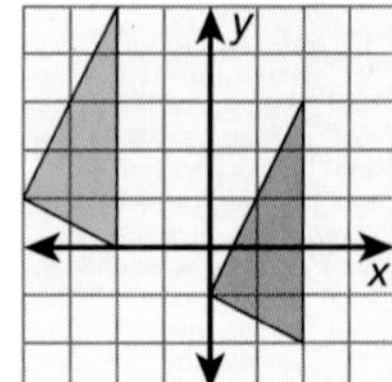

Independent Practice

In Exercises 9–12, match the graph with the ordered pair that describes the translation. Then describe the translation verbally.

a. 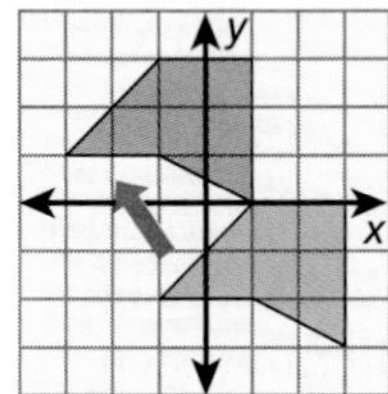

b. 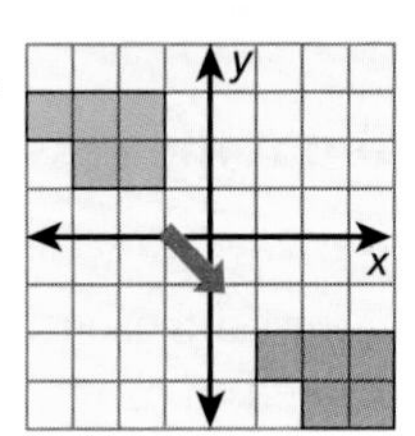

c. 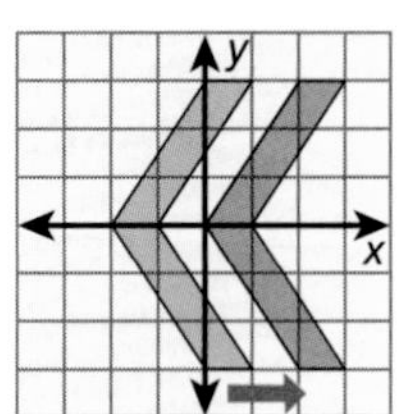

d. 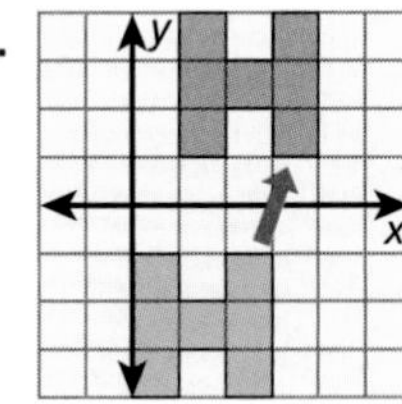

9. $(x + 2, y)$

10. $(x - 2, y + 3)$

11. $(x + 5, y - 5)$

12. $(x + 1, y + 5)$

13. Draw a trapezoid whose vertices are $A(-3, 3)$, $B(-1, 4)$, $C(2, 3)$, and $D(-2, 1)$. Then translate the trapezoid as indicated.

a. 3 units to the left and 6 units down

b. 4 units to the right and 2 units down

c. 5 units to the right and 3 units up

d. 1 unit to the left and 2 units up

14. The motion rule for the translation in Exercise 13a is $(x - 3, y - 6)$. Write the motion rule for the other three translations.

15. Find the area of the trapezoid in Exercise 13. Then use the result to find the area of each of the translated trapezoids.

In Exercises 16–19, copy the figure and translate it twice, as indicated by the arrows. Is the final image a translation of the original figure? If so, describe the single translation that will produce the final image.

16.

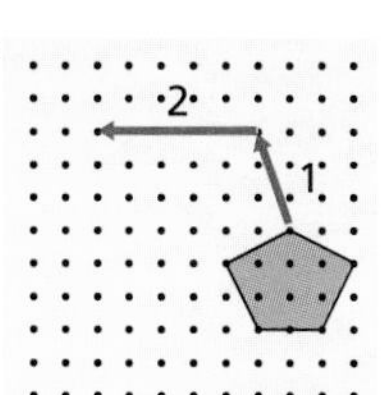

17.

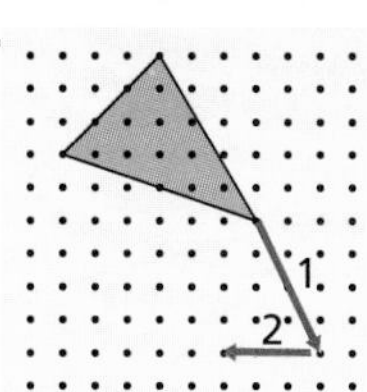

18.

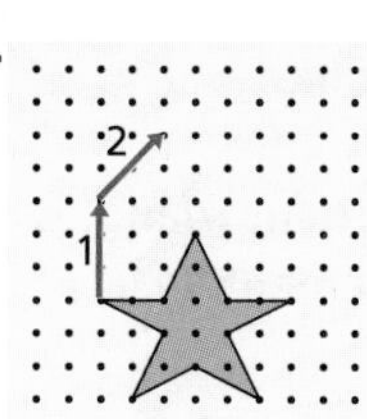

19.

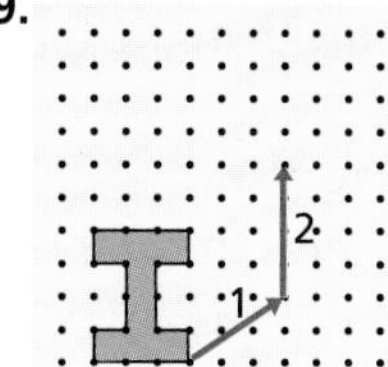

In Exercises 20–23, use the points $A(0, 1)$, $B(3, 4)$, and $C(1, 5)$. Write the vertices of the image of $\triangle ABC$ after it has been translated by the given motion rule.

20. $(x, y + 7)$ **21.** $(x - 1, y + 1)$ **22.** $(x + 2, y - 3)$ **23.** $(x + 3, y + 4)$

24. Draw a triangle in a coordinate plane, and translate the triangle. What can you observe about the line segments that connect the vertices of the original triangle to the corresponding vertices of the image?

Flying in Alabama **In Exercises 25 and 26, use the map at the right. The degree markings indicate the longitude (vertical lines) and latitude (horizontal lines).**

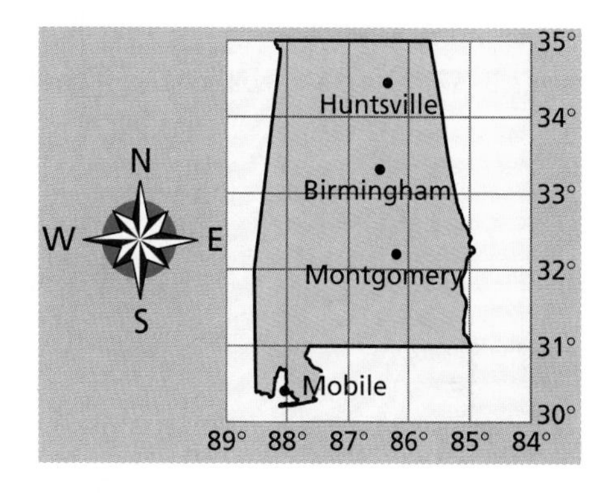

25. You leave Mobile and fly 2.9° north and 1.5° east. Are you flying to Birmingham or Montgomery? Explain.

26. You are flying from Huntsville to Mobile. Describe the translation that will take you there.

Integrated Review

Making Connections within Mathematics

Paper Folding **In Exercises 27–29, fold a piece of paper as shown at the right. Cut out a trapezoid and unfold the paper.**

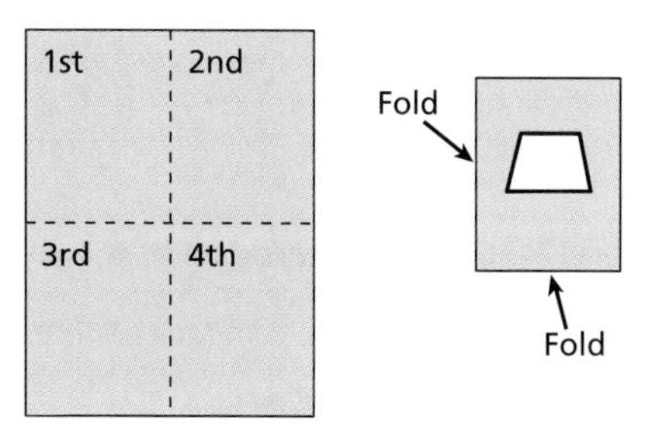

27. How are the first and second trapezoids related?

28. How are the first and fourth trapezoids related?

29. How are the second and fourth trapezoids related?

Exploration and Extension

30. *Animation* Create your own animation of at least 6 screens using the three types of transformations. Then explain the transformation that is used from each screen to the next.

Mid-Chapter SELF-TEST

Take this test as you would take a test in class. The answers to the exercises are given in the back of the book.

In Exercises 1–4, use the congruent triangles at the right to complete the statement. (11.2)

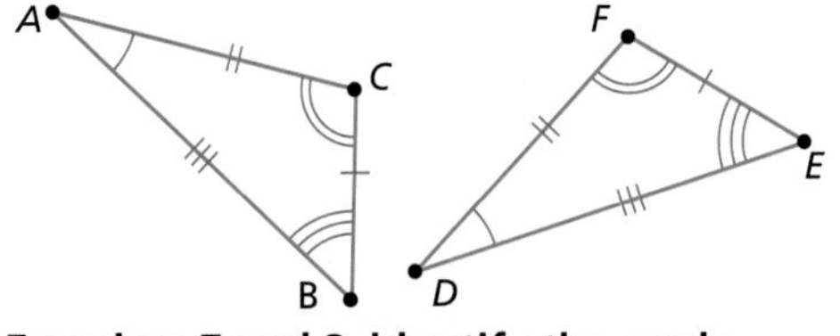

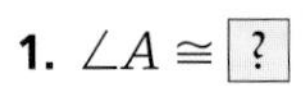

1. $\angle A \cong$ [?] **2.** $\angle F \cong$ [?]

3. $\overline{AB} \cong$ [?] **4.** [?] $\cong \overline{DF}$

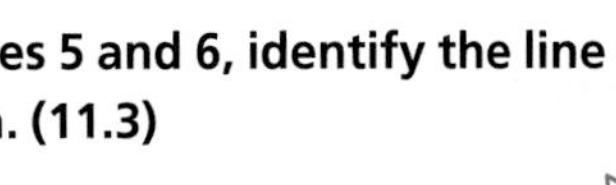

In Exercises 5 and 6, identify the line of reflection. (11.3)

5.

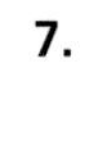

6.

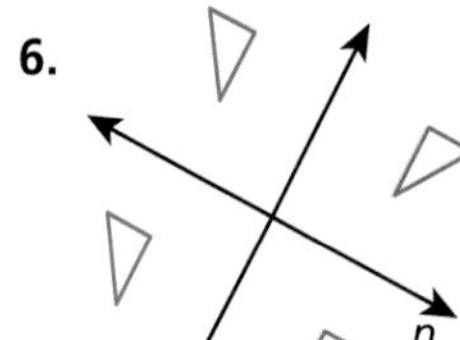

In Exercises 7 and 8, identify the angle and direction of the rotation. (11.4)

7.

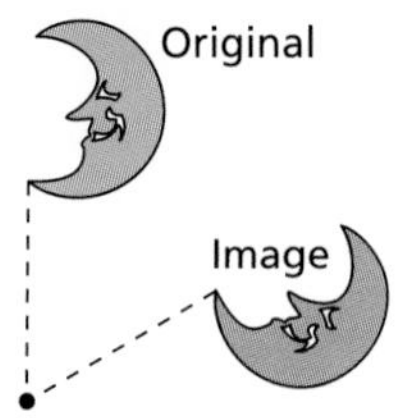

8.

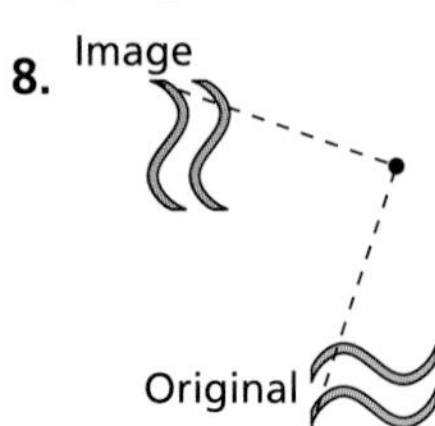

In Exercises 9–12, write a verbal description of the translation. (11.5)

9.

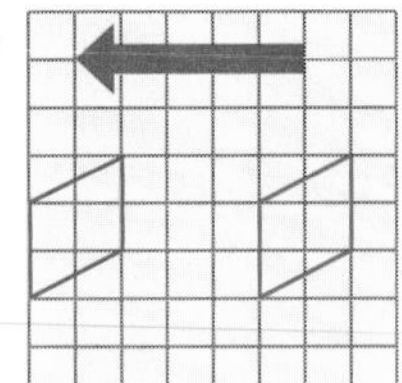

10.

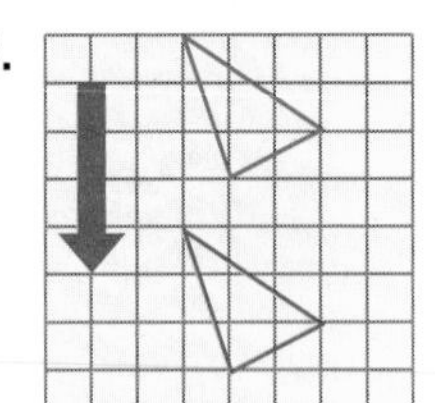

11.

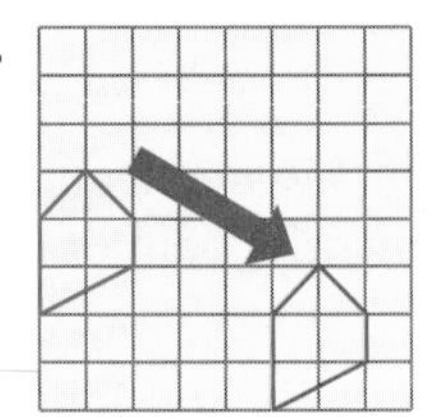

12.

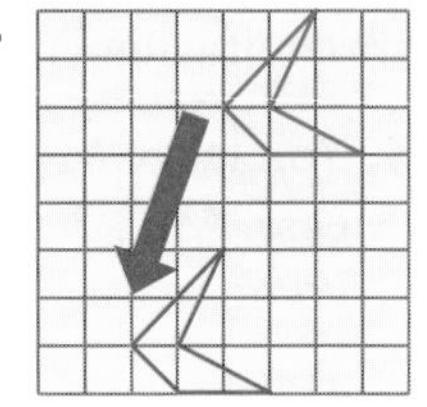

In Exercises 13–16, match the area expression with its polygon. (11.1)

a. $\frac{1}{2}$(Base)(Height) **b.** (Base)(Height) **c.** $(\text{Base})^2$ **d.** $\frac{1}{2}$(Base 1 + Base 2)(Height)

13.

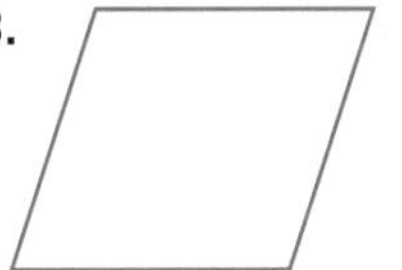

14.

15.

16.

In Exercises 17–20, use the carousel at the right.

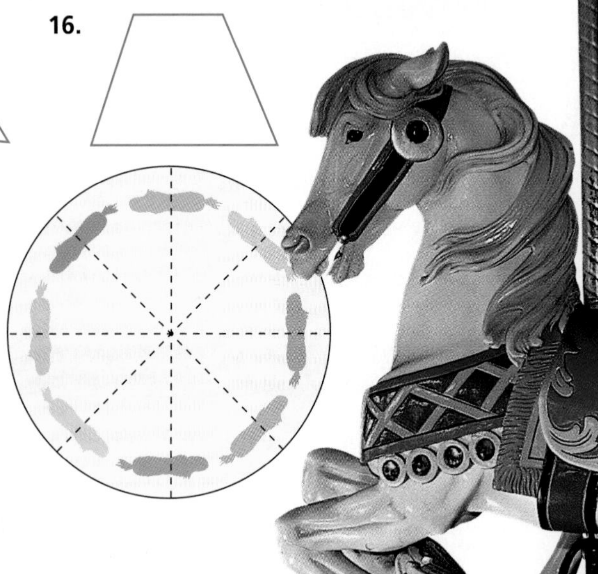

17. Suppose all 8 horses are congruent. Describe the symmetry of the carousel.

18. Suppose every other horse is congruent. Describe the symmetry of the carousel.

19. Suppose every fourth horse is congruent. Describe the symmetry of the carousel.

20. The carousel rotates 15° per second. How long does it take to make one complete revolution?

LESSON INVESTIGATION 11.6

Exploring Similarity

Materials Needed: ruler, protractor, and calculator

Example — *Comparing Photo Sizes*

Photo 2 is a reduction of Photo 1. Use a ruler to measure the width and height of each photo. Then compare the ratios of widths and heights.

Solution Photo 1 has a width of 7.6 centimeters and Photo 2 has a width of 4.5 centimeters. The ratio of these widths is

$$\frac{\text{Photo 1 width}}{\text{Photo 2 width}} = \frac{7.6 \text{ cm}}{4.5 \text{ cm}} \approx 1.7.$$

Photo 1 has a height of 9.1 centimeters and Photo 2 has a height of 5.4 centimeters. The ratio of these heights is

$$\frac{\text{Photo 1 height}}{\text{Photo 2 height}} = \frac{9.1 \text{ cm}}{5.4 \text{ cm}} \approx 1.7.$$

From these four measurements, it appears that the ratios are the same. ■

Photo 1

Exercises

In Exercises 1–8, use a protractor and ruler to measure the indicated line segment or angle in both photos. Then find the ratio of the Photo 1 measurement to the Photo 2 measurement.

1. $\overline{AH}$
2. $\overline{BD}$
3. $\overline{CG}$
4. $\overline{EH}$
5. $\angle ABC$
6. $\angle AHE$
7. $\angle BAH$
8. $\angle GCD$

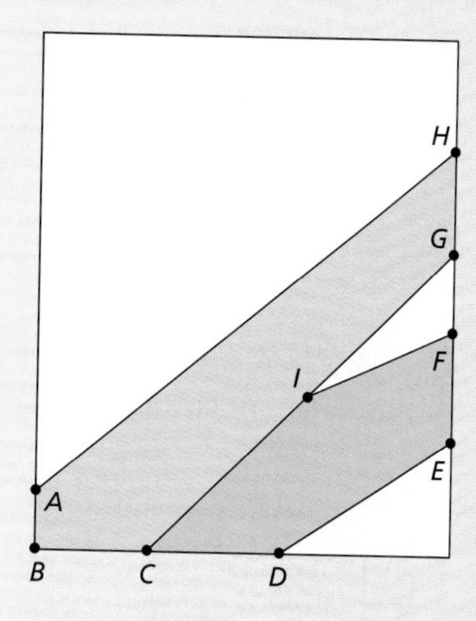

Photo 2

9. What do you conclude about the ratios of segments and angles in the photographs?

11.6 Exploring Similarity

What you should learn:

 How to recognize similar figures

 How to use properties of similar figures

Why you should learn it:

You can use properties of similar figures to solve real-life problems, such as finding the height of a sign.

Goal 1 Recognizing Similar Figures

In the *Investigation* on page 511, you discovered that the two photographs are **similar.** That is, they have the same rectangular shape, but are not necessarily the same size. Here are some other examples of similar and nonsimilar figures.

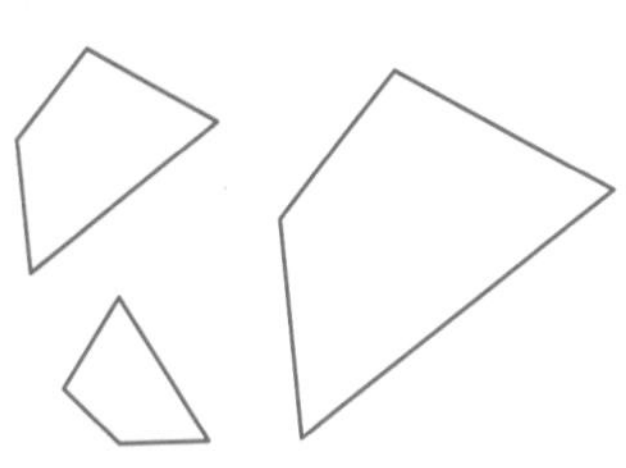

Similar

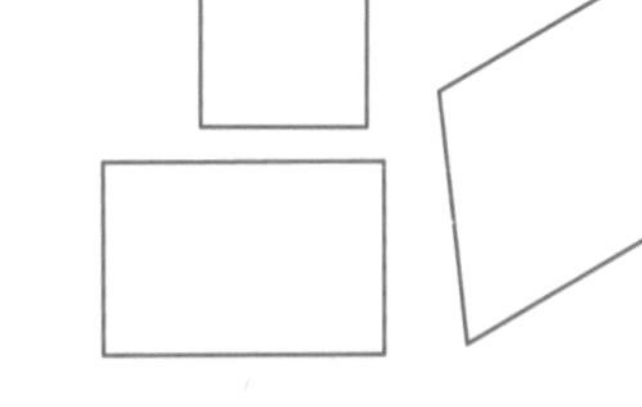

Nonsimilar

LESSON INVESTIGATION

Investigating Similar Triangles

Group Activity Plot the following points on a coordinate plane: $A(2, 4)$, $B(4, 0)$, $C(2, -1)$, $D(4, 8)$, $E(8, 0)$, and $F(4, -2)$. Draw $\triangle ABC$ and $\triangle DEF$. These triangles are similar. Use a protractor to measure each angle. How do the measures of the angles compare?

Compute the length of each side. (You can use the Pythagorean Theorem.) Find the ratios of the lengths of corresponding sides. What can you conclude?

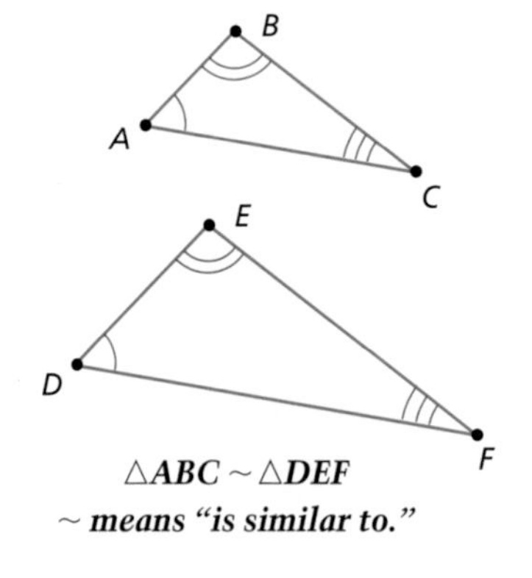

$\triangle ABC \sim \triangle DEF$

$\sim$ means "is similar to."

Corresponding Angles

$m\angle A = m\angle D$

$m\angle B = m\angle E$

$m\angle C = m\angle F$

Corresponding Sides

$\frac{DE}{AB} = \frac{EF}{BC} = \frac{DF}{AC}$

In this investigation, you may have discovered the following properties of similar triangles.

Properties of Similar Triangles

1. Two triangles are similar if their corresponding angles have the same measures.
2. If two triangles are similar, the ratios of corresponding sides are equal. This common ratio is the **scale factor** of the one triangle to the other triangle.

Goal 2 Using Properties of Similar Figures

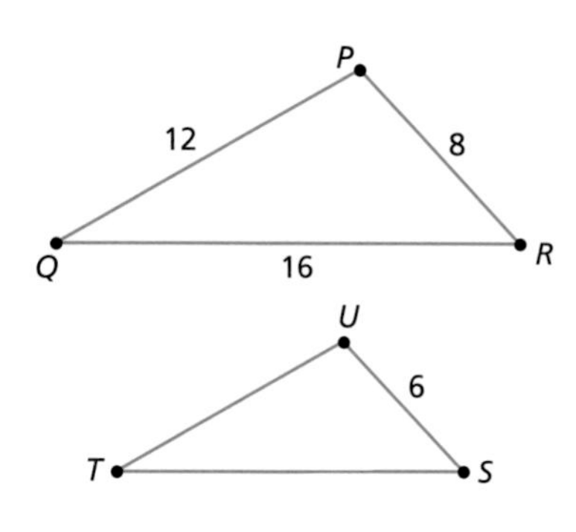

Example 1 *Using Properties of Similar Triangles*

In the figure at the left, $\triangle PQR \sim \triangle UTS$. Find the lengths of $\overline{ST}$ and $\overline{TU}$.

Solution Because the triangles are similar, you know that the ratios of the lengths of corresponding sides are equal. You can use these ratios to write proportions that allow you to solve for the missing lengths.

$$\frac{US}{PR} = \frac{TU}{QP} \qquad \frac{US}{PR} = \frac{ST}{RQ}$$

$$\frac{6}{8} = \frac{TU}{12} \qquad \frac{6}{8} = \frac{ST}{16}$$

$$12 \cdot \frac{6}{8} = 12 \cdot \frac{TU}{12} \qquad 16 \cdot \frac{6}{8} = 16 \cdot \frac{ST}{16}$$

$$9 = TU \qquad 12 = ST$$

■

Real Life
Sign Design

U.S. Medical Corps

Example 2 *Using Properties of Similar Figures*

The symbol at the left represents the United States Medical Corps. You are given a copy of a symbol that is 4 inches tall and 5 inches wide, and are asked to create a sign that has a large Medical Corps symbol. If the large symbol is 6 feet wide, how tall should it be?

Solution Let H be the height of the large symbol.

$$\frac{\text{Large-symbol height}}{\text{Small-symbol height}} = \frac{\text{Large-symbol width}}{\text{Small-symbol width}}$$

$$\frac{H}{4 \text{ inches}} = \frac{6 \text{ feet}}{5 \text{ inches}}$$

$$\frac{H}{4 \text{ inches}} = \frac{72 \text{ inches}}{5 \text{ inches}}$$

$$H = 57.6 \text{ inches}$$

The large symbol should be 57.6 inches or 4.8 feet tall. ■

Communicating about MATHEMATICS

SHARING IDEAS about the Lesson

Extending the Example In Example 2, if the large symbol is 9 feet tall, how wide should it be?

EXERCISES

Guided Practice

CHECK for Understanding

1. ***Writing*** In your own words, state what it means for two figures to be similar.
2. ***The Shape of Things*** Give some examples of similar figures in your classroom.
3. Which two figures are similar?

   a. b. c.

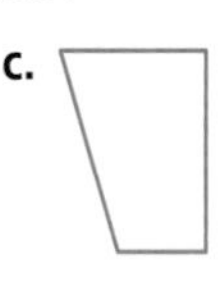

4. Which two figures are similar?

   a. b. c.

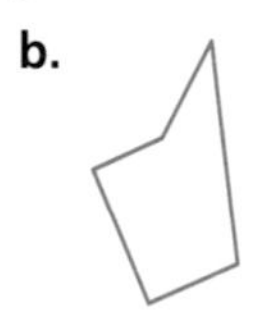

In Exercises 5–8, $\triangle HIJ \sim \triangle KLM$, as shown at the right.

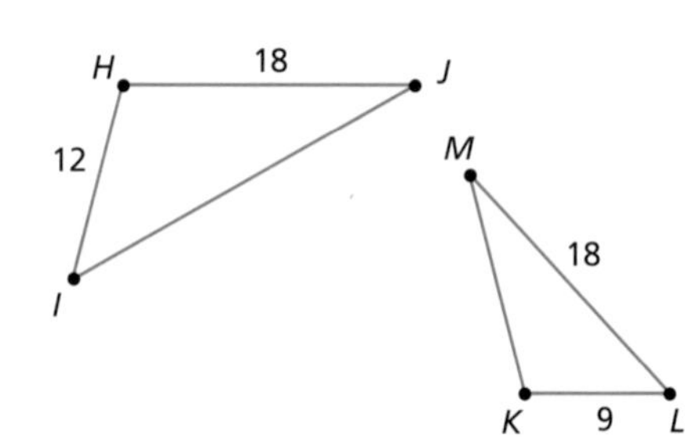

5. Write three proportions for $\triangle HIJ$ and $\triangle KLM$.
6. Find the scale factor of $\triangle HIJ$ to $\triangle KLM$.
7. Find KM and IJ.
8. $m\angle H = m\angle$ [?].

Independent Practice

9. Which two figures are similar?

   a. b. c.

10. Which two figures are similar?

    a. b. c.

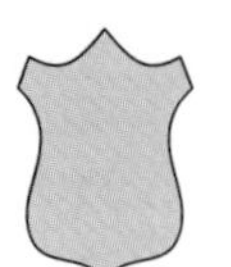

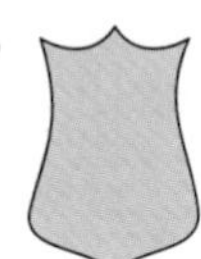

In Exercises 11–14, quadrilaterals *QRST* and *WXYZ* are similar, as shown at the right.

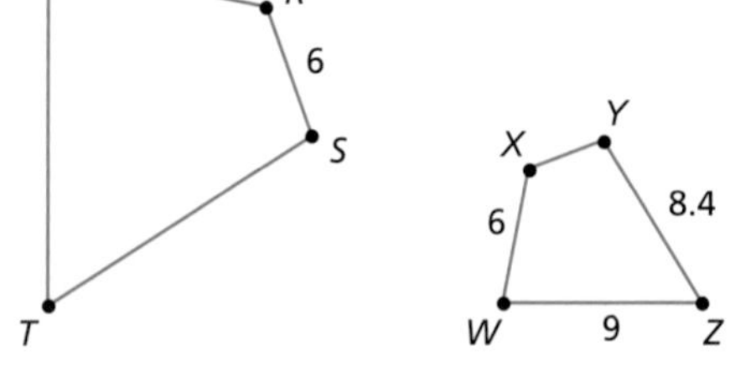

11. Write four equal ratios for $QRST$ and $WXYZ$.
12. Find the scale factor of $QRST$ to $WXYZ$.
13. Find the following.

    a. QT b. ST c. XY
14. $m\angle T = m\angle$ [?].
15. ***Paper Folding*** Can you fold an $8\frac{1}{2}$-inch by 11-inch piece of paper in half so that each half is similar to the original? Explain.
16. ***Paper Folding*** Can you fold a 4-inch by 8-inch piece of paper in half so that each half is similar to the original? Explain.

In Exercises 17–19, determine the scale factor (large to small) of the similar figures. Then solve for x.

17. x 15 12 16

18.

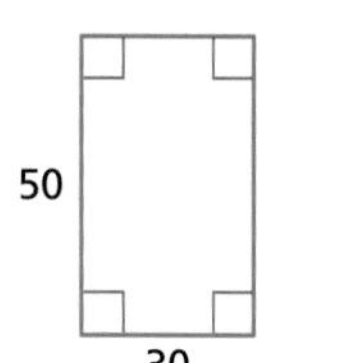

19.

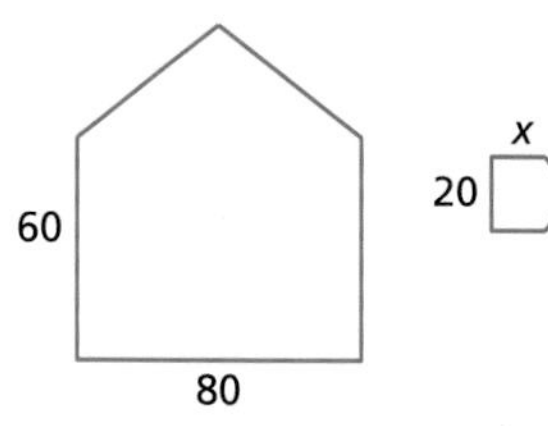

True or False? **In Exercises 20–23, is the statement true or false? Explain.**

20. All congruent figures are similar.

21. All similar figures are congruent.

22. All squares are similar.

23. All right triangles are similar.

24. *What Do You Think?* One of your classmates says that the quadrilaterals $ABCD$ and $EFGH$ are similar. What do you think? Explain your reasoning.

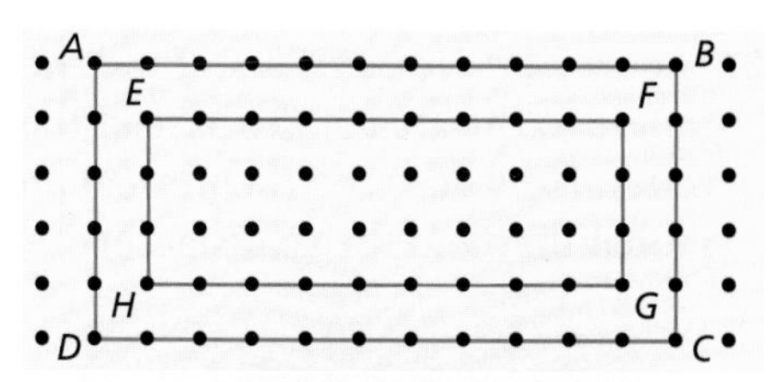

25. *Drawing Triangles* Draw two isosceles right triangles that are not congruent. Are the triangles similar? Explain.

Giant Redwoods **In Exercises 26–28, use the following information.**

A 5-foot person is standing in the shadow of a giant redwood tree. The tips of the person's shadow and the tree's shadow coincide. The person's shadow is 4 feet long and the tree's shadow is 200 feet long.

26. Draw a diagram of the situation. Let a ray of sunlight be the third side of the triangle.

27. Explain why the two triangles are similar.

28. Find the height of the giant redwood tree.

The world's largest tree in volume is a 275-foot Sierra redwood known as the General Sherman Tree. The tree is between 2200 and 2500 years old.

Integrated Review

Making Connections within Mathematics

Proportions **In Exercises 29–31, match the proportion with the pair of similar figures. Then solve for x.**

a. 5 2 4 8 4 x

b.

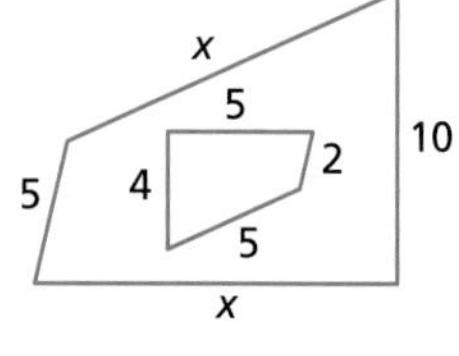

c.

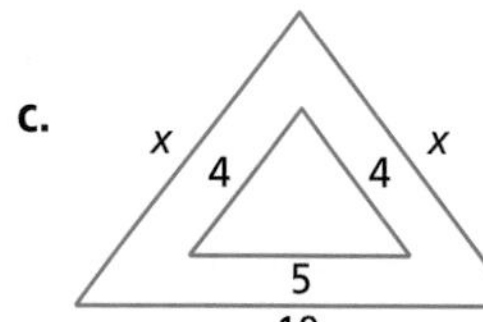

29. $\frac{2}{5} = \frac{5}{x}$

30. $\frac{4}{x} = \frac{5}{10}$

31. $\frac{8}{4} = \frac{x}{5}$

Exploration and Extension

Experimental Probability **In Exercises 32–35, use a box, such as a shoe box, or a shoe-box lid. Draw a rectangle that is similar to (but smaller than) the base of the box. Cut the rectangle out and place it in the box.**

32. Toss a piece of unpopped popcorn (or a pebble) into the box 50 times. Record the number of times it lands on the paper rectangle. Let n be this number.

33. Divide n by 50. This is the experimental probability of landing on the paper rectangle.

34. Find the scale factor of the paper rectangle to the base of the box.

35. *Theoretical Probability* Find the ratio of the area of the paper rectangle to the area of the base of the box. This is the theoretical probability. How does the theoretical probability relate to the experimental probability and the scale factor?

Mixed REVIEW

In Exercises 1–4, use the Pythagorean Theorem to solve for *x*. Round your result to 2 decimal places. (9.3)

1. x, 8, 9

2. x, x, 128

3. 12, x, 11

4. x, 12, 18.06

5. Which triangles in Exercises 1–4 are similar? **(11.6)**

6. In $\triangle ABC$, $BC = 5$, $AC = 7$, and $AB = 10$. Which angle is smallest? Which is largest? **(10.9)**

In Exercises 7–12, solve the percent problem. (7.8)

7. What is 15% of 55?

8. What is 115% of 77?

9. 66 is 150% of what number?

10. 80 is 40% of what number?

11. 53.76 is what percent of 112?

12. 31.92 is what percent of 84?

In Exercises 13–16, evaluate the expression when $x = -6$, $y = 2$, and $z = -3$.

13. $\left|\frac{x}{z}\right|$

14. $|z| - |x|$

15. $(|x| - |y|) \cdot z$

16. $\frac{1}{x}(|y \cdot z|)$

In Exercises 17–20, you have a bag that contains 22 yellow marbles, 20 red marbles, and 14 green marbles. You choose 1 marble from the bag. What is the probability that it is the indicated color? (8.8)

17. Yellow

18. Red

19. Green

20. Red or Green

11.7 Problem Solving Using Similar Figures

What you should learn:

Goal 1 How to use similar figures to solve real-life problems

Goal 2 How to compare perimeters and areas of similar figures

Why you should learn it:

You can use properties of similar figures to solve real-life problems, such as finding the dimensions of a poster.

Goal 1 Using Similar Figures

When two figures are similar, remember that corresponding angles are congruent and corresponding sides are proportional.

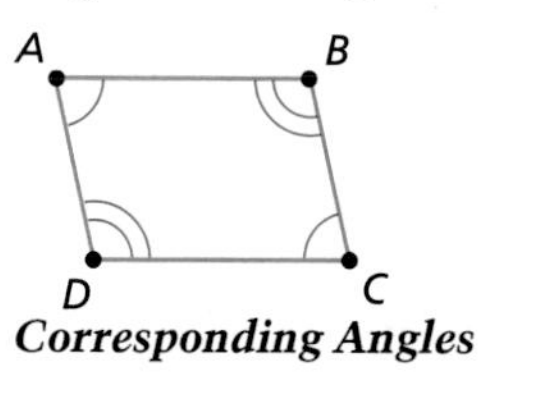

Corresponding Angles

$\angle A \cong \angle E$, $\angle B \cong \angle F$
$\angle C \cong \angle G$, $\angle D \cong \angle H$

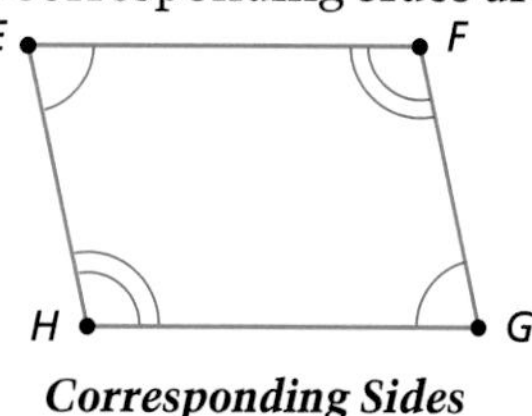

Corresponding Sides

$$\frac{AB}{EF} = \frac{BC}{FG} = \frac{CD}{GH} = \frac{AD}{EH}$$

Example 1 Using Properties of Similar Figures

You are designing a poster to advertise the next meeting of the Space Club. You begin by drawing the sketch shown at the left. You want the actual poster to have a base of 2 feet. Describe the measures of the actual poster.

Solution The angle measures of the actual poster are the same as the angle measures of the sketch. The actual poster has a base (and top) of 2 feet or 24 inches. You can find the length of the slanted sides of the parallelogram by using a proportion.

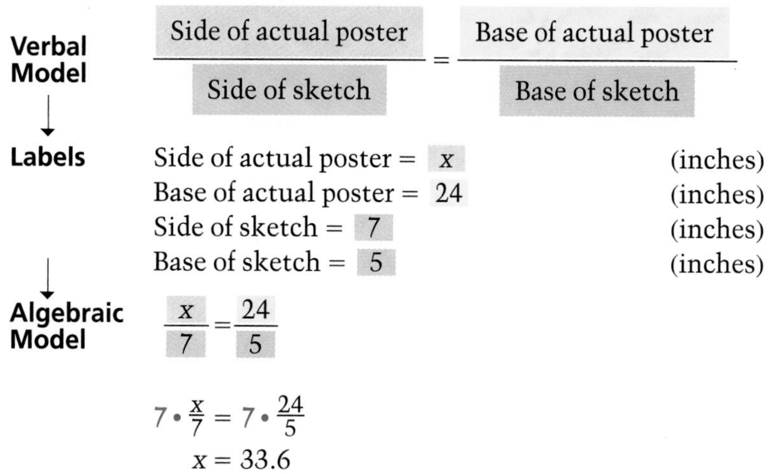

Verbal Model
$$\frac{\text{Side of actual poster}}{\text{Side of sketch}} = \frac{\text{Base of actual poster}}{\text{Base of sketch}}$$

Labels

Side of actual poster = x (inches)
Base of actual poster = 24 (inches)
Side of sketch = 7 (inches)
Base of sketch = 5 (inches)

Algebraic Model
$$\frac{x}{7} = \frac{24}{5}$$

$$7 \cdot \frac{x}{7} = 7 \cdot \frac{24}{5}$$
$$x = 33.6$$

The length of each slanted side should be 33.6 inches. ■

Goal 2 Comparing Perimeters and Areas

Example 2 *Comparing Perimeters and Areas*

Compare the perimeter and area of the poster sketch in Example 1 to the actual poster.

Solution The perimeters of the sketch and poster are

$$\begin{aligned} \text{Perimeter} &= AB + BC + CD + AD \\ &= 5 + 7 + 5 + 7 \\ &= 24 \text{ inches.} \end{aligned}$$

Perimeter of sketch

$$\begin{aligned} \text{Perimeter} &= 24 + 33.6 + 24 + 33.6 \\ &= 115.2 \text{ inches.} \end{aligned}$$

Perimeter of poster

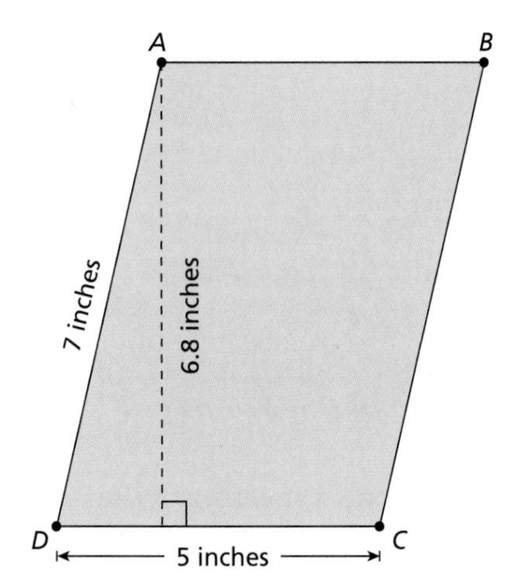

The ratio of the perimeter of the actual poster to the perimeter of the sketch is

$$\frac{\text{Perimeter of Actual Poster}}{\text{Perimeter of Sketch}} = \frac{115.2 \text{ inches}}{24 \text{ inches}} = 4.8.$$

To find the area of the sketch, you need to measure its height, as shown at the left. The area is

$$\begin{aligned} \text{Area of sketch} &= (\text{Base})(\text{Height}) \\ &= (5)(6.8) \\ &= 34 \text{ square inches.} \end{aligned}$$

Area of sketch

In the *Communicating* feature below, you are asked to find the area of the actual poster. ■

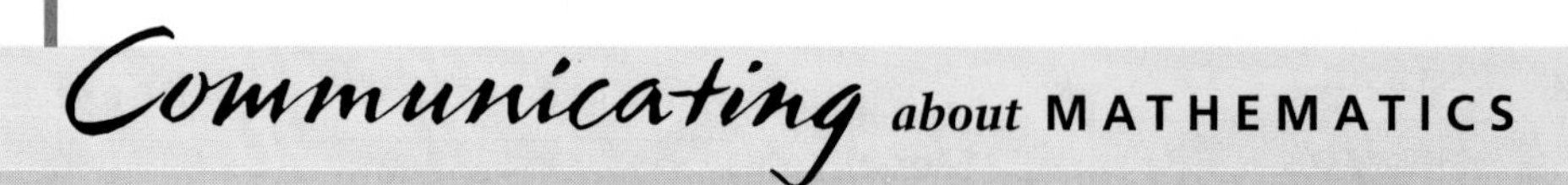

SHARING IDEAS about the Lesson

Comparing Perimeter and Area

A. Find the height of the actual poster in inches. Explain.

B. Find the area of the actual poster in square inches.

C. Which of the following ratios is *not* equal to 4.8? What can you conclude?

1. $\frac{\text{Base of Poster}}{\text{Base of Sketch}}$ **2.** $\frac{\text{Height of Poster}}{\text{Height of Sketch}}$ **3.** $\frac{\text{Area of Poster}}{\text{Area of Sketch}}$

4. $\frac{\text{Perimeter of Poster}}{\text{Perimeter of Sketch}}$ **5.** $\frac{\text{Length of Slanted Side of Poster}}{\text{Length of Slanted Side of Sketch}}$

EXERCISES

Guided Practice

CHECK for Understanding

Model Car **You are designing a model car that is a scale replica of a full-sized car. The license plate on the car is 6 inches by 12 inches. The model's license plate has a length of $\frac{1}{2}$ inch.**

1. Find the height of the model's license plate.
2. Find the scale factor of the car to the model.
3. Find the perimeter and area of each license plate.
4. Find the ratio of the car plate's perimeter to the model plate's perimeter. Compare the result to the scale factor.
5. Find the ratio of the car plate's area to the model plate's area. How does this ratio compare to the scale factor?

The world's smallest running car is powered by a tiny electric motor. The white objects in the photo are grains of rice.

Independent Practice

6. *Estimation* The actual painting is 26 inches high. Use the reduction below to estimate the width of the actual painting.

7. *Like Father, Like Son* The scale factor of a baby lizard to its parent is 1 to 16. The parent is 24 centimeters long. How long is the baby?

8. *Estimation* The blueprint below is drawn with a scale factor of $\frac{1}{16}$ inch to 1 foot. Estimate the perimeter and area of the actual room.

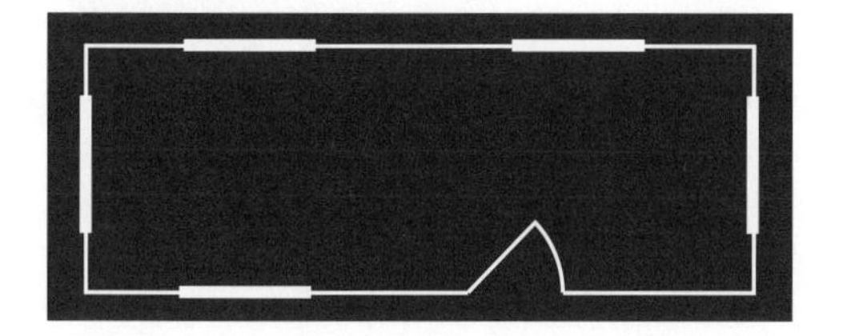

9. *USA Map* The continental United States is about 2500 miles wide. Use the map to estimate its height.

Keepers of the Secret **In Exercises 10–14, use the photo at the right.**

Keepers of the Secret, at the right, was painted by Julie Kramer Cole. The actual painting is 16 inches wide. The painting contains four owls. Two are clearly visible. Can you find the other two?

10. Find the height of the actual painting.

11. Find the scale factor of the actual painting to the photo.

12. Find the perimeter and area of the actual painting. Then find the perimeter and area of the photo (use inches).

13. Find the ratio of the painting's perimeter to the photo's perimeter. How does this ratio compare to the scale factor?

14. Find the ratio of the painting's area to the photo's area. How does this ratio compare to the scale factor?

Integrated Review

Making Connections within Mathematics

Similar Figures **In Exercises 15 and 16, use the fact that the figures are similar to solve for *x*. Then find the perimeter and area of each figure.**

15.

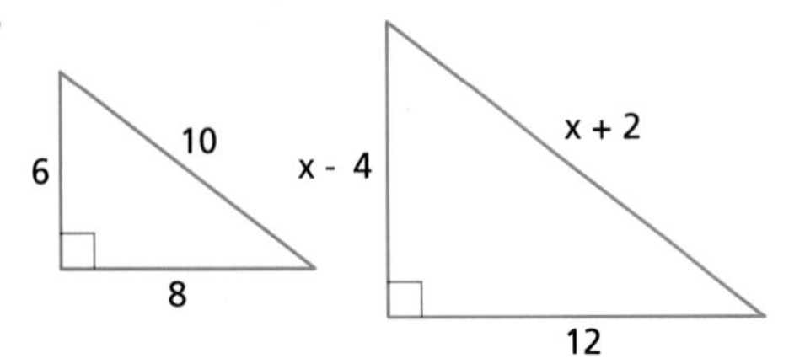

16.

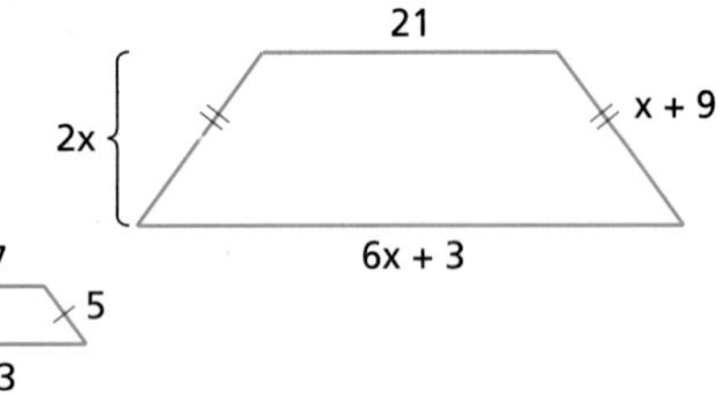

Exploration and Extension

Triangle Grid **In Exercises 17–20, copy the triangular grid at the right on isometric dot paper. Then decide whether it is possible to outline four different sizes of similar figures with the indicated shape, none overlapping. Illustrate the similar figures with a sketch.**

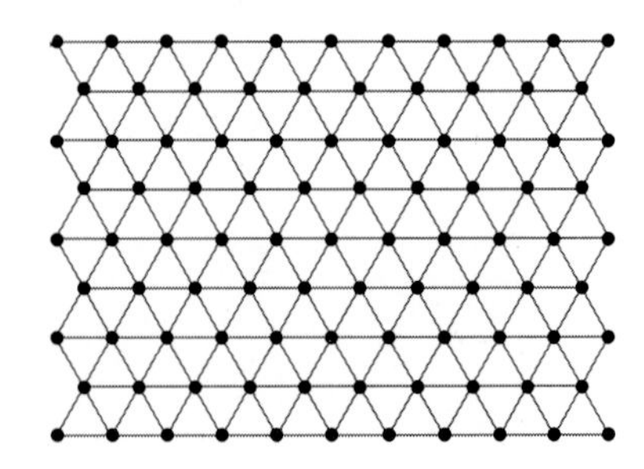

17. Triangles

18. Trapezoids

19. Hexagons

20. Parallelograms

LESSON INVESTIGATION 11.8

Exploring Ratios with Right Triangles

Materials Needed: protractor, ruler, paper, calculator

Example — *Finding Ratios of Side Lengths*

Draw a right triangle whose acute angles measure 25° and 65°. Measure the side lengths, a, b, and c, to the nearest millimeter. Then use a calculator to compute the ratios $\frac{a}{b}$, $\frac{a}{c}$, and $\frac{b}{c}$.

Solution Begin by drawing a right angle, as shown below. Then choose any point on the horizontal ray and label it point A. With your protractor, measure and draw an angle of 25°. Use the ruler to measure the side lengths. For the triangle below, the lengths are $a \approx 35$ millimeters, $b \approx 75$ millimeters, and $c \approx 83$ millimeters. The ratios of the side lengths are

$$\frac{a}{b} \approx \frac{35}{75} \approx 0.467, \quad \frac{a}{c} \approx \frac{35}{83} \approx 0.422, \quad \frac{b}{c} \approx \frac{75}{83} \approx 0.904.$$

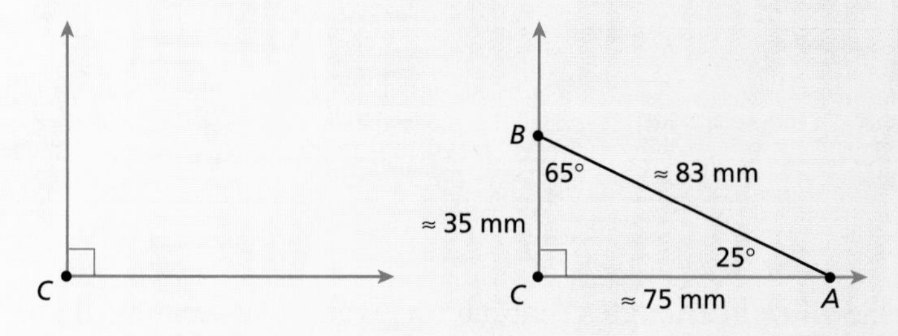

■

Exercises

Group Activity **In Exercises 1–3, each person in your group should draw right triangles that have the given angle measures and complete the table.**

	$m\angle C$	$m\angle A$	$m\angle B$	a	b	c	$\frac{a}{b}$	$\frac{a}{c}$	$\frac{b}{c}$
1.	90°	40°	50°	?	?	?	?	?	?
2.	90°	45°	45°	?	?	?	?	?	?
3.	90°	30°	60°	?	?	?	?	?	?

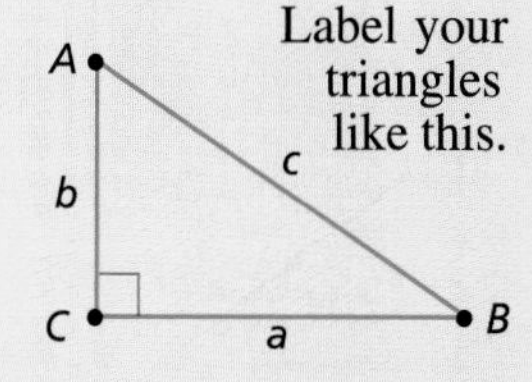

In Exercises 4 and 5, discuss the questions with others in your group.

4. Are all the triangles drawn for Exercise 1 similar to each other? What about Exercises 2 and 3?

5. What can you observe about the ratios found by each person in your group?

11.8 Trigonometric Ratios

What you should learn:

Goal 1 How to find trigonometric ratios

Goal 2 How to use the Pythagorean Theorem to find trigonometric ratios

Why you should learn it:

You can use trigonometric ratios to solve real-life problems, such as comparing heights of a hot-air balloon.

Goal 1 Finding Trigonometric Ratios

In the ancient Greek language, the word "trigonometry" means "measurement of triangles." A **trigonometric ratio** is a ratio of the lengths of two sides of a right triangle. The three basic trigonometric ratios are **sine, cosine,** and **tangent,** which are abbreviated as *sin, cos,* and *tan.*

Trigonometric Ratios

$\sin A = \dfrac{\text{Side opposite } \angle A}{\text{Hypotenuse}} = \dfrac{a}{c}$

$\cos A = \dfrac{\text{Side adjacent to } \angle A}{\text{Hypotenuse}} = \dfrac{b}{c}$

$\tan A = \dfrac{\text{Side opposite } \angle A}{\text{Side adjacent to } \angle A} = \dfrac{a}{b}$

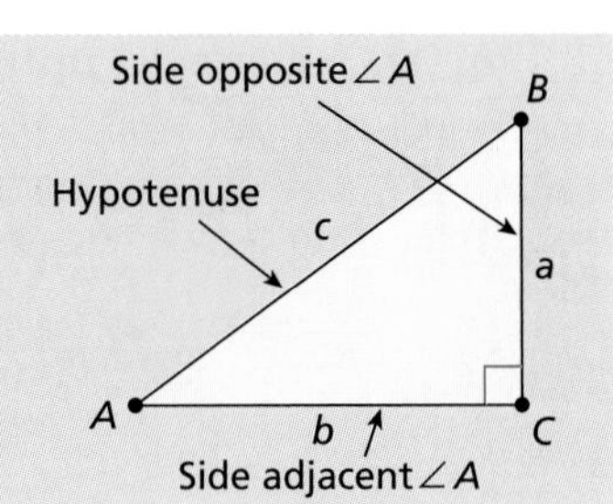

Because all *right* triangles that have a given measure for $\angle A$ are similar, the value of a trigonometric ratio depends only on the measure of $\angle A$. It does not depend on the triangle's size.

Example 1 *Finding Trigonometric Ratios*

For the triangle at the left, find the sine, cosine, and tangent of $\angle P$ and $\angle Q$.

P, 3, 5, R, 4, Q

Solution The length of the hypotenuse is 5.

For $\angle P$, the length of the opposite side is 4, and the length of the adjacent side is 3.

$\sin P = \dfrac{\text{opp.}}{\text{hyp.}} = \dfrac{4}{5}$

$\cos P = \dfrac{\text{adj.}}{\text{hyp.}} = \dfrac{3}{5}$

$\tan P = \dfrac{\text{opp.}}{\text{adj.}} = \dfrac{4}{3}$

For $\angle Q$, the length of the opposite side is 3, and the length of the adjacent side is 4.

$\sin Q = \dfrac{\text{opp.}}{\text{hyp.}} = \dfrac{3}{5}$

$\cos Q = \dfrac{\text{adj.}}{\text{hyp.}} = \dfrac{4}{5}$

$\tan Q = \dfrac{\text{opp.}}{\text{adj.}} = \dfrac{3}{4}$ ■

Goal 2 Using the Pythagorean Theorem

Example 2 *Using the Pythagorean Theorem*

Draw an isosceles right triangle. Then use the triangle to find the sine, cosine, and tangent of 45°.

Hypotenuse $\sqrt{2}$
Opposite side 1
Adjacent side 1
45°

Solution Begin by drawing an isosceles right triangle. Each acute angle of such a triangle measures 45°. Because all isosceles right triangles are similar, it doesn't matter what size triangle you draw. To make the calculations simple, you can choose each leg to have a length of 1. Using the Pythagorean Theorem, you can write

$c^2 = a^2 + b^2$ *Pythagorean Theorem*

$c^2 = 1^2 + 1^2$ *Substitute for a and b.*

$c = \sqrt{2}$. *Square Root Property*

Now, using $a = 1$, $b = 1$, $c = \sqrt{2}$, you can find the sine, cosine, and tangent as follows.

$$\sin 45° = \frac{\text{opp.}}{\text{hyp.}} = \frac{1}{\sqrt{2}} \approx 0.707$$

$$\cos 45° = \frac{\text{adj.}}{\text{hyp.}} = \frac{1}{\sqrt{2}} \approx 0.707$$

$$\tan 45° = \frac{\text{opp.}}{\text{adj.}} = \frac{1}{1} = 1$$

Communicating about MATHEMATICS

SHARING IDEAS about the Lesson

Comparing Sines of Acute Angles Use the following triangles to find the sine of 10°, 20°, 30°, 40°, 50°, 60°, 70°, and 80°. Organize your results in a table. What conclusions can you make about the values of the sine of an angle?

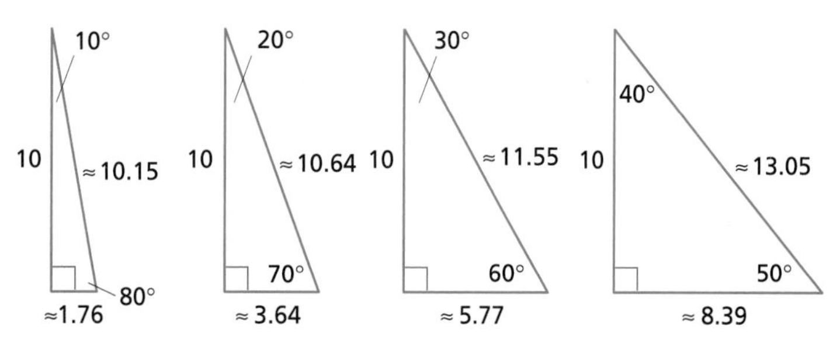

EXERCISES

Guided Practice

CHECK for Understanding

In Exercises 1–3, match the trigonometric ratio with its definition.

a. $\frac{\text{Side opposite } \angle Q}{\text{Hypotenuse}}$ **b.** $\frac{\text{Side opposite } \angle Q}{\text{Side adjacent } \angle Q}$ **c.** $\frac{\text{Side adjacent } \angle Q}{\text{Hypotenuse}}$

1. $\tan Q$ **2.** $\cos Q$ **3.** $\sin Q$

4. Sketch a 40°-50°-90° triangle. Use a ruler to approximate the sine, cosine, and tangent of 40°.

5. Sketch a 40°-50°-90° triangle that is larger than the one in Exercise 4. Approximate the sine, cosine, and tangent of 40°. Do you get the same results as in Exercise 4?

In Exercises 6–11, use $\triangle XYZ$ at the right to find the trigonometric ratio.

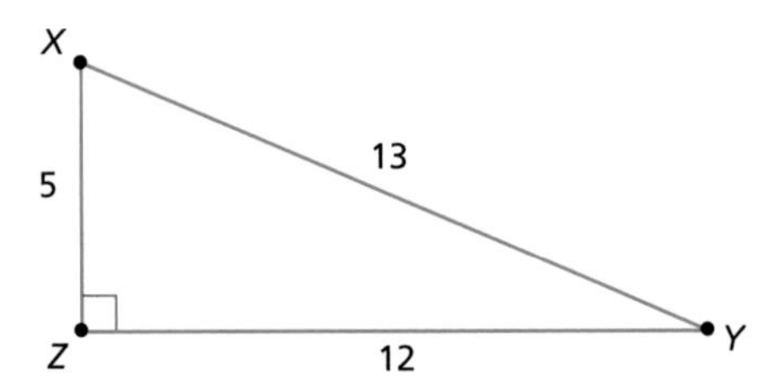

6. $\sin X$ **7.** $\sin Y$

8. $\cos X$ **9.** $\cos Y$

10. $\tan X$ **11.** $\tan Y$

Independent Practice

In Exercises 12–17, use $\triangle DEF$ at the right to find the trigonometric ratio.

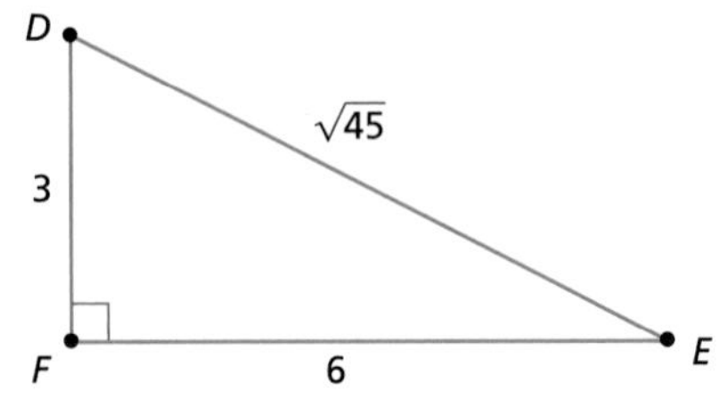

12. $\sin D$ **13.** $\sin E$

14. $\cos D$ **15.** $\cos E$

16. $\tan D$ **17.** $\tan E$

In Exercises 18–21, solve the triangle for its unlabeled angle and side. Then write six trigonometric ratios that can be formed with the triangle.

18.

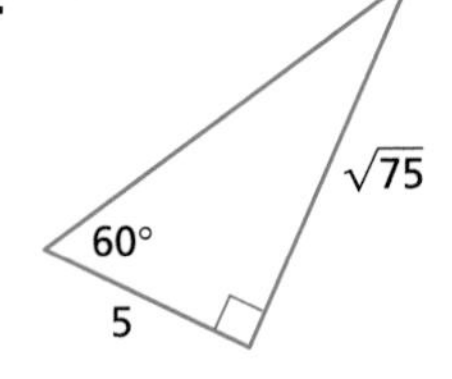

19.

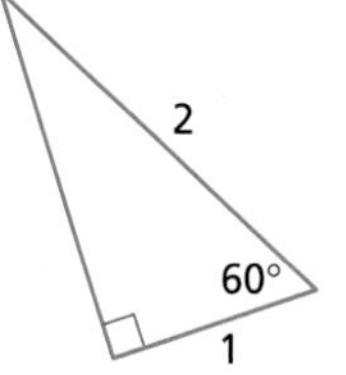

20.

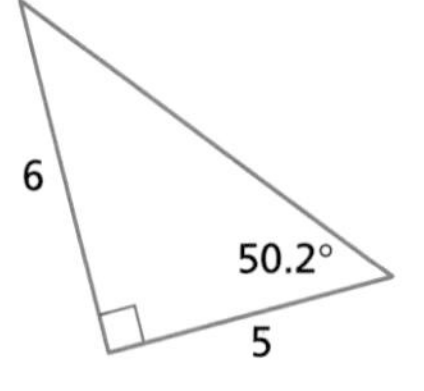

21.

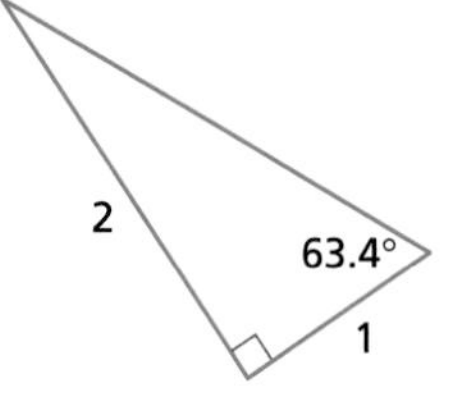

In Exercises 22 and 23, draw a right triangle, $\triangle ABC$, that has the given trigonometric ratios.

22. $\tan A = \frac{15}{8}, \cos B = \frac{15}{17}$

23. $\sin A = \frac{2}{\sqrt{13}}, \cos A = \frac{3}{\sqrt{13}}$

In Exercises 24–26, use the triangles and data that you recorded in *Communicating about Mathematics* on page 523.

24. Copy and complete the table.

x	10°	20°	30°	40°	50°	60°	70°	80°
$\sin x$	?	?	?	?	?	?	?	?
$\cos x$	?	?	?	?	?	?	?	?

25. Describe the relationship between the sine and cosine values in the table.

26. From the table, which angle has the property that $\sin A = \cos A$? Draw a triangle that illustrates your answer. Explain your reasoning.

Ballooning **In Exercises 27–29, use the following information.**

A balloon is released and travels straight upward. You are standing 20 meters from the point of release of the balloon.

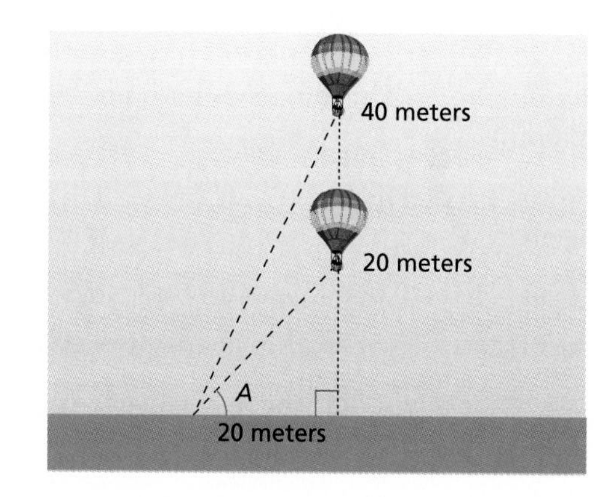

27. Write the tangent ratio for each height of the balloon shown in the diagram.

28. Does $\tan A$ increase or decrease as $m\angle A$ increases? Explain.

29. How high is the balloon when $\tan A = \frac{1}{2}$?

Integrated Review

Making Connections within Mathematics

30. *Estimation* Which of the following best approximates the measure of the peak of the triangular front of the pyramid on a dollar bill?

a. 24° **b.** 44° **c.** 64° **d.** 84°

31. *Estimation* Which of the following best approximates the measure of the base angles of the triangular front of the pyramid on a dollar bill?

a. 8° **b.** 28° **c.** 48° **d.** 68°

Exploration and Extension

In Exercises 32–35, each of the triangles is similar to one of the triangles at the bottom of page 523. Write a proportion that allows you to solve for *x*. Then solve for *x*.

32.

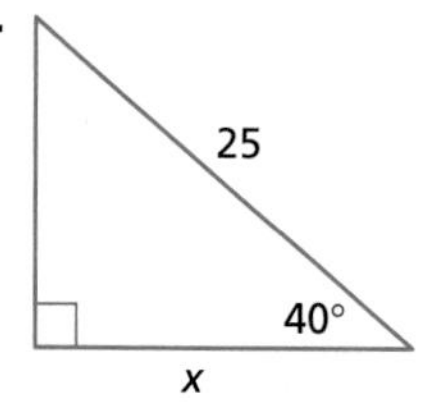

33.

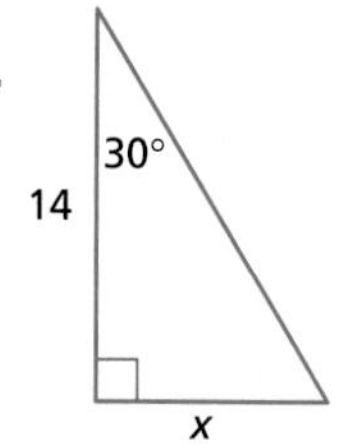

34.

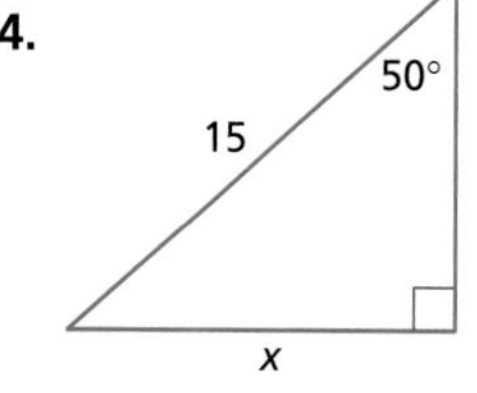

35.

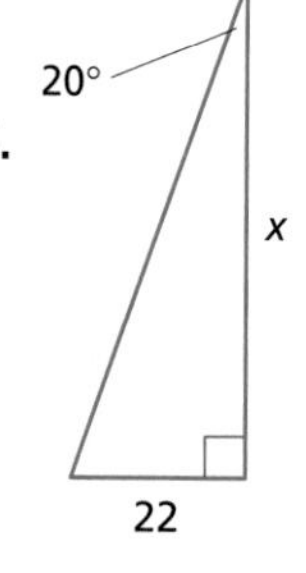

USING A CALCULATOR

Solving Right Triangles

Scientific calculators are able to evaluate the sine, cosine, and tangent of an angle.

Example — *Using a Scientific Calculator*

Use a scientific calculator to evaluate the sine, cosine, and tangent of 30° and 60°. Then use the triangle at the right to evaluate the same trigonometric ratios. Compare your results.

Solution Before beginning, make sure your calculator is set in *degree mode.* (Scientific calculators have another angle measure mode called *radian mode.* This mode will not produce the correct values for this example.)

30°	Display	60°	Display
30 [sin]	0.5	60 [sin]	0.866025
30 [cos]	0.866025	60 [cos]	0.5
30 [tan]	0.577350	60 [tan]	1.732051

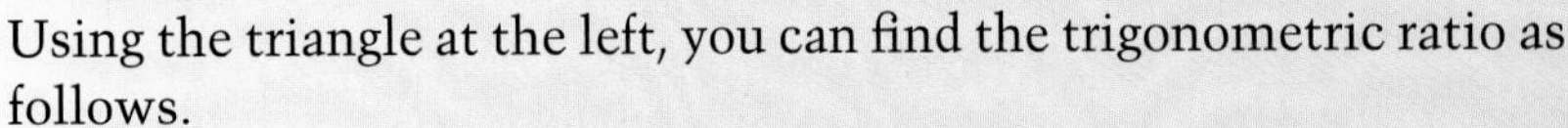

Using the triangle at the left, you can find the trigonometric ratio as follows.

$\sin 30° = \frac{1}{2} = 0.5$ $\quad \sin 60° = \frac{\sqrt{3}}{2} \approx 0.866$

$\cos 30° = \frac{\sqrt{3}}{2} \approx 0.866$ $\quad \cos 60° = \frac{1}{2} = 0.5$

$\tan 30° = \frac{1}{\sqrt{3}} \approx 0.577$ $\quad \tan 60° = \frac{\sqrt{3}}{2} = 1.732$

Exercises

1. Use a scientific calculator to evaluate the sine, cosine, and tangent of 15° and 75°. Then use the triangle at the right to evaluate the same trigonometric ratios. Compare your results.

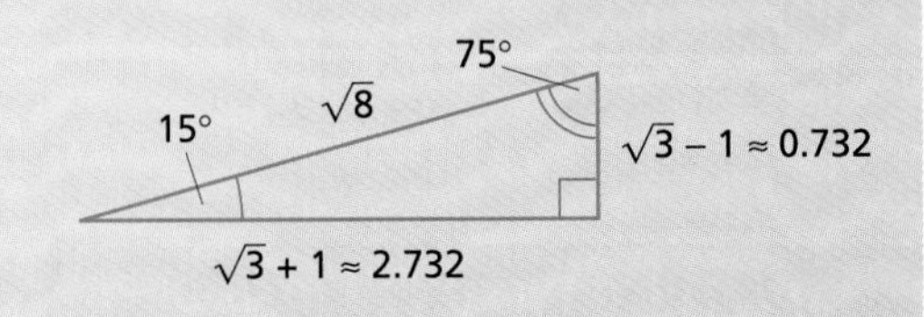

2. ***Discovering a Trigonometric Property*** Choose any acute angle, A. Then use a scientific calculator to complete the table. What can you conclude? Write your conclusion as a rule that applies to all acute angles.

A	$\sin A$	$(\sin A)^2$	$\cos A$	$(\cos A)^2$	$(\sin A)^2 + (\cos A)^2$
?	?	?	?	?	?

11.9 Problem Solving Using Trigonometric Ratios

What you should learn:

Goal 1 How to use trigonometric ratios to solve right triangles

Goal 2 How to use trigonometric ratios to solve real-life problems

Why you should learn it:

You can use trigonometric ratios to solve real-life measurement problems, such as finding the height of a building.

Goal 1 Solving Right Triangles

In Lesson 9.3, you learned how to use the Pythagorean Theorem to solve a right triangle. In that situation, you were given the lengths of two sides of the triangle and asked to find the length of the third side.

In this lesson, you will learn how to solve a right triangle when you are given the length of only one of the sides and the measure of one of the acute angles.

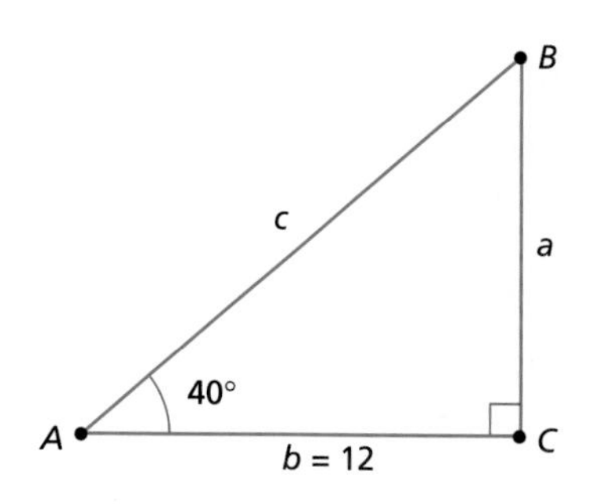

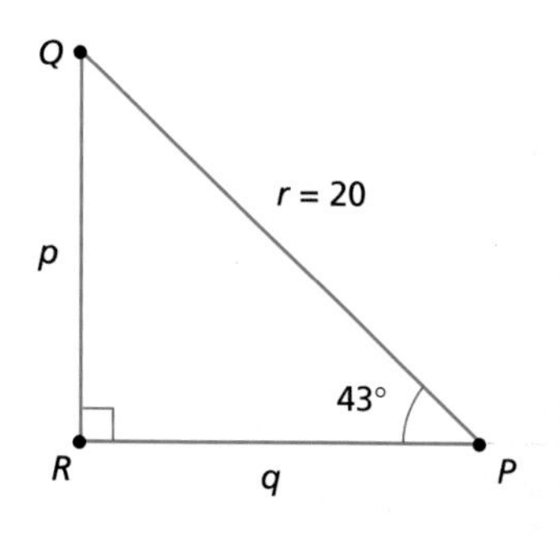

Example 1 *Solving a Right Triangle*

a. In $\triangle ABC$ at the left, find a.

b. In $\triangle PQR$ at the left, find q.

Solution

a. To find a, you can write the trigonometric ratio for the tangent of A.

$\tan A = \frac{a}{b}$	*Definition of tangent of A*
$\tan 40° = \frac{a}{12}$	*Substitute for A and b.*
$0.8391 \approx \frac{a}{12}$	*Use a calculator.*
$12(0.8391) \approx 12\left(\frac{a}{12}\right)$	*Multiply each side by 12.*
$10.07 \approx a$	*Simplify.*

The length of $\overline{BC}$ is about 10.07.

b. To find q, you can write the trigonometric ratio for the cosine of P.

$\cos P = \frac{q}{r}$	*Definition of cosine of P*
$\cos 43° = \frac{q}{20}$	*Substitute for P and r.*
$0.7314 \approx \frac{q}{20}$	*Use a calculator.*
$20(0.7314) \approx 20\left(\frac{q}{20}\right)$	*Multiply each side by 20.*
$14.63 \approx q$	*Simplify.*

The length of $\overline{RP}$ is about 14.63. ■

Goal 2 Solving Real-Life Problems

Real Life
Construction

Example 2 *Using Trigonometric Ratios*

You are restoring a historical building. To estimate the height of the roof's peak, you stand 30 feet back from the building and measure the angle to the peak to be 52=×″. How high is the peak? Use your result to find the area of the side of the building.

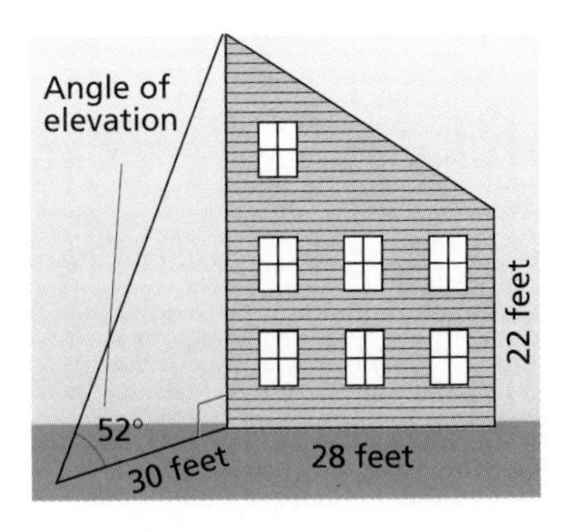

Solution Let h represent the height of the building (in feet).

$\tan 52° = \frac{h}{30}$ *Definition of tangent*

$1.280 \approx \frac{h}{30}$ *Use a calculator.*

$30(1.280) \approx 30 \cdot \frac{h}{30}$ *Multiply each side by 30.*

$38.4 \approx h$ *Simplify.*

The height of the building is about 38.4 feet. To find the area of the side, use the formula for the area of a trapezoid. (The height of the building is one of the bases of the trapezoid.)

$$\text{Area} = \frac{1}{2}(\text{Base 1} + \text{Base 2})(\text{Height})$$
$$\approx \frac{1}{2}(38.4 + 22)(28)$$
$$\approx 845.6$$

The area of the side is about 845.6 square feet.

Gregory and Carolynn Schipa are the owners of The Weather Hill Restoration Company, which specializes in restoring historical buildings.

Communicating about MATHEMATICS

SHARING IDEAS about the Lesson

Measuring Angles A carpenter's level can be used to draw horizontal lines. Explain to your partner how you could use a carpenter's level and a protractor to estimate the angle of a roof.

EXERCISES

Guided Practice

CHECK for Understanding

In Exercises 1–4, use the triangle at the right.

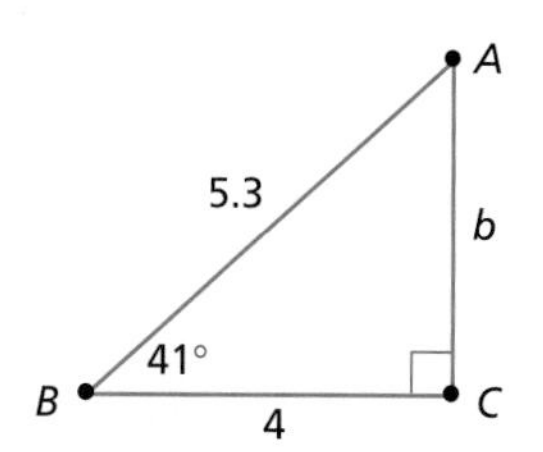

1. Use $\tan 41°$ to solve for b.
2. Use $\sin 41°$ to solve for b.
3. Use the Pythagorean Theorem to solve for b.
4. *It's Up to You* Which of the methods in Exercises 1–3 do you prefer? Explain your reasoning.

Flagpole **In Exercises 5 and 6, use the diagram at the right.**

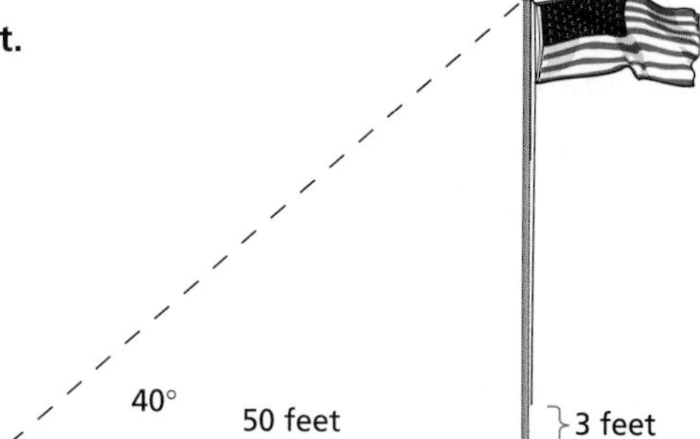

5. You stand 50 feet from the flagpole. The angle from the point where you stand to the top of the flagpole is 40°. How tall is the flagpole?
6. The rope used to raise and lower the flag is secured three feet above the ground. What is the minimum length of the rope? (*Hint:* The rope must be tied in a loop.)

Independent Practice

In Exercises 7–10, find the value of the trigonometric ratio.

7. $\cos 33°$
8. $\tan 74°$
9. $\sin 44°$
10. $\cos 80°$

In Exercises 11–14, find the length of the labeled side. Round your results to two decimal places.

11. 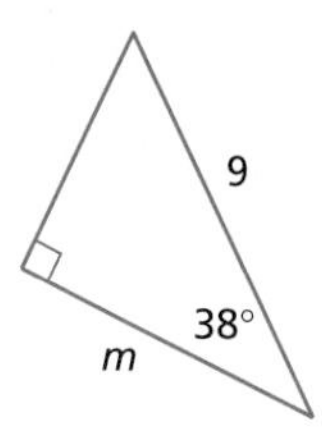

12. 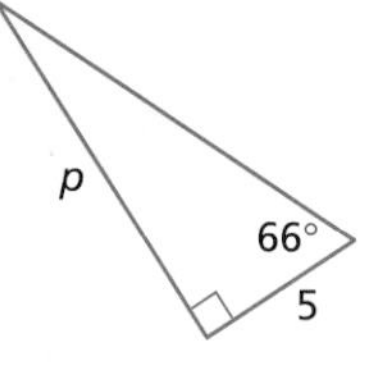

13. 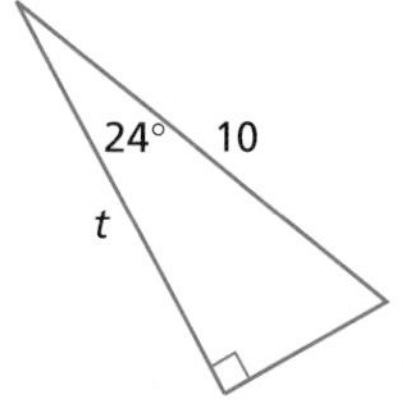

14. 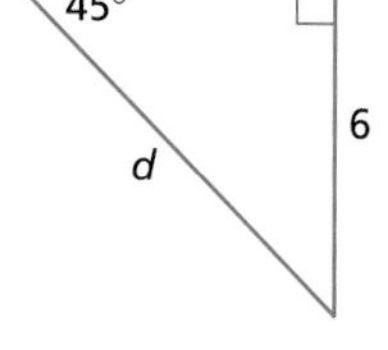

In Exercises 15–18, solve the right triangle for all labeled sides and angles. Round your results to two decimal places.

15. 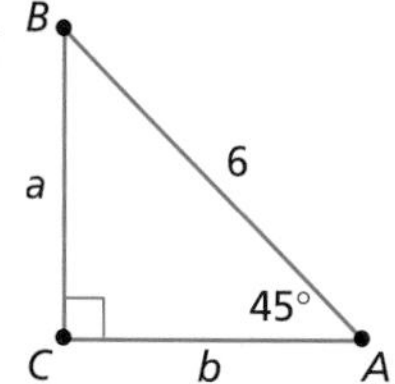

16. 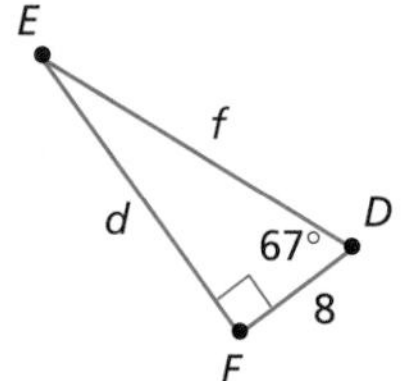

17. 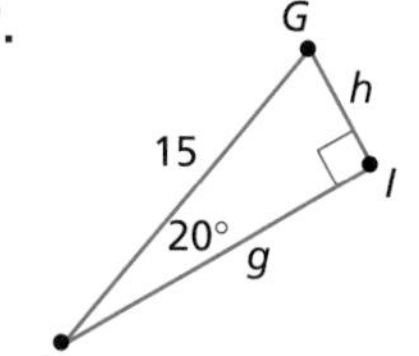

18. 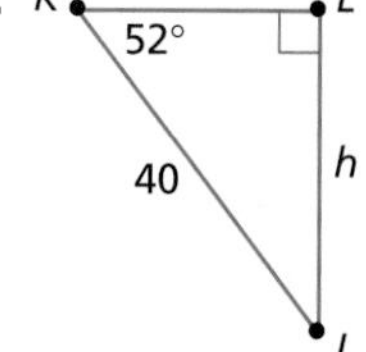

19. *Loading Dock* The angle of depression between the road and the loading ramp is 8°. Find the height, h, of the loading dock.

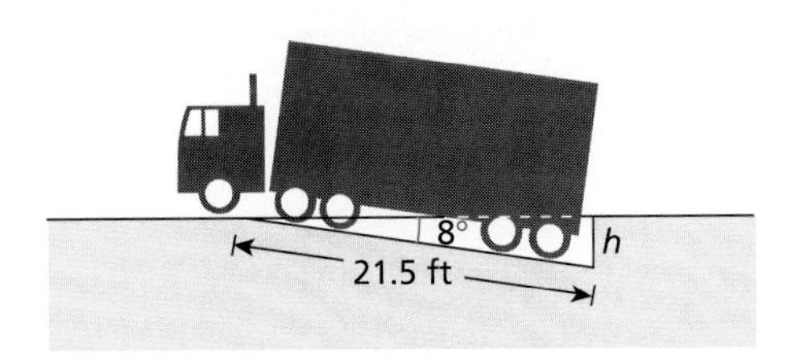

20. *Lighthouse* You are standing on a path 10 meters from a lighthouse. The angle of elevation to the top is 50°. How tall is the lighthouse?

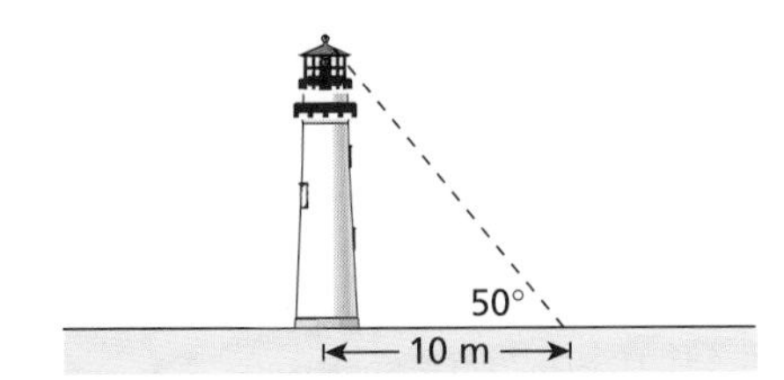

Giraffes **In Exercises 21–24, use the following information.**

You are standing 22 meters from a mother giraffe and her baby. The angle of elevation to the mother's head is 11° and the angle to the baby's head is 5°.

21. How tall is the mother?
22. How tall is the baby?
23. What is the difference in heights of the two giraffes?
24. Is the difference in height equal to $22 \sin(11°–5°)$? Explain.

Integrated Review

Making Connections within Mathematics

Television Channels **In Exercises 25–28, use the graph at the right.** ***(Source: Nielsen Media Research)***

25. What percent of the possible channels are watched by people who receive 36 channels?
26. What percent of the possible channels are watched by people who receive 58 channels?
27. What percent of the possible channels are watched by people who receive 80 channels?
28. If the people who receive 80 channels watched the same percent as those who receive 36 channels, how many channels would they watch?

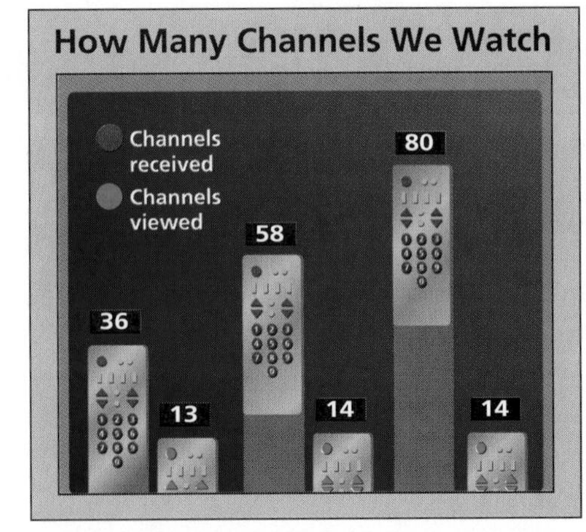

Studies show that people don't watch more channels when more are available.

Exploration and Extension

29. *You Be the Teacher* Imagine that you are teaching this course and are writing a test that covers the geometry in Chapters 10 and 11. Write a test that has 20 questions and covers the important concepts in the two chapters. In your test, include at least one multiple-choice question, at least one question that requires a calculator, and at least one question that requires a paragraph answer.

Chapter Summary

What did you learn?

Skills

1. Find the area and perimeter of a polygon. **(11.1)**
2. Determine whether two figures are congruent. **(11.2)**
3. Reflect a figure in a line. **(11.3)**
   - Connect reflections and line symmetry. **(11.3)**
4. Rotate a figure about a point. **(11.4)**
   - Connect rotations and rotational symmetry. **(11.4)**
5. Translate a figure in a plane. **(11.5)**
   - Translate a figure in a coordinate plane. **(11.5)**
6. Recognize similar figures. **(11.6)**
   - Use properties of similar figures. **(11.6)**
   - Compare perimeters and areas of similar figures. **(11.7)**
7. Find trigonometric ratios. **(11.8)**
   - Use trigonometric ratios to solve right triangles. **(11.9)**

Problem-Solving Strategies

8. Use geometric figures to model and solve real-life problems. **(11.1–11.9)**

Exploring Data

9. Use tables and graphs to solve problems. **(11.1–11.9)**

Why did you learn it?

Congruence, similarity, and transformations can be used to answer questions about many real-life situations. For instance, in this chapter you learned how to analyze the symmetry of clothing patterns, how to use transformations to create computer animations, how to use properties of similar figures to plan a sign or a poster, and how to use trigonometric ratios to indirectly measure the height of a building.

How does it fit into the bigger picture of mathematics?

Before studying this chapter, you already knew that part of geometry is measuring lengths, angles, perimeters, and areas. In this chapter, you learned that geometry is more than measuring. It is also about comparing figures to determine whether they are congruent or similar. For instance, if one figure can be slid, flipped, or rotated so that it fits exactly on top of another figure, then the two figures are congruent. Knowing that figures are congruent or similar is important because it can save you problem-solving time. For instance, if you have already measured the angles of one figure, you don't need to remeasure the angles of a figure that you know is similar (because you know the similar figure must have the same angle measures).

Chapter REVIEW

In Exercises 1–4, find the perimeter and area of the polygon. (11.1)

1.

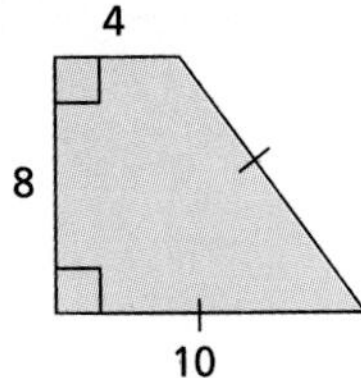

2.

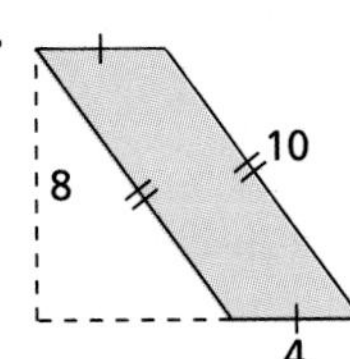

3.

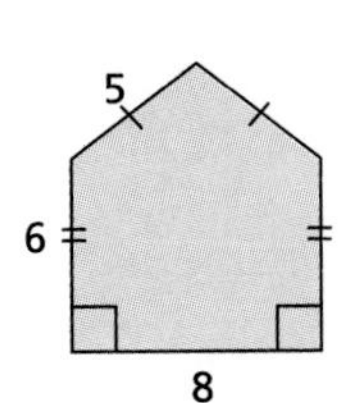

4.

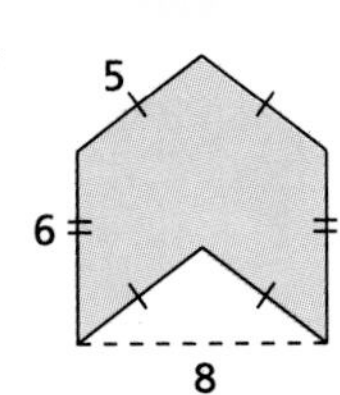

In Exercises 5–10, use the fact that $\triangle PQR \cong \triangle UYS$ to complete the statement. (11.2)

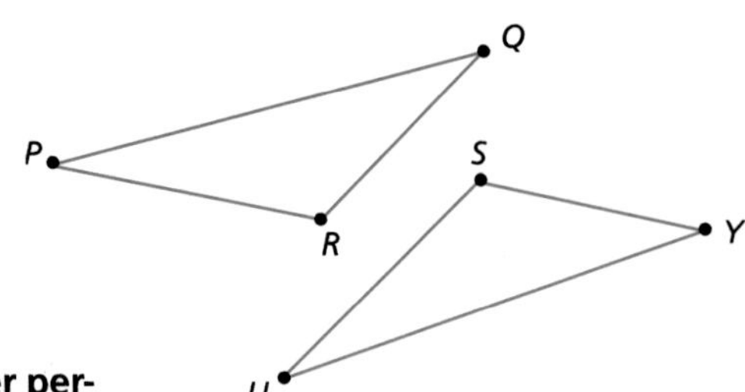

5. $\overline{PQ} \cong$?

6. $\angle S \cong$?

7. $\angle Q \cong$?

8. $\overline{SU} \cong$?

9. $\angle U \cong$?

10. $\overline{QR} \cong$?

In Exercises 11–14, find the vertices of the figure after performing the indicated transformation. (11.3–11.5)

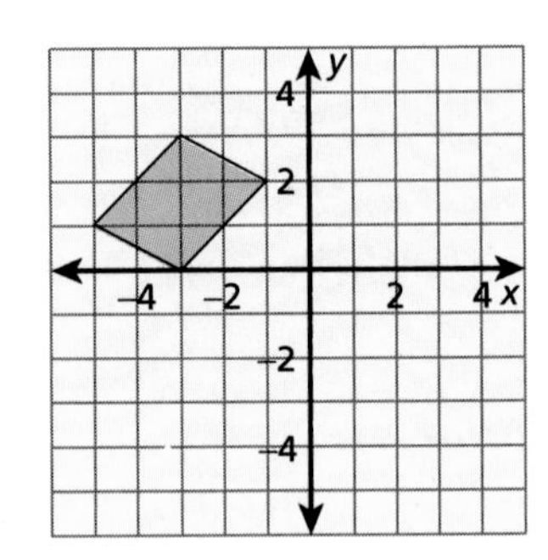

11. Reflect the figure in the y-axis.

12. Rotate the figure counterclockwise 90° about the origin.

13. Translate the figure 6 units to the right and 5 units down.

14. Reflect the figure in the x-axis, then translate it 2 units to the right.

Ferris Wheel **In Exercises 15–17, use the Ferris wheel at the right. (11.3–11.5)**

15. Describe the rotations of the wheel that will transform the spokes onto themselves.

16. When the wheel rotates 45°, are the seats rotated or translated? Explain your reasoning.

17. With the seats, does the Ferris wheel have any lines of symmetry? Without the seats, does it have any lines of symmetry? Explain.

In Exercises 18–21, use the fact that $\triangle UVW \sim \triangle XYZ$. (11.6)

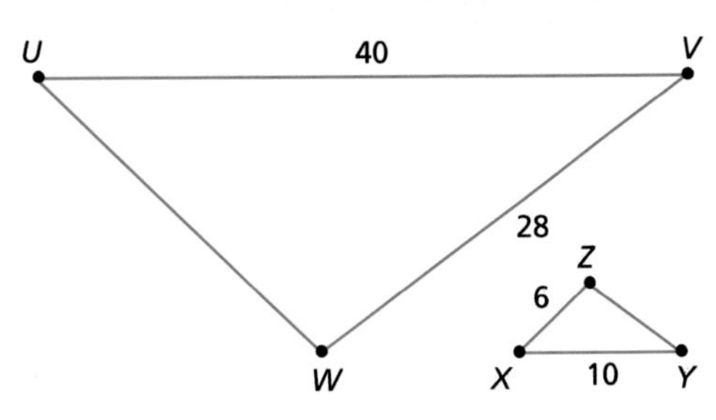

18. Write three proportions for $\triangle UVW$ and $\triangle XYZ$.

19. Find the scale factor of $\triangle XYZ$ to $\triangle UVW$.

20. Find UW and YZ.

21. Find the ratio of the perimeter of $\triangle XYZ$ to the perimeter of $\triangle UVW$. Is this ratio equal to the scale factor?

In Exercises 22–24, complete the statement using *always, sometimes,* or *never.* (11.6)

22. Two congruent figures are ? similar.

23. Two similar figures are ? congruent.

24. A square and a trapezoid are ? similar.

Dollar Bills **In Exercises 25–28, use the drawing of the dollar bill. (11.7)**

25. The actual width of a dollar bill is 6.6 centimeters. Use the scale drawing at the right to find the actual length.

26. Find the scale factor of an actual dollar bill to the drawing.

27. Find the area of the bill and its drawing.

28. Find the ratio of the area of an actual bill to the area of the drawing. How does this relate to the scale factor?

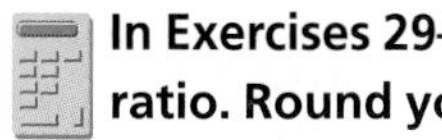

In Exercises 29–34, use $\triangle MNO$ to find the trigonometric ratio. Round your result to two decimal places. (11.8)

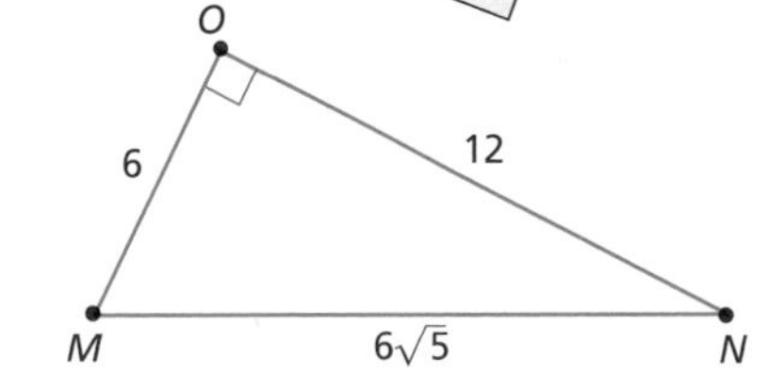

29. $\sin M$ **30.** $\sin N$

31. $\cos M$ **32.** $\cos N$

33. $\tan M$ **34.** $\tan N$

In Exercises 35–38, find the value of *x*. (11.9)

35.

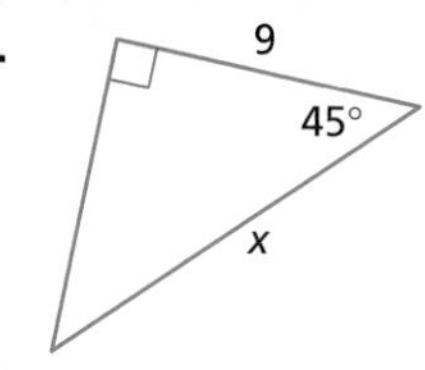

36.

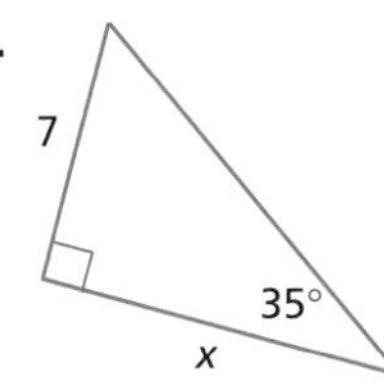

37.

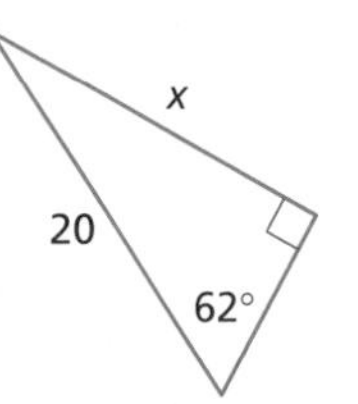

38.

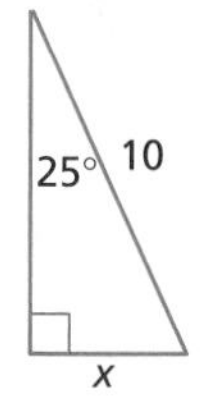

Drawing a Diagram **In Exercises 39–41, use the following information. (11.9)**

A helicopter is hovering over a lifeboat that is floating in the ocean. You are in another boat that is 300 feet from the lifeboat. From your location, the angle of elevation to the helicopter is 42°.

39. Draw a diagram of the situation.

40. *Without* using a calculator, estimate the height of the helicopter.

41. Use trigonometry and a calculator to find the height of the helicopter.

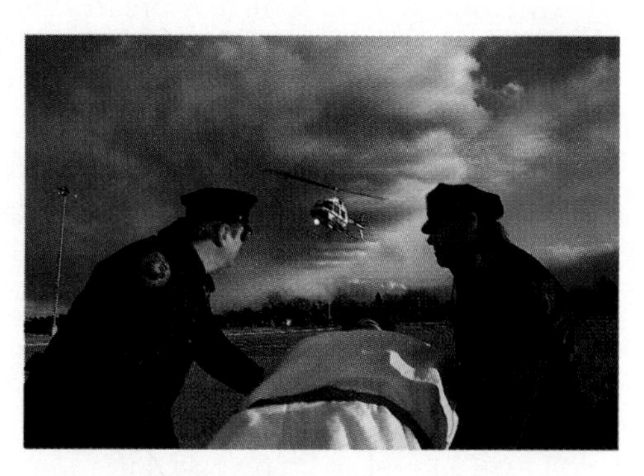

Because of their ability to hover, land, and take off in small areas, helicopters are often used in rescue operations.

Voyage to Mars **In Exercises 42–60, use the figure to find the indicated measure. (11.1, 11.2, 11.6, 11.8)**

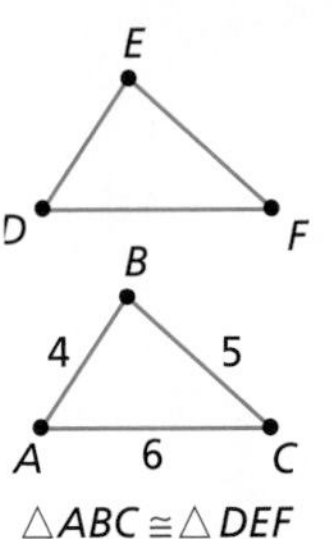

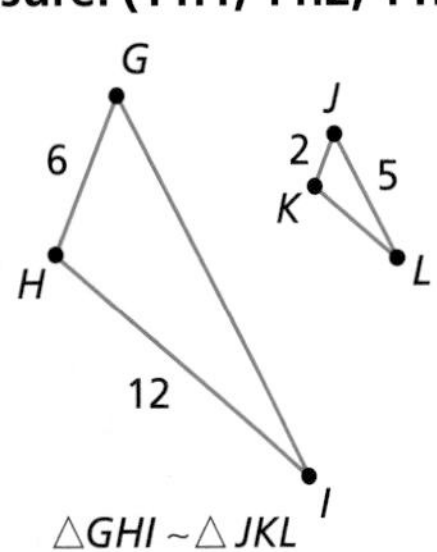

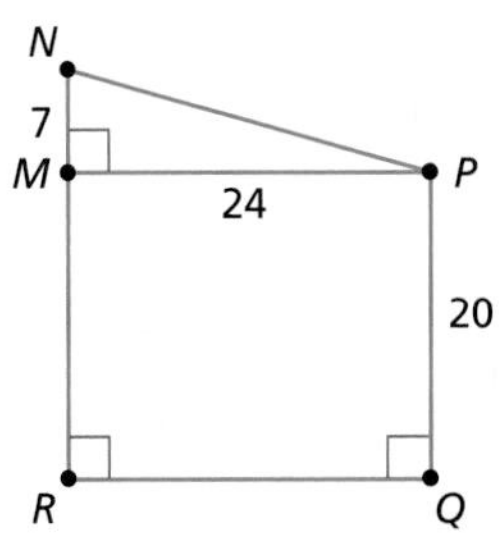

42. **a:** Length of $\overline{EF}$

43. **c:** Length of $\overline{DF}$

44. **d:** Perimeter of $\triangle DEF$

45. **e:** Perimeter of $MPQR$

46. **f:** Length of $\overline{KL}$

47. **g:** Perimeter of $\triangle GHI$

48. **h:** Perimeter of $\triangle JKL$

49. **i:** Length of $\overline{NP}$

50. **j:** Perimeter of $\triangle MNP$

51. **k:** Area of $\triangle MNP$

52. **l:** Tangent of $\angle P$

53. **m:** Tangent of $\angle N$

54. **n:** Sine of $\angle P$

55. **o:** Length of $\overline{QR}$

56. **r:** Cosine of $\angle P$

57. **s:** Area of $MPQR$

58. **t:** Perimeter of $NPQR$

59. **u:** Area of $NPQR$

60. **w:** Scale factor of $\triangle GHI$ to $\triangle JKL$

61. *Voyage to Mars* Use the letter codes found in Exercises 42-60 to decode the following messages. Each message is related to a CD that was developed by the Planetary Society, Time Warner Interactive Group, and Russia's Institute for Space Research. (Some of the messages are names of people who worked on the project and the other messages are items that are contained on the CD.) The CD will be part of a space capsule that will be sent to Mars in 1995.

I. 6 5 $\frac{24}{25}$ $\frac{7}{24}$ 480 5 33 5 $\frac{7}{25}$

II. 56 564 15 25 96 11 $\frac{24}{7}$ 88 $\frac{24}{25}$ $\frac{24}{25}$ 25 $\frac{7}{24}$

III. 3 5 $\frac{24}{25}$ 24 4   96 11 88   3 24 $\frac{24}{25}$ $\frac{7}{24}$ 15 480

IV. 480 96 5 $\frac{24}{25}$ 96 $\frac{24}{25}$ 88 84

V. 96 11 88 $\frac{24}{7}$ 5 $\frac{24}{25}$ 96 25 5 $\frac{7}{25}$

6 11 $\frac{24}{25}$ 24 $\frac{7}{25}$ 25 6 $\frac{7}{24}$ 88 480

VI. $\frac{7}{24}$ 25 96 96 $\frac{7}{24}$ 88   33 $\frac{24}{25}$ 88 88 $\frac{7}{25}$ $\frac{24}{7}$ 5 $\frac{7}{25}$

This painting is by artist Kelly Freas. It first appeared on the cover of a science fiction novel in 1954. The title of the painting is hidden in VI.

This photo was taken on Mars during the Viking II mission.

Chapter TEST

In Exercises 1–3, find the area and perimeter of the figure.

1.

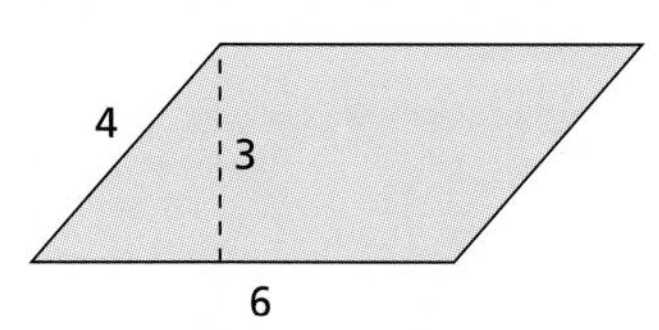

2.

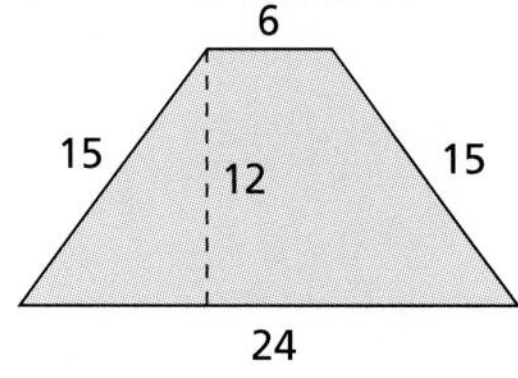

3.

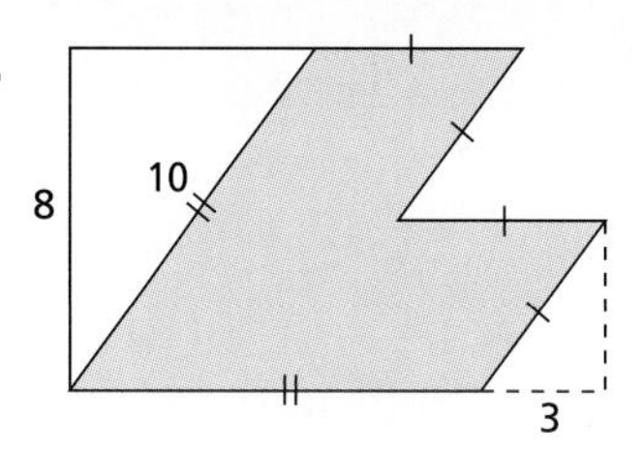

In Exercises 4–7, write the coordinates of the image of $\triangle MNP$.

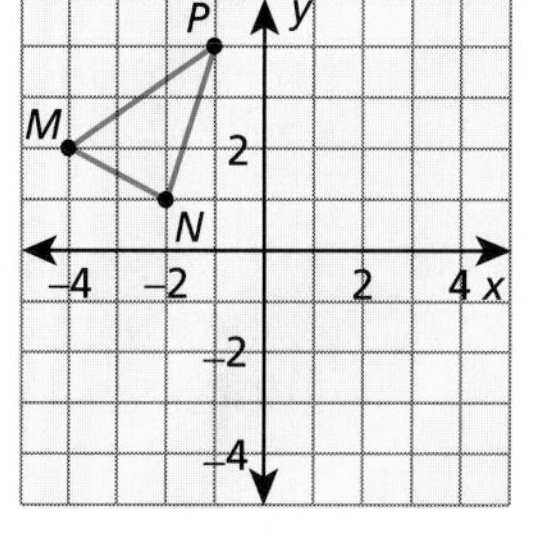

4. Reflect $\triangle MNP$ in the y-axis.

5. Rotate $\triangle MNP$ clockwise 90° about the origin.

6. Rotate $\triangle MNP$ clockwise 180° about the origin.

7. Translate $\triangle MNP$ 2 units to the right and 3 units down.

In Exercises 8–11, state whether the blue lion is congruent to the red lion. If it is, describe the transformation that will map the blue lion to the red lion.

8.

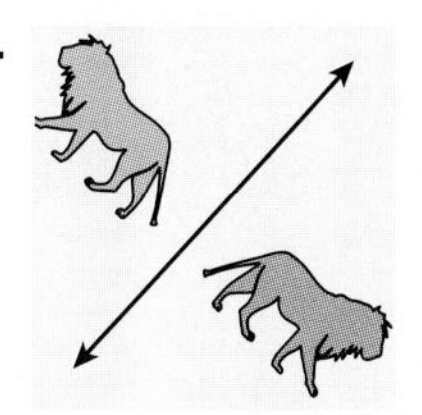

9.

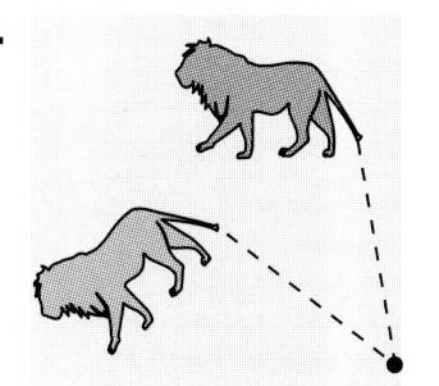

10.

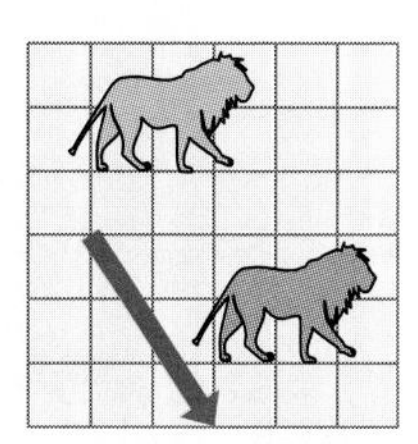

11.

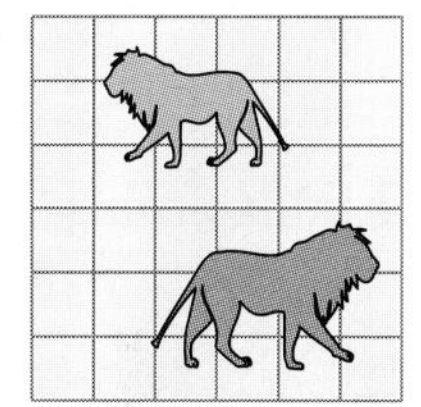

In Exercises 12 and 13, use the diagram at the right.

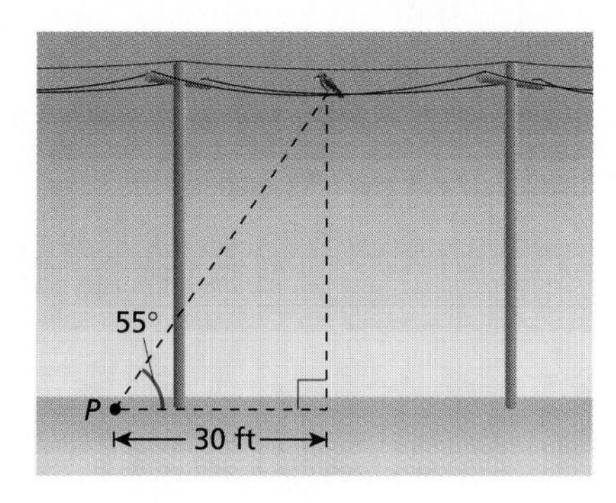

12. How high is the bird?

13. Find the distance between the bird and point P.

In Exercises14 and 15, you are creating a small model of a sculpture that is 78 inches high and 24 inches wide.

14. If the model is 6 inches wide, how tall should it be?

15. If the model is 6 inches high, how wide should it be?

In Exercises 16–20, use the triangles at the right.

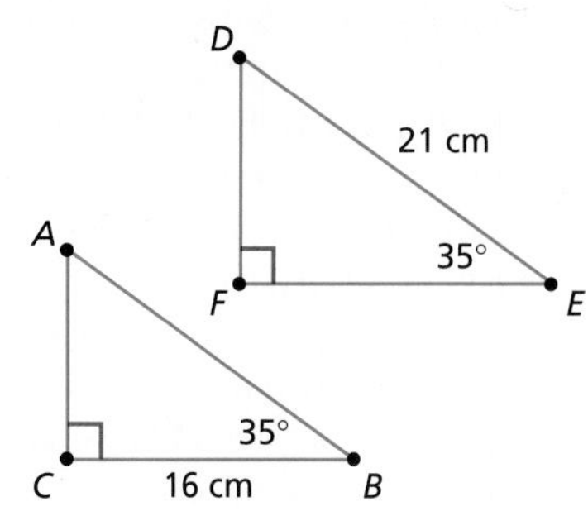

16. Find tan 35°.

17. Use the result of Exercise 16 to find AC.

18. Find cos 35°.

19. Use the result of Exercise 18 to find EF.

20. Are $\triangle ABC$ and $\triangle DEF$ congruent or similar? Explain.

CHAPTER 12

Measurements in Geometry

LESSONS

To draw plans for buildings or manufactured items, drafters use triangles with their T squares and drawing boards. Product designer Sava Cvek created the triangles shown here to make his drawing tasks easier. Two of the triangles have "wheels" so they can be moved into position easily and the other is an "adjustable triangle." The scale drawing is Sava Cvek's plan for the adjustable triangle.

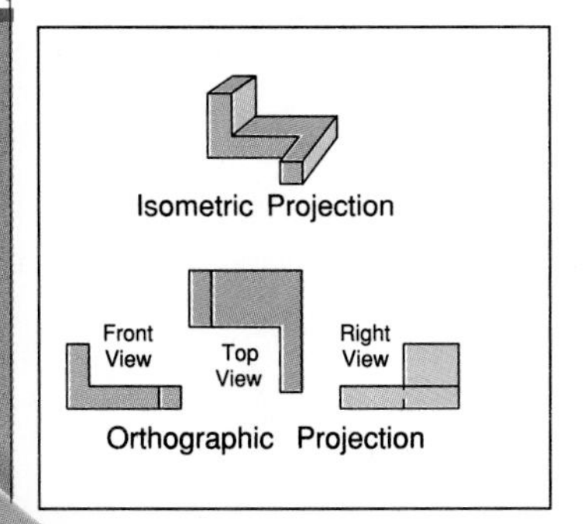

This chapter introduces you to measurement formulas for three-dimensional objects, thus it contains many two-dimensional drawings of three-dimensional shapes. Most of these drawings are *isometric projections* which look realistic, but are not accurate drawings of the sides of the objects.

To provide information about the exact size and shape of a three-dimensional object, scale drawings are made of its top, side and front. These drawings are called *orthographic* (right angle) *projections.*

LESSON INVESTIGATION 12.1

Exploring the Circumference of a Circle

Materials Needed: can or paper-towel tube, ruler, graph paper

As shown at the right, the **diameter** of a circle is the distance across the circle through its **center.** The **circumference** of a circle is the distance around the circle. In this investigation, you will explore the relationship between the diameter and circumference of a circle.

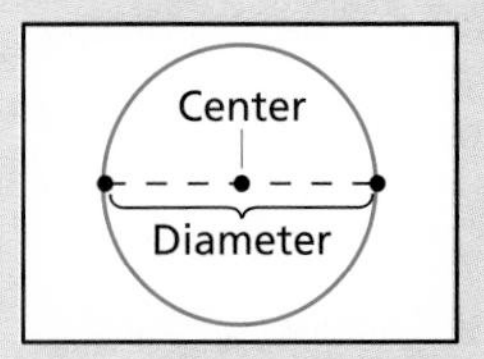

Example — *Comparing Diameter and Circumference*

Mark a point on the rim of a can. Place the marked point on a piece of paper and mark the paper. Then roll the can *without slipping* so it makes exactly one turn and make another mark on the paper. Measure the distance between the marks. Find the ratio of this distance to the diameter of the can.

Solution The measurements you obtain depend on the size of the can. The can shown below has a diameter of 2.5 inches. When it is rolled around once, it rolls 7.9 inches, which means that the circumference of the can is 7.9 inches. For this can, the ratio of the circumference to the diameter is

$$\frac{\text{Circumference}}{\text{Diameter}} = \frac{7.9 \text{ inches}}{2.5 \text{ inches}} = 3.16.$$

In the exercises below, you are asked to repeat this experiment with other circular objects.

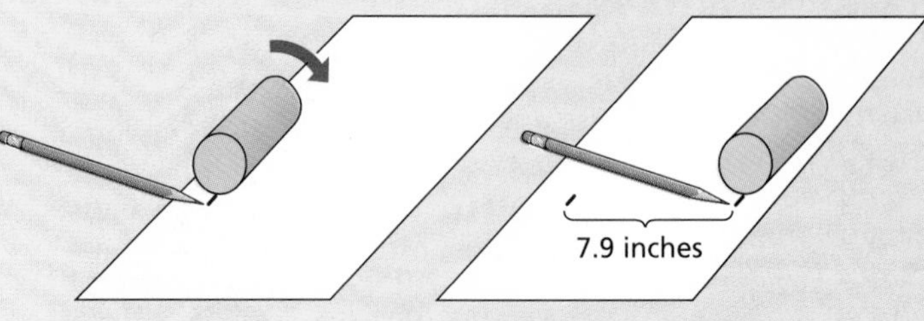

■

Exercises

1. ***Group Activity*** Each person in your group should use a different can, paper-towel tube, or other circular object. If possible, each person should use an object with a different diameter. Repeat the experiment described in the example. Compare the ratios obtained in your group.
2. In your group, discuss the result obtained in Exercise 1. Did everyone obtain about the same ratio as was obtained in the example? If not, the people who obtained different ratios should repeat their experiments. Write a statement about the relationship between the circumference and diameter of a circle.

12.1 Circle Relationships

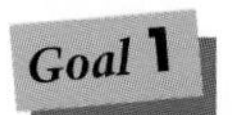
What you should learn:

Goal 1 How to find the circumference of a circle

Goal 2 How to find the area of a circle

Why you should learn it:

You can use the circumference and area of a circle to solve real-life problems, such as finding the circumference of a jar lid.

Goal 1 The Circumference of a Circle

As shown at the right, the **diameter** of a circle is the distance across the circle through its **center.** The **radius** of a circle is the distance from the center to any point on the circle. The **circumference** of a circle is the distance around the circle.

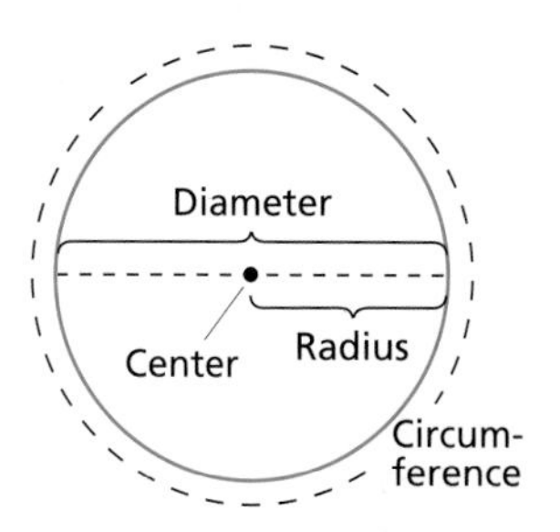

In the investigation on page 538, you may have discovered a special relationship that is true of *all circles.* The ratio of the circumference of any circle to its diameter is about 3.14. This special number is denoted by the Greek letter *pi,* which is written as π. Pi is an irrational number. To four decimal places, $\pi \approx 3.1416$.

The Circumference of a Circle

Let d be the diameter of a circle and let r be its radius. The circumference, C, of the circle is

$$C = \pi d \qquad \text{or} \qquad C = 2\pi r.$$

Example 1 *Finding the Circumference of a Circle*

Real Life
Packaging

Find the circumference of the jar lid at the left.

Solution Because the jar lid has a diameter of 10 centimeters, it follows that its circumference is

$C = \pi d$ *Formula for circumference*

$\approx 3.14(10)$ *Substitute for π and d.*

$= 31.4$ *Simplify.*

The circumference of the jar lid is about 31.4 centimeters. If you need a more accurate measurement for the circumference, you can use a scientific calculator, as shown on page 543. ■

Goal 2 Finding the Area of a Circle

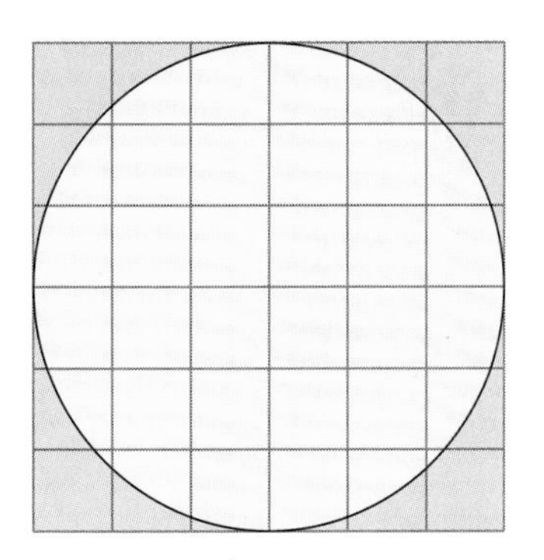

The circle at the left has a radius of 3 units. To estimate the area of the circle, you can reason that each blue corner region has an area of about 2 square units. Because the area of the square is 6^2 or 36 square units, you can reason that the area of the circle must be about $36 - 4(2)$ or 28 square units. The following formula tells you that the *exact* area of the circle is 9π or about 28.3 square units.

The Area of a Circle

Let r be the radius of a circle. The area, A, of the circle is

$A = \pi r^2$.

Example 2 ***Finding the Area of a Circle***

Find the area of the presidential seal at the left.

4.75 in.

Solution The diameter of the seal is 4.75 inches. This implies that the radius is

$r = \frac{1}{2}(4.75) = 2.375$ inches.

Using this measurement, you can find the area of the seal.

$A = \pi r^2$ *Formula for area of a circle*

$\approx 3.14(2.375)^2$ *Substitute for π and r.*

≈ 17.7 *Simplify.*

The area of the seal is about 17.7 square inches. ■

Communicating about MATHEMATICS

SHARING IDEAS about the Lesson

Parts of Circles Each of the circles below has a diameter of 4 inches, and is divided into congruent parts. Find the area of the blue portion of each circle.

A. 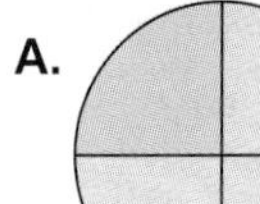 **B.** 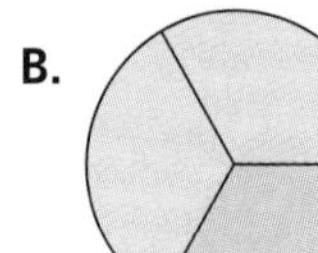 **C.** 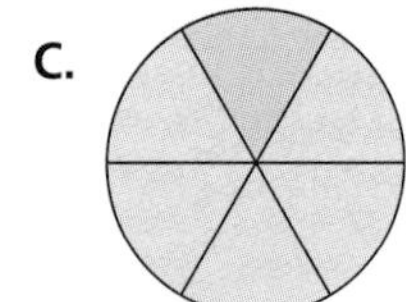

EXERCISES

Guided Practice

CHECK for Understanding

1. How does the formula for the circumference of a circle relate to the Lesson Investigation on page 538?
2. ***Estimation*** Earth has a radius of about 4000 miles. Estimate its circumference at the equator.

In Exercises 3–6, use the figure at the right.

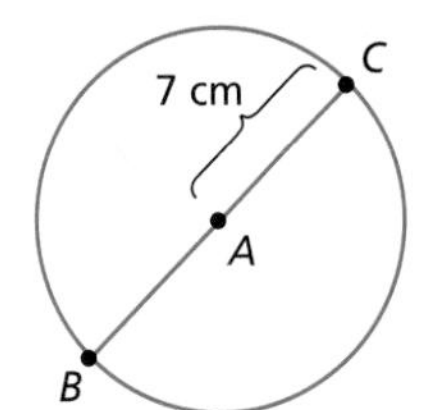

3. Name the diameter.
4. Name a radius.
5. Find the circumference.
6. Find the area.

Independent Practice

In Exercises 7–10, find the circumference and area of the figure. Use 3.14 for π. Round your result to one decimal place.

7. 
$d = 2.4$ cm

8. 
$d = 11.8$ in.

9. 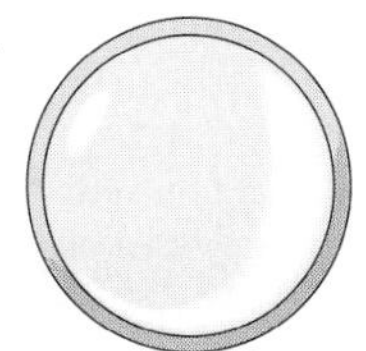
$r = 4$ in.

10. 
$r = 8$ in.

In Exercises 11–14, find the radius and diameter of the figure. Use 3.14 for π. Round your result to one decimal place.

11. 
$C = 4.4$ ft

12. 
$A = 415.48$ in.^{2}

13. 
$A = 17.7$ in.^{2}

14. 
$C = 59.69$ mm

In Exercises 15–18, find the area of the blue portion of the figure. Use 3.14 for π. Round your result to one decimal place.

15. 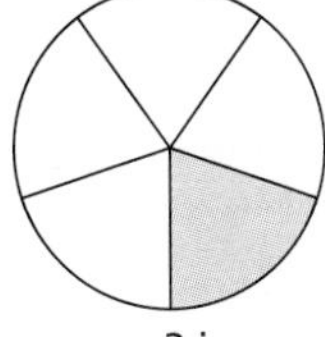
$r = 3$ in.

16. 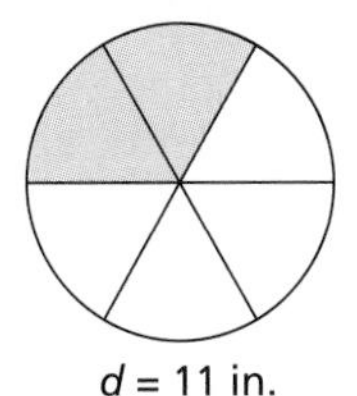
$d = 11$ in.

17. 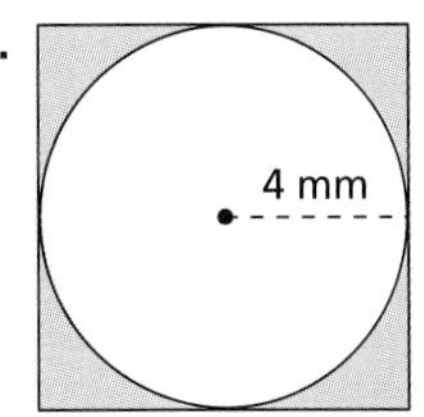

18. 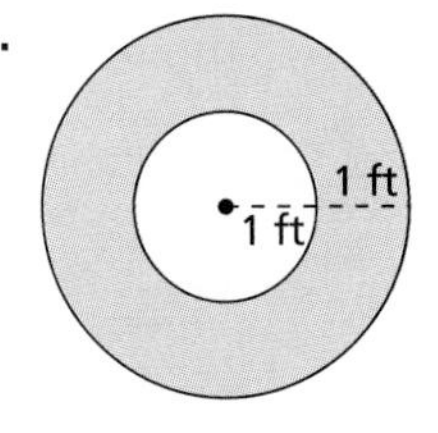

Patterns **In Exercises 19 and 20, create a table. Then describe the pattern.**

19. Find the circumferences of circles whose radii are 1, 2, 3, 4, 5, and 6.

20. Find the areas of circles whose radii are 1, 2, 3, 4, 5, and 6.

21. If the diameter of a circle is doubled, will the area and circumference of the circle double? Explain.

22. Instead of $C = \pi d$, some people prefer to use the formula $C = 2\pi r$ for the circumference of a circle. Which formula do you prefer? Why?

Pizza **In Exercises 23–25, a pizza with a 12-inch diameter is cut into pieces that each have an area of 14.14 square inches.**

23. Find the area of the entire pizza.

24. How many pieces make up the pizza?

25. If the pizza had been cut into 6 pieces, what would have been the area of each?

Washington, D.C. **In Exercises 26–28, use the map at the right.**

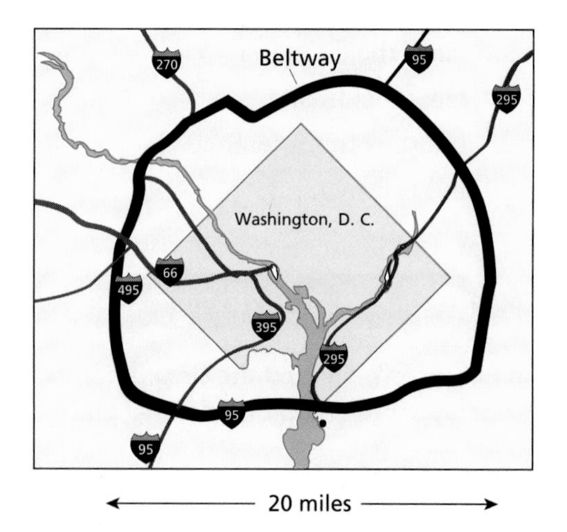

26. The road around Washington, D.C., is called the *Beltway.* Estimate the length of a trip on the Beltway around the entire city.

27. Estimate the area inside the Beltway.

28. *Estimation* About what percent of the region inside the Beltway is Washington, D.C.?

Integrated Review

Making Connections within Mathematics

29. *Intersecting Circles* Two circles of the same size can have no points in common, 1 point in common, 2 points in common, or they may coincide (have all points in common). Draw these four cases.

30. *Intersecting Circles* Draw the different ways that 3 circles of the same size can intersect.

Exploration and Extension

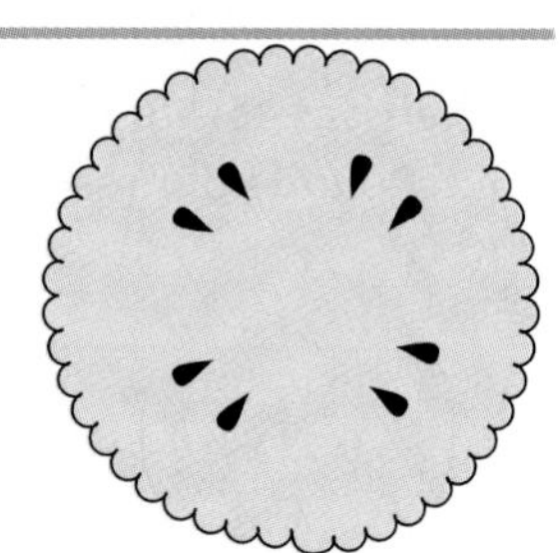

31. *Pie Puzzle* Using four straight cuts, you can cut a pie into eight congruent pieces. Can you cut a pie into eight congruent pieces using only three straight cuts? (You can rearrange the pieces after each cut.) If you can, show how.

32. *Pie Puzzle* Using only two straight cuts and one curved cut, can you cut a pie into eight pieces so that each piece has about the same area? (You can't rearrange the pieces after cutting.) Explain.

USING A CALCULATOR

Circumference and Area of a Circle

Many calculators have a special key for π. The following example shows how to use this key to calculate the circumference and area of a circle.

Example — *Calculating Circumference and Area*

Use a scientific calculator to find the circumference and area of the circle at the right. Round your result to two decimal places.

r = 2.6 in.

Solution The radius of the circle is 2.6 inches. This implies that the diameter is 2(2.6) or 5.2 inches.

$C = \pi d = 5.2\pi$ *Circumference*

$A = \pi r^2 = \pi(2.6)^2$ *Area*

The keystrokes for evaluating the circumference and area are as follows.

Keystrokes	Display	Conclusion
5.2 [×] [π] [=]	16.3362818	$C \approx 16.34$ in.
[π] [×] 2.6 [x^2] [=]	21.23716634	$A \approx 21.24$ in.^{2}

■

Exercises

In Exercises 1–4, find the circumference and area of the indicated circle. Round your results to two decimal places.

1. $r = 1.7$ cm **2.** $r = 5.5$ ft **3.** $d = 3.9$ in. **4.** $d = 10.4$ m

5. A circle has a circumference of 10 inches. Find the diameter and radius of the circle. Round your results to two decimal places.

6. A circle has an area of 6 square meters. Find the radius and diameter of the circle. Round your results to two decimal places.

In Exercises 7–10, find the area of the blue portion of the circle. Round your result to two decimal places.

7.

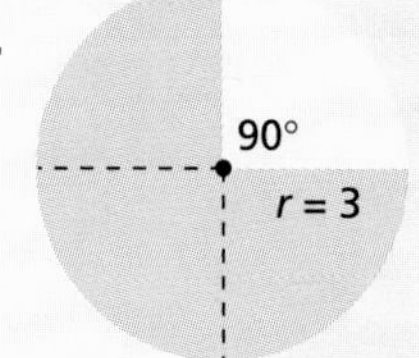

8.

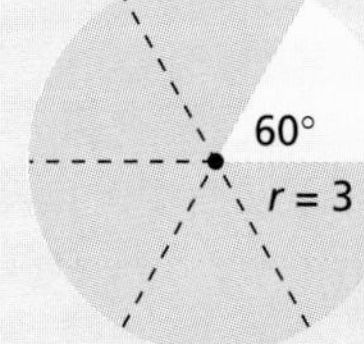

9.

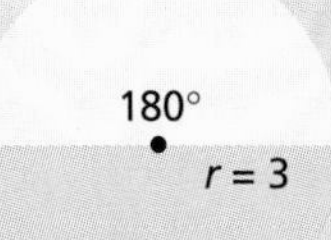

10.

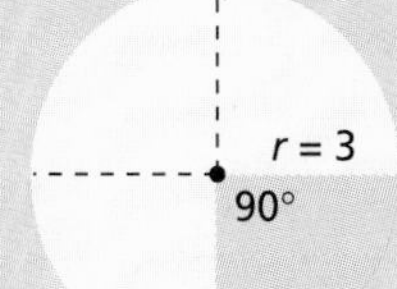

11. Which of the following is the best approximation of π? Explain.

a. 3.14 **b.** 3.1416 **c.** $\frac{22}{7}$ **d.** $\frac{355}{113}$

12.2 Polyhedrons and Other Solids

What you should learn:

How to identify parts of a polyhedron

How to identify parts of cones, cylinders, and spheres

Why you should learn it:

You can use polyhedrons and other solids as models for real-life objects, such as a stick of pepperoni or an apple.

Goal 1 Identifying Parts of Polyhedrons

A **polyhedron** is a solid that is bounded by polygons, which are called **faces.** The segments where the faces meet are **edges,** and the points where the edges meet are **vertices.** Two common types of polyhedrons are **prisms** and **pyramids.**

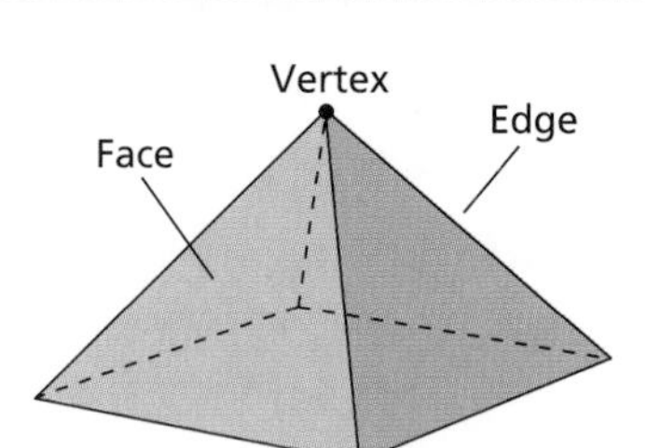

Pyramid with 5 vertices, 5 faces, and 8 edges

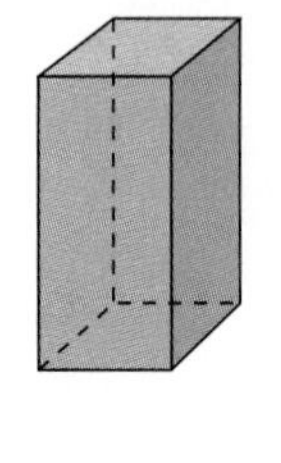

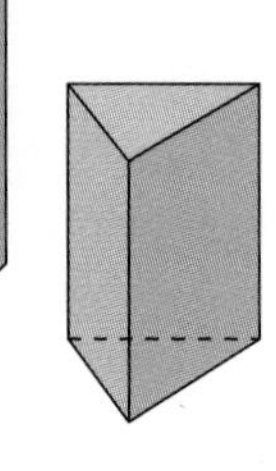

Prisms

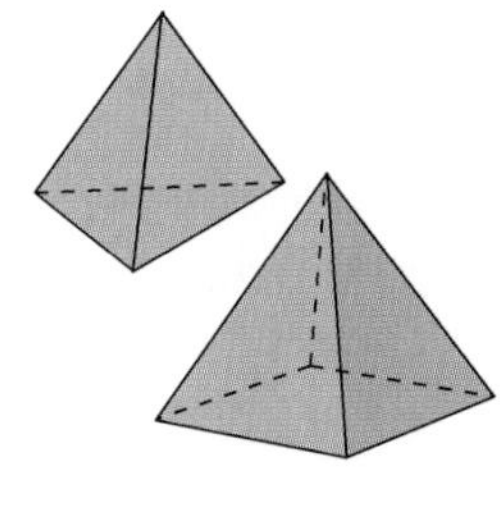

Pyramids

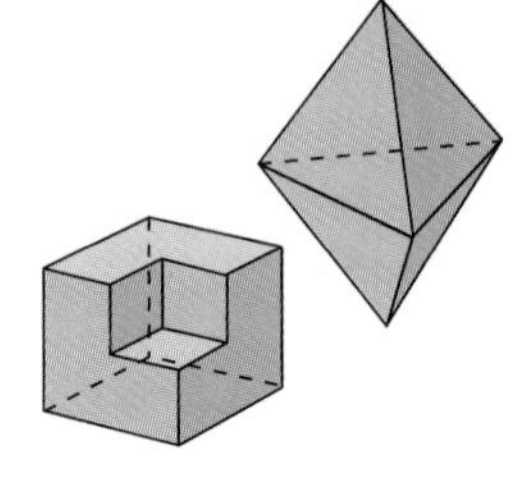

Other polyhedrons

A **net** is a pattern that can be folded to form a solid. The next example shows nets for a prism and a pyramid.

Example 1 *Identifying Parts of Polyhedrons*

Describe the polyhedron that results from folding each net.

a.

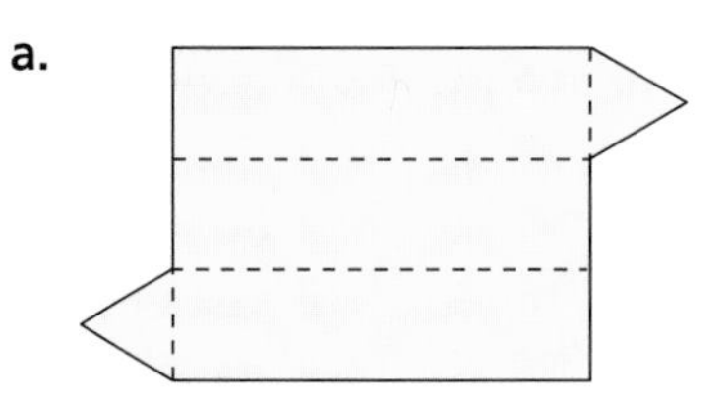

b.

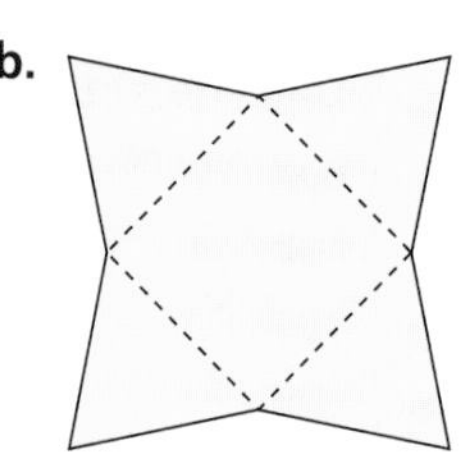

a.

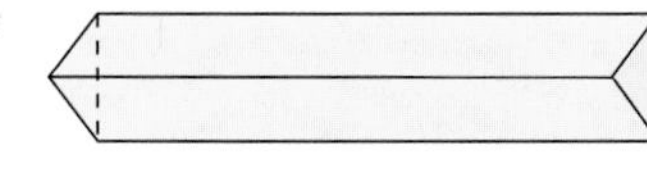

b.

Solution

a. When the net is folded, it forms a prism. The prism has 5 faces, 9 edges, and 6 vertices, as shown at the left.

b. When the net is folded, it forms a pyramid. The pyramid has 5 faces, 8 edges, and 5 vertices, as shown at the left. ■

Goal 2 Identifying Parts of Other Solids

Three other types of common solids, a sphere, a cylinder, and a cone, are shown below.

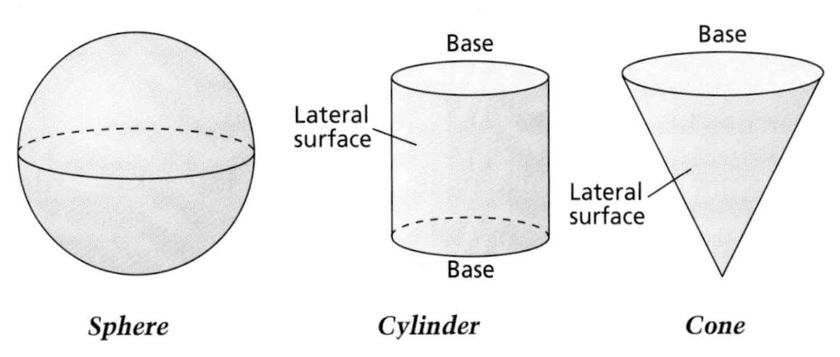

Sphere *Cylinder* *Cone*

Real Life
Food Shapes

Example 2 *Identifying Solids*

Describe the approximate shape of each food before it is cut. Then describe the shape of the cut portions.

a.

b.

Solution

a. Before it is cut, the pepperoni is approximately a cylinder. The shape would be more like a cylinder if the ends were flat instead of rounded. After it is cut, most pieces are almost exact cylinders. The circular bases of each cut piece are large and the lateral surface is small.

b. Before it is cut, the apple is approximately a sphere. The cut piece is not a sphere—it is called a wedge. ■

Communicating about MATHEMATICS

SHARING IDEAS about the Lesson

Describing Solids In Example 2, you can see that a cylinder can be cut to form smaller cylinders. Is this true of other types of solids? For instance, can a prism be cut to form two prisms?

EXERCISES

Guided Practice

CHECK for Understanding

In Exercises 1–5, match the solid with its name.

a. Pyramid **b.** Sphere **c.** Cone **d.** Prism **e.** Cylinder

1. 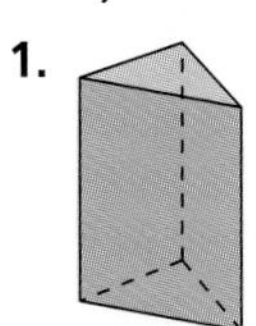

2. 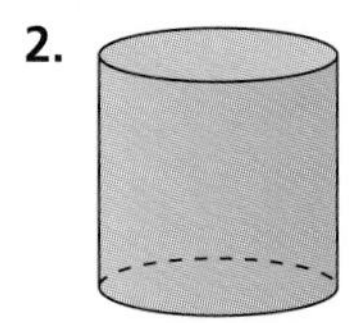

3. 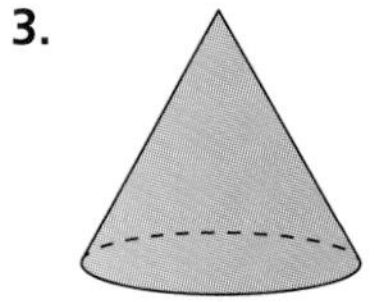

4. 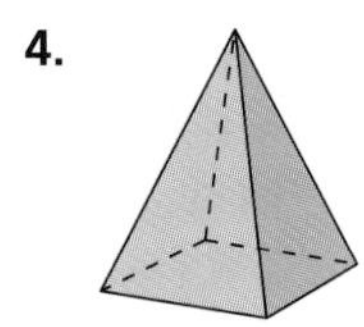

5. 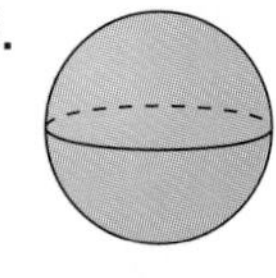

In Exercises 6–9, identify the parts of the solid.

6. 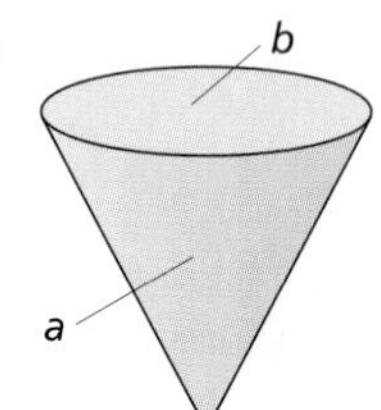

7. 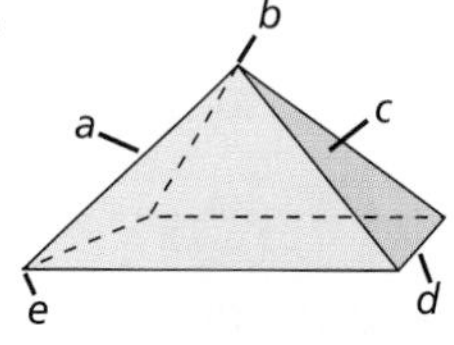

8. 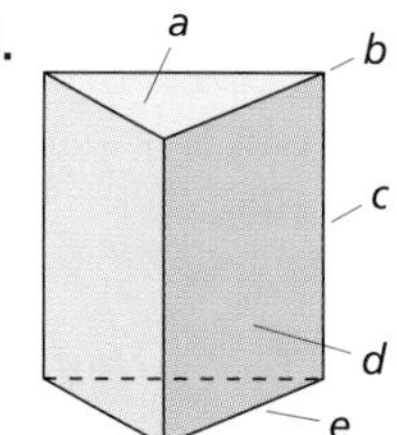

9. 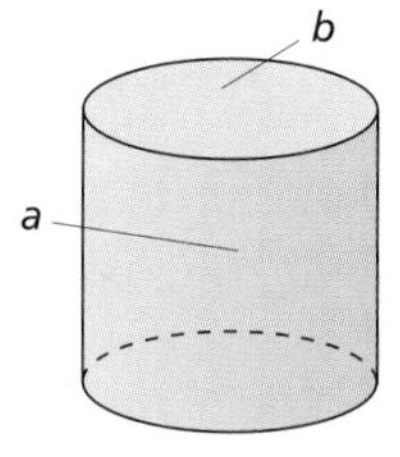

Independent Practice

In Exercises 10–13, identify the solid.

10. 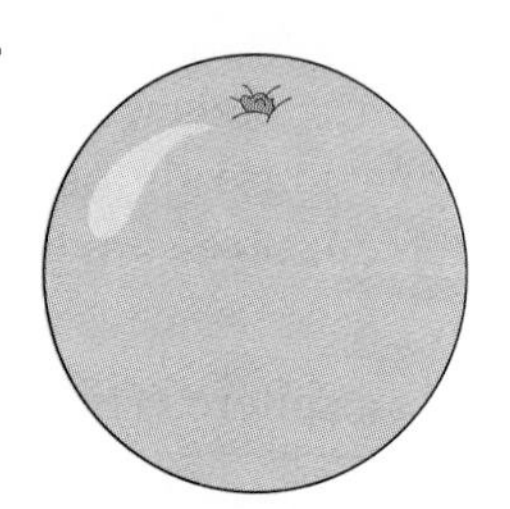

11. 

12. 

13. 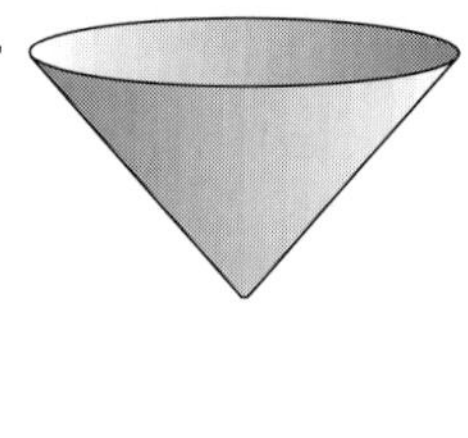

14. How are prisms and cylinders alike?

15. How are pyramids and cones alike?

Nets **In Exercises 16–19, which of the nets can be folded to form a cube?**

16. 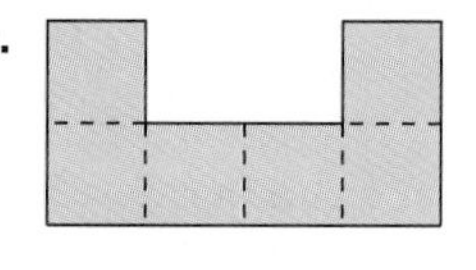

17. 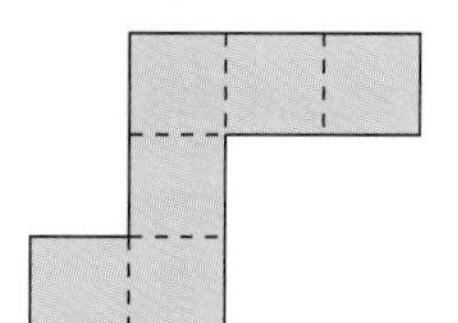

18. 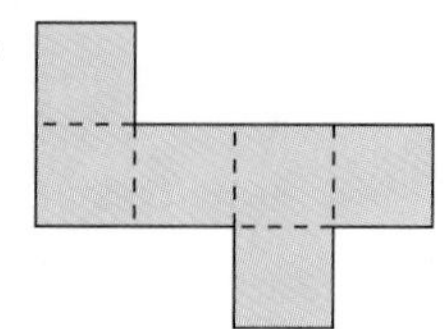

19. 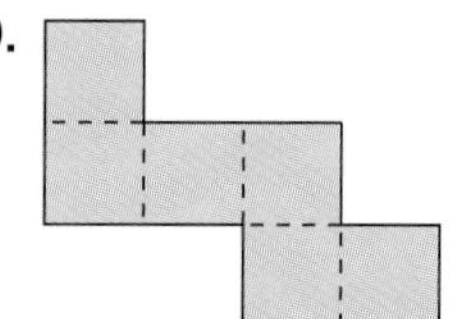

20. ***Think about It*** Sketch other nets that can be folded to form a cube. How many different nets are possible?

Visualizing Solids **In Exercises 21–26, identify the solid formed by folding the net.**

21.

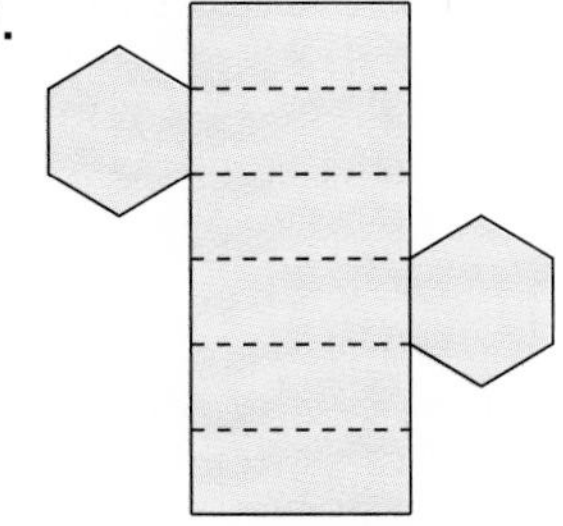

22.

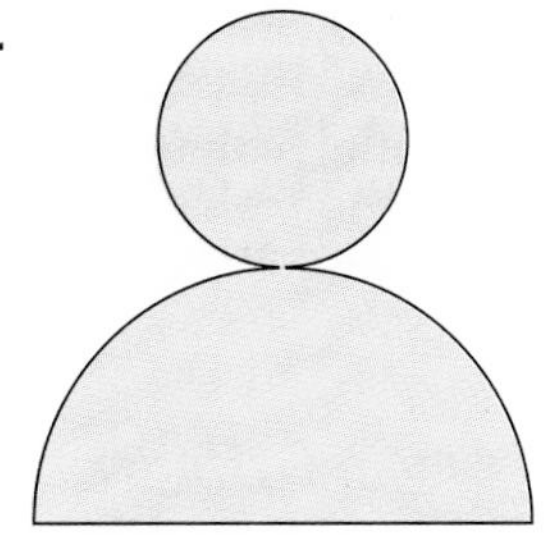

23.

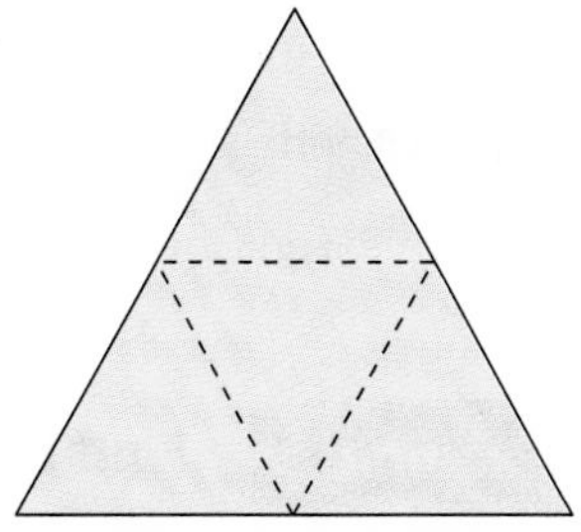

24.

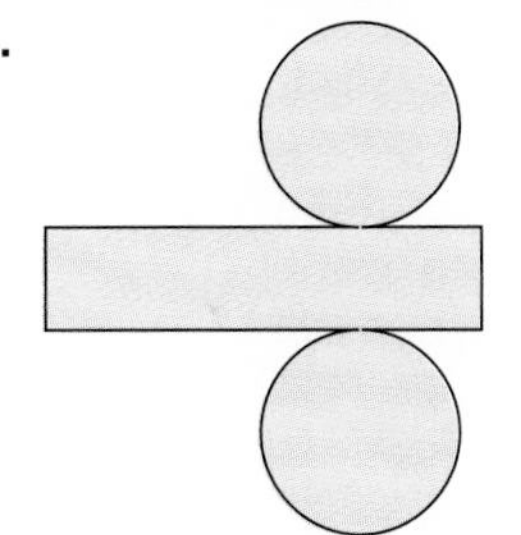

25.

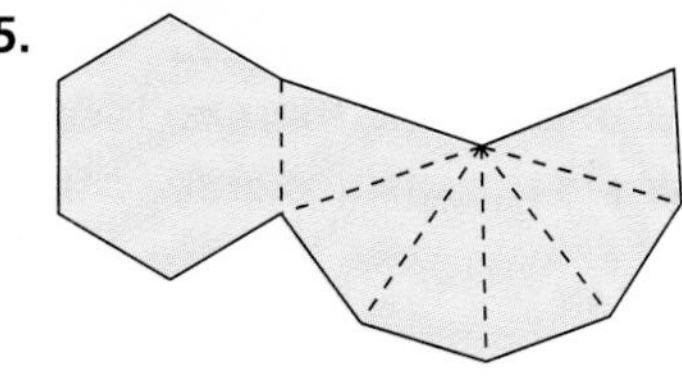

26.

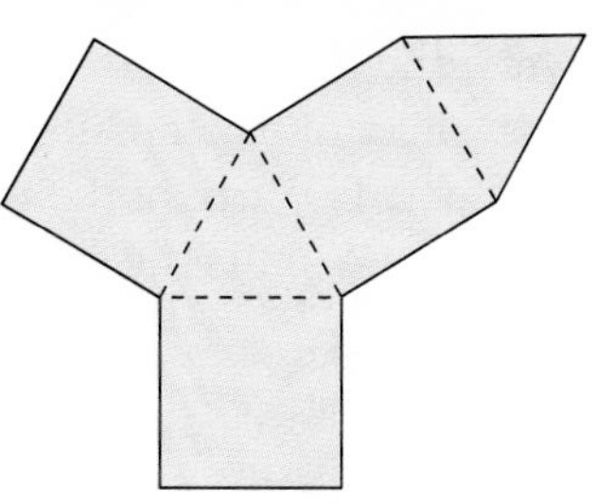

Integrated Review

Making Connections within Mathematics

Perimeter **In Exercises 27–30, find the perimeter (or circumference) and the area of the figure.**

27.

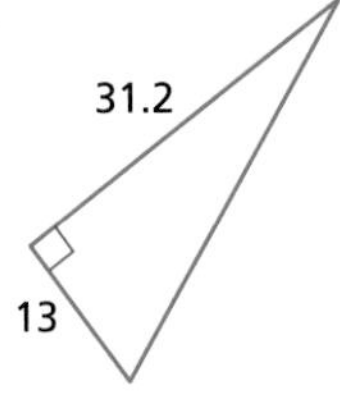

28.

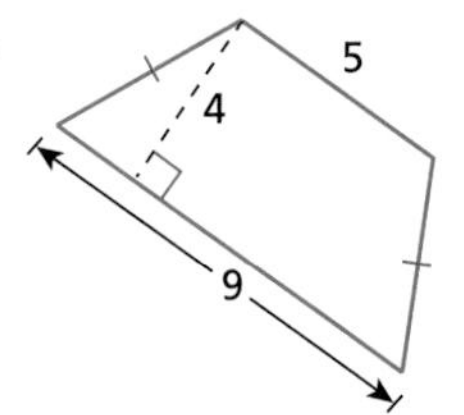

29.

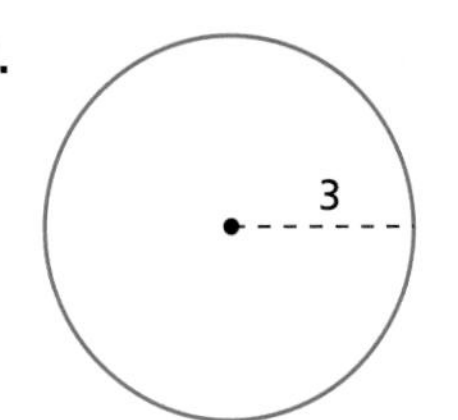

30.

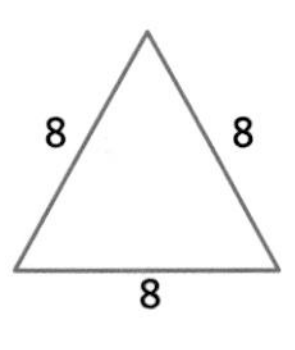

Exploration and Extension

31. *Polyhedron Earth* In the photo at the right, a map of the world has been drawn on an *icosahedron.* Five faces meet to form the top vertex. How many faces does the solid have?

32. Each face of an *octahedron* is an equilateral triangle. From its name, how many faces does an octahedron have?

33. An icosahedron and an octahedron each have faces that are equilateral triangles. Sketch a pyramid whose faces are all equilateral triangles. This type of pyramid is called a *tetrahedron.* How many faces does it have?

LESSON INVESTIGATION 12.3

Exploring the Surface Area of a Cylinder

Materials Needed: paper, compass, scissors, ruler, calculator, transparent tape

In this investigation, you will explore the surface area of a cylinder.

Example — *Find the Surface Area of a Cylinder*

Draw the three pieces shown below on an $8\frac{1}{2}$-inch by 11-inch sheet of paper. Cut the pieces out and tape them to form a cylinder. The area of the three pieces is the *surface area* of the cylinder. What is the surface area?

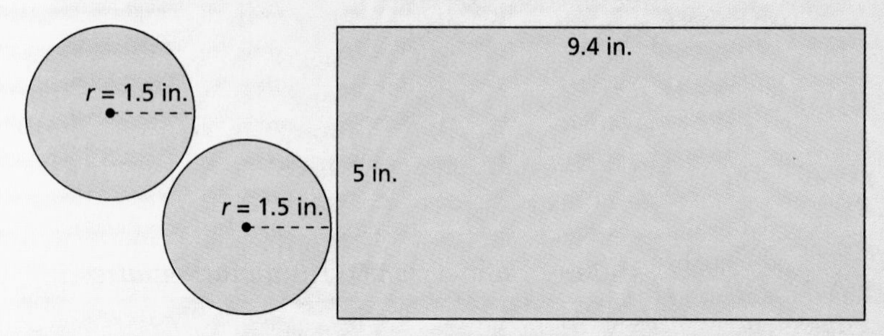

Solution The circles each have an area of $\pi \cdot 1.5^2$ or about 7.07 square inches. The area of the rectangle is $5 \cdot 9.4$ or 47 square inches. The total area of the three pieces is

$$2(\pi \cdot 1.5^2) + (5 \cdot 9.4) \approx 61.14 \text{ square inches.}$$

After the pieces are cut and taped together they form a cylinder, as shown at the right. ■

Exercises

1. ***Group Activity*** With other members of your group, use an $8\frac{1}{2}$-inch by 11-inch sheet of paper to create a cylinder, as shown in the example. All three pieces must fit on a single piece of paper. Your goal is to create a cylinder that has the largest possible surface area. After creating your cylinder, find its surface area. Then compare your group's results with the results of other groups.
2. In Exercise 1, suppose you had been asked to create the "biggest" possible cylinder. What different interpretations could be made of the word "biggest"?
3. When you positioned the rectangle and circles on the piece of paper, which did you draw first: the rectangle or the circles? If you drew the circles first, how did you decide how to make a rectangle that fit the circles? If you drew the rectangle first, how did you decide how to make circles that fit the rectangle?

12.3 Exploring Surface Area of Prisms and Cylinders

What you should learn:

Goal 1 How to find the surface area of a prism and a cylinder

Goal 2 How to use surface area to answer questions about real life

Why you should learn it:

You can use surface area to answer questions about real life, such as comparing the amount of material used to make cereal containers.

Goal 1 Finding Surface Area

The **surface area** of a polyhedron is the sum of the areas of its faces. The figures below show how to find the surface area of a prism and a cylinder. Notice that in each case you obtain the surface area by adding the area of the vertical "sides" to the areas of the top and bottom.

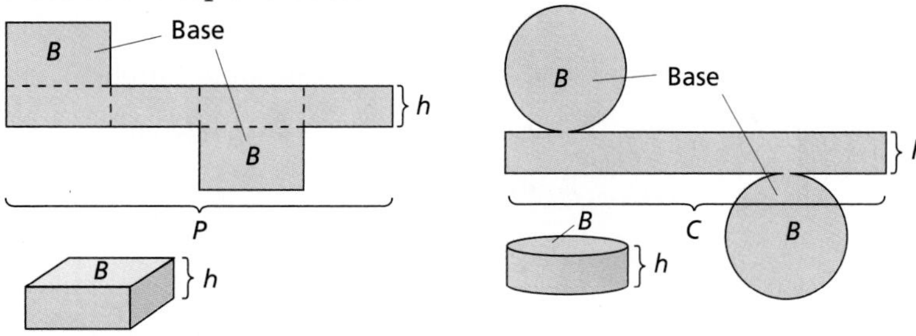

Surface Area of Prism and Cylinder

Prism: The surface area is $S = 2B + Ph$, where B is the area of a base, P is the perimeter of a base, and h is the height of the prism.

Cylinder: The surface area is $S = 2B + Ch$, where B is the area of a base, C is the circumference of a base, and h is the height of the cylinder.

Example 1 *Finding Surface Area*

Find the surface area of the cylinder at the left.

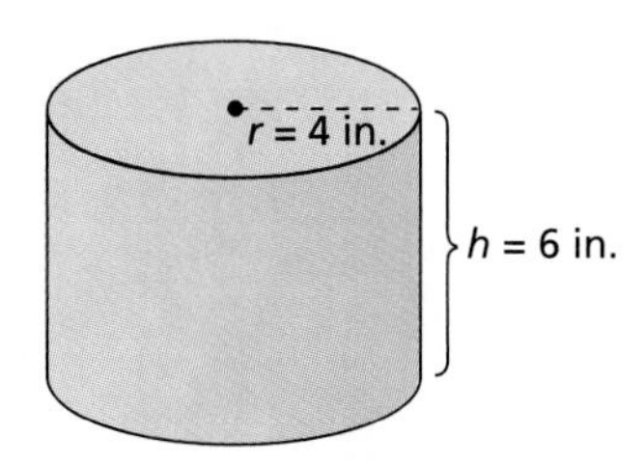

Solution The radius of each base is 4 inches, which means that each base has a surface area of $B = \pi r^2$ or 16π square inches. The circumference of each base is $C = 2(4)\pi$ or 8π inches, and the height is $h = 6$ inches. The surface area is

$S = 2B + Ch$	*Formula for surface area*
$= 2(16\pi) + (8\pi)(6)$	*Substitute for B, C, and h.*
$= 32\pi + 48\pi$	*Simplify.*
$= 80\pi$	*Simplify.*
≈ 251.3	*Use a calculator.*

The surface area is about 251.3 square inches. ■

Goal 2 Solving Real-Life Problems

Example 2 *Comparing Surface Areas*

Real Life
Package Design

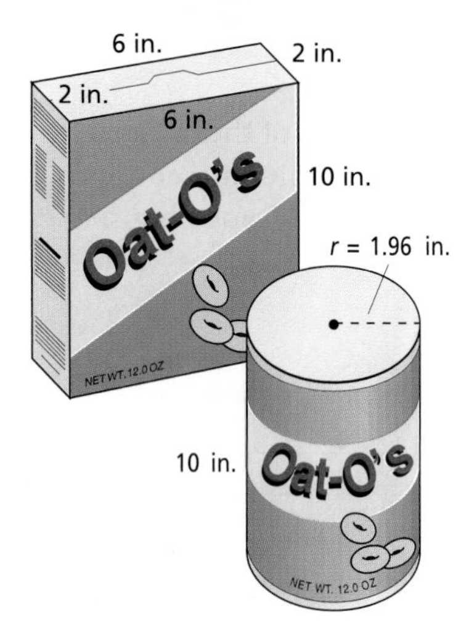

The two cereal containers at the left hold about the same amount of cereal. Which container uses less material?

Solution You can compare the amount of material needed to make each container by comparing their surface areas. The container with the smaller surface area uses less material. The area of each base of the prism is $2 \cdot 6$ or 12 square inches. Thus, the surface area of the prism is

$S = 2B + Ph$	*Surface area of prism*
$= 2(12) + (16)(10)$	*Substitute.*
$= 184$ square inches.	*Simplify.*

The area of each base of the cylinder is $\pi 1.96^2$ or about 12.07 square inches. Thus, the surface area of the cylinder is

$S = 2B + Ch$	*Surface area of cylinder*
$\approx 2(12.07) + (2 \cdot 1.96 \cdot \pi)(10)$	*Substitute.*
≈ 147.3 square inches.	*Simplify.*

Thus, the cylindrical package has less surface area, which implies that it uses less material. ■

Communicating about MATHEMATICS

SHARING IDEAS about the Lesson

Comparing Surface Areas Each of the following containers holds the same amount of liquid. Just from looking at the containers, which do you think has the least surface area? Find the surface area of each. Did you guess correctly?

A.

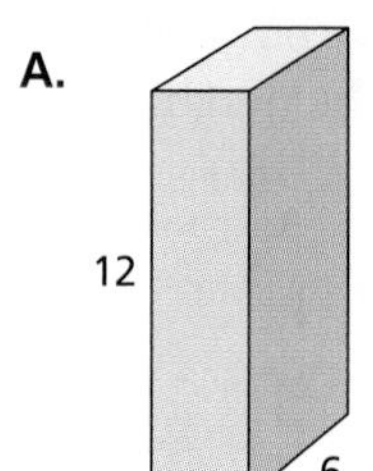

B.

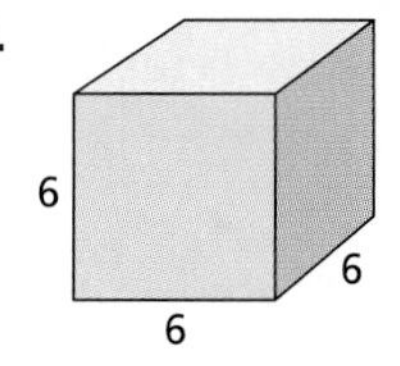

C.

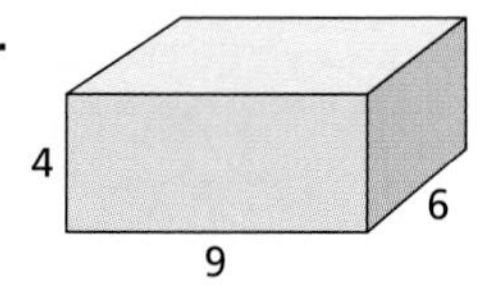

EXERCISES

Guided Practice

CHECK for Understanding

In Exercises 1–4, use the following figures.

A.

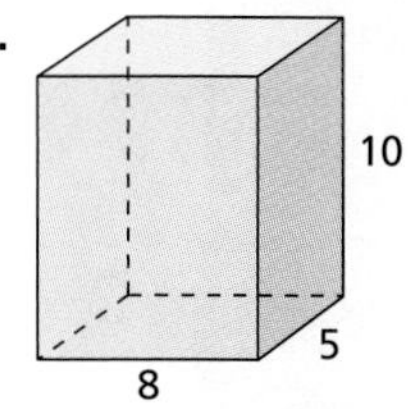

B.

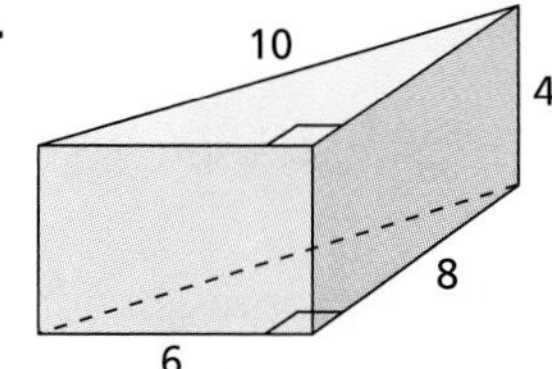

C.

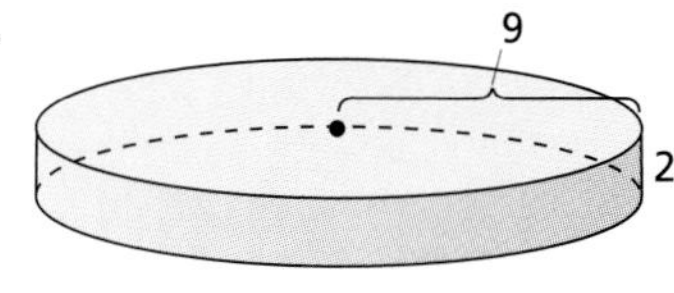

1. Identify each figure.

2. Find the surface area of Figure A.

3. Find the surface area of Figure B.

4. Find the surface area of Figure C.

Independent Practice

In Exercises 5–12, find the surface area of the solid.

5.

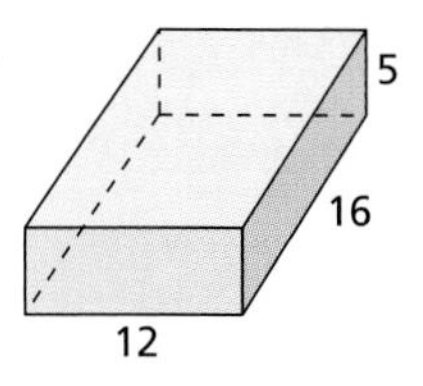

6.

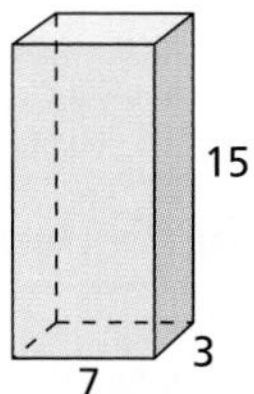

7.

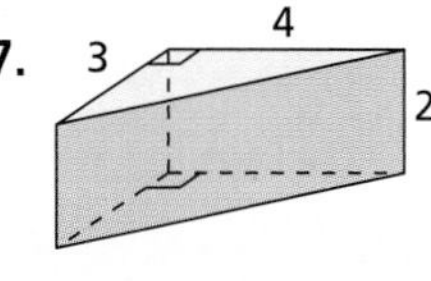

8.

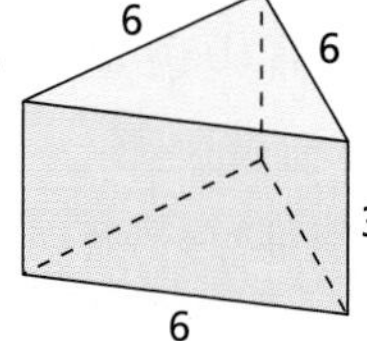

9.

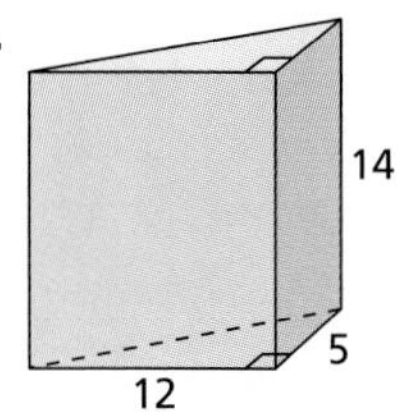

10.

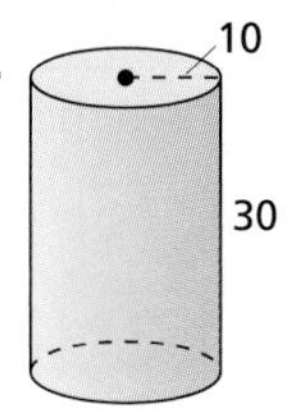

11.

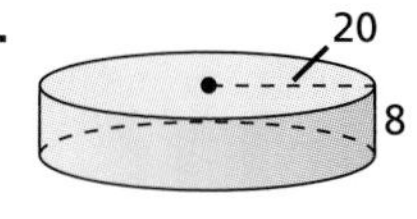

12.

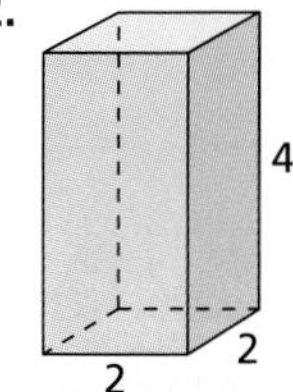

13. Draw a cube. Use dashed lines and shading to make the cube appear three dimensional. If each edge of the cube is 4 inches long, what is its surface area?

14. *Guess, Check, and Revise* Draw a prism that has a surface area of 52 square units. (There is more than one correct answer.)

In Exercises 15 and 16, use the cube at the right.

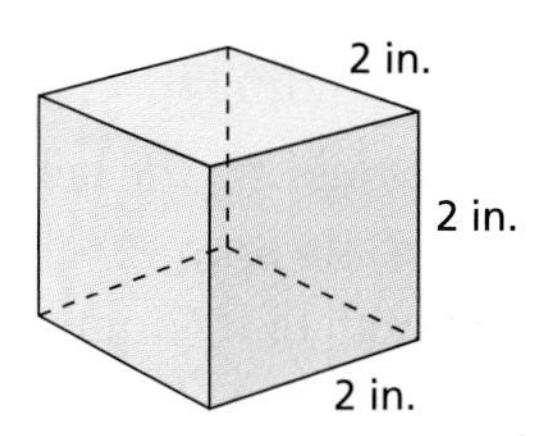

15. Find the surface area of the cube.

16. Imagine that the cube is cut into eight congruent smaller cubes. Find the surface area of each.

Gift Wrapping **In Exercises 17 and 18, consider a gift box that measures 45 centimeters by 27 centimeters by 6 centimeters.**

17. Which sheet of wrapping paper should you choose to wrap the gift? Explain.

a. 67 cm, 50 cm

b.

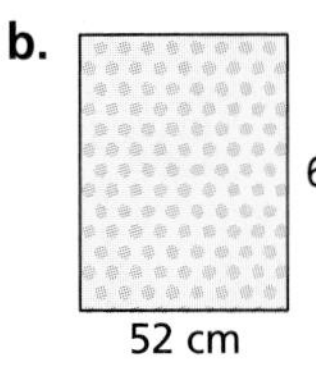

66 cm, 52 cm

c.

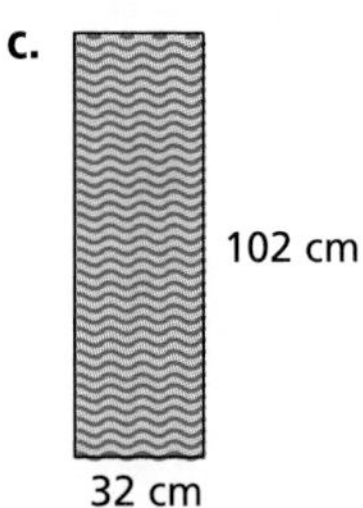

102 cm, 32 cm

18. Describe the smallest rectangular piece of wrapping paper that could be used to wrap the box. Compare its area with the surface area of the box.

Totem Poles **In Exercises 19 and 20, use the following.**

A totem pole is to be carved out of a cylindrical log. The log is 22 feet long and has a diameter of 4 feet.

19. What is the surface area of the cylindrical pole?

20. Will the surface area of the totem pole be greater than or less than the surface area of the log? Explain.

Native American tribes in the Pacific Northwest carved their family and clan emblems on totem poles.

Integrated Review

Making Connections within Mathematics

21. *Estimation* Which best estimates the surface area of this book?

a. 100 in.^{2} **b.** 170 in.^{2} **c.** 220 in.^{2} **d.** 300 in.^{2}

22. *Estimation* Which best estimates the surface area of a 12-ounce soda pop can?

a. 100 in.^{2} **b.** 170 in.^{2} **c.** 220 in.^{2} **d.** 300 in.^{2}

Exploration and Extension

True or False? **In Exercises 23–26, is the statement true or false? Explain your answer by drawing any necessary diagrams.**

23. Doubling the width of a shoe box doubles the base area of the box.

24. Doubling the height of a shoe box doubles the surface area of the box.

25. Doubling the radius of a cylinder doubles the surface area of the cylinder.

26. Doubling the height of a cylinder doubles the area of its curved surface.

27. *Think about It* What happens to the surface area of a shoe box if each of its dimensions is doubled? Give an example.

Mixed REVIEW

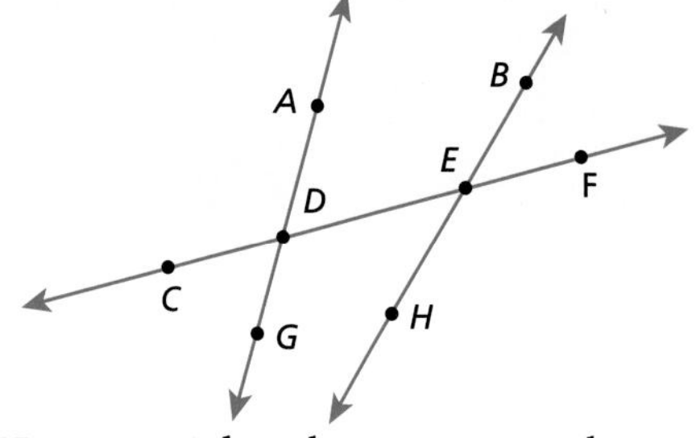

In Exercises 1–7, use the figure at the right. (10.1–10.2)

1. How are $\angle CDA$ and $\angle DEB$ related?
2. Which angle is congruent to $\angle GDE$?
3. Is $\overrightarrow{DA}$ a segment, ray, or line?
4. List 2 other names for $\overleftrightarrow{BH}$.
5. Is $\angle DEB$ acute, right, obtuse, or straight?
6. If $m\angle ADE = 60°$, what is $m\angle CDG$?
7. If $m\angle ADE = 60°$, what is $m\angle ADC$?

In Exercises 8–11, find the area of the indicated circle. (12.1)

8. $r = 8$
9. $r = 13$
10. $d = 8$
11. $d = 13$

In Exercises12–15, find both solutions. (9.1)

12. $x^2 = 16$
13. $169 = y^2$
14. $201.64 = p^2$
15. $m^2 = \frac{25}{9}$

Milestones DESCRIBING MOTION

When the two great mathematicians, Isaac Newton (1642–1727) and Gottfried von Leibniz (1646–1716), developed calculus, they were foreshadowing the development of animation and video games. Just as each frame of a cartoon shows a slight change of movement from the previous one, calculus describes the paths of moving objects by breaking the movements down into smaller and smaller increments.

Even though Walt Disney (1901–1966) did not invent animation, his name has become synonymous with it. At 16 he helped make cartoon ads for movies. In 1923, he moved to Los Angeles but could not find movie work. So, he set up a studio in his garage and in 1928 made the classic *Steamboat Willie* cartoon.

- *Think of the minute hand of a clock. Describe the path its tip makes in one hour.*
- *Think of the hour hand of a clock. Describe the path its tip makes in one hour.*

Mickey Mouse with his creator, Walt Disney

12.4 Exploring Volumes of Prisms

What you should learn:

How to find the volume of a prism

How to use the volume of a prism to solve real-life problems

Why you should learn it:

You can use the volume of a prism to solve real-life problems, such as designing a television studio.

Study Tip...

The formula for the volume of a rectangular prism, $V = lwh$, is a special case of the formula for a general prism. For instance, in the rectangular prism below, notice that the area of each base is $B = lw$, which means that the volume is $V = Bh = lwh$.

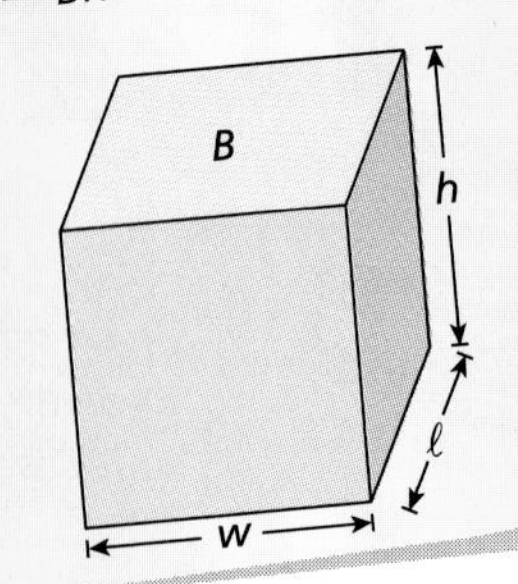

Goal 1 Finding the Volume of a Prism

The **volume** of a solid is a measure of how much it will hold. The standard measures of volume are **cubic units** such as cubic inches, cubic centimeters, and cubic feet. Other measures of volume, such as liters, are discussed on page 560.

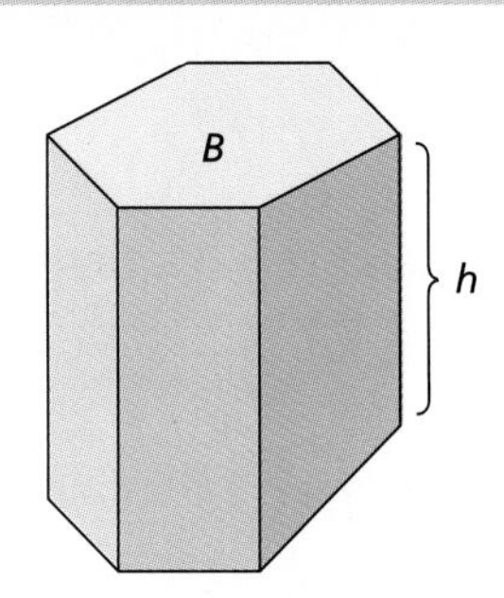

The Volume of a Prism

1. The volume of a prism is the product of its height and the area of its base. That is, $V = Bh$ where B is the area of a base and h is the height of the prism.
2. The volume of a rectangular prism (a prism whose sides are all rectangles) is the product of its length, width, and height. That is, $V = lwh$, where l is the length, w is the width, and h is the height of the prism

Example 1 *Prisms with the Same Volume*

How many different shapes of rectangular prisms can be formed with 64 cubes, each of which is 1 cubic inch?

Solution You need to find all the ways that 64 can be factored into three positive integers. After trying different combinations, you can discover that there are 7 different ways, 6 of which are shown below. (The prism that is 1 by 1 by 64 is not shown.)

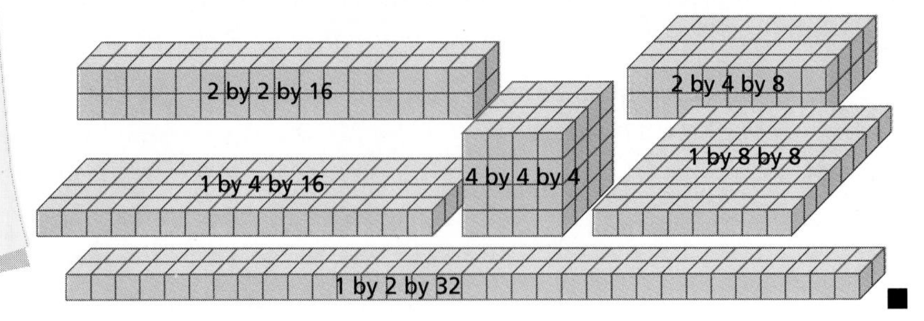

■

Goal 2 Solving Real-Life Problems

Real Life
Room Design

After Lisa Johnson-Smith graduated from Temple University in Philadelphia with a major in journalism, she accepted a job in Washington, D.C., as the host and associate producer of Black Entertainment Television's (BET) new talk show called **Teen Summit.**

Example 2 *Finding the Volume of a Room*

You are designing a studio room for a television talk show. You want the room to have enough seats for an audience of 100. In addition to the audience, the room will have up to 20 staff members and guests. A building code suggests that the room have about 480 cubic feet of air per person. Describe some of the possible room dimensions.

Solution For each of the 120 people to have 480 cubic feet of air space, the room should have a volume of at least

$120(480) = 57{,}600$ cubic feet. *Volume of room*

Typical room heights range between 7 and 12 feet. If the studio is 12 feet high, then the area of its base should be $\frac{1}{12}(57{,}600)$ or 4800 square feet. One possible solution would be to have a base that is 60 feet by 80 feet, as shown below.

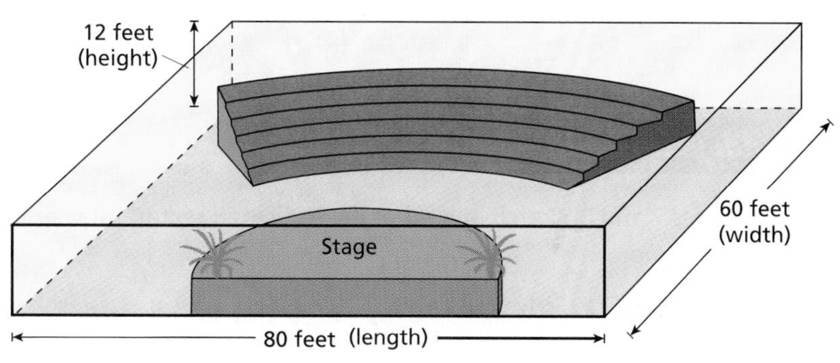

Volume of Room = (length)(width)(height) = 57,600 cubic feet ■

Communicating about MATHEMATICS

Cooperative Learning

SHARING IDEAS about the Lesson

Designing a Studio Example 2 describes only one of many possible designs for the studio. Work with a partner to sketch some other possible designs. Describe their dimensions, advantages, and disadvantages. Remember that you can change the room's height. Compare the floor space for different room heights.

EXERCISES

Guided Practice

CHECK for Understanding

1. Which of these prisms have the same volume? Explain.

a. 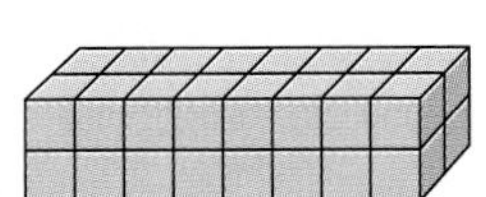

b. 

c. 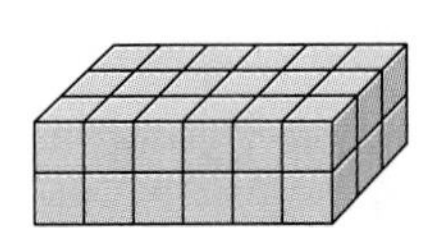

d. 

2. In your own words, state the formula for the volume of a rectangular prism.

In Exercises 3–6, state whether the figure is a prism. If the figure is a prism, find its volume.

3. 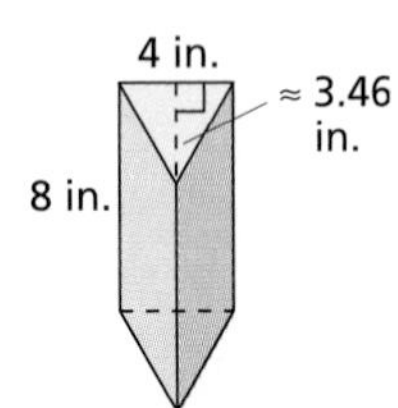

4. 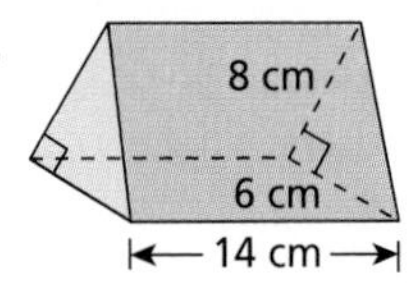

5. 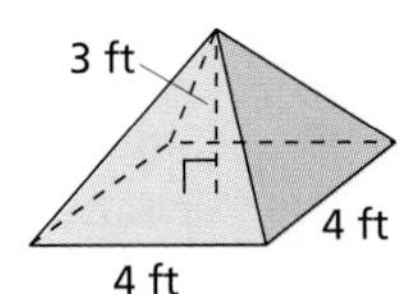

6. 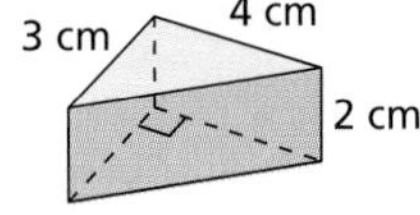

Independent Practice

In Exercises 7 and 8, sketch and label the indicated rectangular prism. Then find its volume.

7. Length: 2 in., Width: 3 in., Height: 5 in.

8. Length: 5 in., Width: 4 in., Height: 6 in.

In Exercises 9–12, find the volume of the prism.

9. 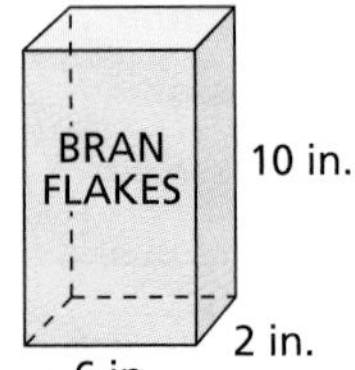

10. 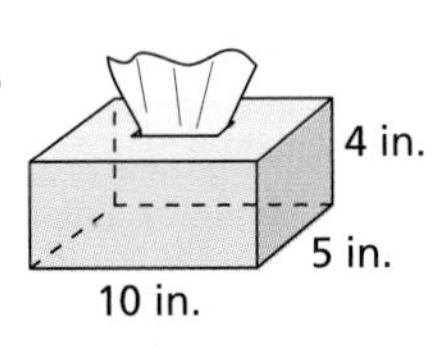

11. 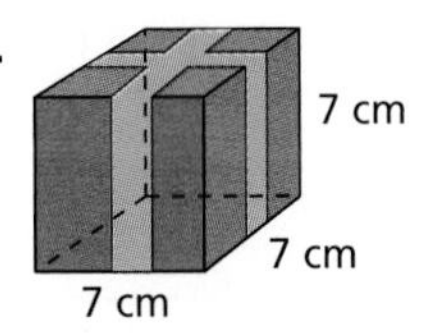

12. 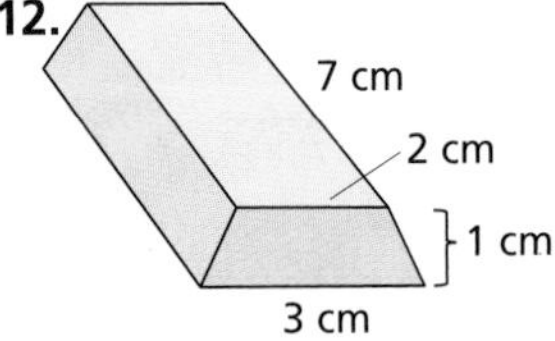

In Exercises 13–16, solve for *x*.

13. $V = 24 \text{ in.}^3$

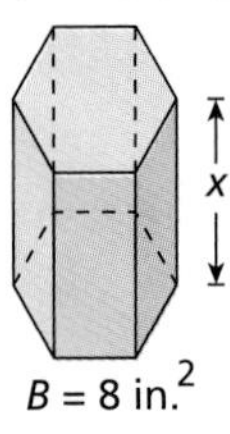

14. $V = 32 \text{ cm}^3$

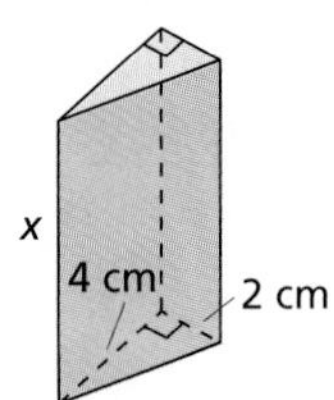

15. $V = 120 \text{ cm}^3$

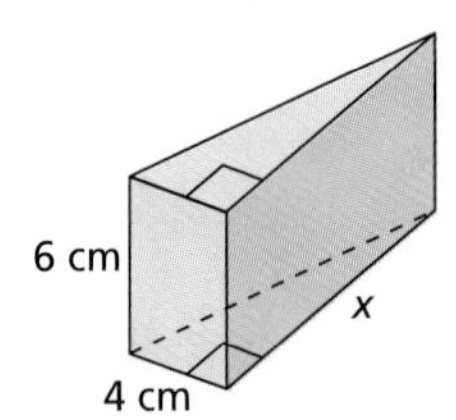

16. $V = 50 \text{ ft}^3$

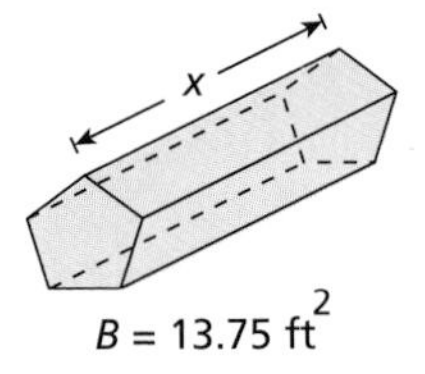

17. *Visualizing Solids* You are given 36 1-inch cubes. Describe the different shapes of rectangular prisms you can build using all the cubes. Of these, which has the greatest surface area?

18. Sketch a rectangular prism that is 2 by 3 by 4. Find its surface area, S, and volume, V.

a. If you double one dimension, how do S and V change?

b. If you double two dimensions, how do S and V change?

c. If you double all three dimensions, how do S and V change?

Designing a Tent **In Exercises 19–22, use the diagram of a tent at the right.**

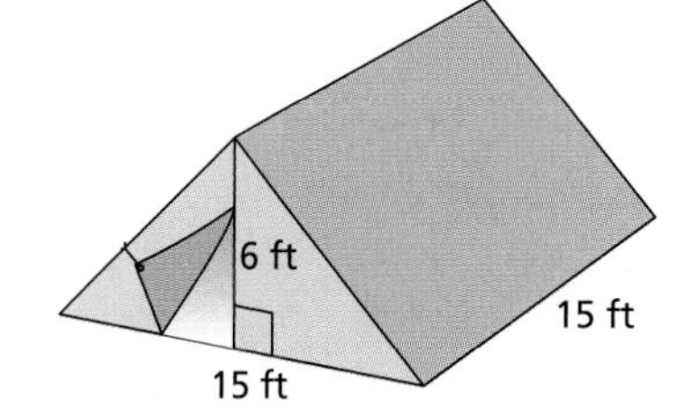

19. How much canvas is used to create the tent? (All sides, including the bottom, are made of canvas.)

20. How many cubic feet of air does the tent hold?

21. Design a different tent that has the same volume but uses less canvas. Which design do you prefer? Explain.

22. *Swimming Pools* Which two pools hold the same amount of water?

a.

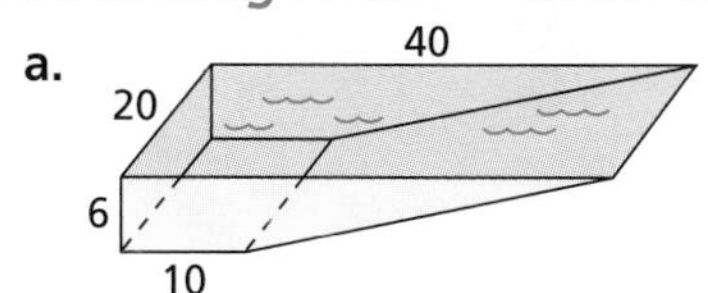

b.

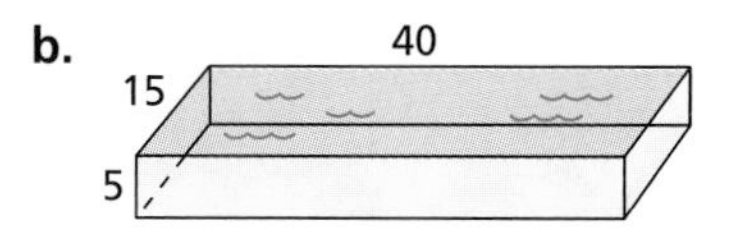

c.

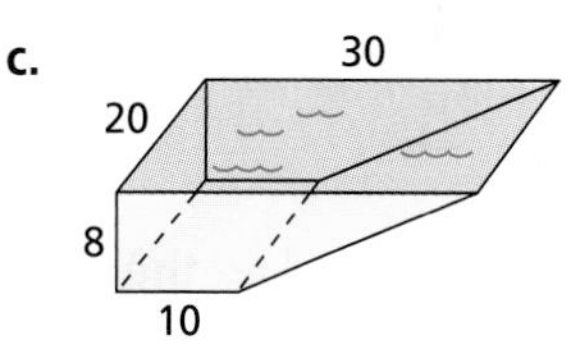

Integrated Review

Making Connections within Mathematics

Coordinate Geometry **In Exercises 23 and 24, plot the points in a coordinate plane. Connect the points and shade the figure so that it looks like a prism. Label the prism's dimensions and find its volume.**

23. $(2, 0), (1, 1), (-2, 0), (-1, 1), (1, -4), (2, -5), (-1, -4), (-2, -5)$

24. $(1, 2), (4, 5), (7, 2), (4, 0), (1, -3), (7, -3)$

Exploration and Extension

Nets **In Exercises 25–27, copy the net and use dashed lines to show how to fold the net to form a prism. Then find its surface area and volume.**

25.

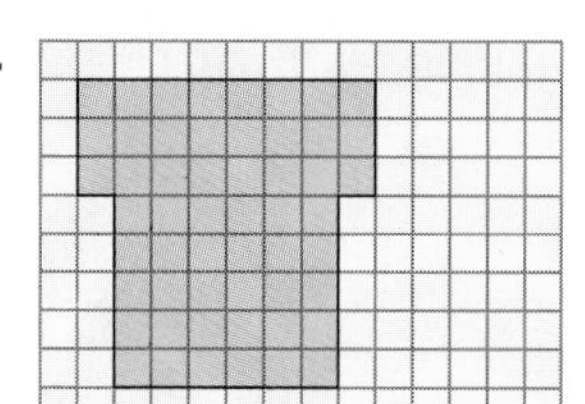

26.

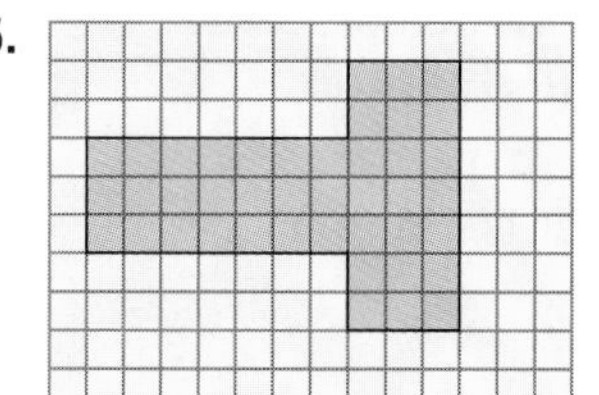

27.

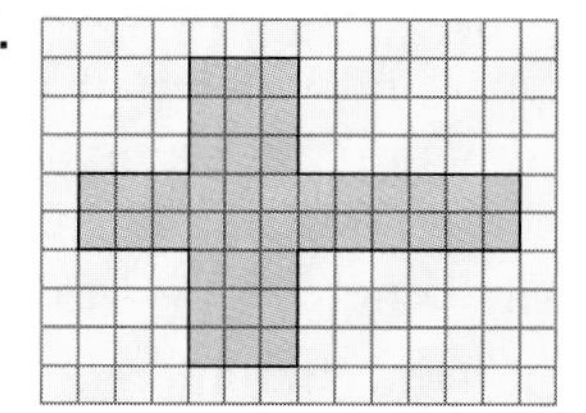

28. *It's Up to You* In Exercises 25–27, is it easier to find the prism's surface area or its volume? Explain your reasoning.

Mid-Chapter SELF-TEST

Take this test as you would take a test in class. The answers to the exercises are given in the back of the book.

In Exercises 1 and 2, find the circumference and area of the circle. (12.1)

1.

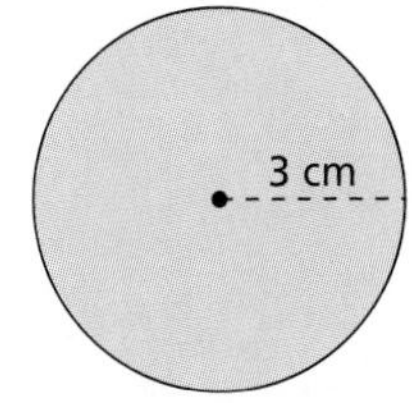

2.

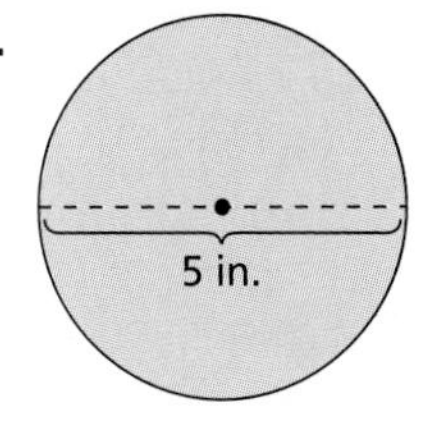

In Exercises 3 and 4, find the area of the blue region (12.1)

3.

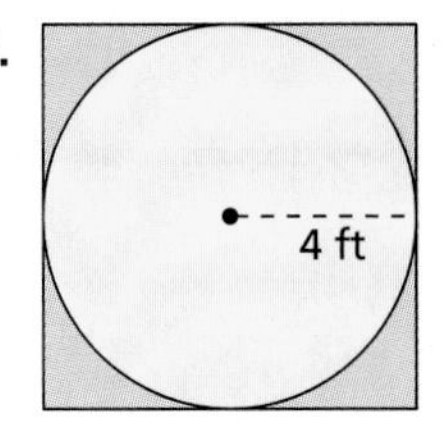

4.

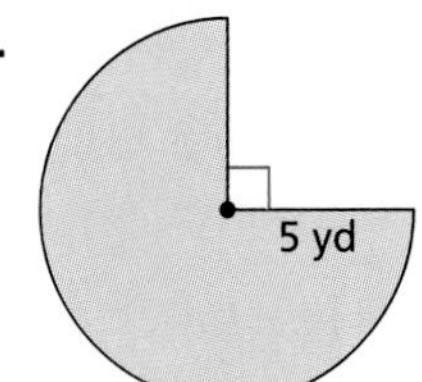

In Exercises 5–9, match the name with a part of the solid at the right. (12.2–12.3)

5. Face

6. Vertex

7. Edge

8. Base

9. Lateral surface

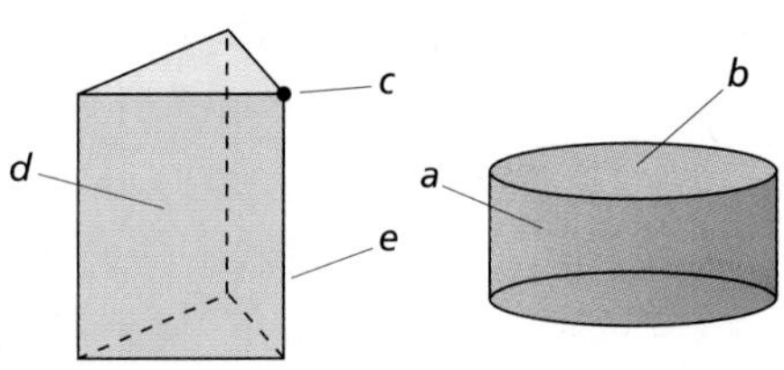

In Exercises 10–13, find the surface area of the solid. (12.3)

10. Diameter: 24 mm
Height: 2 mm

11. Diameter: 14 mm
Length: 49 mm

12.

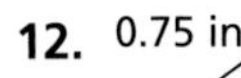

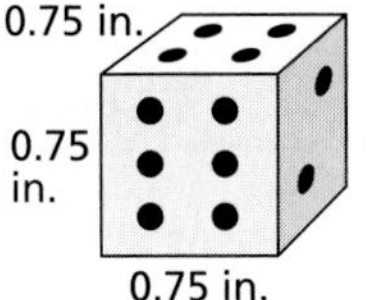

13.

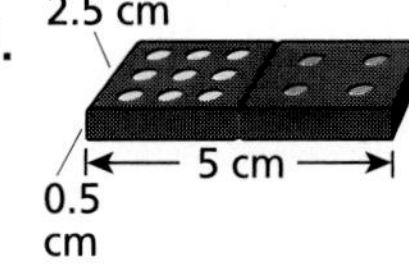

In Exercises 14–17, find the volume of the prism. (12.4)

14.

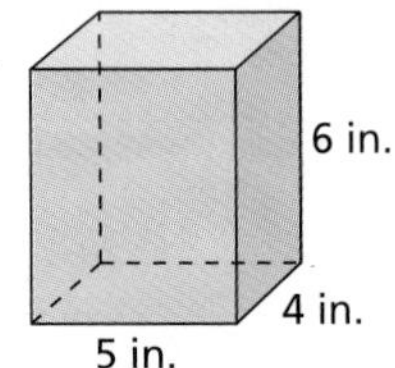

15.

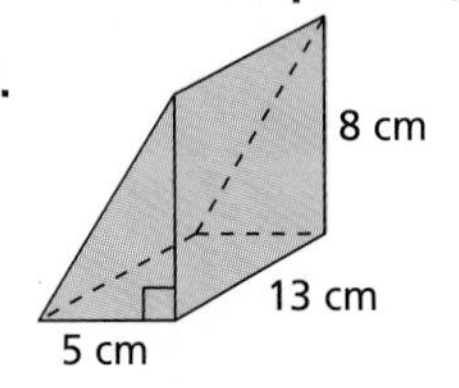

16.

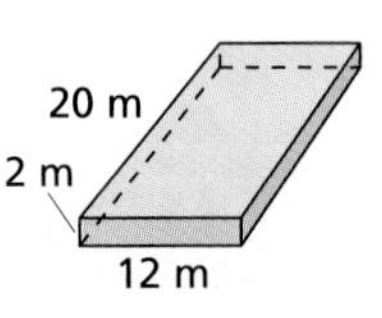

17.

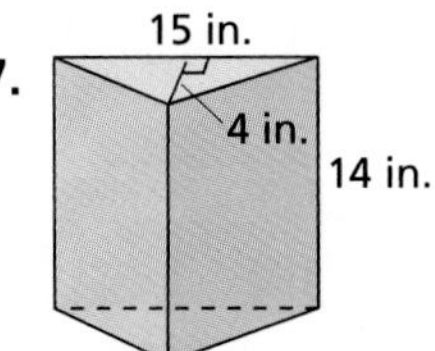

In Exercises 18–20, use the following information.

The Large Electron-Positron Collider, on the border of France and Switzerland, is the largest scientific instrument in the world. It can accelerate particles to nearly the speed of light. The collider is circular, with a diameter of 5.41 miles.

18. Find the circumference of the collider.

19. Find the area of the land inside the collider.

20. The collider's tunnel has a radius of 1.91 feet. What is the tunnel's circumference?

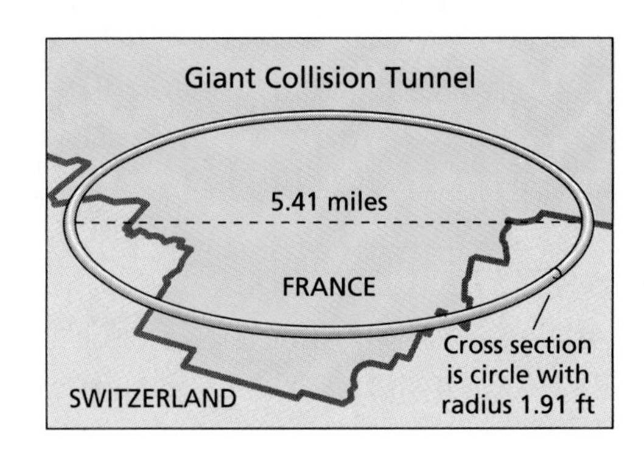

12.5 Exploring Volumes of Cylinders

What you should learn:

How to find the volume of a cylinder

How to use the volume of a cylinder to solve real-life problems

Why you should learn it:

You can use the volume of a cylinder to solve real-life problems, such as comparing the amount of soda pop in different sizes of containers.

Goal 1 Finding the Volume of a Cylinder

Each stack of pennies at the right forms a cylinder. As the number of pennies increases, the height and volume of the stack increase.

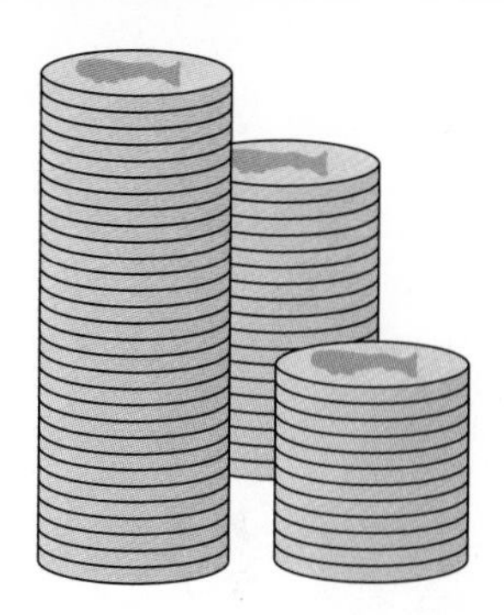

You can find the volume of a cylinder in the same way you find the volume of a prism. That is, you multiply the height of the cylinder by the area of its base.

> **The Volume of a Cylinder**
> The volume of a cylinder is the product of its height and the area of its base. That is, $V = Bh$, where B is the area of a base and h is the height.

Example 1 *Finding Volumes of Cylinders*

a. $r = 3$ m, $h = 5$ m

b. $r = 2$ m, $h = 7$ m

Which cylinder appears to have the greater volume? After finding the volumes, does it surprise you that the green cylinder has about 60% more volume than the blue one?

Find the volume of each cylinder at the left.

Solution

a. The area of the base of this cylinder is $\pi(3^2)$ or about 28.27 square meters. The volume of the cylinder is

$V = Bh$	*Volume of a cylinder*
$\approx (28.27)(5)$	*Substitute for B and h.*
≈ 141.4	*Simplify.*

The volume is about 141.4 cubic meters.

b. The area of the base of this cylinder is $\pi(2^2)$ or about 12.57 square meters. The volume of the cylinder is

$V = Bh$	*Volume of a cylinder*
$\approx (12.57)(7)$	*Substitute for B and h.*
≈ 88.00	*Simplify.*

The volume is about 88.0 cubic meters. ■

Goal 2 Solving Real-Life Problems

1 liter = 33.8 fluid ounces

You have learned that cubic units are the standard units for volume. There are, however, many other commonly used units for volume. Some examples are liters, gallons, quarts, and fluid ounces. To compare volumes that are measured in different units, it helps to change one or more of the measures so that each volume is measured in the same units. ■

Real Life
Unit Prices

$1.79

$1.99

Example 2 *Comparing Volumes*

You are shopping for soda pop, the 2-liter bottle or the 6-pack. Which container has the greater volume? Which is the better buy?

Solution One way to compare the two volumes is to find how many fluid ounces are in the 2-liter bottle. Using the fact that 1 liter is equal to 33.8 fluid ounces, it follows that the 2-liter bottle has a volume of 67.6 fluid ounces. Because the six-pack contains 6(12) or 72 fluid ounces, you can conclude that it has the greater volume.

To decide which is the better buy, you can find the unit price of each container.

$$\text{Unit price} = \frac{\$1.79}{67.6 \text{ fl oz}} \approx \$0.026 \text{ per fl oz} \qquad \textit{Bottle}$$

$$\text{Unit price} = \frac{\$1.99}{72 \text{ fl oz}} \approx \$0.028 \text{ per fl oz} \qquad \textit{Six-pack}$$

The bottle is a slightly better buy. ■

Communicating about MATHEMATICS

SHARING IDEAS about the Lesson

Changing Units of Measure Use a ruler to help you approximate the volume (in cubic inches) of a 12-ounce soda pop can. Then use your result to complete the following.

A. 1 fluid ounce = ? cubic inches.

B. 1 cubic inch = ? fluid ounce.

What do you notice about the two values?

EXERCISES

Guided Practice

CHECK for Understanding

In Exercises 1–4, use the cylinder at the right.

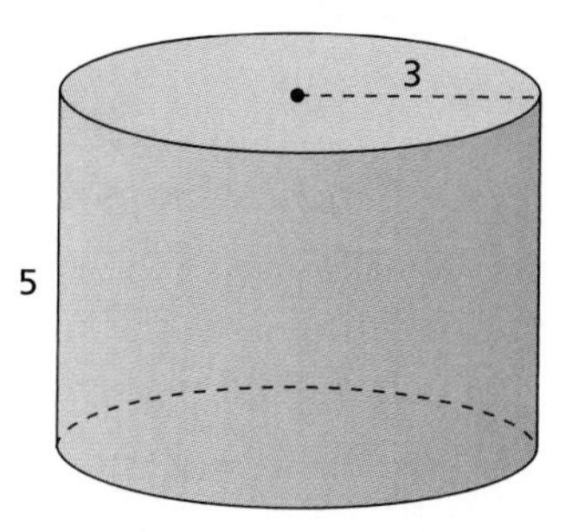

1. Find the area of the base, B.
2. What is the height, h, of the cylinder?
3. State the formula for the volume of the cylinder.
4. Find the volume of the cylinder.

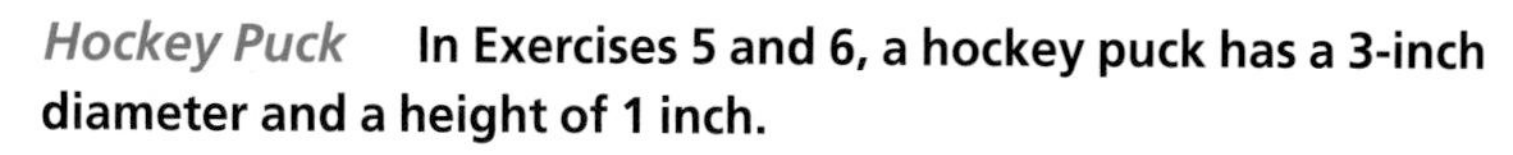

Hockey Puck **In Exercises 5 and 6, a hockey puck has a 3-inch diameter and a height of 1 inch.**

5. What is the area of its base?
6. What is the volume of the hockey puck?

Independent Practice

In Exercises 7–10, find the volume of the cylinder.

7. 2 in.; 6 in.

8. 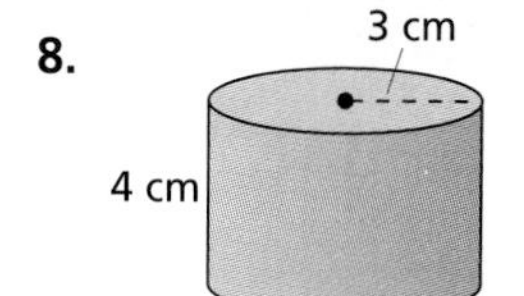

9. 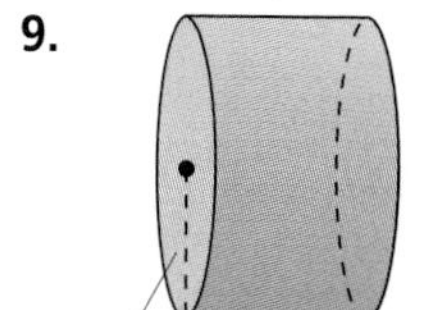

10. 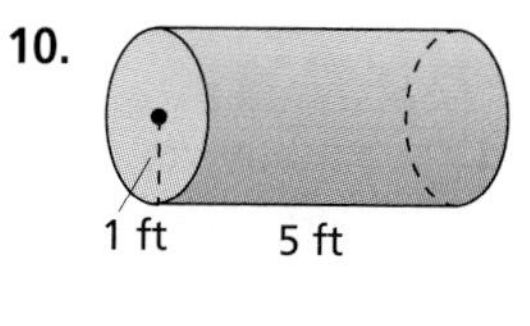

In Exercises 11–14, find the height or the radius of the base.

11. 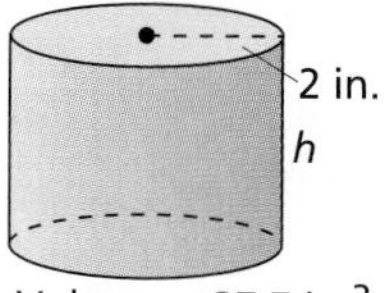

Volume = 37.7 in.^{3}

12. 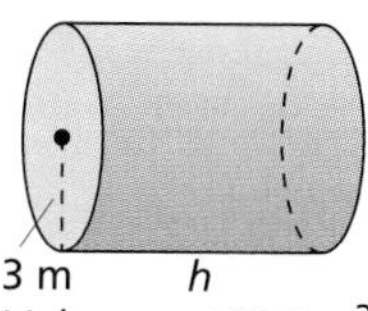

Volume = 197.9 m^3

13. 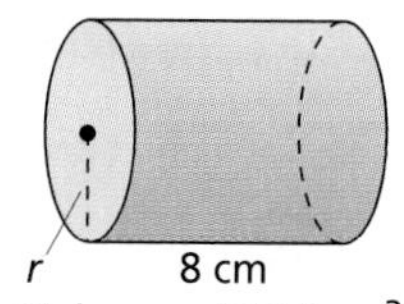

Volume = 290.5 cm^3

14. 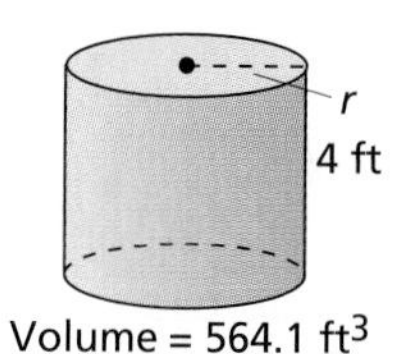

Volume = 564.1 ft^3

In Exercises 15–18, Cylinder A has a 6-inch diameter and a height of 4 inches. Cylinder B has a 4-inch diameter and a height of 6 inches.

15. Without doing any calculations, do you think the volume of cylinder A is greater than, less than, or equal to the volume of cylinder B?
16. Sketch cylinder A and find its volume.
17. Sketch cylinder B and find its volume.
18. Does the cylinder with the greater volume also have the greater surface area? Explain.
19. *Snare Drum* A snare drum has a 14-inch diameter and a height of 5 inches. What is the volume of the snare drum?
20. *Bass Drum* A bass drum has a 20-inch diameter and a height of 14 inches. What is the volume of the bass drum?

Engine Size **In Exercises 21–24, use the following information.**

A cylinder in a V8 engine has a diameter of 4.00 inches. The height of the cylinder is 3.42 inches.

21. What is the volume of the cylinder?

22. When the piston is fully extended, the height of the cylinder changes from 3.42 inches to 0.42 inch. What is the volume at this point?

23. The change in the volume times the number of cylinders is the size of the engine in cubic inches. Find the size of the engine.

24. What is the size of the engine in liters? (Hint: 1 liter = 61.02 cubic inches.)

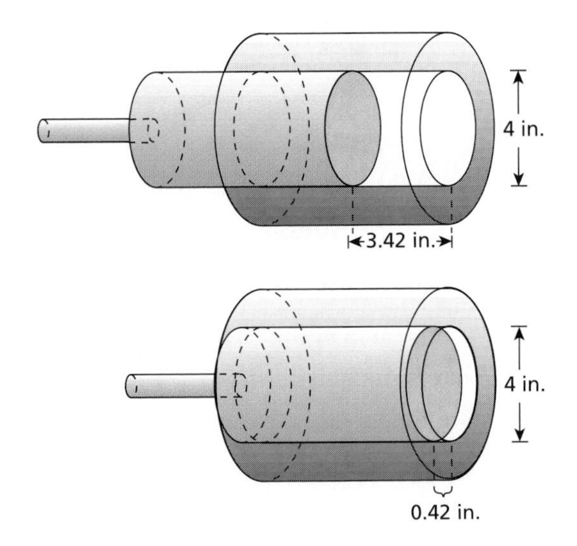

Integrated Review

Making Connections within Mathematics

Nets **In Exercises 25–27, find the volume and surface area of the solid formed by the net.**

25.

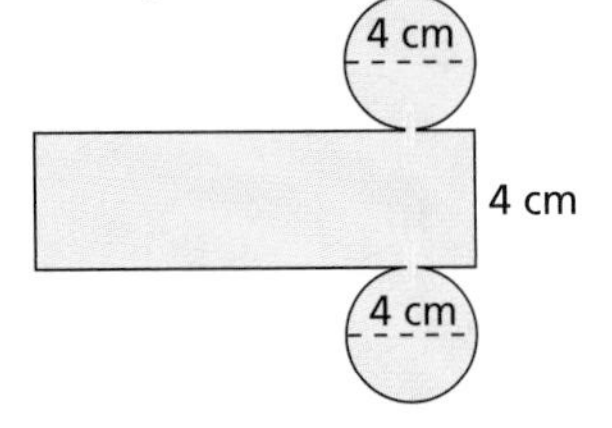

26.

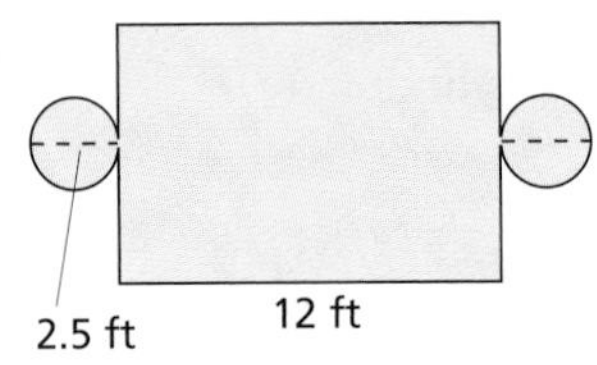

27.

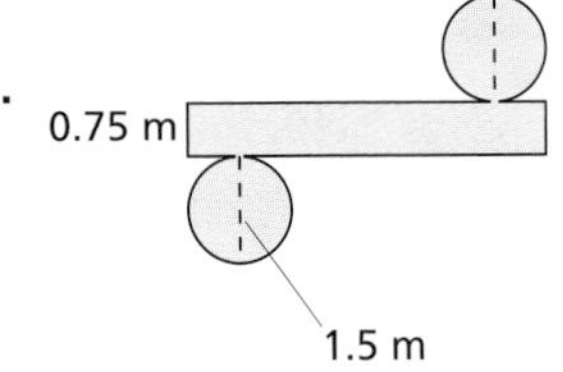

Exploration and Extension

Ringette **In Exercises 28–30, use the following.**

Ringette is a sport similar to ice hockey that is played in Canada, Europe, and the northern United States. The object of the game is to shoot a rubber ring into a net. The volume of the ring is the volume of the cylinder with the outer diameter minus the volume of the cylinder with the inner diameter.

28. Find the volume of the cylinder with the inner diameter.

29. Find the volume of the cylinder with the outer diameter.

30. Find the volume of the ring.

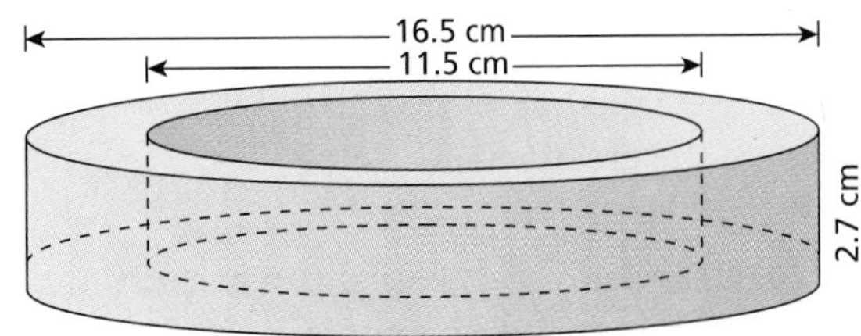

LESSON INVESTIGATION 12.6

Exploring Volume

Materials Needed: thin cardboard, compass, scissors, ruler, unpopped popcorn, paper clips

In this investigation, you will explore the concept of volume.

Example — *Exploring the Volume of a Cone*

Use a compass to draw a circle with a 4-inch radius on a piece of cardboard. Cut it out. Draw a line segment from the circle to its center and cut along the segment. Overlap the cardboard to form an open cone and fix the cardboard with paper clips. Fill the open cone level with unpopped popcorn. (Don't pile the corn above the rim of the open cone.) How many pieces of corn did you use?

Solution The steps used to create and fill the open cone are shown below. The number of pieces of corn depends on the shape of the open cone.

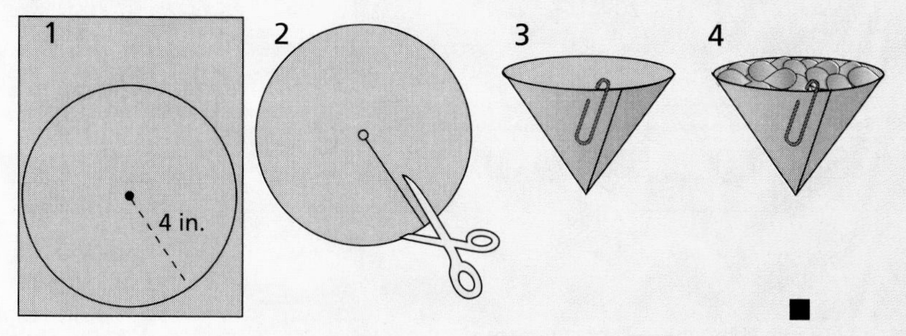

■

Exercises

1. ***Group Activity*** With other members of your group, create the open cone described in the example. Alter the size of the open cone as shown below. For each size, fill the open cone and count the number of pieces of corn. Record your results in a table.

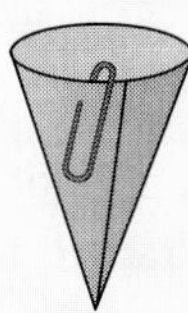
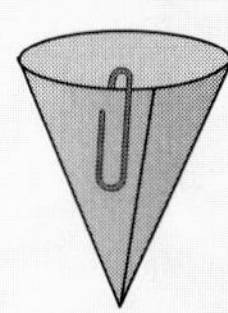
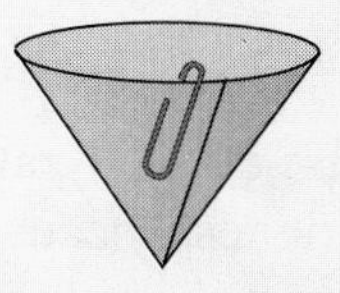
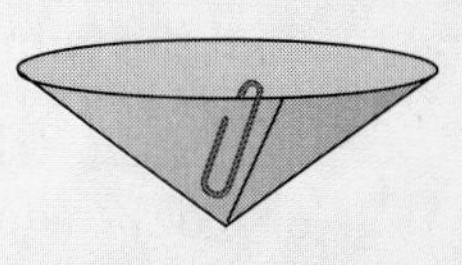

2. Describe the dimensions of the open cone that holds the most popcorn. Compare your group's results with the results of other groups.

12.6 Exploring Volumes of Pyramids and Cones

What you should learn:

Goal 1 How to find the volume of a pyramid and a cone

Goal 2 How to use the volume of a pyramid and a cone to solve real-life problems

Why you should learn it:

You can use the volume of a pyramid or a cone to solve real-life problems, such as finding the volume of a rocket.

Goal 1 Volumes of Pyramids and Cones

To discover the volume of a pyramid, consider a prism that has the same height and base as the pyramid. Fill the pyramid with sand and pour the sand into the prism. The prism will be filled with exactly three pyramids of sand. This suggests that the pyramid has one-third the volume of the prism. The same relationship is true of a cone and a cylinder.

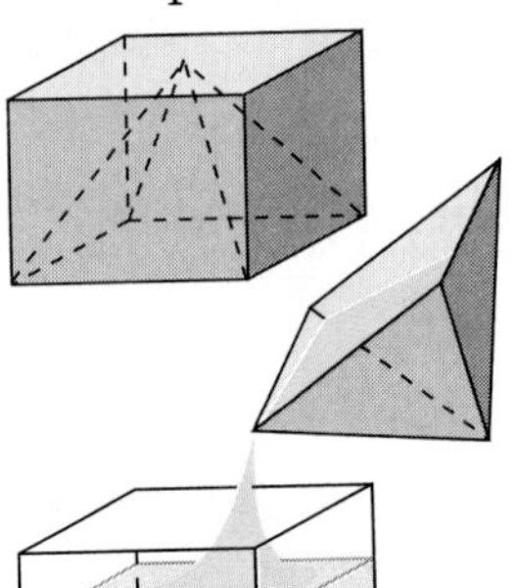

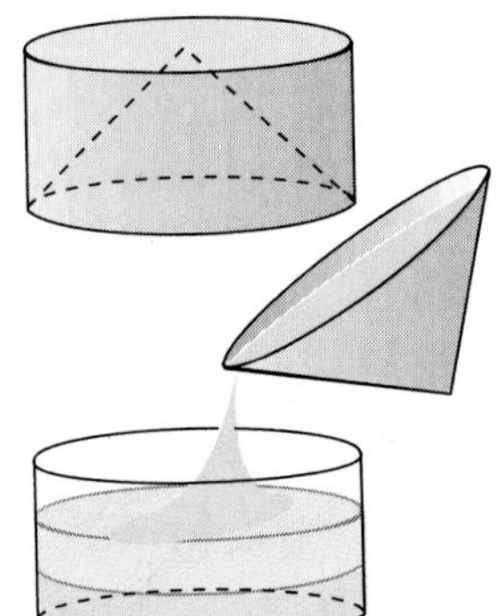

The Volume of a Pyramid or a Cone

The volume of a pyramid or a cone is one-third the product of its height and the area of its base. That is, $V = \frac{1}{3}Bh$, where B is the area of a base and h is the height of the pyramid or cone.

Example 1 *Finding Volumes of Pyramids*

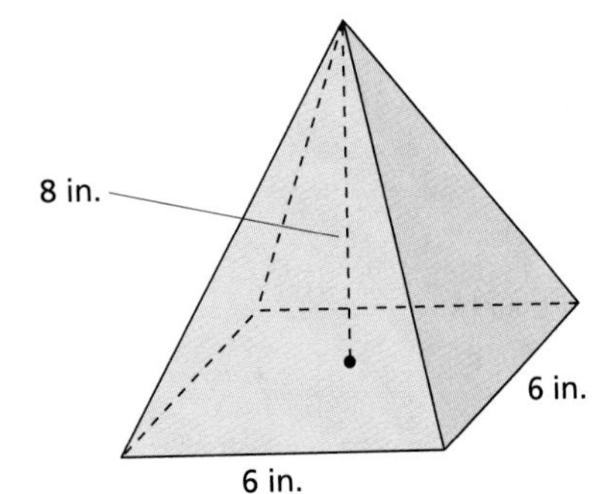

Find the volume of the pyramid, which has a square base.

Solution The base of the pyramid is square with an area of $6 \cdot 6$ or 36 square inches. Because the height of the pyramid is 8 inches, it follows that the volume of the pyramid is

$$V = \frac{1}{3}Bh \quad \textit{Volume of a pyramid}$$
$$= \frac{1}{3}(36)(8) \quad \textit{Substitute for B and h.}$$
$$= 96 \quad \textit{Simplify.}$$

The pyramid has a volume of 96 cubic inches. ■

Goal 2 Solving Real-Life Problems

Example 2 *Finding a Volume*

You are designing a rocket, as shown at the left. The rocket is made by placing a cone on top of a cylinder. What is the total volume of the rocket?

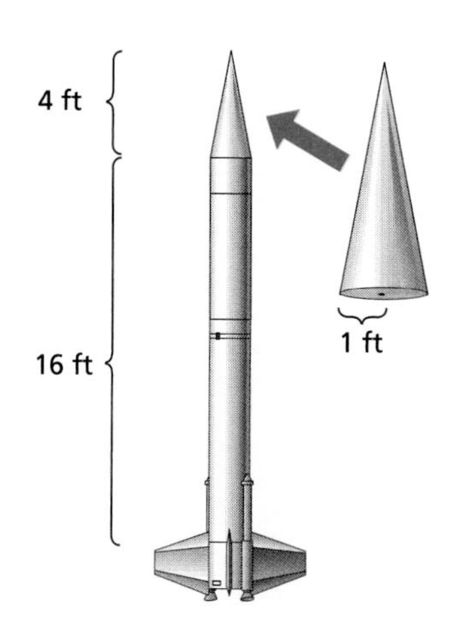

Solution The volume of the rocket is the sum of the volumes of the cone and the cylinder.

$V = \frac{1}{3}Bh$	*Volume of a cone*
$= \frac{1}{3}(\pi r^2)(h)$	*Volume of base is πr^2.*
$= \frac{1}{3}(\pi \cdot 1^2)(4)$	*Substitute for r and h.*
$= \frac{4}{3}\pi$	*Simplify.*
≈ 4.19	*Use a calculator.*

$V = Bh$	*Volume of cylinder*
$= (\pi r^2)(h)$	*Volume of base is πr^2.*
$= (\pi \cdot 1^2)(16)$	*Substitute for r and h.*
$= 16\pi$	*Simplify.*
≈ 50.27	*Use a calculator.*

The volume of the rocket is about 4.19 + 50.27, or about 54.5 cubic feet. ■

Communicating about MATHEMATICS

SHARING IDEAS about the Lesson

Building a Tetrahedron The net below is made of 4 equilateral triangles. It can be folded to form a 4-sided figure called a **tetrahedron.** Make a tetrahedron out of cardboard or stiff paper. Then explain how you can find its volume.

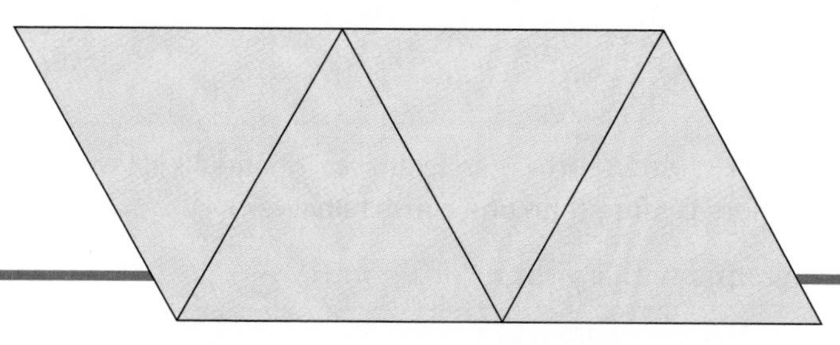

EXERCISES

Guided Practice

CHECK for Understanding

1. How do the bases of pyramids and cones differ?
2. The cylinder and the cone have the same height and radius. Explain how their volumes compare.

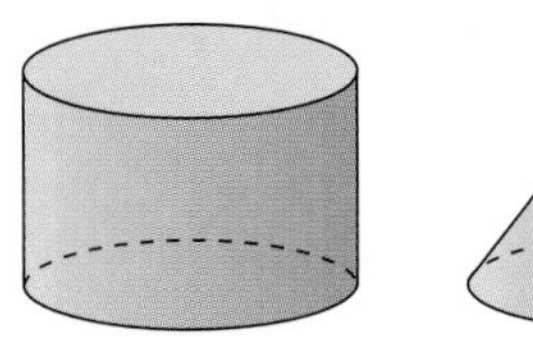

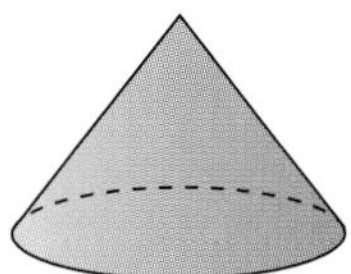

In Exercises 3 and 4, find the volume of the solid.

3. 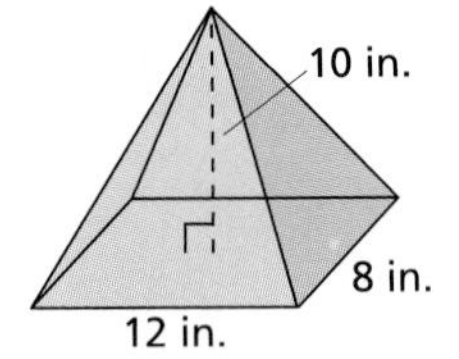

4. 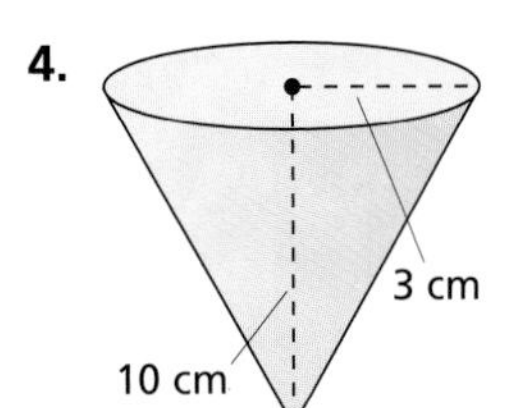

In Exercises 5–8, use the figure at the right.

5. What is the radius of the cone?
6. What is the height of the cone?
7. Find the volume of the cone.
8. Explain how to find the volume of the blue portion of the cube. Then find it.

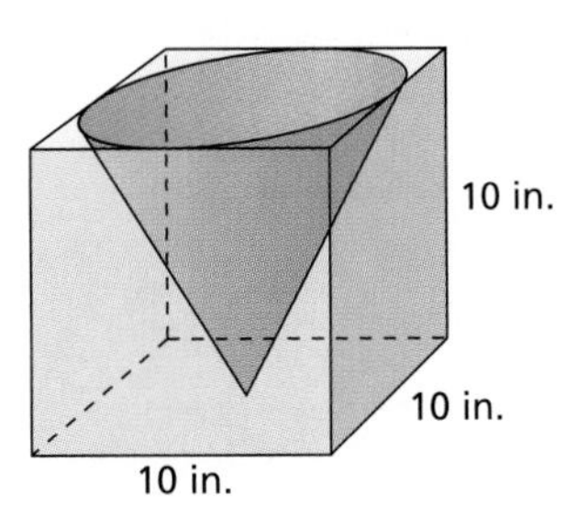

Independent Practice

In Exercises 9–16, find the volume of the solid.

9. 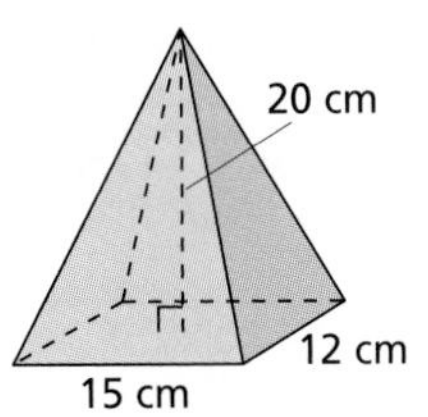

10. 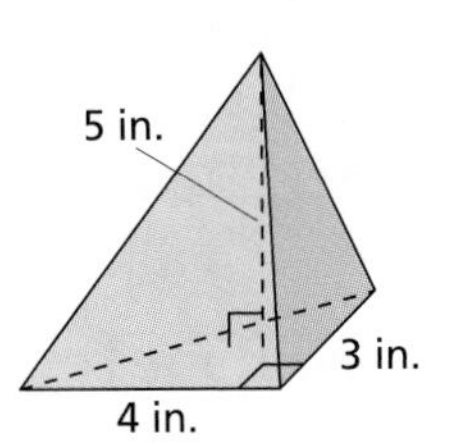

11. 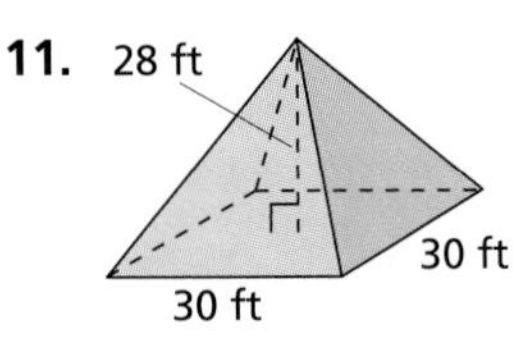

12. 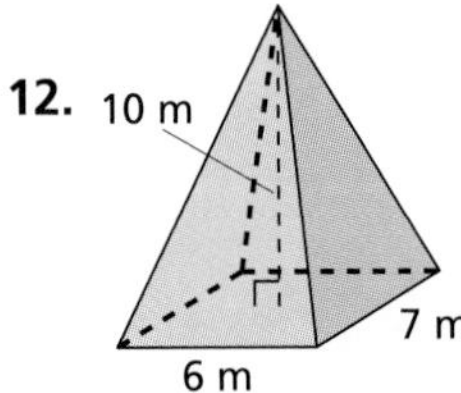

13. 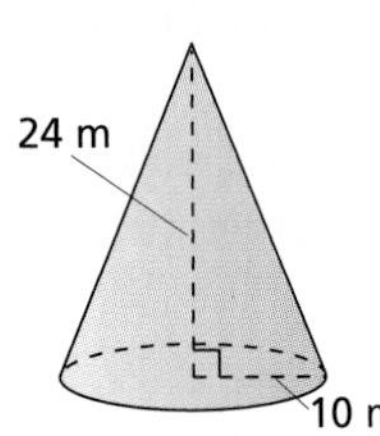

14. 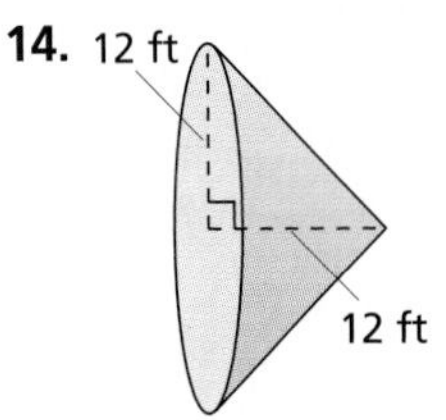

15. 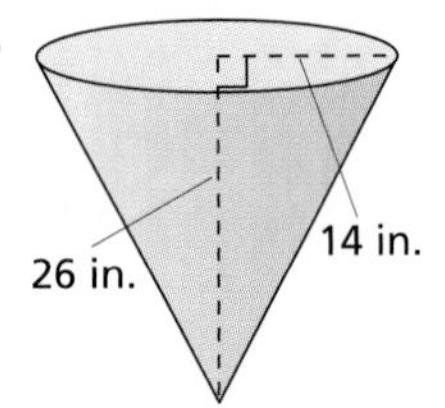

16. 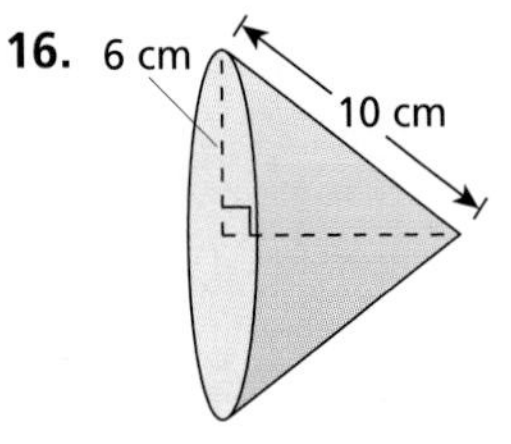

Guess, Check, and Revise **In Exercises 17 and 18, draw the indicated figure. (There is more than one correct answer.)**

17. A pyramid with a volume of 24 mm^3

18. A cone with a volume of $24\pi \text{ ft}^3$

In Exercises 19–22, find the volume of the blue region.

19.

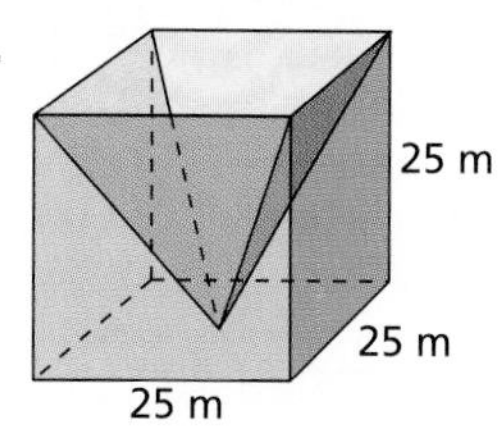

20.

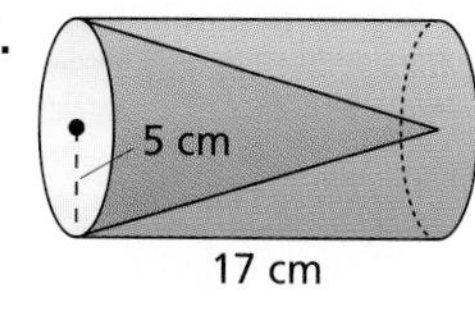

21.

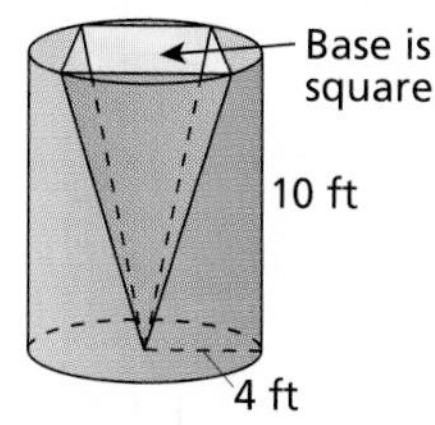

22.

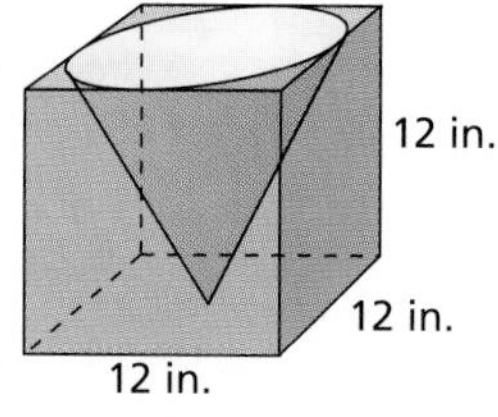

A tornado is a funnel (cone-shaped) cloud. The winds in a tornado can whirl around at more than 200 miles per hour. This tornado has a base diameter of about 200 feet and a height of about 300 feet.

Tornadoes **In Exercises 23–25, use the photo and caption information.**

23. Describe the location of the *base* of the tornado.
24. What is the area of the base, B?
25. What is the volume of the tornado?
26. *Think about It* Which has a greater effect on the volume of a cone, doubling the radius or doubling the height? Explain your reasoning.
27. Copy and complete the table. Round your results to 2 decimal places.

Solid	Base Area	Height	Volume
Pyramid	121 ft^2	7.5 ft	?
Pyramid	486 cm^2	?	3199.5 cm^3
Pyramid	?	5.6 in.	630 $in.^3$

28. Copy and complete the table. Round your results to 2 decimal places.

Solid	Radius	Height	Volume
Cone	3.5 m	7 m	?
Cone	?	9.4 cm	629.99 cm^3
Cone	10.2 mm	?	2179 mm^3

Integrated Review

Making Connections within Mathematics

Fresh Air **In Exercises 29–32, use the graph at the right.** ***(Source: National Academy of Sciences)***

29. What does one small block represent?
30. How much fresh air per minute circulates around a first-class passenger?
31. How much more fresh air per minute circulates around a pilot than around an economy class passenger?
32. How much fresh air per minute would circulate around 80 seats in economy class?

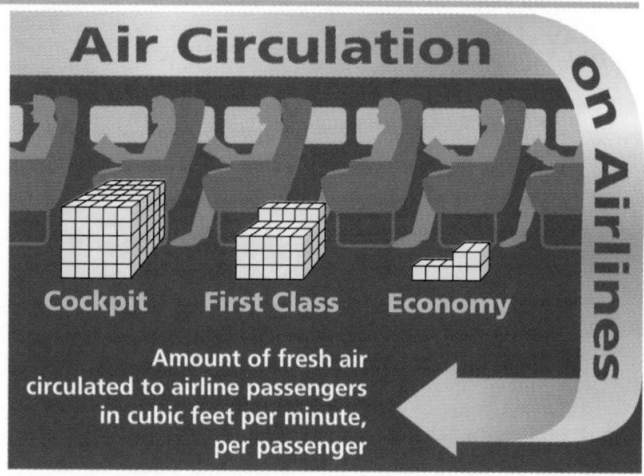

Exploration and Extension

Making a Paper Pyramid **In Exercises 33–36, use the figure. (The figure is not drawn to scale.)**

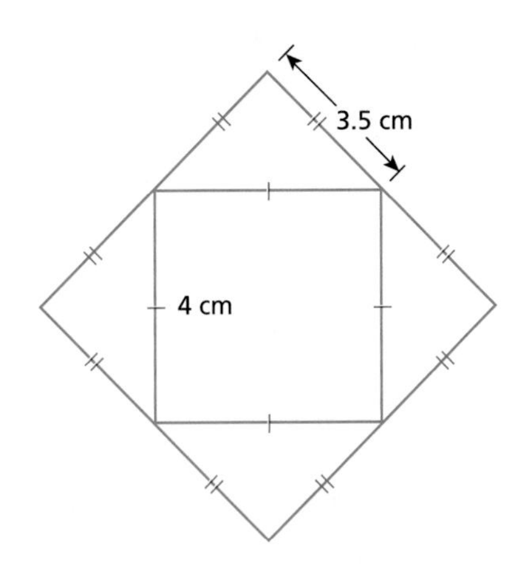

33. Use a ruler to draw a full-scale copy of the figure on stiff paper. Cut the figure out. Then fold and tape the figure to form a pyramid.

34. Use a ruler to measure the height of the pyramid. Then find the volume of the pyramid.

35. Explain how to find the surface area of the pyramid.

36. Find the surface area of the pyramid.

Mixed REVIEW

In Exercises 1–4, find the volume of the figure. (12.4–12.6)

1. **2.**

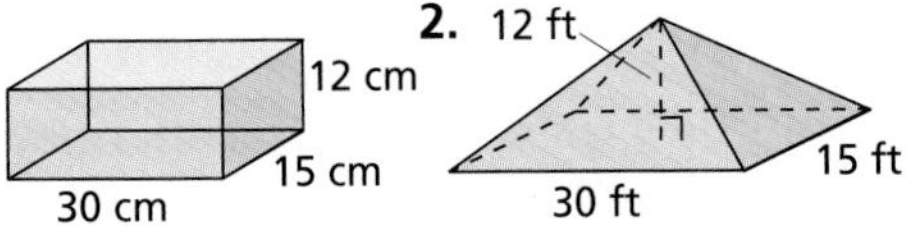

3.

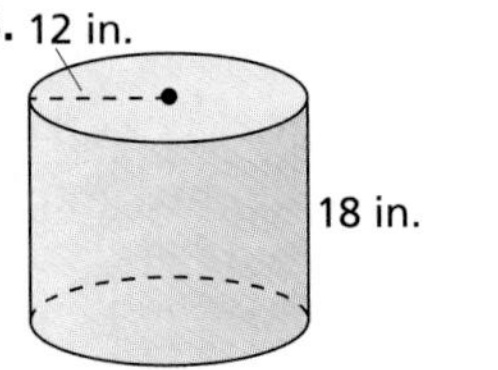

4.

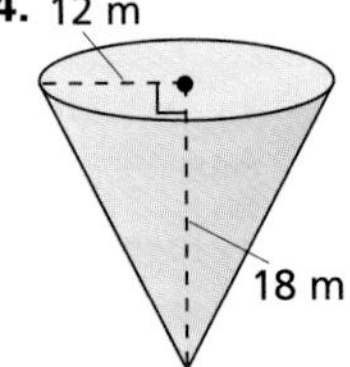

In Exercises 5–8, use Pascal's Triangle on page 372 to answer the question. (8.7–8.8)

5. How many ways can you select 3 people from a group of 6?

6. How many ways can you select 4 birds from a flock of 5?

7. How many ways can you select 3 animals from a herd of 7?

8. How many ways can you select 2 goldfish from 5 goldfish?

In Exercises 9–14, solve the equation. (4.5)

9. $4x + 3 = 2$

10. $2b + 5 = 7 - 5b$

11. $\frac{3}{8}r + 2 = \frac{7}{8}r$

12. $0.6 + 0.2y = 0.5y$

13. $\frac{4}{3} - \frac{5}{7}s = \frac{3}{2}s$

14. $42 + 3m = -4m$

In Exercises 15–20, solve the inequality. (9.7)

15. $24 \leq -2y$

16. $13z + 26 > 0$

17. $7 - 6t > 4t$

18. $r - \frac{6}{5} < \frac{5}{6}$

19. $-\frac{33}{2}p \geq \frac{11}{4}$

20. $\frac{11}{8}p \leq -\frac{33}{4}$

12.7 Exploring Volumes of Spheres

What you should learn:

How to find the volume of a sphere

How to use the volume of a sphere to solve real-life problems

Why you should learn it:

You can use the volume of a sphere to solve real-life problems, such as finding the volume of a natural gas storage tank.

Goal 1 Finding Volumes of Spheres

To discover the volume of a sphere, consider a sphere that is cut into two halves, called **hemispheres.** An open cone is fitted inside one of the hemispheres. The open cone is then filled with sand, which is poured into the hemisphere. The hemisphere will be filled with exactly two cones of sand. Because the volume of the cone is $\frac{1}{3}Bh$, which is $\frac{1}{3}(\pi r^2)r$ or $\frac{1}{3}\pi r^3$, it follows that the volume of the sphere is $4(\frac{1}{3}\pi r^3)$ or $\frac{4}{3}\pi r^3$.

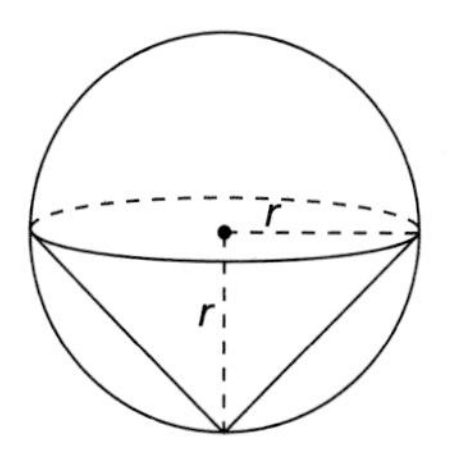

Volume of a Sphere

The volume of a sphere is four-thirds times π times the cube of its radius. That is, $V = \frac{4}{3}\pi r^3$, where r is the radius.

Example 1 *Finding the Volume of a Sphere*

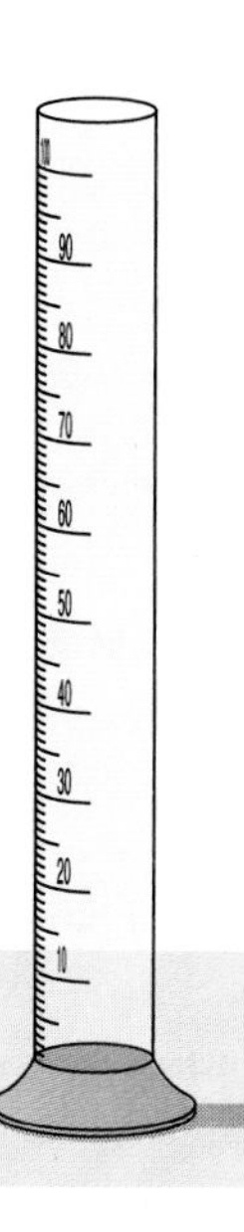

The cylinder at the left is used in chemistry to measure volumes in cubic centimeters (1 cubic centimeter is equal to 1 milliliter). Cut a small hole in a Ping-Pong ball, fill the ball with water, and pour the water into a cylinder. How high in the cylinder will the water come?

Solution The radius of the Ping-Pong ball is 1.9 centimeters, which implies that its volume is as follows.

$V = \frac{4}{3}\pi r^3$ *Volume of a sphere*

$= \frac{4}{3}\pi(1.9)^3$ *Substitute for r.*

≈ 28.73 *Use a calculator.*

The volume is about 28.73 cubic centimeters. This means that the water will be just below the 29 milliliter mark. ■

Goal 2 Solving Real-Life Problems

Spherical storage tanks, such as those at the right, are sometimes used to store natural gas and other gases. The advantage of the spherical design is that the pressure of the gas is evenly distributed on all parts of the tank's surface.

Real Life
Industry

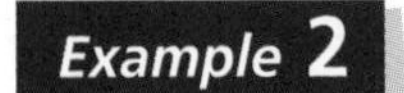

Example 2 *Finding the Volume of a Sphere*

You are designing a spherical storage tank for natural gas. The radius of the tank is 18 feet. How many cubic feet of natural gas will the tank hold? If you double the radius, will the tank hold twice as much?

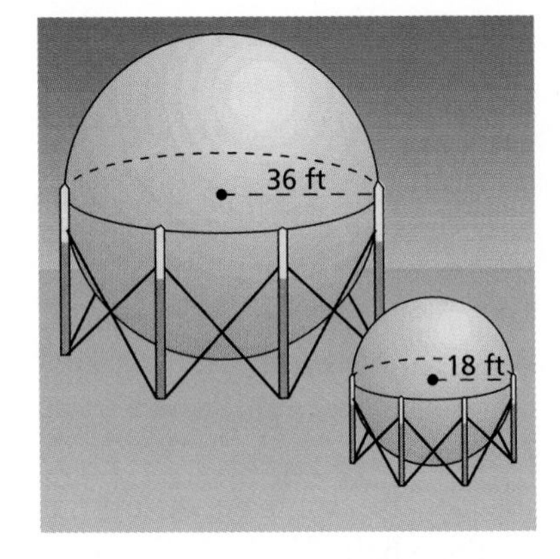

Solution

$V = \frac{4}{3}\pi r^3$	*Volume of a sphere*
$= \frac{4}{3}\pi(18)^3$	*Substitute for r.*
$= 7776\pi$	*Simplify.*
$\approx 24{,}429$	*Use a calculator.*

The tank holds about 24,429 cubic feet of gas. If you doubled the radius, the volume would be

$$V = \frac{4}{3}\pi(36)^3 = 62{,}208\pi \approx 195{,}432 \text{ cubic feet.}$$

This volume is 8 times as much (not twice as much) as the volume of the original tank. ■

Communicating about MATHEMATICS

Cooperative Learning

SHARING IDEAS about the Lesson

Guess, Check, and Revise Work with a partner. You are designing a spherical storage tank that is to contain 30,000 cubic feet of natural gas. Use the guess, check, and revise problem-solving strategy to find the radius of the tank.

EXERCISES

Guided Practice

CHECK for Understanding

In Exercises 1–5, use the figure at the right.

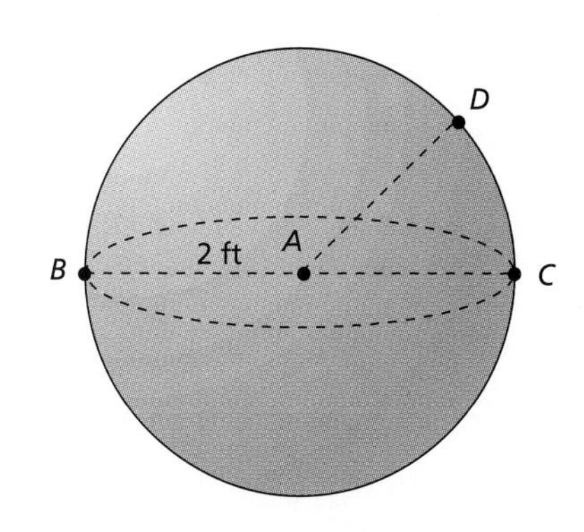

1. Name a radius.
2. Is $\overline{AD}$ a radius? Explain.
3. Name a diameter. What is its measure?
4. What is one half of a sphere called?
5. Find the volume of the sphere.

Independent Practice

In Exercises 6–13, find the volume of the ball.

6. $r = 1.5$ in.

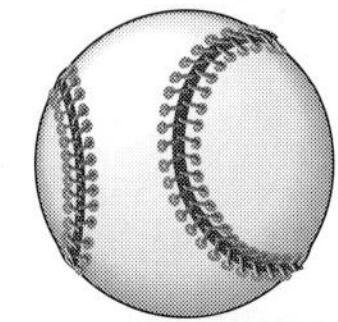

7. $r = 12$ cm

8. $d = 8.6$ in.

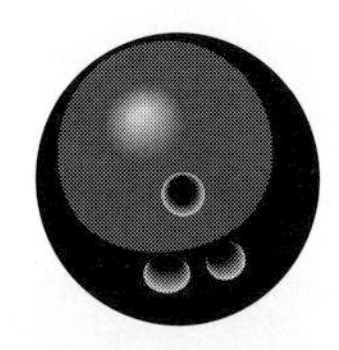

9. $d = 1.68$ in.

10. $d = 2.25$ in.

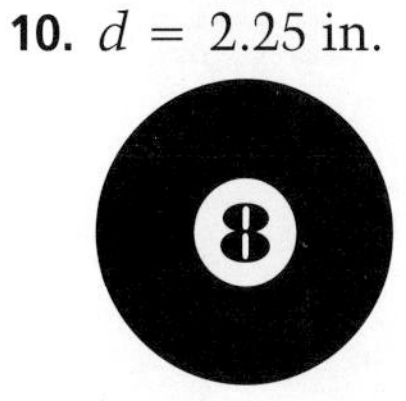

11. $r = 11$ cm

12. $d = 8.25$ in.

13. $r = 1.25$ in.

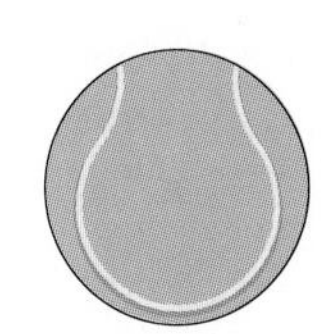

14. From your knowledge of the balls in Exercises 6–13, order them from smallest to largest. Given 1 inch = 2.54 centimeters, write each volume in cubic centimeters. Did you order the balls correctly? Explain.

In Exercises 15–18, find the volume of the blue portion of the figure.

15.

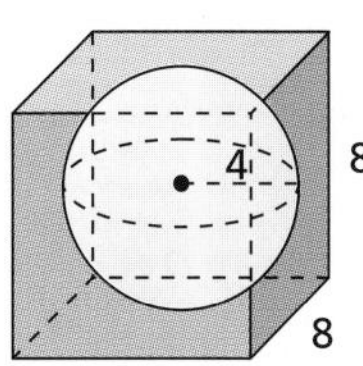

16.

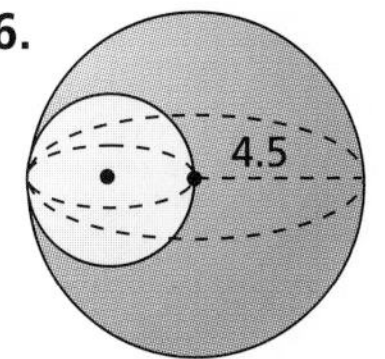

17.

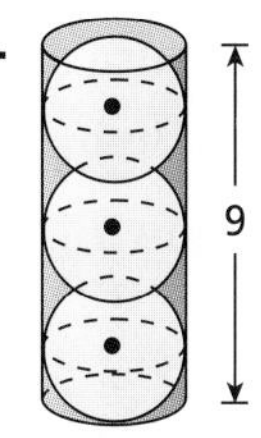

18.

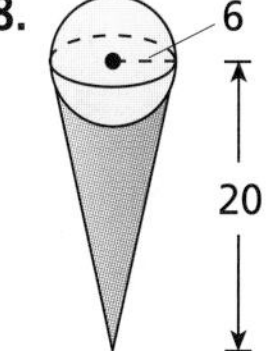

19. Which is larger, the circumference or height of the cylinder in Exercise 17? Explain.

Guess, Check, and Revise **In Exercises 20–23, find the radius of the sphere with the given volume.**

20. 1436.76 cm^3 **21.** 659.58 ft^3

22. $32.52\pi \text{ cm}^3$ **23.** $2929.33\pi \text{ in.}^3$

24. What happens to the volume of a sphere when the radius doubles? Triples? Quadruples? Describe the pattern.

25. *Visualization* Describe the intersection of a sphere and a plane that cuts through the center of the sphere.

Earth and Moon **In Exercises 26–28, use the following.**

Earth has a diameter of about 8000 miles. The volume of the moon is $\frac{1}{50}$ that of Earth.

26. Find the volume of Earth.

27. Find the volume of the moon.

28. What is the moon's diameter?

Sunsphere **In Exercises 29 and 30, use the photo caption.**

29. What is the volume of the sphere?

30. What is the distance from the bottom of the sphere to the ground?

This 266-foot tower, called the Sunsphere, was built for the 1982 World's Fair in Knoxville, Tennessee. The diameter of the sphere is about 89 feet.

Integrated Review

Making Connections within Mathematics

Volume **In Exercises 31–34, find the volume of the solid. Explain your reasoning.**

31.

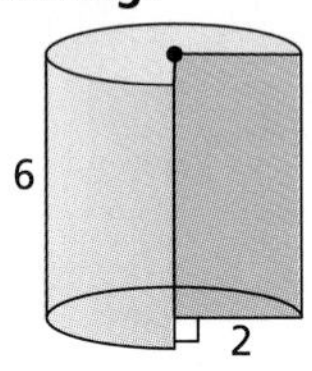

32.

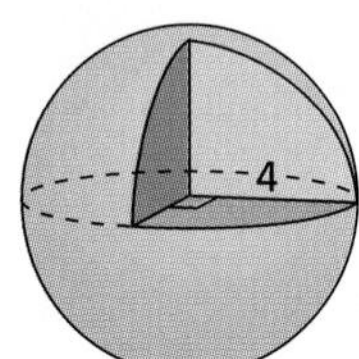

33.

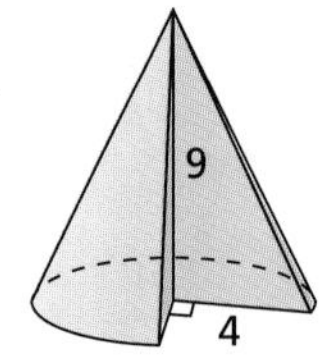

34.

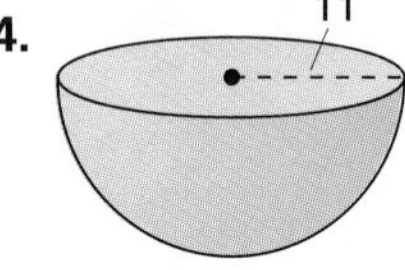

Exploration and Extension

Surface Area **In Exercises 35 and 36, use the figure at the right, which shows a flattened baseball covering.**

35. From the figure, which of the following is the correct formula for the surface area of a sphere? Explain.

a. $S = 2\pi r^2$ **b.** $S = 3\pi r^2$ **c.** $S = 4\pi r^2$ **d.** $S = 5\pi r^2$

36. Find the surface area of a baseball.

37. Use the information in Exercises 26–28 to find the surface area of Earth.

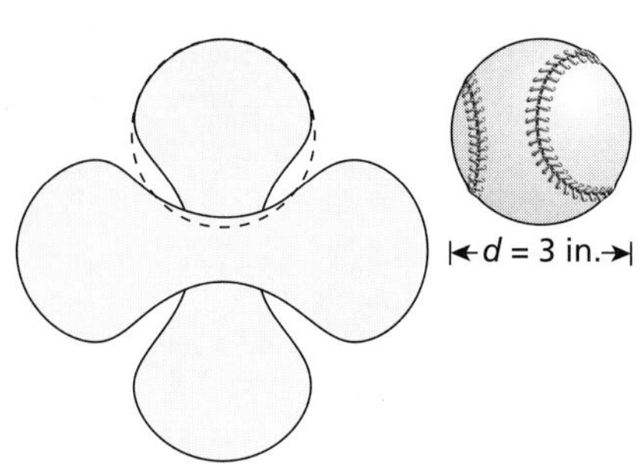

12.8 Exploring Similar Solids

What you should learn:

Goal 1 How to explore ratios of measurements of similar figures

Goal 2 How to use ratios of measurements of similar figures

Why you should learn it:

You can use ratios of measurements of similar figures to solve real-life problems, such as finding the surface area or volume of a building.

Baby mammals are not geometrically similar to their parents (their proportions are different), but baby reptiles are geometrically similar to their parents.

Goal 1 Exploring Measures of Similar Solids

Two solids are similar if they have the same shape and their corresponding lengths are proportional. Here are two examples.

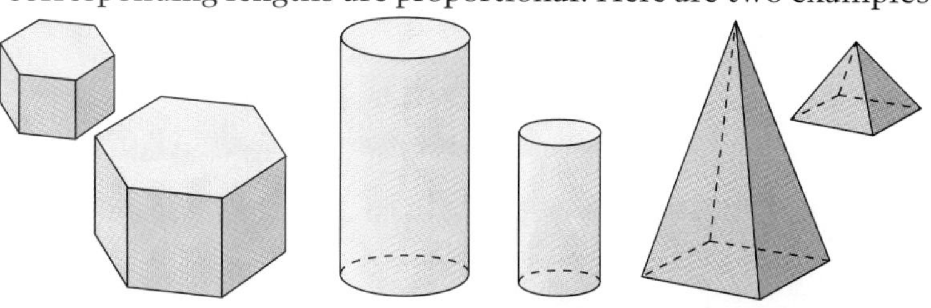

LESSON INVESTIGATION

Investigating Ratios of Similar Solids

Group Activity Use wooden cubes to build cubes that have edge lengths of 1, 2, 3, 4, and 5 inches. Call the 1-inch cube the *original* cube, and the other cubes the *new* cubes. Find the surface area and volume of each cube. Record your results in a table, like that shown below. (In the table, S is the surface area and V is the volume.) Describe the pattern for the ratios in your table.

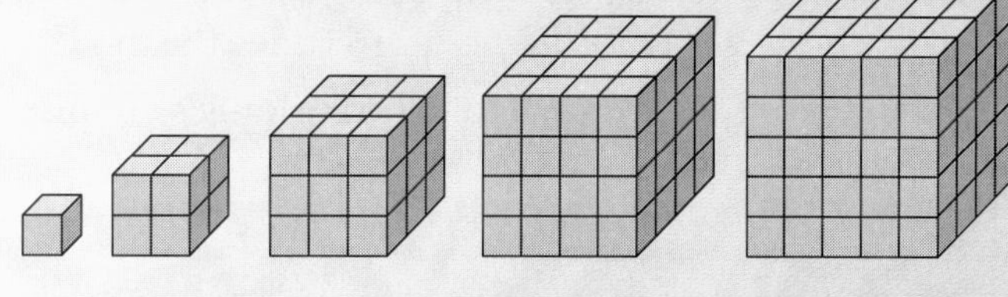

Edge	$\frac{\text{New edge}}{\text{Orig. edge}}$	S (in.^{2})	$\frac{\text{New S}}{\text{Orig. S}}$	V (in.^{3})	$\frac{\text{New V}}{\text{Orig. V}}$
1 in.	1	6	1	1	1
2 in.	2	24	4	8	8
3 in.	?	?	?	?	?
4 in.	?	?	?	?	?
5 in.	?	?	?	?	?

Goal 2 Comparing Ratios of Similar Solids

In the investigation on page 573, you may have discovered the following relationships between measures of similar solids.

Ratios of Measures of Similar Solids

1. If two solids are similar with a scale factor of k, then the ratio of their surface areas is k^2.
2. If two solids are similar with a scale factor of k, then the ratio of their volumes is k^3.

Example 1 *Comparing Surface Areas and Volumes*

You are building a scale model of a building. In your model a length of $\frac{1}{8}$ inch represents a length of 1 foot in the building.

a. What is the scale factor of the building to the model?

b. What is the ratio of the surface area of the building to the surface area of the model?

c. What is the ratio of the volume of the building to the volume of the model?

Architects often build scale models of buildings they are designing. A common scale factor is $\frac{1}{8}$ inch to 1 foot.

Solution

a. To find the scale factor, find the ratio of 1 foot to $\frac{1}{8}$ inch.

$$\frac{1 \text{ foot}}{\frac{1}{8} \text{ inch}} = \frac{12 \text{ inches}}{\frac{1}{8} \text{ inch}} = 12 \cdot \frac{8}{1} = 96$$

The scale factor is 96. That is, the building's lengths are 96 times as large as the model's lengths.

b. Using a scale factor of 96, the building's surface area is 96^2 or 9216 times the model's surface area.

c. The building's volume is 96^3 or 884,736 times the model's volume. ■

Communicating about MATHEMATICS

SHARING IDEAS about the Lesson

Scale Models In Example 1, a rectangular room in the model is 2-inches by 2-inches by 1-inch. Describe two different ways to find the surface area and volume of the actual room in the building.

EXERCISES

Guided Practice

CHECK for Understanding

In Exercises 1–3, decide whether the figures are similar. Explain.

1.

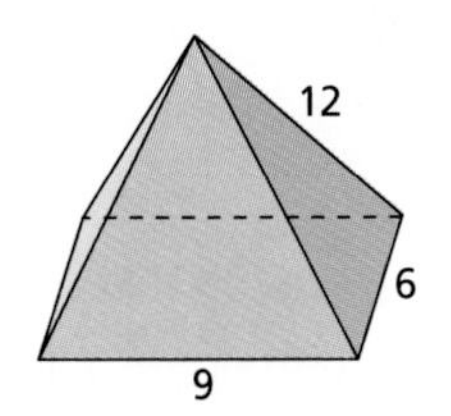

2.

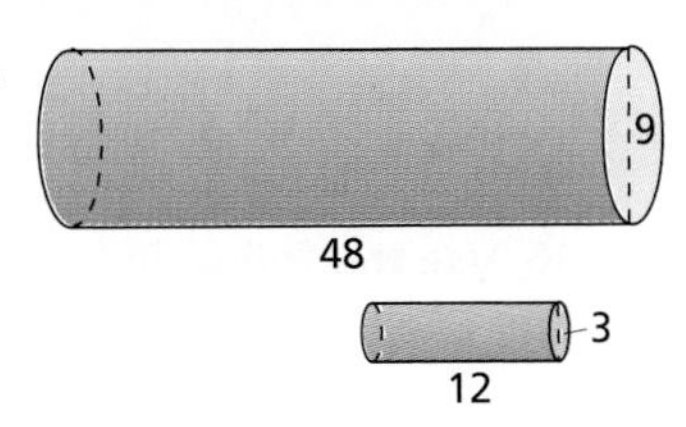

3.

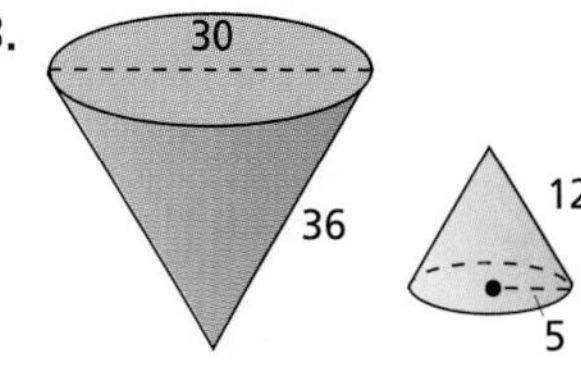

In Exercises 4–8, use the rectangular prisms at the right.

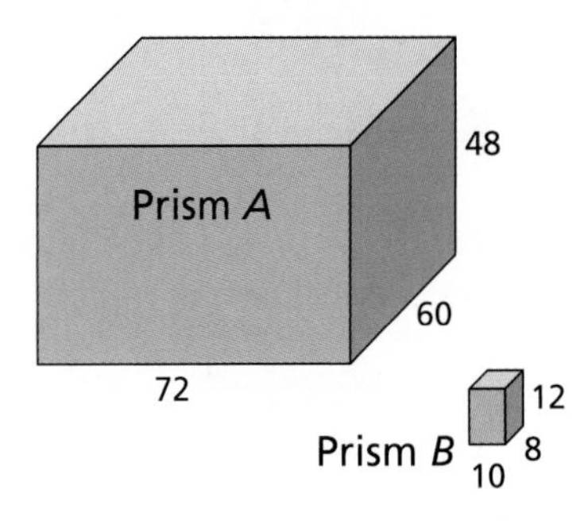

4. Explain why the prisms are similar.

5. Find the scale factor of Prism A to Prism B.

6. Find the surface area of Prism B.

7. Use a ratio of measures of similar solids to find the surface area of Prism A. Check your answer.

8. How are the two volumes related? Explain.

Independent Practice

In Exercises 9 and 10, match the solid with a similar solid.

9.

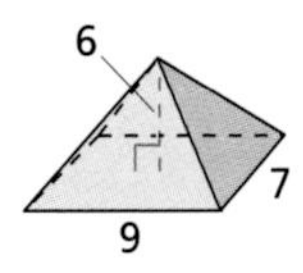

a.

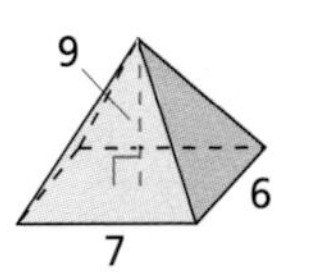

b.

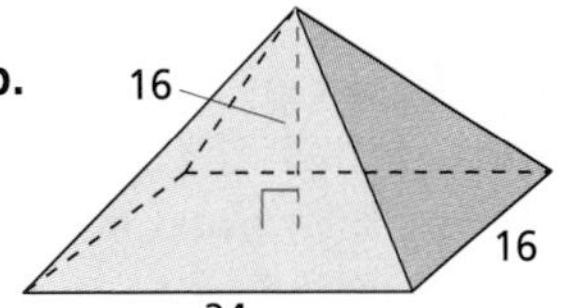

c.

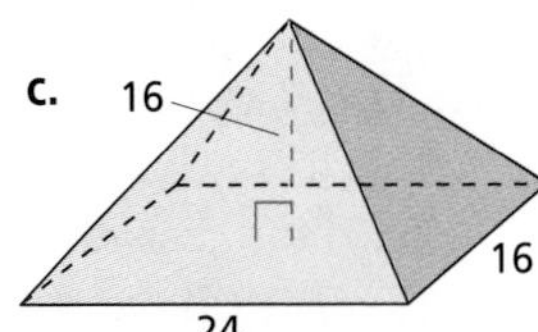

10.

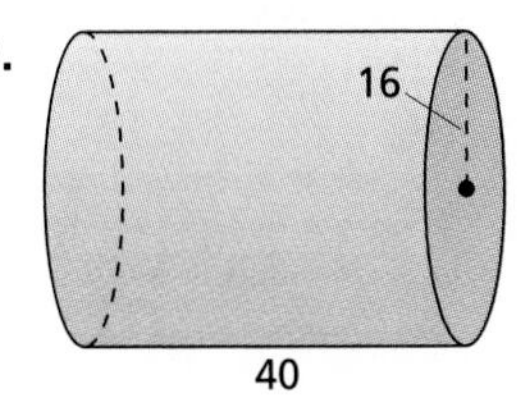

a.

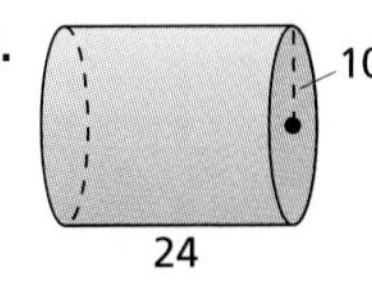

b.

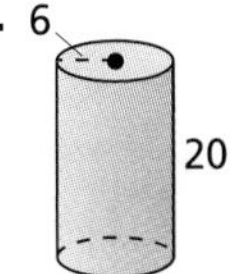

c.

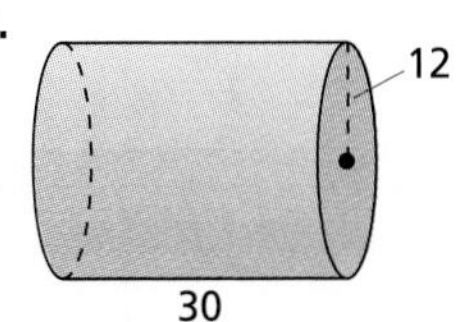

In Exercises 11 and 12, use a proportion to solve for x and y in the similar solids.

11.

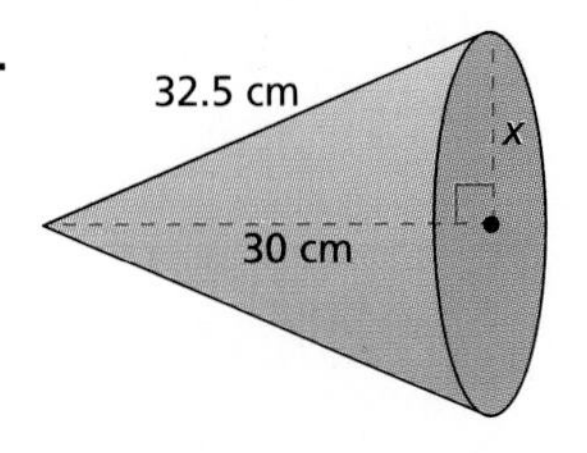

12.

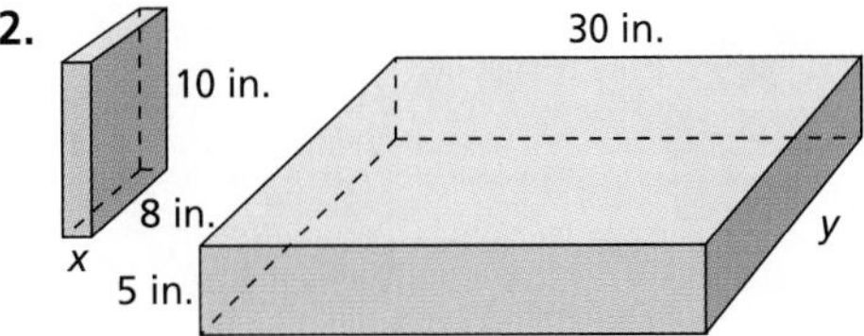

13. Copy and complete the table. The scale factor is the ratio of lengths of Solid A to Solid B.

Scale Factor	Solid A Length	Solid A Surface Area	Solid A Volume	Solid B Length	Solid B Surface Area	Solid B Volume
3	15 ft	400 ft^2	300 ft^3	?	?	?
$\frac{1}{10}$	2 m	46 m^2	50 m^3	?	?	?
7.5	?	11,137.5 cm^2	?	3 cm	?	162 cm^3
$\frac{1}{16}$	4 in.	?	64 $in.^3$	?	24,576 $in.^2$	?

Designing a Tunnel **In Exercises 14–16, use the following.**

You are designing a tunnel through a mountain. In your model, a length of $\frac{1}{12}$ inch represents a length of 1 foot in the actual tunnel.

14. Find the scale factor of the actual tunnel to the model.

15. Find the ratio of the surface area of the actual tunnel to the surface area of the model.

16. Find the ratio of the volume of the actual tunnel to the volume of the model.

Logical Reasoning **In Exercises 17–20, complete the statement using *always, sometimes,* or *never.***

17. Two spheres are ? similar.

18. A cube is ? similar to a pyramid.

19. Two cones are ? similar.

20. A cylinder is ? similar to itself.

Integrated Review

Making Connections within Mathematics

21. *You Be the Teacher* Your classmate says that because the two solids at the right have the same surface area and volume, they must be similar. Do they have the same surface area and volume? Are they similar? How would you explain your answer to your classmate?

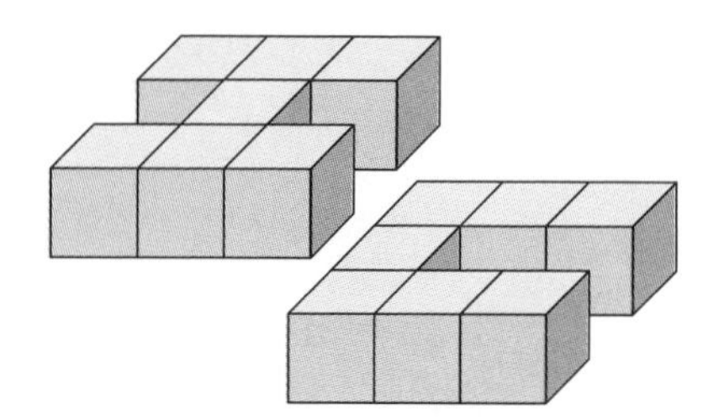

Exploration and Extension

Drawing Solids **In Exercises 22–25, draw and label a solid that is similar to the given solid. (There are two possible answers.)**

22. Scale factor: $\frac{1}{2}$

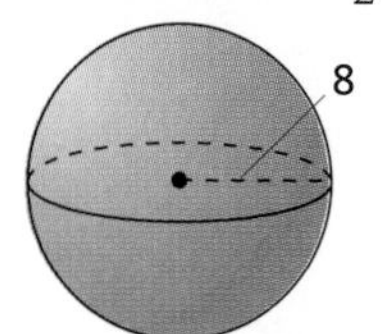

23. Scale factor: $\frac{1}{3}$

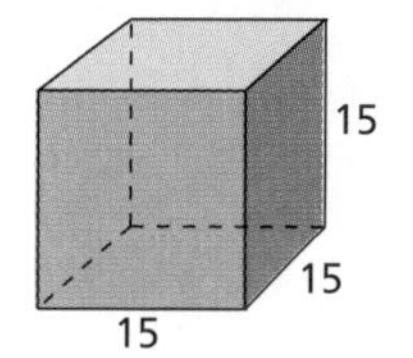

24. Scale factor: 2

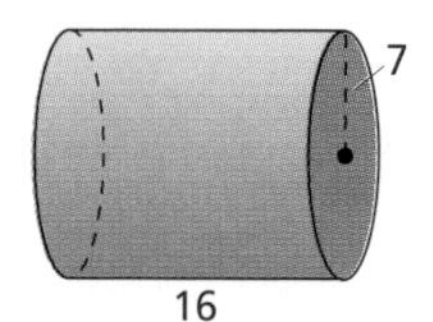

25. Scale factor: $\frac{2}{3}$

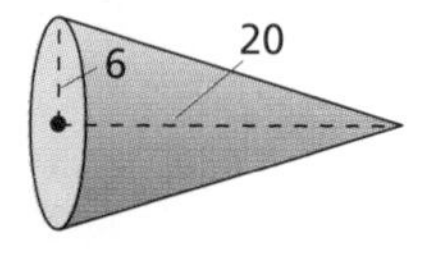

12

Chapter Summary

What did you learn?

Skills

1. Find the circumference of a circle.	**(12.1)**
2. Find the area of a circle.	**(12.1)**
3. Identify different types of solids.	**(12.2)**
▪ Identify faces, edges, and vertices of polyhedrons.	**(12.2)**
4. Find the surface area of a solid.	
▪ Find the surface area of a prism and a cylinder.	**(12.3)**
5. Find the volume of a solid.	
▪ Find the volume of a prism.	**(12.4)**
▪ Find the volume of a cylinder.	**(12.5)**
▪ Find the volume of a pyramid and a cone.	**(12.6)**
▪ Find the volume of a sphere.	**(12.7)**
6. Compare the surface areas and volumes of similar solids.	**(12.8)**

Problem-Solving Strategies

7. Model and solve real-life problems.	**(12.1–12.8)**

Exploring Data

8. Use tables and graphs to solve problems.	**(12.1–12.8)**

Why did you learn it?

From the variety of the real-life examples in this chapter, you can see that measurements of circles and solids are used in almost all walks of life. For instance, in this chapter, you saw how to find the area of one side of the presidential seal. You also learned how surface area can be used to measure the amount of material needed to manufacture different types of containers, how volume can be used to help design a television studio, how volume can be used to decide which type of container is a better buy, how to find the volume of a rocket, and how volume can be used to find how much gas can be stored in a spherical storage tank.

How does it fit into the bigger picture of mathematics?

In this chapter, you studied many different formulas. Some of these formulas are used often enough that you should memorize them. For instance, it is helpful to remember that the circumference of a circle is $C = \pi d$ and the area of a circle is $A = \pi r^2$. In addition to memorizing formulas, we hope that this chapter has helped build your sense of what surface area and volume are actually measuring. For instance, you can think of the surface area as a measure of the material needed to cover the solid and you can think of volume as a measure of the material needed to fill the solid.

In Exercises 1–8, match the expression with its formula. (12.1–12.7)

a. $V = \frac{1}{3}Bh$ b. $S = 2B + Ph$ c. $V = Bh$ d. $V = \pi r^2h$

e. $A = \pi r^2$ f. $S = 2B + Ch$ g. $V = \frac{4}{3}\pi r^3$ h. $C = \pi d$

1. Circumference of a circle
2. Area of a circle
3. Surface area of a prism
4. Surface area of a cylinder
5. Volume of a prism
6. Volume of a cylinder
7. Volume of a pyramid or a cone
8. Volume of a sphere

In Exercises 9 and 10, find the circumference and area of the circle. (12.1)

9.

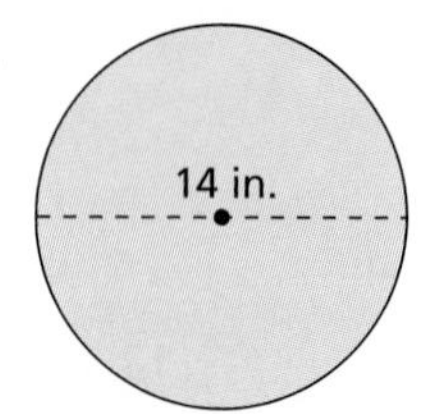

10.

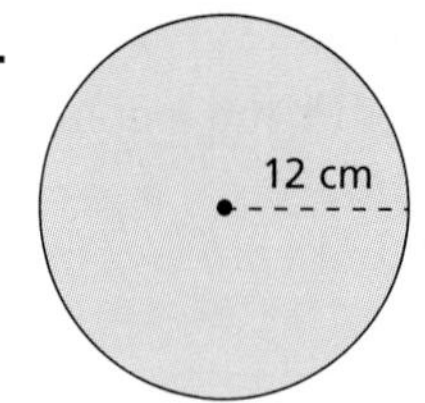

11. Are the blue areas equal? Explain. **(12.1)**

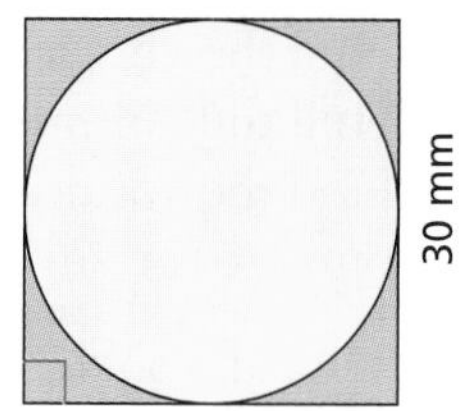

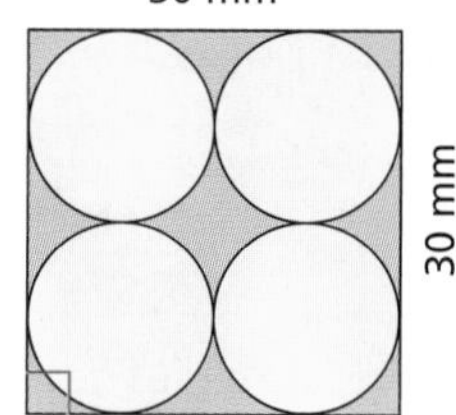

In Exercises 12–15, describe the solid that will result from folding the net. (12.2)

12.

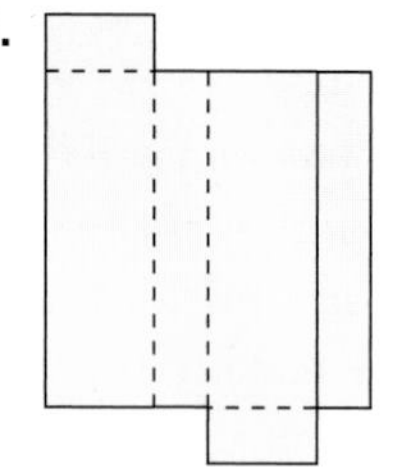

13.

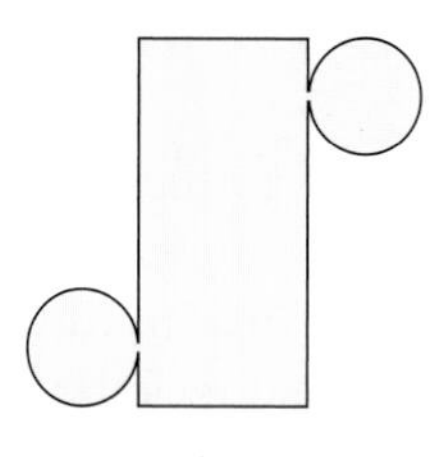

14.

15.

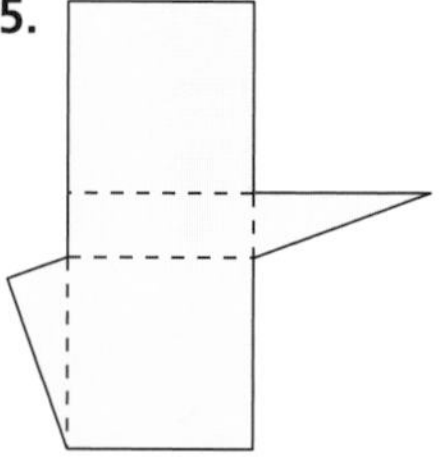

16. Sketch a prism that has isosceles triangles as its bases.

In Exercises 17–20, use the figure at the right. (12.2)

17. Identify the solid.
18. How many faces does the solid have?
19. How many vertices does the solid have?
20. How many edges does the solid have?

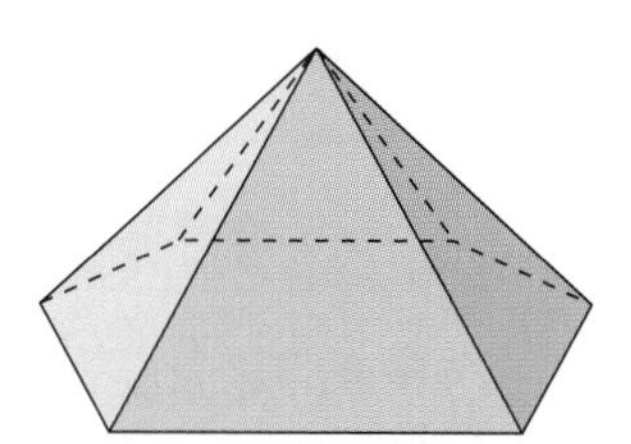

In Exercises 21–24, find the surface area of the solid. (12.3)

21.

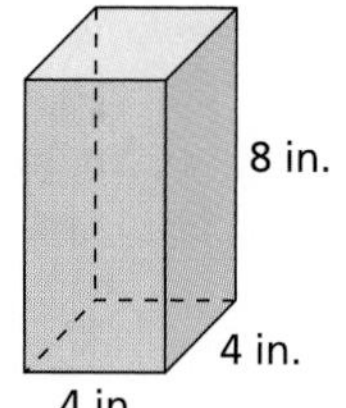

22.

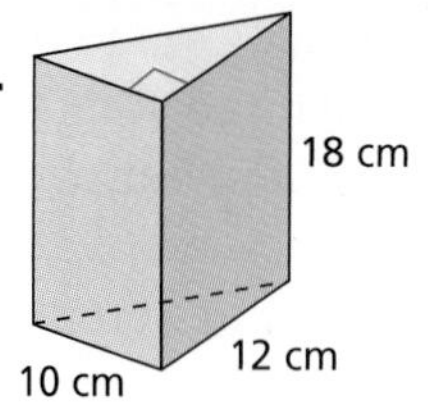

23.

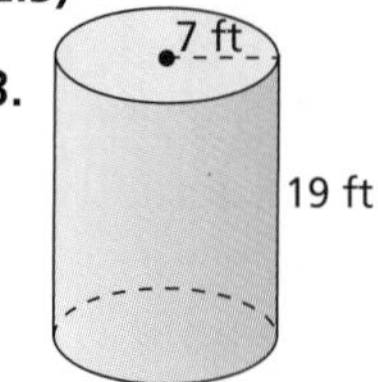

24.

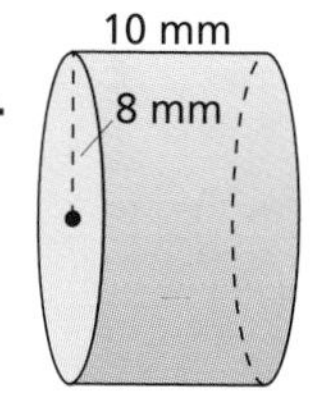

In Exercises 25–32, find the volume of the solid. (12.4–12.7)

25.

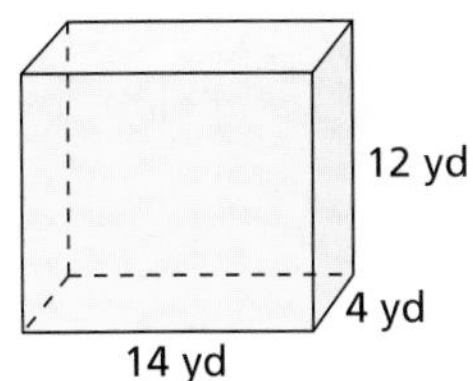

26.

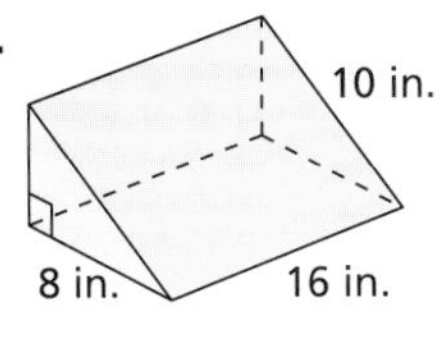

27.

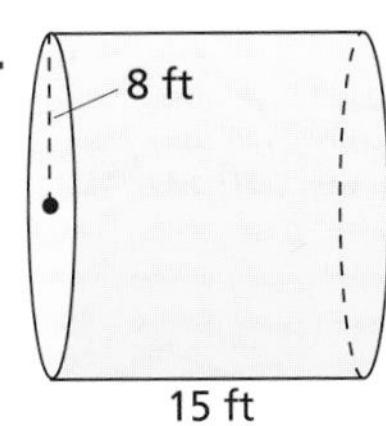

28.

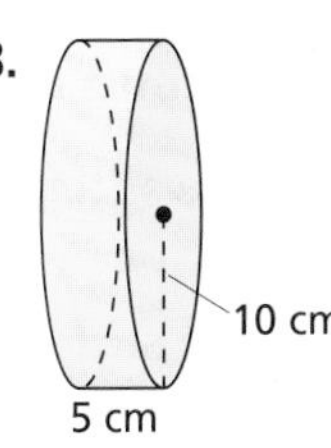

29.

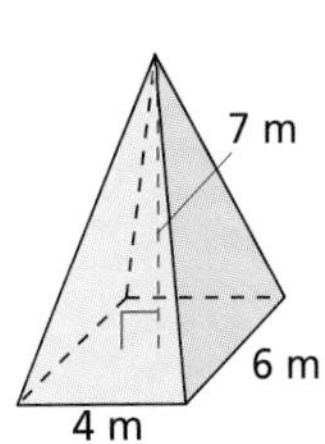

30.

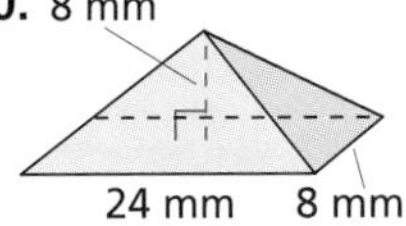

31.

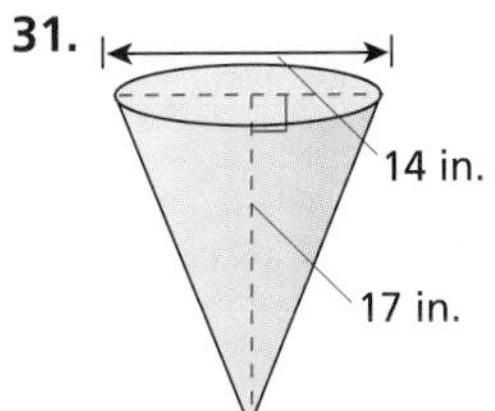

32.

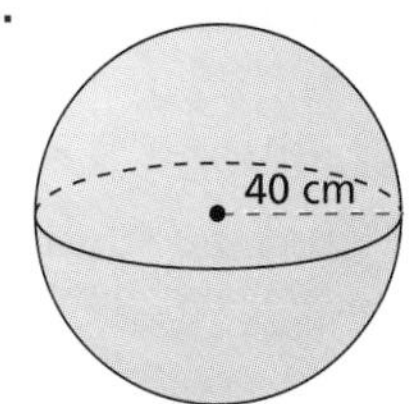

Estimation **In Exercises 33–35, find the volume of the solid to approximate the volume of the fish. (12.5–12.6)**

33.

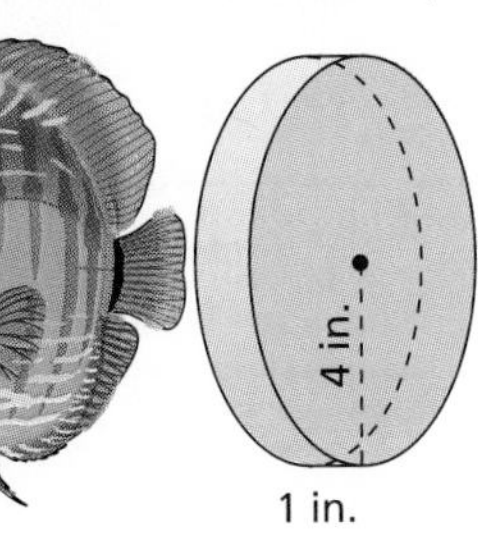

34.

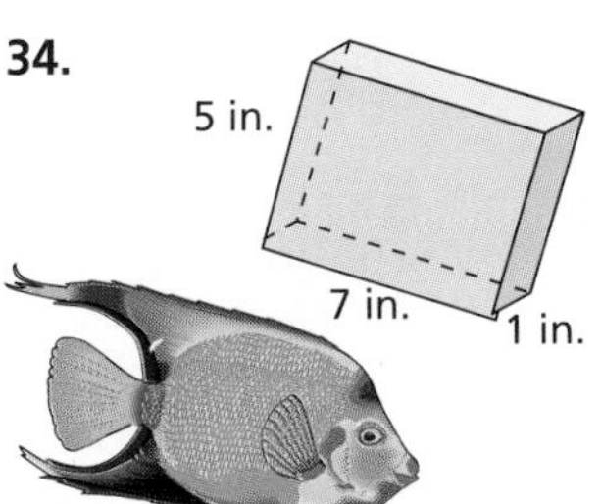

35.

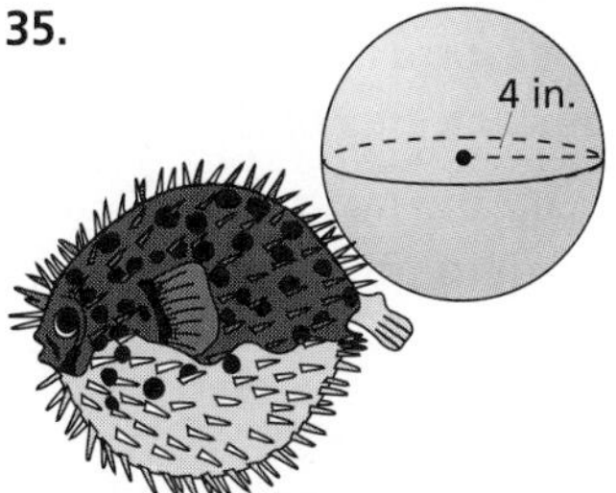

In Exercises 36 and 37, decide whether the solids are similar. If so, find the scale factor. (12.8)

36.

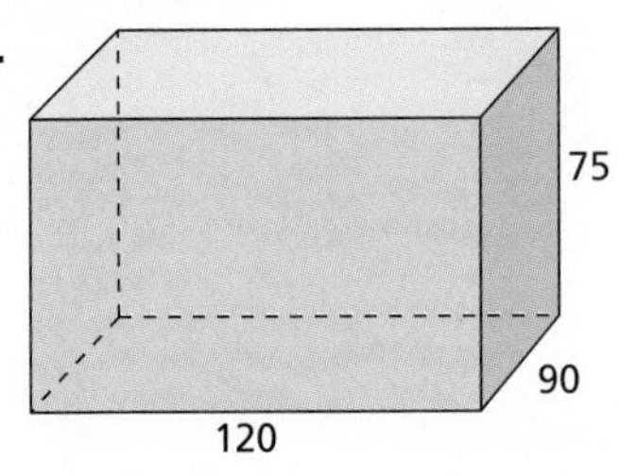

37.

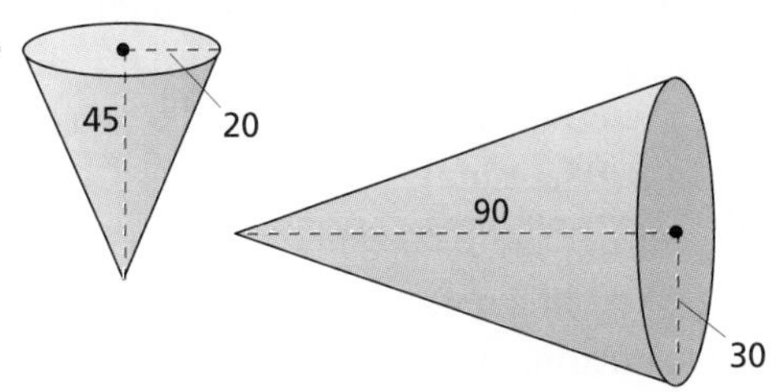

In Exercises 38–41, use the similar cylinders at the right. (12.8)

38. Find the scale factor of the large cylinder to the smaller cylinder.

39. What is the radius of the small cylinder?

40. Find the ratio of the surface area of the large cylinder to the surface area of the small cylinder.

41. How many of the small cylinders will hold as much water as the large cylinder? Explain.

9 cm
54 cm
18 cm

Designing a Model Train **In Exercises 42–48, you are designing model-train cars with shapes like those shown at the right.**

42. Find the surface area and volume of the refrigerator car.

43. Find the surface area and volume of the gondola car.

44. Find the surface area and volume of the tank car. (Hint: The surface area of a sphere of radius r is $S = 4\pi r^2$.)

45. Find the surface area and volume of the passenger car.

46. An actual refrigerator car is 60 feet long. Find the scale factor of an actual train to the model train.

47. Find the ratio of the surface area of the actual train to the model train. What is the surface area of the actual refrigerator car?

48. Find the ratio of the volume of the actual train to the model train. What is the volume of the actual refrigerator car?

Refrigerator Car

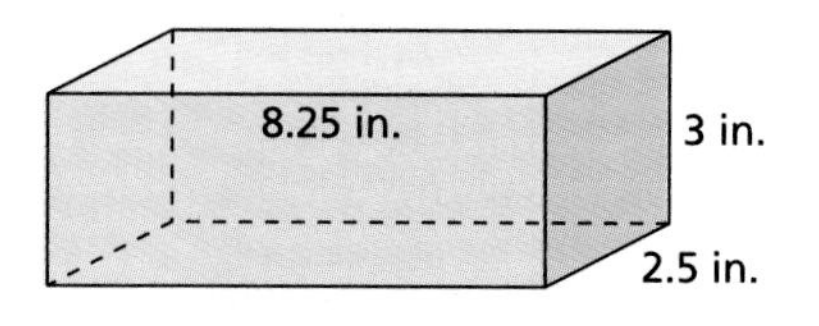

Gondola Car

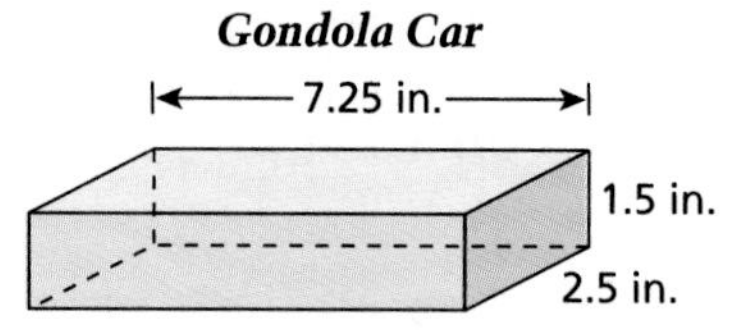

Tank Car

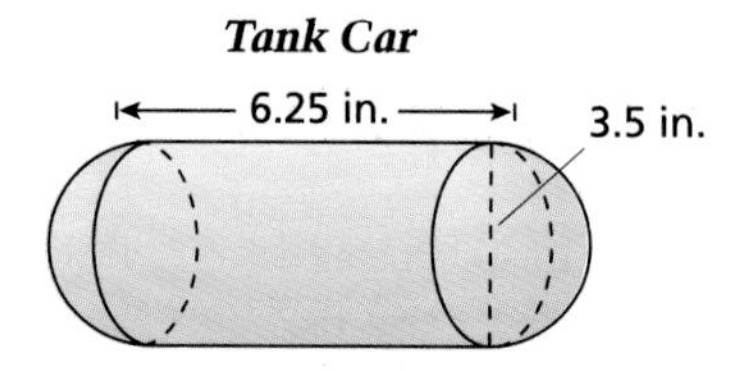

Passenger Car

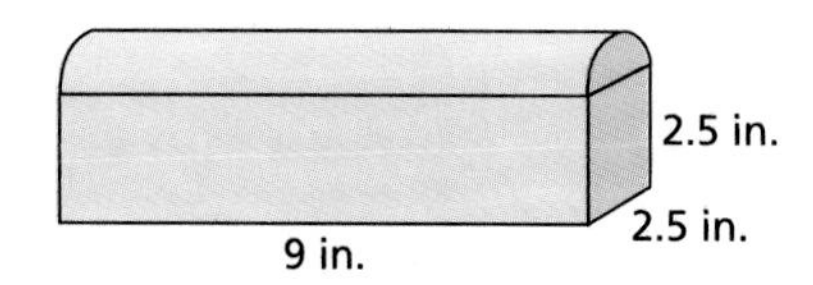

Designing a Tunnel **In Exercises 49–52, use the following information.**

You work for a construction company. You are assigned to design and supervise the construction of a straight railroad tunnel through a mountain. The tunnel needs to be 30 feet high for train clearance and 20 feet wide. The tunnel will be 13,200 feet long.

49. The tunnel can be modeled as half of a cylinder on top of a rectangular prism, so that cross sections of the tunnel are arches. Sketch the tunnel and label your sketch.

50. How much dirt and rock will be removed from the tunnel?

51. Find the surface area of the inside of the tunnel. (Do not include the area of each end.)

52. Suppose that each cross section was a half circle with 30-foot radius. Would this design require more or less removal of dirt and rock? Explain.

The longest railroad tunnel in the United States is the Moffat Tunnel. This 6.2-mile tunnel cuts through the Rocky Mountains in Colorado.

Chapter TEST

In Exercises 1–3, use the prism at the right.

1. How many faces does the prism have?

2. How many vertices does the prism have?

3. How many edges does the prism have?

In Exercises 4–7, use the cylinder at the right.

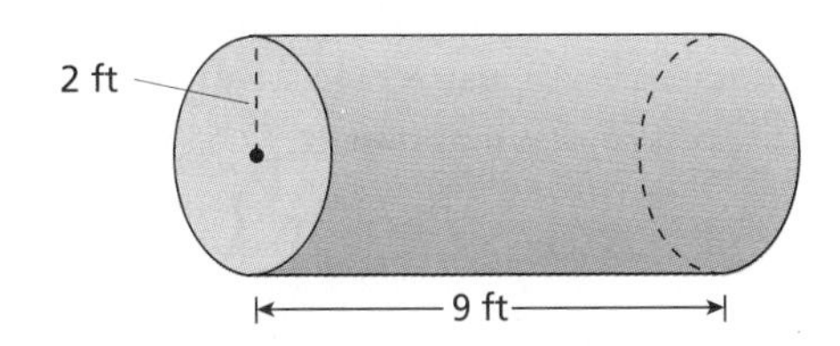

4. Find the circumference of one of the bases.

5. Find the area of one of the bases.

6. Find the surface area of the cylinder.

7. Find the volume of the cylinder.

In Exercises 8–11, find the indicated measure of the solid.

8. Volume

4 cm
3 cm
4 cm

9. Volume

6 ft
3 ft
4 ft
5 ft

10. Surface area

3 in.
2 in.
4 in.

11. Surface area

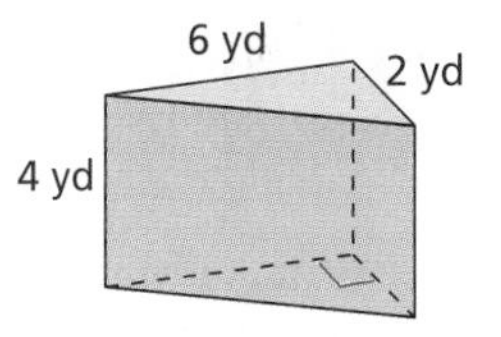

In Exercises 12–16, use the similar solids at the right.

12. Find the surface area of Figure A.

13. Find the volume of Figure A.

14. Find the scale factor of Figure A to Figure B.

15. Find the surface area of Figure B.

16. Find the volume of Figure B.

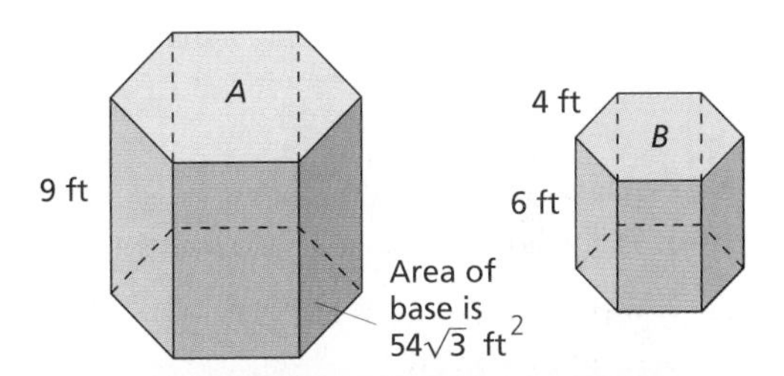

In Exercises 17–19, use the diagram below.

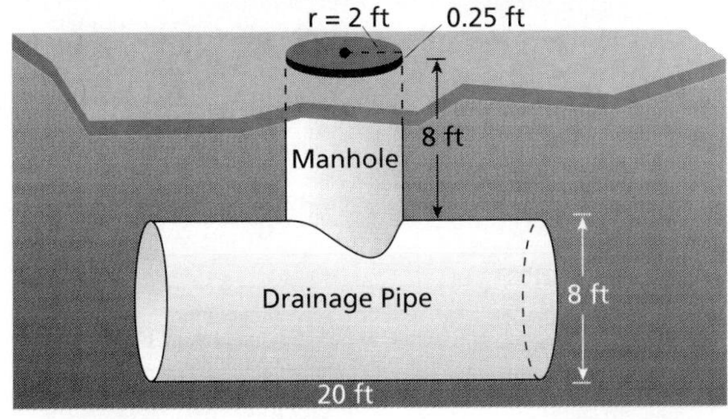

17. What is the surface area of the manhole cover?

18. Approximate the volume of the manhole.

19. What is the volume of the 20-foot cylinder?

The moon's diameter is about 2160 miles.

20. What is the volume of the moon?

Cumulative REVIEW ▪ *Chapters* 7–12

In Exercises 1–4, evaluate the expression. Then simplify, if possible. (7.1, 7.2, 7.4, 7.5)

1. $\frac{1}{2} + \frac{1}{3}$ **2.** $\frac{3}{4} \cdot \frac{8}{9}$ **3.** $\frac{7}{12} - \frac{5}{12}$ **4.** $\frac{12}{5} \div \frac{3}{5}$

In Exercises 5–8, solve the equation. (7.1, 7.2, 7.4, 7.5)

5. $y + \frac{3}{4} = \frac{1}{2}$ **6.** $\frac{1}{3}m = \frac{1}{6}$ **7.** $\frac{4}{5} = 8a$ **8.** $\frac{2}{7} = b - \frac{3}{2}$

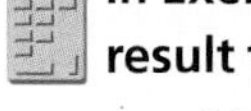

In Exercises 9–12, use a calculator to evaluate the expression. Round your result to two decimal places. (7.1–7.5)

9. $\frac{31}{40} \cdot \frac{26}{51}$ **10.** $\frac{76}{55} + \frac{7}{35}$ **11.** $\frac{93}{95} - \frac{54}{73}$ **12.** $\frac{9}{92} \div \frac{26}{70}$

In Exercises 13–16, write the fraction that represents the portion of the figure's area that is blue. Then write the fraction as a percent. (7.6)

13. 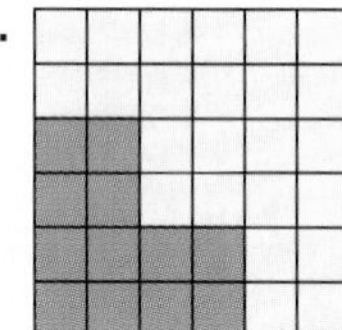

14. 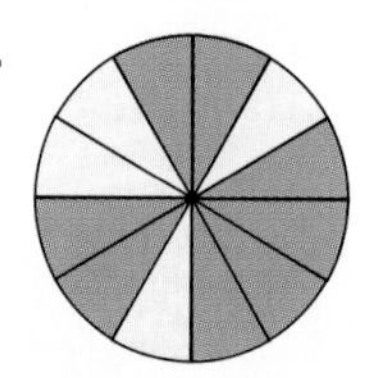

15. 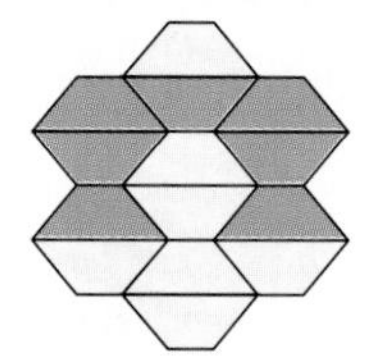

16. 

Job Security **In Exercises 17 and 18, use the circle graph. (7.7, 7.8)** *(Source: Delotte and Touche Trade Survey)*

17. Write each percent as a simplified fraction.

18. Five hundred people participated in the survey. How many people answered in each category?

In Exercises 19–22, write the verbal phrase as a rate or ratio. State whether it is a rate or a ratio. (8.1)

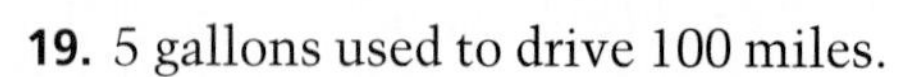

19. 5 gallons used to drive 100 miles.

20. Drove 25 kilometers in 40 minutes.

21. 18 people out of 24 people attended.

22. 99 chicks out of 100 chicks survived.

In Exercises 23–26, solve the proportion. (8.2)

23. $\frac{3}{4} = \frac{x}{32}$ **24.** $\frac{18}{5} = \frac{3}{y}$ **25.** $\frac{z}{6} = \frac{5}{9}$ **26.** $\frac{24}{w} = \frac{3}{5}$

In Exercises 27 and 28, use the table at the right that shows the number of points Tyrone "Muggsy" Bogues scored in each season in the National Basketball Association. (8.6)

27. Find the percent decrease in points from the 1989–90 season to the 1990–91 season.

28. Find the percent increase in points from the 1990–91 season to the 1991–92 season.

Season	Points
1989–90	763
1990–91	568
1991–92	730
1992–93	808

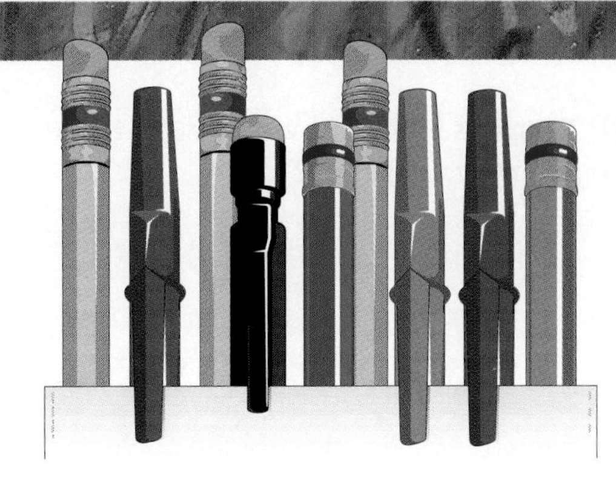

Probability **In Exercises 29–30, you have 3 regular pencils, 2 colored pencils, 1 mechanical pencil, and 3 ballpoint pens. (8.7, 8.8)**

29. How many different ways could you choose one pencil and one pen?

30. If you randomly pick one writing instrument, what is the probability that it is a pencil?

In Exercises 31–34, solve the equation and decide whether the solution is rational or irrational. (9.1–9.2)

31. $x^2 = 121$

32. $84 = y^2$

33. $504 = 4n^2$

34. $-8 = m^2 - 17$

In Exercises 35–37, a and b are the lengths of the legs of a right triangle, and c is the length of the hypotenuse. Find the missing length. (9.3)

35. $a = 7, b = 24$

36. $a = 5, c = 14.87$

37. $b = 13, c = 15.26$

In Exercises 38–43, solve the inequality. Then graph the solution on a number line. (9.5–9.7)

38. $x + 14 \leq 9$

39. $\frac{2}{3} < -4y$

40. $-17 \leq -12n + 19$

41. $5(1 + 2p) < 13$

42. $2(7 - x) > 4x$

43. $3(5 + x) \leq \frac{1}{4}(20 + 8x)$

In Exercises 44–47, decide whether the triangle can have the given side lengths. Explain. (9.3, 9.8)

44.

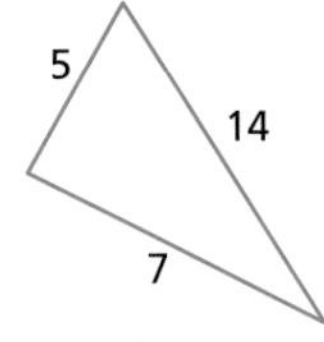

45.

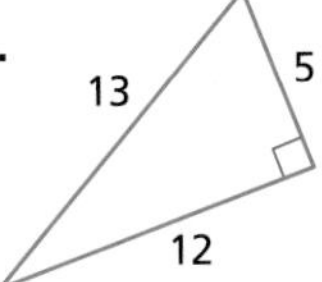

46.

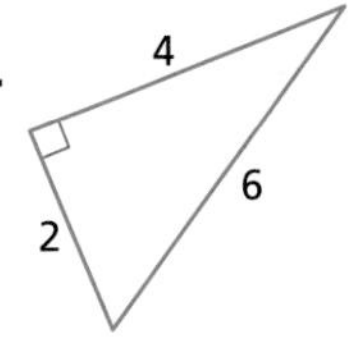

47.

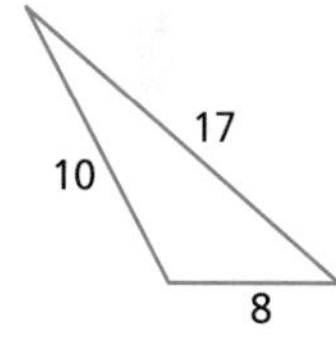

48. *Television* If you have a 25-inch diagonal television screen and its width is 16.7 inches, what is its length? **(9.4)**

In Exercises 49–56, use the figure at the right. (10.1–10.3)

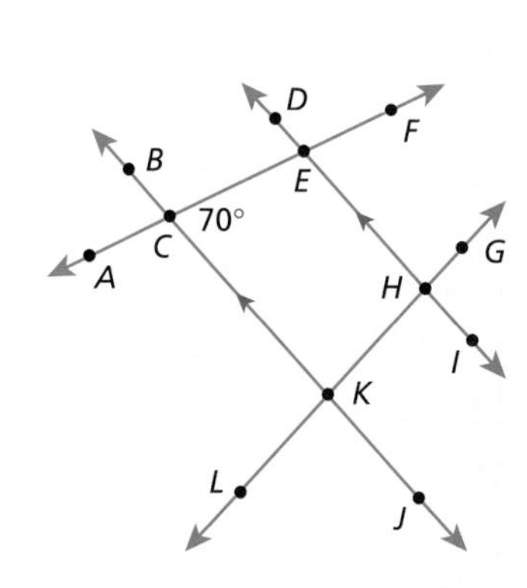

49. Write two other names for the line $\overleftrightarrow{EH}$.

50. List five line segments that have C as an endpoint.

51. Write another name for $\overrightarrow{KH}$.

52. Write three other names for $\angle JKG$.

53. List 2 pairs of vertical angles.

54. List 4 angles whose measure is 70°.

55. List 2 pairs of congruent corresponding angles.

56. List 2 pairs of noncongruent corresponding angles.

In Exercises 57–60, use a protractor to measure the angle. Is it acute, obtuse, right, or straight? (10.2)

57. 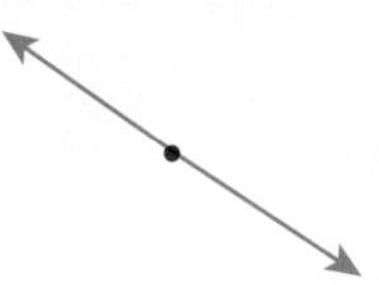

58. 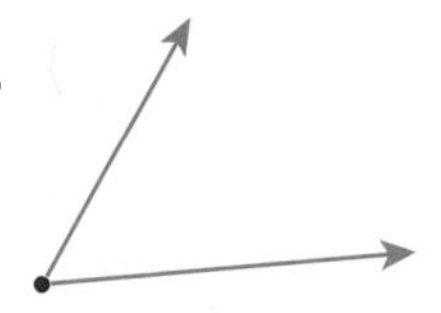

59. 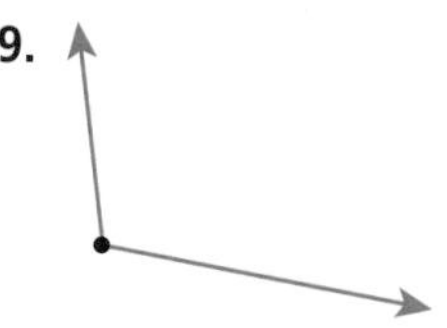

60. 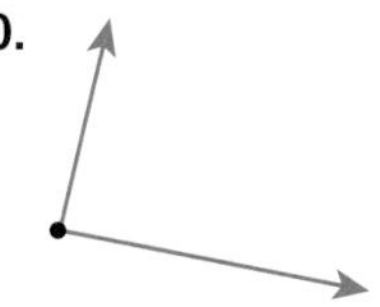

In Exercises 61–64, identify the polygon. (Be specific.) Then find *x*. (10.6, 10.8)

61. 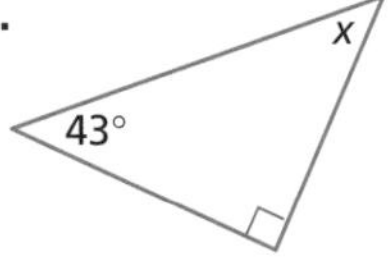

62. 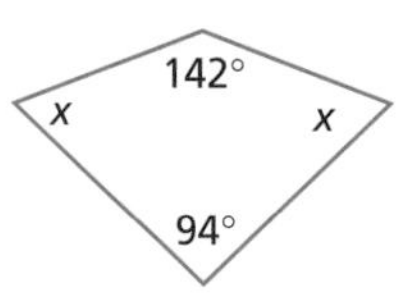

63. 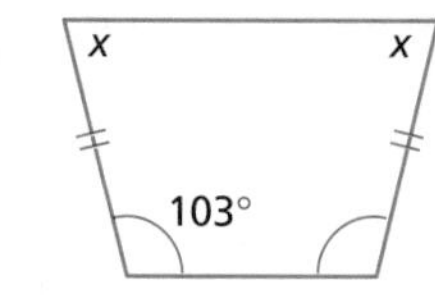

64. 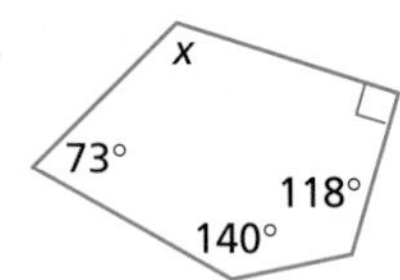

In Exercises 65 and 66, identify any symmetry of the figure. (10.4)

65. 

66. 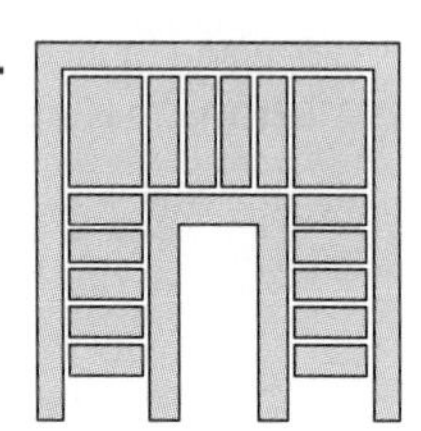

In Exercises 67 and 68, order the angles from the smallest to the largest. (10.9)

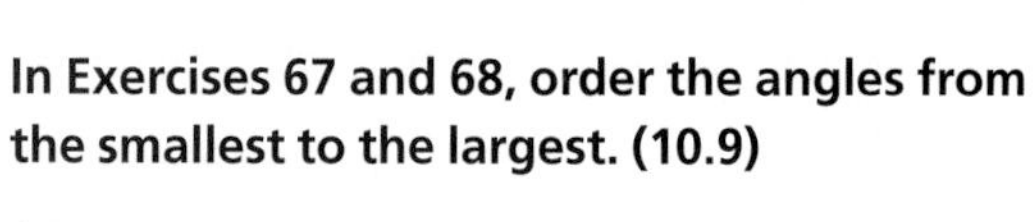

67. 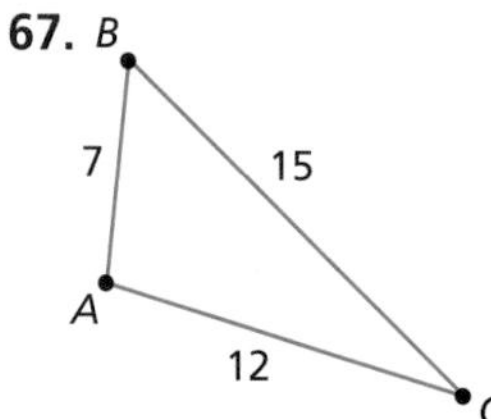

68. 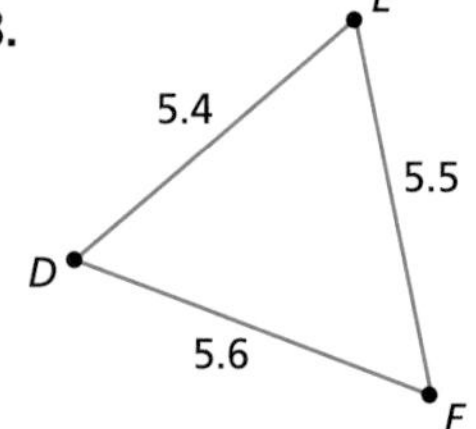

In Exercises 69–72, find the area and perimeter of the polygon. (11.1)

69. 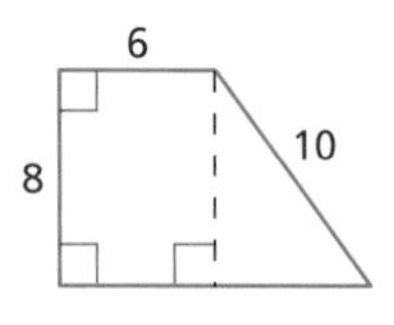

70. 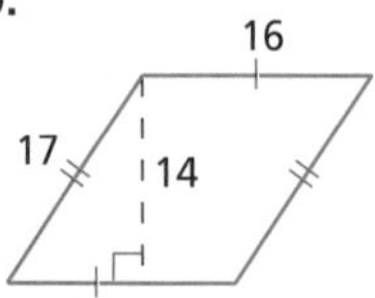

71. 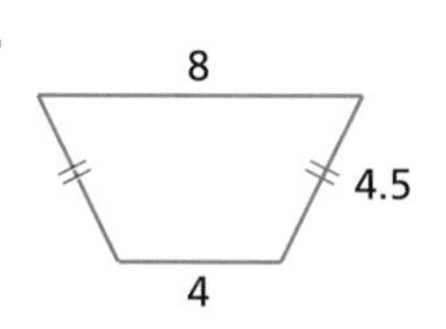

72. 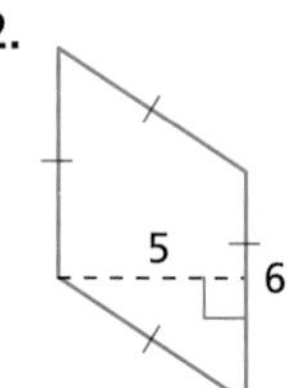

In Exercises 73–75, use the figure at the right, where $\triangle ABC$ is reflected in line ℓ. (11.3)

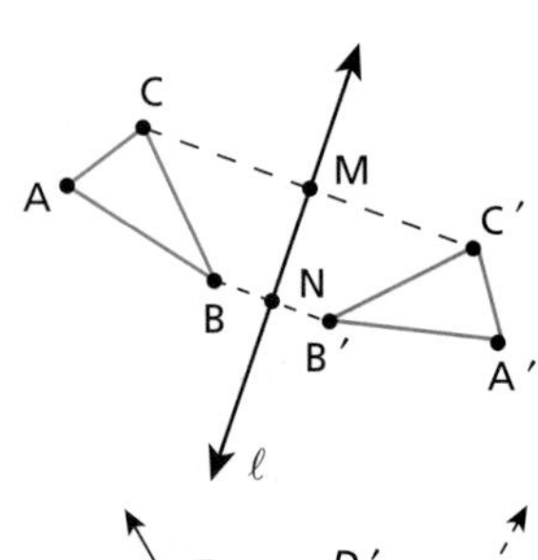

73. Is $\triangle ABC$ congruent to $\triangle A'B'C'$?

74. Is the length of $\overline{BB'}$ half the length of $\overline{CC'}$? Explain.

75. Is $CM = C'M$? Explain.

In Exercises 76–78, use the figure at the right, where $\triangle DEF$ is rotated about point *O*. (11.4)

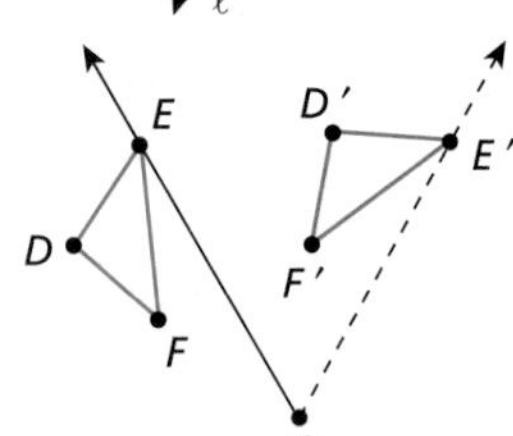

76. Is the rotation clockwise or counterclockwise?

77. Is $\triangle DEF$ congruent to $\triangle D'E'F'$?

78. Is $DD' = FF'$? Explain.

Indirect Measurement **In Exercises 79 and 80, you are measuring the width of the river shown in the diagram. To begin, you place stakes at points *A* and *B*. Then you measure $\overline{AB}$ and $\angle B$. (11.8)**

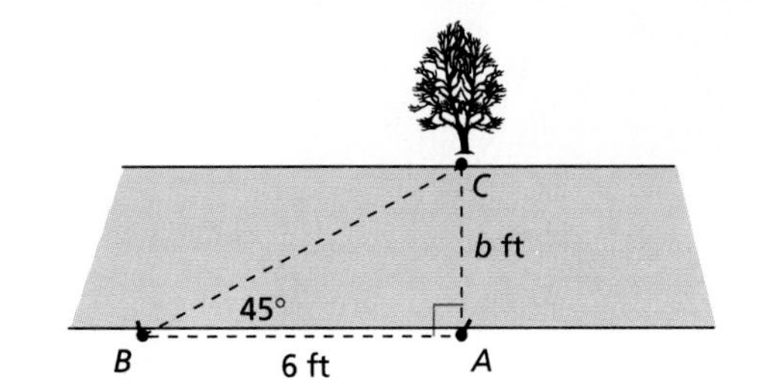

79. Write the trigonometric ratio for the tangent of $\angle B$.

80. Find the width of the river.

In Exercises 81–84, find the circumference and area of the circle. Round your results to two decimal places. (12.1)

81. 

$r = 9$ mm

82. 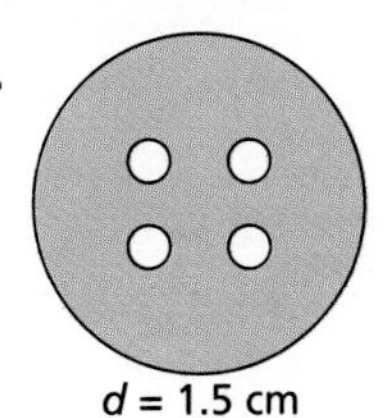
$d = 1.5$ cm

83. 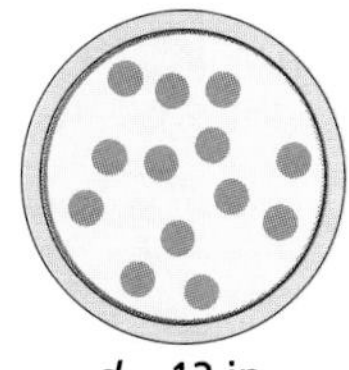
$d = 12$ in.

84. 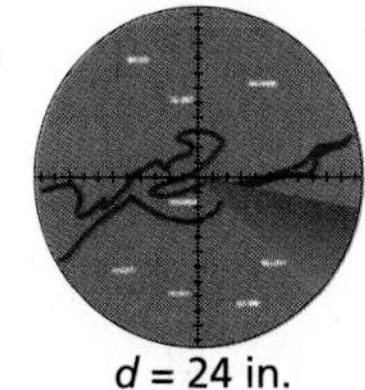
$d = 24$ in.

In Exercises 85–88, find the surface area and volume of the solid. Round your results to two decimal places. (12.3–12.5)

85. 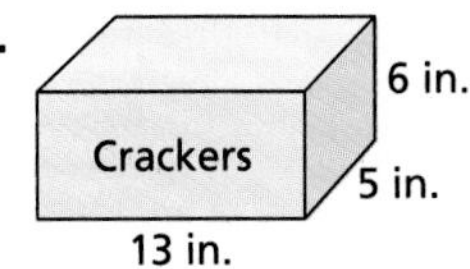

86. 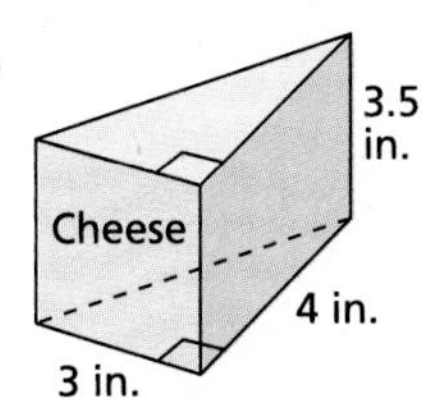

87. 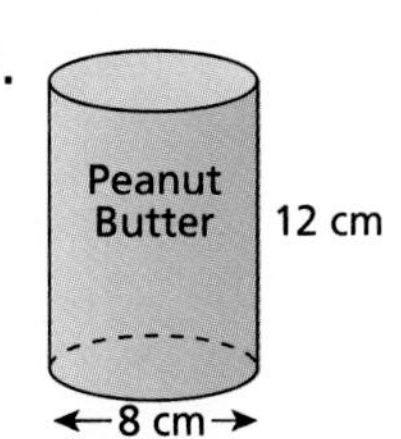

88. 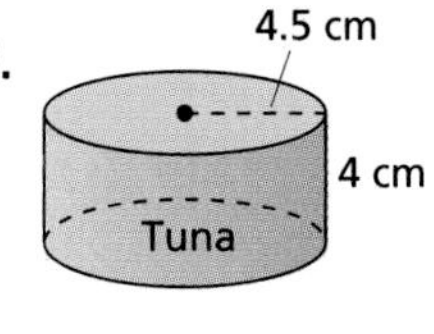

In Exercises 89–92, find the volume of the solid. Round your result to two decimal places. (12.6, 12.7)

89. 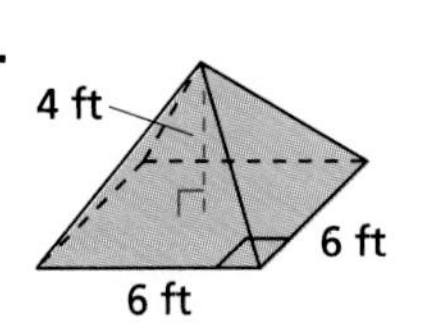

90. 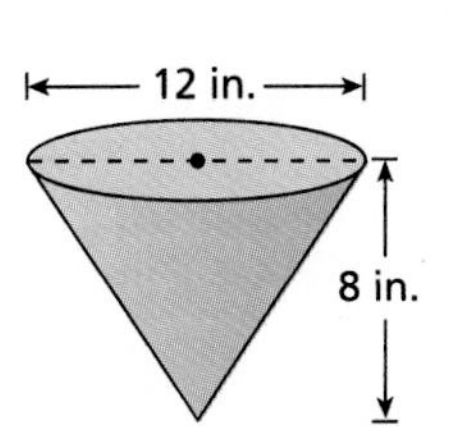

91. 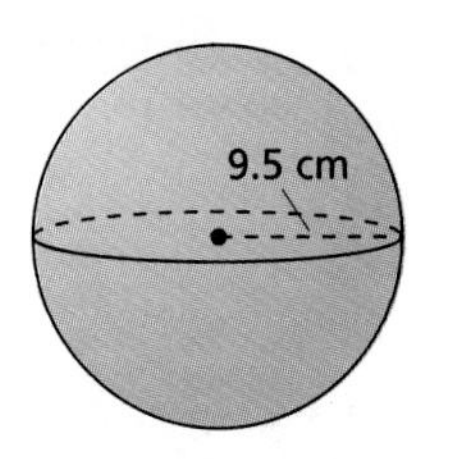

92. 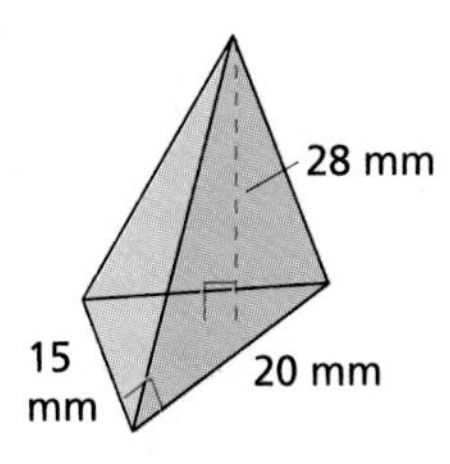

93. *Writing* Trace the polygon at the right on a piece of paper. Then draw a larger, similar polygon. Find the side lengths in centimeters, angle measures, and areas of each polygon. Write a statement that compares these measures. **(11.6)**

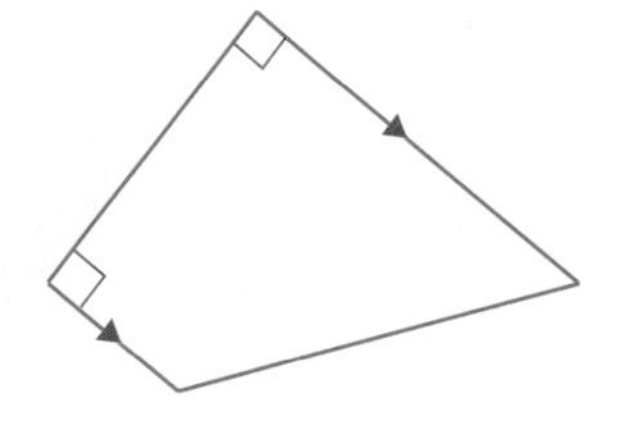

CHAPTER 13

Exploring Linear Equations

LESSONS

Increased air pollution from motor-vehicle emissions has led to the passage of federal and state clean air laws which mandate alternatives to gasoline driven vehicles in the near future. The Aztec, a solar electric car developed by students at Massachusetts Institute of Technology, can travel 135 miles between charges at a top speed of 50 miles per hour.

Real Life

Advanced Transportation

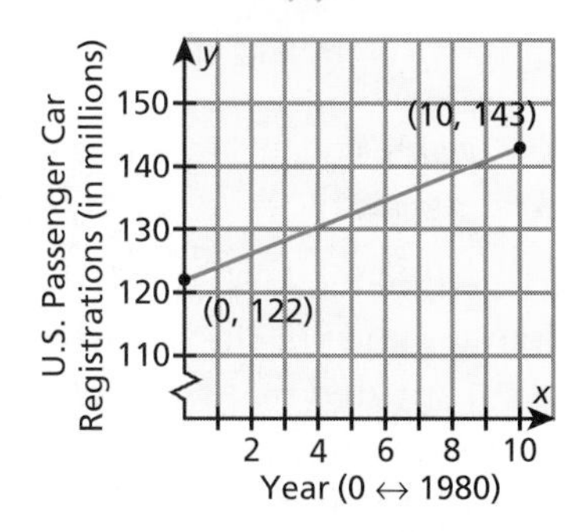

$$\text{Slope} = \frac{\text{Change in } y\text{-values}}{\text{Change in } x\text{-values}}$$

$$= \frac{143 - 122}{10} = 2.1$$

The *slope* of the graph of a nonvertical line describes how the line slants from left to right. In this chapter, you will learn how in real-life situations, slope is used to describe a constant, or average, rate of change; for example, the change in motor vehicle registrations over time.

The linear graph at the left shows the number of United States passenger car registrations from 1980 through 1990. Since the line slants upward, its slope is positive. The slope tells us that the average increase in car registrations from 1980 through 1990 was about 2.1 million cars per year.

13.1 Linear Equations in Two Variables

What you should learn:

Goal 1 How to find solutions of a linear equation in two variables

Goal 2 How to organize solutions of real-life problems using linear equations

Why you should learn it:

You can use linear equations in two variables to solve real-life problems, such as comparing Fahrenheit and Celsius temperature scales.

Goal 1 Solutions of Linear Equations

In this chapter, you will study **linear equations** in two variables. Here are some examples.

$$y = 2x + 1 \qquad C = \pi d \qquad A = 1.06P$$

In a linear equation, variables occur only to the first power. Equations such as $A = \pi r^2$ and $V = s^3$ are not linear.

On page 136, you learned that an ordered pair (x, y) is a **solution** of an equation involving x and y if the equation is true when the values of x and y are substituted into the equation. Most equations involving two variables have many solutions. For instance, three solutions of $y = x + 3$ are shown below.

Equation	Solution (x, y)	Check by Substituting
$y = x + 3$	$(1, 4)$	$4 = 1 + 3$
$y = x + 3$	$(2, 5)$	$5 = 2 + 3$
$y = x + 3$	$(3, 6)$	$6 = 3 + 3$

To find solutions of linear equations in two variables, choose a value for one of the variables, substitute that value into the equation, and then solve for the other variable.

Example 1 *Finding Solutions of Linear Equations*

List several solutions of $2x + y = 10$.

Solution Begin by choosing values of x. Substitute each value into the equation and solve the resulting equation for y.

Choose an x-Value	Substitute for x	Solve for y	Solution
$x = 0$	$2(0) + y = 10$	$y = 10$	$(0, 10)$
$x = 1$	$2(1) + y = 10$	$y = 8$	$(1, 8)$
$x = 2$	$2(2) + y = 10$	$y = 6$	$(2, 6)$
$x = 3$	$2(3) + y = 10$	$y = 4$	$(3, 4)$

■

A **table of values** like that shown at the left can help you organize solutions that you have found.

x-Value	y-Value	Solution
0	10	(0, 10)
1	8	(1, 8)
2	6	(2, 6)
3	4	(3, 4)

Goal 2 Solving Real-Life Problems

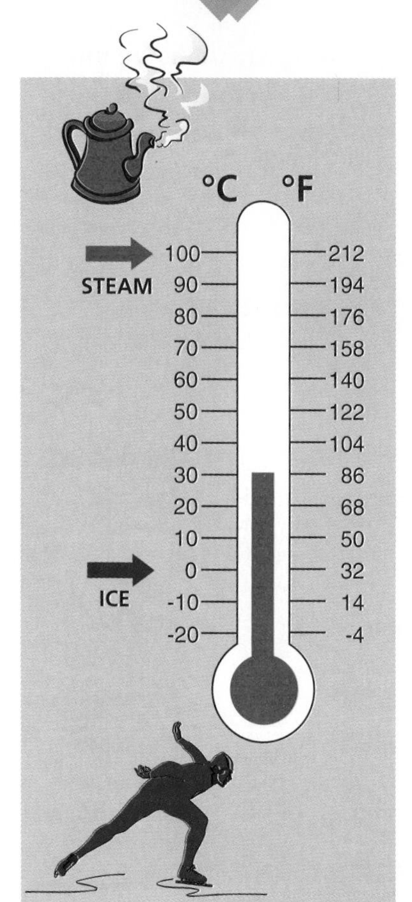

Example 2 *Problem Solving: Organizing Solutions*

The relationship between Fahrenheit temperature F and Celsius temperature C is given by the linear equation

$$F = \frac{9}{5}C + 32.$$

You are designing a poster to help people convert between the two temperature scales. Which temperatures would you use as samples on the poster?

Solution There are several possible temperatures that would be good to use. For instance, on the Celsius scale, water freezes at 0° and boils at 100°. On the Fahrenheit scale, 68° is a comfortable temperature and −4° is very cold.

To find values of F that correspond to values of C, use the following steps.

C-Value	Substitute	Solve for F	Solution
$C = 0$	$F = \frac{9}{5}(0) + 32$	$F = 32$	$(0, 32)$
$C = 100$	$F = \frac{9}{5}(100) + 32$	$F = 212$	$(100, 212)$

To find values of C that correspond to values of F, use the following steps.

F-Value	Substitute	Solve for C	Solution
$F = 68$	$68 = \frac{9}{5}C + 32$	$C = 20$	$(20, 68)$
$F = -4$	$-4 = \frac{9}{5}C + 32$	$C = -20$	$(-20, -4)$

After finding several other temperatures, you could represent them with a poster like that shown at the left. ■

Communicating about MATHEMATICS

Cooperative Learning

SHARING IDEAS about the Lesson

Mental Math Work with a partner. Here are two rules that people use to mentally change between Fahrenheit and Celsius temperatures. Apply the rules to the temperatures in Example 2. How are the two rules related to each other?

A. To change from Celsius to Fahrenheit, divide by 5, multiply by 9, and add 32.

B. To change from Fahrenheit to Celsius, subtract 32, divide by 9, and multiply by 5.

EXERCISES

Guided Practice

CHECK for Understanding

1. Which of the following are solutions of $2x + 3y = 7$? Explain.

   a. $(1, 2)$ b. $(2, 1)$

   c. $(5, -1)$ d. $(4, -1)$

2. Which of the following equations are linear? Explain.

   a. $A = \pi r^2$ b. $C = 2\pi r$

   c. $r + \frac{1}{2}t = 30$ d. $100 - 6p = S$

In Exercises 3 and 4, complete the table of values showing solutions of the equation.

3. $y = x + 5$ 4. $2x - y = 4$

x	−3	−2	−1	0	1	2	3
y	?	?	?	?	?	?	?

Independent Practice

In Exercises 5–8, decide whether the ordered pair is a solution of $7x - y = 5$.

5. $(0, -5)$ 6. $(2, 1)$ 7. $(-1, 12)$ 8. $\left(\frac{1}{2}, -\frac{3}{2}\right)$

In Exercises 9–12, find several solutions of the linear equation. Use a table of values to organize your results.

9. $x + y = 6$ 10. $x + 2y = 13$ 11. $6x + 2y = 24$ 12. $y = \frac{1}{3}x + 2$

Describing Patterns **In Exercises 13 and 14, use the table to decide whether the relationship between *x* and *y* is linear. Explain your reasoning.**

13.

x	−3	−2	−1	0	1	2	3
y	15	13	11	9	7	5	3

14.

x	−3	−2	−1	0	1	2	3
y	1	2	4	7	11	16	22

In Exercises 15 and 16, write the sentence as a linear equation. Then list several solutions.

15. The difference of 6 times a number and 4 times another number is 12.

16. The sum of half a number and twice another number is 54.

Geometry **In Exercises 17–20, match the linear equation with the figure. Then list several solutions.**

a.

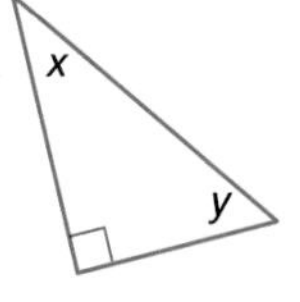

b.

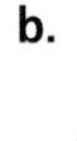

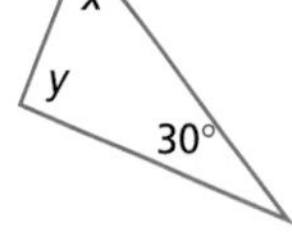

c.

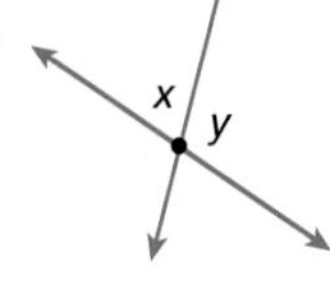

d.

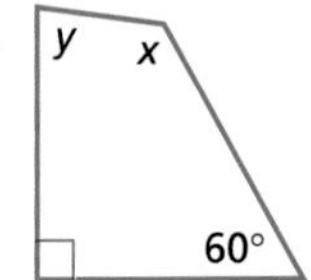

17. $x + y = 150°$ 18. $x + y = 210°$ 19. $x + y = 180°$ 20. $x + y = 90°$

Unit Conversions **In Exercises 21–23, use the equation $y = 2.54x$ which relates a centimeter measurement, y, to an inch measurement, x.**

21. How long, in centimeters, is a 12-inch ruler?

22. How long, in inches, is a 100-centimeter ruler?

23. Create a table to help convert between inches and centimeters.

Children's Books **In Exercises 24–27, use the following information.**

For 1987 through 1992, the amount A (in billions of dollars) spent each year on children's books in the United States can be modeled by the linear equation $A = 0.87 + 0.2t$ where $t = 1$ corresponds to 1987. ***(Source: Book Industry Group, Inc.)***

24. How much was spent in 1987?

25. How much was spent in 1992?

26. How much more was spent each year?

27. Use the graph to estimate the amount spent in 1996. Use the algebraic model to confirm your answer.

In Exercises 28 and 29, determine whether the equations have the same solutions. Explain your reasoning.

28. $3x + 5y = 16$

$12x + 20y = 64$

29. $9x - 2y = 18$

$18x - 4y = 30$

30. ***Think about It*** Does a linear equation in two variables have a *finite* or an *infinite* number of solutions? Write a paragraph explaining your answer.

Integrated Review

Making Connections within Mathematics

31. ***Tables and Solutions*** Complete the table of values for the equation $x + y = 4$.

x	-3	-2	-1	0	1	2	3
y	?	?	?	?	?	?	?

32. ***Scatter Plots and Patterns*** Construct a scatter plot for the data in Exercise 31. Describe the pattern.

Exploration and Extension

33. ***Finding a Pattern*** Consider the ordered pairs $\{(1, 3), (2, 5), (3, 7), (4, 9), (5, 11)\}$.

a. In each ordered pair, how are x and y related?

b. How can you express this relationship as a linear equation?

LESSON INVESTIGATION 13.2

Exploring Graphs of Linear Equations

Materials Needed: graph paper, measuring tape

Example — *Gathering and Plotting Data*

Group Activity Each person in the group should record the measurements indicated in the figure below. (You can use inches or centimeters, but make each pair of measurements in the same units.) Use the measurements to write the following four ordered pairs.

(Span, Height) (Foot, Forearm)

(Thumb, Wrist), (Wrist, Neck)

Plot the four ordered pairs in a coordinate plane like that shown at the right. Include the two lines. Everyone in the group should plot his or her measurements on the same coordinate plane.

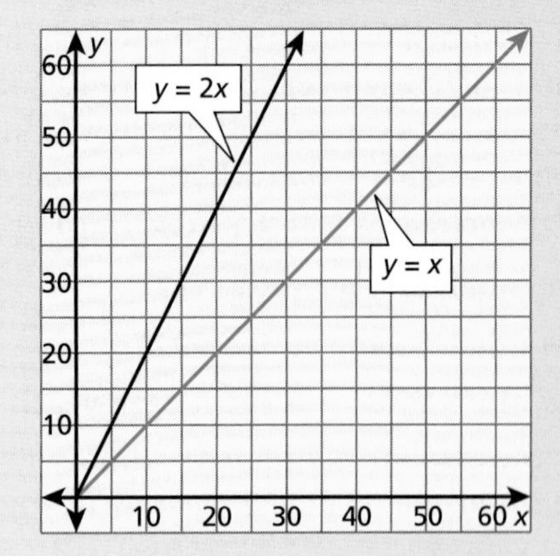

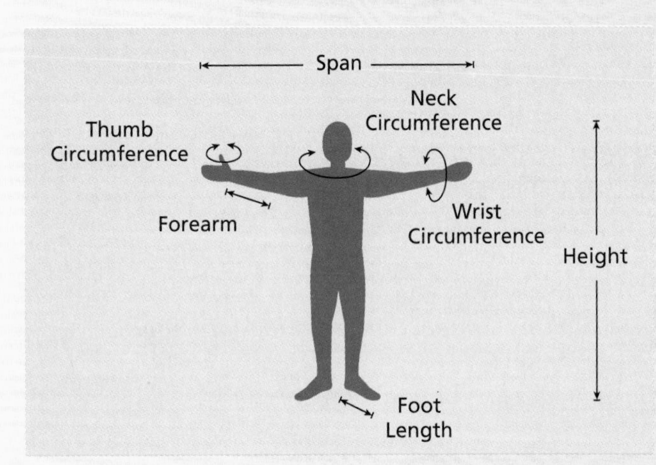

■

Exercises

1. *Group Activity* Discuss any patterns you find in the points that your group plotted.
2. Which of the points tended to fall on or near the line that is labeled $y = x$? How do the x- and the y-coordinates of points on this line compare?
3. Which of the points tended to fall on or near the line that is labeled $y = 2x$? How do the x- and the y-coordinates of points on this line compare?

13.2 Exploring Graphs of Linear Equations

What you should learn:

Goal 1 How to use a table of values to sketch the graph of a linear equation

Goal 2 How to recognize graphs of horizontal and vertical lines

Why you should learn it:

You can use graphs of linear equations to help you recognize relationships between two variables, such as the relationship between the year and the enrollment in a class.

Goal 1 Graphing Linear Equations

LESSON INVESTIGATION

Investigating Graphs of Linear Equations

Group Activity Use the techniques you studied in Lesson 13.1 to find several solutions of the equation $y = 6 - x$. Use x-values of $-3, -2, -1, 0, 1, 2, 3, 4$, and 5. Organize the nine solutions in a table of values. Then plot all nine solutions in a coordinate plane. What do you notice about the points?

In this investigation, you may have discovered that all the solution points lie on a line.

Graph of a Linear Equation

The **graph** of an equation is the graph of all its solutions. The graph of every linear equation is a line.

Example 1 *Graphing a Linear Equation*

Sketch the graph of $y = 2x - 2$.

Solution Begin by making a table of values.

x-Value	Substitute	Solve for y	Solution
$x = -2$	$y = 2(-2) - 2$	$y = -6$	$(-2, -6)$
$x = -1$	$y = 2(-1) - 2$	$y = -4$	$(-1, -4)$
$x = 0$	$y = 2(0) - 2$	$y = -2$	$(0, -2)$
$x = 1$	$y = 2(1) - 2$	$y = 0$	$(1, 0)$
$x = 2$	$y = 2(2) - 2$	$y = 2$	$(2, 2)$
$x = 3$	$y = 2(3) - 2$	$y = 4$	$(3, 4)$

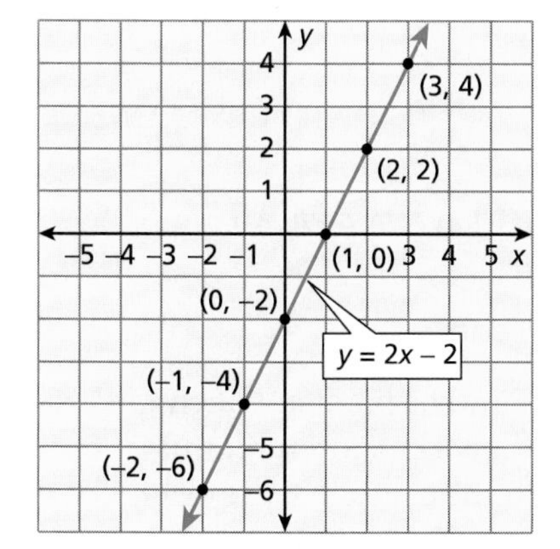

Plot the solutions in a coordinate plane. Finally, draw a line through the points. The line is the graph of the equation. ■

Goal 2 Horizontal and Vertical Lines

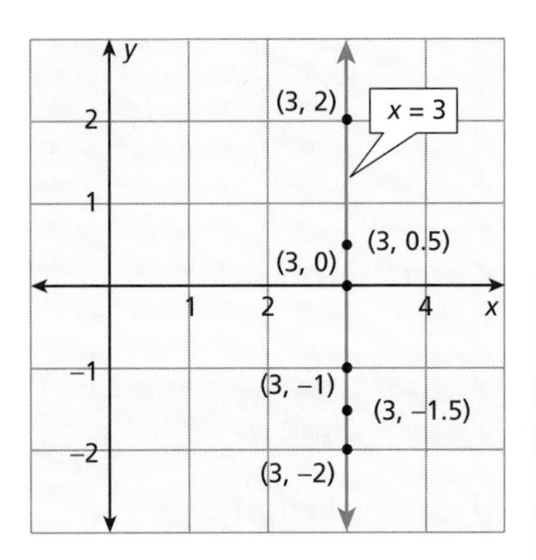

Some linear equations, such as $x = 3$, have just one variable. All solutions of this equation are of the form $(3, y)$. For instance, the points $(3, -2)$, $(3, -1.5)$, $(3, -1)$, $(3, 0)$, $(3, 0.5)$, and $(3, 2)$ are solutions of the equation. If you plot these points, you will notice that they lie on a vertical line.

Graphs of Linear Equations in One Variable

1. The graph of the equation $x = a$ is a vertical line that passes through the point $(a, 0)$.
2. The graph of the equation $y = b$ is a horizontal line that passes through the point $(0, b)$.

Real Life *Education*

Example 2 *Plotting Data*

The enrollments of two classes, Class A and Class B, for 1985 through 1995 are shown in the table. Let $t = 0$ represent 1985. Plot the data and describe the pattern.

t	0	1	2	3	4	5	6	7	8	9	10
A	36	36	36	36	36	36	36	36	36	36	36
B	31	32	33	34	35	36	37	38	39	40	41

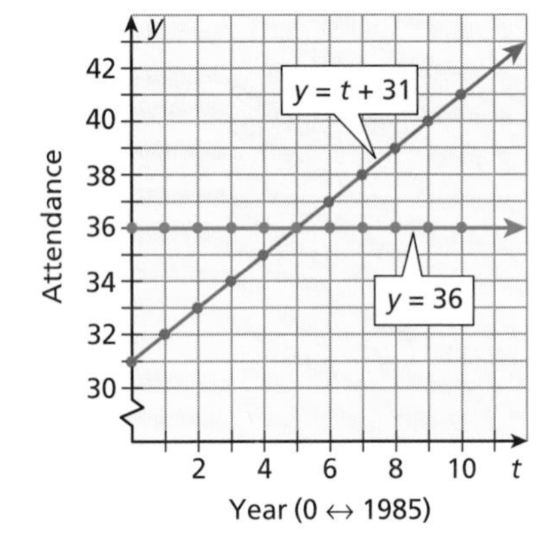

Solution Begin by writing the data as ordered pairs. Then plot the points in a coordinate plane, as shown at the left. From the graphs, you can see that each class's enrollment is represented by a line. For Class A, the line is horizontal because the enrollment did not change. For Class B, the line is not horizontal—it slopes upward because the enrollment is increasing. The enrollment for each class can be modeled by a linear equation.

$y = 36$ *Class A*

$y = t + 31$ *Class B* ■

Communicating about MATHEMATICS

SHARING IDEAS about the Lesson

Finding Points of Intersection Find the point at which the two graphs in Example 2 intersect. Then interpret this point of intersection verbally in the context of the example.

EXERCISES

Guided Practice

CHECK for Understanding

1. *Writing* In your own words, explain how to sketch the graph of an equation.

In Exercises 2–4, match the equation with the description of its graph.

a. Horizontal line **b.** Vertical line **c.** Slanted line

2. $2x + 3y = 8$ **3.** $y = 4$ **4.** $x = -2$

In Exercises 5–8, sketch the graph of the equation.

5. $y = 5$ **6.** $x = -4$ **7.** $y = 3x - 1$ **8.** $x + y = 8$

Independent Practice

In Exercises 9–12, match each equation with its graph.

a.

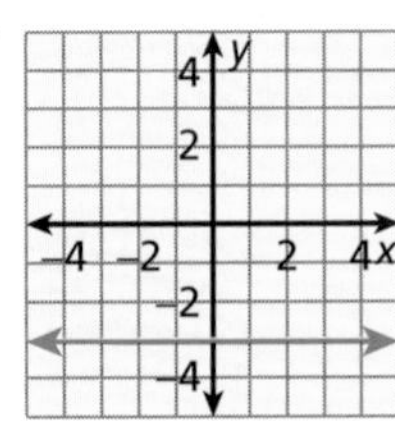

b.

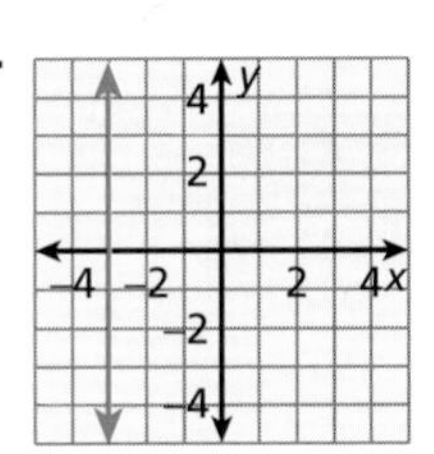

c.

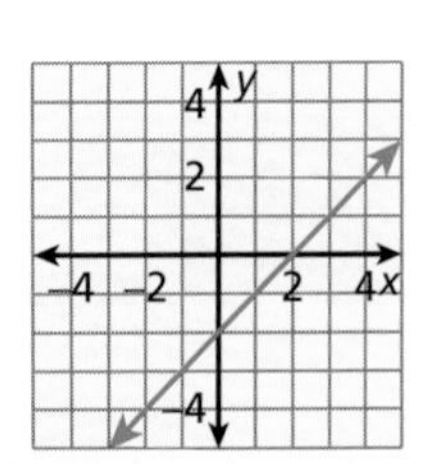

d.

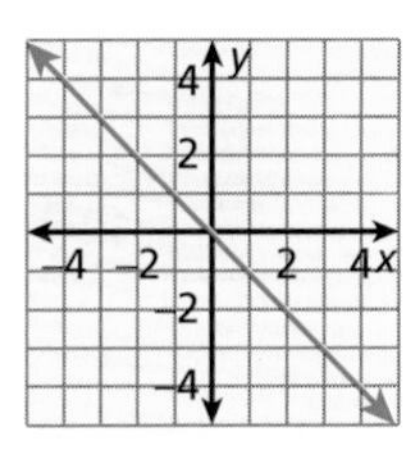

9. $y = x - 2$ **10.** $y = -x$ **11.** $y = -3$ **12.** $x = -3$

In Exercises 13–18, decide whether the ordered pair is a solution to the equation. If not, find a solution.

13. $(0, 4)$; $x - y = 4$ **14.** $(2, 5)$; $y = 7 - x$ **15.** $(-2, 4)$; $y = 8 - 2x$

16. $(1, -1)$; $y - 3x = 4$ **17.** $(-2, 9)$; $y = -5x - 1$ **18.** $(1, -3)$; $y = -3$

In Exercises 19–24, sketch the graph of the equation.

19. $y = x + 4$ **20.** $y = 2x - 6$ **21.** $y = -1$

22. $x = \frac{3}{2}$ **23.** $y = \frac{x}{3}$ **24.** $y = \frac{1}{2}x - 5$

Points of Intersection **In Exercises 25–27, graph both equations on the same coordinate plane. Then find the point of intersection of the two lines.**

25. $x + y = -2$
$y = x - 4$

26. $y = 6x + 14$
$y = 8 + 3x$

27. $x = -5$
$y = -4$

28. The point $(2, 5)$ lies on the graph of $y = cx + 1$. What is the value of c?

29. The point $(3, 4)$ lies on the graph of $y = cx - 8$. What is the value of c?

30. *Geometry* Sketch the graphs of the equations on the same coordinate plane. Which two lines are parallel?

a. $y = 2x + 3$ **b.** $y = x + 3$ **c.** $y = 2x - 1$

31. In the coordinate plane, the graphs of $x = 0$ and $y = 0$ have special names. What are these names?

32. *Population* For 1985 through 1989, the population P (in millions) of Oklahoma is given in the table. Let $t = 0$ represent 1985. Plot the data and describe the pattern. Is the pattern linear? Explain. ***(Source: U.S. Bureau of the Census)***

t	0	1	2	3	4
P	3.27	3.24	3.21	3.18	3.15

The Oklahoma state bird is the scissor-tailed flycatcher. It can catch insects in midair.

33. *Temperature* The average daily temperature of Oklahoma City (in degrees Fahrenheit) from January through August is given in the table. Create a table so that the data can be graphed on a coordinate plane. Then plot the data and describe the pattern. Is the pattern linear? Explain. ***(Source: U.S. National Oceanic and Atmospheric Administration)***

Jan.	Feb.	Mar.	Apr.	May	June	July	Aug.
35.9	40.9	50.3	60.4	68.4	76.7	82.0	81.1

Integrated Review

Making Connections within Mathematics

Graphing Polygons **In Exercises 34–37, sketch the graphs of the equations on the same coordinate plane. What figure is formed by the graphs?**

34. $x = 3, x = -6, y = 1, y = -8$

35. $y = 5, y = -5, y = 3x + 4, y = 3x - 4$

36. $y = -6, y = x, y = -x$

37. $y = -1, y = 4, y = 2x + 7, y = 7 - 2x$

Exploration and Extension

Finding an Equation **In Exercises 38–41, create a table of values that represents the given graph. Describe the relationship between x and y. Then write an equation that represents the relationship.**

38.

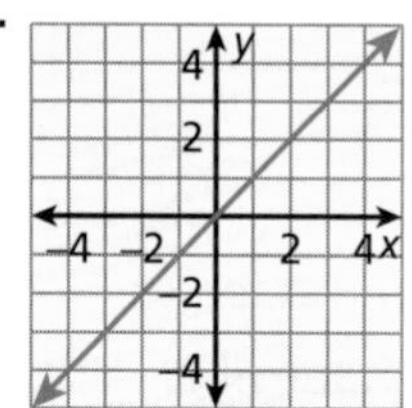

39.

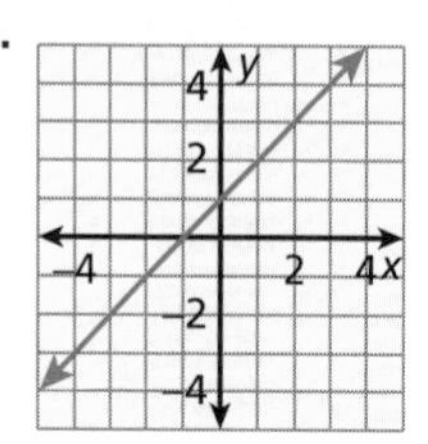

40.

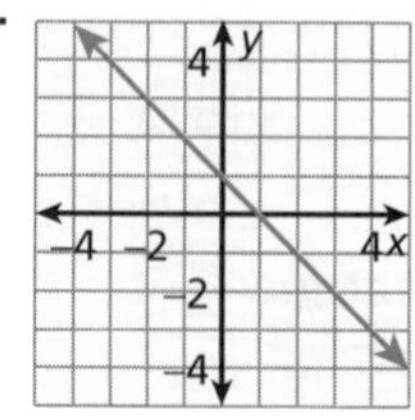

41.

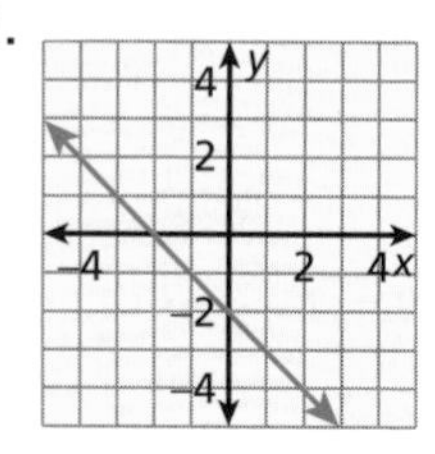

USING A GRAPHING CALCULATOR
Graphing Linear Equations

A graphing calculator can be used to sketch the graph of an equation. The following steps show how to use a Texas Instruments TI-82, a Casio *fx-7700G,* and a Sharp EL-9300C to sketch the graph of $y = 1.5x - 2$, as shown at the right. With each calculator, you must be sure the equation is in the form

$y =$ (expression involving x).

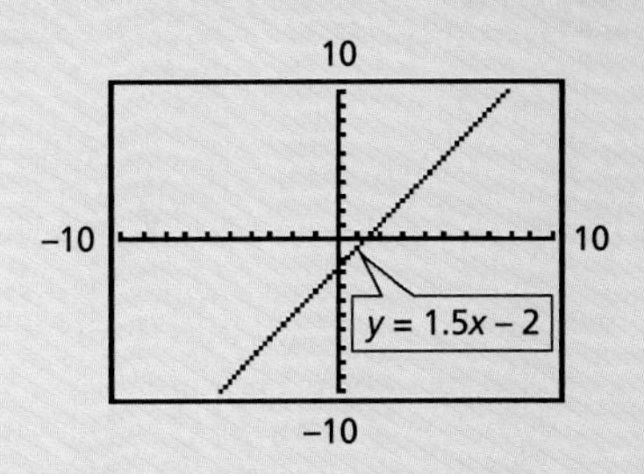

You must also set the range by entering the least and greatest x and y values that you want to be used and the scale (units per tick mark). *

TI-82

[WINDOW] (Set range.)
Xmin = −10
Xmax = 10.
Xscl = 1
Ymin = −10
Ymax = 10
Yscl = 1
[Y=] 1.5 [X, T, θ] [−] 2
:Y_1 = 1.5X − 2
:Y_2 =
:Y_3 =
:Y_4 =
[GRAPH]
[CLEAR] (Clear Screen)

Casio *fx-7700G*

[RANGE] (Set range.)
Xmin = −10
max = 10
scl = 1
Ymin = −10
max = 10
scl = 1
[EXE] [RANGE]
[SHIFT] [F5] (Cls) [EXE]
[GRAPH] 1.5 [X, θ, T] [−] 2
Graph Y = 1.5X − 2
[EXE]
[SHIFT] [F5] (Cls) [EXE]

Sharp EL-9300C

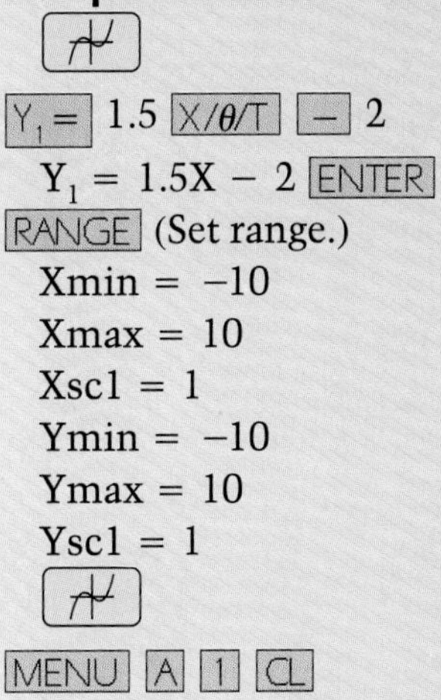

[graph key]
[Y_1=] 1.5 [X/θ/T] [−] 2
Y_1 = 1.5X − 2 [ENTER]
[RANGE] (Set range.)
Xmin = −10
Xmax = 10
Xscl = 1
Ymin = −10
Ymax = 10
Yscl = 1
[graph key]
[MENU] [A] [1] [CL]

Exercises

In Exercises 1–4, sketch the graph. (Use the range shown above.)

1. $y = 2x - 3$ **2.** $y = 0.5x + 4$ **3.** $y = x - 5$ **4.** $y = 0.75x - 2$

In Exercises 5–8, sketch the graph. (Use the indicated range.)

5. $y = x + 20$
Xmin = −10
Xmax = 10
Xscl = 1
Ymin = −5
Ymax = 35
Yscl = 5

6. $y = 2x - 30$
Xmin = −10
Xmax = 10
Xscl = 1
Ymin = −60
Ymax = 10
Yscl = 5

7. $y = 10x + 200$
Xmin = 0
Xmax = 100
Xscl = 10
Ymin = 0
Ymax = 1300
Yscl = 100

8. $y = -75x + 2000$
Xmin = 0
Xmax = 20
Xscl = 2
Ymin = 0
Ymax = 2100
Yscl = 100

* Keystrokes for other calculators are listed in *Technology—Keystrokes for Other Graphing Calculators* found at the end of this text.

13.3 Exploring Intercepts of Graphs

What you should learn:

Goal 1 How to find intercepts of lines

Goal 2 How to use intercepts to sketch quick graphs

Why you should learn it:

You can use intercepts of lines to help solve real-life problems, such as finding the time and distance of a subway trip.

Goal 1 Finding Intercepts of Lines

An **x-intercept** of a graph is the x-coordinate of a point where the graph crosses the x-axis. A **y-intercept** is the y-coordinate of a point where the graph crosses the y-axis. In the graph of $y = x + 2$ at the right, the x-intercept is -2 and the y-intercept is 2.

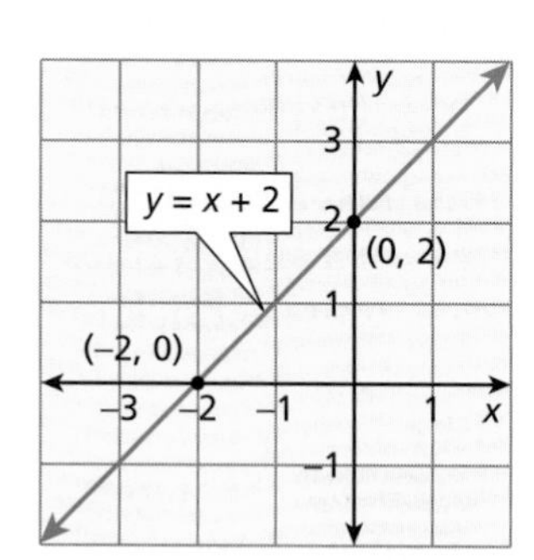

Finding Intercepts of Lines

1. To find an x-intercept of a line, substitute $y = 0$ into the equation and solve for x.
2. To find a y-intercept of a line, substitute $x = 0$ into the equation and solve for y.

Example 1 *Finding Intercepts of a Line*

Find the intercepts of the line given by $y = \frac{3}{2}x - 3$.

Solution

a. To find the x-intercept, let $y = 0$ and solve for x.

$y = \frac{3}{2}x - 3$	*Rewrite original equation.*
$0 = \frac{3}{2}x - 3$	*Substitute 0 for y.*
$3 = \frac{3}{2}x$	*Add 3 to each side.*
$2 = x$	*Multiply each side by $\frac{2}{3}$.*

The x-intercept is 2. The graph contains the point (2, 0).

b. To find the y-intercept, let $x = 0$ and solve for y.

$y = \frac{3}{2}x - 3$	*Rewrite original equation.*
$y = \frac{3}{2}(0) - 3$	*Substitute 0 for x.*
$y = -3$	*Simplify.*

The y-intercept is -3. The graph contains the point $(0, -3)$. ■

Goal 2 Sketching Quick Graphs

To sketch the graph of a linear equation, you only need to find two solution points. The intercepts are good points to use.

Study Tip...
After sketching a quick graph of a line, you can check your graph by finding and plotting a third solution. If the third solution does not lie on the line, then you know that at least one of the three points was plotted incorrectly.

Sketching a Quick Graph of a Line

To sketch a quick graph of a linear equation, find two solutions of the equation. Any two solutions can be used, but the intercepts are often convenient. Plot the two solutions and draw a line through the two plotted points.

Example 2 *Sketching a Quick Graph*

Sketch the graph of $x + y = 4$.

Solution Begin by finding the intercepts. To find the x-intercept, let $y = 0$ and solve for x.

$x + y = 4$	*Rewrite original equation.*
$x + 0 = 4$	*Substitute 0 for y.*
$x = 4$	*Simplify.*

The x-intercept is 4. The graph contains the point (4, 0). To find the y-intercept, let $x = 0$ and solve for y.

$x + y = 4$	*Rewrite original equation.*
$(0) + y = 4$	*Substitute 0 for x.*
$y = 4$	*Simplify.*

The y-intercept is 4. The graph contains the point (0, 4). Plot the points (4, 0) and (0, 4), as shown at the left. Then draw a line through the points. ■

(0, 4)
$x + y = 4$
(4, 0)

Communicating about MATHEMATICS

SHARING IDEAS about the Lesson

Real Life
Transportation

Interpreting Intercepts Your distance (in miles) from home is represented by y, and the time (in minutes) you traveled is represented by x. The relationship between y and x is modeled by the equation $y = 5 - \frac{1}{2}x$. Find the two intercepts of the graph of this equation.

A. Which intercept tells how many miles you are from home?

B. Which intercept tells how many minutes you traveled?

EXERCISES

Guided Practice

CHECK for Understanding

1. Find the x-intercept of $y = 2x - 1$. Explain each step.
2. Find the y-intercept of $5x + 3y = 9$. Explain each step.

In Exercises 3–6, find the intercepts of the line. Then sketch a quick graph.

3. $x + y = 5$ **4.** $x - y = 5$ **5.** $y = \frac{5}{4}x + 3$ **6.** $-7x + 3y = -21$

Independent Practice

In Exercises 7–10, identify the intercepts of the graph.

7. 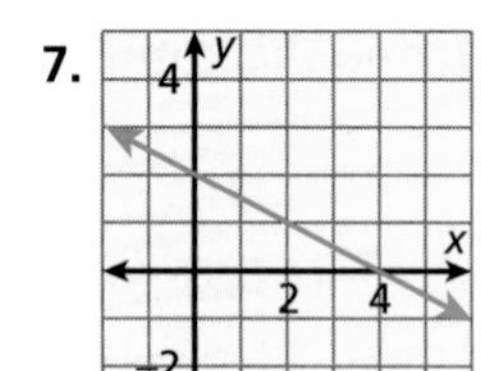

8. 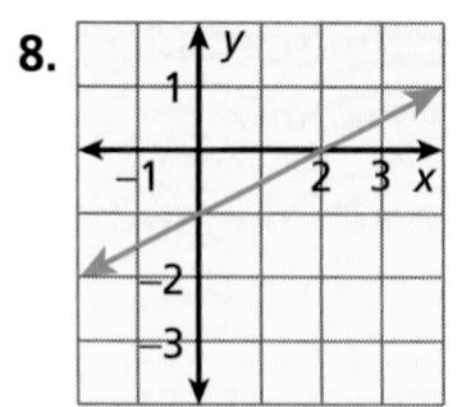

9. 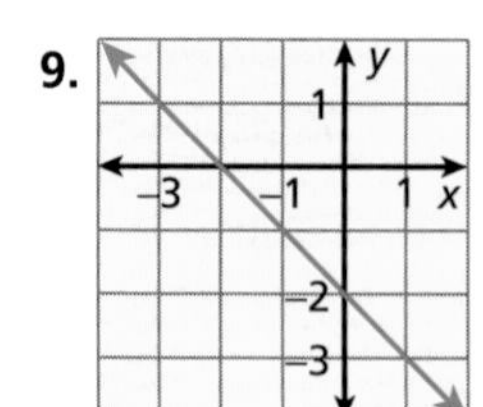

10. 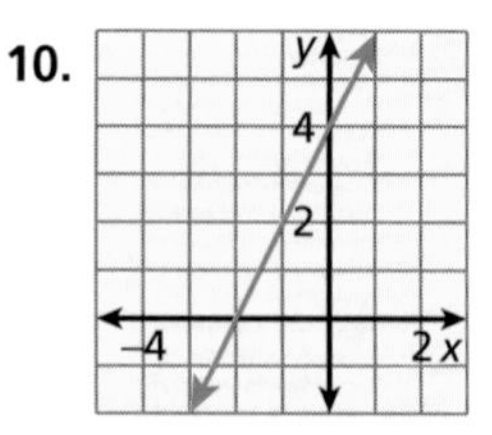

In Exercises 11–14, sketch a line having the given intercepts.

11. x-intercept: -1
y-intercept: 5

12. x-intercept: -5
y-intercept: -2

13. x-intercept: 2
y-intercept: 6

14. x-intercept: 4
y-intercept: -5

In Exercises 15–18, match the equation with its graph.

a. 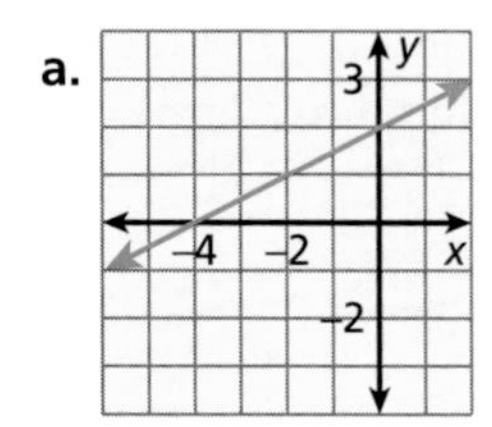

b. 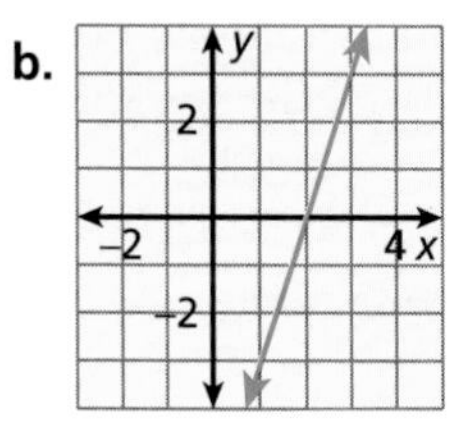

c. 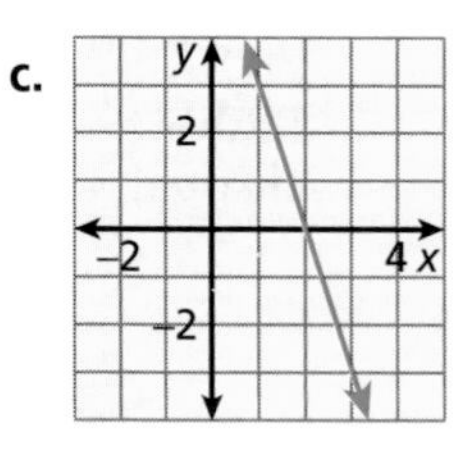

d. 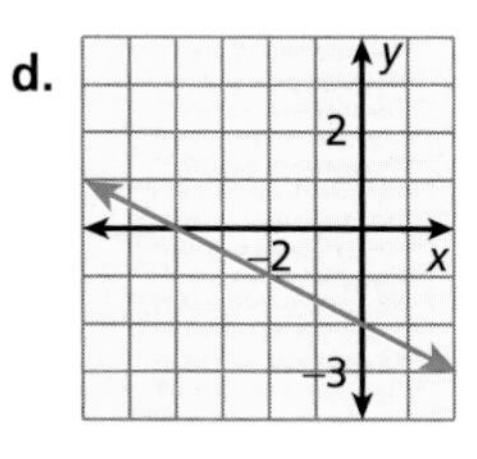

15. $y = \frac{1}{2}x + 2$ **16.** $y = -\frac{1}{2}x - 2$ **17.** $3x - y = 6$ **18.** $3x + y = 6$

In Exercises 19–26, sketch a quick graph of the line. Then create a table of values and compare the values with the points on the line.

19. $y = -3x + 6$ **20.** $y = 4x - 8$ **21.** $x - y = 1$ **22.** $x + y = -3$

23. $3x - 4y = 24$ **24.** $x + 5y = 5$ **25.** $y = -\frac{3}{2}x + 4$ **26.** $y = \frac{4}{3}x + 6$

In Exercises 27 and 28, use a calculator to find the intercepts of the line. Round your results to two decimal places.

27. $y = -3.64x + 2.18$ **28.** $y = 1.85x - 14.302$

29. *Error Analysis* A friend in your math class has sketched a quick graph of the line $4y = -6x + 8$ as shown at the right. Is the graph correct? If not, what did your friend do wrong?

At age 13, Stephen Lovett started a car cleaning service, cleaning and waxing cars after school and on weekends. By age 18, he had five employees.

Business **In Exercises 30 and 31, use the following information.**

You own a car wash business and have discovered that the less you charge for a car wash, the more car washes you sell. Over an eight-week period, you try several prices for 1 week each, as shown in the table. p is the price of a car wash and x is the number of car washes you sold that week.

p	$12	$11	$10	$9	$8	$7	$6	$5
x	400	450	500	550	600	650	700	800

30. Draw a graph of the data.
31. Your total car wash income is

    Income = (Price)(Number sold).

    Make a table showing your total income for the different prices. Which price would you charge? Explain.

32. *Think about It* Not every graph of a line has two distinct intercepts. Give two examples where the graph of a line has only one intercept.

Integrated Review

Making Connections within Mathematics

Coordinates **In Exercises 33–36, the point is on the line $9x + 6y = 36$. Find the missing coordinate.**

33. (2, ?)
34. (6, ?)
35. (?, 9)
36. (?, -6)
37. *Sketching a Graph* Use the points in Exercises 33–36 to sketch the graph of $9x + 6y = 36$. Then name 4 other points on the graph.

Exploration and Extension

Technology **In Exercises 38–40, consider the equation $24x + 54y = 216$.**

38. Rewrite the equation in the form that can be used by a graphing calculator.
39. Use a graphing calculator to sketch the graph. Use a range that makes both intercepts visible.
40. What are the intercepts of the graph?

Mixed REVIEW

In Exercises 1–4, find the circumference and area of the indicated circle. (12.1)

1. $d = 7$ ft **2.** $d = 32$ in. **3.** $r = 8$ cm **4.** $r = 24$ yd

In Exercises 5–8, find the surface area and volume of the solid. (12.3–12.5)

5.

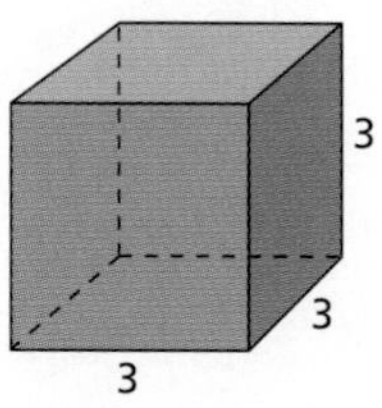

6.

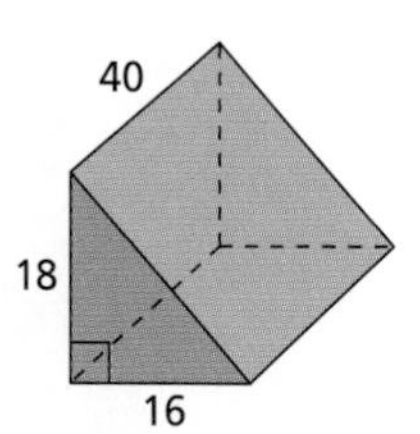

7.

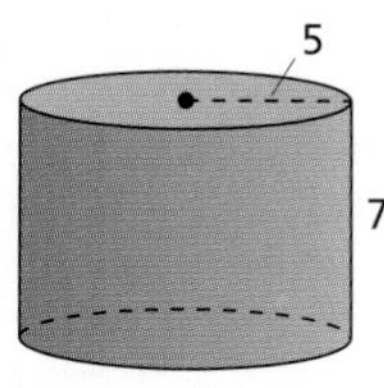

8.

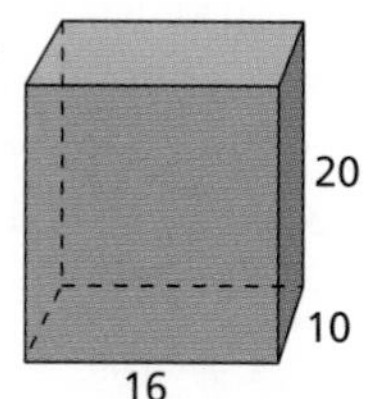

In Exercises 9–12, find the *x*- and *y*-intercepts. (13.3)

9. $y = x - 12$ **10.** $2x = y + 4$ **11.** $3x + 3y = 9$ **12.** $\frac{2}{3}y = \frac{2}{5}x + 6$

In Exercises 13 and 14, is $\triangle ABC$ a right triangle? Explain. (9.4)

13. $A(1, 2), B(5, 2), C(0, 2)$ **14.** $A(1, -1), B(1, 0), C(5, -1)$

Career Interview

Design and Construction Manager

Howard Haywood works for the Massachusetts Bay Transit Authority (MBTA), the oldest transit system in the United States. He is responsible for administering all MBTA construction—tunnels, subway systems, rail systems, bridges, etc.

Q: ***What led you to this career?***
A: I began my career in general construction.
Q: ***What math do you use on your job?***
A: Basic math, algebra, geometry, and calculus. However, I feel that the most important math concepts I use are the problem solving skills such as the ability to decide what steps to take, what math operations to use, how to evaluate the results, and how to determine if the solution is reasonable.
Q: ***What would you like to tell kids about school?***
A: When I was in fifth grade through junior high, I thought that I was a better reader than math student. Then one of my teachers explained to me that math is just like reading. It is a language that has logical steps and a logical progression. Once I tried to "read" math and comprehend it, I found it wasn't so hard.

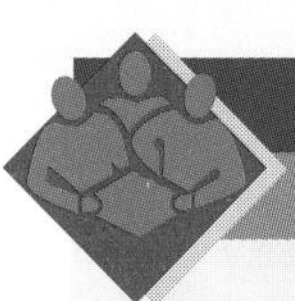

LESSON INVESTIGATION 13.4

Exploring the Slope of a Line

Materials Needed: geoboards or dot paper

Example — *Exploring Slope*

Use a geoboard or dot paper to model the indicated lines.

a. The line slopes up 2 units for each 3 units to the right.

b. The line slopes down 4 units for each 3 units to the right.

c. The line slopes up 3 units for each 3 units to the right.

Solution You can use the hypotenuse of a right triangle to model each line. The vertical side of the triangle models the number of units moved up or down. The horizontal side models the number of units moved to the right.

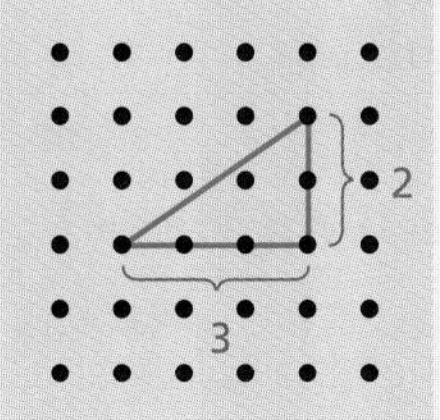

Slopes up 2 units for each 3 units to the right.

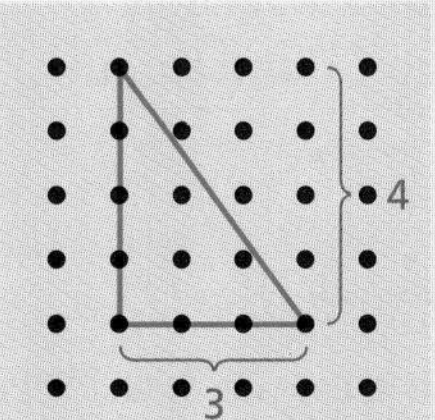

Slopes down 4 units for each 3 units to the right.

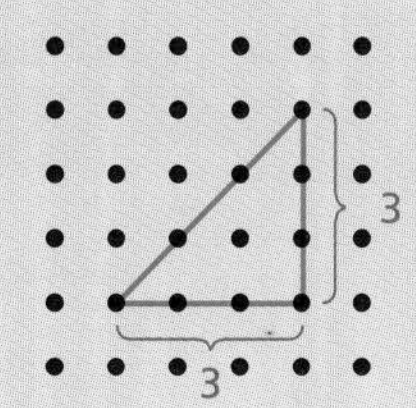

Slopes up 3 units for each 3 units to the right. ■

Exercises

In Exercises 1–4, describe the slope of the hypotenuse.

1. 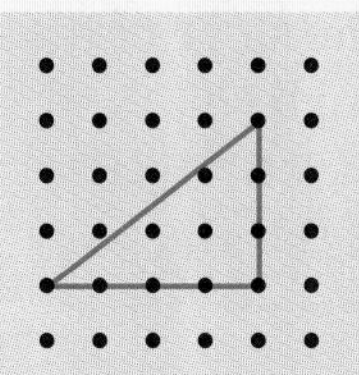

2. 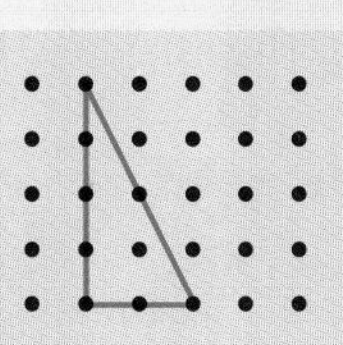

3. 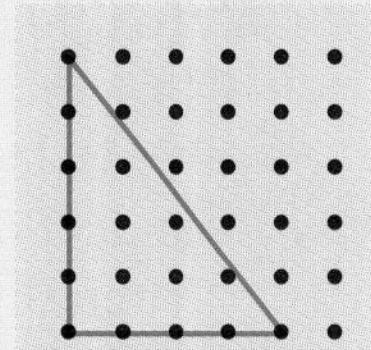

4. 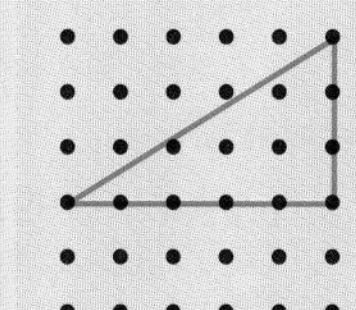

In Exercises 5–8, use a geoboard or dot paper to model the line.

5. The line slopes up 4 units for each 5 units to the right.

6. The line slopes down 4 units for each 5 units to the right.

7. The line slopes down 1 unit for each 4 units to the right.

8. The line slopes up 2 units for each 4 units to the right.

13.4 Exploring Slope

What you should learn:

Goal 1 How to find the slope of a line

Goal 2 How to interpret the slope of a line

Why you should learn it:

You can use the slope of a line to solve real-life problems, such as describing the steepness of a hill.

Goal 1 Finding the Slope of a Line

To find the slope of a line, choose two points on the line. Call the points (x_1, y_1) and (x_2, y_2). The **slope** is the ratio of the change in y to the change in x.

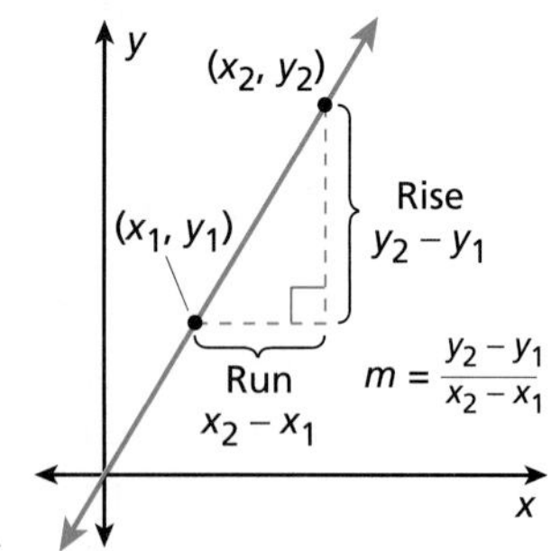

Change in $y = y_2 - y_1 =$ "Rise"

Change in $x = x_2 - x_1 =$ "Run"

These expressions are read as "y sub 2 minus y sub 1" and "x sub 2 minus x sub 1."

> **Finding the Slope of a Line**
>
> The slope m of the nonvertical line passing through the points (x_1, y_1) and (x_2, y_2) is $m = \frac{y_2 - y_1}{x_2 - x_1} = \frac{\text{Rise}}{\text{Run}}$.

Study Tip...

When you are using the formula for slope, it doesn't matter which point you represent with (x_1, y_1). For instance, in Example 1a, the point (1, 2) is represented with (x_1, y_1) and (3, 5) is represented with (x_2, y_2). Try switching the points and applying the formula. You will obtain the same slope.

Example 1 *Finding the Slope of a Line*

a. To find the slope of the line through (1, 2) and (3, 5), let (1, 2) be (x_1, y_1), and let (3, 5) be (x_2, y_2). Then the slope is

$$m = \frac{5 - 2}{3 - 1} = \frac{3}{2}.$$

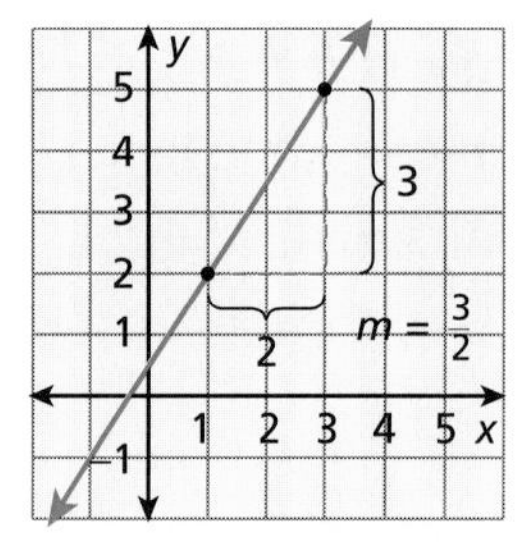

b. To find the slope of the line through (2, 5) and (5, 1), let (2, 5) be (x_1, y_1), and let (5, 1) be (x_2, y_2). Then the slope is

$$m = \frac{1 - 5}{5 - 2} = \frac{-4}{3} = -\frac{4}{3}.$$

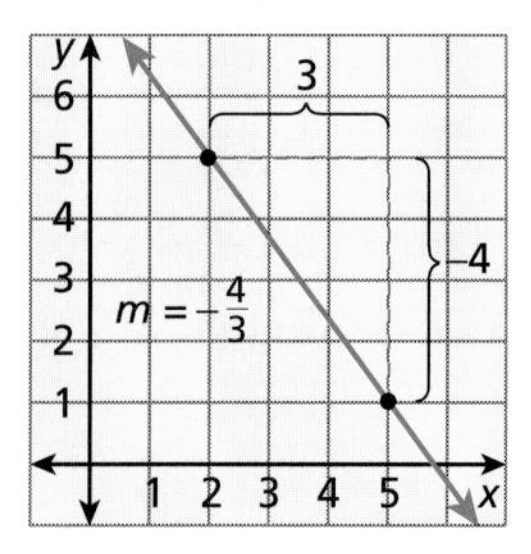

■

Goal 2 Interpreting Slope

The slope of a line tells you whether the line rises to the right, falls to the right, or is horizontal. (Slope is not defined for vertical lines.) If you imagine that you are walking *to the right* on the line, a positive slope means you are walking uphill, a negative slope means you are walking downhill, and a zero slope means you are walking on level ground.

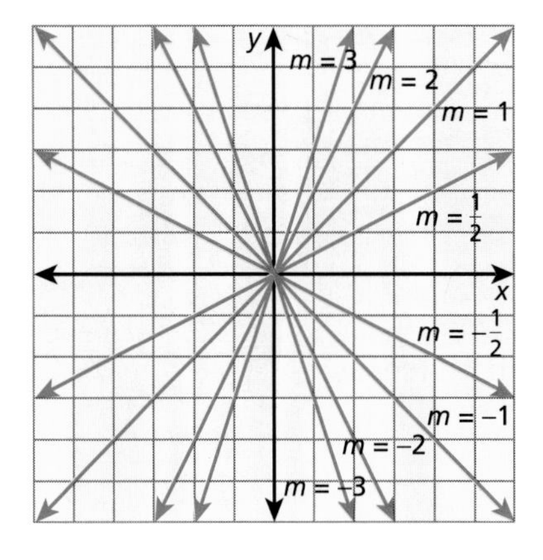

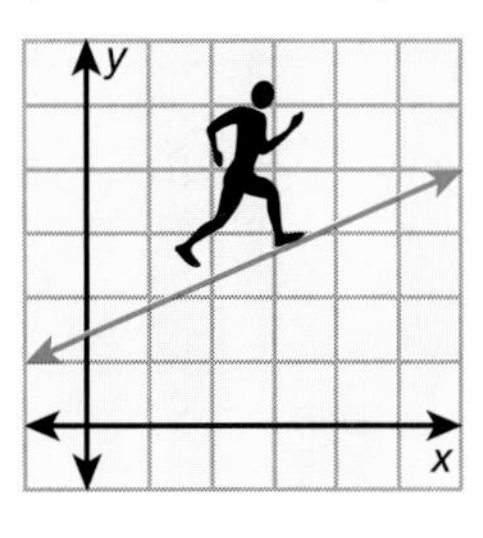

Positive Slope

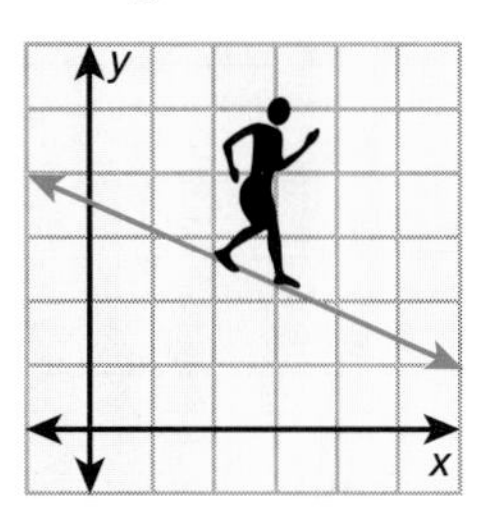

Negative Slope

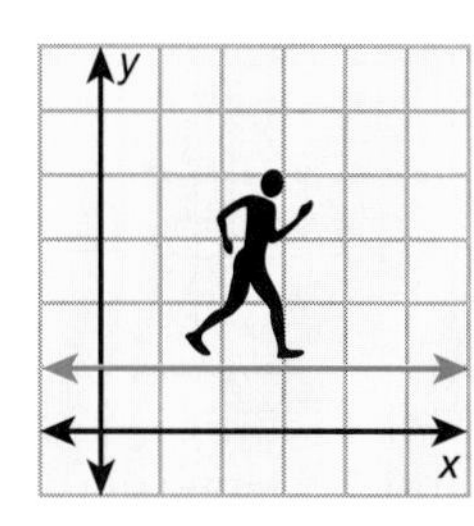

Zero Slope

The slope of a line also tells you how steep the line is. For instance, a line with a slope of 3 is steeper than a line with a slope of 2. Similarly, a line with a slope of -3 is steeper than a line with a slope of -2.

Example 2 *Comparing the Slopes of Two Lines*

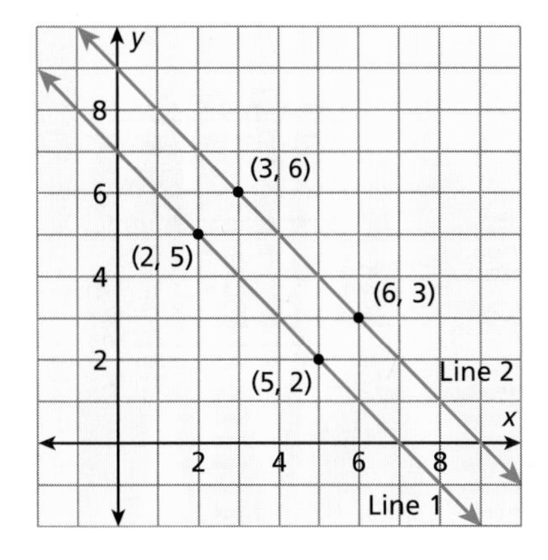

Compare the slopes of the lines through the indicated points.

Line 1: (2, 5), (5, 2) **Line 2:** (3, 6), (6, 3)

Solution

Line 1: $m = \frac{2-5}{5-2} = \frac{-3}{3} = -1$

Line 2: $m = \frac{3-6}{6-3} = \frac{-3}{3} = -1$

Each line has a slope of -1. From the graphs of the lines at the left, you can see that they are parallel. This is another use of slope—lines with the same slope are parallel. ■

Communicating about MATHEMATICS

SHARING IDEAS about the Lesson

Finding the Slope of a Line The points $(-1, 1)$, $(0, 3)$, $(1, 5)$, $(2, 7)$, and $(3, 9)$ all lie on a line. Which two points would you use to find the slope? Can you use any two? Explain.

EXERCISES

Guided Practice

CHECK for Understanding

In Exercises 1–4, find the slope of the line.

1. 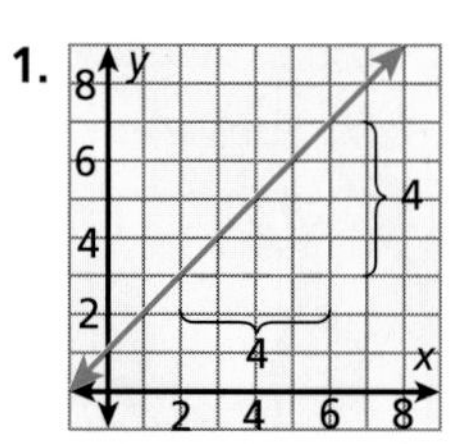

2. 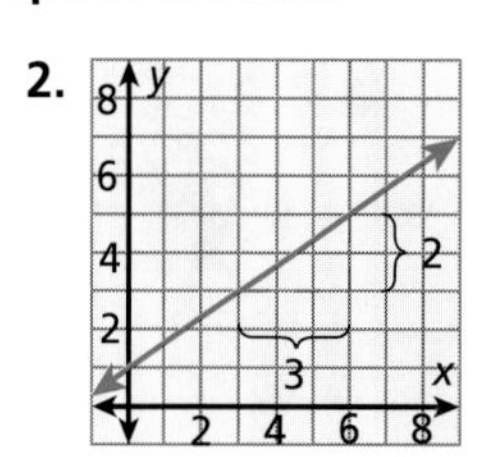

3. 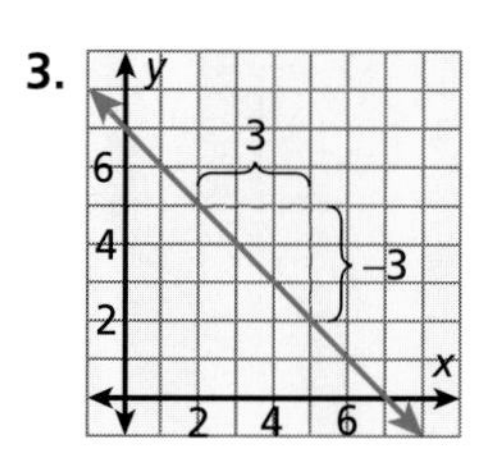

4. 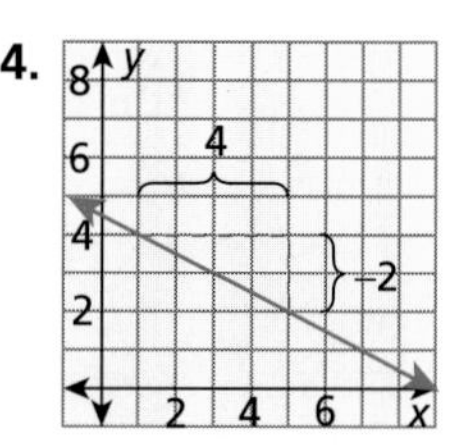

5. Find the slope of the line through $(-1, 3)$ and $(4, 2)$.

6. Sketch a line with a slope of 3 and another with a slope of 4. Which is steeper?

Independent Practice

In Exercises 7 and 8, which slope is steepest?

7. $m = \frac{5}{2}, m = 3, m = 0, m = 5$

8. $m = -1, m = -6, m = -4, m = -\frac{17}{4}$

In Exercises 9–12, find the slope of the line.

9. 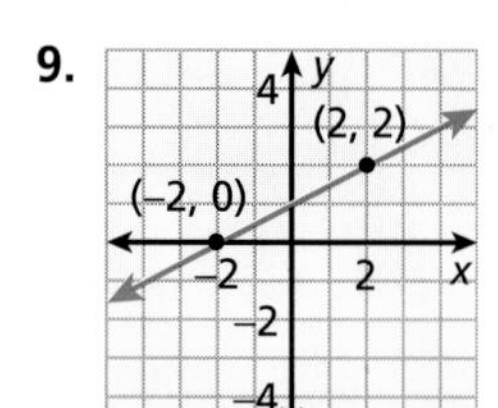

10. 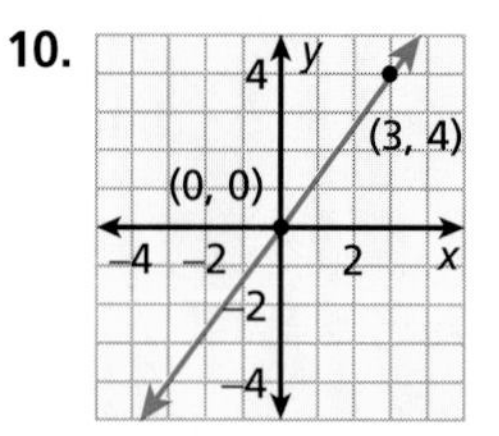

11. 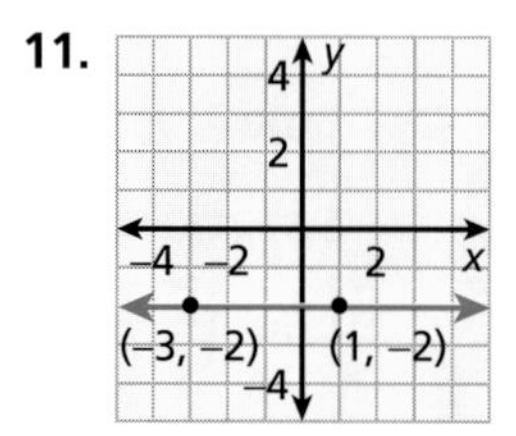

12. 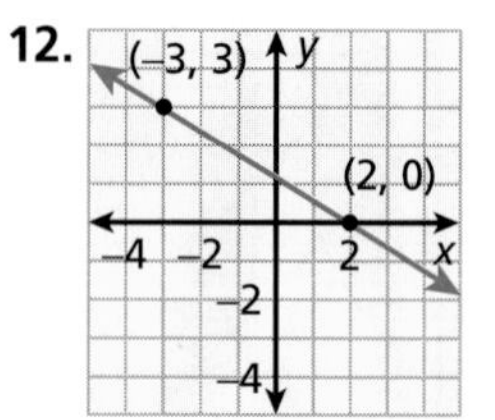

In Exercises 13–16, plot the points. Then find the slope of the line through the points.

13. $(2, 5), (0, 5)$

14. $(3, 4), (4, 3)$

15. $(1, -2), (-1, -6)$

16. $(0, -1), (1, -7)$

In Exercises 17–20, find the slope. (Assume a left-to-right orientation.)

17. 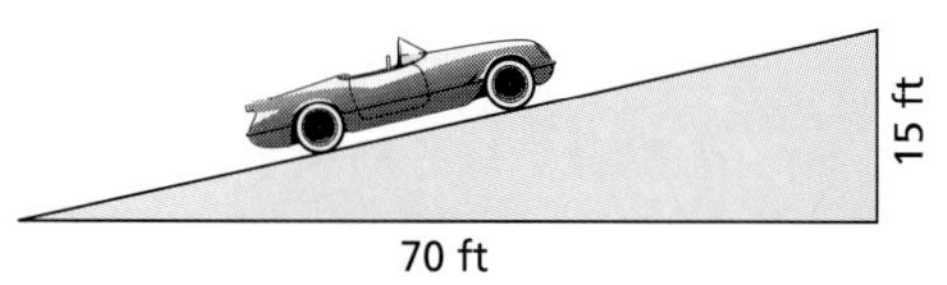

18. 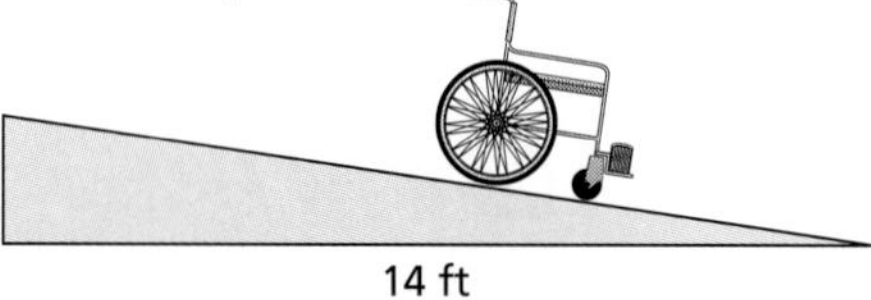

19. 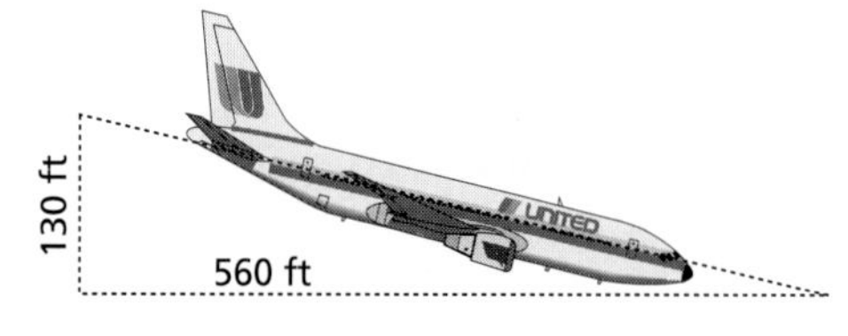

20. 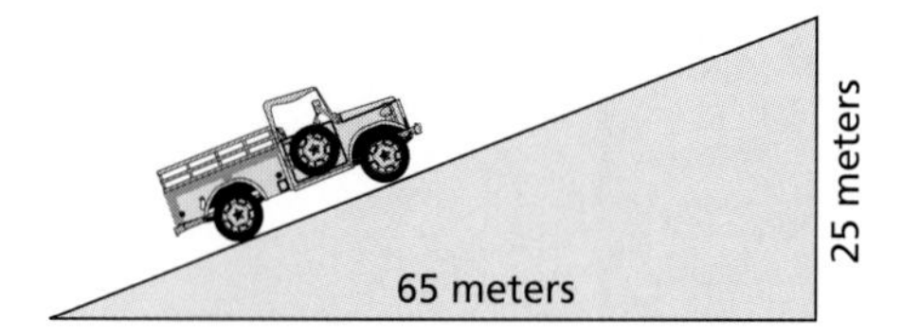

Geometry **In Exercises 21–24, find the slope of the hypotenuse.**

21. 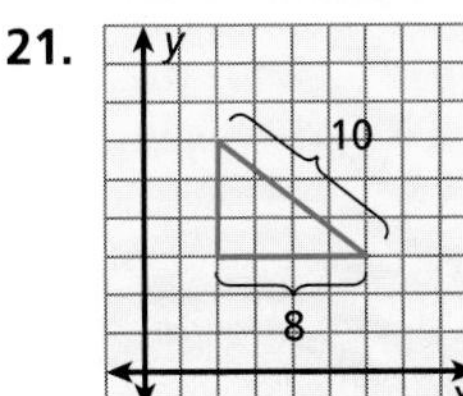

22. 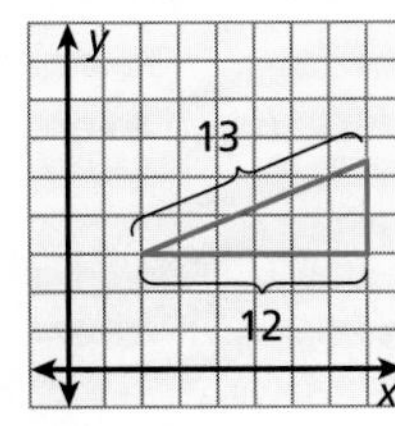

23. 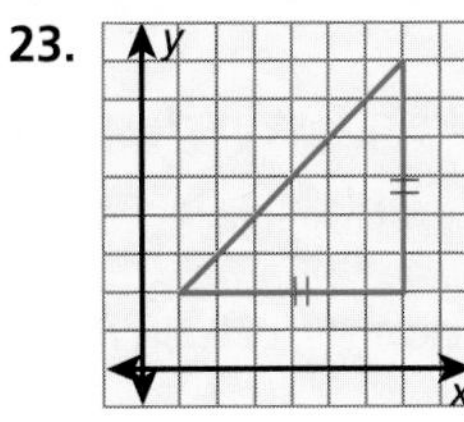

24. 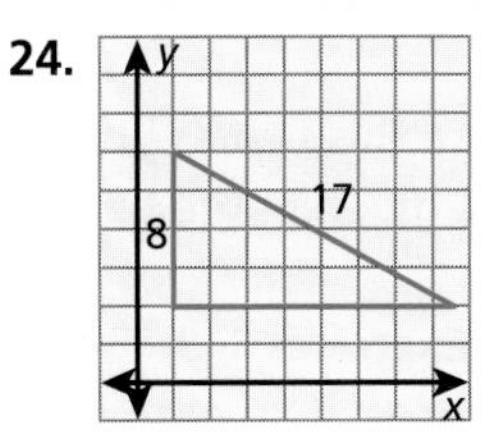

Geometry **In Exercises 25–28, find the slopes of $\overleftrightarrow{AB}$ and $\overleftrightarrow{CD}$. Are the lines parallel? Explain.**

25. $A(3, 3)$, $B(1, -2)$, $C(-4, 4)$, $D(-3, -1)$

26. $A(1, 1)$, $B(0, -2)$, $C(-5, 1)$, $D(-3, -2)$

27. $A(2, 3)$, $B(0, -2)$, $C(4, 3)$, $D(6, 8)$

28. $A(-2, -2)$, $B(2, 6)$, $C(-1, -4)$, $D(-5, 4)$

Stairs **In Exercises 29–31, find the slope of each set of stairs.**

29. 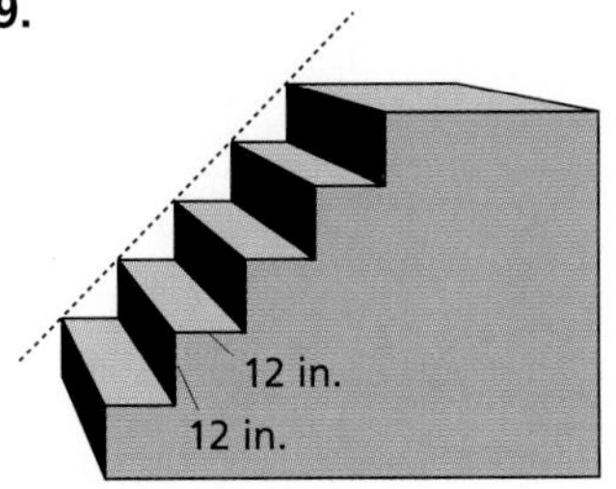

30. 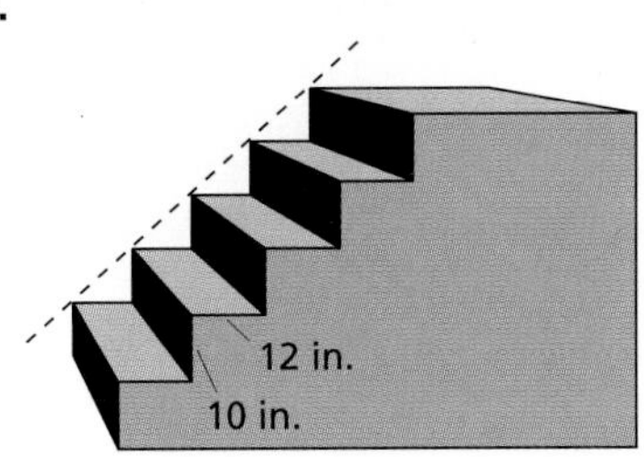

31. 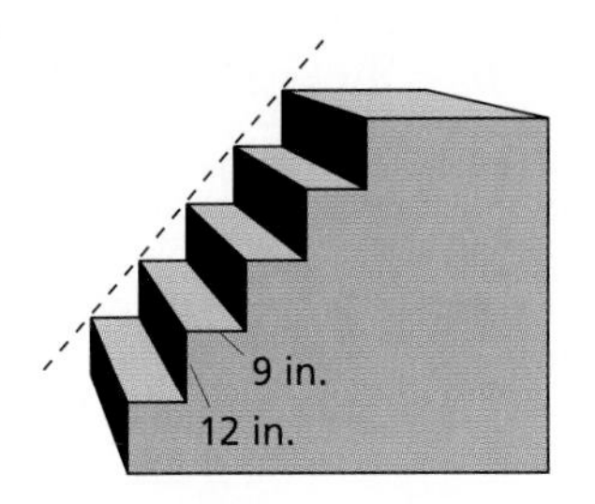

32. *Designing Stairs* You are helping to write a building code that is concerned with public safety. What is the steepest slope for a set of stairs that you think would be safe? Illustrate your answer with a scale drawing.

Integrated Review

Making Connections within Mathematics

Patterns **In Exercises 33 and 34, find the slope of the line through each pair of points. Then describe the pattern.**

33. a. $(-6, 3), (7, 16)$ **b.** $(1, 0), (-2, -9)$ **c.** $(-1, 1), (-4, -14)$ **d.** $(0, -5), (1, 2)$

34. a. $(0, -4), (2, -2)$ **b.** $(1, 5), (-1, -3)$ **c.** $(1, -2), (2, 7)$ **d.** $(0, 8), (-1, -8)$

Exploration and Extension

Misleading Graphs **In Exercises 35 and 36, use the graphs at the right.**

35. Estimate the slope of each graph.

36. Which of the graphs is more misleading? Explain your reasoning.

a. 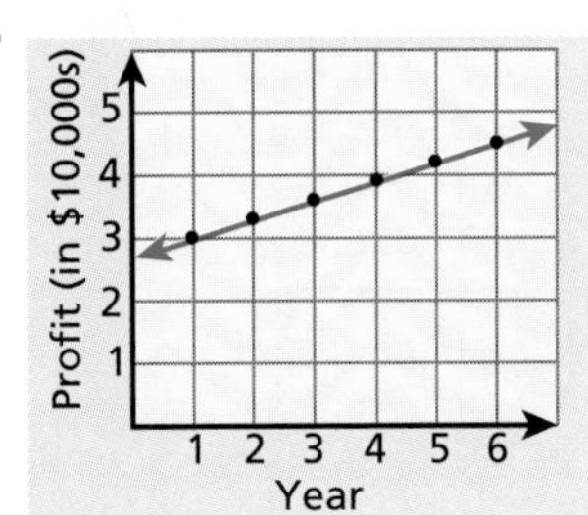

b. 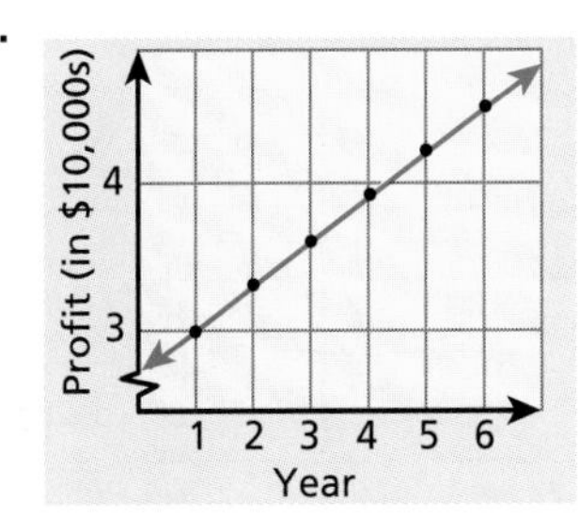

Mid-Chapter SELF-TEST

Take this test as you would take a test in class. The answers to the exercises are given in the back of the book.

In Exercises 1–3, use the equation $y = 28x$, which relates ounces, x, to grams, y. (13.1)

1. A magazine weighs 10 ounces. What is its measure in grams?
2. A textbook has a measure of 700 grams. What is its weight in ounces?
3. Which measure is greater, 1 pound (16 ounces) or 1 kilogram (1000 grams)? Explain.

In Exercises 4–6, decide whether the ordered pair is a solution of the equation $3x + 4y = 28$. (13.1)

4. $(0, 7)$
5. $\left(9, -\frac{1}{4}\right)$
6. $(8, 1)$

In Exercises 7 and 8, use the equation $2x + 5y = 42$. (13.2)

7. Copy and complete the table.

x	1	6	?	?
y	?	?	0	4

8. Use the table of values in Exercise 7 to sketch the graph of $2x + 5y = 42$.

In Exercises 9–11, find the x- and y-intercepts. (13.3)

9. $8x + 2y = 32$
10. $4x + 5y = 20$
11. $12x + 8y = 24$

In Exercises 12–15, find the slope of the line that passes through the points. (13.4)

12. $(2, 9), (4, 12)$
13. $(5, 2), (3, 6)$
14. $(7, 5), (3, 2)$
15. $(0, 7), (2, 10)$
16. Which of the lines in Exercises 12–15 are parallel? **(13.4)**

In Exercises 17–20, decide whether the graph is correct or incorrect. If it is incorrect, sketch the correct graph. (13.3, 13.4)

17. $y = 2x + 1$

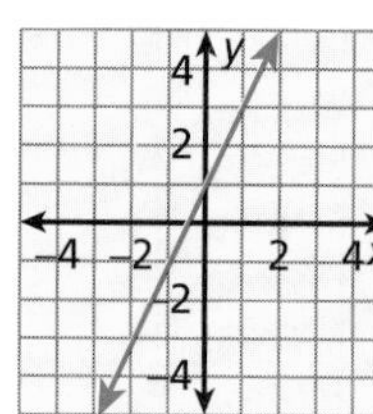

18. $y = 2x - 1$

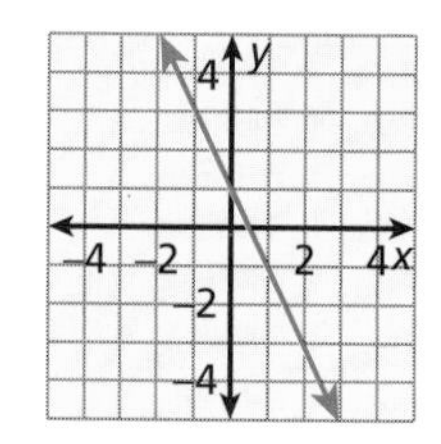

19. $y = -x - 1$

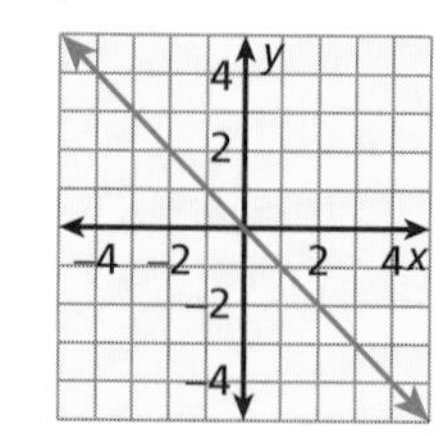

20. $y = \frac{3}{2}x$

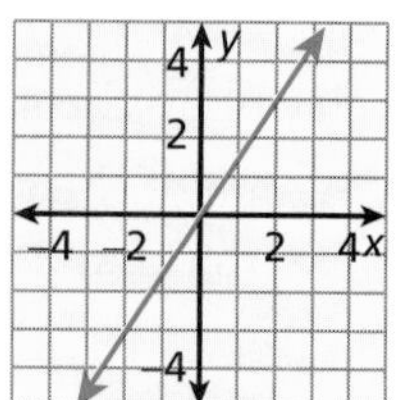

LESSON INVESTIGATION 13.5

Exploring Slope and y-intercepts

Materials Needed: graphing calculator

Example 1 *Comparing Graphs of Linear Equations*

Use a graphing calculator to compare the graphs of the following equations.

a. $y = 2x + 3$ **b.** $y = -x + 3$ **c.** $y = \frac{1}{2}x + 3$

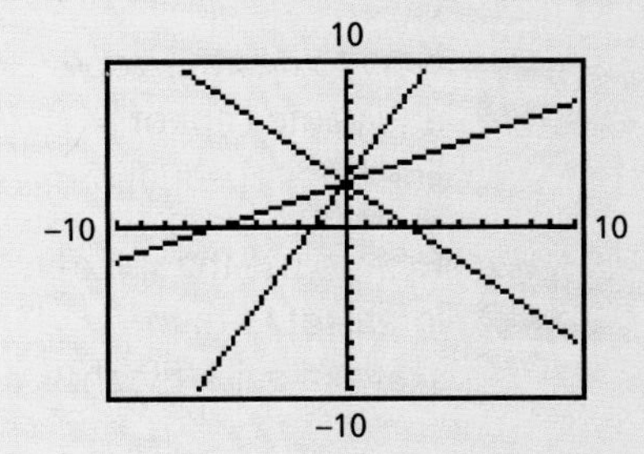

Solution Enter each equation into a graphing calculator. Then graph all three equations on the same calculator screen, as shown at the right. From the screen, you can see that all three lines have the same y-intercept. That is, each crosses the y-axis at the same point (0, 3).

Example 2 *Comparing Graphs of Linear Equations*

Use a graphing calculator to compare the graphs of the following equations.

a. $y = 2x - 3$ **b.** $y = 2x - 1$ **c.** $y = -2x + 3$

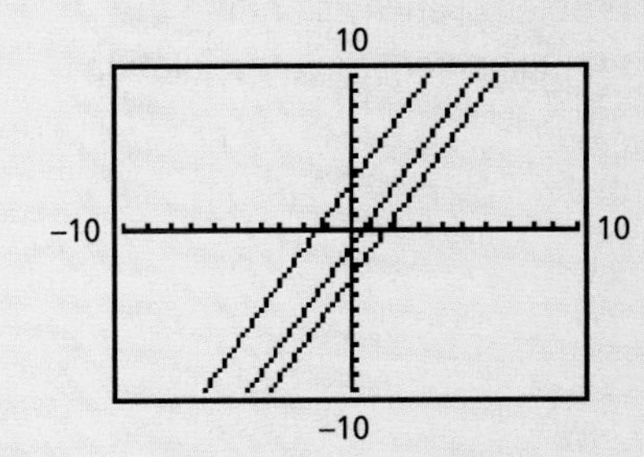

Solution Enter each equation into a graphing calculator. Then graph all three equations on the same calculator screen, as shown at the right. From the screen, you can see that all three lines are parallel. Each has a slope of 2.

■

Exercises

1. Use a graphing calculator to compare the graphs of the following. What can you conclude?

   a. $y = -x - 4$ **b.** $y = x - 4$ **c.** $y = -\frac{1}{2}x - 4$

2. Use a graphing calculator to compare the graphs of the following. What can you conclude?

   a. $y = -x + 5$ **b.** $y = -x - 2$ **c.** $y = -x + 1$

3. Without sketching the graph of $y = -x - 4$, find its slope and y-intercept. (Use the patterns you found in Exercises 1 and 2.) Use a graphing calculator to confirm your answers.

13.5 The Slope-Intercept Form

Goal 1 How to find the slope and y-intercept of a line from its equation

Goal 2 How to use the slope-intercept form to sketch a quick graph

Why you should learn it:

You can use the slope-intercept form of a line to solve real-life problems, such as finding the annual increase in a population.

Goal 1 Using the Slope-Intercept Form

In the *Lesson Investigation* on page 609, you may have discovered that there is a quick way to find the slope and y-intercept of a line. For instance, the line given by $y = 2x + 3$ has a slope of 2 and a y-intercept of 3.

$$y = 2x + 3$$

Slope is 2 *y-intercept is 3.*

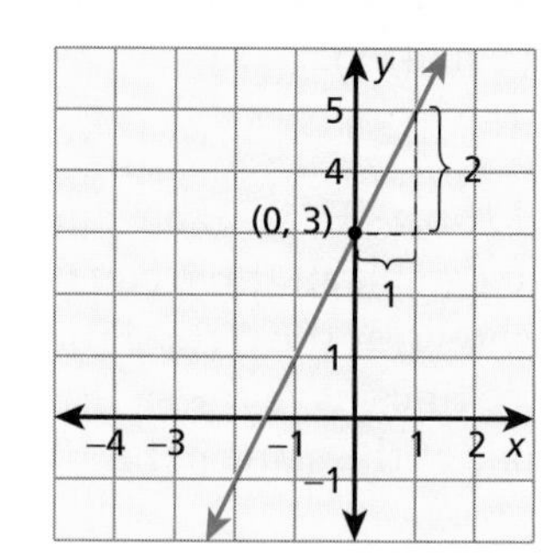

The Slope-Intercept Form of the Equation of a Line

The linear equation

$$y = mx + b$$

is in **slope-intercept form.** The slope is m. The y-intercept is b.

Study Tip...

To use the slope-intercept form, you must first be sure that the linear equation is written in the form

$$y = mx + b.$$

For instance, you can write $x + y = 2$ in slope-intercept form by subtracting x from each side to obtain

$$y = -x + 2.$$

Example 1 *Using the Slope-Intercept Form*

a. The line given by

$$y = x - 4$$
$$y = 1x + (-4)$$

Slope *y-intercept*

has a slope of 1 and a y-intercept of -4.

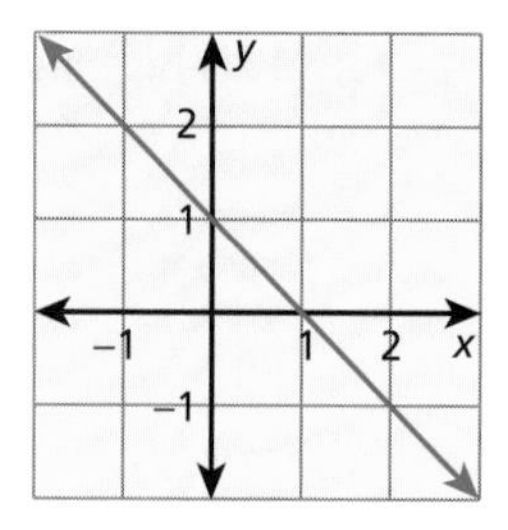

b. The line given by

$$y = -x + 1$$
$$y = (-1)x + 1$$

Slope *y-intercept*

has a slope of -1 and a y-intercept of 1.

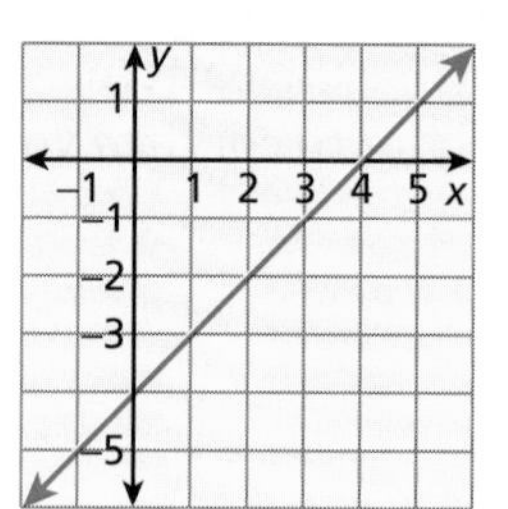

■

Goal 2 Sketching Quick Graphs

In Lesson 13.3, you learned to sketch a line quickly using the x-intercept and y-intercept. The slope and y-intercept can also be used to sketch a quick graph.

Study Tip...

When sketching a quick graph of a line, there are two ways to measure the rise and the run. If the rise and run are both positive or both negative, then the slope is positive. If the rise and run have different signs, then the slope is negative.

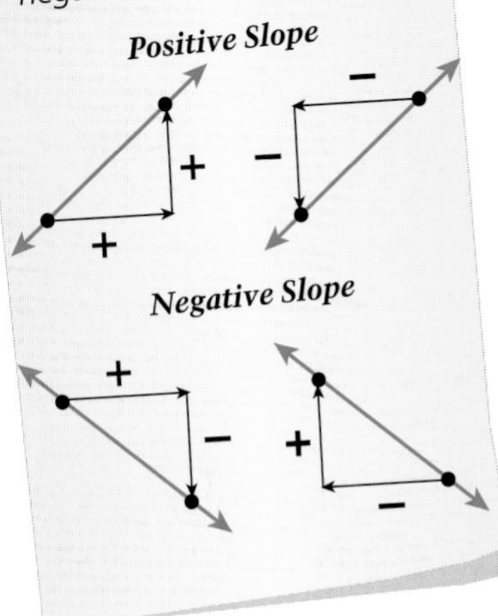

Example 2 *Sketching a Quick Graph*

Sketch a quick graph of $y = \frac{1}{2}x + 2$.

Solution Because this equation is in slope-intercept form, you can conclude that the slope is $\frac{1}{2}$ and the y-intercept is 2. To sketch the graph, first plot the y-intercept, (0, 2). Locate a second point on the line by moving 2 units to the right and 1 unit up. Then draw the line through the two points.

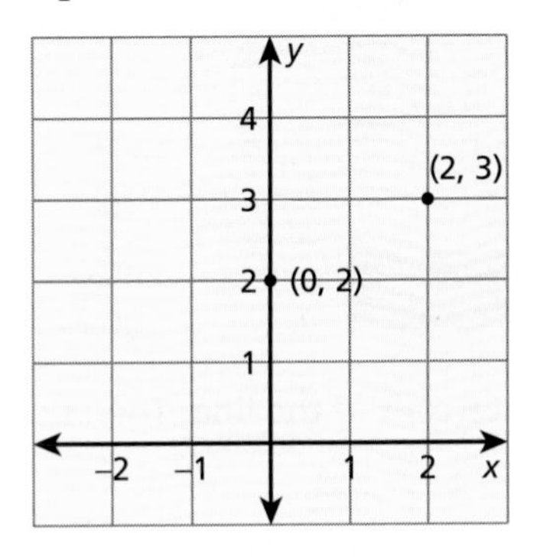

Plot the y-intercept. Then use the slope to find a second point.

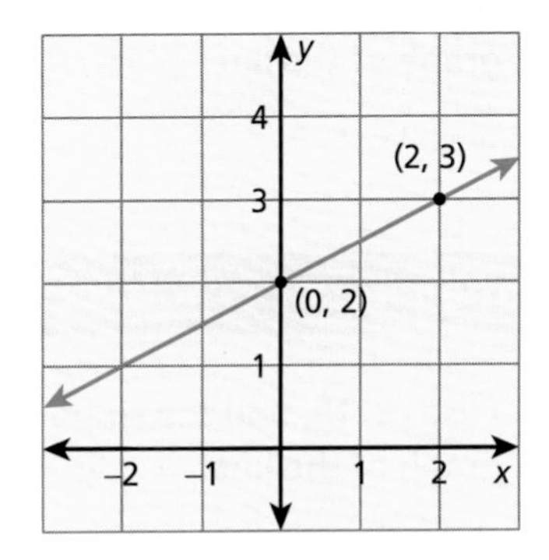

Draw the line that passes through the two points. ■

Communicating about MATHEMATICS

SHARING IDEAS about the Lesson

Real Life **Population**

Interpreting Slope From 1980 to 1992, the population P (in thousands) of Virginia can be modeled by

$P = 85t + 5350$ *Population Model*

where $t = 0$ represents 1980. Copy and complete the table, which shows the population for each year. How is the slope related to the numbers in the table? ***(Source: U.S. Bureau of the Census)***

t	0	1	2	3	4	5	6	7	8	9	10	11	12
P	?	?	?	?	?	?	?	?	?	?	?	?	?

EXERCISES

Guided Practice

CHECK for Understanding

1. *Writing* In your own words, explain why $y = mx + b$ is called the *slope-intercept form* of the equation of a line.
2. Explain how to write $2x + y = 5$ in slope-intercept form.
3. Which of the following is the equation of the line at the right? (There may be more than one correct equation.)

   a. $y = 2x + 3$ b. $y = 2x - 3$ c. $y = -2x + 3$

   d. $y = -2x - 3$ e. $2x + y = 3$ f. $2x - y = 3$

In Exercises 4–7, find the slope and *y*-intercept of the line. Then sketch a quick graph of the line.

4. $y = -4x + 5$
5. $y = \frac{1}{4}x - 1$
6. $-3x + y = 9$
7. $x + 2y = 16$

Independent Practice

In Exercises 8–15, find the slope and *y*-intercept of the line. Then sketch a quick graph of the line.

8. $y = x - 3$
9. $y = -x + 3$
10. $y = -\frac{2}{3}x + 2$
11. $y = 3x$
12. $6y = 24x + 30$
13. $5x + 2y = 0$
14. $x + \frac{1}{2}y = 1$
15. $5x - 10y = 35$

In Exercises 16–19, match the equation with its graph.

a.

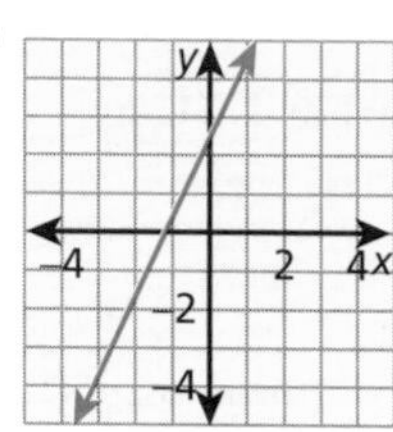

b.

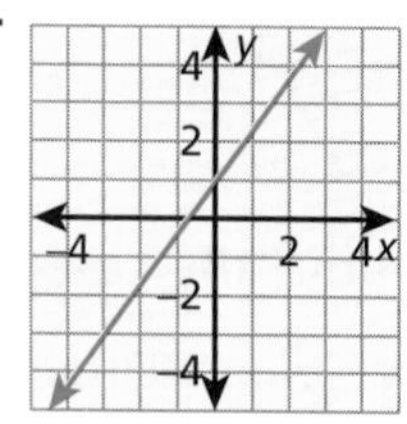

c.

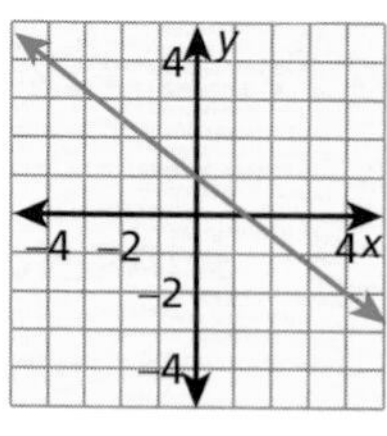

d.

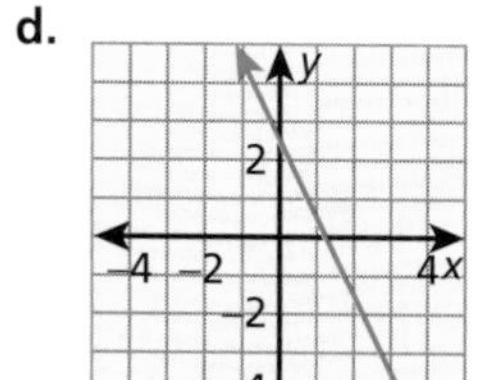

16. $y = -2x + \frac{5}{2}$
17. $y = 2x + \frac{5}{2}$
18. $y = \frac{4}{3}x + 1$
19. $y = -\frac{3}{4}x + 1$

Reasoning **In Exercises 20–23, decide whether the statement is true or false. Explain.**

20. The line $2x + 8y = 40$ has a slope of $\frac{1}{4}$ and a y-intercept of 5.
21. The line $3y = 6x + 5$ has a slope of 6 and a y-intercept of 5.
22. The line $-20x + 15y = 15$ has a positive slope.
23. If the slope of a line is 0, then the line is horizontal.

Population Model **In Exercises 24–26, use the following information.**

From 1984 to 1992, the population P (in thousands) of North Carolina can be modeled by

$$P = 82.3t + 5830.4$$

where $t = 4$ represents 1984.

24. What is the slope and y-intercept of the population model?

25. How much did North Carolina's population increase each year? Does this rate of change correspond to the slope of the model or to the y-intercept of the model?

26. Sketch a quick graph of the model.

Charlotte is the most populated city in North Carolina. It is the state's capital and was first settled in the mid-1700s.

In Exercises 27–30, write the equation of the line in slope-intercept form.

27.

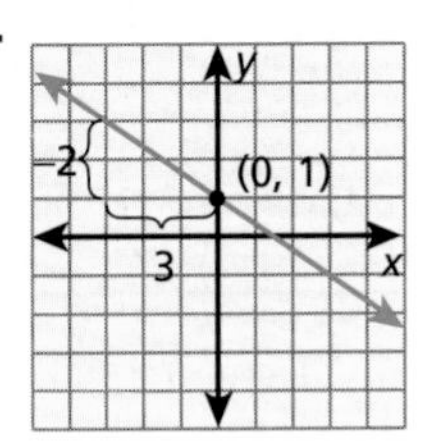

28.

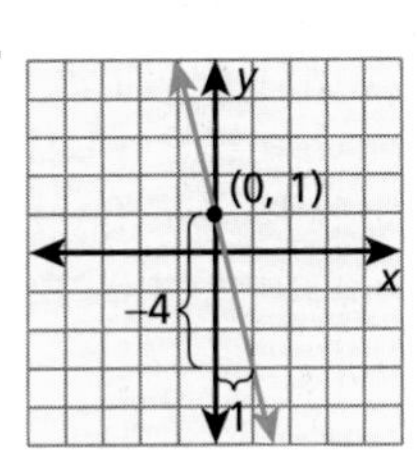

29.

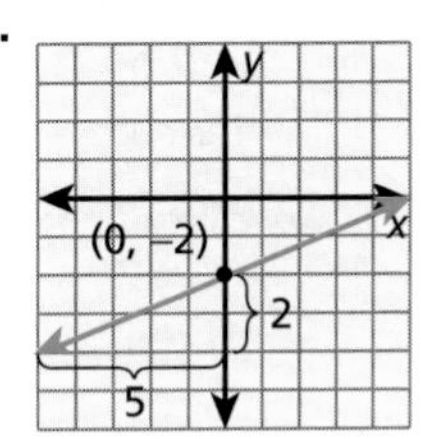

30.

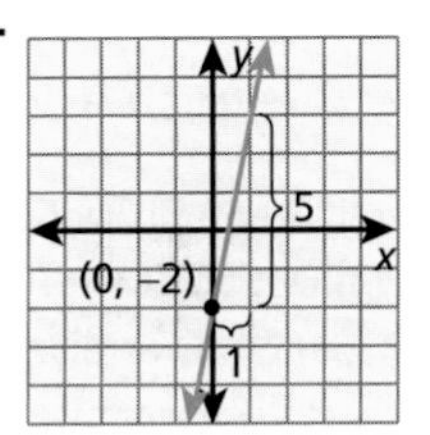

Integrated Review

Making Connections within Mathematics

Coordinate Pairs **In Exercises 31–34, match the equation of the line with the pair of points that lie on the line. Explain your reasoning.**

a. $(4, 2), (-2, -1)$ **b.** $(-2, 0), (4, -3)$ **c.** $(3, 7), (-1, -1)$ **d.** $(-1, 4), (3, -4)$

31. $y = 2x + 1$

32. $y = -2x + 2$

33. $y = \frac{1}{2}x$

34. $y = -\frac{1}{2}x - 1$

Exploration and Extension

Parallel or Perpendicular? **In Exercises 35–38, use the following information to decide whether the lines are parallel, perpendicular, or neither.**

You know that two lines are *parallel* if they have the same slope. For two lines to be *perpendicular,* their slopes must be negative reciprocals of each other. For instance, the lines $y = 2x + 1$ and $y = -\frac{1}{2}x + 1$ are perpendicular because $-\frac{1}{2}$ is the negative reciprocal of 2.

35. $y = 3x - 2$
$y = 3x - 6$

36. $y = -\frac{1}{3}x + 2$
$y = 3x - 1$

37. $4x + 16y = 24$
$3x + 12y = 24$

38. $y = x + 1$
$x + y = 1$

13.6 Problem Solving with Linear Equations

What you should learn:

Goal 1 How to use graphs of linear equations to solve real-life problems

Goal 2 How to use scatter plots to create graphical models

Why you should learn it:

You can use graphs of linear equations to solve real-life problems, such as analyzing the ticket sales necessary to raise a given amount of money.

Goal 1 Using Graphs of Linear Equations

Throughout this chapter, you have studied several real-life situations that can be modeled with linear equations. Two quantities that can be modeled with a linear equation are said to have a **linear relationship.**

For instance, suppose tomatoes cost \$0.65 a pound. The cost A (in dollars) of p pounds of tomatoes can be modeled by $A = 0.65p$. From the graph at the right, you can see that the relationship between A and p is linear. Each time you buy 1 more pound, the cost increases by \$0.65.

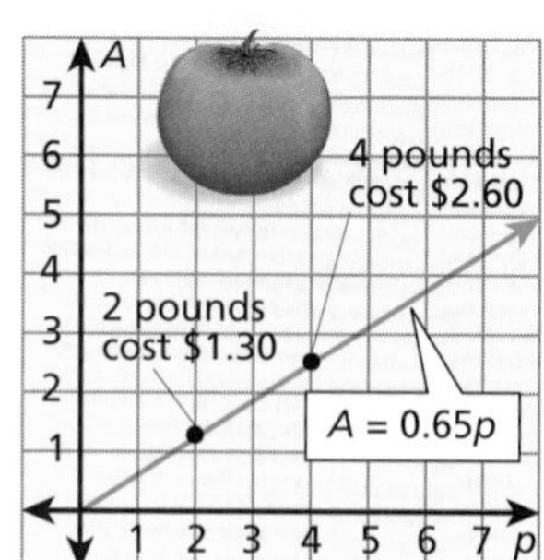

Example 1 *Interpreting Linear Models*

Real Life **Ticket Prices**

You are planning a dinner to raise money for a volunteer fire department. You plan to charge \$8 per adult and \$4 per child, and want to raise \$1200. Use a graph to analyze the following model.

Verbal Model \$8 • Number of Adults + \$4 • Number of Children = \$1200

Labels Number of adults = x

Number of children = y

Algebraic Model $8 \cdot x + 4 \cdot y = 1200$

Solution One way to begin is to create a table of values. You can use the table of values to sketch a graph, as shown at the left. Notice that there are many ways to raise \$1200. For instance, you could sell 150 adult tickets, or you could sell 100 adult tickets and 100 tickets for children.

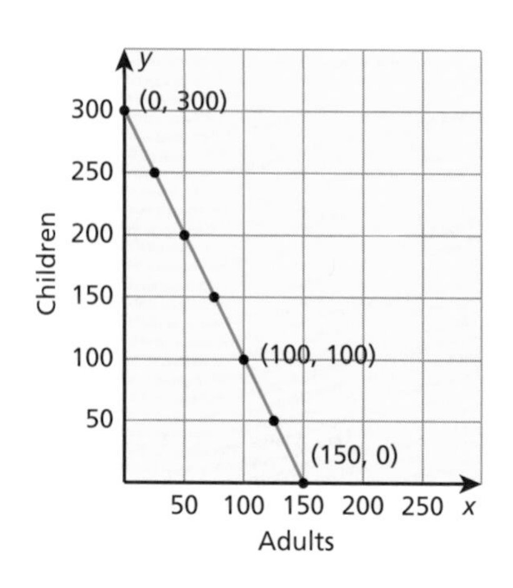

x	0	25	50	75	100	125	150
y	300	250	200	150	100	50	0

■

Goal 2 Using Scatter Plots

Example 2 *Using a Scatter Plot*

Real Life **Travel**

The amount of money A (in billions of dollars) spent by U.S. citizens in Latin America from 1985 through 1992 is given in the table. The year is represented by t with $t = 1$ corresponding to 1985. Draw a scatter plot of the data and use the result to estimate the amount spent in 1994.

Year	1985	1986	1987	1988	1989	1990	1991	1992
t	1	2	3	4	5	6	7	8
A	4.0	4.3	4.8	5.2	5.1	5.4	5.8	6.9

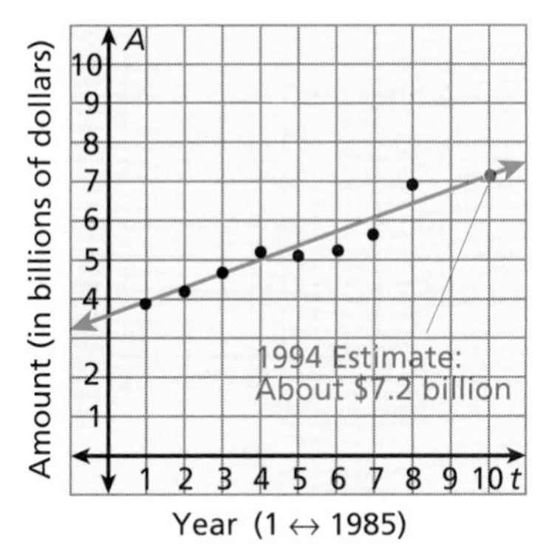

Solution Begin by writing the data as ordered pairs: (1, 4.0), (2, 4.3), (3, 4.8), (4, 5.2), (5, 5.1), (6, 5.4), (7, 5.8), and (8, 6.9). Then sketch a scatter plot, as shown at the left. From the scatter plot, it appears that variables almost have a linear relationship. Sketch the linc that you think best fits the points. Then, use the line to estimate the amount spent in 1994. Using the point (10, 7.2), you can estimate that U.S. citizens spent about $7.2 billion in Latin America in 1994. ■

Communicating about MATHEMATICS

Cooperative Learning

SHARING IDEAS about the Lesson

Using a Scatter Plot The scatter plot shown below shows the population (in thousands) of three different states from 1980 through 1992 with $t = 0$ corresponding to 1980. Use the scatter plot to estimate the population of each state in 1995 and 1998. ***(Source: U.S. Bureau of the Census)***

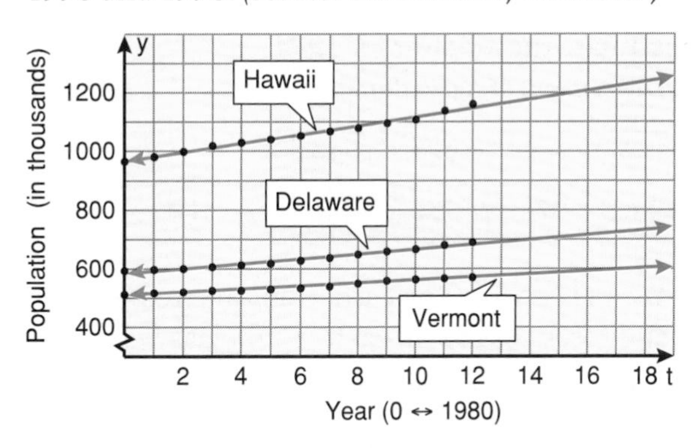

EXERCISES

Guided Practice

CHECK for Understanding

Which Is Linear? **In Exercises 1 and 2, you have two companies. In Company A, the number of employees is increasing by 10% each year. In Company B, the number of employees is increasing by 10 each year. Let n be the number of employees and let t be the year, as indicated in the graph at the right.**

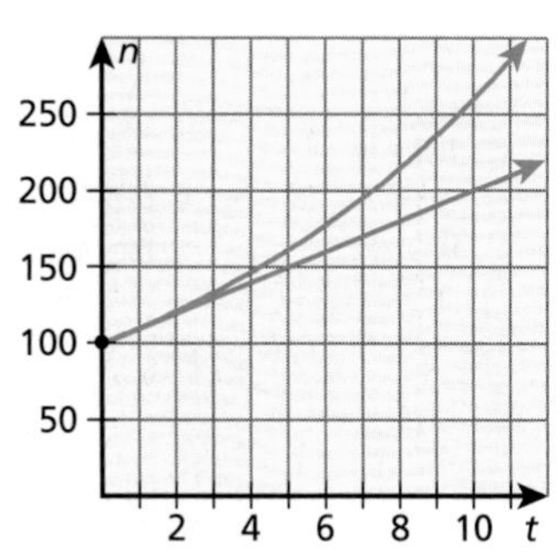

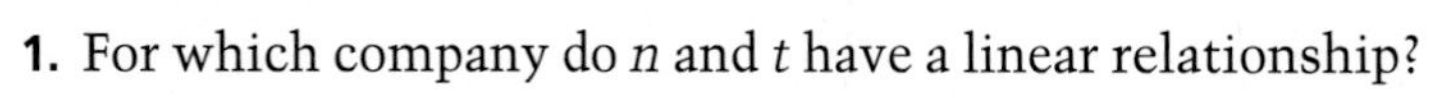

1. For which company do n and t have a linear relationship?
2. Each company began with 100 employees. How many employees will each company have in 10 years?

Tennis and Racquetball **In Exercises 3–6, your company makes tennis rackets that sell for \$50, and racquetball rackets that sell for \$30. Your daily goal is to sell \$1800 worth of rackets.**

\$30 • [Number of racquetball rackets sold] + \$50 • [Number of tennis rackets sold] = \$1800

3. Write an algebraic model to represent the verbal model.
4. Find several solutions of the model. Organize the solutions with a table.
5. Sketch a graph for the data in Exercise 4.
6. Interpret the intercepts of the graph in a real-life context.

Independent Practice

7. *Fruit Consumption* For 1985 through 1989, the average amount of fresh fruit A (in pounds) consumed by an American in a year is given in the table. Use a scatter plot to estimate the amount for 1994. ***(Source: U.S. Department of Agriculture)***

Year	1985	1986	1987	1988	1989
A	86.8	93.1	97.5	97.4	98.8

8. *Railroad Employees* From 1986 through 1991, the number of railroad employees, E (in 1000s), is given in the table. Use a scatter plot to estimate the number (in 1000s) for 1993. ***(Source: Association of American Railroads)***

Year	1986	1987	1988	1989	1990	1991
E	276	249	236	228	216	206

9. *Flowers* Your class is selling carnations and roses. Carnations cost \$1.50 each and roses cost \$3.00 each. Your class wants to earn \$600.
   a. Write a verbal and an algebraic model to represent the sales.
   b. Create a table of values and graph the model.
   c. Interpret the intercepts of the graph in a real-life context.

10. *Baseball Attendance* For 1984 through 1991, the attendance A (in thousands) at major league games is given in the table. Draw a scatter plot of the data and use the result to estimate the attendance in 1994. ***(Source: The National League of Professional Baseball Clubs)***

Year	1984	1985	1986	1987	1988	1989	1990	1991
A	45,262	47,742	48,452	53,182	53,800	55,910	55,512	57,820

11. *Aluminum Cans* From 1972 through 1992, the number of cans produced from one pound of aluminum is shown in the graph at the right. ***(Source: The Aluminum Association)***

a. Estimate the number produced in 1980 and in 1987.

b. What does this graph tell you about aluminum cans produced during this 20-year period?

c. Describe the pattern. Do you think the pattern could continue for another 20 years? Explain your reasoning.

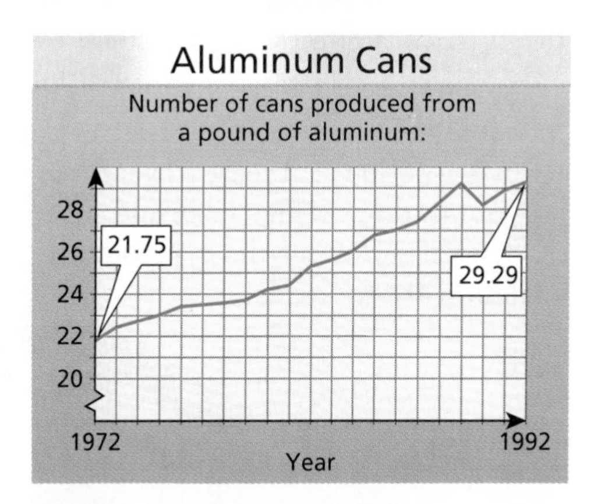

12. *Chemistry* A pan of water that is 68°F is heated until it boils at 212°F. The graph at the right shows the time required to heat the water.

a. How many minutes did it take the water to boil?

b. Write a real-life interpretation of the slope by completing the following. "The water temperature increased at a rate of [?] degrees per minute."

c. Write a real-life interpretation of the y-intercept. "When the pan was put on the stove the water temperature was [?] degrees"

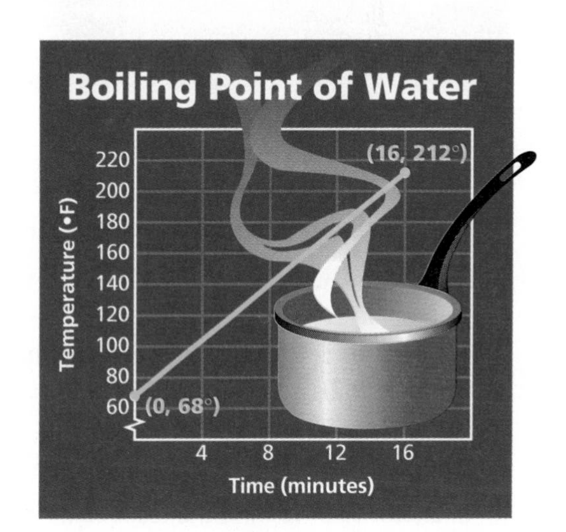

Integrated Review

Making Connections within Mathematics

Visualizing Equations **In Exercises 13–16, match the equation with its graph.**

a. 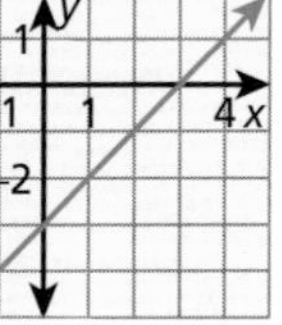

b. 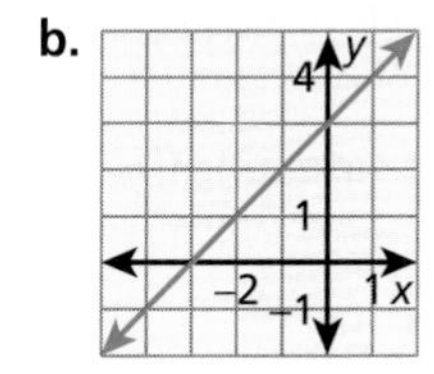

c. 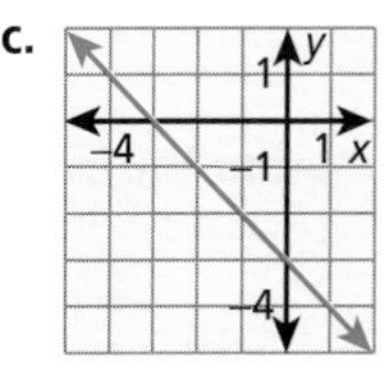

d.

13. $x + y = 3$ **14.** $x - y = 3$ **15.** $x + y = -3$ **16.** $x - y = -3$

Exploration and Extension

17. ***Mass Transportation Puzzle*** Eight different cities in the United States that use subways as part of their mass transportation system are hidden in the graph at the right. Find each city, then state the y-intercept and slope of its line.

18. ***Research Project*** Use your school's library or some other reference source to find data about the population of a state or country over a period of at least 10 years. Organize your findings with a table and a graph. Then use your results to predict the population at a future date.

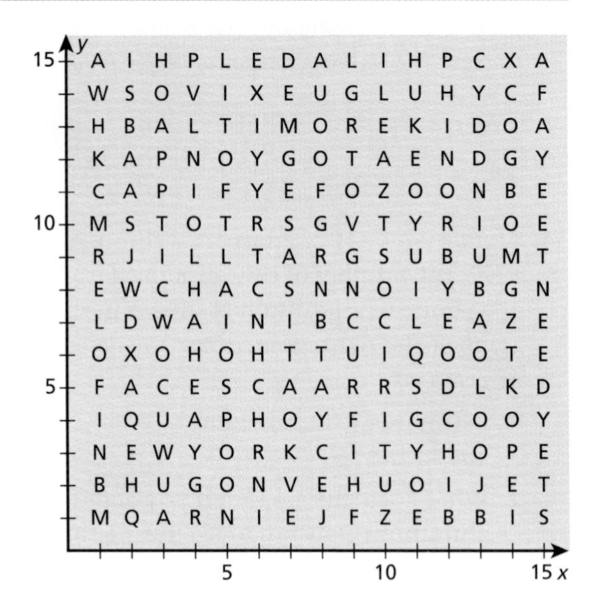

Mixed REVIEW

In Exercises 1–4, find the slope and y-intercept of the line. (13.5)

1. $y = 13x + 2$ **2.** $y = 4x + 21$ **3.** $3y = 12 + 3x$ **4.** $x = 2y + 4$

In Exercises 5–8, decide whether the numbers are rational or irrational. Explain. (9.2)

5. $\frac{2}{3}$ **6.** $\sqrt{529}$ **7.** $\sqrt{599}$ **8.** $\sqrt{\frac{81}{256}}$

In Exercises 9–12, find the least common multiple of the numbers. (6.4)

9. 24 and 36 **10.** 312 and 210 **11.** 111 and 55 **12.** 176 and 264

In Exercises 13–16, solve the percent equation. (8.4)

13. What is 18% of 32? **14.** 15 is 45% of what number?

15. 72 is what percent of 36? **16.** 17 is what percent of 40?

In Exercises 17–20, decide whether the numbers can be side lengths of a triangle. (9.8)

17. 12, 24, 30 **18.** 21, 26, 46 **19.** 13, 14, 29 **20.** 18, 18, 38

13.7 Graphs of Linear Inequalities

What you should learn:

How to check whether an ordered pair is a solution of linear inequality

How to sketch the graph of a linear inequality

Why you should learn it:

You can use graphs of linear inequalities to solve real-life problems, such as analyzing the ticket sales of a fund-raising dinner.

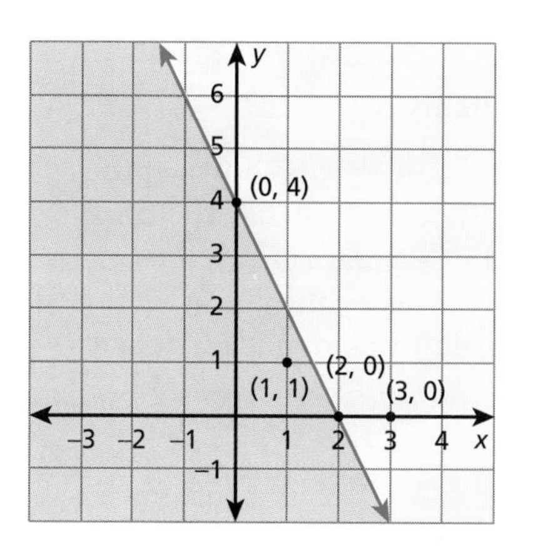

Goal 1 Solutions of Linear Inequalities

In this lesson, you will study **linear inequalities** in x and y. Here are some examples.

$$y < mx + b, \quad y \le mx + b, \quad y > mx + b, \quad y \ge mx + b$$

An ordered pair (x, y) is a **solution** of a linear inequality if the inequality is true when the values of x and y are substituted into the inequality. For instance, $(1, 7)$ is a solution of $y > x + 5$ because $7 > 1 + 5$ is a true statement.

Example 1 *Checking Solutions of Inequalities*

Check whether the ordered pairs are solutions of $y \le -2x + 4$.

a. $(1, 1)$ **b.** $(2, 0)$ **c.** $(3, 0)$

Solution

	(x, y)	Substitute	Conclusion
a.	$(1, 1)$	$1 \overset{?}{\le} -2(1) + 4$	$(1, 1)$ is a solution.
b.	$(2, 0)$	$0 \overset{?}{\le} -2(2) + 4$	$(2, 0)$ is a solution.
c.	$(3, 0)$	$0 \overset{?}{\le} -2(3) + 4$	$(3, 0)$ is not a solution.

The graph of all solutions of $y \le -2x + 4$ is shown at the left. The blue region represents the solutions. For instance the points $(1, 1)$ and $(2, 0)$ lie in the blue region but the point $(3, 0)$ lies in the yellow region. From the graph, can you identify other solutions of the inequality? ■

LESSON INVESTIGATION

■ Investigating Graphs of Linear Inequalities

Group Activity Sketch the graph of $y = x + 2$. Then test several points above the line and below the line. Which group of points contains solutions of the inequality $y \ge x + 2$? What can you conclude about the graph of a linear inequality?

Goal 2 Graphing Linear Inequalities

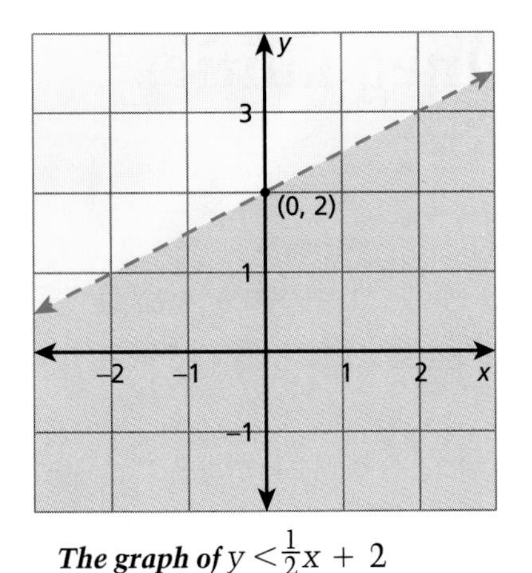

The graph of $y < \frac{1}{2}x + 2$

In the investigation on page 619, you may have discovered that the graph of the linear inequality $y \geq x + 2$ is a half-plane. This observation can be generalized to describe the graph of *any* linear inequality.

Sketching the Graph of a Linear Inequality

The graph of a linear inequality is a **half-plane** that consists of all points on one side of the line that is the graph of the corresponding linear equation.

For the inequality symbols $\geq$ and $\leq$, the points on the line are part of the graph. For the inequality symbols $>$ and $<$, the points on the line are not part of the graph (this is indicated by a dashed line).

Example 2 *Graphing a Linear Inequality*

Real Life
Ticket Sales

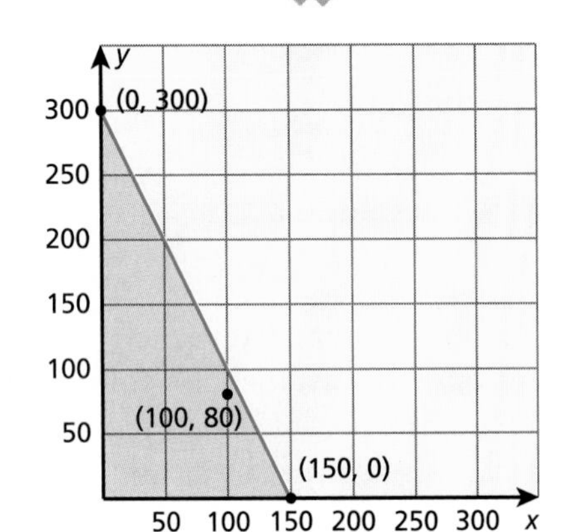

In Example 1 on page 614, the linear model $8x + 4y = 1200$ represented the different ways that you could sell \$8 adult tickets and \$4 tickets for children to earn \$1200 at a fund-raising dinner. Sketch the graph of

$$8x + 4y \leq 1200.$$

What do the solutions of this inequality represent?

Solution The graph of this linear inequality is shown at the left. The solutions represent the different ways that you could sell tickets to earn an amount that is *less than or equal to* \$1200. For instance, the point (100, 80) is a solution. With this solution you would have sold 100 adult tickets and 80 tickets for children for a total of

$$8(100) + 4(80) = \$1120.$$

■

Communicating about MATHEMATICS

SHARING IDEAS about the Lesson

Extending the Example Describe some other solutions of the inequality in Example 2. Which of the solutions represents the greatest number of ticket sales? Is this solution reasonable in the context of the example?

EXERCISES

Guided Practice

CHECK for Understanding

In Exercises 1–4, does the inequality's graph use a solid or dashed line? Explain.

1. $x + y \geq 10$ **2.** $3x + 7y > 42$ **3.** $12x - 17y < 200$ **4.** $-9x + 20y \leq 150$

In Exercises 5–8, use the graph of $y > 2x - 4$ at the right.

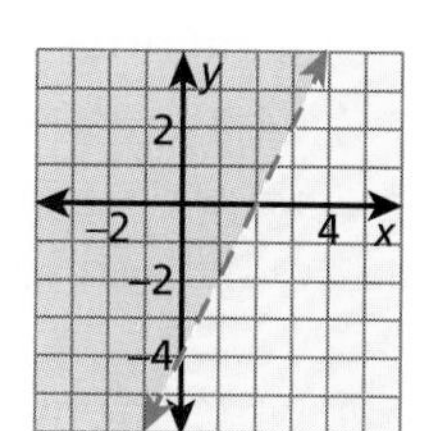

5. Is (2, 3) a solution?

6. Is (3, 2) a solution?

7. Name five other solutions of the inequality.

8. Find the slope and y-intercept of the dashed line.

In Exercises 9–12, sketch the graph of the inequality.

9. $y \leq x + 3$ **10.** $y \geq \frac{1}{2}x - 1$ **11.** $x - y > 2$ **12.** $x > -2$

Independent Practice

In Exercises 13–16, is the ordered pair a solution of $4x + 6y \leq 48$? Explain.

13. (5, 5) **14.** (10, −2) **15.** (−2, 10) **16.** (6, 4)

In Exercises 17–20, match the inequality with its graph.

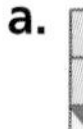

a. 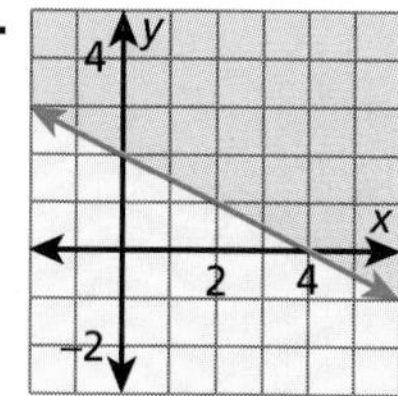

b. 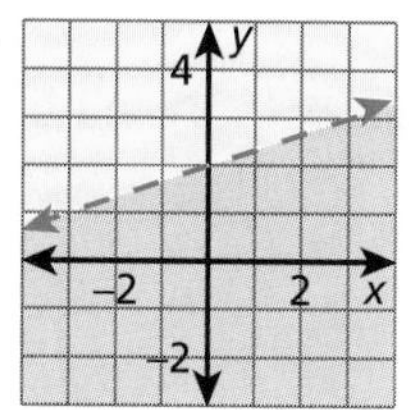

c. 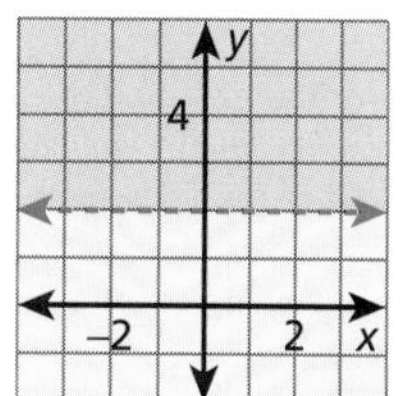

d. 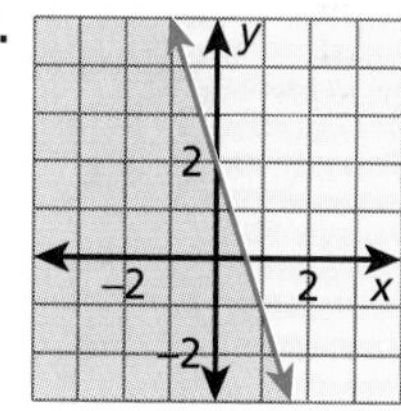

17. $y < \frac{1}{3}x + 2$ **18.** $y > 2$ **19.** $-3x - y \geq -2$ **20.** $x + 2y \geq 4$

In Exercises 21–24, graph the inequality. Then list several solutions.

21. $y \leq \frac{1}{4}x + 1$ **22.** $y > -2x - 2$ **23.** $x + y < 25$ **24.** $4x + 3y \geq 9$

In Exercises 25 and 26, use the following statement.

The sum of twice a number and five times another number is less than 30.

25. Which of the following inequalities represents the sentence?
a. $7y < 30$ **b.** $2x + 5y < 30$ **c.** $2x + 5y > 30$ **d.** $2x + 5x < 30$

26. Graph the inequality in Exercise 25. Then list several solutions.

Completing a Marathon **In Exercises 27–30, use the information in the photo caption and the following.**

A marathon is a long-distance race that covers 26.2 miles.

You are in a marathon race (26.2 miles). To finish the marathon, you realize that you must walk part of the way. Let x represent the number of miles you walk and y represent the number of miles you run.

27. You're not sure you can finish the marathon. Which of the following inequalities best describes your situation? Explain.

a. $x + y = 26.2$ **b.** $x + y \geq 26.2$

c. $x + y \leq 26.2$ **d.** $x + y < 26.2$

28. Sketch a graph of the correct inequality in Exercise 27.

29. Which of the inequalities in Exercise 27 can be interpreted as saying that "you are sure that you cannot finish the marathon."

30. *Problem Solving* Write a real-life interpretation of the other two inequalities in Exercise 27.

In Exercises 31–34, write an equation of the inequality.

31.

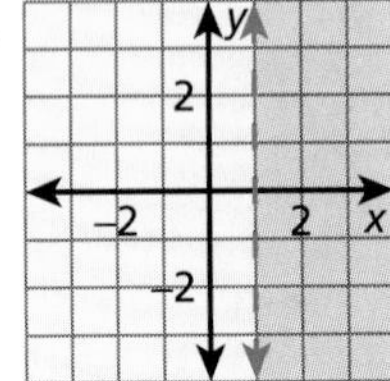

32.

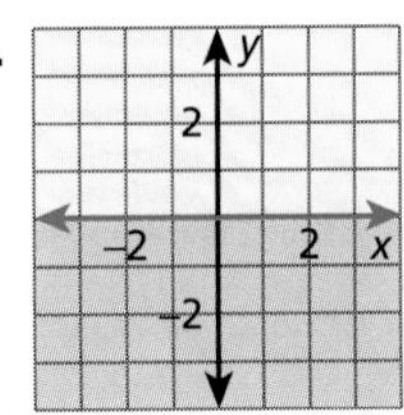

33.

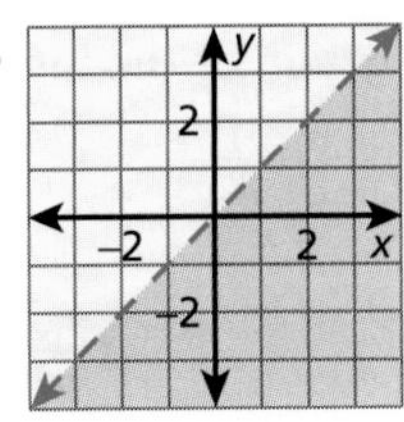

34.

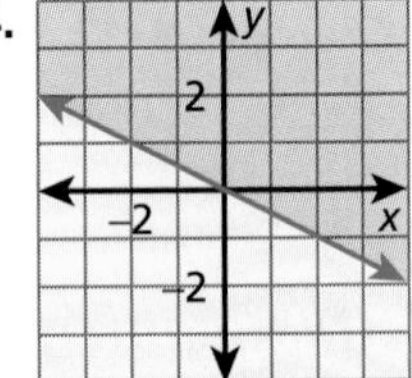

Integrated Review

Making Connections within Mathematics

Error Analysis **In Exercises 35 and 36, find the error in the graph, then correct it.**

35. $-2x + y > 2$

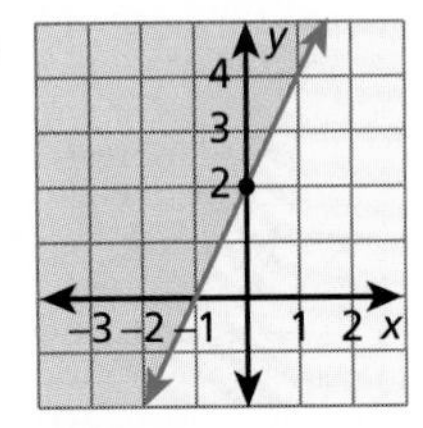

36. $9x - 3y \geq 12$

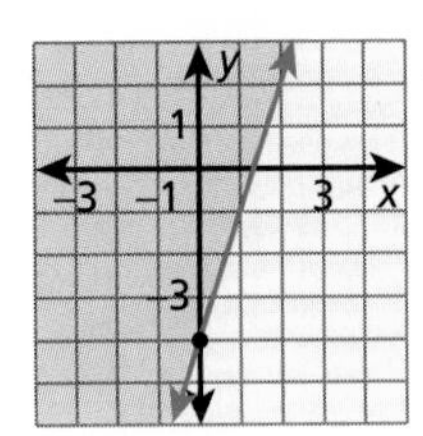

Exploration and Extension

Geometry **In Exercises 37–40, graph each inequality on the same coordinate plane. Identify the geometric figure formed by the set of points that are solutions of all the inequalities.**

37. $\begin{cases} x \geq -3 \\ x \leq 3 \\ y \geq -2 \\ y \leq 2 \end{cases}$

38. $\begin{cases} y < -x + 4 \\ x \geq 0 \\ y \geq 0 \end{cases}$

39. $\begin{cases} y < x + 3 \\ y > x \\ y \geq -1 \\ y < 3 \end{cases}$

40. $\begin{cases} y < -\frac{1}{2}x + 4 \\ y > \frac{1}{2}x - 4 \\ x > -2 \\ x < 4 \end{cases}$

13.8 The Distance and Midpoint Formulas

What you should learn:

How to find the distance between two points

How to find the midpoint between two points

Why you should learn it:

You can use the distance and midpoint formulas to solve real-life problems, such as finding a location for a telephone pole.

Goal 1 Using the Distance Formula

Suppose you were asked to find the distance between the points $A(1, 3)$ and $B(5, 6)$. How would you do it? One way is to draw a right triangle that has the line segment $\overline{AB}$ as its hypotenuse. The right angle occurs at the point $C\,(5, 3)$. As shown at the left, the lengths of the legs of the right triangle are

$$a = 6 - 3 = 3 \quad \text{and} \quad b = 5 - 1 = 4.$$

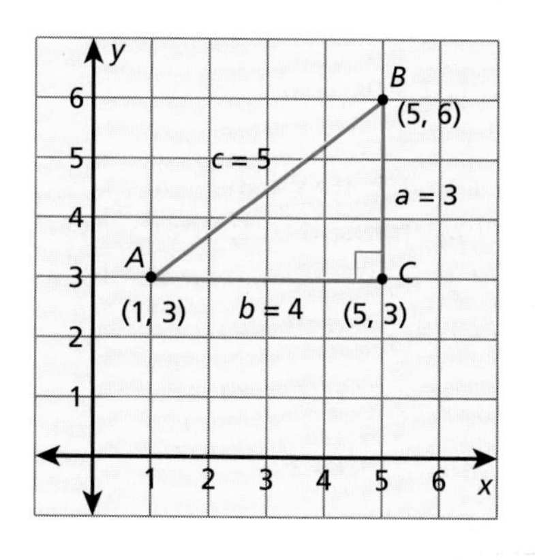

Using the Pythagorean Theorem, you can find the length of the hypotenuse.

$c^2 = a^2 + b^2$ *Pythagorean Theorem*

$c^2 = (3)^2 + (4)^2$ *Substitute for a and b.*

$c^2 = 25$ *Simplify.*

$c = 5$ *Square root principle*

By performing this process with two general points (x_1, y_1) and (x_2, y_2), you can obtain the **Distance Formula.**

The Distance Formula

The distance, d, between the points (x_1, y_1) and (x_2, y_2) is

$$d = \sqrt{(x_2 - x_1)^2 + (y_2 - y_1)^2}.$$

Example 1 *The Distance between Two Points*

Find the distance between (2, 3) and (5, 8).

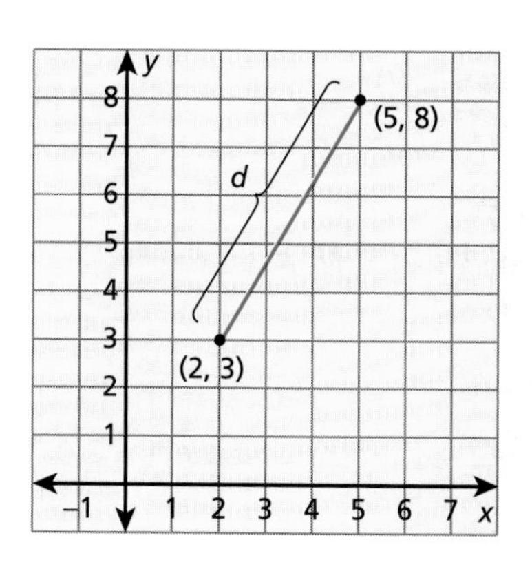

Solution Let $(x_1, y_1) = (2, 3)$ and $(x_2, y_2) = (5, 8)$.

$$\begin{aligned} d &= \sqrt{(x_2 - x_1)^2 + (y_2 - y_1)^2} \\ &= \sqrt{(5 - 2)^2 + (8 - 3)^2} \\ &= \sqrt{3^2 + 5^2} \\ &= \sqrt{34} \\ &\approx 5.83 \end{aligned}$$

The distance between the points is about 5.83 units. ■

Goal 2 Using the Midpoint Formula

LESSON INVESTIGATION

Investigating Midpoints

Group Activity The **midpoint** of A and B is the point that lies on the line segment AB, halfway between the two points. Plot the following pairs of points. Then find the coordinates of each pair's midpoint. Describe a general procedure for finding the coordinates of the midpoint.

a. $A(1, 2), B(5, 6)$ **b.** $A(0, 3), B(6, 5)$

c. $A(2, 1), B(-4, 3)$ **d.** $A(3, 7), B(7, 3)$

In this investigation, you may have discovered the following formula for the midpoint between two points.

The Midpoint Formula

The **midpoint** between (x_1, y_1) and (x_2, y_2) is $\left(\frac{x_1 + x_2}{2}, \frac{y_1 + y_2}{2}\right)$.

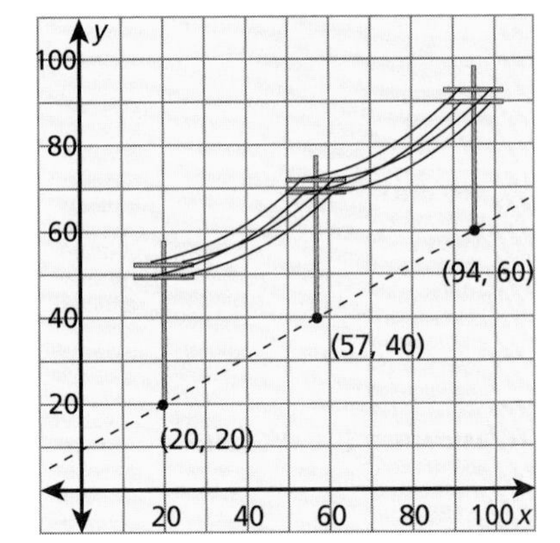

Example 2 *Using the Midpoint Formula*

You work for a telephone company. You are asked to locate the coordinates of a telephone pole that is supposed to lie halfway between two other poles, as shown at the left. Describe the location of the third pole. (The units are measured in feet.)

Solution Let $(x_1, y_1) = (20, 20)$ and $(x_2, y_2) = (94, 60)$.

$$\text{Midpoint} = \left(\frac{x_1 + x_2}{2}, \frac{y_1 + y_2}{2}\right) \quad \textit{Midpoint Formula}$$

$$= \left(\frac{20 + 94}{2}, \frac{20 + 60}{2}\right) \quad \textit{Substitute for coordinates.}$$

$$= (57, 40) \quad \textit{Simplify.}$$

The coordinates of the third pole are (57, 40). ■

Communicating about MATHEMATICS

SHARING IDEAS about the Lesson

Extending the Example In Example 2, find the distances between the poles. What can you conclude?

EXERCISES

Guided Practice

CHECK for Understanding

In Exercises 1 and 2, find the distance between the points. Then find their midpoint. Check your results with a graph.

1. (5, 4), (2, 0)
2. (−1, −3), (−1, −7)

Geometry **In Exercises 3 and 4, use the figure at the right.**

3. Find the perimeter of the figure.
4. Show that the diagonals have the same midpoint.

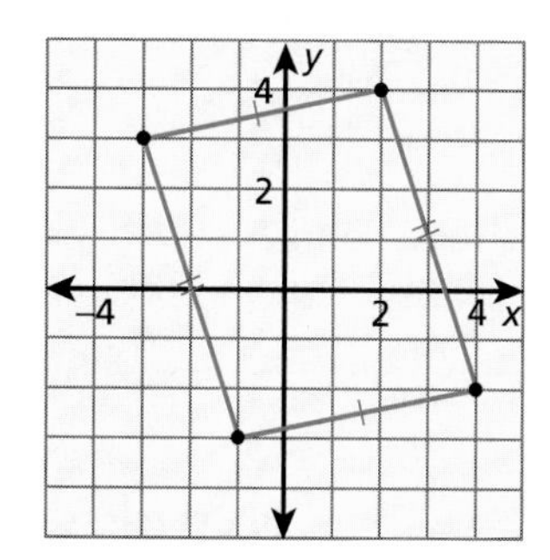

Independent Practice

In Exercises 5–8, use the graph to estimate the distance between the points. Then use the Distance Formula to check your estimate.

5. 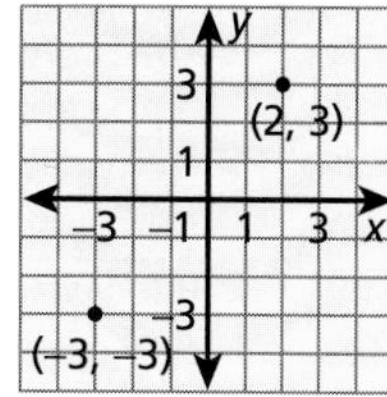

6. 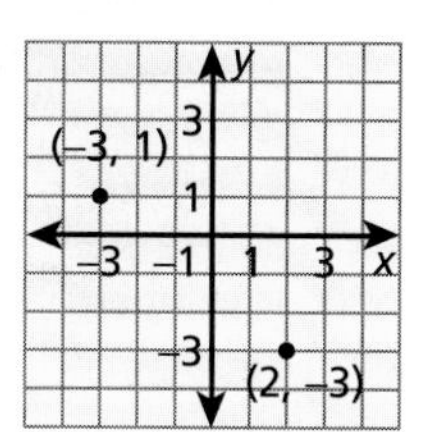

7. 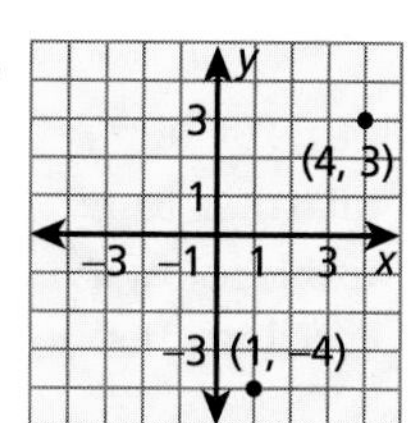

8. 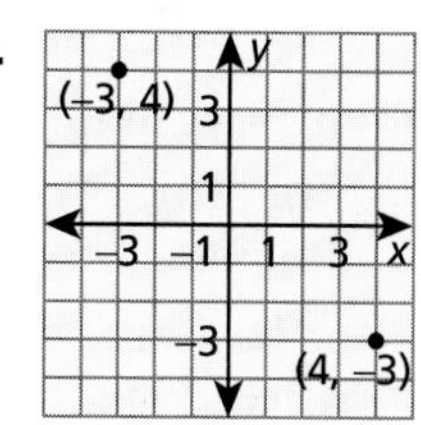

In Exercises 9–12, use the graph to estimate the midpoint of the two points. Then use the Midpoint Formula to check your estimate.

9. 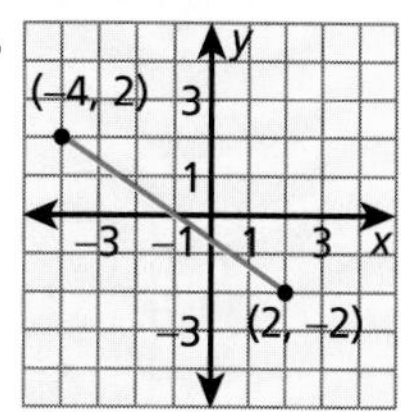

10. 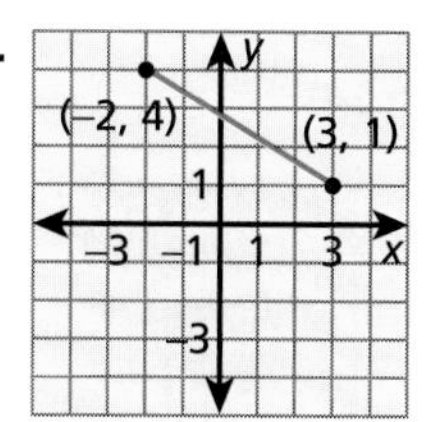

11. 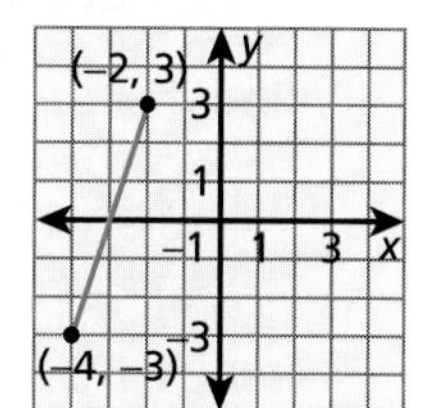

12. 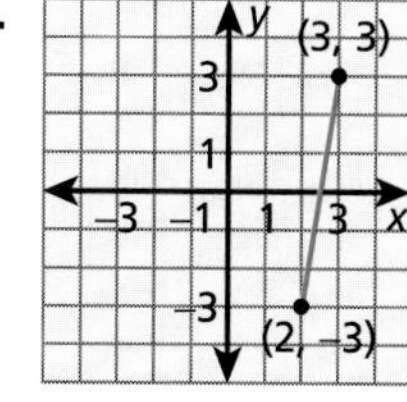

In Exercises 13–16, find the perimeter of the polygon to two decimal places.

13. 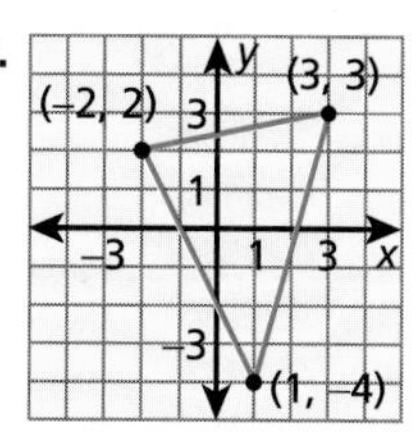

14. 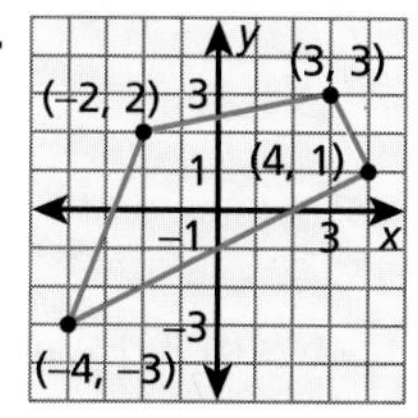

15. 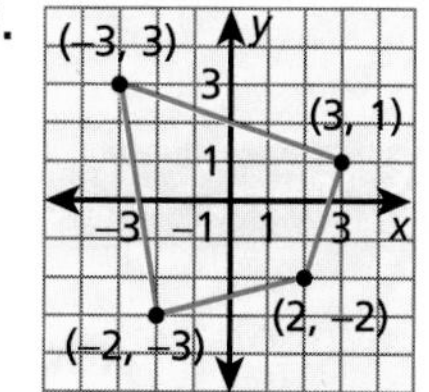

16. 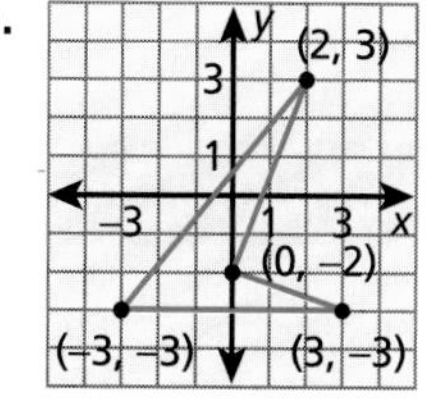

In Exercises 17 and 18, decide whether $\triangle ABC$ is a right triangle. Explain.

17. $A(0, 3)$, $B(2, 5)$, $C(2, 1)$
18. $A(1, 4)$, $B(2, 1)$, $C(5, 7)$

Circles **In Exercises 19 and 20, the labeled points are endpoints of a diameter of the circle. Find the center and radius of the circle.**

19.

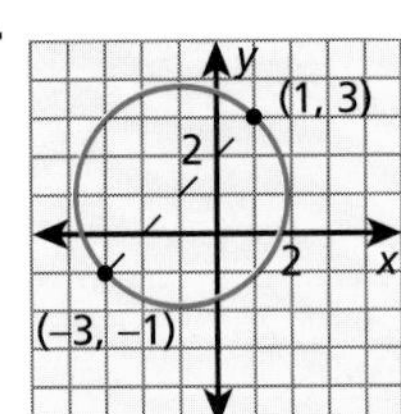

20.

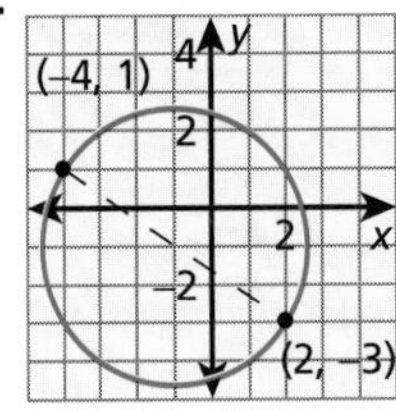

Parallelograms **In Exercises 21 and 22, show that the diagonals of parallelogram *ABCD* have the same midpoints.**

21.

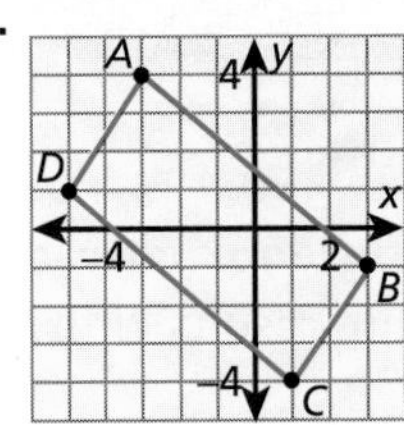

22.

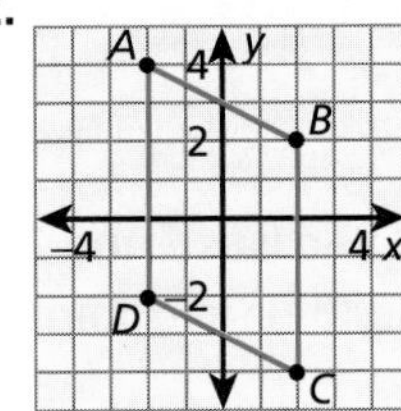

In Exercises 23 and 24, the expression represents the distance between two points. What are the points?

23. $\sqrt{(5 - 3)^2 + (6 - 1)^2}$

24. $\sqrt{(0 - 4)^2 + (-8 + 2)^2}$

Planning a Trip **In Exercises 25 and 26, you live in Louisville, Kentucky, which has the latitude-longitude coordinates (46.2N, 72.6W), and are planning a trip to Wichita, Kansas, which has the latitude-longitude coordinates (37.4N, 97.2W).**

25. Use the map to estimate the distance between the two cities.

26. You are planning a stop halfway between the two cities. What are the latitude-longitude coordinates of the halfway point? If you average 50 miles per hour, how long will it take you to reach the halfway point?

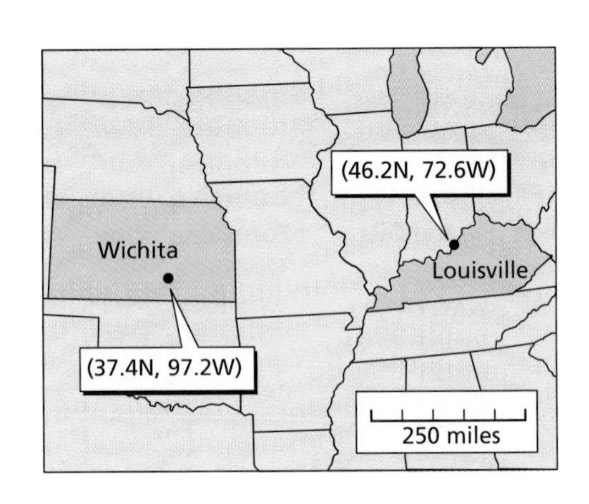

Integrated Review

Making Connections within Mathematics

Geometry **In Exercises 27–30, find the area of the figure.**

27.

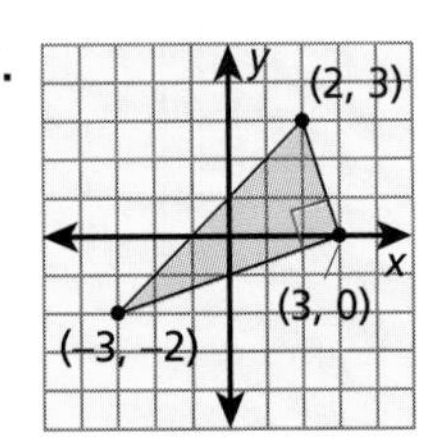

28.

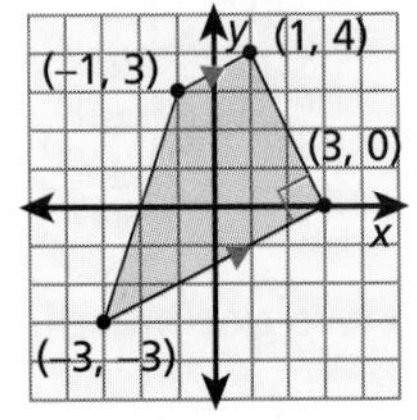

29.

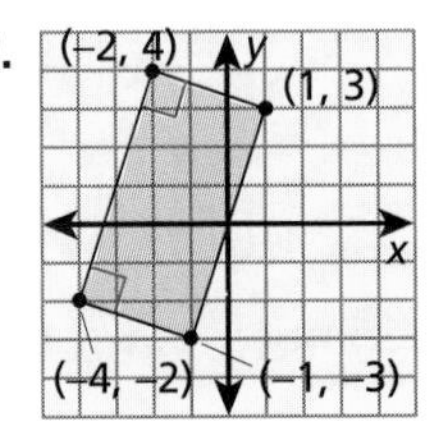

30.

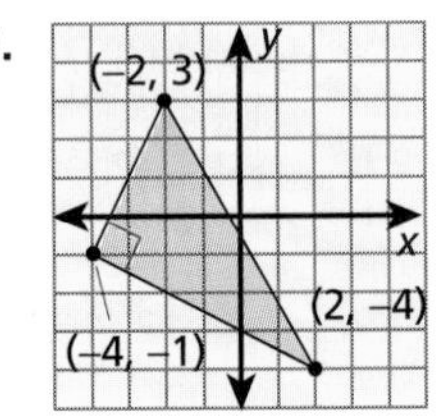

Exploration and Extension

In Exercises 31–34, *C* is a midpoint of $\overline{AB}$. Find the coordinates of *B*. Explain your reasoning and illustrate your answer with a graph.

31. $A(-4, 2)$, $C(-4, -1)$

32. $A(5, 3)$, $C(1, 3)$

33. $A(3, -2)$, $C(1, 0)$

34. $A(2, 5)$, $C(0, -1)$

Chapter Summary

What did you learn?

Skills

1. Find solutions of a linear equation in two variables. **(13.1)**
   - Organize solutions in a table of values. **(13.1)**
2. Use a table of values to graph a linear equation. **(13.2)**
3. Recognize graphs of horizontal and vertical lines. **(13.2)**
4. Find the intercepts of a line. **(13.3)**
   - Use intercepts to sketch a quick graph of a line. **(13.3)**
5. Find the slope of the line passing through two points. **(13.4)**
6. Use the slope-intercept form of the equation of a line. **(13.5)**
   - Use the slope and y-intercept to sketch a quick graph. **(13.5)**
7. Sketch the graph of a linear inequality. **(13.7)**
8. Find the distance between two points. **(13.8)**
9. Find the midpoint between two points. **(13.8)**

Problem-Solving Strategies

10. Model and solve real-life problems. **(13.1–13.8)**
    - Organize solutions of real-life problems **(13.1)**

Exploring Data

11. Use tables and graphs to solve problems. **(13.1–13.8)**
    - Use scatter plots to predict future trends. **(13.6)**

Why did you learn it?

Linear equations can be used to model the relationships between many real-life quantities. For instance, in this chapter you saw how linear equations can be used to model the attendance at college football games, model a trip on a subway, model the populations of states, and model the amount of money earned at a fund-raising dinner. If you take Algebra 1 next year, you will study many other ways that linear equations are used to model real life. In fact, you will learn that a linear equation is the most widely used type of model in all of mathematics.

How does it fit into the bigger picture of mathematics?

The combination of algebra (equations) and geometry (graphs) that you studied in this chapter is called *analytic geometry*. Analytic geometry was developed about 350 years ago, and it has proved to be a very useful way to study both algebra and geometry. In this chapter you saw that you learn a lot about a linear equation by sketching its graph. For instance, from the graph you can find the intercepts and the slope—both of which have important real-life interpretations. As you continue your study of mathematics, remember that the old saying "a picture is worth a thousand words" applies to algebra as well as to other parts of life.

Chapter REVIEW

In Exercises 1–4, is the ordered pair a solution of $5x + 2y = 40$? (13.1)

1. $(4, 10)$ **2.** $(7, 2)$ **3.** $(9, -2)$ **4.** $(-2, 25)$

In Exercises 5–8, create a table of values for the equation. Then use the table to sketch a graph of the equation. (13.1, 13.2)

5. $y = \frac{1}{2}x - 4$ **6.** $y = -\frac{5}{4}x + 2$ **7.** $y = -4x + 6$ **8.** $6x + 4y = 16$

In Exercises 9–12, decide whether the line is *horizontal, vertical,* or *slanted.* (13.2)

9. $y = -10x$ **10.** $x = -4$ **11.** $y = 10$ **12.** $12x - 6y = 54$

In Exercises 13–16, sketch the line that has the given intercepts. (13.3)

13. x-intercept: -4, y-intercept: 7 **14.** x-intercept: 1, y-intercept: 4 **15.** x-intercept: 5, y-intercept: -2 **16.** x-intercept: -6, y-intercept: -1

In Exercises 17–20, use intercepts to sketch a quick graph of the line. (13.3)

17. $y = 2x + 6$ **18.** $y = -\frac{1}{3}x - 1$ **19.** $6x + y = 9$ **20.** $3x - 2y = 10$

In Exercises 21–24, estimate the slope of the line. (10.4)

21.

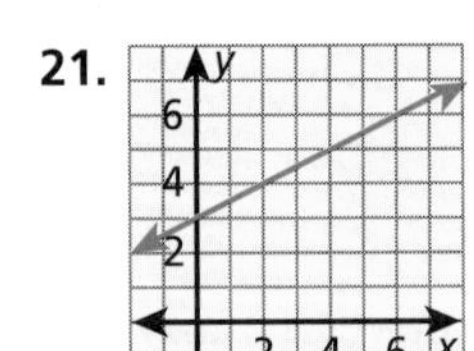

22.

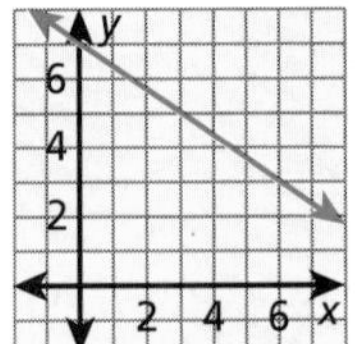

23.

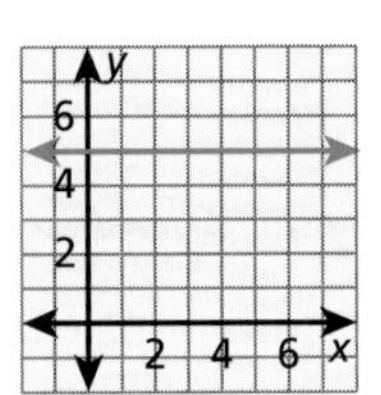

24.

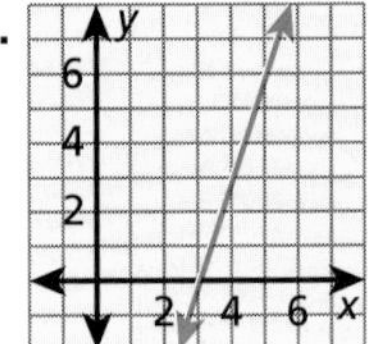

In Exercises 25–28, find the slope of the line through the points. (13.4)

25. $(5, 0), (-1, 6)$ **26.** $(-2, -3), (1, 9)$ **27.** $(-4, 2), (3, 4)$ **28.** $(0, 9), (8, -1)$

29. Which equation is in slope-intercept form? **(13.5)**

a. $6y = 5x - 18$ **b.** $x = \frac{6}{5}y + \frac{18}{5}$ **c.** $y = \frac{5}{6}x - 3$ **d.** $5x - 6y = 18$

30. Explain how to use the slope-intercept form to sketch a quick graph of a line. **(13.5)**

In Exercises 31–34, find the slope and y-intercept of the line. Then sketch a quick graph of the line. (13.5)

31. $y = -\frac{3}{5}x$ **32.** $y = 4x + 5$ **33.** $7x - 6y = 24$ **34.** $x + 5y = 10$

In Exercises 35–38, find the equation of the line. (13.1–13.5)

35. **36.**

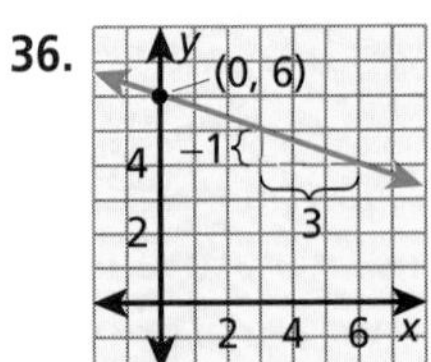

37.

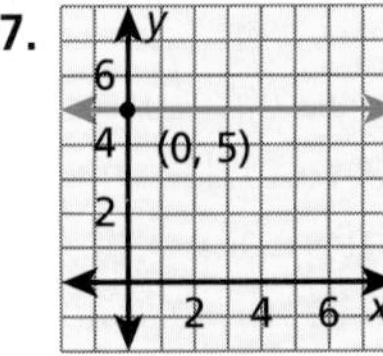

38.

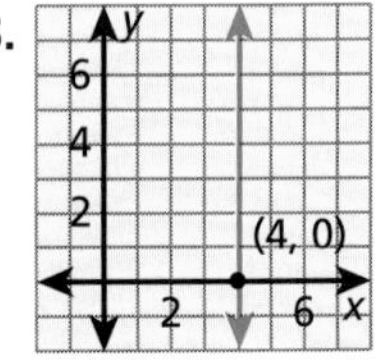

Nursery **In Exercises 39–42, consider the following.**

You own a nursery that sells shrubs and trees to landscapers. You sell rhododendrons for $40 and dogwood trees for $60 each. One day during planting season, you sell $2500 worth of rhododendrons and dogwood trees.

39. Write a verbal model that describes the sale of rhododendrons and dogwood trees.

40. Use the verbal model to write an algebraic model. Let x represent the number of rhododendrons and y represent the number of dogwoods.

41. You sold 31 rhododendrons. How many dogwood trees did you sell?

42. Graph the linear model. Then list several other solutions.

In Exercises 43–48, decide whether the ordered pair is a solution of the inequality. Then decide whether a solid line or dashed line is used to graph the inequality. (13.7)

43. $y < 8x - 12$; $(4, 10)$ **44.** $y \geq -4x + 9$; $(2, 2)$ **45.** $y \leq \frac{11}{2}x + 6$; $(-4, -12)$

46. $-x + \frac{3}{4}y \leq 0$; $(6, 8)$ **47.** $-2y < 3x + 1$; $(1, -3)$ **48.** $3x - 7y < -21$; $(-1, 3)$

In Exercises 49–52, sketch the graph of the inequality. (13.7)

49. $y > -2\frac{1}{2}$ **50.** $y < -x - 1$ **51.** $y \geq 7x + 6$ **52.** $2x - 3y \geq -15$

In Exercises 53–56, find the midpoint and length of the line segment. (13.8)

53.

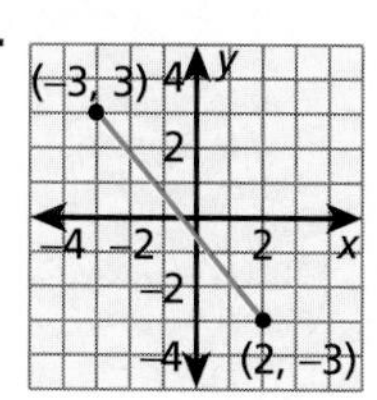

54.

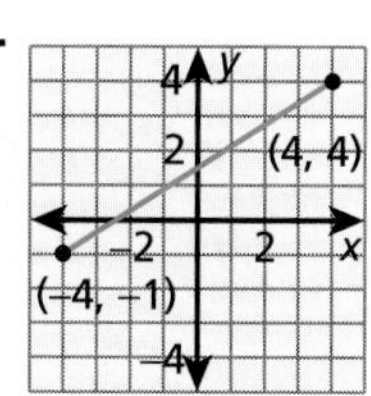

55.

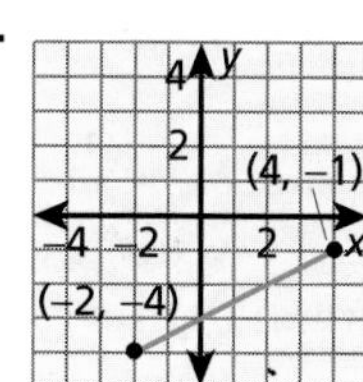

56.

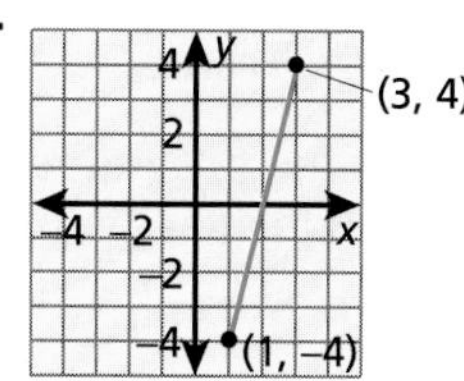

Geometry **In Exercises 57–60, use the figure at the right. (13.8)**

57. Find the length of each side of $\triangle ABC$.

58. Use the converse of the Pythagorean Theorem to show that $\triangle ABC$ is a right triangle.

59. Find the perimeter of $\triangle ABC$.

60. Find the area of $\triangle ABC$.

61. Consider a circle whose center is $(1, 2)$. Exactly two of the points, $(4, 5)$, $(-2, 6)$, and $(4, -1)$ are on the circle. Which two are they? Explain your reasoning.

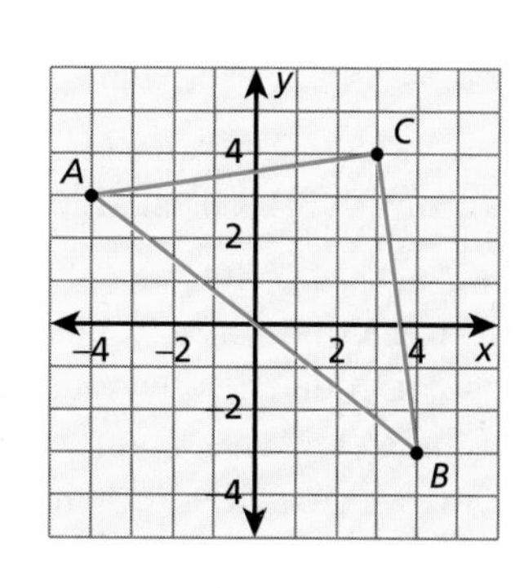

FOCUS ON *Advanced Transportation*

Riding the Commuter Rail **In Exercises 62–66, use the coordinate map and the following information.**

You live in the suburb of a city that has a commuter rail system. There are two stations in the suburb. The stations, your house, and the homes of two friends, Felicia and Heather, are shown on the coordinate map.

y
Station 1
Your house
15
Felicia
10
Station 2
5
Heather
5
10
x

62. Which station is nearer to your house?

63. Which station is nearer to each of your friends?

64. A new station is being built exactly halfway between Station 1 and Heather's house. What are the coordinates of the new station?

65. During the next school year, each of you plans to use the nearest station. Will any of you go to the new one? Explain.

66. Could the new station have been located so that it would be closer to each of your houses than the two existing stations are? Explain.

Bay Area Mass Transportation **In Exercises 67–71, use the following information.**

The cable car system in San Francisco, California, is part of the city's mass transit system. The other part of the system consists of buses and an electric rail system called *BART* for *Bay Area Rapid Transit.*

67. You are riding a cable car whose speed is 9 miles per hour. The distance, d (in miles), you will travel in t hours is modeled by $d = 9t$. Sketch a graph of this model.

68. Use the model in Exercise 67 to find how far you will travel on a 10-minute cable car ride.

69. You are riding a BART car whose speed is 50 miles per hour. The distance, d (in miles), you will travel in t hours is modeled by $d = 50t$. Sketch a graph of this model.

70. Use the model in Exercise 69 to find how far you will travel on a 10-minute BART car ride.

The cable car system in San Francisco was built in 1873.

71. *Guess, Check, and Revise* Riding the bus to your work costs \$0.75 per ride. Taking a taxi to work costs \$3.25 per ride. Let x represent the number of times you rode the bus and let y represent the number of times you took a taxi. During the week, you took 10 rides and spent \$12.50. How many times did you ride the bus? How many times did you take a taxi?

Chapter TEST

In Exercises 1–4, consider the equation $2x + 3y = 6$.

1. Complete the following table of values for the equation.

x	-3	-2	-1	0	1	2	3
y	?	?	?	?	?	?	?

2. From your table of values, name the intercepts of the graph.

3. Sketch a quick graph of the line.

4. Use the graph to find the slope of the line.

In Exercises 5–7, decide whether the ordered pair is a solution of $x + 3y = 13$.

5. $(8, 3)$

6. $(1, 5)$

7. $(-2, 6)$

8. Find the slope of the line that passes through $(3, 5)$ and $(-3, 4)$.

9. Find the slope and y-intercept of the line $y = -2x + 4$. Then sketch the line.

In Exercises 10 and 11, find the slope of the line.

10.

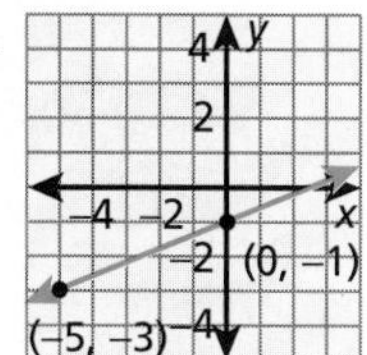

11.

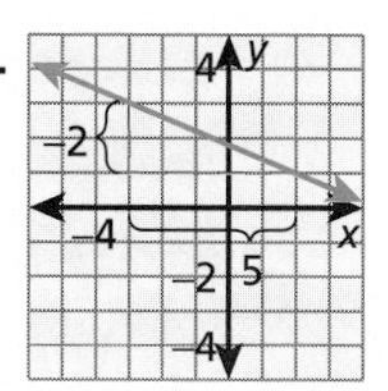

In Exercises 12 and 13, write an equation of the line.

12.

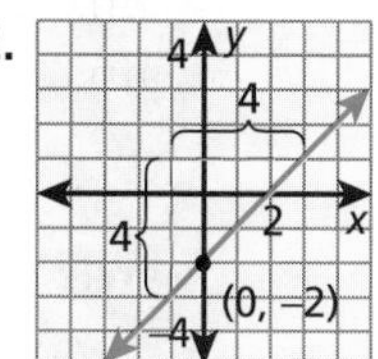

13.

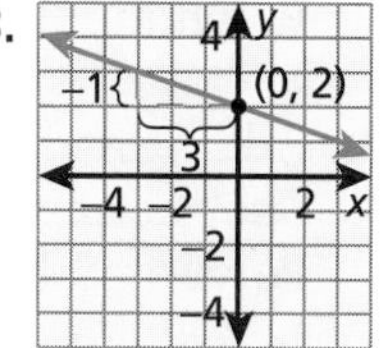

14. Sketch the graph of $y < 3x$.

15. Sketch the graph of $y \geq 2x + 2$.

16. Find the distance between $(3, 0)$ and $(-1, 2)$.

17. Find the midpoint between the points $(-3, -2)$ and $(5, 6)$.

18. A circle in a coordinate plane has $(0, 0)$ and $(6, 8)$ as endpoints of a diameter. Find the area of the circle. Sketch your answer.

In Exercises 19 and 20, use the following.

The speed of an automobile, starting from rest, is shown in the graph at the right. At the end of 12 seconds, the automobile has reached a speed of 48 miles per hour.

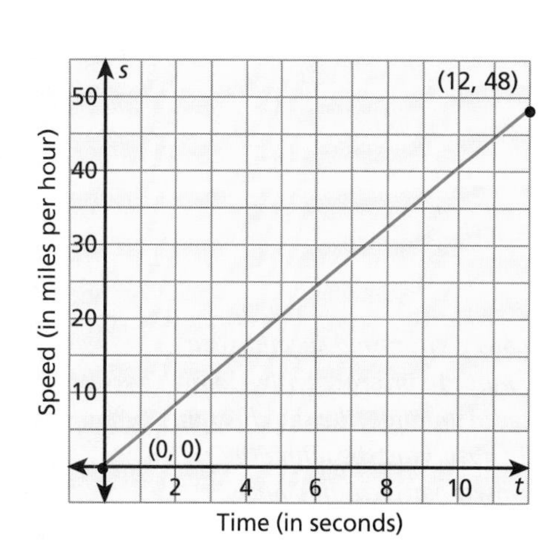

19. Let s represent the speed (in miles per hour) and let t represent the time (in seconds). Which of the following models is correct?

a. $s = 48t$ **b.** $s = 12t$ **c.** $s = 4t$

20. Use the graph to estimate the speed of the automobile at the end of 8 seconds. Then use the correct model found in Exercise 19 to check your result.

CHAPTER

14

Exploring Data and Polynomials

LESSONS

Zoos give animal specialists a chance to interact with endangered animals like the cheetah. By observing the animal's growth patterns and muscle tone, it is possible to study the role diet and exercise play in maintaining the animal's health and well being.

Real Life
Animal Care
Mouse
Lion
Elephant
Rhinoceros
Tortoise
2
10
12
22
50
70
100
Advances in veterinary medicine have greatly improved animal care, led to the preservation of endangered species, and increased the health and life expectancy of livestock and pets.
The box plot shows the average life expectancy of certain animals.
Drawing and interpreting box-and-whisker plots is just one of the topics you will explore in this chapter.

14.1 Measures of Central Tendency

What you should learn:

Goal 1 How to find measures of central tendency

Goal 2 How to use measures of central tendency to solve real-life problems

Why you should learn it:

You can use measures of central tendency to solve real-life problems, such as describing the typical age of the people who attend a party.

Goal 1 Measures of Central Tendency

A **measure of central tendency** is a number that can be used to represent a group of numbers. Bar graphs (or histograms) can help you see numbers that are representative.

LESSON INVESTIGATION

Investigating Central Tendencies

Group Activity The three sets of data below show the ages of people at three different parties. For each party, state a single number that is most representative of the ages of the people. Explain how you chose your number and why you think it is representative.

Party 1: 13, 13, 13, 13, 13, 14, 14, 14, 14, 14
Party 2: 11, 12, 13, 13, 13, 13, 13, 13, 13, 13
Party 3: 10, 11, 12, 13, 14, 15, 16, 17, 18, 19

In this investigation, you may have chosen one or more of the most common measures of central tendency. They are called the **mean** (or average), **median,** and **mode.**

Example 1 *Finding Measures of Central Tendency*

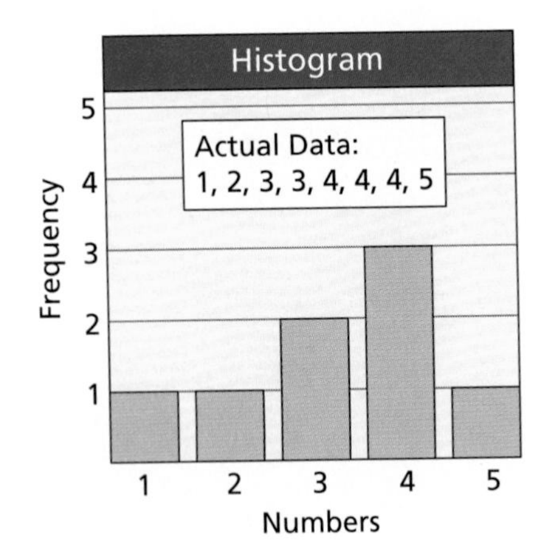

Find the mean, median, and mode of the data.

a. The *mean* of a group of numbers is the average of the numbers. For the numbers shown in the histogram at the left, the mean is

$$\text{Mean} = \frac{1 + 2 + 3 + 3 + 4 + 4 + 4 + 5}{8} = \frac{26}{8} = 3.25.$$

b. The *median* of a group of numbers is the middle number (or average of the two middle numbers) when the numbers are listed in order. Because the two middle numbers of the actual data are 3 and 4, the median is 3.5.

c. The *mode* of a group of numbers is the number that occurs most often. For the data at the left, the mode is 4. ■

Goal 2 Using Measures of Central Tendency

The three measures of central tendency for a group of numbers can be exactly the same, almost the same, or very different. Here are some examples.

> **Study Tip...**
> *To find the median or the mode of a collection of numbers, you should first order the numbers from smallest to largest.*

Group 1: 10, 11, 11, 12, 12, 12, 13, 13, 14
Group 2: 10, 11, 11, 12, 12, 12, 13, 13, 15
Group 3: 10, 10, 10, 12, 13, 13, 14, 14, 84

In Group 1, the mean, median, and mode are each 12. In Group 2, the mean is 12.1 and the median and mode are 12. In Group 3, the mean is 20, the median is 13, and the mode is 10.

Example 2 *Interpreting Central Tendencies*

Real Life **Weather**

The number of days of rain or snow per month in Boise, Idaho, and Lewiston, Idaho, is given in the table. Which measure of central tendency would you use to describe the typical number of days of rain or snow per month in each city?

Month	J	F	M	A	M	J	J	A	S	O	N	D
Boise	12	11	10	8	9	7	2	2	3	7	10	12
Lewiston	13	9	10	8	10	10	4	4	4	9	10	11

Boise, once called the* City of Trees, *is both the capital and the largest city in Idaho.

Solution The means, medians, and modes are as follows.

Boise: Mean = 7.75 Median = 8.5 Mode (none)
Lewiston: Mean = 8.5 Median = 9.5 Mode = 10

Boise does not have a mode because there is no number that occurs most often (2, 7, 10, and 12 each occur twice). Thus, the mode is not a good measure of the typical number of rain or snow days. Either of the other two measures could be used as a representative of the number of rain or snow days. You could say that Boise typically has 8 days of rain or snow per month and Lewiston typically has 9 days. ■

Communicating about MATHEMATICS

SHARING IDEAS about the Lesson

Comparing Salaries A fast-food restaurant employs 16 people. Thirteen of the people make $4.75 per hour. The two assistant managers each make $10 per hour. The manager makes $16 per hour. What number best describes the typical hourly salary? Explain.

EXERCISES

Guided Practice

CHECK for Understanding

1. Name the measure of central tendency associated with each phrase.
   a. middle **b.** most often **c.** average

In Exercises 2–5, use the histogram at the right.

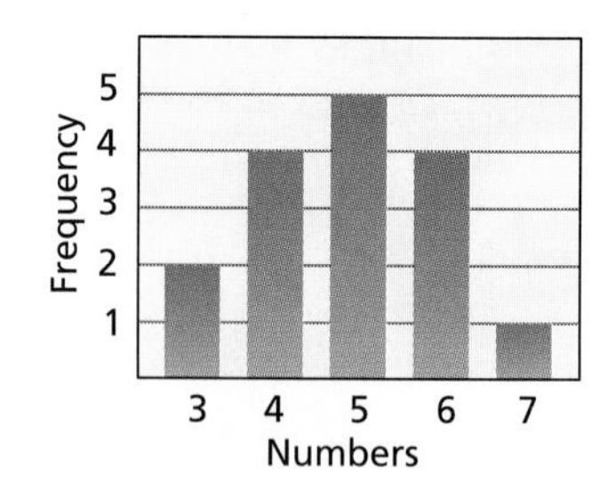

2. Describe and order the data that is represented.
3. Find the mean of the numbers.
4. Find the median of the numbers.
5. Find the mode of the numbers.

Independent Practice

In Exercises 6–9, find the mean, median, and mode of the data.

6. 85, 86, 90, 90, 91, 92
7. 52.8, 53.6, 53.9, 54, 54.5, 54.8, 55.1
8. 34, 35, 36, 36, 37, 37, 38, 38, 39, 40
9. 55, 56, 57, 58, 59, 60, 60, 60, 62

In Exercises 10–12, use the line plot at the right.

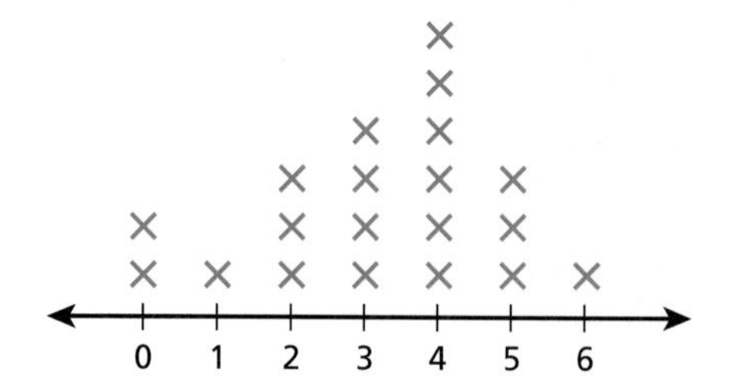

10. How many numbers are represented in the line plot? List them in increasing order.
11. Find the mean, median, and mode of the numbers.
12. *Problem Solving* Describe a real-life situation that can be represented by the line plot.

Mean, Median, or Mode? **In Exercises 13–15, which measure of central tendency best represents the data? Explain your reasoning.**

13. The salaries of 240 people in your neighborhood
14. The number of daily calories you consume for 30 days
15. The number of pairs of shoes owned by each person in your neighborhood

Wildlife Preservation **In Exercises 16–18, use the picture graph, which shows the ages of the wolf population in a state park.**

16. How many 3-year-old wolves are in the park?
17. Find the mean, median, and mode of the ages.
18. Which measure of central tendency best represents the age of the wolf population? Explain.

Chemistry **In Exercises 19 and 20, use the following information and data.**

At a neighborhood swimming pool, you measure the pH level of the water each day for ten days. (The pH level of pure water is 7. Lower pH levels indicate that the water is acidic.)

Data: 7.1, 6.9, 6.7, 6.5, 6.3, 6.9, 6.8, 6.5, 6.3, 6.1

19. Find the mean, median, and mode of the data.

20. The data is listed in order of the days the measurements were taken. Did the water tend to become more acidic or less acidic? On which day was a chemical added to change the pH level?

In Exercises 21 and 22, list six numbers that have the indicated measures of central tendency. (There are many correct answers.)

21. The mean is 15, the median is 15, and the mode is 15.

22. The mean is 10, the median is 9, and the mode is 8.

23. *House Prices* Explain the following statement in your own words. "In 1992, the median sales price of houses in Seattle, Washington, was $141,300." ***(Source: National Association of Realtors)***

24. *Test Scores* On your first four tests of the grading period, you had scores of 95, 89, 91, and 93. What score must you get on your final test to raise your average (mean) to 93?

Integrated Review

Making Connections within Mathematics

Histograms and Line Plots **In Exercises 25–27, use the following data.**

Data: 1, 3, 5, 4, 2, 2, 3, 4, 6, 1, 4, 7, 2, 5, 4, 1, 2, 4, 6, 5

25. Construct a histogram.

26. Construct a line plot.

27. *Probability* You are randomly selecting a number from the given data. Find the probability of choosing the indicated number.

a. 1 **b.** 2 **c.** 3 **d.** 4 **e.** 5 **f.** 6 **g.** 7 **h.** 8

Exploration and Extension

Experimental Probability **In Exercises 28–31, use a pair of six-sided dice.**

28. Toss the dice 50 times and record the totals.

29. Create a histogram of the data.

30. Find the mean, median, and mode of the data.

31. From your results, estimate the probability of tossing totals of 2, 3, 4, 5, 6, 7, 8, 9, 10, 11, and 12 with a pair of dice.

14.2 Stem-and-Leaf Plots

What you should learn:

Goal 1 How to organize data with a stem-and-leaf plot

Goal 2 How to use two stem-and-leaf plots to compare two sets of data

Why you should learn it:

You can use stem-and-leaf plots to help interpret real-life data, such as deciding which part of the United States has a more urban population.

World Cup ***Being able to organize data is important for businesses like the Miami clothing company started by Richard Gonzales. His company won the rights to use the official World Cup soccer logo and mascot for T-shirt designs.***

Goal 1 Using Stem-and-Leaf Plots

A **stem-and-leaf plot** is a technique for ordering data in increasing or decreasing order.

Example 1 *Making a Stem-and-Leaf Plot*

Use a stem-and-leaf plot to order the following data.

Unordered Data: 5, 10, 14, 44, 32, 35, 14, 28, 8, 13, 11, 25, 30, 15, 20, 9, 29, 20, 23, 40, 19, 31, 32, 32, 43, 42, 25, 37, 11, 8, 4, 43, 7, 24

Solution The numbers vary between 5 and 44, so you can let the *stem* represent the tens digits and let the *leaves* represent the units digits. Begin by creating an unordered stem-and-leaf plot as shown on the left. Then order the leaves to form an ordered stem-and-leaf plot as shown on the right.

Unordered Plot

Stem	Leaves
4	4 0 3 2 3
3	2 5 0 1 2 2 7
2	8 5 0 9 0 3 5 4
1	0 4 4 3 1 5 9 1
0	5 8 9 8 4 7

Ordered Plot

Stem	Leaves
4	0 2 3 3 4
3	0 1 2 2 2 5 7
2	0 0 3 4 5 5 8 9
1	0 1 1 3 4 4 5 9
0	4 5 7 8 8 9

4 | 3 represents 43

You can use the ordered stem-and-leaf plot to order the data.

Ordered Data: 4, 5, 7, 8, 8, 9, 10, 11, 11, 13, 14, 14, 15, 19, 20, 20, 23, 24, 25, 25, 28, 29, 30, 31, 32, 32, 32, 35, 37, 40, 42, 43, 43, 44 ■

When you make a stem-and-leaf plot, you should include a key that allows people to tell what the stem and leaves represent. For instance, in Example 1, by knowing that 4 | 3 represents 43, you know that the stem represents the tens digit and the leaves represent the units digits.

Goal 2 Using Double Stem-and-Leaf Plots

Example 2 *Organizing Data*

Real Life
Social Studies

The data below shows the percent of each state's population that is urban (lives in a city). Which group of states is more urban? ***(Source: U.S. Bureau of Census)***

East of Mississippi: AL (60%), CT (79%), DE (73%), FL (85%), GA (63%), IL (85%), IN (65%), KY (52%), MA (84%), MD (81%), ME (45%), MI (71%), MS (47%), NC (50%), NH (51%), NJ (89%), NY (84%), OH (74%), PA (69%), RI (86%), SC (55%), TN (61%), VA (69%), VT (32%), WV (36%)

West of Mississippi: AK (68%), AR (54%), AZ (88%), CA (93%), CO (82%), HI (89%), IA (61%), ID (57%), KS (69%), LA (68%), MN (70%), MO (69%), MT (53%), ND (53%), NE (66%), NM (73%), NV (88%), OK (68%), OR (71%), SD (50%), TX (80%), UT (87%), WA (76%), WI (66%), WY (65%)

During much of the history of the United States, its population was mostly rural (lived in the country or in towns of less than 2500). Now, however, the population is mostly urban (lives in cities or towns of 2500 or more).

Solution A *double* stem-and-leaf plot can help you answer the question. The leaves representing the western states point to the left of the stem and the leaves representing the eastern states point to the right of the stem. From the double plot below, you can see that the western states tend to have more urban populations.

West		East
3	9	
9 8 8 7 2 0	8	1 4 4 5 5 6 9
6 3 1 0	7	1 3 4 9
9 9 8 8 8 6 6 5 1	6	0 1 3 5 9 9
7 4 3 3 0	5	0 1 2 5
	4	5 7
	3	2 6

0 | 7 | 1 represents 70% and 71% urban population

Communicating about MATHEMATICS

Cooperative Learning

SHARING IDEAS about the Lesson

Group Project Use the data in Example 2 to draw a double stem-and-leaf plot that represents the rural (country) populations of the states. Compare your plot to that in Example 2.

EXERCISES

Guided Practice

CHECK for Understanding

In Exercises 1 and 2, list the data represented by the stem-and-leaf plot.

1.

3	1 2 4 4 6
2	0 3 5 7
1	0 1 5 5 6 9
0	2 2 8

3 | 1 represents 31

2.

8 4 3 1	7	0 3 4 6
5 5 0	6	1 4 7
9 7 3 2	5	2 3 5 8 9
4 2 0	4	3

8 | 7 | 0 represents 7.8 and 7.0

3. Use a stem-and-leaf plot to order the following set of data.

18, 6, 52, 41, 43, 8, 29, 24, 33, 30, 2, 55, 28, 32
8, 21, 5, 2, 38, 10, 54, 65, 17, 29, 34, 50, 32, 1

Independent Practice

In Exercises 4–6, list the data represented by the stem-and-leaf plot. Then draw a histogram for the data.

4.

6	4 6 6 8
5	0 2 3
4	1 8
3	5 7 9
2	0 1 1 3

6 | 4 represents 64

5.

12	4 5 6 7
11	0 3 8 8 9
10	1 2 7
9	0 1
8	4 4 5 8

12 | 4 represents 124

6.

17	2 3 4 6 8
16	1 2 3 7
15	1 9 9
14	3 5 9
13	5 6

17 | 2 represents 17.2

7. ***Double Stem-and-Leaf Plot*** List the two sets of data represented by the double stem-and-leaf plot at the right.

8. ***Double Bar Graph*** Draw a double bar graph to represent the data given in the double stem-and-leaf plot at the right. (*Hint:* Double bar graphs are described on page 199.)

7 3 2 2	9	1 2
8 4 3 0	8	3 4
7 4 1	7	3 4 7
5 2	6	2 3 5 6
7 6 4	5	6 8 8
5 3 2	4	0 2 3 7
6 4 1	3	1 3 4 5 7 9

7 | 9 | 1 represents 97 and 91

9. ***Indianapolis 500*** The winning speeds (in miles per hour) in the Indianapolis 500 from 1960 through 1994 are given below. Organize the data with an ordered stem-and-leaf plot.

1960 (139), 1961 (139), 1962 (140), 1963 (143), 1964 (147), 1965 (151), 1966 (144), 1967 (151), 1968 (153), 1969 (157), 1970 (156), 1971 (158), 1972 (163), 1973 (159), 1974 (159), 1975 (149), 1976 (149), 1977 (161), 1978 (161), 1979 (159), 1980 (143), 1981 (139), 1982 (162), 1983 (162), 1984 (164), 1985 (153), 1986 (171), 1987 (162), 1988 (145), 1989 (168), 1990 (186), 1991 (176), 1992 (134), 1993 (157), 1994 (161)

Final Grades **In Exercises 10 and 11, use the set of data below, which lists the scores on a test for students.**

93, 84, 100, 92, 66, 89, 78, 52, 71, 85, 83, 95, 98, 99, 93, 81, 80, 79,
67, 59, 90, 85, 77, 62, 90, 78, 66, 63, 93, 87, 74, 96, 72, 100, 70, 73

10. Order the data in a stem-and-leaf plot.

11. Draw a histogram to represent the data.

Temperature **In Exercises 12 and 13, use the data, which lists the record high temperatures of Bismark, North Dakota, as of 1991.** ***(Source: U.S. National Oceanic and Atmospheric Administration)***

Jan., 62°; Feb., 68°; Mar., 81°; Apr., 93°; May, 98°; June, 107°;
July, 109°; Aug., 109°; Sep., 105°; Oct., 95°; Nov., 75°; Dec., 65°

12. Order the data in a stem-and-leaf plot.

13. Draw a histogram to represent the data.

Population **In Exercises 14 and 15, use the data below showing the percent of each state's population that is between 5 and 17 years old.** ***(Source: U.S. Bureau of Census)***

East of the Mississippi: AL(19%), CT(16%), DE(17%), FL(16%), GA(19%), IL(18%), IN(19%), KY(19%), MA(16%), MD(17%), ME(18%), MI(19%), MS(21%), NC(17%), NH(18%), NJ(17%), NY(17%), OH(18%), PA(17%), RI(16%), SC(19%), TN(18%), VA(17%), VT(18%), WV(18%)

West of the Mississippi: AK(22%), AR(19%), AZ(19%), CA(18%), CO(19%), HI(17%), IA(19%), ID(23%), KS(19%), LA(21%), MN(19%), MO(19%), MT(20%), ND(20%), NE(20%), NM(21%), NV(17%), OK(19%), OR(19%), SD(21%), TX(20%), UT(26%), WA(19%), WI(19%), WY(22%)

14. Order the data with a double stem-and-leaf plot.

15. Find the mean, median, and mode for each group of states and for the total collection of states.

Integrated Review

Making Connections within Mathematics

16. *Probability* If you selected one score from the data in Exercise 10, what is the probability that it is greater than 80?

17. *Probability* If you selected one temperature from the data in Exercise 12, what is the probability that it is less than 80°?

Exploration and Extension

18. *Technology* If you have access to a random number generator (on a computer or a graphing calculator), use the number generator to list 40 numbers between 0 and 1. Multiply each number by 100 and round the result to the nearest whole number. Then organize the results with a stem-and-leaf plot. Do your results appear random? Explain your reasoning.

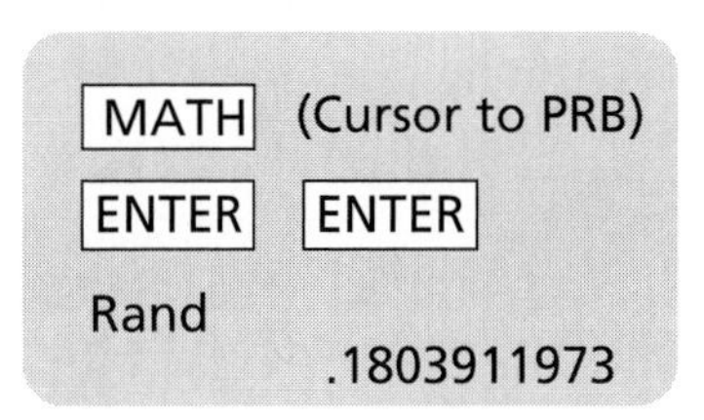

TI-81or TI-82Keystrokes

14.3 Box-and-Whisker Plots

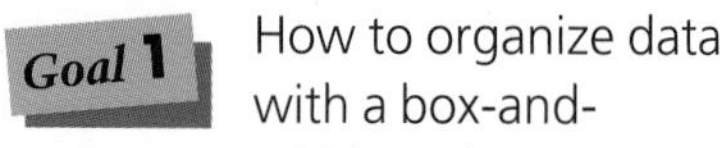

Goal 1 How to organize data with a box-and-whisker plot

Goal 2 How to use box-and-whisker plots to interpret real-life data

Why you should learn it:

You can use box-and-whisker plots to help interpret real-life data, such as describing the age distribution in a state.

Goal 1 Drawing Box-and-Whisker Plots

The median (or **second quartile**) of an ordered collection of numbers roughly divides the collection into two halves: those below the median and those above the median. The **first quartile** is the median of the lower half, and the **third quartile** is the median of the upper half.

1, 5, 6, 12, 14, 18, 20, 26, 27, 29, 30, 31

First quartile $\frac{6+12}{2} = 9$

Second quartile $\frac{18+20}{2} = 19$

Third quartile $\frac{27+29}{2} = 28$

A **box-and-whisker** plot of this data is shown below.

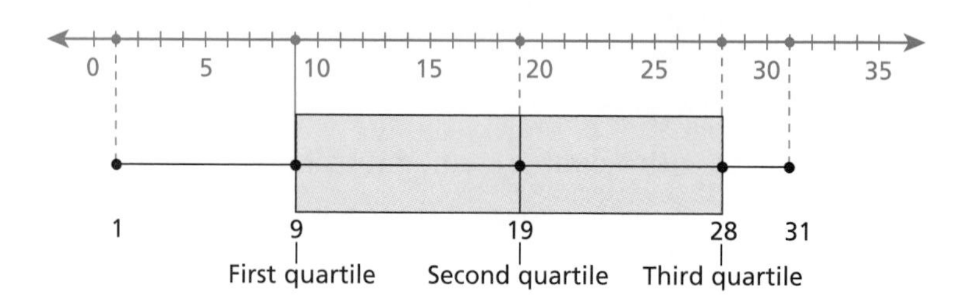

Example 1 *Drawing a Box-and-Whisker Plot*

Draw a box-and-whisker plot for the following data.

22, 65, 23, 19, 42, 62, 38, 29, 50, 46, 28, 36, 25, 40

Solution Begin by writing the numbers in increasing order.

19, 22, 23, 25, 28, 29, 36, 38, 40, 42, 46, 50, 62, 65

Lower half (19, 22, 23, 25, 28, 29, 36) *Upper half* (38, 40, 42, 46, 50, 62, 65)

From this ordering, you can see that the first quartile is 25, the second quartile is 37, and the third quartile is 46. A box-and-whisker plot for the data is shown below.

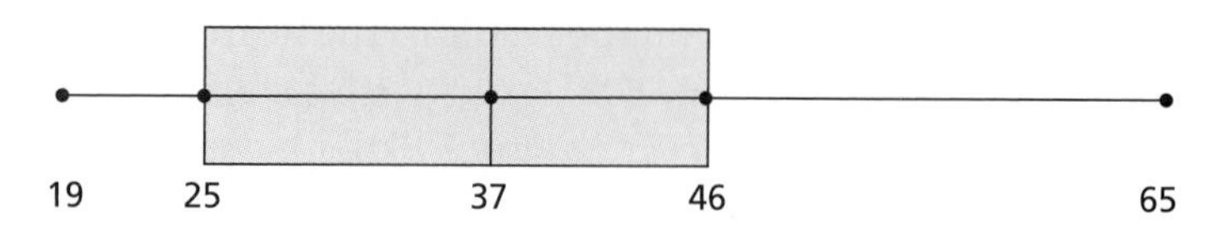

■

> ***Need to Know***
>
> When you draw a box-and-whisker plot, you should label five numbers: the smallest number, the quartiles, and the largest number. These five numbers should be spaced as they would be on a number line.

Goal 2 Using Box-and-Whisker Plots

Real Life
Social Studies

Example 2 *Interpreting Box-and-Whisker Plots*

The box-and-whisker plots below show the age distributions of Alaska's and Rhode Island's populations. What do the plots tell you about the two states? ***(Source: U.S. Bureau of Census)***

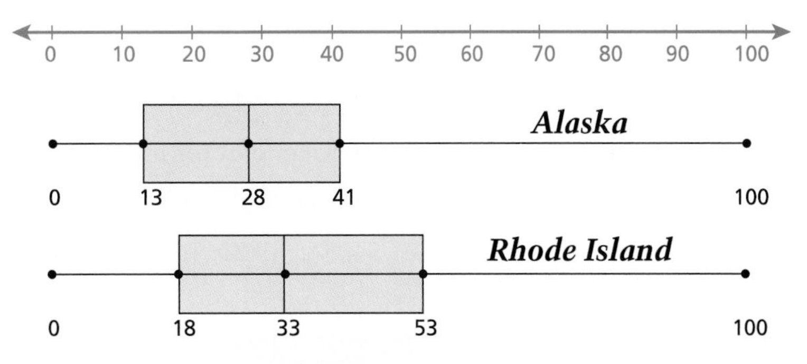

Francesco Schiappa and his great-grandson Joseph represent the four generations of one Rhode Island family.

Solution In Alaska, the median age is 28, which means that about 50% of the population is under 28. The box-and-whisker plot also tells you that 25% of the population is under 13 and 25% is over 41.

Rhode Island's population is quite a bit older. In that state, the median age is 33, which means that about 50% of the population is under 33. The box-and-whisker plot also tells you that 25% of the population is under 18 and 25% is over 53. ■

Communicating about MATHEMATICS

SHARING IDEAS about the Lesson

Extending the Example The box-and-whisker plots below represent the age distributions for Florida and Utah. Which is which? Explain your reasoning.

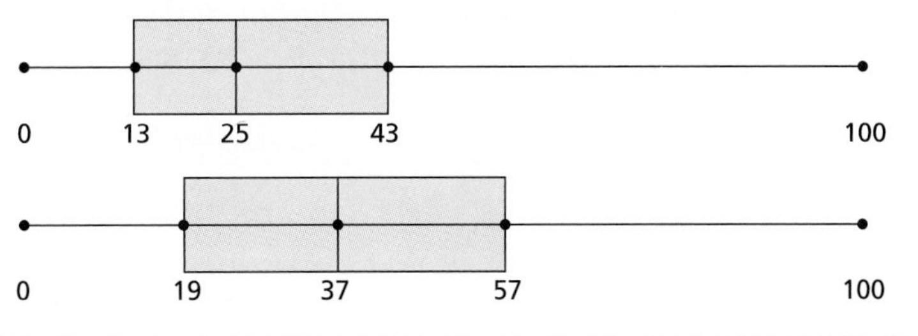

EXERCISES

Guided Practice

CHECK for Understanding

In Exercises 1 and 2, use the box-and-whisker plot at the right.

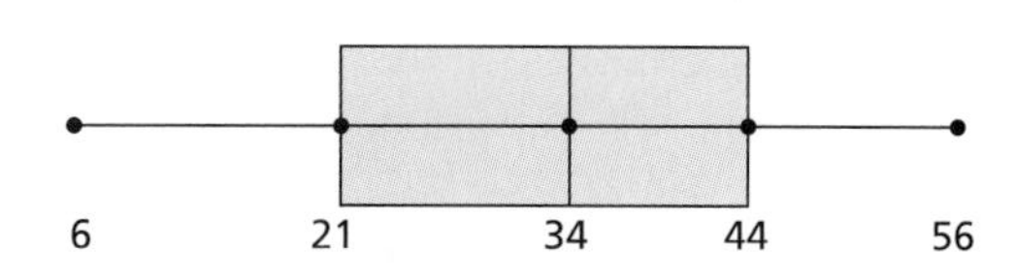

1. Name the smallest and largest numbers.
2. Name the quartiles.

In Exercises 3 and 4, draw a box-and-whisker plot for the data.

3. 3, 4, 8, 10, 13, 17, 21, 26, 29, 31, 32, 36
4. 12, 52, 25, 61, 66, 15, 6, 46, 39, 54, 34, 50, 21, 56, 70, 40

Independent Practice

In Exercises 5–8, use the box-and-whisker plot. There are 20 numbers in the collection, and each number is different.

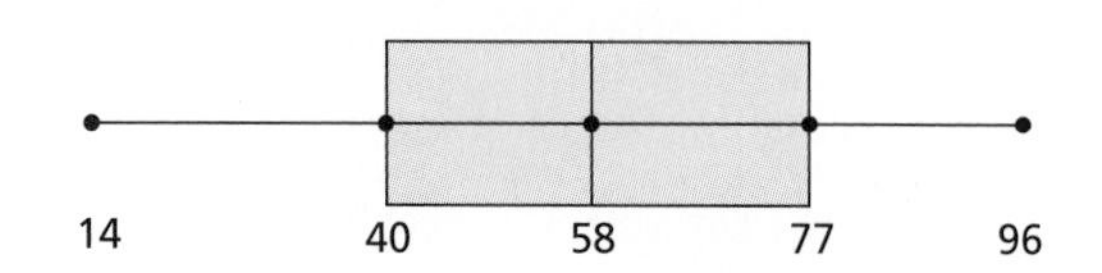

5. Name the smallest and largest numbers.
6. Name the first, second, and third quartiles.
7. What percent of the numbers are less than 40?
8. What percent of the numbers are between 40 and 77?

In Exercises 9 and 10, draw a box-and-whisker plot of the data.

9. 26, 60, 36, 44, 62, 24, 29, 50, 67, 72, 40, 41, 18, 39, 64, 82, 41, 49, 32, 42

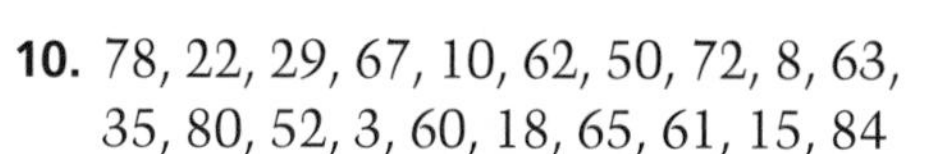

10. 78, 22, 29, 67, 10, 62, 50, 72, 8, 63, 35, 80, 52, 3, 60, 18, 65, 61, 15, 84

11. *Error Analysis* The box-and-whisker plot at the right is supposed to represent the following data. There are four errors. What are they?

    Data: 26, 10, 19, 34, 2, 5, 21, 12, 1, 39, 14, 30, 18, 37, 7, 24

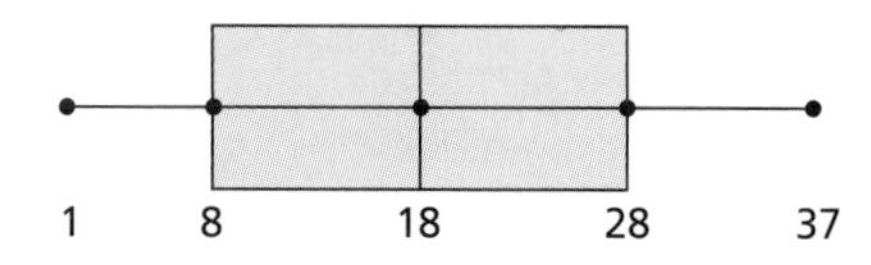

Social Studies **In Exercises 12 and 13, use the box-and-whisker plot, which shows the age distribution for Mississippi.** *(Source: U.S. Bureau of Census)*

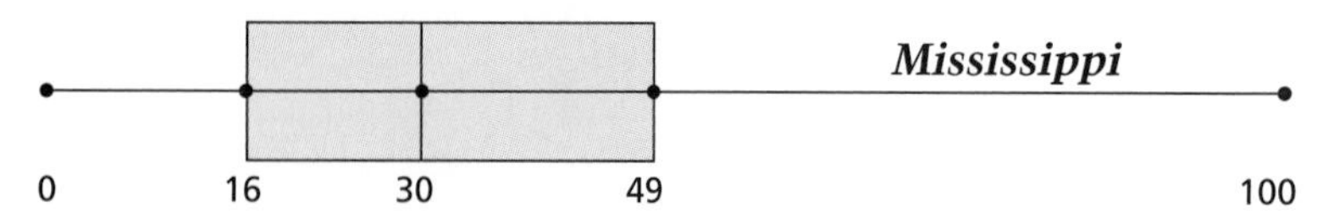

12. Write a description of Mississippi's population.
13. Compare Mississippi's population to the populations discussed on page 643.

In Exercises 14 and 15, use the following data, which lists the price in dollars of several receivers, CD players, and cassette decks in 1994. *(Source: Consumer Reports)*

Scientists band birds in order to study migration patterns.

Receivers:	CD Players:	Cassette Decks:
165, 200, 380, 260, 180,	170, 250, 270, 180, 140,	200, 225, 150, 285, 260,
300, 460, 390, 445, 225,	240, 195, 255, 160, 245,	230, 295, 255, 290, 195,
325, 400, 280, 360	200, 290, 230, 280	265, 280

14. Using the same scale, create a box-and-whisker plot for each stereo component.

15. *Writing* What do the plots tell you about the prices of stereo components? Write your answer in paragraph form.

16. *Think about It* List two different sets of 12 numbers that could be represented by the box-and-whisker plot below.

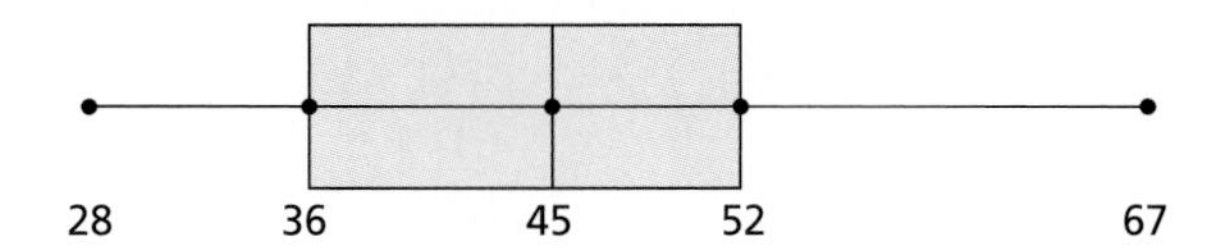

Integrated Review

Making Connections within Mathematics

Percent **In Exercises 17–22, solve the percent equation. Round your result to 2 decimal places.**

17. What is 25% of 78?

18. What is 75% of 130?

19. 16 is what percent of 36?

20. 71 is what percent of 95?

21. 48 is 20% of what number?

22. 84 is 60% of what number?

Exploration and Extension

Finding the Context **In Exercises 23–26, match the description with the most reasonable box-and-whisker plot. Explain your reasoning.**

a. Season scores of a baseball team

b. Weights in pounds of students in 4th-grade class

c. Scores on a 100-point test

d. Ages in years in an algebra class

23.

13 14 15 16 37

24.

62 73 82 89 100

25.

56 62 65 72 107

26.

0 2 4 5 13

USING A GRAPHING CALCULATOR

Box-and-Whisker Plots

Some graphing calculators can be used to sketch box-and-whisker plots. The following steps show how to use a Texas Instruments TI-82 to sketch the box-and-whisker plot shown at the right. The data for the plot is the same as that used in Example 1 on page 642.

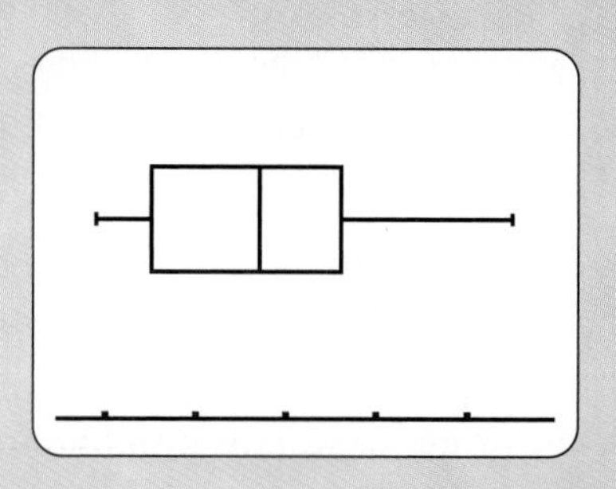

22, 65, 23, 19, 42, 62, 38, 29, 50, 46, 28, 36, 25, 40

TI-82*

To enter data in a list: STAT ENTER (Edit)

L1(1) = 22 ENTER	L1(6) = 62 ENTER	L1(11) = 28 ENTER
L1(2) = 65 ENTER	L1(7) = 38 ENTER	L1(12) = 36 ENTER
L1(3) = 23 ENTER	L1(8) = 29 ENTER	L1(13) = 25 ENTER
L1(4) = 19 ENTER	L1(9) = 50 ENTER	L1(14) = 40 ENTER
L1(5) = 42 ENTER	L1(10) = 46 ENTER	

RANGE
Xmin = 50
Xmax = 70
Xscl = 1
Ymin = 700
Ymax = 1000
Yscl = 50

To set boxplot: 2nd STAT PLOT ENTER (Plot 1)

Choose the following: On, Type: ⊢□⊣, Xlist: L1, Freq: 1

ZOOM 9

To find the minimum, maximum, first quartile, median, and third quartile:

TRACE and ◁ or ▷

Exercises

In Exercises 1–4, use a graphing calculator to sketch a box-and-whisker plot for the data.

1. 15, 19, 37, 15, 25, 33, 30, 27, 31, 37, 37, 14, 28, 34, 24, 25, 35, 18
2. 52, 79, 82, 56, 67, 69, 70, 73, 59, 64, 69, 72, 79, 58, 54, 60, 53, 67
3. 21, 24, 23, 26, 29, 23, 24, 25, 28, 28, 29, 24, 25, 27, 56, 42, 28, 24
4. 39, 52, 36, 45, 42, 43, 37, 43, 46, 42, 38, 34, 36, 45, 49, 34, 45, 48

In Exercises 5–7, create a collection of 16 numbers that corresponds to the box-and-whisker plot. Use a graphing calculator to confirm your answer. (There are many correct collections.)

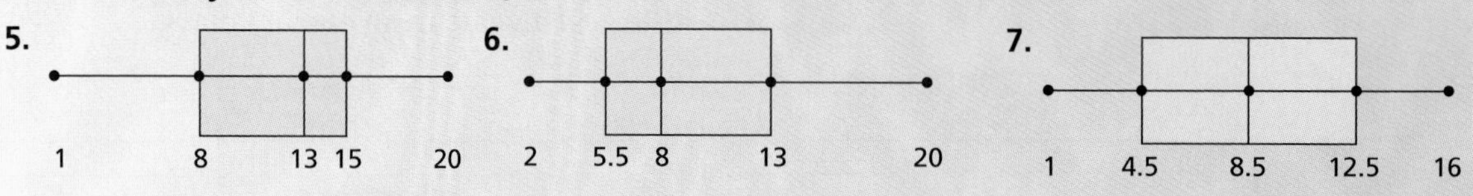

* Keystrokes for other graphing calculators are listed in *Technology—Keystrokes for Other Graphing Calculators* found at the end of this text.

Mixed REVIEW

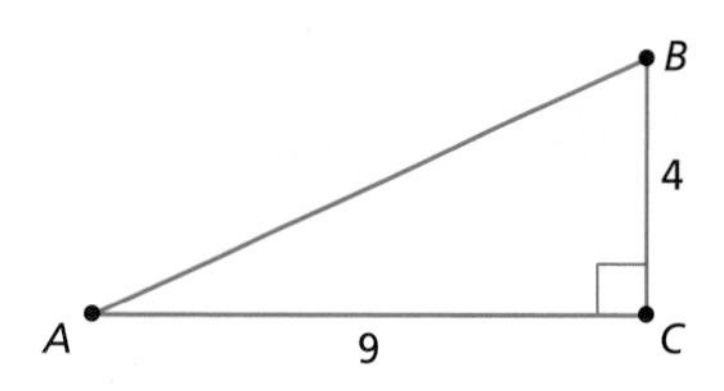

In Exercises 1–8, use $\triangle ABC$ at the right. (11.8)

1. Find the length of the hypotenuse.

2. Find the area of the triangle.

3. Find $\sin A$. **4.** Find $\cos A$. **5.** Find $\tan A$.

6. Find $\sin B$. **7.** Find $\cos B$. **8.** Find $\tan B$.

In Exercises 9–12, use the following data. (14.1, 14.2)

1.5, 1.6, 1.8, 1.8, 1.9, 2.3, 2.4, 2.4, 2.7, 2.7, 2.7, 3.0, 3.0, 3.1

9. What is the mean of the data? **10.** What is the median of the data?

11. What is the mode of the data? **12.** Create a stem-and-leaf plot for the data.

In Exercises 13–16, solve the proportion. (8.2)

13. $\frac{x}{13} = \frac{4}{12}$ **14.** $\frac{b}{6} = \frac{6}{16}$ **15.** $\frac{3}{2} = \frac{15}{r}$ **16.** $\frac{4}{s} = \frac{21}{18}$

Milestones APOTHOCARY MEASUREMENT

1775 1800 1825 1850 1875 1900 1925 1950 1975 2000

Edward Jenner, smallpox vaccine, 1796 · Metric System, 1799 · First United States college of pharmacy, 1821 · Louis Pasteur, anthrax vaccine, 1881 (rabies, 1885) · DPT vaccine, 1891 · Alexander Fleming, penicillin, 1928 · Jonas Salk, polio vaccine, 1953

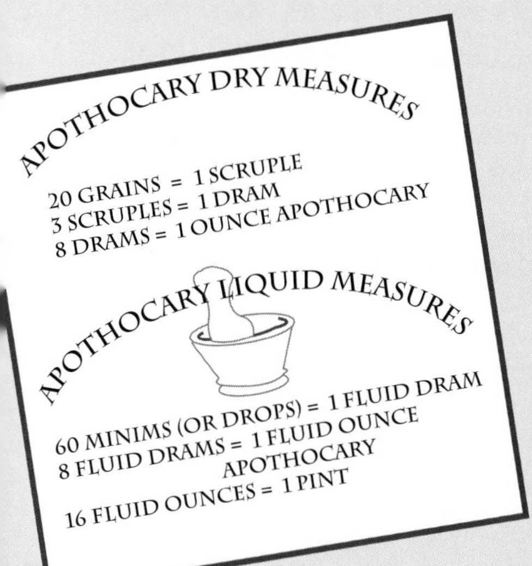

Advances in veterinary medicine have paralleled those in human medicine. The healing effects of certain botanical compounds have been known since pre-history. By the Middle Ages, medicines were dispensed by *apothocaries,* persons who prepared and sold drugs.

For centuries, apothocaries had their own formulas for drugs. In 1542, the first standard table of drugs, or *Pharmacopoeia* (from which we get the terms *pharmacy* and *pharmacist*) was published in Germany. The first United States *Pharmacopoeia* was commissioned by the Continental Congress for the army of the American Revolution in 1778. In 1906, under the Pure Food and Drugs Act, *The United States Pharmacopoeia—The National Formulary* was finally made a legal standard.

- *Today, all medical formulas use metric units. If 1 dry grain = 0.0648 grams, then how many grains are in a 100 mg tablet?*
- *If 1 drop = 0.0616 milliliters, then how many drops are in a 100 mL medicine bottle?*

14.4 Exploring Data and Matrices

What you should learn:

 How to organize data with a matrix

 How to add and subtract two matrices

Why you should learn it:

You can use matrices to help organize real-life data, such as the data that represents income, expenses, and profit for a business.

Goal 1 Using Matrices

A **matrix** is a rectangular arrangement of numbers into rows and columns. For instance, the matrix

$$\begin{bmatrix} 2 & -4 & 5 \\ -3 & 1 & 0 \end{bmatrix}$$

has two rows and three columns. The numbers in the matrix are called **entries.** In the above matrix, the entry in the second row and third column is 0. The entry in the first row and second column is -4.

The plural of *matrix* is *matrices.* Two matrices are **equal** if all of the entries in corresponding positions are equal.

$$\begin{bmatrix} -1 & \frac{3}{2} \\ \frac{1}{4} & 0 \end{bmatrix} = \begin{bmatrix} -1 & 1.5 \\ 0.25 & 0 \end{bmatrix} \qquad \begin{bmatrix} -3 & 2 \\ 4 & 1 \end{bmatrix} \neq \begin{bmatrix} 2 & -3 \\ 1 & 4 \end{bmatrix}$$

You can think of a matrix as a type of table that can be used to organize data.

Real Life **Business**

Example 1 *Writing a Table as a Matrix*

You are opening two fast-food restaurants. The gross income for each of the first four months is shown in the table at the left. Write this table as a matrix. For Restaurant 1, which month had the greatest income? For Restaurant 2, which month had the greatest income?

	Restaurant 1	Restaurant 2
April	\$142,560	\$209,905
May	\$158,430	\$213,413
June	\$162,578	\$206,784
July	\$178,865	\$198,450

Solution The matrix associated with the table has four rows and two columns.

$$\begin{array}{c} \textbf{Income} \\ \text{April} \\ \text{May} \\ \text{June} \\ \text{July} \end{array} \begin{array}{c} \begin{array}{cc} \textbf{Restaurant 1} & \textbf{Restaurant 2} \end{array} \\ \begin{bmatrix} 142{,}560 & 209{,}905 \\ 158{,}430 & 213{,}413 \\ 162{,}578 & 206{,}784 \\ 178{,}865 & 198{,}450 \end{bmatrix} \end{array}$$

Greatest (Restaurant 1: 178,865) — Greatest (Restaurant 2: 213,413)

For Restaurant 1, the greatest income occurred in July. For Restaurant 2, the greatest income occurred in May. ■

Goal 2 Adding and Subtracting Matrices

To **add** or **subtract** matrices, you simply add or subtract corresponding entries, as shown in Example 2.

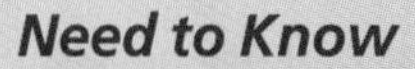

Need to Know

You cannot add or subtract matrices that are different sizes. For instance, you cannot add a matrix that has three rows and two columns to a matrix that has two rows and two columns.

Example 2 *Adding and Subtracting Matrices*

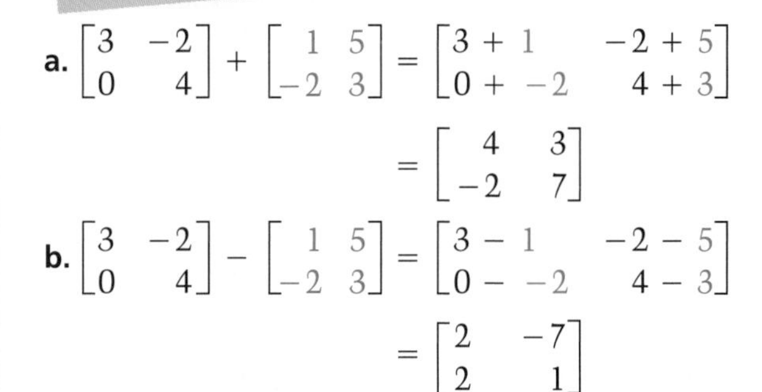

a. $\begin{bmatrix} 3 & -2 \\ 0 & 4 \end{bmatrix} + \begin{bmatrix} 1 & 5 \\ -2 & 3 \end{bmatrix} = \begin{bmatrix} 3+1 & -2+5 \\ 0+-2 & 4+3 \end{bmatrix}$

$= \begin{bmatrix} 4 & 3 \\ -2 & 7 \end{bmatrix}$

b. $\begin{bmatrix} 3 & -2 \\ 0 & 4 \end{bmatrix} - \begin{bmatrix} 1 & 5 \\ -2 & 3 \end{bmatrix} = \begin{bmatrix} 3-1 & -2-5 \\ 0--2 & 4-3 \end{bmatrix}$

$= \begin{bmatrix} 2 & -7 \\ 2 & 1 \end{bmatrix}$ ■

Example 3 *Subtracting Matrices*

Real Life **Business**

Your expenses for the two restaurants described in Example 1 are shown in the table at the left. Use subtraction of matrices to find your monthly profit for each restaurant.

	Restaurant 1	Restaurant 2
April	\$124,792	\$205,316
May	\$138,716	\$197,438
June	\$158,914	\$192,765
July	\$163,047	\$186,652

Solution To find the monthly profit, subtract the expense matrix from the income matrix.

$$\overbrace{\begin{bmatrix} 142{,}560 & 209{,}905 \\ 158{,}430 & 213{,}413 \\ 162{,}578 & 206{,}784 \\ 178{,}865 & 198{,}450 \end{bmatrix}}^{\text{Income}} - \overbrace{\begin{bmatrix} 124{,}792 & 205{,}316 \\ 138{,}716 & 197{,}438 \\ 158{,}914 & 192{,}765 \\ 163{,}047 & 186{,}652 \end{bmatrix}}^{\text{Expenses}} = \overbrace{\begin{bmatrix} 17{,}768 & 4{,}589 \\ 19{,}714 & 15{,}975 \\ 3{,}664 & 14{,}019 \\ 15{,}818 & 11{,}798 \end{bmatrix}}^{\text{Profit}}$$ ■

Communicating about MATHEMATICS

SHARING IDEAS about the Lesson

Extending the Example Create a table that shows the monthly profit for each restaurant in Example 3. Which restaurant had the greater income for the four months? Which had the greater profit? Which restaurant had the better record? Explain your reasoning.

EXERCISES

Guided Practice

CHECK for Understanding

In Exercises 1–3, use the matrix at the right.

$$\begin{bmatrix} 2 & -5 \\ 0 & 3 \\ -4 & 1 \end{bmatrix}$$

1. How many rows and columns does the matrix have?
2. What is the entry in the first row and second column?
3. Describe the position of -4.
4. *True or False?* Are the following matrices equal? Explain.

$$\begin{bmatrix} 0 & \frac{5}{4} & -1 \\ -\frac{8}{5} & 2 & 3 \end{bmatrix} \stackrel{?}{=} \begin{bmatrix} 0 & 1.25 & -1 \\ -1.6 & 2 & 3 \end{bmatrix}$$

In Exercises 5 and 6, find the matrix sum or difference.

5. $\begin{bmatrix} 2 & 3 & 0 \\ -3 & 0 & -1 \end{bmatrix} + \begin{bmatrix} -4 & 1 & -5 \\ -6 & 7 & 2 \end{bmatrix}$

6. $\begin{bmatrix} 2.1 & -1.5 \\ -3.5 & 6.4 \end{bmatrix} - \begin{bmatrix} 1.8 & 4.8 \\ -1.1 & -0.8 \end{bmatrix}$

Independent Practice

In Exercises 7–14, find the sum and difference of the matrices.

7. $\begin{bmatrix} -2 & 3 \\ 1 & -5 \end{bmatrix}, \begin{bmatrix} -1 & 4 \\ -6 & 0 \end{bmatrix}$

8. $\begin{bmatrix} 5 & 2 \\ -2 & 7 \end{bmatrix}, \begin{bmatrix} -8 & 0 \\ -9 & 3 \end{bmatrix}$

9. $\begin{bmatrix} -4 & -5 & 2 \\ 0 & -9 & -3 \end{bmatrix}, \begin{bmatrix} -6 & 3 & 2 \\ 1 & -1 & 4 \end{bmatrix}$

10. $\begin{bmatrix} -4 & 4 \\ 5 & 3 \\ -6 & -2 \end{bmatrix}, \begin{bmatrix} 4 & 7 \\ 2 & -1 \\ -3 & -8 \end{bmatrix}$

11. $\begin{bmatrix} 3 & 5 & -6 \\ -4 & 0 & -6 \\ 4 & 8 & 1 \end{bmatrix}, \begin{bmatrix} -7 & 0 & 3 \\ -1 & -4 & 5 \\ 4 & -2 & 9 \end{bmatrix}$

12. $\begin{bmatrix} 4 & 2 & 8 \\ -2 & 6 & -1 \\ 7 & 9 & -1 \end{bmatrix}, \begin{bmatrix} 12 & -10 & -6 \\ -5 & 0 & 11 \\ -1 & 2 & 3 \end{bmatrix}$

13. $\begin{bmatrix} \frac{1}{3} & \frac{2}{3} & \frac{1}{3} \\ \frac{1}{4} & \frac{1}{4} & \frac{3}{4} \\ \frac{1}{5} & \frac{2}{5} & \frac{3}{5} \end{bmatrix}, \begin{bmatrix} \frac{2}{3} & \frac{2}{3} & \frac{1}{3} \\ \frac{3}{4} & \frac{1}{4} & \frac{1}{2} \\ \frac{1}{5} & \frac{3}{5} & \frac{1}{5} \end{bmatrix}$

14. $\begin{bmatrix} 4.1 & 2.5 & -2.3 \\ 6.8 & 0.4 & -7.3 \\ -4.8 & 4.7 & -5.0 \end{bmatrix}, \begin{bmatrix} -6.3 & 1.5 & 3.6 \\ 2.1 & 4.7 & -1.7 \\ 5.3 & 2.1 & 4.7 \end{bmatrix}$

Mental Math **In Exercises 15 and 16, use mental math to find *a*, *b*, *c*, and *d*.**

15. $\begin{bmatrix} 2a & b \\ c-3 & 2d \end{bmatrix} = \begin{bmatrix} -8 & 3 \\ 1 & -5 \end{bmatrix}$

16. $\begin{bmatrix} -4 & 6 \\ c-1 & -3d \end{bmatrix} = \begin{bmatrix} a+2 & 3b \\ 7 & 6 \end{bmatrix}$

In Exercises 17–20, find two matrices whose sum is the given matrix. Don't use 0 as an element of either matrix. (There are many correct answers.)

17. $\begin{bmatrix} 2 & 5 \\ 6 & 3 \end{bmatrix}$

18. $\begin{bmatrix} 0 & 0 \\ 0 & 0 \end{bmatrix}$

19. $\begin{bmatrix} 1 & 0 \\ 0 & 1 \end{bmatrix}$

20. $\begin{bmatrix} 0 & 1 \\ 1 & 0 \end{bmatrix}$

Pet Store **In Exercises 21–24, use the following information.**

You are opening 2 pet stores. The gross income for each store for the first four months is shown in the left table and the expenses are shown in the right table.

Gross Income	Pet Store 1	Pet Store 2
May	$231,450	$206,210
June	$265,985	$319,754
July	$303,442	$321,615
August	$324,570	$256,419

Expenses	Pet Store 1	Pet Store 2
May	$208,345	$200,926
June	$247,913	$296,575
July	$287,500	$306,480
August	$317,940	$238,212

21. Write each table as a matrix.

22. For each pet store, which month had the greatest income?

23. Find the monthly profit for each store.

24. Which store had a greater profit during the first four months?

Geometry **In Exercises 25–27, use the table at the right, which lists the sides of 4 different triangles.**

25. Write the table as a matrix.

26. Which, if any, of the triangles are right?

27. Find the perimeter of each triangle.

	Side 1	Side 2	Side 3
Triangle 1	5	7	10
Triangle 2	9	12	15
Triangle 3	0.9	4	4.1
Triangle 4	5	6	9.2

Integrated Review

Making Connections within Mathematics

Geometry **In Exercises 28 and 29, the matrix contains information about a rectangle or a circle. Copy and complete the matrix.**

28. Length, Width, Perimeter, Area

$$\begin{bmatrix} 8 & 2 & ? & ? \\ ? & 4 & ? & 40 \\ ? & 8 & 40 & ? \\ ? & 16 & ? & 224 \end{bmatrix}$$

29. Radius, Diameter, Circumference, Area

$$\begin{bmatrix} 1 & ? & ? & ? \\ ? & 6 & ? & ? \\ ? & 10 & ? & ? \\ 7 & ? & ? & ? \end{bmatrix}$$

Exploration and Extension

Magic Squares **In Exercises 30–33, complete the matrix so that every row and column has the same sum. Use the numbers 1–9 only once per matrix.**

30. $\begin{bmatrix} 3 & 8 & 4 \\ ? & ? & ? \\ ? & ? & ? \end{bmatrix}$

31. $\begin{bmatrix} ? & ? & ? \\ 2 & 6 & 7 \\ ? & ? & ? \end{bmatrix}$

32. $\begin{bmatrix} ? & ? & 1 \\ ? & ? & 5 \\ ? & ? & 9 \end{bmatrix}$

33. $\begin{bmatrix} 3 & ? & ? \\ 5 & ? & ? \\ 7 & ? & ? \end{bmatrix}$

Take this test as you would take a test in class. The answers to the exercises are given in the back of the book.

In Exercises 1–3, use the temperatures at the right. (14.1)

80°, 81°, 82°,
83°, 85°, 87°,
87°, 89°, 89°,
87°, 84°, 81°

1. What is the mean of the data?

2. What is the median of the data?

3. What is the mode of the data?

In Exercises 4–6, use the table, which lists the normal monthly rainfall (in inches) in Miami and Jacksonville, Florida. (14.2)

Month	J	F	M	A	M	J	J	A	S	O	N	D
Miami	2.0	2.1	2.4	2.9	6.2	9.3	5.7	7.6	7.6	5.6	2.7	1.8
Jacksonville	3.3	3.9	3.7	2.8	3.6	5.7	5.6	7.9	7.1	2.9	2.2	2.7

4. Create a double stem-and-leaf plot for the data above. Use the whole numbers for the stem and tenths for the leaves.

5. Find the median rainfall for Miami.

6. Find the mean rainfall for Jacksonville.

In Exercises 7–12, use the following temperatures (in degrees Fahrenheit), which are the record highs for May for the indicated cities. (14.3)

Juneau, 82°; Phoenix, 113°; Denver, 96°; Washington, 99°; Wichita, 100°; Boston, 95°; Detroit, 93°; Minneapolis, 96°; Cleveland, 92°; Oklahoma City, 104°; El Paso, 104°; Spokane, 96°

7. Order the data from least to greatest.

8. Find the first quartile.

9. Find the second quartile.

10. Find the third quartile.

11. Draw a box-and-whisker plot of the data.

12. What percent of the data is above the first quartile?

In Exercises 13–15, use the box-and-whisker plot at the right. (14.3)

13. What is the 3rd quartile?

14. What is the greatest number in the data?

15. What is the median of the data?

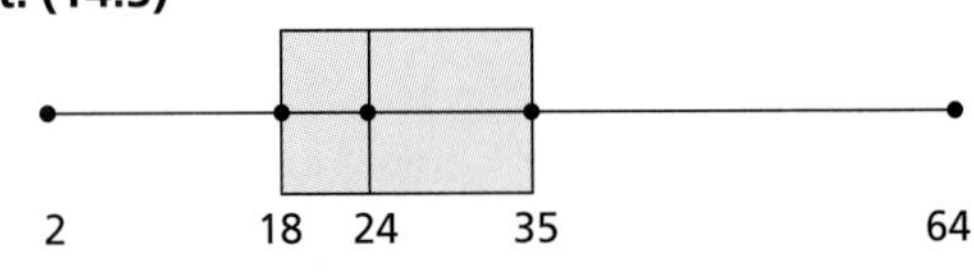

In Exercises 16–18, find the sum or difference. (14.4)

16. $\begin{bmatrix} 2 & 3 \\ 9 & 8 \end{bmatrix} + \begin{bmatrix} 6 & 2 \\ 11 & 0 \end{bmatrix}$

17. $\begin{bmatrix} 4 & 5 \\ 6 & 8 \end{bmatrix} - \begin{bmatrix} 1 & 3 \\ 8 & 21 \end{bmatrix}$

18. $\begin{bmatrix} 21 & 18 \\ 19 & 40 \end{bmatrix} - \begin{bmatrix} 20 & 6 \\ -4 & 16 \end{bmatrix}$

19. The table at the right gives the scores of three students for three tests. Rewrite the table as a matrix.

20. Which student had the highest average for the three tests?

Test	1	2	3
Student 1	74	88	95
Student 2	86	83	81
Student 3	82	71	86

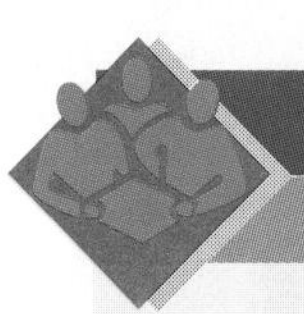

LESSON INVESTIGATION 14.5

Exploring Polynomials

Materials Needed: algebra tiles

In this investigation, you will use algebra tiles like those below to represent expressions called *polynomials.*

This x by x tile represents x^2.

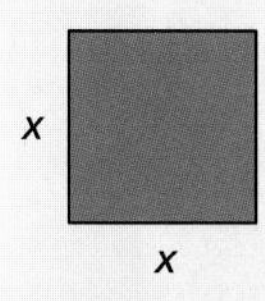

This x by 1 tile represents x.

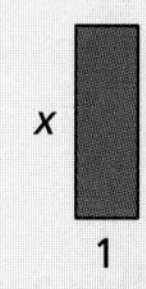

This 1 by 1 tile represents 1.

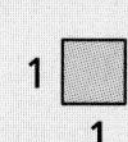

Example — *Exploring Polynomials with Algebra Tiles*

Use algebra tiles to represent the expression $2x^2 + 3x + 5$.

Solution You can represent this expression using two of the large square tiles, three of the rectangular tiles, and 5 small square tiles.

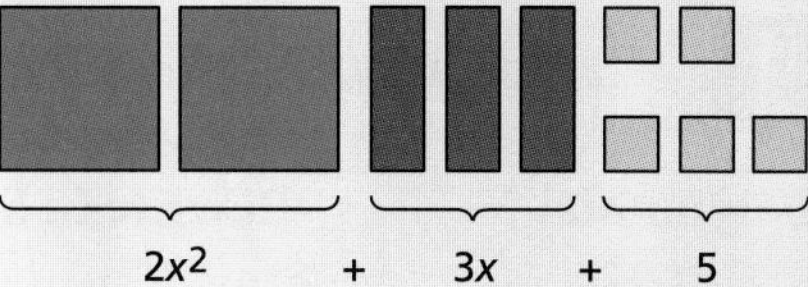

■

Exercises

In Exercises 1–4, write the expression that is represented by the algebra tiles.

1.

2.

3.

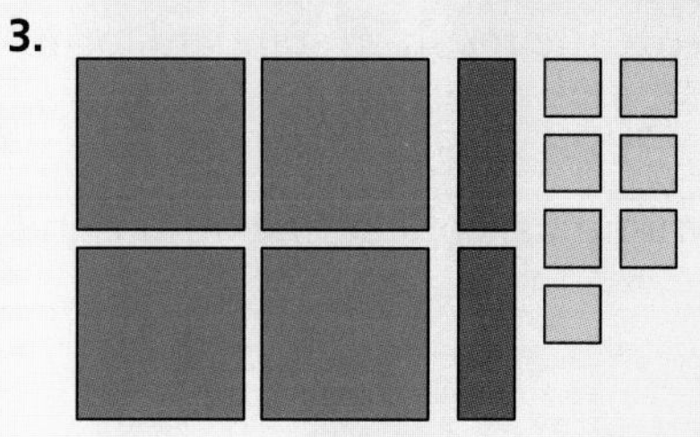

4.

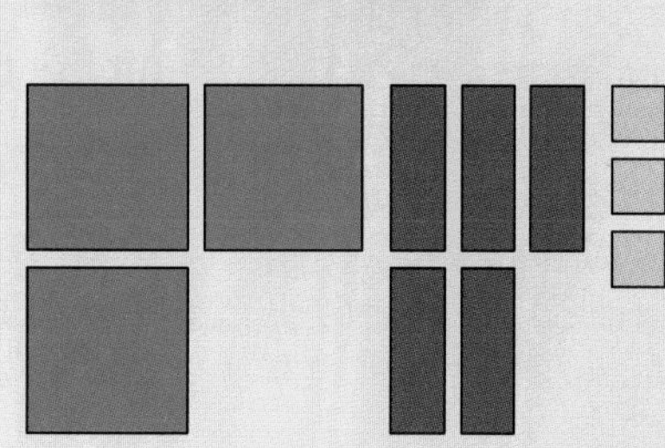

In Exercises 5–8, use algebra tiles to represent the expression. Sketch your result.

5. $3x^2 + 2x + 7$

6. $4x^2 + 5x + 3$

7. $x^2 + 6x + 4$

8. $5x^2 + 7x + 1$

14.5 Exploring Polynomials

What you should learn:

Goal 1 How to identify polynomials and write them in standard form

Goal 2 How to use polynomials to solve real-life problems

Why you should learn it:

You can use polynomials to solve real-life problems, such as finding the time it takes for a wrench to fall from the top of a building.

Goal 1 Identifying Polynomials

A **polynomial** is an expression that has one or more terms of the form ax^n where the coefficient a is any real number and the exponent n is a whole number. Polynomials are identified by the number of terms.

Type of Polynomial	Number of Terms	Example
Monomial	One	$3x^2$
Binomial	Two	$n + 4$
Trinomial	Three	$2y^2 + 4y - 5$

A polynomial is written in **standard form** if the powers of the variable decrease from left to right.

Example 1 *Rewriting in Standard Form*

Original Polynomial	Rewrite in Standard Form
$3m + 4m^3 - 2m^2 + 5$	$4m^3 - 2m^2 + 3m + 5$

■

If two terms have the same variable, raised to the same power, they are called *like terms*. When you combine like terms of a polynomial (by adding their coefficients), you are *simplifying* the polynomial.

Need to Know

The terms of a polynomial are considered to include any minus signs in the polynomial. For instance, the terms of $2x^2 - x + 4$ are $2x^2$, $-x$, and 4. The coefficient of x^2 is 2 and the coefficient of x is -1.

Example 2 *Simplifying Polynomials*

Simplify the polynomial and write the result in standard form.

a. $3x^2 - 4x + x^2 + 5x$ **b.** $5n^2 + 6 + 4n - 7$ ■

Solution To simplify the polynomials, collect like terms, and add their coefficients.

a. $$\begin{aligned} 3x^2 - 4x + x^2 + 5x &= 3x^2 + x^2 - 4x + 5x \\ &= (3 + 1)x^2 + (-4 + 5)x \\ &= 4x^2 + x \end{aligned}$$

b. $$\begin{aligned} 5n^2 + 6 + 4n - 7 &= 5n^2 + 4n + 6 - 7 \\ &= 5n^2 + 4n - 1 \end{aligned}$$ ■

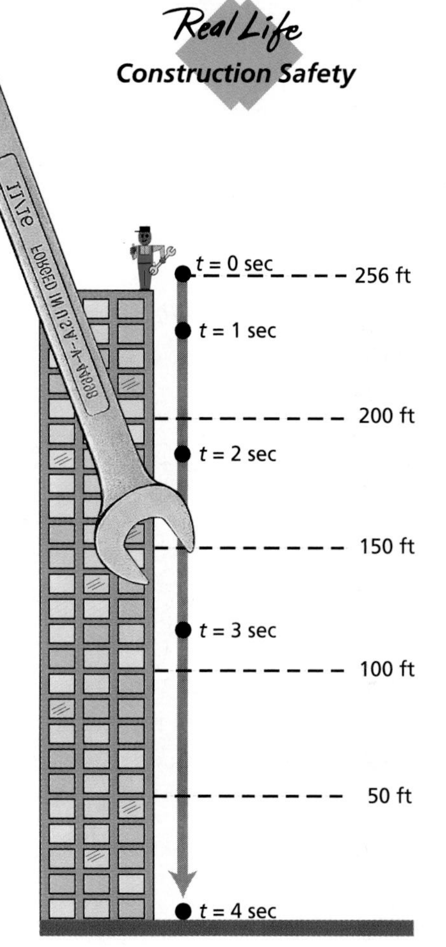

Goal 2 Using Polynomials in Real Life

When you drop a heavy object, does it fall at a constant speed or does it fall faster and faster the longer it is in the air? The answer is that it falls faster and faster.

Example 3 *Using a Polynomial*

A construction worker 256 feet above the ground accidentally drops a wrench from the top of a skyscraper. The height h (in feet) of the wrench after t seconds is given by

$$h = -16t^2 + 256.$$

Find the height when t is 0 seconds, 1 second, 2 seconds, 3 seconds, and 4 seconds. How long does the wrench take to hit the ground?

Solution To find the height, substitute the values of t into the polynomial $-16t^2 + 256$.

Time	Substitute	Height
$t = 0$ (seconds)	$h = -16(0)^2 + 256 = 0 + 256$	256 feet
$t = 1$ (second)	$h = -16(1)^2 + 256 = -16 + 256$	240 feet
$t = 2$ (seconds)	$h = -16(2)^2 + 256 = -64 + 256$	192 feet
$t = 3$ (seconds)	$h = -16(3)^2 + 256 = -144 + 256$	112 feet
$t = 4$ (seconds)	$h = -16(4)^2 + 256 = -256 + 256$	0 feet

Because the height is 0 feet after 4 seconds, you can conclude that the wrench took 4 seconds to fall to the ground. ■

Communicating about MATHEMATICS

SHARING IDEAS about the Lesson

Extending the Example During the first second of fall, the wrench in Example 3 fell 16 feet (from a height of 256 feet to a height of 240 feet). How far did it fall during its second, third, and fourth seconds of fall? Organize your results in a table like that below.

Time	0 to 1	1 to 2	2 to 3	3 to 4
Falling Distance	16 feet	? feet	? feet	? feet

Do your results confirm that the wrench is falling faster and faster? Explain your reasoning.

EXERCISES

Guided Practice

CHECK for Understanding

In Exercises 1–8, is the expression a polynomial? Explain.

1. $y + 1$
2. $3t^{-3}$
3. $4n^{-2} - 7$
4. 5
5. $3x^2 + x^{-1}$
6. $4s^3 - 8s^2 + 12$
7. $\sqrt{5}r^2 - \frac{1}{2}$
8. $2m^4 + m - 4$

In Exercises 9–11, write the polynomial in standard form. Then state its terms.

9. $4x - 2 + 3x^2$
10. $10 - 5r^3 + 4r$
11. $3p - 16p^2 - 12 + p^3$

In Exercises 12–14, simplify the polynomial. Then identify its type.

12. $t^2 + t - 5t + 2t^2$
13. $12 - 6x^3 + 5x^3 - 7$
14. $2n + 1 + 12n - 8$

Independent Practice

In Exercises 15–18, is the expression a polynomial? If it is, state whether it is a monomial, a binomial, or a trinomial.

15. $\frac{1}{2}t^2 - 5t + 3$
16. $9n - \sqrt{2}n^3$
17. $6.2y^4$
18. $\frac{6}{x^2} - 3x^3$

In Exercises 19–21, match the algebra tiles with a polynomial. Then simplify the polynomial and sketch a rearranged version of the tiles that represent the simplified expression.

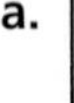

a. 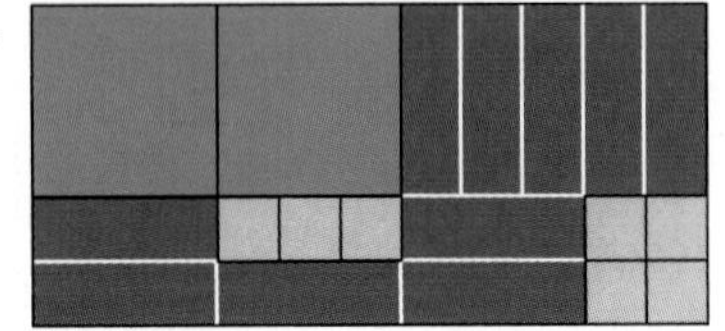

b. 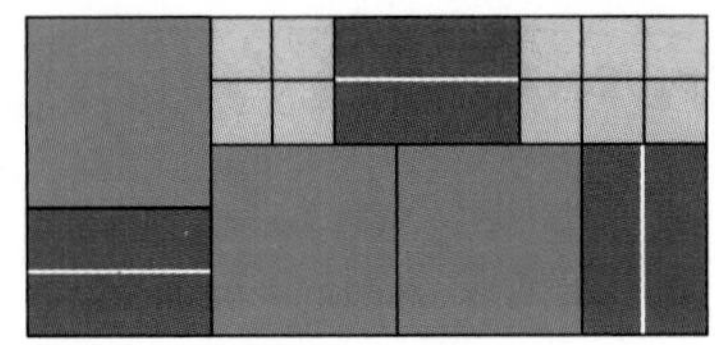

c. 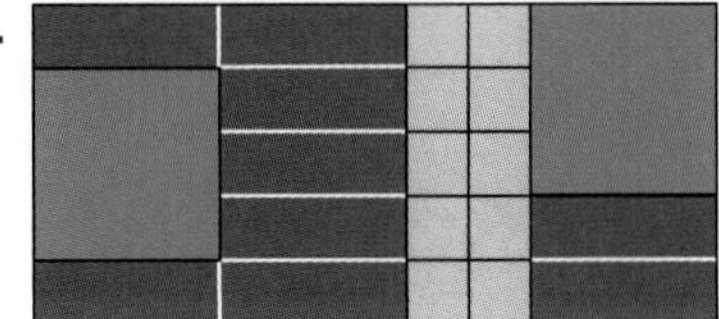

19. $x^2 + 7x + 10 + x^2 + 2x$
20. $2x^2 + 5x + 3 + 5x + 4$
21. $x^2 + 2x + 4 + 2x^2 + 4x + 6$

In Exercises 22–27, write the polynomial in standard form and list its terms.

22. $14m - 10m^2 + 5m^3$
23. $6x^3 - x - 2x^2$
24. $5 - 11y - 8y^3$
25. $9z^2 - 7z + 3 - z^3$
26. $2 - t^4 + t^2 + t$
27. $w + 4w^2 - 3 + 15w^3$

In Exercises 28–33, simplify and write in standard form.

28. $y + 2y^2 - 3y$
29. $x^3 - 3x + 5x - x^3$
30. $8 - 4x^2 + 10x^2 - 11$
31. $15 + 7s^3 - 21 - 3s^2 + s^3$
32. $\frac{4}{3}m - 7 - \frac{2}{3}m + 8$
33. $1.1r^2 - 2.9r + 1.8r^2 + 3.3r$

Look Out Below! **In Exercises 34–36, use the following information.**

You are standing on the Royal Gorge Bridge and accidentally drop your camera. The height h in feet of the camera after t seconds is modeled by

$$h = -16t^2 + 1053.$$

34. Copy and complete the table.

t	0	1	2	3	4	5	6	7	8	9
h	?	?	?	?	?	?	?	?	?	?

35. What is the camera's height after 4 seconds?

36. When will the camera hit the water?

The Royal Gorge Bridge in Colorado is the highest bridge in the world. It is 1,053 feet above the water level.

Integrated Review

Making Connections within Mathematics

Language Skills **In Exercises 37–40, match the term with a phrase.**

a. Three years **b.** Having two modes **c.** Single color photo **d.** Speak many languages

37. Polyglot **38.** Monochrome **39.** Bimodal **40.** Triennium

Mental Math **In Exercises 41–44, evaluate the polynomial for the given values.**

41. $t^2 + 7t - 11;\ t = -1, 0, 2$

42. $-x^2 + 3x + 9;\ x = -2, 0, 5$

43. $2n^3 - 6n + 12;\ n = -3, 0, 3$

44. $-s^3 + 6s^2 - 8;\ s = -1, 0, 4$

Exploration and Extension

45. *Logic Puzzle* Ann, Joy, Lee, Rex, and Ty each have a pet dog. From the clues, determine the name and breed of each person's pet. Copy and use the table to help organize your information. In each box, mark ○ for true and × for false.

Clues:

a. Butch is not Ann's or Rex's dog.

b. Ann and Joy own Daisy and the Great Dane.

c. Max and the collie do not belong to Lee or Ty.

d. Rex's dog and the collie are not Hazel or Duke.

e. Duke and Lee's dog are not boxers.

f. Hazel is a beagle.

g. Duke is not a Great Dane.

h. Two of the dogs are a boxer and a retriever.

	Beagle	Boxer	Collie	Great Dane	Retriever	Butch	Daisy	Duke	Hazel	Max
Ann										
Joy										
Lee										
Rex										
Ty										
Butch										
Daisy										
Duke										
Hazel										
Max										

LESSON INVESTIGATION 14.6

Exploring Polynomial Addition

Materials Needed: algebra tiles

In this investigation, you will use algebra tiles to explore polynomial addition.

Example — *Exploring Polynomial Addition*

Use algebra tiles to represent the polynomials $2x^2 + 3x + 3$ and $x^2 + x + 2$. Then combine the tiles to form one group. What polynomial is represented by the group?

Solution You can represent the two polynomials as follows.

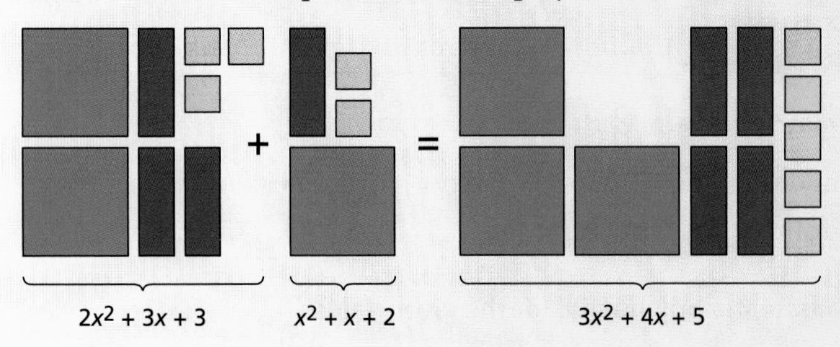

The combined group represents the polynomial $3x^2 + 4x + 5$. This polynomial is called the **sum** of the two original polynomials. It can be written as the *polynomial equation*

$$(2x^2 + 3x + 3) + (x^2 + x + 2) = 3x^2 + 4x + 5.$$ ■

Exercises

In Exercises 1 and 2, write the polynomial equation suggested by the algebra tiles.

1. 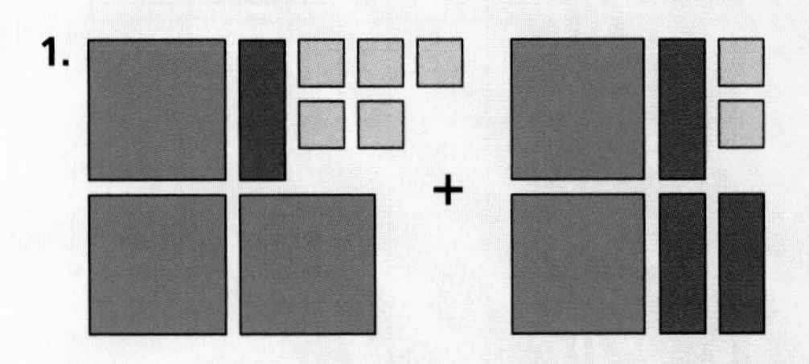

2. 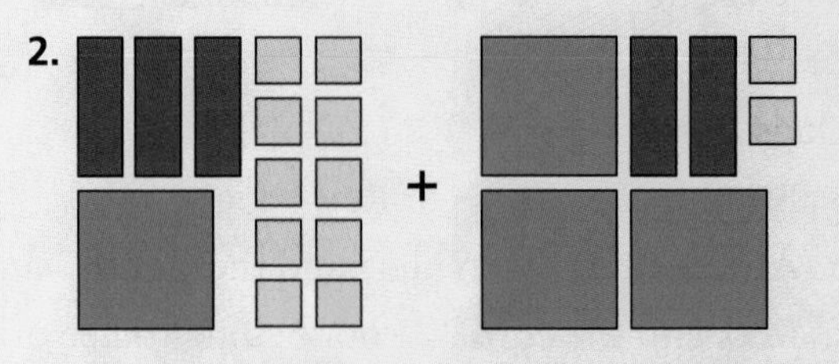

In Exercises 3–6, use algebra tiles to find the sum.

3. $(3x^2 + 2x + 7) + (x^2 + x + 1)$
4. $(4x^2 + 5x + 3) + (2x^2 + 1)$
5. $(x^2 + 6x + 4) + (3x^2 + 2x)$
6. $(5x^2 + 7x + 1) + (4x + 3)$
7. *Writing* In your own words, explain how to add two polynomials.

14.6 Adding and Subtracting Polynomials

What you should learn:

Goal 1 How to add two or more polynomials

Goal 2 How to subtract two polynomials

Why you should learn it:

You can use polynomial subtraction to solve geometry problems, such as finding the area of a region.

Goal 1 Adding Polynomials

In the *Lesson Investigation* on page 658, you may have discovered that you can **add** two polynomials by *combining like terms.* For instance, to add $2x^2 + 3x + 3$ and $x^2 + x + 2$, you can write the following.

$$\begin{aligned}(2x^2 + 3x + 3) + (x^2 + x + 2) &= 2x^2 + 3x + 3 + x^2 + x + 2\\ &= 2x^2 + x^2 + 3x + x + 3 + 2\\ &= 3x^2 + 4x + 5\end{aligned}$$

This technique is called the *horizontal format* for adding polynomials. You can also use a *vertical format.*

$2x^2 + 3x + 3$	*Write each polynomial in standard form.*
$x^2 + x + 2$	*Line up like terms.*
$3x^2 + 4x + 5$	*Add coefficients of like terms.*

The **degree** of a polynomial is its largest exponent. *Any* two polynomials can be added. They do not have to have the same degree.

2nd-Degree Polynomial: $2n^2 - 5n + 3$

3rd-Degree Polynomial: $-n^3 - 4n^2 + 7$

Study Tip...
When you use a vertical format to add two polynomials, be sure that you line up the like terms. Sometimes this means leaving a space. For instance, in Example 1a the polynomial $2n^3 + 3n + 6$ is written with a blank space because there is no n^2-term.

Example 1 *Adding Polynomials*

Add the polynomials.

a. $(-n^3 + 2n^2 - n + 4) + (2n^3 + 3n + 6)$

b. $(x^3 + 5x^2 - 2x + 3) + (2x^2 + 4x - 5)$

Solution You can use either a horizontal or vertical format. The vertical format is shown below.

Line up like terms.

a.
$$\begin{array}{r} -n^3 + 2n^2 - n + 4\\ 2n^3 \qquad\quad + 3n + 6\\ \hline n^3 + 2n^2 + 2n + 10\end{array}$$

b.
$$\begin{array}{r} x^3 + 5x^2 - 2x + 3\\ 2x^2 + 4x - 5\\ \hline x^3 + 7x^2 + 2x - 2\end{array}$$

■

Goal 2 Subtracting Polynomials

To subtract two polynomials, you can use the Distributive Property. For instance, you can subtract $x^2 - x + 3$ from $3x^2 - x + 5$ as follows.

$$(3x^2 - x + 5) - (x^2 - x + 3) = 3x^2 - x + 5 - x^2 + x - 3$$
$$= 3x^2 - x^2 - x + x + 5 - 3$$
$$= 2x^2 + 2$$

Subtract
$$\begin{array}{r} 3x^2 - x + 5 \\ -(x^2 - x + 3) \\ \hline \end{array}$$

Distribute
$$\begin{array}{r} 3x^2 - x + 5 \\ -x^2 + x - 3 \\ \hline 2x^2 \quad\quad + 2 \end{array}$$

A vertical format for this subtraction is shown at the left. Notice that the vertical format requires two steps.

Connections
Geometry

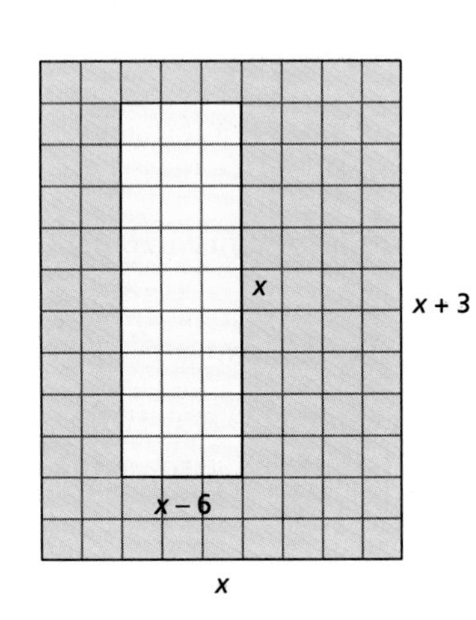

Example 2 *Subtracting Polynomials*

Find an expression that represents the area of the green region at the left.

Solution The area of the larger rectangle is

$$\text{Area} = x(x + 3) = x^2 + 3x.$$

The area of the smaller rectangle is

$$\text{Area} = x(x - 6) = x^2 - 6x.$$

To find the area of the green region, subtract the area of the smaller rectangle from the area of the larger rectangle.

$$(x^2 + 3x) - (x^2 - 6x) = x^2 + 3x - x^2 + 6x$$
$$= x^2 - x^2 + 3x + 6x$$
$$= 9x$$

■

Communicating about MATHEMATICS

SHARING IDEAS about the Lesson

Evaluating Polynomials In Example 2, find the area of the green region when $x = 9$ in the following two ways. Which way do you prefer? Explain.

A. Substitute 9 for x in the expression for the area of each rectangle. Then subtract the area of the smaller from the area of the larger.

B. Substitute 9 for x in the expression for the area of the green region.

EXERCISES

Guided Practice

CHECK for Understanding

In Exercises 1 and 2, use a horizontal format to find the sum or difference.

1. $(3x^2 - 7x + 5) + (3x^2 - 10)$

2. $(n^2 + 8n - 7) - (-n^2 + 8n - 12)$

In Exercises 3 and 4, use a vertical format to find the sum or difference.

3.
$$\begin{array}{r} 2y^3 + y^2 - 4y + 3 \\ +\quad y^3 - 5y^2 + 2y - 6 \\ \hline \end{array}$$

4.
$$\begin{array}{r} 2y^3 + y^2 - 4y + 3 \\ -\quad y^3 - 5y^2 + 2y - 6 \\ \hline \end{array}$$

Independent Practice

Error Analysis **In Exercises 5 and 6, find and correct the error.**

5.
$$\begin{array}{r} -4z^3 + z^2 + \qquad\quad 7 \\ +\quad 3z^3 - \qquad 6z - 5 \\ \hline -z^3 + z^2 + 6z + 2 \end{array}$$

6.
$$\begin{array}{r} 4x^3 + 3x^2 + 4x + 3 \\ -(2x^3 + x^2 + 2x + 1) \\ \hline 2x^3 + 4x^2 + 6x + 4 \end{array}$$

In Exercises 7 and 8, add the polynomials. (Use a horizontal format.)

7. $(-x^2 + 9x - 5) + (6x^2 - 2x + 16)$

8. $(-8a^3 + a^2 + 17) + (6a^2 - 3a + 9)$

In Exercises 9 and 10, subtract the polynomials. (Use a horizontal format.)

9. $(-b^3 + 4b^2 - 1) - (7b^3 + 4b^2 + 3)$

10. $(-5x^3 - 13x + 4) - (-3x^3 + x^2 + 10x - 9)$

In Exercises 11 and 12, add the polynomials. (Use a vertical format.)

11.
$$\begin{array}{r} x^3 + 4x^2 - 9x + 2 \\ +\ -2x^3 + 5x^2 + x - 6 \\ \hline \end{array}$$

12.
$$\begin{array}{r} 2n^4 + 2n^3 - n^2 - 4n + 6 \\ +\quad n^4 + 3n^3 - 3n^2 - 5n + 2 \\ \hline \end{array}$$

In Exercises 13 and 14, subtract the polynomials. (Use a vertical format.)

13.
$$\begin{array}{r} 3t^3 + 4t^2 + t - 5 \\ -\ (t^3 + 2t^2 - 9t + 1) \\ \hline \end{array}$$

14.
$$\begin{array}{r} x^4 + 3x^3 + x^2 + 2x + 5 \\ -(x^4 + 2x^3 + 3x^2 + 4x - 4) \\ \hline \end{array}$$

In Exercises 15–18, perform the indicated operations.

15. $(2x^2 + 9x - 4) + (-8x^2 + 3x + 6) + (x^2 - 5x - 7)$

16. $(-5y^3 - 4y^2 - 1) + (12y^3 + 3y - 11) - (10y^2 + y - 3)$

17. $(z^2 - 5) - (-z^3 + 2z^2 + 3) + (-z^3 - 4z + 8)$

18. $(4x^2 + x - 17) - (x^2 - 15x + 7) - (-7x^2 + x + 6)$

Geometry **In Exercises 19 and 20, find the perimeter of the polygon. Then evaluate the perimeter when $x = 3$.**

19.

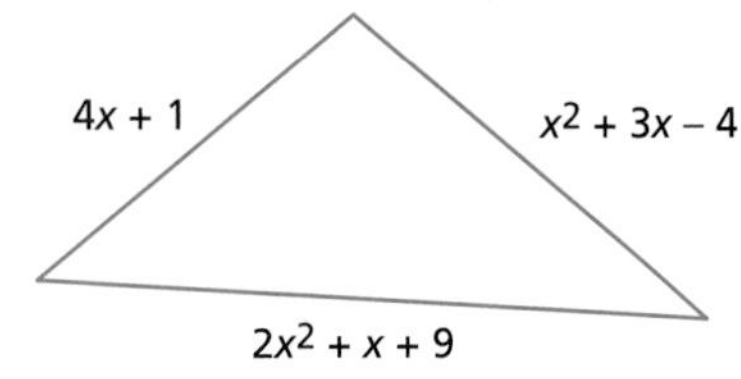

20.

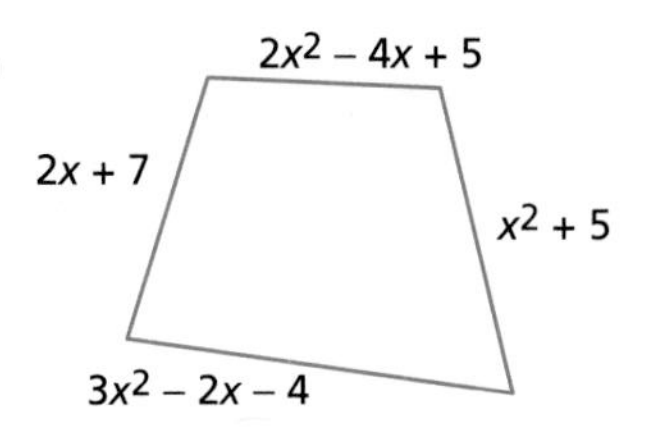

In Exercises 21 and 22, find an expression that represents the area of the green region. Then evaluate the area when $x = 5$.

21.

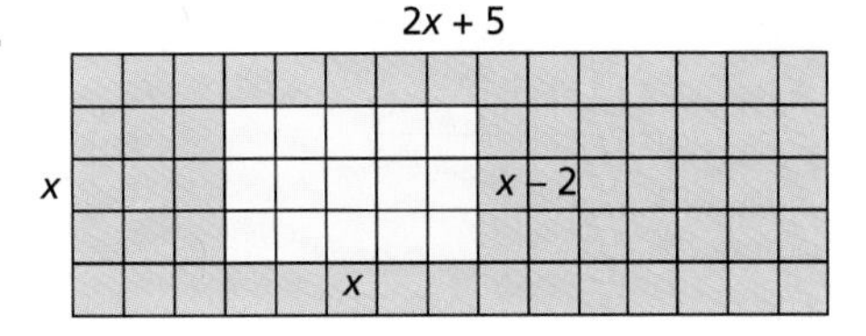

22.

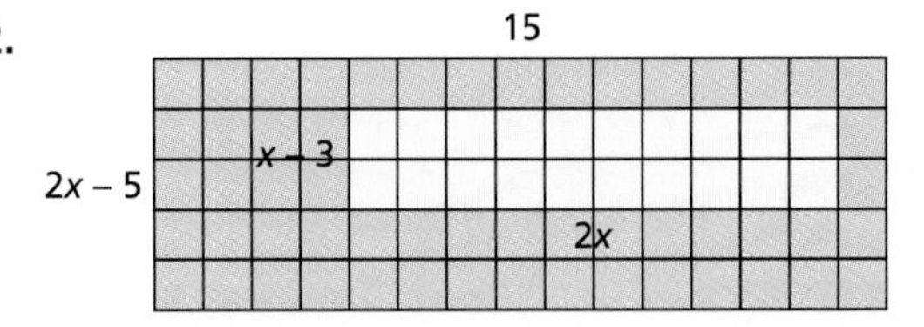

Profit **In Exercises 23–25, use the following information.**

You spend \$500 to start a small business selling a computer software program you wrote. Each program costs you \$2.50 to produce (for the discs and package). You sell each program for \$12.50. Your income and expense for producing and selling x programs are

$$\text{Income} = 12.50x \qquad \text{and} \qquad \text{Expense} = 2.50x + 500.$$

23. Write a polynomial model for the profit you make for selling x programs.

24. How much profit will you make if you sell 200 programs?

25. *Think about It* Will you make twice as much profit if you sell 400 programs? Explain.

Integrated Review

Making Connections within Mathematics

Low-Calorie **In Exercises 26–28, use the figure at the right showing the number of people who bought low-calorie foods and beverages from 1978 to 1993.** *(Source: Caloric Control Council)*

26. How many more people bought low-calorie food and drink in 1993 than in 1978?

27. Estimate the number who bought low-calorie food and drink in 1986. In 1990.

28. Estimate the number of people who will buy low-calorie food and drink in 1995.

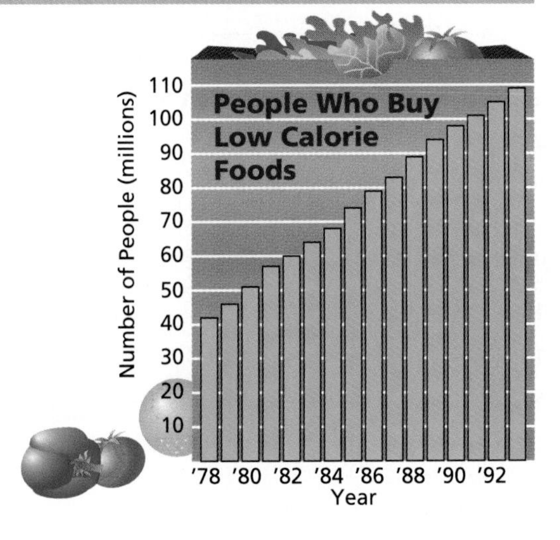

Exploration and Extension

Guess, Check, and Revise **In Exercises 29–34, each of the equations has two solutions. Use a guess, check, and revise strategy to find both solutions.**

29. $x^2 - 3x + 2 = 0$

30. $x^2 - 5x + 6 = 0$

31. $x^2 - 7x + 12 = 0$

32. $x^2 - 9x + 20 = 0$

33. $x^2 - 11x + 30 = 0$

34. $x^2 - 13x + 42 = 0$

35. *Patterns* Use the pattern from Exercises 29–34 to write a polynomial equation that has 7 and 8 as its solutions. Check your result by substituting.

Mixed REVIEW

In Exercises 1–6, use the similar triangles at the right. (11.1)

1. Find the length of the hypotenuse of $\triangle ABC$.
2. Find the length of the hypotenuse of $\triangle DEF$.
3. What is the scale factor of $\triangle DEF$ to $\triangle ABC$?
4. Find the area of each triangle.
5. Find the perimeter of each triangle.
6. Describe the relationship between the scale factor, the perimeters, and the areas.

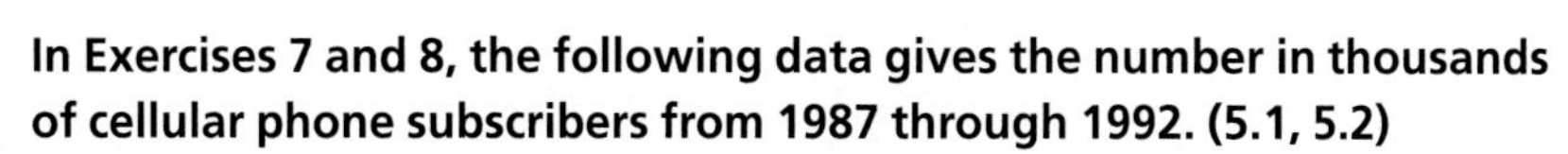

In Exercises 7 and 8, the following data gives the number in thousands of cellular phone subscribers from 1987 through 1992. (5.1, 5.2)

1987: 1,231; 1988: 2,069; 1989: 3,509;
1990: 5,283; 1991: 7,557; 1992: 11,033

(Source: Cellular Telecommunications Industry Association)

7. Draw a picture graph of the data.
8. Draw a bar graph of the data.

In Exercises 9–14, simplify. (7.4, 7.5)

9. $\frac{2}{3} + \frac{1}{6}$

10. $\frac{5}{8} - \frac{3}{4}$

11. $\frac{12}{15} \times \frac{2}{5}$

12. $-\frac{13}{21} \times \left(-\frac{4}{81}\right)$

13. $\frac{16}{21} \div \frac{4}{7}$

14. $\frac{6}{19} \div 2$

In Exercises 15–18, solve the inequality. (9.6)

15. $5y + 13 \geq 28$

16. $3 - 2r < 6$

17. $\frac{1}{2} < \frac{1}{4} - \frac{1}{3}s$

18. $4x + 3 \leq 2x$

In Exercises 19 and 20, add the polynomials. (14.6)

19. $(3x^2 + 2x + 1) + (x^2 + 2x + 3)$

20. $(6x^3 - x^2 + 2) + (3x^2 + 2)$

LESSON INVESTIGATION 14.7

Exploring Polynomial Multiplication

Materials Needed: algebra tiles

In this investigation, you will use algebra tiles to explore polynomial multiplication.

Example — *Exploring Polynomial Multiplication*

Use algebra tiles to represent the product of the polynomials $3x$ and $x + 4$. Then write the polynomial that is represented by the entire group of algebra tiles.

Solution You can represent the product of the two polynomials as follows:

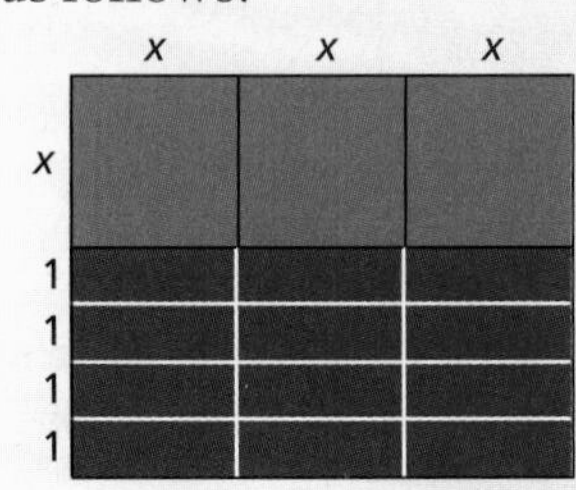

$$\text{Length} = 3x$$
$$\text{Width} = x + 4$$
$$\text{Area} = (\text{Length})(\text{Width}) = 3x(x + 4)$$

The combined group represents the polynomial $3x^2 + 12x$. The result can be written as the *polynomial equation*

$$3x(x + 4) = 3x^2 + 12x.$$

■

Exercises

In Exercises 1–4, write the polynomial equation indicated by the algebra tiles.

1. 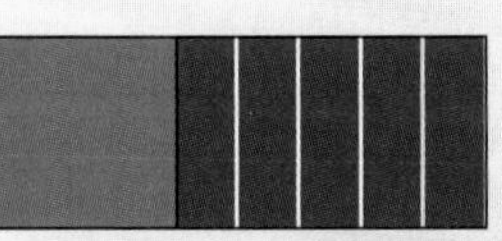

2. 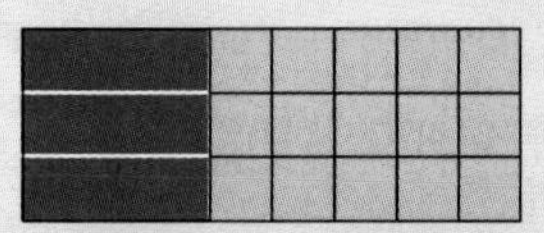

3. 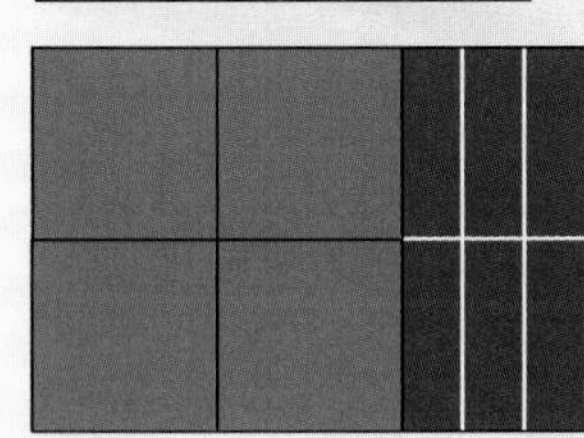

4. 

In Exercises 5–8, use algebra tiles to find the product.

5. $2x(3x + 2)$
6. $x(4x + 3)$
7. $3x(x + 1)$
8. $4x(2x + 3)$

14.7 Multiplying Polynomials

What you should learn:

Goal 1 How to multiply a polynomial by a monomial

Goal 2 How to use polynomial multiplication to solve geometry problems

Why you should learn it:

You can use polynomial multiplication to solve geometry problems, such as finding the area of a region.

Goal 1 Multiplying Polynomials

In this lesson, you will learn how to multiply a polynomial by a monomial. Example 1 reviews the types of polynomials that you already know how to multiply.

Example 1 ***Multiplying Polynomials***

a. $2(3x + 5) = 6x + 10$ *Distributive Property (Lesson 2.1)*

b. $n(n - 3) = n^2 - 3n$ *Distributive Property (Lesson 3.4)*

c. $(y^2)(y^3) = y^5$ *Property of Exponents (Lesson 6.7)*

Each of the products in Example 1 is an example of multiplying a polynomial by a *monomial.* The general rule for finding this type of product is stated below.

Multiplying a Polynomial by a Monomial

To multiply a polynomial by a monomial, multiply each term of the polynomial by the monomial.

Study Tip...

When multiplying polynomials, be sure to check that the signs of the product are correct. One of the most common errors in algebra is to forget to "distribute negative signs."

Incorrect

$-x(2x^2 - 3x + 1)$

$= -2x^3 - 3x^2 + x$

Correct

$-x(2x^2 - 3x + 1)$

$= -2x^3 + 3x^2 - x$

Example 2 ***Multiplying a Polynomial by a Monomial***

a. $3x(x^2 + 2x - 5) = 3x(x^2) + 3x(2x) - 3x(5)$ *Distribute.*

$= 3x^3 + 6x^2 - 15x$ *Simplify.*

b. $n^2(-2n^3 + 4n) = n^2(-2n^3) + n^2(4n)$ *Distribute.*

$= -2n^5 + 4n^3$ *Simplify.*

c. $2b^2(-4b^4 + b + 6) = 2b^2(-4b^4) + 2b^2(b) + 2b^2(6)$

$= -8b^6 + 2b^3 + 12b^2$

d. $-5y(3y^2 + y - 7) = -5y(3y^2) + (-5y)(y) - (-5y)(7)$

$= -15y^3 - 5y^2 + 35y$

Goal 2 Using Polynomial Multiplication

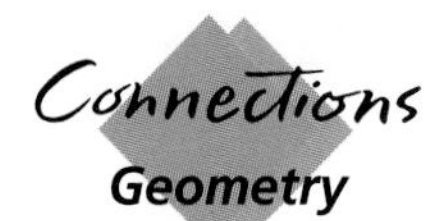

Example 3 *Using Polynomial Multiplication*

The rectangle at the left is divided into five regions. Write an expression for the area of each region. Then write an expression for the area of the entire region.

Solution

Polygon	Expression for Area	Simplify
Rectangle	Area $= x(2x+2)$	$2x^2+2x$
Parallelogram	Area $= x(2x)$	$2x^2$
Trapezoid	Area $= \frac{1}{2}(2x)([x-2]+[x+2])$	$2x^2$
Each triangle	Area $= \frac{1}{2}(2x)(x-1)$	x^2-x

To find an expression for the area of the entire region, you can add the expressions for the areas of the five regions.

$2x^2+2x$	Area of rectangle
$2x^2$	Area of parallelogram
$2x^2$	Area of trapezoid
x^2-x	Area of triangle
$+\ x^2-x$	Area of triangle
$8x^2$	Sum: Area of region

■

Communicating about MATHEMATICS

SHARING IDEAS about the Lesson

Guess, Check, and Revise

A. The area of the entire region in Example 3 can be modeled with the expression

$$\text{Area} = (\text{Length})(\text{Width}) = (3x)(2x+2).$$

Show how you can use the rule for multiplying a polynomial by a monomial to simplify this product.

B. The expression for the area of the entire region obtained in Part A is not the same as that obtained in Example 3. Find a value of x that makes both expressions equal. What is the area of the entire region?

EXERCISES

Guided Practice

CHECK for Understanding

In Exercises 1 and 2, find the product.

1. $4n(2n^2 - 3n + 5)$

2. $-3y^2(y^2 + 2y - 5)$

Geometry **In Exercises 3–7, use the figure at the right.**

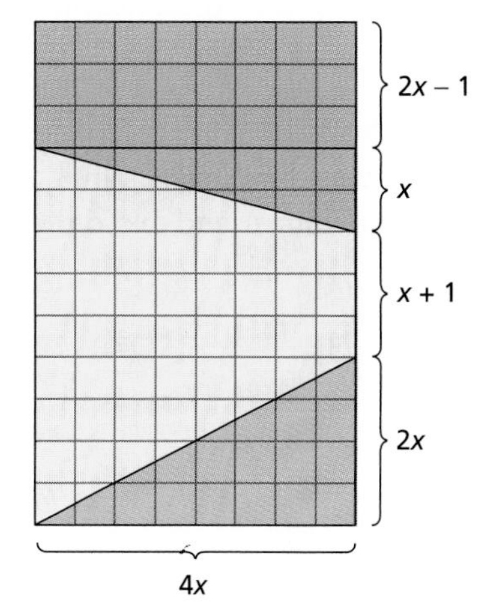

3. Write an expression for the area of each region.

4. Use the results of Exercise 3 to write an expression for the area of the entire region.

5. Write expressions for the length and width of the entire region.

6. Use the result of Exercise 5 to write an expression for the area of the entire region.

7. Compare the expressions obtained in Exercises 4 and 6.

Independent Practice

In Exercises 8–16, multiply.

8. $2y(y^2 + 1)$

9. $4x(x^2 - 2x - 1)$

10. $-2(3x^2 + 6x - 8)$

11. $8t^2(-6t - 5)$

12. $-b^4(8b^2 + 10b - 1)$

13. $2t(-4t^2 + 6t - 2)$

14. $y(-7y^3 + 8y - 4)$

15. $-4z(2z^5 - z^3 + 10)$

16. $n^3(-n^4 + n^3 - n^2 + n - 1)$

Geometry **In Exercises 17–20, use the figure at the right.**

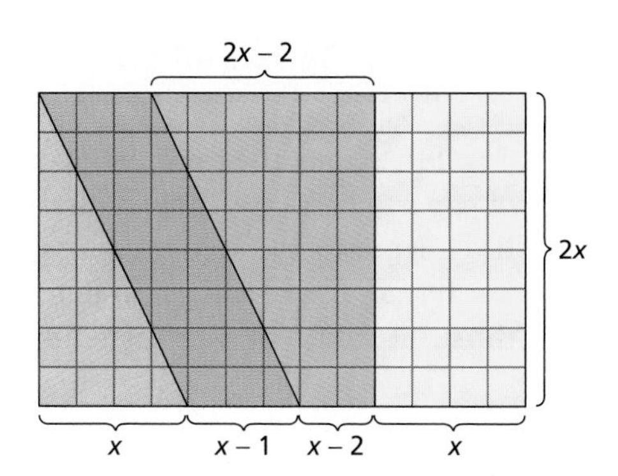

17. Write an expression for the area of each region.

18. Use the results of Exercise 17 to write an expression for the area of the entire region.

19. Use the formula for the area of a rectangle to write an expression for the area of the entire region.

20. Compare the expressions obtained in Exercise 18 and 19.

In Exercises 21 and 22, translate the verbal phrase to an algebraic expression. Then multiply.

21. The product of a number and one more than that number

22. The cube of a number times the difference of the number and 2

Surface Area and Volume **In Exercises 23–26, use the rectangular prism at the right.**

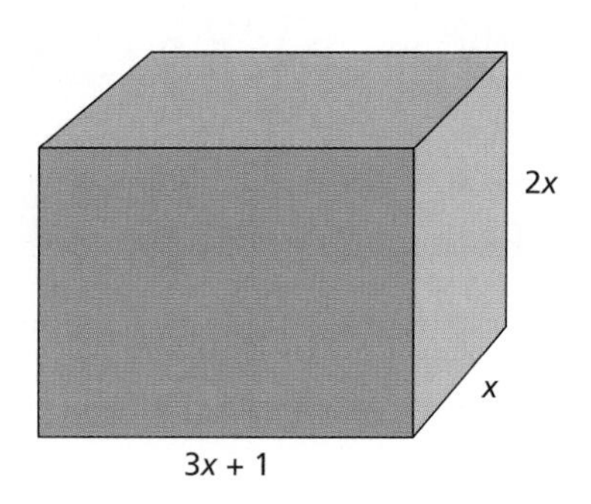

23. Write an expression for the area of the base of the prism.

24. Write an expression for the volume of the prism.

25. Write an expression for the surface area of the prism.

26. The surface area of the prism is 216 square units. What is the volume? Explain your reasoning.

You Be the Teacher **In Exercises 27 and 28, you are helping a friend. Your friend's solution is given. What did your friend do wrong? What could you say to help your friend avoid the error?**

27. $7t^2(-t^3 + 3t^2 - 8t) = 7t^2(-t^3) + 7t^2(3t^2) - 7t^2(-8t)$
$= -7t^6 + 21t^4 + 58t^3$

28. $-3x^3(2x^2 - 5x + 9) = -3x^3(2x^2) - 3x^3(5x) + 3x^3(9)$
$= -6x^6 - 15x^4 + 24x^3$

Integrated Review

Making Connections within Mathematics

In Exercises 29–32, write the expression without using exponents.

29. 4^0 **30.** 2^{-1} **31.** $3^3 \cdot 3^2$ **32.** $5^{-1} \cdot 5^4$

In Exercises 33–36, use the Distributive Property to rewrite the expression.

33. $2(6 + x)$ **34.** $-3(t + 4)$ **35.** $5(2x + 3)$ **36.** $4(3x - 5)$

Exploration and Extension

37. *Puzzle* Copy the four points on a piece of paper. Then show how to draw three line segments through the four points without retracing or lifting your pencil. You must return to the starting point.

38. *Puzzle* Copy the nine points on a piece of paper. Then draw four line segments through the nine points without retracing or lifting your pencil. You do not need to return to the starting point.

LESSON INVESTIGATION 14.8

Exploring Binomial Multiplication

Materials Needed: algebra tiles

In this investigation, you will use algebra tiles to explore multiplication of a binomial by a binomial.

Example — *Multiplying Two Binomials*

Use algebra tiles to represent the product of the binomials $2x + 1$ and $x + 3$. Then write the polynomial that is represented by the entire group of algebra tiles.

Solution You can represent the product of the two polynomials as follows.

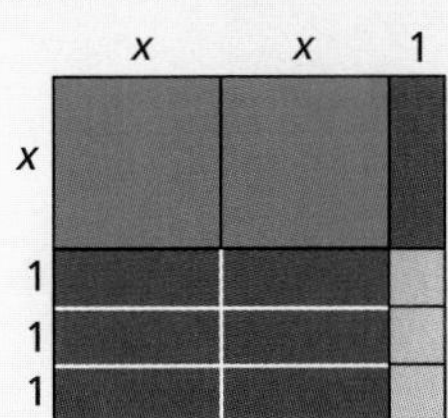

$$\text{Length} = 2x + 1$$
$$\text{Width} = x + 3$$
$$\text{Area} = (\text{Length})(\text{Width})$$
$$= (2x + 1)(x + 3)$$

The combined group represents the polynomial $2x^2 + 7x + 3$. The result can be written as the *polynomial equation*

$$(2x + 1)(x + 3) = 2x^2 + 7x + 3.$$

■

Exercises

In Exercises 1–4, write the polynomial equation indicated by the algebra tiles.

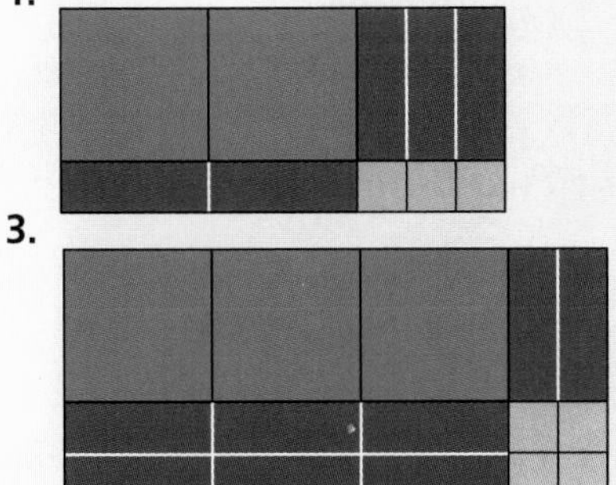

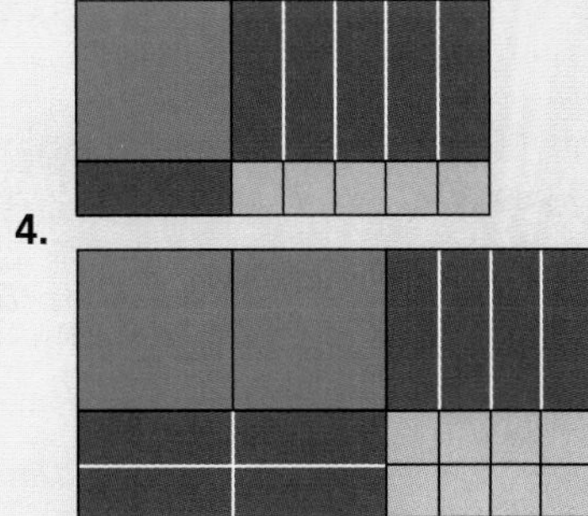

In Exercises 5–8, use or sketch algebra tiles to find the product.

5. $(x + 2)(3x + 1)$ **6.** $(x + 1)(4x + 3)$ **7.** $(x + 1)^2$ **8.** $(x + 2)^2$

14.8 More about Multiplying Polynomials

What you should learn:

How to multiply a binomial by a binomial

How to use polynomial multiplication to solve real-life problems

Why you should learn it:

You can use polynomial multiplication to solve real-life problems, such as finding the dimensions of a picture frame.

Goal 1 Multiplying Two Binomials

In the *Lesson Investigation* on page 669, you learned how to use algebra tiles to model the product of two binomials. In this lesson, you will learn how to use the Distributive Property to multiply two binomials.

Example 1 *Multiplying Two Binomials*

Find the product of $(x + 1)$ and $(2x + 3)$.

Solution In the following solution, notice that the Distributive Property is used in the first step *and* in the second step.

$$\begin{aligned}(x + 1)(2x + 3) &= (x + 1)(2x) + (x + 1)(3)\\ &= (x)(2x) + (1)(2x) + (x)(3) + (1)(3)\\ &= 2x^2 + 2x + 3x + 3\\ &= 2x^2 + 5x + 3\end{aligned}$$

■

Notice in Example 1 that the first use of the Distributive Property is to distribute the binomial $(x + 1)$ over the binomial $(2x + 3)$ to obtain

$$(x + 1)(2x) + (x + 1)(3).$$

Then the Distributive Property is used again to rewrite the expressions $(x + 1)(2x)$ and $(x + 1)(3)$.

Study Tip...

Another way to multiply two binomials is to use a vertical format, as shown below.

$$\begin{array}{r} x + 1\\ \underline{2x + 3}\\ 3x + 3\\ \underline{2x^2 + 2x }\\ 2x^2 + 5x + 3\end{array}$$

Example 2 *Multiplying Two Binomials*

Find the product of $(4x + 2)$ and $(3x + 1)$.

Solution

$$\begin{aligned}(4x + 2)(3x + 1) &= (4x + 2)(3x) + (4x + 2)(1)\\ &= (4x)(3x) + (2)(3x) + (4x)(1) + (2)(1)\\ &= 12x^2 + 6x + 4x + 2\\ &= 12x^2 + 10x + 2\end{aligned}$$

■

Goal 2 Solving Real-Life Problems

Real Life Art

Example 3 *Guess, Check, and Revise*

The area of the picture at the left, including the frame, is 80 square inches. What are the dimensions of the frame?

The next generation may only know the douc langur from photographs. Environmental changes in Southeast Asia have made it an endangered species.

Solution One way to solve the problem is to begin by writing a model for the total area.

Verbal Model Area = Length • Width

Labels
Area = 80 (square inches)
Length = $2x + 7$ (inches)
Width = $2x + 5$ (inches)

Algebraic Model

$$80 = (2x + 7)(2x + 5)$$
$$= (2x + 7)(2x) + (2x + 7)(5)$$
$$= (2x)(2x) + (7)(2x) + (2x)(5) + (7)(5)$$
$$= 4x^2 + 14x + 10x + 35$$
$$= 4x^2 + 24x + 35$$

Using this model, you can use *Guess, Check, and Revise* to find the value of x for which $4x^2 + 24x + 35$ is equal to 80.

Let $x = 1$: $4(1)^2 + 24(1) + 35 = 4 + 24 + 35 = 63$

Let $x = 2$: $4(2)^2 + 24(2) + 35 = 16 + 48 + 35 = 99$

Let $x = 1.5$: $4(1.5)^2 + 24(1.5) + 35 = 9 + 36 + 35 = 80$

Because $x = 1.5$, the frame is 8 inches by 10 inches. ■

Communicating about MATHEMATICS

SHARING IDEAS about the Lesson

The FOIL Method The *FOIL Method* for multiplying two binomials is shown below. Use this method to find the products. Then use the Distributive Property to check your work.

$$(2x + 7)(2x + 5) = 4x^2 + 10x + 14x + 35 = 4x^2 + 24x + 35$$

First: (2x)(2x) Outer: (2x)(5) Inner: (7)(2x) Last: (7)(5)

A. $(2x + 3)(x + 4)$ **B.** $(3x + 5)(x + 2)$

EXERCISES

Guided Practice

CHECK for Understanding

In Exercises 1 and 2, use the Distributive Property to find the product. Explain each step.

1. $(x + 3)(3x + 2)$

2. $(2x + 1)(4x + 5)$

In Exercises 3–6, match the expression with its equivalent expression.

a. $12x^2 + 15x + 3$ **b.** $12x^2 + 52x + 16$ **c.** $12x^2 + 32x + 16$ **d.** $12x^2 + 17x + 6$

3. $(4x + 3)(3x + 2)$

4. $(2x + 8)(6x + 2)$

5. $(12x + 3)(x + 1)$

6. $(6x + 4)(2x + 4)$

Independent Practice

Error Analysis **In Exercises 7 and 8, find and correct the error.**

7.
$$\begin{aligned}&(x + 6)(2x + 5)\\ &= (x + 6)(2x) + 5\\ &= (x)(2x) + 6(2x) + 5\\ &= 2x^2 + 12x + 5\end{aligned}$$

8.
$$\begin{aligned}&(3x + 4)(4x + 3)\\ &= (3x + 4)(4x) + (4x + 3)(4)\\ &= 3x(4x) + 4(4x) + (4x)(4) + 3(4)\\ &= 12x^2 + 16x + 16x + 12\\ &= 12x^2 + 32x + 12\end{aligned}$$

In Exercises 9–17, find the product using the Distributive Property.

9. $(x + 3)(8x + 12)$

10. $(5x + 6)(x + 2)$

11. $(2x + 1)(9x + 7)$

12. $(4x + 5)(5x + 4)$

13. $(10x + 10)(2x + 2)$

14. $(3x + 8)(3x + 2)$

15. $(2x + 3)(4x + 1)$

16. $(2x + 4)(7x + 9)$

17. $(6x + 5)(3x + 5)$

In Exercises 18–20, multiply using a vertical format. Then check the result by using a horizontal format.

18. $(3x + 2)(6x + 8)$

19. $(9x + 6)(3x + 1)$

20. $(x + 10)(4x + 15)$

Geometry **In Exercises 21–24, find the area of the figure. Then evaluate the area when $x = 2$.**

21.

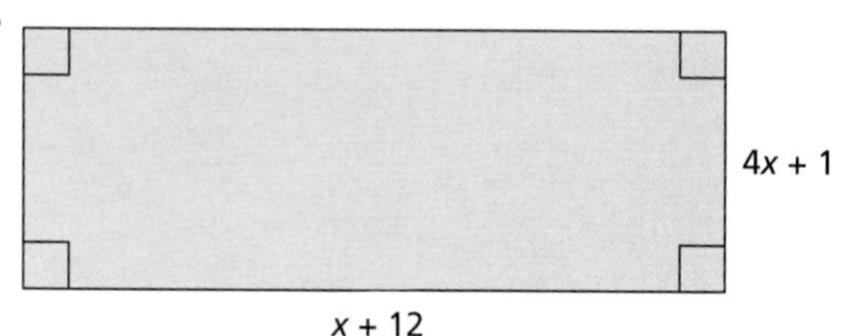

22.

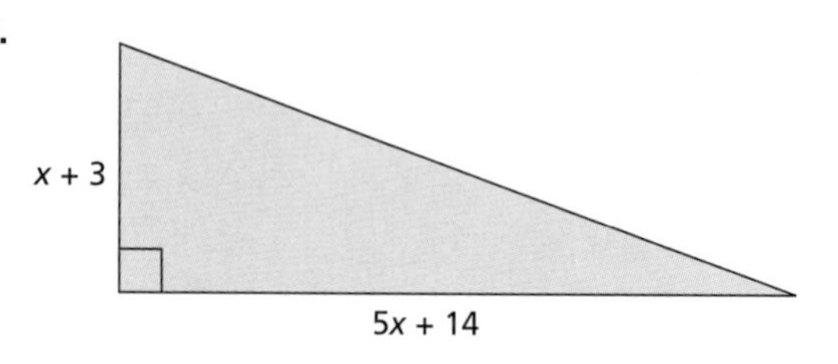

23.

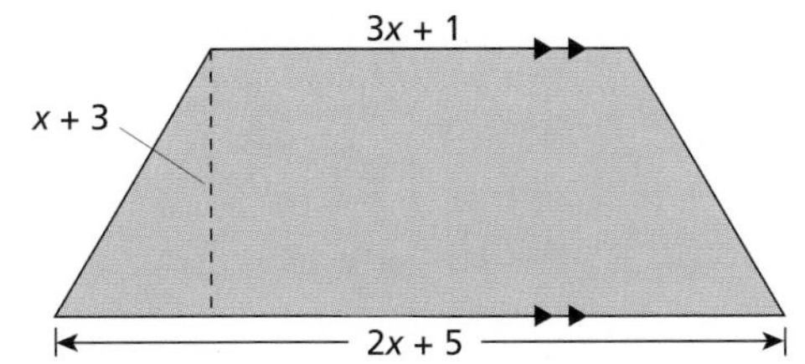

24.

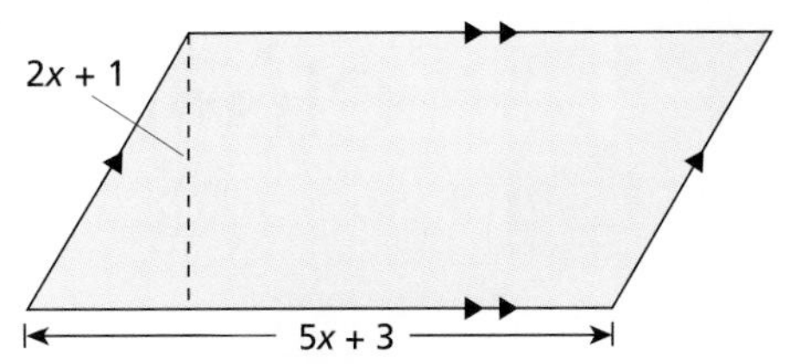

25. Multiply $(x + 15)$ by $(2x + 1)$. Then multiply $(2x + 1)$ by $(x + 15)$. What can you conclude? What property does this illustrate?

Algebra Tiles **In Exercises 26 and 27, use algebra tiles (or a sketch) to model the product. Then use the Distributive Property to confirm your result.**

26. $(1 + 2x)(2 + 3x)$ **27.** $(4 + x)(3 + 2x)$

28. *Designing a Mosaic* You are designing a mosaic that is to have an area of 180 square feet. The mosaic is a rectangle and is made up of rectangular enameled tiles that are each x feet by 1 foot. The mosaic is $(4x + 2)$ feet wide and $(7x + 4)$ feet high. Find the dimensions of the mosaic.

Mexican Art ***About 7.5 million stones make up the Juan O'Gorman mosaic found at Mexico's National Autonomous University. It depicts scenes from the history of Mexico.***

Integrated Review

Making Connections within Mathematics

In Exercises 29–31, multiply the binomials. Then match the result with the product that yields the same result.

a. $(2x + 2)(x + 3)$ **b.** $(x + 5)(3x + 3)$ **c.** $(2x + 10)(x + 3)$

29. $(2x + 6)(x + 5)$ **30.** $(2x + 6)(x + 1)$ **31.** $(3x + 15)(x + 1)$

32. *Estimation* Which of the following best estimates the weight (in pounds) of a 12-fluid-ounce can of soda pop?

a. 1 **b.** 6 **c.** 12 **d.** 20

33. *Estimation* Which of the following best estimates the height (in centimeters) of a 12-fluid-ounce can of soda pop?

a. 1 **b.** 6 **c.** 12 **d.** 20

Exploration and Extension

Binomial Pattern **In Exercises 34–39, multiply.**

34. $(x + 1)(x + 1)$ **35.** $(x + 2)(x + 2)$ **36.** $(x + 3)(x + 3)$

37. $(x + 4)(x + 4)$ **38.** $(x + 5)(x + 5)$ **39.** $(x + 6)(x + 6)$

40. *Binomial Pattern* Each of the products in Exercises 34–39 is an example of a "binomial pattern." Describe the pattern. Then use your description to find the product of $(x + 7)$ and $(x + 7)$. Use the Distributive Property to check your result.

14

Chapter Summary

What did you learn?

Skills

1. Find the mean, median, and mode of a collection of numbers. **(14.1)**
2. Draw a stem-and-leaf plot for a collection of numbers. **(14.2)**
3. Find the quartiles of a collection of numbers. **(14.3)**
   - Draw a box-and-whisker plot for a collection of numbers. **(14.3)**
4. Use matrices to organize data. **(14.4)**
   - Add and subtract matrices. **(14.4)**
5. Identify different types of polynomials. **(14.5)**
   - Simplify polynomials by combining like terms. **(14.5)**
6. Add and subtract polynomials. **(14.6)**
7. Multiply polynomials. **(14.7, 14.8)**
   - Multiply a polynomial by a monomial. **(14.7)**
   - Multiply two binomials. **(14.8)**

Problem-Solving Strategies

8. Use data plots, matrices, and polynomials to model and solve real-life problems. **(14.1–14.8)**

Exploring Data

9. Use tables and graphs to solve problems. **(14.1–14.8)**

Why did you learn it?

In many areas of real life, you will encounter data. Data is much easier to interpret when it is organized. In this chapter, you learned how stem-and-leaf plots and box-and-whisker plots can be used to compare characteristics of different populations, and how matrices can be used to analyze the success of a business. You also studied about polynomials, and you learned that they can be used to model areas of regions.

How does it fit into the bigger picture of mathematics?

If someone were to ask you, "What did you study in math this year?" what would you say? We hope you would say that you were introduced to the two main parts of mathematics: algebra and geometry. We also hope you would say you learned that mathematics is not just a collection of formulas that need to be memorized. Instead, you learned that mathematics is a language that can be used to model and solve real-life problems.

Of course, there is much more to algebra and geometry than is possible to put in this book. You can think of the fourteen chapters in this book as "windows" that give you glimpses of the algebra and geometry that you might be studying in the next two or three years.

Chapter REVIEW

In Exercises 1–7, use the table, which lists the National League teams' home run totals through the first part of the 1994 season. *(Source: USA Today)* **(14.1–14.3)**

Team	Home Run Total	Team	Home Run Total
Cincinnati	38	Houston	44
Los Angeles	42	Philadelphia	32
Atlanta	48	Chicago	36
Colorado	52	St. Louis	36
Montreal	21	New York	49
Pittsburgh	16	San Francisco	35
Florida	34	San Diego	34

With 755 home runs, Henry (Hank) Aaron holds the major league lifetime record. Babe Ruth holds second place with 714.

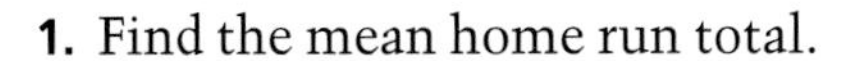

1. Find the mean home run total.

2. Find the median home run total.

3. Find the mode of the home run totals.

4. Which measure of central tendency do you think best represents the data? Explain.

5. Make a stem-and-leaf plot to organize the data.

6. Use the results of the stem-and-leaf plot to draw a histogram for the table.

7. Draw a box-and-whisker plot for the data in the table.

In Exercises 8–10, find the first, second, and third quartiles for the numbers. (14.3)

8. 5, 8, 9, 12, 16, 18, 21, 22, 25, 27

9. 79, 71, 65, 69, 73, 61, 68, 77, 81, 62, 84, 67

10. 35, 67, 95, 21, 16, 100, 47, 82, 50, 0, 89, 71, 31, 54

In Exercises 11–13, use the matrix at the right. (14.4)

$$\begin{bmatrix} 5 & -6 & 8 & 3 & 1 \\ -2 & -7 & 9 & 4 & 2 \\ 10 & 7 & -4 & -8 & 11 \\ 9 & 0 & 6 & -1 & -3 \end{bmatrix}$$

11. How many rows and columns does the matrix have?

12. Identify the entry in the third row and second column.

13. Identify the entry in the second row and fourth column.

In Exercises 14 and 15, find the sum and difference of the matrices. (14.4)

14. $\begin{bmatrix} 0 & -4 \\ 3 & 6 \end{bmatrix}, \begin{bmatrix} 8 & 7 \\ 4 & -6 \end{bmatrix}$

15. $\begin{bmatrix} -1 & 5 & 9 \\ 2 & -8 & 1 \\ -5 & 4 & -3 \end{bmatrix}, \begin{bmatrix} 3 & -1 & 2 \\ 6 & -5 & 8 \\ 4 & -7 & 0 \end{bmatrix}$

16. Give an example of two matrices that cannot be added together.

In Exercises 17–22, simplify the polynomial. Then state whether the result is a monomial, a binomial, or a trinomial. (14.5)

17. $-10x - 7 + 3x + 7$

18. $6z^2 - 4 - 10z + 1$

19. $8 + 3n^2 + 5n - 7n^2$

20. $-12 - 9y + 6 + 2y$

21. $4t^2 - 6t + t^2 + 9t$

22. $8a - a^3 - 8a + 2a^3$

In Exercises 23 and 24, add the polynomials. (14.6)

23. $(6x^2 - 3x - 7) + (x^2 + 3x - 9)$

24. $(-4n^2 + 6n + 2) + (8n^3 - 14)$

In Exercises 25 and 26, subtract the polynomials. (14.6)

25. $(5n^2 + 2n - 11) - (2n^2 - 6n - 8)$

26. $(2x^3 - 4x^2 + x - 1) - (-7x^2 - x)$

In Exercises 27 and 28, find the perimeter of the polygon. (14.6)

27.

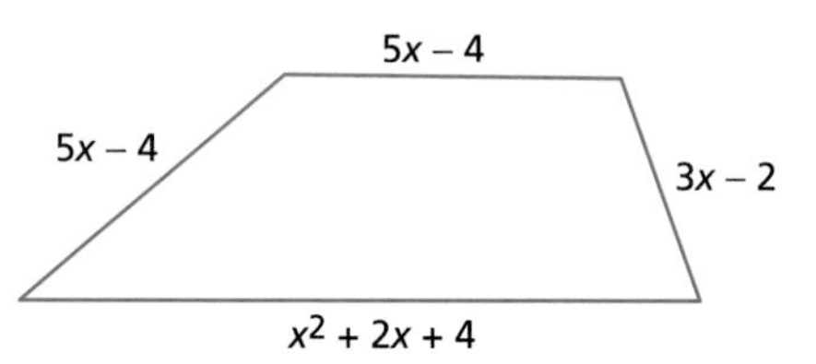

28.

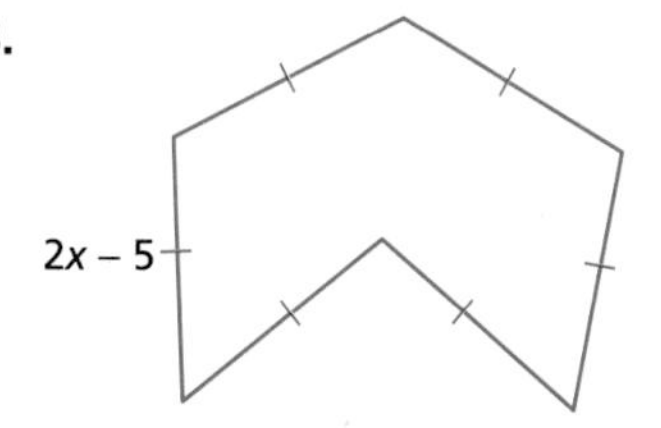

In Exercises 29–34, multiply the polynomials. (14.7, 14.8)

29. $x(x^2 + x)$

30. $(y + 8)(y + 6)$

31. $(2z + 1)(6z + 5)$

32. $n^2(n^3 + 4)$

33. $3t(2t^2 + 4t + 5)$

34. $(4x + 3)(4x + 3)$

Swimming Pool **In Exercises 35–38, use the diagram of the swimming pool at the right. (14.6, 14.7)**

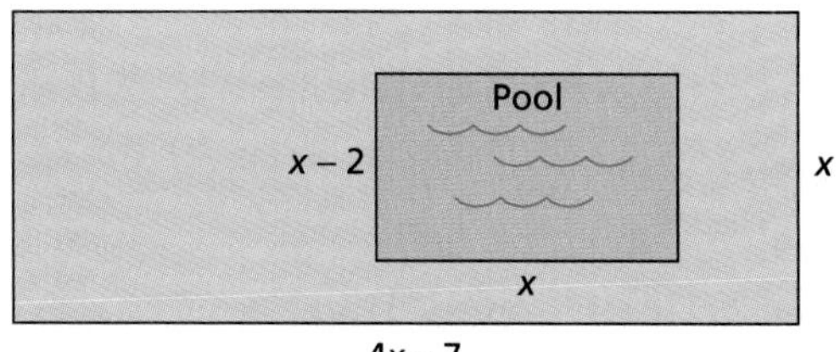

35. Write an expression for the area of the pool.

36. Write an expression for the area of the entire yard, including the pool.

37. Use polynomial subtraction to find an expression for the area of the lawn.

38. The pool is 10 meters by 8 meters. What is the area of the lawn?

In Exercises 39 and 40, write an expression for the area of the entire region. (14.6, 14.8)

39.

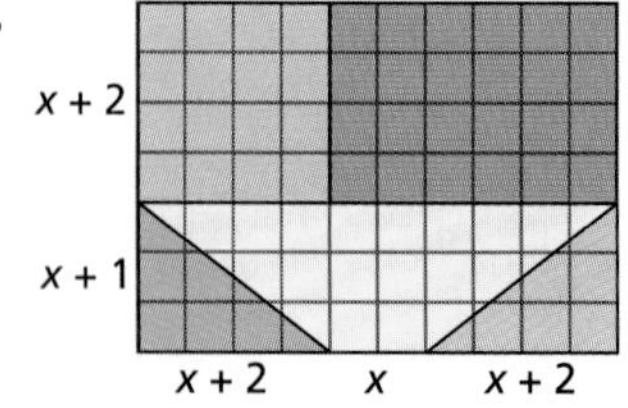

40.

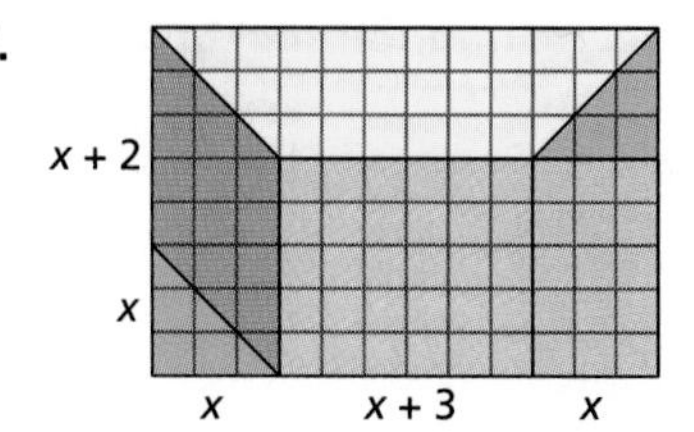

In Exercises 41–58, use the stem-and-leaf plot, the box-and-whisker plot, and the matrix to determine the number for each animal.

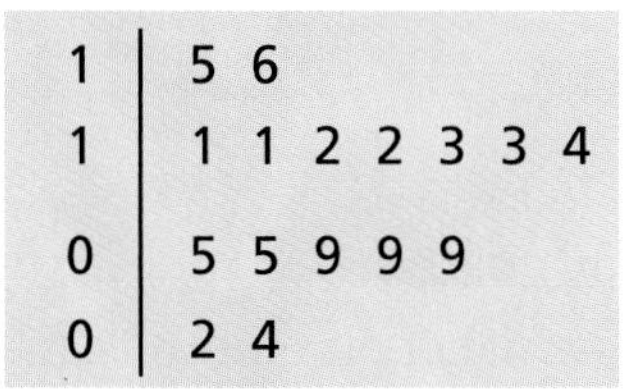

1	5 6
1	1 1 2 2 3 3 4
0	5 5 9 9 9
0	2 4

1 | 5 represents 15

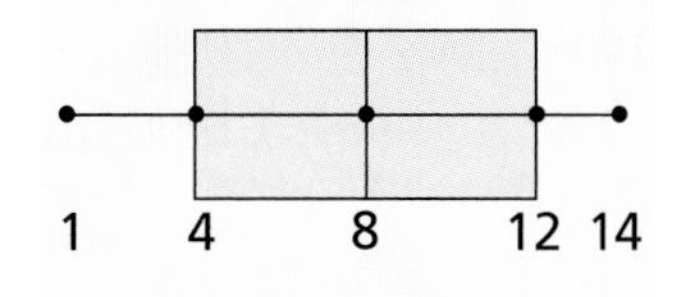

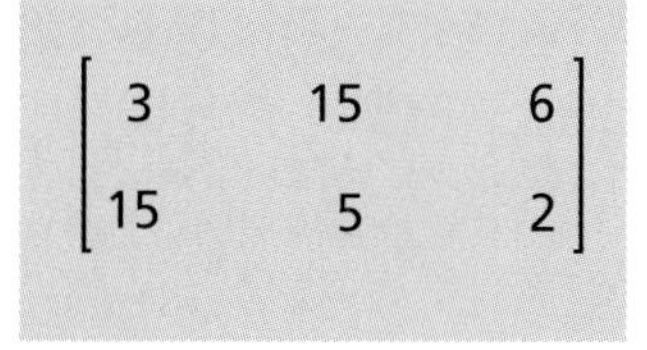

$$\begin{bmatrix} 3 & 15 & 6 \\ 15 & 5 & 2 \end{bmatrix}$$

41. Mean *Rhinoceros*
42. 3rd quartile *Horse*
43. 1st quartile *Bat*
44. Mode *Fish*
45. Maximum *Bear*
46. Median *Platypus*
47. 3rd quartile *Elephant*
48. Minimum *Cat*
49. Median *Giraffe*
50. Maximum *Tiger*
51. 1st quartile *Ostrich*
52. 2nd quartile *Goat*
53. 1st row, 3rd col. *Porcupine*
54. 2nd row, 3rd col. *Rabbit*
55. 2nd row, 2nd col. *Albatross*
56. 1st row, 2nd col. *Dog*
57. 2nd row, 1st col. *Deer*
58. 1st row, 1st col. *Cow*

59. *Animal Facts* Fill in the crossword puzzle (Hint: The answers correspond to the numbers found in Exercises 41–58.)

ACROSS

5. Seabird with 11-foot wingspan
7. Flying mammal
8. Angora breed raised for its wool
11. Mammal with bill and webbed feet
12. Has ears that are 4 ft across
15. Has bones called antlers
16. Weighs $\frac{1}{2}$ lb to 1 lb at birth

DOWN

1. Has keen sense of balance
2. It hops
3. Has stomach with 14 compartments
4. Runs up to 44 mph
6. Has sharp quills
8. Has 18-inch tongue
9. Has no eyelids
10. Has 3 toes on each foot
13. Requires 10–12 gal water per day
14. Adult has a 3-foot long tail
15. First tamed animal

Chapter TEST

In Exercises 1–4, the numbers (in millions) of people camping for the years 1983–1990 are given below. ***(Source: U.S. Bureau of Land Management)***

1983: 84.1	1984: 73.0	1985: 65.3	1986: 95.2
1987: 195.3	1988: 178.7	1989: 173.6	1990: 165.4

1. What is the first quartile?

2. What is the second quartile?

3. What is the third quartile?

4. Create a box-and-whisker plot for the data.

In Exercises 5–7, use {77, 79, 84, 93, 93, 96, 99, 99, 99, 102, 102}

5. What is the mean of the data?

6. What is the median of the data?

7. What is the mode of the data?

In Exercises 8 and 9, find the sum or difference.

8. $\begin{bmatrix} 18 & 12 \\ 4 & 22 \end{bmatrix} + \begin{bmatrix} 6 & 2 \\ 14 & 9 \end{bmatrix}$

9. $\begin{bmatrix} 3 & 18 & 12 \\ 16 & 12 & 15 \\ 12 & 21 & 11 \end{bmatrix} - \begin{bmatrix} 1 & 8 & 4 \\ 18 & 0 & 6 \\ -4 & -21 & 22 \end{bmatrix}$

In Exercises 10 and 11, simplify and write in standard form.

10. $2p^2 - p^3 + p + 2p^3$

11. $3n^3 + 4 - n^2 - n^3 - n$

In Exercises 12 and 13, add or subtract the polynomials and simplify.

12. $(3x^2 + 2x + 5) + (7x^2 - 4x + 2)$

13. $(4p^3 + 6p - 4) - (p^3 - p^2 + 6p - 5)$

In Exercises 14–16, use the stem-and-leaf plot.

14. What is the mean for each group?

15. Construct a box-and-whisker plot for the Group 1 data.

16. Construct a box-and-whisker plot for the Group 2 data.

Group 1		Group 2
0 7 8 5	10	1 4 2 4
3 8 2	9	7 9 2 1 4
6 9 4 1 7	8	7 3 8
4 2 8	7	7 4
7	6	6 2

5|10|1 represents 105 and 101

In Exercises 17 and 18, multiply the polynomials.

17. $2y(3y^2 + 2y + 1)$

18. $(3p + 1)(2p + 3)$

In Exercises 19 and 20, the data in Matrix 1 gives the numbers of dollars deposited into savings accounts at 2 different banks. Matrix 2 shows how much money is in the accounts after 1 year.

19. Write a matrix that shows how much interest was earned in each account.

20. Write a matrix that shows the percent of interest that each account earned.

Deposits		Balances After One Year	
Bank 1	Bank 2	Bank 1	Bank 2
100	150	108	164
450	300	480	325
600	1000	648	1100

Student Handbook

Table of Contents

Reference

THE METRIC SYSTEM

Length

10 millimeters (mm) = 1 centimeter (cm)
10 centimeters = 1 decimeter (dm)
100 centimeters = 1 meter (m)
1000 meters = 1 kilometer (km)
100,000 centimeters = 1 kilometer

Capacity

1000 milliliters (mL) = 1 liter (L)
10 deciliters (dL) = 1 liter

Mass

1000 milligrams (mg) = 1 gram (g)
1000 grams = 1 kilogram (kg)

Area

100 mm^2 = 1 cm^2
100 cm^2 = 1 dm^2
100 dm^2 = 1 m^2
1,000,000 m^2 = 1 km^2

Volume

1000 cm^3 = 1 dm^3
1000 dm^3 = 1 m^3
1 cm^3 = 1 mL
1 dm^3 = 1 L

Converting Units

In the metric system, the units are related by powers of 10.

Table of Units

Prefix	Power of 10	Length	Capacity	Mass
kilo (k)	1000 units	kilometer	kiloliter*	kilogram
hecto (h)	100 units	hectometer*	hectoliter*	hectogram*
deka (dk)	10 units	dekameter*	dekaliter*	dekagram*
——	1 unit	meter	liter	gram
deci (d)	0.1 unit	decimeter*	deciliter*	decigram*
centi (c)	0.01 unit	centimeter	centiliter*	centigram*
milli (m)	0.001 unit	millimeter	milliliter	milligram

*These units are seldom used.

Study Tip...

When you change from a larger unit to a smaller unit, you multiply.

0.24 m = [?] cm

In the Table of Units, there are 2 steps from meters to centimeters, so multiply by 10^2, or 100.

$0.24 \times 100 = 24$

0.24 m = 24 cm

Study Tip...

When you change from a smaller unit to a larger unit, you divide.

3500 mg = [?] kg

In the Table of Units, there are 6 steps from milligrams to kilograms, so divide by 10^6, or 1,000,000.

$3500 \div 1{,}000{,}000 = 0.0035$

3500 mg = 0.0035 kg

Reference

THE CUSTOMARY SYSTEM

Length

12 inches (in.) = 1 foot (ft)
3 feet = 1 yard (yd)
36 inches = 1 yard
5280 feet = 1 mile (mi)
1760 yards = 1 mile

Capacity

1 cup (c) = 8 fluid ounces (fl oz)
2 cups = 1 pint (pt)
2 pints = 1 quart (qt)
2 quarts = 1 half-gallon
4 quarts = 1 gallon

Weight

16 ounces (oz) = 1 pound (lb)
2000 pounds = 1 ton

Area

144 in.^2 = 1 ft^2
9 ft^2 = 1 yd^2
640 acres = 1 square mile

Volume

1728 in.^3 = 1 ft^3
27 ft^3 = 1 yd^3

Time

60 seconds (sec) = 1 minute (min)
3600 seconds = 1 hour (hr)
60 minutes = 1 hour
24 hours = 1 day
7 days = 1 week
360 days = 1 business year
365 days = 1 year
366 days = 1 leap year
10 years = 1 decade
10 decades = 1 century = 100 years

Unit Analysis

When you convert from one unit to another, you can use conversion fractions.

Study Tip...

To change 45 miles per hour to feet per second:

$$\frac{45 \text{ miles}}{\text{hour}} \times \frac{5280 \text{ feet}}{1 \text{ mile}} \times \frac{1 \text{ hour}}{3600 \text{ seconds}} = \left(\frac{45 \times 5280}{3600}\right)\frac{\text{feet}}{\text{second}} = 66 \text{ ft/sec}$$

Study Tip...

To change 120 ounces per square inch to pounds per square foot:

$$\frac{120 \text{ oz}}{\text{in}^2} \times \frac{144 \text{ in.}^2}{1 \text{ ft}^2} \times \frac{1 \text{ lb}}{16 \text{ oz}} = \left(\frac{120 \times 144}{16}\right)\frac{\text{lb}}{\text{ft}^2} = 1080 \text{ lb/ft}^2$$

USING THE TABLES

Table of Squares and Approximate Square Roots

Find 54^2 and $\sqrt{54}$.

Find 54 in the column labeled ***n***. Read across on that line to the columns labeled $\boldsymbol{n^2}$ and $\boldsymbol{\sqrt{n}}$.

n	n^2	$\sqrt{n}$
51	2601	7.141
52	2704	7.211
53	2809	7.280
54	2916	7.348
55	3025	7.416
56	3136	7.483

$54^2 = 2916$ $\sqrt{54} \approx 7.348$

Table of Trigonometric Ratios

Find sin 48°, cos 48°, and tan 48°.

Find 48° in the **Angle** column. Read across on that line to the columns labeled **Sine, Cosine,** and **Tangent.**

Angle	Sine	Cosine	Tangent
46°	0.7193	0.6947	1.0355
47°	0.7314	0.6820	1.0724
48°	0.7431	0.6691	1.1106
49°	0.7547	0.6561	1.1504
50°	0.7660	0.6428	1.1918

$\sin 48° \approx 0.7431$ $\cos 48° \approx 0.6691$ $\tan 48° \approx 1.1106$

TABLE OF SQUARES AND APPROXIMATE SQUARE ROOTS

n	n^2	$\sqrt{n}$	n	n^2	$\sqrt{n}$
1	1	1.000	**51**	2601	7.141
2	4	1.414	**52**	2704	7.211
3	9	1.732	**53**	2809	7.280
4	16	2.000	**54**	2916	7.348
5	25	2.236	**55**	3025	7.416
6	36	2.449	**56**	3136	7.483
7	49	2.646	**57**	3249	7.550
8	64	2.828	**58**	3364	7.616
9	81	3.000	**59**	3481	7.681
10	100	3.162	**60**	3600	7.746
11	121	3.317	**61**	3721	7.810
12	144	3.464	**62**	3844	7.874
13	169	3.606	**63**	3969	7.937
14	196	3.742	**64**	4096	8.000
15	225	3.873	**65**	4225	8.062
16	256	4.000	**66**	4356	8.124
17	289	4.123	**67**	4489	8.185
18	324	4.243	**68**	4624	8.246
19	361	4.359	**69**	4761	8.307
20	400	4.472	**70**	4900	8.367
21	441	4.583	**71**	5041	8.426
22	484	4.690	**72**	5184	8.485
23	529	4.796	**73**	5329	8.544
24	576	4.899	**74**	5476	8.602
25	625	5.000	**75**	5625	8.660
26	676	5.099	**76**	5776	8.718
27	729	5.196	**77**	5929	8.775
28	784	5.292	**78**	6084	8.832
29	841	5.385	**79**	6241	8.888
30	900	5.477	**80**	6400	8.944
31	961	5.568	**81**	6561	9.000
32	1024	5.657	**82**	6724	9.055
33	1089	5.745	**83**	6889	9.110
34	1156	5.831	**84**	7056	9.165
35	1225	5.916	**85**	7225	9.220
36	1296	6.000	**86**	7396	9.274
37	1369	6.083	**87**	7569	9.327
38	1444	6.164	**88**	7744	9.381
39	1521	6.245	**89**	7921	9.434
40	1600	6.325	**90**	8100	9.487
41	1681	6.403	**91**	8281	9.539
42	1764	6.481	**92**	8464	9.592
43	1849	6.557	**93**	8649	9.644
44	1936	6.633	**94**	8836	9.695
45	2025	6.708	**95**	9025	9.747
46	2116	6.782	**96**	9216	9.798
47	2209	6.856	**97**	9409	9.849
48	2304	6.928	**98**	9604	9.899
49	2401	7.000	**99**	9801	9.950
50	2500	7.071	**100**	10000	10.000

TABLE OF TRIGONOMETRIC RATIOS

Angle	Sine	Cosine	Tangent
1°	0.0175	0.9998	0.0175
2°	0.0349	0.9994	0.0349
3°	0.0523	0.9986	0.0524
4°	0.0698	0.9976	0.0699
5°	0.0872	0.9962	0.0875
6°	0.1045	0.9945	0.1051
7°	0.1219	0.9925	0.1228
8°	0.1392	0.9903	0.1405
9°	0.1564	0.9877	0.1584
10°	0.1736	0.9848	0.1763
11°	0.1908	0.9816	0.1944
12°	0.2079	0.9781	0.2126
13°	0.2250	0.9744	0.2309
14°	0.2419	0.9703	0.2493
15°	0.2588	0.9659	0.2679
16°	0.2756	0.9613	0.2867
17°	0.2924	0.9563	0.3057
18°	0.3090	0.9511	0.3249
19°	0.3256	0.9455	0.3443
20°	0.3420	0.9397	0.3640
21°	0.3584	0.9336	0.3839
22°	0.3746	0.9272	0.4040
23°	0.3907	0.9205	0.4245
24°	0.4067	0.9135	0.4452
25°	0.4226	0.9063	0.4663
26°	0.4384	0.8988	0.4877
27°	0.4540	0.8910	0.5095
28°	0.4695	0.8829	0.5317
29°	0.4848	0.8746	0.5543
30°	0.5000	0.8660	0.5774
31°	0.5150	0.8572	0.6009
32°	0.5299	0.8480	0.6249
33°	0.5446	0.8387	0.6494
34°	0.5592	0.8290	0.6745
35°	0.5736	0.8192	0.7002
36°	0.5878	0.8090	0.7265
37°	0.6018	0.7986	0.7536
38°	0.6157	0.7880	0.7813
39°	0.6293	0.7771	0.8098
40°	0.6428	0.7660	0.8391
41°	0.6561	0.7547	0.8693
42°	0.6691	0.7431	0.9004
43°	0.6820	0.7314	0.9325
44°	0.6947	0.7193	0.9657
45°	0.7071	0.7071	1.0000

Angle	Sine	Cosine	Tangent
46°	0.7193	0.6947	1.0355
47°	0.7314	0.6820	1.0724
48°	0.7431	0.6691	1.1106
49°	0.7547	0.6561	1.1504
50°	0.7660	0.6428	1.1918
51°	0.7771	0.6293	1.2349
52°	0.7880	0.6157	1.2799
53°	0.7986	0.6018	1.3270
54°	0.8090	0.5878	1.3764
55°	0.8192	0.5736	1.4281
56°	0.8290	0.5592	1.4826
57°	0.8387	0.5446	1.5399
58°	0.8480	0.5299	1.6003
59°	0.8572	0.5150	1.6643
60°	0.8660	0.5000	1.7321
61°	0.8746	0.4848	1.8040
62°	0.8829	0.4695	1.8807
63°	0.8910	0.4540	1.9626
64°	0.8988	0.4384	2.0503
65°	0.9063	0.4226	2.1445
66°	0.9135	0.4067	2.2460
67°	0.9205	0.3907	2.3559
68°	0.9272	0.3746	2.4751
69°	0.9336	0.3584	2.6051
70°	0.9397	0.3420	2.7475
71°	0.9455	0.3256	2.9042
72°	0.9511	0.3090	3.0777
73°	0.9563	0.2924	3.2709
74°	0.9613	0.2756	3.4874
75°	0.9659	0.2588	3.7321
76°	0.9703	0.2419	4.0108
77°	0.9744	0.2250	4.3315
78°	0.9781	0.2079	4.7046
79°	0.9816	0.1908	5.1446
80°	0.9848	0.1736	5.6713
81°	0.9877	0.1564	6.3138
82°	0.9903	0.1392	7.1154
83°	0.9925	0.1219	8.1443
84°	0.9945	0.1045	9.5144
85°	0.9962	0.0872	11.4301
86°	0.9976	0.0698	14.3007
87°	0.9986	0.0523	19.0811
88°	0.9994	0.0349	28.6363
89°	0.9998	0.0175	57.2900

SYMBOLS

Arithmetic and Algebra

$=$	Is equal to
$\neq$	Is not equal to
$>$	Is greater than
$<$	Is less than
$\geq$	Is greater than or equal to
$\leq$	Is less than or equal to
$\approx$	Is approximately equal to
ab or $a(b)$	a times b
a^n	A number a raised to the nth power
a^{-n}	$\frac{1}{a^n}$
a^0	1
( ), { }, [ ]	Grouping symbols
5 or +5	Positive 5
−5	Negative 5
$\lvert a \rvert$	Absolute value of a number a
$\sqrt{a}$	The principal (positive) square root of a number a
$a : b$ or $\frac{a}{b}$	Ratio of a to b
$P(A)$	Probability of the outcome A
$n!$	n-factorial

Geometry

(a, b)	Ordered pair a, b
$\sim$	Is similar to
$\cong$	Is congruent to
$\triangle ABC$	Triangle ABC
$\overleftrightarrow{AB}$	Line AB
$\overline{AB}$	Segment AB
AB	Measure of segment AB
$\overrightarrow{AB}$	Ray AB
$\angle A$	Angle A
$m\angle A$	Measure of angle A
π	Pi

Right Triangle

$\sin A$	Sine of angle A
$\cos A$	Cosine of angle A
$\tan A$	Tangent of angle A

Reference

FORMULAS

Miscellaneous:

$d = rt$	Distance formula (25)
$F = \frac{9}{5}C + 32$	Temperature conversion to degrees Fahrenheit (123)
$C = \frac{5}{9}(F - 32)$	Temperature conversion to degrees Celsius (589)
$A = P(1 + r)^n$	The balance in a savings account (331)
$d^2 = \frac{3}{2}h$	Distance to the horizon (393)
$0 = -16t^2 + s$	Falling objects (393)
$a^2 + b^2 = c^2$	Pythagorean Theorem (400)

Perimeter:

$P = 4s$	Perimeter of a square (12)
$P = 2(l + w)$	Perimeter of a rectangle (16)
$C = \pi d$ or $C = 2\pi r$	Circumference of a circle (539)

Area:

$A = s^2$	Area of a square (12)
$A = lw$	Area of a rectangle (25)
$A = bh$	Area of a parallelogram (488)
$A = \frac{1}{2}bh$	Area of a triangle (37)
$A = \frac{1}{2}h(b_1 + b_2)$	Area of a trapezoid (45)
$A = \pi r^2$	Area of a circle (540)

Surface Area:

$S = 6s^2$	Surface area of a cube (14)
$S = 2B + Ph$	Surface area of a prism (549)
$S = 2B + Ch$	Surface area of a cylinder (549)
$S = 4\pi r^2$	Surface area of a sphere (572)

Volume:

$V = lwh$	Volume of a rectangular prism (554)
$V = s^3$	Volume of a cube (14)
$V = Bh$	Volume of a prism (554)
$V = \pi r^2 h$	Volume of a cylinder (559)
$V = \frac{1}{3}Bh$	Volume of a pyramid (564)
$V = \frac{1}{3}\pi r^2 h$	Volume of a cone (564)
$V = \frac{4}{3}\pi r^3$	Volume of a sphere (569)

Coordinate Geometry:

$m = \frac{y_2 - y_1}{x_2 - x_1}$	The Slope of a Line (604)
$d = \sqrt{(x_2 - x_1)^2 + (y_2 - y_1)^2}$	The Distance Formula (623)
$\text{Midpoint} = \left(\frac{x_1 + x_2}{2}, \frac{y_1 + y_2}{2}\right)$	The Midpoint Formula (624)

HINTS FOR ESTIMATING

Estimation

When should you estimate rather than find an exact answer?

- An estimate can solve a *yes* or *no* problem.

  Do I have enough money to buy these items?

  Did I really get a 25% savings on my total bill?

- An estimate can be used when you want to know about how long or about how much.

  How much of a tip should I leave?

  About how many miles can I drive on a full tank of gasoline?

- An estimate can be used to eliminate possibilities in multiple-choice situations.

  Choose the best answer.
  $\frac{7}{8}+\frac{13}{14}$ **a.** 1 **b.** 2 **c.** $1\frac{1}{2}$

  <, >, or = ?
  49% of 37 ? 15

- An estimate can be used to check the reasonableness of an answer, especially when using calculators.

  I used a calculator to find $0.17\overline{)1.3685}$ and I got 80.5 as an answer.

  Did I make any mistakes in entering the numbers?

Here are some examples that can help you decide when and how to estimate.

Estimating Sums and Differences

A quick and easy method of estimating some sums is clustering.

Marla's Test Scores

79.1 75.6 82.3
85.0 88.2 69.7

Grades:

A 570+
B 510+
C 450

What score did Marla get?

Example 1 ***Estimating by Clustering***

Estimate.

79.1 + 75.6 + 82. 3 + 85.0 + 88.2 + 69.7

Solution If all the values you want to add or subtract are close to each other, try clustering. Since 79, 75, and 69 are all less than 80, and 82, 85, and 88 are all greater than 80, you can reasonably say all the numbers in the list are close to 80. So, the sum is close to $6 \cdot 80 = 480$. Since the actual sum is 479.9, the estimate is very reasonable. ■

Warning! Clustering is a fast and easy method for estimating sums, but clustering is only valid when the numbers are *really* close to the number you pick! For most estimating situations, you will have to use combinations of the following methods.

Example 2 ***Estimating by Rounding***

Estimate. 52,764 − 36,195 + 7,326

Solution Rounding is often the best method to use for estimating sums and differences. For this problem, round each number to the nearest thousand:

53,000 − 36,000 + 7,000

Regroup the numbers to be able to use mental math to find the total:

53,000 − 36,000 + 7,000 = 53,000 + 7,000 − 36,000

= 60,000 − 36,000 = 24,000

The actual value is 23,895, so the estimate is reasonable. ■

Example 3 ***Front-End Estimation with Adjustments***

Estimate. \$20.00 − (\$3.46 + \$ 8.95 + \$1.33 + \$0.32)

Solution For many addition-subtraction situations, front-end estimating with adjustment produces the most accurate estimates.

Start at the front-end. Estimate the sum using only the front-end (in this case, the dollars) digits:

\$3 + \$8 + \$1 + \$0 = \$12

Look at the rest of the values. \$0.46 + \$0.33 + \$0.32 is a little more than \$1.00. \$0.95 is a little less than \$1.00.

Put them together. The estimated sum is \$12 + \$1 + \$1 =\$14.

So, the estimated value is \$20 − \$14 = \$6.

The actual value is \$5.94, so the estimate is reasonable. ■

******Sales Receipt******

ARNIE'S ART GOODS

\$3.46

\$8.95

\$1.33

\$0.32

Can I buy a poster that costs \$3.95 with the change from a \$20 bill?

Estimating Products and Quotients

Example 1 ***Estimating Products by Rounding***

Estimate. 152.6 • 48.9

Solution Rounding is the most common strategy used in estimating products. For this example, round each value to the nearest 10.

Estimate: 150 • 50 = 7500

Actual value: 7462.14 ■

<, >, or =?

152.6 • 48.9 [?] 8000

Example 2 ***Rounding and Adjusting***

Estimate. **a.** $25.05 \cdot 36.8$ **b.** $34.1 \cdot 63.5$

Solution By rounding, the estimates are **a.** $30 \cdot 40 = 1200$ and **b.** $30 \cdot 60 = 1800$. Notice that in part **a**, both factors are rounded up, so you know the estimate is too large. In part **b**, both factors are rounded down, so you know the estimate is too small.

Actual values: **a.** 921.84 **b.** 2165.35

When you use rounding, you should not be locked into rigid rules. There are many ways to get more accurate estimates. For example, in part **a** above, it helps to think of 25 as $\frac{100}{4}$. So, $25.05 \cdot 36.8 \approx \frac{100}{4} \cdot 36 = 100 \cdot 9 = 900$. How would you adjust the estimate in part **b**? Try some methods of your own.

Choose the best estimates.

$25.05 \cdot 36.8$

a. 500 **b.** 1000 **c.** 1500

$34.1 \cdot 63.5$

a. 1800 **b.** 2400 **c.** 3000

$82\overline{)29,581}$

a. 40 **b.** 400 **c.** 4000

$189.56 \div 0.93$

a. 20 **b.** 200 **c.** 2000

■

Example 3 ***Estimating Quotients***

Estimate. **a.** $82\overline{)29,581}$ **b.** $189.56 \div 0.93$

Solution **a.** If you are using estimating in division only to help you find a "first guess," then use front-end estimating. Think $8\overline{)29} \approx 3^{+}$, so $82\overline{)29,581} \approx 300^{+}$.

If you need a more accurate estimate, consider a combination of techniques. Start off by rounding the divisor to 80. You could round 29,581 up to 30,000. But it would be difficult to divide 30,000 by 80 mentally. Consider using "compatible' or "nice" numbers, that is, numbers you can compute mentally.

Multiples of 8: 8, 16, 24, 32, 40, etc.

Think: 29,581 is closest to 32,000.

So, $82\overline{)29,581} \approx 80\overline{)32,000} = 400$, which is a reasonable estimate when compared to the actual answer of 360 R 61.

b. When the problem involves decimals, you need to consider whether to make the divisor and dividend whole numbers first and then use compatible numbers , or whether you can just estimate from the given problem

$0.93\overline{)189.56} = 93\overline{)18,956} \approx 90\overline{)18,000} = 200$ or $0.93\overline{)189.56} \approx 1\overline{)190} = 190$.

The actual answer is ≈ 203.83, so both estimates are reasonable. ■

Estimating with Fractions and Percents

When multiplying with rational numbers and percents, the easiest and fastest method of estimating is to substitute compatible numbers that are easy to work with mentally.

Example 1 *Estimating Products*

Estimate. **a.** $\frac{3}{7}$ of \$28.98. **b.** 26% of 239.

Solution **a.** \$28.98 is close to \$28. $\frac{1}{7}$ of $\$28 = \$4$.
So, $\frac{3}{7}$ of $\$28.98 \approx \$12.00$.
b. 26% is close to 25%, which equals $\frac{1}{4}$. 239 is close to 240.
So, 26% of $239 \approx \frac{1}{4}$ of $240 = 60$.
The actual answers are **a.** \$12.42 and **b.** 62.14. The estimates are reasonable. ■

When estimating sums and differences of fractions, it is often convenient to use the numbers 0, $\frac{1}{2}$, and 1 as benchmarks.

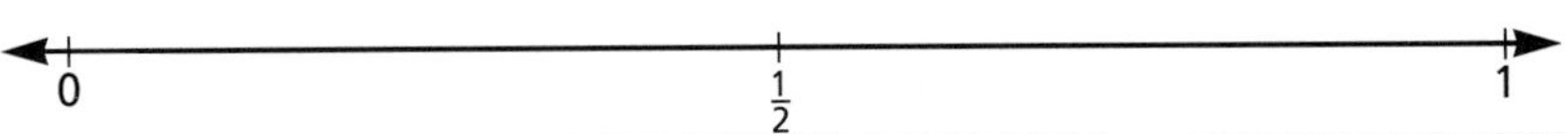

- A fraction is close to 0 if its numerator is very small compared to its denominator.
- A fraction is close to $\frac{1}{2}$ if its numerator is about half its denominator.
- A fraction is close to 1 if its numerator and denominator are about equal.

(You can draw number lines or use fraction strips to help you decide whether a fraction is about 0, about $\frac{1}{2}$ or about 1.)

Example 2 *Estimating Sums and Differences of Fractions*

Estimate. **a.** $\frac{9}{10} + \frac{1}{6} + \frac{2}{5}$ **b.** $8\frac{4}{5} - 1\frac{2}{3} - \frac{7}{9}$

Solution **a.** $\frac{9}{10}$ is about 1, $\frac{1}{6}$ is about 0, and $\frac{2}{5}$ is about $\frac{1}{2}$. So, $\frac{9}{10} + \frac{1}{6} + \frac{2}{5}$ $\approx 1 + 0 + \frac{1}{2} = 1\frac{1}{2}$, which is close to the actual answer of $1\frac{7}{15}$.

b. When you estimate with mixed numbers, you can separate the whole number parts from the fraction parts, or you can consider the numbers as a whole and use front-end estimation. In this problem, all the fraction parts are close to 1, so a first estimate is $9 - 2 - 1 = 6$. Since both of the subtracted numbers were rounded up, and since $\frac{2}{3}$ is not very close to 1, you might consider refining your estimate by rounding $\frac{2}{3}$ down to $\frac{1}{2}$.

So, $8\frac{4}{5} - 1\frac{2}{3} - \frac{7}{9} \approx 9 - 1\frac{1}{2} - 1 = (9 - 1) - 1\frac{1}{2} = 8 - 1\frac{1}{2} = 6\frac{1}{2}$ which is closer to the actual answer of $6\frac{16}{45}$. ■

GLOSSARY

absolute value (101) The distance between the number and 0 on a number line. For example, $|3| = 3$, $|-3| = 3$, $|0| = 0$.

abundant number (279) A natural number with the sum of its factors, except itself, greater than the number.

acute angle (443) An angle whose measure is between 0° and 90°.

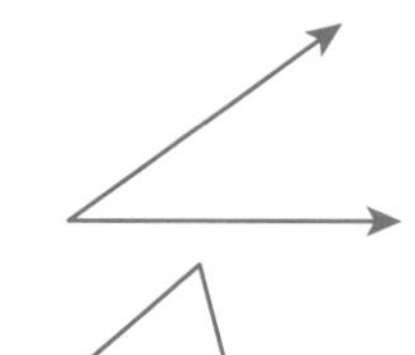

acute triangle (458) A triangle with three acute angles.

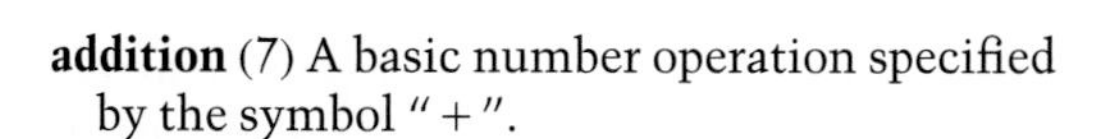

addition (7) A basic number operation specified by the symbol "+".

Addition Property of Equality (65) Adding the same number to each side of an equation produces an equivalent equation. If $a = b$, then $a + c = b + c$.

adjacent side (522) In $\triangle ABC$, side $\overline{AC}$ is adjacent to $\angle A$.

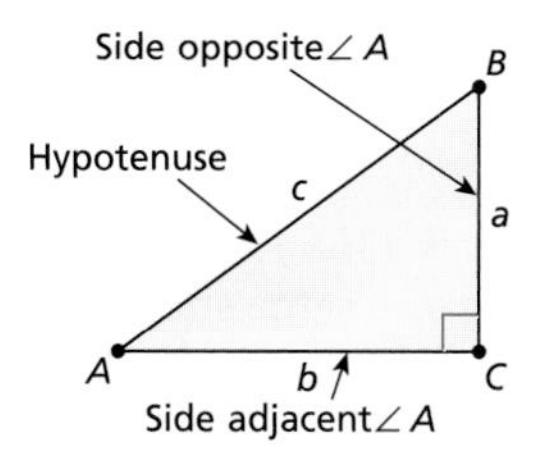

algebraic model (25) An algebraic expression or equation used to represent a real-life situation.

alternate interior angles (451) Angles that lie between lines l and m on opposite sides of a line. For example, angles 3 and 6.

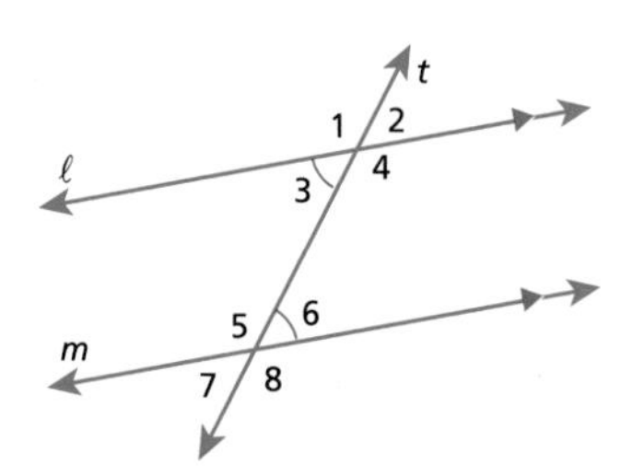

angle (36, 443) A figure consisting of two different rays that begin at the same point. The rays are the *sides* of the angle and the point is the *vertex* of the angle.

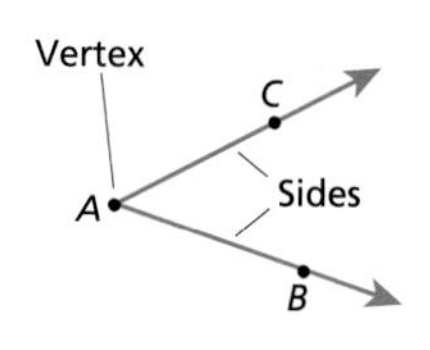

approximate (12) To express a numerical value to different degrees of accuracy. For example, $\sqrt{2}$ can be approximated as 1.41.

area (8) A measure of how much surface is covered by a figure. Areas are measured in square units.

arithmetic mean (127) Another name for the average.

Associative Property of Addition (66) Changing the grouping of the addends does not change the sum.
For example, $(a + b) + c = a + (b + c)$.

Associative Property of Multiplication (66) Changing the grouping of the factors does not change the product. For example, $(ab)c = a(bc)$.

average (127, 634) The sum of the numbers in a list divided by the *number* of numbers in the list.

bar graph (29, 199) A graph that organizes a collection of data by using horizontal or vertical bars to display how many times each event or number occurs in the collection.

base of a percent (358) The number from which a portion is to be found, for example, in the percent equation $\frac{a}{b} = \frac{p}{100}$, the base is b.

base of a power (11) The number or expression that is used as a factor in the repeated multiplication. For example, in the expression 4^6, 4 is the base.

binomial (654, 669) A polynomial that has two terms.

bisect (497) Divide into two equal parts.

box-and-whisker plot (642) A graphical display that uses a box to represent the middle quartiles of a set of data and segments drawn to the extremes at both ends to display the first and fourth quartiles.

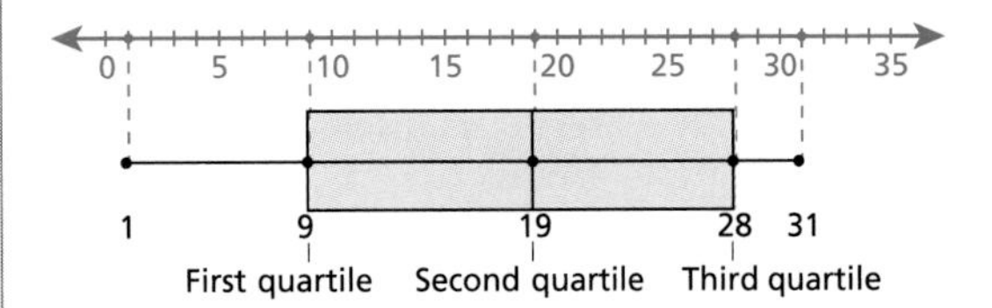

canceling (71, 259) Using slashes to indicate dividing a numerator and a denominator by a common factor.

center of a circle (538, 539) The point inside the circle that is the same distance from all points on the circle.

certain event (229) An event with a probability of 1.

circle (538) The set of all points in a plane that are the same distance from a given point called the center.

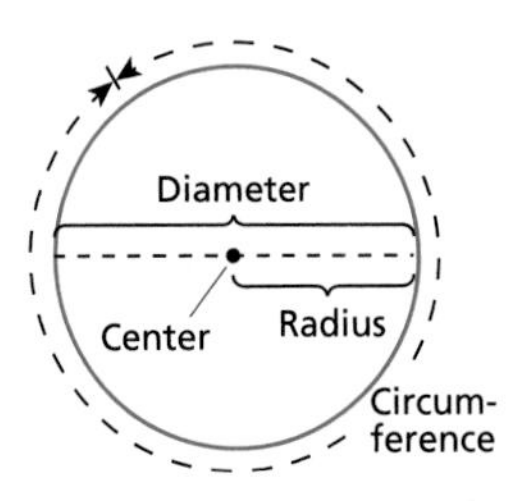

circle graph (300, 323) A graph that displays portions of data collections as parts of a circular region. The parts are often labeled using fractions or percents.

circumference (539) The distance around a circle.

coefficient (110, 654) The numerical factor of an algebraic term. For example, in the term $3x^2$, the coefficient of x^2 is 3.

collecting like terms (55) A procedure for simplifying the sum and difference of like terms in algebraic expressions.

common factor (71, 250) A number that is a factor of two or more numbers. For example, 2 is a common factor of 4 and 6 because 2 is a factor of both 4 and 6.

common multiple (255) A number that is a multiple of two or more numbers. For example, 60 is a common multiple of 5 and 6 because it is a multiple of both 5 and 6.

Commutative Property of Addition (55) Changing the order of the addends does not change the sum. For example, $a + b = b + a$.

Commutative Property of Multiplication (51) Changing the order of the factors does not change the product. For example, $ab = ba$.

complementary angles (155) Two angles whose measures have a sum of 90°.

composite number (245) A natural number that has three or more factors.

concave polygon See *nonconvex polygon.*

conditional equations (59) Equations that are not true for all values of the variables they contain.

cone (545) A solid that has a circular base, a vertex, and a lateral surface.

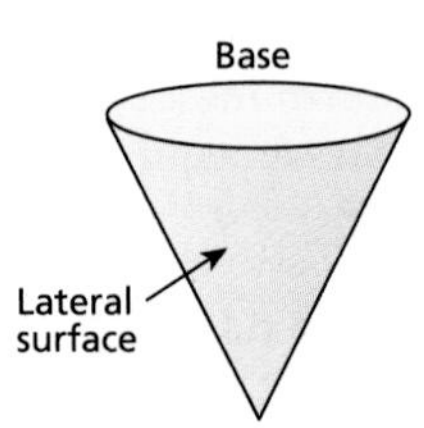

congruent angles (443) Angles that have the same measure.

congruent figures (356) Figures having exactly the same size and same shape.

congruent polygons (467) Two polygons that are exactly the same size and the same shape.

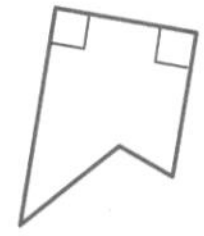

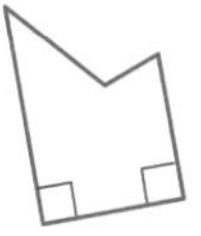

Converse of the Pythagorean Theorem (408) In a triangle, if the sum of the squares of two side lengths is equal to the square of the third side length, then the triangle is a right triangle.

convex polygon (37) A polygon is convex if a segment joining any two interior points lies completely within the polygon.

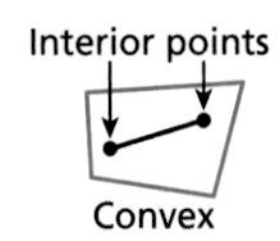

coordinate plane (135) A plane formed by two real number lines called axes that intersect at a right angle; a plane used for locating a point whose coordinates are known.

coordinates (135) An ordered pair of numbers that locate a point on a coordinate graph.

cosine of an angle (522) The cosine of an acute angle in a right triangle is the ratio of the length of the leg adjacent to the acute angle to the length of the hypotenuse. For example, $\cos A = \frac{b}{c}$.

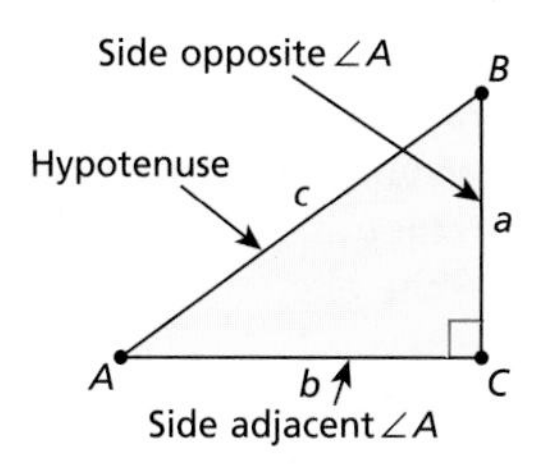

Counting Principle (373) If one event can occur in m ways and another event can occur in n ways, then the two events can occur in mn ways.

Cross-Product Property (349) For two ratios, if $\frac{a}{b} = \frac{c}{d}$, then $ad = bc$.

cross section (487) The exposed surface of a solid cut by a plane.

cube (14) A rectangular prism whose six faces are congruent squares.

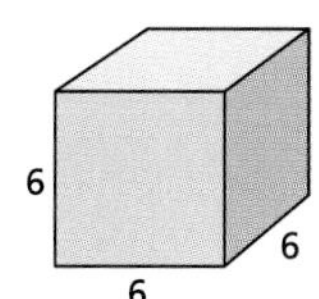

cubic numbers (41) Numbers in the set 1, 8, 27, . . . , n^3.

cubic units (554) Standard measures of volume.

cylinder (545) A solid figure with congruent circular bases that lie in parallel planes.

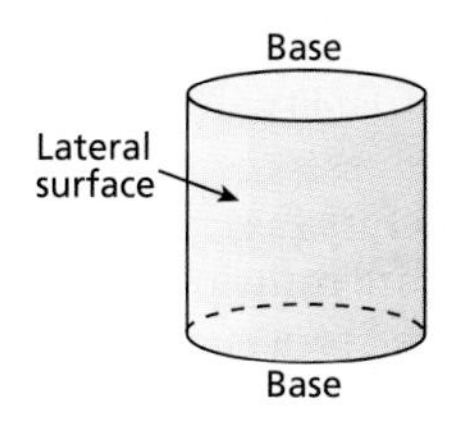

data (28) The facts, or numbers, that describe something.

decagon (34) A polygon that has ten sides.

deficient number (279) A natural number with the sum of its factors, except itself, less than the number.

degree of a polynomial (659) The largest exponent of a polynomial.

diagonal (35) A segment that connects two vertices of a polygon and is not a side.

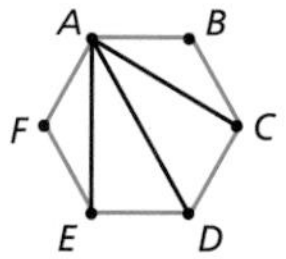

Three diagonals of this hexagon are $\overline{AC}$, $\overline{AD}$, and $\overline{AE}$.

diameter (538) The distance across the circle through its center. The length of the diameter is twice the length of the radius.

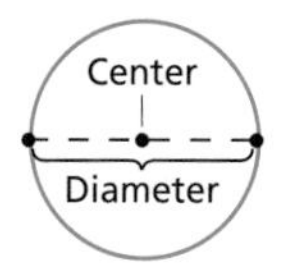

difference (7, 24) The result obtained when numbers or expressions are subtracted. The difference of a and b is $a - b$.

digits (10) The basic symbols used to write numerals. In our base-ten decimal system, the digits are 0, 1, 2, 3, 4, 5, 6, 7, 8, and 9.

dimensions (16) The measure of the magnitude or size of an object. For example, the dimensions of a rectangle are its length and its width.

discount (364) The difference between the regular price and the sale price of an item.

Distributive Property (51) The product of a number and the sum of two numbers is equal to the sum of the two products. For example, $a(b + c) = ab + ac$ and $ab + ac = a(b + c)$.

Dividing Powers Property (270) To divide two powers with the same base, subtract the exponent of the denominator from the exponent of the numerator, that is, $\frac{a^m}{a^n} = a^{m-n}$.

divisible (8, 240) One natural number is divisible by another natural number if the second divides evenly into the first, that is, if there is a 0 remainder after division. For example, 84 is divisible by 2, since 84 ÷ 2 leaves no remainder.

Division Property of Equality (71) Dividing both sides of an equation by the same nonzero number produces an equivalent equation. If $a = b$, then $\frac{a}{c} = \frac{b}{c}$.

edge (14, 544) The segment formed when two faces of a solid figure meet.

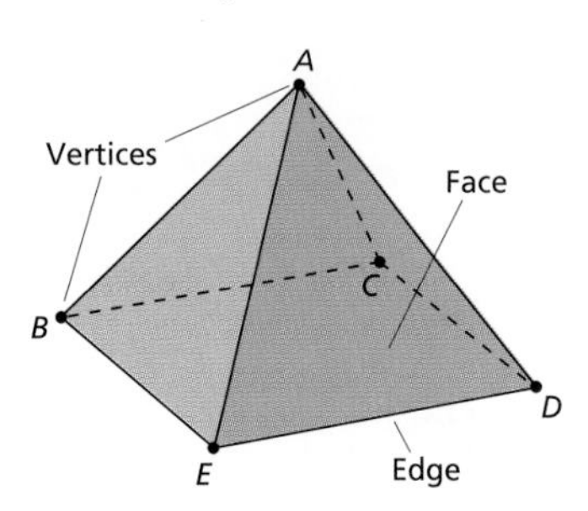

endpoint (34) The point at the end of a line segment or ray.

equally likely outcomes (229) Two or more possible outcomes of a given situation that have the same probability.

equal matrices (648) Two matrices are equal if all the entries in corresponding positions are equal.

equation (59) A statement formed when an equality symbol is placed between two expressions. For example, $3 \times 9 = 27$ and $8 + x = 10$ are equations.

equiangular triangle (458) A triangle in which all three angles have the same measure.

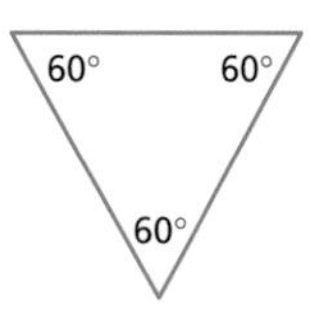

equilateral triangle (170, 457) A triangle in which all three sides have the same length.

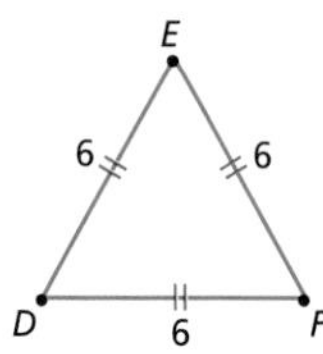

equivalent equations (59) Equations that have the same solutions.

equivalent expressions (51) Expressions that have the same values when numbers are substituted for the variables.

equivalent fractions (259) Fractions that have the same decimal form: $\frac{1}{2}, \frac{2}{4}, \frac{3}{6}$ are equivalent fractions because each is equal to 0.5.

evaluate an algebraic expression (24) To find the value of an expression by replacing each variable in an expression with numbers.

evaluate a numerical expression (17) To perform operations to obtain a single number or value.

even number (10) A whole number that is divisible by 2.

exponent (11) A number or variable that represents the number of times the base is used as a factor. For example, in the expression 2^3, 3 is the exponent.

expression (17, 24) A symbol or combination of symbols that represent a mathematical relationship.

exterior angles (472) Exterior angles are formed when the sides of a polygon are extended.

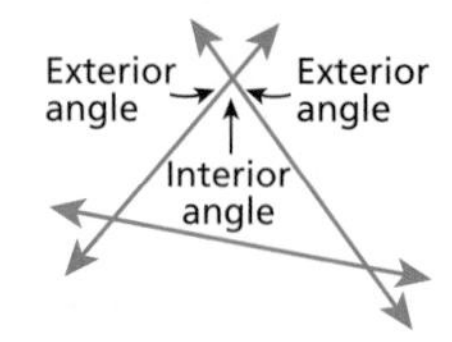

face (14, 544) The flat surface of a solid figure; one of the polygons that make up a polyhedron.

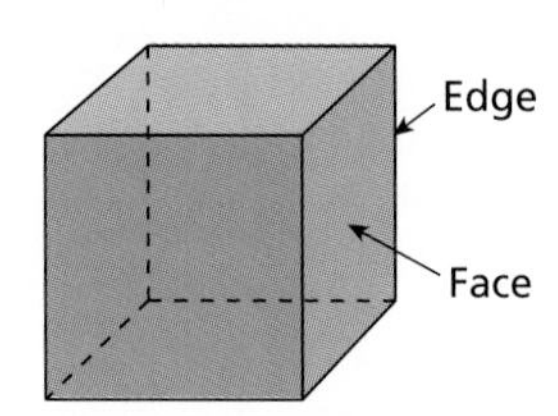

factored form (241) A natural number is factored when it is written as the product of two or more natural numbers.

factorial (376) The product of the numbers from 1 to n is called n-factorial and is denoted by $n!$.

factors (24) Numbers or variable expressions that are multiplied. For example, 4 and x are the factors of $4x$.

favorable outcome (229) An outcome of an experiment for which you are interested in measuring the probability.

Fibonacci Sequence (41) An unending sequence such as 1, 1, 2, 3, 5, 8, 13, 21, ... in which the first two terms are fixed and the other terms are the sum of the two preceding terms.

FOIL pattern (671) A method used to multiply two binomials in a single step. Find the sum of the products of the **F**irst terms, **O**uter terms, **I**nner terms, and **L**ast terms. For example, to find $(2x + 3)(x - 5)$, find

Outer: $(2x)(-5)$ Inner: $(3)(x)$

$$2x^2 + (-10x) + 3x + (-15)$$

First: $(2x)(x)$ Last: $(3)(-5)$

formula (25) An algebraic expression that represents a general way of expressing a relationship in real-life situations.

frequency distribution (200) Organizing data by displaying the number of items or events that occur in an interval.

geometry (34) The study of shapes and their measures.

graph of a linear equation (593) The graph of all its solutions, a line.

graph of a linear inequality (620) The half-plane that consists of all points on one side of the line that is the graph of the corresponding linear equation. For example, this is the graph of the inequality $y > 2x - 4$.

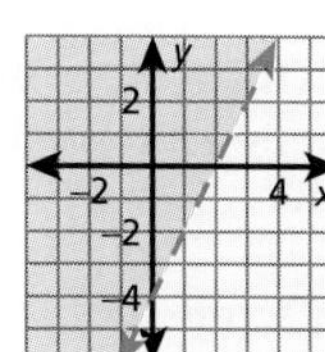

greatest common factor, GCF (250) The largest common factor of two numbers or algebraic expressions. For example, $4a$ is the GCF of $20ab^2$ and $24a^2$.

half-plane (620) In a plane, the region on one side of a line.

height (37, 488, 549) The perpendicular distance between parallel bases of a parallelogram, trapezoid, prism, or cylinder or the distance from a vertex to an opposite side of a triangle, pyramid, or cone.

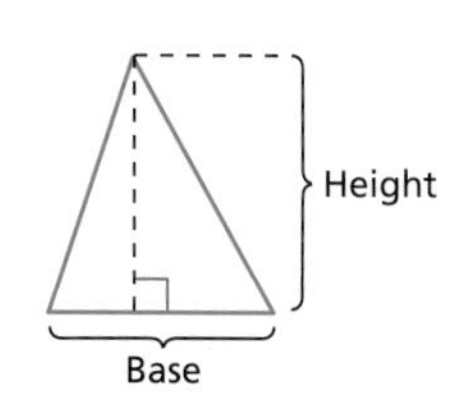

hemisphere (569) One of the two halves of a sphere.

heptagon (34) A polygon with seven sides.

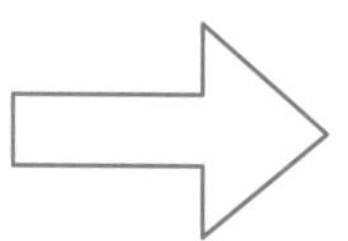

hexagon (34) A polygon with six sides.

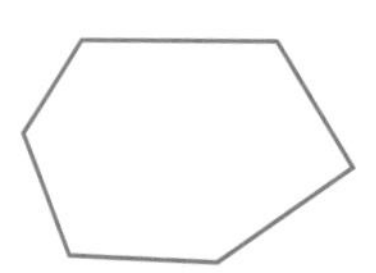

histogram (200) A bar graph in which the bars represent intervals.

hypotenuse (400) The side of a right triangle that is opposite the right angle. It is the longest side of a right triangle.

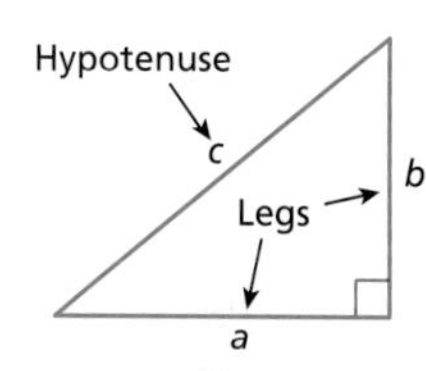

icosahedron (547) A polyhedron with 20 triangular faces.

identities (59) Equations that are true for all values of the variable. For example, the identities $a + 0 = a$ and $a \cdot 1 = a$ are true for all numbers.

image (462, 497, 502, 506) The new figure formed by the transformation of a given figure.

inequality (89) A mathematical sentence that contains a symbol such as $\neq$, $>$, $<$, $\geq$, or $\leq$.

integers (100) The set of numbers . . . –3, –2, –1, 0, 1, 2, 3, ...

interest (331) A percent of the money on deposit (the principal) paid to a lender or depositor for the use of the principal.

interior angle (36, 472) An angle of a polygon formed by two adjacent sides.

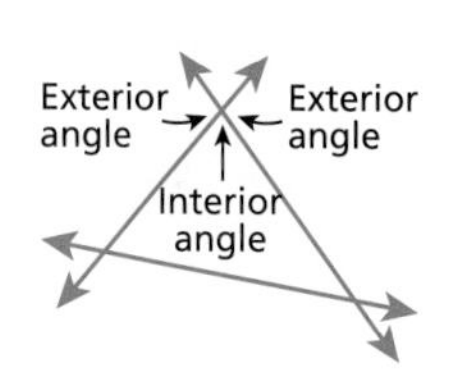

intersecting lines (439) Lines that cross or meet at a common point.

intersection of sets, $\cap$ (162) Contains only the elements found in both of two sets.

inverse operations (131) Operations that undo each other. For example, addition and subtraction are inverse operations, as are multiplication and division.

irrational number (265, 395) A real number that cannot be expressed as the quotient of two integers. For example, the square roots of numbers that are not perfect squares like $\sqrt{2}$, π, and nonterminating nonrepeating decimals like 0.100100010001 ... are all irrational numbers.

isosceles trapezoid (462) A trapezoid whose nonparallel sides have the same length.

isosceles triangle (401, 457) A triangle with at least two sides that have the same length.

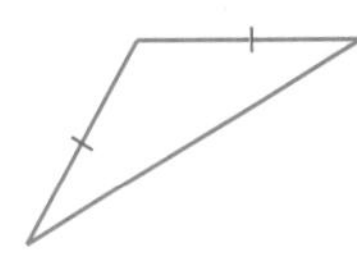

kite (462) A quadrilateral that is not a parallelogram but has two pairs of sides of equal lengths.

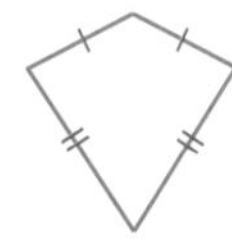

lateral (171) Side.

least common denominator, LCD (299) The least common denominator of two fractions is the least common multiple of their denominators.

least common multiple, LCM (255) The least common multiple of two numbers or algebraic expressions is the smallest of their common multiples. $6a^2b$ is the LCM of $2ab$ and $3a^2$.

Left-to-Right Rule (18) Operations having the same priority in an expression are evaluated from left to right.

leg of a right triangle (400) Either of the two shorter sides of a right triangle.

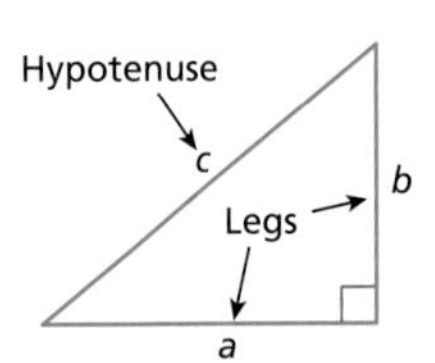

like fractions (294) Fractions that have the same denominator.

like terms (55) Two or more terms in an expression that have the same variable factors raised to the same powers.

linear equation (423, 588) An equation in two variables whose graph is a straight line.

linear inequality (423, 619) An inequality in two variables for which the graph is a half-plane above or below the corresponding linear equation.

linear relationship (614) Two quantities that can be modeled with a linear equation.

line of symmetry (453) A line that divides a figure into two parts, each of which is the mirror image of the other.

line plot (219) A diagram showing the frequency of data on a number line.

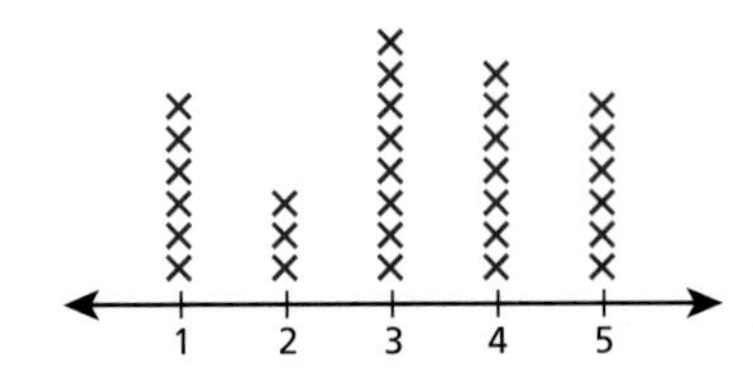

line segment (34, 439) Part of a line consisting of two endpoints and all the points between them.

lowest terms (259) A fraction is in lowest terms if the greatest common factor of the numerator and denominator is 1.

markup (50) The amount a store adds to the wholesale price to get the retail price.

matrix (125, 648) A rectangular array of numbers or data in rows and columns.

mean (127, 634) The average of all the numbers in a set.

measures of central tendency (634) Numbers that can be used to represent a group of numbers. The mean, median, and mode of a distribution.

median (634) The middle number (or the average of the two middle numbers) of a group of numbers listed in order.

midpoint (624) The halfway point on a line segment.

mode (634) The number that occurs most often in a given collection of numbers.

monomial (654) A polynomial with only one term; variable, a number, or a product of variables and numbers.

multiple (255) The product of a given number and any whole number. For example, 8 is a multiple of 1, 2, 4, and 8.

Multiplication Property of Equality (71) Multiplying both sides of an equation by the same nonzero number produces an equivalent equation. If $a = b$, then $ac = bc$.

Multiplying Powers Property (270) To multiply two powers with the same base, add their exponents; that is, $a^m \bullet a^n = a^{m+n}$.

natural number (11, 38) The set of numbers 1, 2, 3, 4, ...

negative correlation (224) Data points on a scatter plot whose y-coordinates tend to decrease as the x-coordinates increase.

negative integer (100) An integer that is less than 0.

negative slope (605) A line in the coordinate plane has a negative slope if it slants downward as you move from left to right.

net (442, 544) A flat pattern that can be folded to form a solid. For example, these nets can be folded to cubes.

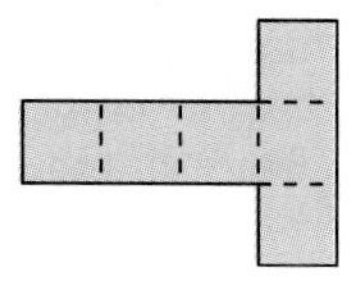

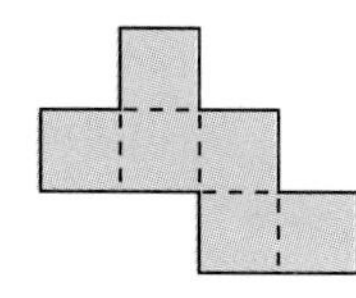

nonagon (34) A polygon with nine sides.

nonconvex polygon (37) A polygon that is not convex because one or more segments joining any two interior points does not lie completely within the polygon.

Not convex

***n*-gon** (34) A polygon with n sides.

nonrepeating decimal (266) A decimal that neither terminates nor repeats.

numerical expression (17) A collection of numbers, operations, and grouping symbols.

obtuse angle (443) An angle that measures between 90° and 180°.

obtuse triangle (458) A triangle with an obtuse angle.

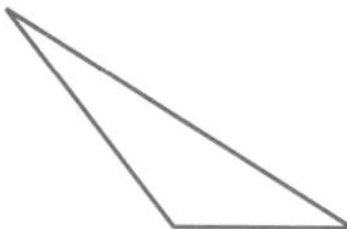

octagon (34) A polygon with eight sides.

octahedron (547) A polyhedron with eight triangular faces.

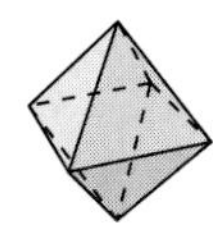

odd number (10) A whole number that is not divisible by 2. The numbers 1, 3, 5, 7, . . . are odd.

opposites (101) Two numbers that have the same absolute value but opposite signs; any two numbers whose sum is 0.

opposite side (522) In $\triangle ABC$, side $\overline{BC}$ is opposite $\angle A$.

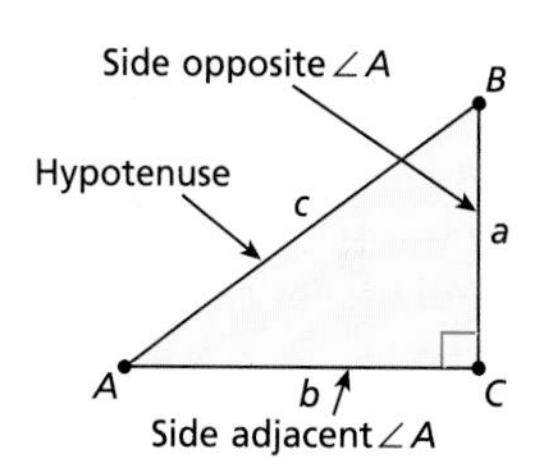

ordered pair (136, 588, 619) A pair of numbers or coordinates used to locate a point in a coordinate plane. The solution of an equation or an inequality in two variables.

Order of Operations (17, 18) A procedure for evaluating an expression involving more than one operation.
1. First do operations that occur within grouping symbols.
2. Then evaluate powers.
3. Then do multiplications and divisions from left to right.
4. Then do additions and subtractions from left to right.

origin (135) The point of intersection in the coordinate plane of the horizontal axis and the vertical axis. The point (0, 0).

outcome (229) A possible result of an event; for example, obtaining heads is an outcome of tossing a coin.

parallel lines (439) Lines in the same plane that do not intersect.

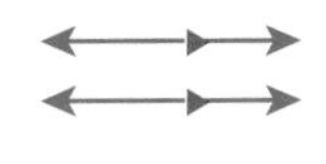

parallelogram (185, 462) A quadrilateral with opposite sides parallel.

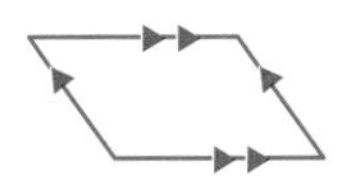

pentagon (34) A polygon with five sides.

percent, % (268, 318) A way of expressing hundredths; that is, a fraction whose denominator is 100. *Percent* means "per hundred."
5% (5 percent) equals $\frac{5}{100}$.

Percent Equation (358) The statement "a is p percent of b" is equivalent to the equation $\frac{a}{b} = \frac{p}{100}$.

percent of increase or decrease (367) An indication of how much a quantity has increased or decreased.

perfect number (279) A natural number with the sum of its factors, except itself, equal to the number. For example 6 is a perfect number since $1 + 2 + 3 = 6$.

perfect square (12) A number whose square root can be written as an exact decimal.

perimeter of a polygon (16) The distance around a figure; the sum of the lengths of the sides.

perpendicular lines (443) Two lines that meet at a right angle.

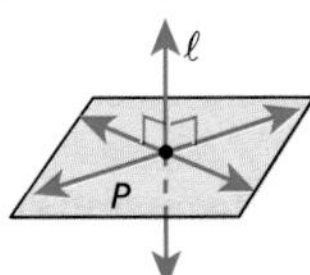

pi (539) The number that is the ratio of the circumference of a circle to its diameter. It is represented by the Greek letter π and is approximately equal to 3.1416.

plotting a point (100, 135) Locating a point that corresponds to a number on a number line or that corresponds to an ordered pair in the coordinate plane.

polygon (34) A closed plane figure that is made up of straight line segments that intersect at their endpoints.

polyhedron (544) A solid that is bounded by polygons, which are called faces.

polynomial (653, 654) A monomial or the sum or difference of monomials.

population (230) A group of people (or objects or events) that fit a particular description.

portion (318) A fraction that compares the measure of part of a quantity to the measure of the whole quantity.

positive correlation (224) Data points on a scatter plot whose y-coordinates tend to increase as the x-coordinates increase.

positive integer (100) The numbers 1, 2, 3, 4, . . . are positive integers.

positive slope (605) A line in the coordinate plane has a positive slope if it slants upward as you move from left to right.

possible outcomes (229) All the different ways an event can turn out.

power (11) An expression such as 4^2 that has a base (4) and an exponent (2).

prime factorization (245) Expression of a composite number as a product of prime factors. The prime factorization of 18 is $2 \cdot 3 \cdot 3$.

prime number (245) A natural number that has exactly two factors, itself and 1. The numbers 2, 3, 5, 7, 11, 13, and so on, are prime numbers.

principal (331) The amount of money loaned by a bank to a borrower or the amount of money on deposit in a bank.

prism (544) A polyhedron that has two parallel, congruent faces called bases.

probability of an event (228) A measure of the likelihood that the event will occur.

product (7, 24) The result obtained when numbers or expressions are multiplied. The product of a and b is $a \bullet b$ or ab.

proportion (349) An equation stating that two ratios are equal. If a is to b as c is to d, then $\frac{a}{b} = \frac{c}{d}$.

protractor (443) A measuring device that can be used to approximate the measure of an angle.

pyramid (441, 544) A space figure whose base is a polygon and whose other faces are triangles that share a common vertex.

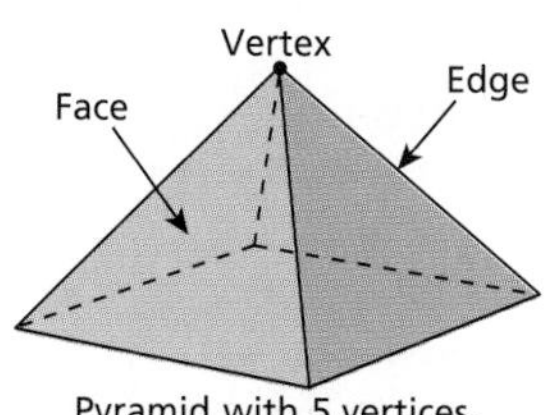

Pyramid with 5 vertices, 5 faces, and 8 edges

Pythagorean Theorem (400) For any right triangle, the sum of the squares of the lengths of the legs, a and b, equals the square of the length of the hypotenuse, c. $a^2 + b^2 = c^2$

Pythagorean triple (401) A set of three natural numbers that represent the sides of a right triangle.

quadrant (135) In the coordinate plane, one of the four parts into which the axes divide the plane.

quadrilateral (34, 462) A polygon with four sides.

quartile (642) The first, second, and third quartiles roughly divide a collection of ordered numbers into four equal groups.

quotient (7, 24) The result obtained when numbers or expressions are divided. The quotient of a and b is $\frac{a}{b}$.

radical (12) The square root symbol, $\sqrt{}$.

radius of a circle (539) A segment that has the center as one endpoint and a point on the circle as the other endpoint.

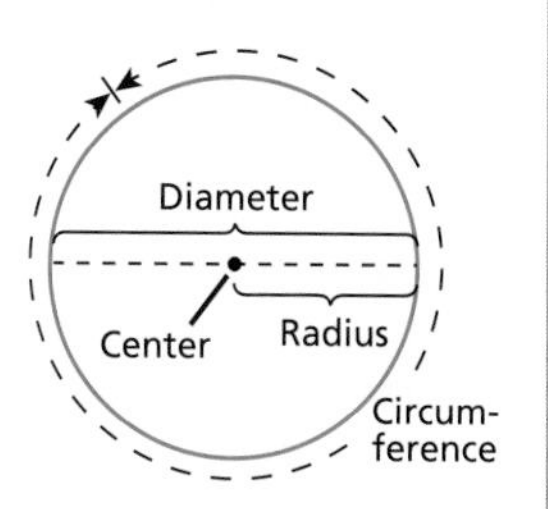

random sample (230) A sample in which every person (or object or event) in the population has an equal chance of having the characteristics of the larger group.

rate (344) The relationship $\frac{a}{b}$ of two quantities a and b that have different units of measure.

ratio (345) The relationship $\frac{a}{b}$ of two quantities a and b that have the same unit of measure.

rational number (265, 395) A number that can be written as the quotient of two integers.

raw data (218) Unorganized data from an experiment or survey.

ray (439) Part of a line that has one endpoint and extends forever in only one direction.

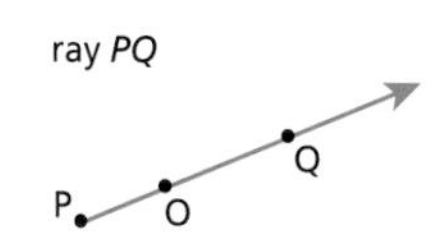

real numbers (62, 395) The set of all rational numbers and irrational numbers together.

reciprocal (130) The reciprocal of a nonzero number a is $\frac{1}{a}$. The product of two reciprocals is 1.

Reciprocal Property (349) For ratios $\frac{a}{b}$ and $\frac{c}{d}$, if $\frac{a}{b} = \frac{c}{d}$ then $\frac{b}{a} = \frac{d}{c}$.

rectangle (8) A parallelogram that has four right angles.

rectangular prism (551) A prism whose bases are rectangles.

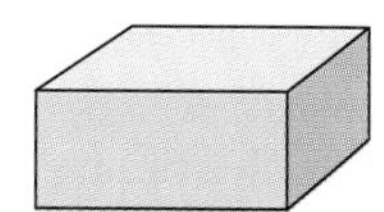

reflection (497) A transformation that flips a figure about a line onto its mirror image on the opposite side of the line.

reflection line (497) In a reflection, a line that is perpendicular to and bisects each segment that joins an original point to its image.

regular polygon (36, 468) A polygon with each of its sides having the same length and each of its angles having the same measure.

relatively prime (251) Two natural numbers are relatively prime if their greatest common factor is one.

repeating decimal (38, 395) A decimal in which a digit or group of digits repeats forever. Repeating digits are indicated by a bar.

$0.3333\ldots = 0.\overline{3}$

$1.47474747\ldots = 1.\overline{47}$

retail price (50) The amount you pay for any item in a store.

rhombus (462) A parallelogram with four sides of equal length.

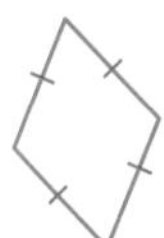

right angle (400) An angle whose measure is 90°.

right triangle (400, 458) A triangle that has a right angle.

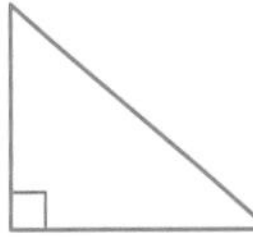

rotation (502) A transformation that turns a figure a given angle and direction about a point.

rotational symmetry (454) A figure has rotational symmetry if it coincides with itself after rotating 180° or less about a point.

round a number (12) To replace a number by another one of approximately the same value that is easier to use.

round-off error (178) The error produced when a number is rounded in a computation.

sample space (229) The set of all possible outcomes of an event.

scale factor (512) In two similar polygons or two similar solids, the **scale factor** is the ratio of corresponding linear measures.

scalene quadrilateral (462) A quadrilateral whose four sides all have different lengths.

scalene triangle (457) A triangle whose three sides all have different lengths.

scatter plot (224) The graph of a collection of ordered pairs (x, y).

scientific notation (274) A short form of writing numbers whose absolute values are very large or very small. A number is written in scientific notation if it has the form $c \times 10^n$, where c is greater than or equal to 1 and less than 10, and where n is an integer.

segment See *line segment.*

sequence (3) An ordered list of numbers.

set (162) A well-defined collection of objects.

side of a polygon (34) One of the straight line segments that make up the polygon.

similar figures (356, 512) Figures that have the same shape but not necessarily the same size.

similar solids (573) Solids that have the same shape with their corresponding lengths proportional.

similar triangles (350) Two triangles that have the same angle measures. For example, triangle ABC is similar to triangle DEF.

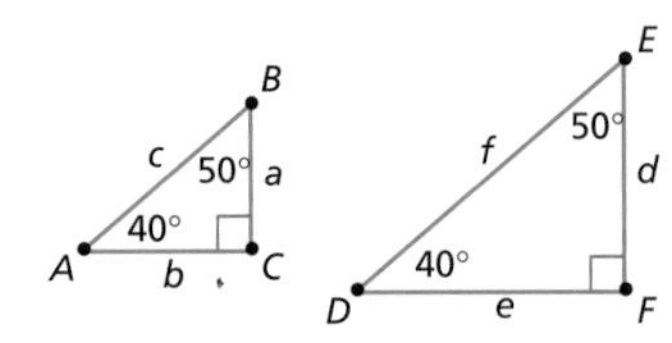

simplest form (259) A fraction (or mixed number) is in simplest form if the fraction (or fraction-part of the mixed number) is less than 1 and in lowest terms (reduced form).

simulation (377) An experiment that models a real-life situation.

sine of an angle (522) The sine of an acute angle in a right triangle is the ratio of the length of the leg opposite the acute angle to the length of the hypotenuse. For example, $\sin A = \frac{a}{c}$.

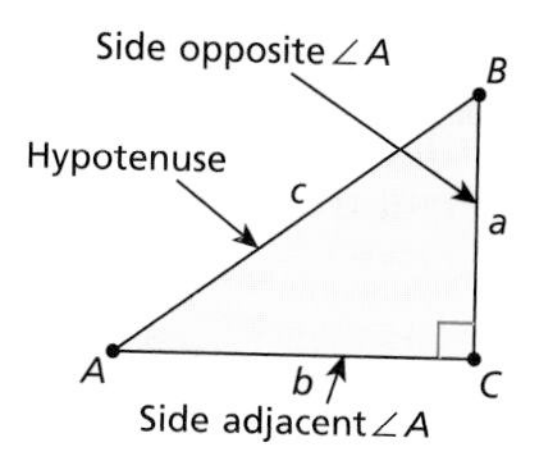

slope (603) The ratio of the difference in y-coordinates to the difference in x-coordinates for any two points on the graph of a linear equation. Also defined as the rise over the run.

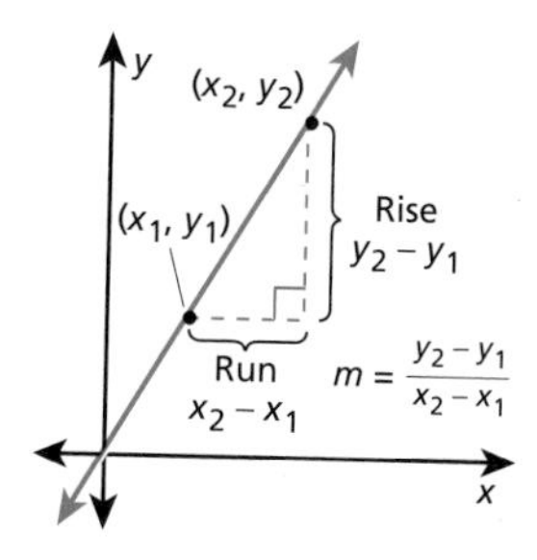

slope-intercept form of an equation (610) A linear equation in x and y that has the form $y = mx + b$, where m is the slope of the line and b is the y-intercept.

solution (59, 89, 136, 619) A number (or ordered pair of numbers) that produces a true statement when substituted for the variable(s) in an equation or an inequality.

solving an equation (59) Finding all the values of the variable that make the equation true.

solving an inequality (89) Finding all the values of the variable that make the inequality true.

solving a right triangle (401) Determining the measures of all six parts; that is, the lengths of the three sides and the measures of the three angles.

spreadsheet (75) A computer program that creates tables.

square (12) A rectangle with four sides of equal length.

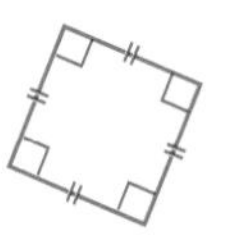

square number (11) The set of numbers 1, 4, 9, $\ldots, n^2$.

square root of a number (12, 390) The number that when squared will produce the given number. Both 7 and –7 are square roots of 49, because $7^2 = 49$ and $(-7)^2 = 49$.

square root symbol See *radical.*

standard form of a polynomial (654) A polynomial written so that the powers of the variable decrease from left to right.

statistics (218) A branch of mathematics that organizes large collections of data in ways that can be used to understand trends and make predictions.

stem-and-leaf plot (638–639) A method of organizing data in increasing or decreasing order.

straight angle (443) An angle whose measure is 180°.

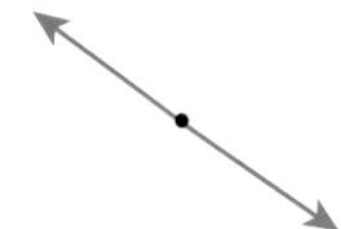

substituting (24) Replacing a variable in an expression by a number.

Subtraction Property of Equality (65) Subtracting the same number from both sides of an equation produces an equivalent equation. If $a = b$, then $a - c = b - c$.

sum (7, 24) The result obtained when numbers or expressions are added. The sum of a and b is $a + b$.

Sum of Opposites Property (105) The sum of any two opposites is zero.

supplementary angles (67) Two angles whose measures have a sum of 180°.

surface area (14, 548) The sum of the areas of all the faces of a solid figure.

tangent of an angle (522) The tangent of an acute angle in a right triangle is the ratio of the length of the leg opposite the acute angle to the length of the leg adjacent to that acute angle. For example, $\tan A = \frac{a}{b}$.

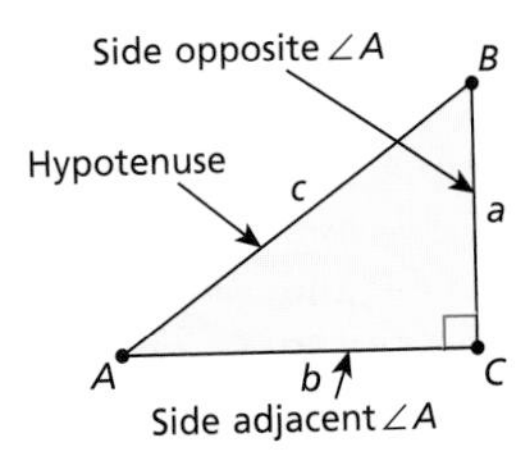

terminating decimal (38) A decimal, that contains a finite number of digits, for example, 0.5.

terms of an expression (24, 116, 654) The terms of an expression are separated by addition signs. In the expression $3x + (-2)$, the terms are $3x$ and -2.

tetrahedron (547, 565) A pyramid with four triangular faces.

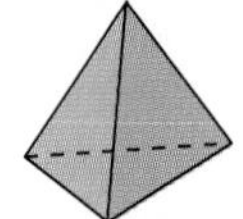

time line (196) A graph on a number line of the dates of several occurrences.

transformations in geometry (502) An operation such as a reflection (flip), rotation (turn), or translation (slide) that maps or moves a figure from an original position (preimage) to a new position (image).

translation (506) A transformation that slides each point of a figure the same distance in a given direction.

trapezoid (45, 462) A quadrilateral with only one pair of parallel sides.

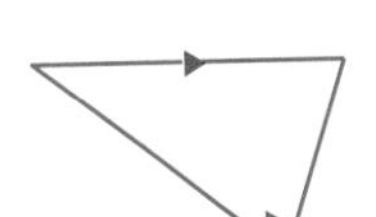

tree diagram (245, 373) A diagram that shows all the prime factors of a number or the possible outcomes of an event.

triangle (34) A polygon with three sides.

Triangle Inequality (90, 425) The sum of the lengths of any two sides of a triangle is greater than the length of the third side of the triangle.

triangular numbers (39) The set of numbers $1, 3, 6, \ldots, \frac{n(n+1)}{2}$.

triangular prism (551) A prism whose bases are triangles.

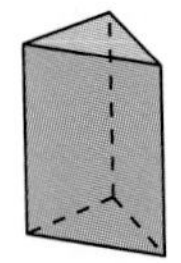

trigonometric ratio (522) A ratio of the lengths of two sides of a right triangle; for example: sine, cosine, and tangent.

trinomial (654) A polynomial that has only three terms.

two-step equation (149) An equation whose solution involves two transformations.

union of sets, ∪ (162) All the elements found in either or both of two sets.

unit analysis (72) A technique that can be used to decide the unit of measure assigned to a product or quotient.

unlike fractions (299) Fractions that do not have the same denominator.

variable (24, 29) A symbol, usually a letter, that is used to represent one or more numbers in an algrebraic expression, for example, x is a variable in the expression $8x + 19$.

Venn diagram (162, 265) A diagram that shows the relationships among sets of numbers or objects.

vertex (34, 443, 544) The point at the corner of an angle, plane figure, or solid figure.

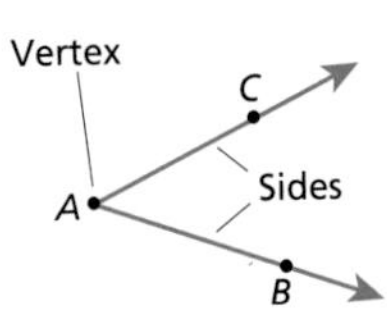

vertical angles (448) Angles whose sides form two pairs of opposite rays.

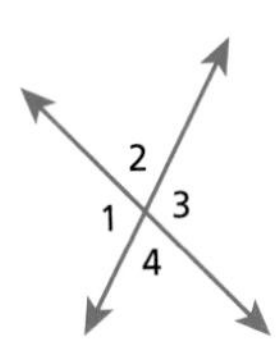

Vertical angles:
∠1 and ∠3, ∠2 and ∠4

volume (14, 554) The measure of the amount of space that an object occupies, or how much it will hold.

whole number (38) Any of the numbers 0, 1, 2, 3, 4, ...

wholesale price (50) The amount that a store paid for an item.

x-axis (135) The horizontal number line in the coordinate plane.

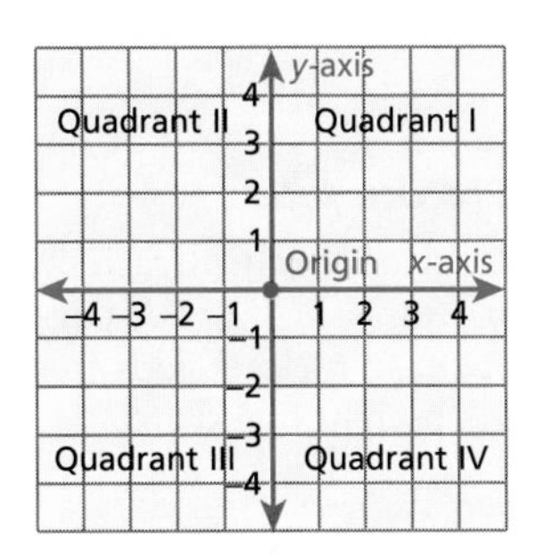

x-coordinate (135) The first number of an ordered pair, the position of the point relative to the horizontal axis.

x-intercept (598) The x-coordinate of the point where a graph crosses the x-axis; the value of x when $y = 0$.

y-axis (135) The vertical number line in the coordinate plane.

y-coordinate (135) The second number of an ordered pair, the position of the point relative to the vertical axis.

y-intercept (598) The y-coordinate of the point where a graph crosses the y-axis; the value of y when $x = 0$.

Zero Property of Addition (105) When zero is added to any number, the sum is the same number. $a + 0 = a$ and $0 + a = a$

GLOSSARY

Reference

SELECTED ANSWERS

CHAPTER 1

1.1 Communicating About Mathematics p. 3

To get the next number: **1.** Add 1 to preceding number, add 2 to preceding number, add 3 . . . etc.; 8, 12, 17, 23, 30 **2.** Add two preceding numbers; 8, 13, 21, 34, 55 **3.** Name the next prime number; 7, 11, 13, 17, 19 **4.** Subtract 1 from 2 times preceding number; 9, 17, 33, 65, 129

1.1 Guided Practice p. 4

Answers vary. **1.** Mathematics is the study of numbers and their uses. **2.** Phone numbers, house numbers, auto license plates **3.** Heights of people in feet and inches, weights of packages in pounds and ounces, speeds of cars in miles per hour, liter, yard, second **4.** Hat sizes: 7, $7\frac{1}{8}$, $7\frac{1}{4}$, $7\frac{3}{8}$, $7\frac{1}{2}$, etc. Each number is $\frac{1}{8}$ more than preceding number.

1.1 Independent Practice pp. 4–5

To get next number: **5.** Name next consecutive odd number; 9,11,13 **7.** Add 2, add 3, add 4, etc. to preceding number; 15, 21, 28 **9.** Add 1 to both numerator and denominator of preceding number; $\frac{5}{6}, \frac{6}{7}, \frac{7}{8}$ **11.** Add $\frac{3}{2}$ to preceding number; 8, $\frac{19}{2}$, 11 **13.** Multiply preceding number by 3; 162, 486, 1458 To get the next letter: **15.** Name second consecutive letter; I, K, M **17.** In every odd-numbered position, except the first: Name the letter of the alphabet that precedes the letter in the preceding odd-numbered position. In every even-numbered position, except the first: Name the letter of the alphabet that follows the letter in the preceding even-numbered position. W, D, V **19.** Write the initial letter in the name of the next counting number; S, E, N **21.** 50, 47, 44, 41, 38, 35

23.

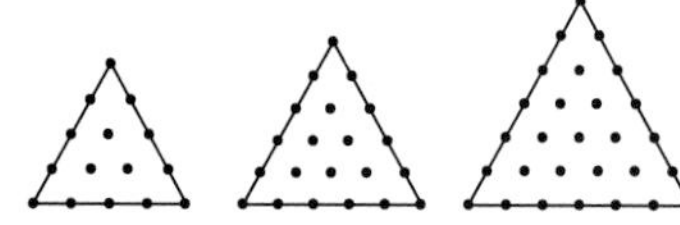

25. You. The first number is minutes and the second number is seconds; so you ran the race in 2 minutes and 39.4 seconds, while your friend ran the race in 2 minutes and 41.8 seconds.
27. They decrease.

1.1 Integrated Review p. 5

29. 18 **31.** 150 **33.** 2.5

1.2 Communicating About Mathematics p. 8

Answers vary. Number of seats in an auditorium section, total cost of identical items **A.** The product of 3 and 6 is 18. $3 \times 6 = 18$ **B.** The product of 5 and 7 is 35. $5 \times 7 = 35$ **C.** The product of 6 and 5 is 30. $6 \times 5 = 30$

1.2 Guided Practice p. 9

1. Addition, subtraction, multiplication, division
2. $+$, $-$, $\times$ or • or ( ), $\div$ or / **3.** See page 8.
4. Answers vary. A cash register adds the prices of various products bought in a store.

1.2 Independent Practice pp. 9–10

5. The product of 6 and 8 is 48. **7.** The sum of 3 and 14 is 17. **9.** The difference of 111 and 56 is 55. **11.** The product of 2 and 54 is 108. **13.** 682 **15.** 719 **17.** 213 **19.** 2.3 **21.** 905.43 **23.** 0.151 **25.** 1 **27.** $\frac{1}{3}$ **29.** 112 **31.** 17 **33.** 41.83 **35.** $\frac{4}{3}$ **37.** 8 **39.** 50,076 **41.** 17.9 **43.** 131.8 **45.** $2+2+2=6$ **47.** $12-6=6$ **49.** 36 million **51.** 572 million **53.** 70

1.2 Integrated Review p. 10

55. $\frac{1}{2}$ **57.** $\frac{3}{2}$ or $1\frac{1}{2}$ **59.** 0.5 **61.** $0.\overline{6}$

1.3 Communicating about Mathematics p. 12

To check, square each answer. **A.** 2.45 **B.** 3.16 **C.** 2.55 **D.** 11.84

1.3 Guided Practice p. 13

1. base, exponent **2.** 3 raised to the 4th power is 81. **3.** 4 **4.** 7 **5.** 9 **6.** The square root of 36 is 6.

1.3 Independent Practice pp. 13–14

7. 6 raised to the 4th power is 1296. **9.** The square root of 1.21 is 1.1. **11.** 12^2, 144 **13.** 3.4^3, 39.304 **15.** $\left(\frac{1}{5}\right)^4$, $\frac{1}{625}$ **17.** 13 **19.** 10.82 **21.** 2.35

23. 8 **25.** 2.1 **27.** 81 **29.** $<$ **31.** $<$ **33.** 2^{10}
35. True, by meaning of power **37.** False; $\sqrt{5} \approx 2.24$, $\sqrt{2} + \sqrt{3} \approx 3.15$ **39. a.** 6 **b.** 49 in.^{2}
c. 294 in.^{2}, $6\,(7 \text{ in.})^2$ **d.** No, $6\,(14^2) \neq 2[6(7^2)]$
41. a. Each is 20.5 ft. **b.** 82 ft
c. No, $4 \times \sqrt{2(420.25)} \neq 2\,(82)$

1.3 Integrated Review p. 15

43. 2.59 **45.** 318.07 **47.** 26.20 **49.** 5

1.3 Mixed Review p. 15

To get the next number: **1.** Add 2 to preceding number; 10, 12, 14 **3.** Add 5 to previous number; 21, 26, 31 **5.** Raise 2 to the power of the preceding number's position and add it to the preceding number; or add 2 times the difference of the two preceding numbers to the preceding number; or raise 2 to the power of the number's position and subtract 1; or add 1 to the product of 2 and the preceding number. 31, 63, 127 **7.** 1417 **9.** 111 **11.** 2604 **13.** 86 **15.** 78,125 **17.** 0.04 **19.** 4.69

1.4 Communicating About Mathematics p. 19

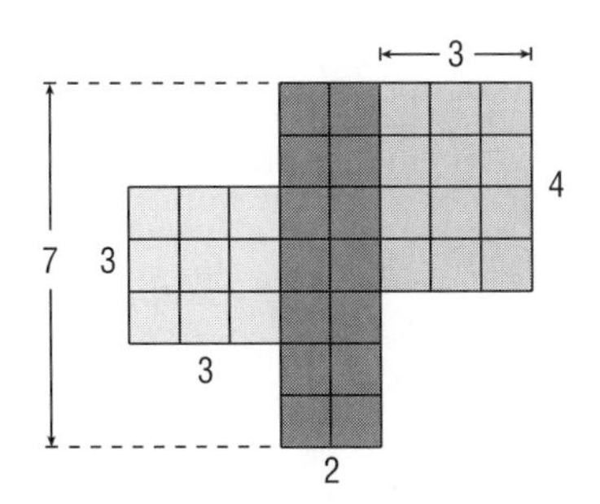

$3^2 + 2 \times 7 + 3 \times 4 = 9 + 14 + 12 = 35$ square units
Yes, order of operations is necessary to always get the same results.

1.4 Guided Practice p. 20

1. To always get the same results. Examples vary. Example: The total number of pencils in 3 boxes of 12 pencils each and 2 boxes of 10 pencils each is $3 \times 12 + 2 \times 10$ or 56, not 380.
2. Evaluating powers, multiplications and divisions, and additions and subtractions; to always get the same results. **3. a.** 6 **b.** 24 **c.** 19
4. a. $3 \times (4+8) - 2 = 34$ **b.** $(7-3) \div 2 \times (8+2) = 20$

1.4 Independent Practice pp. 20–21

5. 9 **7.** 19 **9.** 29 **11.** 52 **13.** 8 **15.** 113 **17.** 42 **19.** 79 **21.** 2 **23.** 113 **25.** 659 **27.** False; $(6+21) \div 3 = 9$ **29.** True **31.** False; $(6+3^2) \div 3 = 5$ **33.** False; $7 + 7 \bullet (2+6) = 63$ **35.** $24 \div 8 + 9 = 12$ **37.** $36 \div (6+12) = 2$ **39.** 2(\$25) + 3(\$20) + 2(\$25), \$160

1.4 Integrated Review p. 21

41. $2 \times 3 + 2^2 + 3 \times 4$, 22 units^2

1.4 Mid-Chapter Self-Test p. 22

1. Answers vary. Height, weight, area To get the next number: **2.** Add 3 to the previous number; 12, 15, 18 **3.** Subtract 9 from the previous number; 63, 54, 45 **4.** Name next larger perfect square; or add 3, add 5, add 7, etc., to the previous number; 16, 25, 36 **5.** Add 1 to the denominator of the previous number; $\frac{1}{4}, \frac{1}{5}, \frac{1}{6}$

6.

7. 7 8 9

8. The product of 12 and 4; 48 **9.** The product of 176 and 12; 2112 **10.** The quotient of 15 and 3; 5 **11.** The quotient of 369 and 41; 9 **12.** 3 cubed; 27 **13.** The square root of 256; 16 **14.** \$52 **15.** \$12 **16.** 20 in. **17.** $2\frac{1}{2}$ in. $\times$ $2\frac{1}{2}$ in. **18.** 64, $6\frac{1}{4}$ in.^{2}

1.5 Communicating About Mathematics p. 25

Descriptions vary. **A.** 24 mi **B.** $18\frac{1}{3}$ mi

1.5 Guided Practice p. 26

1. See page 24. **2.** Find its numerical value. **3.** n **4.** 3 **5.** $4+n$ **6.** 7 **7.** 18 **8.** 5

1.5 Independent Practice pp. 26–27

9. 9 **11.** 48 **13.** 57 **15.** 42 **17.** 25 **19.** 3 **21.** 5 **23.** 19 **25.** 84 **27.** 6 **29.** 125 **31.** 3 **33.** 4 **35.** 12 **37.** b **39.** c **41.** 4 **43. a.** A flying distance, no; it is a distance along a straight line. **b.** 11 mph **45.** $43\frac{3}{4}$ mi

1.5 Integrated Review p. 27

47. 12 **49.** 7 **51.** 100 **53.** 3

1.6 Communicating About Mathematics p. 29

Answers vary.

1.6 Guided Practice p. 30

1. Answers vary. Table, bar graph, line graph
2. Answers vary.

1.6 Independent Practice pp. 30–31

3. Unified Team **5.** Germany
7.

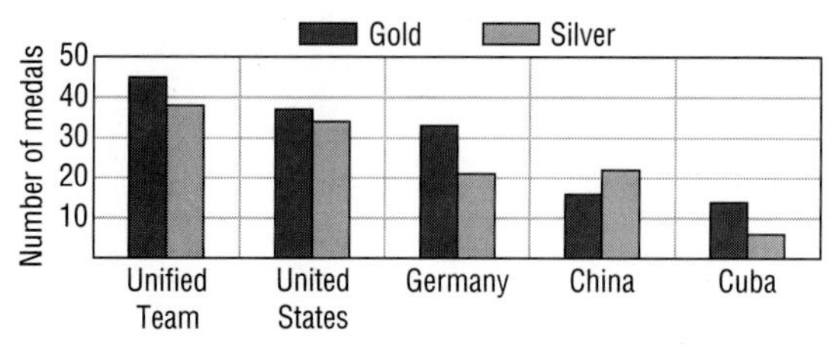

A bar graph is preferable because there are no connections among the numbers of medals won by the various teams.

9. 1988, 1976

11.

Side	1	2	3	4	5	6	7
Perimeter	4	8	12	16	20	24	28
Area	1	4	9	16	25	36	49

13.

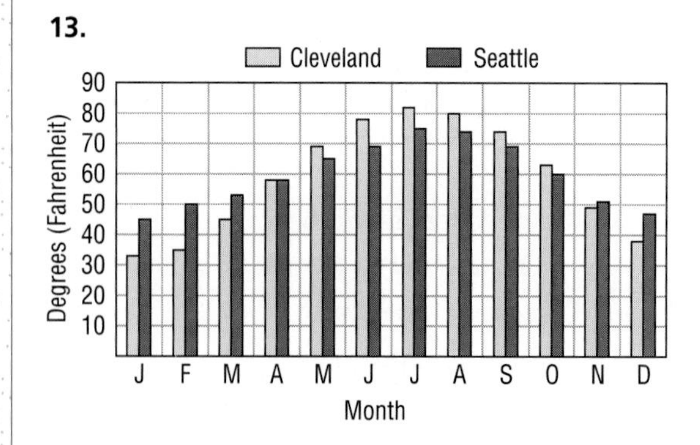

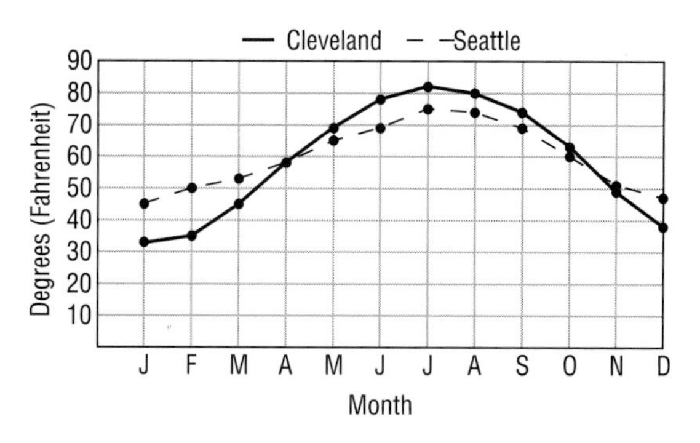

15. Answers vary. **17.** Answers vary.

1.6 Integrated Review p. 31

19. 82.7 **21.** 2.67 **23.** 15.132

1.6 Mixed Review p. 32

1. 85.2 **3.** 1.44 **5.** 1.29 **7.** Addition
9. Subtraction **11.** 10 **13.** 81 **15.** 14 **17.** 5

1.7 Communicating About Mathematics p. 35

14, 20

1.7 Guided Practice p. 36

1. Answers vary. (See page 34.) **2.** G: nonagon, E: octagon, O: quadrilateral, M: decagon, E: heptagon, T: triangle, R: hexagon, Y: pentagon **3.** 2
4. Answers vary.

1.7 Independent Practice pp. 36–37

5. Quadrilateral **7.** Hexagon **9.** No, all sides not segments **11.** Yes, octagon **13.** Answers vary.

15.

Sides	3	4	5	6
Angle measure	60°	90°	108°	120°
Sum of angle measures	180°	360°	540°	720°

17. 144° **19.** The area of a triangle is one half the product of the base and height, $A = \frac{1}{2}bh$. **21.** Yes
23. No, all sides are not segments

1.7 *Integrated Review p. 37*

25. 50 **27.** c **29.** d **31.** a

1.8 *Communicating About Mathematics p. 39*

28, 36

1.8 *Guided Practice p. 40*

1. 0, 1 **2.** $n=1$: $0.\overline{09}$, $n=2$: $0.\overline{18}$, $n=3$: $0.\overline{27}$, $n=4$: $0.\overline{36}$; the repeating decimal is $0.09n$.

1.8 *Independent Practice pp. 40–41*

3.

n	1	2	3	4
$192 \div n$	192	96	64	48

5. Answers vary.

n	1	2	3	4	5	6	7	8	9
$\frac{n}{3}$	$0.\overline{3}$	$0.\overline{6}$	1	$1.\overline{3}$	$1.\overline{6}$	2	$2.\overline{3}$	$2.\overline{6}$	3

7.

n	0	1	2	3	4	5	6
$\frac{n^2}{2}$	0	0.5	2	4.5	8	12.5	18

9. 18, 187, 1876, 18765, 187654, 1876543; each number, after the first, is 10 times the preceding number plus 1 less than the units digit of the preceding number. **11.** 50.2, 50.4, 50.6, 50.8, 51, 51.2; each number, after the first, is 0.2 more than the preceding number.

13. 1 [+] 8 [=] [=] [=] [=] [=] [=] [=] [=]; 33, 41, 49, 57

15. 5 [×] 3 [=] [=] [=] [=] [=] [=] [=] [=]; 405, 1215, 3645, 10935 On some *TI* calculator models you need to reverse the order of the keys for multiplication, that is, enter 3 [×] 5.

17. 1008 [÷] 10 [=] [=] [=] [=] [=] [=] [=] [=]; 1.008, 0.1008, 0.01008, 0.001008

19. 64

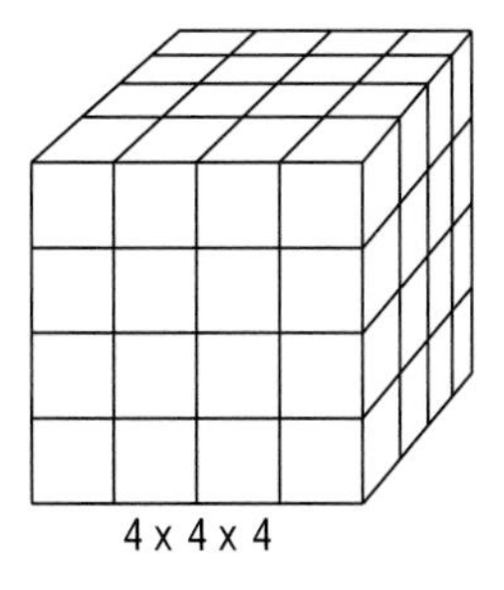

1.8 *Integrated Review p. 41*

21. 14, 17, 20, 23 **23.** 16, 32, 64, 128 **25.** $\frac{5}{6}, \frac{6}{7}, \frac{7}{8}, \frac{8}{9}$ **27.** 3, 7, 4, 8

1.8 *Chapter Review pp. 43–46*

To get the next number: **1.** Add 15 to the preceding number; 75, 90, 105 **3.** Add 4, add 6, add 8, etc., to the preceding number; 30, 42, 56 **5.** Name the next perfect cube; 125, 216, 343 **7.** 1, 10, 19, 28, 37, 46 **9.** d **11.** a **13.** Answers vary. **15.** 34.5 mph **17.** 74.95 **19.** 251 **21.** 54 **23.** 4992 **25.** $3\times4=12$ or $4+4+4=12$ **27.** 13^2, 169 **29.** $\left(\frac{1}{2}\right)^3, \frac{1}{8}$ **31.** $\left(\frac{5}{9}\right)^3, \frac{125}{729}$ **33.** 17 **35.** 1.44 **37.** ≈ 256.8 ft **39.** Answers vary. **41.** 16 **43.** 24 **45.** 8 **47.** 40 **49.** 13 **51.** 6 **53.** $\frac{1}{2}$ **55.** 0 **57.** 64 **59.** 1 **61.** 245 mi **63.** 1991 **65.** 54,800 **67.** Arizona, Maryland, Washington **69.** 1.8 million **71.** Yes, pentagon **73.** No, not all sides are segments **75.** Yes, decagon **77.** The area is $\frac{1}{2}$ the product of the height and the sum of the bases; $A=\frac{1}{2}h(b_1+b_2)$. **79.** Each number, after the first, is 105 more than the preceding number.

n	1	2	3	4	5	6	7	8	9
$105n$	105	210	315	420	525	630	735	840	945

81. 15309 ÷ 3 [=] [=] [=] [=] [=] [=]; 189, 63, 21

83. $P=I-E$ **85.** All of them **87.** 11 miles

89. The weight is the sum of the trailer's weight and the product of 7 and a rover's weight.

91. 64,500 lb

CHAPTER 2

2.1 *Communicating About Mathematics p. 52*

$4(3x+2)=4(3 \bullet 5+2)=68$, $4(3x+2)=12x+8=12 \bullet 5+8=68$; preferences vary.

2.1 *Guided Practice p. 53*

1. See page 51. **2.** b and c **3. a.** Yes, $3\times5=15$ **b.** No, $4\times9=36$ **c.** No, $2\times11=22$ **4.** Answers vary. $3\times5=5\times3$

2.1 *Independent Practice pp. 53–54*

5. 2, $x+4$; $2(x+4)$; $2x+8$ **7.** 3, $2x+2$; $3(2x+2)$; $6x+6$

9. $4x+8$

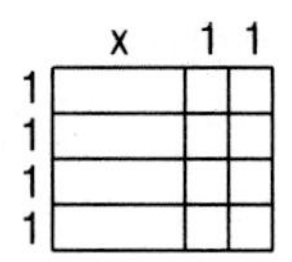

11. $10x+6$

13. $72+63$ **15.** $4x+36$ **17.** $5y+100$ **19.** $x+12$
21. $ab+4a$ **23.** $rs+rt$ **25.** $24+40+48$
27. $3y+yz$ **29.** $12s+12t+12w$
31. $6m+6n+6r+6t$ **33.** $3(6.69)=20.07$; $3.63+16.44=20.07$ **35.** $525(38.79)= 20{,}364.75$; $5874.75+14{,}490= 20{,}364.75$ **37.** $3x+18$
39. a. Total annual pay $=$ $12\times$Total monthly pay
b. $55,200

2.1 Integrated Review p. 54

41. 112 **43.** 88 **45.** 180 **47.** 50 **49.** 184 **51.** 150

2.2 Communicating About Mathematics p. 56

x	1	2	3	4	5
Perimeter, $13x$	13	26	39	52	65

Each perimeter, after the first, is 13 more than the preceding perimeter.

2.2 Guided Practice p. 57

1. See page 55. **2.** Examples vary. $2x+3x=5x$
3. Two numbers can be added in either order.
4. When simplifying or evaluating an expression
5. Yes, have like terms **6.** No, no like terms
7. No, no like terms **8.** Yes, 5 and 2(8) are like terms.

2.2 Independent Practice pp. 57–58

9. $3a$ **11.** $9x+9$ **13.** $12y+10$ **15.** $8a+2b$
17. $14r+2s+5$ **19.** $2x+4y+20z$ **21.** Answers vary. $8x+2x+5+6x$ **23.** b^2+3b **25.** $2x^2$
27. $7x+9$ **29.** $6a+6b$ **31.** $16xy$ **33.** $2x^2y$
35. $5x+y$, 15 **37.** $5x+4y$, 30 **39.** x^2+2xy, 24
41. The expressions are not equivalent. 1st perimeter: $5x+4$, 2nd perimeter: $6x+4$. If the expressions were equivalent, then the perimeters would be the same for all values of x. But, other than 0, there are no such values.

43.

$18x$;

x	1	2	3	4	5
Perimeter, $18x$	18	36	54	72	90

Each perimeter, after the first, is 18 more than the preceding perimeter.
45. a. $3(3x)+9x$ or $18x$ **b.** $58.50

2.2 Integrated Review p. 58

47. The perimeter, P, of the nth figure is $P=n+2$; 12, 17.

2.3 Communicating About Mathematics p. 60

$290+x=425$; $x=135$ Answer: $135 million

2.3 Guided Practice p. 61

1. See page 59. **2.** Examples vary. $4(x+2)=4x+8$ **3.** Examples vary. $x+1=3$
4. Substitute the number in the original equation, simplify, and see if you get an identity.

2.3 Independent Practice pp. 61–62

5. b **7.** d **9.** 6 **11.** 34 **13.** $33-x=24$, 9
15. $8x=56$, 7 **17.** Yes **19.** 3 **21.** No, $23\neq7$
23. No, $4\neq36$ **25.** 11 **27.** 14 **29.** 52 **31.** 5
33. 7 **35.** 48 **37.** Identity; true for all values of x
39. Conditional equation; true only for $x=8$
41. 755.625 lb **43.** 620.25 lb **45.** 10, $x-7=3$
47. 48, $12+t=60$

2.3 Integrated Review p. 62

49. 2 **51.** $9x=18$, 2 **53.** $9x=18$, 2 **55.** $9x=0$, 0

2.3 Mixed Review p. 63

1. 19 **3.** 28 **5.** 10 **7.** 9 **9.** $4r+2$, 18 **11.** $3r+6s$, 42 **13.** $3r^2+2s$, 58 **15.** 2648.8 **17.** 31,383
19. 15.7

2.4 Communicating About Mathematics p. 66

Answers vary.

■ 2.4 Guided Practice p. 67

1. Answers vary. $x=4$, so $x+3=4+3$ and $x+3=7$. **2.** $x=3$, so $x+5=3+5$ and $x+5=8$. **3.** Subtract 24 from both sides. **4.** Add 16 to both sides. **5.** Subtract 72 from both sides. **6.** Add 68 to both sides. **7.** Assoc. Prop. of Mult. **8.** Comm. Prop. of Mult. **9.** Comm. Prop. of Add. **10.** Assoc. Prop. of Add.

■ 2.4 Independent Practice pp. 67–68

11. 34, 34, 86 **13.** 62, 62, 49 **15.** 81 **17.** 139 **19.** 575 **21.** 138 **23.** 2.2 **25.** 13.35 **27.** Added 193 to both sides, simplified. **29.** 380.61 **31.** 3.370 **33.** 8761 **35.** 106° **37.** 4.8 **39.** 78 **41.** 80 **43.** 1040 **45.** 9730 feet

■ 2.4 Integrated Review p. 68

47. > **49.** = **51.** $\frac{1}{2}$ **53.** $1\frac{3}{4}$ **55.** 1.5

■ 2.4 Mid-Chapter Self-Test p. 69

1. $5x+15$ **2.** $2a+4b+8$ **3.** a, d **4.** e, g **5.** Distributive, \$22.20 **6.** $12a$ **7.** $3x+8$ **8.** $9x+21$ **9.** $7x+8$, 36 **10.** a and c **11.** $2(3+x+2)$ or $2x+10$, $3(x+2)$ or $3x+6$ **12.** 13 **13.** 80 **14.** 34 **15.** $x+13-13=28-13$, $x=15$ **16.** $19+4=m-4+4$, $23=m$ **17.** $7+y-7=11-7$, $y+7-7=11-7$, $y=4$ **18.** 10.6 **19.** 13.7 **20.** 11.6

■ 2.5 Communicating About Mathematics p. 72

A. 150 miles **B.** 200 kilometers **C.** 10 dollars **D.** 280 miles **A–D.** Situations vary. **A.** Driving a truck at 50 mph for 3 hours **B.** Riding on a train that is traveling at 80 km per hour for 2.5 hours **C.** Buying 4 pounds of meat at \$2.50 per pound **D.** Buying 10 gallons of gas for a car that uses a gallon of gas for every 28 miles driven

■ 2.5 Guided Practice p. 73

1. See page 71. **2.** Answers vary. If $\frac{n}{3}=6$, then $3 \bullet \frac{n}{3}=3 \bullet 6$ and $n=18$. **3.** Answers vary. If $3n=6$, then $\frac{3n}{3}=\frac{6}{3}$ and $n=2$. **4.** Divide by 6, or multiply by $\frac{1}{6}$ **5.** Multiply by 3 **6.** Answers vary.

■ 2.5 Independent Practice pp. 73–74

7. The product of 2 and x is 4; 2 **9.** The quotient of b and 2 is 3; 6 **11.** 4 **13.** 8 **15.** 30 **17.** 72 **19.** 25 **21.** 40 **23.** 100 **25.** 125 **27.** 2.1 **29.** 524 **31.** 25.6 **33.** 7.5

35.

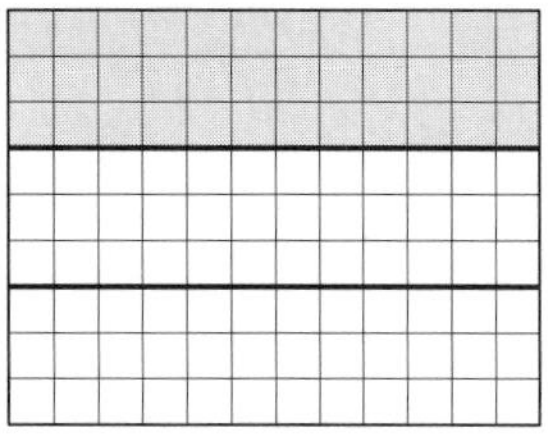

37. 3 **39.** 42 **41.** 9225 **43.** 2312 **45.** 5 **47.** 19 **49.** $4f=28$, 7 **51.** $\frac{b}{12}=2$, 24 **53.** $5x=3.75$, 0.75 **55. a.** $4700=50L$ **b.** 94 ft **57.** $24\times15.75=t$, 378

■ 2.5 Integrated Review p. 74

59. 2 hr 27 min **61.** $\frac{7}{12}$ yd = 21 in. = 1 ft 9 in.

■ 2.6 Communicating About Mathematics p. 77

3; 10 cassettes and 0 discs, 0 cassettes and 8 discs, 5 cassettes and 4 discs

■ 2.6 Guided Practice p. 78

1. $t-20°$ **2.** $s+5$ **3.** $0.25m+1.40$ **4.** $5n-15$ **5.** Write a verbal model, assign labels to it, then write an algebraic model. **6.** Numbers and variables

■ 2.6 Independent Practice pp. 78–79

7. b **9.** f **11.** a **13.** $32-x$ **15.** $\frac{x}{23}$ **17.** $12x+9$ **19.** $3x-4$ **21.** $18(5x)$ **23.** $\frac{b}{2+y}$ **25.** $x(y+13)$ **27.** $m+4$ **29.** $3m$ **31.** $p-3$ **33.** $3p$ **35.** $a-8$ **37.** $\frac{3}{4}a$ **39. a.** $1.5c+b$ **b.** \$8.50 **c.** \$8 **41.** 8 million **43.** 40.4 million

■ 2.6 Integrated Review p. 80

45. $16a-65$, 15 **47.** $8a$, 40 **49.** $11a+62$, 117

■ 2.6 Mixed Review p. 80

1. Subtraction **3.** Multiplication **5.** 4 **7.** $\frac{2}{3}$ **9.** 10 **11.** 0 **13.** $5f=10$, 2 **15.** $6g=12$, 2 **17.** $6s=18$, 3 **19.** $3h=9$, 3

2.7 Communicating About Mathematics p. 82

Answers vary.

2.7 Guided Practice p. 83

1. Equation, 16 **2.** Expression, $2x+6$ **3.** Equation, 9 **4.** Equation, 21 **5.** $c-21=84$, 105 **6.** $10x=75$, 7.5

2.7 Independent Practice pp. 83–84

7. c **9.** a **11.** b **13.** $d+9=20$, 11 **15.** $\frac{p}{12}=4$, 48 **17.** $\frac{y}{45}=5$, 225 **19.** The sum of a number and 15 is 33. **21.** The product of 7 and a number is 56. **23.** The sum of 11 and a number is 23. **25.** Larger number = smaller number + 251, Larger number: 420, smaller number: s; $420=s+251$, $169=s$; Smaller number is 169. **27.** $25=2(w+8)$, $4.5=w$ **29.** This week's sales

2.7 Integrated Review p. 84

31. 4 **33.** 4 **35.** 3.2 **37.** 5

2.8 Communicating About Mathematics p. 86

Answers vary.

2.8 Guided Practice p. 87

1. d, a, f, c, b, e **2.** The solution of the equation may not be the answer to the question.

2.8 Independent Practice pp. 87–88

3. Points needed = Points obtained + Final score **5.** $460=(89+85+92+97)+x$ **7.** Yes **9.** Respectively: 29, 8 + 11, x (miles) **11.** 10 **13.** Sales commission = Commission rate × Annual sales **15.** $x=\frac{1}{20}\cdot 150{,}000$ **17.** \$24,000 **19.** Each shirt sells for \$9.50; You sell 17 T-shirts; \$161.50

2.8 Integrated Review p. 88

21. 75 **23.** 5400 **25.** 21 cm **27.** ≈ 7.21 ft

2.9 Communicating About Mathematics p. 90

A. $11>x>1$ **B.** $8>x>2$

2.9 Guided Practice p. 91

1. See page 89. **2.** True **3.** Yes **4.** No **5.** Yes **6.** No **7.** $x\le 3$ **8.** $x\ge 3\frac{1}{3}$ **9.** $x>14$ **10.** Answers vary. No more than 10 people are allowed on the elevator.

2.9 Independent Practice pp. 91–92

11. 1, $3\frac{1}{2}$ **13.** 46, 1000 **15.** $2\frac{1}{4}$, 2 **17.** $x<6$ **19.** $y\le 6$ **21.** $b\ge 0$ **23.** $t>4$ **25.** $x\ge 1$ **27.** $y\ge 4$ **29.** $x>2.4$ **31.** $k\ge 2$ **33.** $x>20.9$ **35.** $c+5\ge 19.36$, $c\ge 14.36$ **37.** $2x<42$, $x<21$ **39.** $20\ge\frac{m}{6}$, $120\ge m$ **41.** The sum of d and 11 is less than 52. **43.** The product of 3 and h is less than or equal to 60. **45.** 17 is greater than the difference of c and 31. **47.** Yes **49.** No **51.** Yes **53.** Yes **55. a.** Hot Wheelers **b.** 1.15 hours

c.

[Last-place time in minutes] − Minutes < [First-place time in minutes]

$85-m<69$

$16<m$

d.

[Cruisin' Kids' time in minutes] − Minutes ≤ [Brave Bikers' time in minutes]

$76-m\le 71$

$5\le m$

e. ≈ 17.4 mph

2.9 Integrated Review p. 92

57. > **59.** < **61.** >

2.9 Chapter Review pp. 94–96

1. $27+30$ **3.** $4x+24$ **5.** $11m$ **7.** $30x^2+12x$ **9.** $8x^2+4x$ **11.** $20w+22$ **13.** $15a+16b$ **15.** $5x+4y$ **17.** Yes **19.** No **21.** 8 **23.** 51 **25.** 9 **27.** 56 **29.** 4 **31.** Answers vary. **33.** $n+71$ **35.** $\frac{n}{8}$ **37.** $n-16=19$ **39.** $7n=84$ **41.** $\frac{n}{6}=3$, 18 **43.** $n-15\le 3$, $n\le 18$ **45.** 48 **47.** 52 **49.** 14.9 **51.** 21 **53.** 144 **55.** 7.26 **57.** $x>9$ **59.** $y<37.4$ **61.** $x>6$ **63.** $x<40$ **65.** $x>55$ **67.** $2x+26$, $11x+8$ **69.** Price of CD player: \$135; Dollars saved each week: \$14.50 − \$5; Number of weeks: n **71.** $n\ge\approx 14.2$ **73.** \$1110.1 million **75.** \$1789.5 million, \$2289.1 million, \$2749.3 million, \$3057 million, \$3518 million **77.** 1991, 1992, 1993, 1994 **79.** Answers vary. **81.** Answers vary; $13\times 36=468$ in.^{2}

CHAPTER 3

3.1 Communicating about Mathematics p. 101

A. True, distance cannot be negative. **B.** False, 0 is a counterexample. **C.** True, $6 > 4$ **D.** False, -6 lies to the left of -4 on the number line. **E.** True, 0 is neither positive nor negative. **F.** False, $-6 \leq -4$ but $6 \geq 4$.

3.1 Guided Practice p. 102

1. All of them **2.** 0, 1, 2, 3, 4 **3.** 1, 2, 3, 4 **4.** 1 **5.** -1 **6.** 0 **7.** Answers vary. 5 and -5 **8.** 5, -5 **9.** To the right, to the left **10.** Answers vary. In computing how far below par in golf.

3.1 Independent Practice pp. 102–103

11.

−3 −2 −1 0 1 2 3 4

13.

−6 −5 −4 −3 −2 −1 0 1

15.

−8 −7 −6 −5 −4 −3 −2 −1 0 1

17. $<$ **19.** $>$ **21.** $>$ **23.** $-1, 1$ **25.** 3, 3 **27.** -20, 20 **29.** 100, 100 **31.** -250 **33.** 25 **35.** -17 **37.** -15 **39.** $-6, -3, 0, 4, 5$ **41.** $-4, -2, -1, 0, 2$ **43.** $-5, -4, -2, 4, 6$ **45.** 40 **47.** 75 **49.** 45

51.

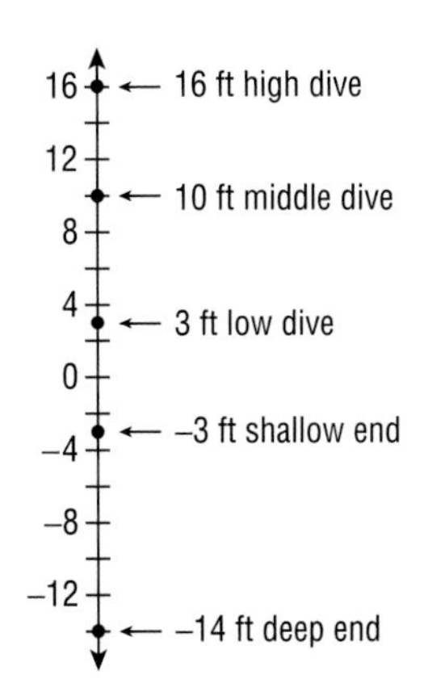

53. 30 ft

3.1 Integrated Review p. 103

55. 21 **57.** 60 **59.** 8 **61.** 11

3.2 Communicating about Mathematics p. 106

A. $-25 + 15 = -10$ **B.** $-15 + (-10) = -25$ **C.** $-40 + 40 = 0$

3.2 Guided Practice p. 107

1. $4 + 3 = 7$ **2.** $2 + (-2) = 0$ **3.** $7 + (-5) = 2$ **4.** $-7 + (-5) = -12$ **5.** $5 + (-2) = 3$ **6.** $1 + (-6) = -5$ **7.** $4 + (-4) = 0$ **8.** $-3 + (-5) = -8$ **9.** One rule is for integers with the same sign, and a second rule is for integers with different signs. Since 8 and -11 have different signs, the second rule should be used to add them. **10.** Positive, $|b| > |a|$

3.2 Independent Practice pp. 107–108

11. $11 + 15 = 26$ **13.** $-13 + (-13) = -26$ **15.** $10 + (-10) = 0$ **17.** $-13 + 13 = 0$ **19.** $13 + 0 = 13$ **21.** $0 + 15 = 15$ **23.** $2 + (-9) = -7$ **25.** $-16 + 12 = -4$ **27.** 7, 6, 5, 4, 3. Sums decrease by 1. **29.** $-4, -2, 0, 2, 4$. Sums increase by 2. **31., 33.** Answers vary. **35.** 3 **37.** -3 **39.** d; -2, \$2 debt **41.** b; 45, 45-yard line **43.** Answers vary. **45.** a and d, b and c

3.2 Integrated Review p. 108

47. 16 **49.** 9 **51.** 3 **53.** 8

3.3 Communicating about Mathematics, p. 110

An overall profit; the graph shows more profit above the line for zero than below it, so the sum of the profits and losses will be positive.

3.3 Guided Practice p. 111

1.

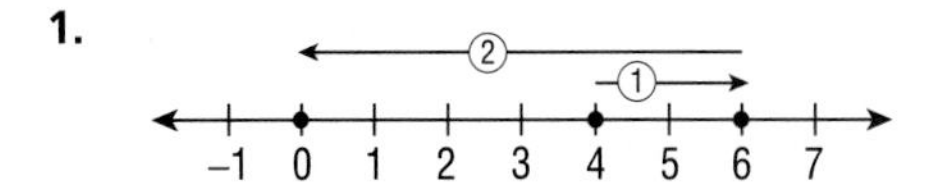

$4 + 2 + (-6) = 0$

2.

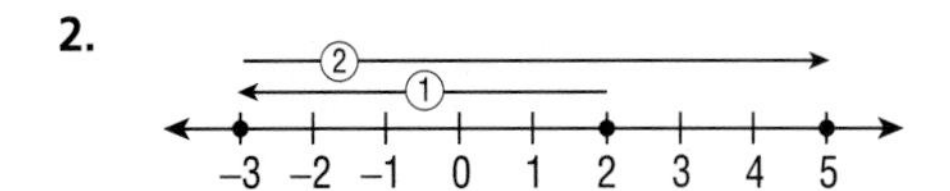

$2 + (-5) + 8 = 5$

3. $-3x, 5x, 7$; $2x + 7$ **4.** Positive; $a + c$ is positive because $|c| > |a|$, and b is positive.

3.3 Independent Practice, pp. 111–112

5. $4+(-5)+6=5$ **7.** $-7+1+(-8)=-14$ **9.** $-8+12+(-1)=3$ **11.** $-12+(-4)+(-8)=-24$ **13.** $5+(-6)+(-13)=-14$ **15.** $-7+(-6)+2+(-7)=-18$ **17.** 26 **19.** -71 **21.** -45

23. Negative, $|-237|>|122+69|$ **25.** $3x+9$, 15 **27.** x, 2 **29.** $3x$, 6 **31.** $25x$, 50 **33.** $10x+8$, 28 **35.** $2x+(-5)$, -1 **37.** $<$ **39.** $>$ **41.** $<$ **43.** No, $6+(-3)+(-4)+8<10$ **45.** 33,000 ft **47.** 34,000 ft

3.3 Integrated Review p. 112

49. $=$ **51.** $<$

3.3 Mixed Review p. 113

1. 6 **3.** 55 **5.** 12 **7.** 3 **9.** $=$ **11.** $<$ **13.** 16 **15.** 20 **17.** 2 **19.** 10

3.4 Communicating About Mathematics p. 116

"There isn't no way" means "there is a way." Two negative conditions applied to a noun and two negative conditions applied to a number both result in a positive condition.

3.4 Guided Practice p. 117

1. $5-(-2)=5+2=7$ **2.** $rs-rt$ **3.** All. Subtracting a quantity gives the same result as adding its opposite. **5.** No. The equation simplifies to $3x+12=3x-12$, which cannot be true for any value of x.

3.4 Independent Practice pp. 117–118

7. $11-6=5$ **9.** $13-18=-5$ **11.** $23-(-8)=31$ **13.** $-10-7=-17$ **15.** $-5-(-5)=0$ **17.** $-5-5=-10$ **19.** $0-27=-27$ **21.** $0-(-61)=61$ **23.** 4, -6 **25.** -1, -11 **27.** 0, 0 **29.** $3x+(-2x)+16$; $3x$, $-2x$, 16 **31.** $7a+(-5b)$; $7a$, $-5b$ **33.** $3x-17$ **35.** $4y-2$ **37.** $3b$ **39.** $-5a-4$ **41.** $-2m+8$ **43.** $-4x$ **45.** -8 **47.** 40 **49.** -56 **51.** The [+/-] key refers to the sign of a number, usually negative; the [−] key refers to the operation of subtraction and is used between two numbers. **53.** $x=2$, $y=3$ **55.** $x=-2$, $y=-3$ **57.** 258°

3.4 Integrated Review p. 118

59.

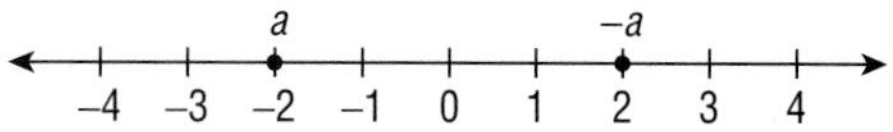

61.

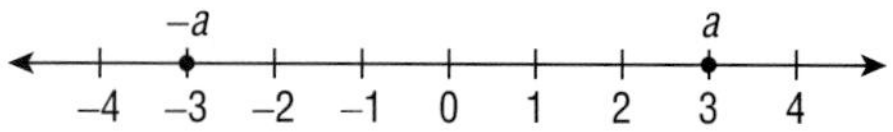

63. 37 **65.** 9

3.4 Mid-Chapter Self-Test p. 120

1.

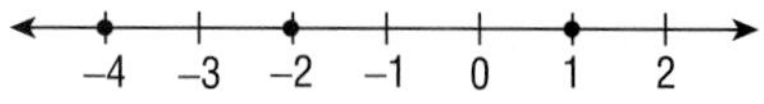

2.

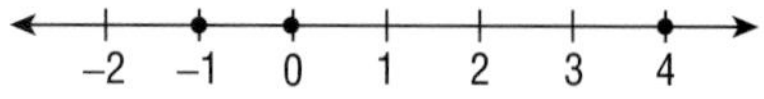

3.

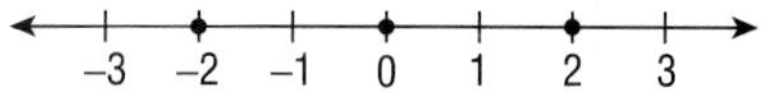

4. -7, 7 **5.** 5, 5 **6.** -3, -2, 3, 4 **7.** 8 **8.** 10 **9.** -5 **10.** 10 **11.** 6 **12.** 10 **13.** 6 **14.** -12 lb **15.** C's, $|-7|$ or 7 is the biggest change. **16.** $6a-6b$, 24 **17.** $4a-2b$, 20 **18.** -1 **19.** 0 **20.** Your score

3.5 Communicating About Mathematics p. 123

About -31°F

3.5 Guided Practice p. 124

1. $3(-4)$, -12 **2.** $3(-x)$, $-3x$ **3.** $4(-8)$, -32 **4.** Positive, negative, negative **5.** x^2 **6.** $(-4)^2$

3.5 Independent Practice pp. 124–125

7. $3 \cdot 9=27$ **9.** $-4 \cdot (-6)=24$ **11.** $5 \cdot (-11)=-55$ **13.** $(-7)(-9)=63$ **15.** $(-10)(3)=-30$ **17.** $(-20)(0)=0$ **19.** $-7x$ **21.** $14a$ **23.** -16 **25.** 2 **27.** -27 **29.** 64 **31.** Negative **33.** 8.4 **35.** -21 **37.** 8130 **39.** 344 **41.** -2 **43.** -2 **45.** -9 **47.** -2 **49.** 1 **51.** -2 **53.** 100°C; 273.15K; -40°C, 233.15K; -10°C, 263.15K

3.5 Integrated Review p. 125

55. Values vary. Either is -5, other is 6.

3.6 Communicating About Mathematics p. 127

Lists vary. 0, 1, 2, 3, 4, −4, −6, −8, −10, −12

3.6 Guided Practice p. 128

1. Positive **2.** Negative **3.** Negative **4.** Positive **5.** Multiply 3 by −2 to see if you get −6. **6.** Negative **7.** Positive **8.** Positive

3.6 Independent Practice pp. 128–129

9. 27 **11.** −6 **13.** −32 **15.** 53 **17.** 0 **19.** −34 **21.** −98 **23.** 79 **25.** −1, −2, −3, −4, −5. Quotients decrease by 1. **27.** 0, 0, 0, 0, 0. Quotients are always 0. **29.** 1 **31.** $1\frac{1}{2}$ **33.** −49 **35.** −24 **37.** 48 **39.** −9 **41.** 36 **43.** −1 **45.** −6 **47.** 44.648 seconds **49.** Yes, see example for Exercise 48.

3.6 Integrated Review p. 129

51. Always **53.** Sometimes **55.** −9 **57.** 23 **59.** 88 **61.** 5 **63.** 72 **65.** $3x-12$ **67.** $-4x+12$

3.6 Mixed Review p. 130

1. 12 **3.** 3 **5.** −32 **7.** −21 **9.** $3b=4.2$ **11.** Every number, after the first, is −2 times the preceding number. 32, −64, 128. **13.** Every number, after the first, is $-\frac{1}{2}$ times the preceding number. $-\frac{1}{8}, \frac{1}{16}, -\frac{1}{32}$. **15.** 27 **17.** 9 **19.** 3

3.7 Communicating About Mathematics p. 132

Answers vary.

3.7 Guided Practice p. 133

1. See pages 65 and 71. **2.** Addition; −4 **3.** Addition (of 8) or Subtraction (of 8); −14 **4.** Multiplication; −35 **5.** Multiplication (of $-\frac{1}{5}$) or Division (by −5); −7 **6.** Answers vary. In computing how far above or below par in golf.

3.7 Independent Practice pp. 133–134

7. Yes **9.** No, $s=4$ **11.** Yes **13.** −13 **15.** 24 **17.** −9 **19.** −12 **21.** −63 **23.** −17 **25.** −14 **27.** $-\frac{1}{2}$ **29.** $x-20=-4$, 16, the difference of 16 and 20 is −4. **31.** $51=-3a$, −17, 51 is the product of −17 and −3. **33.** −1217 **35.** −16 **37.** 1288 **39.** b **41.** e **43.** c **45.** a; 25,370; 25,370 ft **47.** $P=5n-(800+20+200+70)$

3.7 Integrated Review p. 134

49. 57 **51.** 37

3.8 Communicating About Mathematics p. 136

x	3	2	1	0	−1
y	0	1	2	3	4
(x, y)	(3, 0)	(2, 1)	(1, 2)	(0, 3)	(−1, 4)

x	−2	−3
y	5	6
(x, y)	(−2, 5)	(−3, 6)

The graph is a straight line.

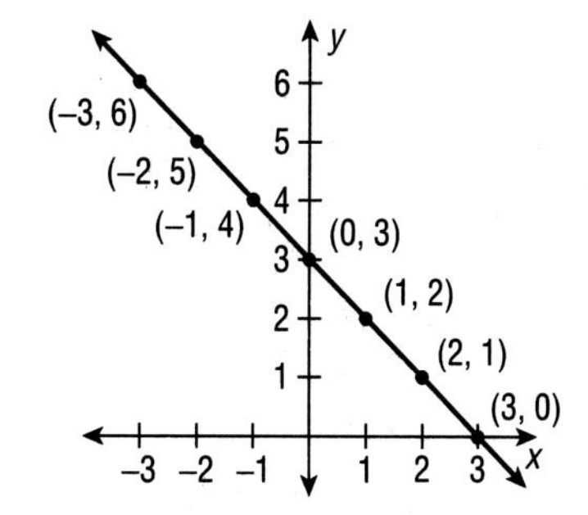

3.8 Guided Practice p. 137

1. Draw two lines that intersect at a right angle. Label the vertical line with a letter, usually *y*, and label the horizontal line with another letter, usually *x*. Label points to the right of and up from their intersection with positive numbers, and label points to the left of and down from their intersection with negative numbers. **2.** Substitute the values of *x* and *y* into the equation and see if the resulting equation consists of the same value on each side of the equal sign. If this occurs, then the ordered pair is a solution. **3.** (−4, 4), II **4.** (2, 3), I **5.** (3, 2), I **6.** (0, 0), none **7.** (−3, −2), III **8.** (−5, −4), III **9.** (2, −5), IV **10.** (0, −4), none

3.8 Independent Practice pp. 137–138

11. J, IV **13.** I, I **15.** L, III
17.–24.

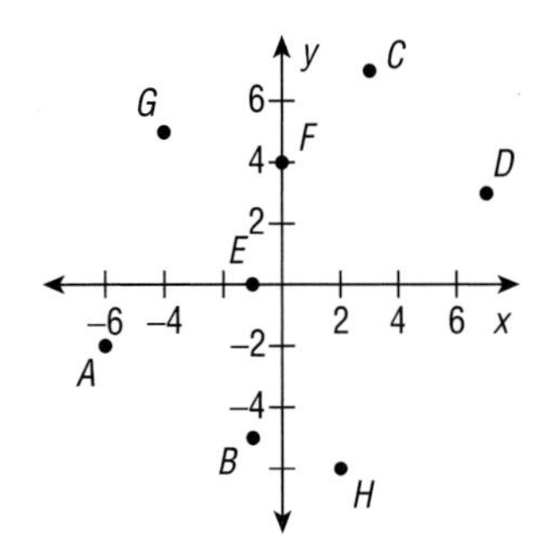

17. III **18.** III **19.** I **20.** I **21.** none **22.** none
23. II **24.** IV **25.** III **27.** II
29.

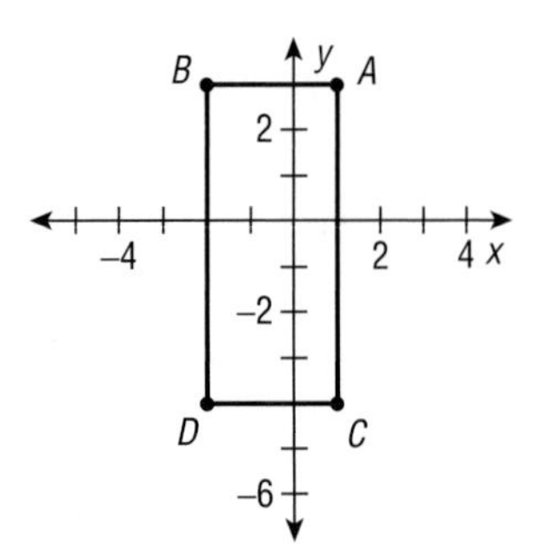

21 units^2, 20 units

31. $3+7=10$; (1, 4), (2, 5), (3, 6) **33.** $8+(-2)=6$; (1, 5), (2, 4), (3, 3)

35.

x	−1	0	1
y	5	0	−5
(x, y)	(−1, 5)	(0, 0)	(1, −5)

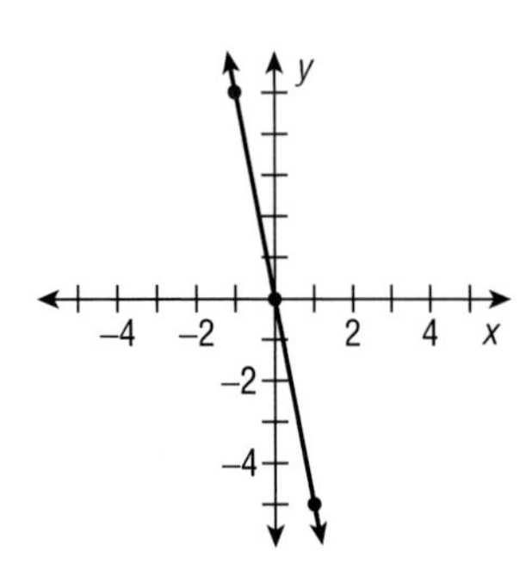

The graph is a straight line.

37.

x	0	2	4
y	3	4	5
(x, y)	(0, 3)	(2, 4)	(4, 5)

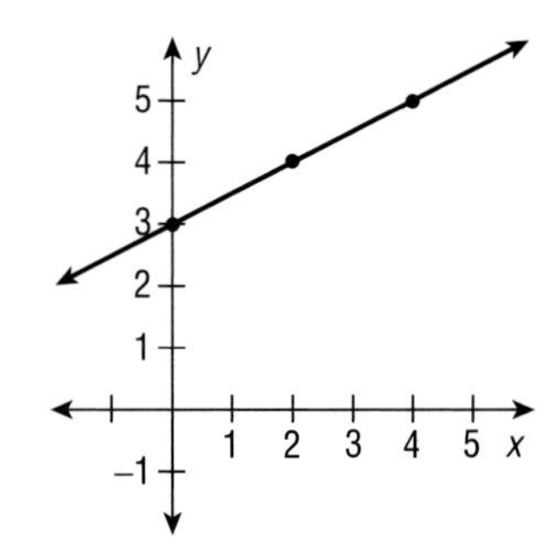

The graph is a straight line.

39.

x	0	1	2
y	−2	1	4
(x, y)	(0, −2)	(1, 1)	(2, 4)

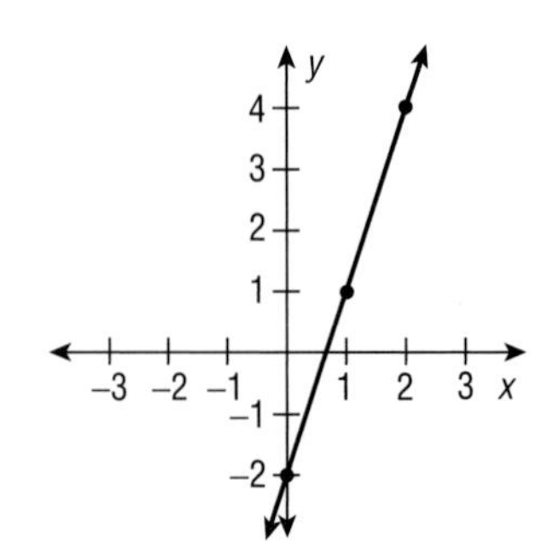

The graph is a straight line.

41. 34 **43.** No; when prices go up, sales will usually go down.

3.8 Integrated Review p. 138

45. −7 **47.** −5 **49.** $5x-18$

3.8 Chapter Review pp. 140–141

1.

–8 –7 –6 –5 –4 –3 –2 –1 0 1

3.

–3 –2 –1 0 1 2 3 4 5 6

5. $-7, -3, 0, 5, 8$ **7.** -365 **9.** 17 **11.** 1
13. -42 **15.** -41 **17.** -2 **19.** 33 **21.** $3x-6$
23. $4x-8$ **25.** $5x-6$ **27.** $24x+12$ **29.** 2
31. Investment 1 is the stock purchase because a bank savings account does not lose money as Investment 1 did at 6 months; so Investment 2 is the bank savings account. **33.** Positive
35. Positive **37.** -21 **39.** 24 **41.** -4 **43.** -36
45. 0.5 **47.** $x-(-8)=7, -1$ **49.** $4x=-28, -7$
51. -15 **53.** 14 **55.** -3 **57.** 56
59.–62.

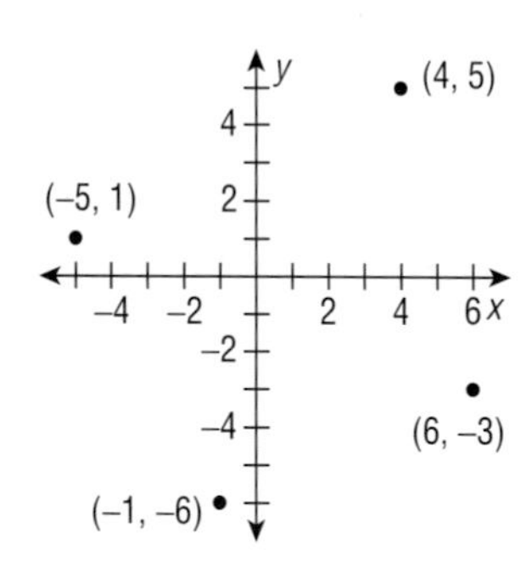

59. II **60.** III **61.** I **62.** IV
63.

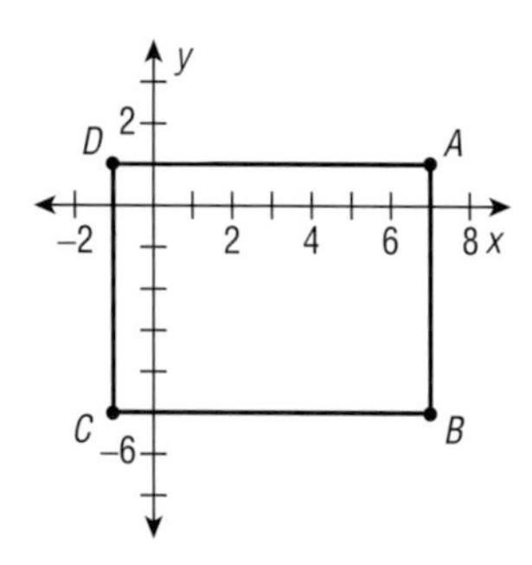

Rectangle; 28 units, 48 units^2
65. No
67.

x	-1	0	1
y	0	1	2
(x, y)	$(-1, 0)$	(0, 1)	(1, 2)

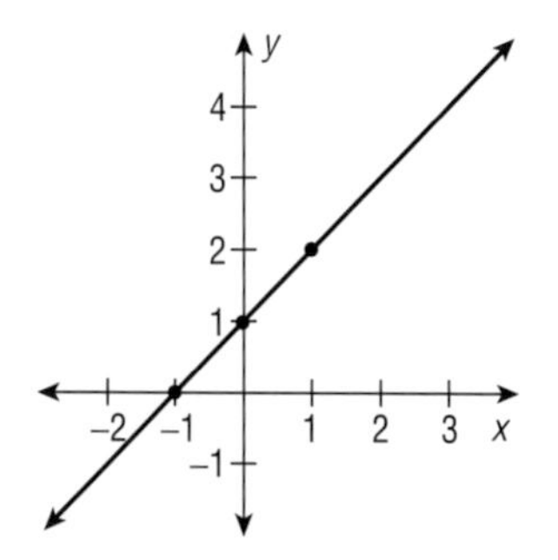

The graph is a straight line.

69.

x	-1	0	1
y	2	0	-2
(x, y)	$(-1, 2)$	(0, 0)	$(1, -2)$

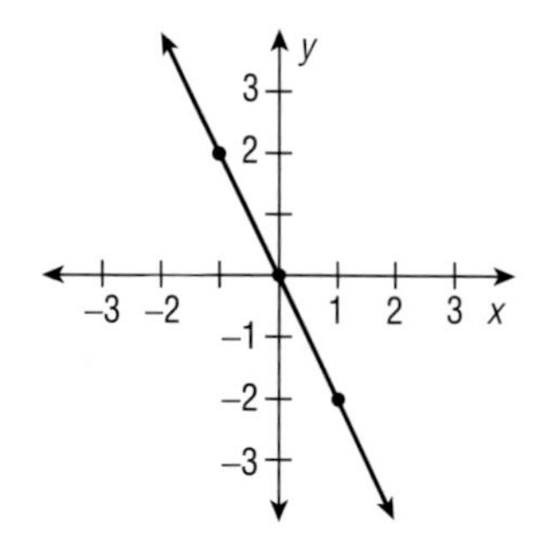

The graph is a straight line.

71. Always. The algebraic expression $-n$ has the same meaning as the verbal expression *the opposite of n.* **73.** Sometimes. The statement is false when $n<0$. **75.** Gymnast 3 **77.** -5 **79.** -2
81. 7 **83.** -8 **85.** -9 **87.** 1 **89.** -4 **91.** -1
93. -7 **95.** 8

■ *3.8 Cumulative Review for Chapters 1–3 pp. 144–145*

To get the next number: **1.** Subtract 2 from the preceding number; 12, 10, 8 **3.** Add 2 to both the numerator and the denominator of the preceding number; $\frac{9}{10}, \frac{11}{12}, \frac{13}{14}$ **5.** To get the next letter, name the letter that is 3 positions earlier in the alphabet than the position in the alphabet of the preceding letter; N, K, H **7.** 65,536 **9.** 6.93
11. 11,441.56 **13.** 70 **15.** 280 **17.** -1 **19.** -8
21. 72 **23.** 13 **25.** No, all sides not segments
27. Yes, hexagon **29.** $3x+3y+9, 30$ **31.** $2z+6y$, 36 **33.** $8x+z+y$, 34 **35.** $10z-9y-25, -1$
37. $18x$, 54 **39.** $zy+2z+|y|$, 40

41. –4 –3 –2 –1 0 1 2 3 4 5

43. –4 –3 –2 –1 0 1 2 3 4 5

45.–48.

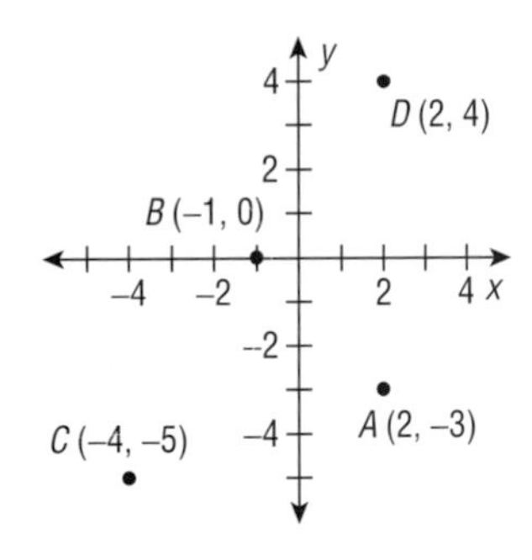

49. 9 **51.** -8 **53.** 10 **55.** -96 **57.** $n<1$ **59.** $x \geq 7$ **61.** $c>2$ **63.** $q \leq 75$ **65.** $-12=n+9$, -21 **67.** $7n>91, n>13$ **69.** Figure 1: $(-1, 4)$, $(1, 4)$, $(1, 2)$, $(-1, 2)$; 8 units. Figure 2: $(-3, 3)$, $(-1, 3)$, $(-1, -1)$, $(-3, -1)$; 12 units. Figure 3: $(-1, 2)$, $(2, 2)$, $(2, -1)$, $(-1, -1)$; 12 units. Figure 4: $(2, 4)$, $(5, 4)$, $(5, -1)$, $(2, -1)$; 16 units. 36 units^2 **71.** About 64.6 mph
73.

x	0	2	4	6	8	10	12
y	6	5	4	3	2	1	0

CHAPTER 4

4.1 Communicating about Mathematics p. 150

14 hours

4.1 Guided Practice p. 151

1. Add 4 to each side. **2.** Subtract 3 from each side. **3.** Subtract 2 from each side. **4.** 8 was not divided by 3. **5.** b **6.** d **7.** a **8.** c

4.1 Independent Practice pp. 151–152

9. 3 **11.** -1 **13.** -13 **15.** 8 **17.** -20 **19.** -45 **21.** 65 **23.** 1 **25.** $3n+7=34, 9$ **27.** $\frac{1}{4}n-2=5, 28$ **29.** $21+7x=-14, -5$ **31.** $57-9n=129, -8$ **33.** $3x+17=38, 7$ **35.** Both of you will leave the museum, walk 2 blocks west to the subway, and ride the subway downtown. Then you and your friend will part at your apartment. **37.** $7\frac{1}{2}$ hours

4.1 Integrated Review p. 152

39. 66 miles **41.** 216 meters

4.2 Communicating about Mathematics p. 154

$300=5n-(250+2n)$, $183.\overline{3}$, 184

4.2 Guided Practice p. 155

1. Combine like terms, subtract 8 from each side, divide each side by 2. **2.** Combine like terms, subtract 7 from each side, $(-1)x=-x$, divide each side by -1. **3.** Combine like terms, subtract 11 from each side, divide each side by -8.

4.2 Independent Practice pp. 155–156

5. No, -3 **7.** No, -7 **9.** Yes **11.** -6 **13.** -13.5 **15.** 8 **17.** -8 **19.** 4 **21.** -1 **23.** $4y-y-5=-29, -8$ **25.** $(5x+15)+(x+20)+(4x-5)=180$, 15 **27.** 288

4.2 Integrated Review p. 156

29. $7x$; $-14, -7, 0, 7, 14$; each number, after the first, is 7 more than the preceding number. **31.** $12x-3$; $-27, -15, -3, 9, 21$; each number, after the first, is 12 more than the preceding number. **33.** $-11x$; $22, 11, 0, -11, -22$; each number, after the first, is 11 less than the preceding number.

4.3 Communicating about Mathematics, p. 159

32 or 33, 1991 or 1992

4.3 Guided Practice p. 160

1. 3 **2.** -4 **3.** $-\frac{1}{5}$ **4.** Not possible **5.** $\frac{1}{7}, \frac{1}{7}, -4$ **6.** $-4, -4, -48$ **7.** 1967 **8.** 1995 **9.** 1940 **10.** 1925

4.3 Independent Practice p. 160–161

11. Divide each side by 3, multiply each side by $\frac{1}{3}$. **13.** Divide each side by 9, multiply each side by $\frac{1}{9}$. **15.** 8 **17.** 64 **19.** $\frac{2}{3}$ **21.** 5 **23.** -1 **25.** 7 **27.** 36 **29.** -5
31. a.

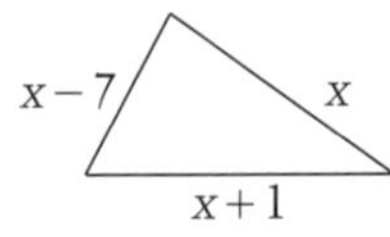

b. 5 cm, 12 cm, 13 cm **33.** CT: ≈ 3,287,133; UT: ≈ 1,722,867

4.3 Integrated Review p. 161

35. about \$20,700 **37.** about \$25,000

4.3 Mixed Review p. 162

1. 27 **3.** 3

5.

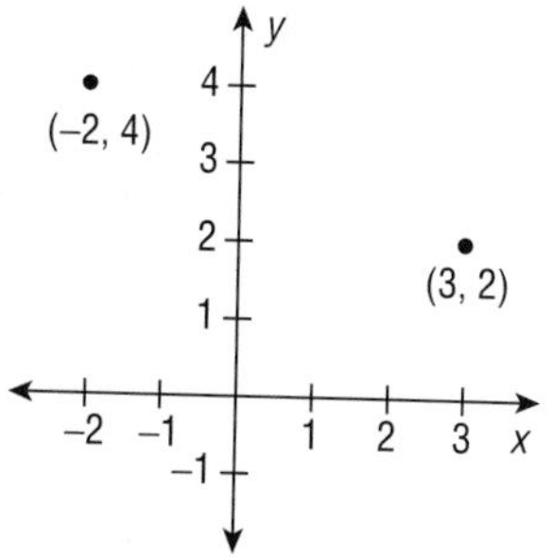

7. $8x-12$ **9.** 7 **11.** 2.6 **13.** 30

4.4 Communicating about Mathematics p. 164

For every \$10,000 increase in salary, there is a corresponding \$300,000 increase in sales; \$1,700,000.

4.4 Guided Practice p. 165

1.–3. Answers vary. **4.** Distributive Property, combine like terms, add 12 to each side, simplify, divide each side by 10, simplify.

4.4 Independent Practice pp. 165–166

5. $2(x+2) \neq 2x+2$, 0 **7.** $-2x-4x \neq -2x$, $-\frac{2}{3}$ **9.** 7 **11.** -3 **13.** 8 **15.** -6 **17.** -80 **19.** -10 **21.** 0 **23.** Answers vary. **23. a.** $\frac{1}{3}x-2=6$, $\frac{1}{3}x=8$, $x=24$ **b.** $x-6=18$, $x=24$ **25.** $12(x-5)=48.9$ **27.** $5x+5+2(3x-6)=180$, 17 **29. a.** $(96+80+14)x+24(x+40)=3370+800$ **b.** \$15

4.4 Integrated Review p. 166

31. 2, -2 **33.** 3, -3 **35.** b

4.4 Mid-Chapter Self-Test p. 167

1.

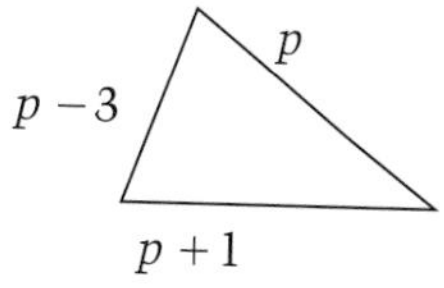

3 cm, 6 cm, 7 cm

2. $\frac{2}{3}$ **3.** $-\frac{21}{8}$ **4.** $\frac{5}{16}$ **5.** 1 **6.** 7 **7.** -4 **8.** 3 **9.** 4 **10.** -1 **11.** 2 **12.** 5 **13.** -6 **14.** 4 **15.** 5 **16.** $2x+3=21$, 9 **17.** $16x-30=-2$, $1\frac{3}{4}$ **18.** $17=2x+x+x+(-4)$, $5\frac{1}{4}$ **19.** 1005 miles **20.** 417 miles

4.5 Communicating about Mathematics p. 170

x	0	1	2	3	4	5	6
Rectangle's perimeter: $4x+6$	6	10	14	18	22	26	30
Triangle's perimeter: $3x+9$	9	12	15	18	21	24	27

$x > 3$

4.5 Guided Practice p. 171

1. Answers vary. **2.** The sides of an equilateral triangle are equal in length. **3.–6.** Answers vary.

4.5 Independent Practice pp. 171–172

7. d **9.** a **11.** 2 **13.** -3 **15.** 15 **17.** -4 **19.** 1 **21.** -6 **23.** $x+x+12=3x+5$, 7 **25.** 10 **27.** 5, 24 **31.** ≈ 40.24 seconds

4.5 Integrated Review p. 172

33. First row: 6, -1; second row: 0.4; third row: -2

4.6 Communicating about Mathematics p. 174

Answers vary.

4.6 Guided Practice p. 175

1. Answers vary. **2.** 60 in., $\frac{1}{3}$ in. per yr, n yr, 56 in., $\frac{4}{3}$ in. per yr, n yr **3.** $60+\frac{1}{3}n=56+\frac{4}{3}n$ **4.** 4 **5.** In 4 years

4.6 Independent Practice pp. 175–176

7.

[Temp. in Santa Fe] − [Santa Fe rate] × [Time for temp. to be the same]

= [Temp. in Minot] + [Minot rate] × [Time for temp. to be the same]

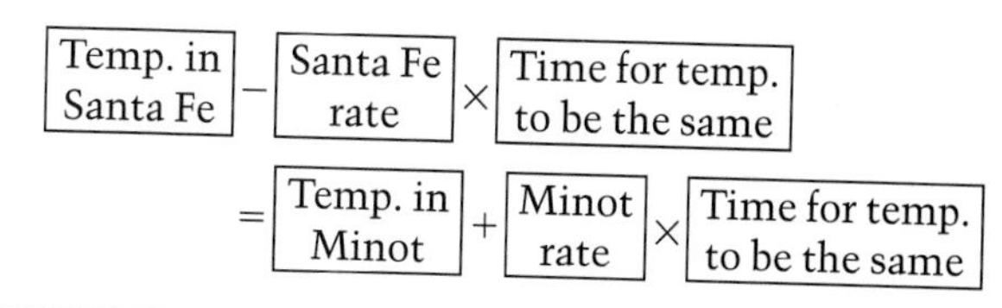

9. $86-3h=56+2h$ **11.** 6 hours **13.** Club 1 is cheaper for less than 20 hours, Club 2 is cheaper for more than 20 hours. **15.** Hexagon: 8, 48; square: 12, 48

■ *4.6 Integrated Review pp. 176–177*

17. 12 **19.** 9 **21.** -8 **23.** $-7h$

■ *4.6 Mixed Review p. 177*

1. 7 **3.** 2 **5.** 2 **7.** 4 **9.** $6n=3, \frac{1}{2}$ **11.** $5x=1-x, \frac{1}{6}$ **13.** 17 **15.** $x \geq \frac{1}{2}$ **17.** $y>3$ **19.** $r \geq 24$

■ *4.7 Communicating about Mathematics p. 179*

When a call is 22 minutes, the two companies charge the same amount. Company A should be chosen when the calls are shorter than 22 minutes, and Company B should be chosen when the calls are longer than 22 minutes.

■ *4.7 Guided Practice p. 180*

1. a, because it is in decimal form **2.** 7.72 **3.** 1st: exact, 2nd: approximate **4.** b, a person's weight varies at least a tenth of a pound each day.

■ *4.7 Independent Practice pp. 180–181*

5. Rounding was done too early (in 2nd step). **7.** 1.67 **9.** -2.38 **11.** 5.06 **13.** -7.08 **15.** 2.77 **17.** -5.81 **19.** -1.00 **21.** \$5.99

23.

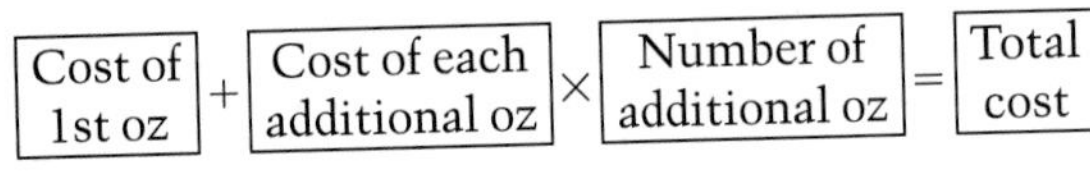

25.

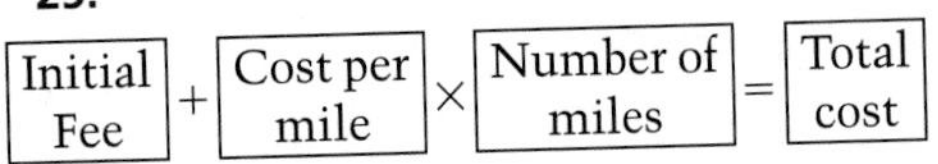

■ *4.7 Integrated Review p. 181*

27. 67.65 **29.** 5.39 **31.** 2.13 **33.** 0.59

■ *4.8 Communicating about Mathematics p. 183*

Explanations may be given to support "yes" or "no."

■ *4.8 Guided Practice p. 184*

1. Answers vary. **2.** b The area of a square is the square of a side. **3.** c (also b) The perimeter of a rectangle is the sum of twice the length and twice the width. **4.** a The area of a triangle is one-half the product of the base and height. **5.** b The perimeter of a square is four times a side. **6.** c (also b) The area of a rectangle is the product of the base and height.

■ *4.8 Independent Practice pp. 184–185*

7. Regular: All sides are congruent and all angles are congruent. Octagon: A polygon with 8 sides. **9.** 128 in. **11.** 3; 14 units, 15 units **13.** 5; 18 units **15.** 8; 25 units **17.** $112\frac{3}{11}^\circ, 67\frac{8}{11}^\circ$ **19.** 165 in.^{2} **21.** 90 ft

■ *4.8 Integrated Review p. 185*

23. 16

■ *4.8 Chapter Review pp. 187–190*

1. Adding 3 to a number **3.** Dividing a number by -4.2 **5.** 6 **7.** 20 **9.** 16 **11., 13.** Answers vary. **15.** 6 **17.** 9 **19.** 30 **21.** $57=6x+5x+x+9, 4$ **23.** The reciprocal of a number is equal to 1 divided by that number. **25.** $\frac{1}{7}$ **27.** -2 **29.** 32 **31.** 11 **33.** -9 **35.** 7 **37.** -2 **39.** 14 **41.** 1987 **43.** \$129.13 billion **45.** Subtract $6n$ from each side, add 39 to each side, divide each side by 6. **47.** 9 **49.** 4 mph for 1 hour is 4 miles. **51. a.** 8.39 **b.** -13.66 **c.** 26.00 **53.** 4.83 **55.** 2.64

57. [Price of game] + [Tax rate] × [Price of game] = [Money saved]

59. 442 cm^2 **61.** 12 ft **63.** 10 ft, 10 ft **65.** 390 ft^2 **67.** 200 feet per second **69.** CHEETAH **71.** MONKEY **73.** RHINOCEROS **75.** JAGUAR **77.** ORANGUTAN **79.** Answers vary.

CHAPTER 5

■ *5.1 Communicating About Mathematics p. 196*

Answers vary.

■ *5.1 Guided Practice p. 197*

1. Tokyo, Stockholm, London, Toronto, Los Angeles, Mexico City

2.

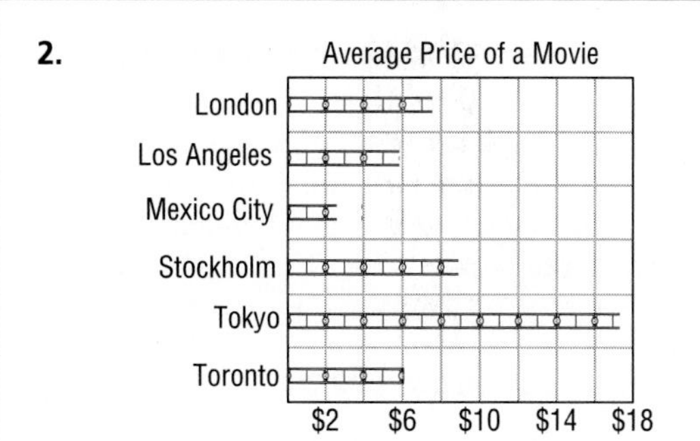

3. 10 **4.** 1950–1970

■ *5.1 Independent Practice pp. 197–198*

5.

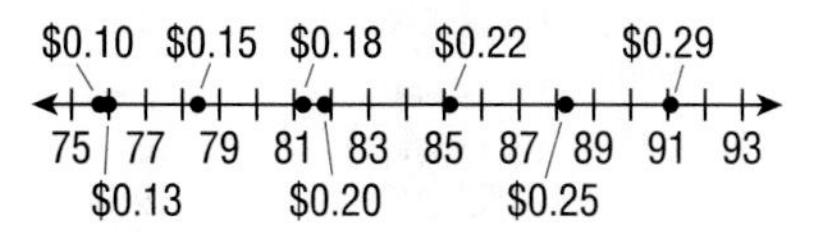

7. 10 million **9.** 20 million
11.

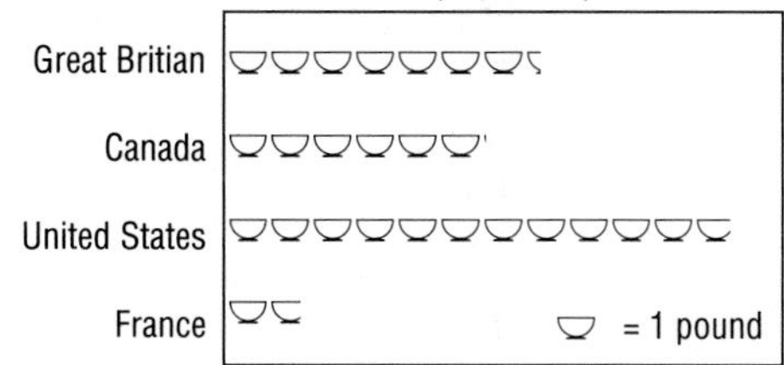

13. 1837 **15.** Becomes royal province, declares independence, becomes ninth state

■ *5.1 Integrated Review p. 198*

17. 700 **19.** 47,500 **21.** 16.64

■ *5.2 Communicating About Mathematics p. 200*

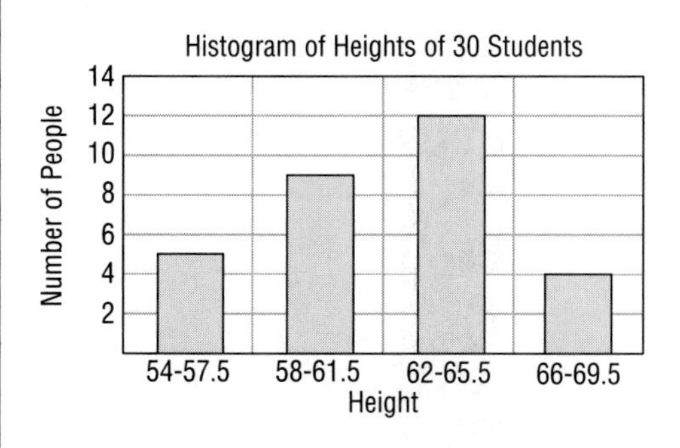

The histogram in the text gives a fairer representation of the data. The more the data is condensed, the less detail is observable; for example, the decrease in frequency from 58–59.5 to 60–61.5 before the final increase in frequency from 60–61.5 to 62–63.5 cannot be observed in the more condensed histogram.

■ *5.2 Guided Practice p. 201*

1. See page 200. **2.** To make a histogram **3.** Own a car, become rich **4.** Own a home, 48 **5.** 62–63.5, 68–69.5 **6.** 56–57.5 and 66–67.5, or 58–59.5 and 64–65.5

■ *5.2 Independent Practice pp. 201–202*

7. Simple, 1 question for 1 period of time **9.** Yes, include 2 more bars **11.** 3–5 and 6–8; 12–14 and 15–17 **13.** 15–17 **15.** Answers vary but may include the following conclusion. For the last three age brackets, there are large increases from 1995 to 2005.

■ *5.2 Integrated Review p. 202*

17.

Perimeter	Number of squares
4	16
8	9
12	4
16	1

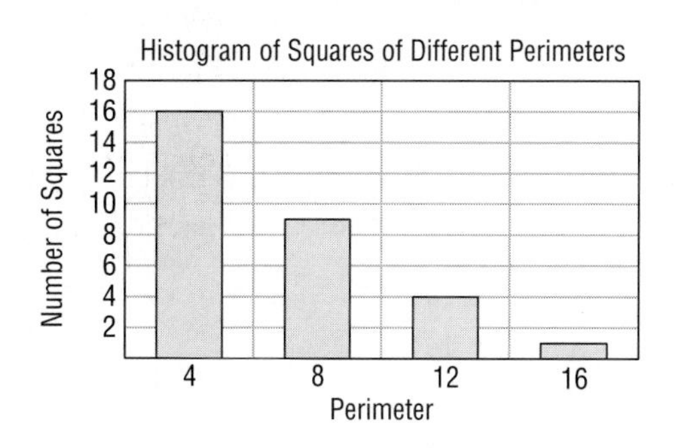

■ *5.3 Communicating About Mathematics p. 204*

$S = (n-2)\,180°$, 1080°; draw an 8-sided figure, measure each angle, and add the measures of the angles.

5.3 Guided Practice p. 205

1. 47 **2.** 17 **3.** 19 **4.** 13

5.3 Independent Practice pp. 205–206

5. Horizontal: 5 years, vertical: 40 million tons **7.** 150 million tons, 195 million tons **9.** Total waste, recycled waste **11.** Total waste equals the sum of the other three quantities.
13.

Base	2	3	4	5
Height	2	2	2	2
Area	2	3	4	5

15.

17. Slightly slower, explanations vary

5.3 Integrated Review p. 206

19. 16 **21.** 13 **23.** c

5.3 Mixed Review p. 207

1. Triple bar **3.** 24 million
5. Answers vary.

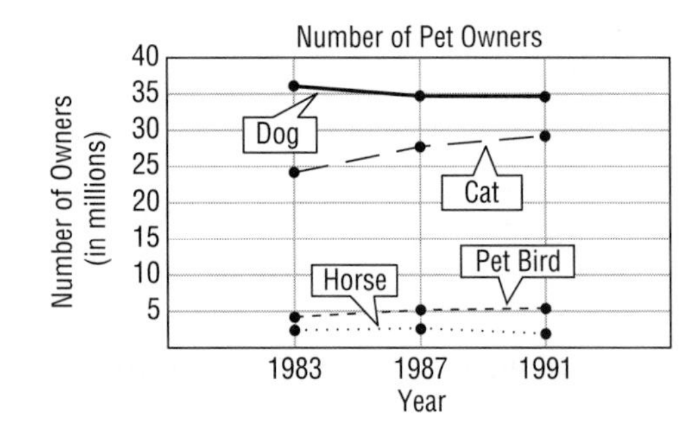

7. $1\frac{1}{2}$ **9.** $-6\frac{1}{2}$

5.4 Communicating About Mathematics p. 210

Answers vary. Because two kinds of people (men and women) choose among the same five colors, only a double or stacked bar graph or a double line graph would make sense.

5.4 Guided Practice p. 211

1. 2.4 **2.** Retail Salesperson **3.** 426 **4.** Answers vary.

5.4 Independent Practice pp. 211–212

5.

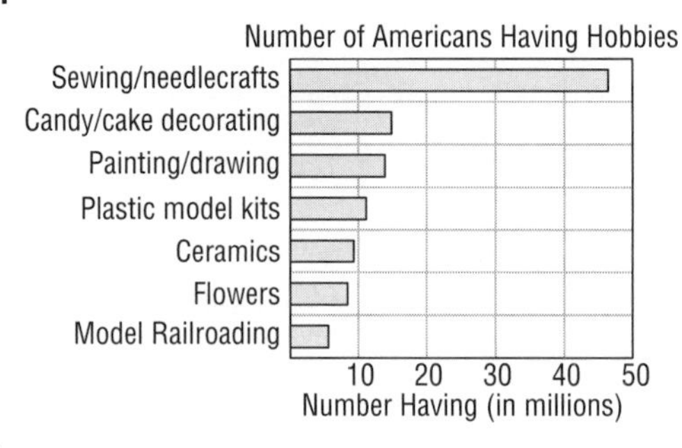

7.

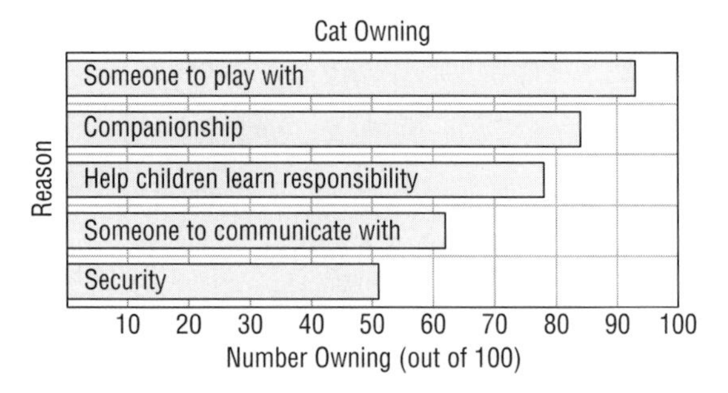

9. a. 0.8 million, 1.3 million, 1.8 million
b. 2.4 million **c.** 1940–1950 **d.** 1980–1990
e. It shows clearly how it has changed over the years.

5.4 Integrated Review p. 212

11.

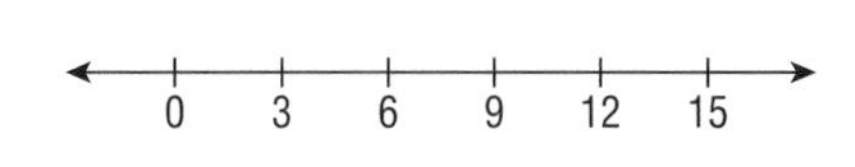

13.

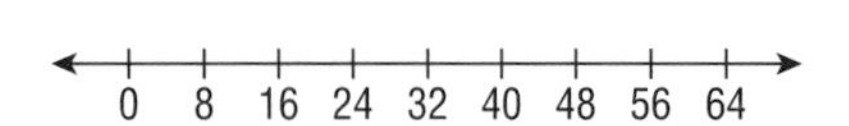

5.4 Mid-Chapter Self-Test p. 213

1. Line **2.** Travel, Hotels, and Resorts **3.** Gas **4.** Travel, Hotels, and Resorts; Gas **5.** Apparel and Footwear **6.** Gas
7.

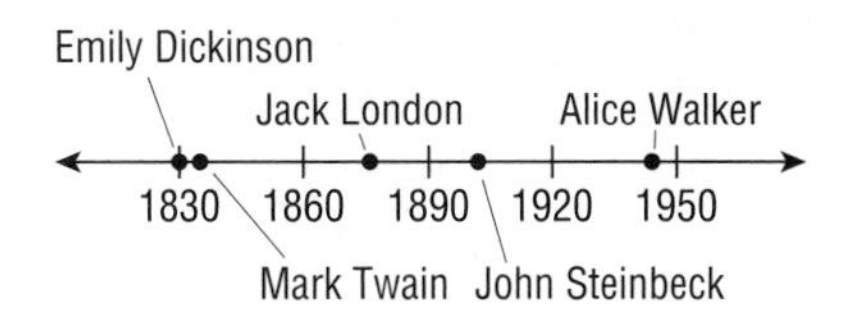

8. Answers vary.

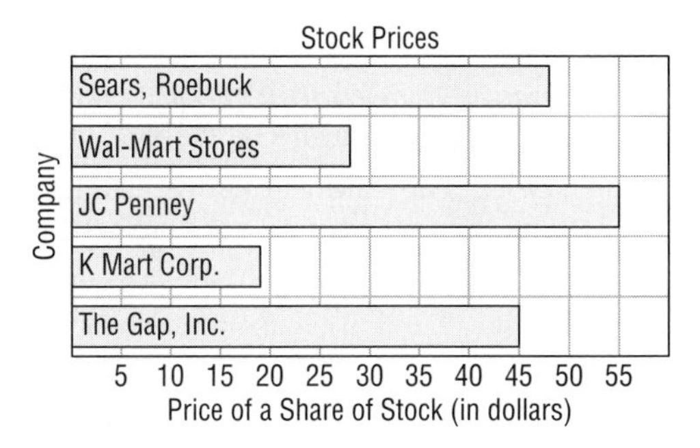

9. Answers vary.

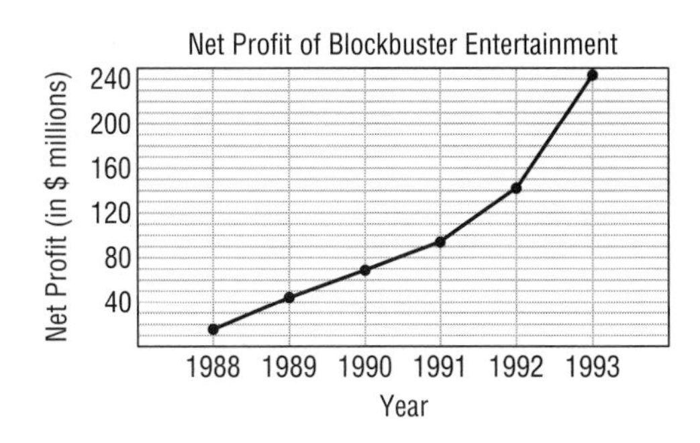

10. Bar **11.** Talking on the telephone **12.** 75 million **13.** Boredom

5.5 Communicating About Mathematics p. 215

Answers vary.

5.5 Guided Practice p. 216

1. The one with the broken vertical scale
2. The one with the unbroken vertical scale

5.5 Independent Practice pp. 216–217

3. There appears to be 2 times as many flutists as drummers. **5.** Yes, the broken vertical scale makes it appear that there are 2 times as many flutists instead of about $1\frac{1}{3}$ times as many flutists. **7.** The pay in December 1993 appears to be 4 times as much as the pay in February 1991. **9.** The broken vertical scale makes it appear that the pay in December 1993 is 4 times as much pay instead of about 1.1 times as much pay.
11. Answers vary. **13.** Answers vary.

5.5 Integrated Review p. 217

15. True, $100\% - 48\% = 52\%$ **17.** False, $100\% - 58\% < 50\%$

5.6 Communicating About Mathematics p. 219

Answers vary.

5.6 Guided Practice p. 220

1. Answers vary. **2.** 6 **3.** Decreases

5.6 Independent Practice pp. 220–221

5. No. Explanations vary.
7.

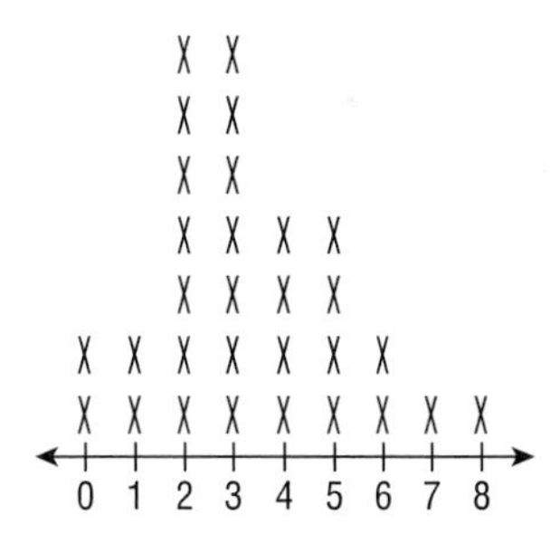

9. a. 6, 30 **b.** $3\frac{2}{15}$ **11. a.** Each of 3 tan figures:1, each of 2 purple figures:2, each of 3 green figures:4, 1 brown figure:6, 1 blue figure:8
b.

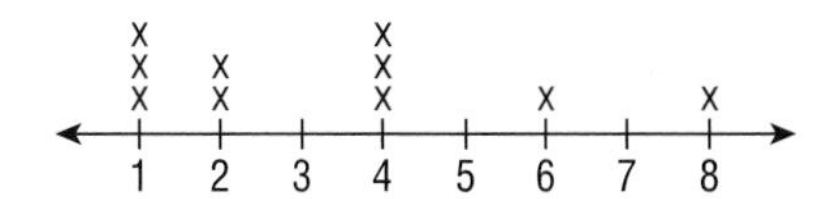

c. 33

SELECTED ANSWERS

5.6 Integrated Review p. 222

13. -5 **15.** 36 **17.** -3

5.6 Mixed Review p. 222

1. The 2nd graph, because the vertical scale on it has a bigger break. **3.** With a line graph having an unbroken vertical scale. **5.** $\frac{1}{2}$ **7.** -1 **9.** 41 **11.** 134 in. **13.** 284 in.^{2}

5.7 Communicating About Mathematics p. 225

Explanations vary. **A.** Positive **B.** No **C.** Negative **D.** Positive

5.7 Guided Practice p. 226

1. Positive **2.** Some duplication **3.** Yes **4.** Yes, 1 hit can bring in up to 4 runs.

5.7 Independent Practice pp. 226–227

5. Negative, descriptions vary **7.** No, descriptions vary **9.** Positive, explanations vary **11.** No, explanations vary **13.** Have a negative correlation **15.** 28,000 ft **17.** Have a positive correlation, cable TV subscribers tend to increase as the years increase **19.** Estimates vary. 79 million

21.

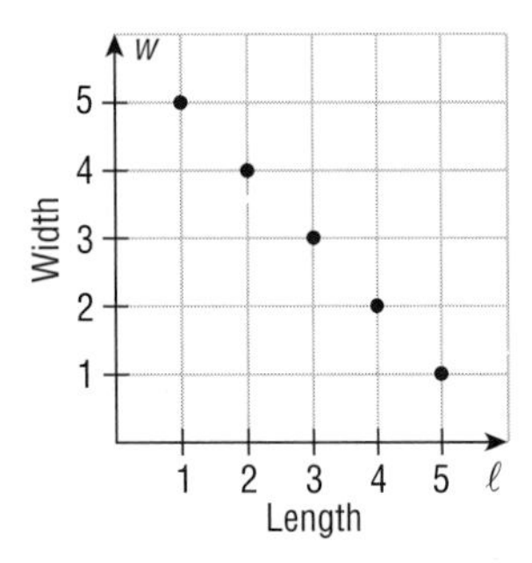

Negative correlation

5.7 Integrated Review p. 227

23. A and B, C and D **25.** 20, 24

5.8 Communicating About Mathematics p. 230

$\approx$0.002, 0.185, $\approx$0.182, $\approx$0.027, 0.005

5.8 Guided Practice p. 231

1. See page 229. **2.** See page 229. **3.** Yes, 8 out of 10 times it will rain. **4.** Answers vary.

5.8 Independent Practice pp. 231–232

5. $\frac{1}{4}$ **7.** $\frac{1}{12}$ **9.** $\frac{4}{7}$ **11.** Answers vary. **13.** $\frac{1}{2}$ **15.** Answers vary. **17.** Answers vary **19.** 0.37, 0.02, 0.415 **21.** 0.16, $1-0.84=0.16$

5.8 Integrated Review p. 232

23. 0.53 **25.** 0.57 **27.** 0.29 **29.** 0.45 **31.** 0.17

5.8 Chapter Review pp. 234–236

1. 1 million **3.** About 0.5 million **5.** About 3.3 lb per person **7.** The number of buyers under 14 appears to be 3 times the number of buyers 14 to 17.

9.

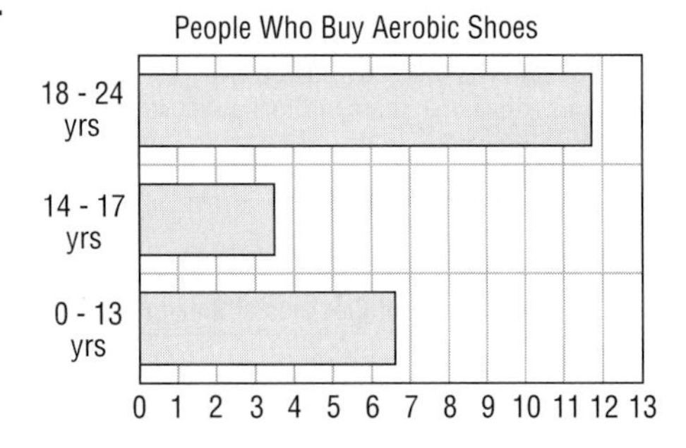

11. About 1640 **13.** Any two of the last three shown on the time line

15.

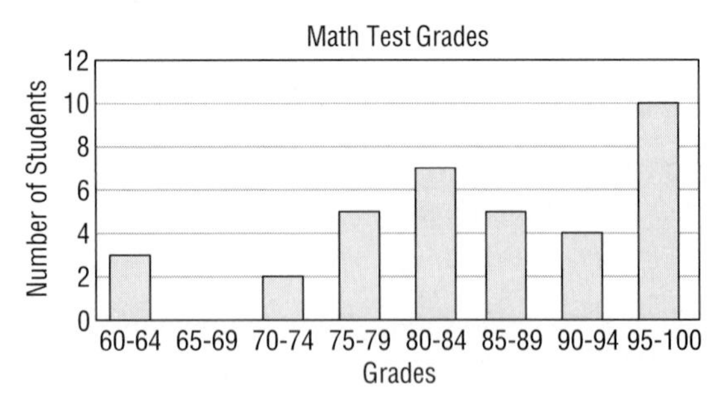

Explanations vary.

17.

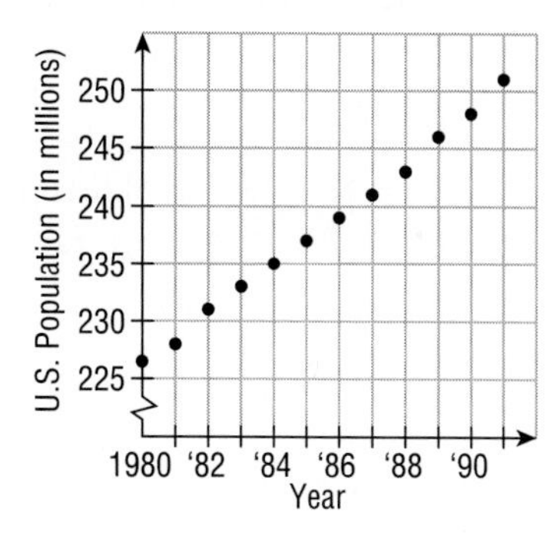

The population is slowly and steadily increasing.

19.

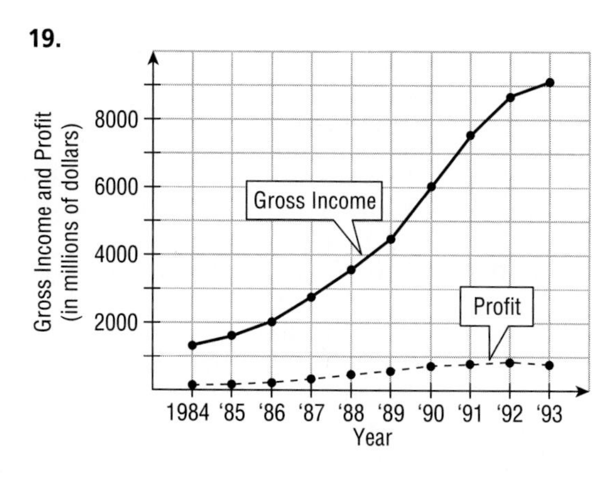

21. 1993 **23.** 5 lb **25.** About 38 lb

CHAPTER 6

■ *6.1 Communicating about Mathematics p. 241*

Use divisibility tests for 2, 5, 10, and 20 (use divisibility tests for 4 and 5) to determine whether two-dollar, five-dollar, ten-dollar, and twenty-dollar bills, respectively, can be given in exchange for the check. Then divide by 2, 5, 10, and 20 to determine the number of two-dollar, five-dollar, ten-dollar, and twenty-dollar bills, respectively, that can be given in exchange for the check.

■ *6.1 Guided Practice p. 242*

1. See page 240. **2.** Divisible only by 3 and 5 **3.** 1, 2, 3, 4, 6, 9, 12, 18, 36 **4.** The side lengths are the factors. **5.** False; 6 is divisible by 3, but is not divisible by 9. **6.** True, $10=2\times5$ **7.** True, $6\times12=72$ **8.** False; the factors of 9 are 1, 3, and 9.

■ *6.1 Independent Practice pp. 242–243*

9. Divisible by all **11.** Divisible by 2, 3, 4, 6, 8 **13.** Divisible by 2 **15.** Divisible by none **17.** 2, 5, and 8 **19.** 0 or 9 **21.** 4 **23.** 360. 2^3 is divisible by 2, 4, and 8; 3^2 is divisible by 3 and 9; 5 is divisible by 5; 2×3 is divisible by 6. So $2^3\times3^2\times5$ is divisible by 2, 3, 4, 5, 6, 8, and 9; and $2^3\times3^2\times5=360$. **25.** 1, 2, 3, 6, 9, 18 **27.** 1, 2, 3, 6, 7, 14, 21, 42 **29.** 1, 2, 5, 10, 25, 50 **31.** 1, 2, 3, 4, 6, 8, 9, 12, 18, 24, 36, 72 **33.** $2\times3\times4$ or $2\times2\times6$ **35.** $2\times3\times35$, $2\times5\times21$, $2\times7\times15$, $3\times5\times14$, $3\times7\times10$, or $5\times6\times7$ **37.** $3\times4\times5$ **39.** No, $15=3\times5$ and 3 is not a factor of 350. **41.** 40 ft $\times$ 40 ft **43.** Because 8 is divisible by 4 **45.** A natural number is divisible by 20 if (a) its last digit is 0 and (b) its next-to-last digit is even (or (b) the number formed by its last two digits is divisible by 4).

■ *6.1 Integrated Review p. 243*

47. $\frac{1}{5}$ **49.** 0

■ *6.2 Communicating about Mathematics p. 246*

Every natural number can be written as products of pairs of numbers. When no two numbers in a pair are identical, there must be an even number of factors; when two numbers in a pair are identical, there must be an odd number of factors. Conversely, if a natural number has an odd number of factors, then there must be two numbers in a pair written as a product that are identical; so the natural number must be a square.

■ *6.2 Guided Practice p. 247*

1. Multiples of 2 other than 2, multiples of 3 other than 3, multiples of 5 other than 5 **2.** Prime **3.** 31, 37 **4.** 1

■ *6.2 Independent Practice pp. 247–248*

5. Prime, exactly two factors **7.** Composite, $35=5\cdot7$ **9.** $2^2\cdot3$ **11.** $2^2\cdot3\cdot5$ **13.** $2^2\cdot3^2$ **15.** $2^2\cdot3\cdot7$ **17.** 2^5 **19.** $2^3\cdot3^2$ **21.** $(-1)\cdot3\cdot3\cdot3$, $(-1)\cdot3^3$ **23.** $3\cdot3\cdot x\cdot x\cdot x$, $3^2\cdot x^3$ **25.** $2\cdot2\cdot2\cdot a\cdot a\cdot a\cdot b\cdot b$, $2^3\cdot a^3\cdot b^2$ **27.** $(-1)\cdot3\cdot3\cdot5\cdot m\cdot n\cdot n\cdot n$, $(-1)\cdot3^2\cdot5\cdot m\cdot n^3$ **29.** 120 **31.** -585 **33.** 1, 2, 4, 8 **35.** 1, 2, 4, 8, 16, 32 **37.** No, by the answers to Exercises 33.–36.

39. Answers vary. $20=3+17$, $22=3+19$, $24=5+19$, $26=3+23$, $28=5+23$, $30=7+23$, $32=3+29$, $34=3+31$, $36=5+31$, $38=7+31$, $40=3+37$ **41.** 1, 2, 3, 5, 6, 9, 10, 15, 18, 25, 30, 45, 50, 75, 90, 150, 225, 450

43.

45.

6.2 Integrated Review p. 248

47. $\frac{1}{3}$

6.3 Communicating about Mathematics, p. 251

A. 1, yes **B.** 45, no **C.** 2, no
Yes, unless they are the same number. Each prime number has only itself and 1 as factors, so only 1 is the greatest common factor of two different prime numbers.

6.3 Guided Practice p. 252

1. 1, 2, 3, 6; 6 **2.** 1, 2, 4, 8; 8 **3.** 1, 5; 5 **4.** 1; 1 **5.** 1 **6.** Yes

6.3 Independent Practice, pp. 252–253

7. 4 **9.** 30 **11.** 72 **13.** 128 **15.** $2yz$ **17.** $3rz$ **19.** 16 and 20, 16 and 28 **21.** 21 and 42, 21 and 63 **23.** No **25.** No **27.** 15, 16; yes, greatest common factor is 1. **29.** 42, 26; no, greatest common factor is not 1. **31.** 6 **33.** 19 **35.** False **37.** True; no, a prime number has only itself and 1 as factors. **39.** True; no, a prime number has only itself and 1 as factors.

6.3 Integrated Review p. 253

41. $2^2 \cdot 31$ **43.** $2^2 \cdot 3^2 \cdot 5^2$

6.3 Mixed Review p. 254

1. Yes: 2, 3, 4, 6, 9; no: 5, 8, 10 **3.** $\frac{1}{12}$ **5.** $\frac{1}{6}$ **7.** $3 \cdot 29$ **9.** $2^2 \cdot 19$ **11.** 6 **13.** -7

6.4 Communicating about Mathematics p. 256

A. 120. $6=2 \cdot 3$, $8=2^3$, $5=5$; least common multiple is $2^3 \cdot 3 \cdot 5=120$. **B.** 12. $3=3$, $4=2^2$, $6=2 \cdot 3$; least common multiple is $2^2 \cdot 3=12$. **C.** 30. $10=2 \cdot 5$, $6=2 \cdot 3$, $15=3 \cdot 5$; least common multiple is $2 \cdot 3 \cdot 5=30$.

6.4 Guided Practice p. 257

1. 4, 12, 24 **2.** 12, 24, 36 **3.** 12 **4.** See page 255, as in Example 1, as in Example 2 **5.** c **6.** b **7.** a **8.** $4a^2b^4$

6.4 Independent Practice pp. 257–258

9. 3, 6, 9, 12, 15, 18, 21; 7, 14, 21. 21 **11.** 6, 12, 18, 24; 8, 16, 24. 24 **13.** 8, 16, 24, 32, 40; 10, 20, 30, 40. 40 **15.** 10, 20, 30, 40, 50, 60, 70, 80, 90, 100, 110, 120, 130; 26, 52, 78, 104, 130. 130 **17.** 3, 6, 9, 12, 15, 18, 21, 24, 27, 30, 33, 36; 4, 8, 12, 16, 20, 24, 28, 32, 36; 18, 36. 36 **19.** 5, 10, 15, 20; 10, 20; 20. 20 **21.** $90=2 \cdot 3^2 \cdot 5$, $108=2^2 \cdot 3^3$, $2^2 \cdot 3^3 \cdot 5=540$. 540 **23.** $125=5^3$, $500=2^2 \cdot 5^3$, $2^2 \cdot 5^3=500$. 500 **25.** $135=3^3 \cdot 5$, $375=3 \cdot 5^3$, $3^3 \cdot 5^3=3375$. 3375 **27.** $144=2^4 \cdot 3^2$, $162=2 \cdot 3^4$, $2^4 \cdot 3^4=1296$. 1296 **29.** $7s^2t=7 \cdot s^2 \cdot t$, $49st^2=7^2 \cdot s \cdot t^2$, $7^2 \cdot s^2 \cdot t^2=49s^2t^2$. $49s^2t^2$ **31.** $3m^4n^4=3 \cdot m^4 \cdot n^4$, $7m^6n^2=7 \cdot m^6 \cdot n^2$, $3 \cdot 7 \cdot m^6 \cdot n^4=21m^6 \cdot n^4$. $21m^6 \cdot n^4$ **33.** Their product, answers vary. **35.** 24 **37.** 6 **39.** 5 and 7 **41.** 1 and 36, 4 and 9, 4 and 36, 9 and 36, 36 and 36
43. 14

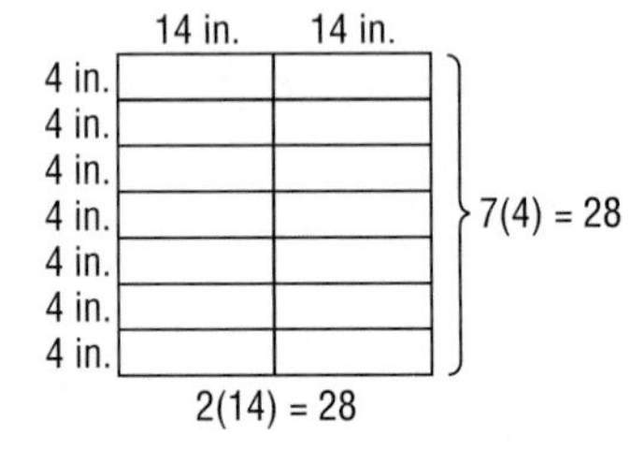

The LCM of 4 and 14 is 28, so each side of the square region is 28 in. Then two 14's and seven 4's are needed for the sides of the square region; $2 \times 7=14$. **45.** A: 6, B: 7

6.4 Integrated Review p. 258

47. 4, 12, 24, 40, 60, 84 To get the next number, add 8, add 12, add 16, etc., to the preceding number.

6.5 Communicating about Mathematics p. 260

Multiplying a fraction by 1 does not change the value of the fraction.

6.5 Guided Practice p. 261

1. $\frac{3}{6}, \frac{1}{2}$ **2.** $\frac{5}{8}, \frac{5}{8}$ **3.** $\frac{4}{6}, \frac{2}{3}$ **4.** $\frac{7}{24}, \frac{7}{24}$ **5.** $\frac{2}{3}, \frac{8}{12}; \frac{3}{4}, \frac{9}{12}; \frac{3}{4}$
6. $\frac{4}{6}, \frac{16}{24}; \frac{5}{8}, \frac{15}{24}; \frac{4}{6}$

6.5 Independent Practice pp. 261–262

7. $2, \frac{7}{10}$ **9.** $3, \frac{3}{14}$ **11.** $5, \frac{2}{15}$ **13.** $18, \frac{2}{3}$
15. $\frac{a}{4b}$ **17.** $\frac{1}{6z}$ **19.** $\frac{3z}{4}$ **21.** $\frac{2}{3pq}$ **23.** A fraction is in simplest form if the only common factor of the numerator and denominator is 1. **25.** Answers vary. $\frac{2}{4}, \frac{3}{6}, \frac{4}{8}$ **27.** Answers vary. $\frac{5}{11}, \frac{15}{33}, \frac{20}{44}$
29. $<$ **31.** $>$ **33.** $<$ **35.** $=$ **37.** $\frac{6}{18}, \frac{5}{12}, \frac{3}{6}, \frac{2}{3}, \frac{8}{10}, \frac{1}{1}$
39. $\frac{1}{500}$ To increase the shutter speed means to leave it open for a shorter period of time, so the fraction would have to be smaller than $\frac{1}{250}$; $\frac{1}{500} < \frac{2}{500} = \frac{1}{250}$. **41.** $\frac{20}{80}, \frac{1}{4}$

6.5 Integrated Review p. 262

43. 2, 60 **45.** 25, 300 **47.** c

6.5 Mid-Chapter Self-Test p. 263

1. Yes: 2, 3, 5, 6, 10; no: 4, 8, 9 **2.** Yes: 2, 4, 5, 8, 10; no: 3, 6, 9 **3.** Yes: 12, 16, 17; no: 11, 13, 14, 15, 18, 19, 20 **4.** 1, 2, 4, 7, 8, 14, 28, 56 **5.** $2^4 \cdot 5$
6. $2^2 \cdot 11$ **7.** $3 \cdot 5 \cdot 7$ **8.** 12 **9.** 3 **10.** 5 **11.** 65
12. 42 **13.** $18x^2$ **14.** $\frac{1}{5}$ **15.** $\frac{5}{34}$ **16.** $\frac{y}{3}$ **17.** 2
18. 28 **17., 18.** The length of the square's side is the LCM of the lengths of the tile's sides.
19. 6-in.: 7, 14-in.: 3 **20.** Least common multiple

6.6 Communicating About Mathematics p. 266

A. $\frac{2}{3}$ **B.** $\frac{5}{3}$ **C.** $\frac{9}{4}$ **D.** $\frac{2}{9}$

6.6 Guided Practice p. 267

1. Rational, $\frac{-3}{1}$ **2.** Irrational **3.** Rational, $\frac{13}{5}$
4. Rational, $\frac{7}{1}$ **5.** Rational, $\frac{2}{5}$ **6.** $0.36, \frac{36}{100}, \frac{9}{25}$
7. $0.75, \frac{75}{100}, \frac{3}{4}$ **8.** $0.64, \frac{64}{100}, \frac{16}{25}$ **9.** $0.05, \frac{5}{100}, \frac{1}{20}$
10. Write the whole number as a fraction with the same denominator as the fraction, then add.

6.6 Independent Practice pp. 267–268

11. $\frac{5}{1}$ **13.** $\frac{1}{4}$ **15.** $\frac{7}{6}$ **17.** $-\frac{13}{8}$ **19.** Rational; 0.6, terminating **21.** Irrational; 2.8284 . . ., nonrepeating **23.** Rational; $0.5\overline{3}$, repeating **25.** Rational; 3.5, terminating **27.** $\frac{8}{10}, \frac{4}{5}$
29. $\frac{84}{100}, \frac{21}{25}$ **31.** $\frac{45}{99}, \frac{5}{11}$ **33.** $\frac{21}{9}, \frac{7}{3}$ **35.** e **37.** c
39. f **41.** $0.\overline{09}, 0.\overline{18}, 0.\overline{27}, 0.\overline{36}, 0.\overline{45}, 0.\overline{54}$; each time after the first is $\frac{1}{11}$ or $0.\overline{09}$ more than the preceding term. **43.** $\frac{18}{5}$ in., $3\frac{3}{5}$ in., 3.6 in.
45. $\frac{20}{3}$ in., $6\frac{2}{3}$ in., $6.\overline{6}$ in.

6.6 Integrated Review p. 268

47. True, every integer can be written with a denominator of 1. **49.** False, $\frac{1}{2}$ is a counter-example.

6.7 Communicating about Mathematics p. 270

$$3^4 \cdot 3^5 = \underbrace{3 \cdot 3 \cdot 3 \cdot 3}_{3^4} \cdot \underbrace{3 \cdot 3 \cdot 3 \cdot 3 \cdot 3}_{3^5} = 3^9$$

6.7 Guided Practice p. 271

1. See page 269. **2.** $\frac{1}{4}$ **3.** $\frac{1}{5^2}$ or $\frac{1}{25}$
4. 1 **5.** $\frac{1}{x^3}$ **6.** See page 270. **7.** p^7 **8.** $r^2 \cdot s^3$
9. $\frac{m^6}{n^4}$ **10.** x^2

6.7 Independent Practice pp. 271–272

11. $\frac{1}{9}$ **13.** 1 **15.** $\frac{1}{t^4}$ **17.** $\frac{3}{s^2}$ **19.** 36 **21.** x^{15}
23. 7 **25.** a^{12} **27.** 0.026 **29.** 5032.844
31. 3 **33.** 6 **35.** 16 **37.** $7^{10} \cdot 7^{-4}$ **39.** $=$ **41.** $>$
43. $10^0, 10^1, 10^2, 10^3$ Each number after the first is 10 times the preceding number. $10^4, 10^5, 10^6$
45. 2^7 **47.** c **49.** 10^{-5}

6.7 Integrated Review p. 272

51. $3^{-2} = \frac{1}{3^2}, \frac{1}{9}$ **53.** $7^2 \cdot 7^3 = 7^{2+3}, 7^5$ **55.** Positive
57. Positive **55., 57.** When the exponent is even, the answer is positive.

6.7 Mixed Review p. 273

1. $0.0\overline{37}$, repeating **3.** $0.\overline{3}$, repeating **5.** $1.\overline{5}$, repeating **7.** 1.389 **9.** -1 **11.** 0.002 **13.** 0.313
15. -3.145

16.–18.

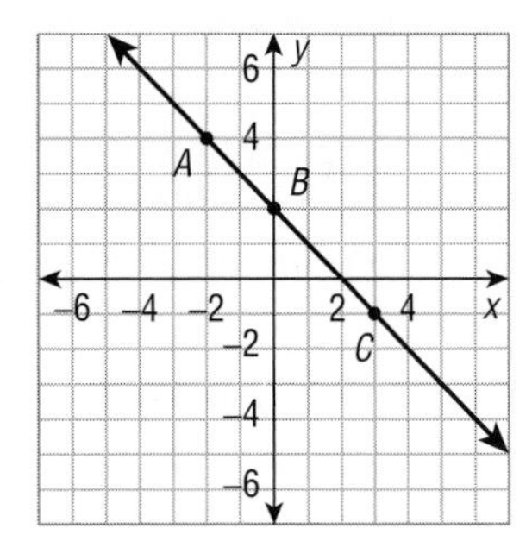

16. II **17.** None **18.** IV **19.** Answers vary. (1, 1)

6.8 Communicating about Mathematics p. 275

A. No, 1.24×10^{-2} **B.** Yes **C.** No, 5×10^{3}

6.8 Guided Practice p. 276

1. b **2.** 5 **3.** -3 **4.** 625,000 **5.** 0.0000087
6. a: Canada, b: China; China has more people and $9>7$.

6.8 Independent Practice pp. 276–277

7. 5×10^{3} **9.** 4.1×10^{-4} **11.** 3.261×10^{7}
13. 1.2×10^{-8} **15.** 0.0057 **17.** 25,000
19. 62,000,000,000 **21.** 0.000000363 **23.** Yes
25. No, 2.56×10^{9} **27.** No, 7.91×10^{-2}
29. 4.96×10^{6}, 4,960,000 **31.** 1.8×10^{-6}, 0.0000018 **33.** 1×10^{9}, $9>8$ **35.** 6×10^{12}
37. Numbers in table should be 20,600,000; 18,900,000; 6,300,000; 4,000,000; 4,000,000; 3,000,000. Answers vary. **39.** 4.704×10^{17}

6.8 Integrated Review p. 277

41. 1600 m^2

6.9 Communicating about Mathematics p. 116

For equivalent rational numbers, $\frac{a}{b}$, the points (b, a) are collinear.

6.9 Guided Practice p. 281

1. 1, 2, 4, 7, 14, 28; $1+2+4+7+14=28$
2. 3, 9, 19, 33, 51, 73 **3.** $A(-1, -2)$, $B(1, 2)$ $C(2, 4)$, $D(3, 6)$ **4.** Each y-coordinate is twice the x-coordinate, 22.

6.9 Independent Practice pp. 281–282

5.

n	1	2	3	4	5	6
n^2+1	2	5	10	17	26	37

7.

n	1	2	3	4	5	6
2^{n-1}	1	2	4	8	16	32

9. To get the next number, square the position number of the number and subtract 1. 24, 35, 48
11. To get the next number, multiply together the two preceding numbers. 32, 256, 8192
13. To get the next number after the first three, add the preceding three numbers. 44, 81, 149

15. 15, 21

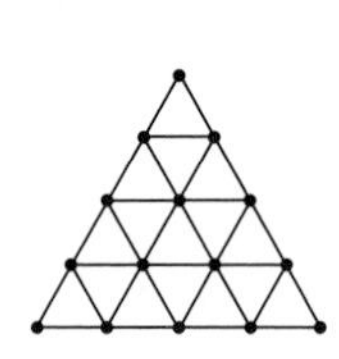
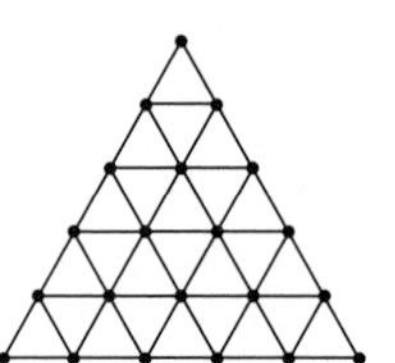

17. 35, 51

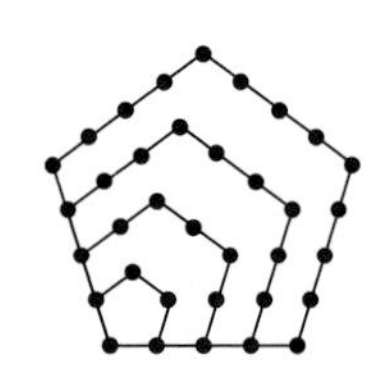
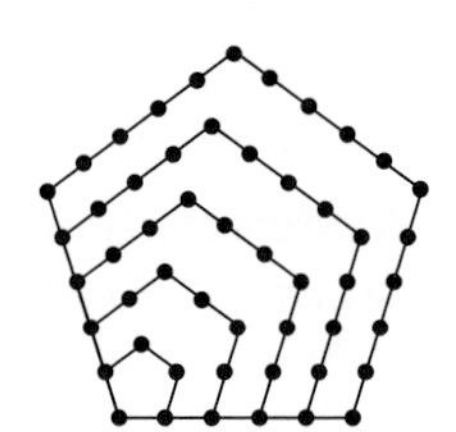

19. 3 and 17 **21.** 47 and 53, 41 and 59, 29 and 71, 17 and 83, 11 and 89, 3 and 97 **23.** 3 and 5, 5 and 7, 11 and 13, 17 and 19, 29 and 31, 41 and 43, 59 and 61, 71 and 73 **25.** To get the next twin primes, add 30 to each prime in the preceding twin primes. 101 and 103, yes

6.9 Integrated Review p. 282

27. 1, 2, 4, 5, 10, 20 **29.** 1, 2, 3, 4, 6, 8, 9, 12, 18, 24, 36, 72

6.9 Chapter Review pp. 284–285

1. Number is divisible by 2, 4, 5, 8, 10
3. Number is divisible by 2, 3, 6, 9 **5.** Number is divisible by 5 **7.** Number is divisible by 2, 3, 4, 6, 8, 9 **9.** Composite; 1, 3, 5, 15 **11.** Prime
13. Composite; 1, 2, 23, 46 **15.** Composite; 1, 2, 4, 8, 16, 32, 64 **17.** $2^4 \cdot 5$ **19.** $2^3 \cdot 3 \cdot 5$
21. $(-1) \cdot 3^3 \cdot 5$ **23.** $2^3 \cdot 5 \cdot x^4$ **25.** $2^2 \cdot 3 \cdot a^6 \cdot b^2$

27. $(-1) \cdot 3^4 \cdot s \cdot t^5$ **29.** 5, 15 **31.** 24, 2160 **33.** $2xy^3$, $4x^4y^8$ **35.** $2y^4z^4$, $42y^5z^5$ **37.** $\frac{1}{7}$ **39.** $\frac{4}{23}$ **41.** $\frac{ab}{9}$ **43.** $\frac{2}{3z^7}$ **45.** $>$ **47.** $=$ **49.** Rational; $0.1\overline{6}$, repeating **51.** Rational; $-.0.625$, terminating **53.** Rational; 1.2, terminating **55.** Irrational; 6.5574 . . ., nonrepeating **57.** $\frac{1}{4}$ **59.** $\frac{1}{49}$ **61.** $\frac{1}{x^4}$ **63.** $\frac{1}{27}$ **65.** $-100a^6$ **67.** $\frac{2n^4}{m^2}$ **69.** 39,000,000 **71.** 0.000946 **73.** 1.2×10^9 **75.** 4.5×10^{-4} **77.** 3.75×10^7 **79.** 1.0×10^0 **81.** To get the next number, add 1 to the preceding denominator. $\frac{1}{6}, \frac{1}{7}, \frac{1}{8}$ **83.** To get the next number, subtract 1 from the absolute value of the numerator of the preceding number, add 1 to the absolute value of the denominator of the preceding number, and keep the negative sign. $-\frac{3}{5}, -\frac{2}{6}, -\frac{1}{7}$ **85.** Yes, 2 **87.** 0.69, 0.16, 0.15; sum is 1. **89.** $\frac{1}{1}, \frac{1}{2}, \frac{1}{4}, \frac{1}{8}, \frac{1}{16}$ **91.** 1 beat, $\frac{1}{2}$ beat, $\frac{1}{4}$ beat **93.** It adds $\frac{1}{2}$ the length of time. **95.** 2, 1, 1 occurs four times. **97.** Answers vary.

6.9 Cumulative Review for Chapters 1-6 pp. 288–291

1. To get the next polygon, rotate it 90° clockwise.

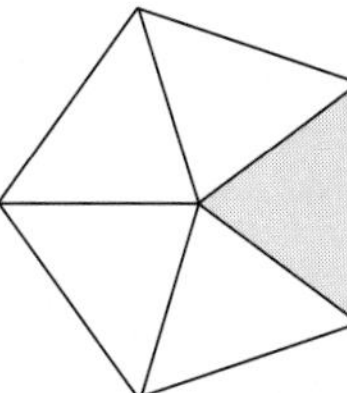

5(4 + 2x), 40

3. 78,125 **5.** 9.17 **7.** 22.63 **9.** 4.22 **11.** -9 **13.** -180 **15.** 69 **17.** b **19.** f **21.** a **23.** $5x + 5y + 5z$, 35 **25.** $-2y - 6x + 3z$, 19 **27.** $-7 + xz + xy$, -25 **29.** $-18 = x - 9$, -9 **31.** $-8y = -104$, 13 **33.** $5 \geq \frac{a}{13}$, $a \leq 65$ **35.** $-5 < z + 9$, $z > -14$

37.

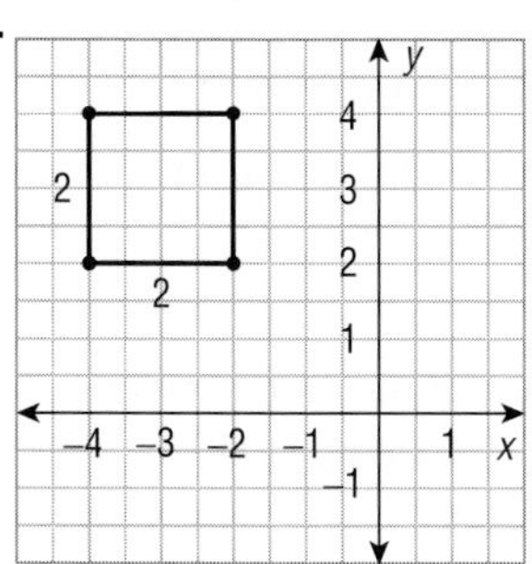

8 units, 4 units²

39.

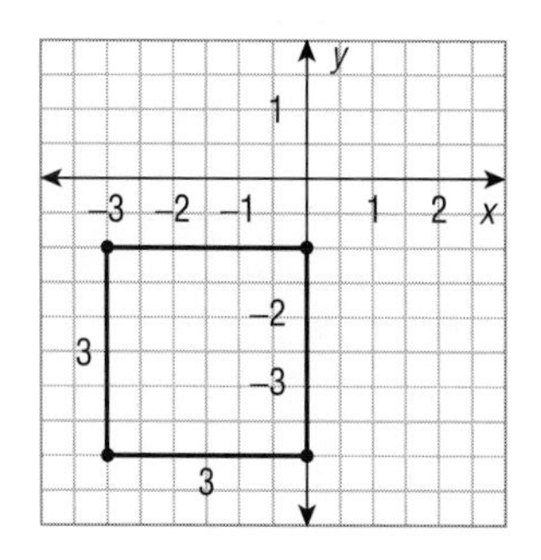

12 units, 9 units²

41. 4, 1; 6, 2; 5, 3 **43.** No; $3(0 + 5 + 0 + 1 + 2 + 4) + (3 + 9 + 2 + 1 + 3) = 54$, $60 - 54 = 6 \neq 9$ **45.** Yes; $3(0 + 8 + 2 + 5 + 6 + 3) + (3 + 3 + 2 + 6 + 1) = 87$, $90 - 87 = 3$ **47.** Addition **49.** $\frac{1}{5}$ **51.** -4 **53.** 4 **55.** 28 **57.** 12 **59.** $(2x - 10°) + (3x - 6°) + (x + 4°) = 180°$, 32°; 36°, 54°, 90° **61.** $5x + 2$ **63.** $5n - 10$ **65.** 3 **67.** -6 **69.** $-\frac{9}{10}$ **71.** $\frac{1}{2}$ **73.** 1.93 **75.** 6 in.² **77.** Yes **79.** 1940's **81.** There appears to be a positive correlation between 35 mph and 45 mph because the gas mileage increases as the speed increases. There appears to be a negative correlation between 45 mph and 75 mph because the gas mileage decreases as the speed increases. **83.** 27 mpg

85. Number of Radio Stations in the U.S. in 1991

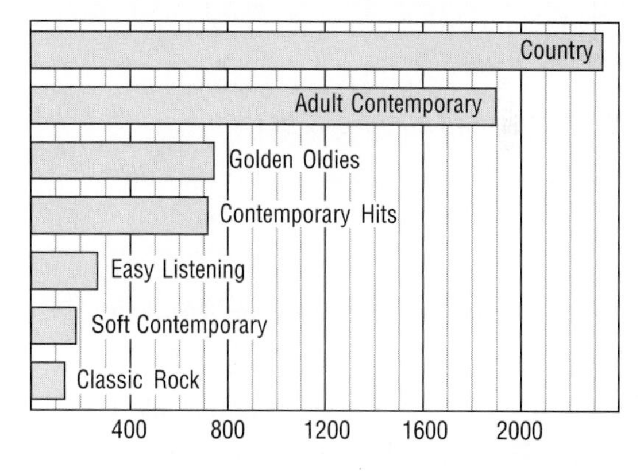

87. 0 **89.** $\frac{3}{4}$ **91.** 2, 36 **93.** 108, 7776 **95.** $3ab$, $36a^2b^4$ **97.** $\frac{5}{9}$, 0.556 **99.** $\frac{26}{27}$, 0.963 **101.** $\frac{1}{9}$ **103.** $\frac{1}{x^2}$ **105.** 1 **107.** $\frac{1}{10,000}$ **109.** 5.2×10^{-5}

CHAPTER 7

7.1 Communicating about Mathematics p. 295

A. $\frac{8}{7}$ **B.** 3 **C.** $\frac{1}{2}$ **D.** $\frac{1}{2}$

7.1 Guided Practice p. 296

1. $\frac{1}{4}+\frac{2}{4}=\frac{3}{4}$ **2.** $\frac{3}{3}-\frac{1}{3}=\frac{2}{3}$ **3.** See page 294. See answer to Exercise 1. **4.** See page 295. See answer to Exercise 2.

7.1 Independent Practice pp. 296–297

5. $\frac{5}{6}$ **7.** $-\frac{15}{15}$, -1 **9.** $-\frac{6}{11}$ **11.** 5 **13.** $\frac{5x}{2}$ **15.** $-\frac{5a}{5}$, $-a$ **17.** $\frac{7}{z}$ **19.** $\frac{3}{5b}$ **21.** $\frac{2}{3}$ **23.** $-\frac{15}{5}$, -3 **25.** $-\frac{14}{4}$, $-\frac{7}{2}$ **27.** $\frac{2}{3}$ **29.** 0.57 **31.** -0.33 **33.** $\frac{3}{8}$, $\frac{7}{8}$, $\frac{11}{8}$, $\frac{15}{8}$ Every fraction after the first is $\frac{4}{8}$ greater than the preceding fraction. $\frac{19}{8}$, $\frac{23}{8}$, $\frac{27}{8}$ **35.** Incorrect. $\frac{3}{5}+\frac{1}{5}=\frac{3+1}{5}=\frac{4}{5}$ **37.** $\frac{1}{4}+\frac{1}{2}=\frac{3}{4}$ **39.** New numbers, in order, are $3\frac{1}{2}$, $1\frac{1}{3}$, 4, 1, $\frac{1}{2}$, 2, $\frac{2}{3}$, 4, 4, $\frac{1}{2}$. **41.** $2\frac{7}{12}$ ft

7.1 Integrated Review p. 297

43. $4\frac{2}{3}$ in.

7.2 Communicating about Mathematics p. 300

$\frac{3}{4}$, $\frac{1}{2}$; $1\frac{1}{4}$ units^2

7.2 Guided Practice p. 301

1. $\frac{2}{3}$, $\frac{1}{2}$; $\frac{2}{3}+\frac{1}{2}=\frac{7}{6}$ **2.** $\frac{5}{8}$, $\frac{1}{2}$; $\frac{5}{8}+\frac{1}{2}=\frac{9}{8}$ **3.** $\frac{11}{15}$; 15 **4.** $\frac{5}{10}$, $\frac{1}{2}$; 10 **5.** $\frac{a}{6}$; 6 **6.** $\frac{9}{2t}$; $2t$

7.2 Independent Practice pp. 301–302

7. $\frac{9}{12}$, $\frac{3}{4}$ **9.** $-\frac{13}{12}$ **11.** $-\frac{5}{15}$, $-\frac{1}{3}$ **13.** $\frac{23}{40}$ **15.** $\frac{3x}{6}$, $\frac{x}{2}$ **17.** $\frac{20+9x}{10x}$ **19.** $-\frac{10}{9t}$ **21.** $\frac{5}{3mn}$ **23.** 1.62 **25.** 0.72 **27.** $7\frac{2}{3}$ **29.** To get the next fraction, add 1 to the absolute values of both the numerator and the denominator and change the sign of the fraction to its opposite. **31.** $\frac{7}{12}$ **33.** $\frac{23}{50}$ **35.** $\frac{11}{100}+\frac{7}{25}+\frac{7}{25}+\frac{9}{50}+\frac{3}{20}=\frac{11}{100}+\frac{28}{100}+\frac{28}{100}+\frac{18}{100}+\frac{15}{100}=\frac{100}{100}=1$

7.2 Integrated Review p. 302

37. $1\frac{1}{2}$ **39.** 2 **41.** 4

7.3 Communicating about Mathematics, p. 304

Explanations and descriptions vary.

7.3 Guided Practice p. 305

1. Write each fraction as a decimal rounded to three decimal places. Add the decimals and round the sum to two decimal places. **2.** 0.67, it is $0.00\overline{3}$ greater. **3.** -0.37 **4.** 0.33

7.3 Independent Practice, pp. 305–306

5. 0.86 **7.** 1.64 **9.** $\frac{17}{36}+\frac{14}{25}\approx 0.472+0.56=1.032\approx 1.03$

11.

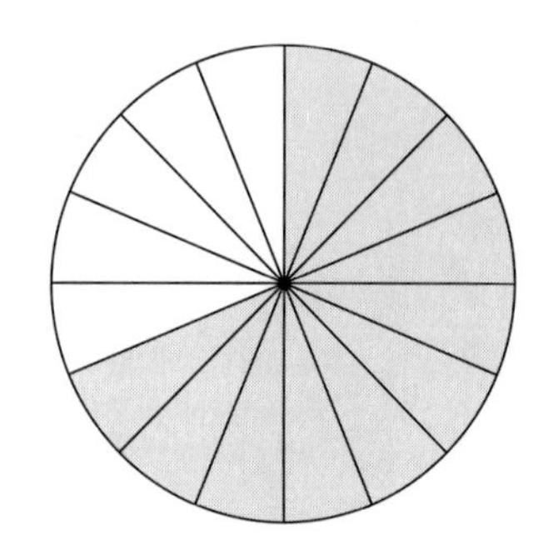

13.

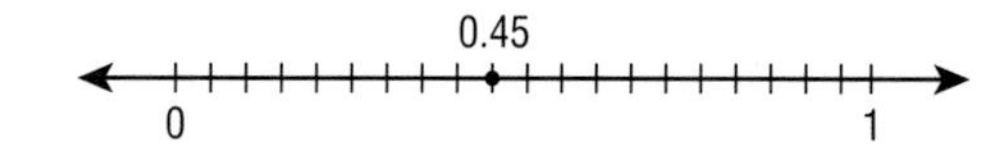

15. $0.658+0.495=1.153\approx 1.15$
17. $0.486y-0.341y=0.145y\approx 0.15y$
19. $1-(0.375+0.323+0.12)=1-0.818=0.182\approx 0.18$ **21.** $2.75+3.125-1.9+4.7=8.675\approx 8.68$
23. 0.8 **25.** 0.19

7.3 Integrated Review p. 306

27. 0.475 **29.** -0.279 **31.** $>$ **33.** $<$ **35.** $<$

7.3 Mixed Review p. 307

1. $\frac{2}{3}$ **3.** 14 **5.** -0.13 **7.** 0.09 **9.** 14 **11.** 60 **13.** $\frac{10}{20}$, $\frac{1}{2}$ **15.** $\frac{25}{100}$, $\frac{1}{4}$

7.4 Communicating about Mathematics p. 309

$371\frac{2}{3}$ ft^2; $(20\frac{1}{2}\times 24)-\frac{361}{3}=492-120\frac{1}{3}=371\frac{2}{3}$

7.4 Guided Practice p. 310

1. See page 308. **2.** $\frac{12}{35}$ **3.** 1 **4.** $\frac{5}{18}$ **5.** 4 **6.** $\frac{2}{3}$ unit by $\frac{4}{5}$ unit, $\frac{8}{15}$ units² **7.** $\frac{6}{7}$ unit by $\frac{3}{6}$ unit, $\frac{3}{7}$ units² **8.** $\frac{3}{4}$ unit by $\frac{5}{6}$ unit, $\frac{5}{8}$ units² **9.** $\frac{7}{8}$ unit by $\frac{3}{4}$ unit, $\frac{21}{32}$ units²

7.4 Independent Practice pp. 310–311

11. $-\frac{16}{27}$ **13.** $\frac{16}{5}$ **15.** $\frac{23}{2}$ **17.** $\frac{1}{9}$ **19.** $\frac{56y}{3}$ **21.** $-\frac{8}{5}$ **23.** $\frac{a}{3}$ **25.** $\frac{9y^3}{49}$ **27.** 2 in.² **29.** 0.260 **31.** -3.354 **33.** $\frac{1}{4}, \frac{2}{3}, \frac{1}{6}$ **35.** $\frac{1}{5}, \frac{5}{8}, \frac{1}{8}$ **37.** 30 ft **39.** 15 ft **41.** 6 ft **43.** $\frac{12}{100}$ or $\frac{3}{25}$ **45.** $\frac{30}{100}$ or $\frac{3}{10}$

7.4 Integrated Review p. 311

47. $\frac{5}{24}$ **49.** $\frac{1}{19}$ **51.** $\frac{6}{55}$

7.5 Communicating about Mathematics p. 313

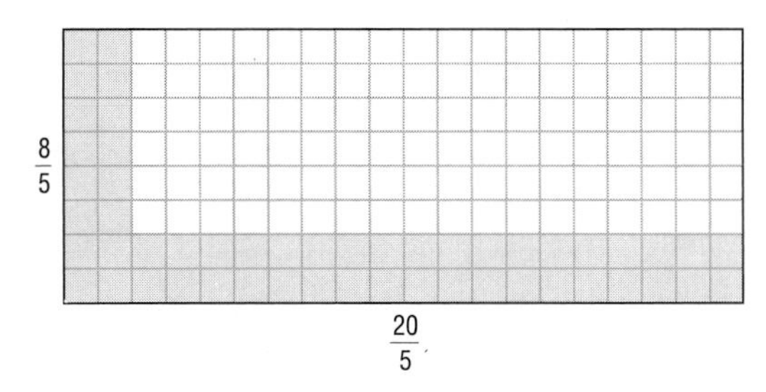

$(\frac{8}{5} \cdot \frac{20}{5}) \div \frac{4}{25} = \frac{32}{5} \cdot \frac{25}{4} = 40$

7.5 Guided Practice p. 314

1. 5 **2.** $-\frac{3}{2}$ **3.** $\frac{1}{7}$ **4.** $\frac{t}{4}$ **5.** The reciprocal of 3 is $\frac{1}{3}$, not $\frac{3}{1}$. $\frac{5}{6} \div 3 = \frac{5}{6} \cdot \frac{1}{3} = \frac{5 \cdot 1}{6 \cdot 3} = \frac{5}{18}$ **6.** Multiply the denominators also, not just the numerators. $\frac{-4}{3} \div \frac{3}{2} = \frac{-4}{3} \cdot \frac{2}{3} = \frac{-4 \cdot 2}{3 \cdot 3} = -\frac{8}{9}$ **7.** $\frac{3}{5}$ **8.** $\frac{27}{2}$ **9.** $\frac{2n}{9}$ **10.** $\frac{7x}{8}$

7.5 Independent Practice pp. 314–315

11. 4 **13.** $\frac{5}{7a}$ **15.** Multiply by the reciprocal of $\frac{3}{5}$, not by the reciprocal of $-\frac{3}{2}$. $-\frac{3}{2} \div \frac{3}{5} = -\frac{3}{2} \cdot \frac{5}{3} = -\frac{5}{2}$ **17.** Multiply by the reciprocal of $\frac{1}{3}$, not by $\frac{1}{3}$. $\frac{1}{3} \div \frac{1}{3} = \frac{1}{3} \cdot 3 = 1$ **19.** $\frac{1}{12}$ **21.** $\frac{1}{20}$ **23.** 3 **25.** 6 **27.** 3, 4.5, 6, 7.5 Each number after the first is $\frac{3}{2}$ or 1.5 more than the preceding number. The result gets larger and larger. **29.** $-\frac{18}{5}$ **31.** -7 **33.** $\frac{8}{15}$ **35.** $-\frac{x}{15}$ **37.** $\frac{4n}{5}$ **39.** $\frac{5}{6}$ **41.** $3.20 per hour, multiply the answer by $3\frac{3}{4}$. **43.** 12 **45.** $\frac{3}{8}x = \frac{3}{40}, \frac{1}{5}$ **47.** $\frac{7}{2}x = \frac{63}{4}, \frac{9}{2}$

7.5 Integrated Review p. 315

49. miles **51.** minutes

7.5 Mid-Chapter Self-Test p. 316

1. $\frac{4}{11}$ **2.** $\frac{4}{6}, \frac{2}{3}$ **3.** $\frac{43}{50}$ **4.** $\frac{4}{10}, \frac{2}{5}$ **5.** $-\frac{1}{2}$ **6.** $\frac{2}{5}$ **7.** $\frac{7}{20}$ **8.** $-\frac{1}{3}$ **9.** $-\frac{11}{20}$ **10.** 1 **11.** $\frac{5}{2}$ **12.** $-\frac{1}{5}$ **13.** $15\frac{7}{15}$ in., $12\frac{4}{5}$ in.² **14.** $\frac{7}{16}$ units² **15.** 28 in. **16.** $1\frac{3}{4}$ ft **17.** $20 **18.** $15 **19.** $\frac{8}{11}$ **20.** b

7.6 Communicating about Mathematics p. 319

A.–C. 25%; can visually divide the 100 small squares into fourths; $\frac{1}{4}$

7.6 Guided Practice p. 320

1. 24%, 24 percent; $\frac{83}{100}$, 83 percent; $\frac{49}{100}$, 49% **2.** 50% **3.** 45% **4.** Models vary.

7.6 Independent Practice pp. 320–321

5. 36% **7.** 34% **9.** a **11.** 5% **13.** 28% **15.** 45% **17.** 60% **19., 21.** Models vary. **23.** 100% − 30% = 70% **25.** c, others each have $33.\overline{3}$% shaded. **27.** c **29.** a **31.** b

7.7 Communicating About Mathematics p. 323

A. 15% **B.** 23% **C.** 45%

7.7 Guided Practice p. 324

1. 36%, 0.36 **2.** 44%, 0.44 **3.** 12%, 0.12 **4.** 30%, 0.3 **5.** See page 322. 0.48 **6.** See page 322. 4.5% **7.** See page 323. 31.25% **8.** Examples vary. A 150% increase in the price of some jewelry.

7.7 Independent Practice pp. 324–325

9. 0.36 **11.** 1.15 **13.** 0.897 **15.** ≈0.147 **17.** 25% **19.** 70% **21.** 140% **23.** 0.9% **25.** > **27.** = **29.** b **31.** d **33.** $\frac{13}{25}$ **35.** $\frac{3}{50}$ **37.** $\frac{8}{5}$ **39.** $\frac{51}{50}$ **41.** 6.25% **43.** ≈78.8% **45.** 162.5%

47. ≈333.3% **49.** 38% **51.** ≈33.3% **53.** 15%
55. 49% + 21% + 13% + 12% + 4% + 1% = 100%
57.

A	$\frac{2}{5}$	0.4	40%
B	$\frac{7}{20}$	0.35	35%
C	$\frac{1}{8}$	0.125	12.5%
D	$\frac{3}{40}$	0.075	7.5%
F	$\frac{1}{20}$	0.05	5%

59. 87.5%

■ *7.7 Integrated Review p. 325*

61. $\frac{1}{5}$ of 895 ≈ $\frac{1}{5}$ of 900 = 180 **63.** $\frac{2}{3}$ of \$14.95 ≈ $\frac{2}{3}$ of \$15 = \$10 **65.** 83% of 320 ≈ $\frac{4}{5}$ of 300 = 240

■ *7.7 Mixed Review p. 326*

1. $\frac{1}{4}$ **3.** 10 **5.** 9% **7.** 25% **9.** 136% **11.** $1\frac{1}{2}$
13. $\frac{5}{18}$ **15.** $-1\frac{1}{2}$ **17.** It is equally likely to rain and not to rain. **19.** $\frac{1}{4}a = \frac{2}{3}$, $2\frac{2}{3}$

■ *7.8 Communicating about Mathematics p. 328*

A.

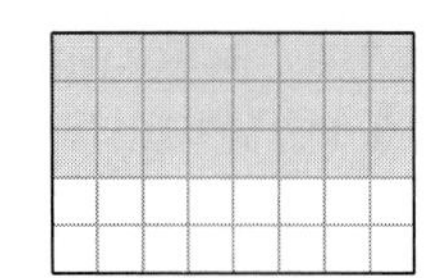

24 squares

B.

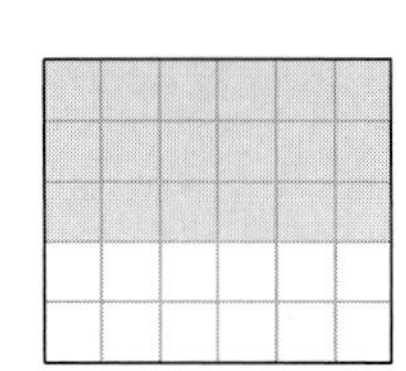

18 squares

C.

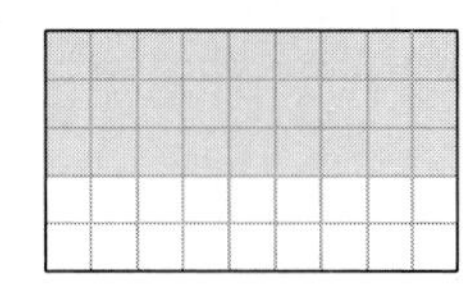

27 squares

To check, divide the number of shaded squares by 40, 30, and 45, respectively.

■ *7.8 Guided Practice p. 329*

1. See page 327. **2.** 171 **3.a.** 4.8 **b.** 32 **c.** 32 **d.** 46

■ *7.8 Independent Practice pp. 329–330*

5. 0.8, 228 **7.** 3.4, 17 **9.** 2.5, 115 **11.** 0.065, 52 **13.** c, 72 **15.** b, 40

17.

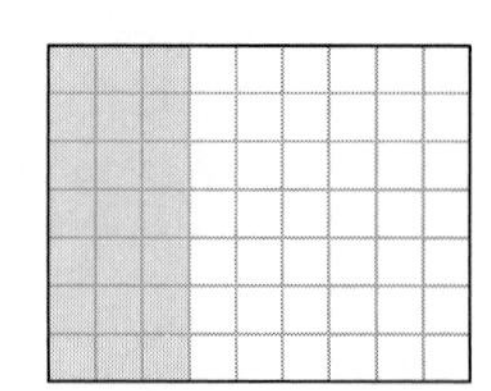

19. 24 **21.** 312 **23.** 46 in., 112 in.^{2} **25.** 69 in., yes, $2(1.5 \times 16 + 1.5 \times 7) = 1.5 \times 2(16 + 7)$ **27.** The width, length, and perimeter of the blue rectangle are each 50% of the width, length, and perimeter of the large rectangle. **29.** The width of the blue rectangle is 50% of the width of the large rectangle, the length of the blue rectangle is 75% of the length of the large rectangle, and the perimeter of the blue rectangle is ≈66.7% of the perimeter of the large rectangle. **31.** 9.78 million
33. 9.291×10^{18}

■ *7.8 Integrated Review p. 330*

35. 0.125, others = 1.25 **37.** 0.06, others = 6

■ *7.9 Communicating about Mathematics p. 333*

a. 328, 82% of 400 = 0.82 × 400 = 328 **b.** 349, 87.2% of 400 = 0.872 × 400 = 348.8 ≈ 349 **c.** 384, 96% of 400 = 0.96 × 400 = 384 **d.** 301, 75.2% of 400 = 0.752 × 400 = 300.8 ≈ 301 **e.** 152, 38% of 400 = 0.38 × 400 = 152 **f.** 229, 57.2% of 400 = 0.572 × 400 = 228.8 ≈ 229

■ *7.9 Guided Practice p. 334*

1. \$20.93 **2.** \$29.11 **3.** \$4.37 **4.** ≈59%

■ *7.9 Independent Practice pp. 334–335*

5. \$4.84 **7.** \$52 **9.** 70.92% **11.** 8.96% **13.** 0.8% **15.** \$370.13 **17.** \$117.65 **19.** \$40.30 **21.** \$9.75 **23.** ≈52.3% **25.** ≈29.7%

7.9 Integrated Review *p. 335*

27. ≈ 185 **29.** ≈ 151

7.9 Chapter Review *pp. 337–339*

1. $\frac{6}{9}, \frac{2}{3}$ **3.** $\frac{11}{26}$ **5.** $-\frac{63}{16}$ **7.** 10 **9.** $-\frac{2}{8t}, -\frac{1}{4t}$ **11.** $\frac{-9-4y}{xy}$ **13.** $\frac{45m}{4}$ **15.** $-\frac{4}{5}$ **17.** $\frac{3}{7}$ **19.** $\frac{3}{5}$ **21.** $-\frac{7}{20}$ **23.** $-\frac{2}{35}$ **25.** $\frac{35}{3}$ **27.** $-\frac{19}{6}$ **29.** $5\frac{1}{2}$ in. **31.** 3 cm **33.** 8 in.^{2} **35.** -0.43 **37.** -0.16 **39.** 0.31 **41.** 0.17 **43.** 1.21 **45.** $-0.68x$ **47.** $\frac{1}{4}+2n=\frac{1}{2}, \frac{1}{8}$ **49.** $\frac{3}{8}$, 37.5% **51.** $\frac{5}{11}$, $\approx$45.5% **53.** 74.5% **55.** 87.5% **57.** $\frac{46}{100}, \frac{23}{50}$ **59.** $\frac{180}{100}, \frac{9}{5}$ **61.** 13 **63.** 280 **65.** b **67.** d **69.** $11\frac{1}{5}$; $\approx$58.3%, $\approx$23.8%, $\approx$17.9% **71.** Subtract only the numerators, not the denominators too; and use a common denominator, which was not done. $\frac{7}{20}-\frac{2}{10}=\frac{7}{20}-\frac{2}{10}\cdot\frac{2}{2}=\frac{7}{20}-\frac{4}{20}=\frac{3}{20}$ **73.** Multiply the denominators also, not just the numerators. $\frac{-2}{9}\div\frac{9}{5}=\frac{-2}{9}\cdot\frac{5}{9}=-\frac{10}{81}$ **75.** $\frac{23}{50}$ **77.** 1 **79.** \$1335 **81.** $\frac{-1}{-2}$, others $=-\frac{1}{2}$ **83.** 14%, 77%; 14%, 76%; 15%, 73%; 24%, 67%; 36%, 54%; 39%, 51%; 40%, 51%; 67%, 27%; 70%, 24%; 85%, 12%.

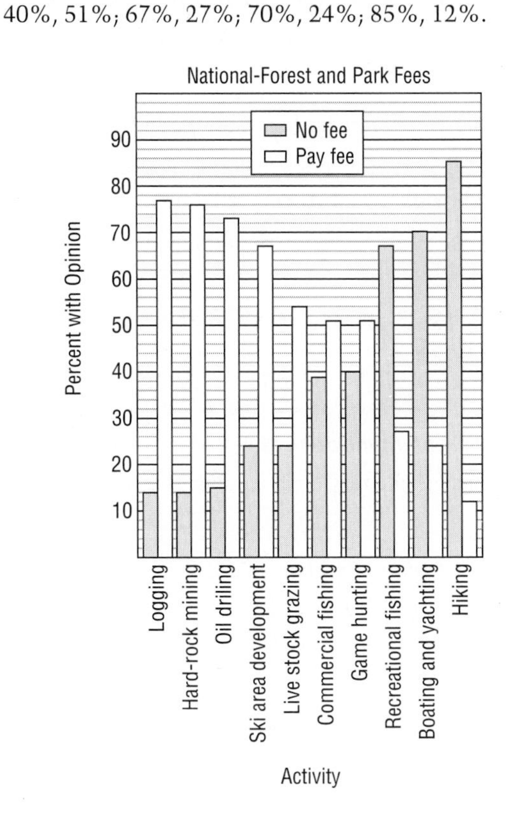

85. $\approx$1.1% **87.** 78% **89.** 54 m **91.** $\approx$188.6 m

CHAPTER 8

8.1 Communicating about Mathematics *p. 345*

A. Rate, 2500 people per year **B.** Ratio, 2.5 **C.** Rate, \$1500 per year

8.1 Guided Practice *p. 346*

1. $\frac{a}{b}$ is a rate, read "*a* per *b*," if *a* and *b* have different units of measure; examples vary, 50 miles per hour is one. $\frac{a}{b}$ is a ratio, read "*a* to *b*," if *a* and *b* have the same units of measure; examples vary, 20 feet to 1 foot is one. **2.** Rate, 25 miles per hour **3.** Ratio, 5 to 2 **4.** Ratio, 1 to 3 **5.** Rate, \$9.33 per pound **6.** \$5 per hour **7.** c **8.** $\frac{4}{5}, \frac{5}{4}, \frac{4}{9}, \frac{9}{4}, \frac{5}{9}, \frac{9}{5}$

8.1 Independent Practice *pp. 346–347*

9. Ratio, 8 to 9 **11.** Rate, 22 points per game **13.** $\frac{4\text{ doctors}}{5\text{ doctors}}$, is a ratio because the units are the same. **15.** $\frac{1600\text{ miles}}{3\text{ days}}$, is a rate because the units are different. **17.** $\frac{24\text{ inches}}{18\text{ inches}}=\frac{4}{3}$ **19.** $\frac{120\text{ seconds}}{300\text{ seconds}}=\frac{2}{5}$ **21.** $\frac{640¢}{400¢}=\frac{8}{5}$ **23.** $\frac{200\text{ centimeters}}{300\text{ centimeters}}=\frac{2}{3}$ **25.** 1500 tickets per hour **27.** b, 17.9¢ per apple < 19.7¢ per apple **29.** a, 21.9¢ per ounce < 23.0¢ per ounce **31.** $\approx$ 41.2 mph **33.** $\frac{15}{8}, \frac{7}{8}$; ratio of perimeters **35.** $\frac{12}{17}, \frac{3}{8}$; ratio of perimeters

8.1 Integrated Review *p. 347*

37. 32 ft **39.** 3 hours

8.2 Communicating about Mathematics *p. 350*

A. $\frac{x}{5}=\frac{3}{30}, \frac{1}{2}$ **B.** $\frac{4}{n}=\frac{12}{30}$, 10 **C.** $\frac{5}{3}=\frac{m}{5}$, $8\frac{1}{3}$ **D.** $\frac{4}{7}=\frac{6}{y}$, $10\frac{1}{2}$

8.2 Guided Practice *p. 351*

1. No, $\frac{2000}{80{,}000}\neq\frac{3000}{100{,}000}$ **2.** Examples vary. **3.** 1 **4.** 15 **5.** 24 **6.** 54 **7.** Similar; $\frac{a}{d}=\frac{b}{e}, \frac{a}{d}=\frac{c}{f}, \frac{b}{e}=\frac{c}{f}$ **8.** Not similar

8.2 Independent Practice pp. 351–352

9. Yes, $1 \times 12 = 4 \times 3$ **11.** Yes, $4 \times 36 = 9 \times 16$ **13.** $\frac{4}{3}$ **15.** $\frac{10}{7}$ **17.** 20 **19.** 18 **21.** $\frac{x}{6} = \frac{8}{9}, \frac{16}{3}$ **23.** $\frac{3}{8} = \frac{m}{24}$, 9 **25.** $\frac{5}{6} = \frac{12}{s}, \frac{72}{5}$ **27.** 23.33 **29.** 4 **31.** $d = 12, f = 10$ **33.** $d = 1\frac{2}{3}, e = 1\frac{2}{3}$ **35.** 680

8.2 Integrated Review p. 352

37. $\frac{3}{2}, \frac{8}{3}, \frac{15}{4}, \frac{24}{5}$. To get the next number, add 5, add 7, add 9, etc., to the numerator of the preceding number and add 1 to the denominator, $\frac{35}{6}, \frac{48}{7}$.

8.3 Communicating about Mathematics p. 354

$\frac{x}{5} = \frac{55}{1}$, 275 m

8.3 Guided Practice pp. 355–356

2. b **3.** $\frac{h}{5} = \frac{15}{3}$ **4.** 25 ft **5.** 12 **7.** 264,000 **9.** 125,714 **11.** a **13.** 440,640 **15.** $\frac{25}{w} = \frac{250}{1}$

8.3 Integrated Review p. 356

17. 4 **19.** 12.5

8.3 Mixed Review p. 357

1. 1 **3.** 2 **5.** $\frac{1}{2}$ **7.** 3 **9.** 12 **11.** 0.83 **13.** 2.06 **15.** Rate, 7 points per quarter **17.** Rate, 2 feet per 5 seconds

8.4 Communicating about Mathematics p. 359

Yes, explanations vary.

8.4 Guided Practice p. 360

1. Write the percent equation, divide to obtain decimal form, multiply each side by 160. **2.** Write the percent equation, divide to obtain decimal form, multiply each side by 100. **3.** 25, 80% **4.** 50,8 **5.** what, 120 **6.** Yes, $\frac{15}{6} = \frac{250}{100}$

8.4 Independent Practice pp. 360-361

7. 73.33% **9.** 7.2 **11.** 68 **13.** 121 **15.** 79.44% **17.** 50 **19.** 100 **21.** $\frac{45}{150} = \frac{p}{100}$

23. 52%

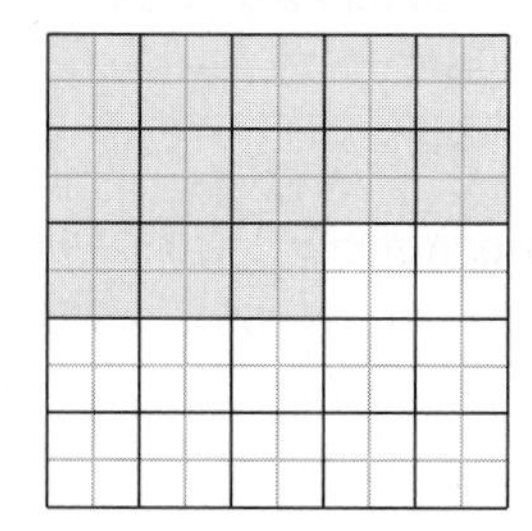

25. 60

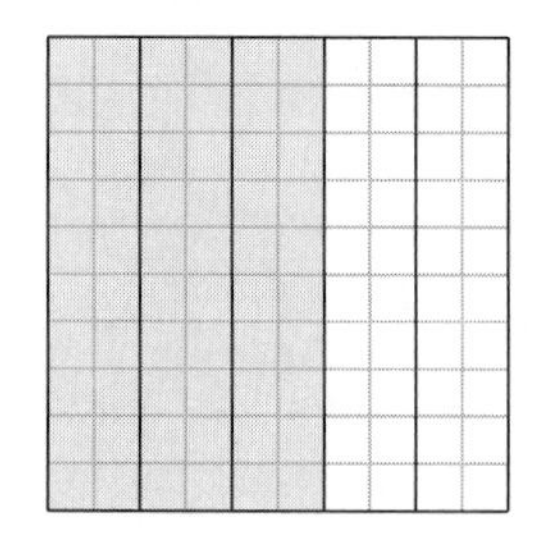

27. 5% **29.** ≈34.6%

8.4 Integrated Review p. 361

31. 0.62 **33.** 0.1824 **35.** 12.34% **37.** 94.32%

8.4 Mid-Chapter Self-Test p. 362

1. Rate, 50 meters per 9 seconds **2.** Rate, 3 lures per fisherman **3.** Ratio, $\frac{14}{1}$ **4.** Rate, 5 days per week **5.** Yes; $\frac{a}{d} = \frac{b}{e}, \frac{a}{d} = \frac{c}{f}, \frac{b}{e} = \frac{c}{f}$ **6.** No **7.** Yes **8.** No **9.** Yes **10.** 6.3 **11.** 24% **12.** 12 **13.** 16 **14.** $\frac{27}{7}$ **15.** 25.92 **16.** 15 **17.** 630 ft **18.** 13,630 **19.** $13\frac{1}{2}$ **20.** $11\frac{1}{4}$

8.5 Communicating about Mathematics p. 364

The base is the regular price, not the sale price.

8.5 Guided Practice p. 365

1.a. \$54.50 **b.** 30% **2.** 100% − 25% = 75% $= \frac{75}{100}, \frac{29.99}{b} = \frac{75}{100}$, solve for b. (\$39.99)

8.5 Independent Practice pp. 365–366

3. 45 **5.** 1,000,000 **7.** \$3000 **9.a.** 39% **b.** ≈ 16% **c.** ≈ 10% **11.** ≈ 50.2% **13.** ≈ 13.5% **15.** ≈23.5 units^2

8.5 Integrated Review p. 366

17. 475 **19.** 54 **21.** 15 **23.** 90

8.6 Communicating about Mathematics p. 368

Decrease: 8¢, 7¢, 2¢, 3¢, 4¢, 1¢, 2¢; Percent: ≈ 19.5, ≈ 21.2, ≈ 7.7, 12.5, ≈ 19.0, ≈ 5.9, 12.5; 85–86, 86–87

8.6 Guided Practice p. 369

1. 175% **2.** ≈ 65.9% **3.** Increase, ≈ 11.4% **4.** Decrease, ≈8.0%

8.6 Independent Practice pp. 369–370

5. Increase, 20% **7.** Decrease, 20% **9.** Increase, ≈122.2% **11.** Increase, ≈17.7% **13.** Decrease, ≈20.7% **15.** Each number after the first is 200% of the preceding number; 16, 32, 64 **17.** Each number after the first is 20% of the preceding number; 25, 5, 1 **19.** True, $2n = n + n$ **21.** False, $80 - 0.20(80) = 64 \neq 60$ **23.** Answers vary. **25.** 60% Increase, 60% Decrease, 500, 300 **27.** ≈76.5% **29.** 7.5%, 7.2%

8.6 Integrated Review p. 370

31. a **33.** 180 **35.** 5.6

8.6 Mixed Review p. 371

1. $\frac{3}{8}$ **3.** $\frac{1}{4}$ **5.** 16 **7.** 6

9.

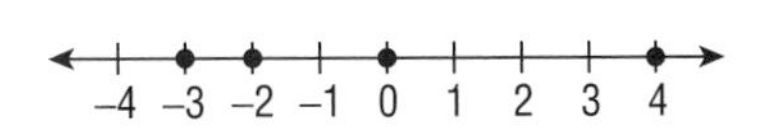

11. –1.01 **13.** 40 **15.** $12x + 20y, -20$ **17.** $71\frac{1}{9}$ **19.** 106.72

8.7 Communicating about Mathematics p. 374

A. 15; ab, ac, ad, ae, af, bc, bd, be,bf, cd, ce, cf, de, df, ef **B.** 20; abc, abd, abe, abf, acd, ace, acf, ade, adf, aef, bcd, bce, bcf, bde, bdf, bef, cde, cdf, cef, def **C.** 21; ab, ac, ad, ae, af, ag, bc, bd, be, bf, bg, cd, ce, cf, cg, de, df, dg, ef, eg, fg

8.7 Guided Practice p. 375

1. 8 **2.** You can choose a book for the first position in 4 ways, then you can choose a book for the second position in 3 ways, then you can choose a book for the third position in 2 ways, then you can choose a book for the fourth position in only 1 way. $4 \cdot 3 \cdot 2 \cdot 1 = 24$

3. 12

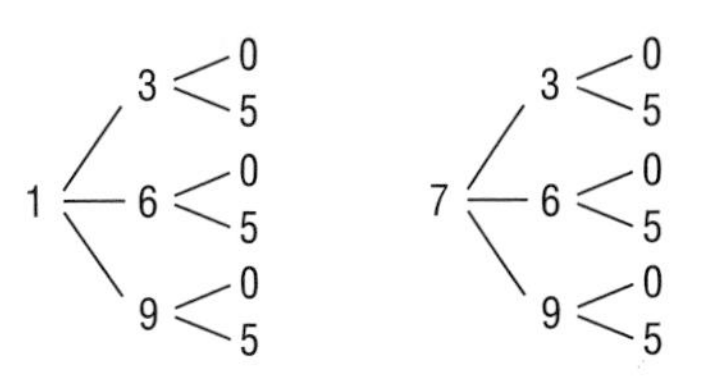

4. 21, go to the 7th row and count in 2 places from the 1.

8.7 Independent Practice pp. 375–376

5. 30 **7.** $\frac{1}{20}$ **9.** 2, 6, 24. To get the next number, multiply the preceding number by 3, by 4, by 5, etc.; 5040.

11. 120 **13.** 70 **15.** 84 **17.** $\frac{5}{18}$ **19.** $\frac{1}{3}$

8.8 Communicating about Mathematics p. 379

A. $\frac{1}{26}$ **B.** $\frac{1}{676}$ **C.** $\frac{1}{10}$ **D.** $\frac{1}{260}$

8.8 Guided Practice p. 380

1. 10 **2.** $\frac{3}{5}$ **3.** $\frac{3}{10}$ **4.** $\frac{1}{10}$

8.8 Independent Practice pp. 380–381

5. 40 **7.** $\frac{1}{40}$ **9.** $\frac{1}{20}$ **11.** $\frac{3}{40}$ **13.** $\frac{1}{12}$ **15.** $\frac{1}{4}$ **17.** $\frac{1}{6}$

19. $\frac{1}{312}$ **21.** $\frac{1}{78}$ **23.** $\frac{1}{26}$ **25.** $\frac{1}{10}$ **27.** $\frac{1}{1000}$

8.8 Integrated Review p. 381

29. $\frac{7}{94}$ **31.** $\frac{15}{47}$

8.8 Chapter Review pp. 383–386

1. f **3.** e **5.** a **7.** Ratio, $\frac{7}{3}$ **9.** Rate, $\frac{46 \text{ miles}}{1 \text{ hour}}$

11. $\frac{27}{50}$, ratio **13.** $\frac{13 \text{ miles}}{2 \text{ days}}$, rate **15.** $\frac{5}{2}$ **17.** $\frac{12}{11}$

19. $\frac{56}{11}$ **21.** $\frac{18}{5}$ **23.** 12.5 **25.** 87.5% **27.** 176

29. 6.45 **31.** Increase, 40% **33.** Increase, 30%

35. Increase, 5% **37.** $\frac{1}{8} = \frac{a}{2}, 2 \cdot \frac{1}{8} = 2 \cdot \frac{a}{2}, \frac{2}{8} = a, \frac{1}{4} = a$

39. $\frac{55}{b} = \frac{44}{100}$ **41.** 25% **43.** $\frac{1}{18}$ **45.** $\frac{1}{9}$ **47.** $\frac{1}{6}$ **49.** $\frac{1}{9}$

51. $\frac{1}{18}$

53.

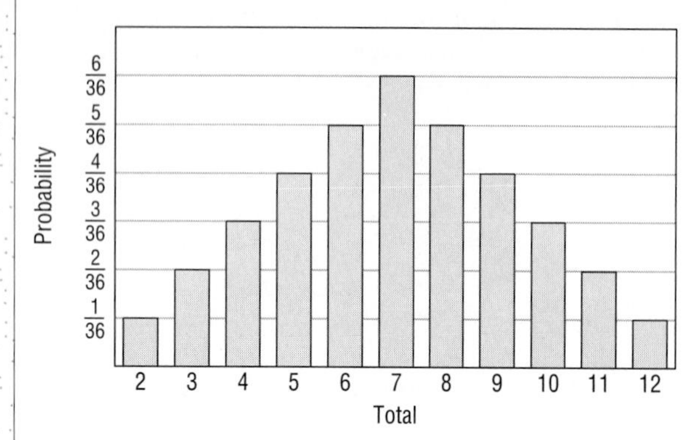

55. $\frac{1}{676}$ **57.** $\frac{1}{67,600}$ **59.** 10 ft **61.** $\frac{8}{32}, \frac{1}{4}$ **63.** \$8.25 **65.** ≈28.2% **67.** 6; abcde, abcdf, abcef, abdef, acdef, bcdef **69.** 17 million **71.** $\frac{39}{50}$ **73.** 288 **75.** $66\frac{2}{3}$ **77.** 32 **79.** 128 **81.** $\frac{1}{128}$

CHAPTER 9

9.1 Communicating about Mathematics p. 391

A. 4.7 **B.** 3.2 **C.** 7.5

9.1 Guided Practice p. 392

1. $2^2 = 4$ and $(-2)^2 = 4$ **2.** 7, −7; yes **3.** $\sqrt{5}$, $-\sqrt{5}$; no **4.** 1.1, −1.1; yes **5.** $\frac{5}{2}, -\frac{5}{2}$; yes **6.** 3.742, −3.742

9.1 Independent Practice pp. 392–393

7. $\sqrt{14}, -\sqrt{14}$ **9.** 8 − 8 **11.** 0.6, −0.6 **13.** $\frac{8}{3}, -\frac{8}{3}$ **15., 17.** Draw a square with 16 tiles. **19.** 3, −3 **21.** 4.690, −4.690 **23.** 5, −5 **25.** 9, −9 **27.** 4.5 **29.** 5.6 **31.** $\sqrt{25} = x$, 5 **33.** $a^2 + 6 = 14$, $\sqrt{8}$ and $-\sqrt{8}$ **35.** ≈3.87 mi **37.** ≈1.12 seconds

9.1 Integrated Review p. 393

39. c

9.2 Communicating about Mathematics p. 396

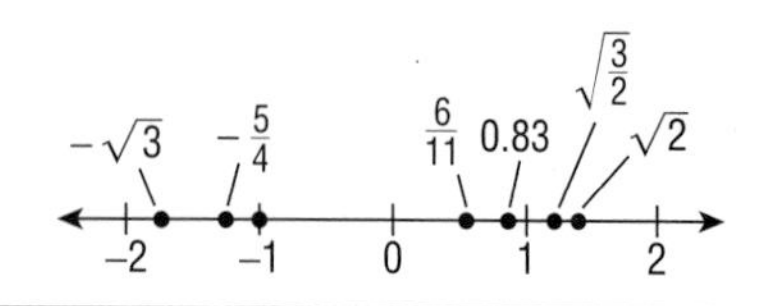

9.2 Guided Practice p. 397

1. See page 395. **2.** See page 395. **3. a.** Yes, the decimal terminates. **b.** Yes, the decimal repeats. **c.** No, the decimal neither terminates nor repeats. **4.** Rational **5.** Irrational **6.** Rational **7.** Rational

8.

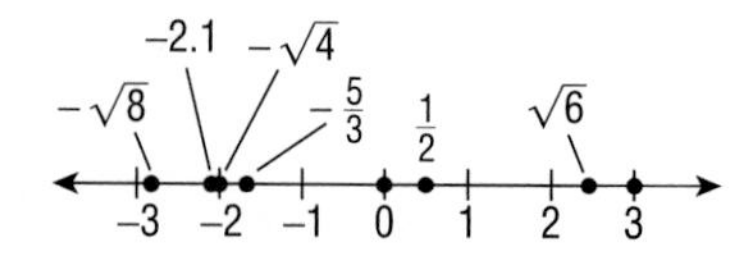

9.2 Independent Practice pp. 397–398

9. Rational, it is the quotient of two integers. **11.** Irrational, its decimal form neither terminates nor repeats. **13.** Rational, its decimal form terminates. **15.** Irrational, its decimal form neither terminates nor repeats. **17.** sometimes, real numbers consist of rational numbers and irrational numbers. **19.** never, all integers are rational numbers. **21.** $\sqrt{2} + 2$, no; decimal neither terminates nor repeats. **23.** 6, yes; can be represented as the quotient of two integers. **25.** f **27.** b **29.** e

31.–36.

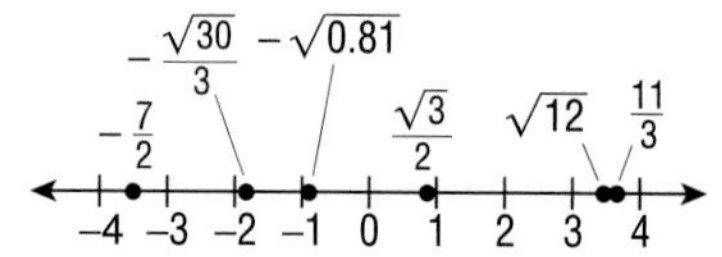

37. $<$ **39.** $=$ **41.** $>$ **43.** No. The area of the shaded square is half the area of the larger square, or 8. So each side length of the shaded square is $\sqrt{8}$. $\sqrt{8}$ is an irrational number, so its decimal form neither terminates nor repeats. **45.** Answers vary.

9.3 Communicating about Mathematics p. 401

Answers vary.

9.3 Guided Practice p. 402

1. $m^2 + n^2 = t^2$ **2.** No, explanations vary. **3.** 4 **4.** 9.2

9.3 Independent Practice pp. 402–403

5. Never **7.** Sometimes **9.** ≈ 13.04 **11.** 9 **13.** ≈ 9.98 **15.** Possible. **17.** Not possible.

19. Not possible. **21.** Not possible. **23.** ≈ 8.49 **25.** 36 **27.** ≈ 8.944 ft **29.** ≈ 5.66 ft **31.** 3rd sides: 6, ≈ 12.8

9.3 Integrated Review p. 403

33.

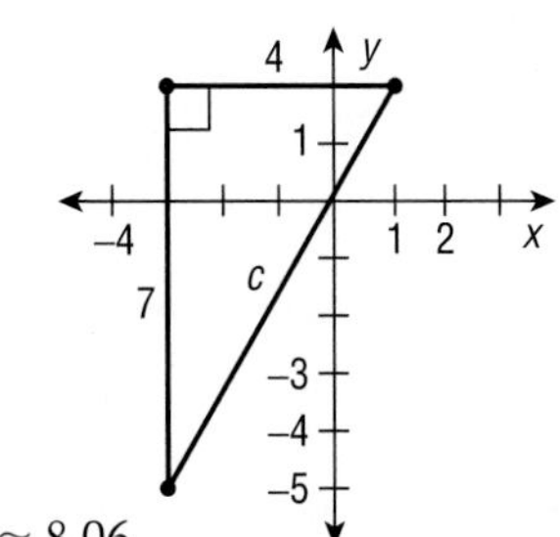

Yes; 4, 7, ≈ 8.06

35. **37.**

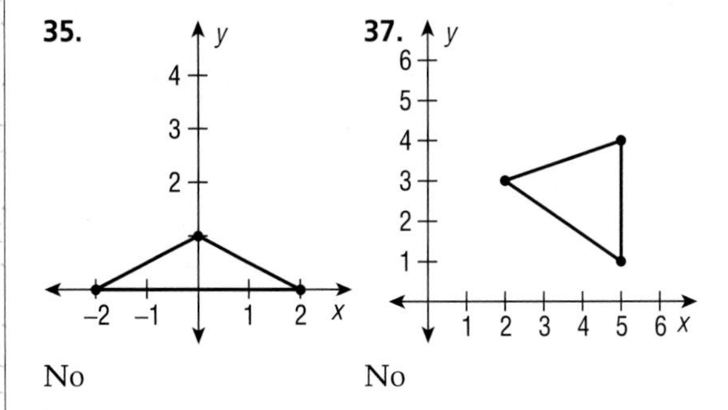

No No

9.3 Mixed Review p. 404

1. ≈ 8.25 units **3.** 5.6 units **5.** 5 **7.** ≈ 92.79 **9.** $x \geq 7$ **11.** $n > \frac{8}{5}$ **13.** 3 **15.** 2

9.4 Communicating about Mathematics p. 406

AC	*BC*	*AB* computed	*AB* measured
$6\frac{1}{4}$ in.	$5\frac{3}{8}$ in.	≈ 8.243 in.	$8\frac{1}{4}$ in.
15.8 cm	13.6 cm	≈ 20.847 cm	20.9 cm

9.4 Guided Practice p. 407

2. No. Use the Pythagorean Theorem to find the length of the other leg. $a^2 + 45^2 = 75^2$, $a = 60$. Then the perimeter is $45 + 75 + 60 = 180$ meters, and the area is $\frac{1}{2}(60 \cdot 45) = 1350$ meters^2.

9.4 Independent Practice pp. 407–408

3. 90 units, 180 units^2 **5.** 8 units, 4 units^2 **7.** No. Since $6^2 + 8^2 = c^2$ and $c = 10$, the number of blocks is 10 and the number of feet is 5000 ft, which is less than 5280 feet. **9.** 12 ft; $\frac{21}{t} = \frac{16+12}{16}$, $t = 12$ **11.** $2\frac{1}{2}$ ft per second

9.4 Integrated Review p. 408

13. 15, 20, 25; 18, 24, 30; 21, 28, 35

9.4 Mid-Chapter Self-Test p. 409

1. 4, -4 **2.** 11, -11 **3.** 0.7, -0.7 **4.** 0.6, -0.6 **5.** ≈ 3.16 units **6.** 8.9 units **7.** ≈ 5.7 units **8.** ≈ 7.4 units **9.** Irrational, its decimal form neither terminates nor repeats. **10.** Rational, its decimal form terminates. **11.** Irrational, its decimal form neither terminates nor repeats. **12.** Rational, its decimal form terminates. **13.** 12.37 **14.** 5 **15.** 90.51 **16.** 35.36 **17.** 21 in. **18.** 70 in. **19.** 210 in.^{2} **20.** False

9.5 Communicating about Mathematics p. 412

A. $b \leq 2$

–3 –2 –1 0 1 2 3

B. $x \leq -5$

–7 –6 –5 –4 –3 –2 –1 0 1

C. $y < -4$

–6 –5 –4 –3 –2 –1 0 1 2

9.5 Guided Practice p. 413

1. c **2.** d **3.** b **4.** a **5.** $x < 15$, $15 > x$ **6.** $x \geq 0$, $0 \leq x$ **7.** $x > -3$, $-3 < x$ **8.** $x \leq -11$, $-11 \geq x$ **9.** $x < -7$

–9 –8 –7 –6 –5 –4 –3 –2 –1 0

10. $x \geq 3$

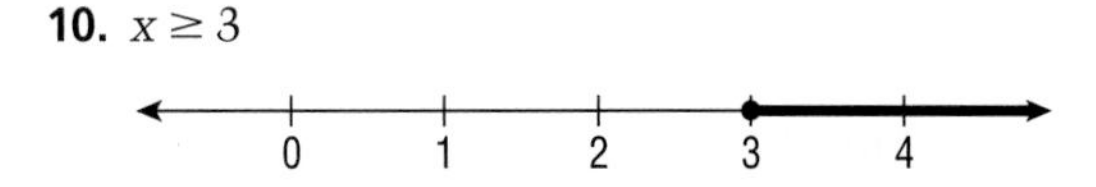

9.5 Independent Practice p. 413–414

11.

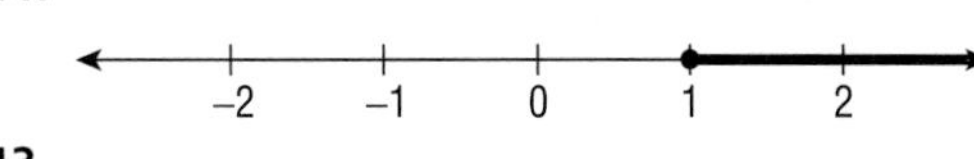

13.

4 5 6 7 8 9

15. $x \geq -5$ **17.** $x \leq -1$ **19.** $x > \sqrt{2}$
21. $x < -\sqrt{3}$ **23.** $x \geq -1$ **25.** $n > 1$
27. $z > -9$ **29.** $-20 \geq x$, the set of all real numbers less than or equal to -20. **31.** $17 > s$, the set of all real numbers less than 17.
33. $t + 17 > 24, t > 7$ **35.** $n + 5 \leq -9, n \leq -14$
37. d, 0 mph and 48.71 mph were both attained by the *Sunraycer.* **39.** $T \geq -453$ **41.** b, a number of students must be a whole number.

■ 9.5 Integrated Review p. 414

43. 59 **45.** 12 **47.** 18

■ 9.5 Mixed Review p. 415

1. 400 ft **3.** ≈ 10.1 ft **5.** Proportions vary, $\frac{y}{10.1} = \frac{60}{1.5}$ **7.** $x < 2$ **9.** $x \geq -3$ **11.** $2^3 \cdot 3^2 \cdot 5$
13. $3^3 \cdot 7$ **15.** $2^4 \cdot 23$ **17.** $\approx 11.1\%$ **19.** $\approx 44.4\%$

■ 9.6 Communicating about Mathematics p. 417

Answers vary.

■ 9.6 Guided Practice p. 418

1. $<$ **2.** $\leq$ **3.** $>$ **4.** $\leq$ **5.** Yes **6.** No **7.** No
8. Yes **9.** $b > 6$ **10.** $x \geq -56$ **11.** $h > -25$
12. $f \geq -72$

■ 9.6 Independent Practice p. 418–419

13. The direction of the inequality symbol was not reversed, as it should have been. **15.** The left side was not divided by -0.4, as it should have been. **17.** b **19.** a **21.** $n \geq \frac{5}{2}$ **23.** $y > 36$
25. $a < -\frac{1}{4}$ **27.** $p \geq -10$ **29.** $n > -42$
31. $x \geq 3$ **33.** $y \leq -10$ **35.** $x < 6$ **37.** 9
39. $1\frac{1}{2}$ hours

■ 9.6 Integrated Review p. 419

41. Yes. Both are 25%. **43.** No. The percents for the other reasons may overlap with each other but they may not overlap with the percent for holiday travel.

■ 9.7 Communicating about Mathematics p. 421

At least 7 ounces

■ 9.7 Guided Practice p. 422

1. $x \leq 5$; add 2 to each side, divide each side by 3.
2. $y < -10$; subtract 2 from each side, multiply each side by -5 and reverse the inequality.
3. Substitute several numbers into the original inequality, some that you think will yield true statements and some that you think will yield false statements. You try to get false statements as well as true ones, and your check cannot be complete.
4. $y \geq 3$

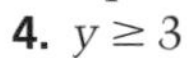

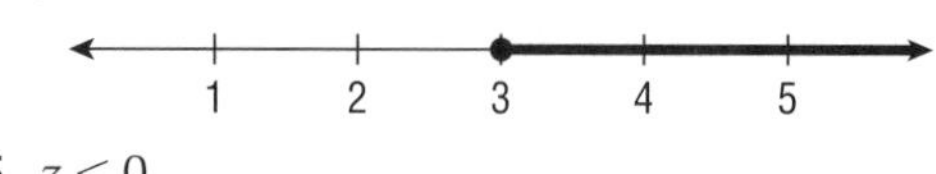

5. $z < 0$

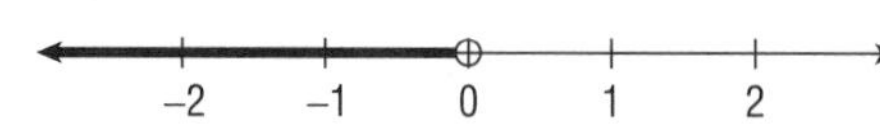

■ 9.7 Independent Practice p. 422–423

7. $(\frac{1}{6})(6y) < (\frac{1}{6})(-4), y < -\frac{2}{3}$ **9.** Never
11. Sometimes **13.** c **15.** b **17.** $x > 3$
19. $a \leq \frac{5}{3}$ **21.** $m \leq 4$ **23.** $x \leq 4$ **25.** $x \geq -3$
27. $n + n + 1 + n + 2 > 18, n > 5$ **29.** $x \leq -4$
31. Less than 7
33.

Number of tickets bought	4	8	12	16	20
Total cost in dollars	11	12	13	14	15

Number of tickets bought	24	28	32	36	40
Total cost in dollars	16	17	18	19	20

■ 9.7 Integrated Review p. 423

35. Yes **37.** No; ordered pairs vary, (0, 0)

■ 9.8 Communicating about Mathematics p. 426

C, $85 + 53 \not> 138$

■ 9.8 Guided Practice p. 427

1. No

6 5

11

2. c, by the Triangle Inequality 3. b, $4+6 \not> 10$ 4. Greater than 2, less than 8 5. Greater than 2, less than 20 6. Greater than 4, less than 36

9.8 Independent Practice p. 427–428

7. No, $1+3 \not> 5$ 9. Yes, $6+8>10$, $6+10>8$, $8+10>6$ 11. 5 cm, 11 cm; 7 in., 25 in.; 11 ft, 31 ft; 15 m, 75 m; 125 cm, 325 cm 13. No 15. Yes 17. No 19. c 21. a 23. No, the sum of the two shorter lengths is not greater than the longest length. 25. If the 6's and the 10's are correct, the 8's are incorrect.

9.8 Integrated Review p. 428

27. $x>8$ 29. $x<5$ 31. 30

9.8 Chapter Review p. 430–431

1. 7, −7 3. $\sqrt{35}$, $-\sqrt{35}$ 5. 4, −4 7. 11, −11 9. Rational, *F* 11. Irrational, *A* 13. Rational, *E* 15. Irrational, *C* 17. 16 19. ≈ 8.06 21. $x \le -5$, the set of all real numbers less than or equal to −5. 23. $m<13$, the set of all real numbers less than 13. 25. $z \ge 8$, the set of all real numbers greater than or equal to 8. 27. $p<-\frac{1}{9}$, the set of all real numbers less than $-\frac{1}{9}$. 29. $x \le \frac{5}{4}$ 31. $x<-3$ 33. $x \le \frac{9}{4}$ 35. $x<-5$ 37. $y \le 2$ 39. Yes, $6^2+8^2=10^2$ 41. Yes; $2+7>8$, $2+8>7$, $7+8>2$ 43. $5t^2=125$; 5, −5 45. $x-7>11$, $x>18$ 47. $-4m \le 52$, $m \ge -13$ 49. $2(5-x)<x+11$, $x>-\frac{1}{3}$ 51. $x^2=380.25$, 19.5 in. 53. 32 and 50 55. $\sqrt{32}$ 57. *d*, $x \le 4$, she can play at most 4 games 59. 40 61. 12 63. 3,078,000 in.^{2}; 21,375 ft^2 65. Solar thermal, the furnace is the generator.

9.8 Cumulative Review for Chapters 7–9 pp. 434–435

1. $\frac{6}{7}$ 3. $\frac{5}{12}$ 5. 1.53 7. 0.62 9. $\frac{25}{18}$ 11. $28n$ 13. $20\frac{1}{6}$ in., $24\frac{19}{24}$ in.^{2} 15. 30% 17. 80% 19. $\frac{3}{5}$, 60%, 0.6 21. Rate, 28 gal per min 23. Ratio, $\frac{8}{1}$ 25. $2800 27. 36 29. $\frac{275}{8}$ 31. 34 33. 76% 35. 70 37. 60 39. 12 41. Increase, ≈3.83% 43. 26 45. 2600 47. 11 49. ≈8.31 51. $b \ge 14$ 53. $n<-\frac{16}{3}$ 55. Yes

CHAPTER 10

10.1 Communicating about Mathematics p. 440

367 mi, 363 mi, 357 mi; 4; no, two consecutive "spokes" from Howell must be traveled and there are only these four most efficient routes to take.

10.1 Guided Practice p. 441

1. b 2. a 3. d 4. c 5. $\overrightarrow{RS}$, $\overline{AB}$, $\overleftrightarrow{PQ}$ 6. Answers vary.

10.1 Independent Practice pp. 441–442

7. $\overleftrightarrow{CD}$, $\overleftrightarrow{CE}$, $\overleftrightarrow{DE}$, $\overleftrightarrow{DF}$, $\overleftrightarrow{EF}$ 9. $\overrightarrow{EB}$, $\overrightarrow{EF}$, $\overrightarrow{EI}$, $\overrightarrow{EH}$, $\overrightarrow{ED}$ ($\overrightarrow{EC}$) 11. $\overleftrightarrow{AG}$ and $\overleftrightarrow{CF}$, $\overleftrightarrow{BH}$ and $\overleftrightarrow{CF}$ 13. Yes, they have the same endpoints. 15. 5 17. *A,B,C,D,E* 19. $\overleftrightarrow{CB}$, $\overleftrightarrow{BE}$ 21. Draw 3 parallel lines. 23. Draw 2 parallel lines, each intersected by a 3rd line.

25.

Figure	a	b	c	d
Number of lines of intersection	0	2	1	3

27. No, only the bases of windows on the same wall (and not side by side) or on opposite walls of the building are parallel. 29. 53 ft

10.1 Integrated Review p. 442

31. $\overleftrightarrow{AB}$ 33. $\overleftrightarrow{AB}$, $\overleftrightarrow{BC}$, $\overleftrightarrow{CD}$, $\overleftrightarrow{DA}$, $\overleftrightarrow{AC}$, $\overleftrightarrow{BD}$ 35. To get the next number, add 2, add 3, add 4, etc., to the preceding number. The numbers are found in the third diagonal column of Pascal's triangle.

10.2 Communicating about Mathematics p. 444

A. 72° B. 90° C. Answers may include heptagon—$51\frac{3}{7}°$, nonagon—40°, decagon—36°

10.2 Guided Practice p. 445

1. *X* 2. $\overrightarrow{XW}$, $\overrightarrow{XZ}$ 3. ∠*X*, ∠*WXY*, ∠*ZXW*, ∠*YXW* 4. ∠*WYZ* 5. Right 6. Obtuse 7. Acute 8. Straight

10.2 Independent Practice pp. 445–446

9. $\angle ABC$, $\angle DBC$, $\angle CBA$, $\angle CBD$ **11.** 2 **13.** 2 **15.** c **17.** b **23.** 35° **25.** 155° **27.** They have the same measure **29.** Straight **31.** Obtuse **33.** Acute **35.** Right

10.2 Integrated Review p. 446

37. $6\frac{2}{3}°$, $28\frac{1}{3}°$ **39.** b

10.3 Communicating about Mathematics p. 449

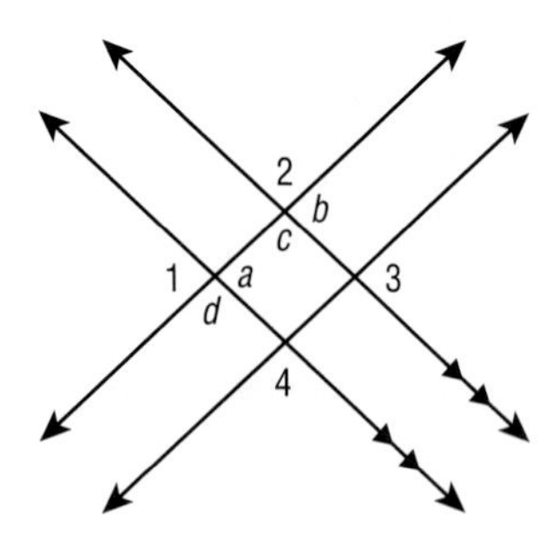

$\angle 1 \cong \angle a, \angle a \cong \angle b$, and $\angle b \cong \angle 3$; so $\angle 1 \cong \angle 3$.
$\angle 2 \cong \angle c, \angle c \cong \angle d$, and $\angle d \cong \angle 4$; so $\angle 2 \cong \angle 4$.
$m\angle 2 = 96°$, $m\angle 3 = 84°$, $m\angle 4 = 96°$.

10.3 Guided Practice p. 450

1. m and n **2.** $\angle 1$ and $\angle 3$, $\angle 2$ and $\angle 4$, $\angle 5$ and $\angle 7$, $\angle 6$ and $\angle 8$ **3.** $\angle 1$ and $\angle 5$, $\angle 2$ and $\angle 6$, $\angle 3$ and $\angle 7$, $\angle 4$ and $\angle 8$ **4.** 55°. When two parallel lines are intersected by a third line, the corresponding angles are congruent; so $\angle 6 \cong \angle 2$.
5. 55°. Vertical angles are congruent; so $\angle 2 \cong \angle 4$.
6. 55°. Vertical angles are congruent; so $\angle 6 \cong \angle 8$.
7. 45°, 135°, 135°; vertical angles are congruent and the sum of the measures of pairs of angles is 180°. **8.** 4

10.3 Independent Practice p. 450

9. l and n **11.** $\angle 1$, $\angle 4$, $\angle 9$, $\angle 12$ **13.** $\angle 2$, $\angle 3$, $\angle 10$, $\angle 11$ **15.** One example: $\angle 1$ and $\angle 5$
17. $\angle 1 \cong \angle 2$, $\angle 2 \cong \angle 3$, and $\angle 3 \cong \angle 4$; so $\angle 1 \cong \angle 4$.
19. $\overleftrightarrow{BC}$, $\overleftrightarrow{DF}$, $\overleftrightarrow{GJ}$, and $\overleftrightarrow{KO}$; $\overleftrightarrow{AK}$, $\overleftrightarrow{CL}$, $\overleftrightarrow{FM}$, and $\overleftrightarrow{JN}$
21. 10 **25.** $m\angle 2 = m\angle 3 = 117°$, $m\angle 1 = m\angle 4 = 63°$, $m\angle 6 = m\angle 7 = 34°$, $m\angle 5 = m\angle 8 = 146°$ **27.** 4; angles are 35°, 55°, 90°, and 180°.

10.3 Integrated Review p. 451

29. Answers vary. **31.** Answers vary.

10.3 Mixed Review p. 452

5. Triangles and a square **7.** $\overleftrightarrow{AB}$ and $\overleftrightarrow{BC}$, $\overleftrightarrow{BC}$ and $\overleftrightarrow{CD}$, $\overleftrightarrow{CD}$ and $\overleftrightarrow{DA}$, $\overleftrightarrow{DA}$ and $\overleftrightarrow{AB}$ **9.** $\angle 8$, $\angle 14$
11. 135° **13.** Vertical congruent angles

10.4 Communicating about Mathematics p. 454

The photo has rotational symmetry.

10.4 Guided Practice p. 455

1. Answers vary. **2.** Answers vary. **3.** 1 line of symmetry **4.** 4 lines of symmetry; rotational symmetry at 45°, 90°, 135°, and 180° in either direction. **5.** 1 line of symmetry **6.** 3 lines of symmetry, rotational symmetry at 120° in either direction

10.4 Independent Practice pp. 455–456

7. Rotational symmetry at 60°, 120°, and 180° in either direction **9.** 1 line of symmetry
11. Possible. Answers vary. **13.** Possible. Answers vary. **15.** Shade the small square in the 5th row and 4th column. **17.** Shade the small square in the 2nd row and 5th column. **19.** No
21. The lion with the lion's reflection has 1 line of symmetry. **23.** The top of each wax cell has 6 lines of symmetry and has rotational symmetry at 60°, 120°, and 180° in either direction.
25. DECK **27.** Answers vary. Examples: BOOK, DICE; HE DIED.

10.4 Integrated Review p. 456

29. $C(-2, -1)$, $D(2, -3)$

10.5 Communicating about Mathematics p. 458

8; 3 isosceles right, 2 isosceles acute, 2 scalene obtuse, 1 scalene right

10.5 Guided Practice p. 459

1. c; all sides have different lengths. d; one angle is obtuse. **2.** a; at least two sides have the same length. d; one angle is obtuse. **3.** c; all sides have different lengths. f; one angle is right. **4.** c; all sides have different lengths. g; all three angles are acute. **5.** a; at least two sides have the same length. b; all three angles have the same measure. e; all three sides have the same length. g; all three angles are acute. **6.** c; all sides have different lengths. g; all three angles are acute. **7.** a; at least two sides have the same length. b; all three angles have the same measure. e; all three sides have the same length. g; all three angles are acute. **8.** a; at least two sides have the same length. f; one angle is right.

10.5 Independent Practice pp. 459–460

9. Answers vary. **11.** Answers vary. **13.** Answers vary. **15.** Scalene right **17.** Isosceles obtuse **19.** Scalene obtuse **21.** Equilateral, acute, equiangular, isosceles **23.** Scalene acute **25.** Isosceles right **27.** ≈ 10.24 units, $4\frac{1}{2}$ units^2; isosceles right **29.** ≈ 9.66 units, 4 units^2; isosceles right **31.** An equilateral triangle is equiangular and vice versa. **33.** The triangle is rigid. You can adjust the rectangle's sticks to form a different shape of quadrilateral. **35.** Triangles will not collapse. **37.** Isosceles obtuse **39.** Scalene right

10.5 Mid-Chapter Self-test p. 461

1. S **2.** $\overleftrightarrow{PQ}$, $\overleftrightarrow{NO}$ **3.** P **4.** Up **5.** $\angle DAB$, $\angle DAF$, $\angle CAE$ **6.** $\angle FAE$, $\angle EAD$, $\angle DAC$, $\angle CAB$ **7.** $\angle FAC$, $\angle EAB$ **8.** $\angle FAB$ **9.** Vertical **10.** Corresponding **11.** Corresponding **12.** $m\angle 2 = m\angle 4 = m\angle 6 = m\angle 8 = 62°$, $m\angle 1 = m\angle 3 = m\angle 5 = m\angle 7 = 118°$ **13.** 6 lines of symmetry; rotational symmetry at 60°, 120°, and 180° in either direction **14.** 5 lines of symmetry; rotational symmetry at 72° and 144° in either direction **15.** 1 line of symmetry **16.** Scalene obtuse **17.** Equilateral, acute, equiangular, isosceles **18.** Scalene right **19.** Isosceles acute **20.** No

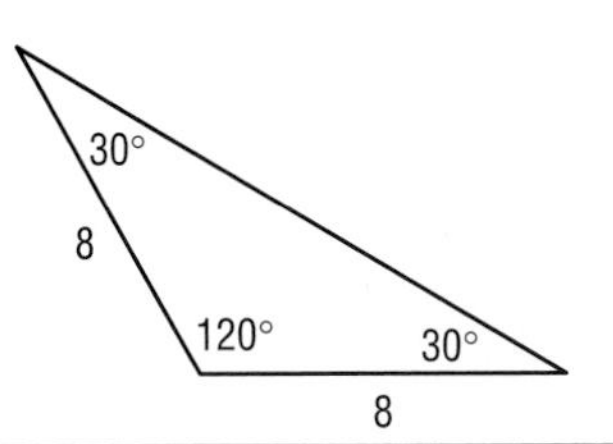

10.6 Communicating about Mathematics p. 463

Answers vary.

10.6 Guided Practice p. 464

1.a. Parallelograms that include rectangles, squares, and rhombuses **b.** Kites, nonspecial quadrilaterals **c.** Trapezoids that include isosceles trapezoids **d.** Rhombuses that include squares **e.** Scalene quadrilaterals that include some trapezoids that are not isosceles **f.** Kites; isosceles trapezoids and some trapezoids that are not isosceles; and parallelograms that include rectangles, squares, and rhombuses **2.** Answers vary. **3.** A rhombus is a parallelogram and a kite is not. **4.** A square is a rectangle with sides of equal length.

10.6 Independent Practice pp. 464–465

5. Trapezoid **7.** Square **9.** Rhombus **11.** Scalene quadrilateral **13.** sometimes; a parallelogram is only one type of quadrilateral **15.** always; a segment joining any two interior points lies completely within a trapezoid **17.** $x = 6$ in. **19.** $x = 6$ ft, $y = 3$ ft **21.** Draw any parallelogram. **23.** Not possible **25.** f **27.** e, f **29.** a **31.** b **33.** Drawings vary. **35.** The sum of the measures of the angles of any convex quadrilateral is 360°. **37.–38.**

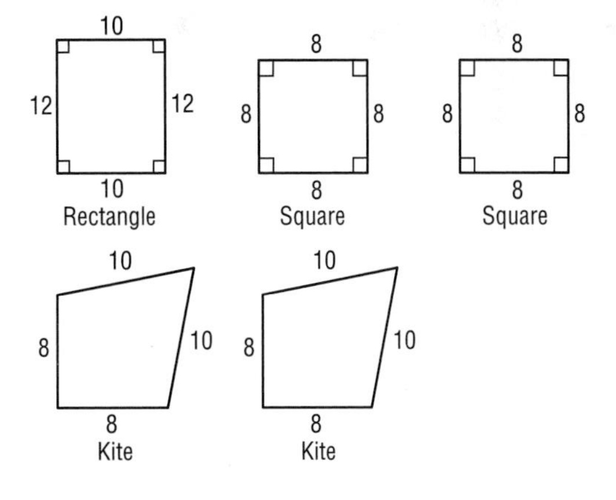

39. The 2 side walls that are kites, the back wall and bottom that are squares

SELECTED ANSWERS

10.6 Integrated Review p. 465

41. Kite **43.** Parallelogram

10.6 Mixed Review p. 466

1. Scalene quadrilateral **3.** Kite
5. 3.14 **7.** $\frac{2}{3}$ **9.** $\frac{5}{6}$ **11.** 9.2×10^{-4} **13.** 4.6×10^{-7}
15. $\frac{3}{5}$ **17.** $\frac{7}{9}$ **19.** $\frac{3}{16}$

10.7 Communicating about Mathematics p. 468

Examples

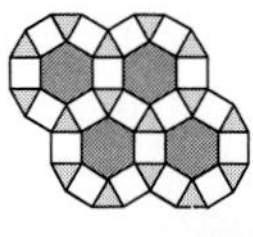

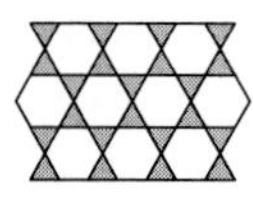

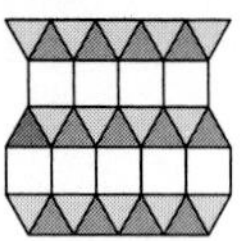

10.7 Guided Practice p. 469

1. See top of page 467. **2.** a and c **3.** b and c **4.a.** Isosceles triangle, no **b.** Regular hexagon, yes **c.** Rectangle, no **d.** Rhombus, no

10.7 Independent Practice pp. 469–470

5. d **7.** a **9.a.** Equilateral, equiangular, regular **b.** Equilateral, equiangular, regular **c.** Congruent to *d* **d.** Congruent to *c* **11., 13.** You should have 6 congruent regular triangles. **15.** A polygon is regular if it is equilateral and equiangular. No, see figure for Exercise 10c on page 469.
17. square; square, rectangle

10.7 Integrated Review p. 470

19. 13; 13, 13, 16, 16, 16, 16 **21.** 3; 6, 18, 18, 20

10.8 Communicating about Mathematics p. 473

A.

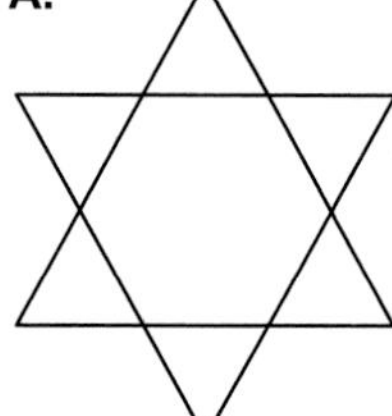

B.

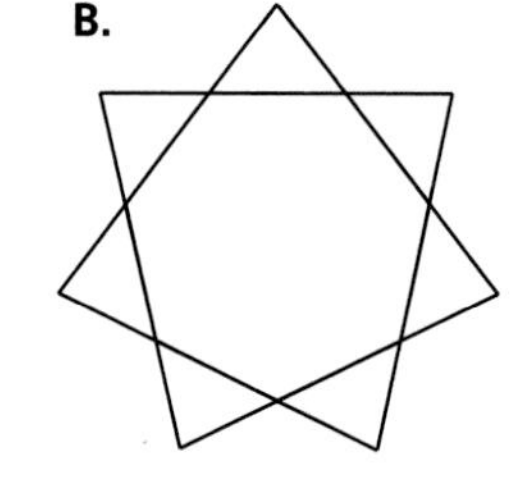

C.

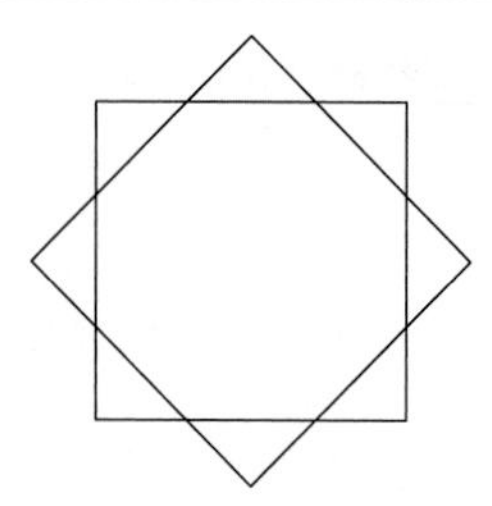

Number of vertex angles in star	Each exterior angle of regular convex polygon	Each vertex angle of star
6	$\frac{370°}{6}=60°$	$180°-2(60°)=60°$
7	$\frac{360°}{7}=51\frac{3}{7}°$	$180°-2(51\frac{3}{7}°)=77\frac{1}{7}°$
8	$\frac{360°}{8}=45°$	$180°-2(45°)=90°$

10.8 Guided Practice p. 474

1. **2.**

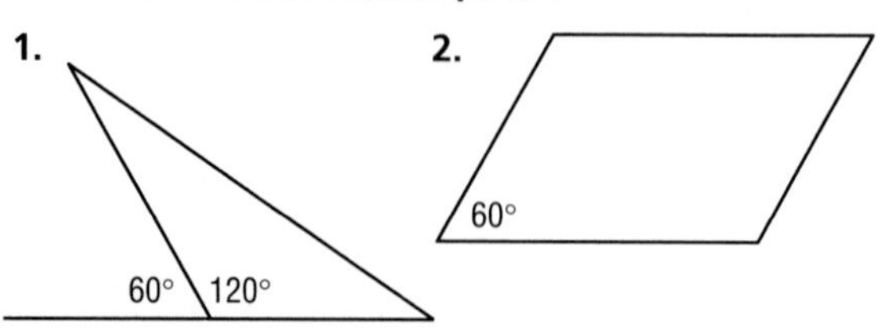

3. 1: 150°, 2: 120°, 3: 90°, 4: 60°, 5: 30° **4.** 1: 120°, 2: 120°, 3: 60°, 4: 60°

10.8 Independent Practice pp. 474–475

5. $\angle ABC$, $\angle BCD$, $\angle CDE$, $\angle DEA$, $\angle EAB$; 540°
7. No **9.** 1: 80°, 2: 45°, 3: 135°, 4: 125°,
11. 1: 90°, 2: 60°, 3: 95°, 4: 85°, 5: 35°, 6: 100°
13. Each interior angle: 144°. Each exterior angle: 36°. **15.** 40°, 70°, 70° **17.** Each is 108°.
19.

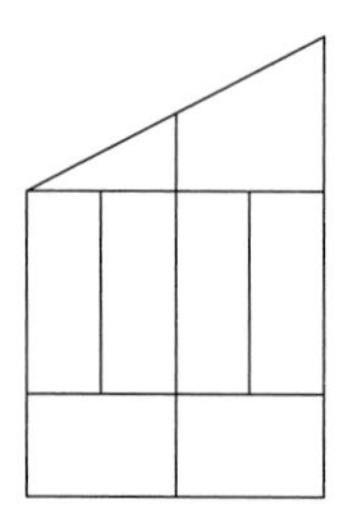

6 rectangles, 1 right triangle, and 1 trapezoid

21. 27°, 63°, 90° **23.** Top one, 140°

10.8 Integrated Review p. 475

25. 60°, 90°, 108°, 120°, $128\frac{4}{7}$°, 135° **27.** 140°

10.9 Communicating about Mathematics p. 478

1. C **2.** A **3.** B In Exercise 2, no two sides are the same length; so no two angles have the same measure; so A matches with Exercise 2. The longest side in Exercise 1 is more than 70% longer than the shorter sides, while the longest side in Exercise 3 is less than 33% longer than the shorter sides; so the largest angle in Exercise 1 is larger than the largest angle in Exercise 3; so C matches with Exercise 1. Then B matches with Exercise 3.

10.9 Guided Practice p. 479

1.a. The longest side is opposite the largest angle and the shortest side is opposite the smallest angle. **b.** The angles opposite the sides of the same lengths have equal measures. **c.** All three angles are congruent and all three sides are congruent. **2.** An isosceles right triangle has a 90° angle and two angles of equal measure. $180° - 90° = 90°$, $\frac{90°}{2} = 45°$ **3.** $\angle B$, $\angle C$, $\overline{AC}$, $\overline{AB}$ **4.** $\angle N$, $\angle O$, $\overline{MO}$, $\overline{MN}$ **5.** $\angle F$, $\angle D$, $\overline{DE}$, $\overline{EF}$ **6.** $\angle R$, $\angle P$, $\overline{PQ}$, $\overline{QR}$ **7.** 40°, 60°, 80°; $\overline{EF}$, $\overline{DE}$ **9.** 15°, 60°, 105°; $\overline{AD}$, $\overline{AG}$ **11.** 45°, 60°, 75°; $\overline{CG}$, $\overline{CH}$ **13.** The hypotenuse, it is opposite the largest angle. **15.** $\angle A$; $AC \approx 6.01$, so $\overline{BC}$ is the shortest side. **17.** $\angle B$; $AC \approx 19.00$, so $\overline{AC}$ is the shortest side. **19.** $\overline{DE}$, $\overline{EF}$; $\angle DEF = 60°$ and $\angle F = 50°$, so $\angle F$ is the smallest angle and $\angle D$ is the largest angle. **21.** $\overline{EF}$, $\overline{DF}$; $\angle D = 48°$ so $\angle D$ is the smallest angle and $\angle E$ is the largest angle. **19., 21.** The shortest side is opposite the smallest angle and the longest side is opposite the largest angle. **23.** 57°, $\angle U$, $\angle TSU$, $\overline{ST}$, $\overline{TU}$ **25.** 8°, $\angle D$, $\angle E$, $\overline{EF}$, $\overline{DF}$ **27.** c. **29.** b. **27., 29.** If a triangle has three sides of the same length, then it has three angles of the same measure; so Exercise 27 matches with c. If no two sides of a triangle have the same length, then no two angles have the same measure; so Exercise 29 matches with b. **31.** 63°

10.9 Integrated Review p. 480

33. 24% **35.** 50's, there was about a 25% decrease.

10.9 Chapter Review pp. 482–484

1. $\overleftrightarrow{AG}$, $\overleftrightarrow{AE}$ **3.** $\overrightarrow{HB}$ **5.** H **7.** Below B, so that $\overline{AB} \parallel \overline{EF}$ and $\overline{HE} \parallel \overline{GF}$ **9.** 50°, acute **11.** 180°, straight **13.** m and n **15.** $\angle 1$, $\angle 4$, $\angle 5$, $\angle 8$ **17.** $m\angle 10$; $m\angle 10 = 80°$, $m\angle 2 = 70°$, and $80° > 70°$ **19.** 1 line of symmetry **21.** 4 lines of symmetry, rotational symmetry at 90° and 180° in either direction **23.** Answers vary. **25.** Scalene obtuse **27.** Isosceles acute **29.** D **31.** F **33.** J **35.** G **37.** B **39.** N **41.** K **43.** Each interior angle: 135°. Each exterior angle: 45°.

45.

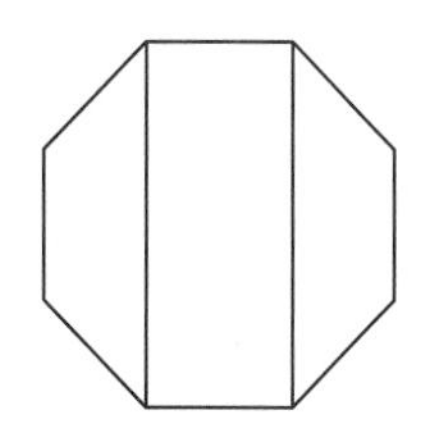

47. $\overline{AC}$, $\overline{BC}$, $\overline{AB}$ **49.** $\angle C$, $\angle B$, $\angle A$ **51.** 50°, 65°, 65° **53.** 90°, 60°, 30° **55.** $\angle A$ and $\angle D$ **57.** 15° **59.** 540° **61.** 1 line of symmetry

CHAPTER 11

11.1 Communicating about Mathematics p. 489

Answers vary.

11.1 Guided Practice pp. 490

1. 24 units², 22 units **2.** 24 units², ≈ 22.8 units **3.** 27.5 units², ≈ 21.1 units **4.** 30 units², ≈ 22.6 units **5.** Nonrectangular parallelogram: 32 units², 26 units. Rectangle: 32 units², 24 units. No, yes

11.1 Independent Practice pp. 490–491

7. $9\sqrt{3} \approx 15.6$ units², $27\sqrt{3} \approx 46.8$ units²; $\frac{1}{2}(6+12)3\sqrt{3}$ or $3(9\sqrt{3})$ **9.** $\frac{1}{2}(10)13.08 \approx 65.40$ units², $8 \times 65.40 \approx 523.20$ units²; 80 units, 38 units

11.

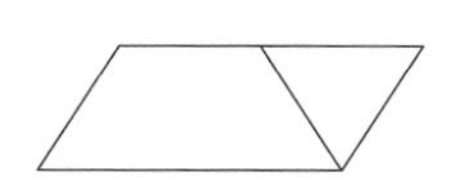

Answers vary.

13.

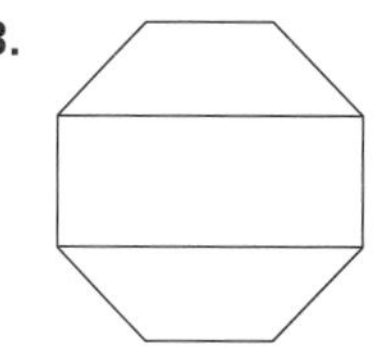

Answers vary.

15. A: Trapezoid, B: Isosceles right triangle, C: Scalene obtuse triangle, D: Kite, E: Parallelogram, F: Square, G: Trapezoid, H: Rectangle, I: Isosceles right triangle, J: Isosceles trapezoid, K: Isosceles right triangle. **17.** 100 units^2 **19.a.** 6 units^2, 6 units^2, 8 units^2, 8 units^2 **c.** 28 units^2 **21.** South Carolina: 30,000 mi^2

11.1 Integrated Review p. 491

23.

height (x)	1	2	3	4	5
area (y)	3	6	9	12	15

For every unit that the height increases, the area increases by 3 units^2.

11.2 Communicating about Mathematics p. 493

A.–D. Drawings vary. **A.** Yes **B.** No **C.** No **D.** No

11.2 Guided Practice p. 494

1. Corresponding sides are congruent and corresponding angles are congruent. Examples vary. **2.** Triangle RST is congruent to triangle XYZ. **3.** Segment RS is congruent to segment XY. **4.** Angle T is congruent to angle Z. **5.** $\overline{RT} \cong \overline{XZ}$, $\overline{ST} \cong \overline{YZ}, \angle R \cong \angle X, \angle S \cong \angle Y$ **6.** Yes, same length **7.** Yes, same measure **8.** No, different sizes **9.** No, different sizes

11.2 Independent Practice pp. 494-495

11. $\overline{MK}$ **13.** $\angle M$ **15.** $\overline{IH}$ **17.** Answers vary. **19.** Answers vary.

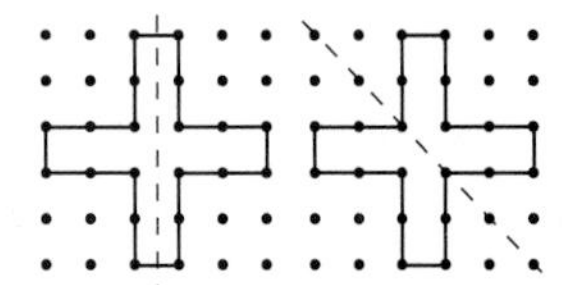

21. Answers vary. **23.** a and c; three prints were made from the same negative, then one of the prints was retouched in order to remove two of the windows.

11.2 Integrated Review p. 495

25. $(1, -3)$ **27.** $(2, 4)$

11.3 Communicating about Mathematics p. 498

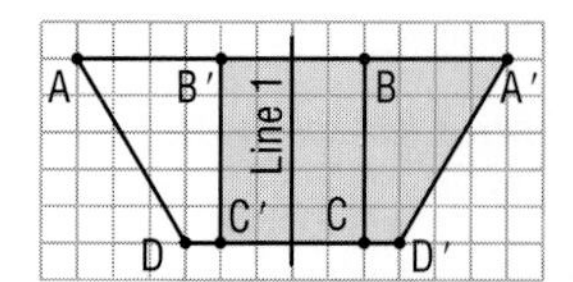

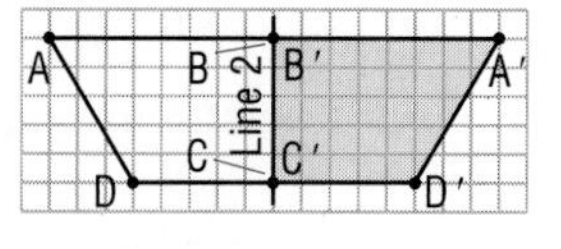

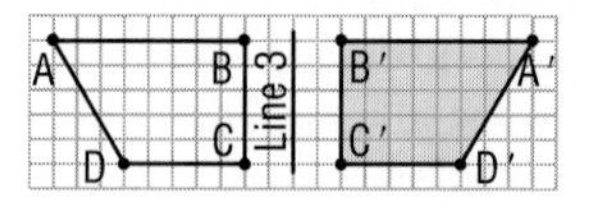

	Area
Original trapezoid	32.5 units^2
New trapezoid 1	45 units^2
New trapezoid 2	65 units^2
New trapezoid 3	65 units^2

	Perimeter
Original trapezoid	$18+\sqrt{34} \approx 23.8$ units
New trapezoid 1	$18+2\sqrt{34} \approx 29.7$ units
New trapezoid 2	$26+2\sqrt{34} \approx 37.7$ units
New trapezoid 3	$36+2\sqrt{34} \approx 47.7$ units

11.3 Guided Practice p. 499

1.

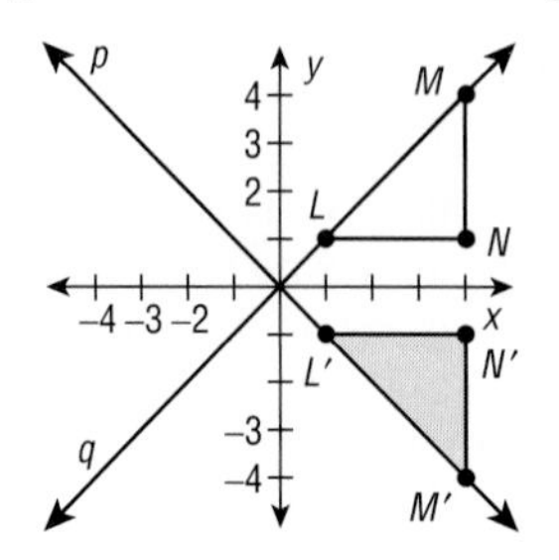

2.

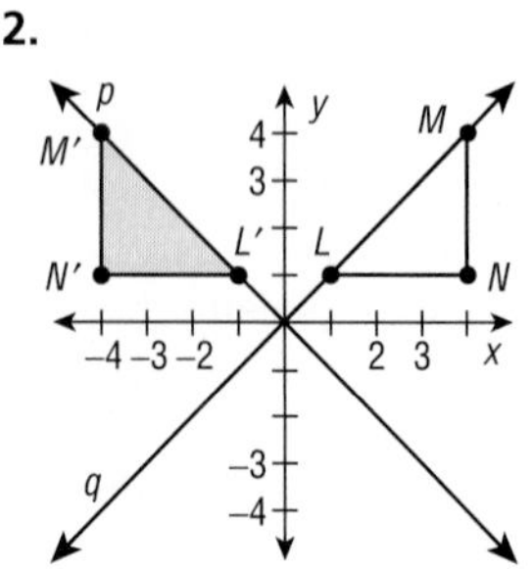

3.

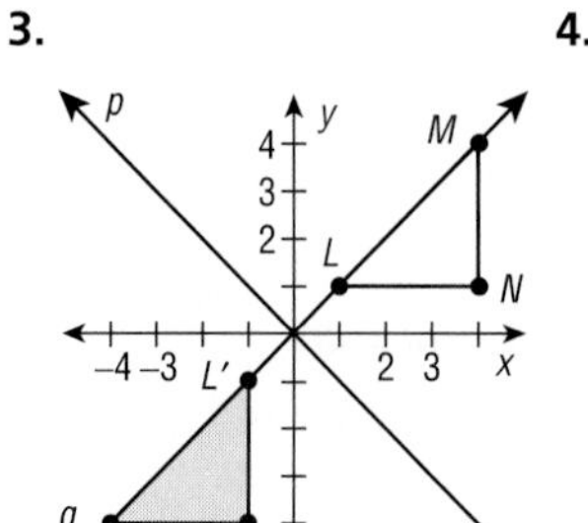

4.

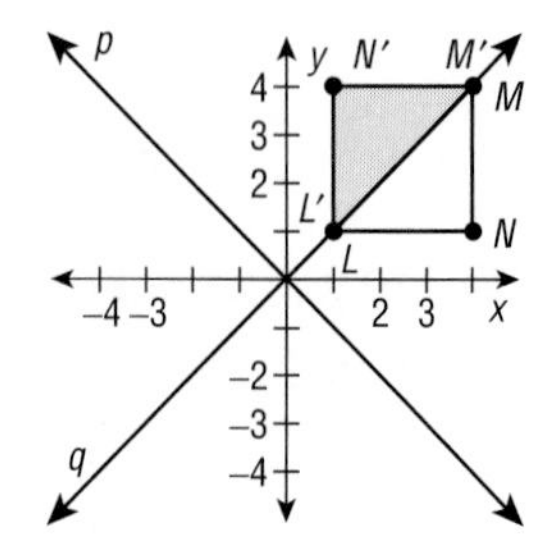

11.3 Independent Practice pp. 499–500

5. $\triangle ONM$ **7.** $\triangle RQP$ **9.** No, your right side is duplicated in a photo. **11.** No **13.** No
15.

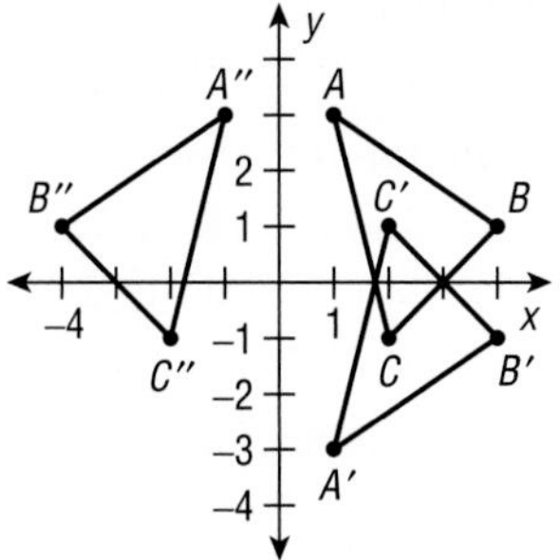

17. Reflection line: Through the origin, leaning to the right at a 45° angle with the x-axis.
19. Reflection line: Through the points $(-1, 0)$ and $(0, -1)$.
21. They are congruent and in the same place.
23. The line $x=4$

11.3 Integrated Review p. 500

25. No **27.** Yes

11.3 Mixed Review p. 501

1.

6
2
12

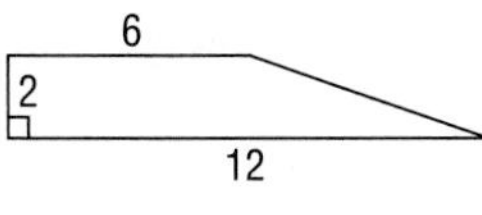

18 units²

3.

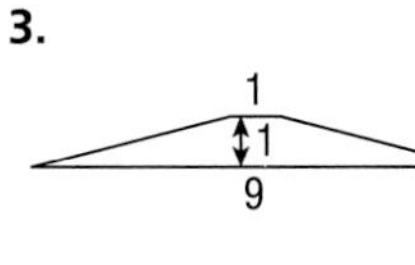

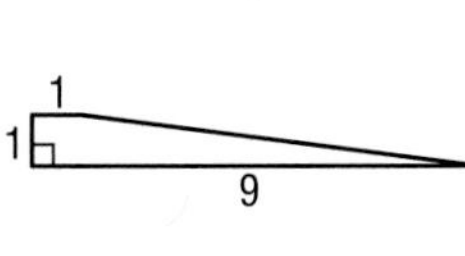

5 units²

5. $\frac{2}{9}$ **7.** $\frac{4}{9}$ **9.** 0.53 **11.** −1.14 **13.** ≈4.6% decrease
15. ≈14.3% decrease

11.4 Communicating about Mathematics p. 503

A. $\frac{1}{4} \times 360° = 90°$ **B.** $1 \times 360° = 360°$
C. $\frac{23}{60} \times 360° = 138°$

11.4 Guided Practice p. 504

1. 60° clockwise or 300° counterclockwise **2.** 90° counterclockwise or 270° clockwise **3.** 120° counterclockwise or 240° clockwise **4.** 60° counterclockwise or 300° clockwise **5.** Rotational symmetry is a special case of a rotation in which the figure fits back on itself. **6.** Answers vary.

11.4 Independent Practice pp. 504-505

7. 180° **9.** 90° **11.** 45° counterclockwise **13.** 60° clockwise **15.a.** 4 **b.** 2 **c.** $\sqrt{20}$ **17.** $(6+\sqrt{20})$ units; the perimeter and area are equal to those of $\triangle ABC$. **23.** $\overline{CF}$ **25.** $\overline{FC}$ **27.** 90° **29.** 30°

11.5 Communicating about Mathematics p. 507

Translation, Rotation, Reflection, Translation, Translation

11.5 Guided Practice p. 508

1. Slide the figure 5 units to the left and 3 units down. **2.** Slide the figure 1 unit to the left and 4 units up. **3.** Slide the figure 5 units to the right and 2 units up. **4.** Slide the figure 5 units to the right and 5 units down. **5.** Reflect the figure in the y-axis. **6.** Slide the figure 5 units to the right. **7.** Rotate the figure 90° clockwise about the origin. **8.** Slide the figure 4 units to the right and 2 units down.

11.5 Independent Practice pp. 508–509

9. Graph c, slide the figure 2 units to the right.
11. Graph b, slide the figure 5 units to the right and 5 units down.
13.

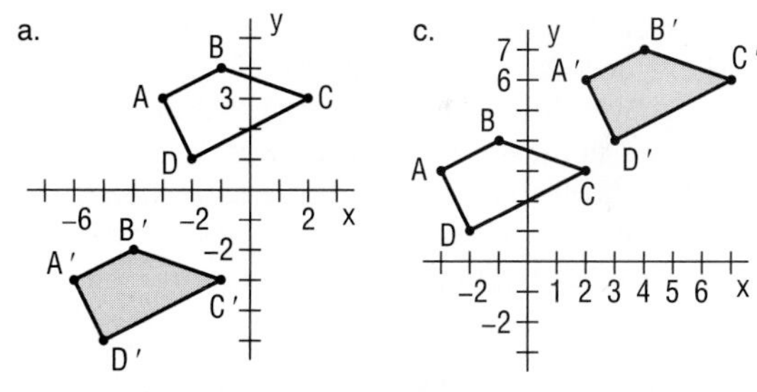

15. Each area is $7\frac{1}{2}$ units². **17.** Yes, slide the figure 1 unit to the left and 4 units down. **19.** Yes, slide the figure 3 units to the right and 6 units up. **21.** $A'(-1, 2)$, $B'(2, 5)$, $C'(0, 6)$ **23.** $A'(3, 5)$, $B'(6, 8)$, $C'(4, 9)$ **25.** Birmingham, Montgomery is only 1.8° north. **27.** Each is the image of the other in a vertical line of reflection. **29.** Each is the image of the other in a horizontal line of reflection.

11.5 Mid Chapter Self-Test p. 510

1. $\angle D$ **2.** $\angle C$ **3.** $\overline{DE}$ **4.** $\overline{AC}$ **5.** n **6.** m **7.** 60° clockwise **8.** 90° clockwise **9.** Slide the figure 5 units to the left. **10.** Slide the figure 4 units down. **11.** Slide the figure 5 units to the right and 2 units down. **12.** Slide the figure 2 units to the left and 5 units down. **13.** b **14.** c **15.** a **16.** d **17.** Rotation symmetry in either direction at 45°, 90°, 135°, 180° **18.** Rotation symmetry in either direction at 90°, 180° **19.** Rotation symmetry in either direction at 180° **20.** 24 seconds

11.6 Communicating about Mathematics p. 513

$11\frac{1}{4}$ ft

11.6 Guided Practice p. 514

1. Answers vary. **2.** Answers vary. **3.** a and c **4.** a and b **5.** $\frac{HJ}{KM}=\frac{HI}{KL}$, $\frac{HI}{KL}=\frac{IJ}{LM}$, $\frac{HJ}{KM}=\frac{IJ}{LM}$ **6.** $\frac{4}{3}$ **7.** $13\frac{1}{2}$, 24 **8.** *K*

11.6 Independent Practice pp. 514–515

9. a and c **11.** $\frac{QR}{WX}=\frac{RS}{XY}=\frac{ST}{YZ}=\frac{QT}{WZ}$ **13.a.** 15 **b.** 14 **c.** $3\frac{3}{5}$ **15.** No; $\frac{5.5}{8.5}\neq\frac{8.5}{11}$ and $\frac{4.25}{11}\neq\frac{8.5}{11}$ **17.** $\frac{5}{4}$, 20 **19.** 4,15 **21.** False; similar figures may not be the same size. **23.** False; the two acute angles of a right triangle can vary. **25.** Yes; corresponding angles have the same measure and the ratios of corresponding sides are equal. **27.** The person and the tree are parts of parallel lines, while the tree's shadow and the ray of sunlight are parts of other lines. When two parallel lines are intersected by a third line, the corresponding angles are congruent; so $\angle 1\cong\angle 2$ and $\angle 3\cong\angle 4$. $\angle 5$ is an angle of both triangles. Since the corresponding angles of the two triangles have the same measures, the two triangles are similar.

11.6 Integrated Review p. 515

29. b, $12\frac{1}{2}$ **31.** a, 10

11.6 Mixed Review p. 516

1. 12.04 **3.** 4.80 **5.** Those in Exercises 1 and 4 **7.** 8.25 **9.** 44 **11.** 48% **13.** 2 **15.** −12 **17.** $\frac{11}{28}$ **19.** $\frac{1}{4}$

11.7 Communicating about Mathematics p. 518

A. 32.64 in., $\frac{24}{5}=\frac{x}{6.8}$ **B.** 783.36 in.^{2} **C.** 3; the ratio of any corresponding lengths = 4.8, while the ratio of the areas ≠ 4.8.

11.7 Guided Practice p. 519

1. $\frac{1}{4}$ in. **2.** 24 **3.** 36 in., $1\frac{1}{2}$ in.; 72 in.^{2}, $\frac{1}{8}$ in.^{2} **4.** 24, it is the same. **5.** 576, it is equal to the square of the scale factor.

11.7 Independent Practice pp. 519–520

7. 1.5 cm **9.** 1563 miles **11.** $\frac{64}{11}$ **13.** $\frac{64}{11}$, it equals the scale factor **15.** 13; 24 units, 24 units^2

11.8 Communicating about Mathematics p. 523

A	sin A
10°	≈0.173
20°	≈0.342
30°	≈0.500
40°	≈0.64

A	sin A
50°	≈0.766
60°	≈0.866
70°	≈0.940
80°	≈0.985

They range from 0 to 1.

11.8 Guided Practice p. 524

1. b **2.** c **3.** a **4.** Answers vary. 0.643, 0.766, 0.839 **5.** Yes **6.** $\frac{12}{13}\approx 0.923$ **7.** $\frac{5}{13}\approx 0.385$ **8.** $\frac{5}{13}\approx 0.385$ **9.** $\frac{12}{13}\approx 0.923$ **10.** $\frac{12}{5}\approx 2.4$ **11.** $\frac{5}{12}\approx 0.417$

11.8 Independent Practice pp. 524–525

15. $\frac{3}{\sqrt{45}}\approx 0.447$ **15.** $\frac{6}{\sqrt{45}}\approx 0.894$ **17.** $\frac{3}{6}=0.5$ **19.** 30°, $\sqrt{3}$; $\sin 60°=\frac{\sqrt{3}}{2}\approx 0.866$, $\cos 60°=\frac{1}{2}=0.5$, $\tan 60°=\frac{\sqrt{3}}{1}=\sqrt{3}\approx 1.732$, $\sin 30°=\frac{1}{2}=0.5$, $\cos 30°=\frac{\sqrt{3}}{2}\approx 0.866$, $\tan 30°=\frac{1}{\sqrt{3}}\approx 0.577$ **21.** 26.6°, $\sqrt{5}$; $\sin 63.4°=\frac{2}{\sqrt{5}}\approx 0.894$, $\cos 63.4°=\frac{1}{\sqrt{5}}\approx 0.447$, $\tan 63.4°=\frac{2}{1}=2$, $\sin 26.6°=\frac{1}{\sqrt{5}}\approx 0.447$, $\cos 26.6°=\frac{2}{\sqrt{5}}\approx 0.894$, $\tan 26.6°=\frac{1}{2}\approx 0.5$

23.

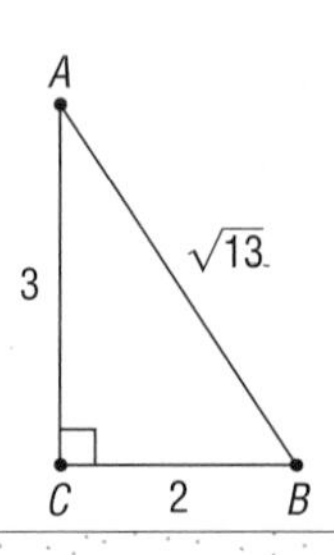

25. $\sin x = \cos (90° - x)$ **27.** 1, 2 **29.** 10 meters

■ 11.8 Integrated Review p. 525

31. d

■ 11.9 Communicating about Mathematics p. 528

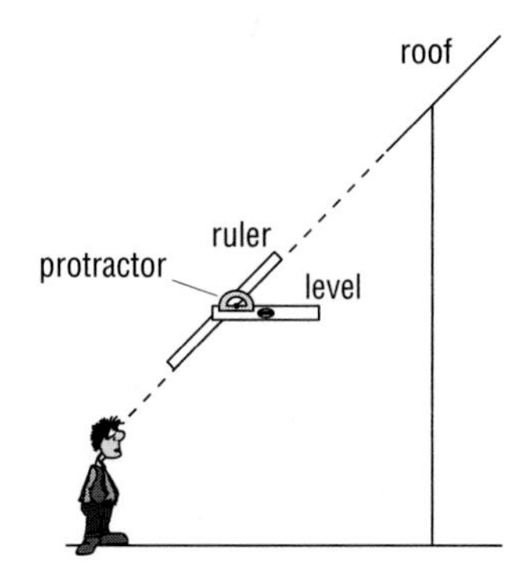

■ 11.9 Guided Practice p. 529

1. ≈3.48 **2.** ≈3.48 **3.** ≈3.48 **4.** Answers vary. **5.** ≈41.95 ft **6.** ≈77.90 ft

■ 11.9 Independent Practice pp. 529–530

7. ≈0.839 **9.** ≈0.695 **11.** ≈7.09 **13.** ≈9.14 **15.** $a \approx 4.24$, $b \approx 4.24$, $m\angle B = 45°$ **17.** $h \approx 5.13$, $g \approx 14.10$, $m\angle G = 70°$ **19.** ≈2.99 ft **21.** ≈4.276 m **23.** ≈2.35 m

■ 11.9 Integrated Review p. 530

25. ≈36% **27.** ≈18%

■ 11.9 Chapter Review pp. 532–534

1. 32 units, 56 units^2 **3.** 30 units, 60 units^2 **5.** $\overline{UY}$ **7.** $\angle Y$ **9.** $\angle P$ **11.** (1, 2), (3, 3), (5, 1), (3, 0) **13.** (5, −3), (3, −2), (1, −4), (3, −5) **15.** 22.5°, 45°, 67.5°, 90°, 112.5°, 135°, 157.5°, 180° in either direction **17.** No, yes. The seats all face one way and make the wheel unsymmetrical. **19.** $\frac{1}{4}$ **21.** $\frac{1}{4}$, yes **23.** sometimes **25.** 15.4 cm **27.** 101.64 cm^2, 21 cm^2 **29.** 0.89 **31.** 0.45 **33.** 2 **35.** ≈12.73 **37.** ≈17.66 **41.** ≈270.12 ft **43.** 6 **45.** 88 **47.** 33 **49.** 25 **51.** 84 **53.** $\frac{24}{7}$ **55.** 24 **57.** 480 **59.** 564

CHAPTER 12

■ 12.1 Communicating about Mathematics p. 540

A. 3.1 in.^{2} **B.** 4.2 in.^{2} **C.** 2.1 in.^{2}

■ 12.1 Guided Practice p. 541

1. Both relate the diameter to the circumference. **2.** ≈25,000 mi **3.** $\overline{BC}$ **4.** $\overline{AB}$, $\overline{AC}$ **5.** ≈44.0 cm **6.** ≈153.9 cm^2

■ 12.1 Independent Practice pp. 541–542

7. 7.5 cm, 4.5cm^2 **9.** 25.1 in., 50.2 in.^{2} **11.** 0.7 ft, 1.4 ft **13.** 2.4 in., 4.7 in. **15.** 5.7 in.^{2} **17.** 13.8 mm^2

Radius	1	2	3
Circumference	6.28	12.56	18.84

Radius	4	5	6
Circumference	25.12	31.40	37.68

For every one-unit increase in the radius, the circumference increases 6.28 units.
21. The circumference doubles; $C_1 = \pi d$, $C_2 = \pi(2d) = 2(\pi d)$. The area quadruples; $A_1 = \pi r^2$, $A_2 = \pi(2r)^2 = 4(\pi r^2)$. **23.** 113.04 in.^{2} **25.** 18.84 in.^{2} **27.** 314 mi^2

■ 12.1 Integrated Review p. 542

29.

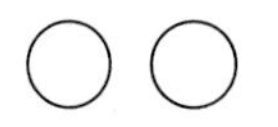
no points

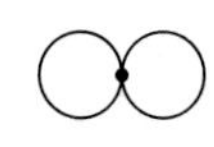
1 point

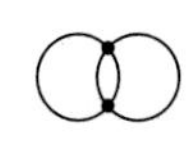
2 points

all points

■ 12.2 Communicating about Mathematics p. 545

True of prisms and pyramids; not true of spheres and cones.

■ 12.2 Guided Practice p. 546

1. d **2.** e **3.** c **4.** a **5.** b **6.** a: Lateral surface, b: Base **7.** a: Edge, b: Vertex, c: Face, d: Edge, e: Vertex **8.** a: Face, b: Vertex, c: Edge, d: Face, e: Edge **9.** a: Lateral surface, b: Base

SELECTED ANSWERS

12.2 Independent Practice pp. 546–547

11. Prism (speaker) **13.** Cone (funnel) **15.** Have base **17.** No **19.** Yes **21.** Prism (with hexagons as bases) **23.** Pyramid (with a triangle as base) **25.** Pyramid (with a hexagon as base)

12.2 Integrated Review p. 547

27. 78 units, 202.8 units^2 **29.** ≈18.8 units, ≈28.3 units^2

12.3 Communicating about Mathematics, p. 550

Least surface area: B. **A.** 252 units^2 **B.** 216 units^2 **C.** 228 units^2

12.3 Guided Practice p. 551

1. A: Prism (with rectangles as bases), B: Prism (with triangles as bases), C: Cylinder **2.** 340 units^2 **3.** 144 units^2 **4.** ≈622.0 units^2

12.3 Independent Practice, pp. 551–552

5. 664 units^2 **7.** 36 units^2 **9.** 480 units^2 **11.** ≈3518.6 units^2 **13.** 96 in.^{2} **15.** 24 in.^{2} **17.** b should be chosen. In c, the area of the paper is too small to cover the box; also, the paper is not wide enough: it needs to be $(27+6)$ cm wide. In a, the paper is not wide enough: it needs to be $(45+6)$ cm wide. **19.** ≈301.6 ft^2

12.3 Integrated Review p. 552

21. c

12.3 Mixed Review p. 553

1. They are corresponding angles. **3.** Ray **5.** Obtuse **7.** 120° **9.** ≈530.9 units^2 **11.** ≈132.7 units^2 **13.** 13, −13 **15.** $\frac{5}{3}$, $-\frac{5}{3}$

12.4 Communicating about Mathematics p. 555

Answers vary.

12.4 Guided Practice p. 556

1. b and c, each volume is 36 units^3 **2.** Volume = length × width × height **3.** Yes, ≈55.36 in.^{3} **4.** Yes, 336 cm^3 **5.** No. **6.** Yes, 12 cm^3

12.4 Independent Practice pp. 556–557

7. 30 in.^{3} **9.** 120 in.^{3} **11.** 343 cm^3 **13.** 3 in. **15.** 10 cm **17.** 1 by 1 by 36, 1 by 2 by 18, 1 by 3 by 12, 1 by 4 by 9, 1 by 6 by 6, 2 by 2 by 9, 2 by 3 by 6, 3 by 3 by 4; 1 by 1 by 36 **19.** ≈603 ft^2 **21.** Answers vary.

12.4 Integrated Review p. 557

23.

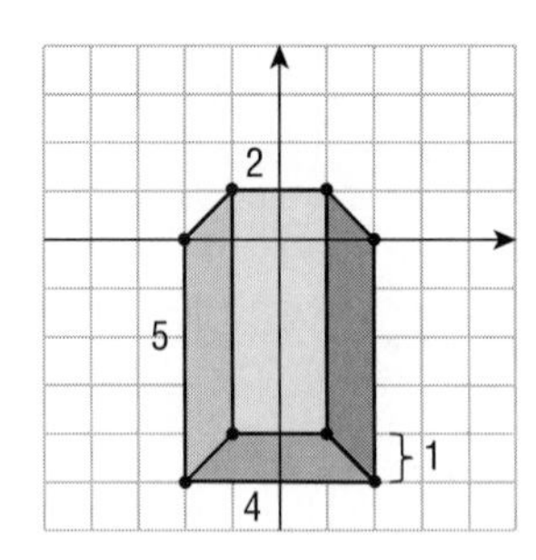

15 units^3

12.4 Mid-Chapter Self-Test p. 558

1. ≈18.8 cm, ≈28.3 cm^2 **2.** ≈15.7 in., ≈19.6 in.^{2} **3.** ≈13.7 ft^2 **4.** ≈58.9 yd^2 **5.** d **6.** c **7.** e **8.** b **9.** a **10.** ≈1055.6 mm^2 **11.** ≈2463.0 mm^2 **12.** 3.375 in.^{2} **13.** 32.5 cm^2 **14.** 120 in.^{3} **15.** 260 cm^3 **16.** 480 m^3 **17.** 420 in.^{3} **18.** ≈17.00 mi **19.** ≈22.99 mi^2 **20.** ≈12.00 ft

12.5 Communicating about Mathematics p. 560

A. ≈1.8408 **B.** ≈0.5432

12.5 Guided Practice p. 561

1. ≈28.274 units^2 **2.** 5 units **3.** $V = Bh$ **4.** ≈141.37 units^3 **5.** ≈7.1 in.^{2} **6.** ≈7.1 in.^{3}

12.5 Independent Practice pp. 561–562

7. ≈75.40 in.^{3} **9.** ≈904.78 yd^3 **11.** ≈3.00 in. **13.** ≈3.40 cm **15.** Greater than **17.** ≈75.40 in.^{3} **19.** ≈769.7 in.^{3} **21.** ≈42.977 in.^{3} **23.** ≈301.59 in.^{3}

12.5 Integrated Review p. 562

25. ≈50.27 cm^3, ≈75.40 cm^2 **27.** ≈1.33 m^3, ≈7.07 m^2

12.6 Communicating about Mathematics p. 565

$V = \frac{1}{3}Bh$

12.6 Guided Practice p. 566

1. The base of a pyramid is a polygon. The base of a cone is a circle. **2.** The cylinder has 3 times the volume of the cone. **3.** 320 $in.^3$ **4.** $\approx$94.25 cm^3 **5.** 5 in. **6.** 10 in. **7.** $\approx$261.80 $in.^3$ **8.** Subtract the volume of the cone from the volume of the cube. $\approx$738.20 $in.^3$

12.6 Independent Practice pp. 566–567

9. 1200 cm^3 **11.** 8400 ft^3 **13.** $\approx$2513.27 m^3 **15.** $\approx$5336.52 $in.^3$ **17.** Answers vary. **19.** $\approx$10,416.67 m^3 **21.** $\approx$395.99 ft^3 **23.** It is 300 ft in the air. **25.** $\approx$3,141,592.7 ft^3 **27.** 302.5 ft^3, 19.75 cm, 337.5 $in.^2$

12.6 Integrated Review p. 567

29. 1 ft^3 of fresh air **31.** 143 ft^3

12.6 Mixed Review p. 568

1. 5400 cm^3 **3.** $\approx$8143.01 $in.^3$ **5.** 20 **7.** 35 **9.** $-\frac{1}{4}$ **11.** 4 **13.** $\frac{56}{93}$ **15.** $y \leq -12$ **17.** $t < \frac{7}{10}$ **19.** $p \leq -\frac{1}{6}$

12.7 Communicating about Mathematics p. 570

$\approx$19.3 ft

12.7 Guided Practice p. 571

1. $\overline{AB}, \overline{AC}, \overline{AD}$ **2.** Yes, it is a segment from the center to the surface. **3.** $\overline{BC}$, 4 ft **4.** A hemisphere **5.** $\approx$33.51 ft^3

12.7 Independent Practice pp. 571–572

7. $\approx$7238.23 cm^3 **9.** $\approx$2.48 $in.^3$ **11.** $\approx$5575.28 cm^3 **13.** $\approx$8.18 $in.^3$ **15.** $\approx$243.92 $units^3$ **17.** $\approx$21.21 $units^3$ **19.** Circumference; $C = 3\pi$ units, $h = 9$, $3\pi > 9$ **21.** $\approx$5.4 ft **23.** $\approx$13.0 in. **25.** It is the largest possible circle when a sphere and a plane intersect. **27.** $\approx$5.4 billion mi^3 **29.** $\approx$369,120.9 ft^3

12.7 Integrated Review p. 572

31. $\approx$56.55 $units^3$, $V = \frac{3}{4}(\pi \cdot 2^2 \cdot 6)$ **33.** $\approx$113.10 $units^3$, $V = \frac{3}{4}(\frac{1}{3}\pi \cdot 4^2 \cdot 9)$

12.8 Communicating about Mathematics p. 574

You can multiply each dimension of the model by 8 to find the dimensions of the actual room in feet (16 ft by 16 ft by 8 ft), then use the dimensions of the actual room to find the surface area and volume of the room. Or you can find the surface area (16 $in.^2$) of the model and multiply by 8^2 to find the surface area of the actual room in square feet, and you can find the volume (4 $in.^3$) of the model and multiply by 8^3 to find the volume of the actual room in cubic feet.

12.8 Guided Practice p. 575

1. Yes. They have the same shape and their corresponding lengths are proportional ($\frac{9}{6} = \frac{6}{4} = \frac{12}{8}$). **2.** No. Their corresponding lengths are not proportional ($\frac{48}{12} \neq \frac{9}{3}$). **3.** Yes. They have the same shape and their corresponding lengths are proportional ($\frac{36}{12} = \frac{15}{5}$). **4.** They have the same shape and their corresponding lengths are proportional ($\frac{72}{12} = \frac{60}{10} = \frac{48}{8}$). **5.** 6 **6.** 592 $units^2$ **7.** 31,312 $units^2$ **8.** Volume of Prism A $= 6^3 \cdot$ Volume of Prism B

12.8 Independent Practice pp. 575–576

9. b **11.** 12.5 cm, 5 cm **13.** Row 1: 5 ft; $44\frac{4}{9}$ ft^2; $11\frac{1}{9}$ ft^3 Row 2: 20 m; 4600 m^2; 50,000 m^3 Row 3: 22.5 cm; 68,343.75 cm^3; 198 cm^2 Row 4: 96 $in.^2$; 64 in.; 262,144 $in.^3$ **15.** 20,736 **17.** always **19.** sometimes

12.8 Integrated Review p. 576

21. Yes, no, they do not have the same shape.

12.8 Chapter Review pp. 578–579

1. h **3.** b **5.** c **7.** a **9.** $\approx$43.98 in., $\approx$153.94 $in.^2$ **11.** Yes, $30^2 - \pi \cdot 15^2 = 30^2 - 4(\pi \cdot 7.5^2)$ **13.** Cylinder **15.** Prism (with triangles as bases) **17.** Pyramid (with a hexagon as base) **19.** 7 **21.** 160 $in.^2$ **23.** $\approx$1143.54 ft^2 **25.** 672 yd^3 **27.** $\approx$3015.93 ft^3 **29.** 56 m^3 **31.** $\approx$872.32 $in.^3$ **33.** $\approx$50.3 $in.^3$ **35.** $\approx$268.1 $in.^3$ **37.** No **39.** 3 cm **41.** 216, $6^3 = 216$ **43.** 65.5 $in.^2$, 27.1875 $in.^3$ **45.** $\approx$120.25 $in.^2$, $\approx$78.34 $in.^3$ **47.** $\frac{921{,}600}{121}$, $\approx$5593.39 ft^2

49.

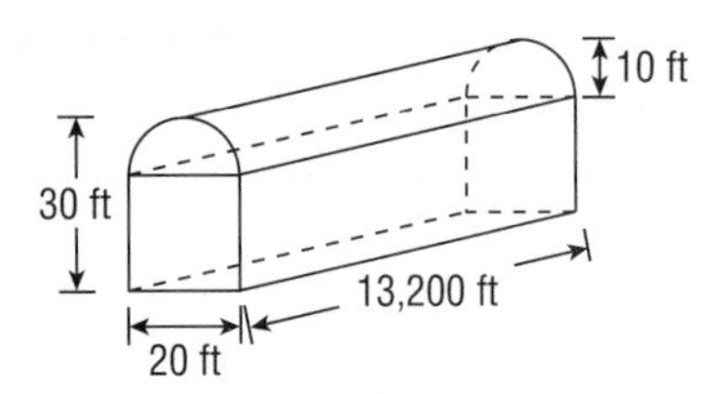

Figure not drawn to scale

51. 1,206,690 ft^2

12.8 Cumulative Review for Chapters 7–12 pp. 582–585

1. $\frac{5}{6}$ **3.** $\frac{2}{12}, \frac{1}{6}$ **5.** $-\frac{1}{4}$ **7.** $\frac{1}{10}$ **9.** 0.40 **11.** 10.24 **13.** $\frac{1}{3}, 33\frac{1}{3}\%$ **15.** $\frac{1}{2}$, 50% **17.** Secure: $\frac{11}{20}$, Somewhat secure: $\frac{7}{25}$, Not secure: $\frac{7}{50}$, Don't know: $\frac{3}{100}$ **19.** 20 mi/gal, rate **21.** $\frac{3}{4}$, ratio **23.** 24 **25.** $\frac{10}{3}$ **27.** ≈25.6% **29.** 18 **31.** 11, −11; rational **33.** $\sqrt{126}, -\sqrt{126}$; irrational **35.** $c=25$ **37.** $a\approx 7.99$ **39.** $y<-\frac{1}{6}$ **41.** $p<\frac{4}{5}$ **43.** $x\le -10$ **45.** Yes; $5^2+12^2=13^2$ **47.** Yes; $8+10>17$, $8+17>10$, $10+17>8$ **49.** $\overleftrightarrow{EI}, \overleftrightarrow{ED}, \overleftrightarrow{DH}, \overleftrightarrow{DI}, \overleftrightarrow{HI}$ **51.** $\overrightarrow{KG}$ **53.** $\angle ACB$ and $\angle KCE$, $\angle BCE$ and $\angle ACK$, and 6 other pairs **55.** $\angle ACB$ and $\angle CED$, $\angle BCE$ and $\angle DEF$, and 6 other pairs **57.** Straight **59.** Obtuse **61.** Scalene right triangle, 47° **63.** Isosceles trapezoid, 77° **65.** 2-line symmetry, 180° rotational symmetry **67.** C, B, A **69.** 72 $units^2$, 36 units **71.** ≈24.19 $units^2$, 21 units **73.** Yes **75.** Yes. In a reflection, the reflection line bisects each segment that joins an original point to its image. **77.** Yes **79.** $\frac{b}{6}$ **81.** 56.55 mm, 254.47 mm^2 **83.** 37.70 in., 113.10 $in.^2$ **85.** 346 $in.^2$, 390 $in.^3$ **87.** 402.12 cm^2, 603.19 cm^3 **89.** 48 ft^3 **91.** 3591.36 cm^3 **93.** The size of the larger similar polygon will vary. The side lengths of the given polygon are 2.9 cm, 2.9 cm, 2.3 cm, and 1.2 cm. The side lengths of the larger similar polygon will vary. The angle measures of the given polygon are 127°, 90°, 90°, and 53°. The angle measures of the larger polygon are the same as those of the given polygon. The area of the given polygon is ≈4.7 cm^2. The area of the larger similar polygon will vary. The statement that compares these measures should say that corresponding side lengths are in the same ratio, corresponding angles are congruent, and the ratio of the areas is equal to the square of the ratio of the side lengths.

CHAPTER 13

13.1 Communicating about Mathematics p. 589

The rules consist of opposite operations done in reverse order.

13.1 Guided Practice p. 590

1. b, $2\cdot 2+3\cdot 1=7$; c, $2\cdot 5+3(-1)=7$ **2.** b, c, and d; variables occur to the first power only. **3.** 2, 3, 4, 5, 6, 7, 8 **4.** −10, −8, −6, −4, −2, 0, 2

13.1 Independent Practice pp. 590–591

5. Yes **7.** No

9.

x	−3	−2	−1	0	1	2	3
y	9	8	7	6	5	4	3

11.

x	−3	−2	−1	0	1	2	3
y	21	18	15	12	9	6	3

13. Yes. For every increase by 1 of x, there is a constant corresponding decrease by 2 of y. **15.** $6x-4y=12$; (2, 0), (0, −3), (4, 3) **17.** b; (100°, 50°), (50°, 100°), (75°, 75°) **19.** c; (100°, 80°), (80°, 100°), (90°, 90°) **21.** 30.48 cm

23.

in.	0.39	0.79	1	1.18	1.57
cm	1	2	2.54	3	4

1.97	2	2.36	1.76	3
5	5.08	6	7	7.62

25. \$2.07 billion **27.** \$2.87 billion **29.** No. When you divide both sides of the second equation by 2, you get $9x-2y=15$. Since the first equation is $9x-2y=18$, the expression $9x-2y$ is equal to two different numbers.

13.1 Integrated Review p. 591

31. 7, 6, 5, 4, 3, 2, 1

13.2 Communicating about Mathematics p. 594

(5, 36) The classes had the same enrollment in 1990.

13.2 Guided Practice p. 595

1. See Example 1 on page 593. **2.** c **3.** a **4.** b

5.

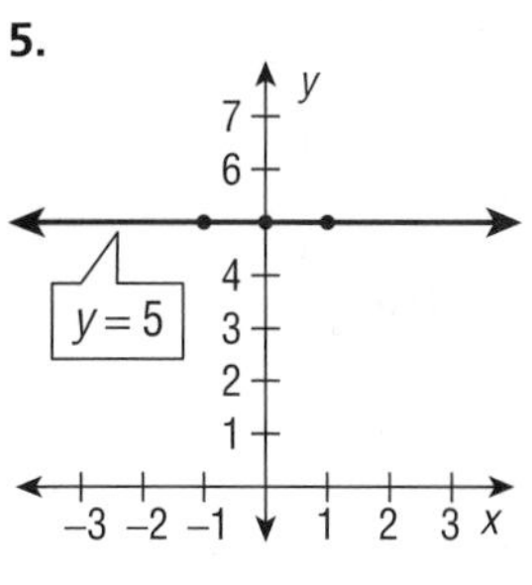

6.

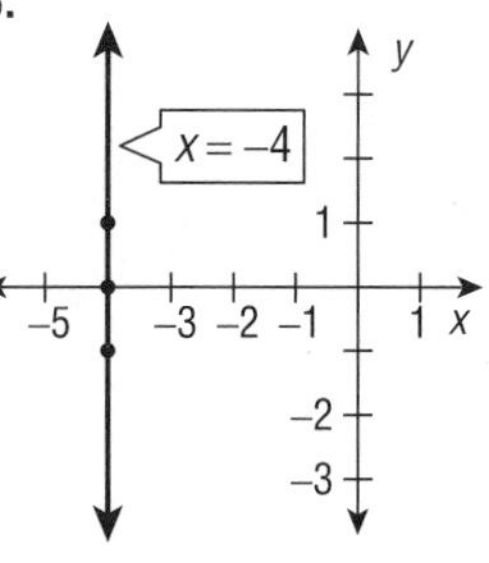

7.

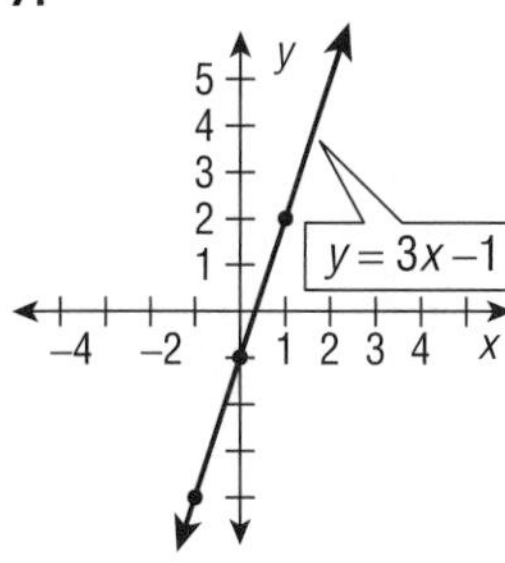

8.

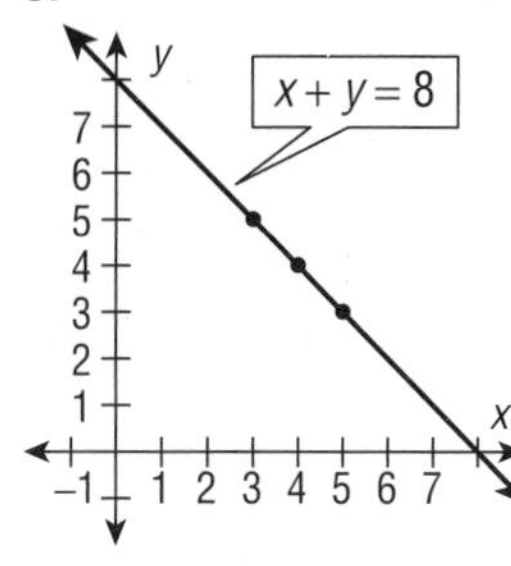

13.2 Independent Practice pp. 595–596

9. c **11.** a **13.** No, $(0, -4)$ **15.** No, $(2, 4)$
17. Yes

19.

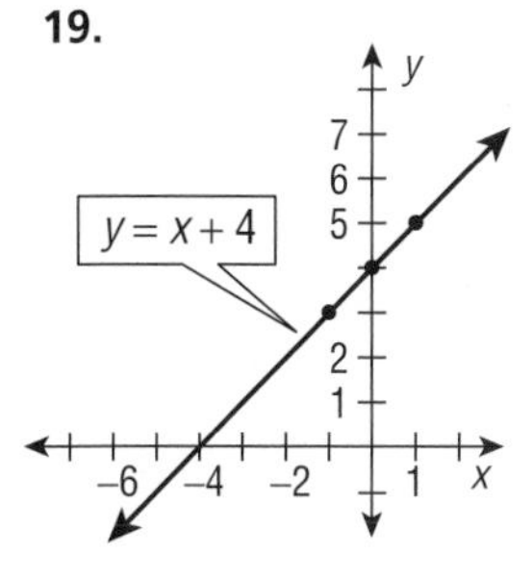

23.

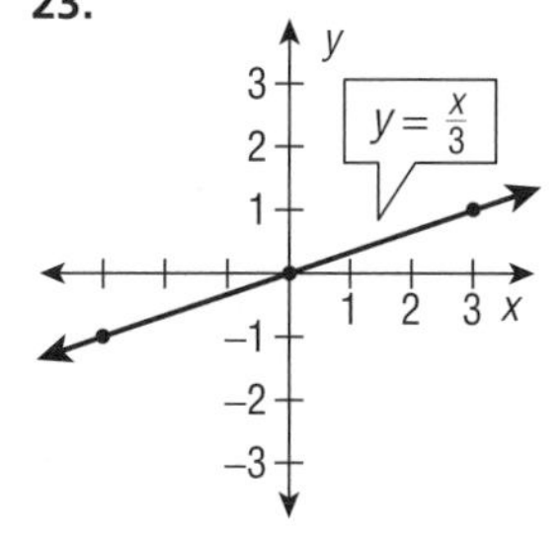

25.

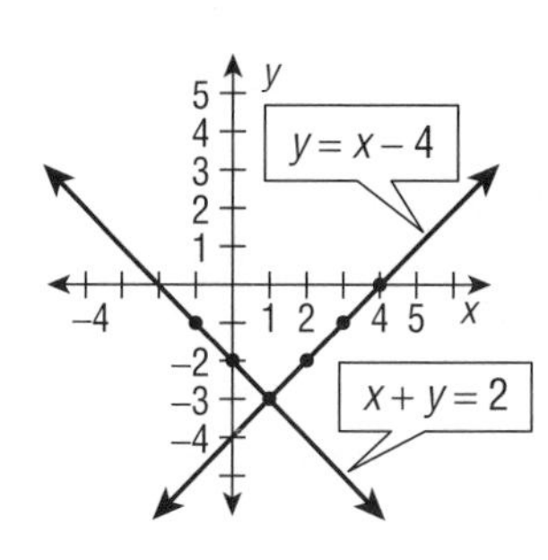

$(1, -3)$

27. $(-5, -4)$ **29.** 4 **31.** y-axis, x-axis

33.

m	1	2	3	4
t	35.9	40.9	50.3	60.4

5	6	7	8
68.4	76.7	82.0	81.1

Temperature (in °F)
Month (1 ↔ January)

For an increase in m, there is an increase in t (with one exception). No, the points do not all lie on the same line.

13.2 Integrated Review p. 596

35.

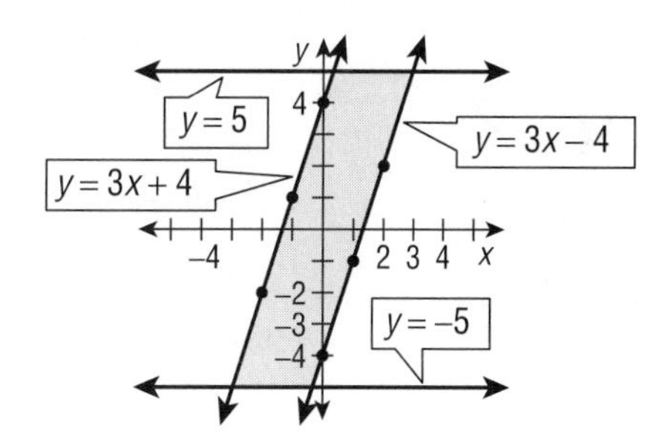

Parallelogram

37.

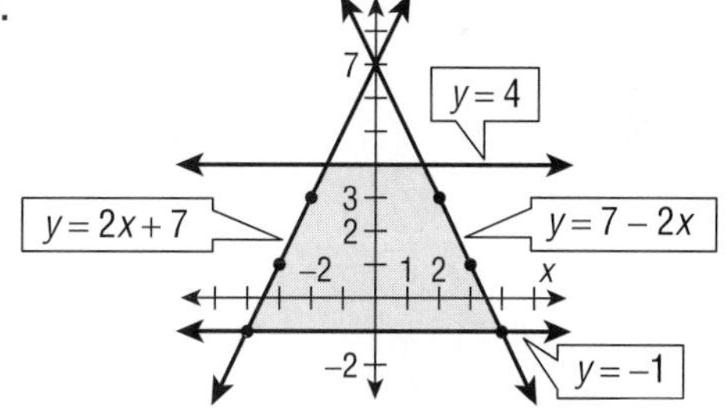

Isosceles trapezoid

13.3 Communicating about Mathematics, p. 599

A. The y-intercept (5) **B.** The x-intercept (10)

13.3 Guided Practice p. 600

1.

$y = 2x - 1$	Rewrite original equation.
$0 = 2x - 1$	Substitute 0 for y.
$1 = 2x$	Add 1 to each side.
$\frac{1}{2} = x$	Divide each side by 2.

2.

$5x+3y=9$	Rewrite original equation.
$5(0)+3y=9$	Substitute 0 for x.
$3y=9$	Simplify.
$y=3$	Divide each side by 3.

3. x-intercept: 5, y-intercept: 5

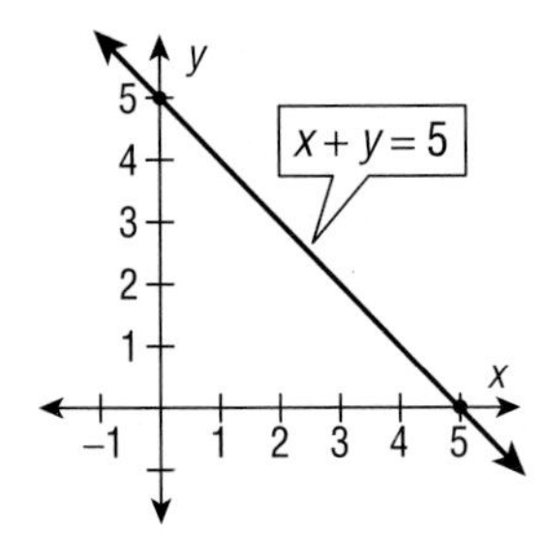

4. x-intercept: 5, y-intercept: -5

5. x-intercept: $-\frac{12}{5}$, y-intercept: 3

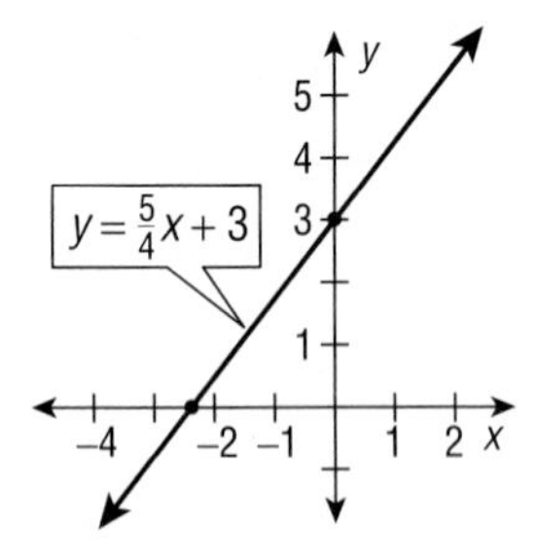

6. x-intercept: 3, y-intercept: -7

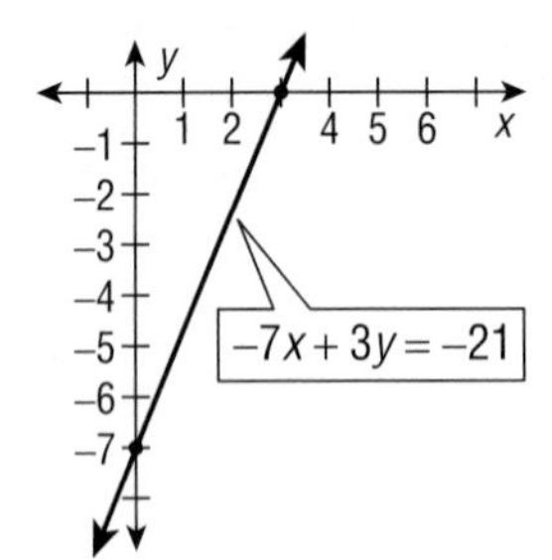

■ *13.3 Independent Practice, pp. 600–601*

7. x-intercept: 4, y-intercept: 2 **9.** x-intercept: -2, y-intercept: -2

11.

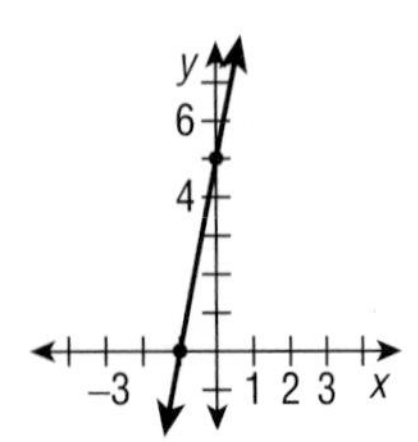

15. a **17.** b

19.

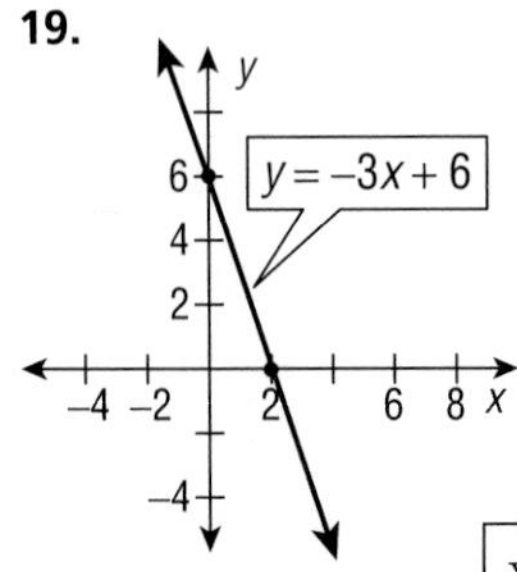

x	0	1	2	3	4
y	6	3	0	-3	-6

23.

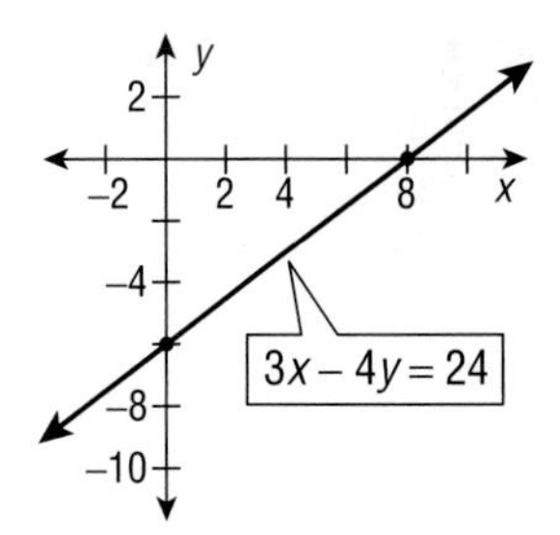

x	8	4	0	-4	-8
y	0	-3	-6	-9	-12

27. x-intercept: 0.60, y-intercept: 2.18

29. x-intercept should be $\frac{4}{3}$.

31.

P	\$12	\$11	\$10	\$9	\$8	\$7	\$6	\$5
j	\$4800	\$4950	\$5000	\$4950	\$4800	\$4550	\$4200	\$4000

■ *13.3 Integrated Review p. 601*

33. 3 **35.** -2

37.

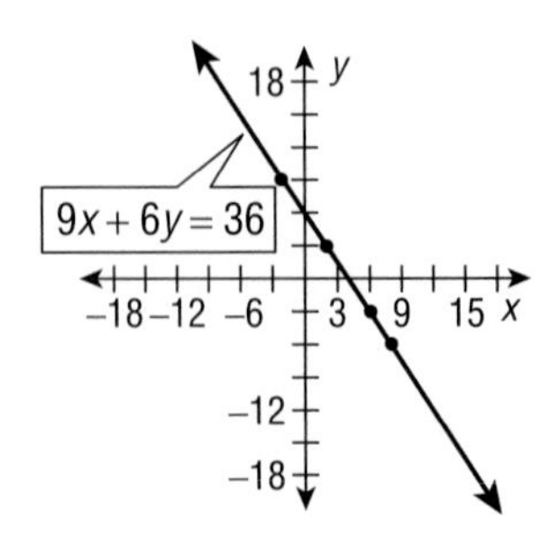

■ *13.3 Mixed Review p.602*

1. ≈ 21.99 ft, ≈ 38.48 ft^2 **3.** ≈ 50.27 cm, ≈ 201.06 cm^2 **5.** 54 units^2, 27 units^3 **7.** ≈ 376.99 units^2, ≈ 549.78 units^3 **9.** 12, -12 **11.** 3, 3 **13.** No; A, B, and C lie on a line.

■ *13.4 Communicating about Mathematics p. 605*

You can use any two points to find the slope. Since the points all lie on the same line, using any two points will produce the same slope.

13.4 Guided Practice p. 606

1. 1 **2.** $\frac{2}{3}$ **3.** -1 **4.** $-\frac{1}{2}$ **5.** $-\frac{1}{5}$ **6.** The line with a slope of 4

13.4 Independent Practice pp. 606–607

7. $m=5$ **9.** $\frac{1}{2}$ **11.** 0

13.

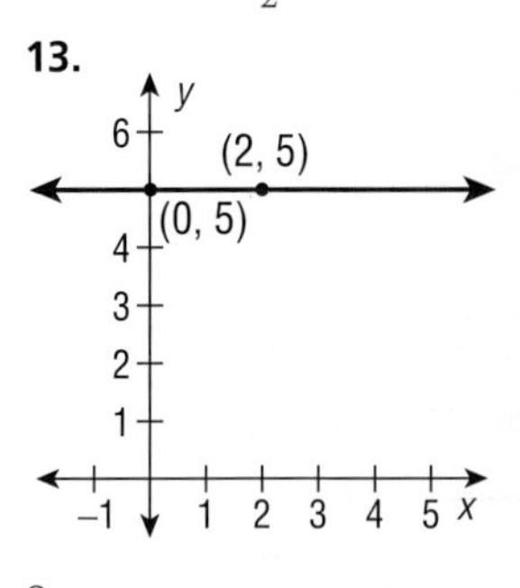

0

15.

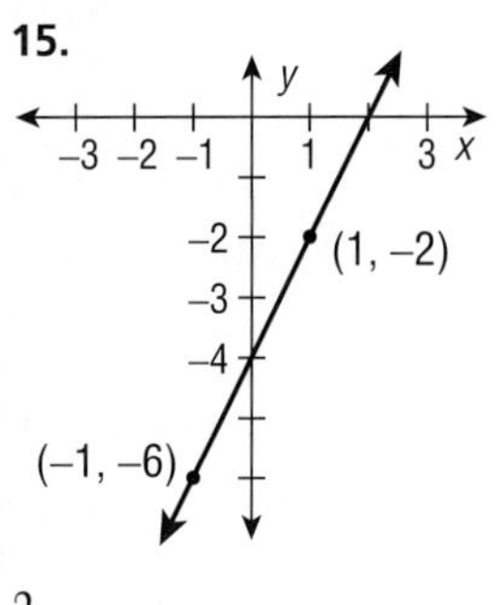

2

17. $\frac{3}{14}$ **19.** $-\frac{13}{56}$ **21.** $-\frac{3}{4}$ **23.** 1

25. No. Slopes are different ($\frac{5}{2}$ and -5).

27. Yes. Slopes are the same ($\frac{5}{2}$ and $\frac{5}{2}$).

29. 1 **31.** $\frac{4}{3}$

13.4 Integrated Review p. 607

33. a. 1 **b.** 3 **c.** 5 **d.** 7 Slopes are consecutive odd numbers.

13.4 Mid-Chapter Self-Test p. 608

1. 280 grams **2.** 25 ounces **3.** 1 kilogram, 1 pound = 448 grams **4.** Yes **5.** No **6.** Yes **7.** 21, 11; 8, 6

8.

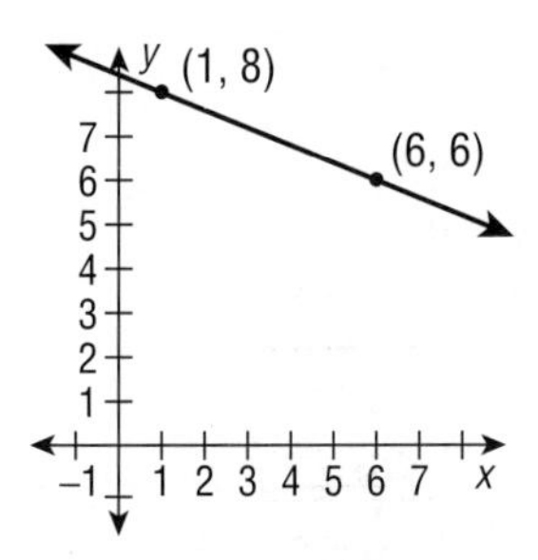

9. 4, 16 **10.** 5, 4 **11.** 2, 3 **12.** $\frac{3}{2}$ **13.** -2 **14.** $\frac{3}{4}$

15. $\frac{3}{2}$ **16.** Those in Exercises 12 and 15

17. Correct **18.** Incorrect. Slope of graph should be 2 and y-intercept should be -1. **19.** Incorrect. x-intercept should be -1 and y-intercept should be -1. **20.** Correct

3.5 Communicating about Mathematics p. 611

t	0	1	2	3	4	5	6	7	8	9	10	11	12
P	5350	5435	5520	5605	5690	5775	5680	5945	6030	6115	6200	6285	6370

For every increase in t of 1, there is a corresponding increase in P of 85, which is the slope.

13.5 Guided Practice p. 612

1. See page 610. **2.** Subtract $2x$ from each side to get $y=-2x+5$. **3.** c and e

4. $-4, 5$

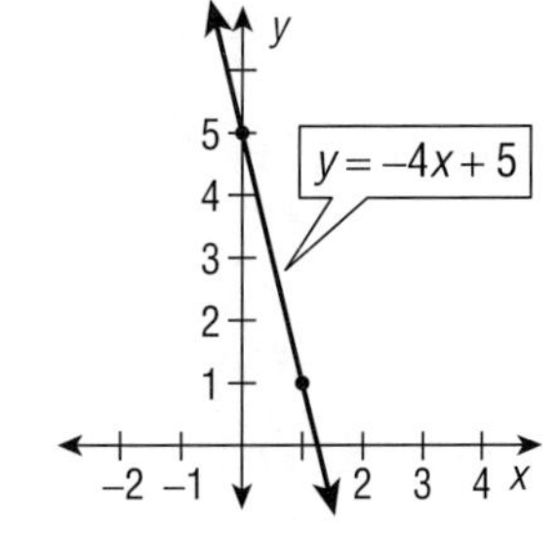

5. 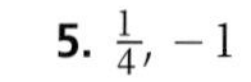 $\frac{1}{4}, -1$

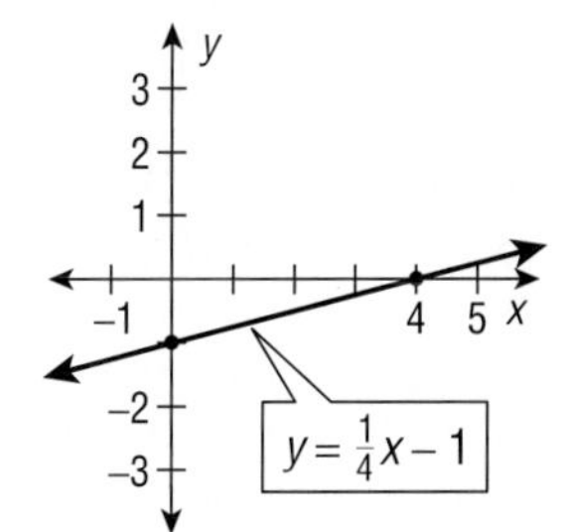

6. 3, 9

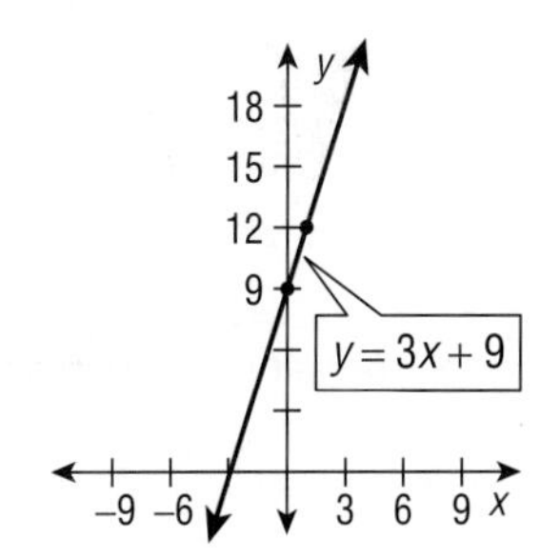

7. $-\frac{1}{2}, 8$

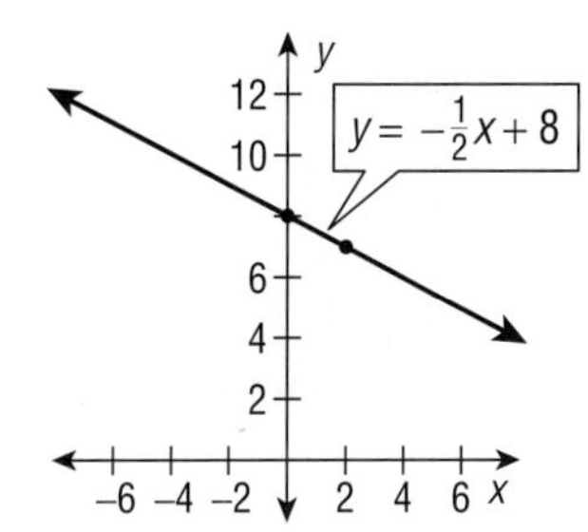

13.5 Independent Practice pp. 612–613

9. $-1, 3$

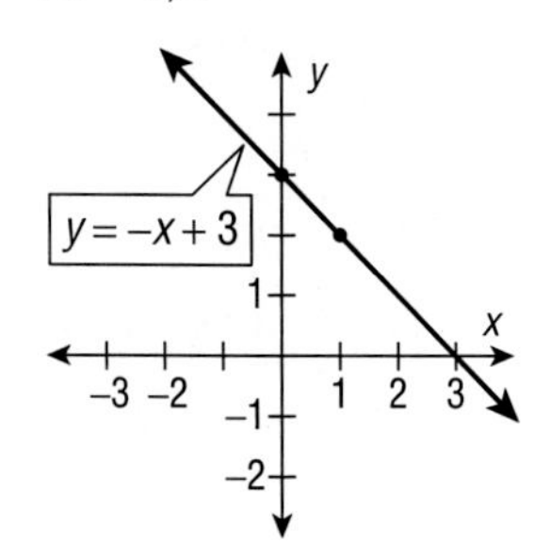

11. $3, 0$

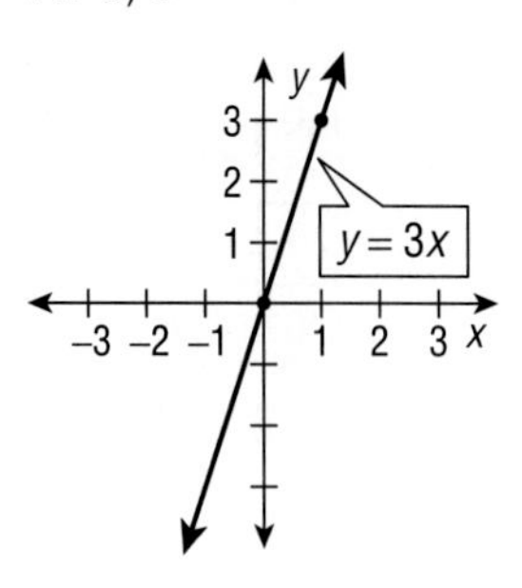

13. $-\frac{5}{2}, 0$

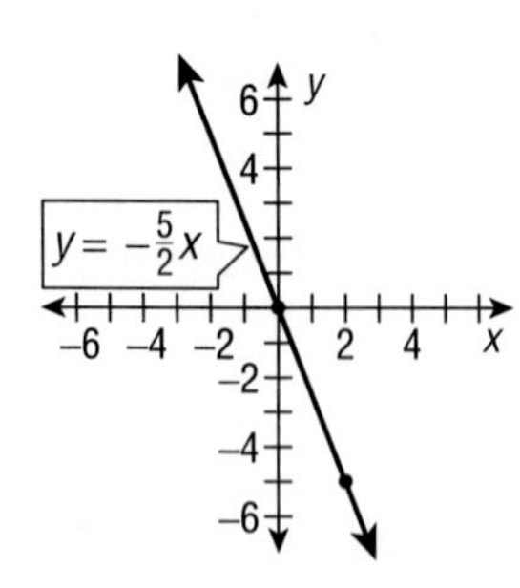

15. $\frac{1}{2}, -\frac{7}{2}$

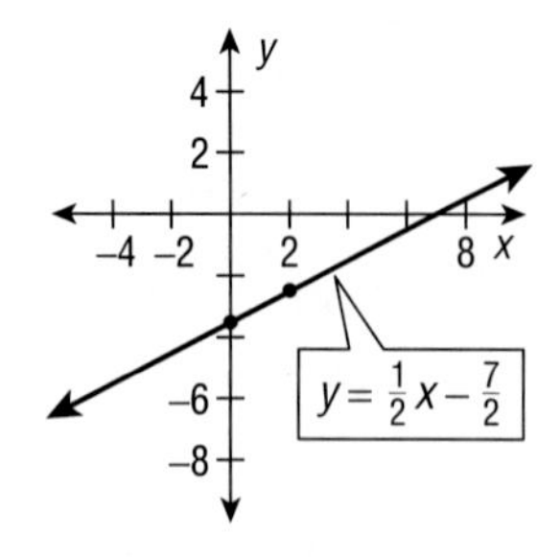

17. a **19.** c **21.** False, slope is 2 and y-intercept is $\frac{5}{3}$. **23.** True. If the slope is 0, $\frac{y_2-y_1}{x_2-x_1}=0$. Therefore, the numerator $y_2-y_1=0$, and $y_2=y_1$. Two points with the same y-coordinates are the same distance from the x-axis, so the line is horizontal. **25.** 82,300; the slope **27.** $y=-\frac{2}{3}x+1$ **29.** $y=\frac{2}{5}x-2$

13.5 Integrated Review p. 613

31. c **33.** a

13.6 Communicating about Mathematics p. 615

Hawaii: 1,190,000 and 1,240,000. Delaware: 700,000 and 730,000. Vermont: 580,000 and 600,000.

13.6 Guided Practice p. 616

1. B **2.** A: 261, B: ≈200 **3.** $30x+50y=1800$

4.

x	10	20	30	40
y	30	24	18	12

5.

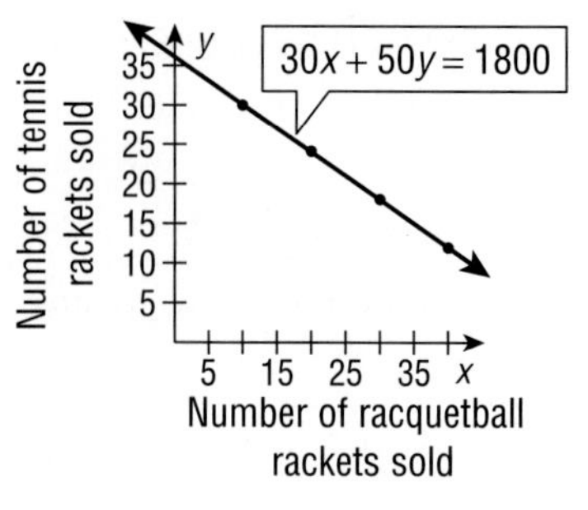

6. If only one kind of racket is sold, selling 60 racquetball rackets or 36 tennis rackets is necessary to reach your goal.

13.6 Independent Practice pp. 616–617

7.

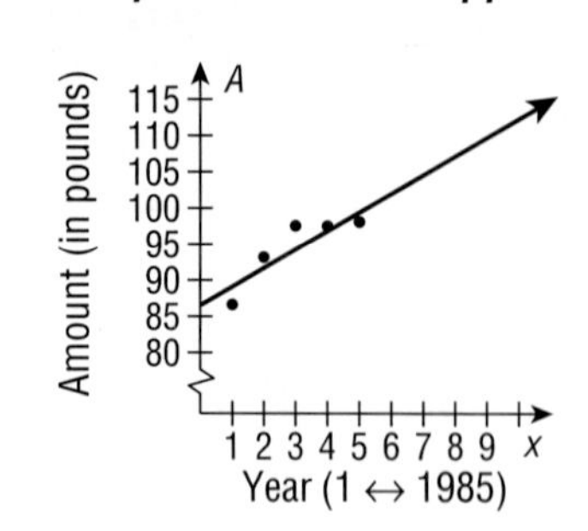

115

9.a.

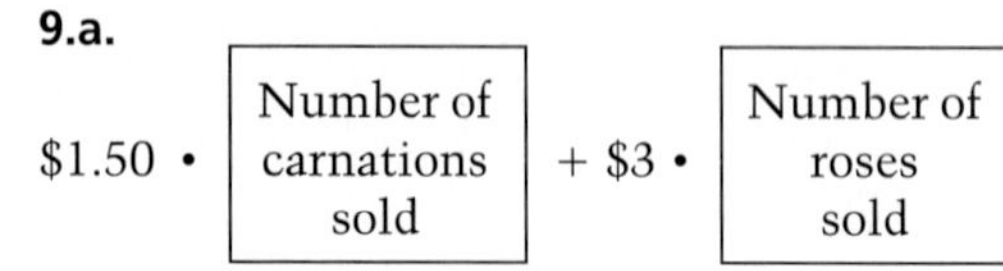

$= \$600$, $1.5x+3y=600$

b.

x	100	120	140	160
y	150	140	130	120

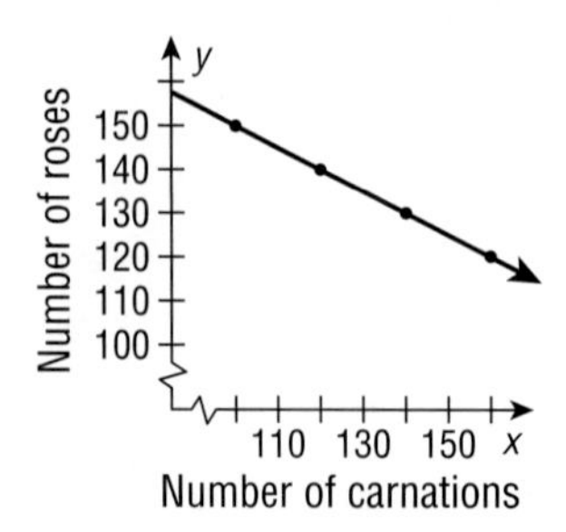

c. If only one kind of flower is sold, selling 400 carnations or 200 roses is necessary to reach your goal. **11.a.** 24.2, 27.4 **b.** The amount of aluminum per can decreased. **c.** The pattern is close to being linear. Answers vary, but the limit would probably be reached before another 20 years.

13.6 Integrated Review p. 617

13. d **15.** c

13.6 Mixed Review p. 618

1. 13, 2 **3.** 1, 4 **5.** Rational. It is the quotient of two integers. **7.** Irrational. Its decimal form neither terminates nor repeats. **9.** 72 **11.** 6105 **13.** 5.76 **15.** 200% **17.** Yes **19.** No

13.7 Communicating about Mathematics p. 132

Answers vary. (0, 300); yes, when no adult tickets are sold you have to sell 300 tickets for children.

13.7 Guided Practice p. 621

1. Solid **2.** Dashed **3.** Dashed **4.** Solid **1.–4.** A solid line consists of points that are on the graph. A dashed line consists of points that are not on the graph. **5.** Yes **6.** No **7.** Answers vary. **8.** 2, -4

9.

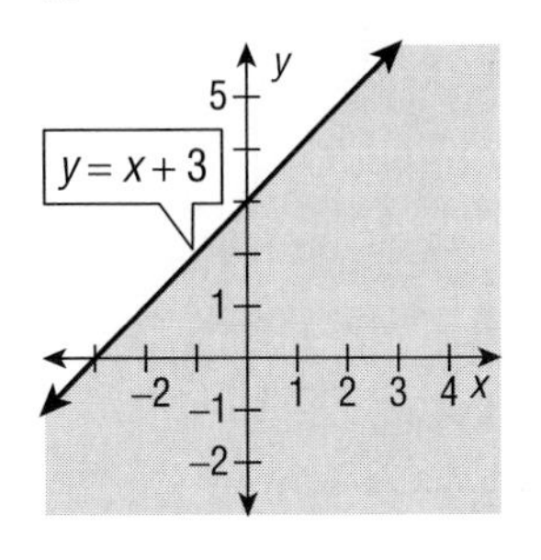

10.

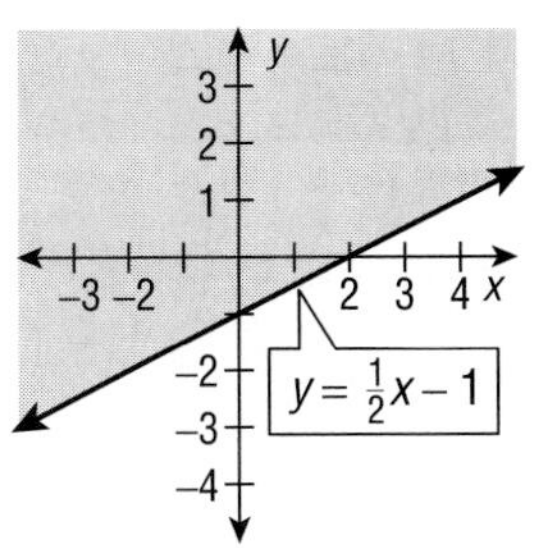

11.

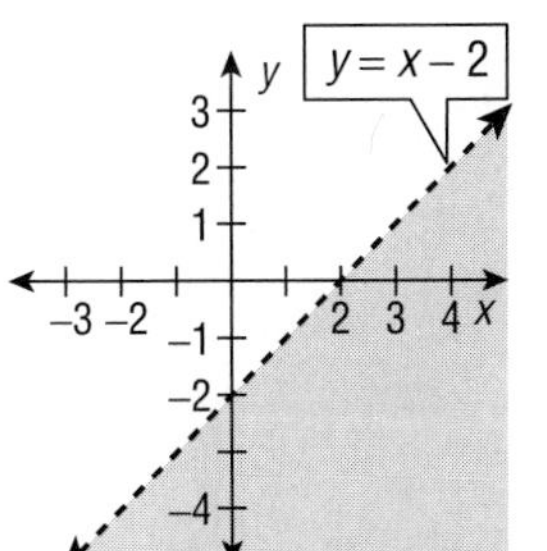

12.

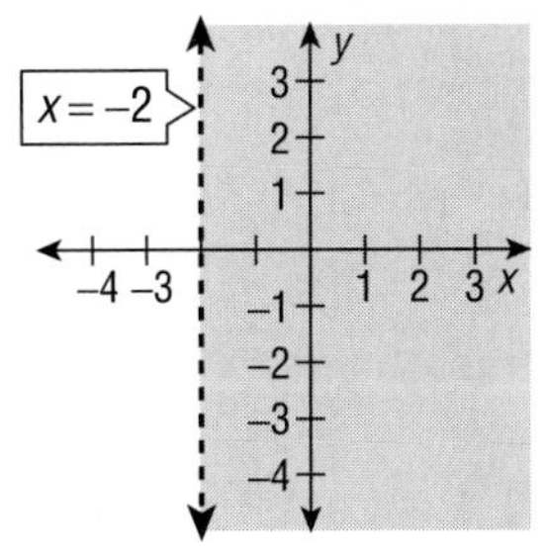

13.7 Independent Practice pp. 621–622

13. No, $4 \cdot 5 + 6 \cdot 5 > 48$ **15.** No, $4(-2) + 6 \cdot 10 > 48$ **17.** b **19.** d

21. **23.**

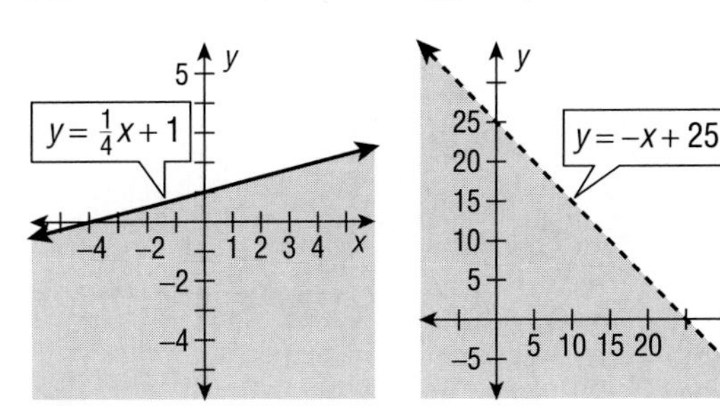

25. b **27.** Inequality c. The sum of the number of miles you walk and the number of miles you run could be equal to 26.2 if you finish the race, or could be less than 26.2 if you do not finish the race. **29.** d **31.** $x > 1$ **33.** $y < x$

13.7 Integrated Review p. 622

35. Line should be dashed.

13.8 Communicating about Mathematics p. 624

$\approx$42.06 ft, $\approx$42.06 ft; the distances are the same.

13.8 Guided Practice p. 625

1. 5, $\left(\frac{7}{2}, 2\right)$ **2.** 4, $(-1, -5)$ **3.** $\approx$22.85
4. $\left(\frac{2+(-1)}{2}, \frac{4+(-3)}{2}\right) = \left(\frac{1}{2}, \frac{1}{2}\right)$, $\left(\frac{-3+4}{2}, \frac{3+(-2)}{2}\right) = \left(\frac{1}{2}, \frac{1}{2}\right)$

13.8 Independent Practice pp. 625–626

5. $\approx$7.8 **7.** $\approx$7.6 **9.** $(-1, 0)$ **11.** $(-3, 0)$ **13.** $\approx$19.09 units **15.** $\approx$19.69 units **17.** Yes, $(\sqrt{8})^2 + (\sqrt{8})^2 = 4^2$ **19.** $(-1, 1)$, $\sqrt{8}$
21. $\left(\frac{-3+1}{2}, \frac{4+(-4)}{2}\right) = (-1, 0)$, $\left(\frac{-5+3}{2}, \frac{1+(-1)}{2}\right) = (-1, 0)$ **23.** (5, 6), (3, 1)
25. 470 mi

13.8 Integrated Review p. 626

27. 10 units^2 **29.** 20 units^2

13.8 Chapter Review pp. 628–629

1. Yes **3.** No

5.

x	-2	-1	0	1	2
y	-5	$-\frac{9}{2}$	-4	$-\frac{7}{2}$	-3

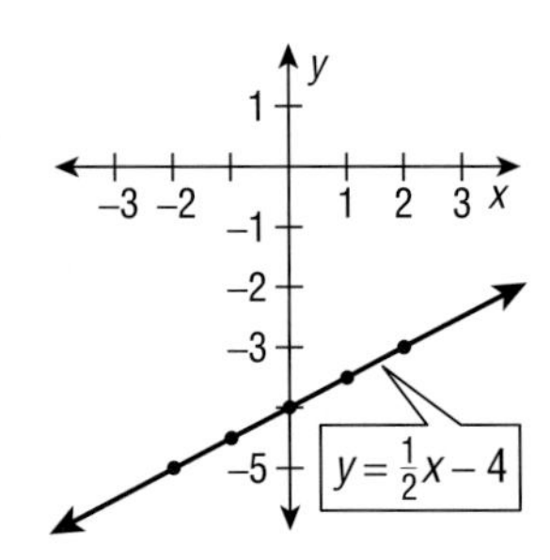

7.

x	-2	-1	0	1	2
y	14	10	6	2	-2

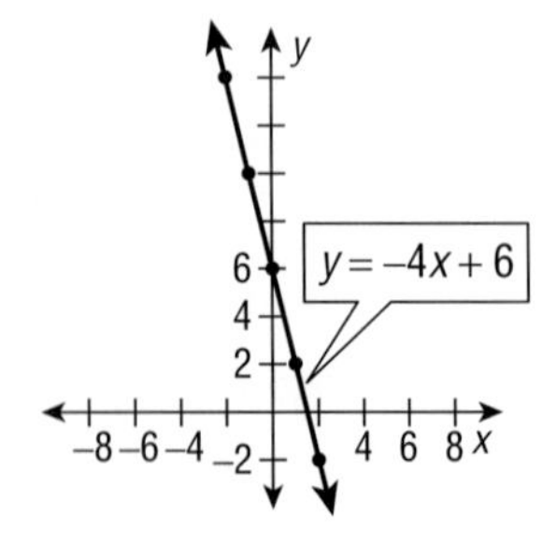

9. Slanted **11.** Horizontal

15. **19.**

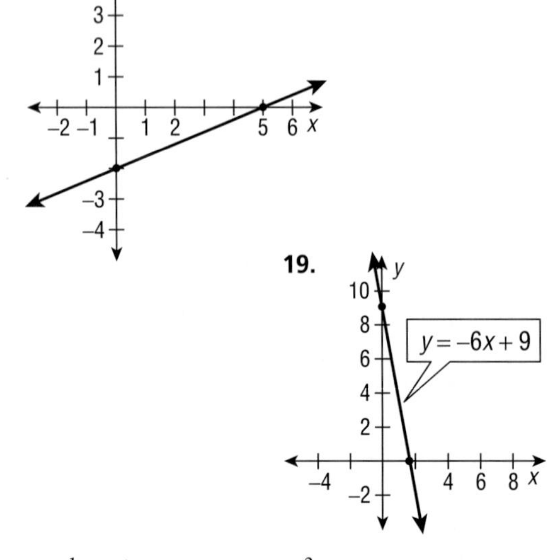

21. $\frac{1}{2}$ **23.** 0 **25.** -1 **27.** $\frac{2}{7}$ **29.** c

33. $\frac{7}{6}$, -4

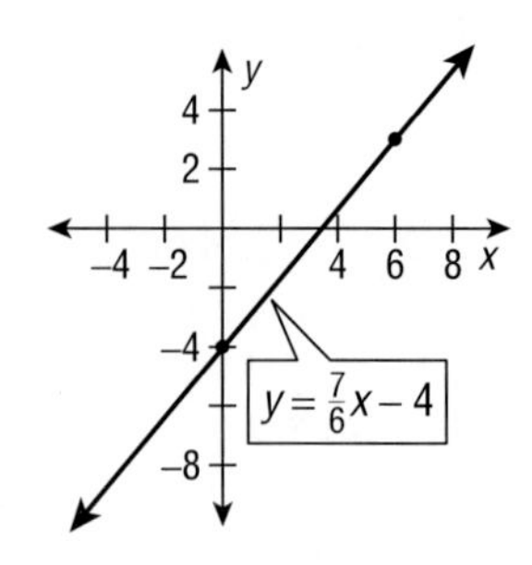

35. $y=\frac{2}{3}x+1$ **37.** $y=5$

39.

$40 • 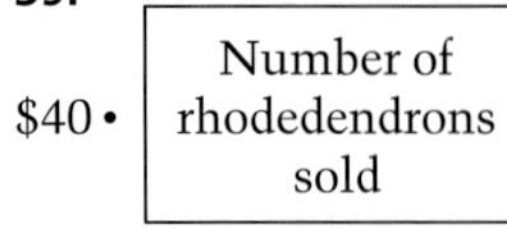 Number of rhododendrons sold + $60 • Number of dogwood trees sold

= $2500

41. 21 **43.** Yes, dashed **45.** No, solid **47.** No, dashed

49.

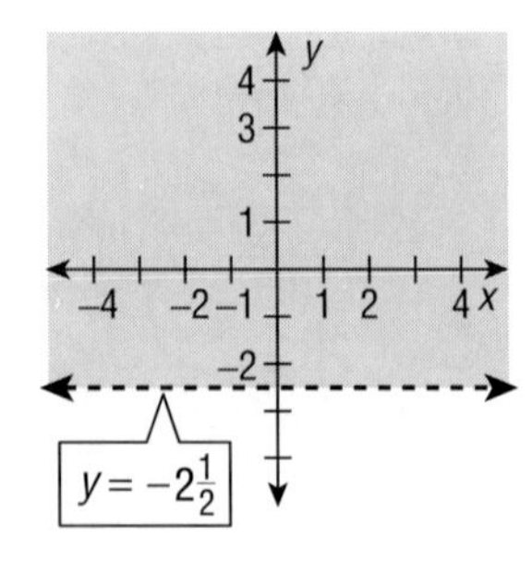

51. $y=7x+6$

53. $(-\frac{1}{2}, 0)$, ≈ 7.81 **55.** $(1, -\frac{5}{2})$, ≈ 6.71

57. $\sqrt{50}$, $\sqrt{50}$, 10 **59.** ≈ 24.14 units **61.** (4, 5) and (4, -1), their distances from (1, 2) are the same ($\sqrt{18}$). **63.** Felicia: Station 1, Heather: Station 2 **65.** Yes, Felicia and Heather; it is nearest.

69.

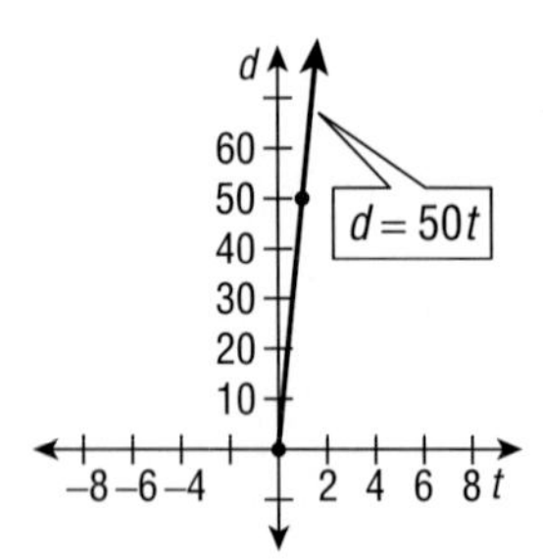

71. 8, 2

CHAPTER 14

14.1 Communicating about Mathematics p. 635

The median or mode, which are both $4.75 per hour; 13 of the 16 salaries are $4.75 per hour.

14.1 Guided Practice p. 636

1.a. Median **b.** Mode **c.** Mean **2.** The numbers 3, 4, 5, 6, and 7 occur from 1 to 5 times each. 3, 3, 4, 4, 4, 4, 5, 5, 5, 5, 5, 6, 6, 6, 6, 7 **3.** $4\frac{7}{8}$ **4.** 5 **5.** 5

14.1 Independent Practice p. 636

7. 54.1, 54, none **9.** $58\frac{5}{9}$, 59, 60 **11.** 3.2, 3.5, 4 **13.** Measures and explanations vary. **15.** Measures and explanations vary. **17.** $2\frac{7}{11}$, 2, 1 **19.** 6.61, 6.6, none **21.** Answers vary. 13, 14, 15, 15, 16, 17 **23.** Answers vary. One wording might be as follows. In 1992 in Seattle, Washington, half the houses cost as much as or more than $141,300 and half the houses cost as much as or less than $141,300.

14.1 Integrated Review p. 637

27.a. $\frac{3}{20}$ **b.** $\frac{1}{5}$ **c.** $\frac{1}{10}$ **d.** $\frac{1}{4}$ **e.** $\frac{3}{20}$ **f.** $\frac{1}{10}$ **g.** $\frac{1}{20}$ **h.** 0

14.2 Communicating about Mathematics p. 639

	6	4 8
0	5	0 3 5
7 7 6 3	4	0 5 8 9
9 5 4 4 2 2 2 1 1 0	3	1 1 5 7 9
9 7 4 0	2	1 6 7 9
8 3 2 2 1	1	1 4 5 5 6 6 9
7	0	

3|4|0 represents 43% western and 40% eastern rural population. This plot represents the opposite to that of Example 2: the eastern states tend to have more rural populations.

14.2 Guided Practice p. 640

1. 2, 2, 8, 10, 11, 15, 15, 16, 19, 20, 23, 25, 27, 31, 32, 34, 34, 36 **2.** Left side: 4.0, 4.2, 4.4, 5.2, 5.3, 5.7, 5.9, 6.0, 6.5, 6.5, 7.1, 7.3, 7.4, 7.8. Right side: 4.3, 5.2, 5.3, 5.5, 5.8, 5.9, 6.1, 6.4, 6.7, 7.0, 7.3, 7.4, 7.6.

3.

6	5
5	0 2 4 5
4	1 3
3	0 2 2 3 4 8
2	1 4 8 9 9
1	0 7 8
0	1 2 2 5 6 8 8

6|5 represents 65.

14.2 Independent Practice pp. 640-641

5. 84, 84, 85, 88, 90, 91, 101, 102, 107, 110, 113, 118, 118, 119, 124, 125, 126, 127

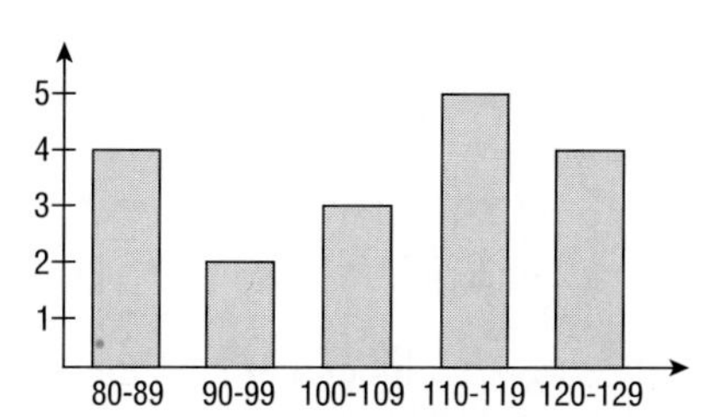

7. Left side: 31, 34, 36, 42, 43, 45, 54, 56, 57, 62, 65, 71, 74, 77, 80, 83, 84, 88, 92, 92, 93, 97. Right side: 31, 33, 34, 35, 37, 39, 40, 42, 43, 47, 56, 58, 58, 62, 63, 65, 66, 73, 74, 77, 83, 84, 91, 92.

9.

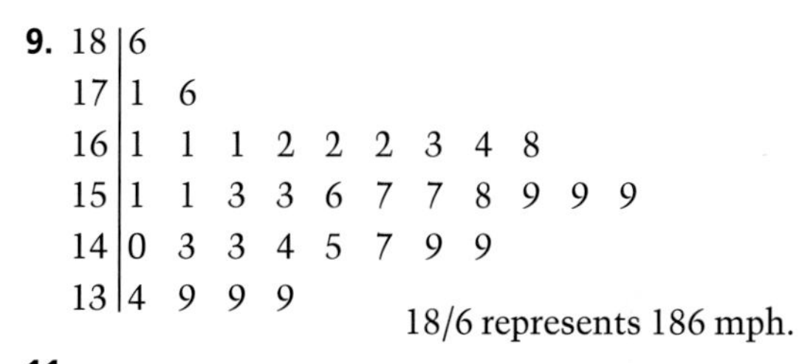

18	6
17	1 6
16	1 1 1 2 2 2 3 4 8
15	1 1 3 3 6 7 7 8 9 9 9
14	0 3 3 4 5 7 9 9
13	4 9 9 9

18/6 represents 186 mph.

11.

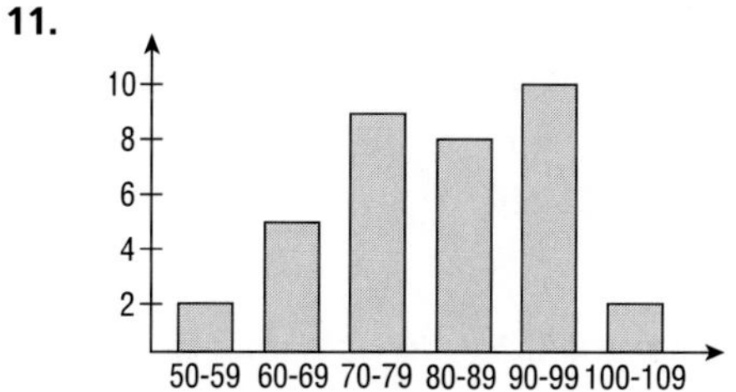

15. East: 17.76%, 18%, none; West: 19.88%, 19%, 19%; Total: 18.82%, 19%, 19%

14.2 Integrated Review p. 641

17. $\frac{1}{3}$

14.3 Communicating about Mathematics p. 643

The bottom plot represents Florida. Many people move to Florida after retiring; so, Florida has a larger percent of older people.

14.3 Guided Practice p. 644

1. 6, 56 **2.** 21, 34, 44
3.

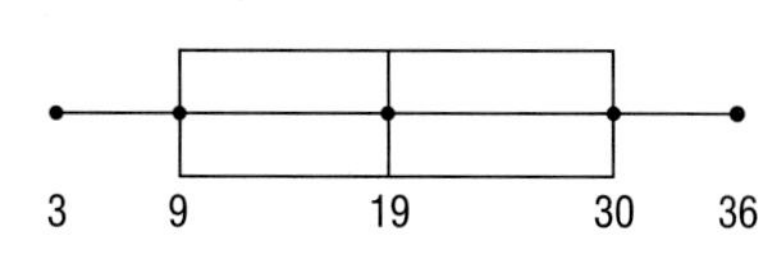

4.

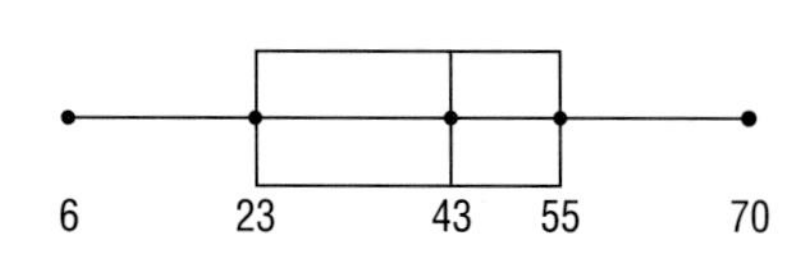

14.3 Independent Practice pp. 644–645

5. 14, 96 **7.** 25%
9.

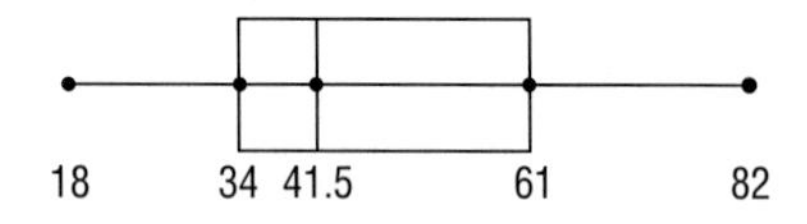

11. 39 is the largest number, not 37. The first quartile is 8.5, not 8. The second quartile is 18.5, not 18. The second quartile should be located $\frac{10}{19.5}$ of the way from 8.5 to 28, not where it is located now. **13.** Mississippi's population is younger than Florida's and Rhode Island's, and is older than Alaska's and Utah's. **15.** Receivers generally cost more than cassette decks, and cassette decks generally cost more than CD players.

14.3 Integrated Review p. 645

17. 19.5 **19.** 44.44% **21.** 240

14.3 Mixed Review p. 647

1. $\sqrt{97}$ **3.** ≈ 0.406 **5.** ≈ 0.444 **7.** ≈ 0.406 **9.** 2.35
11. 2.7 **13.** $\frac{13}{3}$ **15.** 10

14.4 Communicating about Mathematics p. 649

Profit	Restaurant 1	Restaurant 2
April	$17,768	$4,589
May	$19,714	$15,975
June	$3,664	$14,019
July	$15,818	$11,798

Restaurant 2; Restaurant 1; Restaurant 1, profit was greater.

14.4 Guided Practice p. 650

1. 3,2 **2.** -5 **3.** In the third row and first column **4.** Yes, the entries in corresponding positions are equal ($\frac{5}{4} = 1.25$ and $-\frac{8}{5} = 1.6$).

5. $\begin{bmatrix} -2 & 4 & -5 \\ -9 & 7 & 1 \end{bmatrix}$ **6.** $\begin{bmatrix} 0.3 & -6.3 \\ -2.4 & 7.2 \end{bmatrix}$

14.4 Independent Practice pp. 650–651

7. $\begin{bmatrix} -3 & 7 \\ -5 & -5 \end{bmatrix}, \begin{bmatrix} -1 & -1 \\ 7 & -5 \end{bmatrix}$

9. $\begin{bmatrix} -10 & -2 & 4 \\ 1 & -10 & 1 \end{bmatrix}, \begin{bmatrix} 2 & -8 & 0 \\ -1 & -8 & -7 \end{bmatrix}$

11. $\begin{bmatrix} -4 & 5 & -3 \\ -5 & -4 & -1 \\ 8 & 6 & 10 \end{bmatrix}, \begin{bmatrix} 10 & 5 & -9 \\ -3 & 4 & -11 \\ 0 & 10 & -8 \end{bmatrix}$

13. $\begin{bmatrix} 1 & \frac{4}{3} & \frac{2}{3} \\ 1 & \frac{1}{2} & \frac{5}{4} \\ \frac{2}{5} & 1 & \frac{4}{5} \end{bmatrix}, \begin{bmatrix} -\frac{1}{3} & 0 & 0 \\ -\frac{1}{2} & 0 & \frac{1}{4} \\ 0 & -\frac{1}{5} & \frac{2}{5} \end{bmatrix}$

15. $-4, 3, 4, -\frac{5}{2}$

17. Answers vary. $\begin{bmatrix} 1 & 3 \\ 2 & -4 \end{bmatrix}, \begin{bmatrix} 1 & 2 \\ 4 & 7 \end{bmatrix}$

19. Answers vary. $\begin{bmatrix} 3 & 2 \\ -7 & -4 \end{bmatrix}, \begin{bmatrix} -2 & -2 \\ 7 & 5 \end{bmatrix}$

21.

Income	Store 1	Store 2
May	231,450	206,210
June	265,985	319,754
July	303,442	321,615
August	324,570	256,419

Expenses	Store 1	Store 2
May	208,345	200,926
June	247,913	296,575
July	287,500	306,480
August	317,940	238,212

23.

Profit	Store 1	Store 2
May	$23,105	$5,284
June	$18,072	$23,179
July	$15,942	$15,135
August	$6,630	$18,207

25.

	Side 1	Side 2	Side 3
Triangle 1	5	7	10
Triangle 2	9	12	15
Triangle 3	0.9	4	4.1
Triangle 4	5	6	9.2

27. 22, 36, 9, 20.2

14.4 Integrated Review p. 651

29. $2, 2\pi, \pi; 3, 6\pi, 9\pi; 5, 10\pi, 25\pi; 14, 14\pi, 49\pi$

14.4 Mid-Chapter Self-Test p. 652

1. $84\frac{7}{12}°$ **2.** $84\frac{1}{2}°$ **3.** 87°

4.

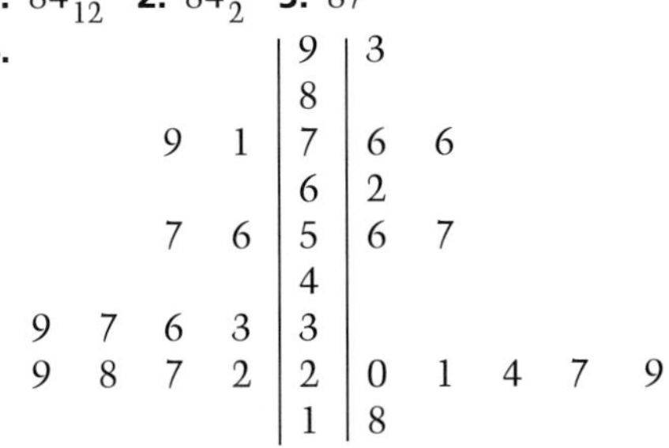

				Stem					
				9	3				
				8					
		9	1	7	6	6			
				6	2				
		7	6	5	6	7			
				4					
9	7	6	3	3					
9	8	7	2	2	0	1	4	7	9
				1	8				

1|7|6 represents 7.1 in. in Jacksonville and 7.6 in. in Miami.

5. 4.25 in. **6.** ≈ 4.28 in. **7.** 82°, 92°, 93°, 95°, 96°, 96°, 96°, 99°, 100°, 104°, 104°, 113° **8.** 94° **9.** 96° **10.** 102°

11.

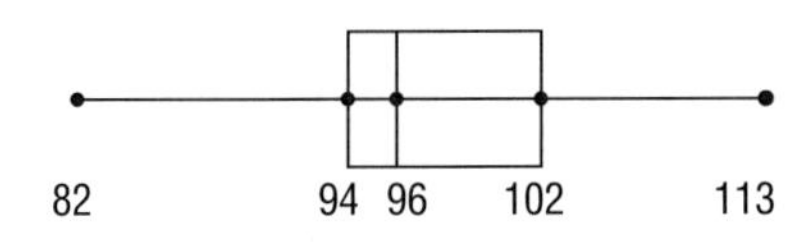

12. 75% **13.** 35 **14.** 64 **15.** 24

16. $\begin{bmatrix} 8 & 5 \\ 20 & 8 \end{bmatrix}$ **17.** $\begin{bmatrix} 3 & 2 \\ -2 & -13 \end{bmatrix}$ **18.** $\begin{bmatrix} 1 & 12 \\ 23 & 24 \end{bmatrix}$

19.

	Test 1	Test 2	Test 3
Student 1	74	88	95
Student 2	86	83	81
Student 3	82	71	86

20. Student 1

14.5 Communicating about Mathematics p. 655

48, 80, 112. Yes. During each second after the first, the wrench falls 32 feet per second faster than it fell during the previous second.

14.5 Guided Practice p. 656

1., 4., 6.–8. Yes. Each coefficient is a real number and each exponent is a whole number. **2., 3., 5.** No. The negative exponent is not a whole number. **9.** $3x^2+4x-2$; $3x^2, 4x, -2$ **10.** $-5r^3+4r+10$; $-5r^3, 4r, 10$ **11.** $p^3-16p^2+3p-12$; $p^3, -16p^2, 3p, -12$ **12.** $3t^2-4t$, binomial **13.** $-x^3+5$, binomial **14.** $14n-7$, binomial

14.5 Independent Practice pp. 656–657

15. Yes, trinomial **17.** Yes, monomial

19. c, $2x^2+9x+10$ **21.** b, $3x^2+6x+10$

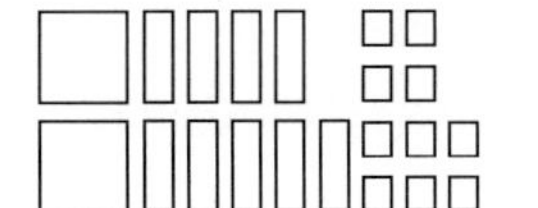

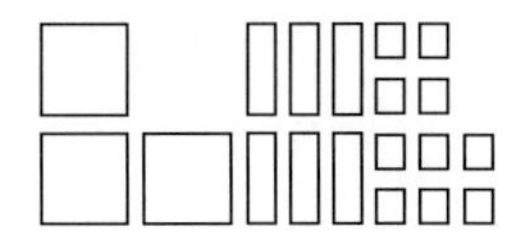

23. $6x^3-2x^2-x$; $6x^3, -2x^2, -x$ **25.** $-z^3+9z^2-7z+3$; $-z^3, 9z^2, -7z, 3$ **27.** $15w^3+4w^2+w-3$; $15w^3, 4w^2, w, -3$ **29.** $2x$ **31.** $8s^3-3s^2-6$ **33.** $2.9r^2+0.4r$ **35.** 797 ft

14.5 Integrated Review p. 657

37. d **39.** b **41.** $-17, -11, 7$ **43.** $-24, 12, 48$

14.6 Communicating about Mathematics p. 660

A. $(x^2+3x)-(x^2-6x)=(9^2+3 \cdot 9)-(9^2-6 \cdot 9)=(81+27)-(81-54)=108-27=81$ units^2
B. $9x=9 \cdot 9=81$ units^2

14.6 Guided Practice p. 661

1. $6x^2-7x-5$ **2.** $2n^2+5$ **3.** $3y^3-4y^2-2y-3$ **4.** y^3+6y^2-6y+9

14.6 Independent Practice pp. 661–662

5.
$$\begin{array}{rrrrr} & -4z^3 & +z^2 & & +7 \\ + & 3z^3 & & -6z & -5 \\ \hline & -z^3 & +z^2 & -6z & +2 \end{array}$$

7. $5x^2+7x+11$ **9.** $-8b^3-4$ **11.** $-x^3+9x^2-8x-4$ **13.** $2t^3+2t^2+10t-6$ **15.** $-5x^2+7x-5$ **17.** $-z^2-4z$ **19.** $3x^2+8x+6$, 57 **21.** $x(2x+5)-x(x-2)$, 60 **23.** $10x-500$ **25.** No; you will make more than twice as much, \$3500.

14.6 Integrated Review p. 662

27. 79 million, 98 million

14.6 Mixed Review p. 663

1. $\sqrt{52}$ **3.** 1.5 **5.** $10+\sqrt{52}, 15+\sqrt{117}$

7.

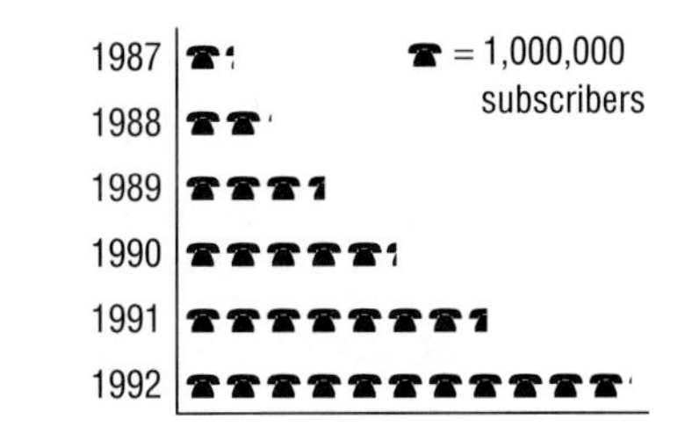

9. $\frac{5}{6}$ **11.** $\frac{8}{25}$ **13.** $\frac{4}{3}$ **15.** $y \geq 3$ **17.** $s < -\frac{3}{4}$
19. $4x^2 + 4x + 4$

■ 14.7 Communicating about Mathematics p. 666

A. $6x^2 + 6x$ **B.** 0 and 3, but for $x = 0$ the figure becomes a point; 72 units^2

■ 14.7 Guided Practice p. 667

1. $8n^3 - 12n^2 + 20n$ **2.** $-3y^4 - 6y^3 + 15y^2$ **3.** Rectangle: $8x^2 - 4x$, small triangle: $2x^2$, trapezoid: $10x^2 + 4x$, large triangle: $4x^2$ **4.** $24x^2$ **5.** $6x$, $4x$ **6.** $24x^2$ **7.** They are the same.

■ 14.7 Independent Practice pp. 667–668

9. $4x^3 - 8x^2 - 4x$ **11.** $-48t^3 - 40t^2$ **13.** $-8t^3 + 12t^2 - 4t$ **15.** $-8z^6 + 4z^4 - 40z$ **17.** Triangle: x^2, parallelogram: $2x^2 - 2x$, trapezoid: $3x^2 - 4x$, rectangle: $2x^2$ **19.** $8x^2 - 6x$ **21.** $n(n+1)$, $n^2 + n$ **23.** $3x^2 + x$ **25.** $22x^2 + 6x$ **27.** $7t^2(-t^3) = -7t^5$, $-7t^2(8t) = -56t^3$

■ 14.7 Integrated Review p. 668

29. 1 **31.** 243 **33.** $12 + 2x$ **35.** $10x + 15$

■ 14.8 Communicating about Mathematics p. 671

A. $2x^2 + 11x + 12$ **B.** $3x^2 + 11x + 10$

■ 14.8 Guided Practice p. 672

1.
$$\begin{aligned}(x+3)(3x+2) &= (x+3)(3x) + (x+3)(2)\\ &= (x)(3x) + (3)(3x) + (x)(2) + (3)(2)\\ &= 3x^2 + 9x + 2x + 6\\ &= 3x^2 + 11x + 6\end{aligned}$$

2.
$$\begin{aligned}(2x+1)(4x+5) &= (2x+1)(4x) + (2x+1)(5)\\ &= (2x)(4x) + (1)(4x) + (2x)(5) + (1)(5)\\ &= 8x^2 + 4x + 10x + 5\\ &= 8x^2 + 14x + 5\end{aligned}$$

3. d **4.** b **5.** a **6.** c

■ 14.8 Independent Practice pp. 672–673

7. Not $+5$, but $(x+6)(5)$; not $+5$, but $(x)(5) + (6)(5)$; not $+5$, but $5x + 30$; answer: $2x^2 + 17x + 30$ **9.** $8x^2 + 36x + 36$ **11.** $18x^2 + 23x + 7$ **13.** $20x^2 + 40x + 20$ **15.** $8x^2 + 14x + 3$ **17.** $18x^2 + 45x + 25$ **19.** $27x^2 + 27x + 6$ **21.** $4x^2 + 49x + 12$, 126 units^2 **23.** $\frac{5x^2 + 21x + 18}{2}$, 40 units^2 **25.** The products are equal, the commutative property of multiplication

27.

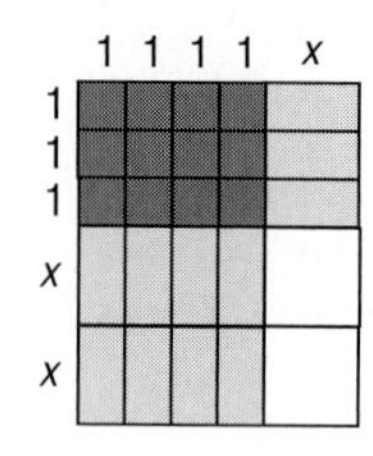

$12 + 11x + 2x^2$

■ 14.8 Integrated Review p. 673

29. $2x^2 + 16x + 30$, c **31.** $3x^2 + 18x + 15$, b **33.** c

■ 14.8 Chapter Review p. 675

1. $36\frac{13}{14}$ **3.** None

5.

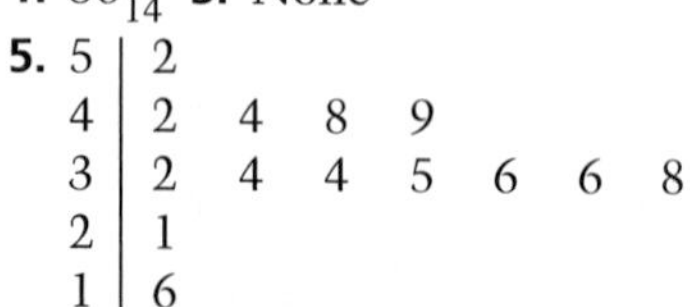

1|6 represents 16 home runs.

7.

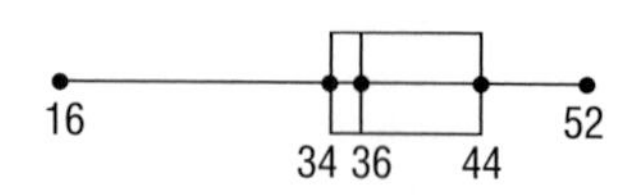

9. 66, 70, 78 **11.** 4, 5 **13.** 4

15. $\begin{bmatrix} 2 & 4 & 11 \\ 8 & -13 & 9 \\ -1 & -3 & -3 \end{bmatrix}$, $\begin{bmatrix} -4 & 6 & 7 \\ -4 & -3 & -7 \\ -9 & 11 & -3 \end{bmatrix}$

17. $-7x$, monomial **19.** $-4n^2 + 5n + 8$, trinomial **21.** $5t^2 + 3t$, binomial **23.** $7x^2 - 16$ **25.** $3n^2 + 8n - 3$ **27.** $x^2 + 15x - 6$ **29.** $x^3 + x^2$ **31.** $12z^2 + 16z + 5$ **33.** $6t^3 + 12t^2 + 15t$ **35.** $x^2 - 2x$ **37.** $3x^2 - 5x$ **39.** $6x^2 + 17x + 12$ **41.** 10 **43.** 7 **45.** 16 **47.** 12 **49.** 8 **51.** 4 **53.** 6 **55.** 5 **57.** 15

CREDITS

Appreciation to the staff at Larson Texts, Inc: who assisted with proofreading the manuscript and preparing and proofreading the art package.

Appreciation to the following art/photo production staff:
Art: Joan Williams
Photographs: Susan Doheny
Cover design: Linda Fishborne.
Cover photos: Y. Gladu/Photo Researchers; inset, DeSciose/The Stock Broker.

ILLUSTRATION CREDITS

Kathy Meisel: 142, 167, 194, 196, 198, 213, 217, 234(b), 260, 297, 365(b), 370, 375(b), 386, 408, 439, 517, 553, 589, 655.
Patrice Rossi: 10, 45, 62, 69, 108, 110, 134, 154, 159, 195, 210, 262, 290, 302, 335, 365(c), 368, 435, 480, 530, 567, 582, 591, 617, 662.

PHOTOGRAPHY CREDITS

CHAPTER 1 **1:** Guy Spangenberg. **2:** Billy Barnes/ Tony Stone Images. **3:** Ken O'Donoghue © D.C. Heath. **5:** Dagmar Fabricius/Stock Boston. **7:** Museum of The History of Science, Oxford University. **14:** Steve Elmore/Stock Market. **21:** Stuart McCall/Tony Stone Images. **22:** Lester Lefkowitz/Tony Stone Images. **25:** NASA. **27:** Phillips-Ramsey. **28:** Jonathan Daniel/Allsport. **31:** George Wuerthner. **32:** Nash Baker © D.C. Heath. **34:** Jeff Foott. **39:** Steve Umland. **41:** Keith Thompson/Tony Stone Images. **43:** Barbara Von Hoffmann/Tom Stack & Associates. **46:** NASA.

CHAPTER 2 **48-9:** John Senzer /Fashion Institute of Technology. **54:** Michael Newman/Photo Edit. **58:** Dan McCoy/Rainbow. **60:** Henry Hilliard. **63:** From The World Book Encyclopedia © 1994 World Book, Inc. By permission of the publisher. **68:** Jonathan Rawle/Stock Boston. **72:** Michael Newman/Photo Edit. **74:** Stephen Dunn/Allsport. **77:** Ken O'Donoghue © D.C. Heath. **82:** Ken O'Donoghue © D.C. Heath. **84:** Tom Pantages. **86:** Darell Lane. **87:** William Johnson/Stock Boston. **92:** David Madison. **94:** Vasamuseet - The Vasa Museum. **95:** David Lissy/Picture Cube. **96:** Lawrence Migdale/Tony Stone Images.

CHAPTER 3 **98-9:** Dance Theatre of Harlem. **101:** George Ranalli/ Photo Researchers. **103:** Bob Daemmrich. **106:** Mark Segal/Tony Stone Images. **108:** Mike Powell/Allsport. **112:** Mark Wagner/ Tony Stone Images. **112:** Mark Wagner/Tony Stone Images. **113:** Anthony Salamone © D.C. Heath. **119:** NASA. **120:** Michael Norton/Adstock Photos. **127:** Bob Daemmrich/Stock Boston. **129:** William R. Sallaz/Duomo. **132:** Paul Avis © D.C. Heath. **134:** AP/Wide World Photos. **141:** Mitchell Layton/Duomo. **142:** Vandystadt Agence de Presse/Allsport. **143:** Michael Newman/Photo Edit. **145:** Daniel L. Feicht/Cedar Point.

CHAPTER 4 **146-47:** Tom & Pat Leeson/DRK. **150:** Barrie Rokeach. **152:** Peabody Museum, Harvard University. Photograph by Hillel Burger. **156:** Courtesy of Nantucket Nectars. **161:** Tom Walker/Stock Boston. **162:** The Bettmann Archive. **164:** David Sams/Stock Boston. **166:** David Young-Wolff/Photo Edit. **171:** Ken O'Donoghue © D.C. Heath. **172:** Jeff Foott. **175:** Tom Bean. **176:** David Young-Wolff/Photo Edit. **179:** Ken O'Donoghue © D.C. Heath. **181:** Duomo. **183:** W. Cody/Westlight. **184:** Paul Avis © D. C. Heath. **188:** Jan Petter Lynaw/Scan Foto. **190:** Jeff Foott.

CHAPTER 5 **192-93:** David Nunuk/Westlight. **196:** United Airlines. **199:** Jeff Watts. **203:** Mark O. Thiessen © National Geographic Society. **204:** From A Treasury of Amish Quilts by Rachel and Kenneth Pellman. © 1991 by Good Books, Intercourse, Pennsylvania. Quilt from the collection of Barbara S. Janos. Used by permission. All rights reserved.. **207:** Anthony Salamone © D.C. Heath. **209:** Janice Fullman/Picture Cube. **210:** Ken Straiton/Stock Market. **214:** Kindra Clineff/Picture Cube. **216:** Ken O'Donoghue © D.C. Heath. **218:** Stephen J. Krasemann/DRK. **219:** Kolvoord/Image Works. **220:** Al Zwiazek/Tony Stone Images. **225:** Tony Stone Images. **226:** Ken O'Donoghue © D.C. Heath. **228:** Bob Daemmrich/Tony Stone Images. **229:** Ken O'Donoghue © D.C. Heath. **230:** Susan Doheny. **235:** J. Pickerell/FPG. **236:** David Young-Wolff/Photo Edit.

CHAPTER 6 **238-39:** Tony Friedkin/FASE. **243:** W. Cody/Westlight. **246:** W. Cody/Westlight. **248:** Jasper Johns *Three Flags* 1958. Encaustic on canvas. 30 7/8x1/2x5 in. (78.4x115.6x12.7 cm.) Collection of Whitney Museum of American Art, 50th Anniversary Gift of the Gilman Foundation, Inc., The Lauder Foundation, A. Alfred Taubman, an anonymous donor, and purchase 80.32. **251:** Wayne Hoy/Picture Cube. **253:** Zigy Kaluzny/Tony Stone Images. **254:** Metropolitan Museum of Art, 56.171.38. **258:** t, Fukuhara, Inc./Weslight; b, Chuck Keeler/Tony Stone Images. **259:** Bruno De Hogues/Tony Stone Images. **262:** Oddo & Sinibaldi/Stock Market. **268:** Ken O'Donoghue © D.C. Heath. **270:** Larry Ulrich/Tony Stone Images. **273:** F. Robert Masini/Phototake. **275:** Lindenmeyr Munroe. **277:** NASA. **279:** Margaret Courtney-Clarke. **280:** Gabe Palmer/Stock Market.**282:** Ken Shung © 1993 The Walt Disney Co. Reprinted with permission of Discover Magazine. **285:** Steve Wilkings/Stock Market. **286:** Michele McDonald. **287:** Frank Siteman/Picture Cube.

CHAPTER 7 **292-93:** Jeff Gnass. **295:** Ian Howarth. **300:** MacDonald Photography/Unicorn Stock Photos. **307:** Anthony Salamone © D.C. Heath. **309:** Courtesy of Four Seasons Sunrooms. **311:** t, Peter Menzel; b, Ken O'Donoghue © D.C. Heath. **313:** Judy Nemeth/Picturesque. **316:** Dan McCoy/Rainbow. **319:** Bridgeman/Art Resource. **323:** Stephen Whalen/Picturesque. **329:** AP/Wide World. **331:** Peter Menzel. **333:** Courtesy of Hispanic Market Connections. **334:** Arthur Tilley/FPG. **339:** Stephen Gorman. **340:** Jeff Gnass. **341:** Bob Daemmrich/Stock Boston.

CHAPTER 8 **342-43:** Paul Avis/Liaison International. **345:** Richard Nowitz. **347:** A. Borrel/Liaison International. **352:** AP/Wide World. **353:** Ken O'Donoghue © D.C. Heath. **354:** Bob Thomason/Tony Stone Images. **355:** Bob Daemmrich. **357:** British Library/The Bridgeman Art Library. **361:** UPI/Bettmann. **362:** Dave Jacobs/Tony Stone Images. **364:** Ken O'Donoghue © D.C. Heath. **374:** Myrleen Ferguson/Photo Edit. **379:** Robert W. Ginn/Picture Cube. **380:** Ken O'Donoghue © D.C. Heath. **381:** Peter Correz/Tony Stone Images. **384**: P.A. Harrington/Peter Arnold, Inc. **385:** Herb Snitzer/Stock Boston. **386:** Ed Lallo/Liaison International. **387:** L.L.T. Rhodes/Tony Stone Images.

CHAPTER 9 **388-89:** Tony Stone Images. **391:** The Woman That Fell from The Sky "Falling Star" by Robert Orduño, Oil on linen, 60"x60". **393:** Susan Lapides. **404:** Chel Beeson © D.C. Heath. **406:** Reuters/Bettmann. **411:** Burk Uzzle. **414:** UPI/Bettmann. **417:** Anne Heimann/Stock Market. **419:** Robert W. Ginn/The Picture Cube. **421:** Ken O'Donoghue © D.C. Heath. **423:** Dan L. Feicht/Cedar Point. **426:** Bohdan Hrynewych/Stock Boston. **432:** t, Alex Bartel/Picture Cube; b, Lionel Delevingne/Stock Boston.

CHAPTER 10 **436-37:** Brian Parker/Tom Stack & Associates. **440:** Peter Yates. **443:** British Museum. **444:** Jim Pickerell/Westlight. **449:** Tony Stone Images. **452:** David Cornwell/Pacific Stock. **454:** Mittet Foto/Tony Stone Images. **456:** l, Mitch Reardon/Tony Stone Image; c, Superstock; r, Art Wolfe/Allstock. **457:** Superstock. **461:** Visuals Unlimited. **463:** Gerry Ellis Nature Photography. **464:** Ken O'Donoghue © D.C. Heath. **467:** Craig Aurness/Westlight. **468:** Vladpans/Leo de Wys. **473:** Rick Strange/The Picture Cube. **475:** Courtesy of Four Seasons Sunrooms. **478:** Telegraph Colour Library/FPG. **484:** l, James Solliday/Biological Photo Service; c, Tom Myers; r, Charles Seaborn/Odyssey.

CHAPTER 11 **486-87:** NASA. **493:** Comstock. **495:** Ken Straiton/Stock Market. **498:** Nancy Sheehan © D.C. Heath. **499:** Richard Hutchings/Photo Edit. **501:** NASA-Ames Research Center. **503:** Steve Gottlieb. **507:** Chris Alan Wilton/Image Bank. **510:** Michele Burgess/The Stock Market. **511:** Joe McDonald/Global Pictures. **513:** Ken O'Donoghue © D.C. Heath. **515:** David Weintraub/West Stock. **519:** Nippondenso Co., Ltd.. **520:** Julie Kramer Cole. **528:** t, Mark Sasahara; b, Ken O'Donoghue © D.C. Heath. **533:** Roger Tully/Tony Stone Images. **534:** t, © Kelly Freas 1954, 1973, 1993, 1996; b, NASA.

CHAPTER 12 **536-37:** Anthony Salamone © D.C. Heath, inset; Ken O'Donoghue © D.C. Heath. **539:** Ken O'Donoghue © D.C. Heath. **540:** The White House. **545:** Ken O'Donoghue © D.C. Heath. **547:** The *Polyhedron Earth* was designed and invented by R. Buckminster Fuller. © 1938, 1967, 1982, & 1992 Buckminster Fuller Institute, Santa Barbara, CA. All rights reserved. Globe distributed by Shasta Visions, Mt. Shasta, CA. (800) 800.3693. Photo © Hugh Barton. **552:** Peabody Museum, Harvard University. Photograph by Hillel Burger. **553:** The Bettmann Archive. **555:** Jules Allen. **560:** Ken O'Donoghue © D.C. Heath. **563:** Ringette Canada. **567:** Sheila Beougher/Liaison International. **570:** Nikolay Zurek/FPG. **572:** Peter Tatiner/Leo de Wys. **573:** Mark J. Barrett/Allstock. **574:** Sava Cvek Associates. **580:** Bruce Benedict/The Stock Broker. **581:** NASA/Tom Stack.

CHAPTER 13 **586-87:** David Leip, MIT Solar Electric Vehicle Team. **596:** H.P. Smith Jr./Academy of Natural Sciences, Philadelphia/VIREO. **601:** Katherine Lambert. **602:** Anthony Salamone © D.C. Heath. **606:** Len Clifford/Natural Selection. **608:** Richard Laird/FPG. **613:** J. Faircloth/Transparencies, Inc. **618:** Richard Wood/The Picture Cube. **622:** Owen Franken/Stock Boston. **629:** Henley & Savage/The Stock Market. **630:** Mark Snyder/Tony Stone Images.

CHAPTER 14 **632-33:** Ron Garrison/Zoological Society of San Diego. **635:** Raymond Barnes/Tony Stone Images. **636:** Jim Brandenburg/Minden Pictures. **637:** Bill Bachmann/Leo de Wys. **638:** Beatriz Terrazas. **639:** Gary Irving/Tony Stone Images. **643:** Jay Paris. **645:** Doug Wechsler/VIREO. **651:** Renee Stockdale/Animals Animals. **655:** John Yurka/The Picture Cube. **657:** Steve Vidler/Leo de Wys. **671:** Art Wolfe. **673:** Robert Frerck/Odyssey Productions. **675:** UPI/Bettmann. **677:** John Cancalosi/Natural Selection.

Reference

INDEX

A

B

C

F

INDEX

INDEX

N

P

INDEX

INDEX

S

Keystrokes for page 223

The following steps show the keystrokes for the *TI-80, TI-81,* the *Casio fx-7700G,* and the *Sharp EL-9300C* that would be used for creating a bar graph for the data given below.

Age of Deer	1	2	3	4	5	6	7	8	9	10	11	12
Number of Deer	18	14	10	7	6	4	3	3	2	1	2	2

TI-80

Window XMIN = 0 YMIN = −1
XMAX = 13 YMAX = 20
XSCL = 1 YSCL = 1

[STAT] [ENTER] (Edit)

L1(1) = 1 L2(1) = 18 L1(7) = 7 L2(7) = 3
L1(2) = 2 L2(2) = 14 L1(8) = 8 L2(8) = 3
L1(3) = 3 L2(3) = 10 L1(9) = 9 L2(9) = 2
L1(4) = 4 L2(4) = 7 L1(10) = 10 L2(10) = 1
L1(5) = 5 L2(5) = 6 L1(11) = 11 L2(11) = 2
L1(6) = 6 L2(6) = 4 L1(12) = 12 L2(12) = 2

[2nd] [STAT PLOT] [ENTER] (Plot1)

Choose the following:
On, Type: [histogram icon], XL: L1, F: L2

[GRAPH]

TI-81

Range Xmin = 0 Ymin = −1
Xmax = 13 Ymax = 20
Xscl = 1 Yscl = 1

[2nd] [STAT], cursor to DATA, [ENTER]

x1 = 1 y1 = 18 x7 = 7 y7 = 3
x2 = 2 y2 = 14 x8 = 8 y8 = 3
x3 = 3 y3 = 10 x9 = 9 y9 = 2
x4 = 4 y4 = 7 x10 = 10 y10 = 1
x5 = 5 y5 = 6 x11 = 11 y11 = 2
x6 = 6 y6 = 4 x12 = 12 y12 = 2

[2nd] [STAT], cursor to DRAW, [ENTER]
(Hist) [ENTER]

Casio *fx-7700G*

Range Xmin = 0 Ymin = −1
Xmax = 13 Ymax = 20
Xscl = 1 Yscl = 1

[MODE] [×] (SD)
[MODE] [SHIFT] 3 (DRAW)
[SHIFT] [Defm] 12 [EXE]

1 [F3] 18 [F1] 7 [F3] 3 [F1]
2 [F3] 14 [F1] 8 [F3] 3 [F1]
3 [F3] 10 [F1] 9 [F3] 2 [F1]
4 [F3] 7 [F1] 10 [F3] 1 [F1]
5 [F3] 6 [F1] 11 [F3] 2 [F1]
6 [F3] 4 [F1] 12 [F3] 2 [F1]

[GRAPH] [EXE]

Sharp *El-9300C*

Range Xdlt = 1
n = 1
Xmin = 1 Ymin = −1
Xmax = 13 Ymax = 20
Xscl = 1 Yscl = 1

[list icon] [MENU], cursor to DEL, 2, [ENTER], 2

X1 = 1 W1 = 18 X7 = 7 W7 = 3
X2 = 2 W2 = 14 X8 = 8 W8 = 3
X3 = 3 W3 = 10 X9 = 9 W9 = 2
X4 = 4 W4 = 7 X10 = 10 W10 = 1
X5 = 5 W5 = 6 X11 = 11 W11 = 2
X6 = 6 W6 = 4 X12 = 12 W12 = 2

[2nd F] [graph icon] [ENTER]

Keystrokes for page 597

The following steps show the keystrokes for the *TI-80, TI-81,* and the *Casio fx-7700G* that would be used to sketch the graph of the equation.

$y = 1.5x - 2.$

TI-80

[WINDOW] (Set range.)
XMIN = −10
XMAX = 10
XSCL = 1
YMIN = −10
YMAX = 10
YSCL = 1

[Y=] 1.5 [X,T] [−] 2
$:Y_1 = 1.5X - 2$
$:Y_2 =$
$:Y_3 =$
$:Y_4 =$

[GRAPH]
[CLEAR] (Clear screen)

TI-81

[RANGE] (Set range.)
Xmin = −10
Xmax = 10
Xscl = 1
Ymin = −10
Ymax = 10
Yscl = 1

[Y=] 1.5 [X|T] [−] 2
$:Y_1 = 1.5X - 2$
$:Y_2 =$
$:Y_3 =$
$:Y_4 =$

[GRAPH]
[CLEAR] (Clear screen)

Casio *fx-7700G*

[RANGE] (Set range.)
Xmin = −10
max = 10
scl = 1
Ymin = −10
max = 10
scl = 1

[EXE] [RANGE]
[SHIFT] [F5] (Cls) [EXE]
[GRAPH] 1.5 [X, θ, T] [−] 2
Graph Y = 1.5 X − 2
[EXE]
[SHIFT] [F5] (Cls) [EXE]